Tectonics, Climate, and Landscape Evolution

Edited by

Sean D. Willett
Department of Earth and Space Sciences
University of Washington
Seattle, Washington 98070
USA

Niels Hovius
Department of Earth Sciences
University of Cambridge
Cambridge CB2 3EQ
UK

Mark T. Brandon
Department of Geology and Geophysics
Yale University
New Haven, Connecticut, 06520
USA

Donald M. Fisher
Department of Geosciences
The Pennsylvania State University
University Park, Pennsylvania 16802
USA

THE
GEOLOGICAL
SOCIETY
OF AMERICA

Special Paper 398

Penrose Conference Series

3300 Penrose Place, P.O. Box 9140 ▪ Boulder, Colorado 80301-9140, USA

2006

Published by The Geological Society of America, Inc.
3300 Penrose Place, P.O. Box 9140, Boulder, Colorado 80301-9140, USA
www.geosociety.org

Printed in U.S.A.

GSA Books Science Editor: Abhijit Basu

Library of Congress Cataloging-in-Publication Data

Tectonics, climate, and landscape evolution / edited by Sean D. Willett ... [et al.]..
p. cm. -- (Special paper ; 398)
Includes bibliographical references and index..
ISBN 0813723981
1. Geology, Structural. 2. Erosion. 3. Climate. I. Willett, Sean D., 1960- II. Special papers (Geological Society of America) ; 398.

QE601.T3934 2006
551.8--dc22

2006040521

Cover: The Shrine of Eternal Waters in the Taroko Gorge of the Liwu River in northeastern Taiwan. Photo courtesy of the Taiwan National Park Service.

10 9 8 7 6 5 4 3 2 1

Contents

Field Studies of Tectonics, Climate, and Landscapes

Geological Society of America
Special Paper 398
2006

Introduction

Sean D. Willett
Department of Earth and Space Sciences, University of Washington, Seattle, Washington 98070, USA
Niels Hovius
Department of Earth Sciences, Cambridge University, Cambridge CB2 3EQ, UK
Mark T. Brandon
Department of Geology and Geophysics, Yale University, New Haven, Connecticut 06520-810, USA
Donald M. Fisher
Department of Geosciences, The Pennsylvania State University, University Park, Pennsylvania 16802, USA

TAIWAN PENROSE CONFERENCE

The Liwu River runs a short course; its channel head at the water divide in Taiwan's Central Range is a mere 35 km from its outflow into the Pacific Ocean. But in those short 35 km, the Liwu has carved one of the world's geographic wonders: the spectacular Taroko Gorge with marble and granite walls soaring nearly 1000 m above the river channel. The Liwu River itself is less impressive. Many visitors have marveled that the little trickle at the base of the gorge has carved such a massive canyon. As tourists marvel at the white marble around the river, however, they may notice the lack of vegetation within the channel, and finally, looking up, realize that the gorge bottom has been scoured by flood waters that reach tens of meters above the channel floor. Indeed, few visitors have seen the Liwu at flood stage, when runoff from typhoon rains causes a rise in the height of the river in the gorge to several hundreds of meters. The rush of turbulent water is able to carry large boulders and abrade and deepen the bedrock channel. The typhoon rains also strip alluvium from the surrounding hillslopes and trigger massive landslides, increasing the load of sediment carried through the gorge.

The Liwu is not a pretty place during floods, and man's perilous coexistence with nature there is symbolized perfectly by the beautiful shrine at the base of the gorge (cover); the shrine is dedicated to the hundreds of construction workers who lost their lives, most in flood-related landslides, while carving the highway through the gorge in the 1950s. As a final ironic act of nature, the shrine itself was twice destroyed by landslides in the 1980s, and rebuilt each time.

Taroko Gorge was a fitting venue for a Penrose Conference in 2003 that addressed the coupled processes of tectonics, climate, and landscape evolution. The young mountains, extreme weather, and dramatic landforms provided an appropriate backdrop to wide-ranging discussions of geomorphic processes, climate and meteorology, sediment generation and transport, the effects of erosion on tectonics, and new analytical and modeling tools used to address these processes and problems. The Penrose Conference also provided an opportunity to assess progress in the field over the past decade since a Chapman Conference on the same topic held at Snowbird, Utah, in 1992 (Merritts and Ellis, 1994). The earlier conference helped motivate a decade of research into the roles of tectonics, surface processes, and climate in creating the topography of Earth. The 2003 conference provided an opportunity to discuss progress made through that decade, and to look ahead to future challenges. Papers in this volume reflect and advance these developments.

COUPLED DYNAMIC SYSTEM

All papers in this volume deal in some form with the physical interactions between tectonics, surface erosion processes, and climate (Fig. 1). The connections in this dynamic system are manifold and many feedback pathways have yet to be explored. However, many of the fundamental relationships between the tectonic growth of topography, erosional destruction of topography, and climatic influence on erosion rates have been identified. Tectonic processes elevate regions of Earth's surface primarily through the isostatic response to crustal thickening (Fig. 1: path *A*). Tectonics also increases relief at multiple length scales through isostatic uplift, faulting, and folding (Fig. 1: path *B*). Increased elevation relative to regional base level increases river channel gradients and thus increases rates of erosion by fluvial incision and transport. In addition, topography at almost any length scale tends to increase orographically localized precipitation. This in turn gives rise to increased river discharge and incision. Insofar as river channels set the local base level for hillslope processes, enhanced

Willett, S.D., Hovius, N., Brandon, M.T., and Fisher, D.M., 2006, Introduction, *in* Willett, S.D., Hovius, N., Brandon, M.T., and Fisher, D.M., eds., Tectonics, Climate, and Landscape Evolution: Geological Society of America Special Paper 398, Penrose Conference Series, p. vii–xi, doi: 10.1130/2006.2398(00).

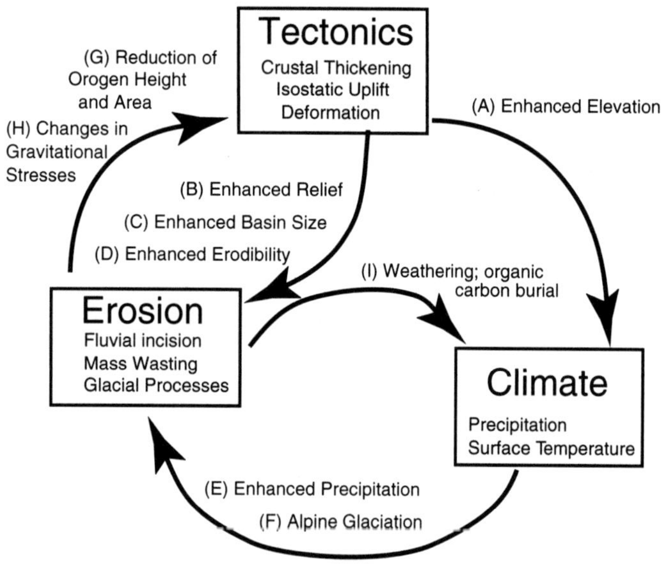

Figure 1. Interactions and feedback pathways for tectonics, climate, and erosional processes.

channel incision leads to increased hillslope failure and sediment supply to channels. Tectonics can directly influence erosion rates at short timescales, as is evident from earthquake triggering of landslides and seismic weakening of rockmass (Fig. 1: path *D*). Another link between tectonics and erosion is the increase in cumulative erosion and sediment yield that occurs as a tectonic region expands, for example, by accretion of new crustal material. As the domain affected by deformation grows, a larger area becomes subjected to high erosion rates and total sediment yield increases (Fig. 1: path *C*).

Climate links surface uplift and erosion rates through several mechanisms, including orographic forcing of precipitation (Fig. 1: path *E*). A second link is provided by the onset or increased efficiency of alpine glaciation, which develops as mountains reach altitudes sufficient to produce and maintain perennial ice (Fig. 1: path *F*). Alpine glaciation is a strongly nonlinear feedback as its onset only occurs when topography grows above the threshold for growing alpine ice (e.g., equilibrium limit altitude).

The feedback of erosion on tectonics is provided by the redistribution of near-surface mass. Tectonic processes are strongly influenced by gravity and surface redistribution of mass will influence gravitational stresses and thus tectonic deformation (Fig. 1: path *H*). The tectonic response can be complex, but the general response, at least of a convergent mountain belt, will be a reduction in the rate of growth of mean elevation and surface area (Fig. 1: path *G*). If erosional fluxes outpace the rate at which tectonics can thicken the crust, mean elevation and orogenic area will be progressively reduced, leading to a progressive decrease in sediment yield and erosional flux.

Mechanisms also exist for erosion to affect climate on a global scale, through influences on the carbon cycle. Weathering of Mg and Ca silicate rocks provides the essential buffer that balances the introduction of CO_2 into the atmosphere and the sequestration of that CO_2 in carbonate rocks (Fig. 1: path I). Erosional refreshing of exposed rock surfaces ensures that chemical weathering can occur at or close to its kinetic limits. Erosion also promotes drawdown of CO_2 by harvesting of life biomass and burial of this material in sedimentary basins, thereby reducing the amount of actively cycling carbon. Together, these two erosion-driven mechanisms are responsible for countering the outgassing of primary CO_2 and stabilizing Earth's climate in a narrow range of conditions suitable for the evolution of life.

RECENT SCIENTIFIC ADVANCES AND THIS VOLUME

The papers collected in this volume reach across fields that have experienced rapid advances in the past decade. Here, we briefly outline some key developments to provide a context for these contributions.

Emergence of Digital Elevation Models

Digital elevation models (DEMs) have become widely available for a large part of the planet and are now established as a common resource for topographic analysis. In addition, computing power and software capabilities have increased dramatically, permitting innovative exploitation of DEMs. Applications range from routine use of digital relief maps as a base for presentation of other data, to mathematically sophisticated analyses to infer physical processes from landscape attributes (Wobus et al., this volume, Chapter 4). Nearly a quarter of the papers in this volume use digital topographic data in some form.

Physics of Geomorphic Processes

As quantitative models of landscape evolution were developed in the 1990s (Willgoose et al., 1991; Chase, 1992; Beaumont et al., 1992; Tucker and Slingerland, 1994), it became clear that progress was limited by the theoretical understanding of the principal surface processes. The representation of tectonic processes and the numerical algorithms used in geodynamic models and landscape evolution models (Braun and Sambridge, 1997; Tucker et al., 2001; Tomkin and Braun, 2002) have grown increasingly sophisticated, but the representation of geomorphic processes has lagged behind. In response, surface processes thought to be important at the landscape scale have been given increasing attention. Much emphasis has been placed on fluvial bedrock incision (Howard et al., 1994; Whipple and Tucker, 1999; Sklar and Dietrich, 2001; Hartshorn et al., 2002). This process is thought to be crucial to landscape evolution because it sets the local base level for hillslope processes, for example, providing the rate control for shallow and deep-seated landslides, thereby governing the timescale of evolution for both channel and hillslope processes. Bedrock landsliding has itself garnered considerable attention in recent years, as it has become clear that this is the principal hillslope process in tectonically active mountain landscapes (Burbank et al., 1996; Hovius et al., 1997). Although simple quantitative paradigms for key erosion processes are now firmly in place, many questions remain about the mechanics, patterns, and rates of these processes, and about the mechanisms and scales of their interactions. In this volume, papers by Gasparini et al. (Chapter 8) and Herman and Braun (Chapter 11) are contributions that quantify and parameterize geomorphic processes, and provide important progress toward building more complete surface processes models.

Critical Orogenic Wedges

The influence of erosion on tectonics is plausible in theory, if somewhat problematic to demonstrate in practice. Critical convergent orogens are a tectonic setting where the response to erosion is predictable, immediate, and large. These orogens are defined by mountain belts that attain a critical topographic slope forming an orogenic wedge with a fixed taper angle (Davis et al., 1983). Under critical conditions, an orogenic wedge is everywhere at or near plastic failure, which implies that deformation is very sensitive to small changes in stress as occur, for example, by mass redistribution by erosion or sediment deposition. The consequence of erosion is thus a reorganization of deformation, internal kinematics, and patterns of uplift to restore the critical taper of a wedge (Willett, 1999). This is a clear example of coupling with feedback, in which erosion affects crustal deformation and rock uplift, which, in turn, affects erosion rates. Under simple conditions, the response of a critical frictional wedge to erosion can be treated analytically (Hilley and Strecker, 2004; Whipple and Meade, 2004). In this volume, the problem of an eroding critical wedge is addressed through the use of analog models (Hoth et al., Chapter 12), analytical solutions (Roe et al., Chapter 13), and numerical models (Stolar et al., Chapter 14).

From Weather to Climate

The influence of climate on erosion processes, landscape evolution, and orogen dynamics is suggested by the close match in some areas of localized precipitation and localized erosion (e.g., Reiners et al., 2003). As advanced landscape evolution models demonstrate important sensitivity to the spatial and temporal distribution of precipitation (Lague et al., 2005), researchers have turned to more complex climate models or even mesoscale meteorological models to drive erosion. In recent decades, vast resources have been invested in the modeling of weather and climate, and leading models deliver detailed constraints on many atmospheric attributes and their changes. However, precipitation in these models varies over a range of timescales and it is not clear how this variability impacts erosion rates over long timescales. It is therefore difficult to define and quantify climate in a way that is meaningful to the evolution of topography. In many ways, this problem parallels the challenge of relating modern topography or modern erosion rates to long-term geologic processes. In this volume, Smith (Chapter 1) reviews a new model that provides a simplified prediction of how stable atmospheric flow interacts with topography to produce precipitation. Barros et al. (Chapter 2) and Anders et al. (Chapter 3) report direct and remotely sensed measurements that resolve patterns of precipitation and their relationship to topography in the Himalaya. Interestingly, they observe that precipitation patterns are correlated with topography on scales of tens of kilometers, suggesting that topography and climate may coevolve through time, even at limited spatial scales. The linear theory of Smith (Chapter 1) for orographic precipitation predicts this short-wavelength correlation between topography and precipitation.

Thermochronometry and Cosmogenic Dating

Advances in low-temperature thermochronometry and exposure dating with cosmogenic isotopes have greatly increased our ability to resolve the timing and rates of erosional processes (von Blanckenburg, 2005; Reiners and Brandon, 2006). New techniques in (U-Th)/He dating of apatite and zircon have increased the temperature range and precision of mineral cooling histories (Farley, 2002). When added to fission-track dating of zircon and apatite and $^{40}Ar/^{39}Ar$ dating of micas and feldspars, rock cooling histories can now be resolved from 350 °C to 70 °C. Exhumation can now be tracked from mid-crustal depth to within a couple of kilometers below the Earth's surface, where they are continued by cosmogenic studies, which provide near-surface measurements of erosion rates. In addition, statistical methods have been developed for evaluating grain age distributions in modern or ancient sediments, providing catchment-wide estimates of (paleo-)erosion rates (Brandon, 1996; Granger et al., 1996). Brewer et al. (Chapter 20) provide an example of this approach, using $^{40}Ar/^{39}Ar$ dating of muscovite from modern sediments to resolve the spatial variability of erosion rates in drainage basins in the Himalaya.

Paleoaltimetry

The ability to estimate the elevation of Earth's land surface back through geologic time has advanced considerably in recent years with the development of stable isotope paleoaltimetry, based on the fractionation of oxygen and hydrogen during Rayleigh distillation in the atmosphere (Poage and Chamberlain, 2001; Rowley et al., 2001; Rowley and Currie, 2006). These new techniques hold great promise to resolve the timing of surface uplift and the formation of high topography. In this volume, Sjostrom et al. (Chapter 19) report isotopic compositions of smectites across the western U.S. and use these results to argue that the Rocky Mountains have been elevated since at least 50 Ma.

REFERENCES CITED

Anders, A.M., Roe, G.H., Hallet, B., Montgomery, D.R., Finnegan, N.J., and Putkonen, J., 2006, this volume, Spatial patterns of precipitation and topography in the Himalaya, *in* Willett, S.D., Hovius, N., Brandon, M.T., and Fisher, D.M., eds., Tectonics, Climate, and Landscape Evolution: Geological Society of America Special Paper 398, Penrose Conference Series, p. 39–53, doi: 10.1130/2006.2398(03).

Barros, A.P., Chiao, S., Lang, T.J., Burbank, D., and Putkonen, J., 2006, this volume, From weather to climate—Seasonal and interannual variability of storms and implications for erosion processes in the Himalaya, *in* Willett, S.D., Hovius, N., Brandon, M.T., and Fisher, D.M., eds., Tectonics, Climate, and Landscape Evolution: Geological Society of America Special Paper 398, Penrose Conference Series, p. 17–38, doi: 10.1130/2006.2398(02).

Beaumont, C., Fulsack, P., and Hamilton, J., 1992, Erosional control of active compressional orogens, *in* McClay, K., ed., Thrust Tectonics: New York, Chapman and Hall, p. 1–18.

Brandon, M.T., 1996, Probability density plots for fission-track grain age distributions: Radiation Measurements, v. 26, p. 663–676, doi: 10.1016/S1350-4487(97)82880-6.

Braun, J., and Sambridge, M.S., 1997, Modelling landscape evolution on geological time scales: a new method based on irregular spatial discretization: Basin Research, v. 9, p. 27–52, doi: 10.1046/j.1365-2117.1997.00030.x.

Brewer, I.D., Burbank, D.W., and Hodges, K.V., 2006, this volume, Downstream development of a detrital cooling-age signal: Insights from ^{40}Ar/^{39}Ar muscovite thermochronology in the Nepalese Himalaya, *in* Willett, S.D., Hovius, N., Brandon, M.T., and Fisher, D.M., eds., Tectonics, climate, and landscape evolution: Geological Society of America Special Paper 398, Penrose Conference Series, p. 321–338, doi: 10.1130/2006.2398(20).

Burbank, D.W., Leland, J., Fielding, E., Anderson, R.S., Brozovic, N., Reid, M.R., and Duncan, C., 1996, Bedrock incision, rock uplift and threshold hillslopes in the northwestern Himalayas: Nature, v. 379, p. 505–510, doi: 10.1038/379505a0.

Chase, C.G., 1992, Fluvial landsculpting and the fractal dimension of topography: Geomorphology, v. 5, p. 39–57, doi: 10.1016/0169-555X(92)90057-U.

Davis, D., Suppe, J., and Dahlen, F.A., 1983, Mechanics of fold-and-thrust belts and accretionary wedges: Journal of Geophysical Research, v. 88, p. 1153–1172.

Farley, K.A., 2002, (U-Th)/He dating: Techniques, calibrations, and applications: Reviews in Mineralogy and Geochemistry, v. 47, p. 819–844.

Gasparini, N.M., Bras, R.L., and Whipple, K.X., 2006, this volume, Numerical modeling of non–steady-state river profile evolution using a sediment-flux-dependent incision model, *in* Willett, S.D., Hovius, N., Brandon, M.T., and Fisher, D.M., eds., Tectonics, climate, and landscape evolution: Geological Society of America Special Paper 398, Penrose Conference Series, p. 127–141, doi: 10.1130/2006.2398(08).

Granger, D.E., Kirchner, J.W., and Finkel, R.C., 1996, Spatially averaged long-term erosion rates measured from in situ–produced cosmogenic nuclides in alluvial sediment: The Journal of Geology, v. 104, p. 249–257.

Hartshorn, K., Hovius, N., Dade, W.B., and Slingerland, R., 2002, Climate-driven bedrock incision in an active mountain belt: Science, v. 297, p. 2036–2038, doi: 10.1126/science.1075078.

Herman, F., and Braun, J., 2006, this volume, A parametric study of soil transport mechanisms, *in* Willett, S.D., Hovius, N., Brandon, M.T., and Fisher, D.M., eds., Tectonics, Climate, and Landscape Evolution: Geological Society of America Special Paper 398, Penrose Conference Series, p. 191–200, doi: 10.1130/2006.2398(11).

Hilley, G.E., and Strecker, M.R., 2004, Steady state erosion of critical Coulomb wedges with applications to Taiwan and the Himalaya: Journal of Geophysical Research, v. 109, p. doi: 10.1029/2002jb002284.

Hoth, S., Adam, J., Kukowski, N., and Oncken, O., 2006, this volume, Influence of erosion on the kinematics of bivergent orogens: Results from scaled sandbox simulations, *in* Willett, S.D., Hovius, N., Brandon, M.T., and Fisher, D.M., eds., Tectonics, Climate, and Landscape Evolution: Geological Society of America Special Paper 398, Penrose Conference Series, p. 201–225, doi: 10.1130/2006.2398(12).

Hovius, N., Stark, C.P., and Allen, P.A., 1997, Sediment flux from a mountain belt derived by landslide mapping: Geology, v. 25, p. 231–234, doi: 10.1130/0091-7613(1997)025<0231:SFFAMB>2.3.CO;2.

Howard, A.D., Seidl, M.A., and Dietrich, W.E., 1994, Modeling fluvial erosion on regional to continental scales: Journal of Geophysical Research, v. 99, p. 13,971–13,986, doi: 10.1029/94JB00744.

Lague, D., Hovius, N., and Davy, P., 2005, Discharge, discharge variability, and the bedrock channel profile: Journal of Geophysical Research–Earth Surface, v. 110, F04006, doi: 10.1029/2004JF000259.

Merritts, D., and Ellis, M.A., 1994, Introduction to special section on tectonics and topography: Journal of Geophysical Research, v. 99, p. 12135–12141, doi: 10.1029/94JB00810.

Poage, M.A., and Chamberlain, C.P., 2001, Empirical relationships between elevation and the stable isotope composition of precipitation and surface waters: considerations for studies of paleoelevation change: American Journal of Science, v. 301, p. 1–15.

Reiners, P.W., and Brandon, M.T., 2006, Using thermochronology to understand orogenic erosion: Annual Review of Earth and Planetary Sciences, v. 34, p. 419–466, doi: 10.1146/annurev.earth.34.031405.125202.

Reiners, P.W., Ehlers, T.A., Mitchell, S.G., and Montgomery, D.R., 2003, Coupled spatial variations in precipitation and long-term erosion rates across the Washington Cascades: Nature, v. 426, p. 645–647.

Roe, G.H., Stolar, D.B., and Willett, S.D., 2006, this volume, Response of a steady-state critical wedge orogen to changes in climate and tectonic forcing, *in* Willett, S.D., Hovius, N., Brandon, M.T., and Fisher, D.M., eds., Tectonics, climate, and landscape evolution: Geological Society of America Special Paper 398, Penrose Conference Series, p. 227–239, doi: 10.1130/2005.2398(13).

Rowley, D.B., and Currie, B.S., 2006, Palaeo-altimetry of the late Eocene to Miocene Lunpola basin, central Tibet: Nature, v. 439, p. 677–681, doi: 10.1038/nature04506.

Rowley, D.B., Pierrehumbert, R.T., and Currie, B.S., 2001, A new approach to stable isotope-based paleoaltimetry: implications for paleoaltimetry and paleohypsometry of the High Himalaya since the Late Miocene: Earth and Planetary Science Letters, v. 5836, p. 1–17.

Sjostrom, D.J., Hren, M.T., Horton, T.W., Waldbauer, J.R., and Chamberlain, C.P., 2006, this volume, Stable isotopic evidence for a pre–late Miocene elevation gradient in the Great Plains–Rocky Mountain region, USA, *in* Willett, S.D., Hovius, N., Brandon, M.T., and Fisher, D.M., eds., Tectonics, climate, and landscape evolution: Geological Society of America Special Paper 398, Penrose Conference Series, p. 309–319, doi: 10.1130/2006.2398(19).

Sklar, L.S., and Dietrich, W.E., 2001, Sediment and rock strength controls on river incision into bedrock: Geology, v. 29, p. 1087–1090, doi: 10.1130/0091-7613(2001)029<1087:SARSCO>2.0.CO;2.

Smith, R.B., 2006, this volume, Progress on the theory of orographic precipitation, *in* Willett, S.D., Hovius, N., Brandon, M.T., and Fisher, D.M., eds., Tectonics, Climate, and Landscape Evolution: Geological Society of America Special Paper 398, Penrose Conference Series, p. 1–16, doi: 10.1130/2006.2398(01).

Stolar, D.B., Willett, S.D., and Roe, G.H., 2006, this volume, Climatic and tectonic forcing of a critical orogen, *in* Willett, S.D., Hovius, N., Brandon, M.T., and Fisher, D.M, eds., Tectonics, climate, and landscape evolution: Geological Society of America Special Paper 398, Penrose Conference Series, p. 241–250, doi: 10.1130/2006.2398(14).

Tomkin, J.H., and Braun, J., 2002, The effect glaciation has on the relief of a fast growing orogen: a numerical modelling study: American Journal of Science, v. 302, p. 169–190.

Tucker, G.E., Lancaster, S.T., Gasparini, N.M., and Bras, R.L., 2001, The Channel-Hillslope Integrated Landscape Development (CHILD) Model, *in* Harmon, R.S., and Doe, W.W., eds., Landscape Erosion and Evolution Modeling: Kluwer Academic/Plenum Publishers, p. 349–388.

Tucker, G.E., and Slingerland, R., 1994, Erosional dynamics, flexural isostasy, and long-lived escarpments: A numerical modeling study: Journal of Geophysical Research, v. 99, p. 12229–12243, doi: 10.1029/94JB00320.

von Blanckenburg, F., 2005, The control mechanisms of erosion and weathering at basin scale from cosmogenic nuclides in river sediment: Earth and Planetary Science Letters, v. 237, p. 462–479, doi: 10.1016/j.epsl.2005.06.030.

Whipple, K.X., and Meade, B.J., 2004, Controls on the strength of coupling among climate, erosion, and deformation in two-sided, frictional orogenic wedges at steady state: Journal of Geophysical Research–Earth Surface, v. 109, F01011, p. doi:10.1029/2003JF000019.

Whipple, K.X., and Tucker, G.E., 1999, Dynamics of the stream-power river incision model: Implications for height limits of mountain ranges, landscape response timescales, and research needs: Journal of Geophysical Research, v. 104, p. 17,661–17,674, doi: 10.1029/1999JB900120.

Willett, S.D., 1999, Orogeny and orography: The effects of erosion on the structure of mountain belts: Journal of Geophysical Research, v. 104, p. 28957–28981, doi: 10.1029/1999JB900248.

Willgoose, G., Bras, R.L., and Rodriguez-Iturbe, I., 1991, A coupled channel network growth and hillslope evolution model, 1, Theory: Water Resources Research, v. 27, p. 1671–1684, doi: 10.1029/91WR00935.

Wobus, C., Whipple, K.X., Kirby, E., Snyder, N., Johnson, J., Spyropolou, K., Crosby, B., and Sheehan, D., 2006, this volume, Tectonics from topography: Procedures, promise, and pitfalls, *in* Willett, S.D., Hovius, N., Brandon, M.T., and Fisher, D.M., eds., Tectonics, Climate, and Landscape Evolution: Geological Society of America Special Paper 398, Penrose Conference Series, p. 55–74, doi: 10.1130/2006.2398(04).

Geological Society of America
Special Paper 398
2006

Progress on the theory of orographic precipitation

Ronald B. Smith[†]
Department of Geology and Geophysics, Yale University, New Haven, Connecticut 06520, USA

ABSTRACT

This paper presents a rèview of recent progress on the theory of orographic precipitation and a discussion of the role of preexisting atmospheric disturbances, especially their strong water vapor fluxes. I also introduce the basic elements of stable moist airflow dynamics and cloud physics, and a new linear theory of orographic precipitation. The theory is tested against two types of data: a single event of Alpine precipitation and the annual climatology of the Oregon coastal ranges. Different methods are used to determine the free "cloud-delay" parameters in the theory, including a statistical analysis of data from conventional rain gauges and isotope analysis of stream samples. The surprising threshold behavior of nonlinear accretion-dominated cloud physics is displayed. Finally, I consider the impact of scale-dependent precipitation patterns on erosion and terrain evolution.

Keywords: mountains, rain, orographic precipitation, clouds.

1. INTRODUCTION

While the influence of terrain on precipitation is widely appreciated, the explanations of the processes involved are often oversimplified. One inaccurate assertion is that mountains can produce precipitation by themselves, acting in normal "climatological" circumstances. This idea is mostly, but not entirely, untrue. Mountains have their most profound influence on precipitation during brief events when significant pre-existing atmospheric disturbances move into mountainous areas. In other words, mountains usually modify, and often amplify, rather than create precipitation. Here are a few examples on both sides of this issue.

First, consider two examples of orographic precipitation that do not require a preexisting disturbance. On the eastern slopes of Hawaii's Big Island, facing the easterly trade winds, bands of precipitation form daily in the summer and move onshore, triggered in part by the airflow blocking effect of the high volcanic mountains (Wang et al., 2000). Four thousand miles to the east, in the summertime Rocky Mountains in Colorado, daily solar heating of the mountain slopes induces thermal updrafts that evolve into precipitating thunderstorms by mid-afternoon (Raymond and Wilkening, 1982; Banta and Barker Schaaf, 1987). These predictable daily occurrences suggest that these mountains require only average seasonal conditions to generate precipitation.

Mountains can also act systematically to suppress precipitation. In regions with steady prevailing winds, the leeward slopes of mountain ranges usually experience descending air in which clouds or precipitation cannot occur. Solar heating effects can also suppress cloud formation. An example is the Mesopotamian valley of Iraq, adjacent to the high Iranian Plateau. In the summer, solar heating of the plateau generates a circulation with a descending air over Mesopotamia, preventing cloud formation and precipitation there (Evans et al., 2004).

The influence of mountains in modifying precipitation from preexisting weather disturbances is very widespread and found in all climate zones. In mid-latitude, cool-season climates, precipitation is generally controlled by the "weather cycle," i.e., the quasi-periodic passing of frontal cyclones. With intervals from 4 to 7 d, these cyclones produce precipitation wherever they occur, but the patterns and amount of precipitation can be strongly modulated by terrain. For example, Pacific Ocean cyclones crossing the mountainous western coast of North America produce copious rainfall events generating floods and landslides. As the cyclones move into the intermountain west and the Rocky Mountains, they produce deep snowfalls that generate mountain glaciers

[†]E-mail: ronald.smith@yale.edu.

Smith, R.B., 2006, Progress on the theory of orographic precipitation, *in* Willett, S.D., Hovius, N., Brandon, M.T., and Fisher, D.M., eds., Tectonics, Climate, and Landscape Evolution: Geological Society of America Special Paper 398, Penrose Conference Series, p. 1–16, doi: 10.1130/2006.2398(01). For permission to copy, contact editing@geosociety.org.

and avalanches (Rauber and Grant, 1986). Between these brief storm events, little precipitation occurs. Later in the spring, when warm-weather episodes melt the mountain snowpack, bursts of river discharge cause further erosion and landscape alteration.

Analogous events can be found in other mid-latitude regions. Nor'easters approaching New England from the south bring strong, moist, easterly airflows over the White and Green Mountains, the Adirondacks, Berkshires, and Catskills. The precipitation patterns are strongly modified by these hills (Passarelli and Boehme, 1983). In Europe, cyclones passing north of the Alps bring snowfall to the northern slopes, while cyclones passing along the Mediterranean Sea bring heavy rains to the southern Alpine slopes. The average or "prevailing" westerly winds have little to do with these events (Buzzi et al., 1998).

Consider now the dry subtropics. Under average conditions in these latitudes, descending air in the Hadley Cell prevents cloud formation. Precipitation is rare. The prevailing dryness is spectacularly reversed when a hurricane forms. When a drifting hurricane hits high terrain, the patterns of precipitation in the cyclone are profoundly altered. Without such disturbances, the mountains produce little precipitation.

A region with frequent hurricanes (i.e., typhoon) impacting high terrain is Taiwan. There, typhoons usually approach from the east (Lin et al., 1999). If the storm center passes south of the Central Mountain Range, the counterclockwise typhoon circulation creates easterlies that hit the eastern slopes. Severe rain rates, flooding, and landslides will occur there (Fig. 1). On the western (lee side) slopes, the air descent is so strong that the air will dry and clouds will disappear. Surprisingly, so intense is the local orographic downslope effect that a dry foehn can exist inside the typhoon.

On other occasions, the typhoon center may pass north of the Central Mountain Range. As the storm with its cyclonic winds moves away toward the west, strong westerlies hit the Central Mountain Range with rain on the western slopes and dry eastern slopes.

In the cool seasons, Taiwan can be impacted by frontal cyclones as well. During the January 2003 GSA Penrose Conference in Taiwan, a strong winter frontal system hit the island while the attendees drove from west to east across the Central Mountain Range. Heavy rain changed to snow as they ascended the mountain roads. Beyond the ridgeline, the skies cleared. A view back toward the west showed the classic foehn wall cloud over the crest, marking the beginning of dry descent. These interesting meteorological events captured the attention of the rock hammer–wielding geologists briefly before they returned to their investigations of faults, metamorphic grades and landforms, seeking clues to the long-term effects of such recurring weather events.

Like Taiwan, the massive Himalayan front in Asia experiences precipitation events in both seasons. In the summer, the monsoon onset vortex brings heavy rains to the foothills of the

Figure 1. Typhoon Toraji over Taiwan on 30 July 2001. This three-band false-color satellite image is from the SeaWiFS sensor (courtesy of National Aeronautics and Space Administration [NASA]). Taiwan is obscured by high clouds, but the coast of China is evident. This storm caused several deaths, numerous landslides, and rapid erosion on the steep slopes of the Central Mountain Range.

Himalayas, while the high Tibetan Plateau is mostly dry (Barros and Lang, 2003). In winter, frontal storms push snow further up into the high terrain.

The dominance of disturbances can also be illustrated in the Southern Hemisphere. Both the New Zealand Southern Alps (Wratt et al., 1996) and the southern Andes lie squarely in the belt of eastward-drifting frontal cyclones. The wet-dry contrast across both ranges is marked, but the precipitation on the windward slopes is intermittent, not continuous. The precipitation events on these ranges are not responding directly to the average prevailing winds, so much as to the impact of discrete moist frontal cyclones.

Even in Antarctica, it is the frontal cyclones translating eastward along the coast that produces moist northerly airflows, carrying water vapor up over the ice sheet where it condenses and precipitates. The orographic lifting enhances the lifting already present in the cyclone to increase the total precipitation rate. The climatological mean winds tell us little about these processes. At low levels, the mean flows near the Antarctic coast are southerly (i.e., downslope), as katabatic winds blow off the continent.

To summarize the connection between meteorological disturbances and orographic precipitation, we should focus on how these disturbances temporarily alter the regional environment. A critical factor is the enhanced horizontal water vapor flux in strong moist winds. As discussed later, orographic precipitation is roughly proportional to the horizontal water vapor flux. The sensitivity to wind speed is amplified by the fact that strong moist winds are better able to climb over high mountains, instead of flowing around. Also important is the increased relative humidity in fronts and cyclones, reducing the amount of lifting needed to produce condensation. Finally, the preexisting rain or snow in an atmospheric disturbance may improve the efficiency with which new orographically generated condensate is converted to precipitation.

The role of weather disturbances in orographic precipitation has implications in climate and terrain reconstructions over geologic time. It implies that wet and dry climates depend on storm tracks more than on mean flows and that brief rain events rather than continuous rains usually control climate, vegetation, and landform processes.

In the following sections, I discuss several key issues in the theory of orographic precipitation. Additional background information can be found in reviews by Smith (1979), Banta (1990), and Barros and Lettenmaier (1994).

2. MOIST AIRFLOW DYNAMICS

The basic thermodynamics of orographic precipitation is firmly established. As air is forced to rise over sloping terrain, the air parcels expand and cool. As the air temperature drops, the relative humidity increases and eventually water vapor must condense. Usually, the condensation occurs on numerous dust particles to form small cloud droplets. After the conversion of cloud droplets to larger hydrometeors (i.e., rain or snow), these particles fall under the influence of gravity to form precipitation (Wallace and Hobbs, 1977).

The starting point for the analysis of orographic precipitation is the study of moist airflow dynamics and the realization that the nature of the terrain-induced ascent depends on the width (a) of the hill (e.g., Smith, 1979). For narrow hills (i.e., $a < 1$ km), the vertical penetration (z) of the forced ascent is limited to about $z = a$. For wider hills, two other factors enter the airflow problem: density stratification and latent heat release.

A measure of the effect of density stratification is the stability frequency (N) given by

$$N^2 = \frac{g}{T}[\gamma - \Gamma], \tag{1}$$

where g is acceleration due to gravity, γ and Γ are the actual and adiabatic lapse rates, respectively. Under typical tropospheric conditions $\gamma = -0.0065$ °C/m, $\Gamma = -0.009.8$ °C/m and $T = 273$ K, so $N \sim 0.01$ s^{-1}. The natural buoyant oscillation period of the air is $T = 2\pi/N \approx 600$ s = 10 min. With air parcels wanting to return to their initial altitude within ten minutes, simple ascent forced by sloping terrain is converted into a more complex gravity wave. Mountain-generated gravity waves have been extensively studied over the past 50 yr (e.g., Smith, 1979; Wurtele et al., 1996). Gravity wave flow fields are characterized by upwind tilting patterns of airflow disturbances, such as those shown in Figure 2. The strength and penetration depth of the forced vertical motion is limited by the phase tilt in these waves. The approximate penetration depth of the vertical motion, for simple hydrostatic flow, is

$$z = U/N. \tag{2}$$

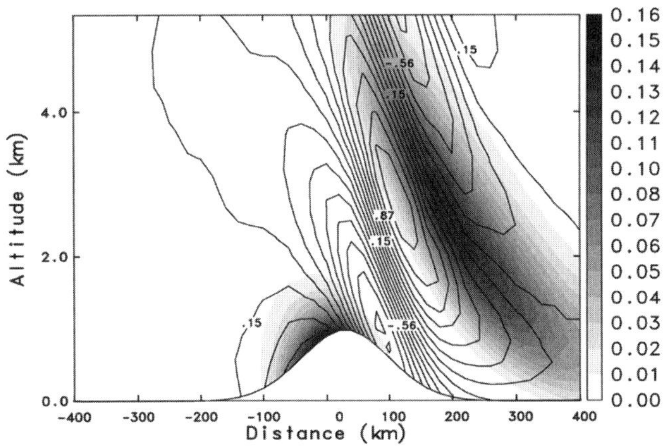

Figure 2. Vertical cross section of moist airflow over an idealized mountain shape. The control parameters are: stability $N = 0.011$ s^{-1}, wind speed $U = 10$ ms^{-1}, mountain width $a = 10$ km, mountain height $h_m = 1000$ m. Airflow is from left to right. The field of vertical velocity is contoured with an interval of 0.086 m/s. The snow mixing ratio is shaded with interval 0.01 g/kg. Note the upwind tilted patterns and the limited vertical penetration of the forced ascent above the windward slope. (Image is from Jiang and Smith, 2003.)

Using wind and stability values, $U = 10$ m/s and $N = 0.01$ s^{-1}, this penetration height is ~1000 m. Air passing over the hill above that level (say at 2000 m) will not feel the terrain forced ascent, or may even descend in the mountain-generated gravity wave (Fig. 2).

For very high terrain, the approaching flow may be decelerated and forced to split and flow around the mountain. This can profoundly reduce the amount of forced ascent. The accepted condition for flow splitting uses scaling laws based on the ambient flow speed (U), the stability frequency (N), and the mountain height (h). Roughly, if the nondimensional mountain height

$$M = hN/U > 1, \quad (3)$$

the air will tend to flow around rather than over the hill. For $U = 10$ m/s and $N = 0.01$ s^{-1}, hills higher than 1 km will exceed this limit and flow splitting will result. The approach to stagnation and flow splitting for a circular Gaussian hill is shown in Figure 3. As M increases toward the critical value ($M = 1.3$ in this case), the minimum wind speed in the windward region drops to zero. Beyond $M = 1.3$, the low-level flow splits and flows around the hill.

The above estimates neglect the effect of latent heat release caused by water vapor condensation during ascent. It has been suggested by several authors that the flow splitting condition (equation 3) might still apply if the adiabatic lapse rate Γ was replaced by the moist adiabatic lapse rate Γ_m, and the dry stability frequency N is replaced by N_m, the moist stability frequency (e.g., Durran and Klemp, 1982; Barcilon and Fitzjarrald, 1985). Typical values of N_m are significantly less than N, often only one-half to one-third as large (e.g., $N_m \sim 0.005$ s^{-1}). This substitution in equation 3 predicts that moist air could rise over hills two or three times as high as dry airflow could. And, using the example above, the penetration of forced ascent could increase to 2 or 3 km. If the depth of the moist layer is $H_w = 3$ km, the deeper penetration of forced ascent will catch a much bigger fraction of the moisture flux across the terrain.

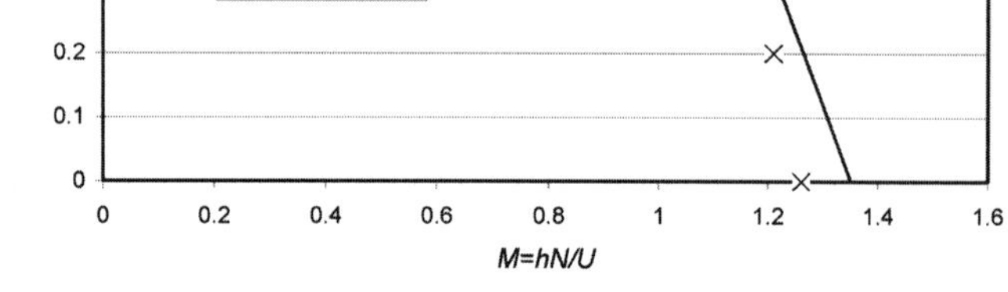

Figure 3. This diagram shows how the wind upstream of a mountain is decelerated depending on the nondimensional mountain height (M). The deceleration ratio, (u/U) was determined from a full three-dimensional nonlinear numerical model of airflow dynamics. The solid curve is for a dry atmosphere. The two sets of symbols repeat the calculation for a moisture-saturated atmosphere. The results collapse to a single curve when M is computed with the moist stability frequency. (Data are from Jiang, 2003.)

The validity of the moist stability substitution has been most carefully studied by Jiang (2003) in the context of orographic precipitation. Jiang showed that the onset of flow nonlinearity and flow splitting could be predicted by equation 3 with N replaced by N_m (Fig. 3). The figure shows the onset of stagnation for dry airflow, and two cases of saturated airflow with different dry stabilities. The critical value of $M = 1.3$ is maintained.

Jiang also showed that substantial regions of dry lee-side descent would occur, but that they did not seriously impact the use of moist stability in flow field computations for the upwind region. This confirmation opens up the possibility of analytic treatments of airflow associated with orographic precipitation (see section 4) without the patching of wet-dry regions proposed by Barcilon and Fitzjarrald (1985).

The influence of latent heating on airflow can take another form when the mountains are high. According to Doyle and Smith (2003), latent heating over a relatively small depth can tune the atmosphere for a nonlinear resonance that leads to strong descent in the lee. This result may solve the old conundrum of how air, heated by water vapor condensation, can be forced to descend on the lee slopes causing foehn.

3. CONDENSED WATER ADVECTION, CONVERSION, FALLOUT, AND EVAPORATION

The second key element in the physics of orographic precipitation is the behavior of the various categories of condensed water. The small cloud droplets, formed by vapor condensation in rising air, will be carried downwind by the ambient flow while undergoing a conversion to larger particles or "hydrometeors" (Uttal et al., 1988; Yau and Rogers, 1989; Yuter and Houze, 2003). Hydrometeors are defined as particles large enough that they have a significant terminal fall speed under the influence of gravity. A typical hydrometeor (i.e., a raindrop or snowflake) has a mass a million times greater than a cloud droplet, so the conversion is a nontrivial step in the formation of precipitation. Current estimates of the time (τ_c) necessary for conversion range from 200 to 2000 s, depending on a wide variety of physical factors, such as temperature, aerosol content, and turbulence.

Once hydrometeors have formed, they fall to the ground with their terminal velocities (V_t). An estimate of $V_t = 2$ m/s is appropriate for snow, while $V_t = 6$ m/s is appropriate for rain. The fallout time (τ_f) depends also on the height from which hydrometeors must fall (H) according to $\tau_f = H/V_t$. With cloud heights varying from 1 to 5 km, values of τ_f can vary widely from 166 s to 2500 s. Colder conditions, with snow, will favor the larger values. Often in mid-latitudes the temperature profile will favor snow aloft, but rain at the surface. The falling snowflakes will melt when they reach the 0 °C level and accelerate to a faster terminal speed.

These conversion and fallout time scales are of primary importance, because the lifting caused by terrain is temporary. After air crosses the ridgeline, descent dominates. Any condensed water that has not converted and precipitated before descent begins is subject to evaporation. A quantitative example can illustrate the point. Consider a ridge with half-width of $a = 10$ km in a wind of $U = 10$ m/s. Under these conditions, the time it takes a parcel to travel from the foothill to the crestline is $T = a/U = 1000$ s. With a cloud-delay time of $\tau_c = \tau_f = 200$ s, the conversion and fallout would have been well completed before descent begins. Conversely, if $\tau_c = \tau_f = 2000$ s, only a small fraction of the condensed water could precipitate. The rest of the condensed water will evaporate on the lee side and return to the vapor state.

The cloud-delay processes also control the amount of precipitation that spills over onto the lee slopes. Lee slope precipitation is significant because it falls into a different watershed than the windward slope precipitation. A few kilometers of downwind shift in precipitation can be amplified to hundreds of kilometers by the terrain-controlled river network.

Certainly the best way to appreciate the role of cloud time scales is to observe a lenticular cloud; one of the most beautiful sights in nature (Fig. 4). In the lenticular cloud, orographic lifting condenses vapor to form cloud droplets or tiny ice crystals, but there is insufficient time for conversion. When the air descends, the warming air evaporates the condensate according to reversible thermodynamics. No precipitation occurs. The air mass has not been dried. Note that the image in Figure 4 can be reversed left to right without changing its basic nature, suggesting this thermodynamic reversibility.

As a prototype physical model of cloud physics, I consider the pair of steady-state condensate advection equations proposed by Smith (2003). The horizontal advection of vertically integrated condensed water is written as

$$\vec{U} \cdot \nabla q_c = S(x,y) - q_c / \tau_c, \tag{4a}$$

$$\vec{U} \cdot \nabla q_f = q_c / \tau_c - q_f / \tau_f, \tag{4b}$$

where τ_c and τ_f are the characteristic time scales for cloud water conversion and hydrometeor fallout. The term S in 4a is the source term, defined as the rate at which supersaturated water vapor is generated by ascent. When S is positive, I assume that the excess vapor is immediately condensed to form small cloud droplets. The cloud water and hydrometeor column densities (q_c, q_f) have units of kgm^{-2}. The rate of conversion of cloud water to hydrometeor (q_c/τ_c) is assumed to be proportional to the cloud water column density with a rate coefficient of τ_c^{-1}. The precipitation term in equation 4b [$P(x,y) = q_f/\tau_f$] is proportional to the hydrometeor column density with a rate coefficient of τ_f^{-1}. I assume that the cloud parameters τ_c and τ_f are constant throughout the region. The symbol $\vec{U}$ is the regionally averaged horizontal wind vector, with eastward and northward components U and V. The vector product of the wind and gradient vectors can be written in Cartesian form $\vec{U} \cdot \nabla q = U \partial q / \partial x + V \partial q / \partial y$.

A simple form for the condensed water source term was proposed by Smith (1979):

$$S(x,y) = \rho q_v \vec{U} \cdot \nabla h(x,y), \tag{5}$$

under the assumptions that the vertical air motion is independent of height, the air is saturated with vapor, and the temperature sounding follows a saturated moist adiabat. The factors ρ and q_v are the air density and specific humidity at Earth's surface. According to equation 5, the condensation rate is proportional to the terrain slope and wind speed. This model is referred to as the raw upslope model, if it is also assumed that conversion and fallout are instantaneous ($\tau_c = \tau_f = 0$) so that equation 4 gives $P(x,y) = S(x,y)$.

In equation 4, as forced ascent drives S positive, the amount of cloud water increases downwind. The conversion term in equation 4 then acts to decrease cloud water and generate rain water. The precipitation term in equation 4b is the final sink to the system. S is negative in downslope regions. Persistent negative S can evaporate cloud water and rain water created upwind and prevent further precipitation.

One special solution of equation 4 can be found by noting that the advection term on the left-hand side scales like q/τ_a, where the advection time scale is $\tau_a = a/U$ and a is the characteristic mountain width. If the cloud conversion is slow (i.e., $\tau_c \gg \tau_a$), equations 4 and 5 give $q_c(x,y) = \rho q_v h(x,y)$ and $q_f = 0$, indicating that the column integrated cloud water is proportional to the mountain height at each point; perhaps illustrated by the lenticular cloud in Figure 4.

4. A LINEAR THEORY OF OROGRAPHIC PRECIPITATION

Several theories of orographic precipitation have been proposed, starting with the early work of Hobbs et al. (1973), Fraser et al. (1973), and Collier (1975). A brief review of existing theories is given by Barstad and Smith (2005). I focus here on the recent linear theory of orographic precipitation by Smith and Barstad (2004). It has several good properties. It (1) is analytically tractable, so that its properties can be easily understood; (2) is applicable to actual complex terrain and arbitrary wind direction, so that it can be tested against real data; (3) reduces to the classical upslope model in equation 5, so that it can be compared with earlier work; (4) includes the basic physical elements: airflow dynamics, condensed water conversion, advection and fallout, and downslope evaporation, leading to a theory of precipitation efficiency; but (5) is limited to modest mountain heights so that $M \ll 1$ (equation 3) and drying ratio DR $\ll 1$ (see below).

Smith and Barstad (2004) pointed out that the airflow dynamics equations (not shown here) and cloud physics equations 4a and 4b can be solved together using Fourier transforms to obtain a single formula relating terrain and precipitation

photo by jane english (www.eheart.com)

Figure 4. Lenticular cloud over Mount Shasta in California. The photo was taken in February 1995 looking northeast over Lake Siskiyou. Note the left-to-right reversible aspect of the image. Photo by Jane English (www.eheart.com).

$$\hat{P}(k,l) = \frac{C_w i\sigma \hat{h}(k,l)}{[1 - imH_w][1 + i\sigma\tau_c][1 + i\sigma\tau_f]}, \quad (6)$$

with vertical wavenumber

$$m(k,l) = \left\{\left[\frac{N_m^{\,2} - \sigma^2}{\sigma^2}\right](k^2 + l^2)\right\}^{1/2},$$

and then

$$P(x,y) = \iint \hat{P}(k,l) e^{i(kx+ly)} dkdl + P_\infty\,. \quad (7)$$

Equation 6 is a transfer function relating the Fourier transforms of the terrain $h(k,l)$ and the precipitation field $\hat{P}(k,l)$. The spatial field of precipitation $P(x,y,)$can be recovered with the inverse Fourier transform in equation 7. The symbols in equations 6 and 7 are described in Table 1. The first bracket in the denominator in equation 6 describes how the source term is modified by airflow dynamics. In this factor, the vertical wavenumber $[m(k,l)]$ controls the depth and tilt of the forced air ascent. The two remaining brackets in equation 6 describe the advection of condensed water during conversion and fallout. The model reduces to the raw upslope model when $H_w = \tau_c = \tau_f = 0$ and the uplift sensitivity factor $C_w = \rho q_v$. Downwind of steep lee slopes, if negative values arise in equation 7, they are truncated with $P = \max(P,0)$. If the background precipitation P_∞ is sufficiently large, and the terrain is small, the truncation is not needed.

TABLE 1. SYMBOLS USED IN THE LINEAR MODEL

Name	Symbol	Typical values
Terrain elevation	$h(x,y)$	1–2000 m
Horizontal wavenumbers	k,l	0.001–0.0000001 m^{-1}
Intrinsic frequency	$\sigma = Uk + Vl$	0.01–0.0001 s^{-1}
Horizontal wind components	U,V	1–50 m/s
Moist stability frequency	N_m	0–0.01 s^{-1}
Water vapor scale height	H_w	1–5 km
Conversion time	τ_c	200–2000 s
Fallout time	τ_f	200–2000 s
Uplift sensitivity factor	C_w	0.001–0.02 kgm^{-3}
Background precipitation rate	P_∞	0–5 mm/h
Vertical wavenumber	$m(k,l)$	0.01–0.0001 m^{-1}

Note: From Smith and Barstad (2004).

Note that the dynamics and cloud-delay factors in the denominator of equation 6 have a similar form. The appearance of $i = \sqrt{-1}$ in each factor causes a phase shift of the pattern relative to equation 5, in addition to the amplitude change. The opposite sign in the dynamic and cloud factors is significant. The negative sign in the dynamics factor gives an upwind shift to the precipitation pattern, while the positive signs in the cloud factors cause a downwind shift (see also section 8). These factors also differ significantly in the way that horizontal wavenumber (k,l) enters the formulae for vertical wavenumber $m(k,l)$ and intrinsic frequency $\sigma(k,l)$. Note that the two cloud time scales, τ_c and τ_f are mathematically analogous.

The properties of the linear model are illustrated in Figure 5 by evaluating it for moist airflow over a triangular ridge with a height of 500 m. Due to the uniform slopes of the triangle ridge, the raw upslope model (equation 5) gives a condensation rate that is constant from ridge base to peak. If this condensed water fell instantly to the ground, the precipitation rate would be 16 mm/hr. On the lee slope, the source function (S) is negative.

When stratified airflow dynamics is included, the condensation pattern is shifted upstream, and its area integral is reduced. Its maximum value of 12 mm/hr occurs near the slope break ~15 km upstream of the crest. When cloud-delay effects are included, the precipitation pattern is shifted downwind. The maximum precipitation rate (3 mm/hr) occurs in the vicinity of the ridge crest. The near coincidence of maximum precipitation and maximum terrain elevation is often seen in nature. Here it arises from a 7 km upwind shift from wave tilt and a 15 km downwind shift due to cloud delays. The cloud delay significantly decreases the total precipitation as cloud droplets and hydrometeors are advected past the crestline to evaporate over the lee slopes.

The efficiency with which vapor can be rained out is a function of mountain scale, as discussed in section 3. The scale dependence of condensation and precipitation is shown in Figure 6. The condensation efficiency is the ratio of source S to the raw upslope value (equation 6). The precipitation efficiency is the ratio of precipitation (P) to condensation source (S). As the scale increases, the vertical motion penetrates more deeply into the atmosphere, increasing the condensation efficiency. Also, as scale increases, more condensed water is able to convert and fall out before being carried to the lee-side region of descent, increasing the precipitation efficiency. Further aspects of the theory are given by Smith and Barstad (2004).

The linear model is easy to apply to real terrain. Required only are the terrain height array $h(x,y)$ and the parameters describing the atmospheric conditions (Table 1). The terrain is transformed, equation 6 is applied, and the inverse transform is computed (equation 7). Two examples are given in Figure 7, based on a 1-km-resolution terrain array. Using the Fast Fourier Transform algorithm, only a few seconds are required to generate these fields on a computer workstation. Both examples, the Olympics in Washington State and the eastern Mediterranean coastal ranges, have terrain with multiple scales. Cloud delays of $\tau_c = \tau_f = 1000$ s were used for both examples.

In the Olympics, a smooth pattern of precipitation begins upwind over the sea. Near the terrain, the pattern responds to the four ridges that verge toward the southwest. The upwind-facing valleys between them receive much less precipitation. The lee slopes in the northeast are dry.

In the Levant, the first mountain range receives heavy precipitation while the second gets much less. Near 33°N, where the first range is much lower, the second range receives more precipitation. The shielding of the leeward by the windward range is surprising, considering that the linear model does not have a complete water vapor budget. If the ranges were separated much more widely, the linear model would predict identical precipitation amounts. The predicted precipitation patterns in these two regions have not yet been compared with rain gauge data.

5. AN ALPINE PRECIPITATION EVENT

One way to study orographic precipitation is to examine particular events (e.g., Browning et al., 1974). The study of special cases works well if the environmental parameters, such as

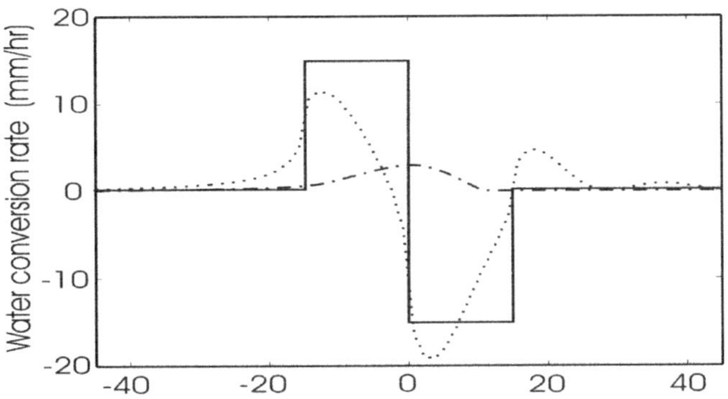

Figure 5. Precipitation rates (mm/h) over a triangle ridge with three different assumptions: solid—precipitation patterns with no airflow dynamics or cloud delays (equation 5), dotted—dynamics only, dashed—dynamics and cloud delays (equations 6 and 7). Parameters are: $a = 15$ km, $U = 15$ m/s, $N_m = 0.005$ s^{-1}, $T_0 = 280$ K, $H_w = 2500$ m, $\tau_c = \tau_f = 100$ s, $A = 500$ m, $P_\infty = 0$. Airflow is from left to right. Note the great reduction in total precipitation caused by airflow dynamics and cloud delays. (Data are from Smith and Barstad, 2004.)

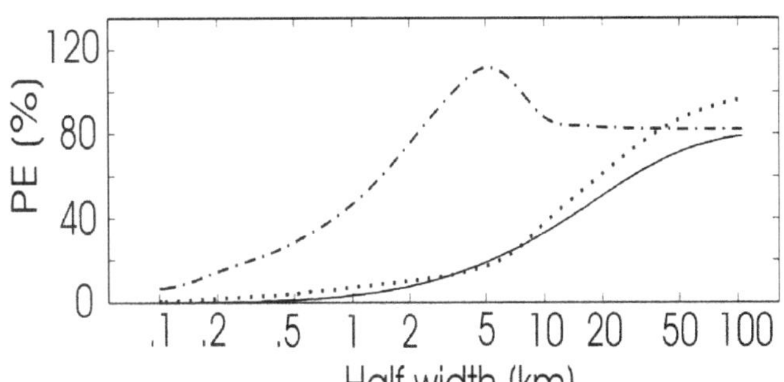

Figure 6. Three efficiency ratios (PE = precipitation efficiency) are plotted as a function of mountain half-width for a Gaussian ridge according to the linear model (equation 6). The condensation efficiency (dashed) is the ratio of the condensation to the raw upslope model (equation 5). The precipitation efficiency (dotted) is the ratio of the precipitation to the condensation. The solid curve is the ratio of precipitation to the raw upslope prediction. (Data are from Smith and Barstad, 2004.)

temperature, lapse rate, wind speed and direction, and background precipitation, can be accurately determined. Also, in real complex terrain, the predicted precipitation fields are highly structured in space and time. Therefore, a dense rain gauge network is required to validate the predicted patterns.

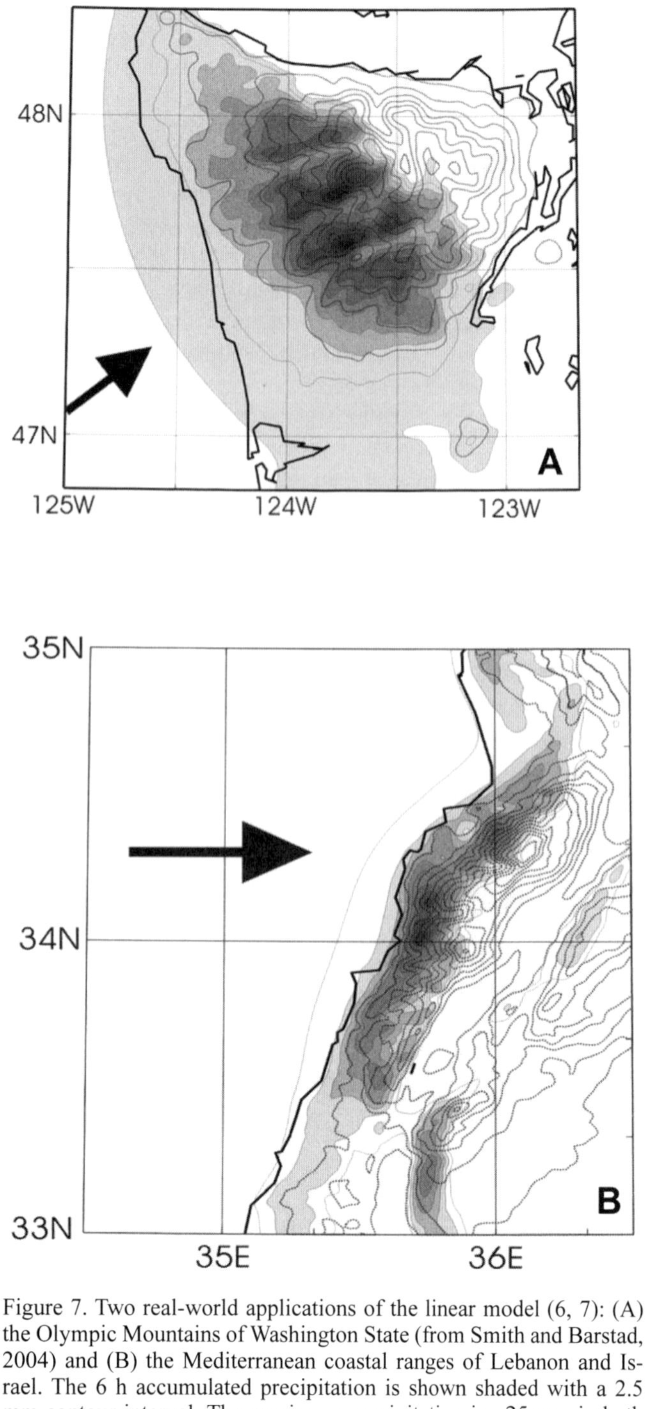

Figure 7. Two real-world applications of the linear model (6, 7): (A) the Olympic Mountains of Washington State (from Smith and Barstad, 2004) and (B) the Mediterranean coastal ranges of Lebanon and Israel. The 6 h accumulated precipitation is shown shaded with a 2.5 mm contour interval. The maximum precipitation is ~25 mm in both cases. The terrain is shown (dotted) with a 200 m contour interval. The coastline is shown with a dark solid line. Large black arrows show the wind direction.

Without any doubt, the most intensely studied event of orographic precipitation in the history of meteorology was the 20 September 1999 Lago Maggiore event during the Mesoscale Alpine Programme (MAP) field phase (Georgis et al., 2003; Bousquet and Smull, 2003; Asencio et al. 2003; Medina and Houze, 2003; Smith et al., 2003; Doyle and Smith, 2003; Rotunno and Ferretti, 2003; Smith, 2003). Readers wishing to compare different scientific approaches to the same natural event have an ideal opportunity here.

The 20 September event was caused by a frontal cyclone transiting Spain, producing an 18 h period of moist southerly airflow against the southern slopes of the Alps (Fig. 8). Using the best available full numerical weather prediction models (COAMPS, MC2, and ECMWF), the water vapor influx and precipitation field were deduced (Smith et al., 2003). These models provide an optimum interpolation between radiosonde stations. The incoming flux of water vapor is illustrated in Figure 9. This figure shows that the depth of the moist layer is ~3 km and that ~2.5 km of lifting must occur if the flow is to cross over the Alpine ridge crest. The two models (COAMPS and MC2) roughly agree in their estimates of incoming flux, adding confidence to the result.

These flux fields were integrated to obtain estimates of the total water budget over a 24 h period (Table 2). The models estimated that ~50×10^{11} kg of water vapor entered Box A (Fig. 8) from the south. The model estimates of precipitation were less

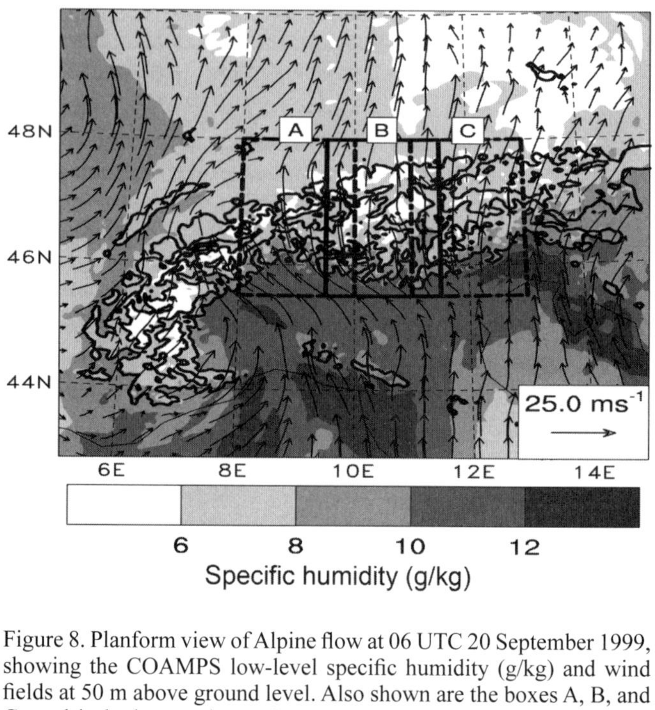

Figure 8. Planform view of Alpine flow at 06 UTC 20 September 1999, showing the COAMPS low-level specific humidity (g/kg) and wind fields at 50 m above ground level. Also shown are the boxes A, B, and C used in budget analyses. Strong moist southerly flow approached the Italian Alps. The dry area in the northeast is caused by the drying action of the Alps. (Data are from Smith et al., 2003.)

certain. They varied from 19 to 12 × 10^{11} kg. The rain gauge observations suggest that the former value is the correct one. The model differences primarily arose from different parameterizations of cloud physics processes. The drying ratio for Box A,

$$\text{DR} = \text{Total precipitation/Vapor influx}, \qquad (8)$$

was estimated to be 35%. Similar values were found for Boxes B and C in Figure 8.

A curiosity in Table 2 (Row e) is that the total precipitation from the raw upslope model (equation 5) is very large, exceeding even the incoming vapor flux. This significant error points out that the processes neglected in the raw upslope model (i.e., airflow dynamics, cloud delays, and water vapor depletion) play essential limiting roles in the precipitation that occurs in the real atmosphere.

It is useful to test the linear theory (equation 6) on this well-documented event. Using the best available parameters, the linear theory was used to predict the precipitation field (Fig. 10), and this was compared against rain gauge data. The cloud-delay values were used as fitting parameters (Barstad and Smith, 2005).

As the conventional error measures, such as bias and root mean square error (RMSE), turned out to favor smooth fields, a more sensitive measure was proposed; the Location Sensitivity Skill (LSS). LSS is a measure of the RMSE skill degradation when the observing station locations are deliberately offset. Only if the skill degradation with station misplacement is significant, can we suppose that the predicted rainfall pattern has correct spatial detail. The results shown in Table 3 indicate that only cloud-delay values in the vicinity of $\tau_c = \tau_f = 1000$ s give suitably low bias and RMSE and sufficiently high LSS. This estimate agrees with the estimates in section 4. Further details and other cases can be found in the original paper.

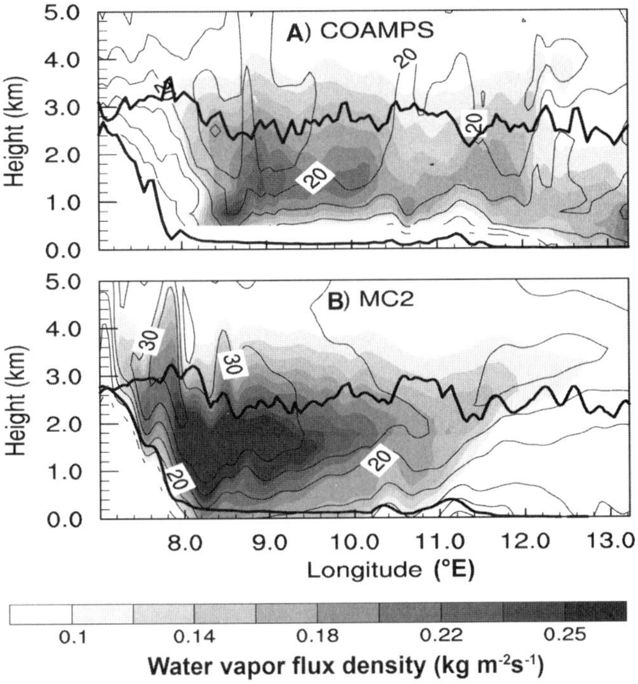

Figure 9. Incoming water vapor flux density (shaded, kg $m^{-2}s^{-1}$) and northward wind speed (contoured, m s^{-1}) through a vertical plane along 45.5°N at 08 UTC on 20 September 1999; (A) COAMPS, (B) MC2. Terrain in the plane of the cross section is shown, along with the projected crestline of the Alpine Massif. The area-integrated water vapor flux is given in Table 2 (Data are from Smith et al., 2003.)

TABLE 2. ALPINE WATER FLUXES, PRECIPITATION AND DRYING RATIOS FOR 20 SEPTEMBER 1999, DERIVED FROM FOUR MODELS

Row	Quantity	COAMPS	MC2	ECMWF forecast	ECMWF analysis
a	WV influx	56	53	42	44
b	WV outflux	33	35	33	29
c	Model precipitation	19	12	17	
d	Actual precipitation	19 (19)	19(19)	19(19)	19(19)
e	Raw upslope precipitation	95	108		
f	DR%*(c/a)	34	23	40	
g	DR%*(d/a)	34	36	45	43

Note: Models and data are for Box A in Figure 8. Water vapor flux and precipitation values are accumulations over 24 h, in units 10^{11} kg. The drying ratio (DR) is given in percent. The ECMWF analysis does not have precipitation. Actual precipitation values are independent estimates from the ETH (Frei and Haeller, 2001, personal commun.), and the University of Vienna (VERA in parentheses). Data are taken from Smith et al. (2003).

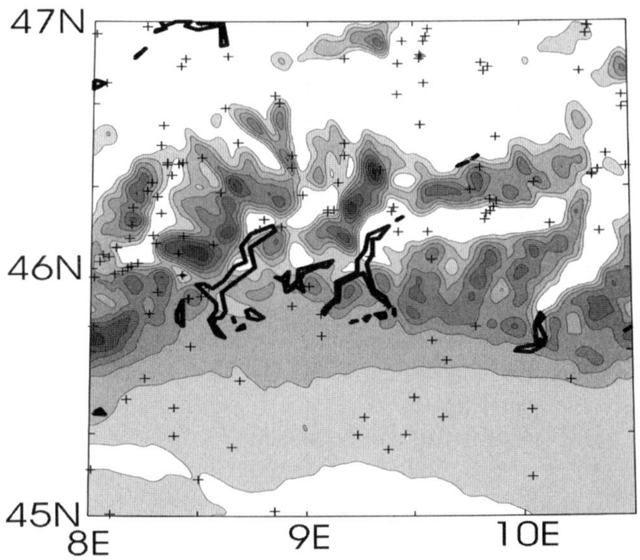

Figure 10. Hourly precipitation simulated by the linear model for 20 September 1999 in the Lago Maggiore area in northern Italy during Mesoscale Alpine Programme (MAP). Crosses indicate station locations. The contour interval is 2.5 mm/h. The Alpine terrain is not shown. (Data are from Barstad and Smith, 2005.)

TABLE 3. PRECIPITATION STATISTICS FOR 20 SEPTEMBER 1999 IN THE LAGO MAGGIORE AREA, COMPARING THE LINEAR MODEL AND RAIN-GAUGE DATA

τ (s)	Average (mm/h)	BIAS (mm/h)	RMSE (mm/h)	LSS_{RMSE}
0	5.4	2.4	10.1	0.04
250	4.1	1.1	5.5	0.00
500	2.7	–0.3	3.7	0.03
750	2.5	–0.5	2.9	0.08
1000	2.6	–0.4	2.5	0.10
1500	2.8	–0.2	2.1	0.09
2000	2.7	–0.3	1.9	0.05
3000	2.3	–0.7	1.8	0.03
4000	1.9	–1.1	1.9	0.02

Note: The measures of goodness for the linear model changes as the characteristic time scale of conversion and fallout of hydrometeors (τ) increase. The observed mean of 77 stations is 3.0 mm/h (from Barstad and Smith, 2005). RMSE is root mean square error; LSS is Location Sensitivity Skill. BIAS is the average error.

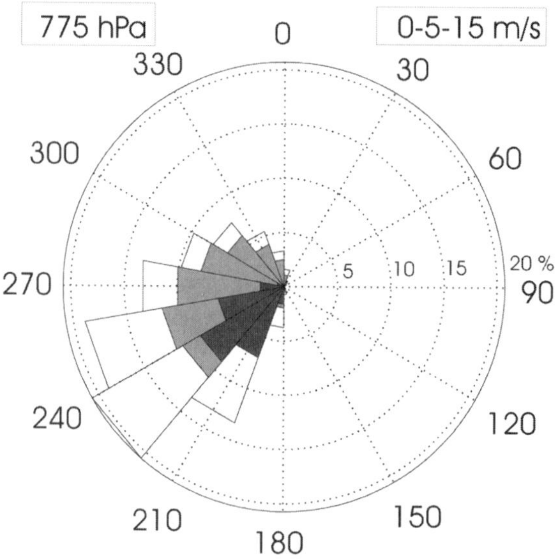

Figure 11. The 775 hPa statistical wind rose at a location near the Oregon coast (44°N, 124°W). Only wind conditions during precipitation are included. Data are from the ERA40 reanalysis for the period 1992–2002. Radial distance is the percent of time with wind direction in the sector. Shading indicates wind speeds greater than 5 and 15 m/s. (Data are from Smith et al., 2005.)

It is useful to recall that southerly flow events, such as the one described above, cannot be extrapolated to understand longer-term climatology. In the Alps, winter northerly flow events and summertime convective events also contribute significantly to the annual average precipitation. In the next section, I consider a region with a narrower range of event types, so that individual events are more consistent with climatology.

6. OREGON CLIMATE GRADIENT

Another approach to the study of orographic precipitation uses average climatological precipitation distributions. For simplicity, I looked for a region where the annual averaged precipitation is composed of a number of fairly similar precipitation events. A reasonable choice for such a study is the coastal region of Oregon; well known for its strong west-to-east, wet-to-dry climate gradient (Bastasch, 1998; Taylor and Hannan, 1999). The small variability from case to case is illustrated in Figure 11. This figure shows that the winds during precipitation events typically blow from the SW quadrant and with wind speeds exceeding 15 m/s. In fact, it seldom rains in western Oregon when the winds are easterly or northerly.

Another important characteristic of the region is that the rainfall is highly correlated with the atmospheric transport of water vapor across the coast (Fig. 12). Almost every pulse of vapor transport has a corresponding pulse of river discharge. This behavior is consistent with the hypothesis that forced ascent causes local precipitation.

Guided by Figures 11 and 12, the linear theory was computed for several representative environments and then averaged with weights corresponding to the frequency and duration of each event type. These climate precipitation fields can be compared

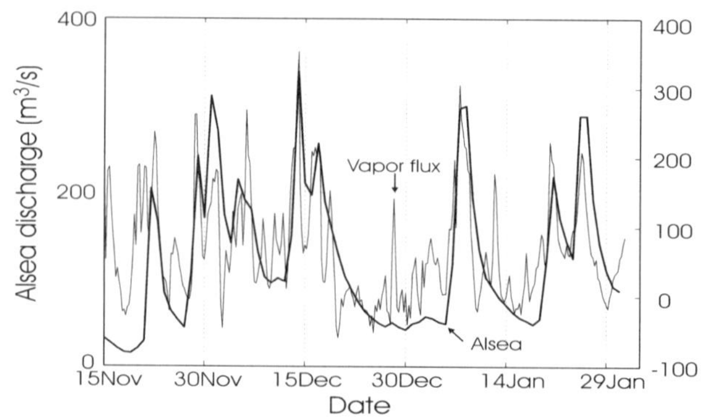

Figure 12. A time series of eastward vertically integrated water vapor flux density across the Oregon coast, computed from ERA40 data at the point 44°N, 124°W (light line: units kg $m^{-1}s^{-1}$). The daily discharge from the Alsea River near Waldport is also shown (dark line: units $m^{-3}s^{-1}$). The period is from 15 November 2001–31 January 2002, including the IMPROVE field project. Both records capture the strong event on 13–14 December 2001. (Data are from Smith et al., 2005.)

against the observed and interpolated annual fields from Daly et al. (1994). The results are shown in Figure 13. The PRISM (Precipitation-elevation Regression on Independent Slopes Model) interpolation scheme used by Daly is probably inaccurate on the smallest scales.

Figure 13 shows four fields related to the Oregon climate transition: the terrain, the annual averaged precipitation, the satellite-derived vegetation pattern, and the precipitation pattern

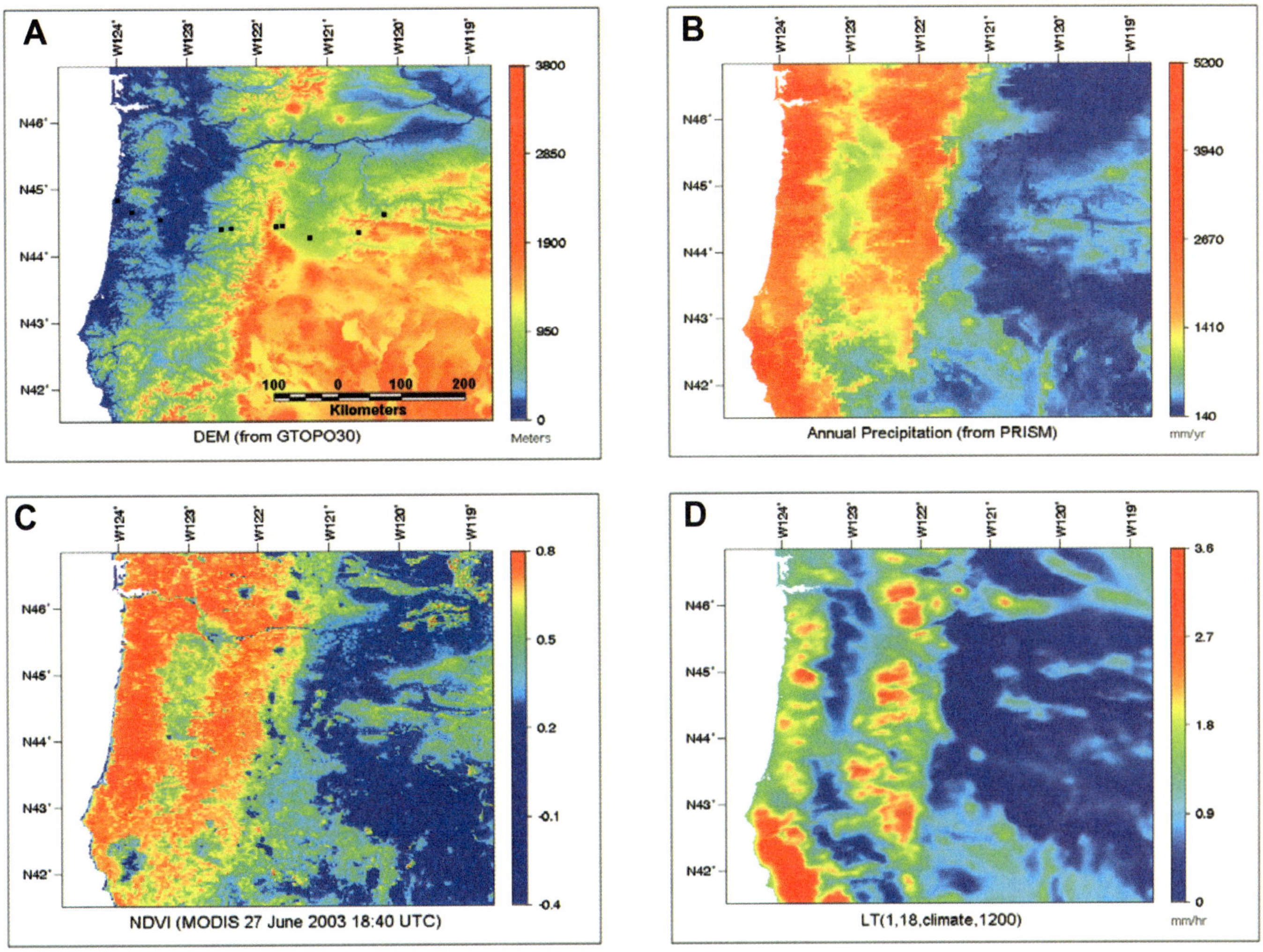

Figure 13. Four-part map of the Oregon climate transition. (A) Terrain elevation (with stream sample sites). (B) Annual precipitation (mm) using the PRISM (Precipitation-elevation Regression on Independent Slopes Model) interpolation algorithm. (C) Normalized difference vegetation index derived from MODIS (Moderate Resolution Imaging Spectroradiometer). (D) Precipitation rate from a linear theory climate run with P_∞ = 1 mm/h and τ = 1200 s. Units are mm/h. All fields are shown with an enhanced pseudo-color scale. Blue indicates low values while red indicates high values. (Data are from Smith et al., 2005.)

derived from the linear model. Annual precipitation amounts are generally greater on the western (i.e., windward) slopes of the coastal and Cascade ranges. East of the Cascades, the precipitation is much reduced. The strong vegetation gradients generally reflect the availability of rainfall and soil moisture.

The model predicted field (Fig. 13D) is computed from equations 6 and 7, using a 1 km terrain map (Fig. 13A) and $\tau_c = \tau_f$ = 1500 s. It captures many of the features of the observed distributions. The terrain and the predicted precipitation fields contain a wide range of physical scales, and thus I found a variety of different balances. The smallest terrain scales (i.e., 1–5 km) have little impact on the precipitation fields, as their fields of forced ascent do not penetrate vertically and any condensate formed is evaporated on the lee slopes before it can convert and precipitate. The intermediate scales (i.e., 5–30 km) produce hydrostatic airflow fields that penetrate to heights given by equation 2. The drift of condensate shifts the maximum precipitation point downstream so that it roughly coincides with the mountain peaks.

On the largest scales (i.e., 30–100 km), the drift is less important and the maximum precipitation occurs on the windward slopes. The strong rain shadow to the east of the Cascades is caused by evaporation in the descending air there.

To deduce optimum cloud-delay times (τ_c, τ_f) for Oregon climatology, I estimated the drying ratio (DR) using stable isotope ratios in dry-season, base-flow stream water samples. As shown in Figure 14, the relative concentration of deuterium to hydrogen

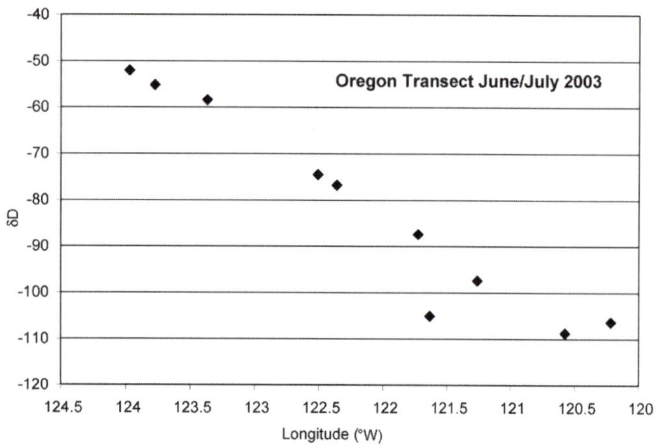

Figure 14. Hydrogen isotope ratios in stream-water samples collected in June–July 2003 as function of longitude. Sample locations (Fig. 13A) span the coastal range and the Cascades in Oregon. (Data are from Smith et al., 2005.)

decreases from west to east, because heavy isotopes are progressively rained out. According to Rayleigh fractionation theory, DR can be expressed in terms of the end-point isotope ratios according to

$$\mathrm{DR} = 1 - (R_p/R_{p0})^{1/(\alpha - 1)}, \tag{9}$$

where R_{p0} and R_p are isotope ratios in precipitation upwind and downwind of the range, and α is the fractionation factor (Dansgaard, 1964; Friedman, 1953; Friedman et al., 1964; Warburton and deFelice, 1986). Using data from Figure 13 in equation 9, I obtained DR = 0.43 ± 0.05.

The connection between cloud delay and drying ratio is shown in Figure 15. The steep decline in DR as cloud delay

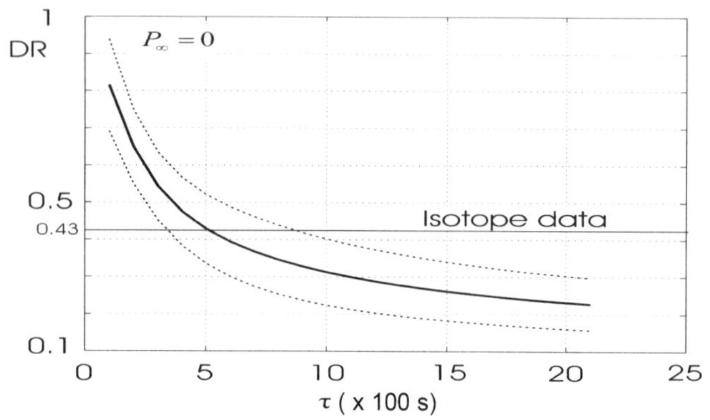

Figure 15. The drying ratio (DR) across western Oregon plotted against the cloud-delay time ($\tau = \tau_c = \tau_f$) using the linear model under an ensemble of meteorological conditions. The background precipitation is set to zero for these runs. The mean DR and the 1σ values are shown. The DR = 0.43 value from the stream-water isotope analysis is also indicated. (Data are from Smith et al., 2005.)

increases is for the reasons described in section 3. With small $\tau = \tau_c = \tau_f$, condensed water is quickly removed by precipitation before lee-side descent begins. If the cloud-delay times are large, most of the condensed water is advected to the lee side where it evaporates. Little total water is removed from the airstream.

According to Figure 15, the isotope data supports a cloud-delay time of 500 s or longer. I took 500 s as a lower bound, because a background precipitation, not included in the model runs, would increase the predicted drying ratio overall. Such values of τ are consistent with section 4. A further discussion of these points can be found in Smith et al. (2005).

7. NONLINEAR CLOUD PHYSICS

The cloud-delay times (τ_c, τ_f) played a central role in the previous discussion but they are, strictly speaking, limited to linear models of cloud physics (equation 4). The τ values are the inverse of the linear rate coefficients required in that model. In reality, there are several nonlinear processes inside clouds (e.g., Yau and Rogers, 1989). The most widely accepted nonlinearities are those associated with collision-coalescence of liquid water drops, riming of snow flakes by supercooled droplets, and accretion of small by large ice crystals. An example of a lightly rimed snowflake is shown in Figure 16. In this photograph, one can see the individual droplets that froze on impact with the snowflake. In these three processes, as the conversion involves collisions between the two types of particles, the rate of conversion of small particles to larger ones is proportional to the product of the two concentrations. To capture this process, Jiang and Smith (2003) have proposed the following system of steady-state equations

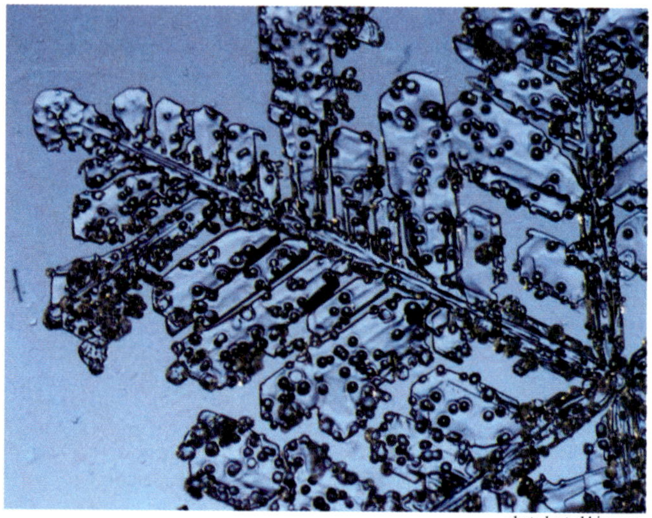

photo by ted kinsman

Figure 16. Photograph of a lightly rimed snowflake. The snowflake first formed by vapor deposition, then fell through a region with small supercooled liquid droplets. The droplets froze upon impact with the crystal. Horizontal scale is ~1 mm. A typical cloud droplet size is ~20 μm. Photo by Ted Kinsman (sciencephotography.com).

$$\vec{U} \cdot \nabla q_c = S - \gamma q_c q_f, \quad (10a)$$

$$\vec{U} \cdot \nabla q_f = \lambda q_c q_f - q_f / \tau_f. \quad (10b)$$

As with equation 4, these equations describe the column densities of cloud particles (q_c) and hydrometeors (q_f). The cloud droplet equation (10a) has a source term S. The hydrometeor equation (10b) has a linear precipitation term in which the fallout time scale appears. These equations differ from equation 4 only in the form of the conversion term. In equation 10, the conversion has a quadratic form in which the product of the two concentrations appears, multiplied by a coefficient γ.

As these partial differential equations (10) are nonlinear, the solution method by Fourier transforms used in section 4 will not apply. Instead, we can remove the spatial variability by integrating horizontally over the volume of air lying above the windward slopes of the mountain (Fig. 17). I restrict my attention to ridge geometry, so that I needed only to integrate in the x direction to obtain average values.

$$Q = \frac{1}{a} \int_{-a}^{0} q dx.$$

A similar procedure could be used with the linear equations (4); an exercise left to the reader. I assumed that no condensed water enters the box from upwind and that the advection out of the box (at the ridge line) is proportional to the wind speed (U). When the nonlinear terms in equation 10 are averaged, new product terms arise of the form $<q_c q_f>$, representing the correlations of cloud and hydrometeor density anomalies within the volume associated with convection, turbulence, or small-scale terrain.

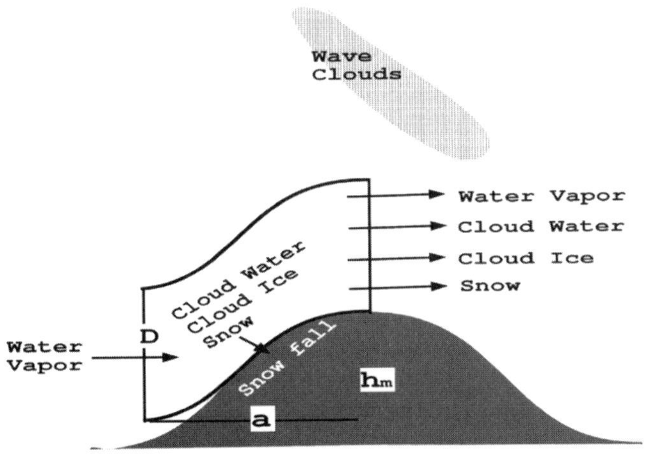

Figure 17. The box model described by equation 11, where a is mountain width and h_m is mountain height. Water vapor enters from the left and is partially condensed by the forced ascent. Condensed water is partitioned between precipitation on the windward slope and the advection of cloud water and hydrometeors over the ridge crest. (Data are from Jiang and Smith, 2003.)

Here I neglect these terms so the average condensate (Q_c,Q_f), equations 10 become

$$0 = S - \gamma Q_c Q_f - (1/\tau_a) Q_c, \quad (11a)$$

$$0 = \gamma Q_c Q_f - (1/\tau_a + 1/\tau_f)\, Q_f, \quad (11b)$$

where S is now the volume averaged source. Integrating equation 5 gives $S = \rho q_v U h_m / a$ (kgm^{-2}s^{-1}). Q_c and Q_f are the averaged condensed water column densities (kgm^{-2}), and $\tau_a = a/U$ is the residence time for air in the test volume. Here, a is the width of the upslope region (Fig. 17).

While nonlinear equations usually require numerical solutions, these particular box-model equations (11) can be solved in closed form. For convenience, I first define a critical condensation rate

$$S_{crit} = \frac{\tau_a^{-1} + \tau_f^{-1}}{\gamma \tau_a}. \quad (12)$$

When $S < S_{crit}$, the physically relevant solution to equation 11 is $Q_c = S\tau_a$ and $Q_f = 0$. This solution describes a state with condensed cloud water only, and with neither hydrometeors nor precipitation. The lenticular cloud in Figure 4 may be an example of this state. When $S > S_{crit}$,

$$Q_c = S_{crit} \tau_a, \quad (13a)$$

and

$$Q_f = (S/S_{crit} - 1)/\gamma\tau_a. \quad (13b)$$

The average upslope precipitation rate is

$$P = Q_f \tau_f^{-1}, \quad (14)$$

so the precipitation efficiency (PE), assuming no lee-side precipitation, is

$$PE = P/S = (1 - S_{crit}/S)/(\tau_f/\tau_a + 1). \quad (15)$$

The threshold behavior in equation 13 and in Figure 18 is a bit of a surprise. The solution to equation 13 tells us that in order for the system to balance strong quadratic accretion-dominated conversion, the source function must be sufficiently strong. In other words, if the source of condensed water is too weak, accretion cannot be sustained. When S is far in excess of the critical value, equations 13 and 14 give

$$P = S/(\tau_f/\tau_a + 1). \quad (16)$$

Equation 16 describes a state with highly efficient conversion of cloud particles to hydrometeors. Hydrometeors are then partitioned between fallout and advection over the crestline, so called spillover. Only when fallout is fast compared to the residence time ($\tau_f \ll \tau_a$) do we achieve the maximum possible upslope precipitation. Equation 16 then reduces to $P = S$.

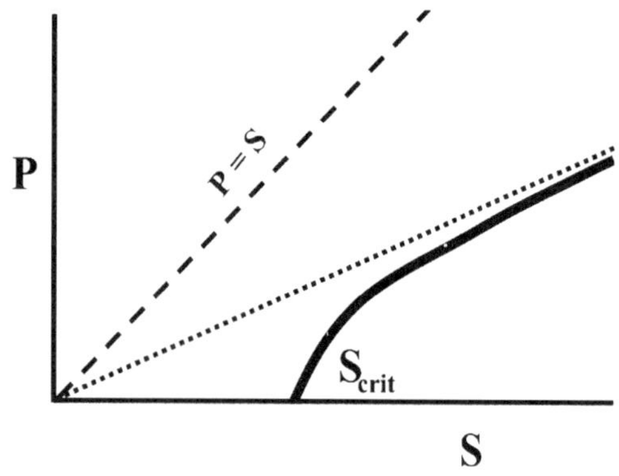

Figure 18. Solution to the nonlinear accretion-dominated box-model system (equation 11). The strength of the condensation source function (*S*) is plotted against the upslope precipitation (*P*). The solid curve is given by equations 13 and 14. The dotted curve is equation 15. The dashed curve (*P* = *S*) represents perfect precipitation. (Theory is after Jiang and Smith, 2003.)

This cloud nonlinearity and threshold property does not necessarily invalidate the linear theory. In real clouds with many competing conversion pathways, abrupt threshold behavior is less likely to occur. Also lumped models, such as our two component representation of condensed water (equation 10), often exaggerate the abruptness of bifurcations. Furthermore, a background precipitation may already surpass the threshold, giving a more linear response to the terrain-forced perturbation. The nonlinear analysis is striking, however, and it suggests at least that the cloud-delay times may vary from case to case, depending on the average rate of water vapor condensation.

Nonlinearities in airflow dynamics and cloud physics can be more fully treated using numerical techniques. Existing community mesoscale models of moist airflow dynamics, run on fast computers, allow repeated simulation experiments to be carried out. To the extent that these models are valid, the runs can be used to determine the sensitivity of precipitation amount and pattern to physical and environmental parameters. Recent work by Colle (2004) investigated the influence of wind speed and shear, moist static stability, freezing level, and barrier height and width.

8. INFLUENCE OF PRECIPITATION ON EROSION AND LANDFORM EVOLUTION

The coupling of precipitation to erosion and landform evolution has been discussed recently by several authors (Burbank et al., 2003; Dadson et al., 2003; Reiners et al., 2003; Molnar, 2003). The idea is an intriguing one as it links processes acting over an enormous range of space and time scales. Millions of orographic precipitation events are needed to alter the terrain.

To motivate the following discussion, consider the shape of the Alpine Massif shown in Figure 8. Along the south side of the Alps, two concave indentations can be seen at latitude 46°N: one at 9°E and the other at 13°E. These two locations experience especially high rainfall, perhaps because of the streams of water vapor that come from the Ligurian and Adriatic Seas or possibly because of the mountain shape itself. This coincidence of rainfall and terrain concavity allows us to speculate that atmospheric focusing of precipitation may have accelerated erosion, causing the terrain to evolve toward its current shape.

With a new scale-sensitive model of orographic precipitation at hand (equation 5), we are in a position to see how airflow dynamics and cloud physics might influence the patterns of erosion. To do this compactly, we must confine our treatment to a linear one-dimensional landscape evolution equation

$$dh(x)/dt = \mathrm{A}\nabla^2 h(x) - \mathrm{B}P(x) + T. \tag{17}$$

According to equation 17, terrain height evolves by diffusion (with coefficient A), by erosion proportional to local precipitation (with coefficient B), and by uniform tectonic uplift (*T*). For simplicity, I considered a situation in which the background precipitation (P_∞) causes a uniform erosion that balances the regional uplift; so that $\mathrm{B}P_\infty = T$. Subtracting this solution from equation 17 gives an evolution equation for the spatially varying part. The terrain evolution governed by equation 17 can be understood most easily by considering the slow changes occurring to a purely sinusoidal landform. Written in complex exponential form, it is:

$$h(x,t) = \mathrm{Re}\left[\hat{h}e^{i(kx+\omega t)}\right], \tag{18}$$

where $\hat{h}$ is the amplitude, k (= $2\pi/\lambda$) is the wavenumber (real), and ω is the complex frequency. If ω is real, the sine wave will maintain its amplitude but shift left or right. If ω is imaginary, the terrain wave will grow or decay with time. If ω is complex, the sine wave will change its amplitude and location as time passes. With equation 18, the perturbation part of equation 17 becomes

$$i\omega\hat{h} = -Ak^2\hat{h} - B\hat{P}. \tag{19}$$

With $\hat{P}$ given by equation 6, equation 19 gives

$$\omega = iAk^2 - \frac{BC_w\sigma}{[1 - imH_w][1 + i\sigma\tau_c][1 + i\sigma\tau_f]}. \tag{20}$$

According to equation 20, the diffusion and precipitation vectors can be added to determine the complex value of ω. The interpretation of ω as a vector in the complex plane will give insight into the coupling (Fig. 19). As expected, the diffusion vector is pure imaginary and points toward the decay field, thus acting to decay or smooth terrain irregularities. The raw upslope precipitation vector ($H_w = \tau_c = \tau_f = 0$), with no gravity waves or cloud delays, points along the negative real axis. Thus, erosion acts to shift the evolving terrain downwind. This is true because the highest precipitation occurs on the steepest upwind slopes, so the erosion rates would be highest there. With weaker erosion on the lee slopes, the entire terrain pattern would move downwind.

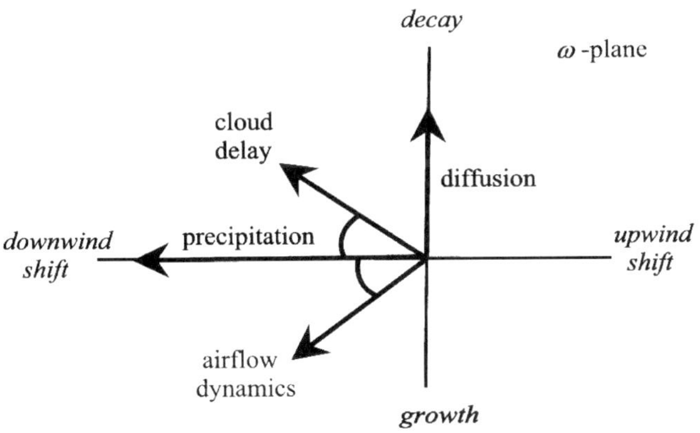

Figure 19. Phase diagram showing the complex ω plane for erosion of a purely sinusoidal terrain. Real and imaginary parts of ω represent the rates of terrain propagation and growth/decay. The pure diffusion vector corresponds to decay of the terrain. If there is a prevailing wind, the erosion due to precipitation causes the terrain to migrate downwind. Other aspects of precipitation physics, e.g., airflow dynamics or cloud physics, modify the erosion patterns slightly and may cause terrain growth or decay.

The effect of atmospheric gravity waves and cloud delays will shorten the precipitation vector in Figure 19 and rotate it. The influence of gravity waves is to rotate the precipitation vector clockwise. Acting alone, this will shift the precipitation further upwind so that erosion in the lower slopes would exceed erosion on the upper slopes and peaks. Terrain growth could occur under this scenario.

Cloud delays rotate the precipitation vector counterclockwise. This effect pushes the precipitation downwind toward the highest terrain. Erosion on the high peaks will decay the terrain.

Linear analysis (equation 20) is certainly oversimplified, and it may mislead us about the real world. It neglects nonlinear factors, such as stream-power erosion on slopes, that are thought to be of central importance in terrain evolution. It does serve to draw our attention to the importance of spatial phase shifts between terrain and precipitation and the fact that these shifts are scale dependent. These phase shifts may play a role in more realistic nonlinear models of terrain evolution.

9. CONCLUSIONS

In this brief review, I have focused on the physics of stable forced airflow lifting and precipitation caused by terrain. Meteorological disturbances such as fronts and cyclones usually play a key role in orographic precipitation by establishing a strong water vapor flux, high relative humidity, and background regional precipitation. Strong, moist airflows are able to rise over high terrain instead of being diverted around. The depth of penetration of the forced ascent, relative to the depth of the moist layer, determines the fraction of the vapor flux that can be condensed. The tilt of the airflow disturbance shifts the condensation pattern upwind. Even more important is the time required for hydrometeor formation and fallout, relative to the transit time for air parcels over the terrain. These cloud delays control precipitation patterns, amount, and spillover. To clarify the physics, I have distinguished between linear and nonlinear processes.

Several other aspects deserve more attention than I can give here: cloud physics theory, observations and parameterizations, convective cloud dynamics, radar and satellite meteorology, and full numerical modeling and data initialization. Advances in these fields will not only help to improve short-term precipitation forecasting related to floods, avalanches, landslides, and water resources, but also teach us more about longer-term processes, such as glacier growth, terrain evolution, and regional climate change.

ACKNOWLEDGMENTS

I very much appreciate the invitation to the GSA Penrose Conference in Taiwan in January 2003. The participants, with their enthusiastic questions and broad insights made the experience memorable. I appreciate discussion with N. Gasparini, Q. Jiang, J. Evans, and I. Barstad. Detailed suggestions by G. Roe and an anonymous reviewer improved the paper. Permission to use photographs by Jane English and Ted Kinsman is appreciated. This work is supported by National Science Foundation grant ATM-0112354 and National Aeronautics and Space Administration (NASA) grant NAG5-9316.

REFERENCES CITED

Asencio, N., Stein, J., Chong, M., and Gheusi, F., 2003, Analysis and simulation of local and regional conditions for the rainfall over the Lago Maggiore Target Area during MAP IOP2b: Quarterly Journal of the Royal Meteorological Society, v. 129, Part B, p. 565–586.

Banta, R.M., 1990, The role of mountain flows in making clouds, *in* Blumen, W., ed., Atmospheric processes over complex terrain: Boston, Massachusetts, American Meteorological Society, Meteorological Monographs, v. 23, no. 45, p. 229–283.

Banta, R.M., and Barker Schaaf, C., 1987, Thunderstorm genesis zones in the Colorado Rocky Mountains as determined by traceback of geosynchronous satellite images: Monthly Weather Review, v. 115, p. 463–476, doi: 10.1175/1520-0493(1987)115<0463:TGZITC>2.0.CO;2.

Barcilon, A., and Fitzjarrald, D., 1985, A nonlinear steady model for moist hydrostatic mountain waves: Journal of Atmospheric Sciences, v. 42, p. 58–67, doi: 10.1175/1520-0469(1985)042<0058:ANSMFM>2.0.CO;2.

Barros, A.P., and Lang, T.J., 2003, Monitoring the monsoon in the Himalayas: Observations in central Nepal, June 2001: Monthly Weather Review, v. 131, p. 1408–1427, doi: 10.1175/1520-0493(2003)131<1408:MTMITH>2.0.CO;2.

Barros, A.P., and Lettenmaier, D.P., 1994, Dynamic modeling of orographically induced precipitation: Reviews of Geophysics, v. 32, no. 3, p. 265–284, doi: 10.1029/94RG00625.

Barstad, I., and Smith, R.B., 2005, Evaluation of an orographic precipitation model: Journal of Hydrometeorology, v. 6, p. 85–99, doi: 10.1175/JHM-404.1.

Bastasch, R., 1998: Waters of Oregon: Corvallis, Oregon, Oregon State University Press, 278 p.

Bousquet, O., and Smull, B.F., 2003, Observations and impacts of upstream blocking during a widespread orographic precipitation event: Quarterly Journal of the Royal Meteorological Society, v. 129, Part B, p. 391–409.

Browning, K.A., Hill, F.F., and Pardoe, C.W., 1974, Structure and mechanism of precipitation and the effect of orography in a wintertime warm sector: Quarterly Journal of the Royal Meteorological Society, v. 100, no. 425, p. 555–562, doi: 10.1256/smsqj.42504.

Burbank, D.W., Blythe, A.E., Putkonen, J., Pratt-Sitaula, B., Gabet, E.,

Oskin, M., Barros, A., and Ojha, T.P., 2003, Decoupling of erosion and precipitation in the Himalayas: Nature, v. 426, no. 6967, p. 652–655, doi: 10.1038/nature02187 D. W.

Buzzi, A., Tartaglione, N., and Malguzzi, P., 1998, Numerical simulations of the 1994 Piedmont flood: Role of orography and moist processes: Monthly Weather Review, v. 126, p. 2369–2383, doi: 10.1175/1520-0493(1998)126<2369:NSOTPF>2.0.CO;2.

Colle, B.A., 2004, Sensitivity of orographic precipitation to changing ambient conditions and terrain geometries: An idealized modeling perspective: Journal of Atmospheric Sciences, v. 61, p. 588–606, doi: 10.1175/1520-0469(2004)061<0588:SOOPTC>2.0.CO;2.

Collier, C.G., 1975, A representation of the effects of topography on surface rainfall within moving baroclinic disturbances: Quarterly Journal of the Royal Meteorological Society, v. 101, p. 407–422, doi: 10.1256/smsqj.42901.

Dadson, S.J., Hovius, N., and Chen, H.G., 2003, Links between erosion, runoff variability and seismicity in the Taiwan orogen: Nature, v. 426, no. 6967, p. 648–651, doi: 10.1038/nature02150.

Daly, C., Neilson, R.P., and Phillips, D.L., 1994, A statistical-topographic model for mapping climatological precipitation over mountainous terrain: Journal of Applied Meteorology, v. 33, p. 140–158, doi: 10.1175/1520-0450(1994)033<0140:ASTMFM>2.0.CO;2.

Dansgaard, W., 1964, Stable isotopes in precipitation: Tellus, v. 16, p. 436–468.

Doyle, J., and Smith, R.B., 2003, Mountain waves over the Hohe Tauern: Influence of upstream diabatic effects: Quarterly Journal of the Royal Meteorological Society, v. 129, no. 588, p. 799–823, doi: 10.1256/qj.01.205.

Durran, D.R., and Klemp, J.B., 1982, The effects of moisture on trapped mountain lee waves: Journal of Atmospheric Sciences, v. 39, p. 2490–2506, doi: 10.1175/1520-0469(1982)039<2490:TEOMOT>2.0.CO;2.

Evans, J., Smith, R.B., and Oglesby, R., 2004, Middle East climate simulation and dominant precipitation processes: International Journal of Climatology, v. 24, p. 1671–1694, doi: 10.1002/joc.1084.

Fraser, A.B., Easter, R.C., and Hobbs, P.V., 1973, A theoretical study of the flow of air and fallout of solid precipitation over mountainous terrain: Part I. Airflow model: Journal of Atmospheric Sciences, v. 30, p. 801–812, doi: 10.1175/1520-0469(1973)030<0801:ATSOTF>2.0.CO;2.

Friedman, I., 1953, Deuterium content of natural waters and other substances: Geochimica et Cosmochimica Acta, v. 4, p. 89–103, doi: 10.1016/0016-7037(53)90066-0.

Friedman, I., Redfield, A.C., Schoen, B., and Harris, J., 1964, The variation of the deuterium content of natural waters in the hydrologic cycle: Reviews of Geophysics, v. 2, p. 177–223.

Georgis, J-F., Roux, F., Chong, M., and Pradier, S., 2003, Triple-Doppler radar analysis of the heavy rain event observed in the Lago Maggiore region during MAP IOP 2b: Quarterly Journal of the Royal Meteorological Society, v. 129, Part B, p. 495–522.

Hobbs, P.V., Easter, R.C., and Fraser, A.B., 1973, A theoretical study of the flow of air and fallout of solid precipitation over mountainous terrain: Part II. Microphysics: Journal of Atmospheric Sciences, v. 30, p. 813–823, doi: 10.1175/1520-0469(1973)030<0813:ATSOTF>2.0.CO;2.

Jiang, Q., 2003, Moist dynamics and orographic precipitation: Tellus, v. 55A, p. 301–316.

Jiang, Q., and Smith, R.B., 2003, Cloud timescale and orographic precipitation: Journal of Atmospheric Sciences, v. 60, p. 1543–1559, doi: 10.1175/2995.1.

Lin, Y.-L., Han, J., Hamilton, D.W., and Huang, C.-Y., 1999, Orographic influence on a drifting cyclone: Journal of Atmospheric Sciences, v. 56, p. 534–562, doi: 10.1175/1520-0469(1999)056<0534:OIOADC>2.0.CO;2.

Medina, S., and Houze, R.A., 2003, Air motions and precipitation growth in Alpine storms: Quarterly Journal of the Royal Meteorological Society, v. 129, Part B, p. 345–371.

Molnar, P., 2003, Geomorphology—Nature, nurture and landscape: Nature, v. 426, no. 6967, p. 612–614, doi: 10.1038/426612a.

Passarelli, R.E., and Boehme, H., 1983, The orographic modulation of pre-warm-front precipitation in southern New England: Monthly Weather Review, v. 111, p. 1062–1070, doi: 10.1175/1520-0493(1983)111<1062:TOMOPW>2.0.CO;2.

Rauber, R.M., and Grant, L.O., 1986, The characteristics and distribution of cloud water over the mountains of northern Colorado during wintertime storms. Part II: Spatial distribution and microphysical characteristics: Journal of Applied Meteorology, v. 25, p. 489–504, doi: 10.1175/1520-0450(1986)025<0489:TCADOC>2.0.CO;2.

Raymond, D.J., and Wilkening, M.H., 1982, Flow and mixing in New Mexico mountain cumuli: Journal of Atmospheric Sciences, v. 39, p. 2211–2228, doi: 10.1175/1520-0469(1982)039<2211:FAMINM>2.0.CO;2.

Reiners, P.W., Ehlers, T.A., Mitchell, S.G., and Montgomery, D.R., 2003, Coupled spatial variations in precipitation and long-term erosion rates across the Washington Cascades: Nature, v. 426, no. 6967, p. 645–647, doi: 10.1038/nature02111.

Rotunno R., and Ferretti, R., 2003, Orographic effects on rainfall in MAP cases IOP2b and IOP8: Quarterly Journal of the Royal Meteorological Society, v. 129, Part B, p. 373–390.

Smith, R.B., 1979, The influence of mountains on the atmosphere: Advances in Geophysics, v. 21, p. 87–230.

Smith, R.B., 1980, Linear theory of stratified hydrostatic flow past an isolated mountain: Tellus, v. 32, p. 348–364.

Smith, R.B., 2003, A linear time-delay model of orographic precipitation: Journal of Hydrology, v. 282, p. 2–9, doi: 10.1016/S0022-1694(03)00248-8.

Smith, R.B., and Barstad, I., 2004, A linear theory of orographic precipitation: Journal of Atmospheric Sciences, v. 61, p. 1377–1391, doi: 10.1175/1520-0469(2004)061<1377:ALTOOP>2.0.CO;2.

Smith, R.B., Jiang, Q.F., Fearon, M.G., Tabary, P., Dorninger, M., Doyle J.D., and Benoit, R., 2003, Orographic precipitation and air mass transformation: An alpine example: Quarterly Journal of the Royal Meteorological Society, v. 9, Part B, p. 433–454.

Smith, R.B., Barstad, I., and Bonneau, L., 2005, Orographic precipitation and Oregon's climate transition: Journal of Atmospheric Sciences, v. 62, p. 177–191, doi: 10.1175/JAS-3376.1.

Taylor, G.H., and Hannan, C., 1999, The climate of Oregon: Corvallis, Oregon, Oregon State University Press, 211 p.

Uttal, T., Rauber, R.M., and Grant, L.O., 1988, Distributions of liquid, vapor, and ice in an orographic cloud from field observations: Journal of Atmospheric Sciences, v. 45, no. 7, p. 1110–1122, doi: 10.1175/1520-0469(1988)045<1110:DOLVAI>2.0.CO;2.

Wallace, J.M., and Hobbs, P.V., 1977, Atmospheric science: An introductory survey: New York, Academic Press, 467 p.

Wang, J.-J., Rauber, R.M., Ochs, H.T., and Carbone, R.E., 2000, The effects of the Island of Hawaii on offshore rainband evolution: Monthly Weather Review, v. 128, p. 1052–1069, doi: 10.1175/1520-0493(2000)128<1052:TEOTIO>2.0.CO;2.

Warburton, J.A., and deFelice, T.P., 1986, Oxygen isotopic composition of central Sierra Nevada precipitation: Part I; Identification of ice-phase water capture regions in winter storms: Atmospheric Research, v. 20, p. 11–22, doi: 10.1016/0169-8095(86)90004-9.

Wratt, D.S., Ridley, R.N., Sinclair, M.R., Larsen, H., Thompson, S.M., Henderson, R., Austin, G.L., Bradley, S.G., Auer, A., Sturman, A.P., Owens, I., Fitzharris, B., Ryan, B.F., and Gayet, J.-F., 1996, The New Zealand Southern Alps experiment: Bulletin of the American Meteorological Society, v. 77, p. 683–692, doi: 10.1175/1520-0477(1996)077<0683:TNZSAE>2.0.CO;2.

Wurtele, M.G., Sharman, R.D., and Datta, A., 1996, Atmospheric lee waves: Annual Review of Fluid Mechanics, v. 28, p. 429–476, doi: 10.1146/annurev.fl.28.010196.002241.

Yau, M.K., and Rogers, R.R., 1989, Short course in cloud physics (3rd edition): Butterworth-Heinemann, 304 p.

Yuter, S.E., and Houze, R.A., 2003, Microphysical modes of precipitation growth determined by S-band vertically pointing radar in orographic precipitation during MAP: Quarterly Journal of the Royal Meteorological Society, v. 129, Part B, p. 455–476.

Manuscript Accepted by the Society 23 June 2005

Printed in the USA

Geological Society of America
Special Paper 398
2006

From weather to climate—Seasonal and interannual variability of storms and implications for erosion processes in the Himalaya

Ana P. Barros[†]
Pratt School of Engineering, Civil and Environmental Engineering Department, Duke University, Durham, North Carolina 27708, USA
Sen Chiao
Division of Engineering and Applied Sciences, Harvard University, Cambridge, Massachusetts 02138, USA
Timothy J. Lang
Department of Atmospheric Sciences, Colorado State University, Fort Collins, Colorado 80523, USA
Douglas Burbank
Department of Geological Sciences, University of California, Santa Barbara, California 93106, USA
J. Putkonen
Department of Earth and Planetary Sciences, University of Washington, Seattle, Washington 98195, USA

ABSTRACT

The extent to which orography may be a product of climate-erosion interactions is largely unknown. One grand challenge is to quantify the precipitation regimes of mountainous regions at the spatial and temporal scales relevant for investigating the interplay of erosion and tectonics in active orogens. In this paper, our objective is to synthesize recent research integrating numerical model simulations, satellite data, and surface observations in the Himalaya to elucidate the role of weather and climate in mountain evolution. We focus on the seasonal and interannual space-time variability of precipitation in the Great Himalayas by studying two preeminent storm regimes in detail—monsoon onsets and depressions in general, and wintertime Western Disturbances (cold season events). High-resolution simulations of heavy precipitation storms for two monsoon onset conditions (1999 and 2001) and one wintertime storm (2000) are used to illustrate the complex patterns of interaction between the mountains and the atmosphere, and to show how these affect the spatial distribution of precipitation. Along with observations from an existing ground-based network, these simulations provide unique insights into the space-time features of seasonal and interannual variability of precipitation. Our analysis indicates that the trajectory of monsoon storms during onset events exerts a strong control on the precipitation amounts and rainfall penetration into the rain shadow. Spatial variability of subsequent storm tracks in any given year helps explain the interannual variability of monsoon precipitation. Both observational data and our simulations define striking spatial variability in precipitation on upwind and downwind flanks of ridges that project into obliquely impinging storms. Specifically, as southeasterly monsoon winds encounter north-south oriented ridges, forced lifting of moist air enhances precipitation on the upwind flanks, whereas less precipitation occurs on downwind flanks. This variability

[†]Email: barros@duke.edu.

Barros, A.P., Chiao, S., Lang, T.J., Burbank, D., and Putkonen, J., 2006, From weather to climate—Seasonal and interannual variability of storms and implications for erosion processes in the Himalaya, *in* Willett, S.D., Hovius, N., Brandon, M.T., and Fisher, D.M., eds., Tectonics, Climate, and Landscape Evolution: Geological Society of America Special Paper 398, Penrose Conference Series, p. 17–38, doi: 10.1130/2006.2398(02). For permission to copy, contact editing@geosociety.org.

is observed at spatial scales as short as ~10 km—a distance equivalent to the spacing of major ridge crests. Because infrequent, singular storm events appear to control the mass input to glaciers, and may determine the frequency and spatial distribution of landslides, these findings provide physically based insight into decoupling high-frequency (seasonal to interannual time scales) from low-frequency (multidecadal to centennial and longer time scales) signals in the interpretation of climate and erosion records in the Himalayas. Furthermore, this research suggests that integrative studies aimed at unraveling the role of climate in landscape evolution must include consideration of storm frequency and intensity along with spatial variability at scales consistent with regional climate forcing.

Keywords: precipitation, orography, landform, climate, erosion.

INTRODUCTION

Mountain belts evolve as the product of tectonics interacting with erosion and weathering processes, which are determined by regional hydrology and hydrometeorology, especially the precipitation regime in the case of humid climates. At steady state, removal of landmass by erosion is compensated by isostatic compensation and tectonically driven rock uplift. In turn, mountains strongly affect climate at regional and even global scale by functioning as obstacles to atmospheric circulation, as elevated heat sources via the production of latent heating of orographic precipitation upwind, and by changing the nature of water and energy exchanges between continental surfaces and the atmosphere via either increased snow cover, elevated plateaus, or the establishment of rain shadow deserts. As the topography rises, precipitation changes (increases upwind, decreases downwind), and, at spatial scales ≥100 km, so does erosion (forming deep valleys with elevated peaks upwind [the Himalayan range], allowing elevated plateaus to form downwind [the Tibetan Plateau]). At the scale of millions of years, spatial and temporal changes in the climate regime, changes in spatial and temporal characteristics of erosion processes, and isostatic rebound and rock uplift interact to determine the evolution and form of mountains (Pinter and Brandon, 1997; Molnar et al., 1993).

One question of interest is whether the critical erosion events occur as pulses of catastrophic magnitude (flash floods and debris flows, landslides) associated with infrequent heavy precipitation events, or whether they can be interpreted in terms of "uniform" along-range average erosion rates over long periods of time (Montgomery and Brandon, 2002). In this paper, we discuss evidence suggesting that the former can be a fundamental erosion mode in the Himalayas, which would explain why small-scale erosion studies that fail to capture the relevant spatial variability of climate forcing might not elucidate the role of erosion in mountain evolution (Burbank et al., 2003). This erosional response, we argue, is related to the regional climate and how the mountain range interacts with the atmospheric circulation. Elsewhere (e.g., the Washington Cascades), average long-term coupling of erosion and climate may be dominant (Reiners et al., 2003). Nevertheless, Hodges et al. (2004) pointed out the close spatial correlation between evidence for active faulting and the maxima of annual precipitation in the Annapurna Range in the central Himalayas. Such a spatial association indicates possible coupling between erosion and deformation around the center of mass of orographic precipitation at geological time scales.

Ice-core records (oxygen isotope ratios and chemical composition) have been increasingly used to characterize the low-frequency variability of precipitation in the Himalayas, and by extension, the South Asian Monsoon (Thompson et al., 2000; Zhao and Moore, 2002; Duan and Yao, 2003). The working hypothesis is that precipitation varies linearly with elevation (either positively or negatively) and that all or the overwhelming majority of the precipitation at glacier locations takes place during the monsoon. We propose in this study that, whereas such data may describe the low-frequency variability of the monsoon at multidecadal to centennial temporal scales, the information content of the records must be deconvolved into high- (seasonal to interannual) and low-frequency modes of variability before attempting a process-based interpretation of the underlying phenomena. Specifically, we show that in the hydrometeorological rain shadow of the central Himalayas (the region of elevated peaks and valleys where most glaciers are located, which lies northward of the precipitation maximum), the contribution of monsoon precipitation to annual accumulation decreases very rapidly with altitude, while the contribution of relatively infrequent wintertime snowstorms increases and dominates spatial and interannual variability. Furthermore, we show that the monsoon contribution to precipitation totals in the hydrometeorological rain shadow is controlled by the strength and path of monsoon depressions originating in the Bay of Bengal, and by monsoon onsets in particular, thus exhibiting high interannual variability, not only in magnitude, but also in spatial distribution. In this context, key issues for the interpretation of centennial-scale climate variations from glacier records become (1) the number and timing of storms that contribute to the observed accumulation, and (2) the definition of non-stationarity and dominant time scales of climate behavior.

These characteristics are important not only from the perspective of elucidating current climate variability, but also for testing the interpretations of climate-tectonics interactions that often rely on average annual precipitation to describe climate controls on landform evolution. A key implication of this study is that, in the range of seasonal to interannual time scales, the atmospheric moisture flux that nurses existing glaciers arrives via chaotic pulse-like events (monsoon depressions, wintertime storms) with high space-time variability, and therefore is fundamentally non-stationary. Such behavior redirects attention to seasonal and interannual variability of storm frequency and intensity and to how these affect the temporal and spatial variability of precipitation as related to climate controls of erosion in the Himalayas.

In this manuscript, we document the spatial features of three storms, including two distinct monsoon onsets (1999 and 2001) and one typical wintertime storm. For this purpose, we rely on numerical simulations using a regional model at very high spatial resolution (1.86 km, unprecedented in this region of the world), which is necessary to resolve the significant cloud features (Barros et al., 2004). The modeling results are presented in the context of extensive data analysis and synthesis from previous diagnostics studies (Barros et al., 2000, 2004; Lang and Barros, 2002, 2004; Barros and Lang, 2003a; Magagi and Barros, 2004). First, we review findings from previous work in order to characterize monsoon weather, and to establish the observational evidence of seasonal and interannual climate variability in the Himalayas. Second, spatial fields of relevant atmospheric variables (wind fields and cloud liquid water) obtained from the numerical simulations are used to illustrate the role of orography in organizing clouds and rainfall on the Himalayan range under different large-scale environmental conditions. We conclude by discussing the implications of our findings for the characterization of erosion processes in the Himalayas.

MONSOON WEATHER IN THE HIMALAYAS

Every summer in South Asia, the thermal contrast between the slowly warming ocean and the rapidly warming land accelerates vertical air flow over the land and thereby drives the landward flow of moisture that underpins the monsoon. The large, high-altitude mass of the Tibetan Plateau greatly strengthens this land-sea contrast regionally, and serves to enhance the intensity of the monsoon. In the Indian subcontinent, the onset of the Southwest (SW) monsoon takes place after a region of strong low-level moisture flux associated with southwesterly cross-equatorial flows off the coast of Somalia (i.e., the Somali jet) becomes well developed (e.g., Krishnamurti et al., 1981; Tang and Reiter, 1984, among others). An elongated area of low pressure, the monsoon trough, becomes established roughly parallel to the Himalayan range, and subsequent monsoon activity is associated with the development of low-pressure systems (mesoscale vortices) along the monsoon trough. Monsoon activity along the south-facing slopes of the Himalayas is associated with the development and westward propagation of monsoon depressions originating from the Bay of Bengal (Das, 1987; Lang and Barros, 2002). Ultimately, the path of Bay of Bengal depressions is determined by the position of the monsoon trough, the Coriolis effect (northward deflection), the origin and strength (vorticity) of the monsoon systems, and their interactions with the land surface as they travel over northern India.

A comprehensive compilation and analysis of observations from a dense hydrometeorological network operating since 1999 in the Marsyandi River basin in central Nepal (Barros et al., 2000; Fig. 1A) revealed the complexity of the relationship between elevation and precipitation in the central Himalayas (Barros et al., 2004), including (1) weak altitudinal gradients of annual rainfall between 1000 and 4500 m; (2) strong ridge-to-ridge zonal (east-west) gradients of monsoon rainfall (1500 mm/5 km); (3) strong ridge-valley gradients during rain storms, especially in the case of deep valleys (elevation differences > 1000 m) and steep hillslopes (70%); and (4) strong altitudinal gradients in rainfall intensity and duration (convective versus stratiform), with longer (shorter) durations and lower (higher) intensities at higher (lower) elevations along the ridges. High elevations (>3000 m above sea level [asl]) can receive up to 40% of their annual precipitation as snowfall during the winter, with the highest-altitude stations (~4000 m asl and above) accumulating the most total winter precipitation (Lang and Barros, 2004; Barros et al., 2004). A comprehensive summary of the observations is presented in Figure 1B. Note that there is a rapid decrease of precipitation amount in the region of the hydrometeorological shadow, and that the spatial gradient associated with this decrease is much stronger for rainfall than for snowfall. Observed cumulative rainfall reaches minimum values well upwind of the topographic divide, thus raising questions with regard to the relative contribution of monsoon vis-à-vis wintertime precipitation to modern glaciers.

At the scale of the Himalayan range, the spatial and temporal variability of precipitation was investigated using NASA-TRMM (National Aeronautics and Space Administration Tropical Rainfall Measurement Mission) and Meteosat-5 (a geostationary satellite maintained by the European Meteorological Satellites Organization, EUMETSAT) analysis data products (Barros et al., 2000, 2004). Scaling analysis of collocated orography and cloudiness fields uncovered two distinct types of orographic controls at different spatial scales: a synoptic-scale control (~300 km) associated with the overall terrain envelope and the major river valleys that cut through the mountains connecting the Indian subcontinent and the Tibetan Plateau; and a mesoscale control characterized by high interannual and spatial variability linked to the topology of the succession of multiscale ridges and valleys along the southern slopes of the Himalaya (5–100 km). Depending on the actual complexity of the terrain, the variance of cloudiness and its scaling behavior reflected the ridge density, as well as its altitudinal range (Barros et al., 2004).

The synoptic and mesoscale dynamics of monsoon and wintertime storms were analyzed based on field campaign data, historical radiosonde records, ECMWF (European Center for Medium-Range Weather Forecasts), and NCEP (National Centers for Environmental Prediction) reanalysis products and high-resolution numerical models (Lang and Barros, 2002, 2004; Barros, 2004; Lang, 2003a). Through these studies, three preeminent weather regimes can be identified that can help to explain the seasonal and interannual variability of precipitation: (1) monsoon onset depressions (Lang and Barros; 2002; Barros and

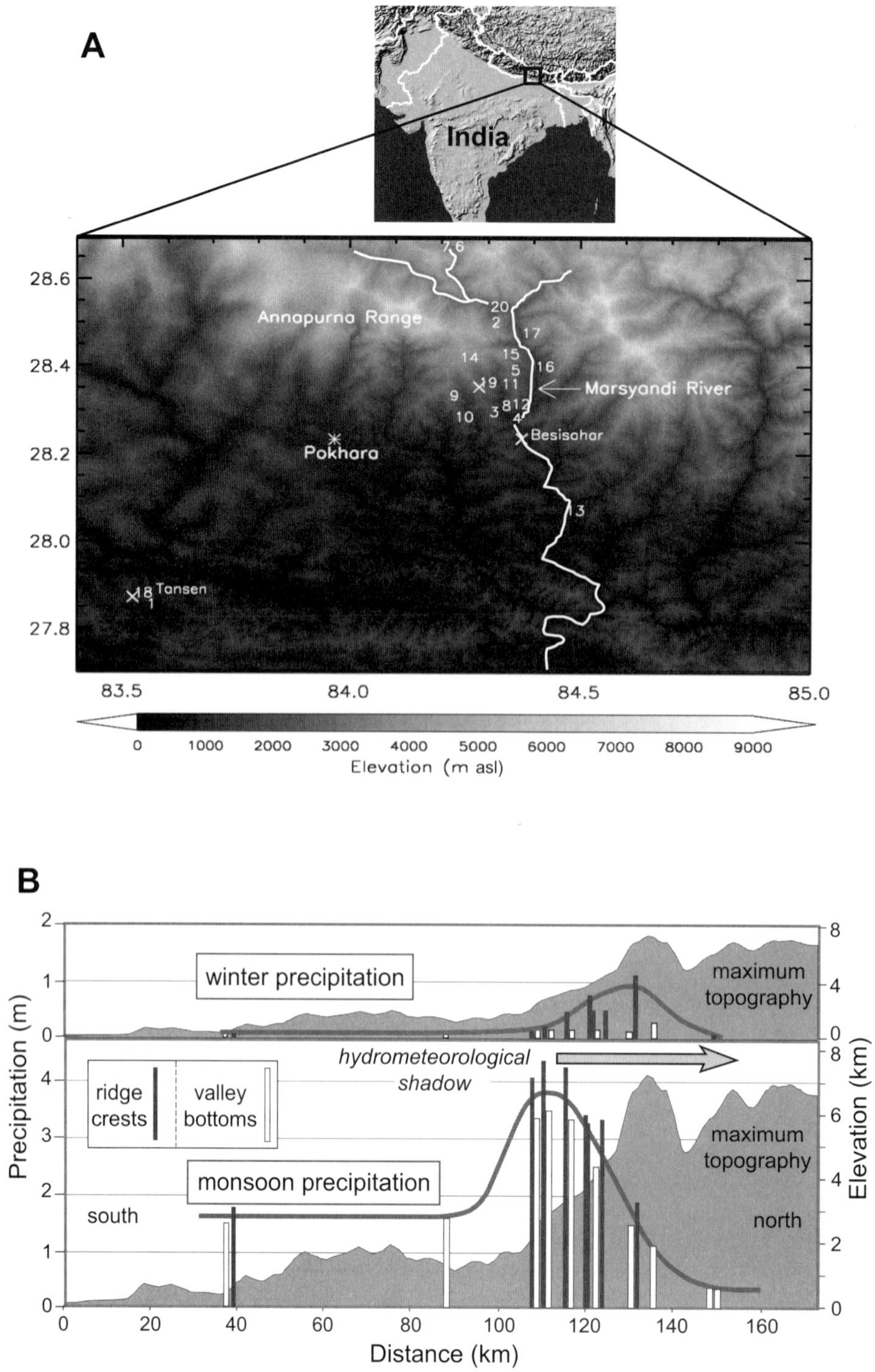

Figure 1. (A) Topography of the Marsyandi River basin. The lower map shows locations of the hydrometeorological stations in the network. The upper map shows the location of the network relative to the Indian subcontinent. (B) Cumulative summer (bottom panel) and winter (top panel) precipitation and maximum topography (averaged along a 60-km-wide swath) are plotted against distance perpendicular to the Himalayan range crest. Precipitation data were collected at 20 stations from 1999 to 2002. The summer monsoon precipitation increases dramatically at all measured elevations as the maximum elevation of the orogen climbs above 2000 m. It then drops off rapidly to the north. Precipitation on the ridge crests is consistently 10%–20% greater than in the valley bottoms. Winter precipitation (top panel) shows a much greater ridge-to-valley contrast, lower precipitation amounts, and a peak of precipitation offset toward the range crest.

Lang, 2003a); (2) mesoscale convection organized by orography (Barros and Lang, 2003a; Barros et al., 2004); and (3) wintertime storms associated with a low-pressure center over the Hindu Kush mountains, the so-called Western Disturbances (Lang and Barros, 2004). We focus on 1 and 3, respectively. Monsoon onset depressions are hereafter referred to as "onset events," and wintertime storms are hereafter referred to as "snow storms" or "cold-season events." The development of convective activity aligned with topographic features is characterized by a strong diurnal cycle, and is particularly strong during the active phases of the monsoon, including onset events (Barros and Lang, 2003a; Barros et al., 2004).

Monsoon onsets consist of heavy, multiday rain events, which are precursors to the subsequent arrival of daily to near-daily rainfall for the rest of the summer season. The onset takes place when strong mesoscale depressions originating from the Bay of Bengal in the first half of June interact with easterly vertical shear forced by the mountains, establishing an asymmetric secondary circulation with strong near-surface upslope (southerly) flow and return flow (northerly) at 600–700 hPa (Lang and Barros, 2002; Barros and Lang, 2003a). Interactions with local topography serve to organize convective activity along the foothills of the Himalayas. At the northerly stations (e.g., 5, 14, and 15 in Fig. 1A), total rainfall during onset events can amount up to 30% of the overall monsoon total (Lang and Barros, 2002, 2004). A climatological analysis of the track of these storms (Fig. 2) shows that they vary widely from year to year (Das, 1987; Lang and Barros, 2002). The southward shift in the year 2000 monsoon track (Fig. 2) caused the 2000 onset rainfall to be up to the 50% less than in 1999. The onset precipitation was even less in 2001, as the onset depression remained south of the range.

As described by Barros and Lang (2003a), monsoon depressions tend to form along the monsoon trough over the Bay of Bengal, and to propagate westward on the northern flank of the

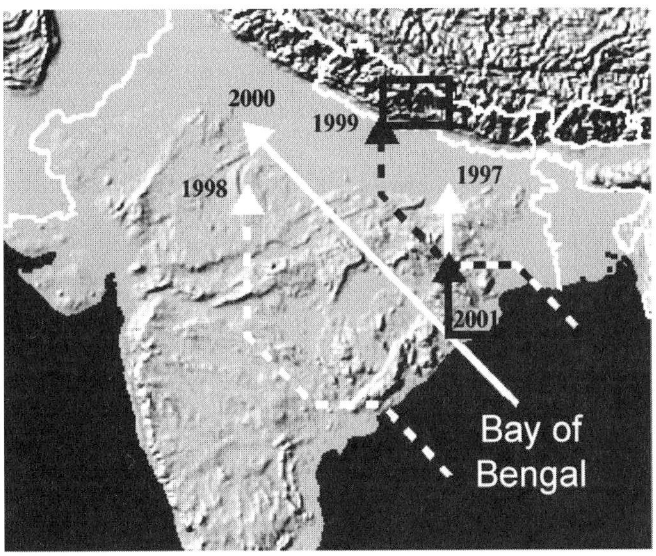

Figure 2. Storm tracks for the Nepal onset depressions (1997–2001). Tracks were determined by the locations of the 850 hPa (~1.5 km) relative vorticity maxima in the European Center for Medium-Range Weather Forecasts (ECMWF) data. The black box surrounds the location of the Marsyandi network. (Image is reproduced from Lang and Barros, 2002.)

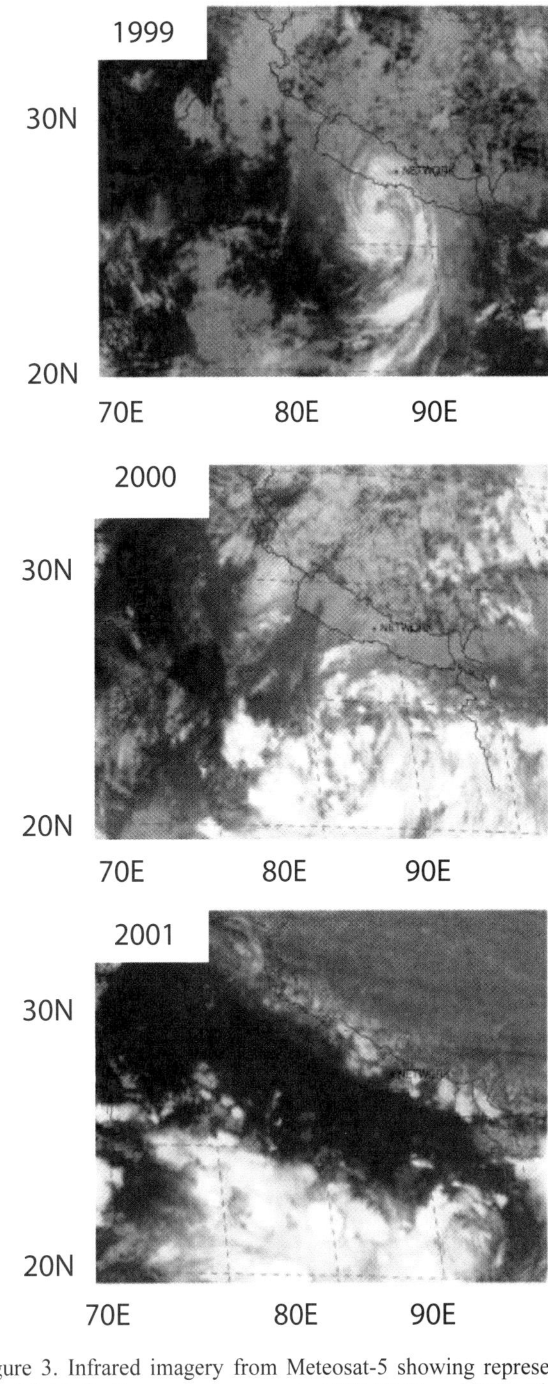

Figure 3. Infrared imagery from Meteosat-5 showing representative cloud fields for three monsoon onsets: 1999, 2000, and 2001. The location of the Marsyandi network is also marked.

westerly jet over the Indian subcontinent. When the depressions follow a trajectory farther away from the Himalayas (as in 2000, Fig. 2), the interactions with the easterly vertical shear along the Himalayas are weakened, thereby decreasing convergence and keeping the depressions away from the Himalayan range and favoring disorganized convection over northern India.

Satellite imagery (Fig. 3) readily depicts cloudiness fields during three monsoon onsets (1999, 2000, and 2001). The climatology of depression strength and motion track influences not only rainfall amounts, but also the spatial variability of precipitation, especially the extent to which rain bands associated with these depressions penetrate northward beyond the leading mountain peaks. Because similar, though generally weaker, depressions occur during the active phases of the monsoon from June to September, this analysis has implications not only for monsoon onsets, but also for the spatial distribution of rainfall over the course of the entire monsoon season.

Lang and Barros (2004) focused their attention on cold-season events. They found that significant snowstorms are associated with terrain-locked low-pressure systems that form when an upper-level disturbance passes over the topographic notch (Fig. 4A) formed by the Himalayas and Hindu Kush mountains (Western Disturbances), causing upper-level SW flow over central Nepal and topographically forced precipitation over extended areas at high elevations. They developed a 30 yr (1973–2002) climatology of these "notch" depressions, which revealed significant interannual variability in central Himalayan winter storms, in particular with regard to the number of such storms per year (fewer than six on average), from years when none occurs (1987, coincidentally a weak monsoon year) to over ten per year.

MODELING STUDIES

Observational data currently available, either ground-based or from satellites, do not provide a multiscale perspective of space-time variability of precipitation on the Himalayan range. Hydrometeorological networks such as the Marsyandi (Fig. 1A) can provide detailed information over the small area covered by the network only. As an indirect measure of precipitation, satellite data present substantial retrieval challenges in regions of complex terrain where precipitation effects are difficult to separate from terrain and snow-cover effects, a difficulty that is compounded with inadequate sampling frequency in either space or time (or both) (Barros et al., 2000; Magagi and Barros, 2004). Model simulations such as those described here allow us to monitor in detail the three-dimensional spatial and temporal evolution of relevant weather systems and their precipitation fields.

Based on analysis and interpretation of observational records (satellite, radiosonde, and rain gauge data), we concluded that onset events provide clear insights on the long-term hydroclimatology of summer monsoon precipitation in the Himalayas. Therefore, we focus on modeling two onset events that are representa-

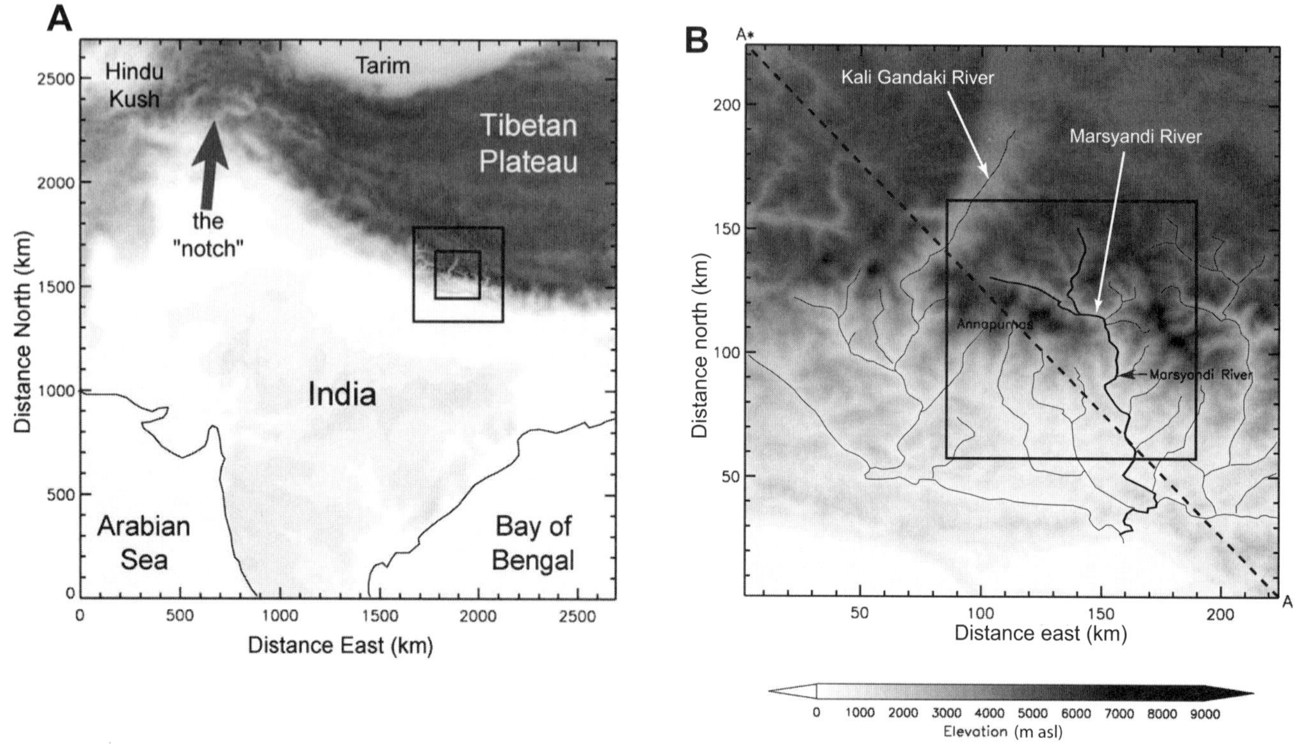

Figure 4. (A) Model computational domain. The grid increments for the three nested meshes are 22.22 km, 3.70 km, and 1.86 km, respectively. (B) Terrain used for the inner domain grid (1.86 km resolution). The rectangular box contains the general area of the hydrometeorological network and current glaciers in central Nepal. The solid line A–A* represents the orientation of the *x-z* cross sections in subsequent analyses.

tive of end members of monsoonal precipitation patterns (1999 and 2001). Furthermore, our research suggests that the character of the ensuing monsoon season correlates with certain characteristics of the monsoon onset event (Barros et al., 2004): a monsoon onset characterized by a well-defined vortex and well-organized convection (e.g., 1999 and 2000) foretells normal or above normal precipitation, with significant penetration of moisture (up to 30% of monsoon total rainfall) into the hydrometeorological shadow in the high Himalayas; whereas a monsoon onset with a weak vortex and widespread disorganized convective activity appears to set the stage for significantly below normal rainfall, especially at high elevations (e.g., 2001). Similarly, we present results for a representative wintertime weather event, specifically a snowstorm in February 2000 (Lang and Barros, 2004). For each case, a description of the synoptic environment of each storm is provided first, followed by a detailed discussion of model results. The analysis of model results will focus on central Nepal (i.e., the network area) where the hydrometeorological network and the glaciers of interest are located.

Description of the Model and Experiment Design

We use a three-dimensional, anelastic, and non-hydrostatic high-resolution model originally introduced by Clark (1979) and Clark and Hall (1991, 1996) in our modeling experiments. Since its original implementation, the model has undergone substantial improvements, and has been particularly successful in studies of small-scale and mesoscale weather phenomena in regions of

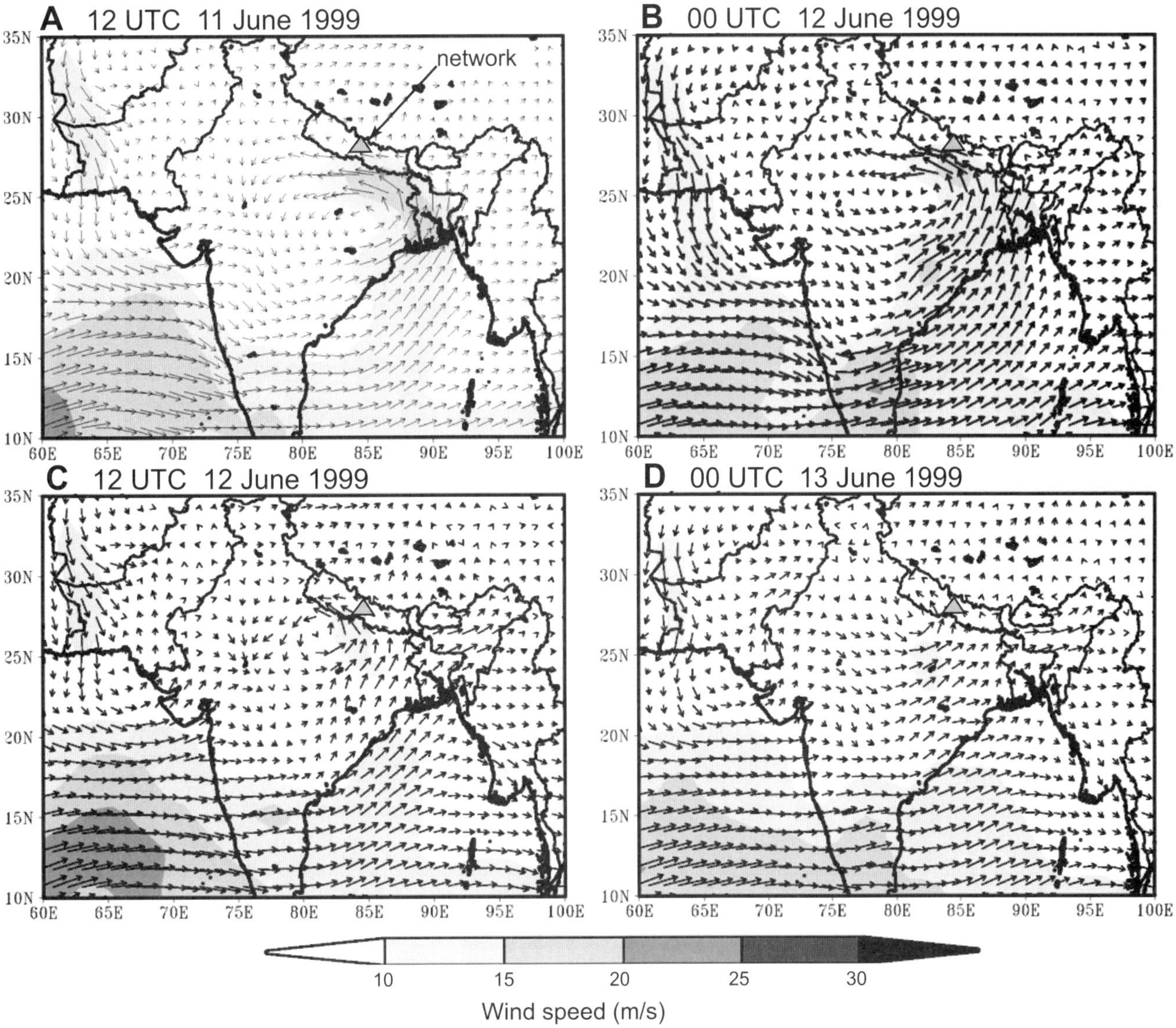

Figure 5. 1999 onset 850 hPa wind streamlines (vectors) and wind speed (shaded) for (A) 12 UTC 11 June, (B) 00 UTC 12 June, (C) 12 UTC 12 June, and (D) 00 UTC 13 June 1999. Gray triangles mark the approximate location of the network. Note that velocities are zero where the terrain height is above 850 hPa (~1.5 km asl).

complex topography, including the Rocky Mountains, the Sierra Madre, and the Himalayas (Clark, 1979; Clark and Hall, 1996; Reinking et al., 2000; Barros and Lang, 2003b; Lang and Barros, 2004, among others).

The model was implemented here using three nested grids. The outer, large-scale domain (22.22 km horizontal resolution) encompassed an area of 2700 × 2700 km, including large portions of the Tibetan Plateau and the Indian subcontinent. The innermost domain (1.86 km horizontal resolution) covered an area of 200 × 200 km, centered on the region where the hydrometeorological network and glaciers are located. The topography used by the model was derived from the Defense Mapping Agency (DMA), United States elevation data, which is available from the Scientific Computing Division of the National Center for Atmospheric Research. Figure 4A displays the topography of the outermost domain with the outline of each of the two inner domains. The model topography for domain 3 (i.e., the inner grid) is representative of the complex local topographic features, with several mountain peaks of the Himalayas higher than 6000 m (Fig. 4B). Time-dependent boundary conditions for model simulations were derived from the European Center for Medium-Range Weather Forecasts (ECMWF) analysis data sets.

In the simulations presented here, the outermost vertical grid increment, Δz, varied smoothly from 50 m at the surface to 200 m at z = 425 m above ground level and increased more gradually to 407 m at 4.15 km above ground level using the stretched vertical coordinate transformation. A modified Kessler (1969) warm rain bulk microphysics parameterization, which predicts the mixing ratios of cloud water (q_c) and rain water (q_r), was used to characterize microphysics. Although this parameterization is lacking in detail necessary to simulate accurate quantitative estimates of precipitation (e.g., ice microphysics are not included), it is adequate to investigate the spatial and temporal evolution of precipitating clouds. Long-wave radiative cooling was described via a combination of the parameterizations of Stephens (1984) and Sasamori (1972). For more information on the implementation of these parameterizations in the model, the reader is referred to Clark et al. (1996). Finally, the version of the model used here does not include land-atmosphere interactions and thus cannot simulate the combined effect of orographically induced ascent

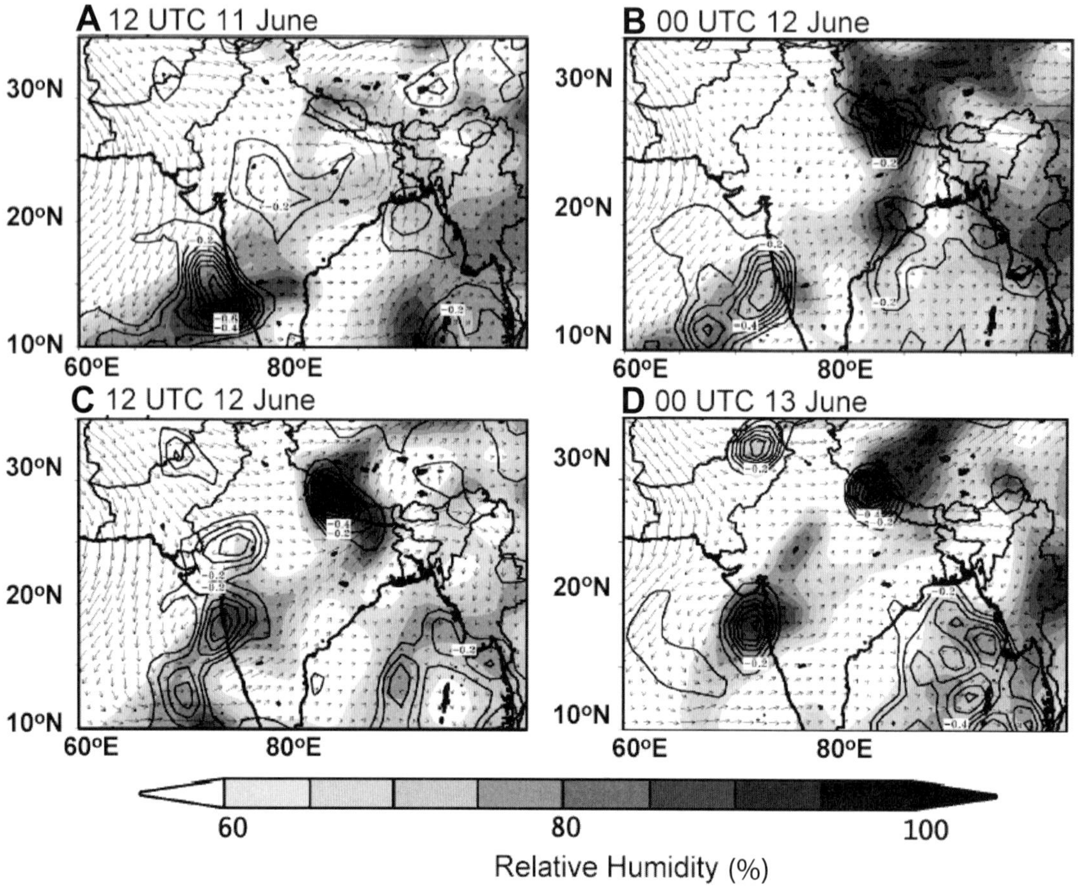

Figure 6. 1999 onset 500 hPa (~5–6 km asl) and relative humidity (RH [%], shaded), horizontal wind streamlines (vectors), and contours of upward vertical wind velocity (0.2 Pa/s ~2 cm/s contour interval; negative velocities are upward; velocity increases from zero at the outermost contour to a maximum at the innermost contour).

and gravity wave dynamics on the stability of the boundary layer, or moisture recycling between the land surface and the lower atmosphere as inferred from analysis of MOHPREX (Monsoon Himalaya Precipitation Experiment) data (Barros and Lang, 2003a).

Overview of Synoptic Conditions

1999 Monsoon Onset

The 1999 monsoon onset in central Nepal was characterized by a low-pressure system (i.e., monsoon depression) that developed in the Bay of Bengal. By 00 UTC on 11 June 1999, this monsoon depression was located near 22.0N, 88.0E, and the low center was ~996 hPa (not shown). Based on 850 hPa winds (~1.5 km in the lower troposphere) from ECMWF objective analysis of model and observations, the evolution of the monsoon depression is depicted in Figure 5 as it moved northwestward toward the central Himalayas at 12 UTC on 11 June (Fig. 5A), 00 UTC on 12 June (Fig. 5B), 12 UTC on 12 June (Fig. 5C), and 00 UTC on 13 June 1999 (Fig. 5D). Note the strong component of flow (>10 m s^{-1}) impinging on central Nepal. Inspection of the vertical structure of horizontal winds, relative humidity distribution, and upward motion fields indicates that the moisture-laden monsoon flow over the mountains extended up to the 500 hPa level (5–6 km in the lower troposphere, Fig. 6), thus implying significant moisture transport to the hydrometeorological network area as well as to the glaciers at higher elevations. The ECMWF analysis fields indicate significant penetration of moisture far into the northern edge of the Himalayan range, consistent with ground observations and IR (infrared) satellite imagery (Barros et al., 2004). Moreover, moisture convergence was enhanced orographically as the incoming air mass reached the mountain slopes (Lang and Barros, 2002; Barros and Lang, 2003a) resulting in the development of a shield-like pattern of mesoscale precipitation with embedded convection, as captured by the TRMM-PR (Tropical Rainfall Measuring Mission- Precipitation Radar) on 12 June 1999 (Fig. 7). The TRMM-PR did not detect precipitation in the northern (rain shadow) area of modern glaciers, and it underestimated rainfall amounts in the network area (Barros et al., 2000). Such systematic underestimation of precipitation at high elevations and in regions of complex terrain in general is a well-known deficiency of satellite precipitation products (TRMM Science Team Working Group Summaries, 2003, available at http://trmm.gsfc.nasa.gov).

2001 Monsoon Onset

The monsoon depression developed in the Bay of Bengal in 12 June 2001 and moved westward toward the Indian subcontinent. The low-pressure center was ~996 hPa before landfall (figure not shown). Associated with this monsoon depression was a southeasterly wind, which brought moist air into the Indian

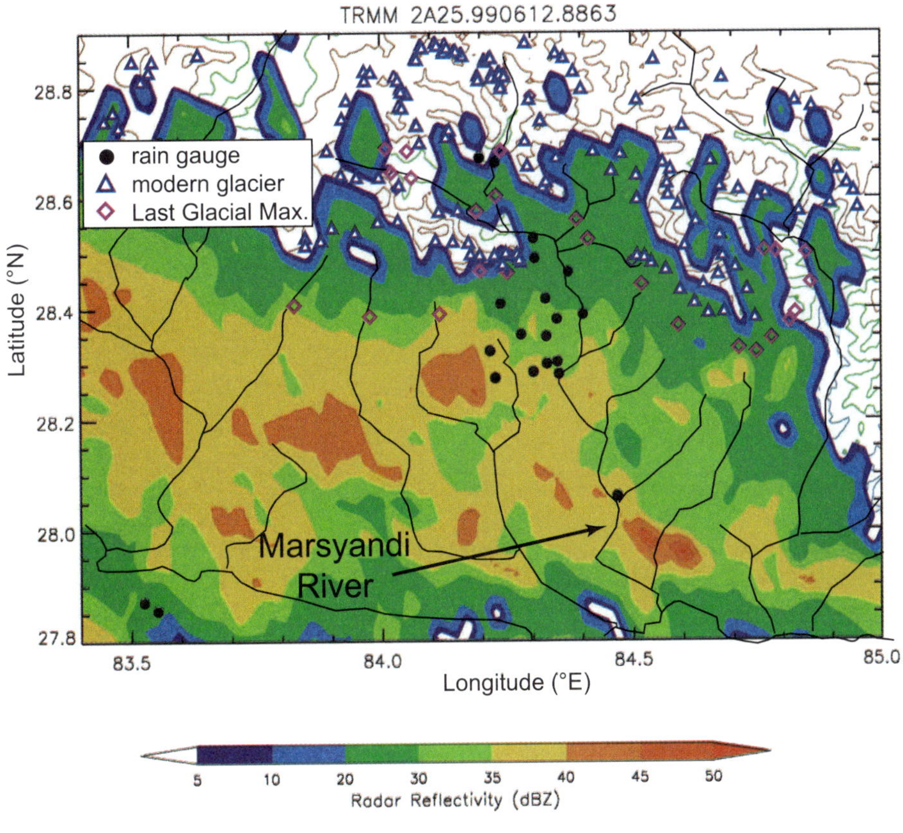

Figure 7. Tropical Rainfall Measurement Mission (TRMM) precipitation radar maximum reflectivity on 12 June 1999. Dark circles indicate positions of hydrometeorological stations shown in Figure 1A. The average location of modern glaciers (triangles) and maximal terminal positions of Late Quaternary glaciers (diamonds) permit assessment of the extent of the remotely sensed precipitation with respect to the glaciers.

subcontinent. On 13 June 2001, the 850 hPa (~1.5 km) height and wind analysis at 00 UTC indicated that the monsoon depression made landfall along the northeast coast of India (Fig. 8A). Twenty-four hours later, by 00 UTC 14 June, the 850 height field clearly showed that the monsoon depression was embedded in the monsoon trough and steered by the flow near the southern periphery of the high-pressure system over the Tibetan Plateau (Fig. 8B). The 500 hPa (~5–6 km) vertical motion and relative humidity fields indicate that strong upward airflow occurred in the western periphery of the monsoon depression (not shown). Subsequently, the monsoon depression moved northwestward away from the central Himalayas.

Although the track of the monsoon depression stayed away from the central Himalayas, significant rainfall (3–6 cm) was observed at the surface network (Barros and Lang, 2003a). The TRMM precipitation radar was able to capture the convection along the periphery of the monsoon depression. TRMM-measured radar reflectivity around 1200 Local Standard Time (LST) 12 June (top panel, Fig. 9) indicates the presence of a strong convective cell near the hydrometeorological network area. By 1200 LST 14 June (Fig. 9, bottom panel), convective activity weakened as the monsoon depression moved further inland, moisture convergence was virtually shut off, and the boundary layer along the Himalayan range became more stable. Once again, negligible precipitation was detected by TRMM at high elevations in the glacier region, as noted previously.

2000 Winter Storm

On 11 February 2000 a snowstorm developed in the Himalayan range, and was detected by ground observations in central Nepal (Lang and Barros, 2004). The 850 hPa height and wind fields from ECMWF analyses on 00 UTC 11 February 2000 (Fig. 10A) show a shallow depression that developed over the western Himalayas when a westerly trough passed over the region between the Himalayas and the Hindu Kush in the northwestern corner of the Indian subcontinent (Fig. 10A). Twelve hours later (Fig. 10B), this system, which drew moisture from the Arabian Sea, intensified as the trough moved eastward. During this period, a strong south-southeasterly low-level jet accompanied the eastward-moving deep trough. Orographic lifting led to the redistribution of the incoming moisture flux and enhancement of precipitation processes over the mountains. At the ground stations in central Nepal, the snow records indicate snow depth accumulations from 5.5 to 22.2 cm (snow density ~0.5; Lang and Barros, 2004). During the life cycle of the system, strong upward motion was observed over the western Himalayas and the Hindu Kush mountains in the ECMWF analysis data set, while strong westerlies over the Indian subcontinent encroached against the Himalayas. The TRMM satellite did not have an overpass concurrently with this event.

Results of Numerical Experiments

1999 Monsoon Onset

A 36 h simulation was run from 00 UTC 12 June to 12 UTC 13 June 1999 for the 1999 monsoon onset event. Time-dependent initial and lateral boundary conditions from ECMWF were imposed at 6 h intervals. Figures 11A and 11B show the simulation results 1.5 km above the ground surface for the outer domain (Δx = 22.22 km) at 12 UTC (18 LST) 12 June, and 00 UTC (06 LST) 13 June 1999, respectively. On 12 UTC 12 June 1999, the

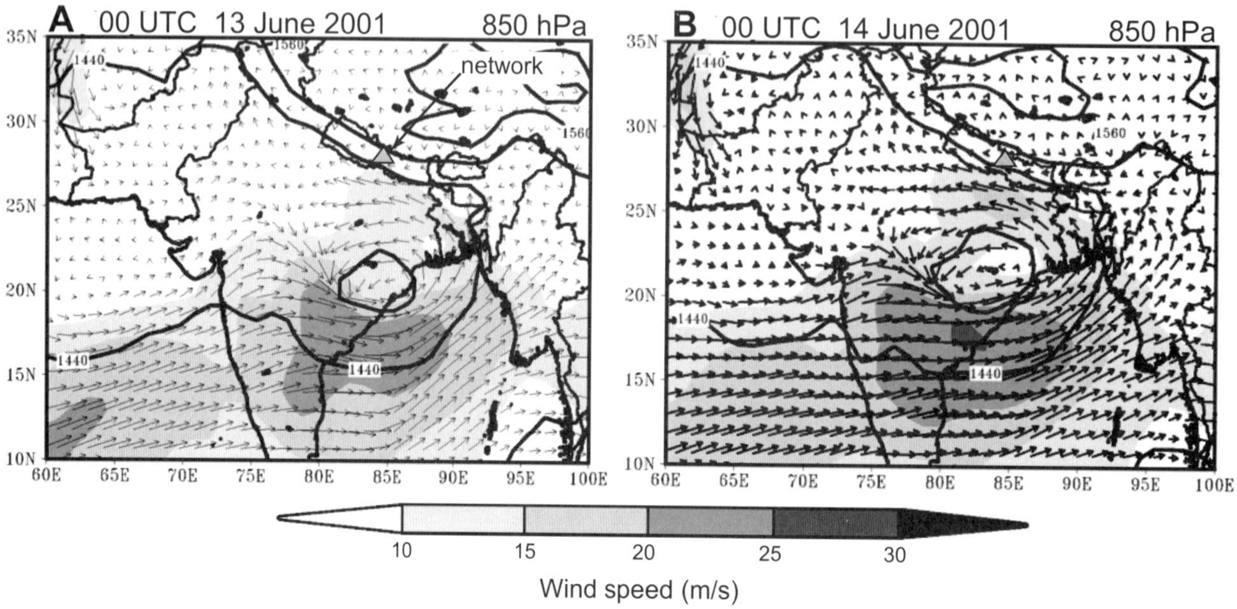

Figure 8. 850 hPa wind fields from European Center for Medium-Range Weather Forecasts (ECMWF) analysis at (A) 00 UTC 13 June, and (B) 00 UTC 14 June 2001, streamlines (vectors), wind speed (shaded), and geopotential heights (labeled contours at 120 m interval). Note that velocities are zero where the terrain height is above 850 hPa (~1.5 km asl).

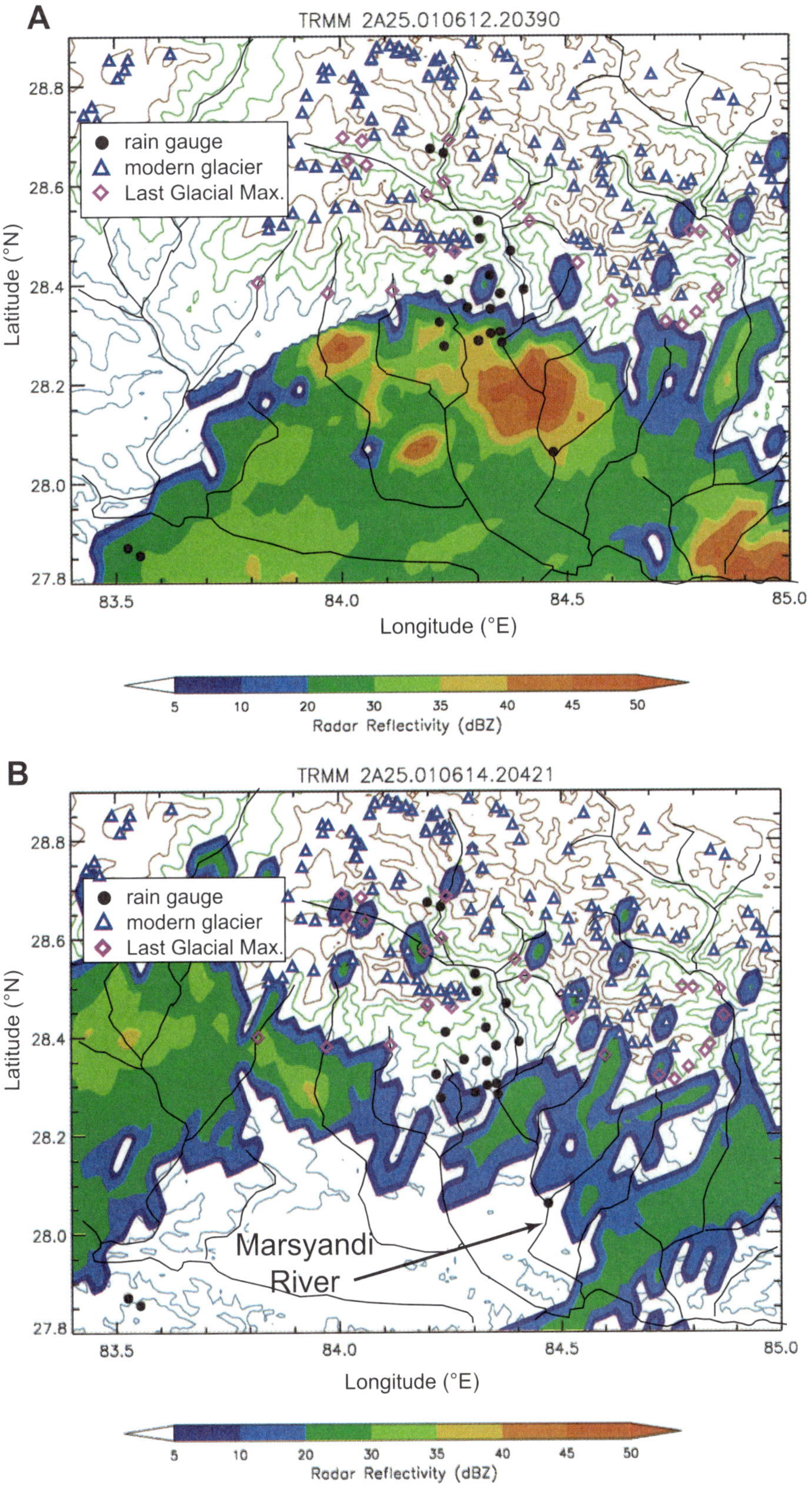

Figure 9. (A) Tropical Rainfall Measurement Mission (TRMM) precipitation radar maximum reflectivity at 06 UTC 12 June 2001. (B) TRMM precipitation radar maximum reflectivity at 06 UTC 14 June 2001. Note that, in comparison to Figure 7, northward moisture penetration is greatly reduced. Symbols are as in Figure 7.

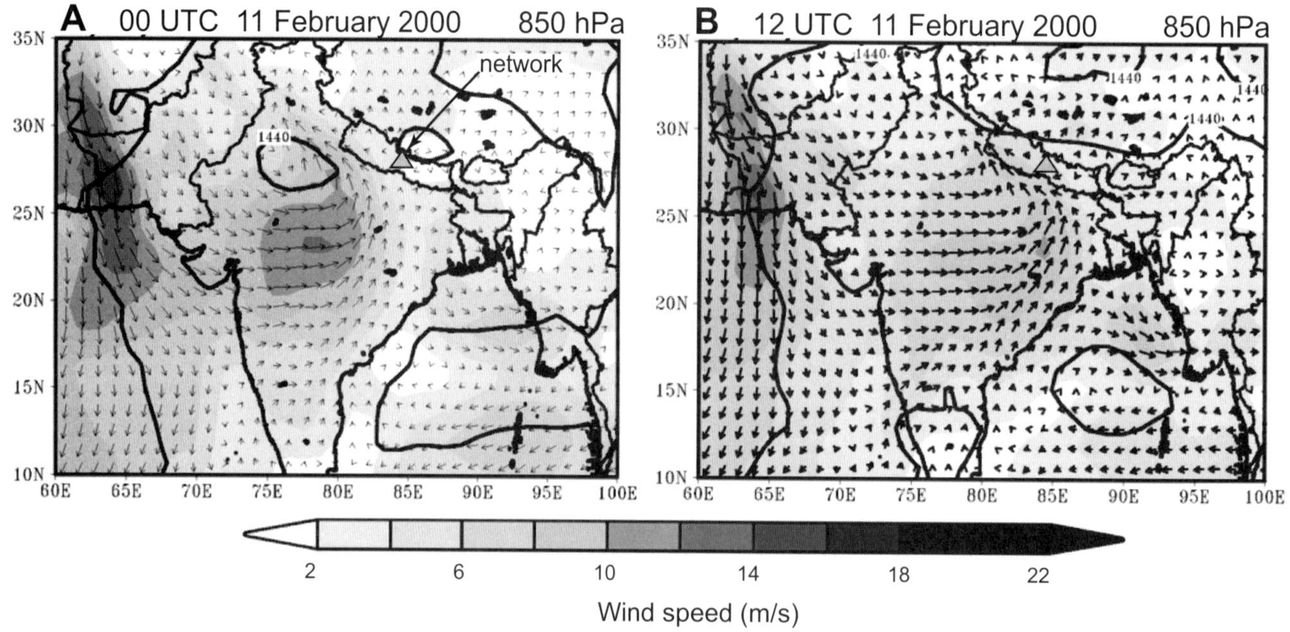

Figure 10. 850 hPa horizontal wind streamlines (vectors) and wind speed (shaded) from European Center for Medium-Range Weather Forecasts (ECMWF) analysis at (A) 00 UTC and (B) 12 UTC 11 February 2000. Gray triangles mark the approximate location of the network. Note that velocities are zero where the terrain height is above 850 hPa (~1.5 km asl).

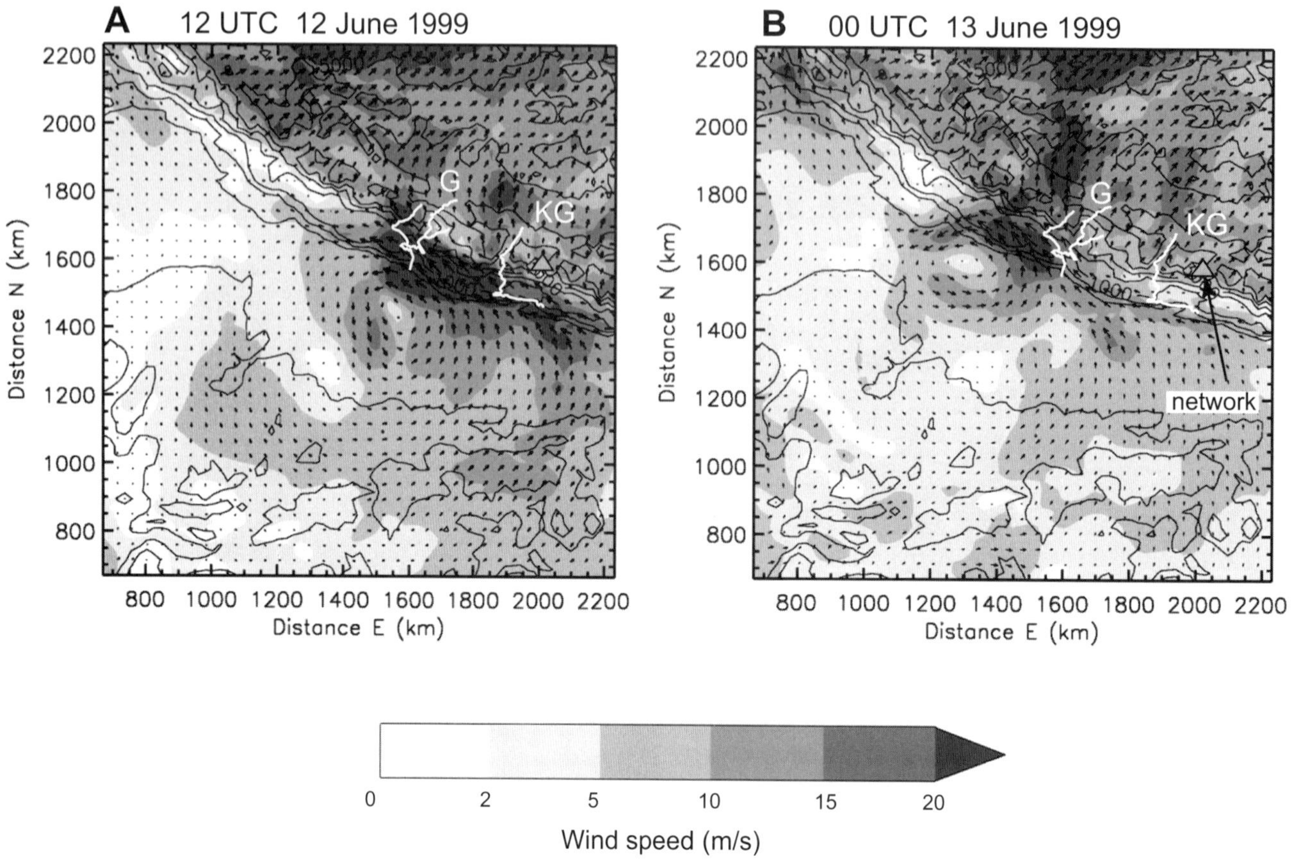

Figure 11. Simulated horizontal wind streamlines (vectors) and wind speed (shaded) 1.5 km above the ground surface for the outer model domain (domain 1, 22.22 km grid resolution) at (A) 12 UTC 12 June 1999, and (B) 00 UTC 13 June 1999. Note the strong northward flow toward the Tibetan Plateau along the Kali Gandaki (KG) and Ghaghra (G) drainages, whereas a region of more stagnant air lies between them. Contours show orography in increments of 1000 m.

model captured correctly the position of the core of the monsoon depression, placed over central Nepal at ~26°N (compare Fig. 3 with Fig. 11A). An impinging wind on the order of 20 m s^{-1} was simulated at that time. The track and the circulation of the monsoon depression are quite well reproduced as it moved northwestward along the Himalayas (Fig. 11B). The moist Froude number Fw is ~0.5 ($Fw = U/N_w h_m$, where U is the basic flow speed near the mountain, N_w is the moist static stability of the incoming airflow, and h_m is the average mountain height. In this case, $U = 20.0$ m s^{-1}, $Nw = 0.0098$ s^{-1}, $h_m = 4000$ m), which indicates that the impinging flow was only partially blocked by the terrain at low levels in the atmosphere (<2 km), consistent with the small area of stagnant flow in Figure 11A between the Ghaghra and Kali Gandaki valleys. Note the jet-like structure of the southerly flow along the Kali Gandaki (KG) and western Ghaghra (G) River valleys, which function as open conduits delivering atmospheric moisture to the northern flanks of the Himalayas. This behavior is consistent with results of covariance analysis of satellite imagery and topography suggesting that trans-Himalayan river valleys determine the multiscaling behavior of cloud fields in the Great Himalayas (Barros et al., 2004). Long-term indirect evidence of the along-valley channelization of atmospheric flows and associated precipitation is also patent in oxygen isotope (e.g., $\delta^{18}O_{mw}$) analysis of stream flow with respect to altitude along the Kali Gandaki River valley (Garzione et al., 2000).

Let us now focus on the simulated mesoscale cloud features during the northwestward movement of the monsoon depression. The simulation (Fig. 12) clearly depicts the distribution of maximum cloud liquid water in the vertical air column for the innermost domain (Δx = 1.86 km) at 12 UTC 12 June (left panel), and 00 UTC 13 June 1999 (right panel). Both windward and lee side of individual mountain peaks received a considerable amount of moisture when the monsoon depression reached the central Himalayas. Note how the organization of clouds fields exhibits a strong association with the spatial arrangement of topography. In particular, the highest liquid water content occurs along the major river valleys and along the windward flanks at middle latitudes (2–4 km) in the High Himalaya.

To gain further insight into the interactions between monsoon depression and orography, we now analyze the distribution of vertical velocity, cloud liquid water, and equivalent potential temperature in the innermost domain along the cross section traversing the Himalaya (section A–A*, Fig. 4B). Orographic forcing leads to the spatial organization of strong and deep updrafts in the foothills of the Himalayan range accompanied by the propagation of orographic gravity waves with updrafts concentrated on upwind slopes (S and SE facing slopes), while downdrafts dominate on top and on the lee side (N and NW facing slopes) of individual terrain peaks at 12 UTC 12 June (Fig. 13C). This topographically driven organization leads to the

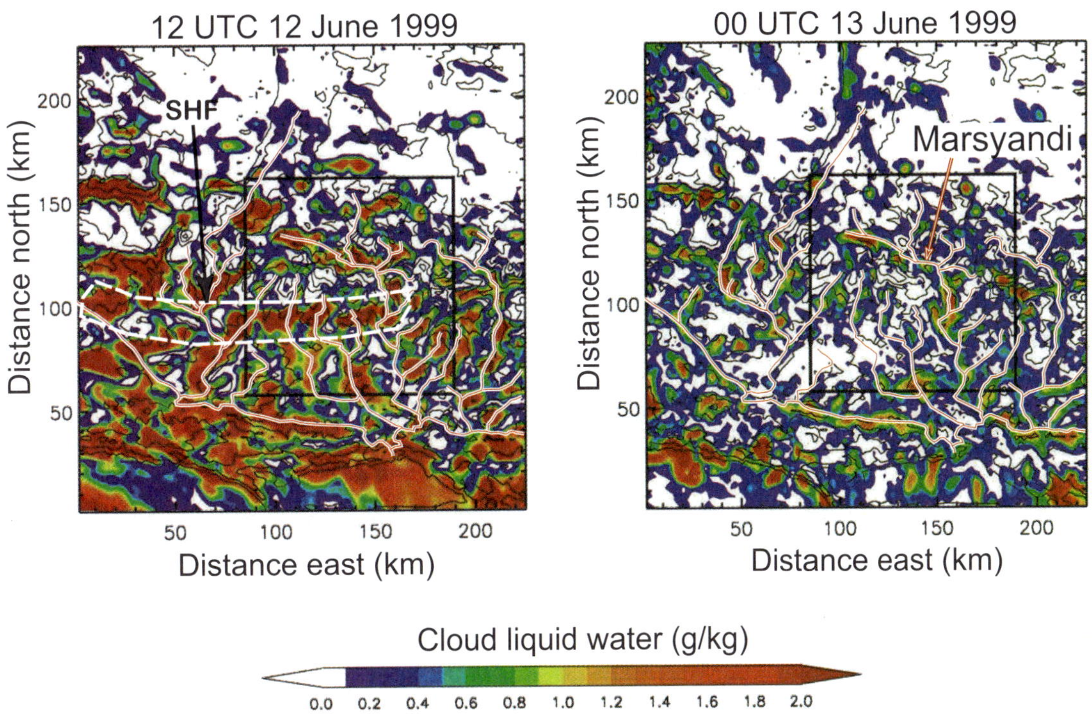

Figure 12. Simulated fields of maximum cloud water in the atmospheric column in the inner domain (domain 3, 1.86 km grid resolution) at 12 UTC 12 June 1999 (left panel), and 00 UTC 13 June 1999 (right panel). Dashed region delineates the southern Himalayan flank (SHF), where rain is intense.

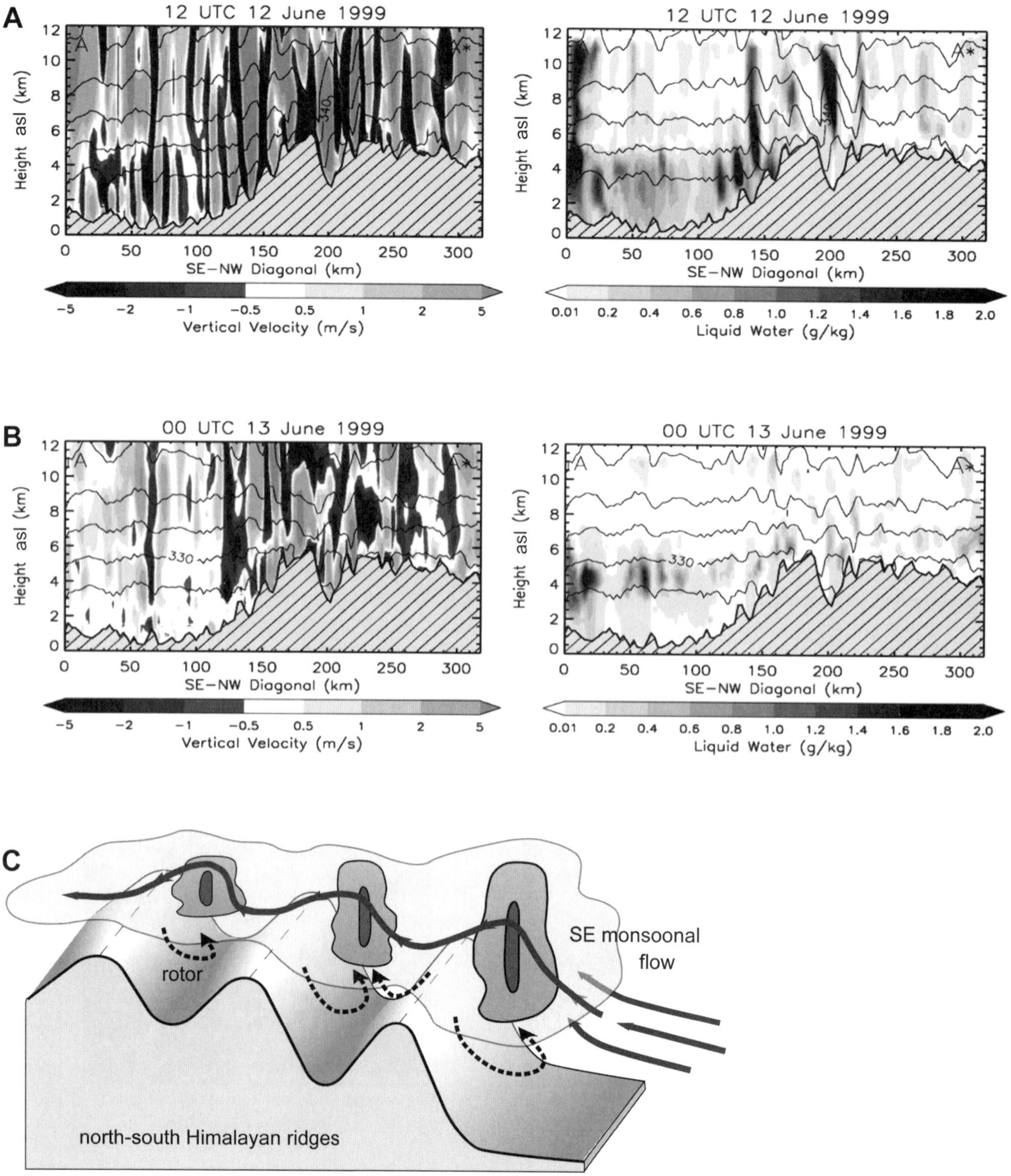

Figure 13. Simulated *x-z* cross section of vertical velocity (m s^{-1}), potential temperature (contour interval 10 K), and cloud liquid water content (g kg^{-1}): (A) 12 UTC 12 June 1999, and (B) 00 UTC 13 June 1999. The cross-section A–A* is depicted in Figure 4B. (C) Cartoon of the orographically induced gravity waves due to SE monsoon flows impinging obliquely on north-south–oriented ridges extending southward from the Himalayan crest. Forced ascending air focuses precipitation on the upwind (W to SW) slopes, with reduced rainfall on downwind slopes and some ridge crests.

formation of cells of high cloud water content, which extends northward into normally dry regions like the Tibetan Plateau at ~12 UTC 12 June (Fig. 13A). Twelve hours later, by 00 UTC 13 June (Fig. 13B), vertical motion weakened on the upwind slopes as the depression moved away toward the west, cutting off the influx of moisture and leading to significant changes in the distribution of cloud water at 00 UTC 13 June.

These results illustrate the role of monsoon depressions as a mechanism for transferring moisture (and latent heat energy) from the Bay of Bengal to the Himalayas and into Tibet. The effectiveness of this transport is controlled by the depression path, and by the number (frequency) of depressions that originate from the Bay of Bengal during the summer monsoon. Over long time scales, and even for different global climate regimes (Bradley, 1991; Kutzbach et al., 1993; Clemens et al., 1991; among others), we propose that the path and number of monsoon depressions that reach the mountain range and penetrate beyond the upwind crest can be used as predictive measures of monsoon intensity. Such events determine precipitation accumulation in existing glaciers (Harper and Humphrey, 2003), the formation of new glaciers, as well as altitudinal-dependent erosion patterns and mechanisms (landslides, river incision rates), thereby potentially influencing landscape evolution (Gabet et al., 2004).

2001 Monsoon Onset

The simulation of the 2001 monsoon onset began at 00 UTC 12 June and ran for 48 h. Model results of wind fields 1.5 km above the ground surface at 00 UTC 13 June and 00 UTC 14 June are shown in Figures 14A and 14B, respectively. The simulation results indicate a cyclonic flow around the northeast coast of India on 00 UTC 13 June (Fig. 14A), a result consistent with the ECMWF analysis (Fig. 8A). By 00 UTC 14 June 2001 (Fig. 14B), the cyclonic circulation remained stationary (Fig. 8B). Overall, the model is able to predict well the location of the monsoon depression over northeast India. The left and right panels in Figure 15 show the maximum cloud water at 00 UTC 13 and 00 UTC 14 June 2001, respectively. Although cells of cloud liquid water formed in the vicinity of individual mountain peaks, large areas in the central Himalayas were not affected by the monsoon depression directly and show clear skies. The southwest to northeast cross sections of simulated vertical

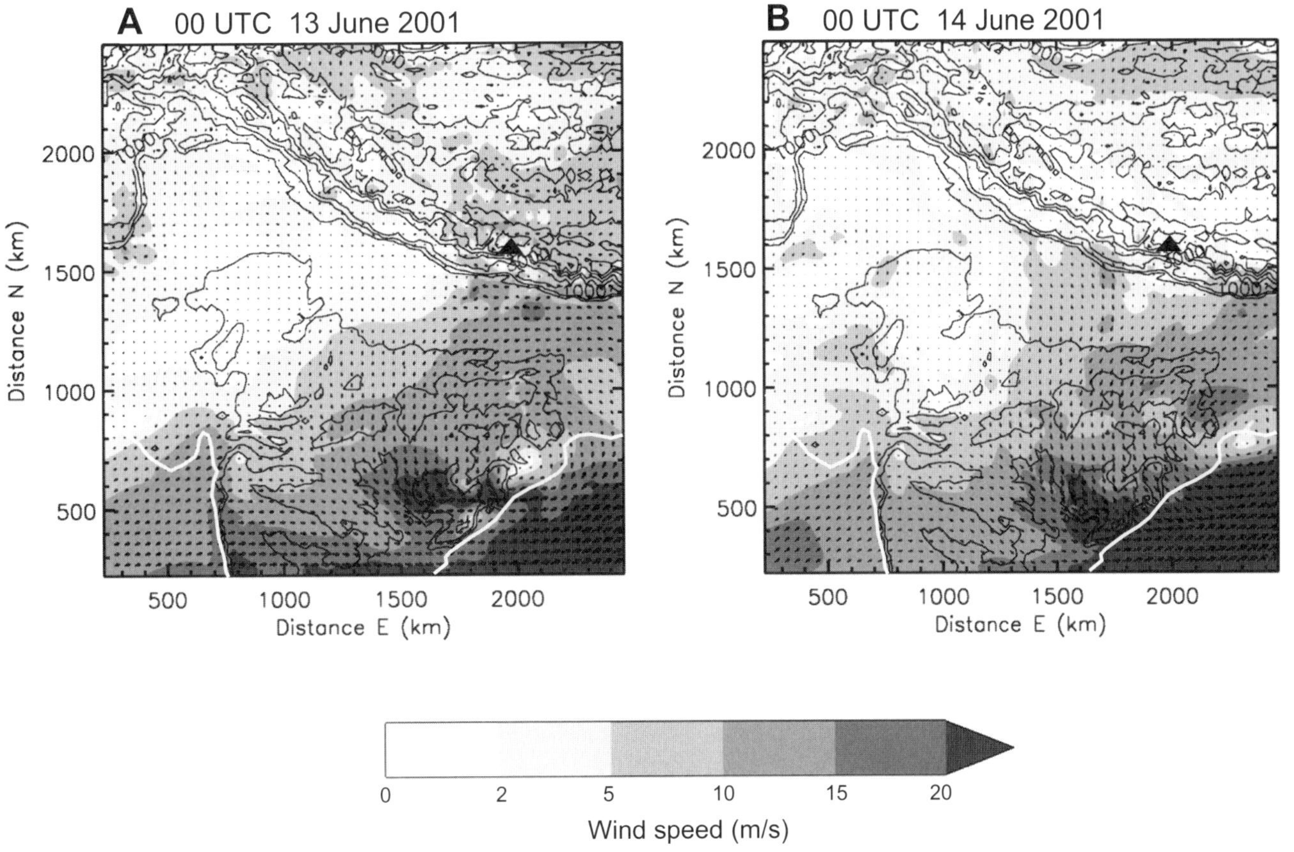

Figure 14. Simulated wind streamlines and wind speed (shaded) 1.5 km above the ground surface for the outer domain (domain 1, 22.22 km grid resolution) at (A) 00 UTC 13 June 2001, and (B) 00 UTC 14 June 2001. Contours show orography in increments of 1000 m. Black triangles mark the approximate location of the network.

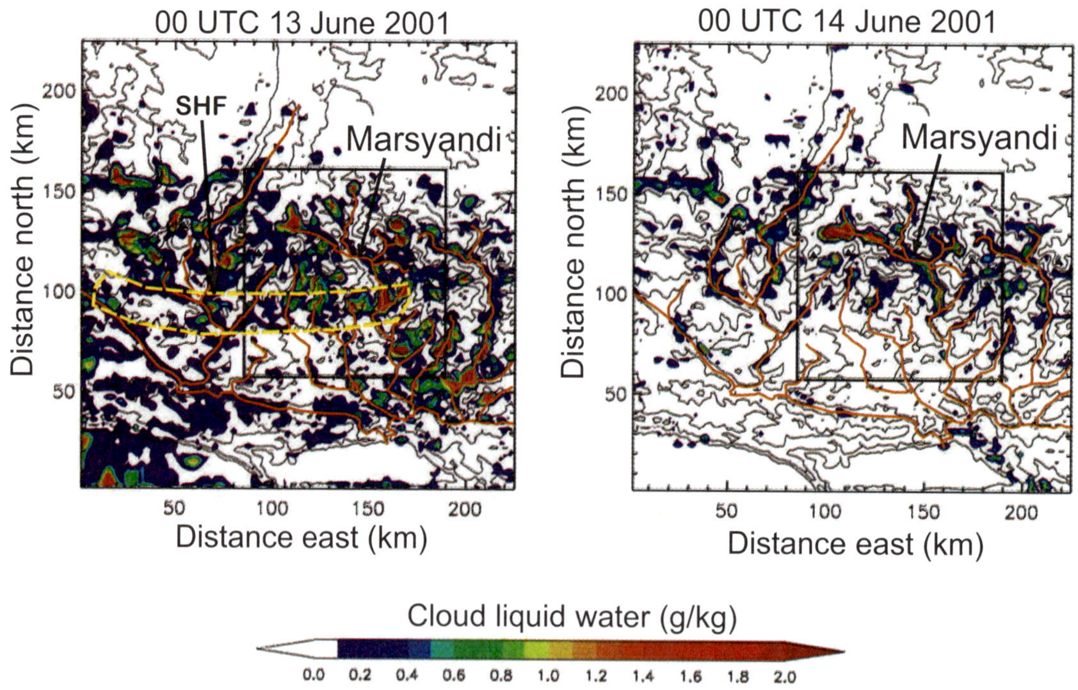

Figure 15. Simulated fields of maximum cloud water in the atmospheric column in the inner domain (domain 3, 1.86 km grid resolution) at 00 UTC 13 June 2001 (left panel), and 00 UTC 14 June 2001 (right panel). Note the significantly reduced precipitation along the southern Himalayan flank (SHF) in comparison to the monsoon onset of June 1999 (Fig. 12).

motion and cloud liquid water (Fig. 16) show that the moisture flux over the complex terrain is much reduced as compared to that in 1999. As a consequence, precipitation is very light over the Himalayan range in general, and on the upwind slopes in particular. The one-to-one correspondence between the magnitude and spatial distribution of the updrafts and the spatial arrangement of successive valleys and ridges shows the importance of orographic gravity waves in organizing local circulations despite weak synoptic forcing. As the monsoon depression moved westward during 14 June 2001, the cloud liquid water reduced dramatically everywhere (Fig. 15, right panel). During that time, a band-like feature of high cloud water content (~2.0 g kg^{-1}) is simulated over the upper Marsyandi Valley, as a result of up-valley easterly flow (Fig. 14B). As shown in Figure 16B, strong vertical motion is simulated at high elevations around 00 UTC 14 June 2001, while it is very weak along the front range. Consequently, cloud water content is very low everywhere due to the lack of sustained moisture influx into the region.

In summary, the numerical simulations predict that orographically organized convection (Fig. 13C) is a persistent feature of weather in the Great Himalayas, as previously inferred from the analysis of satellite imagery over a wide range of scales (Barros et al., 2004). Nevertheless, significant summertime precipitation only occurs when strong and sustained moisture transport into the region is established by environmental conditions associated with strong monsoon depressions (and onsets in particular) on a collision path with the orographic barrier. Moreover, deep penetration of monsoon moisture into the rain shadow north of the high Himalayan summits is predicted to occur only when the monsoon depression impinges directly on the range front, and moisture convergence is channelized along the major river valleys. Therefore, the summertime mass input to the glaciers located at high elevations and sprawling into the northern flanks of the Himalayas and southern Tibet is conditioned on the occurrence of such discrete events.

2000 Winter Storm

The simulation started at 00 UTC 11 February and ran for 36 h. The simulated horizontal wind fields 1.5 km above the ground surface for a portion of the outer model domain at 06 UTC 11 February 2000 (Fig. 17A) and 00 UTC 12 February 2000 (Fig. 17B) show the evolution of a southwesterly low-level jet impinging on the central Himalayas at 10–15 ms^{-1}. This low-level jet is associated with the cyclonic flow around the shallow depression, and turns slightly westward as it approaches the orographic barrier. Although the westward turning of low-level flow was not as significant as in the ECMWF analysis (Fig. 10), the model successfully captures the timing of the storm in central Nepal from 06 UTC to 12 UTC 12 February 2000.

The spatial distribution of maximum cloud water in the innermost model domain is shown in Figure 18. By 06 UTC 11 February (Fig. 18, left panel), scattered clouds are present over

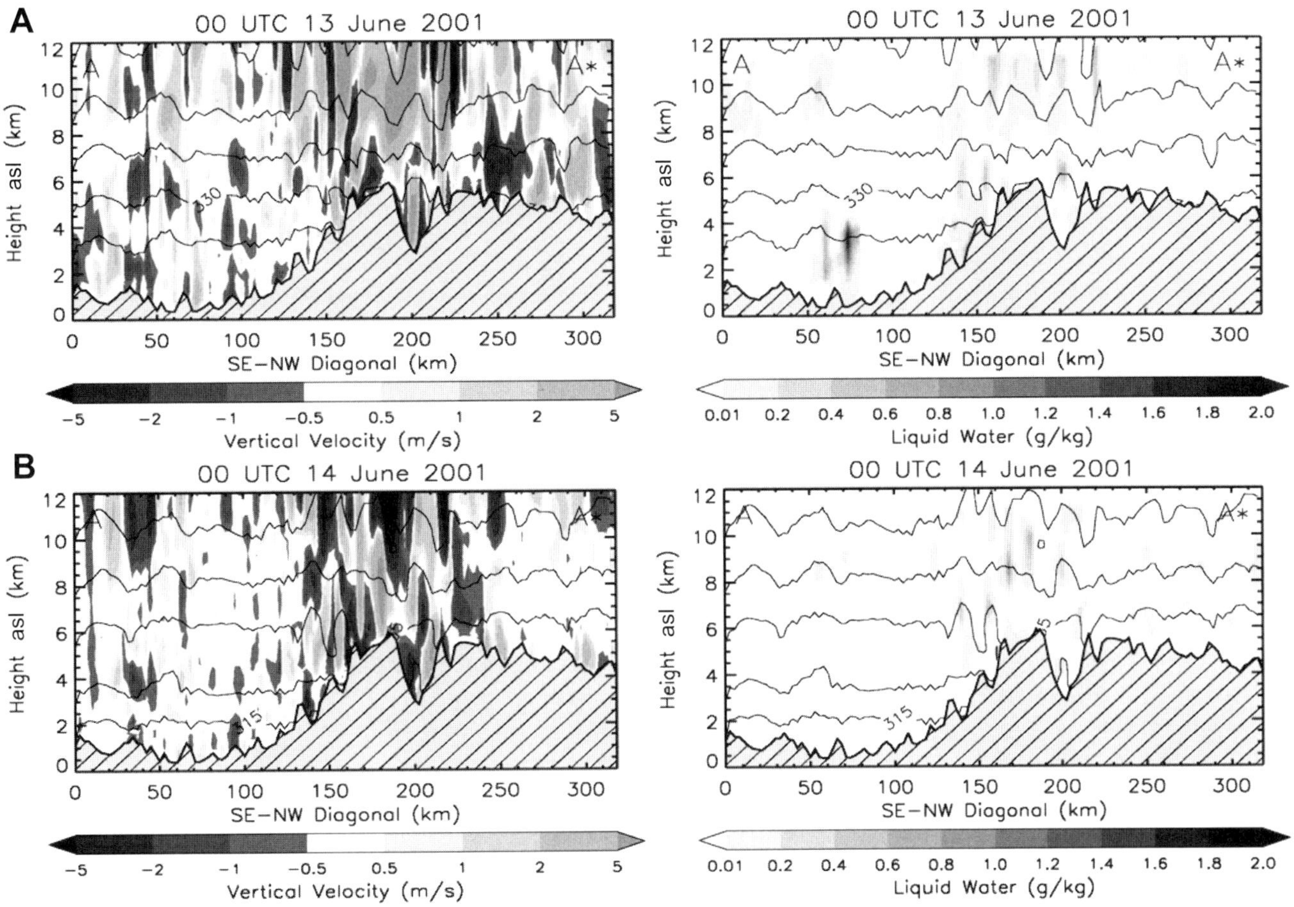

Figure 16. Simulated *x-z* cross section of vertical velocity (m s^{-1}), potential temperature (contour interval 10 K), and cloud liquid water content (g kg^{-1}): (A) 00 UTC 13 June 2001, and (B) 00 UTC 14 June 2001. The cross-section A–A* is depicted in Figure 4B.

the network area, especially over the Annapurnas, with a long linear cloud feature on the windward side, a feature consistent with forced orographic lifting along the Lesser Himalayas in the absence of environmental instability (i.e., no embedded convection). Subsequently, clouds developed around isolated mountain peaks (Fig. 18, right panel) as the trough moved eastward. Note that, in contrast to the monsoon onset events (Figs. 12 and 15), cloud formation along the Marsyandi Valley during the passage of the winter trough is negligible (Fig. 18). Instead, moisture is focused on the surrounding ridges as a result of sustained strong southwesterly flow (Fig. 17) that leads to the development of strong downdrafts in the valley by 00 UTC 12 February.

Cross sections of simulated vertical velocity and cloud water content along the cross-section A–A* (Fig. 4B) are presented in Figure 19. Again, as in the case of monsoon depressions, these results suggest that vertical motion is strongly tied to orographic features, consistent with the generation of orographic gravity waves from the interaction between the impinging air mass and the Himalayan range on 06 UTC 11 February (Fig. 19A) and 00 UTC 12 February 2000 (Fig. 19B), respectively. Note the persistent cloud features in the Lesser Himalayas (Fig. 19), where linear zones of concentrated moisture are predicted (Fig. 18). The vertical cross section of cloud liquid water distribution reveals shallow clouds on 11 and 12 February 2000 (Figs. 19A and 19B), similar to cloudiness fields derived from METEOSAT-IR imagery (see Lang and Barros, 2004, for details). Consequently, precipitation amounts produced by this cold season storm are substantially lower than those attained during the 1999 onset events, though higher than during the 2001 onset. The average number of wintertime storms is on the order of a handful per year, and their cumulative contribution to annual precipitation is significant, especially at high elevations (as much as 100 cm for elevations above 3500 m in the Marsyandi network; Fig. 1B).

OVERVIEW AND DISCUSSION

We have characterized the complexity and space-time intermittency of precipitation in the central Himalayas via a combina-

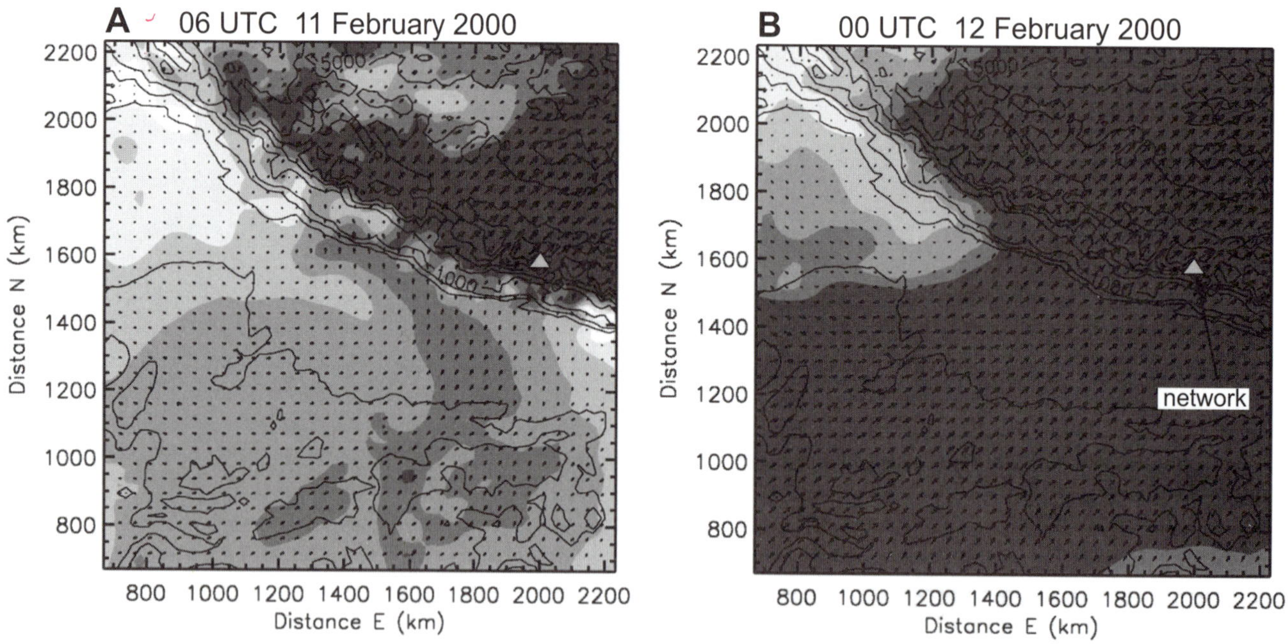

Figure 17. Simulated wind streamlines (vectors) and wind speed (shaded) 1.5 km above the ground surface for the outer domain (domain 1, 22.22 km grid resolution) at (A) 06 UTC 11 February 2000, and (B) 00 UTC 12 February 2000. Contours show orography in increments of 1000 m. Gray triangles mark the approximate location of the network.

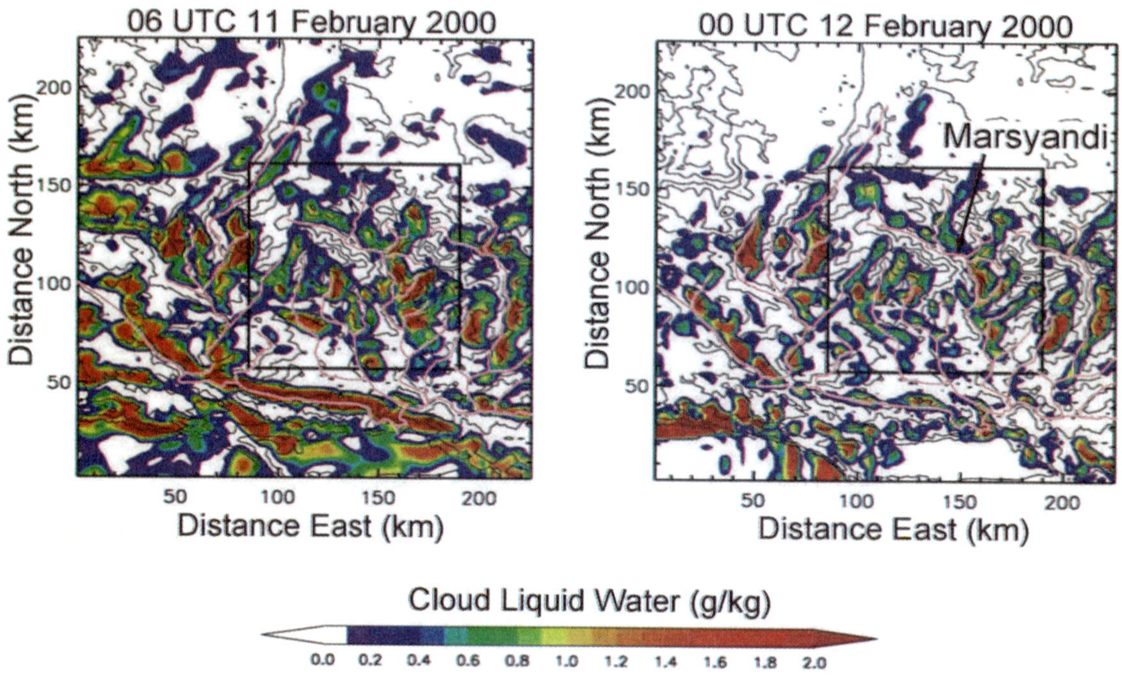

Figure 18. Simulated fields of maximum cloud water in the atmospheric column in the inner domain (domain 3, 1.86 km grid resolution) at 06 UTC 11 February (left panel), and 00 UTC 12 February 2000 (right panel). Note that, in comparison to monsoon simulations (Figs. 12 and 15), much precipitation is focused on ridges, rather than along river valleys.

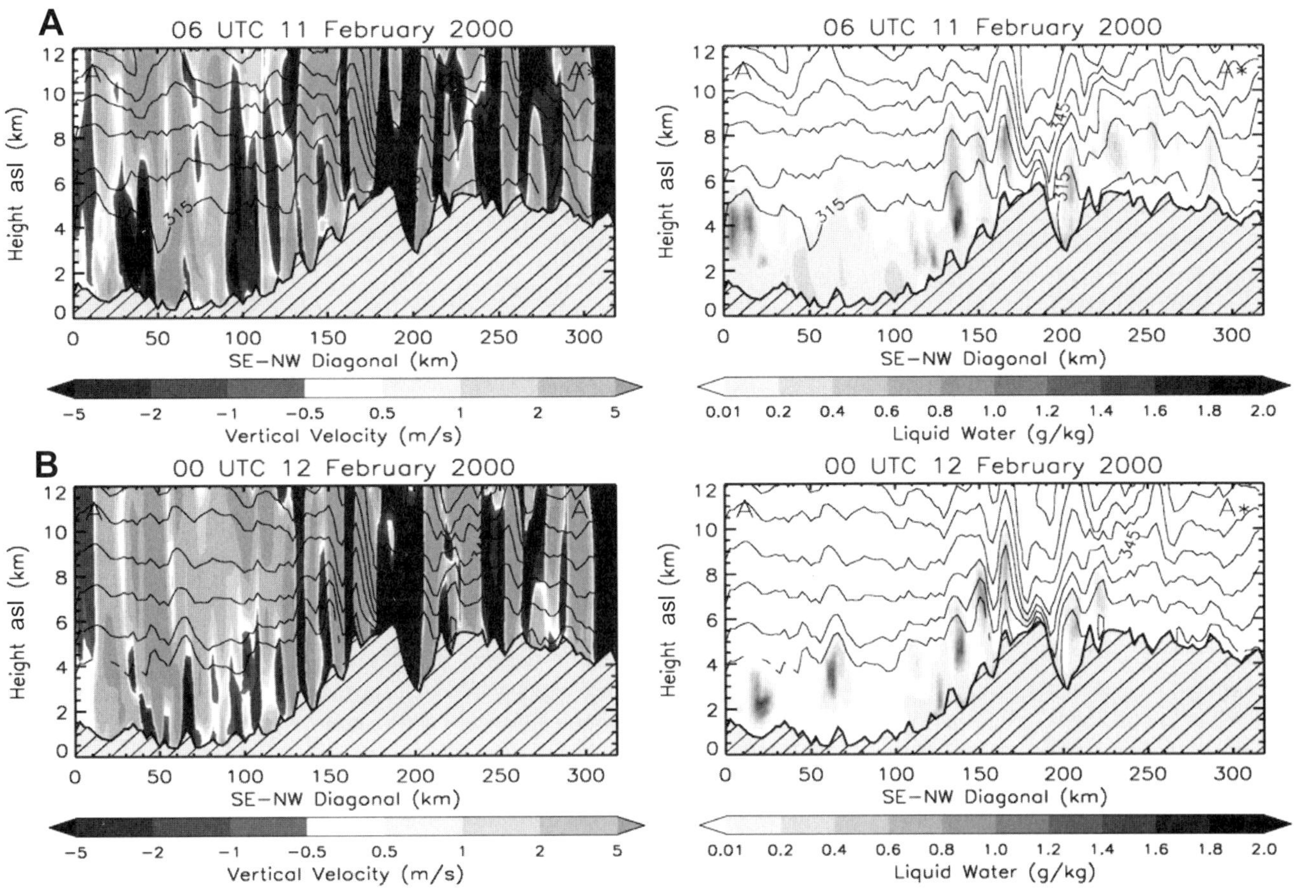

Figure 19. Simulated *x-z* cross section of vertical velocity (m s^{-1}), potential temperature (contour interval 10 K), and cloud liquid water content (g kg^{-1}): (A) 06 UTC 11 February, and (B) 00 UTC 12 February 2000. The cross-section A–A* is depicted in Figure 4B.

tion of ground observations, satellite imagery, and model simulations of dominant weather systems for current climate conditions. The role of orographic land-atmosphere interactions in the spatial organization of precipitation systems is exemplified by three fundamental modes of spatial variability: a synoptic-scale mode linked to global climate controls; a synoptic-scale mode linked to regional climate controls related to monsoon intensity and landform; and finally, a mesoscale mode linked to local orography (Barros et al., 2004). In this work, we show how independent model simulations of representative weather events (monsoon depression and wintertime storms) appear to effectively capture the synoptic and mesoscale modes of variability.

Depending on wind direction, flow over the terrain creates a unique pattern of gravity waves, with vertical motion forming band-like structures oriented roughly parallel to the dominant topographic ridges, which force the waves. Environmental conditions favor the formation of vertically deep waves, with phase lines that tilt into the wind, leading to the simulated phase difference between topography and vertical motion consistent with the theory of Smith (1979). These waves influence cloud heights, as lower cloud tops often are associated with ridges, and higher cloud tops over the intervening slopes. Cloud liquid water content (and by inference rainfall) is maximized where updrafts and the potential temperature gradient are stronger. The simulated spatial patterns of rainfall are consistent with observations from both monsoon onsets, which suggests that that orographic lifting is the key rainfall process (Lang and Barros, 2002). Depending on the synoptic conditions, the gravity wave cells of vertical motion would be expected to modulate the intensity of any incoming convective activity, as well as to prompt the localized release of low-level convective instability in the regions of forced ascent (Barros and Lang, 2003a). This is consistent with the results from Reinking et al. (2000), who showed that orographically produced gravity waves can have important effects on rainfall.

Although a much longer numerical integration is required to fully elucidate the effects of radiative processes and terrain-atmosphere interactions over longer time scales, these model simulations of principal weather systems imply a strong association between rainfall and spatial wind patterns modulated by orography. This association is supported by the analysis of satellite imagery of clouds. Thus, we should expect relative distributions of rainfall on the southern slopes to remain constant from year to year (although absolute amounts will be subject to the interannual variability of the Asian monsoon itself). This inference is further supported by the observations of Lang and Barros (2002), Barros and Lang (2003a), and Barros et al. (2004), who found that relative rainfall distributions remained stable in 1999, 2000, and 2001 in the Marsyandi network. On the northern slopes and in the hydrometeorological rain shadow (Fig. 1B), the temporal scales at which spatial patterns of cumulative precipitation remain stable are at least on the order of decades, consistent with multidecadal variability of the monsoon. Overall, analysis and integration of ground observations, satellite imagery, and model simulations of dominant weather systems for current climate conditions emphasize the need to describe and interpret climate records accounting for chaotic and stochastic variability over a wide range of temporal and spatial scales (frequency, intensity, path distributions of monsoon depressions, and spatial distribution of associated rainfall).

At geologic time scales, a spatially fixed rainfall distribution should have significant effects on erosion and geomorphology, suggesting an interesting long-term feedback between the terrain and atmosphere. For example, persistently greater intensity and magnitude of rainfall on southeasterly flanks of ridges could lead to lower channel slopes, interfluve heights (Whipple et al., 1999), and hillslope angles (Gabet et al., 2004). Furthermore, indirect evidence of spatial consistency between active deformation and heavy rainfall patterns in the Himalayas suggests that this feedback may impact the spatial organization of tectonic activity (Hodges et al., 2004).

Despite recent progress in characterizing the spatial variability of precipitation in major mountain ranges like the Himalayas, the detailed structure of that variability and the fundamental spatial and temporal modes of climate change have remained poorly defined. Similarly, the extent to which orography may be a product of climate-erosion interactions is unknown (Burbank et al., 2003; Reiners et al., 2003; Wobus et al., 2003). Many landscape models either assume a uniform distribution of precipitation (Willett, 1999; Ellis et al., 1999), gradients in precipitation as a function of horizontal distance (Beaumont et al., 2000; Jamieson et al., 2002), or step-wise changes between domains of uniform precipitation (Willett, 1999; Ellis et al., 1999). Some conceptual models assume a homogeneous variation of annual precipitation with elevation, consistent with uniform orographic enhancement of rainfall across a mountain range (Roe et al., 2003). Both our observations and modeling in the Himalaya show strong spatial variability in precipitation at the scale of individual ridges and valleys that serve as obstacles to incoming storms. Our work suggests that annual precipitation averages might be considered appropriate approximations of climate gradients only for mountain ranges oriented perpendicular to the major storm tracks, because moisture is advected directly toward the range crest (i.e., north-south orientation such as occurs for the Rocky Mountains with respect to the prevailing westerlies). Furthermore, the representation of orographic enhancement of rainfall in terms of a linear dependence with elevation neglects the contribution of orography to organize and enhance convective activity, which is responsible for the diurnal cycle of rainfall in mountainous regions, and for spatially persistent heavy precipitation in particular (e.g., the so-called "cloud bursts" in the Himalayas; Barros et al., 2004). Montgomery and Brandon (2002) suggested that frequency in landsliding plays a major role in controlling landscape-scale erosion in tectonically active mountain ranges. Because landsliding activity can be often linked to heavy precipitation (Gabet et al., 2005), the frequency and spatial pattern of orographic convective precipitation must not be ignored.

Simulations using global climate models, with and without the Himalayas and Tibetan Plateau, indicate that the mountains do not have an influence on whether the monsoon takes place (e.g., Zhisheng et al., 2001; Kitoh, 2001; Abe et al., 2003). However, they do show that the presence of the mountains contributes to lengthening the duration of the monsoon; for simulations using different average terrain elevations, more elevated mountain ranges are correlated with earlier monsoon onsets and more intense monsoons (Hahn and Manabe, 1975; Kutzbach et al., 1993; Zhisheng et al., 2001; and many others). We propose that, whereas global climate controls (solar insolation, land-mass distribution, ice-sheet distribution) effectively determine glacial and interglacial oscillations, the characteristics of interannual variability of precipitation for modern climate conditions are reflective of the potential range of variability of climate controls under climate regimes associated with strong monsoon activity (interglacial periods). Under weak monsoon regimes, the effective height of the mountains as an elevated heat source is depressed and constrained to lower elevations in the troposphere. More specifically, during glacial periods, colder sea-surface temperatures and a decrease in land-ocean temperature contrast should lead to a decrease in atmospheric moisture transport during the monsoon, decrease in the moisture-holding capacity of the atmosphere, decrease of the overall accumulation of precipitation, increase in the relative fraction of solid precipitation at lower elevations, and possibly reduced interannual variability in the tropics. This would imply a glacial-age southward shift of the hydrometeorological crest and shortening of the penetration distance of incoming moisture into the hydrometeorological rain shadow. This scenario is consistent with the greater snowline lowering observed for glaciers south of the range crest (Burbank et al., 2003), although the sensitivity of the high-elevation glaciers to seasonal and subseasonal variability of precipitation might remain similar to that for modern climate.

Nevertheless, independently of the monsoon regime, local precipitation amounts and spatial gradients of precipitation should be anchored by their contemporaneous topography. The outstanding questions of how this two-way feedback between mountains and the atmosphere impacts erosion processes, and how erosion and tectonics interact over millions of years to produce modern topography (and hydrometeorology), require an end-to-end integrated study from the dominant spatial scales of precipitation to the orogen scale.

ACKNOWLEDGMENTS

This research was funded over the last five years by the National Aeronautics and Space Administration Tropical Rainfall Measurement Mission (NASA-TRMM) (NAG5-7781) and National Aeronautics and Space Administration Precipitation Measurement Missions (NASA-PMM) (NNG04GP02G), and by the National Science Foundation Continental Dynamics program (EAR-9909647 and EAR-9909498). This work benefited from discussions with K. Hodges, M. Strecker, T.N. Krishnamurti, J. Matsumoto, B. Pratt-Sitaula, N. Humphrey, E. Gabet, and B. Bookhagen.

REFERENCES CITED

Abe, M., Kitoh, A., and Yasunari, T., 2003, An evolution of the Asian summer monsoon associated with mountain uplift simulation with the MRI atmosphere-ocean coupled GCM: Journal of the Meteorological Society of Japan, v. 81, p. 909–933, doi: 10.2151/jmsj.81.909.

Barros, A.P., and Lang, T.J., 2003a, Monitoring the monsoon in the Himalayas: Observations in central Nepal, June 2001: Monthly Weather Review, v. 131, p. 1408–1427, doi: 10.1175/1520-0493(2003)131<1408:MTMITH>2.0.CO;2.

Barros, A.P., and Lang, T., 2003b, Exploring spatial modes of variability of terrain-atmosphere interactions in the Himalayas during monsoon onset: Environmental Sciences and Engineering Program, Division of Engineering and Applied Sciences, Harvard University, Hydrosciences Report Series, no. 03-001, 51 p.

Barros, A.P., Joshi, M., Putkonen, J., and Burbank, D.W., 2000, A study of the 1999 monsoon rainfall in a mountainous region in central Nepal using TRMM products and rain gauge observations: Geophysical Research Letters, v. 27, p. 3683–3686, doi: 10.1029/2000GL011827.

Barros, A.P., Kim, G., Williams, E., and Nesbitt, S.W., 2004, Probing orographic controls in the Himalayas during the monsoon using satellite imagery: Natural Hazards and Earth Systems Science, v. 4, p. 1–23.

Beaumont, C., Munoz, J.A., Hamilton, J., and Fullsac, P., 2000, Factors controlling the Alpine evolution of the central Pyrenees inferred from a comparison of observations and geodynamical models: Journal of Geophysical Research, v. 105, p. 8121–8145, doi: 10.1029/1999JB900390.

Bradley, R.S., 1991, Paleoclimatology: International Geophysics Series, Prentice-Hall, v. 64, 611 p.

Burbank, D.W., Blythe, A.E., Putkonen, J., Pratt, B., Gabet, E., Oskin, M., Barros, A., and Ojha, T.P., 2003, Decoupling of erosion and precipitation in the Himalaya: Nature, v. 426, p. 652–655, doi: 10.1038/nature02187.

Clark, T.L., 1979, Numerical simulations with a three-dimensional cloud model: Lateral boundary condition experiments and multi-cellular severe storm simulations: Journal of Atmospheric Sciences, v. 36, p. 2191–2215, doi: 10.1175/1520-0469(1979)036<2191:NSWATD>2.0.CO;2.

Clark, T.L., and Hall, W.D., 1991, Multi-domain simulations of the time dependent Navier Stokes equations: Benchmark error analyses of some nesting procedures: Journal Computational Physics, v. 92, p. 456–481.

Clark, T.L., and Hall, W.D., 1996, The design of smooth, conservative vertical grids for interactive grid nesting with stretching: Journal of Applied Meteorology, v. 35, p. 1040–1046, doi: 10.1175/1520-0450(1996)035<1040:TDOSCV>2.0.CO;2.

Clark, T.L., Hall, W.D., and Coen, J.L., 1996, Source code documentation for the Clark-Hall cloud-scale model code version G3CH01: Boulder, Colorado, National Center for Atmospheric Research Technical Note NCAR/TN-426+STR.

Clemens, S., Prell, W., Murray, D., Shimmield, G., and Weedon, G., 1991, Forcing mechanisms of the Indian Ocean monsoon: Nature, v. 353, p. 720–725, doi: 10.1038/353720a0.

Das, P.K., 1987, Short and long-range monsoon prediction in India, *in* Fein, J.S., and Stephens, P.L., eds., Monsoons: New York, John Wiley & Sons, p. 549–578.

Duan, K., and Yao, T., 2003, Monsoon variability in the Himalayas under the condition of global warming: Journal of the Meteorological Society of Japan, v. 81, no. 2, p. 251–257, doi: 10.2151/jmsj.81.251.

Ellis, M.A., Densmore, A.L., and Anderson, R.S., 1999, Development of mountainous topography in the Basin Ranges, USA: Basin Research, v. 11, p. 21–41, doi: 10.1046/j.1365-2117.1999.00087.x.

Gabet, E.J., Pratt-Sitaula, B.A., and Burbank, D.W., 2004, Climatic controls on hillslope angle and relief in the Himalayas: Geology, v. 32, p. 629–632, doi: 10.1130/G20641.1

Gabet, E.J., Burbank, D.W., Putkonen, J., Pratt-Sitaula, B.A., and Ojha, T., 2005, Rainfall thresholdings for landsliding in the Himalayas of Nepal: Geomorphology, v. 63, p. 131–143.

Garzione, C.N., Quade, J., DeCelles, P.G., and English, N.B., 2000, Predicting paleoelevation of Tibet and the Himalaya from $\delta^{18}O$ vs. altitude gradients in meteoric water across the Nepal Himalaya: Earth and Planetary Science Letters, v. 183, p. 215–229, doi: 10.1016/S0012-821X(00)00252-1.

Hahn, D.G., and Manabe, S., 1975, The role of mountains in the South Asian Monsoon Circulation: Journal of the Atmospheric Sciences, v. 32, no. 8, p. 1515–1541.

Harper, J., and Humphrey, N.F., 2003, High altitude Himalayan climate inferred from glacial ice flux: Geophysical Research Letters, v. 30, doi: 10.1029/2003GL017329.

Hodges, K.V., Wobus, C., Ruhl, K., Schilgden, T., and Whipple, K., 2004, Quaternary deformation, river steepening, and heavy precipitation at the front of the Higher Himalayan ranges: Earth and Planetary Science Letters, v. 220, p. 379–389, doi: 10.1016/S0012-821X(04)00063-9.

Jamieson, R.A.B., Nguyen, M.H., and Lee, B., 2002, Interaction of metamorphism, deformation and exhumation in large convergent orogens: Journal of Metamorphic Geology, v. 20, p. 9–24, doi: 10.1046/j.0263-4929.2001.00357.x.

Kessler, E., 1969, On the distribution and continuity of water substance in atmospheric circulations: American Meteorological Society, Meteorological Monographs, no. 32, 84 p.

Kitoh, A., 2001, Effect of orography on land and ocean surface temperature: Present and future of modeling global environmental change, *in* Matsuno, T., and Kida, H., eds., Toward integrated modeling: TERRAPUB, p. 427–431.

Krishnamurti, T.N., Ardanuy, P., Ramanathan, Y., and Pash, R., 1981, On the onset vortex of the summer monsoon: Monthly Weather Review v. 109, p. 344–363, doi: 10.1175/1520-0493(1981)109<0344:OTOVOT>2.0.CO;2.

Kutzbach, J.E., Prell, W.L., and Ruddiman, W.F., 1993, Sensitivity of Eurasian climate to surface uplift of the Tibetan Plateau: Journal of Geology, v. 101, p. 177–190.

Lang, J.T., and Barros, A.P., 2002, An investigation of the onsets of the 1999 and 2000 monsoons in central Nepal: Monthly Weather Review, v. 130, p. 1299–1316.

Lang, J.T., and Barros, A.P., 2004, Winter storms in the central Himalayas: Journal of the Meteorological Society of Japan, v. 82, p. 829–844, doi: 10.2151/jmsj.2004.829.

Magagi, R., and Barros, A.P., 2004, Estimation of Latent heating of rainfall during the onset of the Indian monsoon using TRMM-PR and radiosonde data: Journal of Applied Meteorology, v. 43, no. 2, p. 328–349, doi: 10.1175/1520-0450(2004)043<0328:EOLHOR>2.0.CO;2.

Molnar, P., England, P., and Martinod, J., 1993, Mantle dynamics, uplift of the Tibetan Plateau, and the Indian monsoon: Reviews in Geophysics, v. 31, p. 357–396, doi: 10.1029/93RG02030.

Montgomery, D.R., and Brandon, M.T., 2002, Topographic controls on erosion rates in tectonically active mountain ranges: Earth and Planetary Science Letters, v. 201, p. 481–489, doi: 10.1016/S0012-821X(02)00725-2.

Pinter, N., and Brandon, M.T., 1997, How erosion builds mountains: Scientific American, April 1997, p. 74–79.

Reiners, P.W., Ehlers, T.A., Mitchell, S.G., and Montgomery, D.R., 2003, Coupled spatial variations in precipitation and long-term erosion rates across the Washington Cascades: Nature, v. 426, p. 645–647, doi: 10.1038/nature02111.

Reinking, R.F., Snider, J.B., and Coen, J.L., 2000, Influences of storm-embedded orographic gravity waves on cloud liquid water and precipitation: Journal of Applied Meteorology, v. 39, p. 733–759, doi: 10.1175/1520-0450(2000)039<0733:IOSEOG>2.0.CO;2.

Roe, G.H., Montgomery, D.R., and Hallet, B., 2003, Orographic precipitation and the relief of mountain ranges: Journal of Geophysical Research, v. 108, paper no. 2315, doi: 10.1029/2001JB001521.

Sasamori, T., 1972, A linear harmonic analysis of atmospheric motion with radiative dissipation: Journal of the Meteorological Society of Japan, v. 50, p. 505–517.

Smith, R.B., 1979, The influence of mountains on the atmosphere: Advances in Geophysics, v. 21, p. 87–230.

Stephens, G., 1984, Short-wave parameterization revised to improve cloud absorption: Journal of Atmospheric Sciences, v. 41, p. 687–690, doi: 10.1175/1520-0469(1984)041<0687:ASPRTI>2.0.CO;2.

Tang, M., and Reiter, E.R., 1984, Plateau monsoons of the Northern Hemisphere: A comparison between North America and Tibet: Monthly Weather Review, v. 112, p. 617–637.

Thompson, L.G., Yao, T., Mosley-Thompson, E., Davis, M.E., Henderson, K.A., and Lin, P.-N., 2000, A high-resolution millennial record of the south Asian monsoon from Himalayan ice cores: Science, v. 289, p. 1916–1919.

Whipple, K., Kirby, E., and Brocklehurst, S., 1999, Geomorphic limits to climate-induced increases in topographic relief: Nature, v. 401, p. 39–43, doi: 10.1038/43375.

Willett, S.D., 1999, Orogeny and orography—The effects of erosion on the structure of mountain belts: Journal of Geophysical Research, v. 104, p. 28,957–28,981, doi: 10.1029/1999JB900248.

Wobus, C.W., Hodges, K.V., and Whipple, K.X., 2003, Has focused denudation sustained active thrusting at the Himalayan topographic front?: Geology, v. 31, p. 861–864, doi: 10.1130/G19730.1.

Zhao, H., and Moore, G.W.K., 2002, On the relationship between Dasuopu snow accumulation and the Asian summer monsoon: Geophysical Research Letters, v. 29, 2222, doi: 10.1029/2002GL015757.

Zhisheng, A., Kutzbach, J.E., Prell, W.L., and Porters, S.C., 2001, Evolution of Asian monsoons and phased uplift of the Himalaya-Tibetan Plateau since late Miocene times: Nature, v. 411, p. 62–66, doi: 10.1038/35075035.

Manuscript Accepted by the Society 23 June 2005

Printed in the USA

Geological Society of America
Special Paper 398
2006

Spatial patterns of precipitation and topography in the Himalaya

Alison M. Anders[†]
Gerard H. Roe
Bernard Hallet
David R. Montgomery
Noah J. Finnegan
Jaakko Putkonen
Earth and Space Sciences, University of Washington, P.O. Box 351310, Seattle, Washington 98195-1310, USA

ABSTRACT

Spatial variability in precipitation has received little attention in the study of connections between climate, erosion, and tectonics. However, long-term precipitation patterns show large variations over spatial scales of ~10 km and are strongly controlled by topography. We use precipitation rate estimates from Tropical Rainfall Measuring Mission (TRMM) satellite radar data to approximate annual precipitation over the Himalaya at a spatial resolution of 10 km. The resulting precipitation pattern shows gradients across the range, from east to west along the range, and fivefold differences between major valleys and their adjacent ridges. Basin-wide average precipitation estimates correlate well with available measured mean runoff for Himalayan rivers. Estimated errors of 15%–50% in TRMM-derived annual precipitation are much smaller than the spatial variability in predicted totals across the study area. A simple model of orographic precipitation predicts a positive relationship between precipitation and two topographically derived factors: the saturation vapor pressure at the surface and this pressure times the slope. This model captures significant features of the pattern of precipitation, including the gradient across the range and the ridge-valley difference, but fails to predict the east-west gradient and the highest totals. Model results indicate that the spatial pattern of precipitation is strongly related to topography and therefore must co-evolve with the topography, and suggest that our model may be useful for investigation of the relationships among the coupled climate-erosion-tectonic system.

Keywords: orographic precipitation, remote sensing, Himalaya, landscape evolution.

INTRODUCTION

Growing evidence for a strong role of erosion in the tectonic and topographic development of mountain ranges motivates efforts to understand feedbacks between climate, erosion, and tectonics. While spatial gradients in erosional processes and tectonic forcing have prompted much study, most models of erosion consider climate to be spatially uniform. Yet spatial patterns in precipitation are influential in shaping topography through erosion and are sensitive functions of topography at sub–mountain-range scales.

Spatial patterns of precipitation in Nepal (Barros et al., 2000), the Alps (Frei and Schär, 1998), and the Olympic Mountains (Anders et al., 2004; see also Barros and Lettenmaier, 1993) are characterized by remarkable and persistent variation on scales of tens of kilometers. Spatial variability in precipitation confounds the common substitution of drainage basin area for river discharge in modeling patterns of fluvial incision, and the use of this

[†]Present address: Department of Geology and Geophysics, Yale University, P.O. Box 208109, New Haven, Connecticut 06520-8109, USA; alison.anders@yale.edu.
Anders, A.M., Roe, G.H., Hallet, B., Montgomery, D.R., Finnegan, N.J., and Putkonen, J., 2006, Spatial patterns of precipitation and topography in the Himalaya, *in* Willett, S.D., Hovius, N., Brandon, M.T., and Fisher, D.M., eds., Tectonics, Climate, and Landscape Evolution: Geological Society of America Special Paper 398, Penrose Conference Series, p. 39–53, doi: 10.1130/2006.2398(03). For permission to copy, contact editing@geosociety.org.

substitution in areas with strong spatial gradients in precipitation can lead to significant differences in predicted patterns of erosion (Roe et al., 2002, 2003; Finlayson and Montgomery, 2003).

Topography itself has a profound effect on spatial patterns of precipitation both globally and regionally (e.g., Smith, 1979). Mountains influence the flow of air and disturb the vertical stratification of the atmosphere by acting as physical barriers and as sources or sinks of heat (e.g., Barros and Lettenmaier, 1994). At the scale of entire mountain ranges (hundreds of km) the rain-shadow effect has a well-documented and significant influence on precipitation patterns. Enhanced precipitation on the windward side of mountain ranges has been recognized in the geological community as a long-term control on landscape evolution and the geotectonic development of mountain ranges (e.g., Koons, 1989; Beaumont et al., 1992; Hoffman and Grotzinger, 1993;Willett, 1999; Montgomery et al., 2001). However, forced ascent on the windward side is not the only way that the atmosphere interacts with topography; instead of rising over the topography, the air may be blocked or diverted around the range (e.g., Simpson, 1987). Alternately, the flow of air over the topography may excite internal waves, due to the vertical density stratification of the atmosphere (e.g., Durran, 1986). Mountains can also warm the air above them, drawing air from low levels up and triggering condensation or convection on the slopes, as has been observed in thunderstorms on the eastern slopes of the Rocky Mountains (Tripoli and Cotton, 1989). The vertical profiles of temperature and moisture content and the velocity of the incoming air, which vary from storm to storm, as well as characteristics of the topography, including the length, width, and height of the mountain range, determine how the atmosphere interacts with the mountains (e.g., Houze, 1993). Nevertheless, if spatial patterns of precipitation are relatively stable over time and are consistently related to topography in a given region, the pattern of erosion, and therefore the topography, must co-evolve with patterns of precipitation.

Current knowledge about the climatological annual precipitation pattern in the Himalaya is derived from rain gauges, and several interpolations are available (e.g., Leemans and Cramer, 1991, Shrestha, 2000). Additionally, basin-scale orographic precipitation patterns have been studied using rain gauges in the Chenab (Singh et al., 1995) and Kosi basins (Dhar and Rakhecha, 1981).

Orographic precipitation in a limited region of the central Himalaya has been documented with a dense gauge network and a variety of remote sensing techniques to investigate processes controlling precipitation distribution on the scale of 10–20 km during storm events and monsoon seasons (Barros et al., 2000; Lang and Barros, 2002, 2004; Barros and Lang, 2003). These studies have revealed large gradients in seasonal precipitation totals over short (~10 km) spatial scales that are not simply related to elevation (Barros and Lang, 2003). Additionally, they have documented differences in monsoon season (June–September) diurnal precipitation patterns at high (>2000 m) and low elevations that relate to daytime upslope winds switching to weak nighttime downslope winds (Barros et al., 2000; Barros and Lang, 2003). Satellite precipitation radar tracks reveal large regions of stratiform precipitation with embedded convective cells during the monsoon season (Lang and Barros, 2002). A spatial association of weak convective cells and clouds with SW facing ridges during the monsoon season was noted by Barros et al. (2004). Nearly all of the precipitation occurring below 2000 m is rain, while at higher elevation stations snow accounts for 17 ± 11% of annual precipitation totals, with this fraction increasing with elevation (Lang and Barros, 2004). Winter precipitation is associated with Western Disturbances that cause wintertime precipitation over northern India and Kashmir (Lang and Barros, 2004). See Barros et al. (this volume) for modeling studies of monsoon onset and winter storm events illustrating the role of complex topography in shaping cloud patterns.

The long-term pattern of precipitation in mountains generally and the Himalaya in particular is poorly constrained due to a lack of measurements of precipitation on spatial scales of a few tens of kilometers or less, and a lack of measurements extending back over more than a few years or decades. The dearth of information on spatial patterns of precipitation is in part due to the difficulty of measuring precipitation over appropriate spatial and temporal scales. Rain gauges provide information on precipitation, but existing rain-gauge networks, especially in mountainous areas, are generally not dense enough to reveal variability in precipitation over spatial scales of tens of kilometers—scales over which topography and precipitation can vary significantly. Rain gauges themselves are subject to several kinds of errors, including the local disturbance to flow that they create, the difficulty of automatically measuring snow water equivalence, and the fact that gauges are a point measurement and may not represent average precipitation over a larger region (e.g., Groisman and Legates, 1994; Sinclair et al., 1997; Dingman, 2002). In general, rain-gauge networks tend to undersample high elevations relative to lower elevations (e.g., Frei and Schär, 1998; Colle et al., 1999). The extrapolation from a rain-gauge network to a continuous field of precipitation via a statistical algorithm (such as the PRISM method of Daly et al., 2002) relies on an assumed relationship between precipitation and topography that is difficult to justify given the sparse sampling of gauge networks and space-time variability of weather conditions. Finally, the establishment and maintenance of dense gauge networks in remote mountainous regions such as the Himalaya is daunting and to date has only been done over one small area in the central Himalaya (Barros et al., 2000).

The remote sensing of precipitation via radar reflectivity, therefore, provides an attractive approach to defining spatial patterns of precipitation in mountains. The first spaceborne precipitation radar, aboard the Tropical Rainfall Monitoring Mission (TRMM) satellite provides a unique opportunity to assemble data on spatial patterns of precipitation in low-latitude mountain ranges. This instrument allows us, for the first time, to collect spatially continuous information on precipitation in mountains. As such, TRMM is a novel tool worth examining as a means

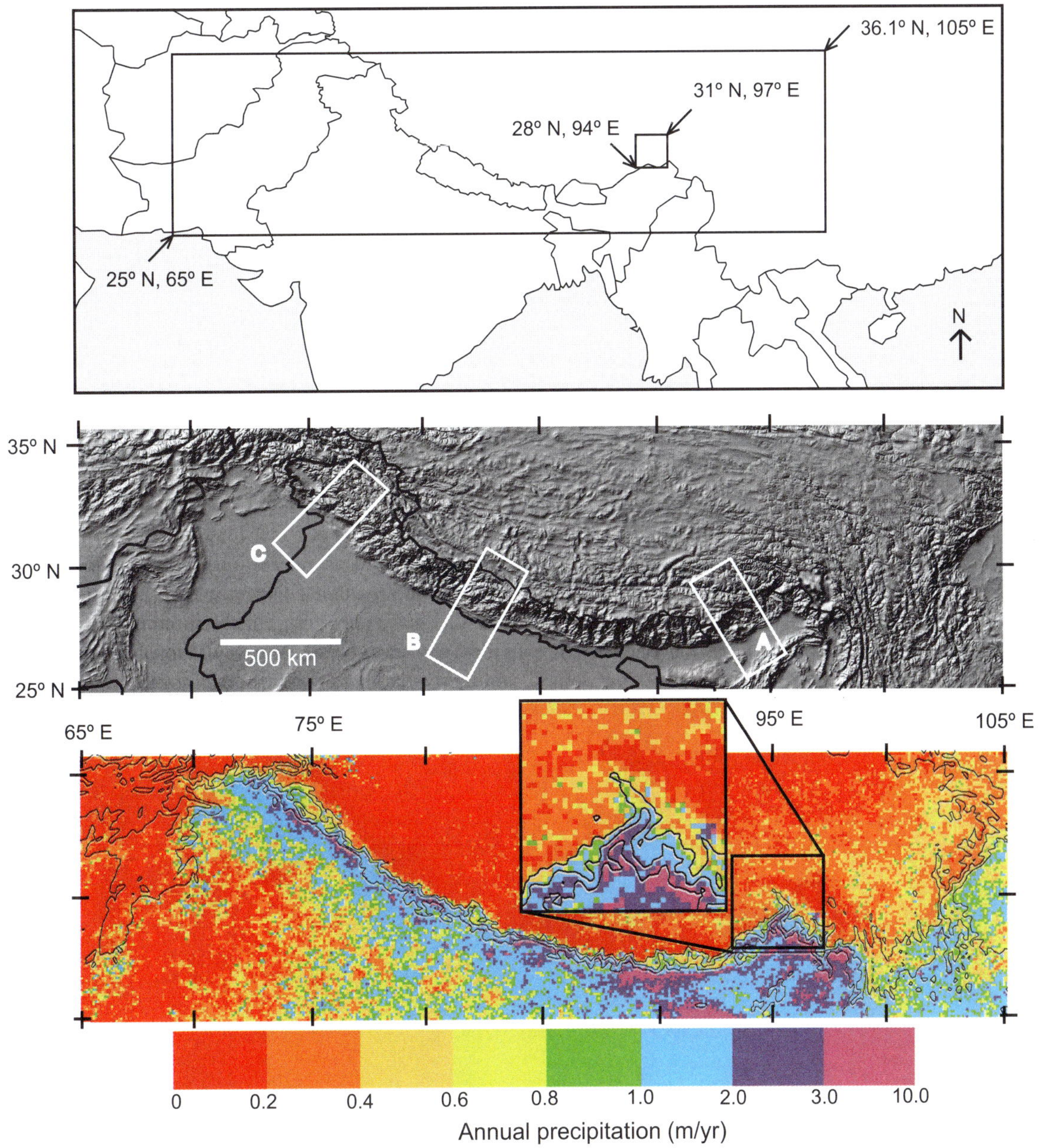

Figure 1. Location map, topography, and Tropical Rainfall Measuring Mission (TRMM) annual precipitation pattern. The study area is indicated in the top panel by the large box, along with a subregion near Namche Barwa at the eastern syntaxis of the range. The topography is shown in shaded relief in the middle panel, and the white boxes indicate the locations of cross sections shown in Figure 3. The lower panel shows the annual precipitation map (m/yr) that we created from four years of the TRMM satellite's precipitation radar precipitation rate estimates for the entire study area; inset is a close-up of the Namche Barwa region. The pattern of precipitation is closely related to topography. At the largest scale, the dry Tibetan Plateau and the wet Indian plains strongly contrast and a subtler gradient from east to west is apparent. At the scale of a few tens of kilometers, precipitation tracks topography, following large valleys north into the Himalaya. In addition, precipitation maxima in the southeastern end of the study area are observed with estimated annual precipitation totals in excess of 9 m/yr.

of understanding spatial patterns of precipitation and erosion in areas sampled by the satellite. Herein, the focus is on using the TRMM satellite to define spatial patterns of annual precipitation in the Himalaya and to compare these patterns with topography in the context of geomorphological applications.

PRECIPITATION PATTERNS IN THE HIMALAYA FROM TRMM

The evaluation of landscape evolution–climate feedbacks in our study area requires a detailed map of precipitation rate

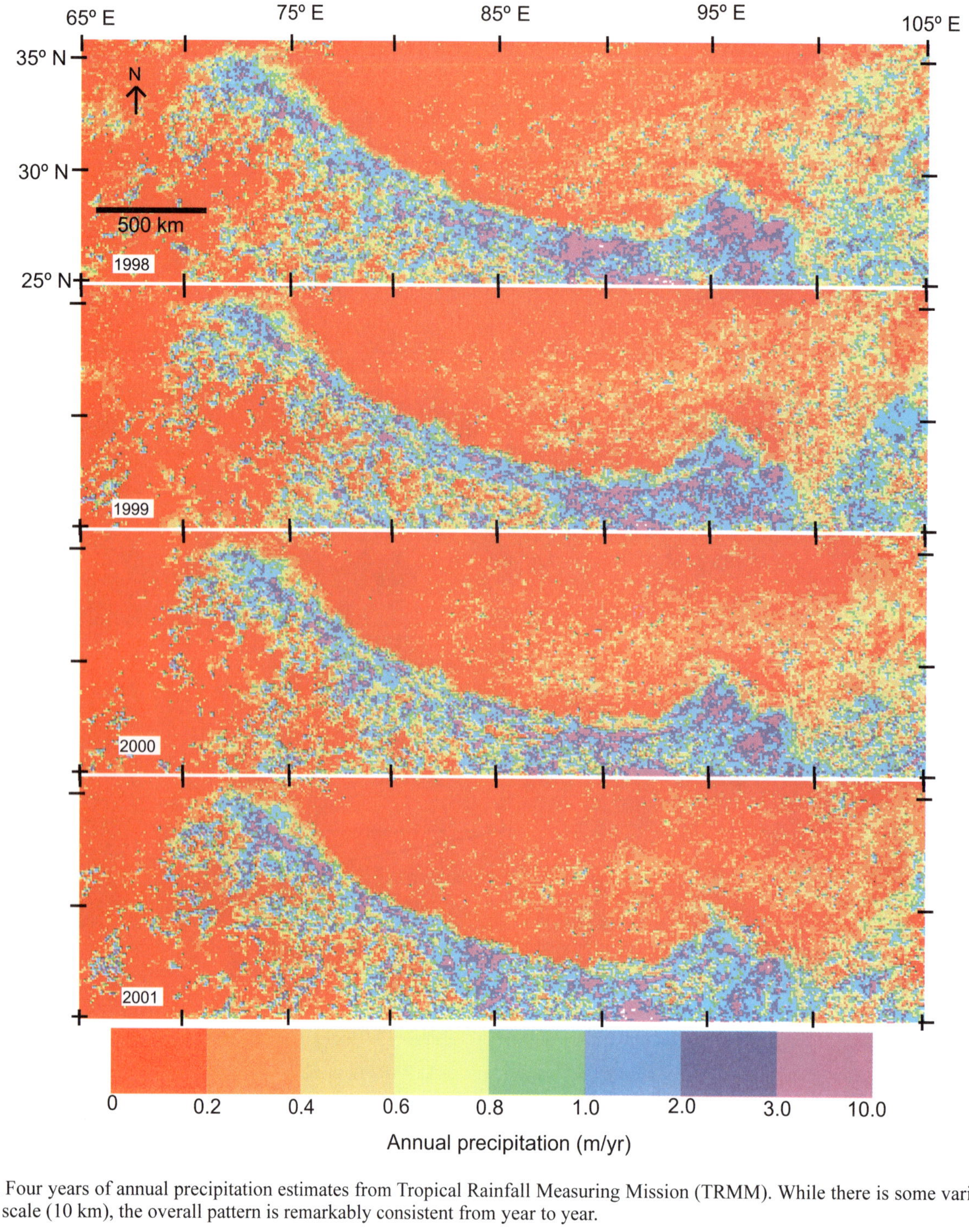

Figure 2. Four years of annual precipitation estimates from Tropical Rainfall Measuring Mission (TRMM). While there is some variability at the pixel scale (10 km), the overall pattern is remarkably consistent from year to year.

over a comparatively wide area (36°N–25°N, 105°E–65°E). The TRMM satellite uses spaceborne radar to provide accurate estimates of near-surface precipitation rates. We obtained the near-surface rain rate estimates from the 2A25 radar profile data (http://daac.gsfc.nasa.gov/hydrology/).

TRMM's precipitation radar operates at a frequency of 13.8GHz and can detect reflectivities down to ~18 dBZ, corresponding to rain rates of ~0.7 mm/h. Snow has a lower radar reflectivity for a given water equivalent than rain, and estimates of the relationship between reflectivity and water equivalent used by the U.S. National Weather Service (http://www.nssl.noaa.gov/teams/watads/public_html/snow/snow.htm) suggest that the ~20 dBZ detection limit would still allow for detection of dry snow with > ~1 mm/h of water equivalent. However, the TRMM algorithms for converting measured reflectivity profiles to near-surface rain rates assume the precipitation is liquid water, which would underestimate the water-equivalence if the precipitation were snow. Additionally, the attenuation estimates are based on liquid precipitation rather than ice. The next generation of satellite precipitation measurement is being developed in the Global Precipitation Measuring Mission and may include a dual-frequency precipitation radar, which will be able to detect lower snow and rain rates and to distinguish between snow and rain using attenuation differences.

As discussed below in our estimates of sampling error, we must balance increasing spatial resolution against a decreasing number of instantaneous estimates of rain rate. This balance influenced our choice to grid our study area (Fig. 1) into 0.1 × 0.1 degree boxes (~10 km × 10 km).

TRMM provides 4 yr of instantaneous rain rate estimates in the study area (1998–2001), which are used to calculate the average rain rates and create a map of average annual precipitation (the climatology). The TRMM satellite orbit was designed to sample every location at different times of the day, over a 46 d cycle. By dividing each year of TRMM estimates into eight 46 d periods, an even sampling of the diurnal cycle was produced, and we avoided bias due to diurnal cycles in precipitation. The average rain rate for each 46 d period was multiplied by the duration of the period to get an estimated volume of precipitation for that time period. These volumes were then summed to obtain an annual precipitation total for each year (Fig. 2). The annual totals from each of the four years studied were averaged to create the average annual climatology (Fig. 1). The total number of samples in a grid box during a 46 d period varied from ~90 to ~400 as a function of latitude. About 95% of all samples were zeros (no precipitation). A small fraction (<0.01%) of the surface rain rate data exceeded 100 mm/h, which is unreasonable for regions larger than 100 km^2 (Kozu et al., 2001). These anomalous data values were considered artifacts and were removed from the analysis.

Figures 1 and 2 present the average annual precipitation climatology and the precipitation totals for each of the four individual years studied. This is the first time that spatially continuous precipitation data at such high spatial resolution (~10 km × 10 km) have been published for the Himalaya, and they reveal striking variability across, along, and within the range, as well as a robust spatial pattern that is consistent between all four individual years. The large-scale features of the precipitation pattern—namely the steep gradient in precipitation from the wet foreland to the arid Tibetan Plateau and the decrease in precipitation from east to west along the range—are clearly illustrated by the TRMM estimates (Fig. 3). In addition, there is remarkable structure at smaller scales, including large precipitation totals following big valleys up from the plains. This is illustrated in the inset portion of Figure 1, which shows a subset of the data near the eastern syntaxis of the range. The zone of high precipitation closely follows the valley of the Tsangpo/Brahmaputra River north, as far north as regions that are within the dry higher Himalaya and southern Tibet, leading to large variations in precipitation between ridges and valleys along the strike of the range. In addition, a double band of high precipitation totals in the central and western portions of the range coincides with the first and second significant rises in topography approaching the Himalaya.

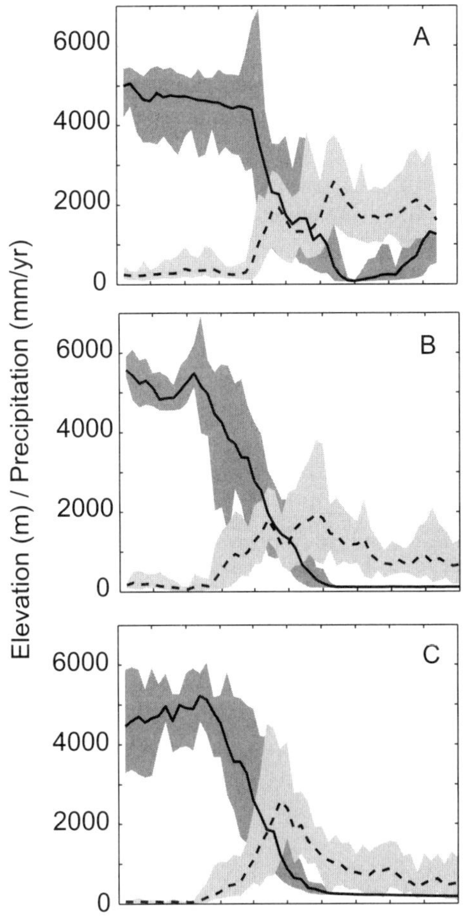

Figure 3. Cross sections of topography and precipitation. The black lines and dark gray shading show the mean elevation (m) and range of elevations along the cross sections indicated in Figure 1. The dashed line is the 4 yr average annual precipitation (mm/yr) estimated with Tropical Rainfall Measuring Mission (TRMM) precipitation radar. Light gray shading indicates the range of precipitation values. The correlation of precipitation and topography is evident in the sharp increase in both along the Himalayan front.

The stationary nature of the spatial pattern of annual precipitation in the four years studied is significant because it supports the notion that spatial patterns of precipitation are stable through time and hence would influence the evolution of topography through the related spatial pattern of erosion. The stability in time of relative differences in precipitation across space is a key requirement for the co-evolution of precipitation patterns with topography. From the point of view of landscape evolution over timescales of 10^4–10^5 yr, we cannot necessarily constrain differences in the magnitude of precipitation in time due to long-term climate variability. But if the spatial pattern of relative precipitation totals is stable, the interactions among precipitation, erosion, and topography can be modeled.

ERROR ESTIMATION

Basin-Wide Average Annual TRMM Precipitation versus Measured Mean Annual Runoff

Evaluating the TRMM precipitation pattern with ground-based measurements is difficult, due to the general lack of rain gauges in the region. Barros et al. (2000) are currently working on the ground validation of the TRMM precipitation radar algorithm on an event basis in a small region in central Nepal. Herein, we wish to evaluate the TRMM-derived annual precipitation pattern at the drainage-basin scale, where data are available. Therefore, we turn to field measurements of river discharge, compiled from several sources by Finlayson et al. (2002) as a check on the integrated pattern of precipitation over drainage basins (Table 1, Fig. 4). The 19 basins were defined using digital topography from SRTM (Shuttle Radar Topography Mission) with a spatial resolution of 1 km. The average TRMM-derived precipitation within each basin was compared to the measured discharge in the basin divided by the basin area (annual mean runoff) (Fig. 4). There is a remarkable correlation

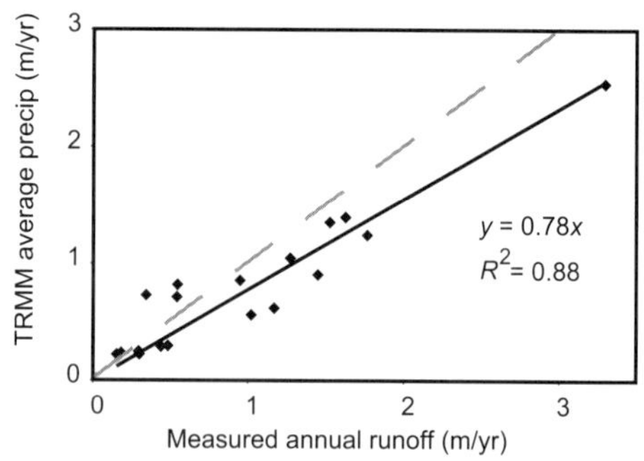

Figure 4. Tropical Rainfall Measuring Mission (TRMM) basin-wide average annual precipitation compared to measured mean runoff. The basin-wide averages of TRMM annual precipitation estimates correlate well with measured mean runoff (annual discharge/basin area) for 19 catchments listed in Table 1. The linear regression forced through the origin is shown as a black line. The 1:1 line is indicated by a dashed line. TRMM estimates generally fall below measured mean runoff values, suggesting that TRMM underestimates precipitation.

TABLE 1. RIVER DISCHARGE MEASURMENT LOCATIONS AND MEAN TROPICAL RAINFALL MEASURING MISSION (TRMM) ANNUAL PRECIPITATION

River and location	Longitude (°E)	Latitude (°N)	Basin area (km^2)	Measured discharge (km^3/yr)	TRMM average annual precipitation (m/yr)
Kali Gandaki—Koketani	84.4	27.8	3452	1.67	0.30
Kali Gandaki—Aval Beni	84.4	27.8	5637	5.35	0.85
Kali Gandaki—Ramdi	84.1	28.1	10,491	15.9	1.35
Kali Gandaki—Kot	84.5	27.9	44,819	19.20	1.40
Seti—Sarang Ghat	84.8	27.8	29,242	8.70	0.23
Seti—Kotre Bazar	84.9	27.8	997	3.29	2.54
Marsyandi—Markichok	85.2	28.0	4164	7.35	1.24
Bhuri—Beni Ghat	87.3	27.6	4994	6.35	1.05
Trisuli—Adamghat	87.2	27.3	5582	8.08	0.90
Trisuli—Betrawati	87.2	26.9	4640	4.73	0.56
Arun—606	74.4	36.0	57,705	19.70	0.73
Arun	74.6	35.6	31,268	13.70	0.29
Arun	72.9	34.8	29,941	8.92	0.22
Tsangpo—179	75.1	32.9	111,449	16.80	0.22
Tsangpo—180	74.4	34.3	170,793	30.70	0.23
Tsangpo—181	89.7	29.4	206,528	60.50	0.25
Jhelum—Baramula	92.0	29.3	12,380	6.70	0.71
Chenab—Alkhnoor	94.6	29.5	21,916	25.60	0.62
Chenab—Panjnab	84.4	27.8	181,226	98.60	0.82

between these quantities in a least-squares linear regression forced through the origin (TRMM average precipitation = 0.77 × measured unit discharge, $R^2 = 0.88$). This correlation is found for rivers with basin areas ranging from 1000–200,000 km^2, mainly along the Himalayan front. We recognize that the annual mean runoff of these rivers is influenced by evaporation and unreturned infiltration, which we do not attempt to model in addition to precipitation. However, the strong correlation between TRMM average precipitation and measured mean runoff indicates that TRMM-derived precipitation estimates provide a reasonable proxy for discharge along the Himalayan front and are an improvement over previous data sets, such as Leemans and Cramer (1991) used by Finlayson et al. (2002). Moreover, it suggests that the spatial pattern of precipitation estimated using TRMM is sufficiently well defined to capture the differences in average precipitation between these 19 basins for use in our purposes, namely, exploring the relationship between precipitation and topography and its role in long-term landscape evolution.

Ground Clutter Error Estimates

In addition to noting the empirical accord between TRMM average annual precipitation and measured mean runoff in a number of rivers, we also constrained the theoretical errors in using TRMM estimates to approximate annual precipitation totals. TRMM estimates are subject to both measurement error and sampling error. The former includes the error in the measurement of radar reflectivity, the attenuation correction to account for the decrease in reflectivity as the beam travels from the rain back to the satellite, and the semi-empirical rain rate–reflectivity relation. Efforts to estimate measurement errors are under way at various ground validation sites, including a site in Nepal that is included in our study area (Barros et al., 2000). A comparison of a limited data set of ~250–300 TRMM precipitation estimates (20–70 rain events) to gauge records at the Nepal calibration site shows a higher probability of detection with better skill scores as well as a higher false alarm rate for low elevation (<2000 m) versus high elevation gauges during the summer monsoon (Barros et al., 2000). This observation of bias at high elevations is not due to snowfall being missed, as these observations were made during the summer. Moreover, the same difference between high and low elevations is seen if only events with rain rates higher than 0.5 mm/h (theoretically within the sensitivity of the satellite) are considered (Barros et al., 2000). Part of this discrepancy could be related to the ground clutter algorithm truncating the profile above precipitation near the surface at high elevations.

As ground clutter artifacts could lead to a systematic, spatially coherent measurement error in mountainous regions, we investigated this component of the measurement error. Ground clutter artifacts are due to radar returns from Earth's surface that are misinterpreted as reflectivity due to water droplets. These commonly result from a radar beam reflecting off of steep slopes. As the incidence angle of the beam increases, the chance for slopes to reflect the beam increases. The ground clutter algorithm used in creating the TRMM near-surface rain rate field guards against ground clutter measurement errors by removing the data less than 1 km above the ground (Meneghini et al., 2000).

In the Himalaya, however, elevation can vary significantly over short spatial scales; several kilometers of relief are expected in some of our 100 km^2 grid boxes, complicating the removal of ground clutter artifacts. The algorithm may be too conservative and cut off reflectivity measurements from precipitation near the ground, or it may erroneously keep ground reflections from isolated peaks. To investigate the potential impact of ground clutter artifacts in the processed surface rain rates, we rely on the assumption that clutter is a function of incidence angle. Specifically, we examined near-surface rain rates as a function of look angle for a subregion of our study area near the eastern syntaxis of the Himalaya (29°N–31°N, 94°E–96°E). For the region as a whole, average rain rate is not a simple function of look angle (Fig. 5). However, the average rain rate over four years is slightly higher for look angles between 13.5 and 17 degrees (the highest look angles) than for look angles from 0 to 13.5 degrees (Fig. 5). This difference cannot be accounted for by either a few very large values of rain rate at high look angles, nor by a greater frequency of rain reported at high look angles. At high look angles, no rain is detected ~97% of the time, identical to that of the data set as a whole and to the subsamples with other look angles.

We examined the spatial pattern of average rain rate derived from samples taken at five classes of look angle and found that for the lowest four groups of look angle, the patterns are remarkably consistent. In the highest class of look angle, representing ~20% of the data, the pattern is somewhat different, with more precipitation measured in the high mountains to the northeast, suggesting some spatial coherence to the difference in rain rate between high and low look angles (Fig. 5). Further work could refine identification of the areas most influenced by this effect and evaluate the total impact on the error of removing the highest look angles from the data set. Nonetheless, for the purposes of this analysis, effects from ground clutter are small compared to the overall pattern, that is, the spatial pattern observed is not dominated by ground clutter artifacts. For the purpose of estimating annual discharge for use in landscape evolution modeling, ground clutter artifacts are negligible.

Sampling Error Estimates

The second type of error, sampling error, is the error in using a discrete number of measurements of rain rate to estimate the average rain rate and compute rainfall totals. When dealing with a small number of samples, sampling error is likely to be much larger than measurement error (Bell and Kundu, 2000). One of the goals of the TRMM mission is to produce monthly precipitation totals in 2° × 2° grids (approx. 40,000 km^2) over the area surveyed by the satellite. Several authors have concluded that these totals have sampling errors of ~10% (e.g., Shin and

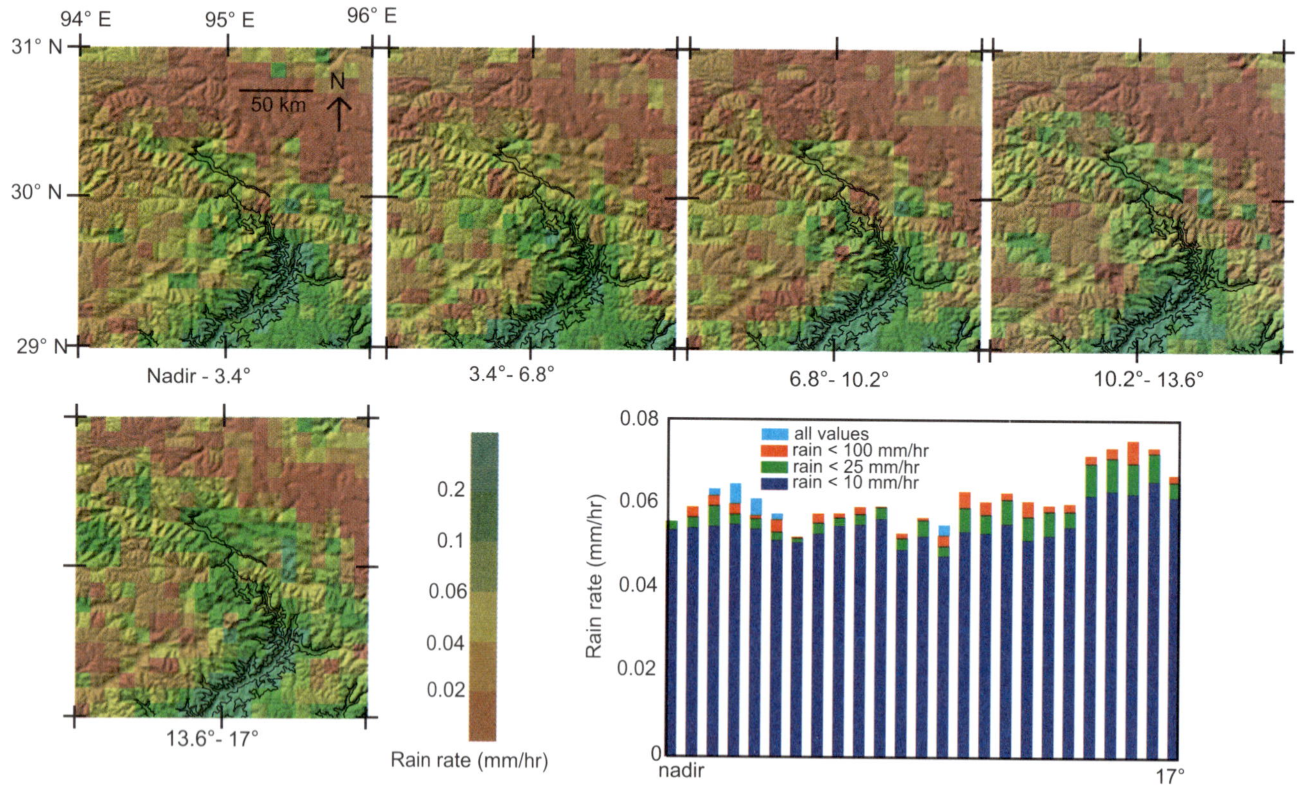

Figure 5. Rain rate as a function of look angle for the Namche Barwa region. The spatial pattern of average rain rate (mm/hr) over the Namche Barwa region is shown as a function of satellite look angle in 5 equal classes from 0 to 17 degrees. For the lowest four classes of look angle, the spatial pattern of rain rate is similar. However, for the highest look angles, higher rain rates are observed up the slopes to the northeast of the river valley than in the other classes, perhaps due to increased ground clutter errors at high look angles. In the lower right, the average rain rate over the Namche Barwa area is shown as a function of look angle, with various cutoffs in the largest values shown as different colors. The cutoff values are in mm/yr and range from 10 to no cutoff. Average rain rate over the domain is not a simple function of look angle, nor do large values of rain rate strongly influence the observed average rain rates.

North, 1988; Bell and Kundu, 2000). We used three independent methods to estimate the sampling errors we would expect for our much higher spatial resolution (100 km^2) precipitation totals with temporal resolution of 4 months to 1 yr (Table 2).

First, two rain-gauge precipitation records from the TRMM ground validation site in Nepal (Khudi, elevation 780 m, Koprung, elev. 3133 m; Barros et al., 2000) were used in a Monte Carlo simulation of sampling error. The year or monsoon-length rain-gauge record was randomly scrambled in day-length intervals and then sampled according to the TRMM overpass cycle to create 100,000 simulated TRMM records. The resulting simulated TRMM estimates of yearly and monsoon precipitation totals were compared to the gauge measurements over these periods to determine the error (standard deviation of the simulated TRMM samples divided by the total gauge-measured precipitation). The error determined from these simulations is 24%–30% for the year and 27%–29% for the monsoon season only (Table 2).

Shin and North (1988) and Bell and Kundu (2000) provided theoretical equations for satellite sampling error, both of which use a Markov model of precipitation such that:

$$R(\tau) = Be^{-|\tau|/\tau_0}, \tag{1}$$

where R is precipitation rate, τ is time, τ_0 is decorrelation time, and B is a constant. Shin and North (1988) proposed an equation based on the orbital parameters of the satellite as well as the decorrelation time of the precipitation and the coefficient of variation, defined as the constant B over the mean rain rate. We used three estimates of the decorrelation time, one of 1.1 h from Bell (1987), a maximum estimate of 10 h from Shin and North (1988), and estimates of 6 h and 5.7 h derived from fitting a Markov model to the Khudi and Koprung gauges in Nepal. We estimated the coefficient of variation using GATE (Global Atmospheric Research program Atlantic Tropical Experiment) data presented in Shin and North (1988), as well as their maximum estimate and one estimate calculated from the Nepal data. This range of parameters gave us nine estimates of yearly and monsoonal sampling error in precipitation totals of 14%–34% and 24%–60%, respectively (Table 1).

Bell and Kundu (2000) provided an alternate model of sampling error. We assume that our sampling grid resolution of

TABLE 2. SAMPLING ERROR ESTIMATES

Monte Carlo simulation			Annual		Monsoon	
			24%		27%	
Shin and North (1988) Estimate						
Decorrelation time (hrs)	1.1 Bell (1987) estimate		6 Nepal data estimate		10 Shin and North (1988) maximum estimate	
	Annual	Monsoon	Annual	Monsoon	Annual	Monsoon
CV = 5 estimate from GATE data	17%	30%	15%	27%	14%	24%
CV = 6.5 estimate from Nepal data	22%	39%	20%	35%	30%	32%
CV = 10 maximum estimate Shin and North (1988)	34%	60%	18%	54%	27%	49%
Bell and Kundu (2000) with rain rate estimate from Nepal						
Decorrelation time (hrs)	1.1 Bell(1987) estimate		6 Nepal data estimate		10 Maximum estimate Shin and North(1988)	
	Annual	Monsoon	Annual	Monsoon	Annual	Monsoon
	58%	52%	41%	37%	18%	17%

~10 km × 10 km is small enough that the entire area is receiving precipitation at the same time, so that Bell and Kundu's (2000) model reduces to

$$\varepsilon \approx \frac{r}{R}\left(\frac{1}{N} - \frac{2\tau_0}{T}\right)^{1/2}, \quad (2)$$

where ε is the sampling error, R is the average rain rate, r is the average rain rate while it is raining, N is the number of samples taken, T is the length of time over which the samples are taken, and τ_0 is the decorrelation time. Using the above-mentioned bounds on decorrelation time and using average rain rate and conditional rain rate from the Nepal data, this method yields error estimates of 18%–58% for yearly totals and 17%–52% for monsoon totals (Table 2).

Given the error estimates determined via the above three methods, we conclude that, even at 10 km resolution, sampling errors are small enough that the precipitation totals are suitable for the purposes we intend of estimating annual river discharge, modeling erosion, and comparing precipitation and topography throughout the Himalaya. At smaller spatial scales, the number of independent samples of precipitation rate decreases, making the errors greater according to all estimates. Therefore, the observed patterns can be used, albeit with caution, to investigate relationships between precipitation patterns and topography at spatial scales of 10 km and larger.

COMPARING PRECIPITATION AND TOPOGRAPHY

In addition to using observed precipitation fields to estimate modern fluvial incision rates, we are interested in understanding how topography and precipitation patterns co-evolve over long time periods. In order to address this question, models of how precipitation relates to topography are needed to determine how precipitation patterns change as topography evolves. Toward this end, we examine the relationship between the TRMM precipitation estimates and characteristics of the modern topography, including elevation and slope. While we note that precipitation patterns can be influenced by topography through many different mechanisms that depend on both topography and the conditions of the air upwind of the mountains, we evaluate a simple theoretical model of orographic precipitation using the observed patterns of precipitation and topography in the Himalaya. The degree to which a simple model is successful in producing a reasonable precipitation pattern gives an indication as to the potential for modeling the co-evolution of precipitation and topography with such a model.

Theoretical Model of Orographic Precipitation

Roe et al. (2002) propose a simple model of orographic precipitation in which precipitation is proportional to the vertically averaged convergence of the moisture flux plus a background term. This model incorporates two effects: the decrease in water vapor with height in the air column, and the change in temperature of the air as it is forced up the topography in the direction of the prevailing winds. Following Roe et al. (2002), the model assumes that orographic precipitation is proportional to two terms. The first is saturation vapor pressure at the surface, which is approximately proportional to the total column moisture content. The second factor is the saturation vapor pressure multiplied by the slope of the topography in the direction of the prevailing wind. Thus:

$$P \propto (\alpha + \beta \times S) \times e_{sat}(T), \quad (3)$$

where P is precipitation, S is slope in the direction of the wind, e_{sat} is saturation vapor pressure, T is surface temperature, and α and β are constants. The two factors in this model favor precipitation at low elevations and in areas with steep slopes facing into the dominant wind at low elevations. We denote surface saturation vapor pressure as *V*p and slope times saturation vapor pressure as *Sv*p.

In order to evaluate the fit of this model to the observations of precipitation and topography in the Himalaya, we must estimate the values of the two terms using the topography. First, the topographic data from GTOPO30 were smoothed to the resolution of the TRMM precipitation estimates (10 km × 10 km cells). We derived a representative surface saturation vapor pressure in each cell by first roughly estimating the air temperature using an average temperature of 30 °C at sea level and a constant lapse rate of 7 °C/km. The saturation vapor pressure (e_{sat}) as a function of temperature can be found using an approximation to the Clausius-Calpeyron relation:

$$e_{sat} = 6.112(mb)e^{\frac{17.67T}{243.5+T}}, \quad (4)$$

where T is temperature in degrees Celsius (e.g., Emanuel, 1994). This gives *V*p as a function of only elevation in the study area. To keep the model very simple, we compute *Sv*p by multiplying *V*p by the slope of the smoothed digital elevation model (DEM) in each grid box, neglecting the direction of the prevailing wind. Aspect could be introduced to this term as a proxy for wind direction. However, in the smoothed DEM of the Himalaya, the majority of steep slopes face the Indian side of the Himalayan arc, which is also the general direction of approach of the major weather systems (i.e., the mesoscale convective systems associated with the summer monsoons), suggesting that the addition of aspect into the model might be a relatively minor refinement. We compared these two topographic variables (*V*p and *Sv*p), as well as other simple topographic variables to the TRMM precipitation estimates using a statistical spatial regression model.

Statistical Relationships between TRMM Precipitation and Topographic Variables

The TRMM precipitation estimates, as well as the topographic variables, are autocorrelated in space; measurements for areas close together in space are likely to be close in value. Thus measurements at different locations are not strictly independent, and a simple regression model may overstate relationships between the variables. Therefore, we used a regression model that accounts for spatial lags of the variables (e.g., Anselin, 1993). It compares the values of chosen topographic variables at each point and the values of these variables at neighboring points to the local measured TRMM precipitation estimate. We used a MATLAB spatial statistics code (Pace and Barry, 1999) to find the ordinary least-squares (OLS) best-fit model of the form above and to evaluate the significance of each variable in this fit.

This model was used to find the best-fit coefficients for our theoretical model of orographic precipitation given the TRMM estimates and GTOPO30 topography. Additionally, we used the spatial regression technique for other topographic variables that may reasonably be related to precipitation, namely, slope (*S*), elevation relative to a reference elevation of 10 km (*E*), and the product of *S* and *E* (*Se*) to compare the fit for these variables versus that for our theoretical model. *E* is used rather than elevation relative to sea level, so that large values of *E* correspond to low elevations, as in *V*p, where large values are found at low elevations. These variables were chosen because they allow for the evaluation of the importance of the nonlinear relationship between *V*p and elevation versus the linear relationship between *E* and elevation, as well as the separation of the effects of slope alone in *S* and slope modified by an elevation term in *Sv*p and *Se*. All the variables were standardized [(value – mean)/standard deviation], as were the TRMM estimates, so that the units and absolute magnitudes of these variables would not be important to the analysis.

The statistical model accounts for the spatial autocorrelation of the topographic variables by computing an average of those variables at the four neighboring points for each point in the domain using a spatial weights matrix *W*. The observed values are regressed on both the topographic variables and the averages of the values at neighboring points (spatial lags). The *m* topographic (t_1, t_2, … t_m) variables being evaluated are put into a matrix, T ($n \times m$):

$$T = \begin{vmatrix} t_1(1) & t_2(1) & \cdots & t_m(1) \\ t_1(2) & t_2(2) & \cdots & t_m(2) \\ \cdots & \cdots & \cdots & \cdots \\ t_1(n) & t_2(n) & \cdots & t_m(n) \end{vmatrix}. \quad (5)$$

The regression model takes the form:

$$T * \begin{vmatrix} a_1 \\ a_2 \\ \cdots \\ a_m \end{vmatrix} + W * T * \begin{vmatrix} b_1 \\ b_2 \\ \cdots \\ b_m \end{vmatrix} = TRMM + error, \quad (6)$$

where TRMM is the $n \times 1$ vector of the TRMM observations over the Himalaya, error is an $n \times 1$ vector of the difference between the estimate and the TRMM value, and the coefficients $a_1 - a_m$ are the OLS coefficients for the topographic variables, while $b_1 - b_m$ are the OLS coefficients for the spatial lags of those variables. We used the MATLAB code to compute the OLS estimates of the model parameters and the likelihood estimates for a hypothesis that each variable and its spatial lag have no effect.

The theoretical model of orographic precipitation that we present in equation 3 yields a good fit to the data in the spatial regression model with a mean error of 0.387. The model is shown in Figure 6 in the same units as the TRMM measurements (m/yr).

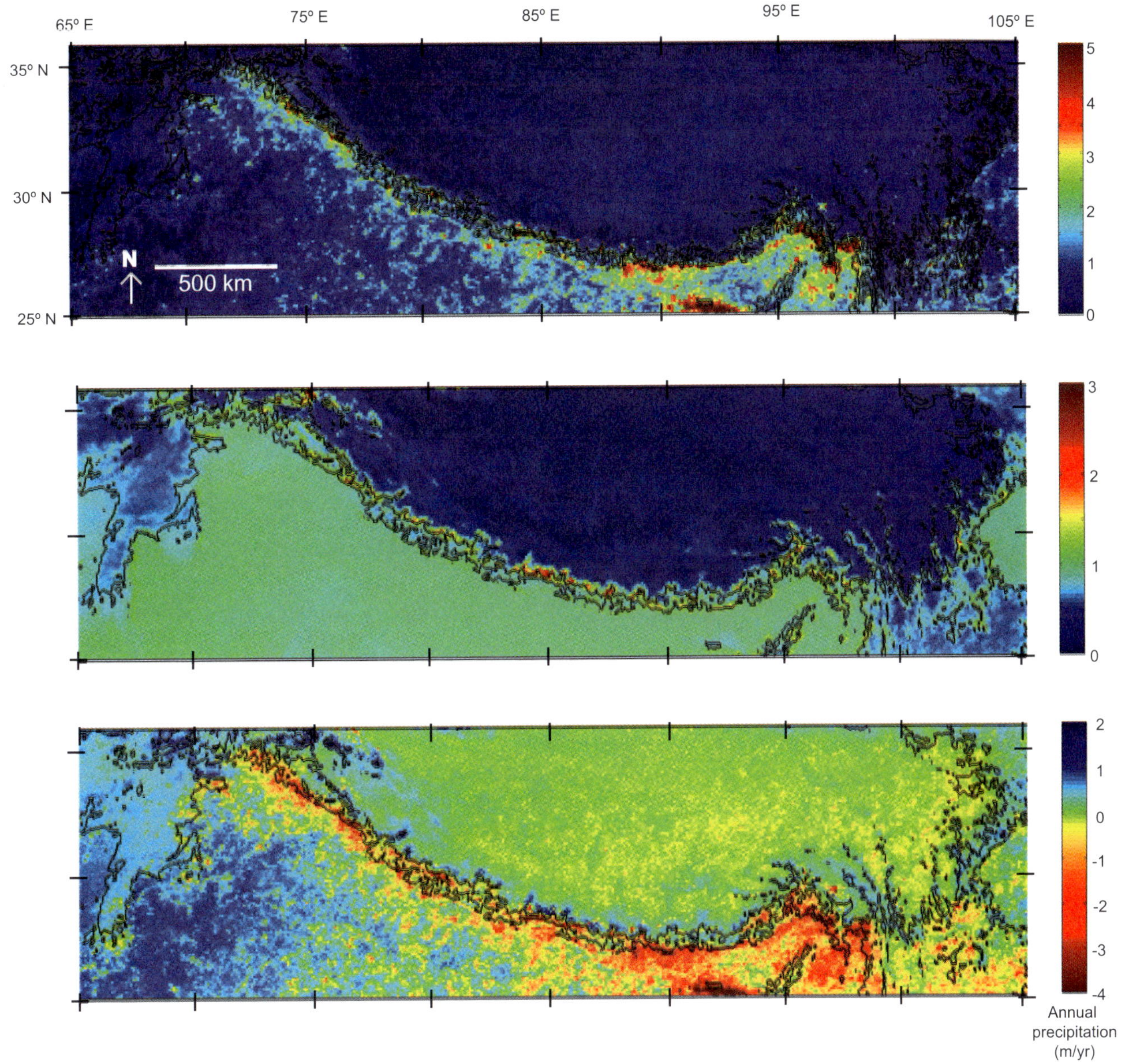

Figure 6. Tropical Rainfall Measuring Mission (TRMM) precipitation pattern, statistical model of precipitation, and error in the model. The TRMM-derived pattern of annual precipitation (m/yr) is shown over the entire study area in the top panel. The ordinary least-squares (OLS) best-fit model of this pattern using *V*p and *Sv*p, shown in the middle panel, captures the gradient across the range and the tracking of precipitation up major valleys. The error in the model in the lower panel, also in m/yr, shows that the large-scale east to west gradient, as well as the smaller-scale maxima in precipitation in the east, are not captured by the model. The solid lines indicate the 1500 and 3500 m topographic contours.

Additionally, we examined the models that resulted from using every combination of 1–5 of the reasonable topographic variables we considered (*V*p, *Sv*p, *S*, *E*, *Se*) and computed for each the mean absolute error and standard deviation of the error for each model. The errors from each model are given in Table 3.

Considering the best models with each number of variables, we find that *V*p is the best 1-parameter model with a mean error of 0.460. The best 2-parameter model is our favored theoretical model: *V*p and *Sv*p, with a mean error of 0.387. In examining the best models with more than 2 parameters, we find that the mean error is not substantially decreased by adding more parameters (0.367 for the best 3-parameter model, 0.360 for a 4-parameter model, and 0.356 for the 5-parameter model) (Table 3). The coefficients associated with the 3, 4, and 5 parameter models are difficult to interpret physically, as they include opposite signs for the coefficients of a variable and the spatial lags of the variable. The relative insensitivity of the mean error to the additional parameters, as well as the un-physical coefficients of these 3-, 4-, and 5-parameter models, indicates that the simpler, physically based, theoretical model not only is a good fit to the data, but also

TABLE 3. ERROR IN STATISTICAL MODELS

Variables	Mean absolute error	Standard deviation of the error
E	0.594	0.923
Vp[†]	0.460	0.806
S	0.716	0.995
Se	0.682	0.955
Svp	0.563	0.837
E, Vp	0.461	0.806
E, S	0.598	0.868
E, Se	0.573	0.836
E, Svp	0.487	0.751
Vp, S	0.458	0.759
Vp, Se	0.431	0.712
Vp, Svp[†]	0.384	0.666
S, Se	0.565	0.834
S, Svp	0.499	0.759
Se, Svp	0.516	0.769
E, Vp, S,	0.452	0.748
E, Vp, Se	0.424	0.707
E, Vp, Svp	0.347	0.664
E, S, Se	0.536	0.813
E, S, Svp	0.463	0.736
E, Se, Svp	0.459	0.730
Vp, S, Se	0.388	0.676
Vp, S, Svp[†]	0.367	0.658
Vp, Se, Svp	0.377	0.663
S, Se, Svp	0.499	0.759
E, Vp, S, Se	0.386	0.671
E, Vp, S, Svp	0.367	0.658
E, Vp, Se, Svp	0.374	0.661
E, S, Se, Svp	0.460	0.729
Vp, S, Se, Svp[†]	0.360	0.656
E, Vp, S, Se, Svp[†]	0.356	0.656

[†]The best model with each number of variables.

provides the best explanation of the observed pattern using the topographic variables we considered.

The coefficients for the OLS best-fit model with the parameters *V*p and *Sv*p are 0.402 and 0.041 on *V*p and *Sv*p, respectively, and 0.114 and 0.452 on their respective spatial lags. The model successfully captures some principal features of the TRMM precipitation pattern, including the dry plateau and the wet foreland at the large scale, and the wet valleys and dry ridges, as well as the double band of precipitation along the central front part of the range. However, as shown in Figure 6, the model fails to capture the extremely high values of precipitation in the eastern end of the range, largely fails to capture the gradient from east to west along the range, and predicts much smaller maximum precipitation totals (~3 m/yr versus 9 m/yr in the TRMM estimates).

In addition to the model for the whole Himalaya, the OLS best-fit coefficients for the preferred model were also found for a small subregion near the eastern syntaxis (Fig. 7); they are 0.014 and 0.216 for *V*p and *Sv*p, respectively, and 0.509 and 0.227 on their spatial lags. As in the case above, the subregion model does capture principal aspects of the variability; the strong valley-ridge difference between the Tsangpo River and the surrounding high areas is reproduced in the model. Also, similar to the larger case, the model does not predict the largest precipitation totals in the TRMM estimates (~3.5 m/yr versus 6 m/yr). That the coefficients

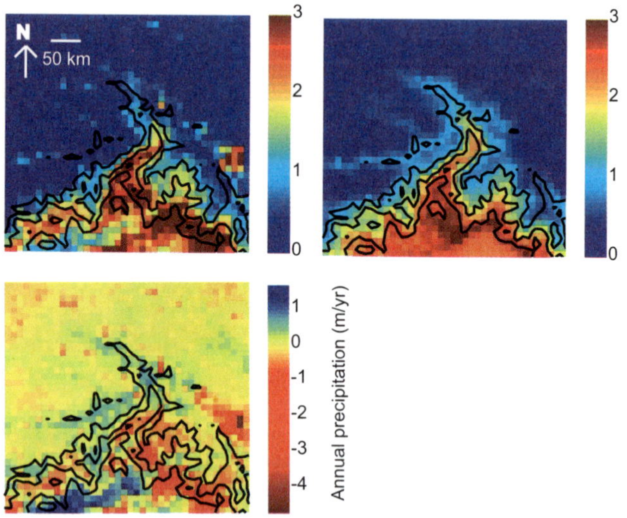

Figure 7. Model of Tropical Rainfall Measuring Mission (TRMM) data for the Namche Barwa subregion. The top left panel shows the TRMM annual precipitation estimates (m/yr). The top right panel shows the ordinary least-squares (OLS) model using *V*p and *Sv*p. The error (model-data) is shown in the bottom panel. As in the case of the entire study area, the model describes principal elements of the precipitation pattern, but misses small-scale (<100 km) details. The solid lines indicate the 1500, 2500, and 3500 m topographic contours.

for this region differ so much from those for the whole study area is likely due to the greater importance of slopes on a regional scale; steep slopes comprise a greater fraction of small region than in the study area as a whole. It also reflects an inherent issue in modeling precipitation: processes involved in orographic precipitation are scale-dependent, and no single set of coefficients should be expected to be generally applicable.

DISCUSSION

The spatial pattern of precipitation as estimated from TRMM in the Himalaya shows considerable variability on scales less than 100 km—spatial variability that is repeated in four separate years of measurements (Figs. 1–3). Similar seasonal gradients are seen in gauge records from the calibration site in Nepal (Barros et al., 2000; Barros and Lang, 2003). Robust gradients in precipitation over spatial scales of tens of kilometers have important implications for both the study of orographic precipitation and for landscape evolution modeling. The fact that the large-scale pattern of precipitation does not change appreciably over time suggests that, at this scale, the pattern strongly reflects topography itself. Topography is obviously constant on this time scale, whereas the characteristics of the incoming air may vary significantly from event to event. At the pixel scale (10 km × 10 km) there are some large interannual differences; we cannot distinguish between those caused by sampling error and those by genuine variation in precipitation from year to year.

Large differences in annual precipitation occurring in areas only tens of kilometers apart along the strike of the mountain range show that the common use of uniform precipitation fields in landscape evolution models does not realistically depict within-range variability in the climatic influence on erosion, even at small spatial scales. Most importantly for the consideration of coupled climatic, erosional, and tectonic processes, the pattern of precipitation in the Himalaya is strongly controlled by the topography, as suggested by the success of our simple theoretical model and supported by remote sensing and modeling of cloud patterns in central Nepal (Barros et al., 2004). A pattern of precipitation that is robust and strongly related to topography must evolve as the topography develops and, hence, must also influence the development of topography through erosion. The co-evolution of precipitation patterns and topography merits further research.

The TRMM precipitation radar is a useful tool for studying precipitation and river discharge patterns in the Himalaya. TRMM can provide reasonable estimates of annual discharge for basins of more than 1000 km^2 in the Himalaya. However, TRMM estimates of average annual precipitation are consistently smaller than measured unit discharge (86% of measured unit discharge) (Fig. 4). Moreover, since they do not account for evapo-transpiration, we would expect that the TRMM estimates for the annual average precipitation in a basin would exceed annual unit discharge. TRMM radar does not detect very low precipitation rates (less than 0.7 mm/h), so, to the extent that events characterized by such precipitation rates contribute to annual total precipitation, TRMM will underestimate this total. Additionally, TRMM's 18 dBZ detection limit does not allow for observation of low-moderate snowfall rates and will underestimate the water equivalence of any snow detected. Snowfall is limited to higher elevations, and there is a tendency for more persistent and lighter rainfall at these same elevations at the Nepal calibration site (Lang and Barros, 2002, 2004). This suggests a bias toward underestimating precipitation at higher elevations. A comparison of TRMM precipitation estimates and measured mean runoff in other regions, especially those with known spatially varying average rain rates, could help constrain the errors associated with TRMM precipitation estimates at high elevations and in complex terrain.

In the Himalaya, ground clutter may have a subtle influence on measurements for look angles greater than 13.5 degrees, but does not greatly impact the observed pattern of TRMM precipitation estimates (Fig. 5). This influence is not seen in more observations of rain at high look angles than at lower look angles, nor as the addition of very high values of rain rate. Rather, the influence seems to be spatially correlated patches with higher average rain rates (Fig. 5). Our preliminary work suggests that a more extensive study of this problem could determine if the combination of measurement and sampling error would be improved by discarding the high look angle measures in some regions.

Three methods of constraining satellite sampling error suggest that for yearly and monsoon precipitation totals in the Himalaya, sampling errors of 15%–50% are likely (Table 2). These errors are not trivial, but neither are they so large as to overwhelm the observed spatial pattern of precipitation; annual precipitation in our climatology varies from a few tenths of a meter to nearly 10 m across the study area. The variability we measured spans nearly two orders of magnitude, far greater than the estimated 15%–50% error. The close correlation of basin-averaged annual precipitation estimates and measured mean runoff supports this conclusion. Moreover, we are primarily interested in the spatial variability in precipitation rather than the magnitude itself, in part because the interannual and longer time-scale variability in this region is largely unconstrained. Provided that spatial patterns similar to those we observed persist over geologically significant time scales, they will be important for questions of landscape (and possibly tectonic) evolution in the mountain range. Such patterns may persist despite changes in the magnitude of precipitation on the whole. Hence, the error estimates and correlation with discharge indicate that TRMM precipitation estimates can be used as a proxy for spatial patterns in precipitation in geomorphic applications, including discharge estimates for use in fluvial incision models and for evaluation of models of precipitation based on topographic variables.

The measurement and sampling errors involved in TRMM estimates of precipitation reported herein are likely to be dependent on conditions specific to the Himalaya. Ground clutter errors may be a function of the relief in the region and formal sampling error estimates require local parameters such as decorrelation time and average rain rate. Consequently, extension of this method of studying spatial patterns of precipitation to other low-latitude regions sampled by the TRMM satellite will require estimation of the local error to determine appropriate temporal and spatial scales for consideration.

A simple, physically based model for precipitation involving a linear combination of *V*p, *Sv*p, and spatial lags of these variables is statistically significant and captures most of the major features of precipitation in the Himalaya at the 10 km scale, including the increased precipitation in large valleys and the double band of high precipitation in the central part of the Himalayan arc (Figs. 6 and 7). The statistical model might be improved by including a preferred wind direction and calculating the topographic slope into the wind rather than just the slope—the current model treats both- north and south-facing slopes equivalently. The relative success of this fairly simple model suggests that the decrease in temperature and saturation vapor pressure with height in the atmosphere and the direct forcing of air up slopes, especially at low elevations, are key controls on precipitation in the Himalaya at scales greater than ten kilometers.

However, the smaller scale (<200 km) maxima in precipitation and the east to west gradient in precipitation along the range are not captured by the statistical model (Fig. 6). This indicates that, not surprisingly, factors other than air temperature and simple upslope motion are important in creating these patterns. In particular, the overland distance that air must travel to the range varies from east to west, and monsoon dynamics, and the associated changes in temperature and humidity, may be important in creating the east to west gradient along the arc. In the small regions with very high precipitation totals, dynamical convergence of air, the triggering of convection, and the motion of convective cells likely increase precipitation locally, in addition to the simple factors included in the statistical model.

The application of our physically based model as a predictive model over evolving topography does require some further research. The difference in the relative magnitudes of the coefficients of the statistical model for the whole region versus the subregion complicates the calibration of our physically based model for use in a predictive sense over evolving topography. An initial step may be to limit the domain considered in the statistical regression to the Himalayan arc itself to obtain parameters specific to this region. The difference between our statistical model and the measured TRMM climatology gives us some sense of the features that would be neglected if we were to model the co-evolution of the precipitation pattern over developing Himalayan topography using our model (Figs. 6 and 7). It remains to be investigated as to whether these features create important feedbacks with the landscape that impact its evolution over time.

Precipitation patterns as measured by TRMM in the Himalaya vary considerably over small spatial scales, as has also been noted for the Alps and the Olympic Mountains of western Washington, USA. Spatial patterns of discharge are certainly important in determining the spatial variability in stream power and fluvial shear stress, and thus an understanding of the spatial patterns of precipitation and discharge is important for estimating spatial patterns of erosion in mountainous landscapes. In addition to the relationship between fluvial incision and erosion, glacial erosion rates are expected to increase with ice discharge (e.g., MacGregor et al., 2000; Tomkin and Braun, 2002), and hence with rates of solid precipitation. Therefore the tendency for more precipitation in certain areas is likely to result in high erosion rates in those areas independent of whether glacial or fluvial processes dominate. There may also be important local relationships between precipitation and erosion. For instance, Reiners et al. (2003) found that long-term erosion rates tracked spatial variation in twentieth-century precipitation across the Washington Cascades. If precipitation itself proves to be a useful predictor of spatial patterns of erosion, the importance of documenting the spatial pattern of precipitation in mountains becomes even more fundamental to our understanding of feedbacks between precipitation and topographic development. Measuring and modeling spatial variability in precipitation will no doubt provide interesting new insights into the continuing study of the interrelationships between erosion, climate, and tectonics.

ACKNOWLEDGMENTS

The data used in this study were acquired as part of the Tropical Rainfall Measuring Mission (TRMM). The algorithms were developed by the TRMM Science Team. The data were processed by the TRMM Science Data and Information System (TDSIS) and the TRMM office; they are archived and distributed by the Goddard Distributed Active Archive Center. TRMM is an international project jointly sponsored by the Japan National Space Development Agency (NASDA) and the U.S. National Aeronautics and Space Administration (NASA) Office of Earth Sciences. We thank Courtney Shumacher for help and advice on obtaining and understanding the TRMM satellite data, Robert Meneghini for information and advice on TRMM and ground clutter, Harvey Greenberg for savvy data processing and graphics, and Ana Barros and Ronald Smith for constructive critiques of the manuscript. This study was partially funded by the National Science Foundation Continental Dynamics Program grant no. 003561.

REFERENCES CITED

Anders, A.M., Roe, G.H., Durran, D.R., Montgomery, D.R., and Hallet, B., 2004, Precipitation and the form of mountain ranges: Bulletin of the American Meteorological Society, v. 85, p. 498–499.

Anselin, L., 1993, Discrete space autoregressive models, *in* Goodchild, M.F., Parks, B.O., and Steyaert, L.T., eds., Environmental modeling with GIS: New York, Oxford University Press, p. 454–469.

Barros, A.P., and Lang, T.J., 2003, Monitoring the monsoon in the Himalayas: Observations in central Nepal, June 2001: Monthly Weather Review, v. 131, p. 1408–1427, doi: 10.1175/1520-0493(2003)131<1408:MTMITH>2.0.CO;2.

Barros, A.P., and Lettenmaier, D.P., 1993, Dynamic modeling of the spatial distribution of precipitation in remote mountainous areas: Monthly Weather Review, v. 121, p. 1195–1214, doi: 10.1175/1520-0493(1993)1212.0.CO;2.

Barros, A.P., and Lettenmaier, D.P., 1994, Dynamic modeling of orographically induced precipitation: Reviews of Geophysics, v. 32, p. 265–284, doi: 10.1029/94RG00625.

Barros, A.P., Joshi, M., Putkonen, J., and Burbank, D.W., 2000, A study of the 1999 monsoon rainfall in a mountainous region in central Nepal using TRMM products and rain gauge observations: Geophysical Research Letters, v. 27, p. 3683–3686, doi: 10.1029/2000GL011827.

Barros, A.P., Kim, G., Williams, E., and Nesbitt, S.W., 2004, Probing orographic controls in the Himalayas during the monsoon using satellite imagery: Naturals Hazards and Earth System Science, v. 4, p. 29–51.

Beaumont, C., Fullsack, P., and Hamilton, J., 1992, Erosional control of active compressional orogens, *in* McClay, K.R., ed., Thrust tectonics: New York, Chapman and Hall, p. 1–18.

Bell, T.L., 1987, A space-time stochastic model of rainfall for satellite remote sensing studies: Journal of Geophysical Research, v. 92, p. 9631–9644.

Bell, T.L., and Kundu, P.N., 2000, Dependence of satellite sampling error on monthly averaged rain rates: Comparison of simple models and recent studies: Journal of Climate, v. 13, p. 449–462, doi: 10.1175/1520-0442(2000)013<0449:DOSSEO>2.0.CO;2.

Colle, B.A., Westrick, K.J., and Mass, C.F., 1999, Evaluation of the MM5 and Eta-10 precipitation forecasts over the Pacific Northwest during the cool season: Weather and Forecasting, v. 14, p. 137–154.

Daly, C., Gibson, W.P., Taylor, G.H., Johnson, G.L., and Pasteris, P., 2002, A knowledge-based approach to the statistical mapping of climate: Climate Research, v. 22, p. 99–113.

Dhar, O.N., and Rakhecha, P.R., 1981, The effect of elevation on monsoon rainfall distribution in the central Himalayas, *in* Lighthill, J., and Pearce, R.P., eds., Monsoon dynamics: New York, Cambridge University Press, p. 253–260.

Dingman, S.L., 2002, Physical hydrology: Upper Saddle River, New Jersey, Prentice Hall, 646 p.

Durran, D.R., 1986, Mountain waves, *in* Ray, P.S., ed., Mesoscale meteorology and forecasting: Boston, American Meteorological Society, p. 472–492.

Emanuel, K.A., 1994, Atmospheric convection: New York, Oxford University Press, 580 p.

Finlayson, D.P., and Montgomery, D.R., 2003, Modeling large-scale fluvial erosion in geographic information systems: Geomorphology, v. 53, p. 147–164, doi: 10.1016/S0169-555X(02)00351-3.

Finlayson, D.P., Montgomery, D.R., and Hallet, B., 2002, Spatial co-incidence of erosional and metamorphic hot spots in the Himalaya: Geology, v. 30, p. 219–222, doi: 10.1130/0091-7613(2002)030<0219:SCORIE>2.0.CO;2.

Frei, C., and Schär, C., 1998, A precipitation climatology of the Alps from high-resolution rain-gauge observations: International Journal of Climatology, v. 18, p. 873–900, doi: 10.1002/(SICI)1097-0088(19980630)18:83.0.CO;2-9.

Groisman, P.V., and Legates, D.R., 1994, The accuracy of United States precipitation data: Bulletin of the American Meteorological Society, v. 75, p. 215–227, doi: 10.1175/1520-0477(1994)075<0215:TAOUSP>2.0.CO;2.

Hoffman, P.F., and Grotzinger, J.P., 1993, Orographic precipitation, erosional unloading, and tectonic style: Geology, v. 21, p. 195–198, doi: 10.1130/0091-7613(1993)021<0195:OPEUAT>2.3.CO;2.

Houze, R.A., 1993, Cloud dynamics: Boston, Academic Press, 573 p.

Koons, 1989, The topographic evolution of collisional mountain belts: A numerical look at the Southern Alps, New Zealand: American Journal of Science, v. 289, p. 1041–1069.

Kozu, T., Kawanishi, T., Kuroiwa, H., Kojima, M., Oikawa, K., Kumagai, H., Okamoto, K., Okumua, M., Nakatsuka, H., and Nizhikawa, K., 2001, Development of precipitation radar onboard the Tropical Rainfall Measuring Mission (TRMM) satellite: IEEE Transactions on Geoscience and Remote Sensing, v. 39, p. 102–116.

Lang, T.J., and Barros, A.P., 2002, An investigation of the onsets of the 1999 and 2000 monsoons in central Nepal: Monthly Weather Review, v. 130, p. 1299–1316, doi: 10.1175/1520-0493(2002)130<1299:AIOTOO>2.0.CO;2.

Lang, T.J., and Barros, A.P., 2004, Winter storms in the central Himalayas: Journal of the Japanese Meteorological Society, v. 82, p. 829–844, doi: 10.2151/jmsj.2004.829.

Leemans, R., and Cramer, W.P., 1991, The IIASA database for mean monthly values of temperature, precipitation and cloudiness of a global terrestrial grid: Laxenburg, Austria, International Institute for Applied Systems Analysis Publication RR-91-18, 62 p.

MacGregor, K.R., Anderson, R.S., Anderson, S.P., and Waddington, E.D., 2000, Numerical simulations of glacial-valley longitudinal profile evolution: Geology, v. 28, p. 1031–1034, doi: 10.1130/0091-7613(2000)028<1031:NSOGLP>2.3.CO;2.

Meneghini, R., Iguchi, T., Kozu, T., Liao, L., Okamoto, K., Jones, J.A., and Kwiatkowski, J., 2000, Use of the surface reference technique for path attenuation estimates from the TRMM precipitation radar: Journal of Applied Meteorology, v. 39, p. 2053–2070.

Montgomery, D.R., Balco, G., and Willett, S.D., 2001, Climate, tectonics and the morphology of the Andes: Geology, v. 29, p. 579–582, doi: 10.1130/0091-7613(2001)029<0579:CTATMO>2.0.CO;2.

Pace, R.K., and Barry, R., 1999, Spatial statistics toolbox 1.1 for Matlab: www.spatial-statistics.com (August 2003).

Reiners, P.W., Ehlers, T.A., Mitchell, S.G., and Montgomery, D.R., 2003, Coupled spatial variations in precipitation and long-term erosion rates across the Washington Cascades: Nature, v. 426, p. 645–647, doi: 10.1038/nature02111.

Roe, G.H., Montgomery, D.R., and Hallet, B., 2002, Effects of orographic precipitation variations on the concavity of steady-state river profiles: Geology, v. 30, p. 143–146, doi: 10.1130/0091-7613(2002)030<0143:EOOPVO>2.0.CO;2.

Roe, G.H., Montgomery, D.R., and Hallet, B., 2003, Orographic climate feedbacks and the relief of mountain ranges: Journal of Geophysical Research, v. 108, doi: 10.1029/2001JB001521.

Shin, K., and North, G.R., 1988, Sampling error study for rainfall estimate by satellite using a stochastic model: Journal of Applied Meteorology, v. 27, p. 1218–1231, doi: 10.1175/1520-0450(1988)027<1218:SESFRE>2.0.CO;2.

Shrestha, M.L., 2000, Interannual variation of summer monsoon rainfall over Nepal and its relation to Southern Oscillation Index: Meteorology and Atmospheric Physics, v. 75, p. 21–28, doi: 10.1007/s007030070012.

Simpson, J.E., 1987, Gravity currents in the environment and laboratory: New York, John Wiley and Sons, 244 p.

Sinclair, M.R., Wratt, D.S., Henderson, R.D., and Gray, W.R., 1997, Factors affecting the distribution and spillover of precipitation in the Southern Alps of New Zealand—A case study: Journal of Applied Meteorology, v. 36, p. 428–442, doi: 10.1175/1520-0450(1997)036<0428:FATDAS>2.0.CO;2.

Singh, P., Ramasastri, K.S., and Kumar, N., 1995, Topographical influence on precipitation distribution in different ranges of Western Himalayas: Nordic Hydrology, v. 26, p. 259–284.

Smith, R.B., 1979, The influence of mountains on the atmosphere: Advances in Geophysics, v. 21, p. 87–233.

Tomkin, J.H., and Braun, J., 2002, The influence of alpine glaciation on the relief of tectonically active mountain belts: American Journal of Science, v. 302, p. 169–190.

Tripoli, G.J., and Cotton, W.R., 1989, Numerical study of an observed orogenic mesoscale convective system: Monthly Weather Review, v. 117, p. 273–328, doi: 10.1175/1520-0493(1989)117<0273:NSOAOO>2.0.CO;2.

Willett, S.D., 1999, Orogeny and orography: The effects of erosion on mountain belts: Journal of Geophysical Research, v. 104, p. 28,957–28,981, doi: 10.1029/1999JB900248.

Manuscript Accepted by the Society 23 June 2005

Printed in the USA

Geological Society of America
Special Paper 398
2006

Tectonics from topography: Procedures, promise, and pitfalls

Cameron Wobus[†]
Kelin X. Whipple
Department of Earth, Atmospheric and Planetary Sciences, Massachusetts Institute of Technology, Cambridge, Massachusetts 02139, USA
Eric Kirby
Department of Geosciences, The Pennsylvania State University, University Park, Pennsylvania 16802, USA
Noah Snyder
Department of Geology and Geophysics, Boston College, Chestnut Hill, Massachusetts 02467, USA
Joel Johnson
Katerina Spyropolou
Benjamin Crosby
Department of Earth, Atmospheric and Planetary Sciences, Massachusetts Institute of Technology, Cambridge, Massachusetts 02139, USA
Daniel Sheehan
Information Systems, Massachusetts Institute of Technology, Cambridge, Massachusetts 02139, USA

ABSTRACT

Empirical observations from fluvial systems across the globe reveal a consistent power-law scaling between channel slope and contributing drainage area. Theoretical arguments for both detachment- and transport-limited erosion regimes suggest that rock uplift rate should exert first-order control on this scaling. Here we describe in detail a method for exploiting this relationship, in which topographic indices of longitudinal profile shape and character are derived from digital topographic data. The stream profile data can then be used to delineate breaks in scaling that may be associated with tectonic boundaries. The description of the method is followed by three case studies from varied tectonic settings. The case studies illustrate the power of stream profile analysis in delineating spatial patterns of, and in some cases, temporal changes in, rock uplift rate. Owing to an incomplete understanding of river response to rock uplift, the method remains primarily a qualitative tool for neotectonic investigations; we conclude with a discussion of research needs that must be met before we can extract quantitative information about tectonics directly from topography.

Keywords: channel geometry, digital elevation models, neotectonics, geomorphology.

1. INTRODUCTION

Across the globe, with few notable exceptions, the steepest landscapes are associated with regions of rapid rock uplift. Given this empirical observation, one might expect that meaningful tectonic information could be extracted from some parameterization of landscape morphology, such as mean topographic gradient. Hillslopes, however, reach threshold slopes wherever erosion rates approach the surface soil production rate (Burbank et al., 1996; Heimsath et al., 1997; Montgomery and Brandon, 2002), limiting their utility as a "fingerprint" of tectonic forcing to relatively low uplift rate environments.

[†]Present address: Cooperative Institute for Research in Environmental Sciences (CIRES), University of Colorado, Boulder, Colorado 80309, USA; cameron.wobus@colorado.edu.

Wobus, C., Whipple, K.X., Kirby, E., Snyder, N., Johnson, J., Spyropolou, K., Crosby, B., and Sheehan, D., 2006, Tectonics from topography: Procedures, promise, and pitfalls, *in* Willett, S.D., Hovius, N., Brandon, M.T., and Fisher, D.M., eds., Tectonics, Climate, and Landscape Evolution: Geological Society of America Special Paper 398, Penrose Conference Series, p. 55–74, doi: 10.1130/2006.2398(04). For permission to copy, contact editing@geosociety.org.

Only the fluvial network consistently maintains its connection to tectonic forcing, and therefore contains potentially useful information about variations in rock uplift rates across the landscape. A number of studies have laid the groundwork for extracting this information by exploring the theoretical response of channels to variations in rock uplift rate, and by analyzing fluvial profiles in field settings where the tectonics have been independently determined (e.g., Whipple and Tucker, 1999, 2002; Snyder et al., 2000; Kirby and Whipple, 2001; Lague and Davy, 2003). Against this theoretical and empirical backdrop, however, there remains some uncertainty as to what can and cannot be learned from an analysis of river profiles, and there still exists no standard method for extracting tectonic information from these data.

In this contribution we attempt to bridge this gap, and discuss the state of the art in our ability to extract tectonic information directly from river profiles. The discussion focuses on the use of digital elevation models (DEMs), which are inexpensive, easily obtained, and can be used to extract much of this information quickly and easily prior to embarking on field campaigns. We discuss the methods employed in delineating tectonic information from DEMs, including data sources, data handling, and interpretation. Case studies from diverse settings are then used to illustrate the utility of DEM analyses in extracting tectonic information from the landscape. We conclude with a discussion of research needs that must be met before we can have a reliable quantitative tool for neotectonics. Throughout the paper, we purposefully restrict our focus to empirical data analysis, discussing theory only as a rudimentary backdrop and in the context of unresolved issues that limit our ability to extract quantitative tectonic information from stream profiles.

2. BACKGROUND

In a variety of natural settings, topographic data from fluvial channels exhibit a scaling in which local channel slope can be expressed as a power-law function of contributing drainage area (e.g., Hack, 1973; Flint, 1974; Howard and Kerby, 1983):

$$S = k_s A^{-\theta}, \quad (1)$$

where S represents local channel slope, A is the upstream drainage area, and k_s and θ are referred to as the steepness and concavity indices, respectively. Equation 1 holds only for drainage areas above a critical threshold, A_{cr}, variably interpreted as the transition from divergent to convergent topography or from debris-flow to fluvial processes (Tarboton et al., 1989; Montgomery and Foufoula-Georgiou, 1993). While many stream profiles will exhibit a single slope-area scaling for their entire length downstream of A_{cr}, segments of an individual profile are often characterized by different values of k_s, θ, or both. These aberrations may appear to be exceptions to the empirical result in equation 1; however, it is actually these variations we wish to exploit to extract tectonic information from the landscape. Much of the remainder of this paper outlines the methodologies employed to extract this information.

As has often been reported, simple models for both detachment- and transport-limited river systems predict power-law relations between channel gradient and drainage area in the form of equation 1 (e.g., Howard, 1994; Willgoose et al., 1991; Whipple and Tucker, 1999). In these models the concavity index, θ, is independent of rock uplift rate, U, assuming U is spatially uniform (Whipple and Tucker, 1999). These models further predict direct power-law relations between the steepness index, k_s, and rock uplift rate (e.g., Howard, 1994; Willgoose et al., 1991). However, although there is strong empirical support for a positive correlation between k_s and U, many factors not incorporated into these simple models can be expected to influence the quantitative relation between k_s and U. Known complexities include: (1) nonlinearities in the incision process (e.g., Whipple et al., 2000; Whipple and Tucker, 1999), including the presence of thresholds (e.g., Tucker and Bras, 2000; Snyder et al., 2003b; Tucker, 2004), and possible changes in dominant incision processes with increasing incision rate (Whipple et al., 2000); (2) adjustments in channel width or sinuosity, herein referred to as channel morphology (e.g., Harbor, 1998; Lavé and Avouac, 2000, 2001; Snyder et al., 2000, 2003a); (3) adjustments in the extent of alluvial cover, bed material grain size, bed morphology, and hydraulic roughness, herein referred to as bed state (e.g., Sklar and Dietrich, 1998, 2001; Hancock and Anderson, 2002; Whipple and Tucker, 2002; Sklar, 2003); (4) changes in the frequency of erosive debris flows (Stock and Dietrich, 2003); and (5) orographic enhancement of precipitation (e.g., Roe et al., 2002, 2003; Snyder et al., 2000, 2003a). Due to the possible influence of each of these complexities, the functional relationship between k_s and U can be expected to vary depending on the geologic setting. While many of these complexities will be important over large length scales, over which rock uplift rates are likely to be nonuniform, we must nonetheless consider these varied feedbacks and nonlinearities if we hope to quantitatively map steepness to rock uplift rates. Additional climatic factors and substrate rock properties also strongly influence k_s, and are difficult to deconvolve from uplift rate signals. Thus at present we do not know how to quantitatively map channel steepness to incision rate (or rock uplift rate at steady state) except, to some extent, through local calibration of incision model parameters, as discussed in the Siwalik Hills example.

Most models predict that profile concavity will be independent of rock uplift rate (if spatially uniform); however, any river response that differs at small and large drainage areas could theoretically induce a change in concavity. For instance, at greater drainage area, one might expect the channel to have more freedom to adjust channel width and sinuosity. Similarly at smaller drainage area, one might expect more variability in the fraction of exposed bedrock, hydraulic roughness, and the relative influence of debris flows. Despite these theoretical considerations, available data suggest little change in the concavity index, θ, of adjusted river profiles as a function of rock uplift rate (e.g., Tucker and Whipple, 2002). Because k_s is a function of U, however (see below, and Snyder et al., 2000; Kirby and Whipple, 2001; Kirby et al., 2003), a downstream change in rock uplift rate may be manifested as a change in profile concavity (e.g., Kirby and Whipple, 2001). Thus, changes in profile concavity can also be exploited in evaluating regional tectonics from topography.

While the qualitative relationships among steepness, concavity, and rock uplift rates can be readily predicted for "adjusted" longitudinal profiles, we note that temporal changes in the climatic and/or tectonic state can complicate these relationships. For example, fluvial systems in a transient state may contain knickpoints caught sweeping through the system in response to base-level fall. If such discontinuities in channel profiles and slope-area scaling are always assumed to reflect spatial variations in rock uplift rate, these profiles may be subject to misinterpretation. However, plan view maps illustrating the spatial distribution of these knickpoints, along with an examination of long profiles and slope-area data, will typically allow these situations to be readily identified, as discussed in section 3.3. Furthermore, such transient profiles, if properly identified, can provide extremely useful information for neotectonic analysis, as described in the San Gabriel Mountains example.

Because the steepness and concavity indices each reflect spatial variations in rock uplift rate, stream profile parameters derived from regressions on natural slope-area data allow us to extract information about regional tectonics. We proceed by discussing the methodologies for extracting these data and delineating breaks in slope-area scaling. We then discuss applications of these methods to deriving tectonic information from longitudinal profile form. We stress that our approach is empirical, and is therefore not tied to any particular river incision model.

3. METHODS

3.1 Data Handling

Digital topographic data suitable for long profile analysis are widely available for sites within the United States, and can be obtained for download directly from the U.S. Geological Survey (USGS) or its affiliated data repositories (http://seamless.usgs.gov/ or http://www.gisdatadepot.com/dem/). For field areas outside of the United States, DEMs can be obtained from local sources or from the National Aeronautics and Space Administration's (NASA) Shuttle Radar Topography Mission (SRTM) (http://www.jpl.nasa.gov/srtm/). DEMs can also be created from stereo pairs of spaceborne satellite imagery (e.g., Advanced Spaceborne Thermal Emission and Reflection Radiometer [ASTER] http://asterweb.jpl.nasa.gov/, or Satellite Probatoire d'Observation de la Terre [SPOT] http://www.spot.com), or from digitized aerial photographs, where available. Note that any DEMs created from remotely sensed data may contain data holes or anomalies due to extreme relief or cloud cover. Depending on the data source, DEMs may also require a projection from geographic coordinates to a format with rectangular, equidimensional pixels throughout the region of interest (c.f. Finlayson and Montgomery, 2003).

Once digital data have been obtained, a variety of methods is appropriate for extracting the requisite stream profile parameters. In practice, any suite of computer scripts that can follow a path of pixels downstream while recording elevation, cumulative streamwise distance, and contributing drainage area data is sufficient for collecting long profile data from a DEM. The methods developed by Snyder et al. (2000) and Kirby et al. (2003) utilize a group of built-in functions in ARC/INFO to create flow accumulation arrays and delineate drainage basins, a suite of MATLAB scripts to extract and analyze stream profile data from these basins, and an Arcview interface for color-coding stream profiles by their steepness and concavity indices in a geographic information system (GIS). While pits and data holes in a DEM usually need to be filled to create flow direction and flow accumulation arrays for basin delineation, profile data should be extracted from the raw DEM matrix to ensure that no data are lost or created at this early stage in the processing.

Once the elevation, distance, and drainage area data are compiled, the next step is to calculate local channel slopes to be used in slope-area plots. If using built-in ARC/INFO functions, slope values should not be extracted from a slope grid computed from a 3 × 3 moving window across the entire DEM: high slopes on channel walls will cause significant upward bias in channel slopes in this case, particularly at large drainage area in narrow bedrock canyons. There are also several problems with using raw pixel-to-pixel slopes from the channel itself (rise/run): (1) many DEMs are created by interpolation of digitized topographic contour maps, grossly oversampling the available data at large drainage area and low channel gradient and leading to bias toward the data at large drainage area in regressions of log S on log A; (2) the algorithms used to convert topographic maps to a raster format often give rise to interpolation errors, which characteristically produce artificial stair-steps associated with each contour crossing of the stream line and tremendous artificial scatter in pixel-to-pixel slope data (Figs. 1 and 2); (3) these stair-steps and the integer format of many DEMs produce multiple flats with zero slope, which cannot be handled in a log-log plot of slope and area (see Fig. 1A); and (4) DEMs with low resolution will often short-circuit meander bends in a river profile, resulting in an overestimate of local channel slope, typically in floodplains at large drainage area.

In order to circumvent many of the problems with raw pixel-to-pixel slopes, raw elevation data can be resampled at equal vertical intervals (Δz), using the contour interval from the original data source as Δz (if known). This step has several benefits: (1) it remains true to the original contour data from which many DEMs are derived; (2) it yields a data set much more evenly distributed in log S–log A space, reducing bias in regression analysis; and (3) it results in considerable smoothing of raw DEM profiles (see Figs. 2A and 2B). As the last two benefits apply even to DEM data not derived from contour maps, we favor the implementation

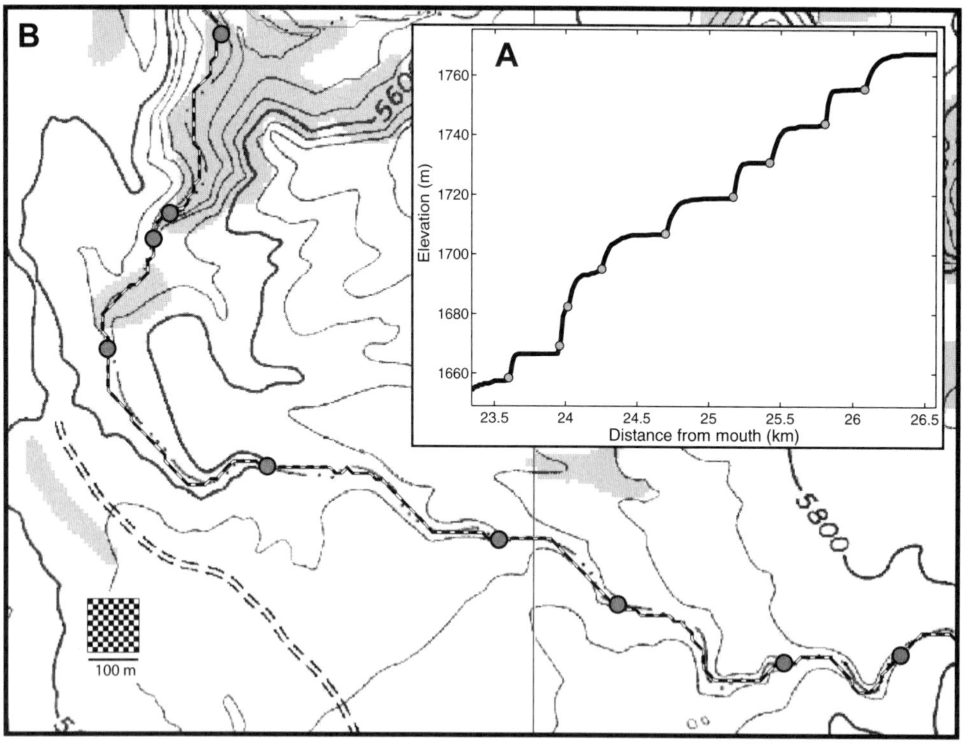

Figure 1. Example of contour extraction method for high-quality U.S. Geological Survey (USGS) 10 m digital elevation model (DEM). (A) Prior to contour extraction, DEM is characterized by abundant artificial data between contour crossings, resulting in a stepped pattern on the profile. Subsampling data at equal vertical intervals (gray dots) produces a smoother, more realistic profile consistent with the source data (see text for discussion). (B) Contour extraction method faithfully reproduces stream profile crossings of contour lines on original USGS topographic map (gray dots). Dashed line—stream line extracted from DEM (compare to dash-dot stream line on the map).

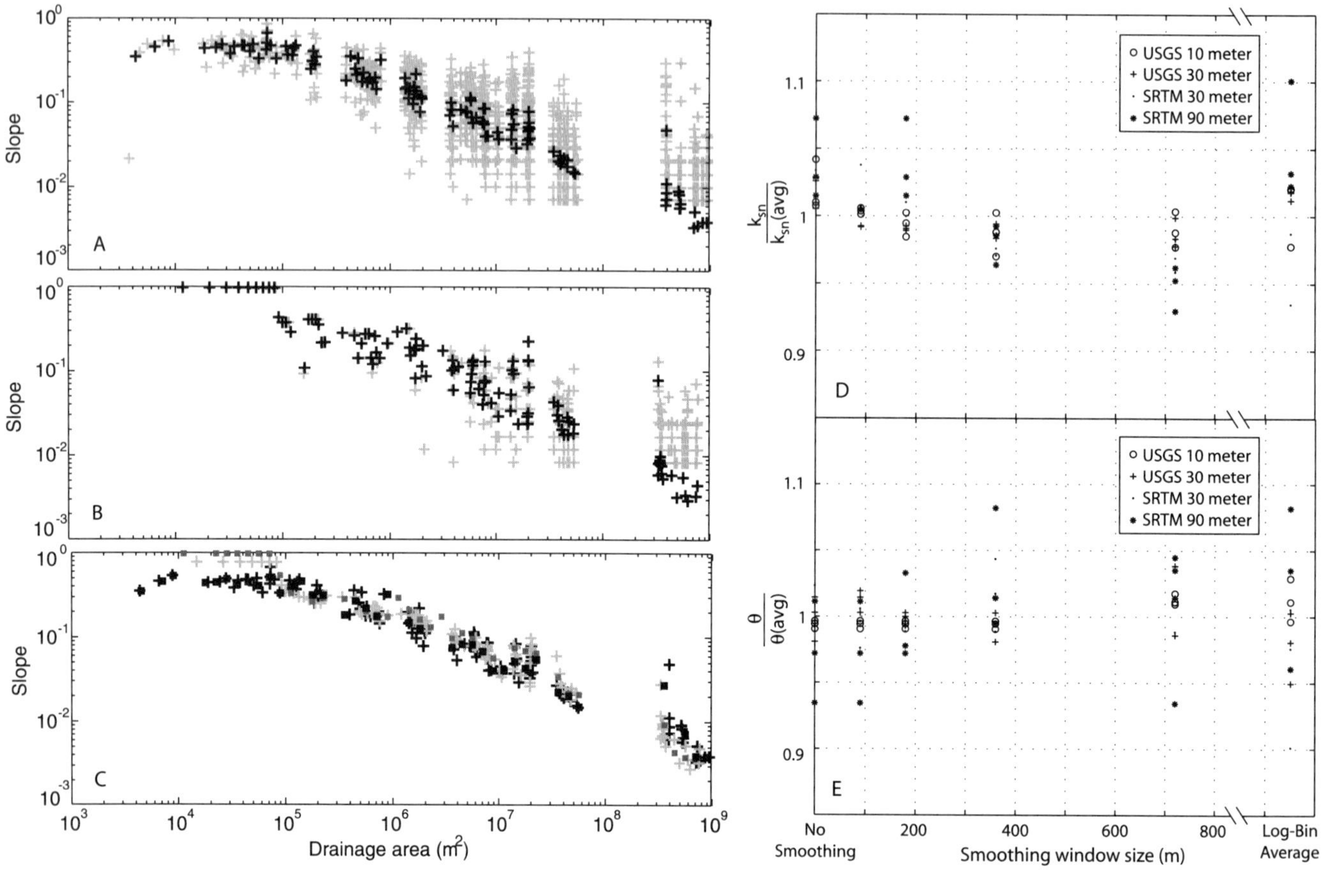

of this resampling in all cases. For high-quality data sources such as the USGS 10-m-pixel DEMs, contour-interval subsampling can be shown to recover faithfully the original contour crossings (Fig. 1B). Profiles extracted from lower-resolution data sources, however, will often exhibit considerable scatter on slope-area plots even after this subsampling (e.g., Fig. 2B); in these cases, additional smoothing can significantly aid interpretation.

As noted by many researchers, slope-area data often exhibit a pronounced break in scaling at $A_{cr} \leq 10^6$ m^2, which in unglaciated environments may represent the transition from debris-flow–dominated colluvial channels to stream-flow–dominated fluvial channels (e.g., Montgomery and Foufoula-Georgiou, 1993) (see Figs. 3A and 3B). This scaling break may be less pronounced in some settings, as a gradual transition from debris-flow–dominated to stream-flow–dominated conditions is reasonably expected (Stock and Dietrich, 2003). Figure 2C between a drainage area of 10^4 and 10^6 m^2 may be an example of this behavior. Regardless of the details at low drainage area, plots of slope-area data at this stage in the processing may reveal a smooth, linear trend below this scaling break, or multiple segments with easily identifiable values of k_s and θ, as illustrated in Figures 3A and 3B. More often, however, slope-area data will exhibit considerable scatter, which may be obscuring natural breaks in scaling along the profile. Further smoothing of the slope data greatly aids identification of scaling breaks without influencing their position, and with predictable effects on the values of k_s and θ (see below). Smoothing methods include using a moving-window average to smooth elevation data prior to calculating channel slopes over a specified vertical interval; regressing on elevation data over a fixed number of elevation points to derive local slope estimates; or averaging the logarithm of raw slopes over log-bins in the drainage area (termed log-bin averaging).

While the position of scaling breaks tends to be insensitive to the choice of smoothing window style and size, steepness indices can be expected to decrease subtly but systematically as the smoothing window grows, and spikes in the data are reduced in magnitude (Fig. 2D). The effects of smoothing on concavity values will also be predictable, but will depend on the relative position of outliers in a particular profile: if the data contain spikes high in the profile, we expect the concavities to decrease with increased smoothing as the regression pivots counterclockwise (flattens); the opposite will be true for data containing spikes near the toe of the channel. Despite these systematic and predictable biases, note that steepness and concavity values will typically fall within ~10% of one another for a wide range of smoothing windows (Figs. 2C–2E). The data in Figure 2 also demonstrate that with appropriate smoothing, data from different sources with greatly varying resolution and quality in fact yield comparable results at long wavelength.

Each smoothing method has its strengths and weaknesses. Log-bin averaging has the advantage that the smoothing window both grows in size with distance downstream and does not produce any averaging of disparate slope values across tributary junctions with a large change in drainage area—the two primary weaknesses of the moving-window averaging approach. However, log-bin averaging alone is susceptible to outliers in particularly rough or low-resolution DEM profiles. Smoothing elevation data with a moving window that grows in size proportional to drainage area may be preferable in some cases.

We stress that investigators must be circumspect about the appropriate scale of observation for DEM analysis. For example, while our algorithms can be shown to recover contour crossings from 10 m DEMs and therefore reproduce the information provided by the original contour map, there may be considerable information missed between these contour crossings. Montgomery et al. (1998) noted up to a fourfold difference in reach-scale channel slopes measured in the field versus those derived from contour maps; Massong and Montgomery (2000) found that slopes derived from field surveys and high-resolution DEMs can be different by up to 100%. In general, we expect the tectonic signals we are interested in to manifest themselves at a scale significantly greater than the contour interval of a topographic map; indeed, Finlayson and colleagues have had success extracting useful information from 30 arc second (~1 km) GTOPO30 data (Finlayson et al., 2002; Finlayson and Montgomery, 2003). However, without extremely high resolution digital topographic data (e.g., laser altimetry), it is clearly inappropriate to extend DEM analysis to geomorphic questions addressed below the reach scale.

3.2 Model Fits

Following data smoothing, we can begin to examine the slope-area data and make decisions about the number of distinct channel segments and the appropriate regression limits for each segment. Many channels can be adequately modeled below A_{cr} with only a single segment, using unique values of k_s and θ (e.g., Fig. 2C). Others may contain multiple segments, reflecting spatial or temporal variations in rock uplift rate, climatic factors, or

←

Figure 2. Effects of smoothing on longitudinal profile data, from San Gabriel Mountains in southern California. (A) Profile data extracted from U.S. Geological Survey (USGS) 10 m digital elevation model (DEM), showing raw profile (light-gray crosses) and data extracted at equal vertical intervals of 12.192 m (40 ft) (black crosses). (B) Profile data extracted from Shuttle Radar Topography Mission (SRTM) 90 m DEM, showing raw profile data (light-gray crosses) and data extracted at 12.192 m intervals (black crosses). Note the considerable reduction in scatter in both cases. (C) Superposition of smoothed data from SRTM 90 m DEM (720 m smoothing window, light-gray crosses), log-bin averaged data for SRTM 90 m DEM (light-gray squares), smoothed data from USGS 10 m DEM (180 m smoothing window, black crosses), and log-bin averaged data for USGS 10 m DEM (black squares). Note general correspondence among all data sets despite variations in resolution. (D) Plot of k_{sn}, normalized to average value of each group, versus smoothing window. (E) Plot of θ, normalized to average value of each group, versus smoothing window. Note that steepness and concavity indices for any data set are consistent within ~10%, regardless of the choice of smoothing window size or style.

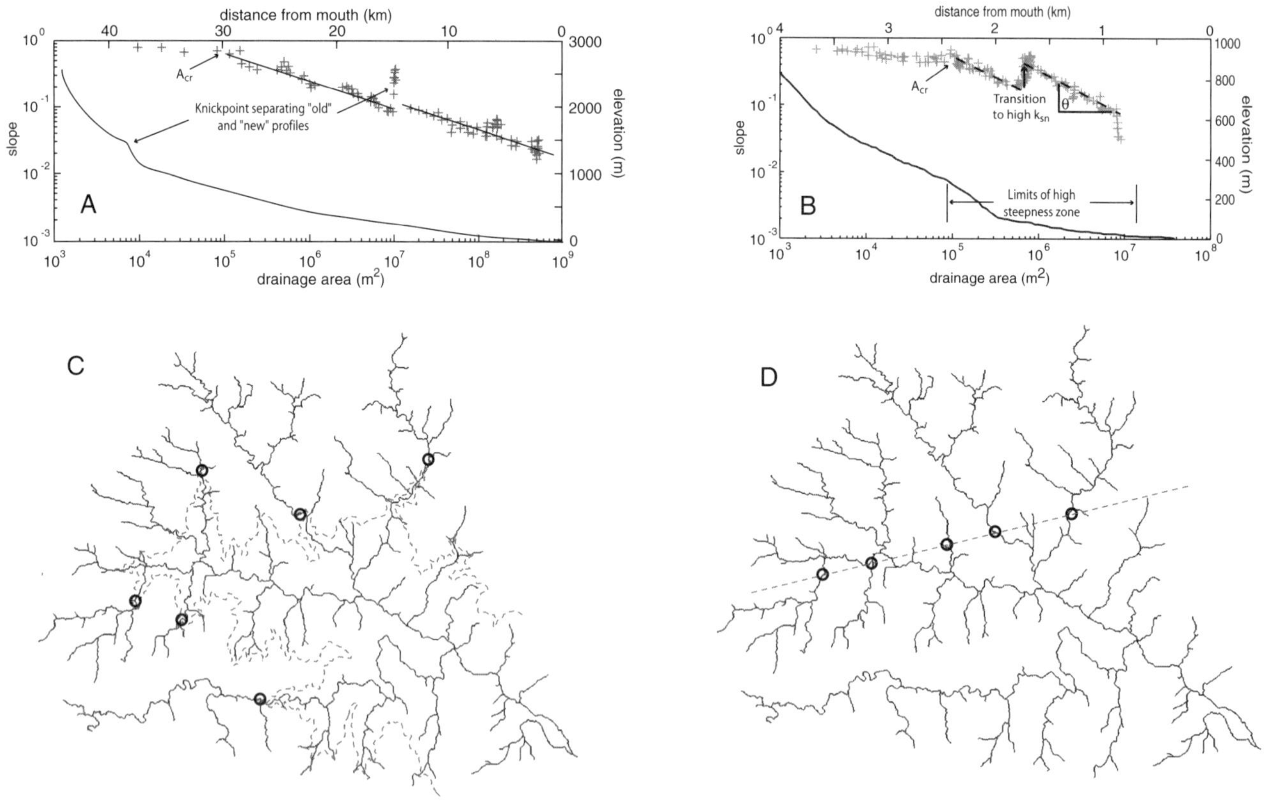

Figure 3. Schematic long profile and map view plots comparing transient and steady-state systems. (A) Transient long profile showing short oversteepened reach separating old and new equilibrium states. (B) Profile crossing from one uplift regime to another, showing channel reaches with constant k_s values separated by a high or low concavity transition zone in between. In this and all subsequent plots, long profile data (elevation versus distance) will be shown as solid lines with linear axes labeled on top and at right; slope-area data (log S vs. log A) will be shown as crosses with logarithmic axes labeled at bottom and at left. A_{cr} marks transition to fluvial scaling (see text); slope of log S vs. log A scaling is the concavity index θ; *y*-intercept is the steepness index k_s. (C) Transient wave of incision propagates through system at a nearly constant vertical rate; knickpoints (white dots) should therefore closely follow lines of constant elevation (dashed line) in plan view. (D) In contrast, knickpoints separating zones of high and low uplift rates (white dots) follow the trend of the accommodating shear zone in map view (dashed line). In many cases, the map pattern of knickpoints may be a better diagnostic of a transient state than long profile form.

the mass strength of rock exposed along the profile (e.g., Fig. 3B). In either case, linear regressions on slope-area data are typically conducted in two ways for each segment to allow intercomparison among different profiles in the basin.

In the first of the two regressions, segments of slope-area data with distinct steepness and/or concavity indices are identified, and are fit with k_s and θ as free parameters, using equation 1 as the regression model. In the second regression, individual segments of slope-area data are fit using a "reference" concavity, θ_{ref}, to determine normalized steepness indices, k_{sn}. A reference concavity is required for interpretation of steepness values, because k_s and θ as determined by regression analysis are, of course, strongly correlated (see equation 1). In practice, θ_{ref} is usually taken as the regional mean of observed θ values in undisturbed channel segments (i.e., those exhibiting no known knickpoints, uplift rate gradients, or changes in rock strength along stream), and can be estimated from a plot superimposing all of the data from a catchment. Reference concavities typically fall in the range of 0.35–0.65 (Snyder et al., 2000; Kirby and Whipple, 2001; Brocklehurst and Whipple, 2002; Kirby et al., 2003; Wobus et al., 2003).

If raw data from a previously completed analysis are not readily available, one can approximately determine the normalized steepness index, k_{sn}, for a reference concavity, θ_{ref}, as follows:

$$k_{sn} = k_s A_{cent}^{(\theta_{ref} - \theta)} \tag{2}$$

and

$$A_{cent} = 10^{(\log A_{max} + \log A_{min})/2}, \tag{3}$$

where k_s and θ are determined by regression, A_{min} and A_{max} bound the segment of the profile analyzed, and A_{cent} is the midpoint value for the segment analyzed. In practice, equation 2 is found to match k_{sn} calculated by regression analysis to within ~10%.

Where the difference between θ and θ_{ref} is large, however, (>0.2), the k_{sn} value is meaningful only over a short range of drainage area near A_{cent}.

The normalized steepness index is analogous to the S_r index proposed by Sklar and Dietrich (1998). Where a regional concavity index is apparent, normalized steepness indices can be shown to correspond closely to S_r indices (see Kirby et al., 2003, Fig. 5B therein). The advantage of the normalized steepness index k_{sn} is that the reference area (A_{cent} here) need not be the same for all channels, or channel segments, analyzed. However, where no typical regional concavity index is apparent, the S_r index may be preferable. Other measures of channel gradient have also been used in tectonic analyses, including the Hack gradient index (e.g., Hack, 1973). Where basin shapes are similar throughout a region, comparison of Hack gradient indices among different channels may be appropriate. However, if we assume that incision rate is related to fluvial discharge, normalized steepness indices may be a more appropriate metric, since contributing drainage area (and thus basin shape) is explicitly incorporated into the analysis as a proxy for fluvial discharge.

3.3 Tectonic Analysis

In the context of extracting tectonic information from longitudinal profiles, the data obtained for steepness and concavity will often yield similar information: a downstream transition between disparate steepness values will typically be bridged by a zone of very high or low concavity (see Fig. 3 and following). This transition zone may be a result of spatially or temporally varying rock uplift rates, temporally varying climatic conditions, or spatially varying rock mass strength. Even abrupt spatial changes in either rock uplift rate or rock properties (across a fault, for instance) may be manifested as a gradual transition in channel gradient (i.e., a high concavity zone) due to gradual downstream changes in sediment size, transport of resistant boulders downstream, or other blurring agents that may diffuse knickpoints in space and time (see Whipple and Tucker, 2002). Moreover, some data-smoothing algorithms have the disadvantage of blurring abrupt changes in channel gradient or elevation (knickpoints). In practice, normalized steepness data are often more useful than concavity data for evaluating regional tectonics, especially for short channel segments, given the sensitivity of θ to scatter in slope data, both real and artificial. However, concavity data are often useful for gross delineation of zones where uplift rates may be systematically changing along a profile (e.g., Kirby and Whipple, 2001; Kirby et al., 2003).

It should be emphasized here that slope-area data are plotted in log-log space, and it is important to be mindful of the compression of data at large drainage area. In particular, when selecting the downstream regression limit for channel segments, small changes in log drainage area are typically associated with large changes in distance along the profile. Large errors may therefore be inherent in any estimates of the distribution of rock uplift rates based on the width of high concavity zones. In addition, one needs to be wary of possible downstream changes in river characteristics not necessarily associated with the underlying tectonics, such as a transition to increasingly alluviated, or even depositional, conditions that may be associated with a rapid decrease in slope and therefore locally high concavity. In other cases, concavity may actually decrease where channels transition to increasingly transport-limited, but incisional, conditions with distance downstream (e.g., Whipple and Tucker, 2002; Sklar, 2003). In other words, some apparent downstream changes in channel steepness or concavity indices may simply be associated with an increase in drainage area. Comparing slope-area data on smaller tributaries that enter orthogonal to the mainstem throughout the zone of interest has proven an effective tool in this regard (see Kirby et al., 2003, Section 6.2 therein; Wobus et al., 2003), as will be illustrated in the Nepal case study here.

Once slope-area data have been extracted and smoothed from each tributary in a basin, it is often useful to superimpose all of the profile data from a catchment on a single plot (e.g., Figs. 4D and 5D). This tool aids in determination of the upper and lower bounds on steepness values in the catchment, segregation of populations with distinct steepness values, and determination of an appropriate reference concavity, as discussed above. With these composite plots, the analysis can be extended from individual tributaries to the regional scale.

The plan view distribution of normalized steepness indices for all the tributaries in a catchment can be an extremely useful tool for delineating tectonic boundaries (e.g., Kirby et al., 2003; Wobus et al., 2003). In tectonic settings containing a discrete break in rock uplift rates, we expect channels with high steepness indices to characterize the high uplift zone, while those with lower steepness indices should characterize the low uplift zone (e.g., Snyder et al., 2000). Channels crossing spatial gradients in rock uplift rate may exhibit readily identifiable knickpoints on a longitudinal profile (z versus x) and slope-area plots. In plan view, the boundary between zones of high and low steepness will also help us to evaluate whether we are seeing a temporally stable break in rock uplift rates (i.e., a fault or shear zone) or a transient condition; the rate of knickpoint migration, and therefore the position of knickpoints in a catchment through time, can be predicted (e.g., Whipple and Tucker, 1999; Niemann et al., 2001) and indicates a spatial distribution of knickpoints very different from that expected in regions of spatially varying uplift rates (Figs. 3C and 3D). In particular, if erodibility is spatially uniform, we expect the vertical rate of knickpoint migration to be constant, suggesting that knickpoints recording a transient condition (due to base-level fall, for example) should lie near a constant elevation (e.g., Niemann et al., 2001). This condition is demonstrated in the San Gabriel Mountains example discussed below.

If regional geologic maps or field observations are available, a superposition of important lithologic contacts is also useful to determine whether regional trends in channel steepness values might be correlative with lithologic boundaries, rather than with a tectonic signal (e.g., Hack, 1957; Kirby et al., 2003, Fig. 9 therein). In such efforts it is critical to recall that lithology is not

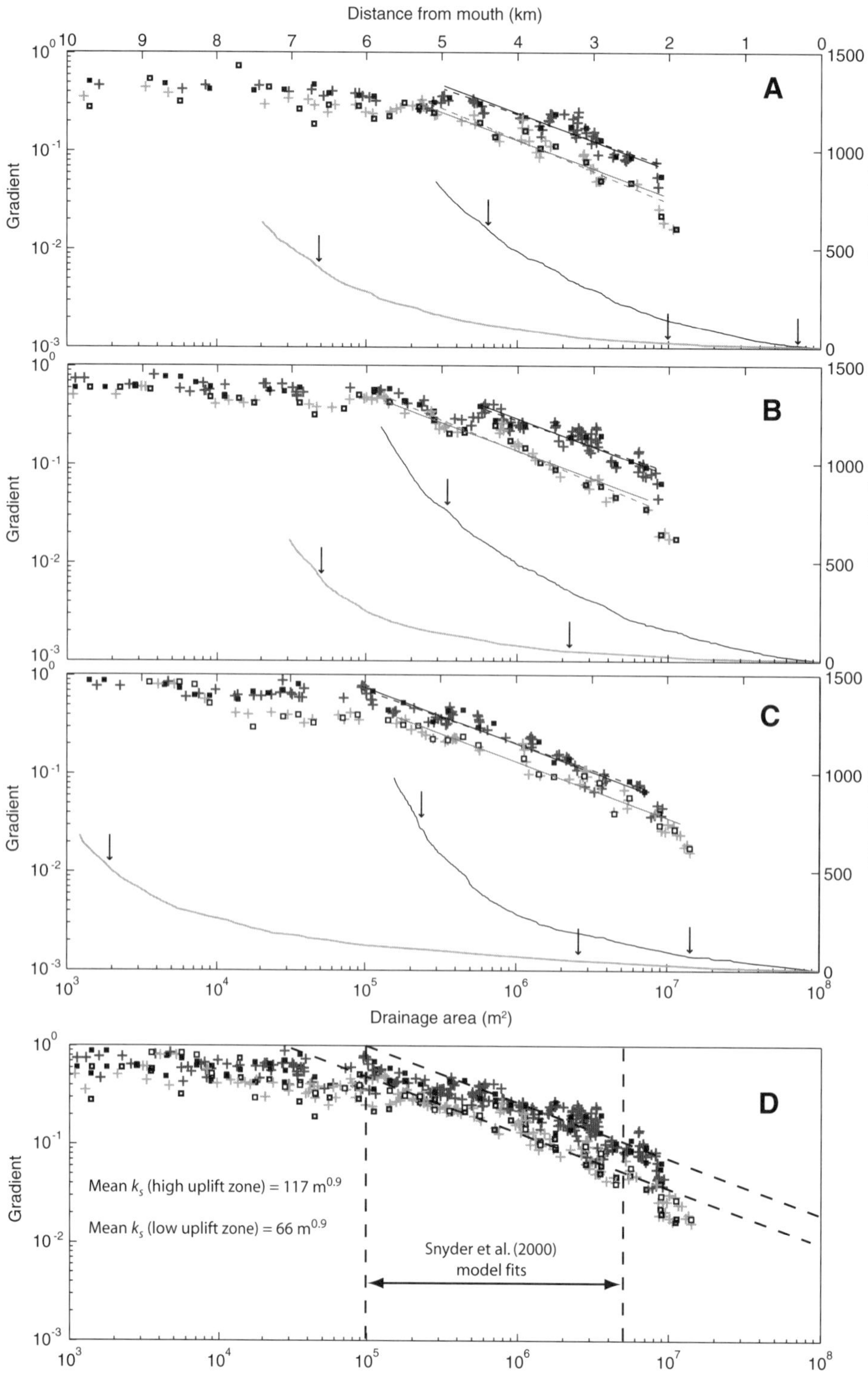

Figure 4. Long profile and slope-area data from a 10-m-resolution digital elevation model (DEM) of the King Range in northern California. Top three plots show pairs of tributaries from high uplift zone (black) and low uplift zone (gray). In all plots, solid lines are fits to data with θ_{ref} = 0.57; dashed lines are fits to data with concavity as a free parameter. Squares are slope-area data using log-bin averaging method; crosses are data using 200 m smoothing window to calculate channel slopes (see text). Arrows above long profiles show regression limits for slope-area fits. (A) Gitchell Creek and Hardy Creek. (B) Shipman Creek and Howard Creek. (C) Big Creek and Juan Creek. (D) Composite slope-area data from all six rivers. Note that in all cases the channels from the high uplift zone are consistently steeper than those in the low uplift zone. Dramatic drop in channel gradient in lowermost segments of the channels reflects a transition to increasingly alluviated conditions, reflecting a decrease in effective rock uplift rate due to Holocene rise in sea level (Snyder et al., 2002), alluviation at channel mouth, or both.

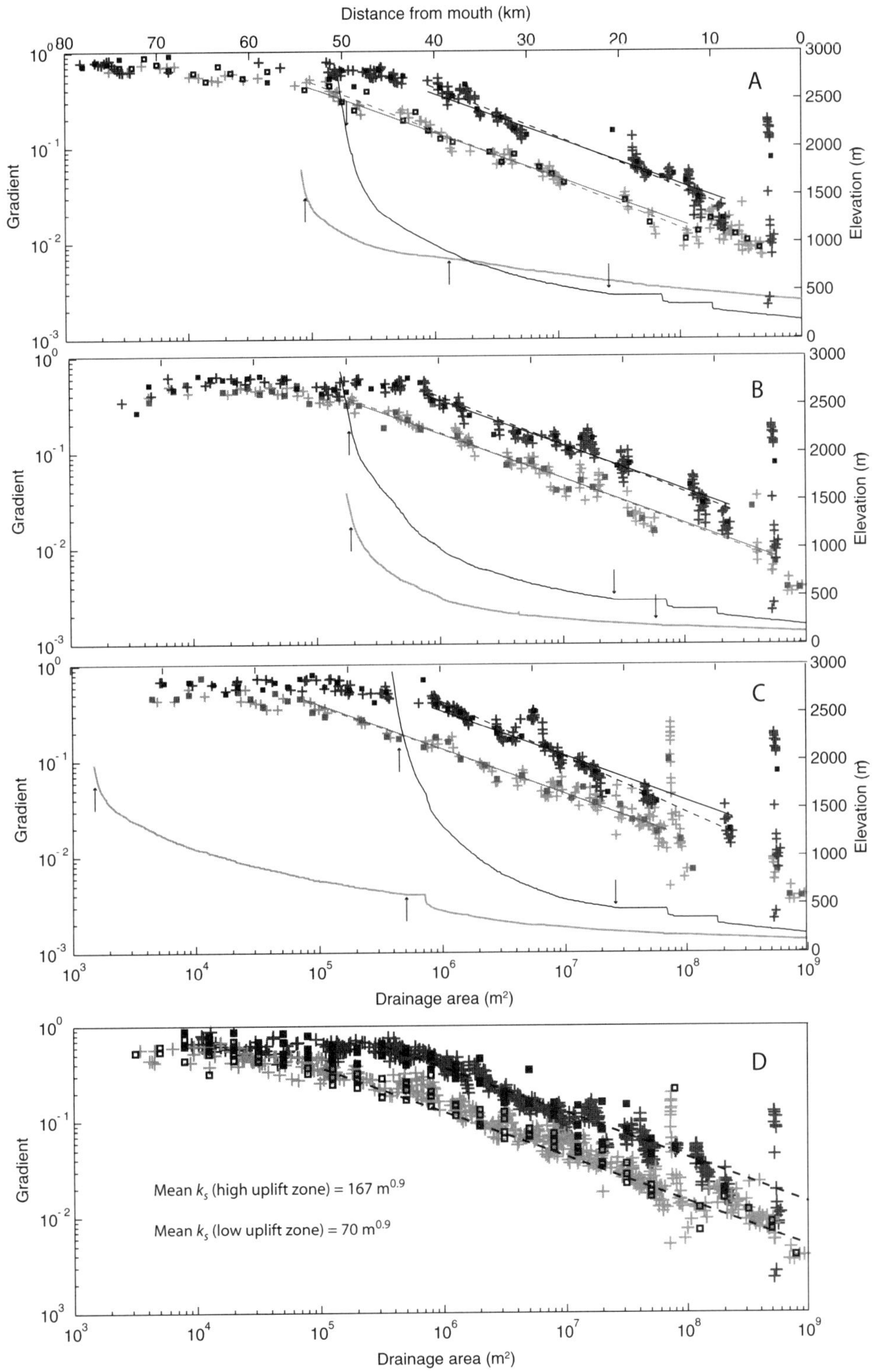

Figure 5. Long profile and slope-area data from a 10-m-resolution digital elevation model (DEM) from the San Gabriel Mountains east of Los Angeles, California. Top three plots show pairs of tributaries from high uplift zone (black) and low uplift zone (gray). In all plots, solid lines are fits to data with $\theta_{ref} = 0.45$; dashed lines are fits to data with concavity as a free parameter. Squares are slope-area data using log-bin averaging method; crosses are data using 400 m smoothing window to calculate channel slopes (see text). Arrows above/below long profiles show regression limits from slope-area data. Rivers shown are: (A) Minegulch and Beartrap; (B) Iron Fork and Little Tujunga; (C) Coldwater and Pacoima. (D) Composite slope-area data from all six rivers. Steps in profiles and spikes in slope-area data are dams and reservoirs in the lower reaches of the drainage basins.

synonymous with rock properties: a competent, well-cemented sandstone can be stronger than a fractured and weathered granite, and the strength of a single unit may vary markedly along strike. However, if depositional or intrusive lithologic boundaries can be demonstrated to correspond to changes in channel gradient, we can often rule out breaks in rock uplift rate as the cause of the channel steepening.

4. CASE STUDIES

Utilizing the methodologies outlined above, we now discuss a series of case studies in which river profile data have been used to extract tectonic information from the landscape. In the first two case studies, both from California, channel steepness values correlate with known variations in rock uplift and exhumation rates as determined from marine terraces, thermochronologic data, and cosmogenic data; however, there is insufficient data available at present to calibrate and uniquely test river incision models. In the third case study, steepness and concavity values derived from stream profiles correlate with the distribution of rock uplift rates above a fault-bend fold in the Siwalik Hills of south-central Nepal, and allow a local calibration of the stream power river incision model. Comparison of two independent calibration methods and application of the calibrated model to predict incision rates across the rest of the landscape allows a semiquantitative verification of the model in this field site. Finally, in the central Nepal Himalaya, breaks in steepness values across the landscape help us to delineate a recently active and previously unrecognized shear zone, the tectonic significance of which is corroborated by $^{40}Ar/^{39}Ar$ thermochronology and structural observations in the field.

4.1 California: King Range and San Gabriel Mountains

In the King Range and San Gabriel Mountains of northern and southern California, respectively, strong spatial gradients in rock uplift and exhumation rates provide excellent opportunities to test the stream profile method in a controlled environment. In the King Range, rock uplift rates near the Mendocino triple junction are quantified from flights of uplifted marine terraces, using radiocarbon dating and correlations with a eustatic sea-level curve (Merritts and Bull, 1989; Merritts, 1996). Based on the record from marine terraces, the field area can be divided into high and low uplift zones, with rock uplift rates varying over approximately an order of magnitude between the two zones (3–4 mm/yr and 0.5 mm/yr, respectively). In the San Gabriel Mountains, a restraining bend on the San Andreas fault creates strong east-west gradients in long-term exhumation rates, as determined from (U-Th)/He and apatite fission-track (AFT) thermochronology. Based on the thermochronologic data, the eastern and western San Gabriel Mountains can be divided into blocks with distinct exhumation histories, with AFT ages ranging from 4 to 64 Ma and rock uplift rates from 0.5 mm/yr in the western block to 2–3 mm/yr in the eastern block (e.g., Blythe et al., 2000; Spotila et al., 2002).

4.1.1 King Range

Using a 30 m USGS DEM of the King Range, Snyder et al. (2000) extracted 21 mainstem river profiles and calculated model parameters for equation 1 from slope-area data. Most of the tributaries in the study area were found to have relatively smooth, concave profiles for much of their length, suggesting a condition in which the rivers have equilibrated with local rock uplift rates. Our data handling methods have been refined and improved over the years since our initial efforts in stream profile analysis (Snyder et al., 2000). That analysis also predates the extensive field work in the region reported in Snyder et al. (2003a). In addition, since that time higher-resolution and higher-quality 10-m-pixel USGS DEMs have become available for the entire King Range study area. We take the opportunity here to revisit the King Range stream profile analysis using better data, refined methods, and field observations. This reanalysis is at once a cautionary tale regarding the uncertainty in best-fit profile concavity indices, and an encouraging example of the robustness of measured channel steepness indices for a reference concavity (k_{sn}), especially within a study area. The two principal conclusions of Snyder et al. (2000) are upheld in this reanalysis: (1) channel steepness increases by a factor of ~1.8 between the low and high uplift rate zones, and (2) there is no statistically significant difference in the concavity index between channels in the low and high uplift rate zones (Fig. 4).

Here we reanalyze only a subset of the 21 drainages studied by Snyder et al. (2000): the largest of the drainages in the high uplift zone (Gitchell, Shipman, Bigflat, and Big Creeks), and the most comparable low uplift zone channels (Hardy, Juan, Howard, and Dehaven Creeks) using 10 m USGS DEMs. We select regression bounds on a case by case basis, rather than simply adhering to the common set of regression limits (0.1 km^2–5 km^2) used by Snyder et al. (2000). The upstream regression limit is still typically near 0.1 km^2, as defined by a kink in the slope-area data, but varies somewhat from drainage to drainage (Fig. 4). The downstream regression limit is set by the sudden reduction in channel gradient associated with the transition to alluviated conditions (typically at ~10^7 m^2; Fig. 4). This transition was mapped in the field (Snyder et al., 2003a) and was likely driven by rapid Holocene sea-level rise (Snyder et al., 2002). Between these regression limits, we find uncorrelated residuals to the model fits.

This revised analysis finds no statistically significant difference in concavity index between the channels sampled in the low and high uplift rate zones, and a regional best-fit mean value of 0.57 ± 0.1 (2σ), considerably higher than, and yet within error of the estimate reported by Snyder et al. (2000), 0.43 ± 0.22 (2σ). Reanalysis of the 30 m DEMs with our original data handling methods (contour extraction but no smoothing) on this subset of drainages confirms that the difference in best-fit concavity results mostly from the change in regression limits, rather than a difference in data quality or data handling methods. Using a reference concavity of 0.45 (for convenience of comparison to data from the other case studies presented here), we find mean k_{sn} values of 66 and 117 $m^{0.9}$ in the low and high uplift rate zones, respectively, yielding a ratio of k_{sn}(high)/k_{sn}(low) of 1.8. Further, the ratio of high uplift zone to low uplift zone average k_{sn} values varies by less than 5% when using all permutations of smoothing, no smoothing, 30 m data, 10 m data, new regression limits, and old regression limits. Thus while the concavity index may be sensitive to the choice of regression limits, steepness indices appear to be robust across a broad range of data quality and user-chosen regression limits. In all cases, we find no statistically significant difference in the concavity index of channels in the high and low uplift rate zones.

The positive relationship between steepness index and rock uplift rate is expected, and is consistent with Merritts and Vincent's (1989) data. However, despite the reanalysis with higher-quality data, a number of conditions limit our ability to quantitatively relate steepness indices to uplift rates in this field setting. First, steepness values in the high uplift zone show considerable variability. Although this may reflect the spatial variations seen in Holocene uplift rates of marine platforms (Merritts, 1996), it does suggest that other variables may influence channel steepness despite the lithologic homogeneity among these drainages. Second, although the uplift rates vary over approximately an order of magnitude, the most reliable estimates of rock uplift rates in the field area are confined to the low and high ends of this range (~0.5 mm/yr and ~3–4 mm/yr). We therefore have only two reliable points to define the functional relationship between steepness index and uplift rate, which does not strongly constrain the range of models that can be fit to the data (e.g., Snyder et al., 2000, 2003a, 2003b).

4.1.2 San Gabriel Mountains

In the San Gabriel Mountains, stream profile data for over 100 streams were extracted from a 10 m USGS DEM of the region, and a composite plot was created from the slope-area data. Most of the tributaries analyzed in the San Gabriel Mountains exhibit smoothly concave profiles, with uniform slope-area scaling below a threshold drainage area of 10^6 m^2 or less (Fig. 5). Among these profiles, concavity indices again show no systematic relationship to rock uplift rates, and a reference concavity of 0.45 was chosen based on the composite slope-area data. Normalized steepness indices in the San Gabriel Mountains range from ~65 to 175 $m^{0.9}$; the highest k_{sn} values (~150–175) are coincident with the youngest cooling ages and highest long-term erosion rates, while the lowest k_{sn} values (~65–80) are coincident with the oldest cooling ages and lowest long-term erosion rates (Fig. 5).

In addition to the adjusted fluvial profiles from which the relationship between steepness and rock uplift rate can be evaluated, a number of rivers in the western San Gabriel Mountains contain abrupt knickpoints that separate upstream and downstream channel segments with distinct steepness indices. The best examples of this are in the Big Tujunga drainage basin, within the slowly uplifting western San Gabriel Mountains (e.g., Blythe et al., 2000; Spotila et al., 2002). In this basin, knickpoints in multiple rivers are found at different points in the basin but nearly constant elevation, suggesting a transient condition as the channels adjust to changing boundary conditions (Fig. 6). Regressions on slope-area data above this knickpoint find k_{sn} values indistinguishable from the regional lower bound (e.g., Fig. 5D), while k_{sn} values below the knickpoint are slightly higher (Fig. 6A). In some of the tributaries, the boundary between these zones is characterized by a significantly oversteepened reach, possibly reflecting a disequilibrium state related to changes in sediment flux during landscape adjustment (e.g., Sklar and Dietrich, 1998; Gasparini, 2003; Gasparini et al., 2005). A preliminary interpretation is that these profiles record a transient response to a recent increase in rock uplift rate—probably during the Quaternary, given the plausible range of rock uplift rates and the height of the knickpoints. A total offset of ~300 m during this period can be inferred from the height of knickpoints and the downstream projection of the less-steep upper-channel segments (Fig. 6A), and may be important both for interpreting young fission-track cooling ages along the lower Big Tujunga (e.g., Blythe et al., 2000), and for understanding the long-term evolution of active fault systems in the Los Angeles region.

In neither the King Range nor the San Gabriel Mountains are we able to quantify the relationship between k_{sn} and U. In the King Range, we do not have enough data to differentiate among various threshold and nonlinear models of k_s versus U and the potential effects of changing bed state on erosional efficiency (e.g., Snyder et al., 2003a); in the San Gabriel Mountains, modern rock uplift rates are not as well constrained and an analysis of (approximately) steady-state channel segments would suffer from similar limitations. Despite these complexities, however, the qualitative results from both field areas are both robust and important: the highest steepness values consistently correspond to the regions with the highest rock uplift and exhumation rates, while the lowest steepness values correspond to the lowest rock uplift and exhumation rates. In both field areas, lithologic differences between the high and low uplift regions are minimal, suggesting that channel steepness is tracking rock uplift rate. If we were to approach either of these field areas without any a priori knowledge of the tectonic setting, a map of steepness indices across the range would provide a great deal of information about the underlying tectonics. A plan view map of knickpoints in longitudinal profiles can also provide constraints on the location and magnitude of recent deformation in the region: in the San Gabriel Mountains, this analysis reveals an apparently recent (ca. 1 Ma?) change in rock uplift rates in the Big Tujunga basin. Importantly, with high-resolution DEMs throughout the United States publicly available, these analyses could be conducted in a matter of hours.

4.2 Siwalik Hills, Nepal

The case study from the Siwalik Hills in central Nepal is similar to the previous examples, in that steepness indices can be compared from zones of different rock uplift rates in a region of spatially uniform lithology. However, the Siwalik Hills provide

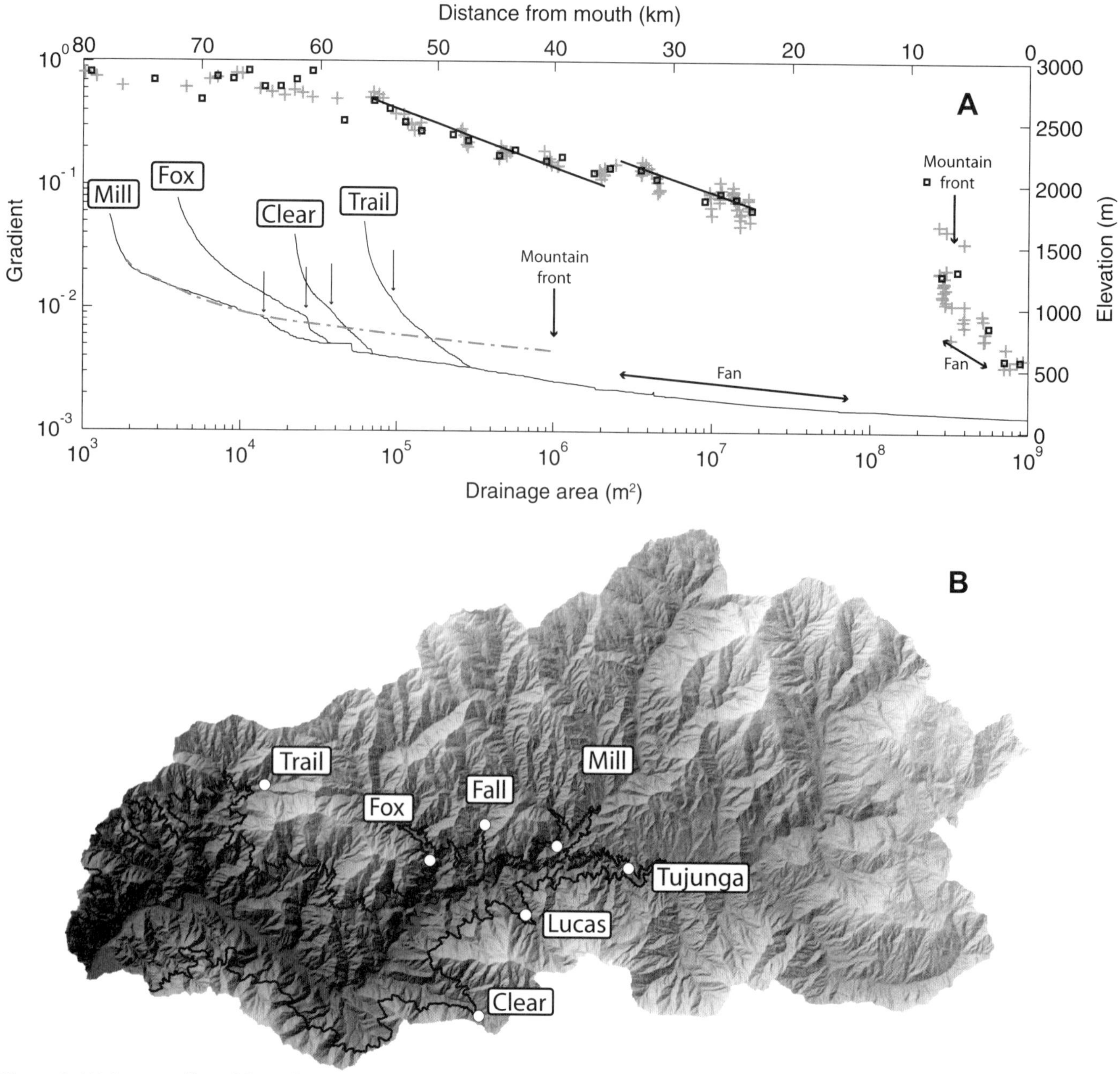

Figure 6. (A) Long profiles of four tributaries to the Big Tujunga River and slope-area data from Trail Canyon, showing apparent transient in long profiles and map pattern. Note that transient is easily seen in *z* vs. *x* plots for Mill, Fox, and Clear Creek, but slope-area plot greatly aids interpretation for Trail Canyon. Dashed line extending from Mill Creek shows projection of upper profile to mountain front. (B) Map pattern of knickpoints (white dots) in Big Tujunga basin. Close correspondence of knicks to 1000 m contour (black line) and irregular pattern in map view again suggests a transient feature (see Fig. 3).

the additional opportunity to examine the topographic signature of rock uplift rate gradients along individual channel profiles, particularly as these gradients affect channel concavities (e.g., Kirby and Whipple, 2001). The Siwalik Hills record modern deformation above a fault-bend fold in the Himalayan foreland, along the Main Frontal Thrust system. Deformation rates inferred from the distribution of Holocene terraces vary from ~4 mm/yr north of the range, to ~17 mm/yr at the range crest, and back to near zero just south of the Main Frontal Thrust. Multiple flights of terraces suggest relatively constant incision rate with time, and data from transverse drainages suggest steady-state profiles in these catchments (Lavé and Avouac, 2000). The fault-bend fold is dissected by drainages oriented both parallel and perpendicular to the strike of the range, and all of the rivers traverse relatively uniform sandstones and siltstones of the Lower and Middle Siwalik Hills (Lavé and Avouac, 2000).

Kirby and Whipple (2001) analyzed 22 channels in the Siwalik Hills, using a 90 m DEM of the region in an effort to assess to what degree strong spatial gradients in rock uplift rate influenced channel concavity and steepness. They argued that systematic changes in concavity indices could be exploited to place bounds on the relationship between channel gradient and incision rate. Here we update those results with analysis of a higher-resolution (30 m) DEM generated from ASTER stereo scenes. Channels parallel to the range crest, and therefore undergoing relatively uniform rock uplift rate, again yield concavity indices ranging from 0.45 to 0.55. Moreover, these strike-parallel drainages exhibit a predictable and quantifiable pattern of k_{sn} values: channels in high uplift settings have the highest steepness indices, while those in low uplift zones have the lowest steepnesses. A plot of steepness coefficients versus uplift rate for these strike-parallel drainages reveals a linear relationship with an intercept statistically indistinguishable from zero (Fig. 7A). This relationship is consistent with a simple detachment-limited stream power model with $n = 1$, where

$$k_s = \left(\frac{U}{K}\right)^{1/n}$$

and K is an erodibility coefficient (e.g., Howard, 1994). Normalized steepness indices, k_{sn}, for $\theta_{ref} = 0.52$ range from ~85 in low uplift regions (~7 mm/yr) to ~200 in high uplift regions (~14 mm/yr).

Although most of the channels contained within uniform uplift regimes have moderate concavities near 0.5, channel segments crossing spatially varying rock uplift rates have anomalously high or low concavities, depending on the direction of flow relative to the gradient in uplift rate. Channels entering the Siwalik anticline from the north have rapidly increasing slopes (negative concavities), as uplift rates increase downstream. Of note is the observation that changes in gradient on these channels span the entire range of increase in gradients on strike-parallel channels (Fig. 7B), suggesting that both the strike-parallel and strike-perpendicular systems exhibit the same manner and degree of response to increasing rock uplift rates. This observation, coupled with the lack of abrupt knickpoints within these channels and the consistency between two independent estimates of the erosion coefficient K (see below), lends support to the hypothesis that these channels have reached a steady-state balance between channel incision and rock uplift.

Using the known range in uplift rates and observed channel geometries across these channel segments, Kirby and Whipple (2001) note that the concavity index provides an independent means of determining n and K. These authors argued that the change in gradient along two tributaries (Dhansar and Chadi Khola) was consistent with n ranging between 0.6 and 1. Reanalysis of the ASTER DEM gives estimated values for n of 0.72 and 0.88, close to the value of $n = 1$ derived in Figure 7A from strike-parallel drainages. The erosion coefficient (K) determined from

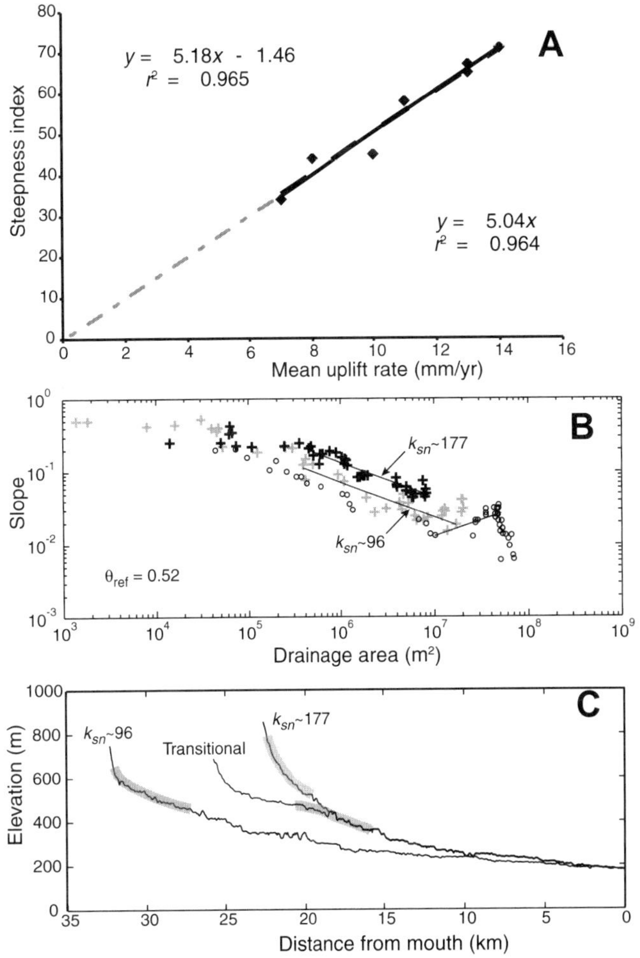

Figure 7. (A) Plot of normalized steepness index (k_{sn}) versus uplift rate (U) for seven strike-parallel channels from the Siwalik Hills in southern Nepal. Data suggest a linear correlation between k_{sn} and U with a zero intercept, consistent with a stream power scaling with $n = 1$. Uplift rates were estimated by Hurtrez et al. (1999) from a fault-bend-fold kinematic model and bedding dips. (B) Example of slope-area data from the Siwalik Hills in southern Nepal. Gray crosses show data from a representative strike-parallel stream in the low uplift rate regime ($U = 7$ mm/yr). Black crosses show data from a representative strike-parallel stream in the high-uplift-rate regime ($U = 13$ mm/yr). Open dots show data from a strike-perpendicular stream crossing the entire anticline, from low rock uplift rate ($U = 4$ mm/yr) to high rock uplift rate ($U = 13$ mm/yr) and back. Negative concavity segment in strike-perpendicular stream can be explained by spatially varying rock uplift rate along the profile (e.g., Kirby and Whipple, 2001). (C) Channel profiles shown in part B. Thick gray lines show approximate extent of slope-area fits used in part B. Negative concavity in transitional profile manifests as a rollover just above junction with the high uplift channel.

the k_s versus U relation (Fig. 7A) is 7.41×10^{-5} $m^{-.04}yr^{-1}$, while K derived independently from the strike-perpendicular channels ranges from 5.25×10^{-5} to 5.76×10^{-5} $m^{-.04}yr^{-1}$. Note that k_{sn} and K are dimensional coefficients, the dimensions of which depend on the ratio of m/n and the value of m, respectively: the estimates reported here assume that $m/n = \theta_{ref}$, and use $\theta_{ref} = 0.52$, and $n = 1$ (i.e., $m = 0.52$). The consistency between estimates derived using both strike-parallel and strike-perpendicular channels suggests that we may be well on our way to deriving quantitative estimates of uplift rates directly from topography in regions with relatively simple tectonics and uniform lithology. For example, a map of the distribution of predicted channel incision rates from calibrated model parameters displays good correspondence with an independent estimate of rock uplift rate in this landscape that has been argued to be in steady state (Hurtrez et al., 1999; Lavé and Avouac, 2000) (Fig. 8).

If this type of analysis can be extended to other field settings, it represents a promising avenue for neotectonic research—if calibration sites can be identified with known uplift rates and similar channel morphologies to a field site of interest, it may be possible to obtain quantitative estimates of uplift rates and their spatial variability at the catchment scale prior to ever setting foot in the field. In the absence of a calibration site, the spatial patterns of uplift rates can still be estimated from the patterns of steepness and concavity identified from stream profiles. The accuracy of our quantitative estimates may decrease with decreasing knowledge of the field site; in particular, downstream changes in channel width (W) that differ from the typical scaling of $W \sim A^{0.4}$ (e.g., Whipple, 2004, and references therein) will not be captured on slope-area plots, and may substantially modify the relationships between uplift rates and steepness indices in ways that must be explored through future work. At a minimum, however, the method provides a quick and easy way to evaluate patterns of rock uplift with a high degree of spatial resolution.

4.3 Central Nepal Himalaya

The preceding case studies illustrate a consistent relationship between steepness and uplift rate in regions where rock uplift rates have been independently determined. The next logical step is to utilize stream profiles for tectonic analysis in regions where modern rock uplift rates are less well understood. The case study discussed here is from central Nepal, where stream profiles are used along with thermochronologic and structural analysis to evaluate two competing models for the tectonic architecture of the Himalaya.

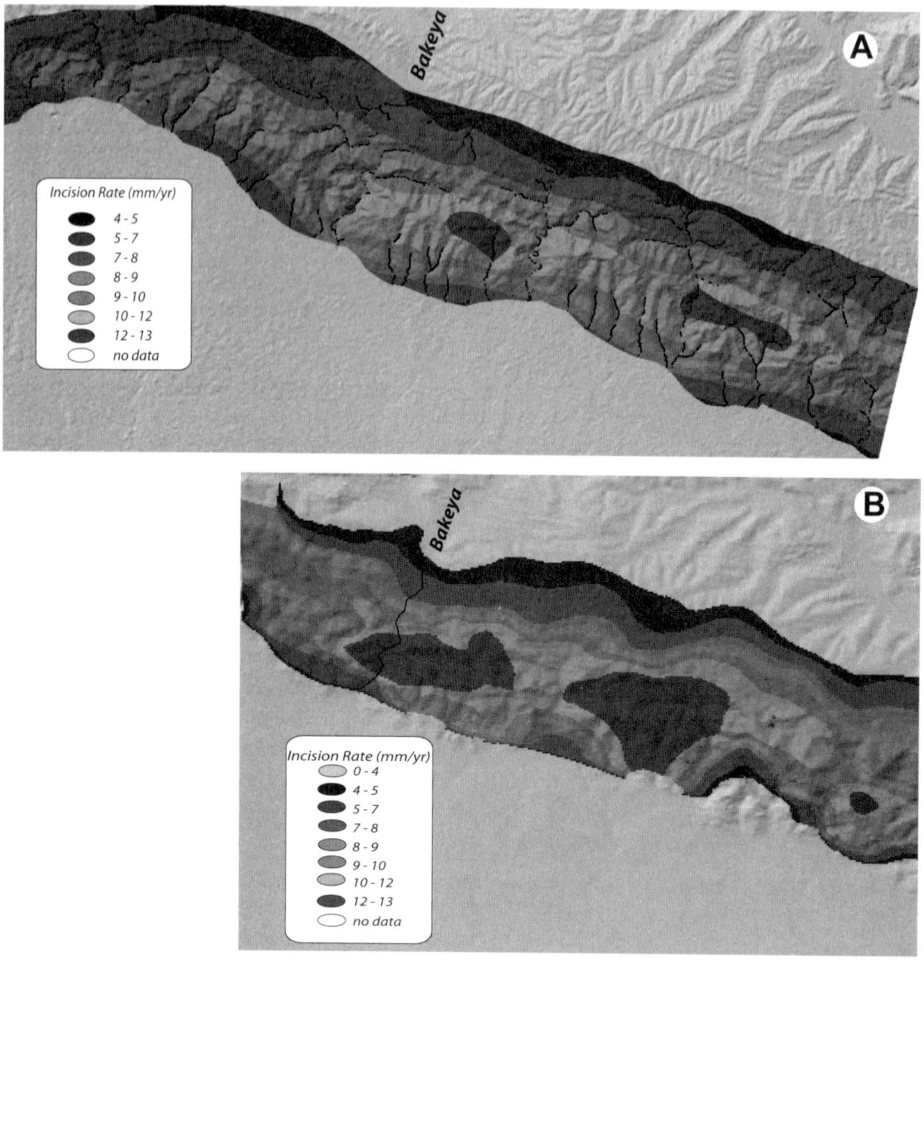

Figure 8. Comparison of incision rate estimates derived from (A) stream profile method, assuming a stream-power incision rule with $n = 1$, and (B) structural study of Hurtrez et al. (1999) and Lavé and Avouac (2000). Note general correspondence of location and magnitude of high incision rate zones from the two methods, suggesting that quantitative estimates of incision rates may be attainable in regions with relatively simple patterns of lithology and uplift. Figure B is adapted from Hurtrez et al. (1999).

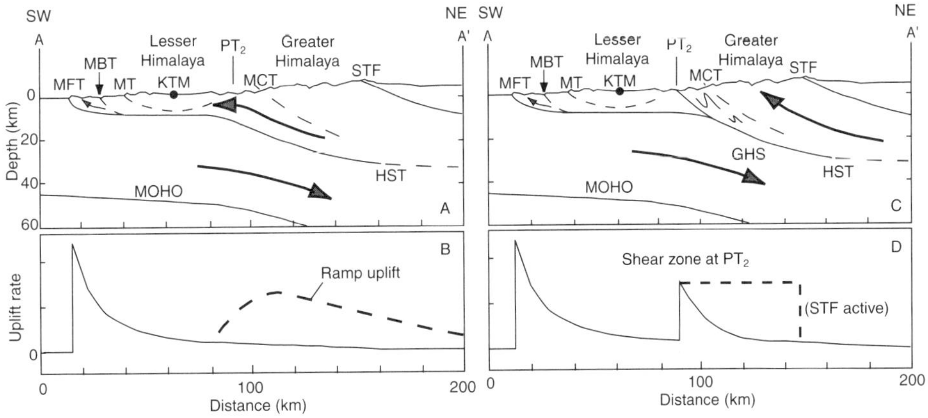

Figure 9. Schematic showing two competing models for Himalayan tectonic architecture and predicted patterns of rock uplift rates for each model. (A) Ramp model predicts uplift gradient spread over 20–30 km, with Main Central Thrust (MCT) carried passively over ramp in Himalayan Sole Thrust (HST). Uplift gradient should be broadly distributed (B), and thermal history for rocks on opposite sides of PT_2 should be similar. Surface breaking thrust model (C) predicts more abrupt break in rock uplift rates centered at PT_2 (D) and distinct thermal histories for rocks on opposite sides of the inferred surface breaking shear zone. MFT—Main Frontal Thrust; MBT—Main Boundary Thrust; MT—Mahabarat Thrust; STF—South Tibetan fault system; HST—Himalayan Sole Thrust; KTM—Kathmandu; PT_2—physiographic transition; GHS—Greater Himalayan sequence. (Adapted from Wobus et al., 2003.)

The High Himalayan peaks in central Nepal are bounded to the south by a sharp topographic break, which can be delineated on a map of hillslope gradients as a WNW-ESE–trending line separating the High Himalaya to the north from the Lesser Himalaya to the south (Wobus et al., 2003; Fig. 1). This physiographic transition—herein referred to as PT_2 for consistency with Wobus et al. (2003) and Hodges et al. (2001)—may be consistent with one of two end-member models for Himalayan tectonics (Fig. 9). In the first model, the topographic transition is sustained entirely by rock uplift gradients due to material transport over a ramp in the Himalayan Sole Thrust. This model predicts similar thermal and deformational histories for rocks on opposite sides of PT_2, and broadly distributed gradients in rock uplift rates from south to north (e.g., Cattin and Avouac, 2000; Lavé and Avouac, 2001). In the second model, surface-breaking thrusts—perhaps including young strands of the Main Central Thrust system (MCT)—remain active at the foot of the High Himalaya. This model predicts disparate thermal and structural histories on either side of PT_2, and more abrupt changes in rock uplift and exhumation rates across discrete structures (e.g., Hodges et al., 2004; Wobus et al., 2003).

As a first step in evaluating which of these models is most relevant to central Nepal, stream profiles were extracted from a 90 m DEM of the region (see Fielding et al., 1994, for description of the data set). Composite plots of slope-area data from all tributaries define regional upper and lower limits of steepness values of 650 and 95 $m^{0.9}$, respectively. In each case, the upper limit is defined by northern tributaries and trunk stream segments, and the lower limit is defined by southern tributaries and trunk stream segments (Fig. 10). This trend is consistent with a decrease from north to south in rock uplift rate, rock strength, or both. Because rock mass quality is typically very similar across these transitions, and because both of the tectonic models considered for the Himalaya include a decrease in rock uplift rates from north to south, we assume that a change in rock uplift rate is the major driving force for the change in k_{sn}. Furthermore, monsoon precipitation appears to be focused just upstream of this major break in steepness indices (Hodges et al., 2004), suggesting that our estimate of the southward decrease in k_{sn} for central Nepal may be conservative as a proxy for rock uplift rate. We focus here on the nature and spatial extent of this steepness transition, to evaluate which of the competing tectonic models is most appropriate for this portion of the Himalaya.

The analysis is similar to that used in the Siwalik Hills (e.g., Kirby and Whipple, 2001), in that tributaries both parallel and perpendicular to the inferred uplift rate gradient are utilized. For the trunk streams perpendicular to the uplift gradient, the width of the high concavity channel segments (the downstream transition from high to low steepness values) provides one estimate of the distance over which rock uplift rates are decreasing downstream (Fig. 10). Note that this should be a maximum estimate of the distance over which uplift rates are changing: downstream adjustments in sediment load across the tectonic boundary may diffuse any uplift rate signal (e.g., Whipple and Tucker, 2002), as will the smoothing algorithm applied in analysis of the data. The smaller tributary channels subparallel to the structural grain typically exhibit smooth profiles without abrupt knickpoints, normal concavities in the range of 0.4–0.6, and uniform steepness values down to the trunk stream junction. Furthermore, either of the tectonic models for central Nepal predicts nearly constant uplift rate within these narrow, east-west trending basins. We therefore infer that these channels record equilibrium profile forms, and use the spatial distribution of high and low steepness tributaries as a second means of estimating the distance over which uplift rates are changing in the system.

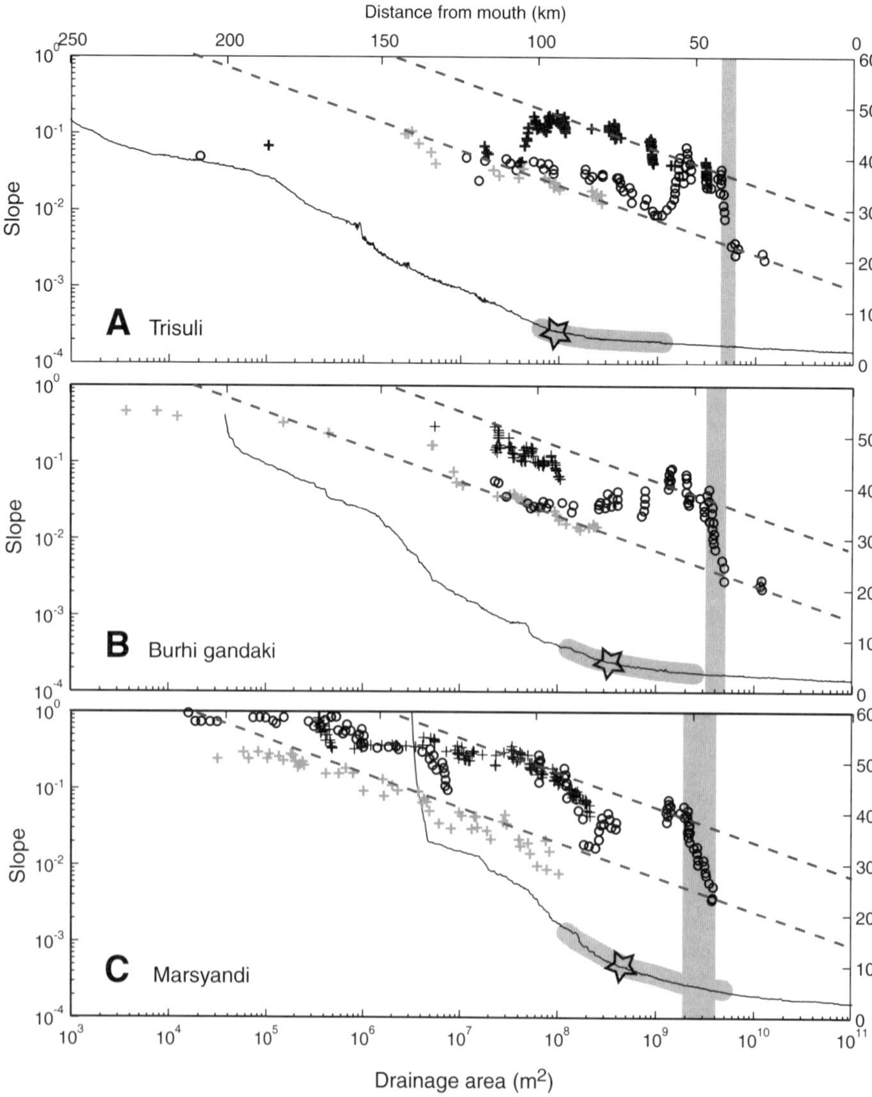

Figure 10. Slope-area plots and trunk stream longitudinal profiles from (A) Trisuli River catchment, (B) Burhi Gandaki catchment, and (C) Marsyandi catchment, all in central Nepal. Smoothing method and contour interval for all cases are 10 km moving average and 30 m, respectively. In each plot, gray crosses show data from representative southern (low uplift zone) tributaries; black crosses show data from representative northern (high uplift zone) tributaries; and open circles show data from trunk streams carving through the range. Dashed lines show approximate regional upper and lower bounds on steepness indices of 650 and 95 $m^{0.9}$, respectively ($\theta_{ref} = 0.45$). Gray shading on slope area and long profile data shows extent of high concavity zone in trunk streams, representing a maximum estimate of the width of rock uplift gradients (see text). Open stars show position of abrupt physiographic transition (PT_2) in central Nepal, as identified from slope maps and satellite photography (see text and Fig. 9). Note that Marsyandi long profile begins on a valley sidewall due to a data gap at the western edge of the catchment.

Slope-area data from the Burhi Gandaki trunk stream allow us to delineate a high concavity zone along ~40 km of streamwise distance, which spans the range in steepness values on a composite plot of slope-area data (Fig. 10B). As discussed above, this represents a maximum estimate of the width of the assumed uplift gradient. Tributary profiles allow us to place more precise constraints on the width of the transition zone: steepness values in strike-parallel streams decrease from ~400 to ~100 over ~20 km in map view (Fig. 11). $^{40}Ar/^{39}Ar$ thermochronologic data from detrital muscovites in the Burhi Gandaki and Trisuli catchments suggest a profound change in time-averaged exhumation rates (or total depth of late Cenozoic exhumation) over an even narrower zone (8–10 km), suggesting that stream profile analysis is capturing a profound tectonic boundary in this setting. Similar results from the Trisuli River suggest that this boundary may continue along strike (Fig. 11; Wobus et al., 2003).

Similar methodologies can be used to delineate the width of the transition zone throughout central Nepal, and additional structural and thermochronologic data can be used to corroborate the stream profile analyses in many cases (Fig. 11). In the Marsyandi

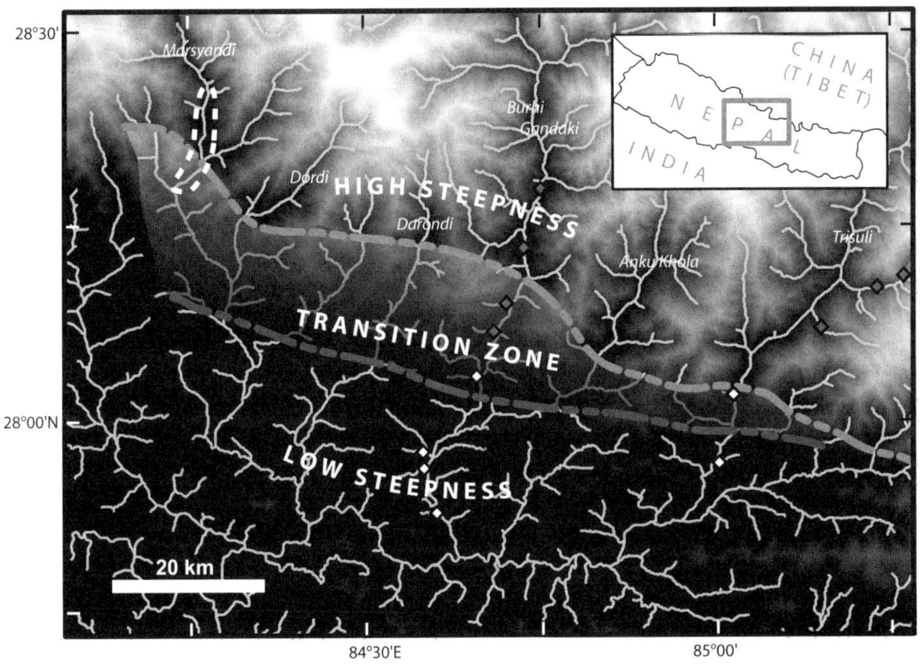

Figure 11. Digital elevation map (DEM) of central Nepal, showing transition zone between high and low uplift rates as determined from stream profile analyses. Diamonds along Trisuli and Burhi Gandaki Rivers show locations of detrital $^{40}Ar/^{39}Ar$ thermochronologic analysis: gray diamonds represent Miocene and younger apparent ages; white diamonds represent Paleozoic and older apparent ages. Note general correspondence of break in cooling history with change in stream profiles, suggesting a long-lived structure accommodating differential exhumation in this region. Dashed white lines in Marsyandi basin show location of intense Quaternary deformation, possibly responsible for accommodating the inferred gradient in rock uplift rates (e.g., Hodges et al., 2004).

trunk stream, a high concavity zone across the topographic break spans a streamwise distance of ~40 km, suggesting a decrease in rock uplift rates over this distance. Tributaries along this segment of the basin allow us to narrow this estimate to between 15 and 20 km (Fig. 11). Detailed structural mapping along the Marsyandi trunk and its tributaries indicates a penetrative brittle shear zone crosscutting all previous fabrics along the upper part of this same river reach. Consistent top-to-the-south kinematics on these north-dipping shear zones suggest that these structures may be accommodating differential motion in this zone, consistent with a model including recently active surface-breaking thrusts in this region (e.g., Hodges et al., 2004).

In the absence of a detailed stream profile analysis to guide structural and thermochronologic studies, sparse data coverage and model uncertainties render geophysical and geodetic data equivocal in discriminating between a surface breaking shear zone and a subsurface ramp in central Nepal. Stream profiles provide an additional piece of data that favors a narrowly distributed rock uplift gradient throughout the study area, and informs our sampling strategy for thermochronologic analyses. Although we are unable to provide an exact estimate of the distance over which rock uplift rates are increasing, stream profiles are a crucial tool in this setting for evaluating a range of tectonic models and identifying sites for future field and laboratory work.

5. DISCUSSION

The correlation between steepness and uplift rate in established tectonic settings, and the ability to delineate temporal and spatial breaks in rock uplift rate in more poorly constrained settings, demonstrate the power of stream profile analysis. However, a number of shortcomings must yet be overcome if stream profile analysis is to become a mainstream tool for neotectonic investigations. In particular, if we wish to derive quantitative estimates of rock uplift rates directly from topography, a great deal of work remains to be done (e.g., Whipple, 2004).

Foremost among our research needs is a continued, systematic dissection of the varied influences on river incision into rock. For example, how does the relative importance of mesoscale processes, such as plucking and abrasion, change with channel slope and incision rate, and how do the rates of these processes differ from simple shear-stress–dependent incision at the bed (e.g., Whipple et al., 2000)? How do changes in bed roughness, sediment flux, and bed cover influence erosion rates (e.g., Sklar and Dietrich, 1998, 2001; Hancock and Anderson, 2002; Sklar, 2003)? What controls changes in bedrock channel width, and how do changes in width influence erosion rates (e.g., Hancock and Anderson, 2002; Montgomery and Gran, 2001; Snyder et al., 2003a; Lague and Davy, 2003)? How do we incorporate critical thresholds for river incision and a stochastic distribution of storms into our erosion models (e.g., Tucker and Bras, 2000; Snyder et al., 2003b; Tucker, 2004)? We have only begun to ask these questions, and more comprehensive field, experimental, and numerical studies must be undertaken before we can hope to have a reliable quantitative tool for neotectonics (e.g.,Whipple, 2004).

Although we cannot yet deconvolve the relative contributions of lithology, adjustments in channel morphology and bed state, climatic variables, and uplift rate on channel steepness, we must continue to incorporate as much of this information as we can into our analysis. Where available, lithologic and climatic information can already be used to qualitatively ascertain the importance of rock uplift rate variations on long profile form. For example, in some settings large breaks in channel steepness across lithologic boundaries may be entirely contained within a single uplift regime; ignoring the effects of lithology would lead to dramatic misconceptions of tectonic signals. In other field areas, a decrease in rock uplift rate may be co-located with an increase in rock strength across major structures, such that the two effects moderate one another in the context of profile steepness. Ultimately, utilizing stream profile analysis to inform detailed structural, thermochronologic, and cosmogenic analyses will continue to be the best approach for neotectonic investigations (e.g., Kirby et al., 2003; Wobus et al., 2003; Hodges et al., 2004). Stream profile analyses cannot be conducted in a vacuum; informing our investigations with as much additional data as possible must remain a priority in any topographic analysis.

Numerical experiments incorporating an orographic forcing of precipitation predict systematic, if minor, changes in profile concavity due to an uneven distribution of precipitation at the range scale (e.g., Roe et al., 2002, 2003). Because precipitation effects are manifested in models of fluvial erosion only through their contribution to river discharge, we typically do not expect fluvial profiles to change abruptly due to spatial gradients in climate. However, the effects of temporal changes in climate may be more pronounced, particularly in settings where river systems take over previously glaciated valleys, or where glacial-interglacial cycling high in a basin creates profound changes in sediment flux within the fluvial part of the system (e.g., Brocklehurst and Whipple, 2002; Hancock and Anderson, 2002). Spatial and temporal changes in climate must both be more fully incorporated into our models if we hope to one day derive quantitative estimates of uplift rates from topography.

The stream profile method measures only changes in channel slope. However, a river's response to changes in uplift rate may include adjustments in a variety of other factors related to dominant incision processes, channel morphology, and bed state (e.g., Whipple and Tucker, 1999, 2002; Sklar and Dietrich, 1998, 2001; Sklar, 2003). Where channels cross boundaries between rocks with considerable differences in rock strength without any change in steepness index, adjustments in channel morphology or bed state may be fully compensating for the change in rock properties (see Montgomery and Gran, 2001; Sklar, 2003). Alternatively, transport-limited, rather than detachment-limited, conditions may be indicated (e.g., Whipple and Tucker, 2002). As discussed in section 2, our ability to quantify many of these feedbacks and internal adjustments is limited. As our understanding of the mechanisms of fluvial response to differential rock uplift improves, these additional degrees of freedom should be incorporated into both modeling and empirical studies of the interrelations among climate, erosion, tectonics, and topography. Until then, we must recognize that adjustments in channel slope are only part of a more complicated equation.

In order to make the connection between topography and tectonics, an assumption is often made that a steady-state balance between uplift and erosion prevails. In many landscapes, this condition will not be met, leading to transient forms such as propagating knickpoints, knickzones, and other disequilibrium channel forms (e.g., Whipple and Tucker, 1999; Stock et al., 2004; Anderson et al., 2005). The severe limitations of the steady-state assumption are well known. However, it is incorrect to presume that model parameters can be constrained only through analysis of steady-state forms. Where data quality is high, deviations from steady-state are often readily discernible through analysis of stream profiles and plan view maps (e.g., Figs. 3 and 6). Recognition of such transient forms is important for at least two reasons. First, they can be useful indicators of tectonic history, such as a sudden base-level fall or a change in differential rock uplift rate at the outlet of a drainage network, as illustrated for the San Gabriel Mountains (Fig. 6). Second, under the right circumstances, profiles with such forms can provide information about two steady-state conditions, when the channel is incising at different, often knowable, rates, above and below the knickpoint, such that an upper steady-state profile is "replaced" by a lower steady-state profile as the knickpoint advances (e.g., Whipple and Tucker, 1999; Niemann et al., 2001). Where sediment flux and channel bed state play important roles, transient response may be complex. For example, these disequilibrium channels may be characterized by transient oversteepened reaches downstream of knickpoints (Gasparini, 2003; Gasparini et al., 2005), which may be mistaken for equilibrium high concavity segments if data quality and resolution are low. It is potential complexities in the transient response of rivers such as this that afford the best opportunity to quantitatively test competing river incision models.

Finally, many DEMs, particularly those with low spatial resolution, contain "bad" data points along a channel due to topographic variability at a smaller spatial scale than the cell size, short-circuiting of sinuous channels in extraction of long profile data, or errors in the algorithm converting an original data source to a raster format. As a result, we often must make a decision as to whether unusual data represent noise or real geological complexity. One potential source of the latter is the influence of large landslides, which may temporarily block fluvial systems causing alternating flats and steps in long profiles. Often the distinction between data noise and geologic complexity is easily made; however, at present there are no quantitative criteria to evaluate whether or not anomalies in stream profiles are real without complementary field investigation. As data quality improves across the globe, part of this difficulty will gradually be overcome. Until then, determining the difference between bad data and geological complexity will typically require field observations to resolve satisfactorily.

6. CONCLUSIONS

Empirical observations and simple models of fluvial erosion suggest a positive correlation between channel gradient and rock uplift rate, which we exploit in the method of stream profile analysis outlined here. Despite our incomplete understanding of the varied processes contributing to fluvial erosion, the stream profile method is an invaluable qualitative tool for neotectonic investigations. In northern and southern California, we show that channel steepness is directly related to rock uplift rate. In the Siwalik Hills, Nepal, changes in steepness and concavity each correlate in a predictable way with rock uplift rate variations, and we can begin constructing a quantitative means of translating topography into tectonics through local calibration of a simple river incision model. Finally, in central Nepal, stream profile analysis provides a crucial discriminator between two models of Himalayan tectonics, and has led to identification of a previously unrecognized shear zone. Our next steps should be focused on refining this promising qualitative tool by incorporating recent advances in process geomorphology into models of stream profile evolution and form. While further numerical studies will be useful in this regard, we are ultimately limited by our lack of empirical data to characterize fluvial response. As such, we should focus on field and laboratory work geared toward understanding variations in fluvial erosion processes and rates with changes in incision rate, bed morphology and bed state, and climate. Only when we can confidently describe all of these feedbacks can we hope to have a reliable quantitative tool for neotectonic analysis.

ACKNOWLEDGMENTS

This work was funded by National Science Foundation (NSF) Tectonics grant EAR-008758 to K.X Whipple and K.V. Hodges that supported the work of C. Wobus, and additional NSF grants to K.X. Whipple that supported the work of B. Crosby (NSF GLD EAR-0208312) and J. Johnson (NSF GLD EAR-0345622). We thank Jerome Lavé for providing access to data and figures for the Siwalik Hills case study, and Bob Anderson and David Montgomery for thoughtful reviews of the original manuscript. Finally, we thank all of the organizers and participants of the Penrose Conference for a fantastic meeting.

REFERENCES CITED

Blythe, A.E., Burbank, D.W., Farley, K.A., and Fielding, E.J., 2000, Structural and topographic evolution of the central Transverse Ranges, California, from apatite fission-track, (U-Th)/He and digital elevation model analyses: Basin Research, v. 12, p. 97–114, doi: 10.1046/j.1365-2117.2000.00116.x.

Brocklehurst, S.H., and Whipple, K.X, 2002, Glacial erosion and relief production in the Eastern Sierra Nevada, California: Geomorphology, v. 42, p. 1–24, doi: 10.1016/S0169-555X(01)00069-1.

Burbank, D.W., Leland, J., Fielding, E., Anderson, R.S., Brozovic, N., Reid, M.R., and Duncan, C., 1996, Bedrock incision, rock uplift and threshold hillslopes in the northwestern Himalayas: Nature, v. 379, no. 6565, p. 505–510, doi: 10.1038/379505a0.

Cattin, R., and Avouac, J.P., 2000, Modeling mountain building and the seismic cycle in the Himalaya of Nepal: Journal of Geophysical Research, v. 105, p. 13,389–13,407, doi: 10.1029/2000JB900032.

Fielding, E., Isacks, B., Barazangi, M., and Duncan, C., 1994, How flat is Tibet?: Geology, v. 22, p. 163–167, doi: 10.1130/0091-7613(1994)022<0163:HFIT>2.3.CO;2.

Finlayson, D.P., and Montgomery, D.R., 2003, Modeling large-scale fluvial erosion in geographic information systems: Geomorphology, v. 53, p. 147–164, doi: 10.1016/S0169-555X(02)00351-3.

Finlayson, D.P., Montgomery, D.R., and Hallet, B., 2002, Spatial coincidence of rapid inferred erosion with young metamorphic massifs in the Himalayas: Geology, v. 30, no. 3, p. 219–222, doi: 10.1130/0091-7613(2002)030<0219:SCORIE>2.0.CO;2.

Flint, J.J., 1974, Stream gradient as a function of order, magnitude, and discharge: Water Resources Research, v. 10, no. 5, p. 969–973.

Gasparini, N., 2003, Equilibrium and transient morphologies of river networks: Discriminating among fluvial erosion models [Ph.D. thesis]: Cambridge, Massachusetts, Massachusetts Institute of Technology, 105 p.

Hack, J.T., 1957, Studies of longitudinal stream profiles in Virginia and Maryland: U.S. Geological Survey Professional Paper 294-B, 97 p.

Hack, J.T., 1973, Stream profile analysis and stream-gradient index: U.S. Geological Survey Journal of Research, v. 1, no. 4, p. 421–429.

Hancock, G.S., and Anderson, R.S., 2002, Numerical modeling of fluvial strath-terrace formation in response to oscillating climate: Geological Society of America Bulletin, v. 114, no. 9, p. 1131–1142, doi: 10.1130/0016-7606(2002)114<1131:NMOFST>2.0.CO;2.

Harbor, D.J., 1998, Dynamic equilibrium between an active uplift and the Sevier River, Utah: Journal of Geology, v. 106, p. 181–198.

Heimsath, A.M., Dietrich, W.E., Nishiizumi, K., and Finkel, R.C., 1997, The soil production function and landscape equilibrium: Nature, v. 388, p. 358–361, doi: 10.1038/41056.

Hodges, K., Hurtado, J.M., and Whipple, K., 2001, Southward extrusion of Tibetan crust and its effect on Himalayan tectonics: Tectonics, v. 20, p. 799–809.

Hodges, K., Wobus, C., Ruhl, K., Schildgen, T., and Whipple, K., 2004, Quaternary deformation, river steepening and heavy precipitation at the front of the Higher Himalayan ranges: Earth and Planetary Science Letters, v. 220, no. 3-4, p. 379–389, doi: 10.1016/S0012-821X(04)00063-9.

Howard, A.D., 1994, A detachment-limited model of drainage basin evolution: Water Resources Research, v. 30, no. 7, p. 2261–2285, doi: 10.1029/94WR00757.

Howard, A.D., and Kerby, G., 1983, Channel changes in badlands: Geological Society of America Bulletin, v. 94, p. 739–752, doi: 10.1130/0016-7606(1983)94<739:CCIB>2.0.CO;2.

Hurtrez, J.E., Lucazeau, F., Lavé, J., and Avouac, J.P., 1999, Investigation of the relationships between basin morphology, tectonic uplift, and denudation from the study of an active fold belt in the Siwalik Hills, central Nepal: Journal of Geophysical Research, v. 104, no. B6, p. 12,779–12,796, doi: 10.1029/1998JB900098.

Kirby, E., and Whipple, K.X, 2001, Quantifying differential rock-uplift rates via stream profile analysis: Geology, v. 29, no. 5, p. 415–418, doi: 10.1130/0091-7613(2001)029<0415:QDRURV>2.0.CO;2.

Kirby, E., Whipple, K., Tang, W., and Chen, Z., 2003, Distribution of active rock uplift along the eastern margin of the Tibetan Plateau: Inferences from bedrock channel longitudinal profiles: Journal of Geophysical Research, v. 108, no. B4, doi: 10.1029/2001JB000861.

Lague, D., and Davy, P., 2003, Constraints on the long-term colluvial erosion law by analyzing slope-area relationships at various tectonic uplift rates in the Siwalik Hills (Nepal): Journal of Geophysical Research, v. 108, no. B2, doi: 10.1029/2002JB001893.

Lavé, J., and Avouac, J.P., 2000, Active folding of fluvial terraces across the Siwalik Hills, Himalayas of central Nepal: Journal of Geophysical Research, v. 105, no. B3, p. 5735–5770, doi: 10.1029/1999JB900292.

Lavé, J., and Avouac, J.P., 2001, Fluvial incision and tectonic uplift across the Himalayas of central Nepal: Journal of Geophysical Research, v. 106, no. B11, p. 26,561–26,591, doi: 10.1029/2001JB000359.

Massong, T.M., and Montgomery, D.R., 2000, Influence of sediment supply, lithology, and wood debris on the distribution of bedrock and alluvial channels: Geological Society of America Bulletin, v. 112, p. 591–599, doi: 10.1130/0016-7606(2000)112<591:IOSSLA>2.3.CO;2.

Merritts, D.J., 1996, The Mendocino triple junction: Active faults, episodic coastal emergence, and rapid uplift: Journal of Geophysical Research, v. 101, p. 6051–6070, doi: 10.1029/95JB01816.

Merritts, D.J., and Bull, W.B., 1989, Interpreting Quaternary uplift rates at the Mendocino triple junction, northern California, from uplifted marine terraces: Geology, v. 17, p. 1020–1024, doi: 10.1130/0091-7613(1989)017<1020:IQURAT>2.3.CO;2.

Merritts, D., and Vincent, K.R., 1989, Geomorphic response of coastal streams to low, intermediate, and high rates of uplift, Mendocino junction region, northern California: Geological Society of America Bulletin, v. 101, p. 1373–1388, doi: 10.1130/0016-7606(1989)101<1373:GROCST>2.3.CO;2.

Montgomery, D.R., and Brandon, M.T., 2002, Topographic controls on erosion rates in tectonically active mountain ranges: Earth and Planetary Science Letters, v. 201, p. 481–489, doi: 10.1016/S0012-821X(02)00725-2.

Montgomery, D.R., and Gran, K.B., 2001, Downstream variations in the width of bedrock channels: Water Resources Research, v. 37, no. 6, p. 1841–1846, doi: 10.1029/2000WR900393.

Montgomery, D.R., and Foufoula-Georgiou, E., 1993, Channel network representation using digital elevation models: Water Resources Research, v. 29, p. 3925–3934, doi: 10.1029/93WR02463.

Montgomery, D.R., Dietrich, W.E., and Sullivan, K., 1998, The role of GIS in watershed analysis, *in* Lane, S.N., Richards, K.S., and Chandler, J.H., eds., Landform monitoring, modelling and analysis: Chichester, England, John Wiley and Sons, p. 241–261.

Niemann, J.D., Gasparini, N.M., Tucker, G.E., and Bras, R.L., 2001, A quantitative evaluation of Playfair's law and its use in testing long-term stream erosion models: Earth Surface Processes and Landforms, v. 26, no. 12, p. 1317–1332, doi: 10.1002/esp.272.

Roe, G.H., Montgomery, D.R., and Hallet, B., 2002, Effects of orographic precipitation variations on the concavity of steady-state river profiles: Geology, v. 30, no. 2, p. 143–146, doi: 10.1130/0091-7613(2002)030<0143:EOOPVO>2.0.CO;2.

Roe, G.H., Montgomery, D.R., and Hallet, B., 2003, Orographic precipitation and the relief of mountain ranges: Journal of Geophysical Research, v. 108, no. B6, 2315, doi: 10.1029/2001JB001521.

Sklar, L., 2003, The influence of grain size, sediment supply, and rock strength on rates of river incision into bedrock [Ph.D. thesis]: Berkeley, University of California, 343 p.

Sklar, L., and Dietrich, W.E., 1998, River longitudinal profiles and bedrock incision models: Stream power and the influence of sediment supply, *in* Tinkler, K.J., and Wohl, E.E., eds., Rivers over rock: Fluvial processes in bedrock channels: Washington, D.C., American Geophysical Union, p. 237–260.

Sklar, L.S., and Dietrich, W.E., 2001, Sediment and rock strength controls on river incision into bedrock: Geology, v. 29, no. 12, p. 1087–1090, doi: 10.1130/0091-7613(2001)029<1087:SARSCO>2.0.CO;2.

Snyder, N., Whipple, K., Tucker, G., and Merritts, D., 2000, Landscape response to tectonic forcing: DEM analysis of stream profiles in the Mendocino triple junction region, northern California: Geological Society of America Bulletin, v. 112, no. 8, p. 1250–1263, doi: 10.1130/0016-7606(2000)112<1250:LRTTFD>2.3.CO;2.

Snyder, N.P., Whipple, K.X, Tucker, G.E., and Merritts, D.M., 2002, Interactions between onshore bedrock-channel incision and nearshore wave-base erosion forced by eustasy and tectonics: Basin Research, v. 14, p. 105–127, doi: 10.1046/j.1365-2117.2002.00169.x.

Snyder, N.P., Whipple, K.X, Tucker, G.E., and Merrits, D.J., 2003a, Channel response to tectonic forcing: Field analysis of stream morphology and hydrology in the Mendocino triple junction region, northern California: Geomorphology, v. 53, p. 97–127, doi: 10.1016/S0169-555X(02)00349-5.

Snyder, N.P., Whipple, K.X, Tucker, G.E., and Merrits, D.J., 2003b, Importance of a stochastic distribution of floods and erosion thresholds in the bedrock river incision problem: Journal of Geophysical Research, v. 39, doi: 10.1029/2001WR001057.

Spotila, J.A., House, M.A., Blythe, A., Niemi, N.A., and Bank, G.C., 2002, Controls on the erosion and geomorphic evolution of the San Bernardino and San Gabriel Mountains, southern California, *in* Barth, A.P., ed., Contributions to crustal evolution of the southwestern United States: Geological Society of America Special Paper 365, p. 205–230.

Stock, G.M., Anderson, R.S., and Finkel, R.C., 2004, Pace of landscape evolution in the Sierra Nevada, California, revealed by cosmogenic dating of cave sediments: Geology, v. 32, no. 3, p. 193–196, doi: 10.1130/G20197.1.

Stock, J.D., and Dietrich, W.E., 2003, Valley incision by debris flows: Evidence of a topographic signature: Water Resources Research, v. 39, no. 4, doi: 10.1029/2001WR001057.

Tarboton, D.G., Bras, R.L., and Rodriguez-Iturbe, I., 1989, Scaling and elevation in river networks: Water Resources Research, v. 25, p. 2037–2051.

Tucker, G.E., 2004, Drainage basin sensitivity to tectonic and climatic forcing: Implications of a stochastic model for the role of entrainment and erosion thresholds: Earth Surface Processes and Landforms, v. 29, p. 185–205, doi: 10.1002/esp.1020.

Tucker, G.E., and Bras, R.L., 2000, A stochastic approach to modeling the role of rainfall variability in drainage basin evolution: Water Resources Research, v. 36, p. 1953–1964, doi: 10.1029/2000WR900065.

Tucker, G.E., and Whipple, K.X, 2002, Topographic outcomes predicted by stream erosion models: Sensitivity analysis and intermodel comparison: Journal of Geophysical Research, v. 107, no. B9, doi: 10.1029/2001JB000162.

Whipple, K., 2004, Bedrock rivers and the geomorphology of active orogens: Annual Reviews of Earth and Planetary Science, v. 32, p. 151–185, doi: 10.1146/annurev.earth.32.101802.120356.

Whipple, K.X, and Tucker, G.E., 1999, Dynamics of the stream-power river incision model: Implications for height limits of mountain ranges, landscape response timescales, and research needs: Journal of Geophysical Research, v. 104, p. 17,661–17,674, doi: 10.1029/1999JB900120.

Whipple, K.X, and Tucker, G.E., 2002, Implications of sediment-flux-dependent river incision models for landscape evolution: Journal of Geophysical Research, v. 107, no. B2, doi: 10.1029/2000JB000044.

Whipple, K.X, Hancock, G.S., and Anderson, R.S., 2000, River incision into bedrock: Mechanics and relative efficacy of plucking, abrasion, and cavitation: Geological Society of America Bulletin, v. 112, no. 3, p. 490–503, doi: 10.1130/0016-7606(2000)112<490:RIIBMA>2.3.CO;2.

Willgoose, G., Bras, R.L., and Rodriguez-Iturbe, I., 1991, A coupled channel network growth and hillslope evolution model: I. Theory: Water Resources Research, v. 27, no. 7, p. 1671–1684, doi: 10.1029/91WR00935.

Wobus, C.W., Hodges, K.V., and Whipple, K.X, 2003, Has focused denudation sustained active thrusting at the Himalayan topographic front?: Geology, v. 31, p. 861–864, doi: 10.1130/G19730.1.

Manuscript Accepted by the Society 23 June 2005

Printed in the USA

Geological Society of America
Special Paper 398
2006

Surface uplift and climate change: The geomorphic evolution of the Western Escarpment of the Andes of northern Chile between the Miocene and present

Florian Kober
Institute of Geology, ETH Zürich, 8092 Zürich, Switzerland, and Institute of Isotope Geology, ETH Zürich, 8092 Zürich, Switzerland
Fritz Schlunegger
Gerold Zeilinger
Heinz Schneider
Institute of Geology, University of Bern, 3012 Bern, Switzerland

ABSTRACT

The Western Escarpment of the Andes at 18.30°S (Arica area, northern Chile) is a classical example for a transient state in landscape evolution. This part of the Andes is characterized by the presence of >10,000 km^2 plains that formed between the Miocene and the present, and >1500 m deeply incised valleys. Although processes in these valleys scale the rates of landscape evolution, determinations of ages of incision, and more importantly, interpretations of possible controls on valley formation have been controversial. This paper uses morphometric data and observations, stratigraphic information, and estimates of sediment yields for the time interval between ca. 7.5 Ma and present to illustrate that the formation of these valleys was driven by two probably unrelated components. The first component is a phase of base-level lowering with magnitudes of~300–500 m in the Coastal Cordillera. This period of base-level change in the Arica area, that started at ca. 7.5 Ma according to stratigraphic data, caused the trunk streams to dissect headward into the plains. The headward erosion interpretation is based on the presence of well-defined knickzones in stream profiles and the decrease in valley widths from the coast toward these knickzones. The second component is a change in paleoclimate. This interpretation is based on (1) the increase in the size of the largest alluvial boulders (from dm to m scale) with distal sources during the last 7.5 m.y., and (2) the calculated increase in minimum fluvial incision rates of ~0.2 mm/yr between ca. 7.5 Ma and 3 Ma to ~0.3 mm/yr subsequently. These trends suggest an increase in effective water discharge for systems sourced in the Western Cordillera (distal source). During the same time, however, valleys with headwaters in the coastal region (local source) lack any evidence of fluvial incision. This implies that the Coastal Cordillera became hyperarid sometime after 7.5 Ma. Furthermore, between 7.5 Ma and present, the sediment yields have been consistently higher in the catchments with distal sources (~15 m/m.y.) than in the headwaters of rivers with local sources (<7 m/m.y.). The positive correlation between sediment yields and the altitude of the headwaters (distal versus local sources) seems to reflect the effect of orographic precipitation on surface erosion. It appears that base-level change in the coastal region, in combination with an increase in the orographic effect of precipitation, has controlled the topographic evolution of the northern Chilean Andes.

Keywords: transient landscape, base-level change, climate change, northern Chile.

Kober, F., Schlunegger, F., Zeilinger, G., and Schneider, H., 2006, Surface uplift and climate change: The geomorphic evolution of the Western Escarpment of the Andes of northern Chile between the Miocene and present, *in* Willett, S.D., Hovius, N., Brandon, M.T., and Fisher, D.M., eds., Tectonics, Climate, and Landscape Evolution: Geological Society of America Special Paper 398, Penrose Conference Series, p. 75–86, doi: 10.1130/2006.2398(05). For permission to copy, contact editing@geosociety.org.

INTRODUCTION

Understanding controls on spatial and temporal variations of surface erosion has become a major research objective in the last few years. This issue has been discussed in detail in the review paper on bedrock rivers and the geomorphology of active orogens by Whipple (2004). Because of the importance of this problem, many studies have been directed toward analyzing (1) the nature of surface erosion in channels and on hillslopes (e.g., Schumm, 1956; Montgomery and Dietrich, 1992; Montgomery et al., 2003), (2) possible feedback mechanisms between erosion in channels and on hillslopes (Tucker and Bras, 1998; Harvey, 2001; Schlunegger et al., 2002), (3) the sensitivity of erosional processes and the resulting sediment flux to climatic and tectonic perturbations (Molnar, 2001; Pratt et al., 2002; Roe et al., 2002), (4) the nature and the magnitudes of parameters that scale with erosional processes for different temporal and spatial length scales (Tucker and Slingerland, 1996, 1997; Hinderer et al., 2002), and (5) possible controls on limits of relief and surface erosion at the local scale and at whole mountain belt scale (Burbank et al., 1996; Whipple et al., 1999; Whipple and Tucker, 1999; Schlunegger and Hinderer, 2003). As suggested by Bull (1991), Rosenbloom and Anderson (1994), Pazzaglia et al. (1998), and Wegmann and Pazzaglia (2002), terraced landscapes represent the simplest examples to address these research objectives, because initial conditions, present form, and ages of different geomorphic elements can possibly be constrained. These landforms represent transient stages toward a steady-state between erosional processes and exogenic/endogenic forces. Transient stages of landscape development can be identified by the presence of knickzones in stream profiles. If digitized, the information extracted from these landscapes can be converted into volumes of eroded rock or depth of local incision per unit time, which, in turn, allows the testing of possible controls and limits on surface erosion (e.g., Whipple et al., 2000; Schlunegger, 2002).

This paper focuses on the landscape evolution of the Arica area, northern Chile, which potentially allows interpreting the tectonic and climatic fingerprints because of the preservation of different steps in the topographic evolution (Mortimer and Saric, 1975; Mortimer, 1980; García, 2002; Wörner et al., 2002). In this region, the landscape is characterized by the presence of flat to gently undulating plains hundreds of km^2 in extent and >1500-m-deep valleys that dissect these landscapes (Fig. 1). Previous studies thought that initiation of incision was the result of (1) eustatic sea-level fall (Herm, 1969; Paskoff, 1980), (2) onset of glaciations in the Pleistocene (which is considered to have increased fluvial runoff) (Wörner et al., 2002), (3) differential tectonic movements (Paskoff, 1980; Mortimer, 1980; Kennan, 2000), (4) enhanced flow of groundwater derived from the Altiplano (Hoke et al., 2002), and (5) the modification of the atmospheric and hydrospheric circulation patterns caused by the rising Andes (Mortimer, 1980; Paskoff, 1980). Our study will use a refined stratigraphic framework to unravel the tectonic and the climatic fingerprints in more detail and highlight the consequential driving forces.

GEOLOGIC SETTING

The Western Escarpment of the Andes of northern Chile is divided into three main longitudinal geological-morphological units that include, from west to east, the Coastal Cordillera, the Western Escarpment, and the Western Andean Cordillera (Fig. 1). The Coastal Cordillera is made up of Upper Jurassic to Cretaceous tholeiitic arc rocks and marine backarc strata, which are partly covered by Oligocene to Miocene alluvial deposits (Wörner et al., 2002) (Figs. 2A and 2B). In the study area, the Coastal Cordillera forms a steep cliff, ~500 m high, that parallels the Andean strike, and that gently dips toward the east. At Arica, the Coastal Cordillera dips beneath the Peruvian Coastal Plain (Fig. 1) (Salas et al., 1966; Tosdal et al., 1984). Hanging paleochannels that are incised into the basement of the Coastal Cordillera are in many places cut at the coastal cliff (Figs. 2A and 2C).

To the east of the Coastal Cordillera starts the Western Escarpment, which includes (from W to E) the Pampa de Diablo, the Oxaya Antiform, and the Pampa de Oxaya (Fig. 1). These pampas form a westward-dipping ramp and are the result of multiple erosional and depositional phases during the Miocene (Mortimer and Saric, 1975; Mortimer, 1980; Wörner et al., 2002). As will be discussed in detail below, we will use the Pampa de Diablo as reference base level prior to the formation of the major valleys. A formation age of 7.5 Ma was assigned to this pampa according to magnetostratigraphic chronologies of the underlying strata (von Rotz, 2003). In the vicinity of the Oxaya Antiform, the Pampa de Oxaya lies at 3000–3600 m above sea level. It dips toward the east, and is characterized by the presence of E-W–aligned relict valleys, 20–100 m deep, separated by ridges with nearly constant curvatures (Fig. 1). These valleys deepen and widen toward the west, indicating a west-directed paleoflow (Wörner et al., 2002). At present, discharge occurs in the opposite direction (i.e., toward the east, Fig. 4A). The substratum of the pampa (Fig. 3) is the ca. 23–19 Ma sequence of ignimbrite sheets and interbedded fluvial sediments that make up the Oxaya Formation (García, 2002). This unit blankets an antecedent mountainous landscape with Mesozoic-Cenozoic plutonic and metasedimentary rocks in its core (eastward migrating evolving arc basement on Fig. 4B). The western border of this primary Oligocene landscape was formed by a reverse fault system (Ausipar fault system, Figs. 4A and 4B) that accommodated crustal shortening (Munoz and Charrier, 1996; García, 2002; Wörner et al. 2002). It appears that uplift along this fault and subsequent erosion resulted in construction of a >400-m-thick sequence of highly concentrated flow deposits (25–23 Ma Azapa Formation) adjacent to this fault (García, 2002).

West of the Oxaya Antiform is the Pampa de Diablo (Fig. 1). This geomorphic feature forms a gently westward-dipping ramp that is cut by valleys with local (e.g., Vitor and Cama-

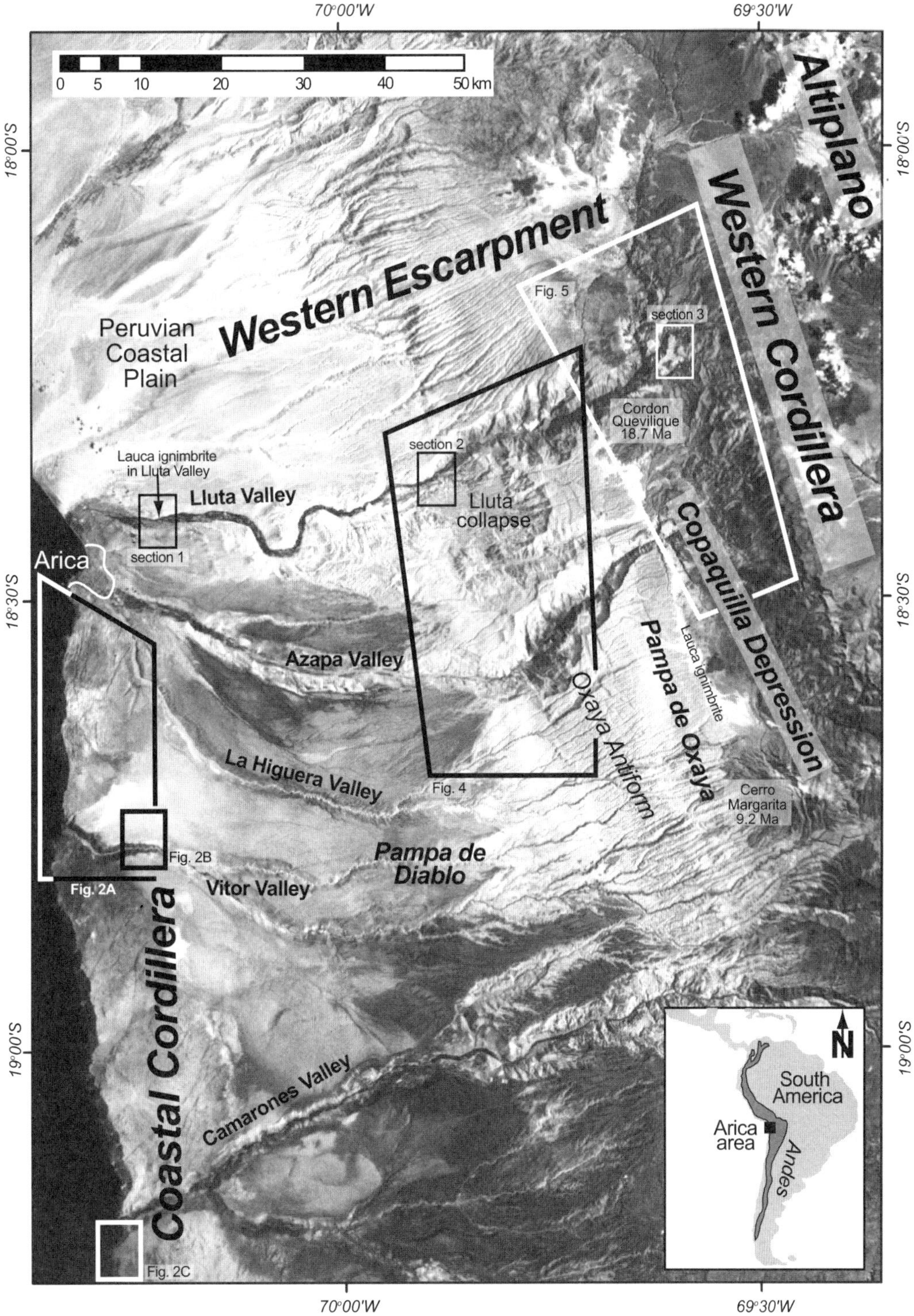

Figure 1. Landsat image and location of study area, the Western Escarpment of the Andes in northern Chile, showing the most important geomorphic units discussed in the text.

rones Valleys, Fig. 1) and distal sources (e.g., Lluta and Azapa Valleys). The Pampa de Diablo is underlain by (in stratigraphic order) the westward-thinning Azapa Formation, the Oxaya Formation, and the Diablo Formation (Fig. 3). This latter unit, deposited between ca. 19 and 7.5 Ma, according to Rb-Sr ages of interbedded ash layers, covering ignimbrites (García, 2002), and new magnetostratigraphic chronologies (von Rotz, 2003), is made up of a large-scale coarsening- and thickening-upward

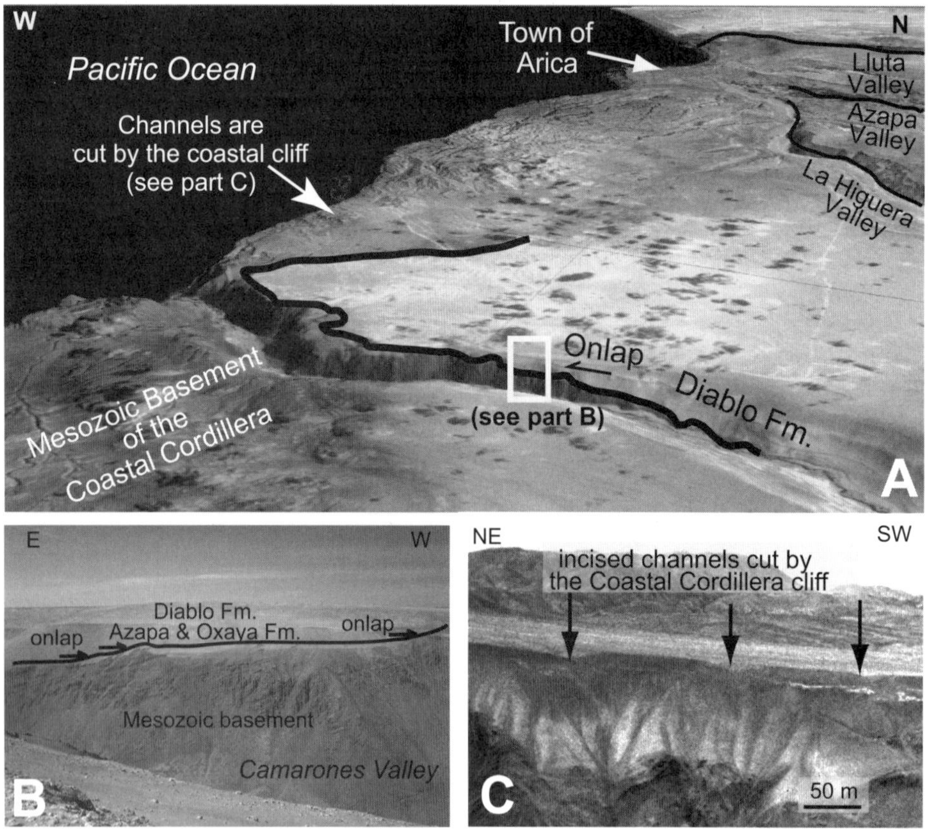

Figure 2. (A) Three-dimensional perspective view of the coastal area and geologic interpretation showing the relationships between the Coastal Cordillera and the overlying deposits. The perspective view which is oriented to the NW was achieved by underlying the Landsat image of Figure 1 with a 50-m-resolution digital elevation model. See Figure 1 for location and scale. Because of accessibility in the field, pictures B and C are taken from valleys farther south (Fig. 1). (B) Onlap situation similar at the Vitor Valley (A), but here in the Camarones Valley (Fig. 1). (C) Detailed situation of cut channels by the Coastal Cordillera similar at the mouth of the Vitor Valley (A), but here at the Camarones River mouth.

sequence with decimeter-thick mudflows at its base and fluvial trough cross-bedded, meter-thick conglomerate beds at its top. In the Arica area, where the Coastal Cordillera is stratigraphically and morphologically below the Peruvian Coastal Plains (Fig. 1), the dispersal systems of the Diablo Formation discharged into the Pacific Ocean (see below for details). South of Arica, however, the fluvio-lacustrine deposits of the Diablo Formation (and Oxaya Formation) onlap, and partly overlap the crystalline basement of the Coastal Cordillera (Fig. 2B). In the Longitudinal Valley some hundreds of kilometers farther south, however, the dispersal systems of the Diablo Formation have had a closed drainage until present (Hartley and Chong, 2002). The occurrence of andesitic and metasedimentary clasts suggests erosion of the Miocene andesitic volcanoes (e.g., the 18.7 Ma Cordon Quevilique, and the 9.2 Ma Cerro Margarita; Fig. 1) and of basement rocks that are exposed in the Western Andean Cordillera (García, 2002; Wörner et al., 2002).

The Copaquilla depression (García, 2002) (Figs. 1 and 5) is made up of a >300-m-thick series of fluvio-lacustrine sediments (e.g., Huaylas Formation) that were deposited between ca. 18 and 7 Ma, according to K-Ar ages of interbedded ignimbrites and tuffs (García, 2002). The Huaylas deposits were sourced from the Western Andean Cordillera (which is made up of the metamorphic basement, Miocene to modern volcanoes, and Miocene to modern fluvio-lacustrine deposits). Locally, the Miocene depositional surface of the Huaylas unit is still well preserved, displaying a fan-shaped surface with a convex curvature. In the Copaquilla depression, the Huaylas Formation is overlain by the 2.7 Ma Lauca ignimbrite (Fig. 5; Wörner et al., 2002), with an erosional unconformity between both units.

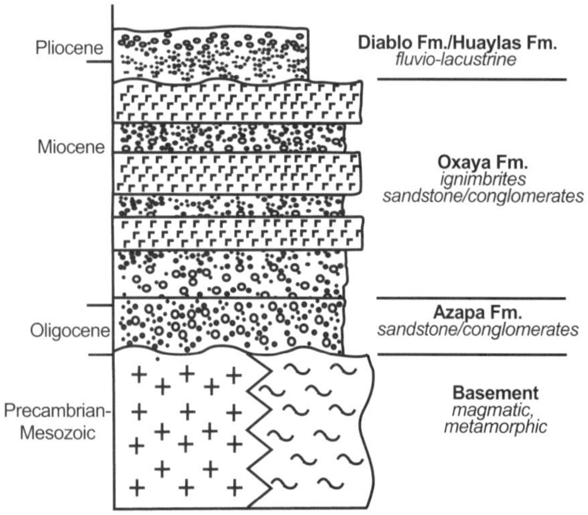

Figure 3. Stratigraphic scheme of the Western Escarpment of the Andes of northern Chile (simplified after Salas et al., 1966; García, 2002; Wörner et al., 2002).

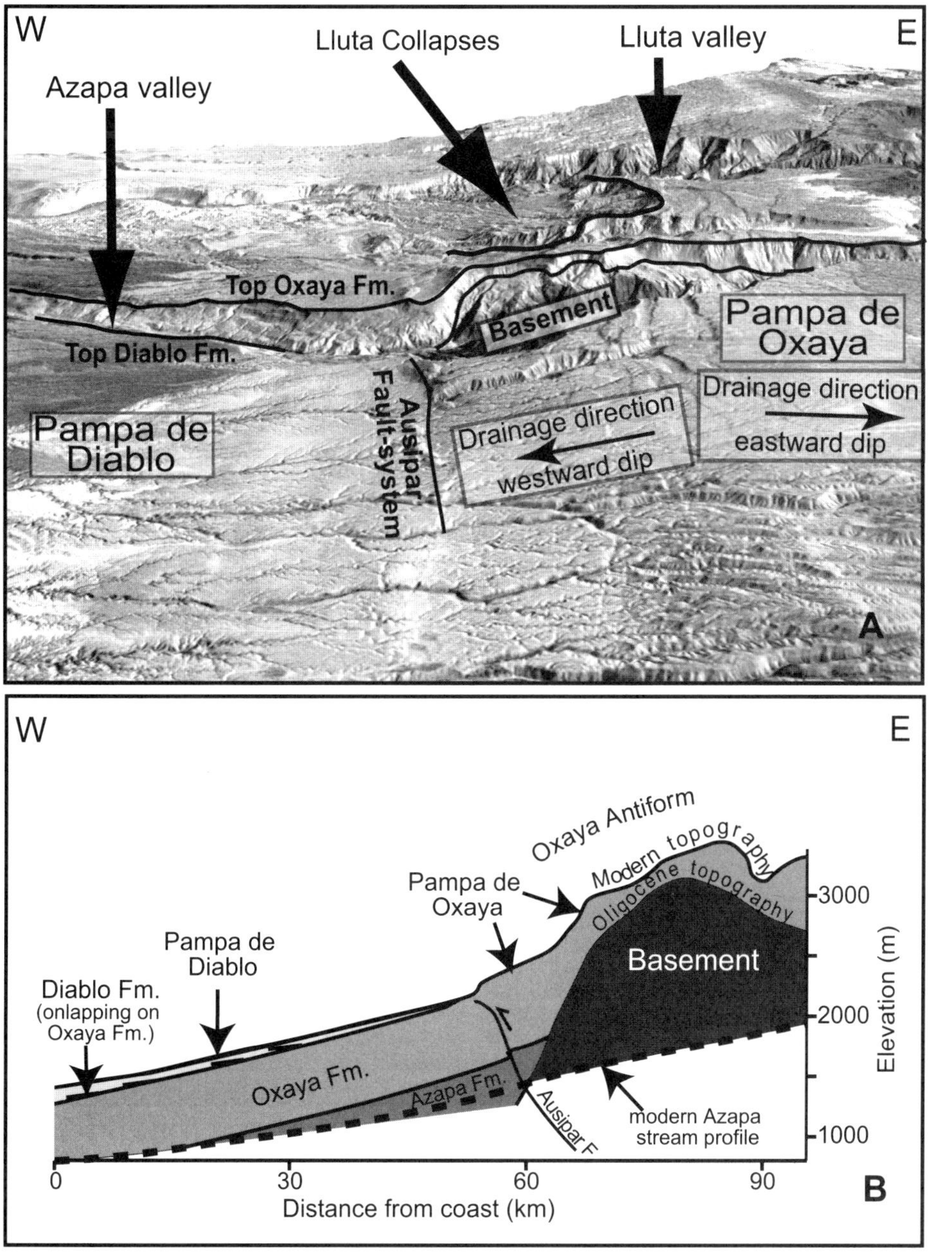

Figure 4. Three-dimensional (3-D) perspective view of the Pampa de Diablo and the Pampa de Oxaya (A), and geomorphic and geologic interpretation that is presented as a cross section (B). See note to Figure 2 for explanation of methodology of 3-D view calculation and scale. The 3-D view is oriented to the N. Valleys here cut ~1000 m deep.

The Lluta collapse (Figs. 1 and 6) is a giant landslide that covers an area of ~600 km^2, which displaced ~50 km^3 of rocks (Wörner et al., 2002; Strasser, 2003). The basal detachment of the Lluta collapse is the top of the Azapa Formation; the slip plane is well exposed on both sides of the Lluta Valley. Because slide blocks are found on both sides of the Lluta Valley, Wörner et al. thought that initiation of landsliding predated the formation of the Lluta Valley. However, a postdating alternative interpretation was proposed by Strasser (2003). The slide blocks are overlain by conglomerates, sandstones, and siltstones, which were deposited in a fluvio-lacustrine environment (Strasser, 2003). The clasts and boulders of the Lluta collapse deposits are Oxaya ignimbrites close to the escarpment, suggesting a local provenance. Farther north in the Lluta Valley, these sediments are interbedded with conglomerates that reveal a distal source (low-grade metasedimentary clasts, and clasts from granites and andesites of the Western Andean Cordillera).

Finally, the Western Cordillera forms the modern magmatic arc system of the central Andes with heights up to 6000 m. This unit is composed of Miocene to Holocene volcanic and sedimentary rocks that were deposited in interarc sedimentary basins on a Precambrian basement (Wörner et al., 2002). The headwaters of the main rivers (Lluta River, Azapa River) charge their water here.

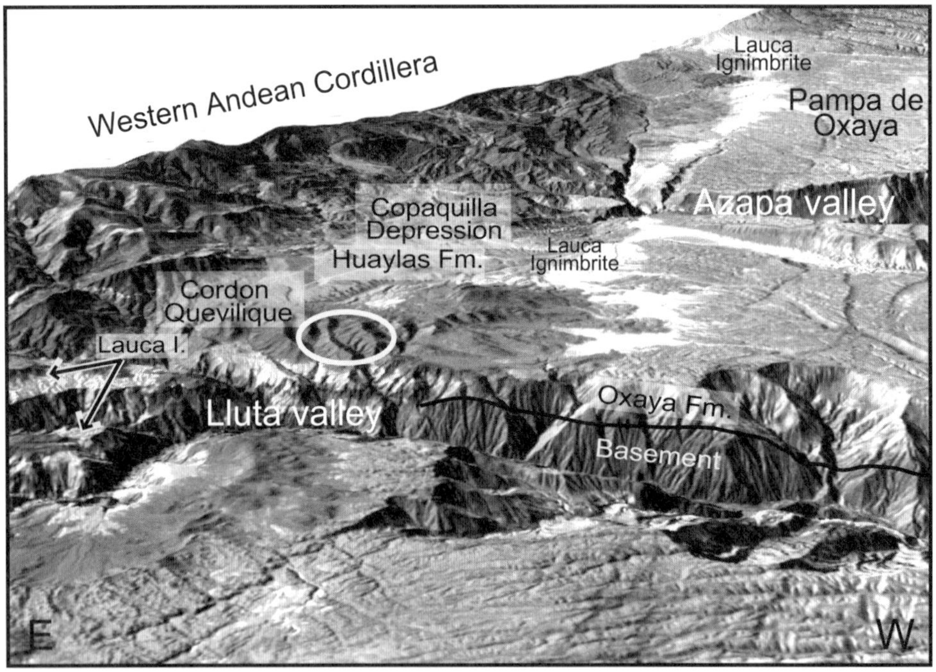

Figure 5. Three-dimensional (3-D) perspective view of the Pampa de Oxaya, the Copaquilla depression, and the Western Cordillera, and geomorphic and geologic interpretations. The view is toward the SE. Here, the Azapa and Lluta Valleys are up to 1500 m deep. See note to Figure 2 for explanation of methodology of 3-D view, calculation, and scale.

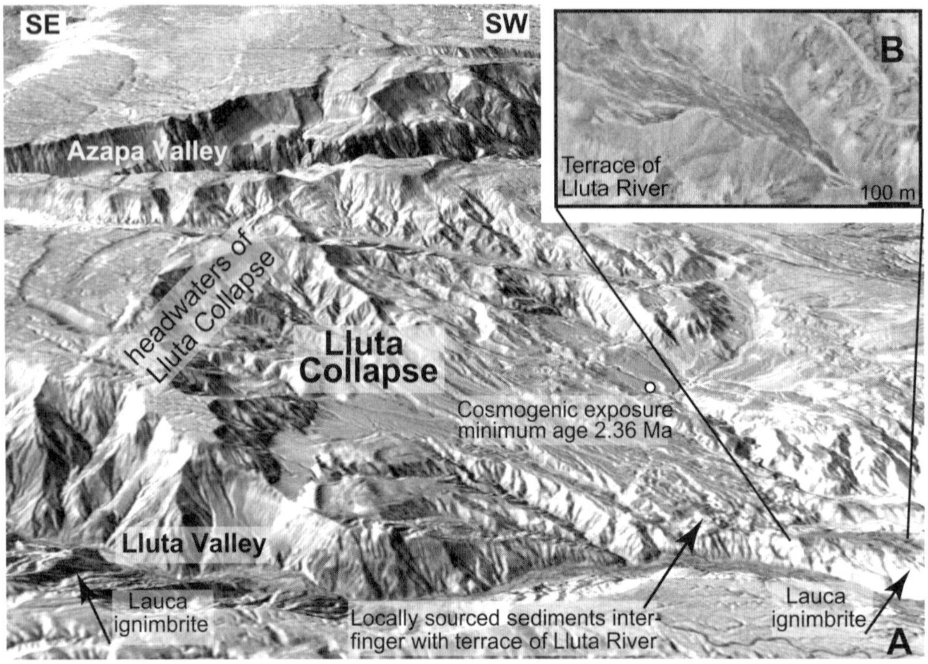

Figure 6. (A) Three-dimensional (3-D) perspective view of the Lluta landslide, with view to the S. See note to Figure 2 for explanation of methodology of 3-D view, calculation, and scale. (B) Detail of parts of the older than 2.4 Ma Lluta River base level, showing the pattern of abundant network of channels and bars.

RECONSTRUCTION OF THE GEOMORPHIC SITUATION PRIOR TO DISSECTION

In order to define the spatial and temporal datum that is needed here as reference for the situation prior to the major phase of valley formation, the top of the Diablo and Huaylas Formations appear the prime target. The Diablo and Huaylas Formations were deposited during the same time interval (Fig. 3). Both units have abundant andesitic clasts, which suggests a distal provenance (i.e., the Western Andean Cordillera; e.g., García, 2002). Because of the isochronous relationship between the Diablo and Huaylas Formations and the similar petrographic composition of the deposits, we interpret that both depositional realms were hydrologically linked. This implies that depositions in the Diablo and the Huaylas areas were presumably controlled by an identical regional base level that is assumed as the Diablo base level in this paper. Furthermore, because the youngest deposits of the Diablo Formation were dated at 7.5 Ma, according to magnetopolarity stratigraphies (von Rotz, 2003), we tentatively assign this late Miocene age to the Diablo base level.

In the Arica area, however, the Diablo base level has its current position ~350 m above the modern sea level, ~10 km inland. Here, this unit is made up of trough cross-bedded conglomerates without any evidence of lacustrine (or playa) deposits, suggesting deposition on a braid plain (García, 2002). This implies that the system feeding the Diablo Formation was hydrologically open and hence must have discharged into the Pacific. Because the rivers feeding the Diablo Formation were open to the Pacific, base-level lowering (possibly either due to surface uplift or due to subsidence of crustal segments; Salas et al., 1966) must have likely initiated incision of the Lluta, Azapa, and all other valleys in the study area. However, this phase of base-level lowering must have terminated at 2.7 Ma at the latest, because a fragment of the Lauca ignimbrite is found in the lower Lluta Valley (see also Wörner et al., 2002) (Fig. 1).

The Diablo base level that is currently ~350 m above sea level appears not only to have controlled aggradation of sediment in the Diablo and Huaylas sedimentary realms, but it probably also affected the large-scale morphometric properties of the Western Andean Cordillera. We observed that in this part of the Andes, hillslopes are smooth and covered by a regolith >1 m thick, and that the hillslopes reveal tangential lower contacts with the depositional surfaces of the Huaylas Formation, which in turn was assigned to be part of the Diablo base level above. Because Huaylas and Diablo depositional surfaces are still well preserved (see Mortimer and Saric, 1975; Mortimer, 1980; Wörner et al., 2002), these landscapes record, to large extents, surface processes during the time of the Diablo base level.

The aggradational nature of the conglomerates of the Diablo and Huaylas Formations, and the smooth hillslopes with a meter-thick regolith cover and tangential lower surfaces suggest that here, the rates of geomorphic processes have been to a large extent transport-limited. Support for this interpretation is given by two sets of results from theoretical models (e.g., Tucker and Slingerland, 1996; Simpson and Schlunegger, 2003). First, a regolith cover and the smooth nature of the hillslopes imply that the dip angles and the curvatures of the hillslopes are controlled by transport rates of regolith (e.g., Tucker and Slingerland, 1997). Second, aggradation of conglomerates in the Diablo and Huaylas depositional realms suggests that the sediment routing systems were at their transport capacity.

The transport-limited stage of sediment flux during deposition of the Diablo and Huaylas Formations implies that at that time, the Western Escarpment of the Andes of northern Chile was presumably in a stage of geomorphic decay. This interpretation is confirmed by theoretical concepts of landscape evolution (e.g., Tucker and Slingerland, 1996, 1997; Whipple et al., 1999). According to these models, geomorphic processes with transport-limited fluxes will result in a decrease in the local and the drainage basin relief, which, in turn, is indicative of the stage of geomorphic decay of a landscape. This situation, however, changed at ca. 7.5 Ma as the lowering of the base level in the Coastal Cordillera caused dissection and headward erosion of the rivers sourced in the eastern parts of the Western Escarpment and the Western Cordillera, respectively. This new stage of landscape rejuvenation will be explored in more detail in the following sections.

HEADWARD EROSION AND CHANGES IN RELIEF

The stream profiles, especially of the distally sourced systems (i.e., the Azapa Valley and the Lluta River), display two segments separated by well-defined knickzones (Fig. 7A). Below these knickzones, the channel gradients of the streams continuously increase in the upstream direction. Similarly, immediately above the knickzones the channels are flat, and their gradients then continuously increase in the upstream direction. Furthermore, the cross-sectional geometries of the valleys reveal distinct trends in the upstream direction that closely correlate with the valley gradients and the distance from the present-day knickzones. Specifically, in the lower reaches where the gradients are flat, the valleys are wider than the river belts (that are made up of braided channels and longitudinal and transverse gravel bars), resulting in the establishment of large floodplains. Here, the rivers flow on alluvial gravels (alluvial channels, e.g., Tucker and Slingerland, 1996). Toward the knickzones, the valleys become narrower, and the gradients increase. In the uppermost 5 km below the center of the knickzones, the rivers incise into bedrock (bedrock channels, e.g., Tucker and Slingerland, 1996), and the valley flanks are nearly vertical. Above the knickzones, the channel gradients decrease, the valleys widen and alluvial channels are occupying the valley floors.

The presence of knickzones that separate the stream profiles into two segments with distinct differences in stream gradients and valley morphologies is interpreted here to reflect that the fluvial systems accommodated the post–7.5 Ma lowering of the base level by headward erosion. According to this interpretation, the knickzones represent the locations of the erosional front, separating an older landscape with morphometric properties that still record the former base level (e.g., the Diablo base level) from a rejuvenated geomorphology that is currently adapting to the modern base level. Note that at present, the knickzones are located in the vicinity of basement-bedrock–Neogene conglomerate transitions (e.g., the Huaylas Fm.), at the interfluves of several tributaries, and on abandoned tectonic features (west-vergent thrust system of Munoz and Charrier, 1996). Therefore, it would be possible to interpret a lithological, tectonic, and/or a discharge control on the establishment of a knickzone in these locations. In this case, the change in bedrock lithologies, the specific drainage pattern configuration in the locations of the present-day knickzones, or, alternatively, enhanced rates of rock uplift in this location have simply resulted in an overall

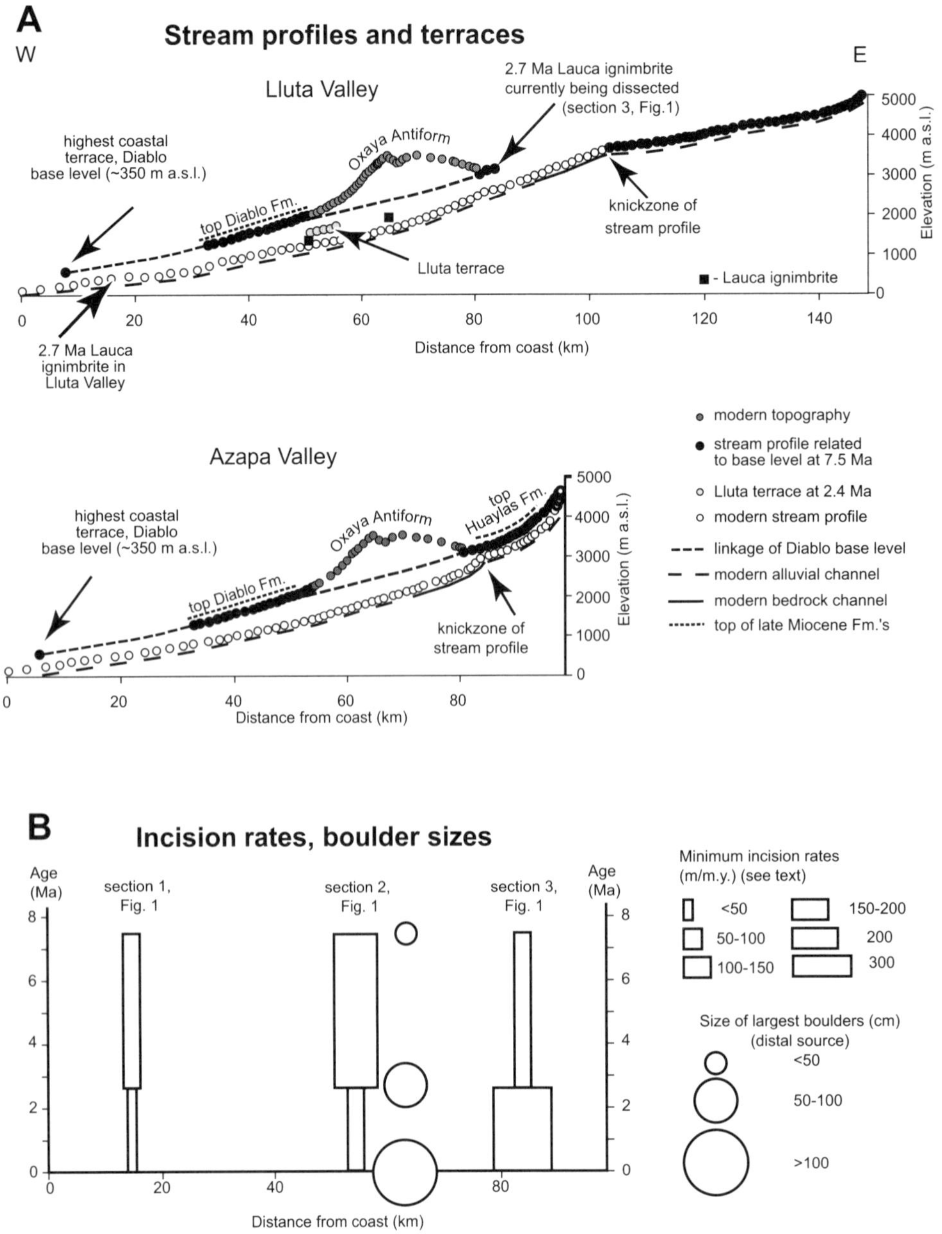

Figure 7. (A) Morphometric information of Azapa Valley, Lluta Valley, and the Diablo base level, and (B) trends in incision rates and sizes of maximum clasts of Lluta River. See text for more information on data acquisition; a.s.l. is above sea level.

reduction of headward erosion rates or in a nearly stationary locus of the knickzone itself.

INCREASE IN FLUVIAL INCISION RATES

Differences in elevations between fluvial and/or strath terraces of known ages are generally used to calculate fluvial incision rates. However, provided that surface uplift and the associated lowering of the base level is accommodated by headward erosion, then differences in elevations between terraces of known ages will only yield information about average local incision rates, and they might differ from effective erosion rates. This is true because during the process of headward erosion, the local incision rates first increase as the erosional front arrives, and then decrease as it passes. This pattern of temporally varying incision rates might not be seen if dated terraces of various elevations are used. Despite these drawbacks in assessing temporal and spatial changes in erosion rates, data are presented here to suggest that minimum local incision rates have increased between ca. 7.5 Ma and present. Wörner et al. (2002) suggested increasing incision rates between ca. 3 Ma and present.

Here, minimum incision rates are estimated for three sectors in the Lluta Valley (see Figs. 1 and 7B). In the lower portion of the Lluta Valley, fluvial dissection was initiated after deposition of the Diablo Formation at ca. 7.5 Ma or younger (as a result of base-level lowering in the Coastal Cordillera as discussed above). At 2.7 Ma at the latest, the modern base level was reached, as indicated by the presence of Lauca ignimbrite fragments in the Lluta Valley some tens of kilometers east of Arica (Fig. 1). In this location, the difference in elevation between the top of the Diablo Formation and the broad valley floor measures ~350 m (Figs. 1 and 7B), yielding rates of fluvial incision of ~70 m/m.y. between 7.5 and 2.7 Ma. The modern Lluta River course is presently further ~20–30 m lower in that location, which results in incision rates toward present of less than 10 m/m.y., likely as a consequence of a more stable sea base level, flatter parts of longitudinal river profiles (most distal), and the long time since the erosional front passed.

In the middle sector of the Lluta Valley, at ~60 km distance from the coast (Fig. 1), alluvial terraces of the Lluta River lie 300 m beneath the top of the Diablo Formation, and 350 m above the modern valley floor (Figs. 6 and 7A). These deposits dip at an angle of ~3.5° in a NNW direction toward the Lluta Valley. Based on a detailed reconstruction of the stratigraphy in this area, Strasser (2003) found that the deposits of the Lluta River terrace (Fig. 6) interfinger with the detritus sourced from the headwaters of the Lluta collapse. We estimated the time when the formation of this terrace system ended using concentrations of cosmogenically produced ^{21}Ne in quartz (see Table 1 for explanation, and Fig. 6A for sample site). This methodology yields a minimum exposure age of 2.36 ± 0.23 Ma for this terrace system. However, in the same location, an ignimbrite that overlies distally sourced conglomerates ~20 m above the present valley floor possibly represents the Lauca ignimbrite. The same ignimbrite is found 260 m above the modern valley floor ~15 km farther east (Fig. 7A). Using this information constrains the resulting incision rates to ~200 m/m.y. between 7.5 and 2.7 Ma, and between 50 and 100 m/m.y. between 2.7 Ma and present. We cannot completely exclude the possibility that the formation of the Lluta River terrace is simply the result of a cut-and-fill process that might have been initiated by the Lluta landslide. However, since evidence of lacustrine sediments in the higher portion east of the Lluta landslide is missing, we do not consider that this landslide significantly altered the base level and hence the lowering process of the Lluta River.

Finally, at ~85 km distance from the ocean (Fig. 1), the Lluta River has cut into 2.7 Ma Lauca ignimbrite deposits (Figs. 1 and 5) that lie ~300 m beneath the Diablo base level, and ~900 m above the modern valley floor. The resulting incision rates are ~60 m/m.y. between 7.5 and 2.7 Ma, and >300 m/m.y. between 2.7 Ma and present.

From the stratigraphic evidence, we conclude that the locus of enhanced incision shifted from the coast to the Western Andean Cordillera (Fig. 7B), and that the time interval between ca. 2.7 Ma and present is characterized by maximum incision rates, especially in the upper catchments (Fig. 7B).

TEMPORAL TRENDS IN STREAM POWER

Flume experiments reveal that flowing water exerts a shear stress on the riverbed that is proportional to the product between stream gradient and water discharge if the flow is unconfined (Bagnold, 1966). This shear stress results in erosion and transport of sediment as bedload if a threshold is exceeded. This critical magnitude depends on the sorting of the bed material, the state of the armoring of the bed by a layer of coarse material, and in particular on the particle sizes of the sediment. Specifically, the critical shear stress has to increase to entrain particles with increasing sizes (especially, if stream gradients decrease during the same time interval).

For the Lluta system, temporal changes of maximum shear stress are estimated at 50 km distance from the ocean (i.e., in the location of the previously described Lluta terrace; Figs. 1 [section 2] and 7B) using the diameters of the largest clasts that had a distal source (i.e., clasts from andesites and granites). Specifically, for each stratigraphic unit that represents the base levels of Figure 7B (conglomerates of Diablo Formation, older than 2.4 Ma terrace of Lluta River, and modern Lluta deposits), we selected at least two >100–500-m^2-large areas where we

TABLE 1. COSMOGENIC ^{21}Ne DATA

Sample	^{20}Ne (× 10^9 atoms/g)	^{21}Ne/^{20}Ne (× 10^{-3})	^{22}Ne/^{20}Ne	^{21}Ne$_{cosmo}$ (× 10^7 atoms/g)	Blank-corrected ^{21}Ne$_{cosmo}$ (× 10^7 atoms/g)	Age (corrected for 70 cm/m.y. erosion) (× 10^6 yr)
CN8b	13.729 ± 0.026	6.35 ± 0.08	0.106 ± 0.004	4.659 ± 0.216	3.874 ± 0.287	2.36 ± 0.23
CN10(blank)	5.622 ± 0.248	4.30 ± 0.06	0.104 ± 0.003			

Note: All concentrations are given and calculated for sampling site. Sample CN8—boulder on abandoned terrace, CN10—boulder in the youngest wash (lower side), assuming a geological blank case. Sample preparation following Ivy-Ochs et al. (1995). Ne data represent cosmogenic-temperature step of 600 °C, higher steps were air. ^{21}Ne$_{cosmo}$ concentration calculated as excess over air. Production rates used after Niedermann (2002) and scaled by modified Lal (1991). Density, ρ = 2.375 gcm^{-3}, attenuation length Λ = 160 gcm^{-2}. Erosion rates are defined by erosion-island-plots of Ne/Be-Be (Kober et al., 2002).

measured the size of ~20–50 clasts. From this data set the maximum diameters of the five largest clasts are presented on Figure 7B. The data reveal a general increase in the maximum clast size from ~50 cm for the 7.5 Ma deposits (i.e., the top of the Diablo Formation), to 50–100 cm for the older than 2.4 Ma Lluta terrace reference, to diameters >100 cm for the modern deposits. In these locations, erosional scours exceeding depths of 1 m are absent and the bars are horizontally stratified, implying that the flow has generally been unconfined. Consequently, the observed general increase in the size of the largest clasts is interpreted here to indicate an augmentation in the magnitudes of maximum discharge events.

Farther west in the Arica area, however, channels that were incised into basement of the Coastal Cordillera are truncated at the coastal steep cliff. Furthermore, they lack any evidence of re-incision (Fig. 2C). This implies, that in the coastal region, precipitation rates fell below the threshold conditions for fluvial incision at 3 Ma at the latest (i.e., at the time when the phase of major base-level lowering was terminated).

COMPARATIVE ESTIMATES OF SEDIMENT YIELDS

The excellent preservation of the relict late Miocene plain—the Diablo landscape—that formed prior to dissection allows us to calculate sediments yields for systems with various sources. Furthermore, because the erosional systems that drain the Arica area have cut through the same lithological architecture, and since incision was initiated for all analyzed systems at the same time (i.e., 7.5 Ma, see above), the comparison of volumes of eroded rocks allows interpretation of possible controls of water discharge on surface erosion rates. Specifically, we aim at identifying whether possible orographic effects of precipitation have had an influence on erosion rates.

Figure 8 illustrates the calculated volumes of rocks that were eroded between 7.5 Ma and present, the present-day sizes of the drainage basins for the systems draining the Arica area, and a resulting erosion rate. The resulting sediment yields are ~15 m/m.y. for systems that are sourced in the Western Andean Cordillera (i.e., the rivers flowing through the Lluta and Azapa Valleys), and <7 m/m.y. for rivers with sources solely on the Western Escarpment. Note that these calculations assume no major increase in the size of the headwaters, which might be expected because of headward erosion. However, catchments rather compete with the systems that discharge to the east (i.e., Altiplano; Mortimer and Saric, 1975). Because the sediment yields represent volumes of eroded rocks that are normalized to the size of the drainage basins, the differences in yields cannot be explained by the various sizes of the headwaters. It appears that the enhanced erosional efficiencies calculated for the rivers with sources in the Western Andean Cordillera simply reflect the observation that modern precipitation rates significantly increase toward the Western Cordillera. As will be argued in the next section, the contrasts in precipitation rates between the coastal region and the Western Andean Cordillera have become enhanced at least between 7.5 Ma and present.

DISCUSSION

It has already been established that fluvial dissection and valley formation was initiated at ca. 7.5 Ma as a result of a major base-level change in the coastal area, and that the fluvial processes have adapted to the new base level by headward erosion. The climate component in valley formation was identified in the comparison between sediment yields calculated for systems with distal and local sources, revealing that the larger yields calculated for the distally sourced systems most likely resulted from the effect of orographic precipitation. Here we will argue that the effect of orographic precipitation has become enhanced between the late Miocene–Pliocene and present. Support for this interpretation is given by the general coarsening trend in the clast size of the conglomerates and the contemporaneous increase in headward erosion rates. These pieces of information imply an increase in effective water discharge. However, a tectonic control cannot be ruled out at the moment (e.g., monoclinal surface uplift proposed by Isacks, 1988, which might have the same effect), so the solution appears nonunique. Nevertheless, if the tectonic component has been the dominant force driving headward erosion, then rates of monoclinal surface uplift should have become enhanced at 3 Ma at the time when incision rates increased (as observed in the Arica area). Existing data, however, do not support such a scenario (Isacks, 1988; Gregory-Wodzicki, 2000). Therefore, the enhanced discharge scenario has to be considered to explain the increase in both the sizes of the largest clasts and incision rates between ca. 3 Ma and present (Fig. 7B), and the overall evolution of the valley systems. This does not mean, however, that the Andes of Chile have become more humid between the late Miocene and present. Indeed, the presence of valleys that are cut by the coastal cliff implies that in the lower portions of the study area, precipitation rates fell below the threshold conditions for fluvial incision of systems with small catchments. Similarly, theoretical considerations about controls on erosion led Molnar (2001) to conclude that an increase in aridity (and hence a decrease in average precipitation rates) might even result in an increase in erosion rates, provided that the frequency and the magnitudes of large floods increase (as supported by the observed increase in the size of the largest clasts), i.e., such an increase might even occur even if average water discharge decreases. The presented data do not allow estimation of whether average precipitation rates decreased. They only indicate a shift toward a more incisive climate, and toward an enhancement of the effect of orographic precipitation as the system evolved.

CONCLUSIONS

The stratigraphic and morphometric data and observations from the Western Escarpment of northern Chile (Arica area) imply that a major phase of valley formation resulted from base-level lowering in the Coastal Cordillera, which was initiated at ca. 7.5 Ma, and which was almost completed at 2.7 Ma at the latest. Additional small-scale, base-level changes, as recorded in

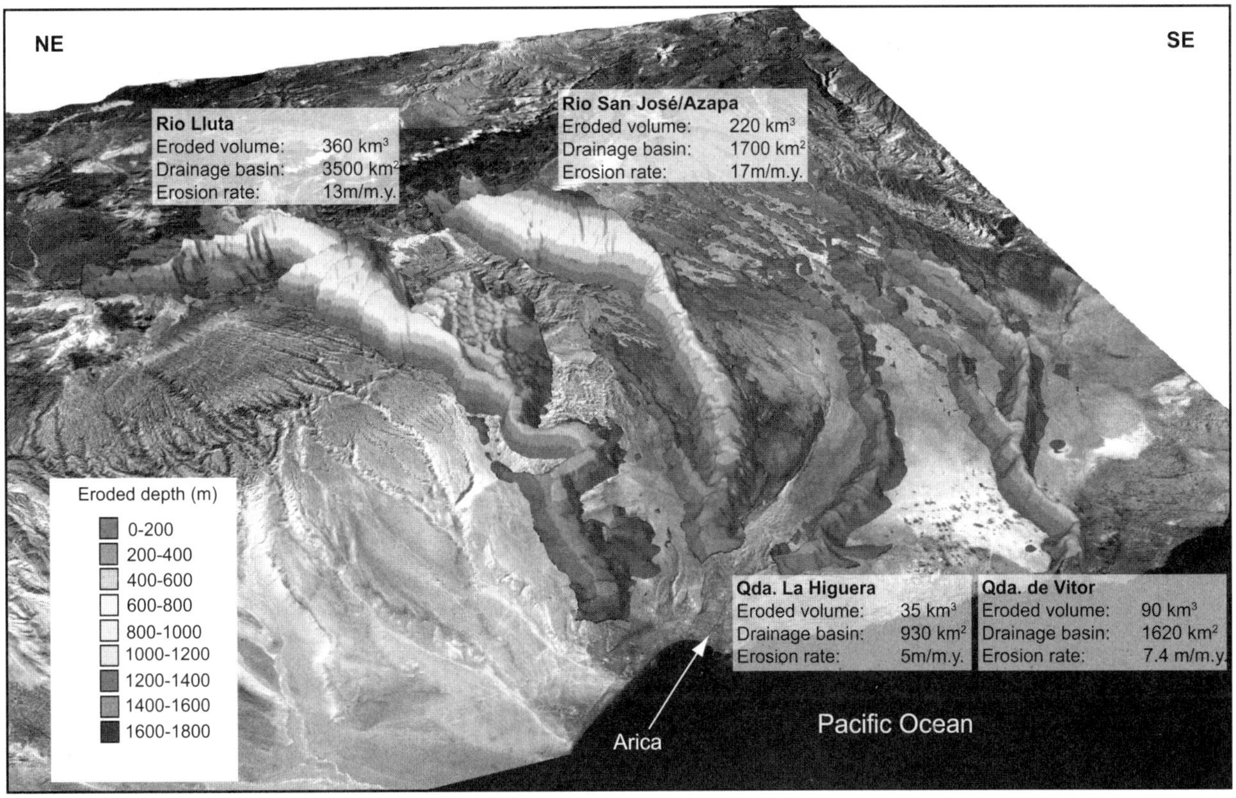

Figure 8. Comparison between volumes of eroded rocks (differential modeling, see text for obtainment), catchment size, and resulting sediment yields of distally and locally sourced rivers. See Figure 1 for scale.

several terraces farther to the north (southern Peru; Ortlieb et al., 1996a) and to the south (e.g., Antofagasta; Ortlieb et al., 1996b), would not result in such prominent incision features. The stratigraphic and geomorphic data also indicate that fluvial incision and valley formation have become enhanced during the last 3 m.y., presumably as a result of enhanced effective water discharge. During the same time interval, however, the Coastal Cordillera and the Atacama Desert became hyperarid (Hartley and Chong, 2002). It appears, therefore, that precipitation became restricted to the higher parts of the Western Escarpment and the Western Cordillera where the major rivers are sourced that deeply cut into the Oxaya and Diablo landscapes. That also highlights the orographic effect on the precipitation pattern due to the raising Andes. The hyperarid conditions that occurred in the lower elevations of the northern Andes of Chile caused the Miocene-Pliocene strata to become covered by crusts of salt and anhydrite that even blanket hillslopes and summits, especially in the region of the Coastal Cordillera. We interpret that this stabilization of surfaces decreased the sediment discharge from the plateaus, e.g., the Pampa de Diablo and Oxaya, to the fluvial transport systems, possibly shifting the rivers into the supply-limited stage. In the case that the fluvial systems are transport-limited, a decrease in sediment discharge—especially at lower elevations—can result in an increase in incision and headward erosion rates.

ACKNOWLEDGMENTS

We thank G. Simpson and G. Hoke for discussions, and R. Wieler, H. Baur, S. Ivy-Ochs, and P. Kubik for support while deducing cosmogenic nuclide data. The paper benefited from reviews by B. Horton, C. Duncan, and S. Willett. This study was funded by the Swiss National Science Foundation (grant 620-576.663).

REFERENCES CITED

Alpers, C.N., and Brimhall, G.H., 1988, Middle Miocene climate change in the Atacama Desert, Northern Chile: Evidence from supergene mineralization at La-Escondida: Geological Society of America Bulletin, v. 100, p. 1640–1656, doi: 10.1130/0016-7606(1988)100<1640:MMCCIT>2.3.CO;2.

Bagnold, R.A., 1966, An approach to the sediment transport problem from general physics: U.S. Geological Survey Professional Paper 422-I, p. I1–I37.

Bull, W.B., 1991, Geomorphic responses to climate change: New York, Oxford University Press, 326 p.

Burbank, D.W., Leland, J., Fielding, E., Anderson, R.S., Brozovic, N., Reid, M.R., and Duncan, C., 1996, Bedrock incision, rock uplift and threshold hillslopes in the northwestern Himalayas: Nature, v. 379, p. 505–510, doi: 10.1038/379505a0.

García, M., 2002, Evolution Oligo-Miocène de l'Altiplano occidental (arc et avant-arc du nord du Chili, Arica): Tectonique, volcansime, sédimentation, géomorphologie et bilan érosion-sédimentation: Géologie Alpine, Mémoire H.S. 40, p. 118.

Gregory-Wodzicki, K.M., 2000, Uplift history of the central and northern Andes: A review: Geological Society of America Bulletin, v. 112, p. 1091–1105, doi: 10.1130/0016-7606(2000)112<1091:UHOTCA>2.3.CO;2.

Hartley, A.J., and Chong, G., 2002, Late Pliocene age of Atacama Desert: Implication for the desertification of western South America: Geology, v. 30, p. 43–46, doi: 10.1130/0091-7613(2002)030<0043:LPAFTA>2.0.CO;2.

Harvey, A.M., 2001, Coupling between hillslopes and channels in upland fluvial systems: Implications for landscape sensitivity, illustrated from the Howgill Fells, northwest England: Catena, v. 42, p. 225–250, doi: 10.1016/S0341-8162(00)00139-9.

Herm, D., 1969, Marines Pliozän und Pleistozän in Nord- und Mittelchile unter besonderer Berücksichtigung der Entwicklung der Mollusken-Faunen: Zitteliana, v. 2, p. 159.

Hinderer, M., Schäuble, H., and Süss, M.S., 2002, Denudation processes at various scales—Toward a new GIS-based global mode: Zentralblatt für Geologie und Paläontologie, Teil 1, no. 3/4, p. 377–397.

Hoke, G.D., Isacks, B.L., and Jordan, T.E., 2002, Contrasting styles of erosion along the western slope of the Central Andean Plateau: Eos (Transactions, American Geophysical Union), AGU Fall Meeting, San Francisco, v. 83, p. 47.

Isacks, B.L., 1988, Uplift of the Central Andean Plateau and bending of the Bolivian Orocline: Journal of Geophysical Research, v. 93, p. 3211–3231.

Ivy-Ochs, S., Schlüchter, Ch., Kubik, P.W., Dittrich-Hannen, B., and Beer, J., 1995, Minimum ^{10}Be exposure ages of early Pliocene for the Table Mountain Plateau and the Sirius Group at Mount Fleming, Dry Valley, Antarctica: Geology, v. 23, p. 1007–1010, doi: 10.1130/0091-7613(1995)023<1007:MBEAOE>2.3.CO;2.

Kennan, L., 2000, Large-scale geomorphology of the Andes: Interrelationships of tectonics, magmatism and climate, *in* Summerfield, M.A., ed., Geomorphology and global tectonics: Chichester, Wiley, p. 167–199.

Kober, F., Schlunegger, F., Ivy-Ochs, S., and Wieler, R., 2002, The dependency of cosmogenic nuclides to climate and surface uplift in transient landscapes: Geochimica et Cosmochimica Acta, v. 66, no. 15A, p. A408.

Lal, D., 1991, Cosmic ray labeling of erosion surfaces: In-situ nuclide production rates and erosion models: Earth and Planetary Science Letters, v. 104, p. 424–439, doi: 10.1016/0012-821X(91)90220-C.

Molnar, P., 2001, Climate change, flooding in arid environments, and erosion rates: Geology, v. 29, p. 1071–1074, doi: 10.1130/0091-7613(2001)029<1071:CCFIAE>2.0.CO;2.

Montgomery, D.R., and Dietrich, W.E., 1992, Channel initiation and the problem of landscape scale: Science, v. 255, p. 826–830.

Montgomery, D.R., Massong, T.M., and Hawley, S.C.S., 2003, Influence of debris flows and log jams on the location of pools and alluvial channel reaches, Oregon Coast Range: Geological Society of America Bulletin, v. 115, p. 78–88, doi: 10.1130/0016-7606(2003)115<0078:IODFAL>2.0.CO;2.

Mortimer, C., 1980, Drainage evolution in the Atacama Desert of northernmost Chile: Revista Geologica de Chile, v. 11, p. 3–28.

Mortimer, C., and Saric, N., 1975, Cenozoic studies in northernmost Chile: Geologische Rundschau, v. 64, p. 395–420.

Munoz, N., and Charrier, R., 1996, Uplift of the western border of the Altiplano on a west-vergent thrust system, northern Chile: Journal of South American Earth Sciences, v. 9, p. 171–181, doi: 10.1016/0895-9811(96)00004-1.

Niedermann, S., 2002, Cosmic-ray-produced noble gases in terrestrial rocks as a dating tool for surface processes, *in* Porcelli, D., Ballentine, C.J., and Wieler, R., eds., Noble gases in cosmochemistry and geochemistry: Reviews in Mineralogy and Geochemistry, v. 47, p. 731–784.

Ortlieb, L., Zazo, C., Goy, J.L., Dabrio, C., and Machare, J., 1996a, Pampa del Palo: An anomalous composite marine terrace on the uprising coast of southern Peru: Journal of South American Earth Sciences, v. 9, p. 367–379.

Ortlieb, L., Zazo, C., Goy, J.L., Hillaire-Marcel, C., Ghaleb, B., and Cournoyer, L., 1996b, Coastal deformation and sea-level changes in the northern Chile subduction area (23°S) during the last 330 ky: Quaternary Science Reviews, v. 15, p. 819–831, doi: 10.1016/S0277-3791(96)00066-2.

Paskoff, R.P., 1980, Late Cenozoic crustal movements and sea level variations in the coastal area of northern Chile, *in* Mörner, N.A., ed., Earth rheology, isostasy, and eustasy: Chichester, Wiley, p. 487–495.

Pazzaglia, F.J., Gardner, T.W., and Merritts, D.J., 1998, Bedrock fluvial incision and longitudinal profile development over geologic time scales determined by fluvial terraces, *in* Tinkler, K.J., and Wohl, E.E., eds., Rivers over rock: Fluvial processes in bedrock channels: Washington, D.C., American Geophysical Union Geophysical Monograph 107, p. 207–235.

Pratt, B., Burbank, D.W., Heimsath, A., and Ojha, T., 2002, Impulsive alleviation during early Holocene strengthened monsoons, central Nepal Himalaya: Geology, v. 30, p. 911–914, doi: 10.1130/0091-7613(2002)030<0911:IADEHS>2.0.CO;2.

Roe, G.H., Montgomery, D.R., and Hallet, B., 2002, Effects of orographic precipitation variations on the concavity of steady-state river profiles: Geology, v. 30, p. 143–146, doi: 10.1130/0091-7613(2002)030<0143:EOOPVO>2.0.CO;2.

Rosenbloom, N.A., and Anderson, R.S., 1994, Hillslope and channel evolution in a marine terraced landscape, Santa Cruz, California: Journal of Geophysical Research, v. 99, p. 14,013–14,029, doi: 10.1029/94JB00048.

Salas, R.O., Kast, R.F., Montecinos, F.P., and Salas, I.Y., 1966, Geologica y recursos minerales del departemento de Arica: Santiago, Instituto de Investigaciones Geologicas Chile, Boletin 21, p. 114.

Schlunegger, F., 2002, Impact of hillslope-derived sediment supply on drainage basin development in small watersheds at the northern border of the central Alps of Switzerland: Geomorphology, v. 46, p. 285–305, doi: 10.1016/S0169-555X(02)00081-8.

Schlunegger, F., and Hinderer, M., 2003, Pleistocene/Holocene climate change, re-establishment of fluvial drainage network and increase in relief in the Swiss Alps: Terra Nova, v. 15, p. 88–95, doi: 10.1046/j.1365-3121.2003.00469.x.

Schlunegger, F., Detzner, K., and Olsson, D., 2002, The evolution towards steady state erosion in a soil-mantled drainage basin: Semi-quantitative data from a transient landscape in the Swiss Alps: Geomorphology, v. 43, p. 55–76, doi: 10.1016/S0169-555X(01)00120-9.

Schumm, S.A., 1956, The role of creep and rainwash on the retreat of badland slopes: American Journal of Science, v. 254, p. 693–706.

Simpson, G., and Schlunegger, F., 2003, Topographic evolution and morphology of surfaces evolving in response to coupled fluvial and hillslope sediment transport: Journal of Geophysical Research., v. 108, 2300. p. ETG 7/1–7/16.

Strasser, M., and Schlunegger, F., 2005, Erosional processes, topographic length-scales and geomorphic evolution in arid climatic environments: the 'Lluta collapse,' northern Chile: International Journal of Earth Sciences, v. 94, p. 433–446, doi: 10.1007/s00531-005-0491-2.

Tosdal, R.M., Farra, A.E., and Clark, A.H., 1984, Cenozoic polyphase landscape and tectonic evolution of the Cordillera Occidental, southernmost Peru: Geological Society of America Bulletin, v. 95, p. 1318–1332, doi: 10.1130/0016-7606(1984)95<1318:CPLATE>2.0.CO;2.

Tucker, G.E., and Bras, R.L., 1998, Hillslope processes, drainage density, and landscape morphology: Water Resources Research, v. 34, p. 2751–2764, doi: 10.1029/98WR01474.

Tucker, G.E., and Slingerland, R., 1996, Predicting sediment flux from fold and thrust belts: Basin Research, v. 8, p. 329–349, doi: 10.1046/j.1365-2117.1996.00238.x.

Tucker, G.E., and Slingerland, R., 1997, Drainage basin response to climate change: Water Resources Research, v. 33, p. 2031–2047, doi: 10.1029/97WR00409.

von Rotz, R., Schlunegger, F., Heller F., and Villa, I., 2005, Assessing the age of relief growth in the Andes of northern Chile: Magneto-polarity chronologies from Neogene continental sections: Terra Nova, doi:10.1111/j.1365-3121.2005.00634.x

Wegmann, K.W., and Pazzaglia, F.J., 2002, Holocene strath terraces, climate change, and active tectonics: The Clearwater River basin, Olympic Peninsula, Washington State: Geological Society of America Bulletin, v. 114, p. 731–744, doi: 10.1130/0016-7606(2002)114<0731:HSTCCA>2.0.CO;2.

Whipple, K.X., 2004, Bedrock rivers and the geomorphology of active orogens: Annual Reviews of Earth and Planetary Science Letters, v. 32, p. 151–185, doi: 10.1146/annurev.earth.32.101802.120356.

Whipple, K.X., and Tucker, G.E., 1999, Dynamics of the stream-power river incision model: Implications for height limits of mountain ranges, landscape response timescales, and research needs: Journal of Geophysical Research, v. 104, p. 17,661–17,674, doi: 10.1029/1999JB900120.

Whipple, K.X., Kirby, E., and Brocklehurst, S.H., 1999, Geomorphic limits to climate-induced increases in topographic relief: Nature, v. 401, p. 39–43, doi: 10.1038/43375.

Whipple, K.X., Snyder, P., and Dollenmayer, K., 2000, Rates and processes of bedrock incision by the Ukak River since the 1912 Novarupta ash flow in the Valley of Ten Thousand Smokes, Alaska: Geology, v. 28, p. 835–838, doi: 10.1130/0091-7613(2000)028<835:RAPOBI>2.3.CO;2.

Wörner, G., Uhlig, D., Kohler, I., and Seyfried, H., 2002, Evolution of the West Andean Escarpment at 18ºS (N. Chile) during the last 25 Ma: Uplift, erosion and collapse through time: Tectonophysics, v. 345, p. 183–198, doi: 10.1016/S0040-1951(01)00212-8.

Manuscript Accepted by the Society 23 June 2005

Printed in the USA

Geological Society of America
Special Paper 398
2006

The application of drainage system analysis in constraining spatial patterns of uplift in the Coastal Cordillera of northern Chile

Anne E. Mather[†]
School of Geography, University of Plymouth, Plymouth, Devon PL4 8AA, UK
Adrian J. Hartley
Department of Geology and Petroleum Geology, Meston Building, King's College, University of Aberdeen, Aberdeen AB9 2UE, UK

ABSTRACT

The Coastal Cordillera of northern Chile is the only subaerial part of the South American continental crust in direct contact with the subducted Nazca plate. Deformation of the cordillera can be directly related to plate coupling at the subduction zone. Knowledge of uplift rate variation along the coast is however limited, mainly due to difficulties in dating and correlating the discontinuous remnants of marine terraces. This paper uses geomorphology and the stream gradient index (SGI) to examine 80 drainages along 200 km of coastline between 21°S and 23°S. Selected catchments are located on similar geology (basaltic andesite and granodiorite) and within the same coastal climate zone. The southernmost part of the transect overlaps with an area of known uplift rates based on dated marine terraces. Characteristics of the SGI index, combined with the geomorphology, enable the coast to be divided into two sectors. Extrapolation of the SGI values from an area of known (sector 1) to unknown uplift rates (sector 2) enables an estimate of relative uplift rates for sector 2. Sector 1 has known uplift rates of 240–384 mm k.y.$^{-1}$. The SGI results suggest sector 2 has overall uplift rates which may be 2–3 times greater than sector 1. The large-scale (70–100 km) differences in apparent differential uplift reflected by the SGI along the Coastal Cordillera are tentatively attributed to aseismic ridge subduction. Smaller-scale (30 km) variations in the SGI associated with the major regional drainages in sector 2 suggest that the latter drainages may be focused in structural lows (with low SGI values) controlled by NE-SW– to E-W–trending reverse faults, although further work is required to clarify this.

Keywords: northern Chile, tectonic geomorphology, stream gradient index.

INTRODUCTION

Geomorphological data have been used to examine Quaternary tectonics in numerous studies. They range from the use of regional planation surfaces (e.g., Coltori and Ollier, 1999) to local marine terraces (e.g., Pillans, 1983; Lajoie, 1986; Merritts and Bull, 1989; Ward and Valensise, 1994). Increasingly, fluvial systems have become a central tool in tectonic geomorphology. Mainly this has been through the use of river terraces (e.g., Rockwell et al., 1984; Van Dissen et al., 1992; Molnar and Gipson, 1994) or profiles (Merritts and Vincent, 1989; Merritts and Hesterberg, 1994). Work on river profiles has indicated that in areas dominated by fluvial bedrock incision (a factor typical in regions of high uplift rates), stream power offers a good explanation of bedrock erosion rates (Seidl and Dietrich, 1992; Seidl et al., 1994). Thus the stream gradient index (SGI)

[†]E-mail: amather@plymouth.ac.uk.

Mather, A.E., and Hartley, A.J., 2006, The application of drainage system analysis in constraining spatial patterns of uplift in the Coastal Cordillera of northern Chile, *in* Willett, S.D., Hovius, N., Brandon, M.T., and Fisher, D.M., eds., Tectonics, Climate, and Landscape Evolution: Geological Society of America Special Paper 398, Penrose Conference Series, p. 87–99, doi: 10.1130/2006.2398(06). For permission to copy, contact editing@geosociety.org.

proposed by Hack (1973), which is based on a surrogate for discharge (stream length) and longitudinal profile gradient has proved popular as an indicator of available stream power that can be derived from cartographic analyses. Work by Merritts and Vincent (1989) demonstrated that the SGI was the most useful indicator of geomorphic response of fluvial systems to tectonics, particularly in lower-order streams. Variations in stream power also occur for reasons other than tectonics, such as base-level changes brought about by river capture or lithologic variations (Mather et al., 2002). It is thus essential to carefully select catchment areas to minimize this nontectonic noise if SGI is to be used as a proxy for uplift. If this can be achieved, the SGI then provides not only a valuable tool for examining patterns of variation in uplift, but also the possibility of extrapolating known uplift data to adjacent areas. It is the aim of this paper to use SGI in conjunction with geomorphological observations and known uplift data to explore patterns of uplift along a 200 km stretch of the northern Chilean coast that is known to be tectonically active but has remained relatively unexplored.

STUDY AREA GEOMORPHOLOGY

The study area covers 200 km of coastline from 21°15′ to 23°S in northern Chile (Fig. 1). From west to east, this area can be divided into four physiographic provinces. These are:

1. A coastal plain. This is a narrow and discontinuous plain present at the foot of a coastal scarp and is defined as the area beneath the 200 m contour line, composed of alluvial fan material and some marine terraces. It extends from 23°30′S to 22°15′S and is most extensively developed between the Mejillones Peninsula and the Coastal Cordillera (Fig. 1). Traced northward, it narrows from 15 km wide to zero at 22°15′S. North of this point some isolated areas of coastal plain development occur between bedrock headlands until 21°30′S, where it is fully developed up to Iquique at 20°15′S.
2. A coastal scarp. The western margin of the cordillera comprises a coastal scarp, a prominent break in slope 700–1000 m high (Fig. 1B), which forms a continuous feature that runs for 900 km slightly oblique to the coastline of northern Chile. The origin of the scarp has been the subject of some debate. Brüggen (1950) and Armijo and Thiele (1990) suggested it represents an active fault, whereas Paskoff (1978, 1980) preferred an origin through marine erosion of a Miocene fault scarp(s). In contrast, Mortimer and Saric (1972) and Hartley and Jolley (1995) used the absence of active fault scarps and the irregular morphology of the scarp to suggest that it represents a weathered paleo-cliffline, developed through marine erosion during post–mid-Miocene uplift. The studied catchments are developed into this feature.
3. The Coastal Cordillera. This range forms a prominent topographic feature 700 km long and 20–50 km wide (Fig. 1B). It is bounded to the east by the 1000-km-long Atacama fault zone (Fig. 1A). Where the fault is active, prominent scarps up to 3 m high can be observed. Elsewhere, where the Atacama fault zone is inactive or comprises a series of splays, the eastern margin of the Coastal Cordillera has a gentle topography with an irregular mountain front embayed by the alluvial fill of the Central Depression. In the study area, the cordillera is typically some 20 km across, with elevations up to 2000 m above sea level (masl), averaging 1000 masl.
4. The Central Depression. The western margin of the Central Depression is represented by the Atacama fault zone. It is 25–100 km wide and has an average elevation of 1000 m in the west that rises to 2000 m in the east. Up to 1100 m of Oligocene to Holocene strata are preserved within the basin of the Central Depression (Hartley et al., 2000). Sedimentary rocks in the Central Depression include alluvial fan, fluvial, lacustrine, playa, and nitrate deposits with interbedded volcanic ashes.

CLIMATE

The Coastal Cordillera forms a topographic barrier between the Pacific Ocean and the Central Depression. The strong temperature inversion along the Pacific Coast between 800 and 1000 m prevents moist Pacific air masses from directly entering the Central Depression, although beneath these elevations they may reach 50–100 km inland. The temperature inversion is caused by cold upwelling waters of the Humboldt Current and the stable location of the Southeast Pacific Anticyclone (Rutllant and Ulriksen, 1979; Trewartha, 1981). Precipitation does occur below the temperature inversion, however the Coastal Cordillera effectively prevents these moist air masses penetrating eastward. While the study area is located in one of the driest areas of the globe, there is a marked transition in moisture availability from the coast to the Central Depression. At Antofagasta, on the coast, average rainfall over the period 1978–1998 was 3.9 mm yr^{-1}. In the Central Depression, only 0.6 mm yr^{-1} of rainfall was recorded for 1974–1991 (Ministerio de Obras Publicas). Precipitation events that affect the Coastal Cordillera are infrequent but do have a significant impact on population centers located at the foot of the coastal scarp. In Antofagasta, seven such events occurred between 1916 and 1999 and resulted in significant mud- and debris-flow deposits (Vargas et al., 2000). Thus although the entire area is hyperarid, the frequency of rainstorm events decreases from the coast across the Coastal Cordillera into the Central Depression. The present-day hyperarid climate of western South America is thought to have commenced between 3 and 4 Ma (based upon geomorphological, sedimentary, and radiometric age dates of the cessation of porphyry copper supergene enrichment profiles; Hartley and Chong, 2002; Hartley, 2003). Prior to this, the sedimentary evidence suggests an arid to semi-arid climate prevailed throughout much of the Tertiary.

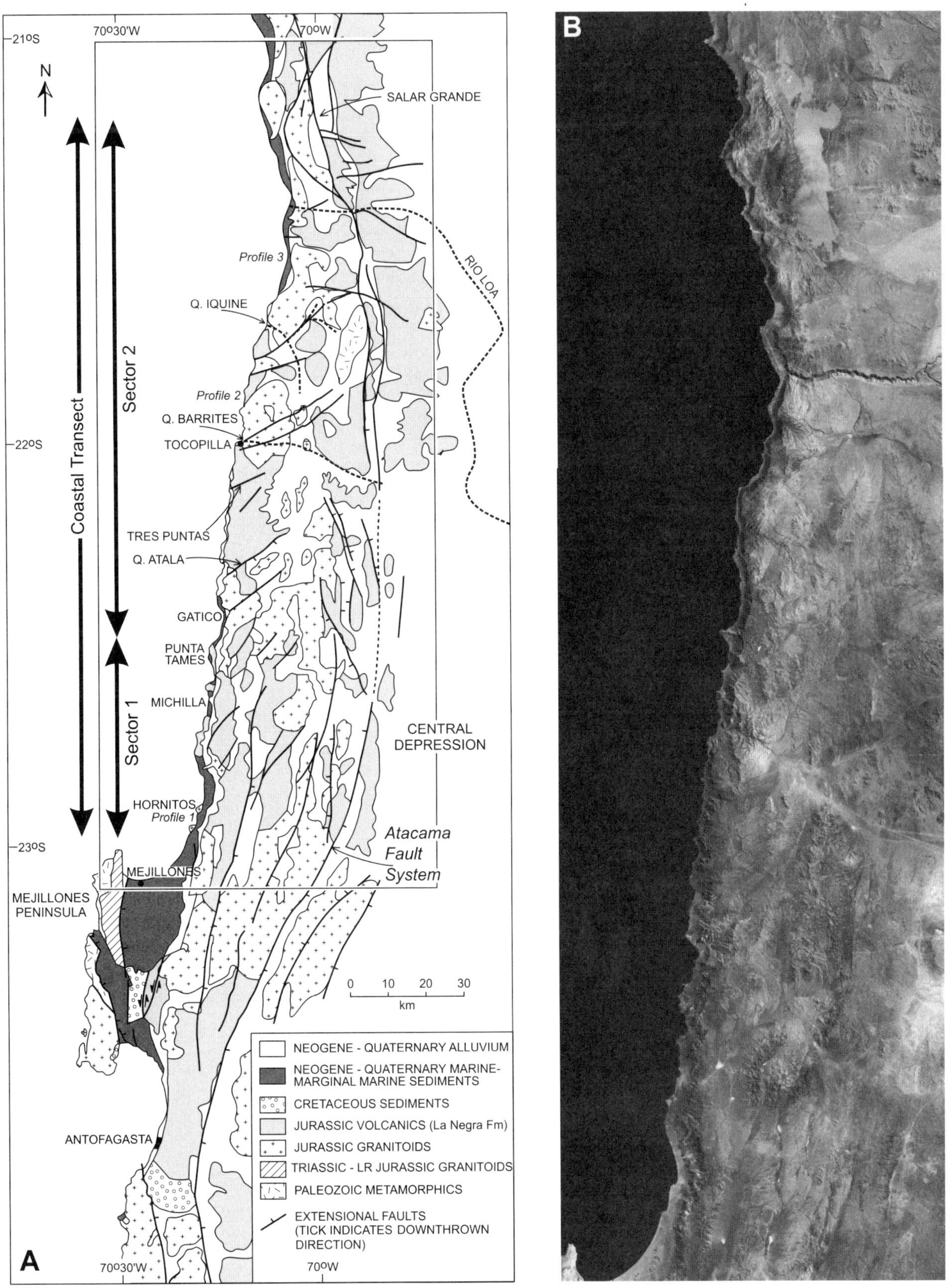

Figure 1. (A) Simplified geology map of the study area in coastal northern Chile with localities mentioned in the text; modified after Boric et al. (1990). (B) Landsat image of boxed area highlighted in A.

BEDROCK GEOLOGY

The catchment geology of the Coastal Cordillera is dominated by Jurassic andesites of the ~10-km-thick La Negra Formation (Garcia, 1967) and associated granitic intrusions (Fig. 1A). In addition, Paleozoic metasediments and Lower Cretaceous conglomerates and limestones are also present (Ferraris and Di Biase, 1978; Boric et al., 1990). These lithologies are cut by a series of predominantly extensional, N-S– to NE-SW–striking faults that dip eastward. These fault trends are picked up in the configuration of the modern coastline (Fig. 1). The eastern boundary of the Coastal Cordillera, the Atacama fault zone, has a long and complex history stretching from the Jurassic to the present day (e.g., Arabasz, 1971; Okada, 1971; Hervé, 1987; Naranjo, 1987; Armijo and Thiele, 1990; Scheuber and Andriessen, 1990; Dewey and Lamb, 1992; Scheuber et al., 1994; Scheuber and González, 1999; González et al., 2003; González and Carrizo, 2003). Since the Oligocene, the Atacama fault zone has accommodated the differential uplift of the Coastal Cordillera relative to the Central Depression. Late Pliocene movement along the fault zone in the study area is considered to have been predominantly dip-slip with a small component of dextral strike-slip (Dewey and Lamb, 1992; González et al., 2003; González and Carrizo, 2003). Uplift and delineation of the cordillera as a separate morphotectonic unit (i.e., distinct from the adjacent Central Depression; Fig. 1) is considered to have taken place in the Oligocene and continues to the present day (Hartley et al., 2000).

TECTONIC SETTING AND QUATERNARY UPLIFT HISTORY

The Coastal Cordillera of northern Chile forms the leading edge of the South American plate and is the only subaerial part of the South American continental crust in direct contact with the subducted Nazca plate. Deformation of the cordillera can therefore be directly related to plate coupling at the subduction zone (e.g., Hartley et al., 2000; Lamb and Davis, 2003; Von Huene and Ranero, 2003). The general seismicity and nature of the Central Andean subduction zone is well known (e.g., Jordan et al., 1983; Cahill and Isacks, 1992; ANCORP Working Group, 1999; Von Huene et al., 1999; Von Huene and Ranero, 2003). The subducted plate beneath northern Chile dips 30° to the east beneath the South American plate, and the Peru-Chile trench is located 70–150 km offshore. The Nazca–South America convergent plate boundary is responsible for some of the largest earthquakes in recorded history (Comte and Pardo, 1991; Tichelaar and Ruff, 1991). In the study area in 1995, a Mw 8.1 interplate earthquake occurred at a depth of 10–50 km along the subduction zone interface (Delouis et al., 1997). Most of the deformation generated by the event occurred in the zone north of Antofagasta to Mejillones (Fig. 1), ranging from 0 at Antofagasta to a maximum of 80 cm of uplift on the southwestern tip of the Mejillones Peninsula (Ortlieb et al., 1996a). This emphasizes the differential nature of individual seismic events in the region, their magnitude, and the direct relationship between deformation at the subduction zone, uplift of the Coastal Cordillera, and drainage response.

Subduction is further complicated by the presence of a series of aseismic ridges on the Nazca plate (Flint et al., 1991; Hartley and Jolley, 1995). These may be responsible for differential uplift of the coastal region as recorded in the Mejillones Peninsula, just to the south of the current study area (Hartley and Jolley, 1995), and also in southern Peru, where the Nazca Ridge has had a marked influence on the topographic development of the adjacent coastline (Macharé and Ortlieb, 1992).

Quaternary records for tectonic uplift in the region are typically less spectacular than individual event figures, because they are time-averaged over 10^5 to 10^6 yr time scales. The classic site within the study area for marine terrace work is in the vicinity of Hornitos, the southernmost part of the transect. Here terraces have been dated using U-series and amino stratigraphic studies of mollusk shells within the coastal terraces (Ortlieb et al., 1996b). This research has revealed remnants of sea-level maxima from oxygen isotope stages (OIS) 5c, 5e, 7, and 9, which give an average uplift rate over the last 330 k.y. of 240 mm k.y.$^{-1}$. Data from Ortlieb et al. (1996b) suggests that the "interval" uplift rates were 250 mm k.y.$^{-1}$ over the last 100–125 k.y.; 240 mm k.y.$^{-1}$ for the interval OIS 5e to 9, and 146 mm k.y.$^{-1}$ for the interval OIS 9–25. These data are the highest-resolution uplift data currently available within the study area. Possible OIS 5c/5e surfaces have been located at elevations of 40 m at Michilla (Fig. 1A) and 15 m at Cobija, just south of Gatico (Fig. 1A; Ortlieb, 1995). This same marine terrace level occurs at 25 m at Hornitos. There is also an unusually high (6–7 m) Holocene terrace, which has been dated using various techniques at ca. 7000 B.P. at Michilla Bay (Ortlieb, 1995). Wells (1988) constructed a Holocene relative sea-level curve in north-central Peru (10°S). Data from this study suggests the maximum sea level at this time was less than 1.5 m above present datum. These data suggest at least 5 m of uplift over the past 7000 yr (around 700 mm k.y.$^{-1}$ uplift) for the Michilla area (Fig. 1A). There is thus the suggestion of uplift rates varying substantially along the coast, yet poor preservation of marine units and difficulties in correlation mean that other surrogates must be used to identify the spatial trends in more detail.

DRAINAGE TYPES

This paper will use the drainages of the Coastal Cordillera to attempt to semiquantify changes in patterns of uplift along the coast. It is thus pertinent to examine the range of drainages present in order to identify those most suitable for analyses in this paper.

The drainage types within the study area can be split into (1) external and (2) coastal. All types are characterized by a steep convex-upward profile (Fig. 2).

External Drainages

External drainages breach the main drainage divide of the Coastal Cordillera and are linked to drainage systems within the

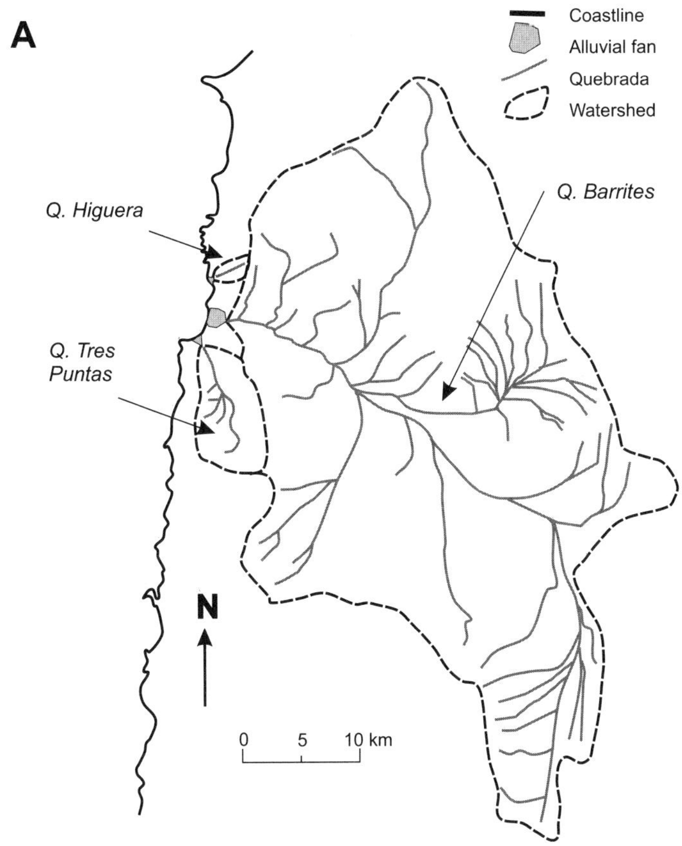

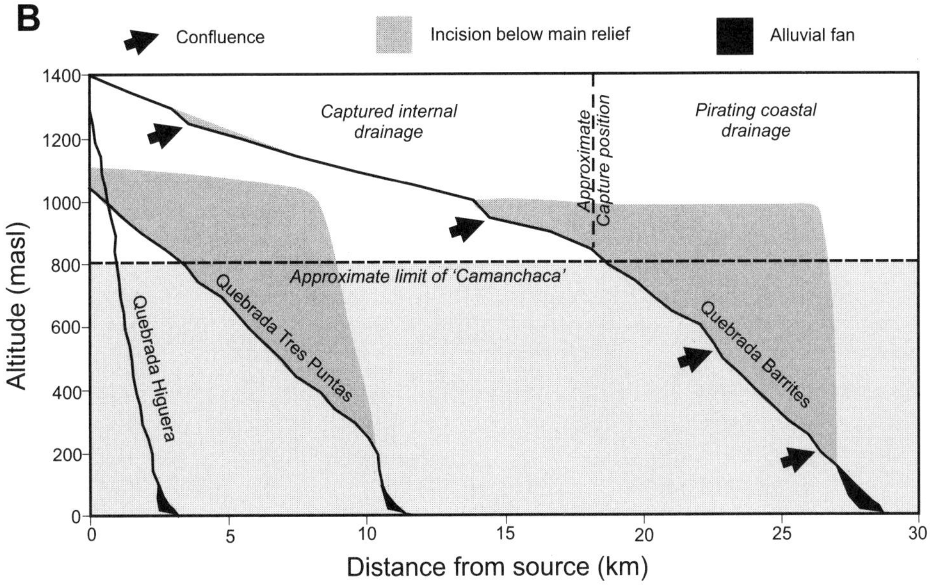

Figure 2. (A) Map showing the relative scales of the different drainages types from Tocopilla (Fig. 1A). The largest catchments are external (Quebrada [Q.] Barrites). The smaller coastal drainages are represented by Q. Tres Puntas and Q. Higuera. (B) Topographic profiles of the range of drainage types shown in A. The line at 800 m shows the approximate limit of the most frequent sea mists (Camanchaca). The gray shaded areas indicate the topographic relief into which the drainages are incised. Image is drawn from 1:50,000 Instituto Geografico Militar de Chile topographic maps; masl—m above sea level.

Central Depression. These systems are thus the largest of those examined and may pass through several climate zones. For example, the Rio Loa (Fig. 1), which is the only perennially flowing river in the area, is some 440 km long and drains from the high Andes, across the Precordillera, and across the Central Depression before crossing the Coastal Cordillera and entering the Pacific Ocean. It experiences winter snowmelt in its upper reaches and hyperaridity in the Central Depression. Smaller external drainage systems, such as the Quebrada Barrites at Tocopilla (Fig. 2), have shorter courses that drain the hyperarid Central Depression and the marginally more humid coastal ranges before entering the Pacific Ocean. To minimize any noise within the data set generated by controls external to the Coastal Cordillera, such as local climate variations, these drainages will not be used in the data collection and analyses for this paper.

Coastal Drainages

Coastal drainages are restricted to the western side of the Coastal Cordillera. The differing spatial scales of these are reflected in Figure 2. They tend to be steeper than the external drainages (Fig. 2) and are all located within the same climate zone (hyperarid, but subjected to the moisture bearing Camanchaca and rare rainfall events—the last one in 1991). These drainages will be used in the data collection and analyses for this paper, because they are developed on similar strength geologies (see next section) and under the same local climate, minimizing any related noise from these sources in the data set.

APPROACH

Marine terraces have only yielded reliable information on uplift rates around Mejillones and Hornitos (Fig. 1A) in the region of Antofagasta. This paper will attempt to use a surrogate for uplift to extrapolate this knowledge of temporal uplift rates and patterns from the best-documented site (Hornitos) along a 200 km stretch of the coastline north of this region (Fig. 1A). The surrogate for uplift that will be utilized is the stream gradient index (SGI) of Hack (1973). The SGI was tested along with other morphologic variables by Merritts and Vincent (1989) in northern coastal California and found to be a reliable indicator of relative uplift rates. The Chilean study area can be considered (for the sake of this paper) to have a relatively uniform geology of granodiorites and basaltic andesites. Within actively eroding areas, such as the bedrock channels, such geologies are difficult to break with a geological hammer, falling into the engineering category "very to extremely strong," with uniaxial compressive strengths in excess of 100 MPa (Mining Life, 2005). Due to the observed field similarities in lithological strength, the area will be treated as one geological unit. To maintain uniformity of climate conditions, only coastal drainages will be used for SGI calculations. All 80 selected catchments feed a system of alluvial fans along the coastal plain and are perpendicular to the coastal scarp. Of these fans, 97% are not fully trenched, so the modern drainages are not linked to modern sea level, but tend to toe out onto the OIS 5c/5e marine terrace.

The approach utilizes the simple equation:

$$\mathrm{SGI} = \Delta h/\Delta l \times L, \tag{1}$$

where $\Delta h/\Delta l$ is the gradient of the reach (in this case an individual reach is taken between each 250 m contour interval) and L is the length of the stream (surrogate for discharge) from the watershed to the middle of a studied reach. All units are in gradient meters. All data were derived manually from 1:50,000 Instituto Geografico Militar de Chile topographic maps. The SGI was calculated for each 250 m fall in channel height, with reach 1 taken as the lowest reach (base of coastal scarp) in all cases and reaches 2 through to 5 occurring in a progressively upstream direction. In addition, information was gathered on stream channel length to give an idea of where the highest SGI values occur in relation to the coast.

RESULTS

The results of the SGI analyses are presented in Figures 3 and 4. Figure 3 presents the data as all SGI values averaged per 5 km of coastline. For simplicity, the data can be separated into two

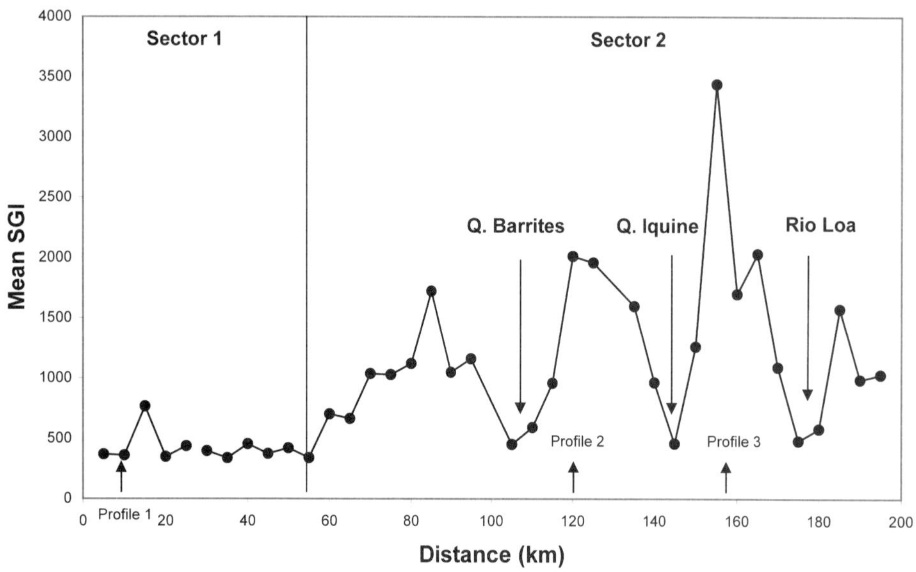

Figure 3. Plot of stream gradient indices (SGI) averaged per 5 km of coastline. The distance is measured from 23°S (0 km on transect) to 21°15′S (200 km on transect). Positions of major drainages are indicated and can be located on Figure 1A. In total, 207 data points were used from 80 coastal drainages of first to third order. Using the SGI values and stream length characteristics, the transect was split into two sectors. See Tables 1 and 2 and text for details. Small arrows indicate the position of topographic profiles in Figure 5.

sectors based on drainage length, SGI profiles, and mean SGI values (Figs. 3 and 4; Table 1) as described below.

Sector 1

Sector 1 extends from 0 to 55 km along the coastal transect. The average stream length is 2.28 km (n = 23). The SGI values calculated for individual reaches throughout the sector have a mean value of 430 and show less variability than sector 2 (Tables 1 and 2). The highest SGI values are commonly found in reach 1, at the base of the coastal scarp (Table 2 and Fig. 4). Aerial photographs and topographic profiles indicate that this is an area of lesser relief within the Coastal Cordillera (Figs. 5A and 6). Arrows on Figure 3 indicate the positioning of the cross sections shown in Figure 5 along the coastal transect.

Sector 2

Sector 2 extends from 55 to 200 km along the coastal transect. It is characterized by longer streams (mean 4.74 km, n = 57), with highest SGI values typically in reaches 1 and 2 (Table 2 and Fig. 4), toward the base of the coastal scarp. Mean SGI per reach is 1195, an order of magnitude higher than sector 1, and values display a greater variability (Tables 1 and 2). Aerial photographs and profiles (Figs. 5B and 7) indicate the prevalence of significant planation surfaces. These are dissected by the drainage systems in this study. Some of the drainages show evidence of rejuvenation and river capture (1 in Fig. 8), and drain across a higher surface (2 on Fig. 7) before dropping off the pronounced coastal scarp onto the coastal plain (3 on Fig. 7). These surfaces may be separated vertically by as much as 200 m (Fig. 5B).

TABLE 1. COMPARISON OF DESCRIPTIVE STATISTICS FOR ALL THE SGI REACH VALUES FOR SECTORS 1 AND 2

	Sector 1	Sector 2
n	47	160
Mean	430	1195
Standard deviation	193	1112
Minimum	125	142
Maximum	1008	8355

DISCUSSION

The SGI and topographic profiles reveal a number of interesting features, which suggest that factors controlling drainage system development in the Coastal Cordillera are operative at different scales. At the largest scale, regional uplift of the Coastal Cordillera is driven by intraplate shear stress generated between the Nazca and South American plates by ridge push and deep mantle flow (Russo and Silver, 1996). Plate convergence vectors suggest that this contractional regime has prevailed from the late Oligocene to the present day (Somoza, 1998; Hartley et al., 2000).

At an intermediate scale of 75–100 km, the SGI and topographic profiles suggest variations in uplift. The region where uplift data are available (sector 1) appears to show the lowest SGI values. Sector 2 contains the highest values. The combination of high SGI values and evidence in aerial photographs of erosionally rejuvenated catchments cutting into and capturing older surface drainage on the coastal scarp (Fig. 8) would appear to indicate high rates of tectonically stimulated geomorphic erosion. This is further supported by the absence or limited development of the coastal plain in this area as a function of net coastal erosion (Fig. 1B). Within this sector, structural control on many

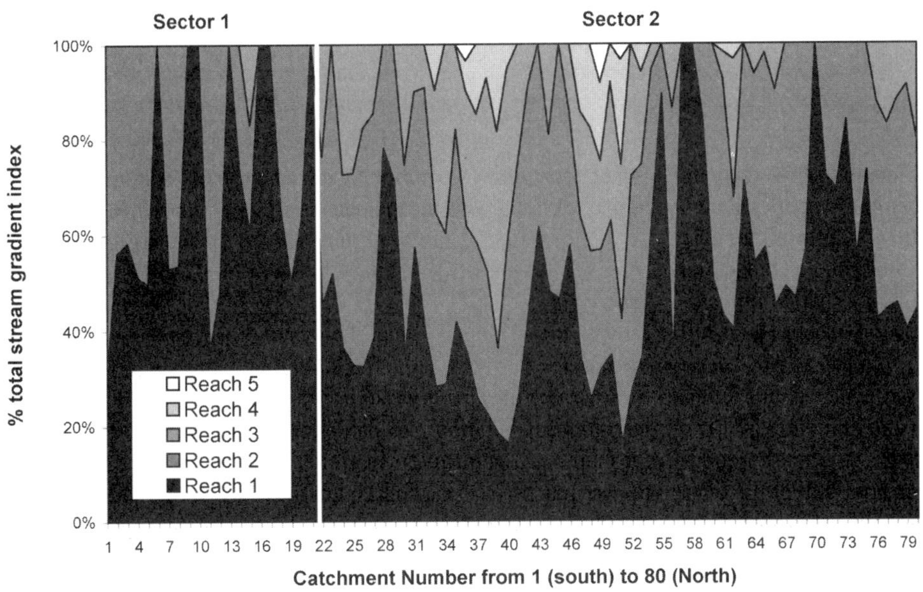

Figure 4. Percentage SGI plot of each of the 80 catchments studied. The percentage scale indicates the relative size of the SGI per drainage reach by catchment, where reach 1 is the base of the coastal scarp in all cases. Numbers on the horizontal axis indicate the catchment sample number with catchment 1 at zero km and catchment 80 at 200 km on Figure 3.

TABLE 2. SECTOR 1 AND SECTOR 2 DESCRIPTIVE STATISTICS FOR THE STREAM GRADIENT INDEX VALUES BY REACH

Reach	*n*	Mean	Standard deviation	Minimum	Maximum
Sector 1					
1	25	456	203	125	1008
2	18	429	187	150	894
3	4	270	14	250	281
4	-	-	-	-	-
5	-	-	-	-	-
Sector 2					
1	55	1465	1110	354	7200
2	52	1357	1278	176	8355
3	36	882	865	142	3250
4	14	528	496	261	2218
5	3	302	32	266	328

Note: Reach 1 is at the coast in all cases, with reaches 2–5 being located progressively inland.

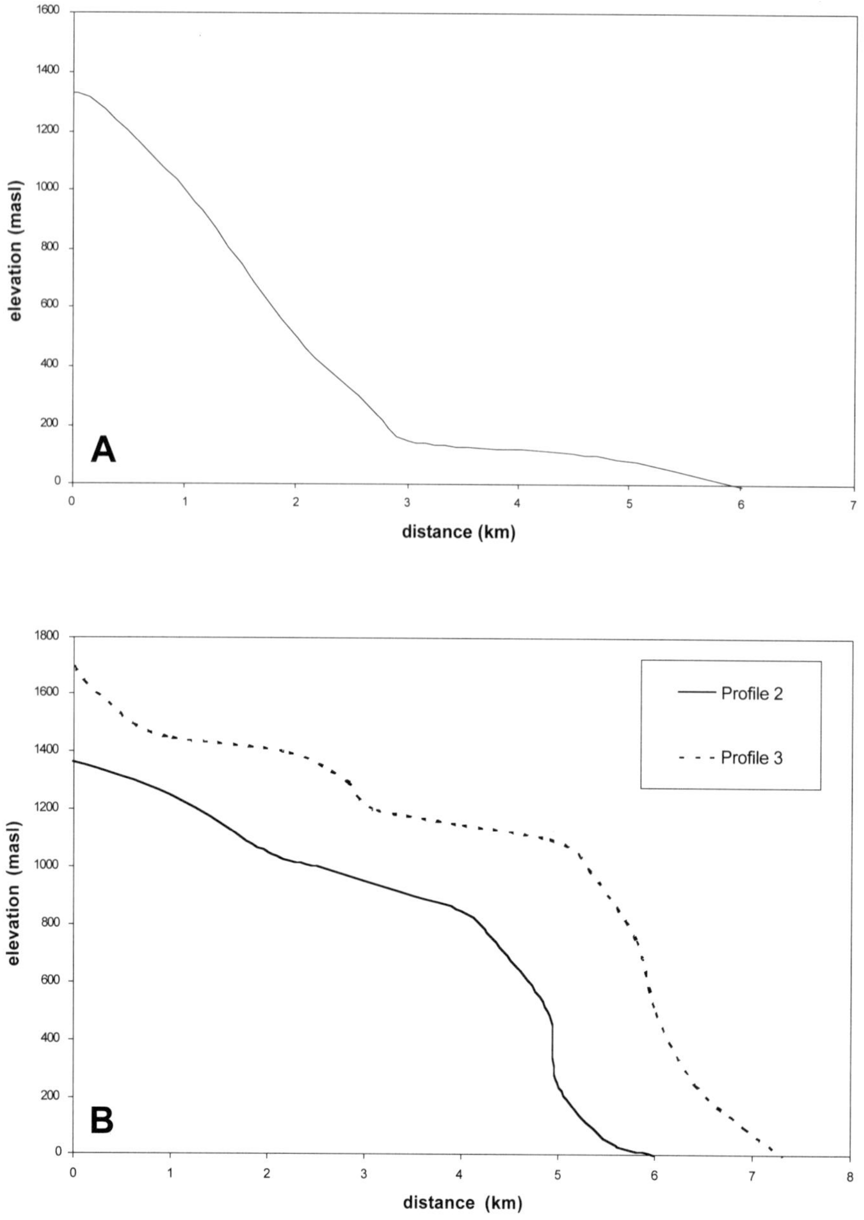

Figure 5. Topographic profiles of each of the sectors taken from 1:50,000 Instituto Geografico Militar de Chile topographic maps. Profile locations are indicated on Figures 1A and 3. (A) Sector 1, profile 1 from Hornitos, also shown as an aerial photograph in Figure 6. (B) Sector 2, profiles 2 (120 km on Fig. 3) and 3 (155 km on Fig. 3, and aerial photograph in Fig. 7). For explanation see text.

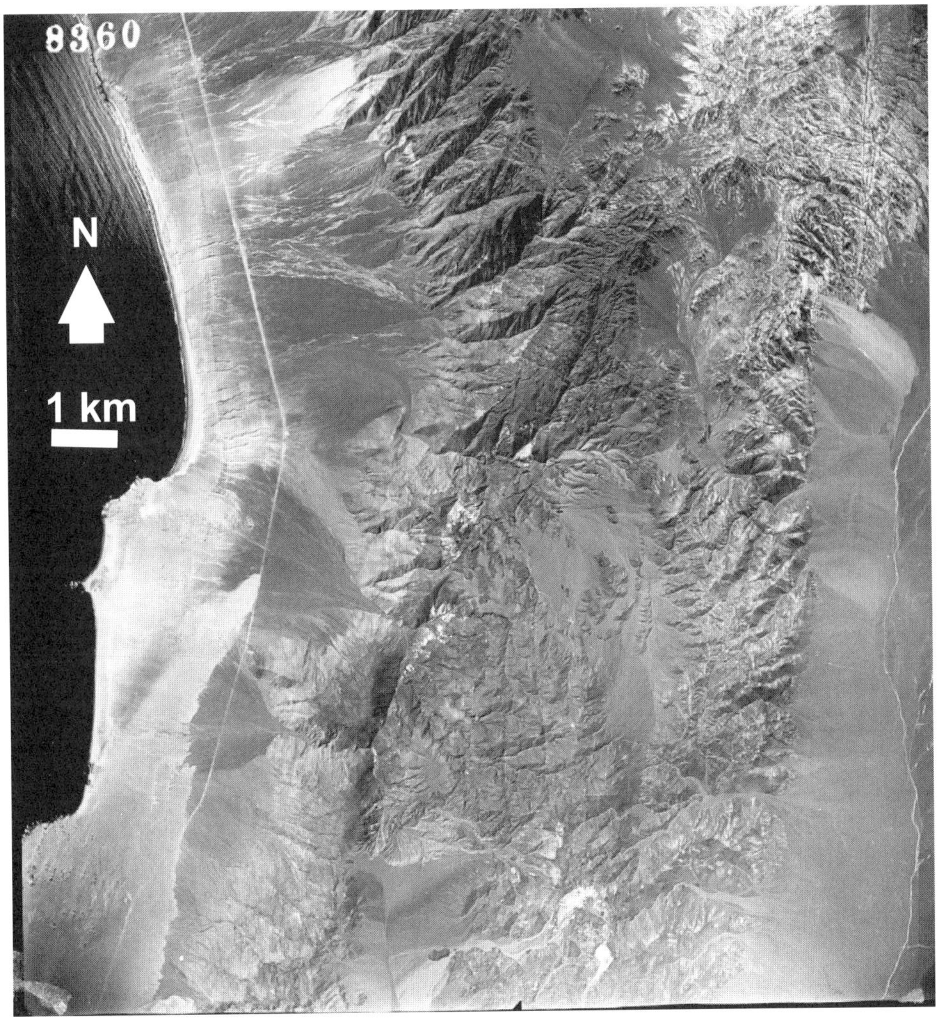

Figure 6. Aerial photograph of the Hornitos area (Figs. 1A and 5A), showing relatively low relief compared with adjacent sectors. Note the wider coastal plain here (3–4 km) and degraded coastal scarp (center of image running N-S).

of the drainage alignments is very clear (Fig. 8). The SGI data and geomorphic evidence indicate that sector 2 has been the most tectonically active part of the coastal transect.

A mechanism that could explain the observed differential uplift along the Coastal Cordillera may be aseismic ridge subduction (Flint et al., 1991; Hartley and Jolley, 1995; Fig. 9). The planation surfaces evident in sector 2 (Figs. 5B and 7) are likely to be marine in origin and of probable Pliocene age, based on known ages of high surfaces at Mejillones (Hartley and Jolley, 1995). The surfaces are located at elevations over 800 m (Fig. 5B) and appear to suggest vertical height offsets of as much as 200 m over some of the low SGI anomalies identified above (Quebrada Iquine on Figs. 1 and 3). Differential uplift of the forearc associated with subduction of irregular seafloor topography is well known, for example that associated with seamounts and the Cocos Ridge off Costa Rica (Wells et al., 1988; Gardner et al., 1992; Fisher et al., 1998), the Nazca Ridge off SW Peru (Macharé and Ortlieb, 1992) and the Mendocino triple junction (Merritts and Vincent, 1989). All of these studies highlighted the coast parallel variation in uplift rate associated with ridge/seamount subduction. Subduction of the so-called Iquique (or Perdido) Ridge offshore northern Chile (Fig. 9) may account for the differential uplift observed in the study area, however, further work is required before the exact causal mechanism can be identified.

At a smaller scale (~30 km), the SGI values show that three of the largest drainages that transect the coastal range occupy areas along the coastal scarp where drainages show the lowest SGI values (Fig. 3). These E-W to NE-SW drainages have the same trend as a series of predominantly southerly dipping reverse faults recently described by Allmendinger et al. (2005) from the Coastal Cordillera between 19° and 21.6°S. These authors noted that these faults significantly affected the topography of the Coastal Cordillera, resulting in the formation of a series of tilted fault blocks that strongly focused drainage. It is likely that the structures identified in the north of the study area by Allmendinger et al. (2005) extend further south to 22°30′S, particularly as the asymmetric nature of the troughs around the SGI anomalies present in Figure 3, with gentler southern sides and steeper northern limbs, could indicate the

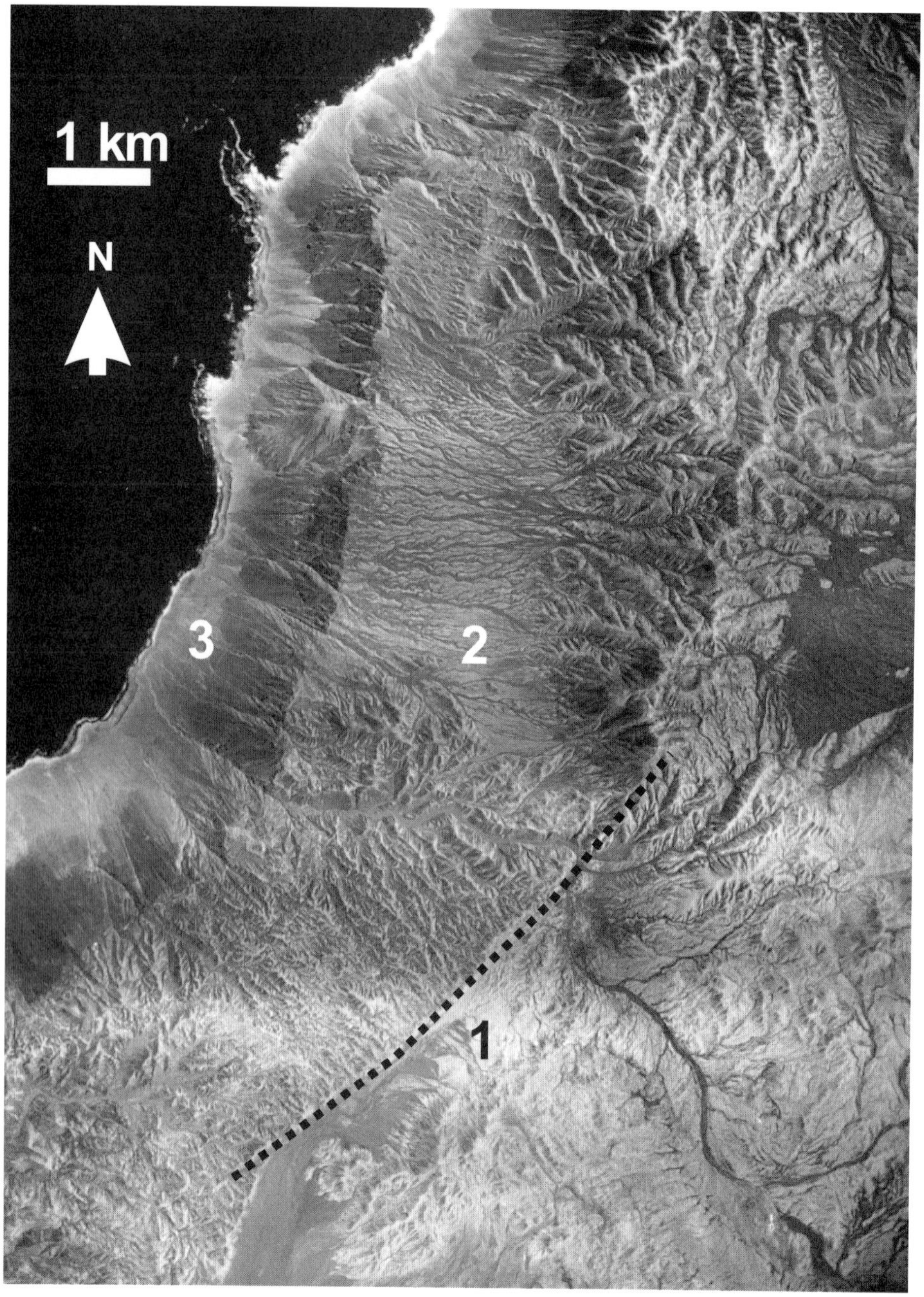

Figure 7. Aerial Photograph of surfaces in sector 2 (Figs. 1A and 5B, profile 3). Main features observed are old ?marine surface (2) and coastal scarp (3). Note how fault (1 and dashed line) parallels coastline and narrow coastal plain (3).

presence of tilted fault blocks. The study by Fisher et al. (1998) of the forearc of Costa Rica noted the development of coast-orthogonal drainage systems generated by differential uplift across fault bounded blocks similar to those seen in the Coastal Cordillera of northern Chile.

It is interesting to postulate about uplift rates in sector 2. Sector 1 uplift rates are of the order of 240–384 mm k.y.$^{-1}$ based on dated 100–125 ka marine terraces (discussed above). Holocene records suggest figures may have been as high as 700 mm k.y.$^{-1}$ for the Holocene locally. SGI values in sector 1, however, are 2–3 times smaller than those in sector 2. This may indicate that more realistic uplift rates for sector 2 may be an order of magnitude higher. Is this realistic? If we compare the data from this study with further up the Pacific coast in northern California, Merritts and Vincent (1989) obtained SGI values of 66–393 in areas of low uplift (400 mm k.y.$^{-1}$); 89–386 in areas of intermediate uplift (1200 mm k.y.$^{-1}$), and values of 114–838 in areas of high uplift (3000–4000 mm k.y.$^{-1}$). These rates were based on late Pleistocene (younger than 72 ka) marine terraces. The overall SGI values for California are much lower than Chile, which will undoubtedly reflect the difference in geology (the Californian study area being

Figure 8. Aerial photograph showing structural control on the drainage alignment of Quebrada Atala (Figs. 1A and ~72 km on Fig. 3 transect). Number 1 indicates where a rejuvenated coastal scarp drainage is capturing an older surface drainage.

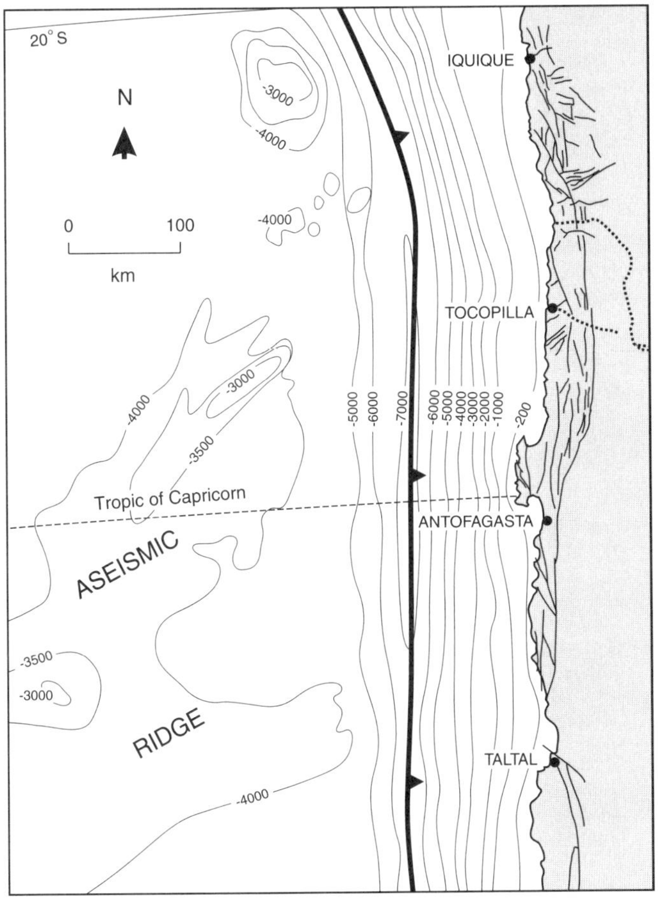

Figure 9. Location of the aseismic ridge currently being subducted beneath the Andean forearc between Iquique and Taltal, together with the outcrop trace of the Atacama fault zone. Note that the landwardly convex trace of the Atacama fault corresponds to the position of the aseismic ridge, possibly suggesting a causal relationship between the two (based on Hartley and Jolley, 1995; Servicio Nacional de Geologia y Minera, Chile, 1982).

predominantly arkosic sandstones, siltstone and shale, and thus of much lower erosional resistance than the andesites and granodiorites of the Chilean study area). In addition, the climate in northern California is much wetter than the Chilean example, with >1000 mm yr^{-1} of precipitation (Merritts and Vincent, 1989). The combination of weaker lithologies and more frequent high-magnitude flood events should make the Californian study area much more adjustable to tectonic change. Thus high SGI values from the Chilean example would be expected, while postulated uplift rates that could range from 480 to 1152 mm $k.y.^{-1}$ over the Late Pleistocene are not unreasonable. A comparison of profile types also indicates that the Chilean examples (typically convex profiles with highest SGI values in the lowermost reaches) most closely approximate the profiles from the high uplift rate areas of Merritts and Vincent (1989).

CONCLUSIONS

SGI values and geomorphic observation have enabled us to examine the spatial patterns of uplift in a 200 km transect of the northern Chilean coast. The results indicate that the area of least-preserved marine terraces corresponds to the most actively uplifting coastal sector in the region studied, and that inferred late Pleistocene to recent uplift rates might have exceeded 1000 mm $k.y.^{-1}$, a rate 2–3 times greater than those currently known for the region. Differential uplift along the Coastal Cordillera is tentatively attributed to aseismic ridge subduction. At a smaller scale, the pattern of SGI values suggests that the major drainages are exploiting structural lows controlled by NE-SW– to E-W–trending reverse faults, although further work is required to clarify this.

ACKNOWLEDGMENTS

Mather would like to thank the Royal Society for providing a travel grant to support this fieldwork. Hartley would like to acknowledge the British Council for support. Both authors would like to thank Guillermo Chong and colleagues at the Universidad Catolica del Norte at Antofagasta for scientific and logistical support, and T. Gardner and an anonymous referee for constructive comments on earlier versions of the manuscript.

REFERENCES CITED

Allmendinger, R.W., González, G., Yu, J., Hoke, G., and Isacks, B.L., 2005, Trench-parallel shortening in the northern Chilean forearc: Tectonic and climatic implications: Geological Society of America Bulletin, v. 117, p. 89–104, doi: 10.1130/B25505.1.

ANCORP Working Group, 1999, Seismic reflection image revealing offset of Andean subduction-zone earthquake locations into oceanic mantle: Nature, v. 397, p. 341–344, doi: 10.1038/16909.

Arabasz, W.J., 1971, Geological and geophysical studies of the Atacama fault zone, northern Chile [Ph.D. thesis]: Pasadena, California Institute of Technology, 275 p.

Armijo, R., and Thiele, R., 1990, Active faulting in northern Chile: Ramp stacking and lateral decoupling along a subduction plate boundary?: Earth and Planetary Science Letters, v. 98, p. 40–61, doi: 10.1016/0012-821X(90)90087-E.

Boric, R., Diaz, F., and Maksaev, V., 1990, Geología y Yacimientos Metaliferous de la Region de Antofagasta: Servicio Nacional de Geología y Minera Chile, Bolletin, v. 40, p. 246 p.

Brüggen, J., 1950, Fundamentos de la Geología de Chile: Santiago, Instituto Geografico Militar, 378 p.

Cahill, T.A., and Isacks, B.L., 1992, Seismicity and shape of the subducted Nazca plate: Journal of Geophysical Research, v. 97, no. 12, p. 17,503–17,529.

Coltori, M., and Ollier, C.D., 1999, The significance of high planation surfaces in the Andes of Ecuador, *in* Smith, B.J., Whalley, W.B., and Warke, P.A., eds., Uplift, erosion and stability: Perspectives on long-term landscape development: Geological Society [London] Special Publication 162, p. 239–253.

Comte, D., and Pardo, M., 1991, Reappraisal of great historical earthquakes in the northern Chile and southern Peru seismic gaps: Natural Hazards, v. 4, p. 23–44, doi: 10.1007/BF00126557.

Delouis, B., Monfret, T., Dorbath, L., Pardo, M., Rivera, L., Comte, D., Haessler, H., Caminade, J.P., Ponce, L., Kausal, E., and Cisteras, A., 1997, The Mw = 8.0 Antofagasta (northern Chile) earthquake of 30 July 1995: A precursor to the end of the 1877 gap: Bulletin of the Seismological Society of America, v. 87, p. 417–445.

Dewey, J.F., and Lamb, S.H., 1992, Active tectonics of the Andes: Tectonophysics, v. 205, p. 79–96, doi: 10.1016/0040-1951(92)90419-7.

Ferraris, F., and Di Biase, F., 1978, Carta Geologica de Chile, Hoja Antofagasta: Carta, Instituto de Investigaciones Geologicas de Chile, v. 30, 48 p., scale 1:250,000.

Fisher, D.M., Gardner, T.W., Marshall, J.S., Sak, P.B., and Protti, M., 1998, Effect of subducting sea-floor roughness on fore-arc kinematics, Pacific coast, Costa Rica: Geology, v. 26, p. 467–470, doi: 10.1130/0091-7613(1998)026 <0467:EOSSFR>2.3.CO;2.

Flint, S., Turner, P., and Jolley, E., 1991, Depositional architecture of Quaternary fan delta deposits of the Andean fore-arc: Relative sea-level changes as a response to aseismic ridge subduction: Special Publication of the International Association of Sedimentologists, v. 12, p. 91–103.

Garcia, F., 1967, Geología del Norte Grande de Chile: Simposium sobre el Geosinclinal Andino: Sociedad Geológica de Chile, v. 3, 138 p.

Gardner, T.W., Verdonck, D., Pinter, N., Slingerland, R., Furlong, K., Bullard, T.F., and Wells, S.G., 1992, Quaternary uplift astride the aseismic Cocos Ridge, Pacific coast of Costa Rica: Geological Society of America Bulletin, v. 104, p. 219–232, doi: 10.1130/0016-7606(1992)104<0219:QUATAC>2.3.CO;2.

González, G., and Carrizo, D., 2003, Segmentación, cinemática y cronologa relativa de la deformacióntardia de la Falla Salar del Carmen, Sistema de Falas de Atacama (23°40′S), norte de Chile: Revista Geológica de Chile, v. 30, p. 223–244.

González, G., Cembrano, J., Carrizo, D., Macci, A., and Schneider, H., 2003, The link between forearc tectonics and Pliocene-Quaternary deformation of the Coastal Cordillera, northern Chile: Journal of South American Earth Sciences, v. 16, p. 321–342, doi: 10.1016/S0895-9811(03)00100-7.

Hack, J.T., 1973, Stream-profile analysis and stream gradient indices: U.S. Geological Survey Journal of Research, v. 1, p. 421–429.

Hartley, A.J., 2003, Andean uplift and climate change: Journal of the Geological Society [London], v. 160, p. 7–10.

Hartley, A.J., and Chong, G., 2002, Late Pliocene age for the Atacama Desert: Implications for the desertification of western South America: Geology, v. 30, p. 43–46, doi: 10.1130/0091-7613(2002)030<0043:LPAFTA>2.0.CO;2.

Hartley, A.J., and Jolley, E.J., 1995, Tectonic implications of late Cenozoic sedimentation from the Coastal Cordillera of northern Chile (22–24°S): Journal of the Geological Society of London, v. 152, p. 51–63.

Hartley, A.J., Chong, G., Turner, P., May, G., Kape, S.J., and Jolley, E.J., 2000, Development of a continental forearc: A Neogene example from the central Andes, northern Chile: Geology, v. 28, p. 331–334, doi: 10.1130/0091-7613(2000)028<331:DOACFA>2.3.CO;2.

Hervé, M., 1987, Movimiento sinistral en el Cretácico Inferior de la Zona de Falla Atacama, al norte de Paposo (24°S), Chile: Revista Geológica de Chile, v. 31, p. 37–42.

Jordan, T.E., Isacks, B.L., Allmendinger, R.W., Brewer, J.A., Ramos, V.A., and Ando, C.J., 1983, Andean tectonics related to geometry of subducted Nazca plate: Geological Society of America Bulletin, v. 94, p. 341–361, doi: 10.1130/0016-7606(1983)94<341:ATRTGO>2.0.CO;2.

Lajoie, K.R., 1986, Coastal tectonics, active tectonics: Washington D.C., National Academy Press, p. 95–124.

Lamb, S., and Davis, P., 2003, Cenozoic climate change as a possible cause for the rise of the Andes: Nature, v. 425, no. 6960, p. 792–797, doi: 10.1038/nature02049.

Macharé, J., and Ortlieb, L., 1992, Plio-Quaternary vertical motions and the subduction of the Nazca Ridge, central coast of Peru: Tectonophysics, v. 205, p. 97–108, doi: 10.1016/0040-1951(92)90420-B.

Mather, A.E., Stokes, M., and Griffiths, J.S., 2002, Quaternary landscape evolution: A framework for understanding contemporary erosion, SE Spain: Land Degradation and Management, v. 13, p. 1–21.

Merritts, D., and Bull, W.B., 1989, Interpreting Quaternary uplift rates at the Mendocino triple junction, northern California, from uplifted marine terraces: Geology, v. 17, p. 1020–1024, doi: 10.1130/0091-7613(1989)017<1020:IQURAT>2.3.CO;2.

Merritts, D., and Hesterberg, T., 1994, Stream networks and long-term surface uplift in the New Madrid seismic zone: Science, v. 265, p. 1081–1084.

Merritts, D., and Vincent, K.R., 1989, Geomorphic response of coastal stream to low, intermediate and high rates of uplift, Mendocino triple junction region, northern California: Geological Society of America Bulletin, v. 101, p. 1373–1388, doi: 10.1130/0016-7606(1989)101<1373:GROCST>2.3.CO;2.

Mining Life, 2005, Miner's Toolbox, Rock Mechanics, Unconfined Compressive Strength Field Index Test: http://www.mininglife.com/Miner/rockmech/UCS_Field_Index.htm (August 2005).

Molnar, P., and Gipson, J.M., 1994, Very long baseline intferometry and active rotations of crustal blocks in the western Transverse Ranges, California: Geological Society of America Bulletin, v. 106, p. 594–606, doi: 10.1130/0016-7606(1994)106<0594:VLBIAA>2.3.CO;2.

Mortimer, C., and Saric, N., 1972, Landform evolution in the coastal region of Tarapaca Province: Revue de Geomorphologie Dynamique, v. 21, p. 162–170.

Naranjo, J.A., 1987, Interpretación de la actividad Cenozoica superior a lo largo de la zona de Falla Atacama, norte de Chile: Revista Geológica de Chile, v. 31, p. 43–55.

Okada, A., 1971, On the neotectonics of the Atacama fault zone region—Preliminary notes on late Cenozoic faulting and geomorphic development of the Coast Range of northern Chile: Bulletin of the Department of Geography, University of Tokyo, v. 3, p. 47–65.

Ortlieb, L., 1995 Late Quaternary coastal changes in northern Chile: International Geological Correlation program (IGCP) Annual Meeting of IGCP project 367 guidebook: ORSTOM, Antofagasta, 175 p.

Ortlieb, L., Barrientos, S., and Guzman, N., 1996a, Coseismic coastal uplift and coralline algae record in northern Chile: The 1995 Antofagasta earthquake case: Quaternary Science Reviews, v. 15, p. 949–960, doi: 10.1016/S0277-3791(96)00056-X.

Ortlieb, L., Zazo, C., Goy, J.L., Hillaire-Marcel, C., Ghaleb, B., and Cournoyer, B., 1996b, Coastal deformation and sea-level changes in the northern Chile subduction area (23 degrees S) during the last 330 ky: Quaternary Science Reviews, v. 15, p. 819–831, doi: 10.1016/S0277-3791(96)00066-2.

Paskoff, R., 1978, Sur l'évolution geomorphologique du grand escarpement côtier du désert Chilien: Geographique Physical Quaterniere, v. 32, p. 351–360.

Paskoff, R., 1980, Late Cenozoic crustal movements and sea-level variations in the coastal area of northern Chile, *in* Mörner, N.A., ed., Earth rheology, isostasy and eustasy: New York, Wiley and Sons, p. 487–495.

Pillans, B., 1983, Upper Quaternary terrace chronology and deformation, South Taranaki, New Zealand: Geology, v. 11, p. 292–297, doi: 10.1130/0091-7613(1983)11<292:UQMTCA>2.0.CO;2.

Rockwell, T.K., Keller, E.A., Clark, M.N., and Johnson, D.L., 1984, Chronology and rates of faulting of Ventura River terraces, California: Geological Society of America Bulletin, v. 95, p. 1466–1474, doi: 10.1130/0016-7606(1984)95<1466:CAROFO>2.0.CO;2.

Russo, R.M., and Silver, P.G., 1996, Cordillera formation, mantle dynamics, and the Wilson cycle: Geology, v. 24, p. 511–514, doi: 10.1130/0091-7613(1996)024<0511:CFMDAT>2.3.CO;2.

Rutllant, J., and Ulriksen, P., 1979, Boundary layer dynamics of the extremely arid northern Chile: The Antofagasta field experiment: Boundary-Layer Meteorology, v. 17, p. 45–55, doi: 10.1007/BF00121936.

Scheuber, E., and Andriessen, P.A.M., 1990, The kinematic and geodynamic significance of the Atacama fault zone, northern Chile: Journal of Structural Geology, v. 12, p. 243–257, doi: 10.1016/0191-8141(90)90008-M.

Scheuber, E., and González, G., 1999, Tectonics of the Jurassic–Early Cretaceous magmatic arc of the north Chilean Coastal Cordillera (22–26°S): A story of crustal deformation along a convergent plate boundary: Tectonics, v. 18, no. 5, p. 895–910, doi: 10.1029/1999TC900024.

Scheuber, E., Bogdanic, T., Jensen, A., and Reutter, K.J., 1994, Tectonic development of the north Chilean Andes in relation to plate convergence and magmatism since the Jurassic, *in* Reutter, K.J., Scheuber, E., and Wigger, P.J., eds., Tectonics of the southern Central Andes: Berlin, Springer-Verlag, p. 121–140.

Seidl, M.A., and Dietrich, W.E., 1992, The problem of channel erosion into bedrock: Catena Supplement, v. 23, p. 101–124.

Seidl, M.A., Dietrich, W.E., and Kirchner, J.W., 1994, Longitudinal profile development into bedrock: Analysis of Hawaiian channels: Journal of Geology, v. 102, p. 457–474.

Servicio Nacional de Geologia y Minera, Chile, 1982, Mapa Geologico de Chile: Santiago, Chile, Servicio Nacional de Geologia y Minera, Chile, scale 1:1,000,000, 1 sheet.

Somoza, R., 1998, Updated Nazca (Farallon)–South America relative motions during the last 40 My: Implications for mountain building in the central Andean region: Journal of South American Earth Sciences, v. 11, p. 211–215, doi: 10.1016/S0895-9811(98)00012-1.

Tichelaar, B.W., and Ruff, L.J., 1991, Seismic coupling along the Chilean subduction zone: Journal of Geophysical Research, v. 96, p. 11,997–12,022.

Trewartha, G.T., 1981, The Earth's problem climates: Madison, Wisconsin, The University of Wisconsin Press, 371 p.

Van Dissen, R.J., Berryman, K.R., Pettings, J.R., and Hill, N.L., 1992, Paleoseismicity of the Wellington–Hutt Valley segment of the Wellington fault, North Island, New Zealand: New Zealand Journal of Geology and Geophysics, v. 35, p. 165–176.

Vargas, G., Ortlieb, L., and Rutllant, J., 2000, Aluviones históricos en Antofagasta y su relación con eventos El Niño/Oscilación del Sur: Revista Geológica de Chile, v. 27, p. 157–176.

Von Huene, R., and Ranero, C.R., 2003, Subduction erosion and basal friction along the sediment-starved convergent margin off Antofagasta, Chile: Journal of Geophysical Research, v. 108, no. B2, 2079, doi: 10.1029/2001JB001569.

Von Huene, R., Weinrebe, W., and Heeren, F., 1999, Subduction erosion along the north Chile margin: Journal of Geodynamics, v. 27, p. 345–358, doi: 10.1016/S0264-3707(98)00002-7.

Ward, S.N., and Valensise, G., 1994, The Palos Verdes terraces, California: Bathtub rings from a buried reverse fault: Journal of Geophysical Research, v. 99, p. 4485–4494, doi: 10.1029/93JB03362.

Wells, L.E., 1988, Holocene fluvial and shoreline history as a function of human and geologic factors in arid northern Peru [Ph.D. thesis]: Stanford, Stanford University, 381 p.

Wells, S.G., Bullard, T.F., Menges, C.M., Drake, P.G., Karas, P.A., Kelson, K.I., Ritter, J.B., and Wesling, J.R., 1988, Regional variations in tectonic geomorphology along a segmented convergent plate boundary, Pacific coast of Costa Rica: Geomorphology, v. 1, p. 239–265, doi: 10.1016/0169-555X(88)90016-5.

Manuscript Accepted by the Society 23 June 2005

Printed in the USA

Geological Society of America
Special Paper 398
2006

Influence of incision rate, rock strength, and bedload supply on bedrock river gradients and valley-flat widths: Field-based evidence and calibrations from western Alpine rivers (southeast France)

G.Y. Brocard†
P.A. van der Beek‡
Laboratoire de Géodynamique des Chaînes Alpines, Université Joseph Fourier, 38 041 Grenoble, France

ABSTRACT

Several process-based models of river incision have been proposed in recent years that attempt to describe fluvial landform development. Although some field tests have been performed, more data are required to test the ability of these models to predict the observed evolution of fluvial landforms. We have investigated several tens of rivers located in the French western Alps that flow across folded sedimentary rocks with strongly contrasting rock strengths. These rivers record significant variations in some of the parameters controlling river incision, notably bedrock lithology, stream power, incision rate, and sediment flux, potentially allowing discrimination between existing models. Variations in incision rates are driven by variations in the amount of disequilibrium introduced in the river profiles during the last glaciation. We use diagnostic indices to investigate transport- and detachment-limited conditions, which include the channel morphology, the occurrence of lithogenic knickpoints, the continuity of alluvial and bedrock reaches, and the slope-area scaling of the river long profile. We observe transitions from detachment-limited to transport-limited conditions with increasing discharge/drainage area and decreasing incision rate. Bedrock strength influences the location of the transition predictably. The formation of transport-limited rivers coincides with the development of a valley flat wider than the active channel, which accommodates variations in bedrock strength, stream power, and incision rate along the transport-limited reaches. We propose and calibrate a model for the development of valley flats along transport-limited rivers and explore some properties of landscape development in mountain ranges controlled by transport-limited rivers.

Keywords: fluvial geomorphology, river incision, landscape evolution, valley width, sediment supply.

†Present address: Department of Geology and Geophysics, University of Minnesota, Minneapolis, Minnesota 55455, USA.

‡E-mail: pvdbeek@ujf-grenoble.fr.

Brocard, G.Y., and van der Beek, P.A., 2006, Influence of incision rate, rock strength, and bedload supply on bedrock river gradients and valley-flat widths: Field-based evidence and calibrations from western Alpine rivers (southeast France), *in* Willett, S.D., Hovius, N., Brandon, M.T., and Fisher, D.M., eds., Tectonics, Climate, and Landscape Evolution: Geological Society of America Special Paper 398, Penrose Conference Series, p. 101–126, doi: 10.1130/2006.2398(07). For permission to copy, contact editing@geosociety.org.

INTRODUCTION

The relief of active mountain belts is the result of the competition between rock uplift and erosion. In nonglaciated areas, the erosion system is controlled by river incision, as streams maintain both the transport of debris generated on mountain slopes and the incision of the bedrock (e.g., Burbank et al., 1996; Benda and Dunne, 1997). Through incision and clearing of colluvium, rivers control the steepness of valley flanks and thus regulate erosion on catchment slopes. At the mountain-range scale, river long profiles control the bulk elevation of the orogen (e.g., Whipple et al., 1999). Because the capacity of a river to incise bedrock and transport sediments depends on its gradient, a positive feedback exists between rock uplift and river incision. Relief is therefore thought to evolve toward a dynamic equilibrium between uplift and erosion; such equilibrium is often assumed and used in neotectonic studies to infer uplift rates from river incision rate measurements, which can be compared with uplift rates obtained by other methods (e.g., Personius, 1995; Burbank et al., 1996; Harbor, 1998; Lavé and Avouac, 2001; Pazzaglia and Brandon, 2001).

The European Alps are a slowly growing orogen, and most of the present-day pattern of uplift is poorly constrained. Our study of fluvial forms in the western Alps therefore started as an assessment of the ability of such analyses to provide useful information on neotectonic activity. Large glaciers have developed in the Alps during the Quaternary glaciations. They have formed in the highest parts of the range and have spread out into the foreland. A study of river long profile development in the western Alps has shown that river incision is mostly triggered by the restoration of graded profiles in rivers that have been severely glacially disturbed (Brocard, 2002; Brocard et al., 2003). Tectonically driven incision is therefore outstripped by the postglacial relaxation of the fluvial system in this area, and the extraction of a neotectonic signature would require highly accurate data on postglacial reequilibration. The southwesternmost part of the Alps, however, has not been occupied by ice. The relief of this area is thought to be close to equilibrium, since river incision rates (Brocard et al., 2003), present-day erosion rates (e.g., Alary, 1998), and thermochronologically derived long-term denudation rates (Seward et al., 1999; Bigot-Cormier et al., 2000; Bernet et al., 2001) are of the same order of magnitude.

This paper presents the equilibrium forms developed by rivers located in a nonglaciated portion of the western Alps (Fig. 1) and their evolution as a function of bedrock erodibility, incision rate, stream power, and sediment flux. Two major and potentially useful markers are described: the river long profile and the valley flat. These features are expected to exhibit detectable variations due to tectonic forcing.

Bedrock river long profiles have received considerable attention in the last decade, because they control the overall relief of mountain ranges. In many numerical models of landscape evolution, the "stream power" incision law (e.g., Howard et al., 1994; Whipple and Tucker, 1999) is used to link river incision rate ($\dot{E}$) to drainage area (A) and gradient (S):

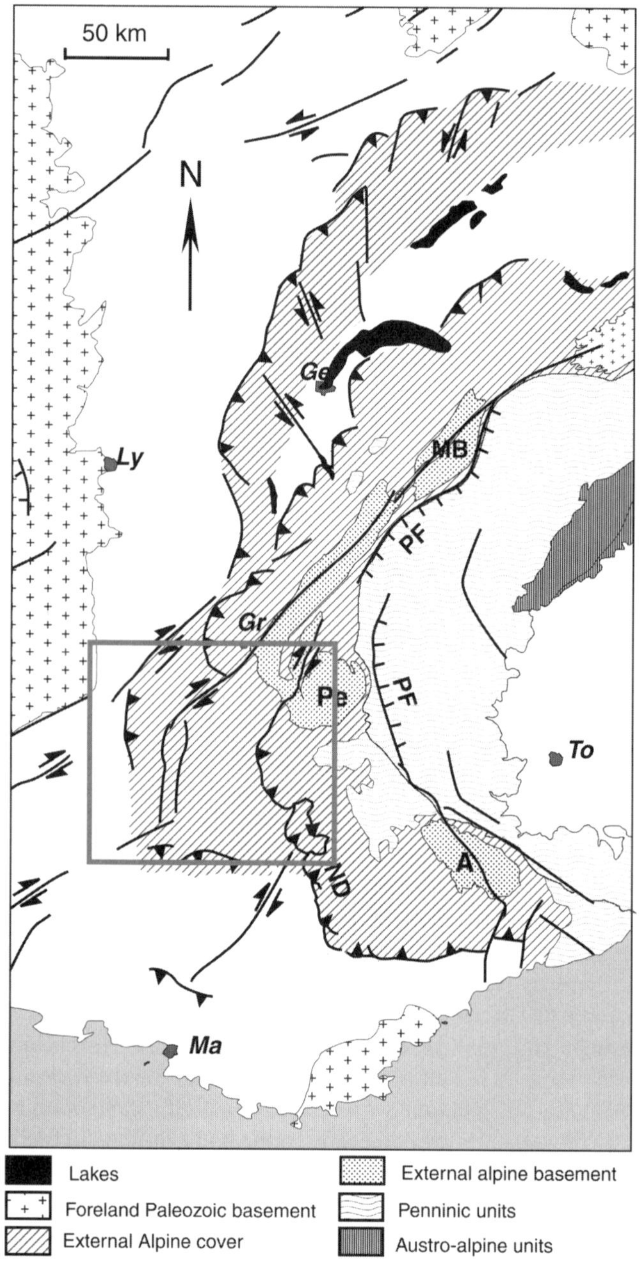

Figure 1. Sketch map of the western Alps, indicating structures that have been active since the late Miocene. PF—Penninic Front; ND—Digne thrust sheet; MB—Mont Blanc Massif; Pe—Pelvoux Massif; A—Argentera Massif. Cities: Ge—Geneva; Gr—Grenoble; Ly—Lyon; Ma—Marseille; To—Torino. Box indicates study area shown in Figure 2.

$$\dot{E}= K_d A^{m_d} S^{n_d} \qquad (1)$$

where K_d is an erosional efficiency factor with dimension [$L^{(1-2m_d)}$ T^{-1}] and m_d and n_d are nondimensional exponents. If relief is in dynamic equilibrium, the uplift rate (U) can be substituted for the incision rate ($\dot{E}$). The stream power law is convenient because it is a process-based erosion law that relates incision to measurable geomorphic parameters. The erosion of bedrock is considered to proceed by abrasion, plucking, and cavitation; the values for the exponents in the law should be determined by the dominant process (e.g., Whipple et al., 2000). River incision in such a detachment-limited model is controlled by the erodibility of the bedrock through the parameter K_d. This simple law has been used extensively, with some authors even inferring patterns of crustal deformation from river profiles (e.g., Kirby and Whipple, 2001; Finlayson et al., 2002).

However, recent work has highlighted the strong influence of bedload transport on incision rate (Sklar and Dietrich, 1998, 2001) and its potential effects on river long profile development (Howard et al., 1994; Sklar and Dietrich, 1998; Whipple and Tucker, 2002). Various models of river incision have been proposed that incorporate bedload control to some degree (e.g., Beaumont et al., 1992; Sklar and Dietrich, 1998). These models predict progressive transitions from detachment-limited rivers, which are well described by the stream power law, toward transport-limited rivers, the incision of which is controlled by the bedload.

A transport-limited incision rule can be derived in a form similar to the detachment-limited stream power law by stating that volumetric transport capacity (Q_{eq}) is a function of stream power, sediment flux is equal to carrying capacity, and incision or deposition rate equals the downstream divergence of sediment flux (e.g., Willgoose et al., 1991):

$$Q_{eq} = K_t A^{m_t'} S^{n_t} \tag{2a}$$

and

$$\dot{E} = \frac{1}{W_c}\frac{d}{dx}Q_{eq} \tag{2b}$$

where K_t is a sediment transport coefficient [$L^{(3-2m_t')}$ T^{-1}], m_t' and n_t are area and slope exponents as in equation 1, and W_c is bankfull channel width. By using the well-known relationships between drainage area and channel length x (e.g., Hack, 1957), $A = K_a x^h$, and between channel width W_c and drainage area (e.g., Leopold and Maddock, 1953), $W_c = K_c A^c$, where K_a and K_c are pre-exponential coefficients with dimensions [$L^{(2-h)}$] and [$L^{(1-2c)}$], respectively, and h and c are dimensionless exponents, a power-law solution for river incision can be found:

$$\dot{E} = K_t A^{(m_t-1)} S^{n_t}, \tag{2c}$$

where $m_t = m_t' - c$. Rivers are expected to behave according to one of these end-member cases (detachment-limited, equation 1; transport-limited, equation 2) or as a hybrid between the two.

Unlike variations in channel profiles or channel width, the potential significance of variations in valley-flat width has received much less attention up to now (notable exceptions being Harbor, 1998; Hancock and Anderson, 2002; Snyder et al., 2003a; Tomkin et al., 2003). The widespread development of valley flats in our study area led us to explore the factors that control valley-flat widths, with the aim of extracting possible tectonic forcing.

In the following, we first present the tectonic, climatic, and lithologic variables that control the incision pattern and fluvial landform development along the studied rivers. Second, we present the diagnostic indices we have used to infer whether river segments are detachment- or transport-limited. These indices are: (1) the channel-bed morphology, (2) the sensitivity of river gradients to bedrock erodibility, (3) the continuity and parallelism of alluvial and bedrock reaches, and (4) the slope-area scaling behavior of the river segments. Third, we investigate the properties of the transition from detachment-limited to transport-limited reaches. We cannot resolve the form of the transition with sufficient resolution to discriminate between simple stream power models and more elaborate sediment-flux–dependent incision models. However, our study area is ideally suited to study the influence of bedrock erodibility and of incision rate on the location of this transition. Fourth, the valley-flat data are exposed. We describe the main physical characteristics of valley flats in our study area. We then explore the influence of bedrock erodibility, river discharge, and incision rate on the valley-flat width. We propose a theoretical framework to account for valley-flat development and test its ability to explain the data. Finally, we discuss the overall accordance between the field data and the available river incision models and evaluate what values for model parameters are implied by our data. We explore what the response of a mixed transport- and detachment-limited river system to uplift rate variations would be and use field examples to demonstrate the implications of such a system for drainage stability.

THE WESTERN ALPS: TECTONICS AND MORPHOLOGY

The western Alps are a collisional orogen that started to emerge above sea level since Eocene times. At the mountain-belt scale, two main units are distinguished (Fig. 1): the Internal or Penninic Alps are composed of highly deformed and metamorphosed rocks of oceanic and continental origin. This unit was deformed and thrusted onto the European passive margin before the Miocene along the Penninic frontal thrust (e.g., Schmid and Kissling, 2000), which currently acts as an extensional detachment fault (Tricart et al., 2001). The External Alps are much less deformed. They were incorporated within the orogenic wedge during the Miocene-Pliocene and are composed of crystalline basement blocks adjacent to foreland fold-and-thrust belts. Our study area is located within the External Alps; it is composed of several tectonic units of the former European passive margin that were inverted during the Pyrenean (Upper Cretaceous) and Alpine (Eocene and Miocene) orogenies. Tectonic units include the Paleozoic basement of the Pelvoux External Crystalline Massif, composed of high-grade metamorphic and igneous rocks, overlain by a Mesozoic sequence of marly and calcareous sediments that are largely detached from the basement by a Triassic

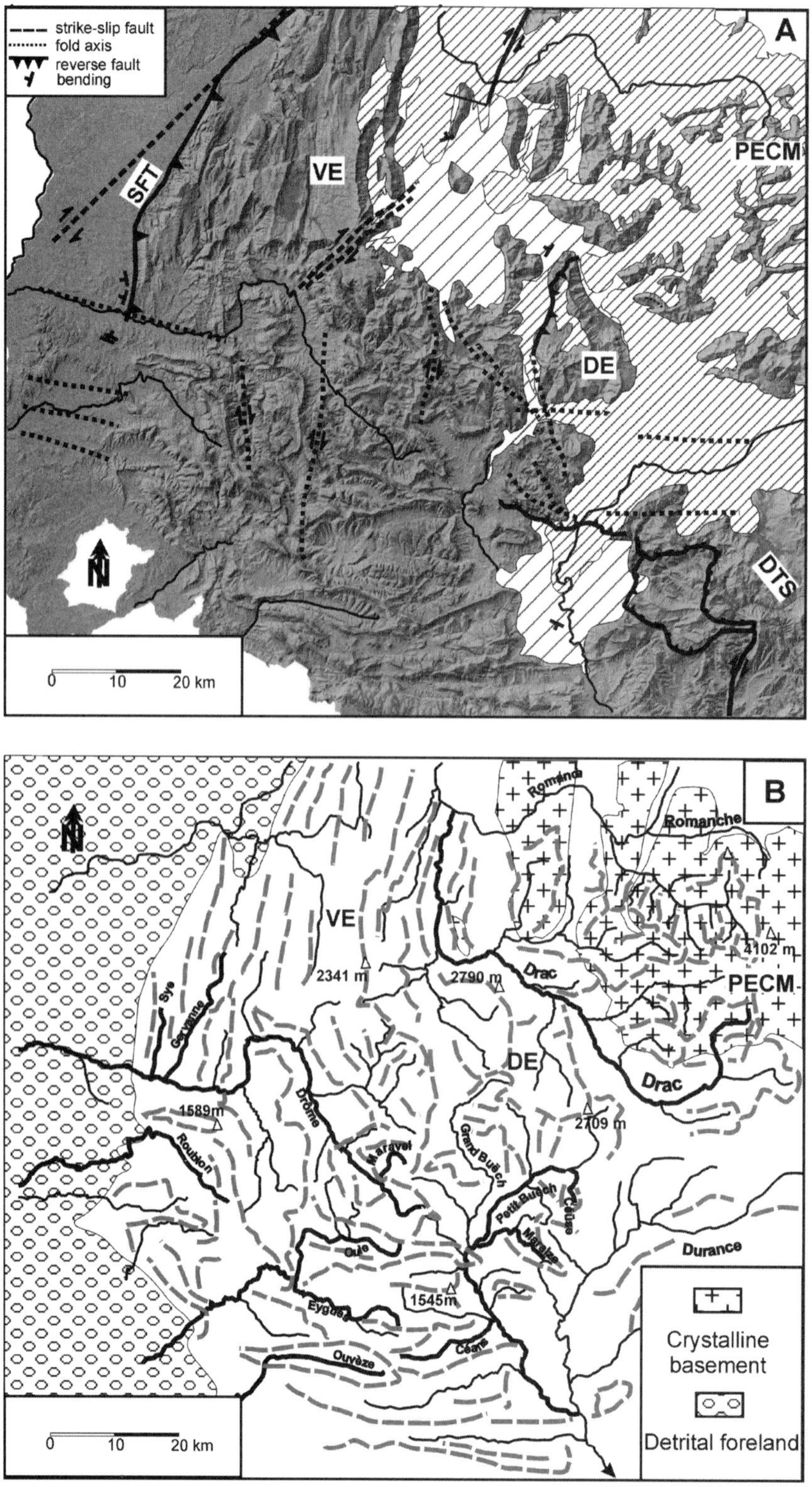

Figure 2. Maps of the study area showing the stream network and the elements that modulate the tectonic and climatic controls on river incision. (A) Shaded-relief map of the 50-m-resolution digital elevation model (DEM) of the study area, with indication of glacier extents during the last glaciation (close hatches— marine isotope stage [MIS] 2; loose hatches—MIS 4) and Pliocene to Quaternary tectonic structures: DTS—Digne thrust sheet; SFT—Subalpine Frontal thrust. (B) Sketch relief of the study area with locations of the rivers quoted in the text. Heavy lines refer to long profiles shown in Figures 4–6. Cross pattern—crystalline rocks of the Pelvoux External Crystalline Massif (PECM); pebble pattern—Alpine foreland; dark-gray dashed lines—mountain crests cored by resistant limestones—Tithonian in the SW of the study area; Barremian in the Vercors Massif (VE); Campanian in the Devoluy Massif (DE).

evaporate layer. The Mesozoic sequence is locally overlain by remnants of the Cenozoic foreland basin, incorporated into the thrust slices by the outward propagation of the orogenic front.

The crystalline Pelvoux Massif is much higher than the surrounding subalpine massifs of the foreland fold-and-thrust belt; most of its peaks are above 3000 m in elevation, and several summits reach 4000 m (Fig. 2). Valleys in the Pelvoux Massif have conspicuously glacial forms; they are very deep (typically 2500 m) with very steep sidewalls. To the west, thick limestone series have produced large gently folded karstic plateaus: the Vercors (peak elevation 2341 m) and Devoluy (2790 m) Massifs. Elsewhere, relief is more subdued, with a suite of marly basins separated by thin crests of limestone that stand ~1000 m above the surrounding valley floors.

The study area has a temperate mid-latitude climate; mean annual precipitation is around 1000 mm yr^{-1} and is evenly distributed throughout the year. To the south, the climate tends toward the Mediterranean type, with dry hot summers, heavy rains in autumn, and damp and fresh winters. To the east, it becomes mountainous, with highest discharges during the spring due to snowmelt. To the north, the climate becomes temperate marine with an increasing influence of Atlantic winds.

Tectonic and Climatic Forcing of River Incision in the Western Alps

During the Quaternary, the driving processes of river incision in the western Alps were continuing rock uplift as well as climatic forcing. The Alps are nowadays moderately active: both long- and short-term uplift and denudation rates are typically on the order of 0.5–1.0 mm yr^{-1} in the western part of the orogen (e.g., Martinod et al., 1996: Jouanne et al., 1998; Bigot-Cormier et al., 2000; Bernet et al., 2001; Tricart et al., 2001). The overall concordance of present-day rock uplift rates, short-term river incision rates, and long-term denudation rates has been interpreted as indicating that the western Alpine relief may tend toward dynamic equilibrium between uplift and erosion (Bernet et al., 2001; Brocard et al., 2003).

During the Quaternary, the Alps experienced widespread glaciations that repeatedly reshaped the landscape. The Alpine glaciers were fed by snowfields located within the Internal Alps, spilled across the External Alps, and spread far onto the northwestern European foreland (Montjuvent, 1978; Mandier, 1984). The ice streams oversteepened the valley long profiles downstream of confluence steps and riegels, and carved long and deep troughs into the softer rock beds. Since the end of the last glaciation, the rivers that flow along these reaches tend to restore graded long profiles. This is achieved by filling of the glacial troughs with sediments and by incising the oversteepened reaches. Most of the glacial trough lakes have been quickly filled with suspended river load (e.g., Chapron, 1999). In many places, however, rivers are still aggrading, as their gradients are not yet steep enough to allow the bedload to overpass the glacial troughs. The incision of oversteepened reaches remains active, as graded profiles have not been achieved even along the largest streams.

The study area (Figs. 1 and 2) covers a bulk area of 5500 km^2. Our morphological analysis is focused on three medium-size rivers and 42 smaller streams (Fig. 2B). The main rivers are the Drac, the Drôme, and the Buëch; morphometric properties for their catchments are given in Table 1. These rivers preserve numerous terraces that have been dated using cosmogenic ^{10}Be to infer incision rates (Brocard et al., 2003; cf. next section here). The Drac River drains the resistant crystalline rocks (granites, gneisses, and amphibolites) of the Pelvoux Massif in its headwaters before flowing northwestward into an isoclinal structure of Mesozoic sediments. The Buëch and Drôme Rivers, as well as the smaller investigated rivers, flow exclusively within the Mesozoic sedimentary cover through interference structures of Late Cretaceous to Miocene folds. Their bedrock (Fig. 3) is composed of very thick (up to 2000 m) marly formations, interbedded with thin (20–80 m) levels of highly resistant massive limestone, in addition to thicker (up to 300 m) but softer rhythmic successions of marl and marly limestone.

Seismic activity, relatively insignificant in most of the western Alps, is practically absent in the study area (see, for

TABLE 1. MORPHOMETRIC DATA FOR THE THREE MAJOR CATCHMENTS IN THE STUDY AREA

	Drac River	Buëch River	Drôme River
Length (km)	125	75	110
Catchment area (km^2)	2095	1473	1645
Mean elevation (m)	1487	1067	786
Maximum elevation (m)	3669	2709	2041
Minimum elevation (m)	180	450	86
Mean annual discharge (m^3 s^{-1})	17	16	19
Peak discharge—decadal flood (m^3 s^{-1})	430	370	340
Maximum measured peak discharge (m^3 s^{-1})	700	557	556

Note: Morphometric data compiled from the Institut Géographique National 50-m-resolution digital elevation model (DEM). Discharge data from Réseau des données sur l'eau du bassin Rhône-Méditerranée-Corse (http://www.rdbrmc.com/debithydro/). Discharge stations: Drac River: Pont de Claix (125 km downstream, at confluence with Romanche River); Buëch River: Laragne (53 km downstream; upstream catchment area 1100 km^2); Drôme River: Saillans (67 km downstream; upstream catchment area 1150 km^2).

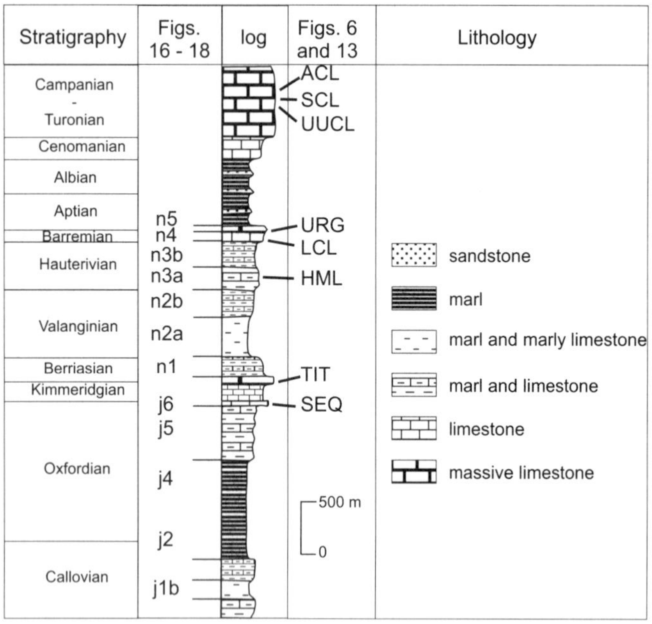

Figure 3. Simplified log of the sedimentary cover units encountered by the nonglaciated rivers and abbreviations used in the study. Simplified stratigraphic groups used in Figures 6 and 13: SEQ—Sequanian (Upper Kimmeridgian) limestone; TIT—Tithonic (Portlandian-Tithonian) limestone, referred to in the text as Tithonian; HML—Hauterivian marly limestone; LCL—Lower Cretaceous (Barremian-Aptian) limestone; URG—Urgonian (Barremian) reef limestone; ACL—Campanian argillaceous limestone; SCL—Campanian siliceous limestone; UUCL—undifferentiated Upper Cretaceous detrital limestone (Turonian to Coniacian). The abbreviations in Figures 15–17 correspond to stratigraphic codes used on the BRGM (Bureau des Recherches Géologiques et Minières) 1:50,000 scale geological maps.

instance, http://sismalp.obs.ujf-grenoble.fr/sismalpuk.html). No post-Miocene deformation features have been identified with certainty so far. However, the highest terrace level of the Drôme River is slightly bent and upwarped in the downstream vicinity of the Subalpine Front (Fig. 2A), the main frontal thrust of the Alps during the Miocene (Brocard, 2002). The terrace deformation suggests that the frontal thrust has maintained a low level of activity throughout the Pliocene-Pleistocene. No differential uplift has been identified upstream of the Subalpine Front within the fold-and-thrust belt, where most of the studied rivers are located. We thus assume that the tectonically driven component of river incision is constant throughout the study area.

In the investigated area, glaciers descending from the Internal Alps during cold periods were not large enough to reach the foreland and were diverted by north-south–trending structures (Fig. 2A). Their influence therefore decreases westward. To the east, rivers like the Drac and its tributaries were partly invaded by ice streams that strongly modified their long profile (Montjuvent, 1973; Brocard et al., 2003). To the west, valleys remained ice free during glaciations. The combination of a constant and relatively slow, tectonically driven component to river incision, together with a rapid and unevenly distributed, climatically driven component allows observation of a wide range of incision rates throughout an area that is lithologically, tectonically, and climatically homogeneous.

Incision Rate Data

Incision rates have been measured along the Drac and Buëch Rivers, which have constructed numerous terraces during the Pleistocene. Of these, the Drac River was most strongly affected by glacial advances, during which a series of glaciers dammed the river valley. Fill terraces were built upstream from these ice dams (Montjuvent, 1973). Since the Last Glacial Maximum, the Drac River has entrenched the fill deposits and underlying bedrock, carving out cut-fill and strath terraces. Cosmogenic ^{10}Be dating of these terraces (Brocard et al., 2003) has shown that incision is controlled by retreating knickpoints: at a site located ~15 km upstream of the major glacier dam, incision began several thousand years after the inception of glacial retreat and occurred at a rate greater than 60 mm yr^{-1} during less than 5 k.y. before dropping to 8–11 mm yr^{-1} over the last 7 k.y. A degraded unstable knickpoint occurs in the present-day long profile of the Drac River (cf. next section); this knickpoint is interpreted as being the remnant of the initial glacially oversteepened reach that has migrated ~55 km upstream during Holocene time. Incision rates downstream of this knickpoint are significantly higher (7.4 ± 1.0 mm yr^{-1}) than upstream (4.6 ± 0.7 mm yr^{-1}).

The Buëch represents an intermediate type of river, the regime and bedload characteristics of which have been modified by the meltwater contribution of large glaciers located in its upper catchment (Brocard et al., 2003). The glacially increased discharge and sediment flux led to the formation of paired terraces (Mandier, 1984). The Buëch River preserves three terrace levels with treads that stand 20, 80, and 190 m above the present-day valley floor, respectively. ^{10}Be dating of these terraces shows that the river's incision rate is roughly constant at 0.8 mm yr^{-1} when integrated over periods longer than the mean duration of the glaciations (Brocard et al., 2003).

The Drôme River catchment and the many smaller river catchments analyzed here have never been glaciated. The glacial-interglacial climatic fluctuations, however, significantly modified their regime and triggered the formation of paired strath terraces along the largest streams (e.g., Drôme, Bès, Gervanne). The treads of these terraces are degraded, which precludes their dating using cosmogenic ^{10}Be. However, they occur in three levels at similar elevations above the river bed as in the Buëch catchment. If we assume that the terraces of the Buëch and Drôme Rivers were abandoned simultaneously, long-term incision rates of the Drôme River can be estimated at ~0.7–0.9 mm yr^{-1}. Terrace levels are parallel to the present-day river profile, suggesting that this rate is homogeneous along its course upstream of the mountain front (Brocard, 2002).

DIAGNOSTIC INDICES AND FIELD EVIDENCE FOR DETACHMENT- AND TRANSPORT-LIMITED RIVERS

Channel Morphology

The morphology and sediment pattern in active river channels can be used to define the river type (e.g., bedrock, alluvial, or mixed bedrock-alluvial; Howard, 1998) and the process of incision (e.g., Pazzaglia et al., 1998; Whipple et al., 2000). As river bed morphology is very sensitive to climatic and anthropogenic changes (e.g., Gautier, 1992; Snyder et al., 2003a), it must be used with care to assess the processes involved in river incision over the time scales required to shape river long profiles and valley flats (Personius, 1995; Wegmann and Pazzaglia, 2002). However, our field observations along the Alpine rivers are consistent with the parameters describing the mode of long profile development exposed hereafter.

Upstream of the knickpoint discussed in the previous section, the Drac River flows over thick fluvio-glacial deposits within a wide and gentle valley, whereas it has incised up to 300-m-deep gorges downstream. Where incising limestones, basalts, and gneisses, the river exhibits a mixed bedrock-alluvial bed. All of these rocks are intensively folded and fractured. Abrasion features are restricted to massive limestone strata and gneisses. Elsewhere, incision appears to be dominated by plucking. Where incising marly levels, however, the river channel is braided and flows over a narrow strath; the bedrock is extensively covered with gravel in these reaches. In its lowermost reaches, the Drac River again flows over alluvial deposits from a large fan that developed at the confluence with the Romanche River after deglaciation.

Rivers in the nonglaciated area are naturally braided and wander within wide valley flats. Their bedrock is usually entirely blanketed with a thin layer of channel sand-and-gravel deposits (Gautier, 1992). However, gravels have been intensively extracted from many river beds during the twentieth century, so that most of the bedload cover has been stripped off the active channels. The thickness of channel deposits is commonly a few meters; this is interpreted as the maximum thickness the river can remove during peak discharge stages (Mackin, 1937; Howard, 1998; Wegmann and Pazzaglia, 2002). In many places, the rivers cross a thin but conspicuous series of Tithonian massive limestone, composed of submarine micritic breccia. Where rivers cross this unit, the valley floor is reduced to the active channel, at the bottom of narrow gorges. In some of these gorges, the rivers further exhibit knickpoints associated with outstanding abrasion features such as potholes.

On the basis of these field observations, the rapidly incising Drac River can be regarded as a detachment-limited, plucking-dominated, mixed bedrock-alluvial river, with restricted transport-limited reaches below its knickpoint. In contrast, the slowly incising rivers of the nonglaciated catchments, as well as the upper Drac River, are alluvial or transport-limited, mixed bedrock-alluvial rivers. The use of other criteria hereafter will confirm this trend over longer time scales.

Bedrock Strength and Slope Variations in Long Profiles

The long profiles of detachment-limited rivers are influenced by the (tensile) bedrock strength, which partly controls the parameter K_d in the stream power law (equation 1), whereas transport-limited long profiles are sensitive to bedload caliber (Snow and Slingerland, 1987; Gasparini et al., 1999; Sklar and Dietrich, 2001). We have investigated the effect of bedrock strength along the Drac River and along the nonglaciated rivers by constructing long profiles of the main rivers in the study area. Elevation and distance data were digitized from 1:25,000 scale topographic maps with contour intervals of 10 m. Along smaller streams, long profiles were extracted using Arcview GIS (geographic information systems) from a 50-m-resolution digital elevation model (DEM) obtained from the Institut Géographique National (cf. http://www.ign.fr/). As grid nodes generally sample points on valley flanks rather than the valley floor, elevation data of long profiles are overestimated and very rough. This technique was only applied where changes in slope in the river profiles are larger, both in length and amplitude, than the high-frequency noise of the DEM data.

In Figure 4, the logarithm of gradient of the Drac River is plotted as a function of the logarithm of downstream distance. Equilibrium long profiles should plot as straight segments in such a diagram (Bishop and Goldrick, 2000). Lithological knickpoints are generated where equilibrium rivers cross zones of hard bedrock. Numerous knickpoints occur along the Drac River; they

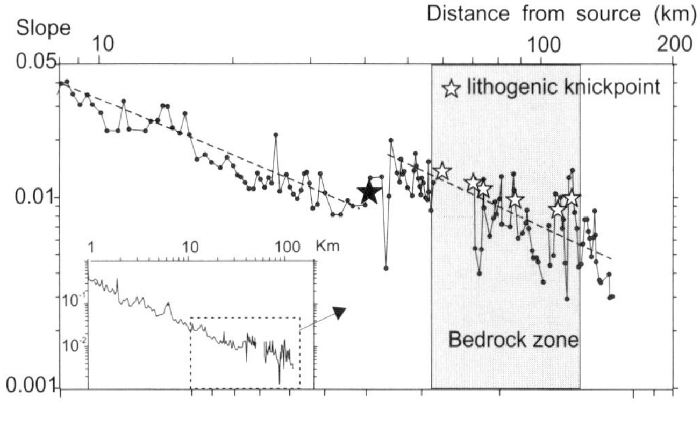

Figure 4. Logarithmic slope-distance plot of the Drac River (modified from Brocard et al., 2003). Open stars—lithogenic knickpoints. Solid star—retreating disequilibrium knickpoint. Data extracted from 1:25,000 scale topographic maps; inset shows complete plot. Best-fit linear regressions upstream and downstream of the knickpoint are:

$$S = 0.34^{+0.33}_{-0.13} D^{-1.01\pm0.12}$$

and

$$S = 1.4^{+7.4}_{-1.2} D^{-1.22\pm0.72},$$

with correlation coefficients $r^2 = 0.85$ and 0.12, respectively.

correlate with more resistant bedrock, such as Mesozoic reef limestone and Paleozoic micaschist. The topographic resolution is not sufficient to construct lines along these reaches and evaluate the degree of equilibrium of these knickpoints. In contrast, a clearly unstable knickpoint is currently propagating toward the Drac headwaters into unconsolidated sediments deposited during the last glaciation (solid star in Fig. 4), separating the river long profile into two segments—a regularly graded upstream reach and a downstream reach with steep and highly variable gradients. Upstream of this knickpoint, bedrock is not exposed in the river bed and incision rates are roughly half those downstream (cf. previous sections).

In the nonglaciated area, the sensitivity of the long profile to variations in bedrock strength appears to depend on stream power. Along small streams (Fig. 5), knickpoints occur where the streams cross massive limestones. In contrast, no knickpoints are observed along streams with discharges comparable to the Drac River, even where they cross hard lithologies, so that their long

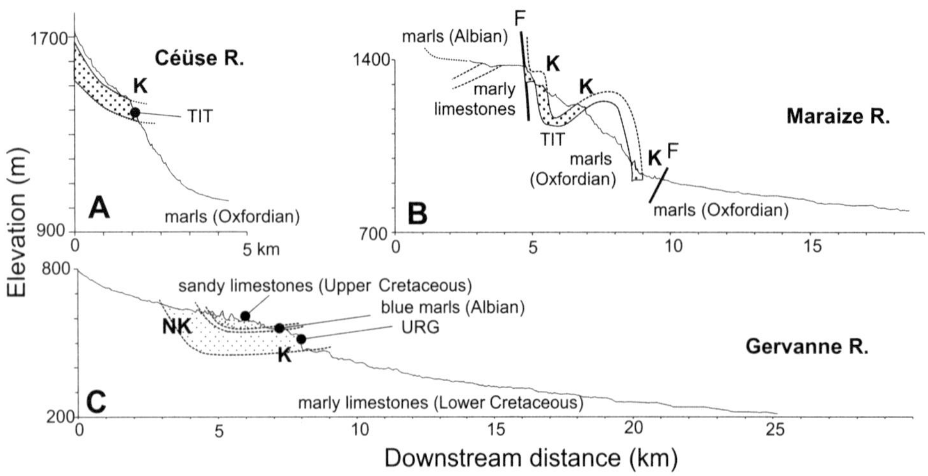

Figure 5. Examples of lithogenic knickpoints along small streams of the nonglaciated area. Horizontal axis—downstream distance (km). Vertical axis—elevation (m). Data extracted from the 50-m-resolution digital elevation model (DEM). (A) Céüse River; (B) Maraize River; (C) Gervanne River. K—knickpoints; NK—reach on resistant rock without a knickpoint. For lithological codes, see Figure 3. F—Fault; URG—Urgonian (Barremian) reef limestone. Note (1) the steep gradients of the Céüse and Maraize Rivers downstream of TIT (Tithonic [Portlandian-Tithonian] limestone) knickpoints, sustained by the Tithonian boulders supplied to the river beds; and (2) the influence of bedrock bedding dip on knickpoint generation along the Gervanne River.

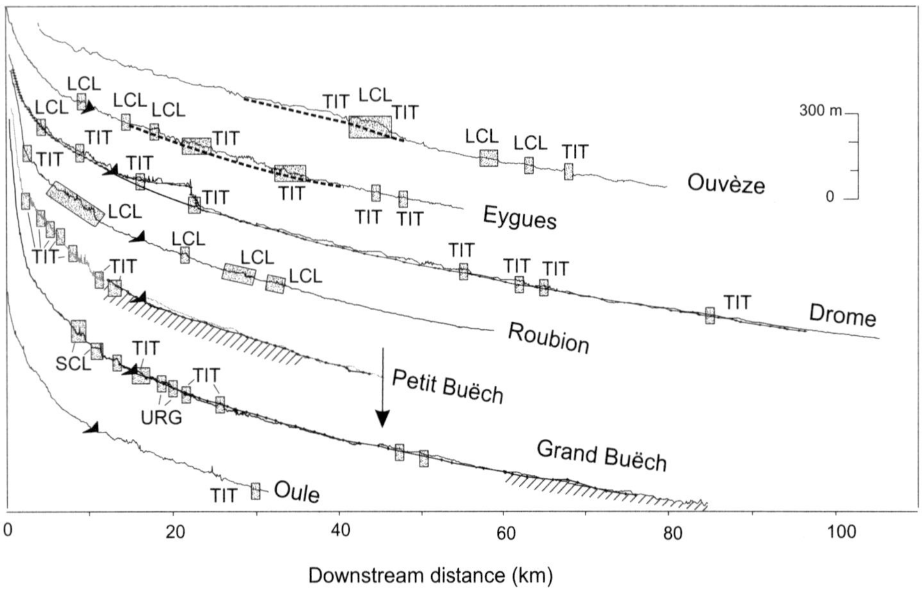

Figure 6. Long-profiles of some of the largest streams of the nonglaciated area. Thick lines with dots—data extracted from 1:25,000 scale topographic maps. Thin continuous lines—data extracted from the 50-m-resolution digital elevation model (DEM). Thick lines—profile shape drawn manually where the profiles extracted from the DEM are too rough. Wide arrowheads—locations where drainage area reaches 30 km². Thin vertical arrow indicates confluence of Petit Buëch and Grand Buëch Rivers. Boxes with dotted pattern indicate reaches of potential knickpoint development, with an indication of bedrock composition; for lithological codes, see Figure 3. Hatched pattern—alluvial reaches where rivers flow across sediment-filled glacial troughs. The large knickpoint in the Drôme River just downstream of 30 km² corresponds to the historic Claps rockslide that blocked the river, leading to widespread alluviation upstream. The base of the alluvial deposits defines a smooth equilibrium profile (Brocard, 2004).

profiles become evenly concave (Fig. 6). Furthermore, the slopes of the large nonglaciated rivers are conspicuously similar and do not show any correlation with rock strength (Fig. 7).

In detail, the long profiles of the nonglaciated rivers locally exhibit very subdued knickzones (e.g., Ouvèze River, Fig. 6) or straight segments that correlate with outcrops of Tithonian limestone. The knickzones expand several kilometers upstream and downstream of the reaches where bedrock is composed of Tithonian limestone. Along the knickzones, the active channels are fed with limestone clasts from the surrounding cliffs. Based on visual observations, it appears that these are significantly coarser than the clasts delivered by other sources. Although other limestone units are resistant enough to generate knickpoints, their denser internal stratification precludes the delivery of clasts as large as those produced by the massive Tithonian limestone breccias.

It is worth noticing that the Tithonian limestones also deliver boulders that control the river gradients where their density is high (e.g., Maraize River, Fig. 5). In such locations along the smaller streams, the active channels are armored with boulders overlying the marly bedrock. The river gradients are considerably steeper than the gradient required to incise the marly levels. The boulders are too large to be transported during flood events and cannot be regarded as a part of the bedload. We would argue that the boulders are almost equivalent to Tithonian bedrock in these reaches.

As a conclusion, the Drac River and the small streams of the nonglaciated area behave as typical detachment-limited rivers in that they are sensitive to bedrock strength. The regular concavity of the nonglaciated river long profiles, in contrast, cannot be explained by the detachment-limited stream power model. Their gradient is insensitive to bedrock changes but appears controlled by bedload caliber, a behavior typical of transport-limited rivers. Note that along the small streams, below the zone of debris-flow scouring, detachment-limited behavior is only observed in limestone-dominated reaches. Along reaches that cross marls and marly limestones, variations in bedrock erodibility do not induce changes in slope. Because the small streams of the nonglaciated area flow for most of their length over marls, detachment-limited behavior is limited to restricted reaches where the rivers cross the most resistant units (see following). Along the lower Drac River, in contrast, transport-limited behavior is restricted to short reaches crossing Aalenian marly limestones.

Parallelism of Alluvial and Bedrock Long Profiles

The Buëch River provides additional evidence for a bedload-controlled gradient along a bedrock river. Along its middle reaches, the river has incised its bedrock almost continuously since the last glaciation (Fig. 8). Its floodplain is underlain by a thin layer of sand-and-gravel deposits blanketing a wide strath. Upstream and downstream of these reaches, the river is reworking fluvio-glacial sediments that fill shallow glacial troughs carved during the last two glaciations. These loose sediments are easily detached and readily incorporated into the bedload, similar to sediments stored along alluvial rivers. The Buëch River thus behaves as an alluvial river along these formerly glaciated reaches. The long profiles of the bedrock and alluvial reaches are strikingly similar; no changes in slope are observed at the

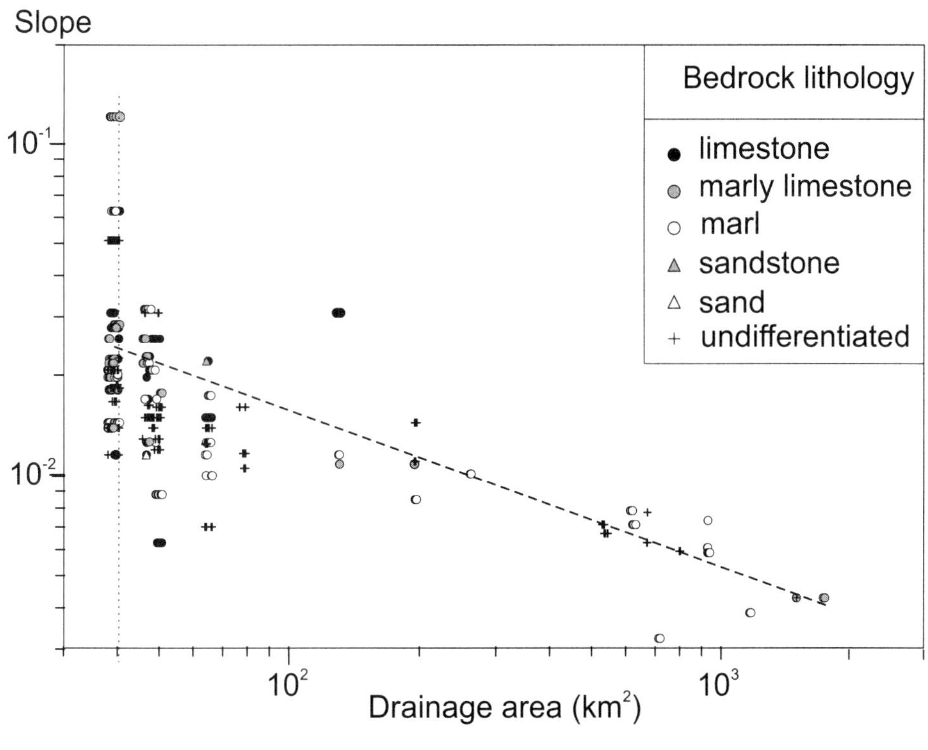

Figure 7. Slope-area plot for the slowly incising rivers of the nonglaciated area for different bedrock lithologies (990 sites in total). Slope extracted from 1:25,000 scale topographic maps, bedrock from 1:50,000 scale geological maps, drainage area extracted from the 50-m-resolution digital elevation model (DEM). Regression line is based on 631 sites located beyond the dotted vertical line that represents the critical area (A_{cr}) of the detachment- to transport-limited transition in Tithonian limestone (40 km^2, cf. Fig. 13). Regression parameters are listed in Table 2.

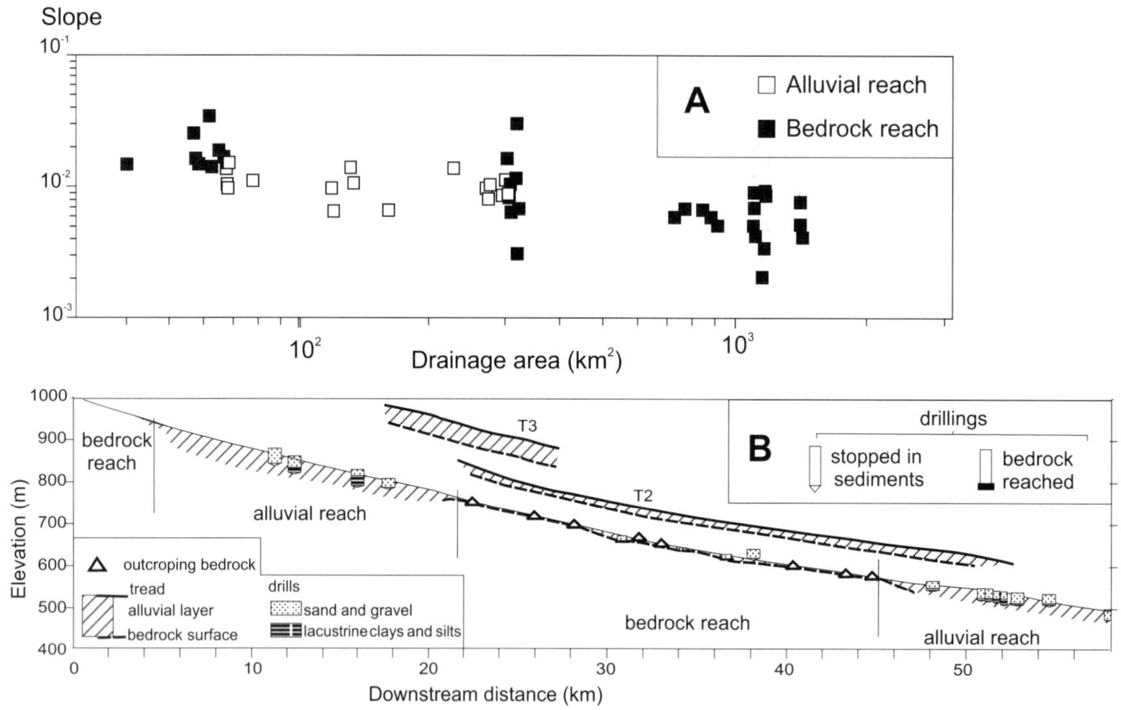

Figure 8. (A) Slope-area plot of the Buëch River, distinguishing bedrock and alluvial reaches. (B) Long profile and alluvial sediment thickness along the Buëch River. Sediment thickness was assessed from the compilation of field observations and archived logs of the underground database of the Bureau des Recherches Géologiques et Minières (BRGM). Drillings are indicated with an indication of whether they reached bedrock or were stopped in alluvial sediments. T2 and T3 are fill terraces abandoned at 60 and 190 ka, respectively (Brocard et al., 2003). Buëch long profile and T2 and T3 tread and strath profiles are projected onto the valley axis.

transitions between alluvial and bedrock reaches (Figs. 8 and 9). Moreover, the modern long profile in the bedrock reach is parallel to both the straths and the treads of the fill-terrace levels (Fig. 8). Therefore, the river gradient is clearly not set by bedrock strength.

Slope-Area Scaling of Long Profiles

Transport- and detachment-limited rivers may be distinguishable on the basis of the slope-area scaling behavior of their long profiles, although this may not in itself provide a diagnostic index (Whipple and Tucker, 2002). For detachment-limited rivers, the relationship between equilibrium slope and drainage area, for constant incision rate and erosional efficiency, can be written as (Whipple and Tucker, 2002):

$$S_d = (\dot{E}_d / K_d)^{(1/n_d)} A^{-\theta_d} \tag{3a}$$

and

$$\theta_d = m_d / n_d, \tag{3b}$$

where S_d is the detachment-limited gradient, $\dot{E}_d$ the detachment-limited incision rate, and θ_d the intrinsic concavity. The pre-exponential constant $(\dot{E}_d / K_d)^{(1/n_d)}$ is known as the steepness index.

The relationship between equilibrium slope and drainage area can be written similarly for transport-limited rivers, assuming that incision rate is equal to the catchment erosion rate (Whipple and Tucker, 2002):

$$S_t = (\beta \dot{E}_t / K_t)^{(1/n_t)} A^{-\theta_t} \tag{4a}$$

and

$$\theta_t = (m_t - 1) / n_t, \tag{4b}$$

where S_t is the transport-limited gradient, $\dot{E}_t$ the transport-limited incision rate, and β the proportion of sediments produced by the watershed that is converted into river bedload. θ_t and $(\beta \dot{E}_t / K_t)^{(1/n_t)}$ are the intrinsic concavity and steepness index of a transport-limited river, respectively.

We have compared the slope-area scaling for the Drac River to that of the slowly incising rivers of the nonglaciated area (Fig. 9; Table 2). Slopes were measured by digitizing the 1:25,000 scale topographic maps, and drainage areas were extracted from the 50-m-resolution DEM. Since lithologies may vary over very short distances along the river profiles, it is difficult to ascribe a specific slope value to any rock formation using the present data. Values of concavity and steepness indices as a function of lithological units are therefore not included in the following discussion.

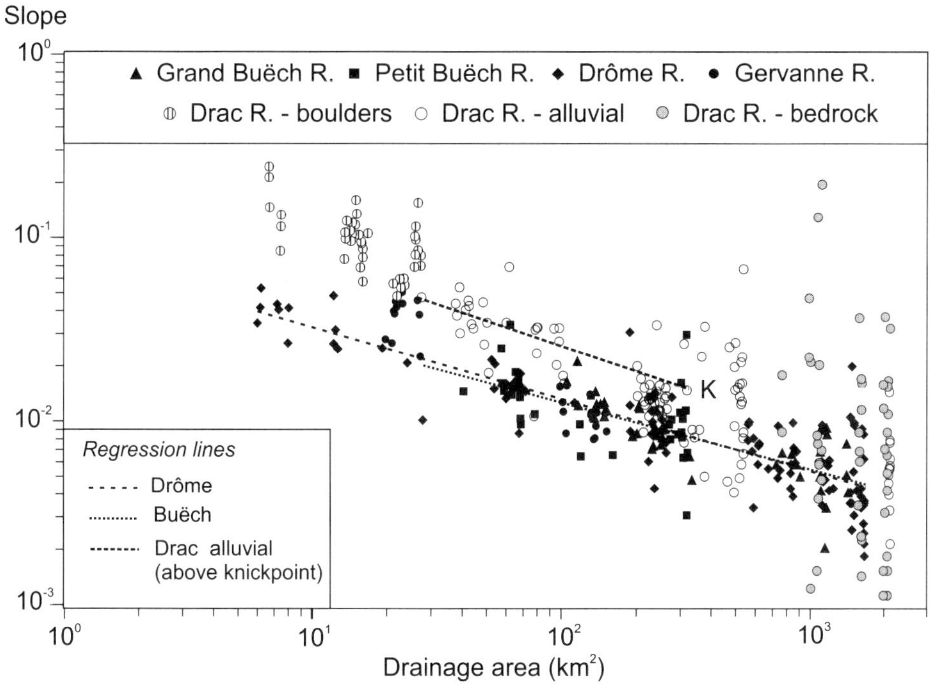

Figure 9. Comparative plot of gradient as a function of drainage area for a rapidly incising river (Drac River), and several slowly incising rivers (Buëch, Drôme, Gervanne), showing that the concavity, slope values, and scatter of slope values are higher along the rapidly incising river. K—disequilibrium knickpoint of the Drac River. Dashed lines—best-fit regression lines for the Drôme and Buëch Rivers, and for the alluvial reach upstream of the knickpoint in the Drac River (see text for discussion). Regression parameters are listed in Table 2.

TABLE 2. SLOPE-AREA RELATIONSHIPS FOR THE MAJOR CATCHMENTS IN THE STUDY AREA

		Drac River (upstream of knickpoint)	Buëch River	Drôme River	Set of 18 transport-limited rivers (Fig. 7)
Intrinsic concavity[a]	$\theta = (m_t - 1) / n_t$	0.56 ± 0.12	0.36 ± 0.08	0.39 ± 0.05	0.42 ± 0.03
Steepness index[b]	$k_s = \left(\frac{\beta E}{K_t}\right)^{1/n_t}$	753^{+6491}_{-676} (17.4 – 72.4)	$9.7^{+38.7}_{-7.8}$ (12.3 – 38.0)	$16.9^{+24.3}_{-9.8}$ (11.2 – 43.7)	$26.7^{+17.0}_{-10.3}$ (12.6 – 39.8)
Correlation coefficient	r^2	0.85	0.49	0.75	0.60
Incision rate	$\dot{E}$ (mm y^{-1})	4.6 ± 0.7	0.8 ± 0.07	0.8 ± 0.1	0.8 ± 0.1
Transport coefficient[c] ($n_t = 1$)	$K_t\ \beta$	$18.0 \pm 12.6 \times 10^{-5}$	$4.5 \pm 2.6 \times 10^{-5}$	$4.8 \pm 3.1 \times 10^{-5}$	$4.4 \pm 2.6 \times 10^{-5}$

[a]The quoted uncertainty on the intrinsic concavity corresponds to the 95% confidence limits

[b]The steepness index is estimated in two different ways: the numbers with error bars correspond to the estimates using the best-fit intrinsic concavity and its 95% confidence limits; the range quoted in parentheses encompasses 95% of the data points for a fixed θ = 0.4.

[c]The transport coefficient is estimated on the basis of a common θ = 0.4 for all rivers.

The intrinsic concavities of the slowly incising rivers are all remarkably similar, and their gradients at any drainage area are clearly lower than those of the rapidly incising Drac River. The intrinsic concavity can be evaluated by linear regression through the profile data only if erosion rate ($\dot{E}$) and erosion or transport coefficient (K_d or K_t) are constant along the rivers. This is clearly not the case for the Drac River, where the retreating knickpoint zone at least is out of equilibrium and where bedrock erodibility is highly variable (Fig. 4). In its headwaters, large crystalline boulders are delivered to the Drac River bed by debris flows and rockfalls from surrounding hillslopes. Gradients in these reaches are, therefore, neither controlled by the bedrock nor by the bedload and are excluded from our analysis. The intrinsic concavity for the Drac River was, therefore, only evaluated upstream of the retreating knickpoint within the alluvial segment to which the postglacial base-level fall has not yet been communicated. This reach should behave similarly to the nonglaciated transport-limited rivers. The intrinsic concavity of the upstream alluvial reach of the Drac River is not significantly different from that of the Buëch and Drôme Rivers, but the steepness index is. For the downstream bedrock reach in the Drac River, gradients are much steeper than for comparable drainage areas in the other rivers, but the large scatter in the data renders values for the intrinsic concavity and steepness index meaningless. We expect such scatter to occur, because of the variable lithologies, the alternation of transport- and detachment-limited reaches, and the fact that we cannot unambiguously demonstrate equilibrium downstream of the knickpoint.

The intrinsic concavity is sensitive to the precipitation gradient within the watershed (e.g., Roe et al., 2002). Latitudinal and longitudinal gradients in mean annual precipitation are low throughout the study area, although they increase to the NW due to northwesterly damp oceanic winds (Fig. 10). The orographic gradients are more important. The mean and maximum elevations of the Drac River catchment are distinctly higher than those of the highest nonglaciated catchments (cf. Table 1). We therefore have to ensure that the higher gradients of the Drac River are not due to orographic effects. To test this hypothesis, we use precipitation records of the meteorological stations located in the study area, obtained from the Météo France database (http://www.meteo.fr/meteonet/temps/france/clim/cli.htm#). The mean annual precipitation record is calculated over varying periods of time, depending on the time span of operation of the stations (~20–80 yr). From the data, we determined the orographic gradient in the Vercors Massif, located at the core of the study area. This massif is particularly suitable for such an analysis because the stations are located in large and shallow valleys that do not affect the orographic gradient at the regional scale. In the higher-relief Pelvoux Massif, in contrast, data are more scattered because rain often falls on the downwind valleys and not on the mountains where it is generated. The observed altitudinal precipitation gradient in the Vercors Massif is close to linear (Fig. 10).

The orographic effect on concavity is taken into account by weighting the contributing area of each DEM cell by its elevation. Latitudinal precipitation gradients can be neglected at the scale of the study area when compared with the altitudinal gradient. The weighting applied to the DEM is:

$$P_i = (0.65\ Z_i + 560) / 1032, \qquad (5)$$

where P_i is the weighting factor for grid cell I, and Z_i is its altitude. The denominator represents the predicted precipitation at the average elevation (726 m) of the study area. Drainage areas were then extracted from the weighted grid using the ArcView flow accumulation function. Resulting slope-area relationships for the Drac and Drôme Rivers are shown in Figure 11.

The figure shows that correcting for orographic effects, in fact, enhances the differences. As average precipitation is higher in the Drac catchment than in the nonglaciated catchments, the slope-area relationship becomes even steeper when using the weighted areas. The orographic weighting has negligible effect on the slowly incising rivers because they do not show large elevation differences. Differences in precipitation thus do not account for the observed differences in long profile gradients.

Of course, present-day records must be considered with caution to explain fluvial forms that develop over 10^4–10^5 yr time scales. We lack proxy records of paleoclimate gradients during the Holocene and Upper Pleistocene. We expect, however, that the orographic gradient would increase during damper climate conditions (e.g., beginning of the Holocene), and decrease when climate is drier (e.g., during the end of the last glaciation). Moreover, mean annual precipitation cannot be used to assess the effect of large-magnitude flood events, which potentially could play an important role in controlling fluvial morphology (e.g., Snyder et al., 2003b). A precise study of this effect from a statistical analysis of river gauging data (e.g., Tucker and Bras, 2000) would, however, require the same extrapolation when applied as a surrogate for paleoclimate.

It is commonly assumed that the intrinsic concavity of transport-limited systems is lower than that of detachment-limited rivers, although there is in fact very little data to support this assumption (cf. Whipple and Tucker, 2002, and references

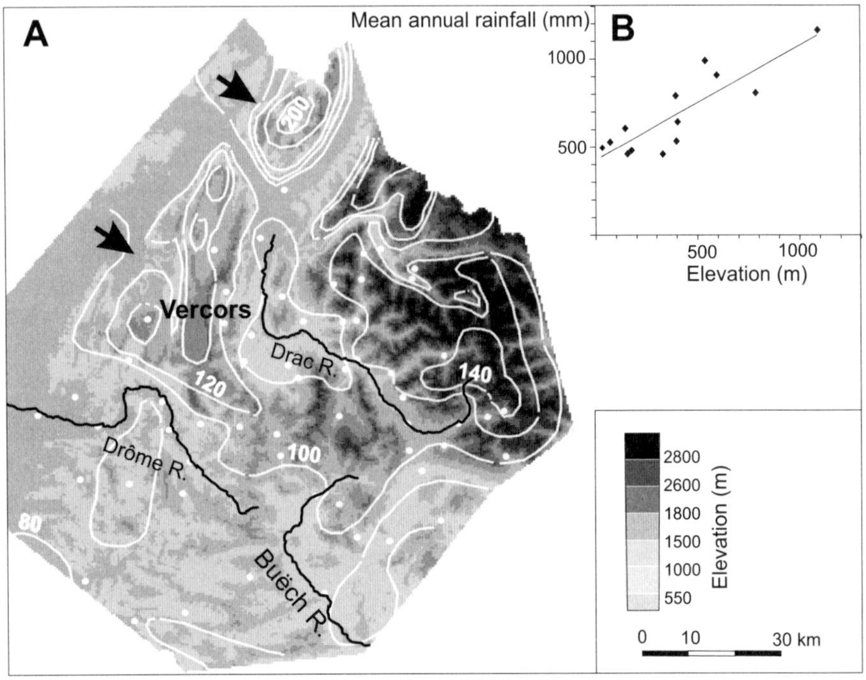

Figure 10. Assessment of the orographic precipitation gradient in the study area. (A) Map of mean annual precipitation (in mm) measured at Météo-France meteorological stations (white dots) overlain on coarse digital elevation model (DEM) topography. Black arrows indicate prevailing moisture transport from the NW. (B) Plot of precipitation values for the Vercors stations as a function of altitude. The regression yields: Precipitation (mm) = 0.65 × elevation (m) + 560, with r^2 = 0.74.

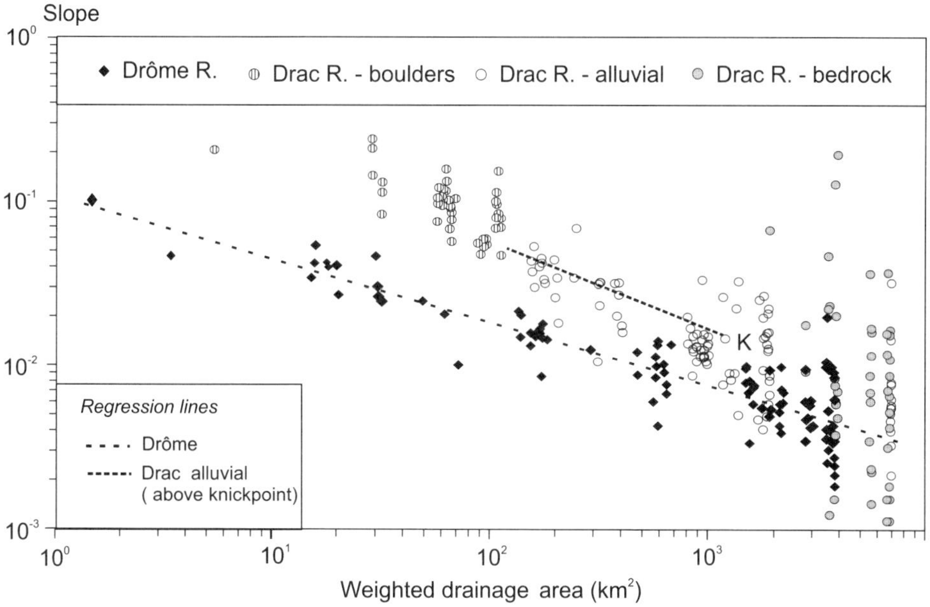

Figure 11. Slope-area plot of the Drac and Drôme Rivers after weighting the drainage areas for orographic precipitation (see text for discussion). The correction exacerbates the difference in gradient between the slowly and rapidly incising rivers; intrinsic concavities for these slope–weighted area plots are 0.39 ± 0.08 ($r^2 = 0.44$) for the Drôme River and 0.60 ± 0.13 ($r^2 = 0.59$) for the alluvial reach of the Drac River. K—knickpoint.

therein). We have argued herein that most of the downstream reach of the Drac River behaves as a detachment-limited system, whereas the Buëch and Drôme Rivers are transport-limited. The slope-area relationships (Fig. 9) do not contradict the hypothesis that the intrinsic concavity of transport-limited systems is lower than that of detachment-limited systems, even though no statistically meaningful intrinsic concavity can be calculated for the downstream reach of the Drac River.

From a comparison of equations 3a and 4a, it can be shown that higher incision rates only favor detachment-limited conditions if $n_d < n_t$ (Howard, 1980; Whipple and Tucker, 2002). The value of n_d has been assessed to be around 2/3 (i.e., shear stress dependent) in the case of plucking-dominated incision, and 5/3 if abrasion prevails (Whipple et al., 2000). Field observations of the Drac River bedrock morphology suggest incision dominated by plucking (see "Channel Morphology"), which implies $n_d \approx 2/3$. Therefore, our data suggest the value of n_t to be >2/3, and possibly close to 1, which is consistent with gravel bedload transport in rivers (cf. Whipple and Tucker, 2002, and references therein).

The slope-area plots of the slowly incising rivers are clustered when compared with the great dispersal of the Drac River values. As uplift and incision rates can be regarded constant in the nonglaciated area, the clustering suggests that rivers in this area are characterized by very similar K_t values (Table 2). This indicates that either the caliber of the bedload evolves similarly along the transport-limited rivers, or that fluctuations in caliber generate only mild variations in gradient. We currently lack detailed bedload caliber data along these streams and are therefore not able to resolve this question in detail. The steeper long profile gradient in the alluvial upper reach of the Drac River appears to be consistent with the larger incision rate and translates into an insignificantly different K_t value for this reach with respect to the slowly incising rivers (Table 2).

THE TRANSITION FROM DETACHMENT-LIMITED TO TRANSPORT-LIMITED CONDITIONS

It has been demonstrated theoretically (e.g., Howard, 1980; Whipple and Tucker, 2002) that a river can change from detachment- to transport-limited behavior along its course if the intrinsic concavities for transport-limited and detachment-limited incision are different. In the case of uniform lithology, uplift rate, sediment flux, and river transport capacity, the river will shift from detachment- to transport-limited behavior downstream if $\theta_d > \theta_t$ (Fig. 12). The critical area where the transition occurs (A_{cr}) can be

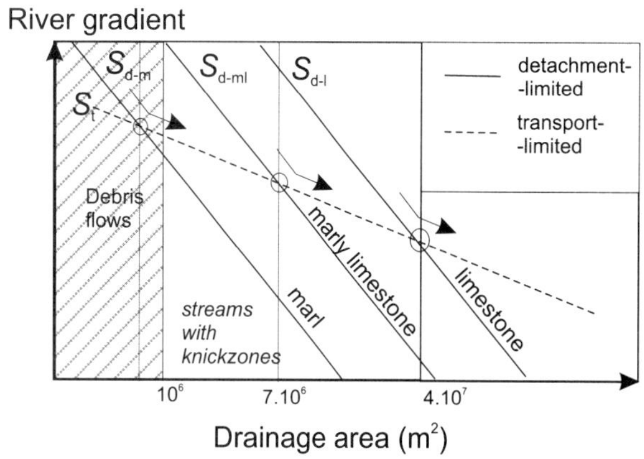

Figure 12. Sketch diagram showing the location of the detachment- to transport-limited transition and the influence of bedrock strength. The equilibrium river gradient corresponds to the greater of the detachment- and transport-limited slopes (S_d and S_t, respectively; the subscripts m, ml, and l refer to marl, marly limestone, and limestone, respectively). The figure assumes $\theta_d > \theta_t$. Inspired by Whipple and Tucker (2002, their Fig. 1).

found by stating $S_d = S_t$ in equations 3 and 4 and solving for area (Whipple and Tucker, 2002):

$$A_{cr} = [K_d^{1/n_d}(K_t / \beta)^{-1/n_t}]^{1/(\theta_t-\theta_d)} \dot{E}^{(1/n_t-1/n_d)/(\theta_t-\theta_d)}. \quad (6)$$

We have shown that the slowly incising rivers switch from detachment-limited conditions (where they cross resistant rocks) upstream to transport-limited conditions downstream. Our study area is particularly suitable to test the influence of bedrock erodibility (K_d) and incision rate ($\dot{E}$) on the critical drainage area (A_{cr}). Other parameters are more difficult to test, but the influence of K_t and β could be evaluated in further studies of these rivers.

Influence of Bedrock Lithology

Slope-area plots should show a kink at the critical drainage area if $\theta_d \neq \theta_t$. However, as lithology varies frequently along the streams we have studied here, the detachment-limited intrinsic concavity θ_d is difficult to evaluate (cf. Fig. 9). Moreover, different rivers will not likely cross the same lithology at the point where the transition occurs, and the transition will thus not be observed directly.

In order to constrain the critical area, we have, therefore, made an inventory of the occurrence or absence of knickpoints where rivers cross the most resistant lithological units. Drainage areas at these locations were extracted from the 50-m-resolution DEM. In total, some 200 knickpoints and 60 gorges without knickpoints were investigated (Fig. 13).

As predicted by the model, the critical drainage area scales with bedrock resistance: the most resistant units (Tithonian and "Urgonian" limestones) show knickpoints, and thus detachment-limited behavior, for the largest areas. In contrast, knickpoints disappear at smaller drainage areas for lesser-resistant units (e.g., Hauterivian marly limestone). The upper threshold of knickpoint disappearance along the entire river system is set by the Tithonian limestone threshold and occurs between 30 and 45 km².

The precise value of the critical area is, however, difficult to assess, because at intermediate drainage areas some rivers demonstrate knickpoints when flowing over resistant units whereas others do not. Variations in the nature and caliber of sediment load, sediment flux, lithological strength, and bedding strike and dip are the most probable factors responsible for this scatter. Preliminary field observations suggest that variations in bedrock bedding and fracturing, together with bedding dip, are the primary cause. The Gervanne River shows a dramatic illustration of this effect (Fig. 5): it does not show a clear knickpoint where it flows over steeply dipping Urgonian limestone, whereas it flows over the same but flat-lying unit in a spectacular knickpoint 3 km downstream. No evidence for possible groundwater sapping has been observed in the lower knickpoint. In addition to these effects, some knickpoints may be buried under ephemeral sediment accumulations or landslides. Finally, stochastic variations in sediment supply and river carrying capacity are expected to cause lateral shifts in the transition along a river (e.g., Whipple and Tucker, 2002). A part of the scatter could therefore also be explained by temporal displacements of the detachment- to transport-limited transition, triggered by high-frequency variations in the climatic forcing of the rivers' transport capacity. An unknown

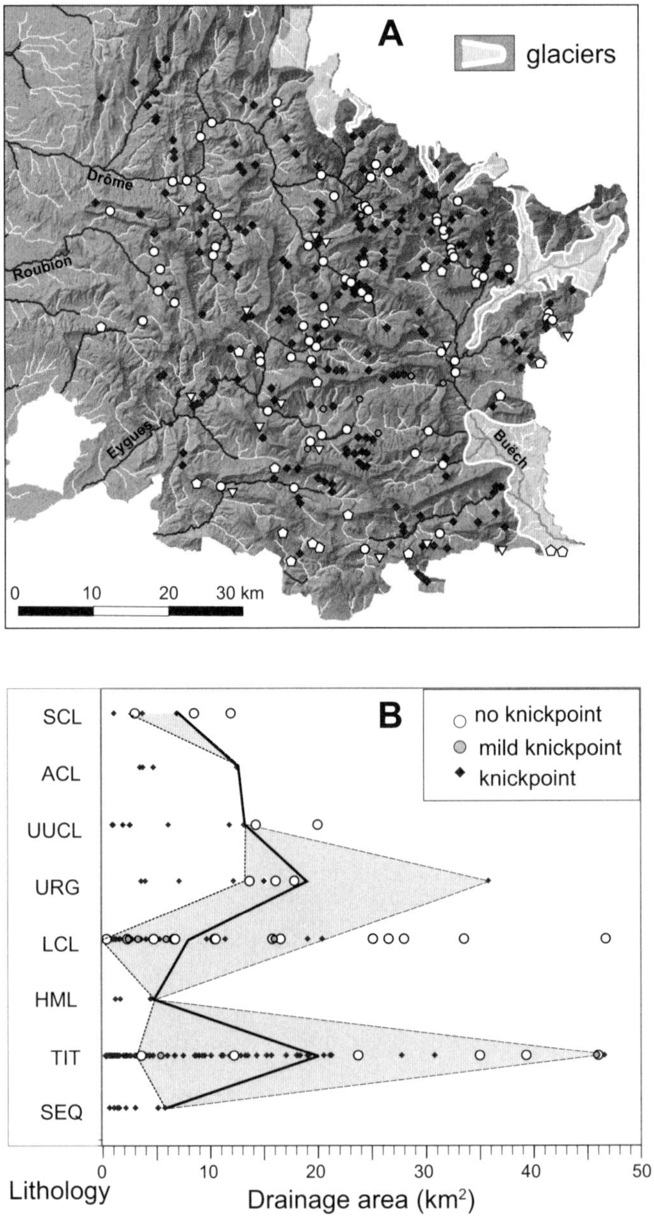

Figure 13. Critical area (A_{cr}) for the detachment- to transport-limited transition for some of the hardest lithological units in the study area. (A) Location of the studied reaches with and without knickpoints. The drainage network is displayed in white for drainage areas larger than 2.5 km² and in black for drainage areas larger than 38 km². Windgaps (pentagons) and captures likely to occur in the near future (inverted triangles) are also indicated. (B) Locations of knickpoints, subdued knickpoints, and reaches without knickpoints, as a function of drainage area and bedrock type (see Fig. 3 for abbreviations). Dashed line—largest area for occurrence of knickpoints; dotted line—smallest area of appearance of non-knickpoint reaches; heavy line—weighted mean area for overlapping data.

number of knickpoints and non-knickpoint reaches could thus be relict, nonequilibrium features.

We have argued herein that it is reasonable to consider that uplift is homogeneous throughout the study area, but this cannot be firmly demonstrated. Thus, differences in uplift rate could also account for part of the dispersion. However, the data do not show any marked and logical spatial trend. If uplift gradients play a role in the dispersion, its effects are hidden by the addition of other processes.

In Figure 12, we have placed the critical area for three types of lithology according to our analysis (Fig. 13), to show how bedrock resistance to erosion controls the area at which streams become transport-limited. The transition in the marly levels is difficult to observe because the critical area is small (<10^6 m^2) and possibly within the debris-flow realm (Stock and Dietrich, 2003). It also appears to be strongly controlled by the bedload characteristics. Marls such as the Oxfordian Black Shales only deliver small blocks of sandstone and vein calcite (typically 1 cm in length) as bedload to the streams. If the upstream watershed is only composed of such marls, badlands develop and streams are detachment-limited (e.g., Howard and Kerby, 1983). When the badlands propagate into the overlying marly limestones, bedload supply and caliber increase and the streams become transport-limited. Notwithstanding these uncertainties, the ~40-fold increase in critical drainage area between the weakest and most resistant lithological units in our study area may be used to constrain variations in erodibility (K_d) between these units from equation 6 if we suppose $n_d = 2/3$ (as inferred previously). To do this requires an estimate for θ_d, however, which we will derive in the following.

Influence of Incision Rate

Using the incision rates inferred from cosmogenic dating (Brocard et al., 2003), the nonglaciated rivers and the Drac River can be used to put loose constraints on the evolution of the critical area (A_{cr}) with incision rate, for some of the main bedrock types encountered (Fig. 14). Values of A_{cr} for the slowly incising rivers are extracted from Figure 13. As explained above, the critical area in marls is not easy to define but it is <1 km^2. Along the Drac River, only the value for marly limestones can be well constrained, as small transport-limited reaches occur on this lithology downstream. It appears to be nearly two orders of magnitude larger than the corresponding critical area for the slowly incising rivers. The probable location of the transition for marls in the Drac River is currently occupied by a glacial trough filled with sediments. The Drac River does not cross Tithonian limestone in its downstream reaches. However, it does cross lesser-resistant lithologies that develop knickpoints. These less-resistant units thus provide a lower limit for the critical area in Tithonian limestone. Altogether the data, although limited, are consistent with a power-law dependence of critical drainage area on incision rate, as predicted by equation 6. From a comparison of A_{cr} for marly limestones, we estimate the dependence to be close to cubic

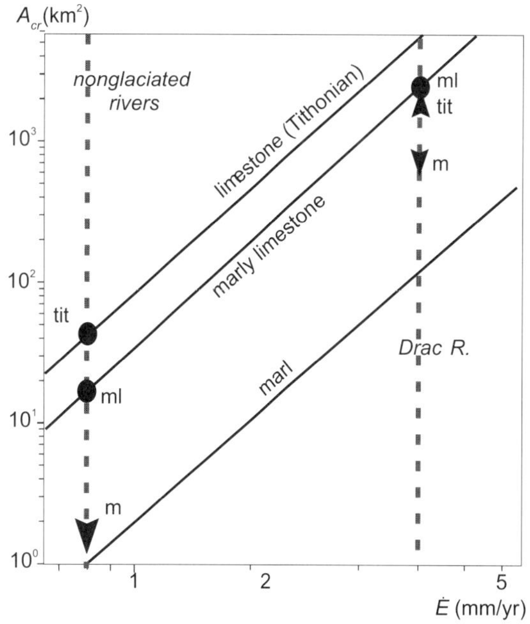

Figure 14. Sketch diagram of the evolution of the critical drainage area (A_{cr}) with incision rate ($\dot{E}$) in the case where $\theta_d > \theta_t$ and $n_d < n_t$. Three different lithologies (*m*—marl; *ml*—marly limestone; *tit*—Tithonian massive limestone) are considered and positioned according to the available field data. See text for discussion. Dots indicate that the critical area is constrained; upward and downward facing arrows indicate minimum and maximum estimates for A_{cr}, respectively.

(Fig. 14), which is consistent with equation 6 for $n_t = 1$, $n_d = 2/3$, $\theta_t \approx 0.4$, and $0.5 < \theta_d < 0.6$.

VALLEY FLAT DEVELOPMENT: A DIAGNOSTIC FEATURE OF TRANSPORT-LIMITED SYSTEMS

Description and Behavior of Valley Flats in Western Alpine Rivers

In our study area, we observe that valley flats wider than the river channel develop systematically as soon as the rivers become transport-limited (i.e., as soon as lithological knickpoints disappear). Along small, slowly incising streams, and along large and intermediate, rapidly incising rivers (i.e., the Drac River and its tributaries), valley flats only develop in marly levels. Within the same stratigraphic unit, the valley flat widens with increasing discharge. It also propagates into more resistant limestone units downstream (e.g., Fig. 15). In the field, we thus find the following suite of fluvial forms with increasing drainage area: (1) rivers showing a series of gorges with knickpoints separating reaches with well-developed valley flats; (2) rivers with a suite of valley flats of varying width, separated by narrow valley-bottom segments with or without knickpoints; and finally (3) rivers without knickpoints, which have a continuous valley flat of highly variable width. The development of a valley flat significantly larger

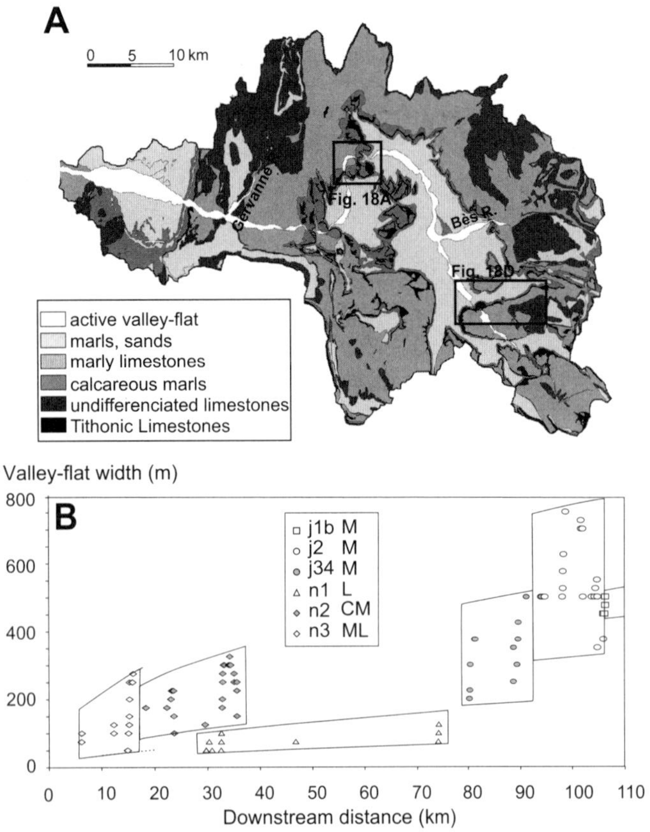

Figure 15. Effect of bedrock lithology on the valley-flat width. (A) Simplified lithological map of the Drôme watershed. The valley flats of the Drôme River and some of its largest tributaries are superposed where wide enough to be represented. Boxes indicate detailed topographic maps of Figure 18. (B) Downstream variations in valley-flat width along the Céans River and corresponding bedrock lithology. Boxes indicate upper and lower limits of calibrated widening rates for each lithology (cf. Fig. 16; Table 3). M—marls; L—limestones, CM—calcareous marls, ML—marly limestones. Codes for stratigraphic units are defined in Figure 3.

than the channel is thus a diagnostic feature of the transport-limited reaches.

The valley flats are occupied by braided channels where these have not been contained by dikes. Meandering reaches are rare and restricted to smaller streams located at the western margin of the field area, near the Alpine foreland (e.g., Roubion River, Fig. 2). The valley flats are generally composed of thin (2–5 m) sand-and-gravel deposits overlaying straths. Low fill-terrace remnants can cover some parts of the straths, as observed along the Buëch and Drôme Rivers, and are composed of sand-and-gravel channel-bed deposits with locally very thick covers of overbank deposits. Previous ^{14}C dating suggests that these fill terraces were emplaced during distinct climatic events since the end of the last glaciation (Rosique, 1996).

Valley-flat widths range from a few tens of meters to several kilometers. The widest valley flat encountered is that of the Drôme River where it flows into the foreland (Fig. 15A); it is 3 km wide, but the fluvial deposits covering it are no more than 12 m thick (Mandier, 1988). The main factors that control valley-flat width in our field area appear to be bedrock lithology, river discharge, and incision rate.

To investigate these controls, we measured the valley-flat width at 3200 sites along 45 rivers from 1:25,000 scale topographic maps. Only the most common lithological units were investigated. We avoided transition zones from one bedrock type to another because of possible boundary effects, as well as valley flats obviously underlain by fill deposits of anomalous thickness and bedrock types that do not favor the preservation of clear valley-flat borders due to intense landsliding. Where proper segments could be found, valley-flat width was measured every 150 m. Drainage area and downstream distance at each site were extracted from the 50-m-resolution DEM.

Influence of Bedrock Lithology

Along all the investigated streams, variations in valley-flat width are strongly controlled by bedrock lithology. Lithological forcing can observed directly in map view (e.g., for the Drôme River, Fig. 15A) or by plotting valley width and lithology along a single river (e.g., Céans River, Fig. 15B). Valley-flat widths vary suddenly where the rivers cross lithological boundaries. The local lithological units are mixes between limestones and clays; valley-flat width correlates positively with clay content in the bedrock. However, there is large scatter in the data: the variability of valley-flat width along a single bedrock segment is in many cases greater than the difference between two successive segments of different lithologies, or even than the difference between the mean widths of two different bedrock segments.

Influence of Drainage Area

The valley-flat width increases with discharge along single streams, as can be clearly observed along reaches crossing a single lithological unit (e.g., Fig. 15B). However, it is necessary to correct for the lithological effect in order to better observe the influence of discharge. To achieve this, we have plotted the valley-flat width as a function of drainage area for different bedrock lithologies, compiling the width data for all rivers. We expect that valley-flat width (W_{vf}) should have a power-law dependence on drainage area (A), as has been observed elsewhere (Snyder et al., 2003a; Tomkin et al., 2003); we will provide arguments for and a derivation of this relationship in a later section:

$$W_v = K_v A^{c_v}, \quad (7)$$

where K_v is a pre-exponential coefficient with dimension $[L^{(1-2c_v)}]$ that we will term the "widening factor" and c_v is a dimensionless exponent.

The data show, as expected, that valley-flat width scales with drainage area (Fig. 16; Table 3). The best-fit values of c_v overlap the interval 0.3–0.4 with 95% confidence whatever the lithology

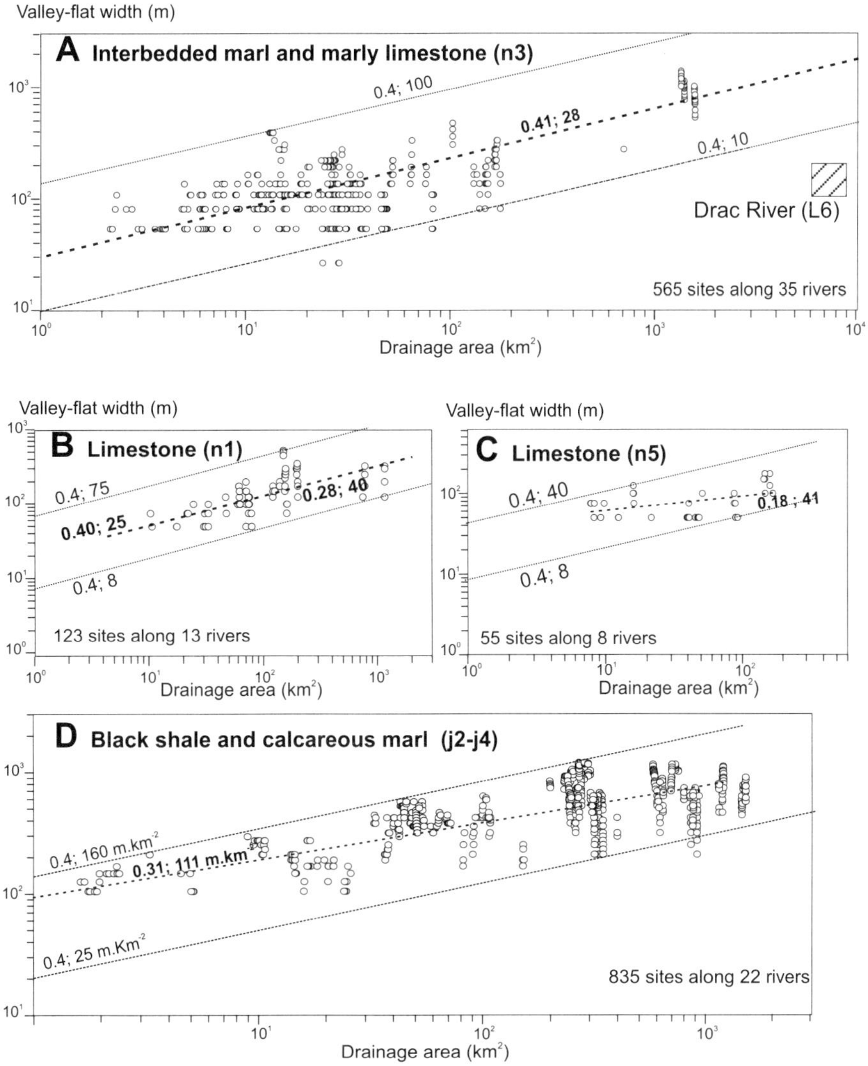

Figure 16. Examples of downstream increase of the valley-flat width in various bedrock types. The first number indicated on the straight lines is the exponent c_v in equation 7, while the second number is the widening factor (K_v). Dashed lines are regressions on the data, continuous lines contain 95% of the data for $c_v = 0.4$. Some of the main types of bedrock lithologies are presented: marly limestone (A); well-bedded (B) and massive (C) limestone; shale and marls (D). Stratigraphic codes j2, j4, n1, n3, and n5 are defined in Figure 3. In A, the data of the slowly incising rivers are compared to those for the rapidly incising Drac River (hatched box) for lithologically comparable stratigraphic units: the Hauterivian (n3) and Aalenian (l6).

TABLE 3. VALLEY-FLAT WIDTH DATA

Stratigraphic unit	Lithology	No. of sites	K_{vf}	c*	r^2	Range of K_{vf} for $c = 0.4$#
c3	sandy limestone	102	47	0.34 ± 0.11	0.22	18 – 130
n6	marls	153	157	0.11 ± 0.06	0.07	24 – 135
n5	massive limestone	55	41	0.18 ± 0.13	0.11	8 – 40
n4	well-bedded limestone	152	47	0.21 ± 0.07	0.14	10 – 75
n3	marly limestone	565	28	0.41 ± 0.03	0.61	10 – 100
n2	marls	226	63	0.29 ± 0.07	0.29	12 – 100
n1	well-bedded limestone	118	21	0.40 ± 0.07	0.42	8 – 75
j6	limestone	22	41	0.18 ± 0.21	0.08	5 – 38
j5	limestone	47	54	0.18 ± 0.12	0.12	4 – 42
j2-j4	black shales	835	111	0.31 ± 0.03	0.46	25 – 160

*Value of the exponent c in Eq. (7) with 95% confidence interval and K_{vf} calculated by regressions on the data set for some of the most common rock units of the field area.

#Lower and upper values of the K_{vf} factor bracketing the valley-width dataset for the same rock units, taking an exponent $c = 0.4$ for the drainage area dependency.

investigated, except for the Aptian marls (n6). The best-resolved values of c_v (e.g., for j2–j4, n1, n3) all lie within the interval 0.3–0.4. There appears to be a tendency for c_v to decrease in the most resistant limestones (e.g., j5, j6, n4, n5), but this trend is not statistically significant.

Minimum and maximum values of the widening factor K_v (taking $c_v = 0.4$) are plotted as dotted lines for a few lithological units (Fig. 16). The valley-flat width increases significantly faster in more erodible lithologies, such as the mid-Jurassic marls (j2–j4: K_v = 25–160 m km$^{-0.8}$), than in the resistant mid-Cretaceous ("Urgonian") limestone (n5: K_v = 8–40 m km$^{-0.8}$). Within the Hauterivian marly limestones (n3), valley-flat width increases at an intermediate rate (K_v = 10–100 m km$^{-0.8}$).

The relatively high scatter of the limestone data sets stems from the higher relative error on width measurements in addition to various other factors. For example, the high scatter for the c3 limestones and n6 marls can be partly explained by non-negligible lateral facies variations in these units. In addition, some valley widths in c3 were measured in the hinge of a syncline located near the foreland that could be actively folding (cf. Fig. 2). The valley width at this locality is unexpectedly large. This can be either explained by the low dip angle of the strata or by very low uplift and incision rates in the synclinal hinge (see below).

Influence of Incision Rate

Valley flats are commonly thought to narrow with increasing incision rate. From field observations, the valley flat was found either to follow such a trend along a single river (e.g., Harbor, 1998) or to not correlate with uplift rate when comparing different rivers (Snyder et al., 2003a). We can test the potential influence of incision rate by comparing the valley-flat width in the most downstream section of the rapidly incising Drac River, where it exhibits transport-limited characteristics, with that of the slowly incising rivers. Because the Drac River cuts through Lower Jurassic series in its downstream reaches, whereas the slowly incising rivers incise Upper Jurassic and Cretaceous rocks, we cannot directly compare data for the same unit. We therefore compare the bedrock underlying the Drac valley-flat (Aalenian marly limestone) with the most similar unit, in terms of lithological composition and bedding thickness, of the nonglaciated area (Hauterivian marly limestone; Fig. 16A). For similar drainage areas, the valley flat of the Drac River is an order of magnitude narrower than the valley flats of the slowly incising rivers. This observation clearly supports the common assumption of a narrowing of valley flats with increasing incision rate, and suggests the relationship between valley-flat width and incision rate to be inversely linear.

A Model for Valley-Flat Development: The Erosion Frequency

From the field data, it appears that the valley-flat width along transport-limited rivers is controlled by the same factors that control the gradient of detachment-limited rivers. In contrast, the gradient of the transport-limited rivers is set by the bedload caliber. By definition, incision rate and bedrock strength do not modulate the transport-limited stream gradient directly; they only act indirectly on the profile shape by affecting the flux and caliber of the slope debris delivered to the streams. The fact that valley flats develop as soon as detachment-limited conditions disappear is a strong indication that the valley flat accommodates most of the variations in bedrock strength and incision rate along transport-limited rivers.

Factors Controlling Valley-Flat Width

We propose that valley-flat width reflects the frequency of strath erosion. Along transport-limited reaches, erosion occurs during stages that are high enough to rework the alluvial layer down to the strath. The braided and meandering streams wander over the strath from one side of the valley flat to another. Over large periods of time, the frequency of erosion of any point of the strath (F) should be independent of the frequency of channel switching and be set primarily by the widths of the active channel (W_c) and valley flat (W_v) (e.g., Tomkin et al., 2003):

$$F = W_c / W_v. \tag{8}$$

The erosion frequency will be controlled by ratio of the actual (transport-limited) incision rate and the potential (detachment-limited) incision rate:

$$F = \dot{E}_t / \dot{E}_d, \tag{9}$$

so that:

$$W_v = (\dot{E}_d / \dot{E}_t) W_c. \tag{10}$$

Combining equation 10 with equations 1 and 2c for detachment- and transport-limited incision, respectively, gives:

$$Wv = \left(\frac{K_d A^{m_d} S^{n_d}}{K_t A^{(m_t - 1)} S^{n_t}} \right) W_c = \left(\frac{K_d}{K_t} \right) A^{(m_d + 1 - m_t)} S^{(n_d - n_t)} W_c. \tag{11}$$

Relationship between Valley-Flat Width and Drainage Area

We did not compare W_v and W_c directly along the rivers, because W_c varies at very short time scales. Instead, we use a relationship that links channel width (W_c) and drainage area (A), in the scope of establishing a time-averaged channel width. This also allows us to investigate the direct relationship between valley-flat width and drainage area described previously.

The active channels of transport-limited streams flow over a bed composed of channel sand-and-gravel. They should thus behave like alluvial rivers (Howard, 1998; Whipple and Tucker,

2002) and their bankfull channel width (W_c) should depend on the river discharge (Q) (e.g., Leopold and Maddock, 1953):

$$W_c = K_a Q^a, \tag{12}$$

where K_a is a dimensional coefficient [$L^{(1-3a)}$] and a is a positive exponent; a has frequently been shown to be close to 0.5 (e.g., Leopold and Maddock, 1953). The drainage area (A) is generally used as a proxy for (Q) through a power-law relationship for basin hydrology:

$$Q = K_q A^b, \tag{13}$$

where K_q is a dimensional coefficient [$L^{(3-2b)}$] and b a positive exponent; b generally lies between 0.7 and 1 (e.g., Talling and Sowter, 1998; Stock and Montgomery, 1999; Snyder et al., 2003a). The relation between bankfull channel width and drainage area leads to the functional form:

$$W_c = K_c A^c, \tag{14}$$

with $K_c = K_a K_q$ and $c = a\,b$. Combining equations 11 and 14 yields:

$$\begin{aligned} W_v &= \left(\frac{K_d A^{m_d} S^{n_d}}{K_t A^{(m_t-1)} S^{n_t}}\right) K_c A^c \\ &= \left(\frac{K_d K_c}{K_t}\right) A^{(c+m_d+1-m_t)} S^{(n_d-n_t)} \end{aligned}. \tag{15}$$

We can now use the power-law relationship between slope and area for transport-limited rivers (equation 4a) to retrieve a relationship between valley width and drainage area as in equation 7:

$$\begin{aligned} W_v &= \left(\frac{K_d K_c}{K_t}\right) A^{(c+m_d+1-m_t)} \left[\left(\frac{\beta \dot{E}}{Kt}\right)^{\frac{1}{n_t}} A^{\frac{(1-m_t)}{n_t}}\right]^{-(n_d-n_t)} \\ &= \left(\frac{K_d K_c}{K_t}\right)\left(\frac{\beta \dot{E}}{K_t}\right)^{\frac{(n_d-n_t)}{n_t}} A^{\left(c+m_d \frac{n_d(1-m_t)}{n_t}\right)} \end{aligned} \tag{16}$$

Whether the relationship between channel and valley width is linear thus depends on the values of m_d, m_t, n_d, and n_t. For the values inferred in the previous section ($n_t = 1$; $n_d \approx 2/3$; $(m_t - 1)/n_t \approx 0.4$; $0.5 \leq m_d/n_d \leq 0.6$) the scaling between W_v and W_c should be close to linear. Recent studies have come up with power-law exponents for valley widths that were either similar to those for channel widths (Snyder et al., 2003a) or significantly larger than these (Tomkin et al., 2003).

We have constrained values of c_v (equation 7) for different lithological units and showed that best-fit values lie between 0.3 and 0.4. These values are similar to or lower than values for c obtained by measuring downstream variations in channel width from 1:25,000 scale topographic maps. For the present-day Drôme River, we determined $c = 0.57$ ($r^2 = 0.66$). The present-day data are not representative of the long-term character of the river, however, because river width has been reduced artificially in many places during the twentieth century (Gautier, 1992; Landon, 1999). Scarce eighteenth century data for the Buëch River in its "natural" state (Gautier, 1992) suggest that $c \approx 0.35$. We therefore consider a value for c of 0.57 as a maximum estimate. At present, we cannot unambiguously decide whether the channel-width and valley-width exponents are the same in our study area. This would require making an inventory of nondisturbed channel widths, and these are difficult to find. Moreover, one could argue that the bedrock strath is only stripped of its sediment cover during peak discharges. Because such floods are triggered by precipitation events on restricted areas, the exponent b that relates discharge to drainage area may decrease with discharge, and the value that is appropriate for scaling bankfull channel width and area may not be the relevant scaling factor for valley width.

According to the model, bedrock erodibility acts on the valley-flat width through the parameter K_d (equation 16). As expected, K_v values are lower for resistant rocks and larger for softer rocks (Fig. 16; Table 3). However, the scatter in the data is too large to propose characteristic values for each lithological unit. The dispersion in the data, like that for the critical area of the detachment- to transport-limited transition, can be ascribed to a variety of factors. In order of decreasing supposed importance we can quote: the volume and caliber of the sediment load, variations in the composition of lithological units at the regional scale, and bedding strike and dip with respect to valley orientation.

Substituting Spatial Changes in Discharge with Temporal Changes in Discharge

Temporal fluctuations in river regime and dynamics are responsible for the formation of paired strath terraces, as observed along the Drôme, Buëch, and Gervanne Rivers, by successive valley-flat enlargement and reduction. These terraces attest to the effects of river carrying capacity, sediment load, and caliber in modulating valley widths. Where paired terraces are preserved, it is possible to estimate the former valley-flat width. The most suitable level for such an analysis is the T2 level of the Drôme River, supposed to be 60–75 k.y. old. We measured the width of the paleo–valley flat represented by this level where possible and extracted drainage areas at these sites from the 50-m-resolution DEM, assuming that drainage areas have not evolved through time. This assumption is reasonable, considering that no significant stream capture has occurred in the catchment since the abandonment of the terrace level.

As for downstream increases in valley-flat width, our model predicts that bedrock strength should modulate the rate at which the valley flat enlarges during cold periods. The ratio of the paleo–valley width over the active valley width should be constant within each bedrock unit. Our analysis suggests that the valley flat was 1.2–2 times wider than present during the last glaciation (Fig. 17). There appears to be some correlation between widening factor and bedrock erodibility, but the scatter of the data does not allow quantitative investigation of bedrock control on widening. Moreover, widening on the Oxfordian marly

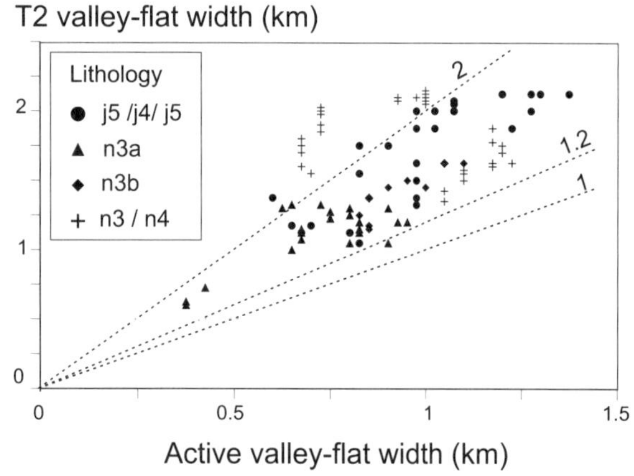

Figure 17. Effects of varying river regime on the valley-flat width: example of the Drôme River. Comparison of the widths of the present-day valley flat with that of the T2 terrace level. Lines indicate some values of the ratio T2 width/current width. Lithology: for stratigraphic significance, refer to Figure 3. J5 and n3 are rhythmic series of interbedded marly limestone and marls (n3a marlier than n3b); n4 is thickly bedded limestone. The / sign indicates that the valley flat covers several lithological units.

levels (j4) has been restricted because the terrace edges encountered Kimmeridgian marly limestones (j5) on both sides of the valley. A similar phenomenon occurred when the paleo–valley flat widened in the Hauterivian marly limestones (n3) until it encountered Barremian limestone (n4). Our estimates of widening factors in these lithologies are therefore minimum estimates.

As is the case for the detachment- to transport-limited transition zones, climate variability during the Pleistocene could partly explain the scatter in valley-flat widening factors, because it should act on river regime and overall discharge. It is also possible that the rock strength itself depends on climate through the probable development of thick permafrosts. The variability of the valley-flat width in a given rock type may, finally, also be explained at some places where the terraces have not evolved over a time long enough to reach the potential valley-flat width. Valley-flat widening in response to changes in discharge and sediment flux is a process that may take many thousands of years (Hancock and Anderson, 2002). This effect may be especially important along the marly levels where the erosion frequency is low, because the valley flat is very wide. Correcting for it requires an assessment of valley-flat maturity. Mature valley flats should exhibit poorly curved borders with less intense and frequent variations in strike than their active channels. Immature reaches are characterized by portions of valley flanks with irregularities that are more similar to that of the active channels.

DISCUSSION

Our field observations in western Alpine rivers lead us to propose a model for fluvial response to variations in bedrock strength, bedload supply, and tectonically or climatically driven incision rates that combines detachment-limited and transport-limited behavior.

We have shown that slowly incising "equilibrium" streams in our study area are detachment-limited in their upstream reaches and show a transition to transport-limited behavior downstream. The Drac River, which incises an order of magnitude more rapidly because it has been pushed out of equilibrium during glaciations, shows an inverse pattern: it is transport-limited upstream of a retreating knickpoint, which represents the present-day limit to where postglacial base-level drop has been communicated up the river, and detachment-limited downstream. Observations and characteristics that permit us to distinguish between transport-limited and detachment-limited behavior include the channel morphology, river gradients, the occurrence or absence of lithological knickpoints, and the development of a valley flat. The latter two characteristics are the most diagnostic: within detachment-limited reaches, rivers react to variations in bedrock strength by adjusting their gradient, whereas in transport-limited reaches, they react by adjusting their valley-flat width.

By mapping out the occurrence of knickpoints, we have constrained the critical area for the transition from detachment-limited to transport-limited behavior and have quantified the influence of rock strength and incision rates on the location of this transition. These data allow us to make estimates of model parameter values that we will outline further in the following. An important aspect of our model is the behavior of transport-limited streams: their gradients are set by bedload supply and caliber, and are not directly influenced by bedrock strength or incision rate. There is, however, an indirect control of these parameters on transport-limited gradients through the dependence of the flux and caliber of bedload that is delivered to the streams on erosion rate and rock type. We have shown that transport-limited rivers respond to variations in rock strength and incision rate by adjusting their valley-flat widths in a predictable and quantifiable manner.

Many of our observations of the interrelationships between bedrock strength, incision rate, discharge, river gradients, and valley widths are corroborated by similar observations in comparable settings, e.g., for streams in northern California (Snyder et al., 2003a) and Oregon (Personius, 1995); the Clearwater River in the Olympic Mountains (Pazzaglia and Brandon, 2001; Wegmann and Pazzaglia, 2002), the Sevier River in Utah (Harbor, 1998), or rivers crossing the Lesser Himalayas of Nepal (Lavé and Avouac, 2001; Attal and Lavé, 2005). Our data set contributes to the global set of observations with the advantage that it was collected in a restricted area with constant tectonic uplift rates but strongly varying incision rates, for rivers that are both in and out of equilibrium, and locally strongly variable but regionally homogeneous bedrock.

The conceptual model we distill from our observations has important implications for river incision models, the coupling between uplift/incision rates and river profiles in equilibrium settings, and the stability of drainage networks. We will concisely explore these consequences now.

Ability of Models to Reproduce the Observed Fluvial Morphology

Models used to study fluvial incision at the orogen scale are often based on a simple detachment-limited description (e.g., Whipple et al., 1999; Willett et al., 2001). Our study of the western Alps shows that many rivers do not behave as detachment-limited bedrock rivers, and would therefore be inappropriately modeled by such a description. Our data indicate that the drainage net is a mixture of detachment- and transport-limited reaches, the pattern of which is controlled by a transition that depends on drainage area, lithology, and incision rate. Such a pattern has recently been proposed from theoretical considerations (Whipple and Tucker, 2002). Our data are consistent with the Whipple and Tucker (2002) model of combined detachment-limited and transport-limited behavior, and are accurate enough to evaluate some parameters of the model.

The steady-state solution to the transport-limited model that we use in equation 4 assumes that the river incision rate is equal to the catchment-wide erosion rate (Whipple and Tucker, 2002). In the nonglaciated catchments, this assumption appears reasonable, since sediment storage is reduced to thin layers of colluvium on the slopes. Moreover, short-term catchment denudation rates, calculated from sediment mass balances integrated over a few decades (Alary, 1998) to centuries (e.g., Brocard, 2004), are not significantly different from long-term river incision rates. Finally, in many places, the valley flanks preserve large pediments that bear witness to former stages of valley development. No changes in slope steepness or flank amplitude can be noticed between pediment surfaces and active valley flanks, suggesting that the relief is in steady state.

From a comparison of equations 3 and 4, it appears that the downstream transition from detachment- to transport-limited conditions requires m_d/n_d to be larger than $(m_t - 1)/n_t$, whereas the tendency toward detachment-limited conditions at higher incision rates requires that n_d be smaller than n_t. Since we suspect plucking to be the dominant incision process in the detachment-limited reaches, their incision should be controlled by shear stress, and n_d should be around 2/3 (Whipple et al., 2000). As argued by Whipple and Tucker (2002), a value for $n_t = 1$ appears to be an appropriate estimate for gravel transport. From a comparison of the critical drainage area (A_{cr}) for different incision rates for a single lithology, we estimate the exponent in equation 6 that describes this scaling to be close to 3, which is consistent with $n_t = 1$, $n_d = 2/3$, $(m_t - 1)/n_t \approx 0.4$, and $0.5 < m_d/n_d < 0.6$, so that m_t should be around 1.4 and m_d between 0.3 and 0.4. Because m_t embeds the exponent on the channel width-area relationship c, which is close to 0.6 where we measured it (but note the problems with measuring present-day channel widths in our study area outlined in "A Model for Valley-Flat Development: The Erosion Frequency"), this requires the sediment flux in the transport-limited rivers to scale approximately with the square of drainage area (i.e., $m_t' \approx 2$ in equation 2a), consistent with the derivation of a sediment transport equation from the Einstein-Brown equation (Willgoose et al., 1991). Finally, we can estimate appropriate K_d values by solving equation 6 for K_d:

$$K_d = \left[A_{cr}^{(\theta_t - \theta_d)} \dot{E}^{(1/n_d - 1/n_t)} (K_t / \beta)^{1/n_t} \right]^{n_d} . \tag{18}$$

Taking $A_{cr} = 10^6$ and 4×10^7 m^2 for marls and limestones, respectively (cf. Fig. 13), $\dot{E} = 0.8$ mm yr^{-1}, $K_t = 4.5 \times 10^{-5}$ m yr^{-1} (cf. Table 2), β = 1, and the other parameters as above, we find that K_d should lie between 1.8 and 4.7×10^{-5} m$^{0.4}$ yr^{-1} for marls and between 1.1 and 3.7×10^{-5} m$^{0.4}$ yr^{-1} for limestone. A roughly twofold variation in erosional resistance between the softest and hardest lithologies in our study area is consistent with our estimates of the widening factors for valley flats in these lithologies (cf. Fig. 16, Table 3), which are linearly dependent on bedrock erodibility in our model. These K_d values could be compared with experimental measures of erodibility or tensile rock strength (e.g., Sklar and Dietrich, 2001; Lavé and Avouac, this volume) in order to test whether we have completely separated lithological effects from other controls on incision in our model.

The above constraints on model parameters, specifically those for detachment-limited incision, are similar to earlier estimates by Stock and Montgomery (1999), Snyder et al. (2000), and van der Beek and Bishop (2003). The latter study, which concentrated on the Upper Lachlan River, a low-relief, low–incision rate catchment on crystalline basement rocks in SE Australia, found that a detachment-limited incision model provided the best fits of five different model formulations tested, for $m_d = 0.3$–0.4, $n_d = 0.7$, and $K_d = 7 \times 10^{-7}$ m$^{0.4}$ yr^{-1}. Although the concordance in exponent values between these independent studies is promising, two discrepancies require an explanation. Firstly, why do detachment-limited conditions appear much more widespread in the Lachlan catchment (as shown by the persistence of lithological knickpoints up to drainage areas of at least 1000 km^2 and the relatively good performance of a simple detachment-limited stream power model in predicting incision), whereas incision rates in the Lachlan catchment are at least two orders of magnitude lower than in the Alpine rivers? Secondly, the two-orders-of-magnitude difference in estimated K_d values between these two studies requires an explanation, even though these values are both within the range of values estimated in different settings by Stock and Montgomery (1999). The difference in lithology underlying both study areas (calcareous sediments in the western Alps; granites and metasediments in the Lachlan catchment) will undoubtedly have an effect on K_d; the direction of change matches our a priori expectations (i.e., the Lachlan catchment is underlain by harder bedrock than the western Alpine rivers). We are not able to quantify the lithological effect at this stage; this would require extending our Alpine study into the crystalline massifs where the glacial disruption of river profiles is ubiquitous. An alternative explanation for large variations in K_d values between areas with different incision rates could be the presence of erosion thresholds and a stochastic distribution of flood events, as proposed by Snyder et al. (2003b). In any case, the significantly lower K_d values for the Lachlan River compared to the western Alpine rivers would promote detachment-limited behavior. Moreover, there is a significant difference in bedload caliber, because the Lachlan River only carries sand. This much finer bedload (in comparison with the western Alpine rivers) will require much lower transport-limited slopes and therefore promote detachment-limited conditions.

Even though our observations are consistent with the model proposed by Whipple and Tucker (2002), they do not exclude other model formulations. In particular, our data appear equally consistent with either "undercapacity" (Beaumont et al., 1992) or "tools" (Sklar and Dietrich, 1998) models, which take the effect of sediment flux on bedrock incision more fully into account. As shown by van der Beek and Bishop (2003), if sediment flux is close to or far from carrying capacity, their end-member cases are equivalent to the transport-limited and detachment-limited stream power models, respectively. In most cases, both these models will also predict downstream transitions from detachment-limited–like to transport-limited–like behavior, but they predict gentler transitions from one to the other, without a clearly defined critical drainage area (van der Beek and Bishop, 2003). Although these models are intuitively more realistic than the simple stream power approach, they are also more difficult to test because they require constraining variables such as sediment fluxes and channel widths. Along most of the investigated rivers, the width of the anastomosing channel belt is very sensitive to factors such as fluctuating anthropogenic pressure (deforestation) and short-term climatic changes (e.g., the Little Ice Age). Moreover, many rivers have been embanked since the Middle Ages (e.g., Bouchayer, 1925; Gautier, 1992). The evaluation of long-term channel widths is thus difficult. Therefore, the transport-versus detachment-limited incision model is, given the present resolution of our data, the simplest model that is both testable and consistent with our data.

However, none of these models currently include the development of a valley flat, where river incision is evenly distributed over a strath that is considerably wider than the bankfull channel. To our knowledge, the only model in which strath formation was explicitly studied is that of Hancock and Anderson (2002). In that model, however, strath widening rates are simply assumed to be constant and independent of rock type or drainage area, whereas our data show that these parameters strongly influence widening rates. From our study, valley flats appear as important features, because they accommodate lithological, climatic, and tectonic forcings, and thus potentially provide crucial information about spatio-temporal variations in uplift rate or climate. Moreover, the geomorphologic imprint of valley flats is not negligible, as they may occupy significant portions of the topography.

Coupling and Decoupling between Uplift/Incision Rates and Fluvial Profile Form

Following the conceptual work of Whipple and Tucker (1999), several authors have attempted to use fluvial profile forms to directly infer incision and uplift rates in equilibrium rivers (e.g., Kirby and Whipple, 2001; Finlayson et al., 2002). These studies use the relationship between river gradient and drainage area predicted by the steady-state formulation of the detachment-limited stream power law (equation 3). There is, however, an obvious problem in the basic assumptions of this approach if rivers become transport-limited. For decreasing uplift and incision rates, there will be a stage where transport-limited slopes are higher than detachment-limited slopes, and river gradients do not decrease further. Instead, valley widening will take over in order to expend the excess incision capacity. Therefore, river gradients may be much less sensitive recorders of rock uplift rates, and rivers may have a much greater degree of freedom in adapting to variable uplift rates, than previously assumed. An indication of this effect can be found in the data presented by Snyder et al. (2000), which show a much weaker correlation between river gradients and uplift/incision rates than predicted by the detachment-limited stream power model (except if $n_d >> 1$, which is theoretically implausible). Similarly, Tomkin et al. (2003) found that no single model fits incision data for the Clearwater River in the Olympic Mountains, which has to adapt to order-of-magnitude variations in long-term uplift rates along its length. Snyder et al. (2003a, 2003b) suggest that a threshold for incision combined with a stochastic distribution of floods may play a role to explain this discrepancy, but their model is not unique. While we have not tested the potential contribution of incision thresholds and discharge stochasticity in our field area, and the observed large variations in K_d values between the western Alpine rivers and the Lachlan River suggest that this effect may play a role (idem previous discussion), our observations clearly suggest that sediment flux and caliber exert an important nontectonic constraint on river profiles.

Stability of Detachment- and Transport-Limited Drainage Networks

The question of drainage stability and the notion of stream piracy have been important issues in large-scale fluvial geomorphology. Bishop (1995) provided a review of the processes involved and outlined the problems associated with both the notion of stream capture and that of drainage stability through the erosion of significant amounts of crustal section. In particular, stream capture requires the head of one stream to retreat across a drainage divide and into the catchment of another stream. Our model suggests the circumstances under which this may be achieved, and, in contrast, under what circumstances the drainage net remains stable through the erosion of different stratigraphic units.

The stream network of the nonglaciated part of our study area is composed of detachment-limited reaches, the gradients of which vary with bedrock strength, and of transport-limited reaches, the gradients of which are insensitive to lithological variation. We have shown that the Tithonian limestones trigger the formation of knickpoints on the course of streams that drain areas up to 40 km^2 (Fig. 13). At equilibrium, both detachment-limited and transport-limited reaches incise at the same rate. However, the bedrock is strongly folded and composed of stratigraphic units with highly contrasting strengths. As the structures are incised, the folded and tilted layers responsible for the formation of knickpoints migrate in plan view with respect to the stream network. The response of the detachment-limited and transport-limited reaches to the displacement of a band of resistant bedrock is different, and is illustrated here by two field examples from the Drôme catchment (Fig. 18).

Several small tributaries of the Drôme River that flowed over marly levels up to a few tens of thousands of years ago have recently encountered Tithonian limestone on their course (Fig.

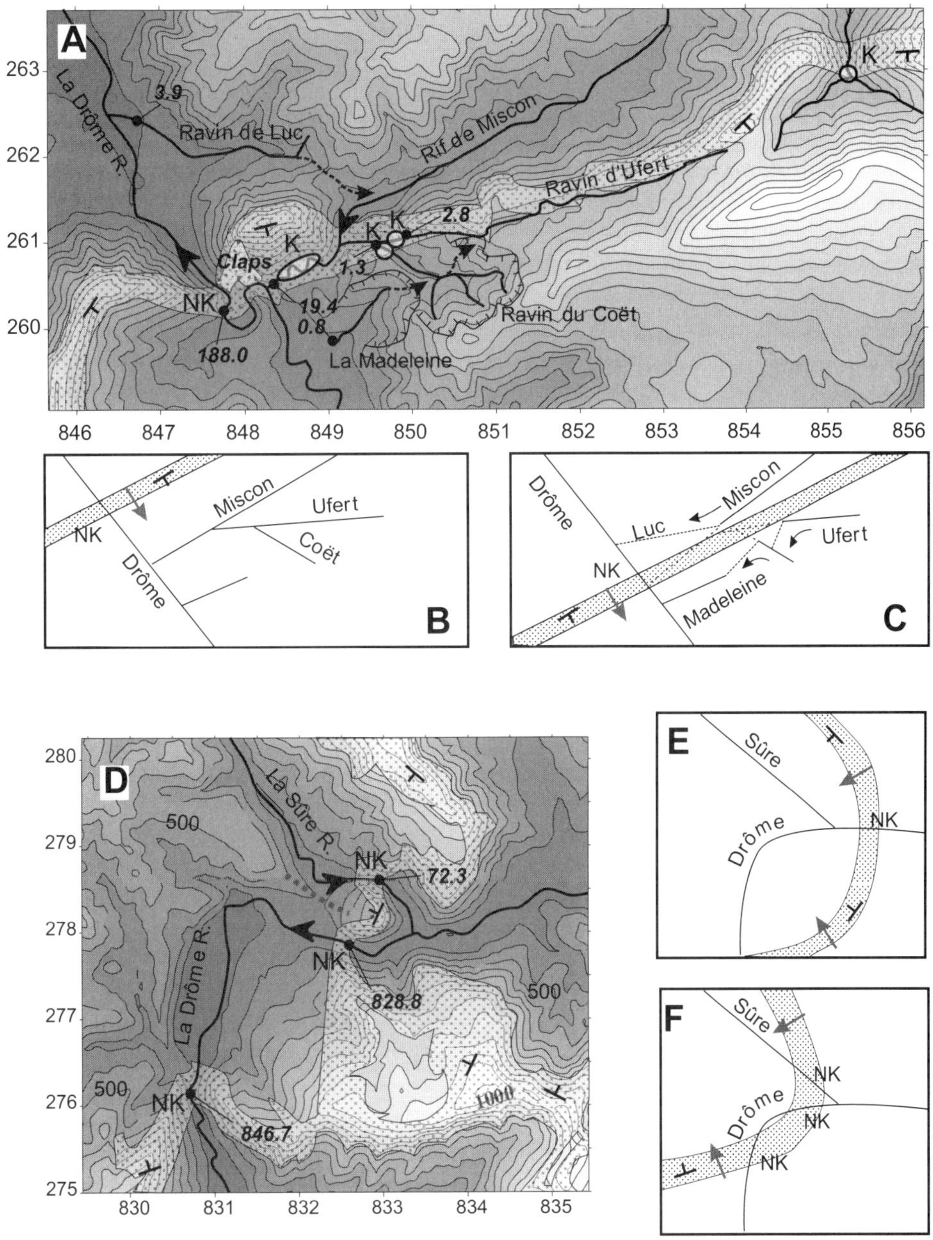

Figure 18. Examples of stream capture in detachment-limited streams (A–C) and drainage stability in transport-limited streams (D–E) in the Drôme River catchment. (A and D) Detailed maps of present-day topography (contour spacing: 100 m) and drainage system, extracted from the 50-m-resolution digital elevation model (DEM). Dotted bands—Tithonian limestone; K—knickpoints; NK—reaches on Tithonian limestone without knickpoints. Dashed lines indicate direction of stream head retreat and inferred future captures. Italic numbers—drainage area in km^2. Axes correspond to the Institut Géographique National Lambert III grid. Locations of these detailed maps are indicated in Figure 15. (B and E) Sketch maps of inferred former drainage system and outcrop of Tithonian limestone. (C and F) Sketch maps of the expected future drainage system. Gray arrows indicate migration of Tithonian limestone outcrops; black arrows expected captures.

18A–B). As these tributaries have drainage areas smaller than the critical area in Tithonian limestone, they have developed pronounced knickpoints where crossing this unit. The knickpoints grow in amplitude as long as the tributaries' incision does not keep pace with that of the main stem. As the knickpoint acts as the base level for the upstream reaches, incision rate is reduced upstream during knickpoint growth, and the streams are uplifted with respect to the surrounding streams. This makes them potential targets for capture by headward retreat of the surrounding stream heads through easily erodible lithologies. According to the valley-flat model, the phenomenon should be enhanced by an increase of the valley-flat width upstream of the knickpoint as incision rate is temporarily lowered. In the example shown in Figure 18A, a suite of captures is about to occur. At the final stage, all tributaries will flow parallel to the structural strike (Fig. 18C). Once the captures have occurred, the crests of Tithonian limestone will exhibit windgaps. Numerous windgaps are preserved throughout the study area and record an intense activity of drainage reorganization (Fig. 13). They are readily distinguishable from common passes in that they appear as notches across the mountain crests and lithological units. They preserve very low axial slopes that can only be achieved by stream erosion. Other ongoing capture sites can be detected in many localities (Fig. 13), and a dramatic capture has occurred in historical times (upper Maraize River; Goguel, 1954).

The transport-limited streams, in contrast, are insensitive to the geological structure. This lack of adaptation is well illustrated by the confluence of the Drôme and Sûre Rivers, where the network is clearly superimposed on the structures (Fig. 18D). A Tithonian limestone layer has recently been excavated by the lower course of the Sûre River. The formation of a knickpoint would soon have triggered the capture of the Sûre River upstream of the limestone layer to the neighboring Drôme River, but the Sûre is a transport-limited river where it reaches the Tithonian limestone and no knickpoint has developed. The hairpin loop of the Sûre River at its confluence with the Drôme is thus a stable feature that will be maintained throughout its incision history.

The drainage network thus presents a threshold for stability that is set by the transition from detachment- to-transport-limited behavior of the most resistant bedrock type. Smaller streams are less stable and continuously adapt to the bedrock structure, while larger streams are fully superimposed. This may explain why examples of small-scale stream capture are ubiquitous, but no unequivocal examples of large-scale river piracy are known (Bishop, 1995).

CONCLUSIONS

The drainage network of the nonglaciated western Alps can be regarded as a mixture of transport-limited and detachment-limited reaches. Evidence for both transport-limited and detachment-limited incision can be found in the field, and other diagnostic features, such as the intrinsic concavity, are in accordance with theoretical predictions for these two models. As expected, the transition from detachment- to transport-limited behavior occurs with increasing discharge. The transition shifts upstream with increasing bedrock erodibility and decreasing incision rate. The exact form of the transition, whether it is sharp or progressive, cannot be evaluated from our field data, because the same bedrock lithology cannot be followed over sufficient lengths along a single stream. A valley flat develops systematically in association with transport-limited segments. The factors that control the detachment-limited river gradient (i.e., bedrock strength and incision rate) are transferred to the valley-flat width, while the transport-limited river gradient is far less sensitive to these factors. Thus, the valley-flat width increases with increasing bedrock erodibility, increasing discharge, and decreasing incision rate. A model for valley-flat development, based on the frequency of strath erosion, has been proposed. It reasonably accounts for most of the observed trends in valley-flat widths. However, the valley-flat width data exhibit a high scatter that could reflect the effects of bedrock heterogeneity and variability in bedload caliber throughout the study area. The map pattern of the dispersion does not show any trend that could correlate with possible previously unrecognized gradients in rock uplift; further calibrations are needed to extract any possible tectonic forcing. They should take into account the orientation of bedrock fabric relative to the river strike, the intrinsic variability of the lithological units within the studied area, and an assessment of the variability of the bedload characteristics. Such investigations are needed to make the analysis of fluvial forms an efficient tool for the detection of neotectonic activity in moderately active orogens.

ACKNOWLEDGMENTS

Funding for this project was provided by the Institut National des Sciences de l'Univers, Centre National de la Recherche Scientifique (INSU-CNRS) through Programme National de Recherche Sols et Erosion project 99PNSE07. Many of the statistical analyses have been performed by Florence Revol. We thank Nicole Gasparini and Jonathan Tomkin for their thorough reviews of the manuscript, and Mark Brandon for editorial handling.

REFERENCES CITED

Alary, C., 1998, Mécanismes et bilans de l'érosion dans un bassin versant méditerranéen aménagé: Le cas de la Durance (S-E France) [Ph.D. thesis]: Marseille, Université Aix-Marseille III, 255 p.

Beaumont, C., Fullsack, P., and Hamilton, J., 1992, Erosional control of active compressional orogens, *in* McClay, K.R., ed., Thrust tectonics: London, Chapman Hall, p. 1–18.

Benda, L., and Dunne, T., 1997, Stochastic forcing of sediment routing and storage in channel networks: Water Resources Research, v. 33, p. 2865–2880, doi: 10.1029/97WR02387.

Bernet, M., Zattin, M., Garver, J.I., Brandon, M.T., and Vance, J.A., 2001, Steady state exhumation of the European Alps: Geology, v. 29, p. 35–38, doi: 10.1130/0091-7613(2001)029<0035:SSEOTE>2.0.CO;2.

Bigot-Cormier, F., Poupeau, G., and Sosson, M., 2000, Dénudations différentielles du massif cristallin externe alpin de l'Argentera (Sud-Est de la France) révélées par thermochronologie traces de fission (apatites, zircons): Compte Rendu de l'Academie des Sciences de Paris, Sciences de la Terre et des Planètes, v. 330, p. 363–370.

Bishop, P., 1995, Drainage rearrangement by river capture, beheading and diversion: Progress in Physical Geography, v. 19, p. 449–473.

Bishop, P., and Goldrick, G., 2000, Geomorphological evolution of the East Australian continental margin, *in* Summerfield M.A., ed., Geomorphology and global tectonics: Chichester, John Wiley and Sons Ltd., p. 225–254.

Bouchayer, A., 1925, Le Drac dans la plaine de Grenoble de 1280 à 1651: Revue de Geographie Alpine, v. 13, p. 115–173.

Brocard, G., 2002, Origine, variabilité spatio-temporelle et signature morphologique de l'incision fluviatile dans les Alpes du Dauphiné (SE France) [Ph.D. thesis]: Grenoble, Université Joseph Fourier, Géologie Alpine, Mémoire Hors Serie, no. 37, 165 p.

Brocard, G., 2004, Le grand lac du Claps de Luc-en-Diois (Drôme): Évaluation, à la lumière d'une analyse morphologique, du volume d'un lac comblé: Bulletin de la Société Géologique de France, v. 175, p. 303–312.

Brocard, G.Y., van der Beek, P.A., Bourlès, D.L., Siame, L.L., and Mugnier, J.-L., 2003, Long-term fluvial incision rates and postglacial river relaxation time in the French western Alps from ^{10}Be dating of alluvial terraces with assessment of inheritance, soil development and wind ablation effects: Earth and Planetary Science Letters, v. 209, p. 197–214, doi: 10.1016/S0012-821X(03)00031-1.

Burbank, D.W., Leland, J., Fielding, E., Anderson, R.S., Brozovic, N., Reid, M.R., and Duncan, C., 1996, Bedrock incision, rock uplift and threshold hillslopes in the northwestern Himalayas: Nature, v. 379, p. 505–510, doi: 10.1038/379505a0.

Chapron, E., 1999, Contrôle climatique et tectonique de la sédimentation lacustre dans l'avant-pays alpin (lac du Bourget) durant le Quaternaire récent [Ph.D. thesis]: Chambéry, Université de Savoie, Géologie Alpine, Mémoire Hors Serie, no. 30, 258 p.

Finlayson, D.P., Montgomery, D.R., and Hallet, B., 2002, Spatial coincidence of rapid inferred erosion with young metamorphic massifs in the Himalayas: Geology, v. 30, p. 219–222, doi: 10.1130/0091-7613(2002)030<0219:SCORIE>2.0.CO;2.

Gasparini, N.M., Tucker, G.E., and Bras, R.L., 1999, Downstream fining through selective particle sorting in an equilibrium drainage network: Geology, v. 27, p. 1079–1082, doi: 10.1130/0091-7613(1999)027<1079:DFTSPS>2.3.CO;2.

Gautier, E., 1992, Recherche sur la morphologie et la dynamique fluviale dans le bassin du Buëch [Ph.D. thesis]: Paris, Université Paris X-Nanterres, 434 p.

Goguel, J., 1954, Une capture subactuelle dans les Hautes Alpes: Compte Rendu Sommaire de la Société Géologique de France, p. 181–182.

Hack, J.T., 1957, Studies of longitudinal stream profiles in Virginia and Maryland: U.S. Geological Survey Professional Paper 294-B, p. 45–97.

Hancock, G.S., and Anderson, R.S., 2002, Numerical modeling of fluvial strath-terrace formation in response to oscillating climate: Geological Society of America Bulletin, v. 114, p. 1131–1142, doi: 10.1130/0016-7606(2002)114<1131:NMOFST>2.0.CO;2..

Harbor, D.J., 1998, Dynamic equilibrium between an active uplift and the Sevier River, Utah: Journal of Geology, v. 106, p. 181–194.

Howard, A.D., 1980, Thresholds in river regimes, *in* Coates, D.R., and Vitek, J.D., eds., Thresholds in geomorphology: Winchester, Massachusetts, Allen and Unwin, p. 227–258.

Howard, A.D., 1998, Long profile development of bedrock channels: Interaction of weathering, mass wasting, bed erosion, and sediment transport, *in* Tinkler, K., and Wohl, E.E., eds., Rivers over rock: Fluvial processes in bedrock channels: Washington, D.C., American Geophysical Union, p. 297–319.

Howard, A.D., and Kerby, G., 1983, Channel changes in badlands: Geological Society of America Bulletin, v. 94, p. 739–752, doi: 10.1130/0016-7606(1983)94<739:CCIB>2.0.CO;2.

Howard, A.D., Dietrich, W.E., and Seidl, M.A., 1994, Modeling fluvial erosion on regional to continental scales: Journal of Geophysical Research, v. 99, p. 13,971–13,986, doi: 10.1029/94JB00744.

Jouanne, F., Génaudau, N., Ménard, G., and Darmendrail, X., 1998, Estimating present-day displacement fields and tectonic deformation in active mountain belts: An example from the Chartreuse Massif and the southern Jura Mountains, western Alps: Tectonophysics, v. 296, p. 403–419, doi: 10.1016/S0040-1951(98)00156-5.

Kirby, E., and Whipple, K.X., 2001, Quantifying differential rock-uplift rates via stream profile analysis: Geology, v. 29, p. 415–418, doi: 10.1130/0091-7613(2001)029<0415:QDRURV>2.0.CO;2.

Landon, N., 1999, L'évolution contemporaine du profil en long des affluents du Rhône moyen—Constat régional et analyse d'un hydrosystème complexe, la Drôme [Ph.D. thesis]: Paris, Université Paris IV—Sorbonne, 2 volumes, 734 p.

Lavé, J., and Avouac, J.-P., 2001, Fluvial incision and tectonic uplift across the Himalayas of central Nepal: Journal of Geophysical Research, v. 106, p. 25,561–25,593, doi: 10.1029/2001JB000359.

Leopold, L.B., and Maddock, T., 1953, Hydraulic geometry of stream channels and some physiographic implications: U.S. Geological Survey Professional Paper 252, 57 p.

Mackin, J.H., 1937, Erosional history of the Big Horn Basin, Wyoming: Geological Society of America Bulletin, v. 48, p. 813–894.

Mandier, P., 1984, Signification dynamique et climatique des formations et terrasses fluviatiles Quaternaires dans les Alpes et leur périphérie: Bulletin de l'Association Française pour l'Etude du Quaternaire, no. 1984(1–3), p. 113–118.

Mandier, P., 1988, Le relief de la moyenne vallée du Rhône au Tertiaire et au Quaternaire, Essai de synthèse paléogéographique: Documents du Bureau des Recherches Géologiques et Minières, no. 151, 3 volumes, 865 p.

Martinod, J., Jouanne, F., Taverna, J., Ménard, G., Gamond, J.F., Darmendrail, X., Notter, J.C., and Basile, C., 1996, Present-day deformation of the Dauphiné (SE France) Alpine and Subalpine massifs: Geophysical Journal International, v. 127, p. 189–200.

Montjuvent, G., 1973, La transfluence Durance-Isère: Essai de synthèse du Quaternaire du bassin du Drac (Alpes françaises): Géologie Alpine, v. 49, p. 57–118.

Montjuvent, G., 1978, Le Drac, morphologie, stratigraphie et chronologie Quaternaires d'un bassin alpin: Paris, Comité National de la Recherche Scientifique, 433 p.

Pazzaglia, F.J., and Brandon, M.T., 2001, A fluvial record of long-term steady-state uplift and erosion across the Cascadia forearc high, western Washington State: American Journal of Science, v. 301, p. 385–431.

Pazzaglia, F.J., Gardner, T.W., and Merritts, D.J., 1998, Bedrock fluvial incision and longitudinal profile development over geologic time scales determined by fluvial terraces, *in* Tinkler, K., and Wohl, E.E., eds., Rivers over rock: Fluvial processes in bedrock channels: Washington, D.C., American Geophysical Union, p. 207–235.

Personius, S.F., 1995, Late Quaternary stream incision and uplift in the forearc of the Cascadia subduction zone, western Oregon: Journal of Geophysical Research, v. 100, p. 20,193–20,210, doi: 10.1029/95JB01684.

Roe, G.H., Montgomery, D.R., and Hallett, B., 2002, Effects of orographic precipitation variations on the concavity of steady-state river profiles: Geology, v. 30, p. 143–146, doi: 10.1130/0091-7613(2002)030<0143:EOOPVO>2.0.CO;2.

Rosique, T., 1996, Morphogénèse et évolution des paléoenvironnements alpins de la fin des temps glaciaires au début de l'Holocène [Ph.D. thesis]: Marseille, Université d'Aix-Marseille I, 288 p.

Schmid, S.M., and Kissling, E., 2000, The arc of the western Alps in the light of geophysical data on deep crustal structure: Tectonics, v. 19, p. 62–85, doi: 10.1029/1999TC900057.

Seward, D., Ford, M., Bürgisser, J., Lickorish, H., Williams, E.A., and Meckel, L.D., 1999, Preliminary results of fission-track analyses in the southern Pelvoux area, SE France, *in* Gosso, G., Jadoul, F., Sella, M., and Spalla, M.I., eds., 3rd Workshop on Alpine Geological Studies: Memorie di Scienze Geologiche, v. 51, p. 25–31.

Sklar, L., and Dietrich, W.E., 1998, River longitudinal profiles and bedrock incision models: Stream power and the influence of sediment supply, *in* Tinkler, K., and Wohl, E.E., eds., Rivers over rock: Fluvial processes in bedrock channels: Washington, D.C., American Geophysical Union, p. 237–260.

Sklar, L., and Dietrich, W.E., 2001, Sediment and rock strength controls on river incision into bedrock: Geology, v. 29, p. 1087–1090, doi: 10.1130/0091-7613(2001)029<1087:SARSCO>2.0.CO;2.

Snow, R.S., and Slingerland, R.L., 1987, Mathematical modelling of graded river profiles: Journal of Geology, v. 95, p. 15–33.

Snyder, N.P., Whipple, K.X., Tucker, G.E., and Merritts, D.J., 2000, Landscape response to tectonic forcing: Digital elevation model analysis of stream profiles in the Mendocino triple junction region, northern California: Geological Society of America Bulletin, v. 112, p. 1250–1263, doi: 10.1130/0016-7606(2000)112<1250:LRTTFD>2.3.CO;2.

Snyder, N.P., Whipple, K.X., Tucker, G.E., and Merritts, D.J., 2003a, Channel response to tectonic forcing: Field analysis of stream morphology and hydrology in the Mendocino triple junction region, northern California: Geomorphology, v. 1315, p. 1–31.

Snyder, N.P., Whipple, K.X., Tucker, G.E., and Merritts, D.J., 2003b, Importance of a stochastic distribution of floods and erosion thresholds in the bedrock river incision problem: Journal of Geophysical Research, v. 108, no. B2, 2117, doi: 10.1029/2001JB001655.

Stock, J., and Dietrich, W.E., 2003, Valley incision by debris flows: Evidence of a topographic signature: American Geophysical Union, Water Resources Research, v. 39, p. 1089, doi: 10.1029/2001WR001057.

Stock, J.D., and Montgomery, D.R., 1999, Geologic constraints on bedrock river incision using the stream power law: Journal of Geophysical Research, v. 104, p. 4983–4993, doi: 10.1029/98JB02139.

Talling, P.J., and Sowter, M.F., 1998, Erosion, deposition and basin-wide variations in stream power and bed shear stress: Basin Research, v. 10, p. 87–108, doi: 10.1046/j.1365-2117.1998.00088.x.

Tomkin, J.H., Brandon, M.T., Pazzaglia, F.J., Barbour, J.R., and Willett, S.D., 2003, Quantitative testing of bedrock incision models for the Clearwater River, NW Washington State: Journal of Geophysical Research, v. 108, 2308, doi: 10.1029/2001JB000862.

Tricart, P., Schwartz, S., Sue, C., Poupeau, G., and Lardeaux, J.-M., 2001, La dénudation tectonique de la zone ultradauphinoise et l'inversion du front briançonnais au sud-est du Pelvoux (Alpes occidentales): Une dynamique miocène à actuelle: Bulletin de la Société Géologique de France, v. 172, p. 49–58.

Tucker, G.E., and Bras, R.L., 2000, A stochastic approach to modeling the role of rainfall variability in drainage basin evolution: Water Resources Research, v. 36, p. 1953–1964, doi: 10.1029/2000WR900065.

van der Beek, P.A., and Bishop, P., 2003, Cenozoic river profile development in the Upper Lachlan catchment (SE Australia) as a test of quantitative fluvial incision models: Journal of Geophysical Research, v. 108, 2309, doi: 10.1029/2002JB002125.

Wegmann, K.W., and Pazzaglia, F.J., 2002, Holocene strath terraces, climate change, and active tectonics: The Clearwater River basin, Olympic Peninsula, Washington State: Geological Society of America Bulletin, v. 114, p. 731–744, doi: 10.1130/0016-7606(2002)114<0731:HSTCCA>2.0.CO;2.

Whipple, K.X., and Tucker, G.E., 1999, Dynamics of the stream-power river incision model: Implications for height limits of mountain ranges, landscape response timescales, and research needs: Journal of Geophysical Research, v. 104, p. 17,661–17,674, doi: 10.1029/1999JB900120.

Whipple, K.X., and Tucker, G.E., 2002, Implications of sediment-flux dependent river incision models for landscape evolution: Journal of Geophysical Research, v. 107, 2039, doi: 10.1029/2000JB000044.

Whipple, K.X., Kirby, E., and Brocklehurst, S.H., 1999, Geomorphic limits to climate-induced increases in topographic relief: Nature, v. 401, p. 39–43, doi: 10.1038/43375.

Whipple, K.X., Hancock, G.S., and Anderson, R.S., 2000, River incision into bedrock: Mechanics and relative efficacy of plucking, abrasion and cavitation: Geological Society of America Bulletin, v. 112, p. 490–503, doi: 10.1130/0016-7606(2000)112<490:RIIBMA>2.3.CO;2.

Willett, S.D., Slingerland, R., and Hovius, N., 2001, Uplift, shortening and steady state topography in active mountain belts: American Journal of Science, v. 301, p. 455–485.

Willgoose, G.R., Bras, R.L., and Rodriguez-Iturbe, I., 1991, A physically based coupled network growth and hillslope evolution model. 1. Theory: Water Resources Research, v. 27, p. 1671–1684, doi: 10.1029/91WR00935.

MANUSCRIPT ACCEPTED BY THE SOCIETY 23 JUNE 2005

Printed in the USA

Geological Society of America
Special Paper 398
2006

Numerical modeling of non–steady-state river profile evolution using a sediment-flux-dependent incision model

Nicole M. Gasparini*
Rafael L. Bras
Department of Civil and Environmental Engineering, Massachusetts Institute of Technology, Cambridge, Massachusetts 02139, USA
Kelin X. Whipple
Department of Earth, Atmospheric and Planetary Sciences, Massachusetts Institute of Technology, Cambridge, Massachusetts 02139, USA

ABSTRACT

We use a numerical model to investigate disequilibrium conditions in detachment-limited river networks. Erosion rates are modeled using two different equations that include sediment flux as a variable for determining incision rates into bedrock. A number of numerical simulations are performed to explore erosion patterns, channel profile shape, and network concavity after an increase in uplift rate across the network. In the case where an increase in sediment flux (relative to carrying capacity) is considered only to decrease incision rates, the main channel has a two-part response to a faster uplift rate; initially a knickpoint steepens channel slopes locally, but at later times channel slopes rise throughout the network. However, in the case where an increase in sediment flux can both enhance and suppress incision rates, the transient network response can be much more dynamic; channel slopes (and also elevations) can both rise and fall, all in response to a single increase in uplift rate. The response varies depending on the magnitude of change in uplift rate and the initial ratio of sediment flux to sediment carrying capacity. In all examples, the lower parts of the network respond quickly to an increase in uplift rates by increasing channel slopes, while the response of erosion rates in the upper parts of the network occurs later. As a result, the change in sediment flux delivered to higher order channels lags the initial changes in the slope of these channels and causes a complex response in erosion rates. These findings highlight that erosion rates at any point in the network respond to changes both downstream and upstream, and therefore variables such as sediment flux that integrate the upstream response can play an important role is shaping the transient morphology of river networks.

Keywords: bedrock incision, sediment flux, transient landscapes, channel profiles.

MOTIVATION

In many regions, bedrock channels form an important link in the landscape by transporting sediment eroded off the hillslopes to the lower alluvial reaches of a drainage network. As sediment moves through these channels, it may play a critical role in determining the rate of fluvial incision into bedrock. Gilbert (1877) discussed the processes responsible for bedrock incision, but only recently has much attention been focused on this problem. A number of field and flume studies have investigated the many

*Present address: Department of Geology and Geophysics, Yale University, P.O. Box 208109, New Haven, CT 06520-8109, USA; nicole.gasparini@yale.edu.

Gasparini, N.M., Bras, R.L., and Whipple, K.X., 2006, Numerical modeling of non–steady-state river profile evolution using a sediment-flux-dependent incision model, *in* Willett, S.D., Hovius, N., Brandon, M.T., and Fisher, D.M., eds., Tectonics, climate, and landscape evolution: Geological Society of America Special Paper 398, Penrose Conference Series, p. 127–141, doi: 10.1130/2006.2398(08). For permission to copy, contact editing@geosociety.org.

different controls on bedrock incision (e.g., Foley, 1980; Gardner, 1983; Wohl, 1993; Wohl and Ikeda, 1997; Hancock et al., 1998; Whipple et al. 2000a; Whipple et al., 2000b; Kim, 2004), and recent studies indicate that sediment load plays a first-order role in controlling bedrock incision. In the East Central Range of Taiwan, Hartshorn et al. (2002) found that abrasion by suspended sediment, along with channel bed geometry and bedding planes of weaker lithologies all affect erosion rates. Flumes studied by Sklar and Dietrich (2001) indicate that bedrock incision is a nonlinear function of bedload sediment flux; low supply rates of sediment encourage erosion, while high supply rates of sediment hinder erosion. Most numerical models use the stream-power model to calculate incision rates (e.g., Howard, 1994; Rosenbloom and Anderson, 1994; Stock and Montgomery, 1999; Whipple and Tucker, 1999; Snyder et al., 2002). However, some studies suggest that the stream-power model may not adequately describe the processes controlling the morphology of bedrock channels (e.g., Howard et al., 1994; Sklar and Dietrich, 1998; Snyder et al. 2003a; Tomkin et al. 2003; van der Beek and Bishop, 2003) because it does not consider variables such as sediment delivery to and transport rates in the channels, thresholds for detachment and transport, local grain size, downstream sorting and changes in channel width. These variables may all play an important role in channel evolution.

With these ideas in mind, we investigate two different bedrock erosion equations which include sediment flux as a variable that influences fluvial incision rates. Using the CHILD numerical landscape evolution model (Tucker et al., 2001a, 2001b), we focus on the transient morphology of bedrock rivers as they respond to an increase in rock uplift rate. All of the examples here consider channels in which erosion is limited by the amount of material that can be detached from the bed, but the detachment rate is influenced by the amount of sediment fluxing through the channel. Under the assumption of spatially uniform uplift rate, all of the incision models produce a similar equilibrium morphology. However, the transient morphologies produced using the different models can be quite different and may be more indicative of the incision processes, which is why we focus on transients here.

Predictions on transitions in the channel network from a low to high uplift rate were also modeled by Whipple and Tucker (2002). In their study, they use the stream-power model to calculate incision rates in detachment-limited channels. In response to a sudden increase in rock uplift rate, the stream-power river incision model predicts an abrupt break in slope from the initial steady-state gradient upstream to the final steady-state gradient downstream. This break in slope is a discrete feature that migrates upstream from the mouth. Whipple and Tucker (1999) called this migrating slope-break a knickpoint, but pointed out it is distinct from classic knickpoints defined as steps in the elevation of the river bed. Throughout this paper, we use the term knickpoint to describe abrupt, migrating slope-breaks, following Whipple and Tucker (1999). Channel slopes remain steep in the wake of the knickpoint, and as a result, the elevation of every point on the channel profile increases from its initial value. Whipple and Tucker (2002) also considered transitions in a transport-limited network in which erosion rates are calculated from continuity of mass using a transport equation without a threshold. In this case, slopes increase more slowly and uniformly throughout the channel; there is no knickpoint in the transport-limited case. In both cases, channel slopes and elevations never exceed their final equilibrium values.

Considering sediment flux in the incision equation results in some important differences in transient river networks from those described by Whipple and Tucker (2002). Even though we only model detachment-limited channels in this study, we find that disequilibrium channels often do not contain knickpoints and that the channel profile can be fairly smooth and resemble steady conditions. In some cases, there are periods during which channel slope decreases after an increase in uplift rate. These previously unpredicted transient states result from using a very simplified sediment transport law (erosion thresholds are ignored) and climate model (a single storm rate, storm duration, and inter-storm duration are used in all cases) and without including the effects of grain-size changes or thresholds. However, even under these conditions, complex transient morphologies result.

If sediment-flux plays a significant role in bedrock erosion, and more importantly, if large amounts of sediment flux can alter the transient response of channels, this will have important implications on studies which use geomorphology as a tool for interpreting the tectonic history of a region. For example, knickpoints and regions of varying steepness in tributaries to the Red River, Yunnan Province, China, led Schoenbohm et al. (2004) to infer that the region has experienced two phases of river incision, although they are not able to state for certain the relative roles of climate and tectonics in driving these phases of river incision. Interpretation of the tectonic history of the Red River area was in part complicated by an observation of very high channel concavity indices associated with the steepest, lower channel segments. This association of unusually high concavity with high steepness zones of transient river profiles may be an indication of the type of transient sediment-flux effects illustrated here. Wobus et al. (this volume) found that some channels in the San Gabriel Mountains, California, have knickpoints which divide the channels into two regions with distinct steepness values (Fig. 6 of Wobus et al., this volume). Interestingly, the San Gabriel knickpoints are not simple slope-breaks as expected from the stream power family of models; they exhibit a marked local over-steepening over a short reach in the immediate vicinity of the slope-break, forming more of a classical knickpoint step in the channel profile. As in the Red River example, this local over-steepening during river transient response to an apparent change in rock uplift rate may be explained by the sediment-flux effects illustrated in the analysis presented here. The apparent transient state of some channels in the San Gabriels, possibly due to a recent increase in rock uplift rates, is supported by the apatite fission-track ages found by Blythe et al. (2000). Along with isotopic and thermochronologic data, numerical modeling studies such as this one have the potential to validate or exclude possible past climatic and tectonic scenarios which produced the landscape that we see today.

EROSION EQUATIONS AND EQUILIBRIUM CONDITIONS

Following Whipple and Tucker (2002), we start by expressing the rate of incision into bedrock (E_d) as

$$E_d = Kf(Q_s)A^m S^n, \quad (1)$$

where K is an erodibility parameter which depends on factors such as lithology, climate, and channel properties (e.g., Stock and Montgomery; 1999) and its units depend on m and n; and $f(Q_s)$ embodies the importance of sediment load in eroding the channel bed (Q_s is the volumetric rate of incoming sediment). A is upstream drainage area; S is channel slope; and m and n are conventionally considered to be positive constants which can be derived from physical channel properties (e.g., Howard et al., 1994; Whipple and Tucker, 1999), although recent studies suggest that in some cases m and n may not be positive (Tomkin et al., 2003; Sklar, 2003). In this study, we will always assume that m and n are positive. The rate of change in elevation $\partial z/\partial t$ is given by:

$$\frac{\partial z}{\partial t} = U - E_d \quad (2)$$

In all cases considered here the uplift rate, U, is spatially uniform. (We do not model lateral advection of rock; uplift is solely in the vertical direction.) When the landscape reaches equilibrium, or steady-state, elevations do not change in time (assuming a spatially uniform vertical uplift rate) leading to the following equilibrium condition:

$$U = E_d \quad (3)$$

Following from Equations 1 and 3, the common expression for equilibrium channel slope (e.g., Howard, 1980; Howard et al., 1994; Moglen and Bras, 1995; Whipple and Tucker, 1999; Snyder et al., 2000) can be written somewhat incompletely as,

$$S = \left(\frac{U}{Kf(Q_s)}\right)^{\frac{1}{n}} A^{-\theta} \quad (4)$$

where

$$\theta = \frac{m}{n}, \quad (5)$$

when U and $f(Q_s)$ are uniform in space; θ is the channel concavity.

Equation 4 is a bit non-conventional because it includes the $f(Q_s)$ term. The two different forms of $f(Q_s)$ are given in later sections. Because $f(Q_s)$ includes channel slope and drainage area, Equation 4 is more complicated than it first appears. However, it is useful to think of channel slope as inversely proportional to both K and the value of $f(Q_s)$ (assuming, as we do, that n is positive).

In all examples we use values of m and n so that $m/n = 0.5$. Channel concavity θ has been studied by numerous researchers (e.g., Hack, 1957; Flint, 1974; Tarboton et al., 1991; Tucker and Whipple, 2002), and 0.5 is an average representative value for many landscapes.

The stream-power incision model does not depend on the sediment flux, and therefore $f(Q_s) = 1$. Given that $f(Q_s) = 1$, Equation 4 describes exactly the equilibrium slope-area relationship for the stream-power model, and channel slopes are expected to increase with uplift rate using the stream-power model (Whipple and Tucker, 1999; Whipple, 2001). We do not illustrate any numerical simulations using the stream-power model. However, the stream-power model is widely used so we will contrast the behavior of other more complex models presented here to that of the stream-power model.

Almost-Parabolic Model

The almost-parabolic model described here is very similar to the relationship described by Whipple and Tucker (2002). It embodies the role of sediment flux in bedrock incision as first described by Gilbert (1877), and recently, by others (e.g., Slingerland et al., 1997; Sklar and Dietrich, 1998, 2001). The basic premise is that sediment carried in the load can enhance erosion as it impacts the channel bed, causing wear. Gilbert (1877) stated that sediment can effectively erode the bed when the sediment load is well below the sediment transport capacity. However, as the sediment load increases (in relation to the carrying capacity) the sediment begins to cover the bed and protect it from erosion. We describe this dual role of sediment in enhancing and inhibiting erosion through the $f(Q_s)$ factor in Equation 1.

Following the model presented by Sklar and Dietrich (1998), Whipple and Tucker (2002) described a parabolic form of $f(Q_s)$ as a function of the ratio of Q_s/Q_c, where Q_c is the sediment transport capacity. In the relationship used by Whipple and Tucker (2002) $f(Q_s)$ increases from 0 to 1.0 as Q_s/Q_c increases from 0 to 0.5 (sediment enhances erosion) and $f(Q_s)$ decreases from 1.0 to 0 as Q_s/Q_c increases from 0.5 to 1.0 (sediment covers the bed). The function used in this study for $f(Q_s)$ is the same as that used by Whipple and Tucker (2002) except we slightly adapt the function so that erosion can still take place when there is there is no sediment load, theoretically through processes other than wear by sediment, such as plucking or solution (e.g., see Wohl, 1993; Hancock et al., 1998; Wohl, 1998; Whipple et al., 2000a, for a discussion of bedrock erosion processes). This assumption is also made partly as a boundary condition to avoid infinite slopes (Whipple and Tucker, 2002).

The above description of the $f(Q_s)$ function translates into the following equations: when $Q_s/Q_c > 0.1$(from Whipple and Tucker, 2002),

$$f(Q_s) = 1 - 4\,(Q_s/Q_c - 0.5)^2 \tag{6}$$

and when $Q_s/Q_c < 0.1$,

$$f(Q_s) = 2.6\,(Q_s/Q_c) + 0.1 \tag{7}$$

Figure 1 illustrates this relationship. We describe the sediment transport capacity using a simple power-law function:

$$Q_c = K_t A^{m_t} S^{n_t} \tag{8}$$

In this simplified form of the sediment transport equation, there is some range in the value used for m_t. (We will always use $n_t = 1$.) Generally, one expects there to be a threshold for sediment entrainment which is sensitive to grain-size variations, and Equation 8 does not contain a threshold. For example, the Meyer-Peter and Müller sediment transport equation (Meyer-Peter and Müller, 1948) can be written as

$$Q_c \propto (\tau - \tau_c)^{3/2} \tag{9}$$

τ or bed shear stress, is often derived as

$$\tau \propto A^{1/3} S^{2/3} \tag{10}$$

(e.g., Tucker and Slingerland, 1997). When Equations 9 and 10 are applied in a landscape evolution model with uniform vertical uplift rates under transport-limited conditions without a threshold, convex channels result (Gasparini, 2003). In order to generate realistic concavity values (that is, concave-upward channels) without including the complicating effects of grain-size variation (e.g., Snow and Slingerland, 1987; Pizzuto, 1995; Sinha and Parker, 1996; Gasparini et al., 2004), the value of m_t can be increased and is considered to be a proxy for downstream changes in grain size (Howard, 1980; Whipple and Tucker, 2002). We illustrate numerical experiments using $m_t = 1.5$ and $m_t = 1.3$.

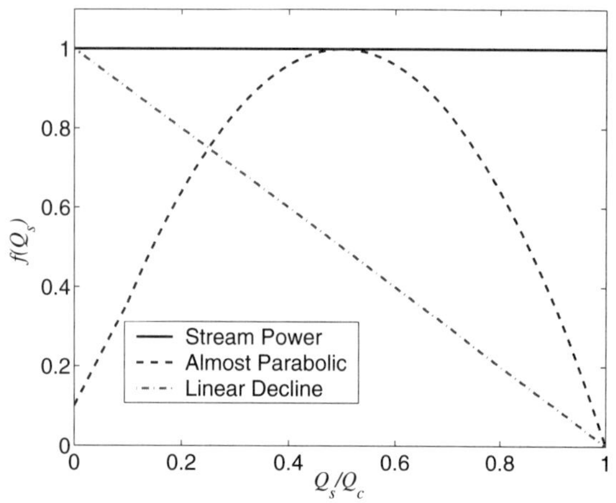

Figure 1. Dependency of three different erosion equations on the ratio of sediment load to sediment carrying capacity.

Substituting Equations 6, 7, and 8 into the equilibrium slope-area relationship (Equation 4) and making the equilibrium assumption that $Q_s = \beta UA$ (β represents the fraction of the sediment load that is bedload, and here is always considered to be 1.0), we can solve for the almost-parabolic slope-area relationship. These results were shown by Whipple and Tucker (2002) and we have added to their solution the slope-area relationship in the linear region of the almost-parabolic model ($Q_s/Q_c < 0.1$).

For the case where $n = 1$ and $n_t = 1$: when $Q_s/Q_c > 0.1$,

$$S = \frac{\frac{\beta U}{K_t} A^{1-m_t}}{1 - \frac{K_t}{4K\beta} A^{m_t - m - 1}} \tag{11}$$

and when $Q_s/Q_c < 0.1$,

$$S = \frac{U}{0.1K} A^{-m} - \frac{26\beta U}{K_t} A^{1-m_t} \tag{12}$$

For the case where $n = 2$ and $n_t = 1$: When $Q_s/Q_c > 0.1$,

$$S = \frac{K_t}{4K\beta} A^{m_t - m - 1} + \frac{\beta U}{K_t} A^{1-m_t} \tag{13}$$

and when $Q_s/Q_c < 0.1$,

$$S = \frac{-13\beta U}{K_t} A^{1-m_t} + 5\sqrt{\left(\frac{2.6\beta U}{K_t} A^{1-m_t}\right)^2 + \frac{0.4U}{K} A^{-m}} \tag{14}$$

Equilibrium channel slopes increase with uplift rate using the almost-parabolic model (Equations 11–14), similarly to the predictions of equilibrium slopes using the stream-power model (e.g., Whipple and Tucker, 1999). When $n = 1$, slopes vary directly with uplift rate (Equations 11 and 12). The equilibrium relationship for Q_s/Q_c is given by

$$\frac{Q_s}{Q_c} = \frac{\beta UA}{K_t A^{m_t} S^{n_t}} \tag{15}$$

In the special case when $n = 1$, so $S \propto U$, the equilibrium value of Q_s/Q_c does not vary with uplift rate and therefore $f(Q_s)$ also does not vary with uplift rate (assuming that $n_t = 1$).

Generally, one might expect $f(Q_s)$ to vary throughout the drainage network. In fact, this is the key point of using an incision rule such as the almost-parabolic model, otherwise an equation with a constant erodibility, such as the stream-power model, applies. However, in the special case where $m_t = 1.5$ (and $m/n = 0.5$), $f(Q_s)$ does not vary downstream. In this case, the equilibrium slope-area relationships for the almost-parabolic model

(Equations 11–14) all predict that $S \propto A^{-0.5}$. In order for $f(Q_s)$ to be constant in space under equilibrium conditions, Q_s/Q_c must also be spatially invariable, as predicted by Equation 15 under these conditions ($m/n = 0.5$ and $m_t = 1.5$).

Linear-Decline Model

As an alternative method to the almost-parabolic model, we explore a generalized version of the under-capacity model described by Beaumont et al. (1992) and Kooi and Beaumont (1994). The linear-decline model assumes that as the sediment load increases (with respect to the sediment transport capacity), more energy is required to transport sediment, so less energy is available to expend on erosion of the bed. Here we use the model described by Whipple and Tucker (2002) (Fig. 1):

$$f(Q_s) = 1 - Q_s/Q_c \tag{16}$$

Using the same equilibrium assumption for sediment load as in the previous section ($Q_s = \beta UA$) and the same sediment transport equation ($Q_c = K_t A^{m_t} S^{n_t}$) the equilibrium slope-area relationships described by Whipple and Tucker (2002) can be derived. For $n = 1$ and $n_t = 1$:

$$S = \frac{U}{K} A^{-m} + \frac{\beta U}{K_t} A^{1-m_t} \tag{17}$$

For $n = 2$ and $n_t = 1$:

$$S = \frac{\beta U}{2K_t} \sqrt{\left(\frac{\beta U}{2K_t} A^{1-m_t} \right)^2 + \frac{U}{K} A^{-m}} \tag{18}$$

Again, equilibrium slopes increase with uplift rate, as they do with the stream-power and almost-parabolic models. Similarly to the almost-parabolic model, when $m_t = 1.5$ and $m/n = 0.5$, the value of $f(Q_s)$ remains constant throughout the network and channel concavity is constant. This is not the case when $m_t \neq 1.5$. Also, when $n = 1$ and $m_t = 1.5$, $f(Q_s)$ does not change with uplift rate.

TRANSIENT RESPONSE TO AN INCREASE IN UPLIFT RATE

In all cases, the equilibrium slope-area relationship predicts that channel slopes will increase with uplift rate. However, this relationship applies only for equilibrium. We now explore network transitions after a step increase in uplift rate throughout the drainage network using both incision rules described above.

The CHILD numerical landscape evolution model is used for all of the studies presented here (Tucker et al., 2001a,b). The experimental set-up is the same for all cases shown. A synthetic square drainage network is used with closed (no-flow) boundaries on all four sides and one corner outlet through which water and sediment can exit the network. The average distance between points is 50 m. Because the CHILD model operates on an irregular mesh, the spacing is not exactly 50 m between every node (the variation is less than a few meters). Water and sediment are assumed to follow the path of steepest descent across the landscape and can only leave the network at a single corner outlet. All points across the network are being raised at the same vertical uplift rate, and the outlet point feeds a downstream boundary of zero elevation. All simulations begin with a network that is in equilibrium with a low uplift rate (erosion balances the uplift rate across the network). The uplift rate is instantaneously increased and remains at the higher rate for the duration of the experiment.

For simplification, hillslope erosion is not considered in these simulations. In locations which only drain themselves, which in natural landscapes would be hillslopes, the lowering rate is calculated using $Q_s = 0$. In some examples, the points with the smallest drainage areas have very high slopes because they have no sediment influx to aid in erosion, but the behavior of these points is not considered in any of our analysis. The lack of hillslopes is addressed further in the discussion section.

Almost-Parabolic Model

In this section, we illustrate two numerical experiments using the almost-parabolic model for bedrock incision. Both begin with the same initial condition: equilibrium with an uplift rate of 0.1 mm/yr ($K = 5 \times 10^{-5}$yr^{-1}, $m = 0.5$, $n = 1$, $K_t = 5 \times 10^{-5}$yr^{-1}, $m_t = 1.5$, $n_t = 1$). In the first experiment, we increase the uplift rate by two times and in the second by four times. As will be illustrated, with these parameters the transient response is sensitive to the magnitude of change in uplift rate. Transient variations in sediment load and $f(Q_s)$ are responsible for the results shown here. However, keep in mind that in equilibrium, $f(Q_s)$ is not expected to change spatially or with uplift rate because we are using parameter values of $m_t = 1.5$, $m = 0.5$ and $n = 1$.

Figure 2 shows the response of main channel elevations (A) and slopes (B) after a twofold increase in uplift rate. The initial response looks very much like a knickpoint traveling up the network as one would expect from the stream-power model (Whipple and Tucker, 2002). (Note that we use the term knickpoint to refer to regions in which slope increases downstream; these regions are sometimes subtle on the channel profile, but are always clear in the slope-area data.) The knickpoint, however, does not increase channel slopes all the way up to their new equilibrium values. After the knickpoint has swept up the main channel, elevations and slopes continue to gradually increase everywhere (Fig. 3A and B), similar to transient conditions in a transport-limited channel (Whipple and Tucker, 2002).

The combination response illustrated by Figures 2 and 3, where initially a knickpoint propagates up the network and later slopes slowly increase throughout the network to their final values, is similar to the mixed-channel response illustrated by Whipple and Tucker (2002). In their example, they chose erosion parameters so that the equilibrium profile of the transport-limited and stream-power detachment limited channels are exactly the

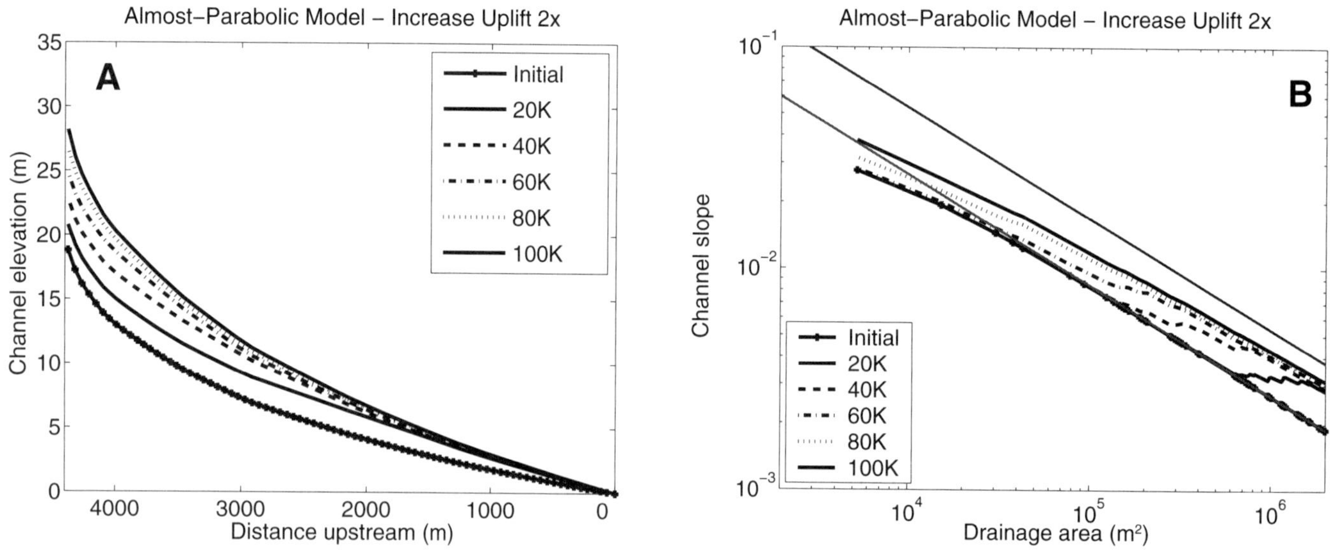

Figure 2. Initial change in main channel profile (A) and slope (B) in response to a twofold increase in uplift rate using the almost-parabolic model. The lines running through the slope-area plot are the equilibrium relationships for the old (lower) and new (upper) uplift rates. See text for parameters used in numerical simulations.

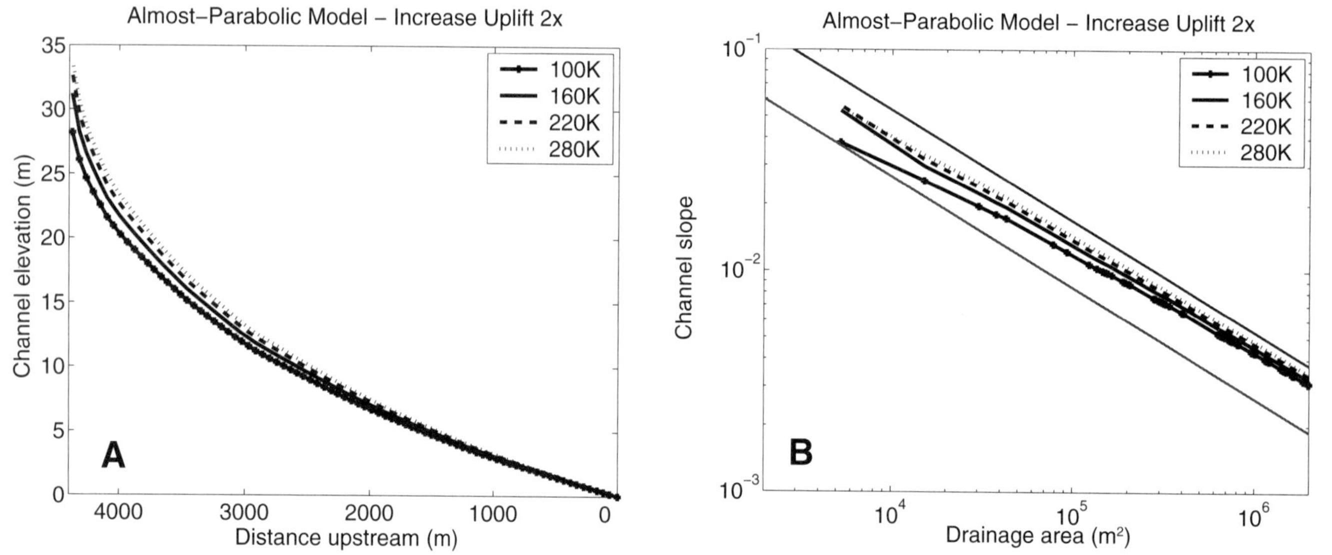

Figure 3. Later change (following the previous figure) in main channel profile (A) and slope (B) in response to a twofold increase in uplift rate using the almost-parabolic model. The lines running through the slope-area plot are the equilibrium relationships for the old (lower) and new (upper) uplift rates.

same. Their results show that the transient response to an increase in uplift rate has tendencies of both types of channels and depends on when changes in sediment load become important and the response switches from stream-power style to transport-limited style. Variations in sediment load also control the response here, as will become clear later.

Note that the initial equilibrium slope-area relationship in Figure 2 deviates in the smaller drainage areas from the predicted slope-area relationship. The deviation in the smaller drainage areas is a boundary condition of the model. The predicted slope-area relationship (plotted in Fig. 2) assumes that $Q_s = \beta UA$, but this is not the case at points which only drain themselves and have no incoming sediment load ($Q_s = 0$). Even though these points have reached equilibrium, their value of $f(Q_s)$ differs from that predicted using $Q_s = \beta UA$, and therefore their slope values also differ from those predicted assuming $Q_s = \beta UA$. This boundary condition affects the predicted slopes in the uppermost reaches of the network in both examples using the almost-parabolic model.

A fourfold increase in uplift rate causes some unexpected transients in the main channel. Following the increase in uplift rate, the knickpoint actually increases slopes above their new equilibrium values (Figs. 4B and 5B). In response to this overshooting of slopes above their new equilibrium values, channel elevations must subsequently be reduced. Slopes actually fall back down to a value lower than their new equilibrium value (Fig. 5B), and finally they must rise again (not shown). This overshooting and undershooting creates a whiplash-like response in channel profiles (Fig. 5A). A larger increase in uplift rate, using these same parameters, results in even more extreme changes in slopes (higher overshooting).

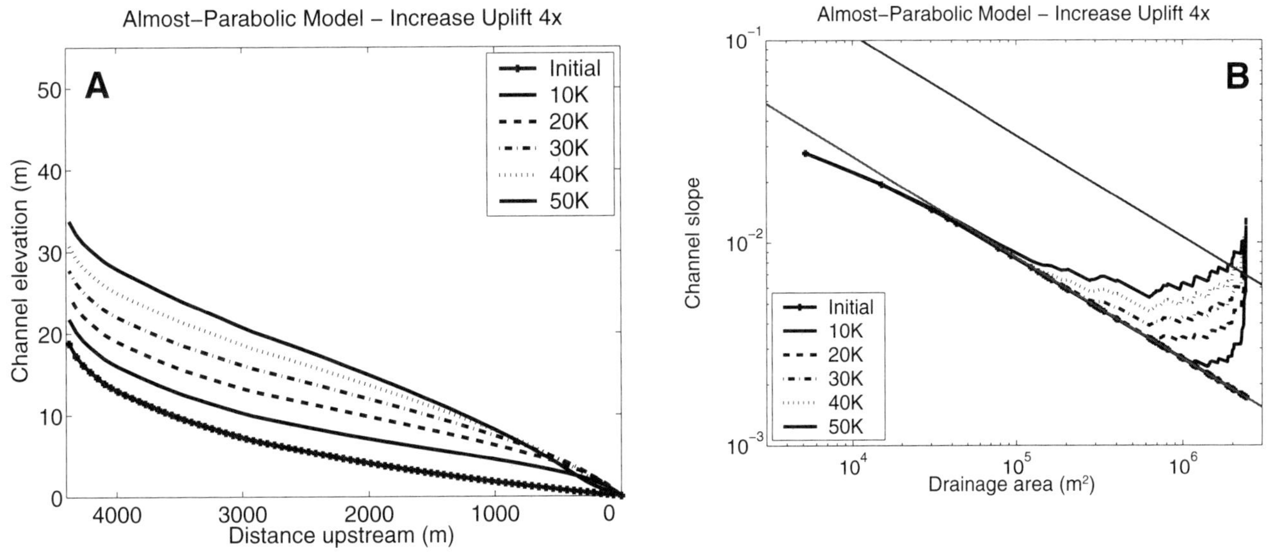

Figure 4. Initial change in main channel profile (A) and slope (B) in response to a fourfold increase in uplift rate using the almost-parabolic model. The lines running through the slope-area plot are the equilibrium relationships for old (lower) and new (upper) uplift rates. See text for parameter values used in numerical simulations.

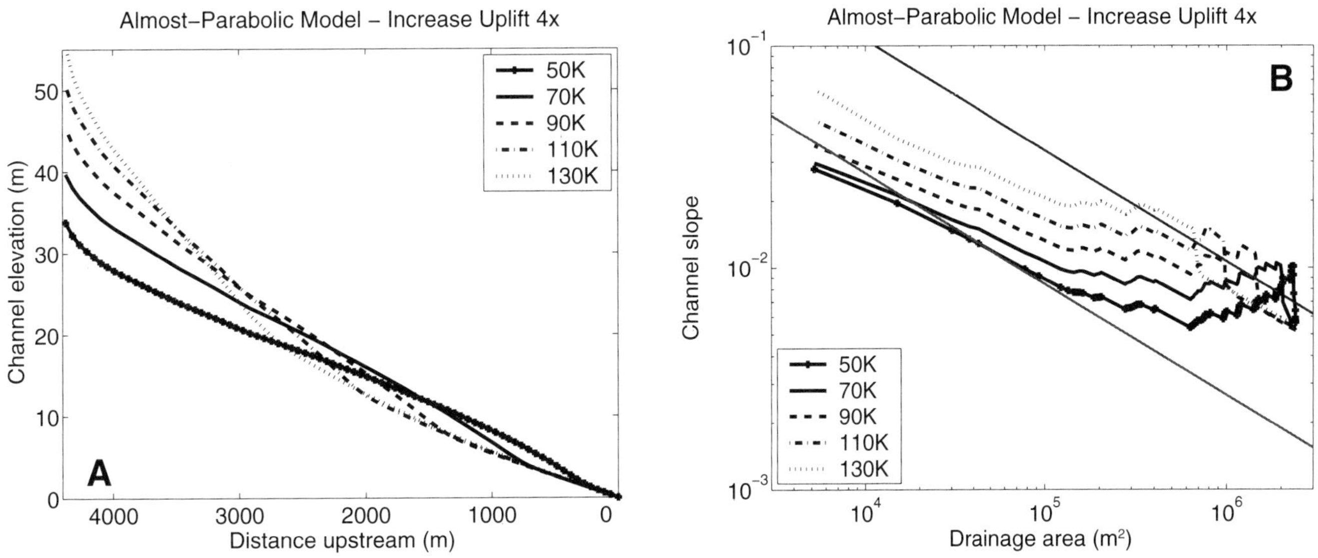

Figure 5. Later change (following the previous figure) in main channel profile (A) and slope (B) in response to a fourfold increase in uplift rate using the almost-parabolic model. The lines running through the slope-area plot are the equilibrium relationships for old (lower) and new (upper) uplift rates.

The complex response in channel slopes produced from a single increase in uplift rate is all the result of changes in sediment load and its effects on erodibility [through $f(Q_s)$]. Initially, the response to an increase in uplift rate is felt only at the outlet. The rest of the points in the network continue to erode at the old equilibrium uplift rate (Whipple and Tucker, 1999). (U_{old} is used to refer to the original smaller uplift rate in this discussion.) Figure 6 illustrates the pattern in erosion rates across the topography; shading is according to the ratio of the erosion rate to the new uplift rate (U_{new}). The illustrations are from the fourfold uplift increase experiment, therefore points which are eroding at the old uplift rate have an erosion ratio of 0.25 (shaded white in this figure). Points eroding at the new equilibrium value have an erosion ratio of 1.0 (shaded dark gray in this figure). Light-gray points have just started to respond to the change in uplift rate, while black points are eroding faster than the new uplift rate. (There are no black points in Figure 6A.) Figure 6A shows that points near the outlet respond first, while the rest of the network continues to erode at the same original rate (U_{old}). (This pattern is the same regardless of the magnitude of change in uplift rate.) As time moves on, erosion rates increase as a wave moving up the network (Figs. 6B and 6C). In Figures 6B and 6C, the black points are eroding faster than the new uplift rate (causing channel elevations to be reduced in Fig. 5A), and their erosion ratio will eventually reduce back to unity.

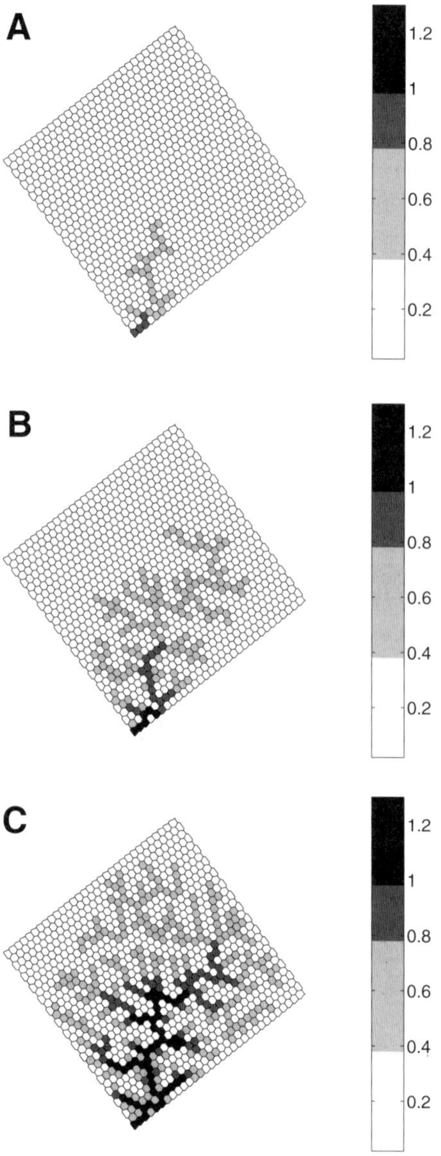

Figure 6. Response of erosion rates across the landscape at 10 k.y. (A), 30 k.y. (B), and 60 k.y. (C) after increasing the uplift rate by fourfold using the almost-parabolic model. Shading is by the ratio of erosion rate to new uplift value, so a value of 0.25 corresponds to the old erosion rate. The black polygons are areas that are eroding faster than the new equilibrium erosion rate. The total illustrated area is 1.6 × 1.6 km².

Because the points in the upper reaches of the network are eroding at the same original value (U_{old}), there is no change in the amount of sediment that they send downstream. Therefore, initially there is no change in the sediment load received by the outlet point. However the outlet point feels the increase in uplift rate and adjusts its slope to erode at the higher rate of U_{new}. Therefore the outlet point adjusts its slopes based on the old sediment load. This can lead to overshooting of slopes, as the profiles in Figures 4 and 5 illustrate.

As stated above, the outlet point needs to erode at the new uplift rate. Initially, the incoming sediment load at the outlet (or any point) does not change, but the slope can adjust, changing the value of Q_c, and furthermore, the value of $f(Q_s)$. This results in the following equation for the interim erosion rates at the outlet:

$$U_{new} = Kf(Q_s)_{trans} A^m S^n_{trans}, \tag{19}$$

where $f(Q_s)_{trans}$ and S_{trans} are the transient values of $f(Q_s)$ and S in response to the uplift increase. (Note that drainage area does not change with a change in uplift rate.) Rearranging Equation 19, we obtain an expression for S_{trans} as a function of $f(Q_s)_{trans}$:

$$S_{trans} = (U_{new}/Kf(Q_s)_{trans})^{1/n} A^{-m/n}. \tag{20}$$

In order to describe the transient channel slope, we need to describe $f(Q_s)_{trans}$. We have already predicted that the value of Q_s at the outlet remains at its old equilibrium value of $\beta U_{old} A$ (initially). In contrast to the sediment flux, channel slope at the outlet responds immediately to an increase in uplift rate, resulting in a larger value of Q_c (Equation 8). As a result, Q_s/Q_c will always decrease in response to an increase in uplift rate, and this can have different effects on $f(Q_s)_{trans}$.

The response of $f(Q_s)_{trans}$ will depend on the initial value of Q_s/Q_c. Figure 7 is a schematic of the expected response of $f(Q_s)_{trans}$ based on the examples from this section. The equilibrium value of Q_s/Q_c in these examples is 0.75 (for both U_{new} and U_{old}). More important than the actual value of Q_s/Q_c is that $Q_s/Q_c > 0.5$. Therefore a decrease in Q_s/Q_c (as expected) will initially cause $f(Q_s)_{trans}$ to rise (first part of solid arrow-line in Fig. 7).

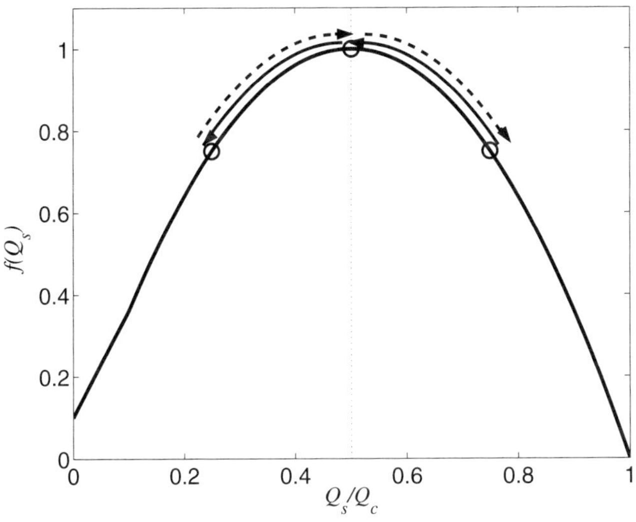

Figure 7. Cartoon example of how an increase in uplift rate can change $f(Q_s)$ using the almost-parabolic model. The solid arrow-lines indicate the initial response of $f(Q_s)$, and the dashed arrow-lines indicate the later response. For the example presented with a twofold increase in uplift rate, the changes in Q_s/Q_c and $f(Q_s)$ are represented by the first solid-arrowed line and the second dashed-arrowed line only. In the twofold uplift increase example $f(Q_s)$ never decreases.

If Q_s/Q_c is reduced to a value smaller than 0.5, the value of $f(Q_s)_{trans}$ will then decline (second part of solid arrow-line in Fig. 7). If Q_s/Q_c is reduced far enough, the value of $f(Q_s)_{trans}$ can decrease below the equilibrium value of $f(Q_s)$ (not shown in Fig. 7), leading to a transient slope value which is greater than the new equilibrium slope value (Equation 20).

The decrease in Q_s/Q_c results in the following expression for the transient value of $f(Q_s)$ when $Q_s/Q_c > 0.1$:

$$f(Q_s)_{trans} = 1 - 4\left(\frac{\beta U_{old} A}{K_t A^{m_t} S^{n_t}_{trans}} - 0.5\right)^2 \tag{21}$$

and when $Q_s/Q_c < 0.1$:

$$f(Q_s)_{trans} = 2.6\left(\frac{\beta U_{old} A}{K_t A^{m_t} S^{n_t}_{trans}}\right) + 0.1\,. \tag{22}$$

Combining Equation 20 with either Equation 21 or 22, we obtain an equation for the predicted value of S_{trans} (for $n = 1$ and $n_t = 1$) when $Q_s/Q_c > 0.1$:

$$S_{trans} = \frac{\dfrac{\beta U_{old}}{K_t} A^{1-m_t}}{1 - \dfrac{U_{new} K_t}{4\beta U_{old} K} A^{m_t - m - 1}} \tag{23}$$

and when $Q_s/Q_c < 0.1$:

$$S_{trans} = \frac{U_{new}}{0.1K} A^{-m} \frac{26\beta U_{old}}{K_t} A^{1-m_t}\,. \tag{24}$$

Equations 23 and 24 predict the initial change in slope before the sediment load starts to increase. These equations apply at any location in the drainage network where the upstream sediment load has not yet adjusted, but the downstream point is now eroding at U_{new}. Because erosion rates do not adjust immediately from U_{old} to U_{new}, Equation 23 applies only at the outlet where the downstream change in erosion rate, or equivalently uplift rate, is instantaneous because it is the boundary condition. Equation 23 is the predicted maximum slope. In actuality, S_{trans} is not always reached because upstream points begin to adjust (increasing the sediment flux at the outlet) before the outlet has time to reach S_{trans} predicted by Equation 23. (Note that similar equations can be derived for $n = 2$, $n_t = 1$ but they are not shown here.)

The transient slope value (from Equation 23 or 24) can be compared with the new equilibrium slope (predicted using U_{new} and Equations 11 or 12), in order to predict cases where the transient slope will overshoot its new equilibrium value. In the case when initially $Q_s/Q_c < 0.1$, the transient slope will always be greater than the new equilibrium slope (compare Equation 12 using U_{new} as the uplift value with Equation 24). This is because $f(Q_s)$ will always initially decrease with an increase in uplift rate when $Q_s/Q_c < 0.1$. When $Q_s/Q_c > 0.1$ initially, whether or not S_{trans} is greater than the new equilibrium slope depends on the initial value of Q_s/Q_c and the magnitude of change in Q_s/Q_c. If $Q_s/Q_c < 0.5$ initially, transient slopes will always over-steepen because $f(Q_s)$ will always initially decrease with an increase in uplift rate. However, when $Q_s/Q_c > 0.5$ initially, the magnitude of change in uplift rate determines whether or not transient slopes over-steepen. This last scenario ($Q_s/Q_c > 0.5$ initially) is the case for the two examples we have presented.

In the case of a twofold increase in uplift rate (Figs. 2 and 3), $f(Q_s)_{trans} = 1$. Because $f(Q_s)_{trans} = 0.75$ initially, S_{trans} is less than the new equilibrium slope value (Equation 20), and as a result, transient slopes only increase. In this example, Q_s/Q_c decreases initially to 0.5, but it never falls below 0.5 or, in other words, goes over the hump in Figure 7. As a result, slopes rise initially, but to a value less than their new equilibrium value (Fig. 2B). Later, when sediment load increases, resulting in an increase in Q_s/Q_c and a decrease in $f(Q_s)$, slopes rise even further (Fig. 3B).

For the fourfold uplift increase case, $f(Q_s)$ is pushed to below its initial value of 0.75, and slopes rise above their new equilibrium value, as was illustrated in Figures 4B and 5B. Changes in $f(Q_s)$ through time at the outlet and two points upstream are illustrated in Figure 8A. As slope increases at the outlet through time (Fig. 9A) $f(Q_s)$ rises and then falls (Fig. 8A); at the same time, the value of Q_s/Q_c is steadily decreasing (Fig. 8B). Once the upper parts of the network start to respond, the sediment flux at the outlet begins to increase. Eventually, the increasing sediment load overtakes the effects of the over-steepening slopes, and Q_s/Q_c starts to increase (Fig. 8B). By this point, erosion rates have already risen above their new equilibrium value, and they now begin to decrease (Fig. 9B).

The slope at the outlet does not reach the predicted value of S_{trans} from Equation 24 (Fig. 9A). Because it takes some time for the slope at the outlet to rise, upstream points begin to adjust while the slope at the outlet is still rising. As soon as the upstream points start to adjust, they send more sediment downstream. Equation 23 predicts the maximum change in slope if the incoming sediment load does not change. In this case, the sediment load begins to increase before the predicted maximum transient slope is reached.

The pattern of change in upstream points is similar to that at the outlet. Changes in $f(Q_s)$ are dampened at the upstream points. Because $f(Q_s)$ does not drop as low in the upstream points, but the slopes still increase, this allows for the erosion ratio to vary more than it does at the outlet. There is some loss of information upstream, which in this case leads to more variability in erosion rates upstream. The outlet point and points near to the outlet feel the boundary condition more closely and are more tied to it, while points higher upstream feel the change later. Upstream points are not bound to immediately adjust their slopes to erode at U_{new}, and therefore a lag in information sent upstream or misinformation through the overshooting of slopes, leads to more freedom in the transient erosion rates upstream.

As stated earlier, when the equilibrium value of $Q_s/Q_c \leq 0.5$, an increase in uplift rate will always result in an initial decrease in $f(Q_s)$ (starting at the top of the hump, or to the left of it, in Fig. 7). We performed a number of numerical experiments where the initial and final values of Q_s/Q_c are less than 0.5 (not illustrated), and the slopes always over-steepened, regardless of the magnitude of increase in uplift rate.

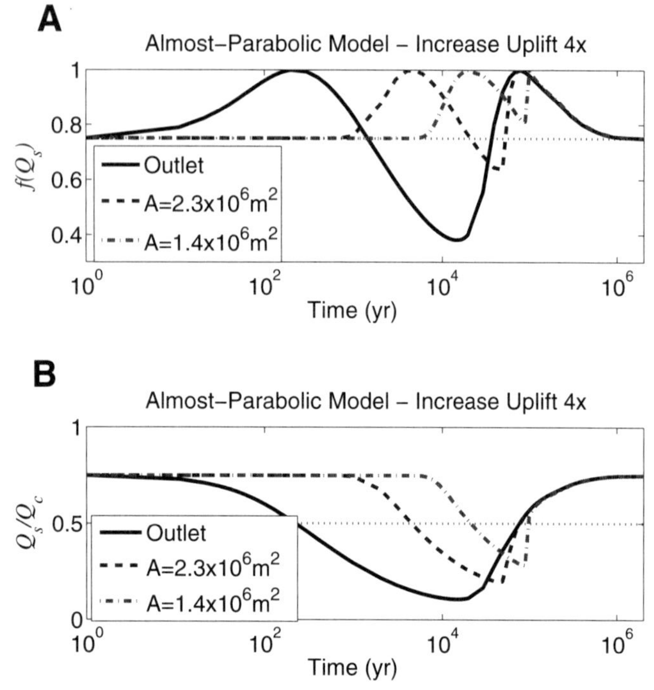

Figure 8. Response over time of $f(Q_s)$ (A) and Q_s/Q_c (B) at three different points after a fourfold increase in uplift rate using the almost-parabolic model. In (A) the thin-dotted line represents the equilibrium value of $f(Q_s)$, which does not change in space and time. In (B) the thin-dotted line ($Q_s/Q_c = 0.5$) represents the value of Q_s/Q_c where $f(Q_s)$ changes from increasing to decreasing. Parameter values for this simulation are given in the text.

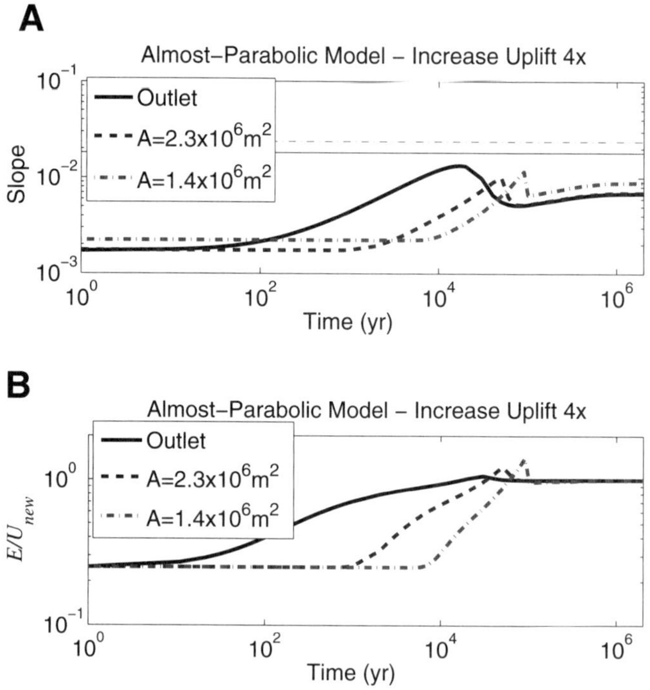

Figure 9. Response over time of channel slope (A) and erosion ratio (B) from the same simulation (fourfold increase in uplift using almost parabolic model) and at the same three points illustrated in Figure 8. Thin horizontal lines in (A) are the predicted transient slopes from Equation 23 for the three different points. Note that the predicted transient slopes for the outlet (area = $2.4 \times 10^6 m^2$) and the next illustrated upstream point (area = $2.3 \times 10^6 m^2$) are essentially the same, so they appear as only one line on this plot

Linear-Decline Model

In this section, we discuss an example of channel network response to an increase in uplift rate using the linear-decline model. Changes in uplift rate are felt first at the outlet and then propagate upstream (in a similar fashion to the almost-parabolic model). There is no case in which the linear-decline model predicts over-steepening of channel slopes during transitions from a low to high uplift rate.

Figure 10 illustrates the initial response in main channel elevations (A) and slopes (B) to a fivefold increase in uplift rate ($K = 1 \times 10^{-6} yr^{-1}$, $m = 1$, $n = 2$, $K = 2 \times 10^{-4} m^{0.4} yr^{-1}$, $m_t = 1.3$, $n_t = 1$). Initially a knickpoint sweeps upstream. However, channel slopes increase to a value less than their new equilibrium value (Fig. 10B). Once the knickpoint passes a point in the channel, its slope (and therefore elevation, since the lower points are also steepening) continues to rise. This leads to a later response in which slopes are rising everywhere throughout the network (Fig. 11B). This combination-type response is similar to the twofold uplift increase example using the almost-parabolic model and

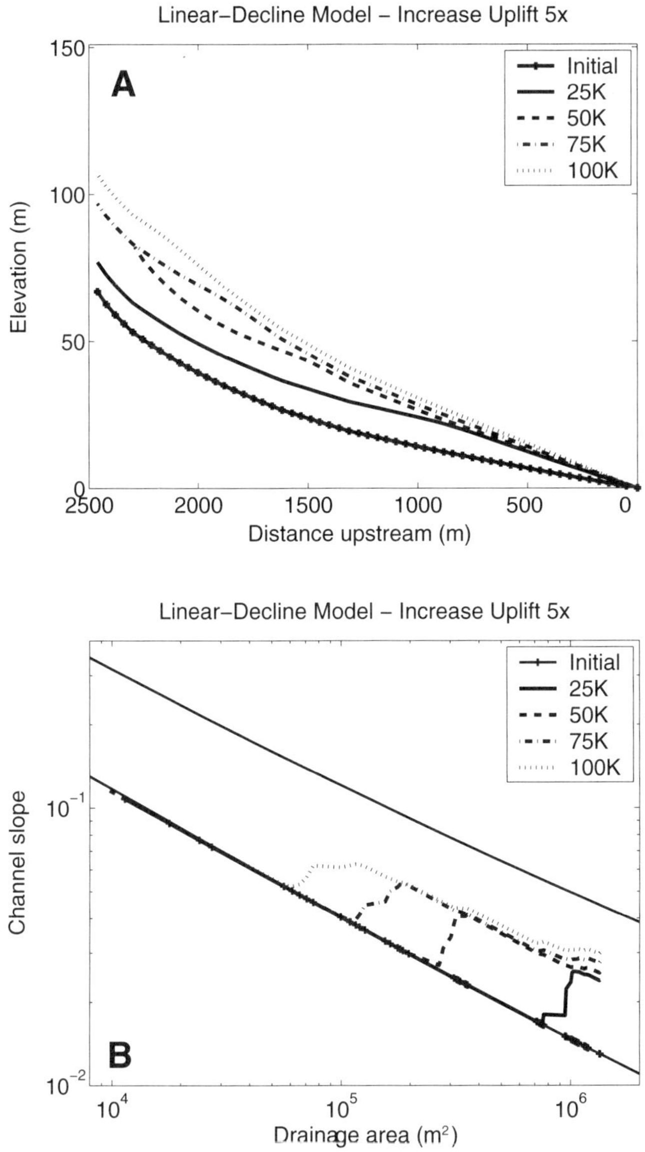

Figure 10. Using the linear-decline model, changes in main channel profile (A) and slope (B) through time in response to a fivefold increase in uplift rate. (The lines running through the slope-area plot are the equilibrium relationships for old (lower) and new (upper) uplift rates.) The temporary shortening of the profile at 50K is due to network rearrangement. See text for model parameters.

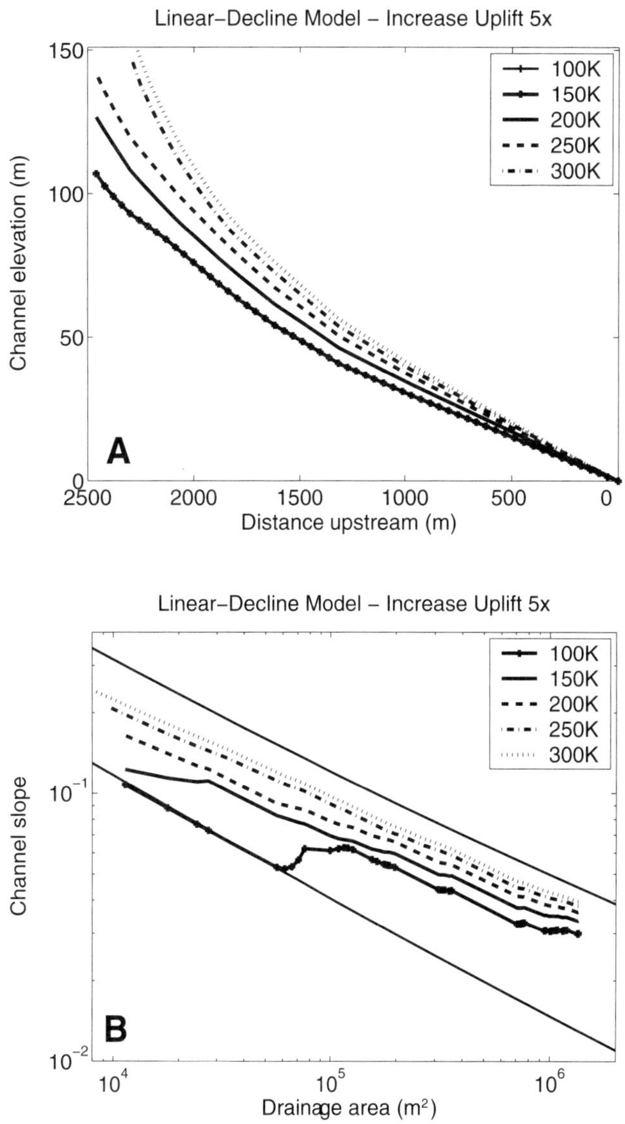

Figure 11. Later changes in time (after Fig. 10) in main channel profile (A) and slope (B) in response to a fivefold increase in uplift rate using the linear-decline model. (The lines running through the slope-area plot are the equilibrium relationships for old (lower) and new (upper) uplift rates.)

the mixed-channel response of threshold channels described by Whipple and Tucker (2002). The response illustrated in Figures 10 and 11 is also due to the lag in response of the sediment flux.

As was the case with the almost-parabolic model, the sediment flux at the outlet can be assumed to stay constant while channel slope rises to accommodate for the increase in uplift rate. Given that the channel slope rises but sediment flux remains the same, the value of Q_s/Q_c declines in response to an increase in uplift rate. With the linear-decline model, $f(Q_s)$ can only increase as Q_s/Q_c declines (Fig. 1 and Equation 16), so channel slopes will never overshoot their new equilibrium value (Equation 20). The expression for the transient slope in the case of $n = 2$, $n_t = 1$ is

$$S_{trans} = \frac{\beta U_{old}}{2K_t} + \sqrt{\left(\frac{\beta U_{old}}{2K_t} A^{1-m_t}\right)^2 + \frac{U_{new}}{K} A^{-m}} \,. \quad (25)$$

This is very similar to the equilibrium expression (Equation 18) except that two of the uplift terms are replaced with U_{old}. Given that $U_{new} > U_{old}$, this equation always predicts that the transient slope will be less than the new equilibrium slope. (A similar equation can be derived for the $n = 1$, $n_t = 1$ case but is not illustrated here.)

The increase in $f(Q_s)$ at the outlet (and two other points upstream) is shown in Figure 12A. Channel slope increases (Fig. 13A) while sediment load remains the same causing Q_s/Q_c to decrease (Fig. 12B). The decrease in Q_s/Q_c results in a rise in $f(Q_s)$ (Fig. 12A). After rising, $f(Q_s)$ remains stable for a period and the outlet slope remains at S_{trans}. Once the sediment load begins to increase, $f(Q_s)$ starts to decline and the slope at the outlet begins to rise again. Because $m_t \neq 1.5$, the equilibrium value of $f(Q_s)$ does not stay constant with uplift rate and eventually decreases below its initial value. This decrease in $f(Q_s)$ happens at later times (Fig. 12A) and does not result in an over-steepening of slopes or erosion rates that are greater than U_{new} (Fig. 13).

In this example, all points reach their transient slope value as predicted by Equation 25 because the transient slope value is between the old and new slope value. Because the outlet reaches S_{trans} before the upper points begin to adjust, the outlet remains at S_{trans} for a period of time before increasing its slope further (plateau of the thick solid line in Fig. 13A).

Figures 12 and 13 illustrate that the response at upstream points lags the response at the outlet. The points upstream from the outlet adjust more freely than the outlet does, and erosion rates vary more upstream in that they both increase and decrease (Fig. 13B). The dash-dot line in Figure 12A illustrates that for a very short period (~50–70 k.y.) $f(Q_s)$ declines while channel slope does not keep pace (Fig. 13A), causing erosion rates to slightly decrease (Fig. 13B).

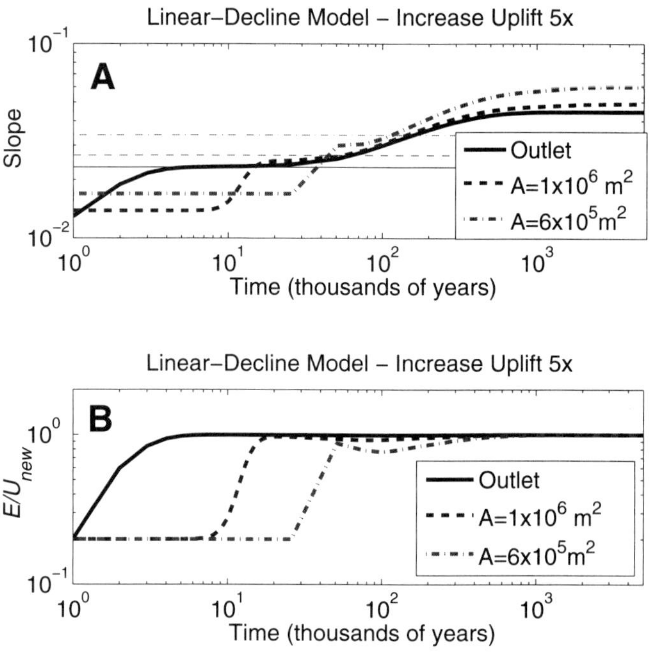

Figure 13. Thick lines represent the change over time of the channel slope (A) and erosion ratio (B) from the same simulation and at the same three points illustrated in Figure 12. Thin horizontal lines in (A) are the predicted transient slopes from Equation 18 for the three different points.

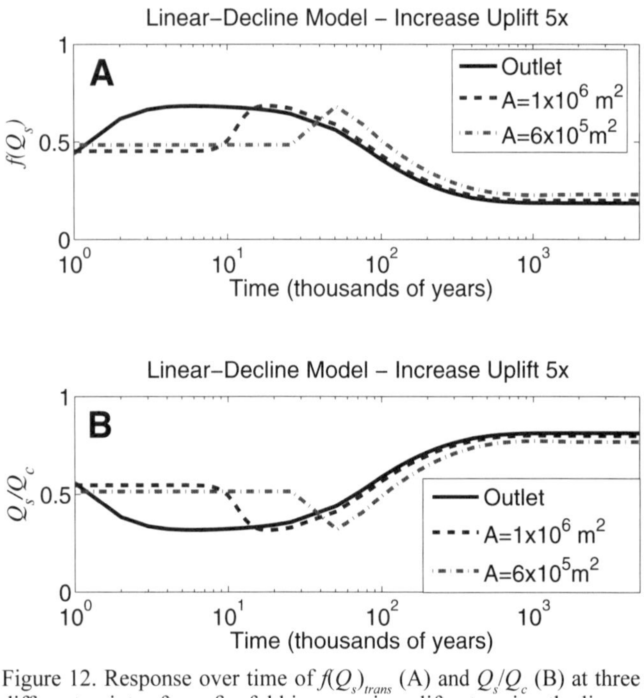

Figure 12. Response over time of $f(Q_s)_{trans}$ (A) and Q_s/Q_c (B) at three different points after a fivefold increase in uplift rate using the linear-decline model. (See text for model parameters.)

DISCUSSION

Using two different sediment-flux–dependent incision models, our results show that the channel response to a change in uplift rate can be more complex than a single knickpoint propagating upstream. The complex response of the channels is a result of different variables (sediment flux and channel slope) responding at different time scales in different parts of the network and embodies the ideas put forth by Schumm (1977). The upper parts of the network respond to an increase in uplift rate after the lower parts of the network. As a result, the increase in sediment flux from upstream lags the initial increase in channel slopes downstream. Because the sediment flux affects the erodibility, changes in the sediment flux cause the erosion rates to change and the downstream parts of the network must readjust their slopes once the sediment flux increases. The key difference between the incision models used in this study and the stream-power model is that the stream-power model is not sensitive to changes upstream because it is only a function of channel slope.

Knickpoints have been observed in many channels and are predicted with the stream-power model, so one might question whether further complications in the erosion equation are necessary. We argue that knickpoints are not the only manner in which bedrock channels respond to transient conditions. For example, Blythe et al. (2000) suggest that there was an acceleration in uplift rate in the San Gabriel Mountains, California, over the past 3 m.y. In these drainages, we have observed many higher-order channels that have knickpoints and exposed bedrock. However, landslides are inundating the upstream parts of the drainages with sediment, and erosion rates appear to be limited by the transport and wear of large boulders. In other words, there are transport-limited reaches of channel upstream from bedrock channels. These transient dynamics could not be predicted using the stream-power model alone.

The unexpected transient results using the almost-parabolic model could have important implications when trying to infer past conditions in the landscape. If a channel is cutting laterally as well as vertically, periods of incision can leave behind a strath terrace. The almost-parabolic model predicts that the channel may go through periods of both rising and falling as the incision rate varies in response to a single step increase in uplift rate. If the periods of different channel incision rates were recorded in fluvial terraces, they could be interpreted as the response to multiple changes in forcing (Hancock and Anderson, 2002). Furthermore, if the decline in channel slopes was recorded in strath terraces, it would probably be interpreted as a decline in uplift rate and not an increase (Pazzaglia et al., 1998). In the example that we illustrated, channel slopes at the outlet decline for a period of 30,000 yr. This is a significant period of time, and therefore its record could remain in the landscape and its meaning could easily be misinterpreted. Note that the period of declining slopes depends on the magnitude of change in uplift rates and the parameter values used in the simulation. We have performed other simulations in which the period of declining slopes can be even longer and leave a more significant signal throughout the network, rather than just predominantly at the outlet.

There are a number of limitations in our model. We do not consider landslides, and this could be an important omission. In essence, our model assumes that sediment delivery from the hillslopes keeps pace with fluvial erosion. If hillslopes adjust over long time scales, this could be a critical oversimplification. For example, Fernandes and Dietrich (1997) showed that hillslopes can take tens of thousands of years to adjust to a twofold increase in base-level downcutting. Even though the channels may take more time to equilibrate than the hillslopes, as the hillslope evolves, it steepens and delivers more sediment to the channel. If the sediment load delivered from the hillslopes to the channel changes over tens of thousands of years, this could potentially change the response of the channel, and the landscape as a whole might take even longer to adjust. Furthermore, episodic inputs of sediment to the channel from landslides or debris flows could leave the channel in a constant state of adjustment as sediment pulses are transported throughout the network. Another potential variable that is missing from this study is the role of thresholds for entrainment and transport of sediment. Snyder et al. (2003a) and Tucker (2004) found that thresholds, along with the stochastic nature of storms and floods, add a highly nonlinear response to the system and play a large role in landscape morphology. Finally, although we have made some improvements by allowing channel erodibility to respond to a change in uplift rate (as opposed to channel slope only), we do not consider other variables, such as grain size, hydraulic roughness, and channel width that could potentially adjust during transients and vary with different equilibrium erosion rates (Stark and Stark, 2001; Chitale, 2003; Gasparini et al., 2004). Snyder et al. (2003b) found that channel width did not vary between regions of different uplift rates, but few studies have been made on the adjustment of channel form to uplift rates, leaving room for further investigation. Currently available channel width data has been discussed by Whipple (2004).

It is difficult to predict what effects further complications might have on our results. In this study, we find that there are some similarities between the two different sediment-flux models. For example, they both predict that there will be periods of time during which the channel has not yet reached steady-state but it appears graded, with relatively uniform concavity throughout and no knickpoints [similar to the hybrid model presented by Whipple and Tucker (2002)]. It is possible that other complications, such as thresholds and sediment-delivery from the hillslopes might distinguish the response of these models further. Even so, it is much easier to find differences between models in idealized numerical settings than it is in natural settings. We suggest that numerical studies such as this one should be used along with other data, such as thermochronologic data, historical landslide data and sediment transport data, in order to fully understand the processes controlling bedrock incision. Furthermore, this study challenges some commonly held assumptions about bedrock channels. Oftentimes, a lack of knickpoints is used as an indicator of steady-state conditions; however, this study suggests that transient conditions do not necessarily imply the presence of knickpoints. This has implications in tectonic-geomorphologic studies which attempt to extract tectonic information from river profile form, and, for instance, is one reason why morphometic parameters, such as the steepness index, cannot be quantitatively inverted for rock uplift rate (e.g., Hurtrez et al., 1999; Snyder et al., 2000; Kirby and Whipple, 2001; Lague and Davy, 2003; Kirby et al., 2003; Wobus et al., 2003; Wobus et al., this volume).

CONCLUSIONS

Transient network response is very sensitive to the applied erosion model. Using the linear-decline model, an increase in uplift rate can result initially in a knickpoint propagating through the network which does not, however, raise slopes to their new equilibrium values; at later times, channel slopes rise slowly throughout the network. When incision rates are calculated using the almost-parabolic model, some cases predict that channel slopes will over-steepen and erosion rates will surpass their new equilibrium values and later decline. As a result, in some cases using the almost-parabolic model, channel elevations can increase and decrease in both space and time. With both models, the complex pattern of elevation changes at a single point results from the lag in response of the sediment flux relative to the response of channel slopes. These results are more complex than previous predictions using the stream-power model, in which knickpoint propagation is the sole method for a network to respond to an increase in uplift rate.

Using both incision models, there are periods in which the channel profile is relatively smooth. At these times, one could easily misinterpret this as an indication of steady-state conditions. Further, declining slopes in response to an increase in uplift rate could leave a very deceptive mark in the landscape. This study highlights that complexities in the total network response, in particular the lag in response of the sediment load, may play an important role in determining the transient morphology of fluvial bedrock networks.

ACKNOWLEDGMENTS

This work was supported by the NASA Earth System Science Fellowship Program and NASA grant NAG5–10684. We thank Rudy Slingerland and an anonymous reviewer for their helpful suggestions. This paper benefited greatly from discussions with Greg Tucker and Daniel Collins.

REFERENCES CITED

Beaumont, C., Fullsack, P., and Hamilton, J., 1992, Erosional control of active compressional orogens, *in* McClay, K.R., ed., Thrust Tectonics: New York, Chapman and Hall, p. 1–18.

Blythe, A.E., Burbank, D.W., Farley, K.A., and Fielding, E.J., 2000, Structural and topographic evolution of the central Transverse Ranges, California from apatite fission-track, (U-Th)/He and digital elevation model analyses: Basin Research, v. 12, p. 97–114, doi: 10.1046/j.1365-2117.2000.00116.x.

Chitale, S.V., 2003, Modeling for width adjustment in alluvial rivers: Journal of Hydraulic Engineering, v. 129, p. 404–407, doi: 10.1061/(ASCE)0733-9429(2003)129:5(404).

Fernandes, N.F., and Dietrich, W.E., 1997, Hillslope evolution by diffusive processes: the timescale for equilibrium adjustments: Water Resources Research, v. 33, p. 1307–1318, doi: 10.1029/97WR00534.

Flint, J.J., 1974, Steam gradient as a function of order, magnitude, and discharge: Water Resources Research, v. 10, p. 969–973.

Foley, M.G., 1980, Bed-rock incision by streams: Geological Society of America Bulletin, v. 91, p. 2189–2213, doi: 10.1130/0016-7606(1980)91<577:BIBSS>2.0.CO;2.

Gardner, T.W., 1983, Experimental study of knickpoint and longitudinal profile evolution in cohesive, homogeneous material: Geological Society of America Bulletin, v. 94, p. 664–672, doi: 10.1130/0016-7606(1983)94<664:ESOKAL>2.0.CO;2.

Gasparini, N.M., 2003, Equilibrium and transient morphologies of river networks: discriminating among fluvial erosion models [Ph.D. thesis]: Massachusetts Institute of Technology, 232 p.

Gasparini, N.M., Tucker, G.E., and Bras, R.L., 2004, Network-scale dynamics of grain-size sorting: Implications for downstream fining, stream-profile concavity and drainage basin morphology: Earth Surface Processes and Landforms, v. 29, p. 401–421, doi: 10.1002/esp.1031.

Gilbert, G.K., 1877, Report on the geology of the Henry Mountains [Utah], Technical report: U.S. Geological Survey, Rocky Mountain Region: Washington, D.C., U.S. Government Printing Office, 160 p.

Hack, J.T., 1957, Studies of longitudinal stream profiles in Virginia and Maryland: U.S. Geological Survey Professional Paper 294-B, p. 45–97.

Hancock, G.S., and Anderson, R.S., 2002, Numerical modeling of fluvial strath-terrace formation in response to oscillating climate: Geological Society of America Bulletin, v. 114, p. 1131–1142, doi: 10.1130/0016-7606(2002)114<1131:NMOFST>2.0.CO;2.

Hancock, G.S., Anderson, R.S., and Whipple, K.X., 1998, Beyond power: Bedrock river incision process and form, *in* Tinkler, K.J., and Wohl, E.E., eds., Rivers over Rock: Fluvial Processes in Bedrock Channels: American Geophysical Union Geophysical Monograph 107, p. 35–60.

Hartshorn, K., Hovius, N., Dade, W.B., and Slingerland, R.L., 2002, Climate-driven bedrock incision in an active mountain belt: Science, v. 297, p. 2036–2038, doi: 10.1126/science.1075078.

Howard, A.D., 1980, Thresholds in river regimes, *in* Coates, D.R, and Vitek, J.D. eds., Thresholds in Geomorphology: Allen and Unwin, p. 227–258.

Howard, A.D., 1994, A detachment-limited model of drainage basin evolution: Water Resources Research, v. 30, p. 2261–2285, doi: 10.1029/94WR00757.

Howard, A.D., Seidl, M.A., and Dietrich, W.E., 1994, Modeling fluvial erosion on regional to continental scales: Journal of Geophysical Research, v. 99, p. 13971–13986, doi: 10.1029/94JB00744.

Hurtrez, J.E., Lucazeau, F., Lave, J., and Avouac, J., 1999, Investigation of the relationship between basin morphology, tectonic uplift, and denudation from the study of an active fold belt in the Siwalik Hills, central Nepal: Journal of Geophysical Research, v. 104, p. 12,779–12,796, doi: 10.1029/1998JB900098.

Kim, J.Y., 2004, Controls over bedrock channel incision [Ph.D. thesis]: University of Glasgow, 453 p.

Kirby, E., and Whipple, K.X., 2001, Quantifying differential rock-uplift rates via stream profile analysis: Geology, v. 29, p. 415–418, doi: 10.1130/0091-7613(2001)029<0415:QDRURV>2.0.CO;2.

Kirby, E., Whipple, K.X., Tang, W., and Chen, Z., 2003, Distribution of active rock uplift along the eastern margin of the Tibetan Plateau: Inferences from bedrock channel longitudinal profiles: Journal of Geophysical Research, v. 108, no. B4, 2217, doi: 10.1029/2001JB000861.

Kooi, H., and Beaumont, C., 1994, Escarpment evolution on high-elevation rifted margins: Insights derived from a surface processes model that combines diffusion, advection, and reaction: Journal of Geophysical Research, v. 99, p. 12191–12209, doi: 10.1029/94JB00047.

Lague, D., and Davy, P., 2003, Constraints on the long-term colluvial erosion law by analyzing slope-area relationship at various tectonic uplift rates in the Siwaliks Hills (Nepal): Journal of Geophysical Research, v. 108, no. B2, 2129, doi: 10.1029/2002JB001893.

Meyer-Peter, R., and Müller, R., 1948, Formulas for bedload transport, *in* Proceedings, 2nd Meeting International Association of Hydraulic Research: Stockholm, International Association of Hydraulic Research, p. 39–64.

Moglen, G.E., and Bras, R.L., 1995, The effect of spatial heterogeneities on geomorphic expression in a model of basin evolution: Water Resources Research, v. 31, p. 2613–2623, doi: 10.1029/95WR02036.

Pazzaglia, F.J., Gardner, T.W., and Merritts, D.J., 1998, Bedrock fluvial incision and longitudinal profile development over geologic time scales determined by fluvial terraces, *in* Tinkler, K.J., and Wohl, E.E., eds., Rivers Over Rock: Fluvial Processes in Bedrock Channels: American Geophysical Union Geophysical Monograph 107, p. 207–236.

Pizzuto, J.E., 1995, Downstream fining in a network of gravel-bedded rivers: Water Resources Research, v. 31, p. 753–759, doi: 10.1029/94WR02532.

Rosenbloom, N.A., and Anderson, R.S., 1994, Hillslope and channel evolution in a marine terraced landscape, Santa Cruz, California: Journal of Geophysical Research, v. 99, p. 14013–14029, doi: 10.1029/94JB00048.

Schoenbohm, L.M., Whipple, K.X., Burchfiel, B.C., and Chen, L., 2004, Geomorphic constraints on surface uplift, exhumation, and plateau growth in the Red River region, Yunnan Province, China: Geological Society of America Bulletin, v. 116, p. 895–909, doi: 10.1130/B25364.1.

Schumm, S.A., 1977, The Fluvial System, John Wiley and Sons, 338 p.

Sinha, S.K., and Parker, G., 1996, Causes of concavity in longitudinal profiles of rivers: Water Resources Research, v. 32, p. 1417–1428, doi: 10.1029/95WR03819.

Sklar, L., 2003, The influence of grain size, sediment supply and rock strength on rates of river incision into bedrock [Ph.D. thesis]: University of California, Berkeley, 342 p.

Sklar, L., and Dietrich, W.E., 1998, River longitudinal profiles and bedrock incision models: Stream power and the influence of sediment supply, *in* Tinkler, K.J., and Wohl, E.E., eds., Rivers over Rock: Fluvial Processes in Bedrock Channels: American Geophysical Union Geophysical Monograph 107, p. 237–260.

Sklar, L., and Dietrich, W.E., 2001, Sediment and rock strength controls on river incision into bedrock: Geology, v. 29, p. 1087–1090, doi: 10.1130/0091-7613(2001)029<1087:SARSCO>2.0.CO;2.

Slingerland, R.S., Willett, S.D., and Hennessey, H.L., 1997, A new fluvial bedrock erosion model based on the work-energy principle: Eos (Transactions, American Geophysical Union), v. 78, Fall Meeting Supplement F299.

Snow, R.S., and Slingerland, R.L., 1987, Mathematical modeling of graded river profiles: Journal of Geology, v. 95, p. 15–33.

Snyder, N.P., Whipple, K.X., Tucker, G.E., and Merritts, D.J., 2000, Landscape response to tectonic forcing: Digital elevation model analysis of stream profiles in the Mendocino triple junction region, northern California: Geological Society of America Bulletin, v. 112, p. 1250–1263, doi: 10.1130/0016-7606(2000)112<1250:LRTTFD>2.3.CO;2.

Snyder, N.P., Whipple, K.X., Tucker, G.E., and Merritts, D.J., 2002, Interactions between onshore bedrock-channel incision and nearshore wave-base erosion forced by eustasy and tectonics: Basin Research, v. 14, p. 105–127, doi: 10.1046/j.1365-2117.2002.00169.x.

Snyder, N.P., Whipple, K.X., Tucker, G.E. and Merritts, D.J. 2003a, Importance of a stochastic distribution of floods and erosion thresholds in the bedrock river incision problem: Journal of Geophysical Research, v. 108, no. B2, 2117, doi: 10.1029/2001JB001655.

Snyder, N.P., Whipple, K.X., Tucker, G.E., and Merritts, D.J., 2003b, Channel response to tectonic forcing: Analysis of stream morphology and hydrology in the Mendocino triple junction region, northern California: Geomorphology, v. 53, p. 97–127, doi: 10.1016/S0169-555X(02)00349-5.

Stark, C.P., and Stark, G.J., 2001, A channelization model of landscape evolution: American Journal of Science, v. 301, p. 486–512.

Stock, J.D., and Montgomery, D.R., 1999, Geologic constraints on bedrock river incision using the stream power law: Journal of Geophysical Research, v. 104, no. B3, p. 4983–4993, doi: 10.1029/98JB02139.

Tarboton, D.G., Bras, R.L., and Rodriquez-Iturbe, I., 1991, On the extraction of channel networks from digital elevation data: Hydrological Processes, v. 5, p. 81–100.

Tomkin, J.H., Brandon, M.T., Pazzaglia, F.J., Barbour, J.R. and Willett, S.D., 2003, Quantitative testing of bedrock incision models for the Clearwater River, NW Washington state: Journal of Geophysical Research, v. 108, no. B6, 2308, doi: 10.1029/2001JB000862.

Tucker, G.E., 2004, Drainage basin sensitivity to tectonic and climatic forcing: implications of a stochastic model for the role of entrainment and erosion thresholds: Earth Surface Processes and Landforms, v. 29, p. 185–205, doi: 10.1002/esp.1020.

Tucker, G.E., Lancaster, S.T., Gasparini, N.M., Bras, R.L., and Rybarczyk, S.M., 2001a, An object-oriented framework for distributed hydrologic and geomorphic modeling using triangulated irregular networks: Computers & Geosciences, v. 27, p. 959–973, doi: 10.1016/S0098-3004(00)00134-5.

Tucker, G.E., and Slingerland, R.L., 1997, Drainage basin responses to climate change: Water Resources Research, v. 33, p. 2031–2047, doi: 10.1029/97WR00409.

Tucker, G.E. and Whipple, K.X., 2002, Topographic outcomes predicted by stream erosion models: Sensitivity analysis and intermodel comparison: Journal of Geophysical Research, v. 107, no. B9, 2179, doi: 10.1029/2001JB000162.

Tucker, G.E., Lancaster, S.T., Gasparini, N.M., and Bras, R.L., 2001b, The channel-hillslope integrated landscape development model (CHILD), *in* Harmon, R.S., and Doe, W.W. eds., Landscape Erosion and Evolution Modeling: Kluwer Academic/Plenum Publishers, p. 349–388.

van der Beek, P., and Bishop, P., 2003, Cenozoic river profile development in the Upper; Lachlan catchment (SE Australia) as a test of quantitative fluvial incision models: Journal of Geophysical Research, v. 108, no. B6, 2309, doi: 10.1029/2002JB002125.

Whipple, K.X., 2001, Fluvial landscape response time: how plausible is steady-state denudation?: American Journal of Science, v. 301, p. 313–325.

Whipple, K.X., 2004, Bedrock rivers and the geomorphology of active orogens: Annual Review of Earth and Planetary Sciences, v. 32, p. 151–185, doi: 10.1146/annurev.earth.32.101802.120356.

Whipple, K.X., Hancock, G.S., and Anderson, R.S., 2000a, River incision into bedrock: Mechanics and relative efficacy of plucking, abrasion, and cavitation: Geological Society of America Bulletin, v. 112, p. 490–503, doi: 10.1130/0016-7606(2000)112<0490:RIIBMA>2.3.CO;2.

Whipple, K.X., Snyder, N.P., and Dollenmayer, K., 2000b, Rates and processes of bedrock incision by the Upper Ukak River since the 1912 Novarupta ash flow in the Valley of Ten Thousand Smokes, Alaska: Geology, v. 28, p. 835–838, doi: 10.1130/0091-7613(2000)028<0835:RAPOBI>2.3.CO;2.

Whipple, K.X., and Tucker, G.E., 1999, Dynamics of the stream-power river incision model; implications for height limits of mountain ranges, landscape response timescales, and research needs: Journal of Geophysical Research, B, Solid Earth and Planets, v. 104, p. 17661–17674, doi: 10.1029/1999JB900120.

Whipple, K. X., and Tucker, G. E., 2002, Implications of sediment-flux dependent river incision models for landscape evolution: Journal of Geophysical Research, B, Solid Earth and Planets, v. 107, no. B2, 2039, doi: 10.1029/2000JB000044.

Wobus, C.W., Hodges, K.V., and Whipple, K.X., 2003, Has focused denudation sustained active thrusting at the Himalayan topographic front?: Geology, v. 31, p. 861–864, doi: 10.1130/G19730.1.

Wohl, E.E., 1993, Bedrock channel incision along Piccanniny Creek, Australia: Journal of Geology, v. 101, p. 749–761.

Wohl, E.E., 1998, Bedrock channel morphology in relation to erosional processes, *in* Tinkler, K.J., and Wohl, E.E., eds., Rivers over Rock: Fluvial Processes in Bedrock Channels: American Geophysical Union Geophysical Monograph 107, p. 133–151.

Wohl, E.E., and Ikeda, H., 1997, Experimental simulation of channel incision into a cohesive substrate at varying gradients: Geology, v. 25, p. 295–298, doi: 10.1130/0091-7613(1997)025<0295:ESOCII>2.3.CO;2.

Manuscript Accepted by the Society 23 June 2005

Printed in the USA

Geological Society of America
Special Paper 398
2006

Changes of bedload characteristics along the Marsyandi River (central Nepal): Implications for understanding hillslope sediment supply, sediment load evolution along fluvial networks, and denudation in active orogenic belts

Mikaël Attal[†]
Jérôme Lavé
Laboratoire de Géodynamique des Chaînes Alpines, Université Joseph Fourier, Grenoble, France

ABSTRACT

Understanding and quantifying fluvial transport and bedrock abrasion processes have become major concerns in modeling landform response to tectonic and climatic forcing. Recent theoretical and experimental investigations have in particular stressed the importance of sediment supply and size in controlling bedrock incision rate. Many studies on the downstream evolution of pebble size have focused on unraveling the respective roles of selective sorting and abrasion, without paying much attention to sediment sources. In order to track sediment supply and characteristics from source to sink in an active tectonic setting, where long-term selective deposition can be excluded, we systematically measured sediment size and lithology on gravel bars along the Marsyandi River and its tributaries (Himalayas of central Nepal), and also in sediment source material from hillslopes (landslides, moraines, terrace deposits). The downstream evolution in lithological distribution is found to be in close agreement with common views on pebble abrasion and present views on denudation in the range: (1) pebbles from the more rapidly uplifted and eroded Higher Himalayan gneissic units are over-represented, due to their major contribution to sediment influx, (2) easily erodible lithologies such as schists, sandstones, and limestone are under-represented relative to resistant rock types like quartzite. More surprisingly, we observe a general downstream coarsening of gravel bar material along the middle and lower Marsyandi River, whereas downstream sediment fining is typical of most river systems. A simple integrative model that tracks pebbles from hillslope to the main stem of the river and includes abrasion coefficients for the different Himalayan lithologies and size distribution of hillslopes sediment supplies accounts for both changing lithologic proportion along the Marsyandi and for the downstream coarsening of gravel bar material. This coarsening mainly results from differences in sediment sources along the Marsyandi Valley, in particular from differences in size distributions of landslide and moraine material. However, the median pebble size of subsurface material in gravel bars is coarser than median size of the blocky material in the source. The choice of the measurement methods and their potential bias are discussed but cannot explain this surprising feature displayed by our measurements. We suspect

[†]E-mail: mattal@ujf-grenoble.fr.

Attal, M., and Lavé, 2006, Changes of bedload characteristics along the Marsyandi River (central Nepal): Implications for understanding hillslope sediment supply, sediment load evolution along fluvial networks, and denudation in active orogenic belts, *in* Willett, S.D., Hovius, N., Brandon, M.T., and Fisher, D.M., eds., Tectonics, Climate, and Landscape Evolution: Geological Society of America Special Paper 398, Penrose Conference Series, p. 143–171, doi: 10.1130/2006.2398(09).

that due to sediment transport modalities in active tectonic settings, the subpavement grain-size distribution on gravel bars is not representative of the average bedload size distribution. Consequently, pebble abrasion is more easily demonstrated by description of pebble lithology than by the downstream evolution of pebble size. Our study also shows, in contrast with previous studies, that experimentally derived abrasion coefficients can account for the downstream evolution of pebbles without calling for additional fining processes. We conclude that the eroded lithology and hillslope sediment source exert a major influence on the downstream evolution of sediment characteristics, on bedload ratio, and probably on bedrock erosion efficiency. These conclusions have important implications in terms of river profile evolution, landscape denudation, internal erosion coupling, and the response of the fluvial network to glacial-interglacial fluctuations.

Keywords: sediment, erosion, transport, fluvial network, active orogen, Himalayas.

1. INTRODUCTION

Recent studies on coupling between tectonics, climate, and erosion (e.g., Koons, 1989; Molnar and England, 1990; Beaumont et al., 1992; Avouac and Burov, 1996; Willett, 1999) have emphasized the key role of the erosional processes and their efficiency in shaping and uplifting mountain ranges. More particularly, the fluvial network has been recognized as having a major control on landscape denudation by setting the local base level of the hillslopes (e.g., Burbank et al., 1996; Whipple et al., 1999). Such observations have spurred many studies on the way rivers incise bedrock. Several heuristic models have been proposed, which can be grouped into three types: the detachment-limited model, which proposes a determinant incising efficiency linked to the stream power (Howard and Kerby, 1983), the transport-limited model (Willgoose et al., 1991), and the mixed-tools model (Sklar and Dietrich, 1998, 2004). The last two models strongly depend on upstream sediment supply and on sediment size. In particular, sediments introduce nonlinear behavior that can strongly affect the transitory regime (Whipple and Tucker, 2002) as well as the late-stage orogen evolution (Baldwin et al., 2003). In high mountain streams or in rivers draining through steep canyons, removal of static boulders on the meter scale or larger can also introduce an additional nonlinear component to river incision processes (Howard et al., 1994) and an eventual feedback between hillslope erosion and fluvial downcutting. However, few recent studies have focused on the evolution of boulders, blocks, and sediments from the hillslopes toward the mountain range outlet, even though pebble fining and abrasion have been pointed out as potential key processes (Howard, 1998; Whipple and Tucker, 2002). In the fluvial geomorphology community, there has been a long-standing debate over the causes of downstream sediment fining in rivers (e.g., Bradley, 1970; Goede, 1975; Knighton, 1982; Brierley and Hickin, 1985; Brewer and Lewin, 1993; Kodama, 1994a; Heller et al., 2001; Surian, 2002). In part this is because two kinds of processes may be acting at the same time: fining by selective transport and fining by pebble abrasion. Observed apparent fining rates in natural rivers were generally found to be much higher than experimental abrasion rates (Kuenen, 1956; Bradley, 1970; Shaw and Kellerhals, 1982; Kukal, 1990; Brewer and Lewin, 1993). Several hypotheses have been proposed to explain this discrepancy: a predominant role of selective transport (Brierley and Hickin, 1985; Paola et al., 1992; Brewer and Lewin, 1993; Surian, 2002), the role of chemical weathering (Bradley, 1970; Jones and Humphrey, 1997) or an underestimation of experimental abrasion rates, most of the experimental devices reproducing too-slow hydrodynamic regimes (Kodama, 1994b).

In an attempt to unravel the interplay between abrasion and other causes of pebble fining, and more generally to identify the role of sediments in fluvial transport and incision in active orogens, we must focus on a setting where only one of these processes is acting. This paper, therefore addresses the case in which abrasion is assumed to be the dominant process acting during fluvial transport. To this aim, it is important to choose a river system that presents neither short- nor long-term depositional sections: in such a setting, both selective sorting and pebble surface weathering during deposition can be expected to be minimal. Such behavior can be observed along rivers draining across actively uplifting and eroding mountain ranges. However, most of the world mountain ranges are relatively narrow and river length between its source and outlet in a depositional area (foreland basin or intermountainous basins) rarely exceeds more than 30–50 km, except for rivers draining large ranges like the Himalayas or the eastern Andes.

In this study, we focused our attention on the downstream evolution of sediment characteristics along a central Nepal Himalayan river, the Marsyandi River. The choice of this 200-km-long river system was dictated both by the necessity to study a sufficiently long river system to observe a significant evolution, as well as by the existence of numerous constraints on both lithologies (Colchen et al., 1986), erosion rates, and thus sediment supply rates from hillslopes (Lavé and Avouac, 2001; Burbank et al., 2003; Pratt-Sitaula et al., 2004). In addition, a previous study on gravel bar material along the Kali Gandaki, a nearby river system, has already indicated important downstream variations in pebble lithologies (Mezaki and Yabiku, 1984). In contrast with previous studies on pebble evolution and particle abrasion rates, we paid particular attention to the sediment evolution from the hillslopes down to the depositional plain at the outlet of the range, characterizing the size distribution of the hillslope sediment sources that feed the fluvial network. Field measurements were conducted in the autumns of 2000 and 2001.

After some brief theoretical considerations on pebble size evolution along a river incising into an actively eroded landscape, we first review the geological and geomorphological setting of the Marsyandi watershed. We then present the methodology used to characterize the sediment sources, i.e., landslides, moraines, terrace deposits, and tributaries, and to measure the size distribution and lithologic composition of gravel bar material along the main stem of the Marsyandi River and its tributaries. Measurements, results, and geomorphic implications are discussed first qualitatively and second in the light of a simple integrative model that takes into account abrasion rates determined for Himalayan lithologies in an experimental device (Attal and Lavé, 2003; Attal, 2003).

2. GENERAL CONSIDERATIONS ON DOWNSTREAM PEBBLE SIZE EVOLUTION IN A UNIFORMLY ERODED LANDSCAPE

In several studies (Kuenen, 1956; Bradley, 1970; Brewer and Lewin, 1993; Kodama, 1994a), downstream fining rate, expressed in % per km, is directly compared to abrasion rates obtained from experimental studies. According to the authors, this comparison allows evaluation of the respective efficiency of sorting and abrasion processes. However, this approach is valid only when there is a unique sediment point source in the headwater, i.e., if sediment supply from tributaries and hillslopes further downstream can be neglected. For rivers draining through an actively eroded region, this comparison is invalid. We will demonstrate this for a simplified linear drainage geometry, i.e., defined by the relation $A = wL/s$, where A is the drainage area, w is the average width of the watershed between the two lateral interfluves, and s and L are the respective average sinuosity and length of the river. This relation is simply a particular case of the more general Hack's law, with an exponent of 1. Each block or rock fragment, after being delivered from the hillslopes to the river network, will be submitted to breaking, crushing, and abrasion that tend to round the fragment and decrease its diameter. We assume that pebbles are mostly reduced in size by abrasion after a few kilometers (Krumbein, 1941; Kuenen, 1956; Pearce, 1971) and that pebbles are abraded following the commonly used Sternberg's law (1875):

$$\frac{dV}{V} = -kdL \quad \text{or} \quad \frac{dD}{D} = -\frac{k}{3}dL, \tag{1}$$

where k is the pebble abrasion coefficient, and D and V are the pebble diameter and volume, respectively. We also assume that the products of abrasion are mostly fine materials that then transit as suspended load (Kuenen, 1956). If we now consider a uniform erosion rate ε and uniform sediment supply from the hillslopes and lateral tributaries with a unique fragment size D_0, we can write for any point along the main river stem:

* for the total sediment flux:

$$Q_s(L) = \varepsilon w L/s, \tag{2}$$

* for the bedload sediment flux:

$$Q_b(L) = \frac{\varepsilon w}{s}\int_0^L e^{-kx}\,dx = \frac{\varepsilon w}{sk}(1-e^{-kL}), \tag{3}$$

* for the mean pebble size:

$$\bar{D}(L) = D_0\frac{\int_0^L e^{\frac{k}{3}x}dx}{\int_0^L e^{-ks}dx} = \frac{3}{4}D_0\frac{\left(1-e^{\frac{4k}{3}L}\right)}{\left(1-e^{-kL}\right)}. \tag{4}$$

At great distances from the river source, or for high values of the erodibility k, an asymptotic behavior is rapidly reached both for the bedload flux and the mean grain size. Moreover, the asymptotic value for the mean pebble size $D = 3/4D_0$ is independent of the erodibility coefficient (Fig. 1A).

Asymptotic behavior arises after a distance of the order of $3/2k$ from the balance between the quantity lost by abrasion and the continuous supply of fresh material from hillslopes. It can be demonstrated that this phenomenon is also observed with a more realistic model, i.e., with a more complex watershed geometry and a complete grain size distribution for the sediment sources, or even with hillslopes delivering several lithologies with different erodibilities: the asymptotic values for the mean grain size are slightly different but still independent of k. A recent study along a U.S. river that drains a homogeneous lithology in the Olympic Mountains indeed shows such seemingly asymptotic

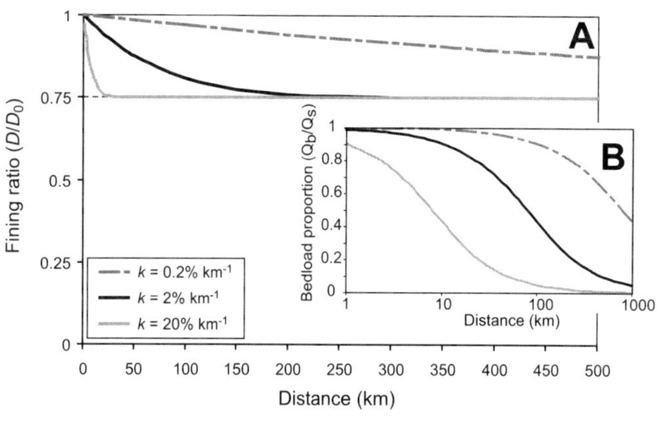

Figure 1. Downstream evolution of fluvial sediment size produced by pebble abrasion following Sternberg's law in a uniformly eroded linear watershed (equations 3 and 4): (a) fining ratio and (b) bedload proportion for different abrasion coefficients k.

behavior (Heller et al., 2001). For rivers draining a uniformly eroded region, the apparent downstream fining rate of bedload is close to 0, even if the abrasion rate of the eroded lithology is very high. Therefore, the downstream size evolution cannot be translated directly in terms of equivalent abrasion rates, except for the upper reaches, where the river length is lower than the critical length $3/2k$ (in the above example

$$\frac{dD}{dL} = \frac{k}{6} D_0).$$

However, in this case, downstream fining can be difficult to demonstrate given the usually large uncertainties in field measurement and the moderate change between upstream and downstream values.

To unravel the role and amplitude of abrasion, it is therefore more pertinent to track the ratio of bedload to total load (Fig. 1B) rather than the downstream evolution of the mean pebble size. However, bedload flux measurements are usually very difficult to monitor for large rivers, in particular for the long periods of time that are necessary to estimate the average flux. Alternatively, it is possible to study a river crossing contrasting lithologies and track the dilution rate of the upstream lithologies in the downstream ones. In light of the simplified model presented above, the downstream evolution of the relative proportions of the different lithologies is expected to be sensitive both to their relative abrasion coefficients but also to the absolute values of these coefficients. The Marsyandi, the setting of which is detailed below, displays such characteristics.

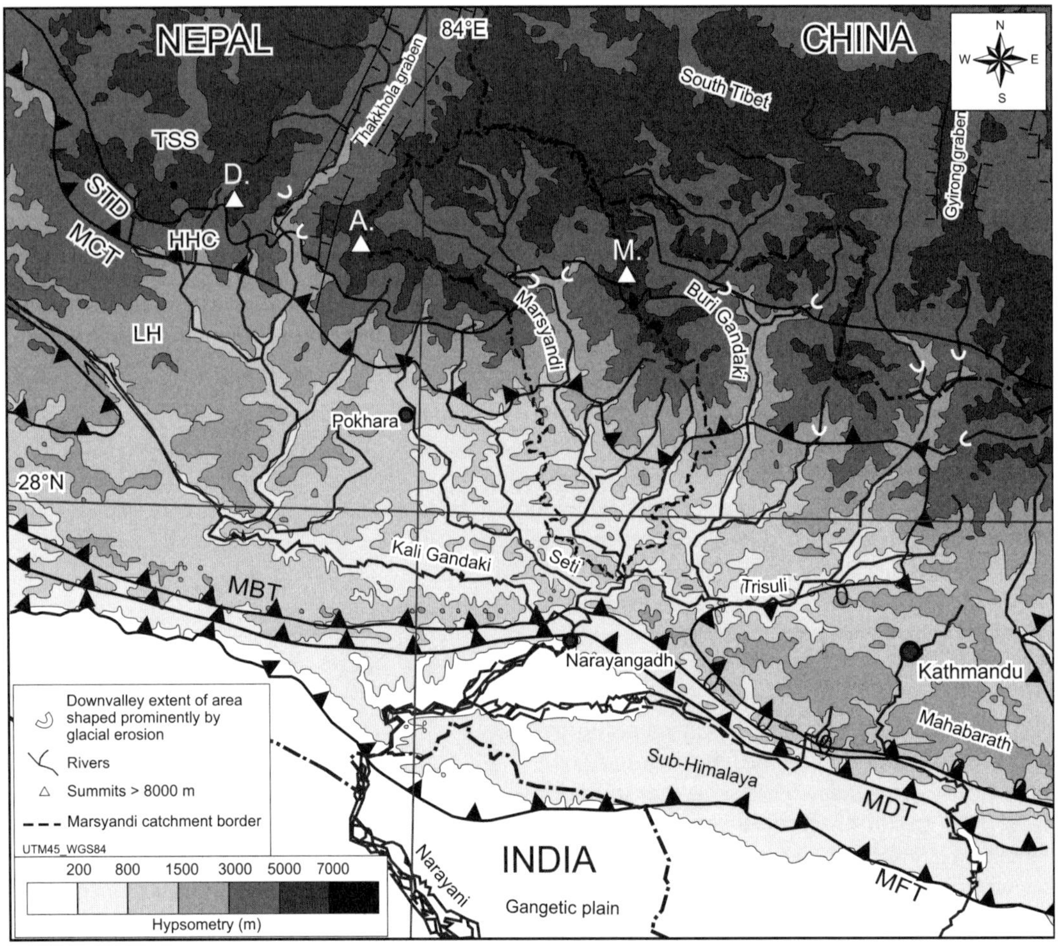

Figure 2. Topographic map of the Narayani basin modified from Lavé and Avouac (2001). Thick dashed line follows the catchment boundary of the Marsyandi River, for which the bedload evolution is characterized in this study. The down-valley extent (white arcuate segments) of areas shaped prominently by glacial erosion (Duncan et al., 1998) defines the beginning of dominant valley shaping by fluvial incision, i.e., domains where bedrock landslides supply most of the coarse river sediments. The major faults are the South Tibetan Detachment (STD), the Main Central thrust (MCT), the Main Boundary thrust (MBT), the Main Dun thrust (MDT), and the Main Frontal thrust (MFT). From west to east, >8000-m-high summits are labeled: D—Dhaulagiri, A—Annapurna, M—Manaslu. Domains are TTS—Tethyan Sedimentary Series, LH—Lesser Himalaya, and HHC—Higher Himalayan Crystalline.

3. LOCATION AND GEODYNAMIC SETTING OF THE MARSYANDI WATERSHED

The Marsyandi River drains across the Himalayan range in central Nepal. Its source is located north of the Annapurnas (Fig. 2). On its upper reaches, it drains to the southeast to skirt round the Annapurnas ridge. It then drains to the south and reaches the Trisuli River after a course of ~170 km. From east to west, the Trisuli, Buri Gandaki, Marsyandi, Seti, and Kali Gandaki join to form the Narayani River system, which drains to the Terai plain. They form one of the most important hydrographic network of the Himalayan range, west of the Kathmandu basin.

The Marsyandi watershed is superimposed on three main structural units (Fig. 2): the Tethyan Sedimentary Series, the Higher Himalayan Crystalline, and the Lesser Himalayan units. These structural units roughly coincide with the main geographic domains across the Himalayas of Nepal (Fig. 2). The Tethyan Sedimentary Series extends through the south Tibetan Plateau and the northern flanks of the Higher Himalayan summits. It consists of a thick stack of Paleozoic and Mesozoic sediments that are slightly metamorphosed and intruded by an early Miocene granitic body, the Manaslu Granite (e.g., Le Fort, 1986; Searle, 1999; Fig. 3). In these units, the dominant lithologies are limestone and schist, but fine sandstone and quartzite levels are frequent (Colchen et al., 1986). The core and southern flank of the Higher Himalayan topography correspond to the crystalline units of the Higher Himalayan Crystalline, which consists mainly of medium- to high-grade aluminous (Formation I) and calcic (Formation II) paragneisses and orthogneisses (Formation III; Fig. 3). To the south, the topography drops abruptly from elevations greater than 6000 m in the Higher Himalaya to around 1000 m in the Lesser Himalaya. The rocks in the Lesser Himalayan units consist of low-grade metasediments (sandstones, phyllites, schists, quartzites of Pre-Cambrian age = mostly Kuncha Formation; Fig. 3) forming a large antiformal duplex structure. Just below the Main Central thrust, the northern part of the anticline is overlain by schists, micaschists, quartzite, and limestones metamorphosed in garnet to kyanite facies (e.g., Pêcher, 1989; Colchen et al., 1986; Schelling 1992). The southern part of the anticline, where metamorphism has been less intense, is overlain by the Mahabarat range, mostly in the eastern part of the Narayani watershed. The Mahabarat units are composed of schists and Cambrian to Eocene Tethyan sediments intruded by Late Cambrian to Ordovician granites. All these Himalayan units and sheets are overriding the Indo-Gangetic plain and have generated thin-skinned tectonic deformation, giving rise to the Siwalik Hills, which form the most frontal Himalayan relief. The Siwalik or sub-Himalayan rocks are composed of easily erodible Neogene sandstones, siltstones, and conglomerates.

The boundaries between the different domains roughly coincide with major faults. These are from north to south: the South Tibetan Detachment, a gently dipping normal fault underlying the Tethyan Sedimentary Series (Burchfiel et al., 1992), the Main Central thrust, a ductile shear zone that separates the Lesser Himalayan units from the Higher Himalayan Crystalline (e.g., Le Fort, 1986), the Main Boundary thrust, which marks the limit between the sub-Himalaya and the Lesser Himalaya, and the Main Dun thrust and the Main Frontal thrust, which correspond to inner and southern thrusts associated with the sub-Himalayan folds.

Currently, the most active tectonic feature appears to be the Main Frontal thrust, which absorbs most of the convergence (~21 ± 1.5 mm/yr) between India and south Tibet, (Lavé and Avouac, 2000). However, important vertical movements also affect the Higher Himalaya, around 100 km north of the Main Frontal thrust. This phenomenon has been inferred to be the consequence of the ramp-flat geometry of the main detachment at depth, the Main Himalayan thrust, on which the main faults of the range connect (Lavé and Avouac, 2001). In the central Himalaya, uplift rates, inferred from fluvial incision rates, display strong variations across the range: they peak at 6–15mm/yr in the frontal Siwaliks, drop to 0–2 mm/yr above the Main Dun thrust (Lavé and Avouac, 2000, 2001), and to 1–2.5 mm/yr across the Mahabarat, then decrease to around <0–1 mm/yr in the Lesser Himalaya, and finally rise again to values ranging between 2 and 5 mm/yr across the Higher Himalayan Crystalline (Lavé and Avouac, 2001; Burbank et al., 2003; Pratt-Sitaula et al., 2004). The south Tibetan region is characterized by a general extensional regime (Armijo et al., 1986) and probably by very low vertical motions. The hydrologic network of the Marsyandi drains across contrasted uplifting and eroding areas, and sediment supply rates from the hillslopes are therefore suspected to vary strongly along Marsyandi course.

In response to the intense tectonic activity and sharp topographic gradients, the Marsyandi region is being actively eroded, and the erosion products are exported through the Marsyandi fluvial network toward the Indo-Gangetic plain and the Bengal Fan.

4. FIELD MEASUREMENTS IN MARSYANDI VALLEY

4.1. Sampling Sites and Geomorphic Objects

To address the issue of the sediment mass transfer in the Marsyandi watershed, we focused on two distinct geomorphic objects: the sources of coarse sediments delivered to the hydrographic network (~20 measurement sites) and the transported bedload exposed on gravel bars along the Marsyandi River and its tributaries (~40 measurement sites; Fig. 3).

Three types of sediment sources were considered, which deliver most of the coarse material producing bedload. First, we surveyed landslide deposits in the Marsyandi Valley and its surroundings. These probably represent the main process of hillslope erosion and the main source of sediment in this kind of active orogenic setting (Burbank et al., 1996; Hovius et al., 1997). Choosing a representative site was the most difficult task of the counting procedure because of the heterogeneity of landslide deposits (Figs. 4B and 4C). We observed that the material usually becomes coarser downhill, probably because large blocks

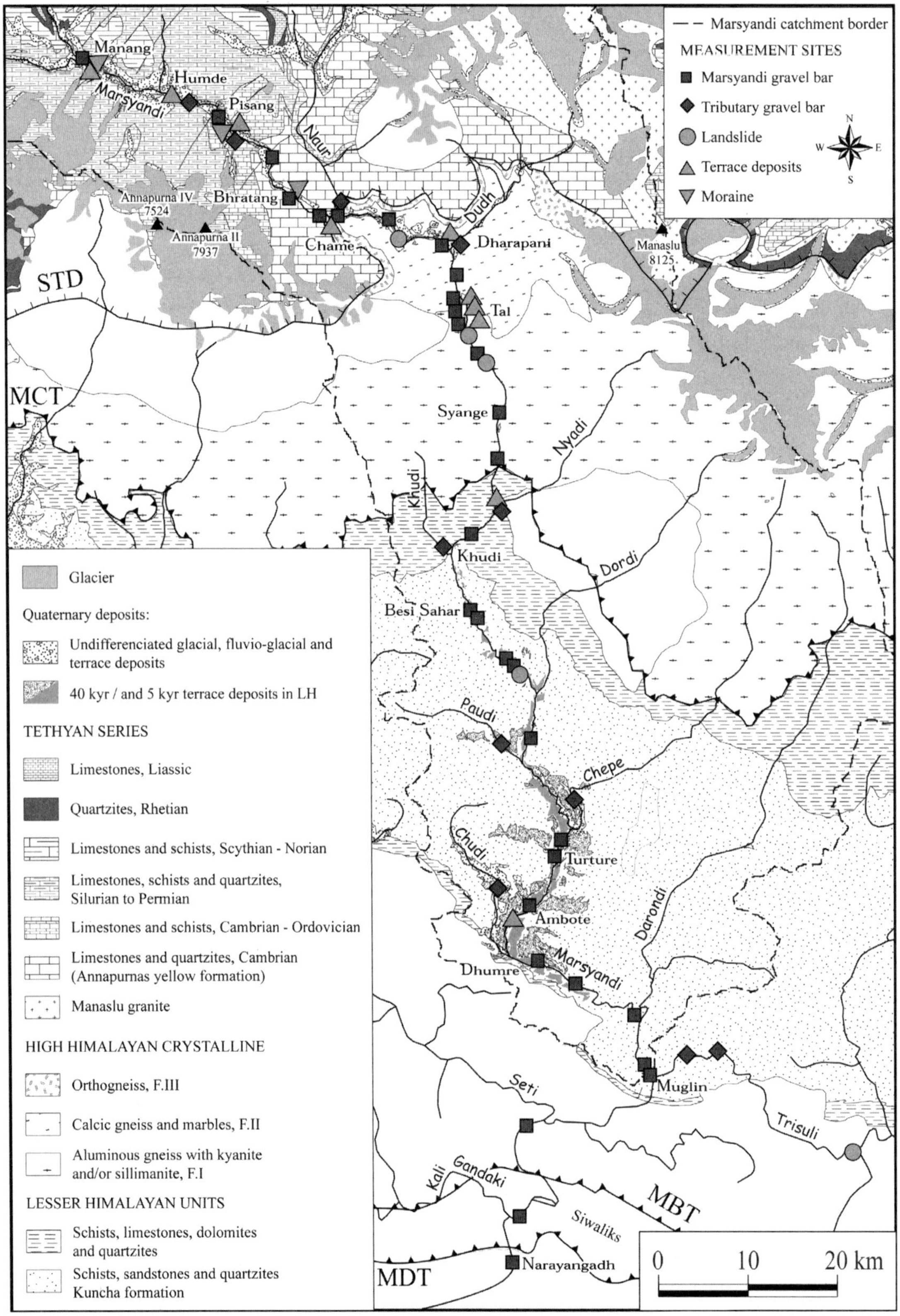

Figure 3. Geological map of the Marsyandi catchment, modified from Colchen et al. (1986). The different geomorphic objects surveyed in this study are also reported on the map: gravel bars along the Marsyandi River and its tributaries, moraines, landslides, and terrace deposits. The major faults are the South Tibetan Detachment (STD), the Main Central thrust (MCT), the Main Boundary thrust (MBT), and the Main Dun thrust (MDT). LH—Lesser Himalaya.

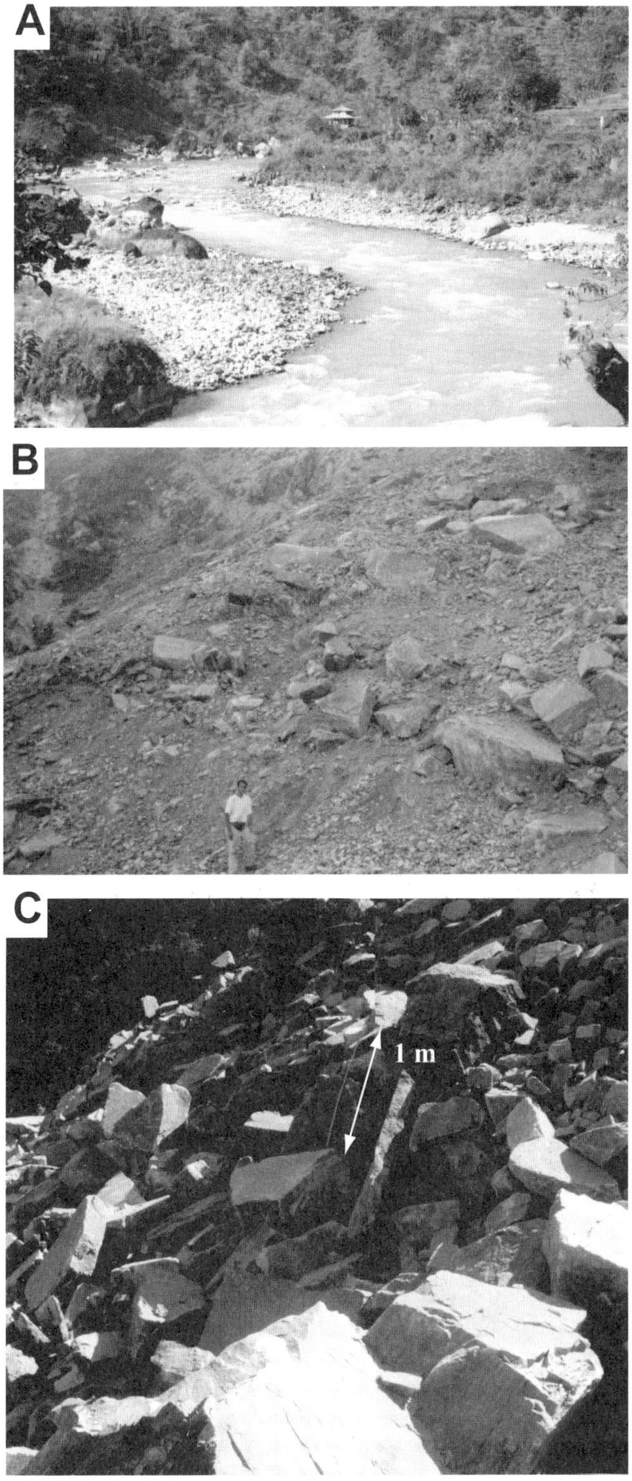

Figure 4. (A) Photo of a gravel bar near BesiSahar showing heterogeneity of the surface material with the presence of pluri-metric static boulders. (B) and (C) Photos of a surveyed bedrock landslide in quartzites: in a cross section (B) produced by recent road clearing, and at surface (C) 20–30 m below this road, showing clearly the large range of particle sizes and the heterogeneity of such deposits, with depletion of fine material at surface resulting from wash out and from segregation processes during landslide movement.

travel greater distances, and it is also much coarser at the surface, because fine material has been washed out and also because segregation processes during landslide movement tend to expel the largest block toward the surface. To get an idea of the size distribution of the whole landslide deposit, it would be necessary in theory to measure the distribution of both surface and core material for different points along the landslide deposits. However, such a procedure is highly time-consuming, and we focus on the core material in the central part of the landslide, which provides an adequate representation of the average size distribution.

In the upper part of the Marsyandi watershed, glacial and periglacial processes deliver different material to the fluvial network. We surveyed both modern and late Pleistocene to Holocene moraines to evaluate this source. In the upper Marsyandi Valley, many remnants of formerly glaciated landscapes can be observed (Fig. 3; Fort, 1993; Lavé and Avouac, 2001) down to the confluence with the Naur Khola; the most extensive deposits can be followed from Manang to Pisang. Despite some controversies about their exact origin and age (Fort, 1993), these deposits probably represent a volume of several cubic kilometers of moraine-type material, and their erosion since the Last Glacial Maximum has delivered a large quantity of sediments to the upper Marsyandi.

Across the Lesser Himalaya, thick terrace deposits also contribute to sediment supply. These deposits correspond mostly to the filling of temporary lakes resulting from the obstruction of the river by a landslide or to the filling of the valley during episodes of alluviation (Lavé and Avouac, 2001; Pratt et al., 2002). An important late Pleistocene deposit is well preserved throughout the Lesser Himalaya from Besi Sahar down to Dhumre along the Marsyandi and most tributaries (Yamanaka and Iwata, 1982; Fig 3). This fill terrace material, the middle part of which has been dated at around 40 ka (Pratt-Sitaula et al., 2004), consists of mixed fluvial and debris-flow units with some organic-rich clay lenses. Its initial volume could have represented more than 7 km^3; around 80% has been eroded along the Marsyandi and its tributaries valleys. If we assume steady erosion of this fill terrace and compare it with present sediment fluxes in Himalayan rivers, corresponding to erosion rates around 1–3 mm/yr (His Majesty's Government of Nepal Undertaking, 1994; Lavé and Avouac, 2001), we end up with a terrace material contribution to the present sediment fluxes of ~2%. The second important terrace deposit in the Lesser Himalaya, between Khudi and Dhumre, corresponds to one or several debris flows that ran from the Higher Himalayan Crystalline through the valley between 5.2 ± 0.15 k.y. B.P. (^{14}C dating of a trunk within the debris flow material north of Besi Sahar, E84°22.8′–N28°14.7′, sample MAR-301 dated by the conventional method, ^{14}C age = 4559 ± 631 yr B.P.) and 4.8 k.y. B.P. (Yamanaka and Iwata, 1982). The Marsyandi River has re-incised the mud-flow deposit, leaving fill-cut terraces, but has not yet reached the bedrock valley floor. The initial volume of the flow could have represented 1.2 ± 0.4 km^3, and 70% has been eroded. With the same assumption as that for the Pleistocene terrace, we also end up with a contribution of around 1%–3% to the present sediment fluxes. We therefore suspect that reworking of terrace deposits represents only a minor source of sediment into the Marsyandi. These two examples highlight the fact that temporary sediment storage in terraces and gravel bars and subsequent reworking do not significantly affect the long-term sediment flux. Downstream fining by selective transport can therefore hardly be effective in an actively eroded mountain range.

Along the Marsyandi and its tributaries, numerous gravel bars can be observed. With the exception of the narrow reaches across the Higher Himalaya, where they can be rare or poorly developed, the gravel bars are usually more than 100 m long and 20 m wide. As for most gravel rivers worldwide, gravel bar material presents clear surface coarsening. We therefore conducted measurements both on surface and subsurface gravels. When measuring subsurface gravels, the surface layer was removed to the thickness of the largest boulders before volumetric sampling. Gravel bars can also display large pebble size variations between their extremities (Fig. 4A). In order to keep consistency between sampling sites, we surveyed (wherever possible) a similar hydrodynamic setting, and we chose the central part of the gravel bars. Two or three sampling points were generally chosen on a line parallel to the river and at least ten meters distant to define the variability of the sediment characteristics across the central part of the bar. Following Parker and Klingeman (1982), we assumed that the material composing the gravel bars is representative of the bedload transported by the river during flooding stages. This issue will, however, be discussed more at length in the final sections.

4.2 Counting and Sampling Procedures

Counting and sampling procedures are inspired from previous studies (see review and analysis by Kellerhals and Bray, 1971). We distinguished surface measurements, i.e., counting on lines or grid nodes, and volume measurements by weighing of a volumetric sample. Distributions obtained from these methods have been shown to be directly reliable (Kellerhals and Bray, 1971; Church et al., 1987; Displas and Sutherland, 1988). At each site, and for each method, two or three measurement stations were utilized. In order to display and compare size distributions from these different sampling or counting procedures, it was also necessary to define a common variable to describe pebble dimension. We choose to consider the *b*-axis of the particles, i.e., their intermediate axis, and to perform geometric corrections when necessary (see more details in Appendix A).

Surface measurements were applied to gravel bar surfaces, landslide surfaces, terrace and moraine cross sections. We used both tape measure and photo counting methods. For the tape measure method, a line of 15 m was posed on the geomorphic object and particles were measured and lithologically identified every 50 cm. For the photo counting, particle size was measured at the 100 nodes of a regular numeric square grid. A comparison of the two methods is detailed in Appendix B.

Volumetric measurements were applied for subsurface measurements of the gravel bar, landslide, terrace, and moraine material: 100–250 kg of material were collected and sieved with 1, 2, and 4 cm square mesh sieves. Material coarser than 4 cm was sorted by lithology and particles were weighed separately. Fractions finer than 4 cm were weighed in classes corresponding to the mesh size. The size distribution for particles coarser than 1 cm was established in the field. The size distribution of the fraction finer than 1 cm was analyzed in the laboratory, using sieves and an optic laser method for fractions larger and finer than 1 mm, respectively.

On landslide and moraine cross sections (Figs. 5A and 5B), both surface and volumetric counting were used and generally provided similar D_{50} values. The upper tail of the distribution curves differ, however, because the surface methods include larger blocks than the volumetric method. The maximum block size on photos, which encompass 5–20-m-wide surface, reaches 2 m (Fig. 5A), but only 40 cm (≤50 kg) in our volumetric sampling. Accordingly, we consider that volumetric sampling provided a reasonable estimate of the median size D_{50} of the whole material, but largely underestimated D_{90} values.

Along the Marsyandi course, five main lithologies were distinguished: limestones (including marbles), schists (including phyllites, sericites, micaschists), crystalline rocks (gneiss and granitic rocks), quartzites, and sandstones. Each structural unit drained by the Marsyandi River contains contrasting proportions of these different lithologies, a context helpful to identify the particle provenance and thus favorable to study abrasion processes, as discussed in section 2. From the source to the confluence with the Kali Gandaki, the river drains mostly limestones across the Tethyan Sedimentary Series, gneisses across the Higher Himalayan Crystalline, schists in the upstream part of the Lesser Himalaya, and sandstones, schists, and quartzites in its downstream part.

Visual identification of entire or freshly broken pebbles, with an eventual test for presence of carbonates with hydrochloric acid was used to define sediment lithology. Gneissic pebble identification was the easiest one, and errors were probably very rare. In contrast, for the other lithologies, the identification of some pebbles was somewhat arbitrary in particular for "intermediate" lithologies such as schistosed sandstones, quartzitic sandstones (Kuncha Formation), or silicified limestones (Annapurnas Formation). However, identification errors or misleading choices concern only a few pebbles within the whole sampled volume at each site. Therefore, the relative error for dominant lithologies reaches only a few percent, but it may be important and eventually reach 100% when the proportion of a lithology is lower than 5%–10%.

4.3 Sources Characteristics

Landslides occur along the entire valley but deliver a very wide range of size distribution to the fluvial network (Fig. 5A). Distribution curves obtained in distinct lithologies present glob-

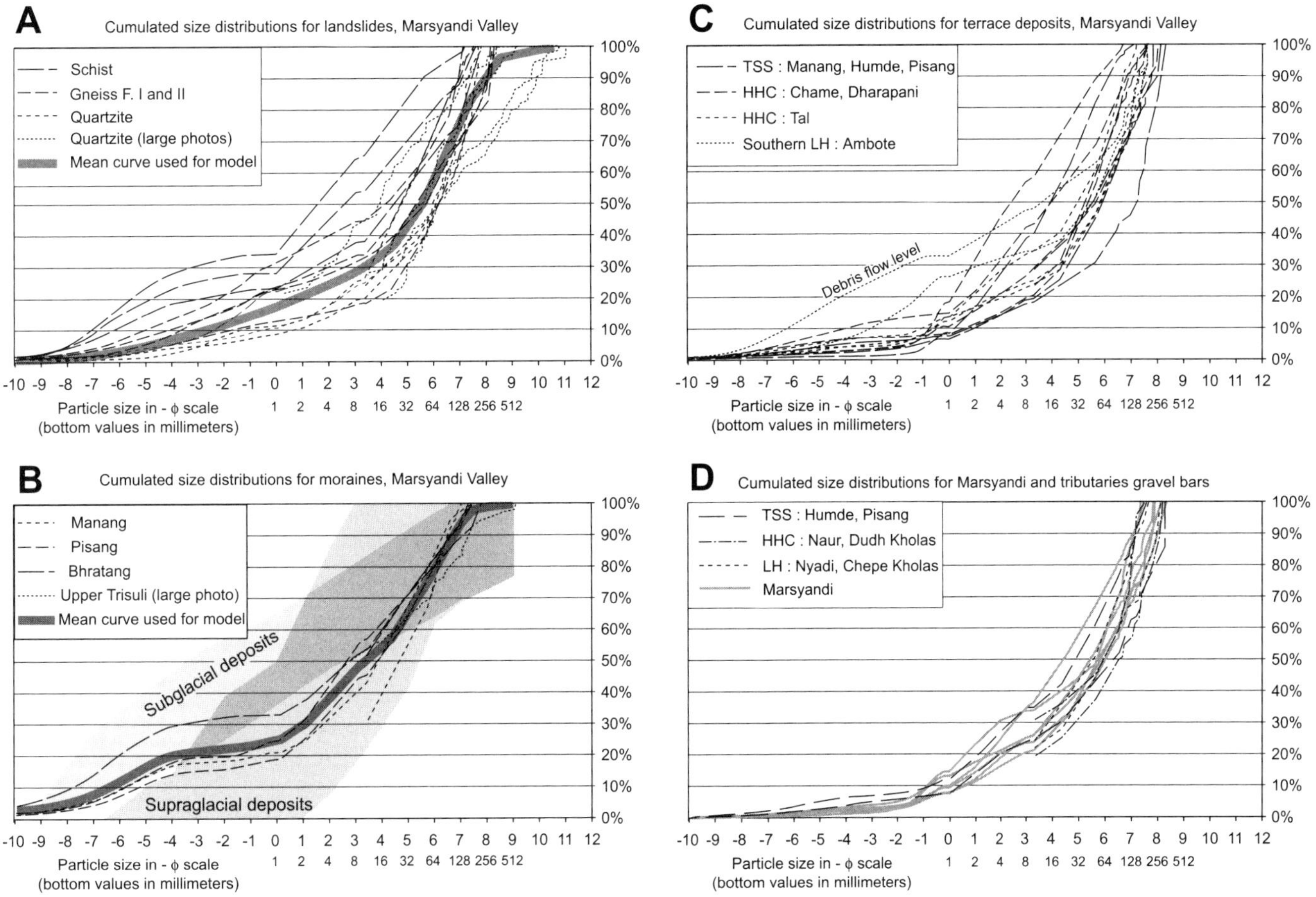

Figure 5. Cumulative size distributions for different sediment sources and gravel bars from subsurface measurements. (A) Bedrock landslides: the size of material provided by landslides is highly variable, with D_{50} ranging between 3 and 74 mm. The size distribution appears to depend on bedrock lithology: landslides initiated in quartzitic and gneissic units deliver coarser material than schist units. An average distribution curve for competent units (thick gray line) was computed without this finest distribution from schist units, but including photo counting to account for the coarse upper tail of the distribution. (B) Moraine deposits provide on average finer material, with D_{50} varying between 6 and 24 mm. Curves obtained from field measurements (mostly dump moraines) fall in the domain covered by supraglacial deposit spectra (gray shading; Campy and Macaire, 1989; Benett and Glasser, 1996). As for landslides, the coarse upper tail of the distribution obtained by photo counting has been included to derive an average distribution curve (thick gray line). (C) Terrace deposits are very heterogeneous, due to their mode of formation. They consist of a succession of sedimentary bodies superimposed during aggradation events. As a consequence, size distribution curves can have very distinct shapes and D_{50} varies between 5 and 147 mm. (D) Gravel bars do not show such dispersion. Marsyandi and tributary gravel bar curves show a similar trend with D_{50} values ranging between 20 and 100 mm. All the curves end at values of D lower than 400 mm (except for photo analysis curves), showing the limitation of the sampling method to take into account coarse material. Domains are TTS—Tethyan Sedimentary Series, LH—Lesser Himalaya, and HHC—Higher Himalayan Crystalline.

ally similar trends. The median size, D_{50}, of the material appears, however, to depend on the nature of rock outcropping in the area (Fig. 5A). D_{50} varies between 37 and 74 mm for gneiss of the Formation I, between 18 and 28 mm for gneiss of the Formation II, between 21 and 74 mm for quartzites, and between 3 and 23 mm for schists. Other parameters, like the degree of fracturation and weathering of the bedrock, probably have a significant role, but were not measured for this study. Silt and sand fractions account for 15%–20% of the volume of landslide deposits, except for a deep-seated landslide in schists, where strong weathering could have substantially increased the proportion of fine material.

Moraines are located in the upper valley above ~2500 m elevation. In contrast to landslides, they provide a large proportion of very fine material (Fig. 5B), around 20%–30% of silt and a very low sand fraction (between 1/16 and 1 mm). D_{50} values range between 6 and 24 mm. The material is mainly composed of limestone fragments; the moraines we surveyed being located upstream of the South Tibetan Detachment. The fine mineralogical structure of the rocks of these units may explain the divergence observed for the silt fraction between our curves and the curves found for supraglacial material derived from crystalline rocks (shaded domains in Fig. 5D; Campy and Macaire, 1989; Benett and Glasser, 1996).

For the two hillslope sources, all the distribution curves derived from volumetric sampling end at values of *D* lower than 30 cm, revealing the limitations of our sampling method in accounting for coarse material. Nevertheless, the distribution derived from photo counting indicates that the upper tail of the curve is limited and that the missing coarsest fraction generally represents less than 10%.

It has to be noted that the distributions we measured on hillslopes overestimate in some way the size of the material that will effectively become bedload material in the fluvial network. Many blocks are fractured, in particular through schistosity plans. After having reached the river channel by sliding or falling along the hillslope and traveling a few kilometers in the river, these blocks would probably have been split several times, and their size would have decreased dramatically.

4.4 Gravel bar Material Characteristics

When compared to source material, gravel bar material displays much less dispersion (Fig. 5D). Most of the 12 complete size distributions (i.e., with the silt to gravel fractions finer than 1 cm) we measured along the Marsyandi River and its tributaries display similar curves. Sand and silt represent only 8–15% of the total volume of gravel bar sediments. Around half of the fine material present in the hillslope sources, plus all the products of blocks and pebble abrasion, are therefore evacuated directly as suspended and wash load, independent of the material traveling as bedload. Size and lithological distributions of gravel bar material have been established for ~30 sites distributed along the Marsyandi River from Manang to the confluence with the Trisuli in Mugling, and for 10 gravel bars on the main tributaries close to their confluence with the Marsyandi. Additional gravel bars were surveyed along the lower Trisuli and on the Narayani close to the Main Dun thrust.

4.4.1 Downstream Evolution of Lithologies

The downstream evolution of the lithological composition for both surface (Fig. 6B) and subsurface (Fig. 6A) pebble material (fraction > 5 cm) is remarkably consistent. From one site to the next, scattering is, however, important and can reach up to 40% for major lithologies and more than 100% for minor ones. This reflects more the poor statistics resulting from reduced sample sizes than the lateral input by the tributaries (Fig. 6A). The general trend for both diagrams conforms to the different geologic units crossed by the Marsyandi. As explained in section 3, the river drains across three main structural units, each of them having its proper lithological characteristics. The lithological composition of the gravel bar material reflects both the influence of sediment supply coming from these structural units and the different erodibilities related to each lithology. To illustrate these two processes, note for example, that the proportion of limestone decreases rapidly downstream from the South Tibetan Detachment, probably in response to a high erodibility of limestone (Kuenen, 1956) as well as an important dilution in gneissic pebbles, amplified by increasing local erosion and sediment supply rates when crossing the Higher Himalaya. In contrast, the steady downstream increase in the proportion of quartzite, up to 50%–60% in the lower Trisuli and Narayani, despite minor proportions of quartzite in the different source units, reflects the much higher resistance to abrasion of quartzite pebbles relative to other lithologies (Kuenen, 1956; Bradley, 1970).

Gravel bar compositions also help to unravel several characteristics of the transported material. The lithologies that are poorly resistant to abrasion are expected to be present in low-order drainage systems, reflecting the composition of local sources, and to diminish in abundance downstream. This is the case for schists and sandstones: the schists represent important lithologies in the upper Tethyan series and in the Lesser Himalayan units. The tributary in Sabche, the Paudi, and Chepe Kholas present a proportion of 15–40% of schists, illustrating the importance of schists in the corresponding local sources. However, their proportion drops to <5% on average along the Marsyandi River. A similar pattern is observed for the proportion of sandstones between the Paudi and Chudi Kholas on the one hand and the main stem of the Marsyandi on the other hand. Finally, some tributaries draining southern Tibet, like the Naur and Dudh Kholas, or the Marsyandi in Chame just below the South Tibetan Detachment trace, display surprising features—for all of them, the proportion of limestone is much lower than expected from the size of the contributing area in calcareous units. For example, the Dudh Khola gravel composition displays 60% of granitic pebbles, 40% of gneissic pebbles, and almost no limestone pebbles, despite the fact that calcareous units represent a third of its contributing area. Such unexpected results could arise either from heterogeneous erosional behavior of the lithologies (limestones eroded mostly in sandy fractions or by chemical erosion, etc.) or more probably from stochastic behavior related to discrete sediment supply events, like landslides or glacial lake outbursts (e.g., Benda and Dunne, 1997).

4.4.2 Downstream Evolution in Grain Size

Downstream profiles of the median size (D_{50}) of gravel bar material (Fig. 7) are characterized by significant scatter, even more pronounced than for the lithological evolution. This scatter can exceed 100% of the median grain size from one site to the next, and even at the scale of a single gravel bar. We suspect that such scatter arises from the insufficient sample size and from the spatial variability in hydrodynamic conditions along and between gravel bars that lead to more- or less-intense local coarsening.

Despite the scatter, the general trend for subsurface samples is clear and opposite to what would be expected in gravel-bed rivers: there is no downstream fining, but downstream coarsening, with three domains for which D_{50} seems to remain more or less uniform. At the boundaries between these domains, a sharp

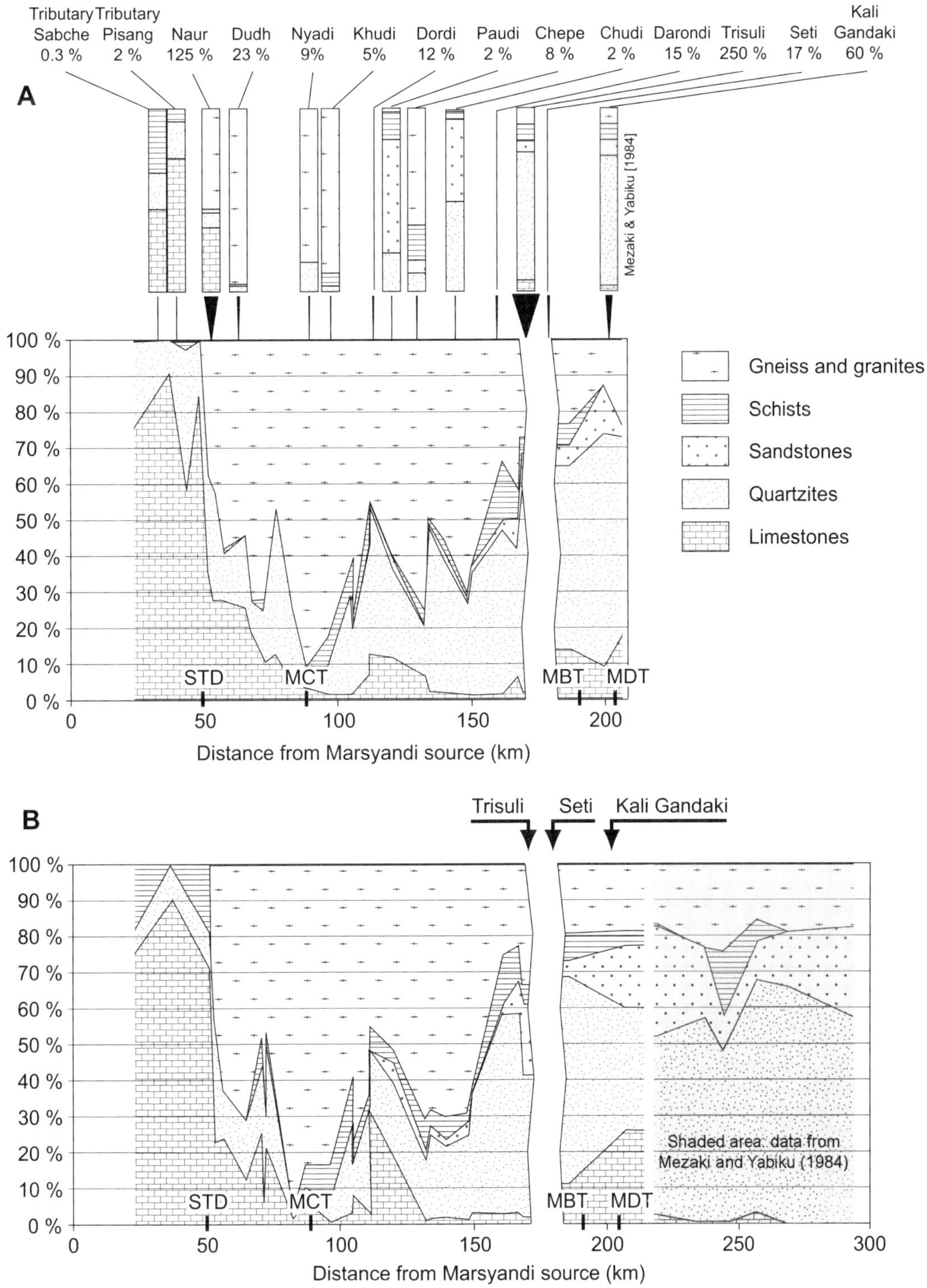

Figure 6. Evolution of the lithologic average content for pebbles larger than 5 cm in gravel bar material along the Marsyandi River (average value issued from the two or three independent measurements on each gravel bar), (A) for the subsurface material and main tributaries, and (B) for the surface material. Important scattering from one point to the next is due to the insufficient size of the sampled volume. The global trend is, however, quite similar for surface and subsurface data: the lithological composition of the gravel bar material reflects both the influence of sediment supply coming from these structural units and the different resistances to abrasion related to each lithology. Resistant rocks (quartzites) are over-represented compared to soft rocks (schist and sandstone). For surface measurements, disparities between Mezaki and Yabiku's (1984) data and ours mostly concern sandstones and limestones and probably arise from distinct criteria for lithology identification (a good fit is obtained for the unambiguous gneissic pebbles). The major faults are the South Tibetan Detachment (STD), the Main Central thrust (MCT), the Main Boundary thrust (MBT), and the Main Dun thrust (MDT).

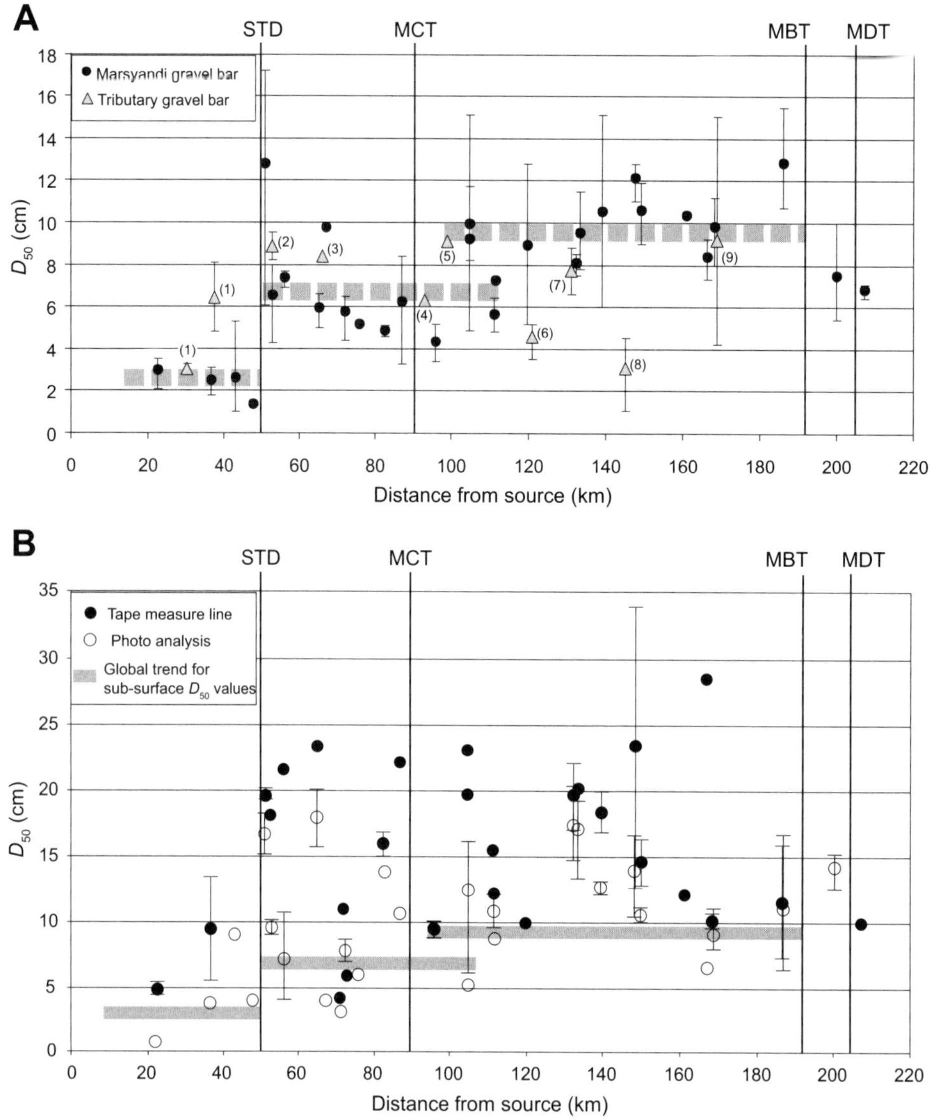

Figure 7. Evolution of the median size of gravel bar material along the Marsyandi River, for (A) the subsurface material and (B) for the surface material. Circle and vertical bars show the respective average value and range of values obtained from the two or three different sampling points on each measurement site. Downstream coarsening appears clearly. Subsurface D_{50} seems to remain roughly constant across the main structural units (trend drawn by thick gray dotted line); increases are abrupt and located roughly at the limit between these units. These trends are less visible for surface D_{50}, due to a larger scatter. Surface values are systematically 1–4 times higher than subsurface ones, showing a coarsening of the surface layer of the bars. Tributaries: (1) unnamed small tributaries, (2) Naur, (3) Dudh, (4) Nyadi, (5) Khudi, (6) Paudi, (7) Chepe, (8) Chudi, (9) Trisuli. The major faults are the South Tibetan Detachment (STD), the Main Central thrust (MCT), the Main Boundary thrust (MBT), and the Main Dun thrust (MDT).

increase of D_{50} occurs downstream. These domains correspond roughly to the main structural units. At this stage, it is thus tempting to attribute such changes to some relationship between sediment sources and geologic or morphotectonic units. We saw, for example, in the previous section, that moraine material is finer than landslide material: the increase in mean pebble size from southern Tibet to the Higher Himalayan gorges could, therefore, result from an increasing supply in landslide material. Similarly, it would be tempting to attribute the increase in pebble size south of the Main Central thrust to an increasing signature of landslides initiated in quartzitic units, which are characterized by slightly coarser size distributions than for landslides initiated in the Higher Himalayan Crystalline gneissic units.

Additional observations can be made on the median pebble size of the gravel bar surface (Fig. 7B). D_{50} values for surface sample are systematically higher than those for subsurface samples. Logically, they follow the same rough downstream coarsening. However, the flat trend observed for subsurface D_{50} across a given structural unit is not observed for surface D_{50}, which varies across a wide range of values, between one to five times subsurface D_{50} values. At the crossing of the South Tibetan Detachment, the difference in surface D_{50} is still visible, whereas no major change occurs going from the Higher Himalayan to the Lesser Himalayan domains. These observations suggest that there is no clear relationship between surface and subsurface size distribution. The vertical coarsening-up observed on gravel bars is not uniform: it varies both along the course of the river and at the scale of a single bar, probably in response to variable local hydrodynamic conditions.

4.4.3 Terrace Material and Temporal Variations in Transported Sediments

Size distribution in terrace deposits display larger dispersion than on gravel bars (Fig. 5C). This likely arises from the fact that terrace deposits correspond to local events in space and time, such as landslide dam filling, alluviation, or debris flows. The amplitude of the event influences the changes of the river hydrodynamic variables and thus the characteristics of the fill. Even within a single terrace deposit, very large vertical and horizontal variations in size distribution of the material can be observed. As a consequence, individual measurement of D_{50} in terrace material varies between 6 and 160 mm.

To show possible temporal variations in pebble characteristics along the Marsyandi, we compared the size distribution in alluvial terrace deposits to that of the closest surveyed gravel bar (Fig. 8). First, we note that the median pebble sizes in terrace material do not strongly differ, on average, from those in the gravel bars, and that they all cluster around 5 cm, except at the Nyadi Khola site, for which a mixed debris-flow–type origin cannot be excluded. Individually, the differences between terrace and gravel bar material are larger than 50%, but, in fact, are of the same order as the scatter observed from one gravel bar to the next. Differences in lithologic compositions between gravel bars and terraces are also within the observed scatter along the Marsyandi channel and do not suggest any major changes in sediment sources. The only noticeable exception is observed in Chame, few kilometers downstream of the South Tibetan Detachment trace, where we previously identified a strong anomaly in both the Marsyandi and Naur Khola gravel bars. Terrace deposit proportions include more lithologies representative of the Tethyan Sedimentary Series than the gravel bars, which are characterized by a marked over-representation of Higher Himalayan Crystalline gneissic elements. As already argued, if the river network is supply-limited by discrete landslides, then important temporal variation can be expected, in particular in the low-order channels. Downstream, in the main stem, we expect that mixing becomes

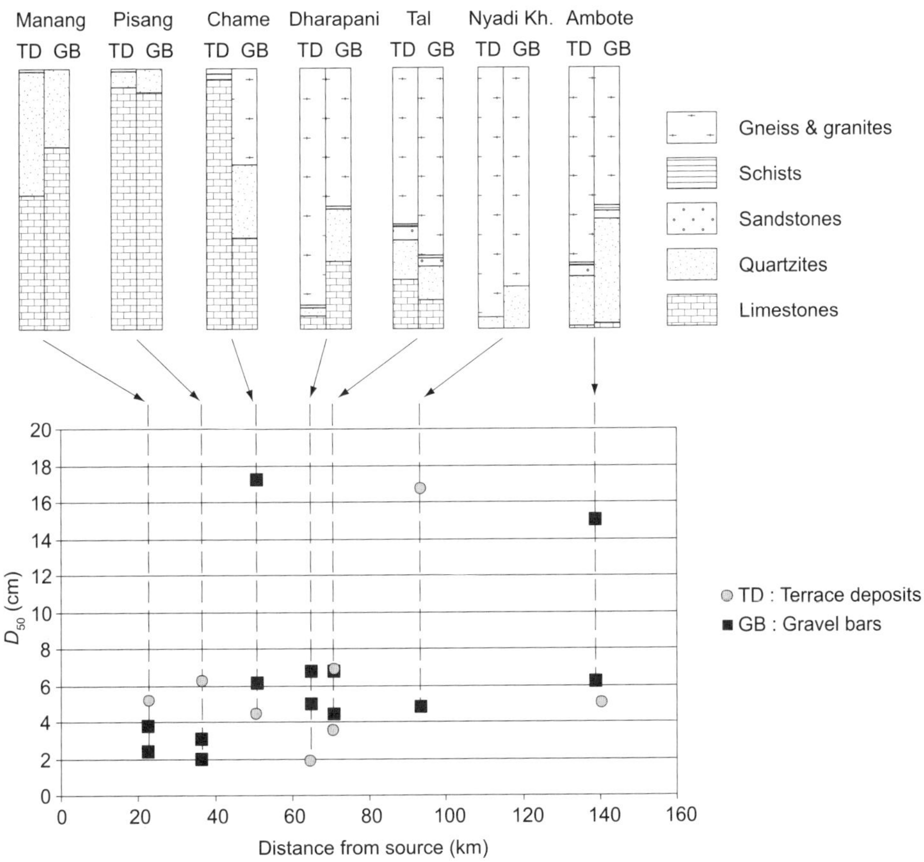

Figure 8. Comparison of median size and lithologic content of terrace deposits and of their closest gravel bars. Terrace deposits provide an insight into the material transported by the river in the past. Some differences can be observed between present-day deposits and fossil ones, but these differences are too small in regard to measurement uncertainties to be interpreted in terms of variations in sediment supply.

sufficiently efficient to dampen these discrete signatures. Alternatively, the variations in pebble characteristics in Chame could reflect important changes in spatial erosion and hydrodynamic regimes, due to the transient response to glacier retreat or advance since the late Pleistocene.

Finally, our measurements on terrace deposits and gravel bars confirm that spatial and temporal scatter, whether due to our measurement procedures, climatic variations, or some stochastic aspect of sediment supply and river transport, has more impact on the gravel size distribution than on the lithologic composition.

5. COMPARISON BETWEEN EXPERIMENTAL AND NATURAL ABRASION

In order to extend in a more quantitative way the data interpretations on gravel bars and sources in the Marsyandi Valley, we conducted experimental measurements on Himalayan lithologies and checked their consistency with our observations by means of a simplified integrative model.

5.1 Experimental Abrasion Rates of Himalayan Lithologies

Between three and ten rounded pebbles were sampled in the field for each of the most representative lithologies of the Marsyandi watershed. Pebbles came from active gravel bars: no obvious weathering rind was observed. Their *b*-axes ranged from 4 to 10 cm, which roughly corresponds to the median pebble size along the Marsyandi. The pebbles were run into a circular flume (Attal and Lavé, 2003; Attal, 2003) designed to characterize pebble abrasion processes and rates. The run was conducted under a turbulent flow velocity of 2.3 m/s and average shear stress of the order of 250 Nm^{-2}. These conditions are close to the conditions prevailing in the Marsyandi River during annual peak discharge across the Lesser Himalaya. Because the dominant lithologies in the Marsyandi River are mostly resistant to abrasion, ~60 kg of alpine crystalline pebbles of similar size were added to 15 kg of pebbles of different Himalayan lithologies in order to obtain a bedload flux of 80 kg $s^{-1}m^{-1}$ or 0.03 m^2s^{-1} by width unit. Weight loss by abrasion was quantified by weighing the dry Himalayan pebbles before and after the run. Traveling distances ranged between 2.3 and 7.0 km, according to run durations of 0.5–1.5 h and a mean pebble velocity of 1.3 m s^{-1}. Abrasion coefficients are thus reported in % loss per kilometer (Table 1).

Weight loss values are relatively uniform (with relative standard deviations lower than 1) for resistant lithologies and for marbles. Quartzite abrasion rates are very low, ~0.15% per km. Granites and calcic gneisses (Formation II) have rates of 0.4% per km, but aluminous gneisses (Formation I) have rates three times higher: 1.4% per km. This difference is probably due to the presence in these gneisses of well-developed weak mica beds, which make the rock less resistant to abrasion. Each of the

TABLE 1. EXPERIMENTAL ABRASION RATES FOR DIFFERENT HIMALAYAN LITHOLOGIES

Structural unit	Lithology	Abrasion rate (%.km^{-1})			
		Mean value	1σ	min	max
TSS lower units (Paleozoic)	Quartzitic sandstone	0.4	0.3	0.16	0.9
TSS lower units (Ordovician)	Limestones	2.6	0.8	1.7	3.4
TSS lower units (Paleozoic)	Annapurna limestone	0.5	0.2	0.3	0.7
TSS lower units (Paleozoic)	Schist	7	8	0.4	20
Manaslu granite	Granites	0.4	0.2	0.28	0.6
HHCrystalline, FII	Calcic gneisses	0.4	0.2	0.12	0.7
HHCrystalline, FI	Aluminous gneisses	1.4	0.6	0.5	2.4
Upper series of LH, northern part of the anticlinorium	Schist, sericites,and micaschist	16	10	4.3	31
Upper series of the LH, northern part of the anticlinorium	Marble	1.7	0.4	1.4	2.2
Lower LH, Kuncha formation	Sandstone	1.7	1.2	0.17	3.8
Lower LH, Kuncha formation	Schists/sandstones	2.5	0.6	1.8	2.9
LH	Quartzite	0.15	0.04	0.1	0.2
Upper series of the LH, southern part of the anticlinorium	Schist	23	16	5.4	45
Siwaliks MDT	Sandstone	6	5	1.4	13
Siwaliks MFT	Sandstone	31	22	16	47

Note: Measurements were conducted in a circular flume (Attal and Lavé, 2003; Attal, 2003) with 4–8 cm pebbles (*b*-axis) sampled along the Marsyandi river. Some lithologies like limestone, gneiss, granite, or marble display roughly homogeneous values, in contrast to schist and sandstone, which present heterogeneous behavior to abrasion. This behavior depends on their degree of strengthening by diagenesis and/or metamorphism, their content in micas and phyllitic minerals and their degree of schistosity. TSS— Tethyan Sedimentary Series; HHC—Higher Himalayan Crystalline; LH— Lesser Himalayan units; MDT—Main Dun Thrust; MFT—Main Frontal Thrust.

calcareous rock types have roughly uniform abrasion rates due to their fine and homogeneous structure at the pebble scale, but these rates are variable if we consider the limestone group as a whole; they vary from 0.5% per km for the Annapurna silicified limestones to 2.6% per km for the black limestones of the Ordovician Pisang Formation. These differences are probably linked to the various degrees of metamorphism and recrystallization of the rocks and to their various content in quartz.

Soft lithologies present an important dispersion in abrasion rates. For sandstone, abrasion rates and dispersion increase with decreasing degrees of metamorphism or diagenesis, because these processes contribute to strengthen rocks. The abrasion rates obtained experimentally are 0.4% for the Tethyan quartzitic sandstones, 1.7% for the Lesser Himlayan sandstones of the Kuncha Formation, 6% and 30%, respectively, for the Tertiary sandstones from the Main Dun thrust zone and from the Main Frontal thrust zone. For schist pebbles, abrasion rates can vary by an order of magnitude according to their degree of weakness; this weakness is linked to their mica and phyllitic mineral content as well as to the intensity and spacing of cleavage planes (bedding, schistosity). Schists display on average very high weight loss values, between 7% and 23% per km. Due to their particular structure and mineralogical composition, schist pebbles are abraded very rapidly and often in a mode that rarely occurs for other lithologies: the pebble splits into two or three pieces of similar size.

For resistant lithologies, the average values are consistent with the results obtained for similar lithologies run in other abrasion experimental devices (Schoklitsch, 1933; Kuenen, 1956), although they are slightly higher, by a factor 1–3, than these previously published abrasion rates. For soft lithologies, our experimental abrasion rates are 4–20 times higher than previously published rates. We suggest that this is due to higher impact velocities and thus to more efficient impact wear in our experimental flume than in the tumbling barrels that were often used in previous studies.

In the following, we will be using our average experimental values, keeping in mind, however, that they cannot account for the large dispersion observed for some lithologies, and also that they possibly correspond to minimum values. Indeed, the pebbles of each lithologic class that we sampled to run in the experimental device could not be fully representative of the hillslope material, because the most-fractured or least-resistant fraction of the source material already disappeared during fluvial transport and abrasion, and therefore is not included in the sampled pebbles.

5.2 Integrative Model for the Marsyandi Watershed

In a given river section, the sediment load size distribution results from the contribution from all the parts of the watershed, modulated by the local erosion rate (or hillslope sediment supply rate) and by size reduction processes. If we assume first that most of the material supplied by the hillslopes can be transported along the fluvial network whatever the initial size (except for very large blocks that represent a minor contribution to total sediment flux), second that abrasion coefficients do not depend on hydrodynamic conditions or sediment flux, and third that we can neglect additional weight loss during the first traveled kilometers due to very efficient abrasion and edge rounding processes of angular pebbles (Krumbein, 1941; Kuenen, 1956; Pearce, 1971), then size reduction can be adequately represented by Sternberg's law (equation 1) with a constant abrasion coefficient, like in section 2. Under these simplifying assumptions, the size distribution for the jth lithology can be written as an integral term on the upstream fluvial network (*fn*):

$$f_j(D) = \left(\iint_{fn} p_j(\vec{x}) f_j^0(\vec{x}, De^{k_j L(\vec{x})/3}) A(\vec{x}) \dot{\varepsilon}(\vec{x}) d\vec{x} \right) \Bigg/ \left(\iint_{fn} A(\vec{x}) \dot{\varepsilon}(\vec{x}) d\vec{x} \right), \quad (5)$$

where $p_j(\vec{x})$ represents the local proportion of the jth lithology, $A(\vec{x})$ the local contributing area, $\dot{\varepsilon}(\vec{x})$ the corresponding average erosion rate, k_j the abrasion coefficient, $L(\vec{x})$ the distance along the fluvial network between the sediment source and the considered river section, and

$$f_j^0(\vec{x}, D)$$

the size distribution of the sediment supply from the hillslope to the river network in x, with

$$\int_D f_j^0(\vec{x}, D) dD = 1.$$

Median size D_{50} is thus defined as the diameter for which:

$$\left(\sum_j \int_{D_{mbl}}^{D_{50}} f_j(D) dD \right) \Bigg/ \left(\sum_j \int_{D_{mbl}}^{+\infty} f_j(D) dD \right) = 1/2, \quad (6)$$

where D_{mbl} represents the smallest gravel size traveling as bedload. The proportion of the jth lithology for pebbles larger than D_m = 5 cm is expressed by:

$$P_j = \int_{D_m}^{+\infty} f_j(D) dD \Bigg/ \sum_j \int_{D_m}^{+\infty} f_j(D) dD. \quad (7)$$

It should be noted that equation 5 is probably not appropriate for suspended load, which, in any event, is not considered here.

In an attempt to invert this integrative model for the abrasion coefficient k_j for each lithology, we thus need three independent data sets. First, we need the size distribution for the sources: we consider the average distribution of the ones we measured along the Marsyandi valley (thick gray line on Figs. 5A and 5B). We initially assume that landslides represent the main source of sediment supply and consider a uniform distribution curve, independent of lithology, which corresponds to the average size

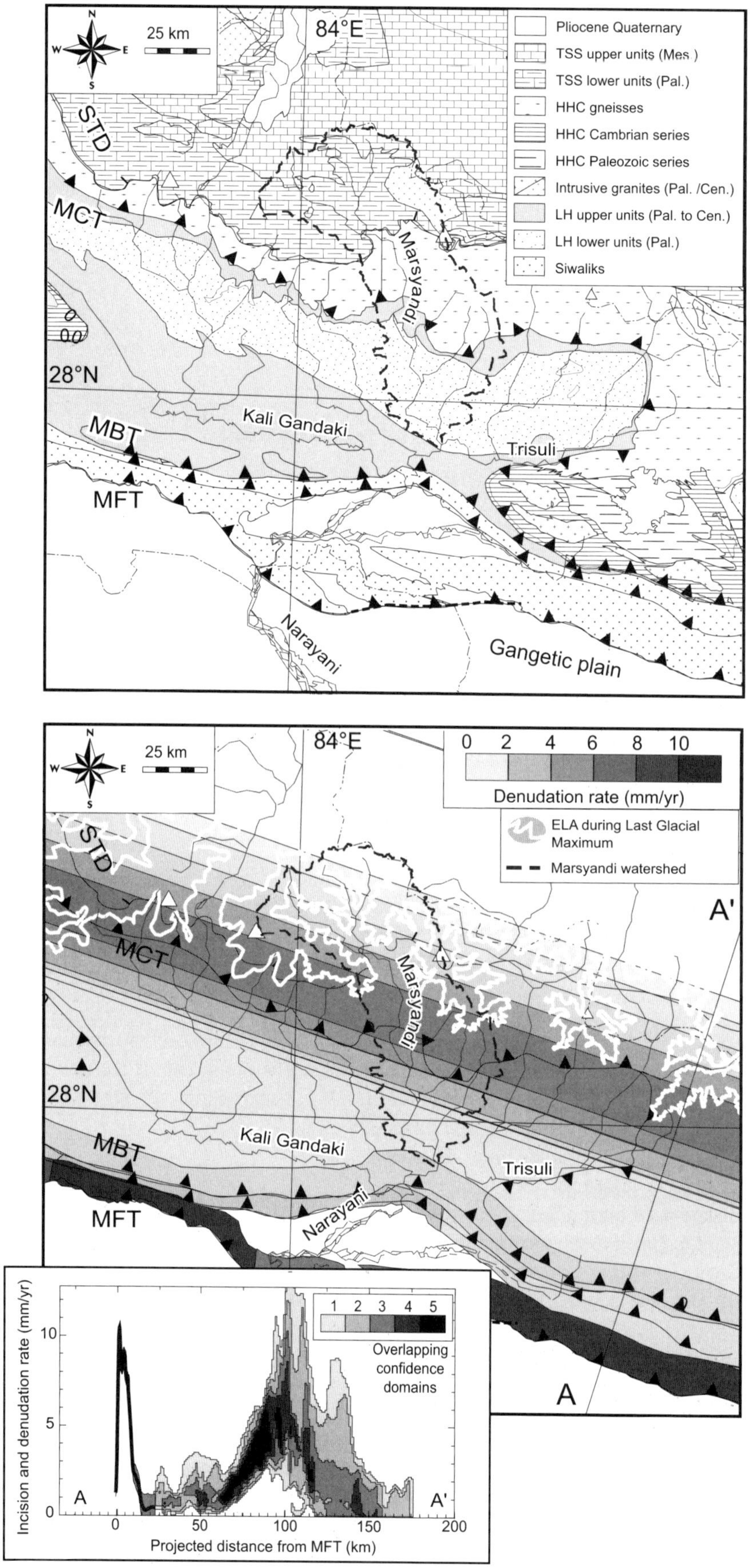

25 km
84°E
Pliocene Quaternary
TSS upper units (Mes.)
TSS lower units (Pal.)
HHC gneisses
HHC Cambrian series
HHC Paleozoic series
Intrusive granites (Pal. /Cen.)
LH upper units (Pal. to Cen.)
LH lower units (Pal.)
Siwaliks
STD
MCT
Marsyandi
28°N
MBT
Kali Gandaki
Trisuli
MFT
Narayani
Gangetic plain
25 km
84°E
0 2 4 6 8 10
Denudation rate (mm/yr)
ELA during Last Glacial Maximum
Marsyandi watershed
STD
A'
MCT
Marsyandi
28°N
MBT
Kali Gandaki
Trisuli
MFT
Narayani
A
Incision and denudation rate (mm/yr)
0
5
10
1 2 3 4 5
Overlapping confidence domains
A
A'
0 50 100 150 200
Projected distance from MFT (km)

TABLE 2. PROPORTION OF THE MAIN LITHOLOGIES IN THE DIFFERENT GEOLOGIC UNITS AND AVERAGE ABRASION COEFFICIENT USED IN THE MODELS

	Granite	Gneiss	Schist, sericite, and micaschist	Quartzite	Sandstone	Limestone and marbles
Abrasion coefficient	0.4	1.3	16	0.15	2.0	2.0
Lithologic proportion (%)						
TSS upper units	-	-	16	13	15	56
TSS lower units	-	-	20	10	10	60
Intrusive granites	100	-	-	-	-	-
HHC gneisses	-	95	-	5	-	-
LH upper units	0	0	55	20	10	15
LH lower units	-	-	55	10	35	-

Note: See Figure 9A for correspondence. TSS— Tethyan Sedimentary Series; HHC—Higher Himalayan Crystalline; LH— Lesser Himalayan units.

distribution of landslide deposits (Fig. 5A). Additional supply by moraine material will be introduced only for the last tested model (section 5.3.2). Second, a lithologic map is required (Fig. 9A): based on previous geologic mapping (Colchen et al., 1986) and corresponding cross sections, rough proportions of each lithology are estimated for the main structural units (Table 2). Finally, a denudation map is required (Fig. 9B): as a first order proxy, we use an erosion map extrapolated from fluvial incision rates in central Nepal, assuming that incision rates are roughly in equilibrium with catchment erosion rates (Lavé and Avouac, 2001). Recent work in the Marsyandi Valley confirmed the gross pattern view of inferred downcutting and erosion rates along Marsyandi Valley (Burbank et al., 2003; Pratt-Sitaula et al., 2004) with highest incision rates across the Higher Himalayan Crystalline, low rates across the Lesser Himalaya, and moderate rates across the Mahabarat and Main Dun thrust fold. In addition, suspended load measurements in east and central Nepal suggest that present denudation rates are roughly similar to long-term denudation rates (Lavé and Avouac, 2001). Finally, it has to be noted that only relative spatial variations in erosion rates are required by equation 5.

Because large uncertainties are associated with these data sets, we did not perform a formal inversion and preferred to explore different end-member models, in order to assess the roles of abrasion coefficients and local hillslope sediment supply rates. To conduct such direct modeling, we extracted the fluvial network from the worldwide 30″ digital elevation model.

Figure 9. Maps of (A) lithology and (B) denudation rates used for the model. Lithologic map is extrapolated from Colchen et al. (1986). Denudation map is derived from terrace record and fluvial Shields stress inversion along five transverse Himalayan rivers (Lavé and Avouac, 2001). Stacked incision curves for these rivers are presented in the lower left angle of the diagram; curves are projected on a N18° profile, roughly perpendicular to the Himalayan structures. Equilibrium line altitude (ELA) during Last Glacial Maximum has been considered in model (10G; section 5.3) as the limit between an upper zone with sources dominated by moraine material and a lower zone with sources dominated by landslides. Domains are TTS—Tethyan Sedimentary Series, LH—Lesser Himalaya, and HHC—Higher Himalayan Crystalline. The major faults are the South Tibetan Detachment (STD), the Main Central thrust (MCT), the Main Boundary thrust (MBT), the Main Dun thrust (MDT), and the Main Frontal thrust (MFT).

5.3 Results of the Different Models

5.3.1 Lithologic Proportion

The results of seven end-member models (Fig. 10A–G, corresponding to models 10A–G, respectively) have been compared to the lithologic data for the Marsyandi (Fig. 10H), i.e., the lithologic composition of the material coarser than 5 cm. First, we considered the relative surface proportion of each lithology outcropping in the contributing area (Fig. 10A), i.e., the pebble distribution produced by a uniformly eroded landscape, in the absence of abrasion during pebble transport. As expected from a rapid look on the geologic map, the Tethyan series, Higher Himalayan Crystalline, and Lesser Himalayan units represent similar areas in the Marsyandi basin: in consequence, their dominant lithologies—limestones, gneisses, and schists, respectively—occur in similar proportions close to the confluence with the Trisuli. This result sharply contrasts with our observations and suggests that pebble abrasion and/or variations in sediment supply have to be incorporated. The second model therefore includes variations in sediment supply (Fig. 10B), but still excludes pebble abrasion. Increasing hillslope erosion in the Higher Himalayan Crystalline unit, according to the denudation map in Figure 9B, produces the observed proportional increase in gneissic pebbles, but is unable to predict the low concentrations in schists, the rapid downstream decrease in limestone where the river crosses the Higher Himalayan Crystalline, and the net increase in quartzite. Including a uniform abrasion coefficient of 2%/km (Fig. 10C) helps to accentuate the decrease in limestone, but does not solve the misfits for the other lithologies. Introducing lithology-dependent abrasion coefficients, as evidenced by our experimental results on Himalayan pebbles, therefore seems necessary (Fig. 10D). The average coefficients permit us to account for most of the observed concentration curves along the Marsyandi, in particular the increase in quartzite and vanishing of schist pebbles, although the latter effect is overestimated by the model. On the other hand, we could

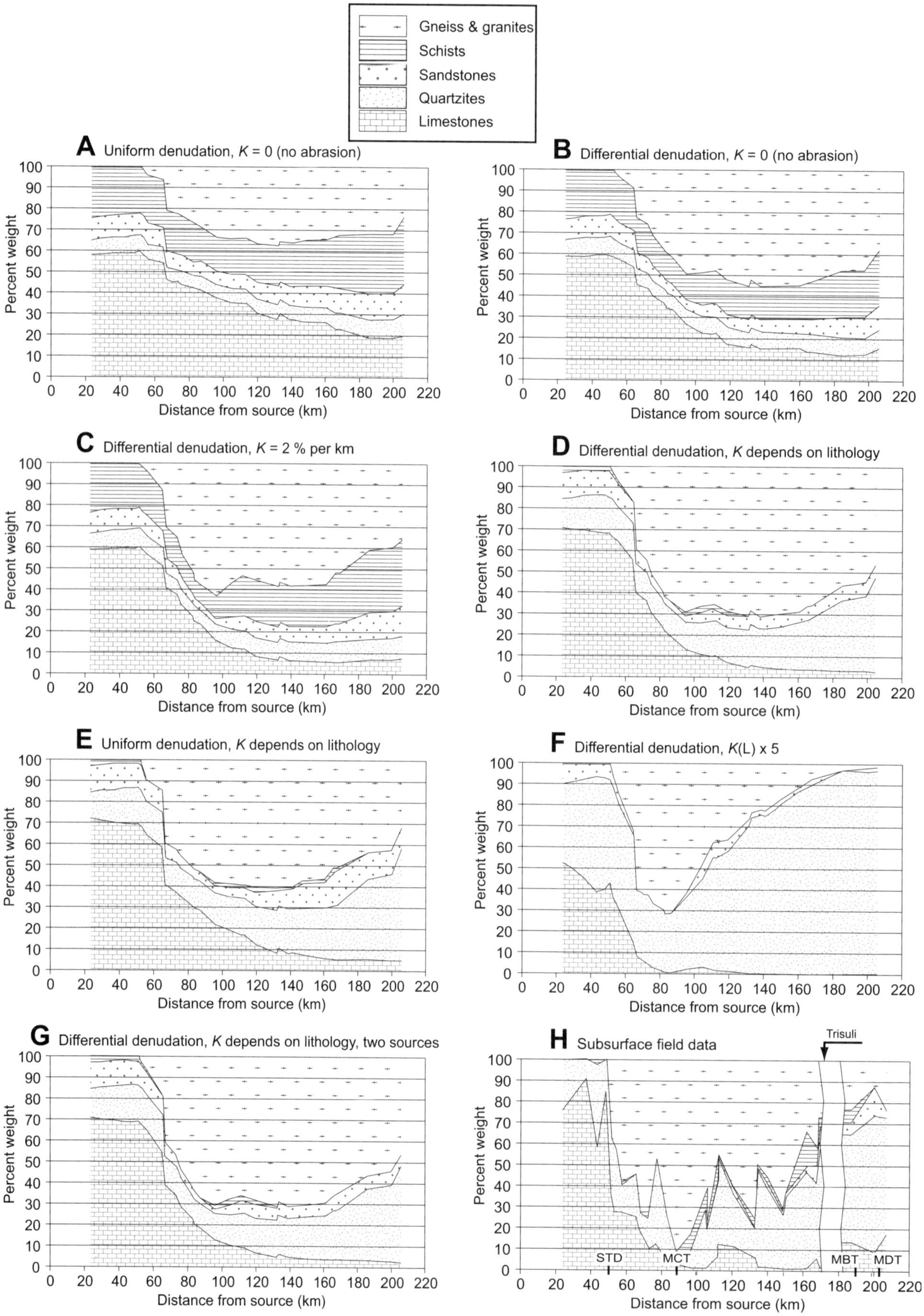

Gneiss & granites
Schists
Sandstones
Quartzites
Limestones
A Uniform denudation, K = 0 (no abrasion)
B Differential denudation, K = 0 (no abrasion)
C Differential denudation, K = 2 % per km
D Differential denudation, K depends on lithology
E Uniform denudation, K depends on lithology
F Differential denudation, K(L) x 5
G Differential denudation, K depends on lithology, two sources
H Subsurface field data
Trisuli
STD
MCT
MBT
MDT
Percent weight
Distance from source (km)

neglect differential hillslope erosion rates and focus on differential abrasion coefficients (Fig. 10E). As for model 10D , most of the concentration curves reasonably fit the observations along the Marsyandi. However, when compared to the data, the decrease in limestone proportions is not sharp enough, the proportion of the gneiss across the Higher Himalayan Crystalline is too low, and the proportion of sandstone in the Lesser Himalaya is too high. Therefore both processes, differential supply rates and abrasion coefficients, are necessary to explain the gross features of the observed lithologic evolution along the Marsyandi.

In section 2, it was argued from a simple analytic model that the downstream evolution of the pebble fraction in the sediment load strongly depends on the abrasion coefficient (Fig. 1B). When two or more lithologies with distinct resistance to abrasion are present in the catchment, it can easily be shown that the downstream evolution in lithological proportion depends not only on the relative values of the abrasion coefficients but also on their absolute values. This prediction can be tested with the integrative model, by assuming the different abrasion coefficients multiplied by factors of 1/5 and 5, respectively. The first case, i.e., a decrease in abrasion efficiency, leads to a figure that is similar to model 10B, except for the schist, the abundance of which is reduced. The second case (Fig. 10F), i.e., an increase in abrasion efficiency, produces a rapid predominance of quartzite pebbles over all other lithologies in the Lesser Himalaya. Both scenarios can thus be rejected: the abrasion coefficients that our model requires to explain the data we collected along the Marsyandi River are roughly constrained, and they are of the same order as the experimental abrasion coefficient we obtained experimentally, although large uncertainties remain.

5.3.2 Size Distributions

When landslide material reaches the fluvial network, we suspect that the finer fraction is rapidly washed out and travels as suspended load, and that the size distribution of the residual material becomes coarser. We have already observed in the size distribution curves (Fig. 5) that gravel bar material is indeed depleted in fine particles in comparison to the hillslope sources. In order to avoid any bias, only fractions traveling as bedload are considered here, and the median size will be computed for the fractions larger than a cut-off value of 1 mm. For each previous end-member model, the predicted downstream evolution of this truncated median size does not fit the observations (Fig. 11). First, all the models predict a uniform median size or a slight downstream decrease, in sharp contradiction with the data along the Marsyandi. Second, all the models underestimate the data trend by a factor of two, except in the Tethyan Sedimentary Series, where they overestimate gravel size.

Most of the highest parts of the landscape, above 4000 m, display important present-day and fossil glacial signatures. Such preservation of formerly glaciated landscapes suggests that the system has not yet returned to equilibrated fluvial landscape. In these areas, the present denudation rates could be different from the long-term rates. Moreover, sediment supply by landslides could represent a minor contribution relative to re-erosion of moraine material and debris produced by periglacial processes. Therefore, let us assume that in the upper part of the watershed, the hillslope sediment supply is better accounted for by the average size distribution of moraine material (thick gray line on Fig. 5B). To do this simply, we consider that the equilibrium line altitude (ELA) during the Last Glacial Maximum (Duncan et al., 1998) defines the limit between hillslope supply dominated by landslides and by moraine-type material (Fig. 9B). Obviously, large glacial valleys have developed far below this line, like in the upper Marsyandi between Manang and Dharapani (Fort, 1993; Lavé and Avouac, 2001), down to probably 2000–2500 m (Fig. 2). However, it is beyond the scope of this paper to account for a precise location of glacial remnants: introducing in model 10G such a simplistic limit only provides a first-order view on the role of source distribution. Along the Marsyandi Valley, the lithologic composition is almost insensitive to this change, and differences in curves of models 10D and 10G are of the order of few percents or less (Fig. 10). In contrast, the downstream evolution of the median grain size reflects the introduction of finer size distributions from moraine material in the upper part of the watershed. The modeled curve displays similar features to the observed data: low median sizes above the South Tibetan Detachment, intermediate sizes across the Higher Himalayan Crystalline, and higher sizes across the Lesser Himalaya. We therefore propose that the downstream coarsening pebble sizes could arise from varying sediment sources along the Marsyandi Valley and that source characteristics can exert a strong control on the evolution of river sediments. Though it provides a more acceptable downstream trend, the model 10G still largely underestimates the median pebble size observed on the gravel bars in the Higher Himalayan Crystalline and Lesser Himalayan units. An explanation for this misfit will be discussed in the section 6.1.

5.3.3 Size Distribution for Individual Lithologies

The consistency of model 10G can be tested in more detail through the size distribution in each lithology. To attenuate the large scattering in the data, we grouped the measurement stations in seven zones according to their morphostructural position along the river (Fig. 12A), and summed their weight into a

Figure 10. Downstream evolution, along the Marsyandi River, of the lithologic content of bedload pebbles larger than 5 cm predicted by different end-member models (A to G), as compared to observed subsurface gravel bar material (H). Model A displays the lithologic surface proportion in the contributing watershed (= uniform erosion and no abrasion); models B to F correspond to end-member models showing the influence of hillslope sediment supply rates and abrasion coefficients. They were computed for a uniform size distribution (landslide-type) of hillslope sediment supply. In contrast, model G also includes hillslope sediment supply of moraine-type material in the upper part of the drainage basin. The major faults are the South Tibetan Detachment (STD), the Main Central thrust (MCT), the Main Boundary thrust (MBT), the Main Dun thrust (MDT), and the Main Frontal thrust (MFT).

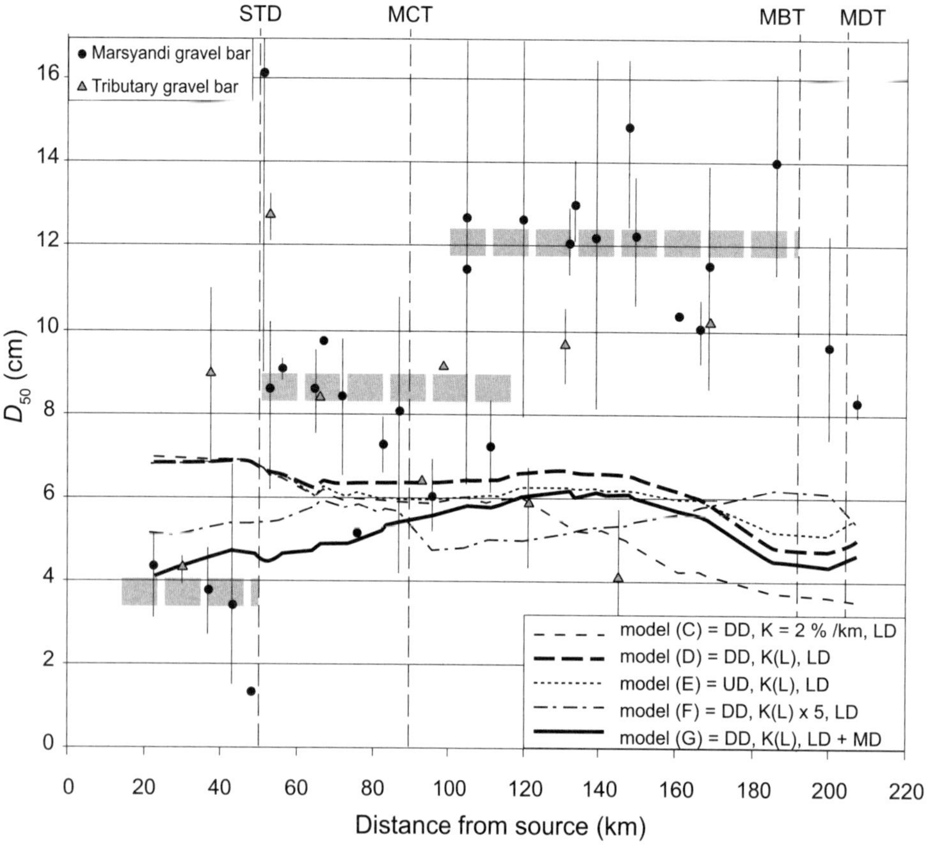

Figure 11. Downstream evolution, along the Marsyandi River, of the median size in the measured subsurface gravel bar material and in bedload as predicted by the different models. D_{50} has been determined by removing particles finer than 1 mm to be directly comparable with modeled bedload D_{50} values (particles finer than 1 mm are considered as suspended load in the model). Vertical bars show the range of values obtained from different sampling points on each measurement site. Modeled curves: DD—differential denudation, UD—uniform denudation, K(L)—abrasion coefficient depending on lithology (experimental values), LD—landslide distribution for source, MD—moraine distribution for source. D_{50} obtained from the model are systematically smaller than field values. The two-sources model is the only one that achieves partial reproduction of the observed downstream coarsening, showing the importance of source characteristics in controlling river sediment characteristics. The major faults are the South Tibetan Detachment (STD), the Main Central thrust (MCT), the Main Boundary thrust (MBT), and the Main Dun thrust (MDT).

single virtual sample. Such a representation will help us to look more at the details and misfit of model 10G. First, as already discussed in section 4.4.1 for the southern Tibetan tributaries Naur and Dudh Khola (Fig. 6A), the gneiss/limestone ratio increases dramatically in that segment (zone 2), much more than predicted by the model. The differences of lithologic composition between gravel bar material and terrace deposit in Chame could suggest, however, that the present distribution in this zone of the upper Marsyandi Valley (zone 2) is a temporary situation. The sudden increase in crystalline pebbles may arise from a complex erosional pattern, in terms of volume and size distribution of the sources, due to the interplay between broadly distributed erosion of formerly glaciated landscapes and localized regressive erosion of the fluvial network that tries to reach equilibrium profiles. Alternatively, the anomalous lithologic composition of the downstream Chame gravel bars could reflect the stochastic behavior of sediment supply. The second misfit concerns the downstream increase of the quartzite/gneiss ratio, underestimated by our model. Several explanations can be proposed: an underestimation of the hillslope erosion rate in the Lesser Himalaya, of the quartzite proportion in Lesser Himalayan units, or more likely of the abrasion coefficient for gneiss. The third misfit concerns the schists, underestimated by our model: this appears to be due to the fact that we both replaced dispersed values of abrasion coefficients (Table 1) by a single average value (a simplifying assumption that can introduce a systematic bias since relation [equation 5] is nonlinear in k_j), and also to a systematic misattribution in the field of micaceous sandstone to schist. Similarly, the attribution of fine sandstone with calcareous cement

Figure 12. (A) Downstream evolution of the size distribution, averaged by zone, for subsurface gravel bar material coarser than 5 cm. Lithologies have been isolated and the contribution of each fraction is expressed in percent of the total weight of the material. Dark-gray columns correspond to field data and light-gray ones to the model 10G, with differential erosion rates, K depending on lithology, and two source distributions. Under-representation of coarsest fractions by field measurement methods appears clearly. (B) Downstream profiles of D_{50}* for crystalline and quartzite pebbles larger than 5 cm in sub-surface gravel bar material along the Marsyandi River. Dark gray lines correspond to the mean values by zone for the field data and light gray lines to modeled data. The major faults are the South Tibetan Detachment (STD), the Main Central thrust (MCT), the Main Boundary thrust (MBT), and the Main Dun thrust (MDT). Zone 1—Tethyan Series; Zone 2—Upper Higher Himalayan Crystalline (between STD and Dudh Khola); Zone 3—Lower Higher Himalayan Crystalline (between Dudh Khola and MCT); Zone 4—Upper Lesser Himalaya (between MCT and Dordi Khola); Zone 5—Middle Lesser Himalaya (between Dordi and Chudi Khola); Zone 6—Lower Lesser Himalaya (between Chudi Khola and the confluence with the Trisuli River); Zone 7—Trisuli River across the Mahabarat (between the confluence with the Marsyandi and Narayangadh).

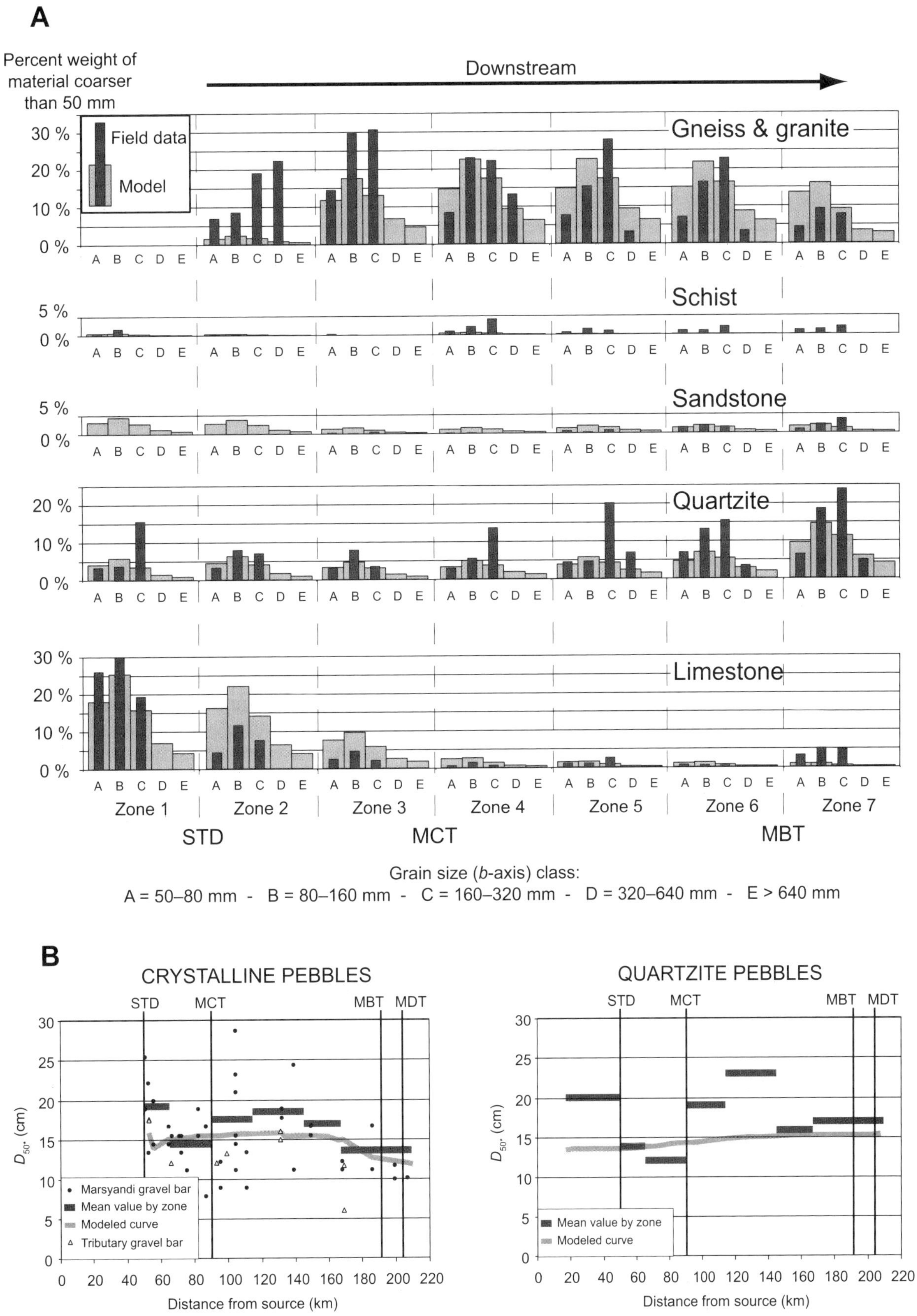
A
Percent weight of material coarser than 50 mm
Downstream
Field data
Model
Gneiss & granite
Schist
Sandstone
Quartzite
Limestone
Zone 1
Zone 2
Zone 3
Zone 4
Zone 5
Zone 6
Zone 7
STD
MCT
MBT
Grain size (b-axis) class:
A = 50–80 mm - B = 80–160 mm - C = 160–320 mm - D = 320–640 mm - E > 640 mm
B
CRYSTALLINE PEBBLES
QUARTZITE PEBBLES
STD
MCT
MBT
MDT
D50* (cm)
Marsyandi gravel bar
Mean value by zone
Modeled curve
Tributary gravel bar
Distance from source (km)

to limestone could explain the complete absence of sandstone in our data along the upper Marsyandi.

Finally, the histograms of Figure 12A show that the observed distribution peaks are systematically displaced toward coarser fraction for quartzite and gneissic pebbles in comparison to the model. However, both lithologies show surprisingly good agreement between modeled and measured D_{50*} (Fig. 12B), where D_{50*} corresponds to the D_{50} value determined for the fraction coarser than 5 cm. This result suggests firstly that the shift in peak distribution in the field data histograms compensates for the absence of the largest pebble size, which is poorly taken into account in the field measurements (see discussion in section 4.2 on procedures bias for the coarsest fraction), and second that most of the misfit observed in Figure 11 between modeled and measured D_{50} is related to a depletion in the gravel fraction (1–50 mm) in bar material compared to the model.

6. DISCUSSION

6.1 Why is Gravel bar Material Coarser than Hillslope Material Supplied to the Fluvial Network?

A major remaining unanswered question is why the gravel bar material is generally coarser than the source material? In the absence of comparable studies that would have stressed a similar problem, we can envisage several hypotheses. The apparent discrepancy might arise first from our data set, second from our model of pebble abrasion, and third from the misleading hypothesis that gravel bar material is representative of the bedload material.

The main weakness of our sampling procedure relates to the insufficient volume we analyzed in each station. As already discussed in section 4.2, the size distribution obtained by volumetric sampling is prone to large uncertainties and leads to a systematic cut-off for the coarsest fraction. The observed distributions are characterized by a systematic absence of boulders larger than 0.4 m (Fig. 5), in contrast with what is predicted by the integrative model. However, a few joint volumetric and surface analyses in sections indicate that D_{50} values are not too sensitive to this cut-off. On gravel bars, the counting procedure cannot in any case be the source of the coarsening effect relative to hillslope material, because the bias would play in the opposite sense. In contrast, for landslide deposits, this bias could lead to an underestimation of the median size of the distributions. However, even distributions from photo counting do not display median sizes larger than 80 mm. Despite a low number of measurement sites in landslide deposits, it would be surprising that all the measurements could be finer than average material from landslide deposits. By preferentially choosing the core of the landslide deposits, we could have missed the coarsest fractions that are concentrated both at the surface and downhill part of the deposit. Clarifying this point would require additional detailed studies of the size distribution of the bulk of a landslide. Finally, it should be noted that the downstream evolution in size distribution and D_{50*} (Fig. 12B) does not indicate any major differences between gneiss and quartzite pebbles, even in the Lesser Himalaya where hillslope supply of gneissic pebbles has stopped. This observation argues against the fact that the coarsening observed in gravel bars across the Lesser Himalaya could result from a major change in source characteristics in the Lesser Himalaya. Two additional observations corroborate this conclusion: first, the size distributions for landslides initiated in Higher Himalayan Crystalline gneissic (Formation I) units and Lesser Himalayan quartzitic units do not appear to differ significantly; and second, the gravel bar material along the tributaries that drain the Lesser Himalayan units do not show coarser material than along the Marsyandi.

In our model, we assumed that abrasion coefficients are independent of particle size. For unimodal size distribution, experimental results (Kuenen, 1956; Attal, 2003) indicate that the abrasion coefficient increases with particle size. However, for mixed pebble size distributions, very preliminary results obtained in our abrasion experimental device suggest that the abrasion coefficient is in fact inversely proportional to particle size (Attal, 2003), probably because shocks with larger impacting pebbles strongly enhance relative weight loss of small impacted pebbles, either by abrasion, crushing, or splitting. In this case, abrasion during transport could lead to a rapid depletion in gravel and small pebble fractions, and therefore to a coarsening of the particles traveling as bedload. This hypothesis, however, needs a more quantitative support from future experimental work.

Up to now, we assumed that subsurface gravel bar material is representative of the bedload material. This hypothesis, supported by flume experiments (e.g., Parker and Klingeman, 1982), is closely tied to the notion of equal mobility transport in gravel rivers. Wilcock and Southard (1989) showed that these results are strongly dependent on the experimental device and that sediment-recirculating flumes lead to distinct distribution of subsurface material and bedload. However, they also suggested that natural systems might behave like a sediment-fed flume with the equal mobility scenario. We are not aware of many field studies on mountain rivers that try to compare size distributions of bedload during peak flow and of gravel bar material at low stage. Habersack and Laronne (2001) made such a comparison for a Swiss mountain river and found that subsurface material was much coarser than bedload at intermediate- to high-flow stages; unfortunately, they did not conduct sampling during the peak discharge. Experimental results, even in recirculating flumes (Wilcock and McArdell, 1997; Wilcock, 1997), indicate that full mobility of all pebble sizes is met when the average shear stress reaches three times the critical shear stress to set the median pebble size in motion. However, this criterion is not completely fulfilled by the coarsest fraction, and even with full mobility, the coarse fraction does not reach an equal mobility, i.e., the bedload transport rate decreases for large pebble size (Wilcock and McArdell, 1997). This last point implies that large pebbles travel at lower velocity than small pebbles. As a consequence, the residence time of the coarse pebbles from hillslopes to the high-order channels has to be longer: at low stage, when all pebbles are at rest on the gravel bars, the coarse fractions are therefore over-represented compared to the instantaneous bedload flux during peak flow or bedload motion conditions (Fig. 13). Along the Marsyandi channel and gravel bars, fluted blocks several meters in size attest that the largest blocks are almost immobile (Fig. 4A). This suggestion of differential travel velocities is equivalent to considering that there is a continuous decrease in velocity from a maximum for small and median sizes to almost 0 for the coarsest size.

More generally, we do not know exactly how gravel bars or channel bottom exchanges subsurface material with bedload via its surface layer (Parker, 1991) and how a gravel bar is built from the bedload material. From experimental results, we can suspect that these exchanges between gravel bar material and bedload depend on hydrodynamic conditions, in particular on fluvial shear stress during transporting conditions (Parker, 1990; Wilcock and McArdell, 1997). At high shear stress conditions, full mobility conditions would favor the equivalence in pebble size distributions between bedload and subsurface gravel bar material, whereas at intermediate shear stress conditions, the subsurface material would be coarser than the bedload. Along the Marsyandi River, the shear stress profile (Lavé and Avouac, 2001) would be consistent with such a view: the shear stress values prevailing across the Lesser Himalaya are much lower than across the Higher Himalayan Crystalline and would lead to coarser median pebble size for gravel bars (Fig. 14). In conclusion, as long as equal mobility conditions have not been fully met during peak flows in mountain rivers, we can cast doubt about the fact that subsurface gravel bar material can be used as a representative proxy of bedload material. In the Marsyandi Valley, the discrepancy observed between sources and gravel bar material could arise from such a biased hypothesis.

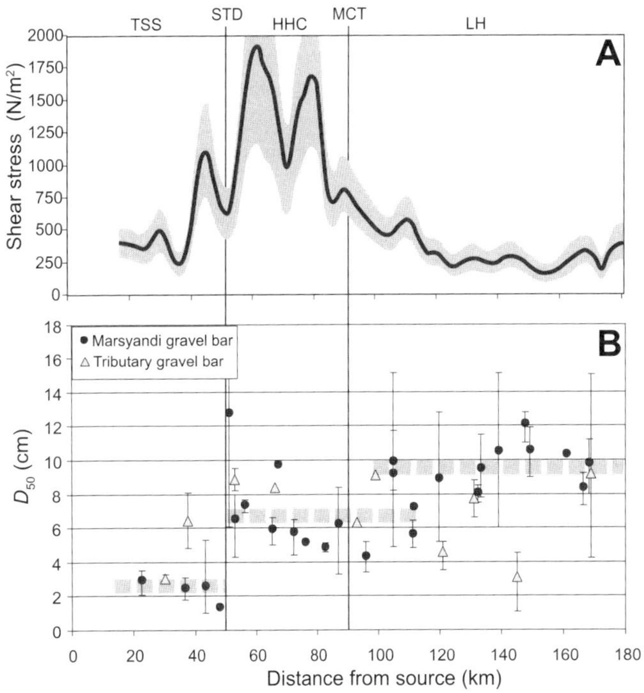

Figure 14. (A) Shear stress profile computed for the 10 yr return peak flow conditions adapted from Lavé and Avouac (2001). Shear stress values display a maximum across the Higher Himalaya: here, full mobility conditions would favor the equivalence of size distribution between bedload and subsurface gravel bar material. In the Lesser Himalaya, low shear stress values may explain the observed coarsening of gravel bar material, in particular the median size of subsurface material (B). The major faults are the South Tibetan Detachment (STD) and the Main Central thrust (MCT). Domains are TTS—Tethyan Sedimentary Series, LH—Lesser Himalaya, and HHC—Higher Himalayan Crystalline.

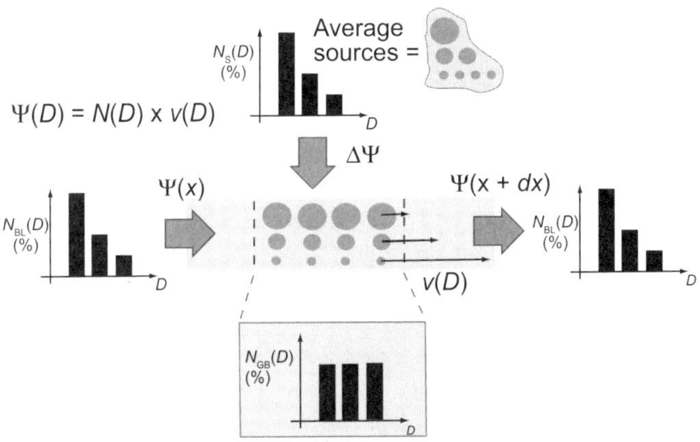

Figure 13. Differential motion diagram: for a given pebble size D, bedload flux $\Psi(D)$ across a river segment is equal to the number $N(D)$ of particles of size D multiplied by their average traveling velocity $v(D)$. If particles move at the same velocity independent of their size, bedload size distribution $N_{BL}(D)$ is equal to the size distribution $N_{GB}(D)$ of the sediment stored on gravel bars and channel bottom at low-flow stage. If large particles move slower than small ones, sediments stored on gravel bars at low flow are coarser than bedload. For the sake of simplicity, no significant abrasion has been considered in this diagram, and, consequently, bedload size distribution $N_{BL}(D)$ is equal to the size distribution $N_S(D)$ of the hillslope sources of material.

6.2 Pebble Abrasion, Sediment Supply, and Transport Modalities Across Active Orogens

In previous studies of pebble size evolution along rivers, experimentally derived abrasion coefficients were not able to explain downstream fining. Several explanations were proposed for this observation: (1) in settings where deposition is occurring, selective transport is a more efficient downstream fining process than abrasion (Brierley and Hickin, 1985; Parker, 1991; Paola et al., 1992; Brewer and Lewin, 1993; Surian, 2002); (2) when long-term gravel storage in gravel bars and terraces occurs, the role of weathering could have a major impact by increasing abrasion coefficients (Bradley, 1970; Jones and Humphrey, 1997); (3) an underestimation of abrasion coefficients in experimental mills has also been advocated, because of miscalculation of traveling distances (Mikos and Jaeggi, 1995) or because experimental devices do not reproduce saltating pebble trajectories (Kodama,

1994b). In the case of active orogens, like the central Himalayas, the two first explanations can be discarded. Even if there is a factor 1–20 above previous estimates (e.g., Kuenen, 1956), the increase in abrasion coefficient by two orders of magnitude required to explain field downstream fining (Kodama, 1994b) is not observed in our experimental measurements, which more or less reproduce the hydrodynamic conditions along the lower Marsyandi during a decadal flood. Thus, our study shows that downstream fining cannot provide a direct and visual estimate of pebble abrasion rate. First, grain size distributions result both from downstream abrasion processes and from the hillslope supply in fresh material: the mean pebble size fining rate is therefore not expected to reflect pebble abrasion rate in a straightforward manner, except in few cases when abrasion coefficients are low (when the asymptotic behavior described in section 2 is not yet reached), or when the sediment source is restricted to the upper part of the watershed. Second, in many morphotectonic settings, as in the Marsyandi Valley, the size distribution of hillslope source material, and/or erosion rates are spatially nonuniform. In addition, the different lithologies exposed in the watershed can have very distinct abrasion coefficients. These complexities make the downstream evolution in grain size ever more difficult to interpret. Finally, it is suspected that the pebble size distribution on gravel bars does not exactly reflect the bedload distribution and could depend on hydrodynamic conditions, and thus lead to erroneous interpretation. We therefore strongly suspect that the size distribution and median pebble size on gravel bars do not constitutes pertinent variables when studying pebble abrasion in active orogens.

Instead, on the basis of the Marsyandi River case study, we propose that the lithologic content of gravel bars bar material represents a much more sensitive tool to unravel pebble abrasion coefficients. Lithologic content is only weakly sensitive to sampling procedure (Figs. 6A and 6B) and to hydrodynamic conditions. It also seems only weakly sensitive to the size distribution of hillslope source material. However, it requires having a rough map of the erosion rates, as well as of the exposed lithologies. Ultimately, these parameters have to be included into an integrative model, and inverse models have to be run to adjust the different abrasion coefficients. According to the Marsyandi case study, the inversion results are sensitive to: (1) the estimate of lithologic content in the different geologic units, and the manner vertical proportions from geologic cross sections are transformed into surface proportions; (2) the lithologic sorting criteria, which has to coincide with those described in regional geologic maps and cross sections; and (3) the estimate of the local hillslope supply rates at a relevant temporal scale, i.e., of the order of the average time required by a pebble to travel from its hillslope source to the outlet of the range.

Following the above procedure in the Marsyandi Valley, but using experimental abrasion coefficients, it can be shown that our observations are fully consistent with a simple model in which abrasion is the only factor of downstream fining, as expected in a uniformly eroding landscape. Even the predicted ratio of bedload to total sediment flux, between 12% and 18% for the major rivers at the outlet of the Himalayan range (Fig. 15), is in reasonable accord with values usually assumed by engineers when planning dam construction in active mountains setting. These ratios obviously represent a maximum, since other hillslope sources delivering mostly fine-grained material (e.g., soil erosion, shallow landslides) have not been included in model sediment budgets, and also because a higher proportion of fine material produced by landslides occurring in schist units is not incorporated (Fig. 5A).

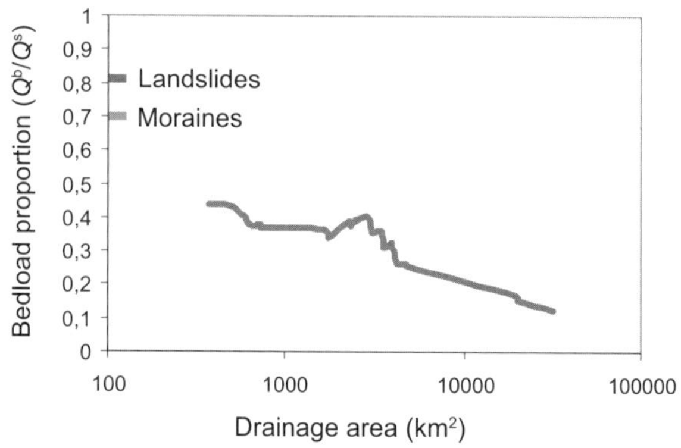

Figure 15. Bedload proportion relative to the total sediment flux predicted by our integrative model (10G) with differential denudation rates, abrasion coefficients from experimental studies, and two distinct hillslope source size distributions. The average proportion of material coarser than 1 mm in landslide and moraine material are also represented.

Finally, according to our results, the best strategy for studying abrasion along the fluvial network in an active orogen is to look at the lithologic composition of the surface material, because it is roughly similar to the subsurface material but easier to study.

Our work also shows that moraines and landslides deliver different sediment size distributions to the river network, as suggested both by direct measurements on sediment sources and by the smaller pebble sizes along the upper Marsyandi reaches. Even landslide deposit characteristics can vary greatly between schist and quartzite sources (Fig. 5A). The sediment supplies from the hillslope can thus be subject to important spatial variations, due to distinct erosion processes, or distinct lithologies. These variations have a direct incidence on the bedload ratio and on the size of the particles transported by the river network, and therefore on the balance between sediment load and river carrying capacity. In recent years, several investigators have argued that sediment flux, and in particular the balance between bedload and carrying capacity, strongly influences the rate and mode of fluvial incision into bedrock (Sklar and Dietrich, 1998, 2001; Howard, 1998; Hancock et al., 1998; Whipple and Tucker, 2002). However, the downstream loss of mass for the bedload fraction is not taken into account by most of the models developed. At the scale of the Marsyandi or Narayani basins, our model results indicate a bedload ratio decrease by a factor ~5 between the source and the mountain front (Fig. 15). This bedload ratio can be crudely approximated by a power law with an exponent –0.2. This downstream decrease in bedload is significant enough to be introduced in transport-limited or mixed-tools fluvial incision models, and in landscape evolution models. Thus, exponent values close to –0.5, as implicitly assumed by Whipple and Tucker (2002) in order to explore the behavior of detachment-limited incision models, appear to be too low in view of our results along the Marsyandi Valley. Moreover, the non-monotonous behavior of the bedload ratio curve, due to change in lithology, sources, and local hillslope supply rates can eventually produce, in a given setting, several channel segments where the river would be alternately more detachment- or transport-limited.

The variations in proportion of fine sediment and median size of the sediment supply between moraines and landslide material may also have a profound impact on river behavior in response to glacial-interglacial climatic changes. The proportions of both types of material are expected to vary strongly with climate and the advance or retreat of glaciers. These changes in load size characteristics, in addition to variations in discharge or sediment flux, could in particular have a nontrivial impact on the building of alluvial or avalanche fans or on terrace formation downstream of formerly glaciated valleys (Hancock and Anderson, 2002; Pratt-Sitaula et al., 2004). On the other hand, a recent study in the San Gabriel Mountains (Lavé and Burbank, 2004) suggests a progressive replacement of shallow erosion processes (dry and wet raveling, soil slip, shallow landslides, etc.) by more deep-seated landslides when erosion and uplift rates increase in active mountains. In such a scenario, the ratio of coarse to fine material and the median size of the coarse material delivered by hillslopes to the fluvial network are expected to increase dramatically around this transition and to affect fluvial downcutting modalities. Similarly, when a landscape is rejuvenated with regressive erosion propagating upstream and triggering more landslides, the increased supply of coarser sediment from the hillslopes can represent a negative feedback to river incision and landscape erosion. If such processes and catena could be confirmed and quantified in more detail, they would represent additional coupling between hillslopes and the fluvial network, which possibly have been underestimated until now.

6.3 Pebble Abrasion Coefficients, Bedrock Erodibility, and Mountain Denudation

The differences in abrasion coefficients obtained experimentally, roughly corroborated by the comparison between integrative model results and data along the Marsyandi, are consistent with the differences of erodibility coefficients derived from river profiles and terrace incision rates along Himalayan rivers (Lavé and Avouac, 2001). For the main structural units, the ratio between the average pebble abrasion coefficients is of the same order as the ratio between average bedrock erodibility coefficients. The abrasion rates and bedrock erosion efficiency are both 10–20 times lower for gneiss than for the Siwalik sandstones, and roughly of the same order for Higher Himalayan Crystalline and Lesser Himalayan units, if we assume that the average bedrock erodibility is partially constrained by the most resistant lithologies outcropping in the channel (Foley, 1980), i.e., quartzites and sandstone for the Lesser Himalaya. This would suggest that fluvial incision rates and river geometry are quite sensitive to the incised lithologies, as has been suggested by Sklar and Dietrich (2001), and that abrasion by bedload would be the dominant process of river incision. Indeed, except if there is a causal relationship between lithologic abrasion coefficient and the degree of fracturation, a mechanism like plucking would not be sensitive to the incised lithologies in the same manner as observed from pebble abrasion. Thus, lithologies have a dual role: first, because they control source characteristics, bedload fraction, pebble size distribution, and abrasion rate during fluvial transport, the lithologies of the contributing area have a long-distance effect on the bedload flux and mobility, i.e., on the coverage effect of the channel bottom and thus on the exposure of bedrock to abrasion and plucking; second, the local lithology directly impacts the detachment rate when bedrock is exposed (local effect).

According to our abrasion data (Table 1), abrasion coefficients in natural lithologies can vary by more than two orders of magnitude. This can have fundamental consequences on tectonic-erosion coupling, on the equilibrium elevation of mountain ranges, and on the rate of topographic decrease for mountains and relief after tectonic cessation. Granitic cores in ranges would help to maintain high elevation, even with subdued uplift. In contrast, schist massifs or thin-skinned tectonic folds composed dominantly of weak sandstone can respond to high uplift rates without developing high topography and a steep fluvial network. These soft lithologies are therefore more able to cope with high uplift rates and to maintain a dynamic equilibrium with high erosion rates. This could explain why mountains characterized by moderate tectonic activity (Alps, Kyrgyz Ranges) can maintain topographies as high as very active schist- and sediment-dominated ranges (Southern Alps, Taiwan Range).

7. CONCLUSIONS

To our knowledge, this contribution represents the first attempt to study jointly hillslope sediment supply and gravel bar material at the scale of a 4000 km^2 basin, and to couple them in an integrative model. Despite questions being left because of uncertainties in measurements, this kind of approach brings new insights on sediment evolution along the fluvial network in active orogens. We highlight below the important implications of our study in the Marsyandi Valley and some avenues for further investigations.

1. Many former studies on downstream evolution of pebble size and lithologic composition have focused on unraveling the respective roles of selective sorting and abrasion, in particular looking at apparent fining rates. However, this study, like Heller et al. (2001), shows that, except in particular settings, the pebble size evolution is not a very

relevant variable: in active orogens, particle size is more sensitive to the size distribution of the local hillslope sediment supply and to their temporal variations than to abrasion processes. Moreover, we strongly suspect that the subsurface grain size distribution on gravel bars is significantly coarser than the average bedload size distribution, in contradiction with the assumption derived from the equal mobility concept in high-energy rivers. In the Marsyandi Valley, the erosion and transport processes have led to a paradoxical downstream coarsening of the gravel bar material. In addition, this material is found to be coarser than the sediments delivered to the fluvial network from the hillslopes. Gravel bar material in active tectonic settings would therefore represent a poor estimator for average downstream evolution of the transported bedload. In contrast, pebble abrasion can be more easily evidenced by the downstream evolution of the lithologic composition of gravel bars, because its measure displays larger variations and is less sensitive to methodological bias and hillslope sediment supply.

2. As expected from theoretical considerations for active orogens, it is not necessary to invoke selective sorting processes to explain the downstream lithologic and size evolution of pebbles along an incising river system. In addition, and in contrast with former studies, the abrasion coefficients required by our simple integrative model of downstream sediment evolution are consistent with experimentally derived abrasion coefficients.
3. The differences in abrasion coefficients are comparable with differences in erodibility coefficients as derived from bedrock incision (Lavé and Avouac, 2001), suggesting that abrasion could be the dominant process in bedrock river incision. Depending on lithology, the abrasion coefficient can vary by more than two orders of magnitude: the eroded lithology could therefore have a major influence on the denudation and tectonic history of active orogens, as well as on postorogenic decay.
4. Finally, our study shows that hillslope sediment supply (landslides, moraines, etc.) may have a major signature in the downstream evolution of pebble size and suspended/bedload ratio. Recent theoretical and experimental investigations have stressed the importance of sediment supply and size in controlling bedrock incision rates. The variations in size distribution from hillslope sediment supply could therefore have important implications on river profile development and on the response of the fluvial network to glacial-interglacial fluctuations. They could also introduce some additional internal coupling between hillslopes and the fluvial network.

ACKNOWLEDGMENTS

We are most grateful to Tank Ojha and "Himalayan Experience" for the help and technical support in the organization of the field surveys. The manuscript benefited from thorough comments and syntax corrections by Pieter van der Beek and Alexander Whittaker. We are indebted to Douglas Burbank for having financially supported our field work through National Science Foundation Continental Dynamic Program (grant ERA-99-09647). Experimental and modeling work conducted at the Laboratoire de Géodynamique des Chaînes Alpines was supported by the Institut National des Sciences de l'Univers programs "coup de pouce" and Programme National de Recherches en Hydrologie.

APPENDIX A: *B*-AXIS AND CORRECTION FACTORS OF PEBBLE DIMENSION

Appendix A.1.

In order to display and compare size distributions from our different sampling and/or counting procedures, we defined a common variable for pebble dimension. We choose to consider the *b*-axis of the particles, i.e., their intermediate axis. For each sampling procedure, we therefore derived a specific correction factor to convert the measured pebble size into the corresponding *b*-axis value. On tape measure line, the *b*-axis was directly measured on particles with a caliper. For the photo determination, we consider that most pebbles are imbricated and that the (*a*-*b*)-axes section is roughly horizontal. In this case, the *b*-axis corresponds to the smallest visible axis on photos. However, pebbles are often partly buried and some (*a*-*b*)-axes section strongly deviate from the horizontal. For these reasons, the *b*-axis tends to be underestimated on photos.

For subsurface samples, we used square mesh sieves. If the (*b*-*c*)-axes section was a circular one, sieving would give direct results. Unfortunately, it is rarely the case and sieving tends to underestimate the *b*-axis value of pebble sizes, due to the fact that particles with the *b*-axis longer than the mesh size can pass through the sieve. To correct the result from this effect, we used measurements of *b*- and *c*-axis realized on tape measure lines. About 1000 pebbles were measured, 15–60 pebbles per site. Maximum value of b/c was 6.2, mean value per site ranged between 1.5 and 2.5. For the whole pebble set, the average value of b/c was 1.9. This value corresponds to an underestimation of the *b*-axis by sieve dimension of 25% (see calculation in Appendix A.2.). No trend appears along the river course, showing that global pebble shape does not change significantly. For subsurface particles coarser than 4 cm, each particle was weighed. The *b*-axis was determined by considering a sphere with a density of 2700 kg/m^3. The error is thus directly linked to the shape of the particle: the *b*-axis is overestimated for an elongated pebble, whereas it is underestimated for a platy pebble. From the measures made on the same 1000 pebbles, the ratio $r_b = b/(abc)^{1/3}$ was calculated, with $(abc)^{1/3}$ corresponding to the diameter of the sphere of equivalent volume, i.e., the *b*-axis deduced from the weight of the pebble. Values for r_b varied between 0.7 (elongated) and 1.8 (platy), the average values for the different stations ranged between 1.0 and 1.2. For the whole pebble set, the average value of r_b was 1.1. Weighing pebbles led thus to an average underestimation of the *b*-axis of 10%. Like for the ratio b/c, the data did not show any significant downstream variations in pebble shape.

All the D_{50} values presented in the text body and figures correspond to an equivalent *b*-axis value, which has been calculated from our field measurements according to the above corrections.

Appendix A.2

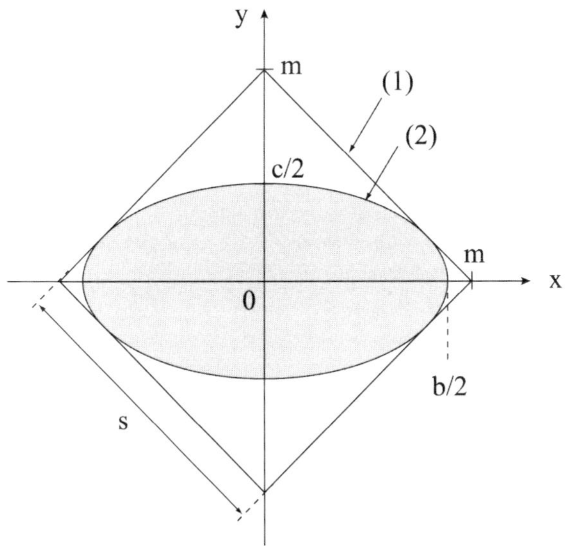

To calculate the maximum *b*-axis value of an elliptic pebble that passes through a square, we consider the parametric equation for a mesh of size *s*:

$$y = -x + m, \tag{8}$$

where $m = \dfrac{s}{\sqrt{2}}$.

The equation describing the pebble is:

$$\frac{x^2}{b^2} + \frac{y^2}{c^2} = \frac{1}{4}, \tag{9}$$

and its ellipse eccentricity $k = \dfrac{b}{c}$.

The parametric equation of the contact points between an elliptic pebble and the mesh is thus given by equations 8 and 9, and leads to a second-order polynomial equation:

$$(4 + 4k^2)\,x^2 - (8k^2m)\,x + (4k^2m^2 - b^2) = 0, \tag{10}$$

The solution for which is $b = \dfrac{2km}{\sqrt{1+k^2}} = \dfrac{k\sqrt{2}}{\sqrt{1+k^2}}s$.

APPENDIX B: COMPARISON OF THE MEDIAN SIZE D_{50} OBTAINED FROM TAPE MEASURE AND PHOTO COUNTING

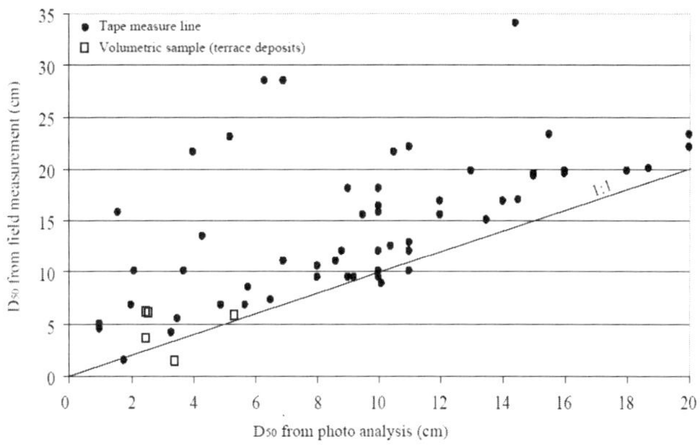

D_{50} values obtained by tape measure lines are systematically larger than D_{50} values obtained by photo analysis. Several factors relative to photo counting contribute to this systematic bias: first, the dimensions of partially hidden pebbles are underestimated; second, if the pebble is not oriented with its longest axis close to the horizontal, then measuring the shortest visible dimension also leads to an underestimation of the intermediate axis of the pebble; third, the area covered by a photo is generally 2 m wide, whereas tape measure lines are deployed on 15 m. The tape measure method therefore allows us to consider coarser particles than the photo method and leads to more representative results for the coarse fraction.

REFERENCES CITED

Armijo, R., Tapponnier, P., Mercier, J.L., and Han, T.L., 1986, Quaternary extension in southern Tibet: Field observations and tectonic implications: Journal of Geophysical Research, v. 91, p. 13,803–13,872.

Attal, M., 2003, Erosion des galets des rivières de montagne au cours du transport fluvial: Étude expérimentale et application aux réseaux hydrographiques d'orogènes actifs [Ph.D. thesis]: Grenoble, France, Université J. Fourier, 279 p.

Attal, M., and Lavé, J., 2003, Pebble and bedrock abrasion during fluvial transport in active orogenic context: Experimental study and application to natural hydrographic networks: European Geophysical Society–American Geophysical Union–European Union of Geosciences Joint Assembly, Nice, France, Geophysical Research Abstracts, v. 5, p. 470.

Avouac, J.P., and Burov, E., 1996, Erosion as a driving mechanism of intracontinental mountain growth: Journal of Geophysical Research, v. 101, p. 17,747–17,769, doi: 10.1029/96JB01344.

Baldwin, J.A., Whipple, K.X., and Tucker, G.E., 2003, Implications of the shear stress river incision model for the timescale of postorogenic decay of topography: Journal of Geophysical Research, v. 108, no. B3, 2158, doi: 10.1029/2001JB000550.

Beaumont, C., Fullsack, P., and Hamilton, J., 1992, Erosional control of active compressional orogens, *in* McClay, K.R., ed., Thrust tectonics: London, Chapman and Hall, p. 1–18.

Benda, L., and, Dunne, T., 1997, Stochastic forcing of sediment routing and storage in channel networks: Water Resources Research, v. 33, p. 2864–2880.

Benett, M.R., and Glasser, N.F., 1996, Glacial geology: Ice sheets and landforms: New York, John Wiley and Sons, 376 p.

Bradley, W.C., 1970, Effect of weathering on abrasion of granitic gravel, Colorado River (Texas): Geological Society of America Bulletin, v. 81, p. 61–80.

Brewer, P.A., and Lewin, J., 1993, In-transport modification of alluvial sediment: Field evidence and laboratory experiments: Special Publication of the International Association of Sedimentologists, v. 17, p. 23–35.

Brierley, G.J., and Hickin, E.J., 1985, The downstream gradation of particle sizes in the Squamish River, British Columbia: Earth Surface Processes and Landforms, v. 10, p. 597–606.

Burbank, D.W., Leland, J., Fielding, E., Anderson, R.S., Brozovic, N., Reid, M.R., and Duncan, C., 1996, Bedrock incision, rock uplift and threshold hillslopes in the northwestern Himalayas: Nature, v. 379, p. 505–510, doi: 10.1038/379505a0.

Burbank, D.W., Blythe, A.E., Putkonen, J.K., Pratt-Sitaula, B.A., Gabet, E.J., Oskin, M.E., Barros, A.P., and Ojha, T.P., 2003, Decoupling of erosion and precipitation in the Himalayas: Nature, v. 426, p. 652–655, doi: 10.1038/nature02187.

Burchfiel, B.C., Zhiliang, C., Hodges, K.V., Yuping, L., Royden, L.H., Changrong, D., and Jiene, X., 1992, The South Tibetan detachment system, Himalayan orogen: Extension contemporaneous with and parallel to shortening in a collisional mountain belt: Geological Society of America Special Paper 269, 41 p.

Campy, M., and Macaire, J.J., 1989, Géologie des formations superficielles: Géodynamique–faciès–utilisation: Paris, Masson, 433 p.

Church, M., McLean, D.G., and Wolcott, J.F., 1987, River bed gravels: Sampling and analysis, *in* Thorne, C.R., Bathurst, J.C., and Hey, R.D., eds., Sediment transport in gravel bed rivers: New York, John Wiley and Sons, p. 43–79.

Colchen, M., Le Fort, P., and Pêcher, A., 1986, Annapurna-Manaslu-Ganesh Himal 1/200,000[e] geological map: Paris, Editions du Centre National de la Recherche Scientifique, 136 p., scale 1:200,000.

Displas, P., and Sutherland, A.J., 1988, Sampling techniques for gravel sized sediments: Journal of Hydraulic Engineering, v. 114, no. 5, p. 484–501.

Duncan, C.C., Klein, A.J., Masek, J.G., and Isacks, B.L., 1998, Comparison of late Pleistocene and modern glacier extents in central Nepal based on digital elevation data and satellite imagery: Quaternary Research, v. 49, p. 241–254, doi: 10.1006/qres.1998.1958.

Foley, M.G., 1980, Bed-rock incision by streams: Geological Society of America Bulletin, v. 91, p. 2189–2213, doi: 10.1130/0016-7606(1980)91<577:BIBSS>2.0.CO;2.

Fort, M., 1993, Etude géomorphologique d'une chaîne de collision intracontinentale: Memoire de Thèse de l'Université de Paris VII, 700 p.

Goede, A., 1975, Downstream changes in the pebble morphometry of the Tambo River, eastern Victoria: Journal of Sedimentology and Petrology, v. 45, p. 704–718.

Habersack, H.M., and Laronne, J.B., 2001, Bed load texture in an alpine gravel bed river: Water Resources Research, v. 37, no. 12, p. 3359–3370, doi: 10.1029/2001WR000260.

Hancock, G.S., and Anderson, R., 2002, Numerical modeling of fluvial strath-terrace formation in response to oscillating climate: Geological Society of America Bulletin, v. 114, p. 1131–1142, doi: 10.1130/0016-7606(2002)114<1131:NMOFST>2.0.CO;2.

Hancock, G.S., Anderson, R., and Whipple, K.X., 1998, Beyond power: Bedrock river incision process and form, *in* Tinkler, K., and Wohl, E.E., eds., Rivers over rock: Fluvial processes in bedrock channels: American Geophysical Union Geophysical Monograph 107, p. 35–60.

Heller, P.L., Beland, P.E., Humphrey, N.F., Konrad, S.K., Lynds, R.M., McMillan, M.E., Valentine, K.E., Widman, Y.A., and Furbish, D.J., 2001, Paradox of downstream fining and weathering-rind formation in the lower Hoh River, Olympic Peninsula, Washington: Geology, v. 29, p. 971–974, doi: 10.1130/0091-7613(2001)029<0971:PODFAW>2.0.CO;2.

His Majesty's Government of Nepal Undertaking, 1994, Nepal Electricity Authority, Engineering Directorate Investment Department: Middle Marsyandi Hydroelectric Project, August (1994), 122 p.

Hovius, N., Starck, C.P., and Allen, P.A., 1997, Sediment flux from a mountain belt derived by landslide mapping: Geology, v. 25, p. 231–234, doi: 10.1130/0091-7613(1997)025<0231:SFFAMB>2.3.CO;2.

Howard, A.D., 1998, Long profile development of bedrock channels: Interaction of weathering, mass wasting, bed erosion, and sediment transport, *in* Tinkler, K., and Wohl, E.E., eds, Rivers over rock: Fluvial processes in bedrock channels: American Geophysical Union Geophysical Monograph 107, p. 237–260.

Howard, A.D., and Kerby, G., 1983, Channel changes in badlands: Geological Society of America Bulletin, v. 94, p. 739–752, doi: 10.1130/0016-7606(1983)94<739:CCIB>2.0.CO;2.

Howard, A.D., Dietrich, W.E., and Seidl, M.A., 1994, Modeling fluvial erosion on regional to continental scales: Journal of Geophysical Research, v. 99, p. 13,971–13,986, doi: 10.1029/94JB00744.

Jones, L.S., and Humphrey, N.F., 1997, Weathering-controlled abrasion in a coarse-grained, meandering reach of the Rio Grande: Implications for the rock record: Geological Society of America Bulletin, v. 109, p. 1080–1088, doi: 10.1130/0016-7606(1997)109<1080:WCAIAC>2.3.CO;2.

Kellerhals, R., and Bray, D.I., 1971, Sampling procedure for coarse fluvial sediments: Journal of Hydraulic Engineering, v. 97, no. 8, p. 1165–1180.

Knighton, A.D., 1982, Longitudinal changes in the size and shape of stream bed material: Evidence of variable transport conditions: Catena, v. 9, p. 25–34, doi: 10.1016/S0341-8162(82)80003-9.

Kodama, Y., 1994a, Downstream changes in the lithology and grain size of fluvial gravels, the Watarase River, Japan: Evidence of the role of abrasion in downstream fining: Journal of Sedimentary Research, v. A64, p. 68–75.

Kodama, Y., 1994b, Experimental study of abrasion and its role in producing downstream fining in gravel-bed rivers: Journal of Sedimentary Research, v. A64, p. 76–85.

Koons, P.O., 1989, The topographic evolution of collisional mountain belts: A numerical look at the Southern Alps, New Zealand: American Journal of Science, v. 289, p. 1041–1069.

Krumbein, W.C., 1941, The effects of abrasion on the size, shape and roundness of rocks fragments: The Journal of Geology, v. XLIX, p. 482–520.

Kuenen, Ph.H., 1956, Experimental abrasion of pebbles: 2. Rolling by current: The Journal of Geology, v. 64, p. 336–368.

Kukal, Z., 1990, The rate of geological processes: Earth-Science Reviews, v. 28, 284 p.

Lavé, J., and Avouac, J.P., 2000, Active folding of fluvial terraces across the Siwaliks Hills (Himalayas of central Nepal): Journal of Geophysical Research, v. 105, p. 5735–5770, doi: 10.1029/1999JB900292.

Lavé, J., and Avouac, J.P., 2001, Fluvial incision and tectonic uplift across the Himalaya of central Nepal: Journal of Geophysical Research, v. 106, p. 26,561–26,592, doi: 10.1029/2001JB000359.

Lavé, J., and Burbank, D, 2004, Erosion rates and patterns in the Transverse Ranges (California): Journal of Geophysical Research, v. 109, doi: 10.1029/2003JF000023.

Le Fort, P., 1986, Metamorphism and magmatism during the Himalayan collision, *in* Coward, M.P., and Ries, A.C., eds, Collision tectonics: Geological Society [London] Special Publication 19, p. 159–172.

Mezaki, S., and Yabiku, M., 1984, Channel morphology of the Kali Gandaki and the Narayani Rivers in central Nepal: Journal of Nepal Geological Society, v. 4, p. 161–176.

Mikos, M., and Jaeggi, M.N.R., 1995, Experiments on motion of sediment mixtures in a tumbling mill to study fluvial abrasion: Journal of Hydraulic Research, v. 33, p. 751–772.

Molnar, P., and England, P., 1990, Late Cenozoic uplift of mountain ranges and global climate change: Chicken or egg?: Nature, v. 346, p. 29–34, doi: 10.1038/346029a0.

Paola, C., Parker, G., Seal, R., Sinha, S.K., Southard, J.B., and Wilcock, P.R., 1992, Downstream fining by selective deposition in a laboratory flume: Science, v. 258, p. 1757–1760.

Parker, G., 1990, Surface-based bedload transport relation for gravel rivers: Journal of Hydraulic Research, v. 28, no. 4, p. 417–436.

Parker, G., 1991, Selective sorting and abrasion of river gravel. I. Theory: Journal of Hydraulic Engineering, v. 117, no. 2, p. 131–149.

Parker, G., and Klingeman, P.C., 1982, On why gravel bed streams are paved: Water Resources Research, v. 18, no. 5, p. 1409–1429.

Pearce, T.H., 1971, Short distance fluvial rounding of volcanic detritus: Journal of Sedimentary Petrology, v. 41, p. 1069–1072.

Pêcher, A., 1989, The metamorphism in the central Himalaya: Journal of Metamorphic Geology, v. 7, p. 31–41.

Pratt, B., Burbank, D.W., Heimsath, A., and Ojha, T., 2002, Impulsive alluviation during early Holocene strengthened monsoons, central Nepal, Himalaya: Geology, v. 30, no. 10, p. 911–914, doi: 10.1130/0091-7613(2002)030<0911:IADEHS>2.0.CO;2.

Pratt-Sitaula, B., Burbank, D.W., Heimsath, A., and Ojha, T., 2004, Landscape disequilibrium on 1,000–10,000 year scales, Marsyandi River, Nepal, central Himalaya: Geomorphology, v. 58, p. 223–241, doi: 10.1016/j.geomorph.2003.07.002.

Schelling, D., 1992, The tectonostratigraphy and structure of the eastern Nepal Himalaya: Tectonics, v. 11, p. 925–943.

Schoklitsch, A., 1933, Über die Verkleinerung der Geschiebe in Flussläufen, Akademie Wissenschaft Wien, Mathematik-naturwissenschaft (Proceedings of the Vienna Academy of Science), Section IIa: Sitzungsber, v. 142, p. 343–366 (in German).

Searle, M.P., 1999, Extensional and compressional faults in the Everest-Lhotse massif, Khumbu Himalaya: Journal of the Geological Society of London, v. 156, p. 227–240.

Shaw, J., and Kellerhals, R., 1982, The composition of recent alluvial gravels in Alberta river beds: Alberta Research Council Bulletin, v. 41, 151 p.

Sklar, L., and Dietrich, W.E., 1998, River longitudinal profiles and bedrock incision models: Stream power and the influence of sediment supply, *in* Tinkler, K., and Wohl, E.E., eds., Rivers over rock: Fluvial processes in bedrock channels: American Geophysical Union Geophysical Monograph 107, p. 237–260.

Sklar, L.S., and Dietrich, W.E., 2001, Sediment and rock strength controls on river incision into bedrock: Geology, v. 29, p. 1087–1090, doi: 10.1130/0091-7613(2001)029<1087:SARSCO>2.0.CO;2.

Sklar, L.S., and Dietrich, W.E., 2004, A mechanistic model for river incision into bedrock by saltating bed load: Water Resources Research, v. 40, W06301, doi:10.1029/2003WR002496.

Sternberg, H., 1875, Untersuchungen über Längen- und Querprofile geschiebführender Flüsse: Zeitschrift für Bauwesen, v. 25, p. 483–506.

Surian, N., 2002, Downstream variation in grain size along an Alpine river: Analysis of controls and processes: Geomorphology, v. 43, p. 137–149, doi: 10.1016/S0169-555X(01)00127-1.

Whipple, K.X., and Tucker, G.E., 2002, Implications of sediment-flux-dependent river incision models for landscape evolution: Journal of Geophysical Research, v. 107, no. B2, doi: 10.1029/2000JB000044.

Whipple, K.X., Kirby, E., and Brocklehurst, S., 1999, Geomorphic limits to climatically induced increases in topography relief: Nature, v. 401, p. 39–43, doi: 10.1038/43375.

Wilcock, P.R., 1997, The components of fractional transport rate: Water Resources Research, v. 33, no. 1, p. 247–258, doi: 10.1029/96WR02666.

Wilcock, P.R., and McArdell, B.W., 1997, Partial transport of a sand/gravel sediment: Water Resources Research, v. 33, no. 1, p. 235–245, doi: 10.1029/96WR02672.

Wilcock, P.R., and Southard, J.B., 1989, Bed load transport of mixed size sediment: Fractional transport rates, bed forms, and the development of a coarse bed surface layer: Water Resources Research, v. 25, no. 7, p. 1629–1641.

Willett, S.D., 1999, Orogeny and orography: The effects of erosion on the structure of mountain belts: Journal of Geophysical Research, v. 104, p. 28,957–28,981, doi: 10.1029/1999JB900248.

Willgoose, G., Bras, R.L., and Rodriguez-Iturbe, J., 1991, A coupled channel network growth and hillslope evolution model, 1. Theory: Water Resources Research, v. 27, p. 1671–1684, doi: 10.1029/91WR00935.

Yamanaka, H., and Iwata, S., 1982, River terraces along the Middle Kali Gandaki and Marsyandi Khola, central Nepal: Journal of the Nepal Geological Society, v. 2, p. 95–111.

Manuscript Accepted by the Society 23 June 2005

Printed in the USA

Geological Society of America
Special Paper 398
2006

Escarpment erosion and landscape evolution in southeastern Australia

Arjun M. Heimsath[†]
Department of Earth Sciences, Dartmouth College, Hanover, New Hampshire 03755, USA
John Chappell
Research School of Earth Sciences, Australian National University, Canberra, ACT 0200, Australia
Robert C. Finkel
Center for Accelerator Mass Spectrometry, Lawrence Livermore National Laboratory, Livermore, California 94550, USA
Keith Fifield
Research School of Physical Sciences and Engineering, Australian National University, Canberra, ACT 0200, Australia
Abaz Alimanovic
Research School of Earth Sciences, Australian National University, Canberra, ACT 0200, Australia

ABSTRACT

Passive margin escarpments are extensively studied around the world, and understanding their evolution continues to present one of the more compelling interdisciplinary challenges tackled by earth scientists today. Escarpments reflect the morphotectonic development of passive margins and can separate regions with different climatic histories, but the inferred rapid rates of escarpment retreat have been at odds with actual measurements of land surface denudation. In this paper we present results from extensive cosmogenic ^{10}Be and ^{26}Al analyses across the escarpment of southeastern Australia to quantify the erosional processes evolving the highland, lowland, and scarp face landscapes. We document new relationships between soil production rates and soil thicknesses for the highland and lowland landscapes and compare these soil production functions with those published in our earlier studies from the highlands and at the base of the escarpment. Both new functions define exponential declines of soil production rates with increasing soil depths, with inferred intercepts of 65 and 42 m/m.y. for the highland and lowland sites, respectively, and slopes of –0.02. Exposed bedrock at both of the new sites erodes more slowly than the maximum soil production rates, at 22 ± 3 and 9 ± 2 m/m.y., respectively, thus suggesting a "humped" soil production function. We suggest that instead of a humped function, lithologic variations set the emergence of bedrock, which evolves into the tors that are found extensively across the highlands and at the crest of the escarpment by eroding more slowly than the surrounding soil-mantled landscape. Compared to soil production rates from previous work using ^{10}Be and ^{26}Al measurements from two different sites, these results show remarkable agreement and specifically quantify a soil production function for the region where soil production rates decline exponentially with increasing soil thickness, with an intercept of 53 m/m.y. and a slope of –0.02. Erosion rates determined from ^{10}Be concentrations from outcropping tors, bedrock, and saprolite from a main spur ridge perpendicular to the escarpment, and sediments from first- and zero-order catchments draining the main ridges, show a clear

[†]E-mail: arjun.heimsath@dartmouth.edu.

Heimsath, A.M., Chappell, J., Finkel, R.C., Fifield, K., and Alimanovic, A., 2006, Escarpment erosion and landscape evolution in southeastern Australia, *in* Willett, S.D., Hovius, N., Brandon, M.T., and Fisher, D.M., eds., Tectonics, Climate, and Landscape Evolution: Geological Society of America Special Paper 398, Penrose Conference Series, p. 173–190, doi: 10.1130/2006.2398(10).

linear decline with elevation, from ~35 m/m.y. near the escarpment base to ~3 m/m.y. at the escarpment crest. This order of magnitude difference in erosion rates may be due to increases in stream incision with distance downslope on the escarpment, or to decreases in precipitation with elevation, neither of which we quantify here. The rates do agree, in general, with our soil production functions, suggesting that the biogenic processes actively eroding soil-mantled landscapes are shaping the evolution of the escarpment despite our observations of block fall and debris-flow processes across the steep regions near the scarp crest. Our results support recent results from studies using low-temperature thermochronology, which suggest that the escarpment is relatively stable after having retreated rapidly immediately following rifting. Differences between our rates of surface erosion caused by processes active today and the scarp retreat rates needed to place the escarpment in its present position need to be explained by future work to untangle the mysteries of escarpment evolution.

Keywords: cosmogenic nuclides, altitudinal transect, tectonic geomorphology, scarp evolution.

INTRODUCTION

Passive continental margins and the escarpments that are typically associated with them have captured the attention of much geomorphic study, helping to build the application of numerical modeling as well as develop applications of new analytical techniques in the field of quantitative geomorphology (e.g., Bierman and Caffee, 2001; Cockburn et al., 2000; King, 1962; Ollier, 1982; Summerfield, 1999; Tucker and Slingerland, 1994; van der Beek and Braun, 1998). Interest in escarpments is well founded, as understanding landscape evolution of a passive margin escarpment forces an integration of climate, tectonic, and surface process studies. Traditional views of escarpment evolution have suggested high rates of escarpment retreat, parallel to the continental margin, in comparison with low rates of denudation in the low-relief highlands as well as across the more highly dissected lowlands.

Despite such long-standing interest in understanding the evolution of passive margin escarpments, relatively few data actually quantify the rates of surface erosion and long-term denudation across these margins to confirm qualitative models of rapid escarpment retreat. Recent work using low-temperature thermochronology place critical long-term constraints on escarpment denudation (e.g., Brown et al., 2002; Cockburn et al., 2000; Persano et al., 2002), while cosmogenic nuclides have defined shorter-term surface erosion rates above (Heimsath et al., 2001), across (Bierman and Caffee, 2001; Cockburn et al., 2000), and below (Heimsath et al., 2000) escarpments in southeastern Australia and Namibia. These data refute the traditional models of escarpment evolution, and instead suggest that relatively rapid rates of postrifting denudation are followed by long periods of lower erosion rates and the development of a long-term steady-state in the topographic form of the escarpment, where relatively slow landscape change is occurring.

We tackle the problem of escarpment evolution by applying the well-developed mass-balance approach initially articulated by Gilbert (1877), quantitatively laid out by Culling (1960, 1965), and eloquently made accessible to a broad audience by Carson and Kirkby (1972). This conceptual framework has been extensively applied recently to quantify soil production (e.g., Heimsath et al., 1997, 1999), landscape evolution (e.g., Dietrich et al., 1995), as well as model the dynamic responses of the land surface to changes in climate and tectonic forcing (e.g., Tucker and Slingerland, 1997; van der Beek and Braun, 1999). Specifically, we are interested in the vertical lowering rate of the land surface. For a bedrock surface, this rate is simply the erosion rate of the surface. For a soil-mantled landscape, this is the rate of conversion of the underlying weathered bedrock to mobile material: the soil production rate. Soil production rates have been hypothesized and recently documented to decline exponentially with increasing soil thickness in a relationship termed the soil production function (Heimsath et al., 1997). Note that soil production rates only equal total erosion rates (including solute loss), i.e., landscape lowering rates relative to a local datum, if local soil thickness is roughly constant over time. This conceptual framework applies to upland, soil-mantled landscapes with no recent history of glaciation, without significant eolian deposition, and only across divergent (convex-up) noses where there is no net soil deposition (Dietrich et al., 1995; Heimsath et al., 1997, 1999). Field verification of this function helps quantify recent landscape evolution models (Dietrich et al., 2003), and the local steady-state soil thickness assumption was verified at one of the field areas used here (Heimsath et al., 2000).

In this paper we present new results from a comprehensive study done across the passive margin escarpment (cf. Great Escarpment; Ollier, 1982) of southeastern Australia. We use two in situ–produced cosmogenic nuclides, ^{10}Be and ^{26}Al, to quantify erosion rates and processes from the low-relief coastal lowlands to the highlands above the escarpment. These erosion rates are compared with two new, well-defined, soil production functions that we present from the highlands at the escarpment crest and from the coastal lowlands. These soil production functions illustrate differences in the processes eroding the highlands in comparison with the coastal lowlands and escarpment base. These 44 new cosmogenic nuclide data are used with two independent data sets published using cosmogenic nuclides and landscape morphology from the base of the escarpment and from the highlands above the escarpment to show that erosion rates decrease with elevation from escarpment base to crest and that soil production rates decrease exponentially with increasing soil thickness. Our findings are consistent with recent studies (e.g., Cockburn et al., 2000; Matmon et al., 2002; Persano et al., 2002; van der Beek et al., 2002) that concluded that passive margin escarpments have been relatively stable and are not retreating as rapidly as suggested by earlier studies of escarpment evolution. To place our results in the context of the extensive work done to understand landscape evolution across the southeastern Australia rift margin, we begin with a brief review.

ERODING AND EVOLVING THE ESCARPMENT

The passive continental margin of southeastern Australia provides an excellent example of a landscape where the processes and rates of evolution have been studied over multiple time scales (e.g., Bishop, 1988; Dumitru et al., 1995; Lambeck and Stephenson, 1985; Nott, 1992; O'Sullivan et al., 1996; Ollier, 1995; Persano et al., 2002; Seidl et al., 1996; Stephenson and Lambeck, 1985; van der Beek and Braun, 1998; Wellman, 1987). This margin is thought to have begun with the rifting between the Australian continent and the Lord Howe Rise to the east ca. 85–100 Ma (Hayes and Ringis, 1973; Weissel and Hayes, 1977). While there has been much debate about the subsequent uplift, erosion, and evolution of the landscape associated with this rift, the large-scale morphology is roughly similar to other passive margins around the world (Matmon et al., 2002), although an important difference exists in that the drainage divide occurs inland of the scarp crest. The highland region of gentle topography, low-relief, and relatively slow erosion rates (e.g., Bishop, 1986; Bishop and Brown, 1992; Bishop et al., 1985; Nott, 1992; Ollier, 1978; Pain, 1985; Wellman, 1987) is separated from the more deeply incised coastal belt by what was once commonly referred to as the Great Escarpment (Ollier, 1982). Modeling of the evolution of the escarpment is extensive (e.g., Lambeck and Stephenson, 1985; van der Beek and Braun, 1998) and depends on field determination of erosion rates, but much modeling of this and other escarpments has been done without empirical constraints on relevant time scales.

Slow rates of Tertiary erosion were deduced for the southeastern highlands, west of the escarpment, in several studies by dividing the age of plateau-forming basalt flows into the elevation difference between the plateau and the channels that had incised through the flows (Bishop, 1985; Bishop et al., 1985; Wellman, 1979, 1987; Wellman and McDougall, 1974). Because of the likely posteruption erosion of the basalt surface, the rates calculated were the minimum rates of incision. All of the studies deduced rates less than 10 m/m.y., with significant uncertainty, for the southeastern highlands, and slight differences depended on the time since basalt emplacement. Young (1983) and Young and McDougall (1982, 1993) extended the examination of river incision through dated basalts to infer slow rates of scarp retreat both west and east of the Great Escarpment, suggesting rates as low as 18 m/m.y.

The approach of determining long-term denudation rates for the region by using low-temperature thermochronology measured across the southeastern highlands, as well as along the present coastline, was pursued relatively early, though often with few data (Dumitru et al., 1991, 1995; Foster and Gleadow, 1991; O'Sullivan et al., 1995, 1996). These studies were motivated, at least in part, by debate over the origin, not the erosion rates, of the southeastern highlands. O'Sullivan et al. (1995, 1996) used apatite fission-track thermochronology (AFTT) to deduce two separate periods of rapid denudation based on distinct periods of rapid cooling recorded in the apatite. They suggested that the initial continental extension associated with the splitting of the Tasman Sea in the mid-Cretaceous resulted in kilometer-scale (>2 km) denudation across much of the present-day southeastern highlands, possibly due to rock uplift following underplating inland from the rift. The second period of cooling recorded in samples from the coastal regions was several tens of millions of years later, associated with the continental breakup during the Late Cretaceous (ca. 80 Ma) through the Paleocene (ca. 60 Ma). These results implied that the highlands that were uplifted and highly eroded 90–100 Ma and the more recently eroded coastal belt arrived at a position near their present morphologies ca. 60 Ma. As Bishop (1986) pointed out, most of the models of the history of the southeast Australian highlands (as opposed to models of the evolution of the continental margin) implied a stable continental divide, which roughly coincided with the Great Escarpment. The inferred tectonic processes leading to the stability of this divide and the southeastern highlands were not stated, but the conclusions regarding very slow Tertiary denudation rates on the highlands were similar to those drawn from a passive isostatic rebound model as presented by Stephenson and Lambeck (1985). Closer examination of O'Sullivan et al.'s research (1995, 1996) shows few data from potentially anomalous areas, and the supposed peak in denudation rates on the coastal strip is not supported by numerous data from other regions (Gleadow et al., 2002; van der Beek et al., 2001). Recent AFTT work (Persano et al., 2002) concludes that the highlands have been tectonically stable with relatively constant denudation rates throughout continental breakup, and suggests erosion rates <10 m/m.y. since the late Paleozoic–early Mesozoic, similar to findings by Bishop and Goldrick (2000) and Moore et al. (1986).

Slow denudation rates are part of the assumed conditions used by Thomas (1989a, 1989b, 1995), Twidale (1971, 1985, 1993), and Twidale and Vidal Romani (1994) to suggest the differential weathering processes leading to the formation of relief in some landscapes, as well as features such as inselbergs and tors. The etching process reported by these authors involves the development of a saprolite (chemically weathered rock that retains its relict structure but has not been physically transported) resulting from the prolonged interaction of groundwater with the underlying bedrock. The weathering front progresses into the bedrock, penetrating into fractures and joints, and weathers the bedrock preferentially into saprolite and core-stones. Saprolite may then be stripped (for reasons and by processes not specified by these authors, but potential mechanisms could include climate change to a wetter, more humid period likely to cause higher rates of erosion) to expose the differentially weathered bedrock, resulting in the formation of relief in the landscape. Depending on the spacing of the bedrock fractures and joints, the exposure of the core-stones may result in the formation of landscape features as diverse as tors, inselbergs, angular turrets, or granite domes.

Seidl et al. (1996) and Weissel and Seidl (1997) focused on the mechanisms for propagation of the escarpment inland following the splitting of the Tasman Sea. They used a mass-balance approach to estimate denudation rates ranging from 0.04 m/m.y. to 12 m/m.y. east (downstream, or below) of the escarpment since the time of rifting. These rates were consistent with the minimum estimates of Dumitru et al. (1991), despite the rough constraints on their measurements and their assumption of initial topography, which was not substantiated. They also inferred escarpment retreat rates (Seidl et al., 1996) by dividing the average distance of many gorge heads from the rift margin (200 km) by the period since rifting (100 m.y.) to get a long-term average rate of 2 km/m.y. This rate agreed roughly with retreat rates cited from other passive continental margins (Ollier and Marker, 1985), though the same simple assumptions were used, and the reasoning was not mechanistic.

Weissel and Seidl's (1998) analyses of cosmogenic nuclide measurements from bedrock rivers determined bedrock incision rates of ~7 m/m.y. from upstream of the edge of the escarpment (which agrees with highland denudation rates of 1–10 m/m.y. for the highlands), and greater than 100 m/m.y. inferred from strath terraces downstream of the escarpment knickpoint. As they point out, their rates were consistent with their models of escarpment propagation across the coastal region with very low rates of erosion from the highland region (~4 m/m.y.), higher rates of bedrock channel incision near the upper edge of the escarpment than from upstream (~20 m/m.y. versus ~7 m/m.y.), and the highest erosion rates across the gorge head (~100 m/m.y.). Importantly, both studies (Seidl et al., 1996, and Weissel and Seidl, 1998) forward a process-based argument for the propagation of the escarpment. They suggested that as long as there is enough fluvial transport power in the streams, the propagation rate of the escarpment is primarily determined by the slope failure mechanisms and the mass wasting from the steep rock slopes at the escarpment head (also note modeling results of Tucker and Slingerland [1994], which suggested that weathering-limited conditions across the escarpment knickpoint set the retreat rate of the escarpment).

In contrast with the extensive research on the evolution and denudation of the southeastern highlands, there is a paucity of data addressing the erosional history of the coastal plain, and work that has been done appears fraught with equivocation. The notable exceptions were the low-temperature thermochronology studies that found a discrete period of rock cooling from rock samples along the coast (O'Sullivan et al., 1995, 1996; Persano et al., 2002). Perhaps one reason there have been few studies similar to those carried out on the highlands is that the retreat of the escarpment has eroded most of the material used in the earlier studies, so they must now use datable basalts to infer the evolution of the highlands from the coastal region. While the highly dissected region adjacent to and below the escarpment suggests high erosion rates, the actual rates have not been well quantified beyond a few recent studies (Heimsath et al., 2000; Persano et al., 2002; Weissel and Seidl, 1998).

Young (1977) did, however, report mantles of Tertiary sediment, duricrust, and basalt on the coastal lowlands at the foot of the escarpment. Young and McDougall (1982) reported further on this work and used datable rock to report K-Ar ages averaging ca. 30 Ma from basalt flows that postdate the silcrete formation on the lowland sites. They used the basalt ages in relation with the silcrete development to conclude that the coastal lowland resulted from erosion following an early Tertiary uplift of the highland. Similarly, based on terrestrial sedimentary deposits examined along the southeastern coast, Nott et al. (1991) concluded that the uplift of the highland region and the development of the coastal lowland occurred by the early Tertiary. They proposed that a late Mesozoic period of deep weathering was followed by an early Tertiary period of uplift and erosion. Subsequently, by the early Miocene (ca. 22 Ma), global sea-level rise may have led to a period of deposition of sediments on the present coastal plain. These aggraded sediments were exposed when the drop in sea level in the late Miocene to Pliocene (ca. 8 Ma) allowed the incision of streams into the coastal landscape. There is considerable dispute surrounding these lowland studies.

FIELD SITES

To quantify erosion rates and processes across the escarpment, we extended the field area that was used to quantify soil production rates with in situ–produced cosmogenic nuclides (^{10}Be and ^{26}Al; Heimsath et al., 2000; Figs. 1 and 2). Results from the Heimsath et al.'s (2000) study were later compared to soil production and erosion rates determined at a site on the highlands (Heimsath et al., 2001), but the highland site was roughly 60 km northwest of the escarpment base site and ~12 km west of the escarpment crest. For the study presented here, we directly connect the coastal lowland with the escarpment by sampling along ridge and river transects (Fig. 3), as well as extensive new sampling on the highlands at the escarpment lip and also on the coastal lowlands. Sampling at the escarpment lip and on the highlands near Brown Mountain was also motivated by the desire to couple our quantification of physical erosion rates with recent and ongoing work into chemical weathering in the same field region (Banfield and Eggleton, 1989; Taunton et al., 2000a, 2000b).

The Nunnock River is a tributary to the Bemboka-Bega River, which drains into the Pacific Ocean at Bega (Fig. 1). The granite cliffs of the escarpment, downslope from the low-gradient headwaters region, and a relatively steep drop off the escarpment

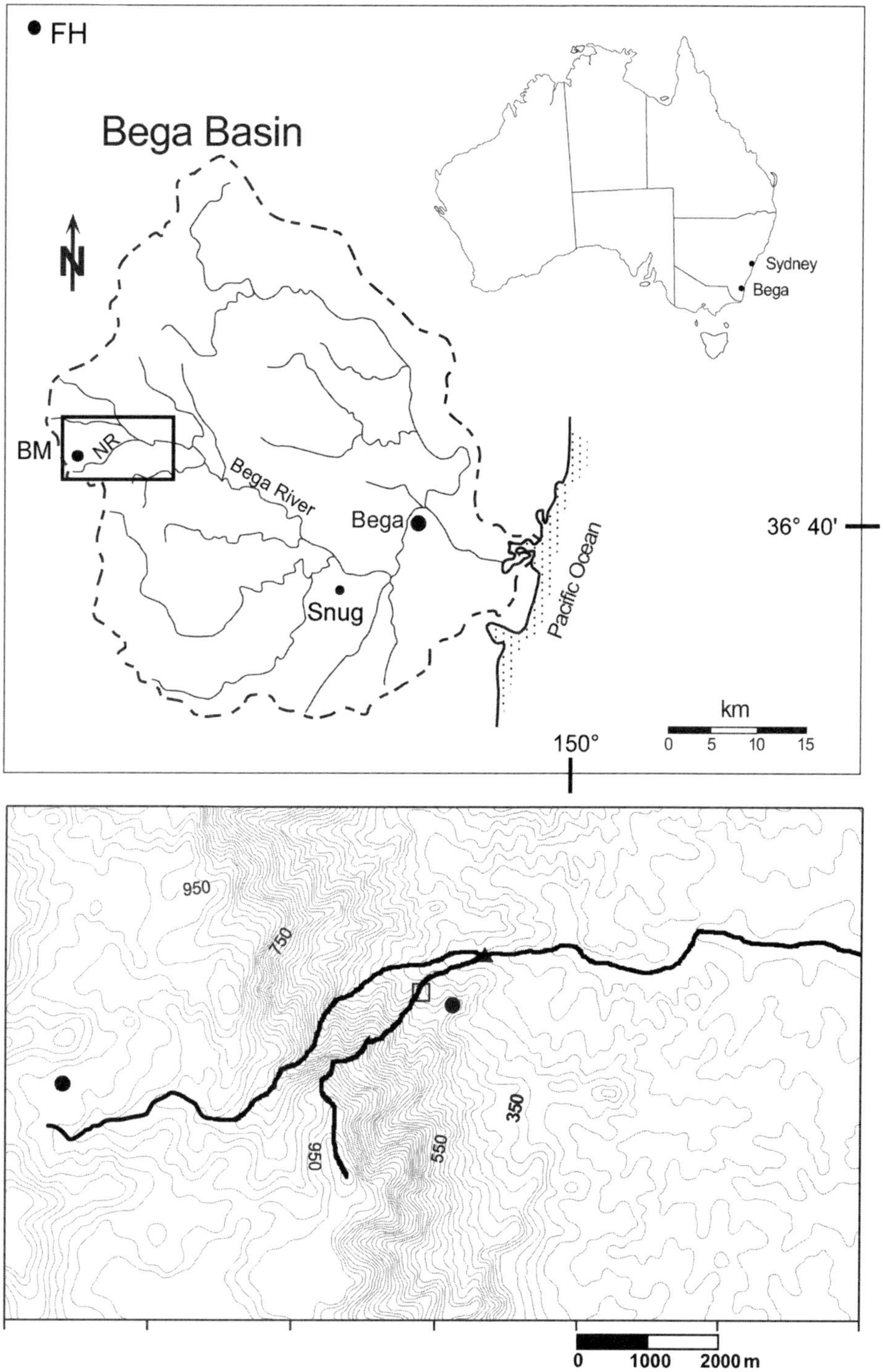

Figure 1. Location map showing the topography across the escarpment face. Outline of the Bega basin shown by dashed line, with Nunnock River shown by NR in the middle of the location rectangle. Topography of the lower panel is with 20 m contour intervals from 1:100,000 digital data (Australian Surveying and Land Information Group) from Bega (sheet 8824) map. Locations used for the Heimsath et al. (2000) study are shown by: black dot on ridge crest at base of escarpment is where Figure 2 photos were taken from; open square shows approximate location of the surveyed area of the Heimsath et al. (2000) study; and black triangle shows stream sediments from Nunnock River used in that study (modified from Heimsath et al., 2000). Approximate locations of the lowland (Snug) and two highland (Brown Mountain [BM] and Frogs Hollow [FH]) sites shown by black dots and labels on map, with the FH site ~20–30 km north of the shown region. Black lines across the escarpment topography show the profiles plotted in Figure 3A–B, where the channel line is extended farther here than in the profile data.

Figure 2. Photographs from the ridge crest tor location used to develop the tor erosion model presented in Heimsath et al. (2000). View is upslope showing the spur ridge used for elevation transect of this study (A), where the "V" visible on skyline is the incision of Nunnock River across the escarpment crest. Note tree cover of the escarpment and the relatively small region of exposed cliffs, directly above middle person's head, which is not visible from below. (B) View downslope to the Pacific Ocean, horizontal skyline roughly 50 km distant, shows drop of the higher-relief escarpment slopes to more gently sloping, lower-relief coastal lowlands.

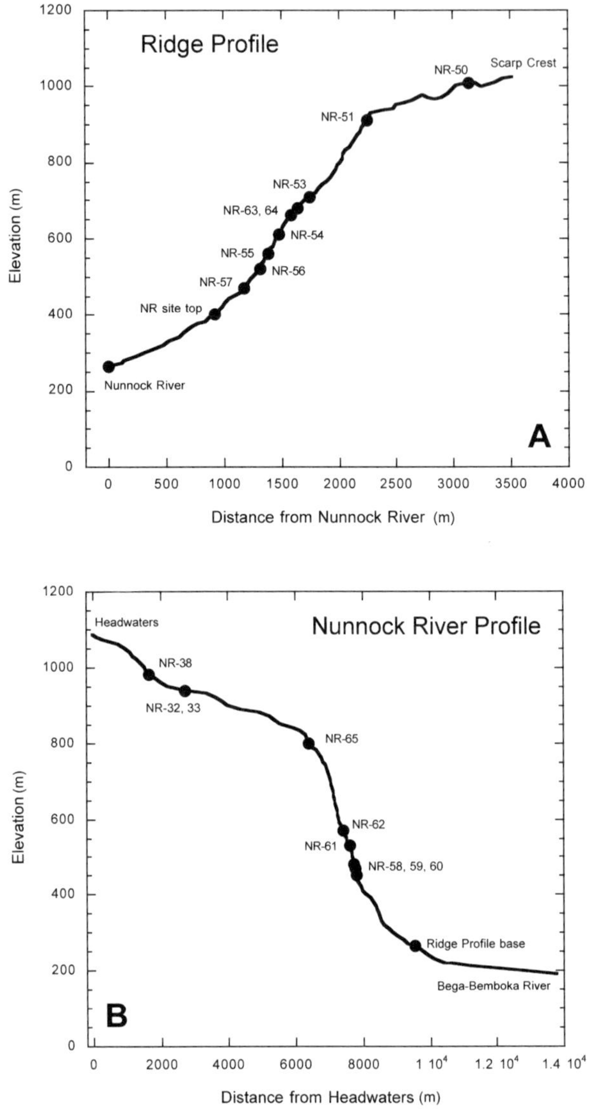

Figure 3. Elevation versus distance profiles for the spur ridge used to collect samples for this study (A), as well as the long profile of Nunnock River (NR) as it traverses the escarpment (B). Note vertical exaggeration for both figures and that the distance axes are not at the same scale for A and B. Data for both profiles collected by hand off the 1:20,000 regional maps using 10 m contour intervals. Ridge profile begins where the spur ridge intersects with Nunnock River, shows the top of the field site of Heimsath et al.'s (2000) study area, and marks sample elevations with the filled black circles. Corresponding sample ID numbers are noted adjacent the circles. Profile ends at the scarp crest. River profile begins in the low-gradient, swampy meadow headwaters of the Nunnock River at the farthest upslope extent of the channel. Catchment area at the channel head is approximately one hectare, and the landscape is soil-mantled, dotted with outcropping tors. Small catchment sediment sample average elevations are shown on the river profile with filled black circles and are marked with the sample ID numbers used in Table 1. Nunnock River changes from sand-bedded, relatively incised channel across the low-gradient highlands to a steeply dropping, bedrock river across the escarpment with obvious fluting and fluvial incision channels. Large boulder deposits are observed beginning near the escarpment base, but are not present in the Bega-Bemboka River.

define the drainage area of Nunnock River. Evidence of debris-flow deposits are found in each of the second-order drainages downslope from the escarpment, though scour to bedrock on the escarpment face is relatively limited. The exposed cliff region across the top of the escarpment is not as dramatic or extensive as those from the Weissel and Seidl (1998) study area, but they are eroding by block failure across their face and thin exfoliation sheet erosion across their tops. There is little sediment accumulation beneath the cliffs, and it appears that the upper region of the escarpment is weathering-limited, as are regions of the channels (the channels are bedrock with the exception of patchy debris-flow cover). Spur ridges are soil-mantled and perpendicular to the escarpment. Slopes are sharply convex across the tops of the regions and steepen downslope to the channel (Heimsath et al., 2000). Between Nunnock River and the coastline, the landscape is composed of rolling, granodiorite hills with very few outcropping tors in comparison with the highland region. The granodiorites are Late Silurian to Early Devonian and are part of the Bega Batholith Igneous Suite (Lewis and Glen, 1995). Relief is <200 m across the coastal lowland and increases from the coast inland to the escarpment, which rises abruptly for ~800 m directly west of the scarp base field site.

Rainfall on the coastal lowlands (~900 mm/yr), and especially near the escarpment, is considerably higher than on the tablelands, though it does fall sporadically throughout the year (Richardson, 1976). Rainfall intensities are reportedly higher near the escarpment (personal commun. with landowners), though this has not been quantified beyond the past 10 yr. Using stations of the Australian Bureau of Meteorology (www.bom.gov.au/climate/averages) closest to the field sites, the average rainfall ranges from 870 mm/yr for the coastal lowland site (Snug, at ~200 m elevation), to over 1200 mm/yr at the escarpment base (Nunnock River, at 400 m), to 690 mm/yr at the escarpment crest (Brown Mountain, at 1000 m), to 720 mm/yr at the highland site (Frogs Hollow, at 900 m). Observed vegetation at the escarpment base reflects this higher rainfall compared with the relatively dry and cold tablelands region, and forest cover is intermediate to wet schlerophyll forest with patches of dense understory, especially in the valley bottoms. Ferns grow widely on the more humid slopes around the Nunnock River field area, but are not observed on the slopes nearer the coast at the Snug site.

The field area of Heimsath et al. (2001) is representative of the low-relief plateau of the highlands above the escarpment, but is northwest of the Nunnock River portion of the escarpment leading to the coastal lowlands focused on here, motivating sampling another highland area around Brown Mountain. This highland site is the same region studied by Banfield and Eggleton (1989) and Taunton et al. (2000a, 2000b). The highlands, in general, are between ~800 and 1500 m above sea level. Bedrock varies between Ordovician metasediments and Devonian granites across the region (Richardson, 1976). Morphology of the highlands directly above the Nunnock River site (i.e., Brown Mountain) is similar and is characterized by low-relief hillslopes (<100 m) that are gently sloping (<25°) and are soil-mantled, but punctuated frequently by tors. These tors are typical examples of the more coherent bedrock that outcrops through a saprolitic granite landscape (e.g., Linton, 1955; Twidale, 1985; Williams et al., 1986). The weathered nature of the soil-mantled saprolite is exposed in numerous road cuts around the field areas. These road cuts also reveal the prevalence of core-stones, some fraction of which are likely to emerge sometime in the future as tors.

The climate of the southeastern highlands is relatively cool, averaging ~18–22 °C in the summer and 3–5 °C in the winter. There are periods of freezing temperatures and snow in the winter, but these rarely last long. Rainfall is not distinctly seasonal and, with the exception of drought periods, does not vary much from ~550–750 mm/yr (McAlpine and Yapp, 1969; Richardson, 1976). Forest cover, where it occurs, is typically dry schlerophyll, with very little understory. Grassland and rolling savanna woodland vegetation are also widespread across this region of the tablelands. Climate changes during the Pleistocene and during the Pleistocene–Holocene transition (Bryant et al., 1994; Chappell, 1991; Dodson, 1977, 1987; Dodson and Wright, 1989; Harrison, 1993; Kershaw et al., 1991) probably affected the biotic communities and therefore the dominant soil production and transport processes. These processes were likely similar to the Nunnock River site since the biota are similar, but the drier and colder climate may reduce the frequency of the bioturbation at the highland sites. Potential impacts of historic agriculture were avoided by selecting protected areas for our work.

Soil production and transport is observed to be primarily due to biogenic processes across all field areas, although there is some evidence of overland flow, especially during storms following fires that remove any vegetative understory. Burrowing wombats were the most obviously disruptive biota and were observed to occupy all parts of the landscape except where soil thickness was less than ~30 cm, and their burrows extend into the weathered bedrock in places. Other observed biogenic processes included burrowing echidnas and shallow burrowing lyre-birds, worms (worm trails extended into the saprolite and worm castings in the soil included quartz crystals that could have been conveyed up from the underlying saprolite), and tree throw (Fig. 4). Ant or termite mounds revealed significant quartz grains and pieces of unweathered granodiorite, but we did not observe ant burrows extending into the saprolite; therefore, we infer that such microfauna contribute to transport rather than production of the soil.

METHODS

In situ–produced cosmogenic nuclides (e.g., ^{10}Be and ^{26}Al) are used increasingly to quantify bedrock erosion (e.g., Lal, 1991; Nishiizumi et al., 1991), soil production (e.g., Heimsath et al., 1997; Small et al., 1999), and spatially averaged rates of erosion (e.g., Bierman and Steig, 1996; Brown et al., 1995; Granger et al., 1996). Concentrations of these nuclides accumulate in materials at or near Earth's surface as cosmic rays bombard atoms, such as Si and O in quartz and other minerals in rock and sediments (Lal and Arnold, 1985; Lal and Peters, 1967). Cosmic-ray production of nuclides is balanced by radioactive decay of the nuclides in surfaces that are not eroding and by removal of the target material in eroding surfaces. Application of cosmogenic nuclide analysis to understanding landscapes is extensively and well-reviewed (Bierman, 1994; Cockburn and Summerfield, 2004; Gosse and Phillips, 2001; Lal, 1988; Nishiizumi et al., 1993) and will only be summarized here.

Nuclide concentrations measured from quartz have been shown to reflect the exposure history of the sample and are dependent on the production and decay rates of the nuclide, as well as

Figure 4. Photos showing the more dramatic of the biogenic processes acting across each of the field areas. Tree throw of the *Eucalyptus* sp. is common and typically shows uprooted granitic saprolite clinging to the roots (A). Here the fan-shaped, light-colored deposit from the uprooted tree was observed to be primarily quartz grains from the saprolite, which were penetrated by the roots prior to uprooting. Wombat burrows (B) commonly penetrate the saprolite, such that the downslope deposits from the burrows are mixtures of saprolite and soil, similar to the observations of gopher burrowing by Gabet (2000). Tape measure for scale shows burrow diameter to be ~30 cm, with the trowel handle at the approximate location of the soil-saprolite interface.

the erosion rate of the sample (Lal, 1988; Lal and Arnold, 1985; Nishiizumi et al., 1986, 1991). As shown by Lal (1991), and neglecting muon effects, the nuclide concentration, C, depends on the production of the nuclide at an exponential depth, z_x, function of the surface production rate, P_0, and its radioactive decay constant, λ,

$$\frac{dC}{dt} = P_0 e^{-\mu z_x} - \lambda C, \tag{1}$$

where μ is the absorption coefficient (equal to material density divided by the mean attenuation length for cosmic rays, Λ, where Λ ~165 g cm^{-2}), and $\lambda = \ln 2/t_{1/2}$, where $t_{1/2} = 1.5 \times 10^6$ yr for ^{10}Be and $t_{1/2} = 7.01 \times 10^5$ yr for ^{26}Al). If erosion, ε, is assumed, based on geomorphic examination, to be constant over the exposure history of the sample, equation 1 can be solved analytically (Lal, 1991), yielding,

$$C = c_0 e^{-\lambda t} + P(H,\theta)\left(\frac{1}{\lambda + \mu\varepsilon}\right) \tag{2}$$

at secular equilibrium, when $t >> (\lambda + \mu\varepsilon)^{-1}$, where $P(H, \theta)$ is the nuclide production rate (atom g^{-1}yr^{-1}) at the soil-bedrock boundary under slope-normal soil depth, H, and on a surface with slope θ, and C_0 is the initial concentration of the nuclide. $P(H, \theta)$ is calculated as a factor of the surface nuclide production rates (e.g., Dunne et al., 1999). Under steady-state conditions, where local soil thickness is relatively constant with time, soil production, ε_s, equals erosion, ε, at the surface, which is also equal to the rate of lowering of the soil-saprolite interface (Heimsath et al., 1997) and can therefore be equated to the local denudation rate. Equation 2 can therefore be solved to yield steady-state soil production or erosion rates. Nuclide concentration accumulated in river or catchment sediments can be interpreted by equation 2 to infer catchment-averaged steady-state erosion rates, assuming relatively short transport times and relatively homogeneous catchment lithology (e.g., Granger et al., 1996; Bierman and Steig, 1996).

We apply this methodology to determine erosion rates in three different ways. Most simply, we sample the outermost centimeter or two of exposed bedrock (tor tops as well as the top of the cliff region of the escarpment) that shows field evidence of eroding by grain-grain spallation or by exfoliation of thin sheets. Concentrations of ^{10}Be and ^{26}Al in these samples yield the inferred steady-state erosion rate of the exposed bedrock surfaces, assuming that samples have had simple exposure histories (e.g., no significant burial or shielding that is now not evident) and have been eroding at nearly constant rates (i.e., not by large block failure or landsliding). For soil production rate determination, we followed the typical geomorphic transect from bedrock exposed at convex ridge crests onto the soil-mantled part of the landscape and dug soil pits to sample the top centimeter or two of saprolite or weathered bedrock directly beneath different depths of the relatively thin (<1 m) soil cover. Concentrations of nuclides from these samples were used with a corrected nuclide production rate that accounts for the shielding of the soil mantle above the sampled layer to yield a soil production rate, which is equivalent to a landscape-lowering rate relative to a local datum (Heimsath et al., 1999). Finally, we collected stream and creek sediments or sediment from zero-order catchments, measured the concentrations of the nuclides, and used the nuclide concentrations to infer a catchment-averaged erosion rate (e.g., Granger et al., 1996). Each of these applications depends on careful geomorphic site selection, such that the numerous assumptions and simplifications detailed in the above studies are met.

Roughly 500 g of bedrock, saprolite, or sediment from each sample were crushed, sieved to separate particles <1 mm size, and chemically purified following the procedure of Kohl and Nishizumi (1992) to yield ~50 g of quartz, from which Be and Al were extracted. For results reported here we only measure concentrations of ^{10}Be, because sample lithology and exposure history were similar to those reported in Heimsath et al. (2000, 2001) and did not warrant the test for steady-state exposure history enabled by measuring ^{26}Al (e.g., Nishiizumi et al., 1993). Samples for ^{10}Be measurement were spiked with a ^{9}Be carrier solution calibrated by a Be atomic absorption standard that differed by <2% from the Be carrier used for the Nishiizumi et al. (1989) analyses. Preparation and chemistry on these samples were done at Dartmouth College as well as at the Research School of Earth Sciences at the Australian National University (ANU). We measured concentrations of ^{10}Be at the Lawrence Livermore National Laboratory, Center for Accelerator Mass Spectrometry (LLNL-CAMS) facility (Davis et al., 1990) as well as at the ANU Nuclear Physics facility and Purdue University's PRIME Lab, and standardized the measurements to the ICN (ICN Biomedical, Inc.) ^{10}Be standards. By doing the chemistry on these samples in two very different labs and measuring the nuclide concentrations on three very different accelerator mass spectrometers, we also have an excellent test for the potential analytical uncertainties in the cosmogenic nuclide methodology used for geomorphic studies.

Production rates for ^{10}Be and ^{26}Al in quartz were based on the sea-level and high-latitude production rates of 6 and 36.8 atoms g^{-1} yr^{-1} (Nishiizumi et al., 1989), rescaled to 5.1 and 31.1 atoms g^{-1} yr^{-1}, respectively (Stone, 2000), and were corrected for latitude and altitude effects (Lal, 1991), as well as for the slope and shielding of the sample (Dunne et al., 1999; Masarik and Vanya, 2000; Nishiizumi et al., 1989). These production rates were used for consistency with previous reports of erosion rates or exposure ages, despite the growing debate over production rates (Clark et al., 1995; Dunai, 2000; Nishiizumi et al., 1996; Stone et al., 1998a) and the potential contribution of muons to nuclide concentrations under moderate and high erosion rates (e.g., Granger and Smith, 2000; Stone et al., 1998b).

RESULTS AND DISCUSSION

The Highlands

Results from the new nuclide analyses are reported here for ^{10}Be concentrations, with the erosion rates or soil production rates determined from them (Table 1). Table 1 does not distinguish between samples prepared at different laboratories or measurements made at different accelerators. The results presented below incorporate these differences and agree so well that we believe the soundness of our methodology is maintained. Seven samples from directly beneath the actively eroding soil mantle define a soil production function for the highland landscape at Brown Mountain (Fig. 5A). Two samples from bedrock exposed at the soil surface yield steady-state erosion rates and suggest that maximum soil production rates occur under some finite soil mantle (Ahnert, 1987; Carson and Kirkby, 1972; Dietrich et al., 1995; Gilbert, 1877), such that the soil production function could have a "humped" form (Cox, 1980). Measured nuclide concentrations were corrected for observed soil depths (Heimsath et al., 1999), and steady-state erosion rates, calculated with equation 2, define an exponential decline of apparent soil production rates with increasing soil depths for the saprolite samples. The variance-weighted best fit to these points, excluding the bedrock points, is

$$\varepsilon(H) = 65 \pm 15e^{-(0.020\pm0.003)H}, \tag{3}$$

where the soil production rate, $\varepsilon(H)$, is the vertical lowering rate in Table 1 in m/m.y., and H is in cm. Comparison of these data with the soil production and exposed bedrock erosion rates reported for the highland site of Heimsath et al. (2001) shows remarkable similarity in rates and form despite the sites being nearly 60 km apart and the samples being collected and analyzed about four years later in a different laboratory (Heimsath et al., 2001, analyses were done at the University of California, Berkeley; Fig. 5B). The apparent soil production function reported by Heimsath et al. (2001) was fit to five samples using both ^{10}Be and ^{26}Al from beneath soil thicknesses from 25 to 65 cm and had a dramatically steeper slope to it, such that the function defined an intercept that was unreasonably high for any rates measured across the highlands:

$$\varepsilon(H) = 143 \pm 20e^{-(0.042\pm0.003)H}. \tag{4}$$

Arguably, the combined soil production data from the highland sites (Fig. 5B) would define a humped soil production function, as first articulated by Carson and Kirkby (1972) and also discussed by Cox (1980). This is discussed below, after presentation of all the soil production data.

Similarities in morphology between the two highlands sites suggest that the conclusions reached by Heimsath et al. (2001) are applicable to the escarpment crest region. Specifically, that extensive outcropping tors are found at both sites and that the weathering profiles visible in the road cuts, and analyzed at a site very close to our Brown Mountain site by Banfield and Eggleton (1989) and Taunton et al. (2000a, 2000b), support the notion of a propagating weathering front into the relatively stable highlands. Their measurements suggest considerable erosion by solute processes, while our cosmogenic nuclide–determined lowering rates are recording a total lowering rate with no differentiation between chemical and physical weathering processes. It is clear from the weathering profiles that solute processes are significant, and it is also clear from the active bioturbation across the landscape that physical soil production and transport are significant. At this point, we cannot resolve the relative roles that both play, though it is clear that they are coupled as Riebe et al. (2001, 2003) found elsewhere, and we are examining the problem with a detailed look at how the extent and rate of weathering varies across the landscape (Burke and Heimsath, 2003).

Average erosion rates from ^{10}Be concentrations from stream sediments taken from small catchments draining to Nunnock River above the escarpment range from 13 ± 1.7 to 20 ± 2.5 m/m.y.

TABLE 1. EROSION RATES FROM COSMOGENIC NUCLIDE CONCENTRATIONS

Sample	Depth† (cm)	Type‡	PROD. factor§	Elev. (m)	Wt. (g)	Be (mg)	^{10}Be (10^6 atoms/g)	±	ε (m/m.y.)	±	% error
Cosmogenic nuclide concentrations from small catchment sediments											
NR-32		Avg	0.9	940	20.89	0.451	0.258	0.012	15.47	1.50	9.7
NR-33		Avg	0.9	940	19.40	0.447	0.251	0.012	15.88	1.55	9.8
NR-38		Avg	1.0	980	21.04	0.434	0.307	0.023	13.43	1.69	12.6
NR-58		Avg	0.9	450	48.79	0.455	0.153	0.011	27.79	3.42	12.3
NR-60		Avg	0.6	478	51.30	0.524	0.082	0.004	29.20	3.04	10.4
NR-61		Avg	0.6	525	47.69	0.542	0.064	0.007	35.87	5.64	15.7
NR-62		Avg	1.2	570	25.89	0.513	0.175	0.011	27.12	3.02	11.1
NR-65		Avg	0.4	800	48.86	0.489	0.092	0.007	20.13	2.51	12.5
Cosmogenic nuclide concentrations from brown mt. saprolites											
NR-34	60	BM	1.5	950	19.80	0.437	0.354	0.056	18.08	4.68	25.9
NR-36	0	BM	1.2	950	22.12	0.445	0.213	0.019	22.21	3.11	14.0
NR-37	25	BM	1.4	965	20.30	0.441	0.186	0.021	31.65	6.77	21.4
NR-39	55	BM	1.0	970	22.38	0.442	0.250	0.020	16.70	2.99	17.9
NR-41	50	BM	0.9	1033	21.24	0.435	0.149	0.018	26.59	5.87	22.1
NR-42	100	BM	0.4	1033	20.33	0.436	0.192	0.016	6.23	1.15	18.4
NR-43	70	BM	0.7	1066	21.69	0.435	0.195	0.024	16.05	3.58	22.3
NR-45	73	BM	0.8	1091	21.32	0.438	0.519	0.052	6.79	1.37	20.1
NR-50		Rdg	1.3	1015	23.70	0.443	0.400	0.017	12.44	1.15	9.2
Cosmogenic nuclide concentrations from ridge crest samples											
NR-51		Rdg	1.0	912	20.98	0.443	0.241	0.012	16.20	1.58	9.8
NR-53		Rdg	1.4	710	22.12	0.445	0.212	0.010	20.84	2.01	9.6
NR-54		Rdg	0.7	606	20.30	0.441	0.078	0.012	32.90	6.60	20.0
NR-55		Rdg	0.6	566	21.04	0.434	0.085	0.005	31.00	3.50	11.3
NR-56		Rdg	0.7	517	22.38	0.442	0.089	0.006	33.56	3.96	11.8
NR-57		Rdg	1.1	468	19.79	0.440	0.110	0.009	36.27	4.76	13.1
NR-59		Rdg	1.1	470	34.68	0.530	0.108	0.011	32.67	5.01	15.3
NR-63		Rdg	1.4	660	49.21	0.508	0.206	0.009	25.33	2.43	9.6
NR-64		Rdg	1.3	680	48.82	0.499	0.210	0.011	19.95	2.03	10.2
Cosmogenic nuclide concentrations from tor samples											
NR-30		Tor	2.0	1125	23.70	0.443	1.963	0.071	3.26	0.28	8.6
NR-31		Tor	2.0	1125	20.98	0.443	1.692	0.064	3.78	0.33	8.8
NR-35		Tor	1.8	950	23.90	0.442	0.396	0.018	14.19	1.35	9.5
NR-40		Tor	1.9	1033	19.79	0.440	0.317	0.021	18.80	2.16	11.5
NR-44		Tor	1.9	1071	20.46	0.438	1.470	0.072	4.18	0.41	9.9
NR-52		Tor	1.5	819	23.90	0.442	0.245	0.010	20.01	1.83	9.1
Cosmogenic nuclide concentrations from lowland site, SNUG											
NR-101	85	Snug	1.2	212	31.88	0.453	0.090	0.020	11.13	2.66	23.9
NR-102	27	Snug	1.1	214	31.15	0.456	0.1	0.01	24.51	2.50	10.2
NR-103	70	Snug	1.1	215	30.24	0.391	0.1	0.1	9.75	1.30	13.3
NR-104	90	Snug	1.1	216	31.28	0.325	0.13	0.02	9.48	1.53	16.2
NR-105	70	Snug	1.2	217	30.61	0.370	0.21	0.04	4.19	0.88	21.0
NR-106	45	Snug	1.1	220	30.94	0.456	0.16	0.01	11.43	0.73	6.4
NR-107	38	Snug	1.2	221	30.20	0.454	0.12	0.01	17.27	1.47	8.5
NR-108	60	Snug	1.2	222	29.63	0.359	0.13	0.01	11.30	0.89	7.9
NR-109	50	Snug	1.1	223	30.67	0.406	0.03	0.02	57.38	6.92	12.1
NR-110	70	Snug	1.2	227	29.68	0.368	0.177	0.027	7.01	1.13	16.2
NR-111	0	Snug	1.1	228	29.97	0.377	0.421	0.081	8.72	1.79	20.6

Note: All errors are propagated to soil production or erosion rate, ε.

†Average soil density: 1.2 g/cm^2.

‡Avg are catchment averaged, Rdg are Spur ridge crests.

§Production factor scales nuclide production for slope, soil depth, elevation, and latitude. Location of site at escarpment base: 36.62 S Lat, 149.5 E Long. ^{26}Al and ^{10}Be concentrations scaled from sea-level production rates of 36.8 and 6 atoms g^{-1} yr^{-1}.

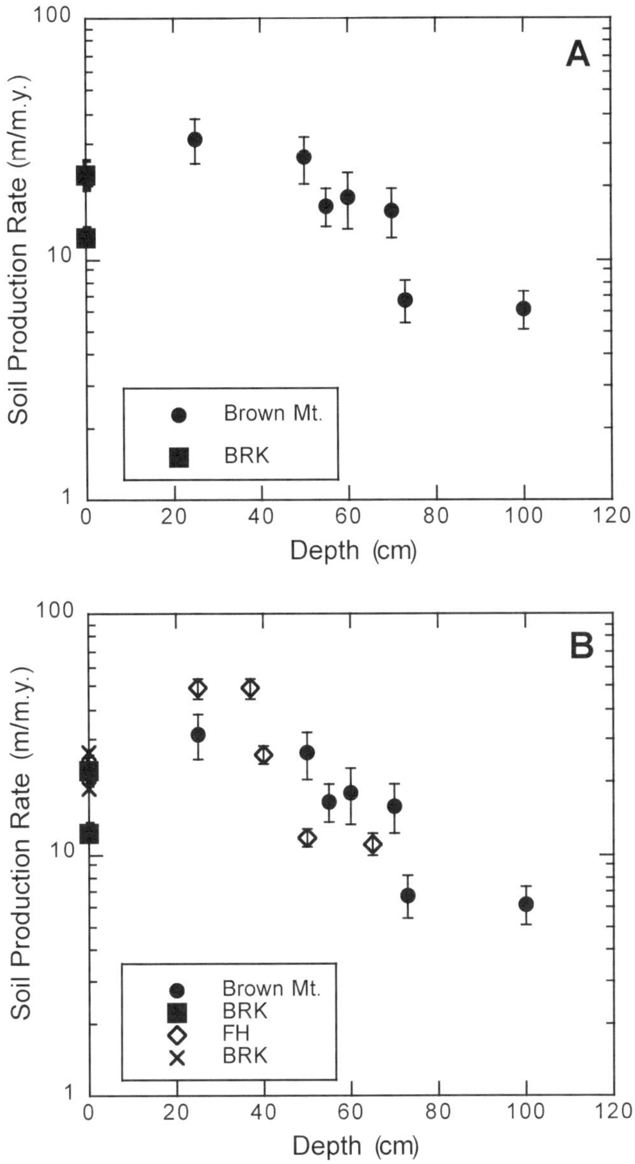

Figure 5. Soil production (m/m.y.) plotted as a function of overlying soil depth (cm). Solid black circles in A are rates derived from ^{10}Be concentrations (Table 1) in samples from the highland site near Brown Mountain (BM), and error bars represent all errors propagated through the nuclide calculations, i.e., uncertainty in Be carrier concentrations, accelerator mass spectrometry, bulk densities, and soil depth measurements, and attenuation length of cosmic rays. See text equation 3 for variance-weighted best fit to the soil production data. Black squares plotted at zero depth are erosion rates of bedrock (BRK) exposed at ground surface. Comparison with the soil production and bedrock erosion rates reported in Heimsath et al. (2001) from both ^{10}Be and ^{26}Al is shown in B, where the open diamonds are soil production rates and crosses at zero depth are exposed bedrock samples at the ground surface from that site, Frogs Hollow (FH). See text equation 4 for the soil production function from Frogs Hollow.

(Table 1; Fig. 3B). These rates are at the low end of the escarpment erosion rates discussed below, but agree well the highland soil production rates shown in Figures 5A and 5B. This is encouraging support for the use of small catchment sediments to quantify average erosion rates for a landscape, as the processes of soil production and transport are the dominant geomorphic forces shaping the highlands. Landslides and extensive gullying are only observed where there is a clear link between overgrazing by livestock and local degradation of the land surface following increases in rainfall. Tor erosion rates are slower, as discussed in Heimsath et al. (2000, 2001), and by quantifying the sediment budget for the catchment, the interpretation of the nuclide concentrations in the sediment samples reflects well the proportional contributions of tor erosion and soil production from the saprolite (Granger et al., 2001; Heimsath et al., 2001). The overlap between average and point specific rates supports the suggestion that the highlands may have reached a steady-state since emerging from the last glacial cycle, when periglacial processes were likely to be dominant, a suggestion that was raised in the "Departures from Steady State: (II) Soil Production" section of Heimsath et al. (2001, p. 180–181).

The Lowlands

Ten samples from the saprolite-soil boundary at the coastal lowland site (Snug) define soil production rates for soil depths between 27 and 90 cm (Fig. 6A). No intermediate soil depths were observed, and the sample from bedrock exposed at the soil surface led to a lower soil production rate than the sample from beneath the shallowest soil, a finding similar to the highland sites. Analyses of the nuclide concentrations followed procedures outlined above, correcting for overlying soil thickness to infer steady-state soil production rates. Using the saprolite soil production rates to determine a variance-weighted best fit for the soil production function, we determine that,

$$\varepsilon(H) = 42 \pm 5e^{-(0.020 \pm 0.002)H}, \tag{5}$$

where the variables are the same as defined previously. With the exception of sample NR-109, which determined an especially high soil production rate and is likely to be an outlier due to local disturbance of the long-term average soil thickness (i.e., if soil thickness had been locally stripped, due to anthropogenic activity or local burrowing, then the long-term nuclide production rates would not be correctly scaled for depth and the inferred soil production rate would be anomalously high), the Snug data complement and extend the function defined at the base of the escarpment (Nunnock River) by Heimsath et al. (2000). The notable difference between the Snug and Nunnock River sites was the absence of intermediate soil thicknesses, and further investigation is required to decided whether this is true across the coastal lowland region, as we observe it to be for the highlands, or if there is the potential for human impact that may have left a plow layer. A plow layer is a surficial layer of soils with near-uniform thickness that would be present if any plowing of the site had homogenized soils down to a finite depth, typically ~30 cm. The soil production function defined from Nunnock River, at the base of the escarpment, was from both ^{10}Be and ^{26}Al concentrations measured in 15 samples of saprolite and bedrock and showed the following very well-defined exponential decline of soil production rates with increasing soil thicknesses (Heimsath et al., 2000):

$$\varepsilon(H) = 53 \pm 3e^{-(0.020\pm0.001)H}. \quad (6)$$

Plotting the two data sets together shows nearly identical relationships despite the offset of the intercept between the two functions (Fig. 6B). This overlap is surprising, as the coastal lowlands have lower relief and are more gently sloping than the soil-mantled slopes adjacent to the escarpment base. Conventional wisdom would suggest that the lowlands and highlands would be eroding more slowly than the escarpment base, yet our cosmogenic nuclide data suggest otherwise.

If we plot soil production rates from all four of the field sites from lowland to highland, the overlap becomes especially clear (Fig. 7). Indeed, using all these data, again with the exception of the exposed bedrock samples that are eroding more slowly (discussed below), to determine a variance-weighted soil production function tightens the relationship of equation 6 such that

$$\varepsilon(H) = 53 \pm 2e^{-(0.022\pm0.001)H}. \quad (7)$$

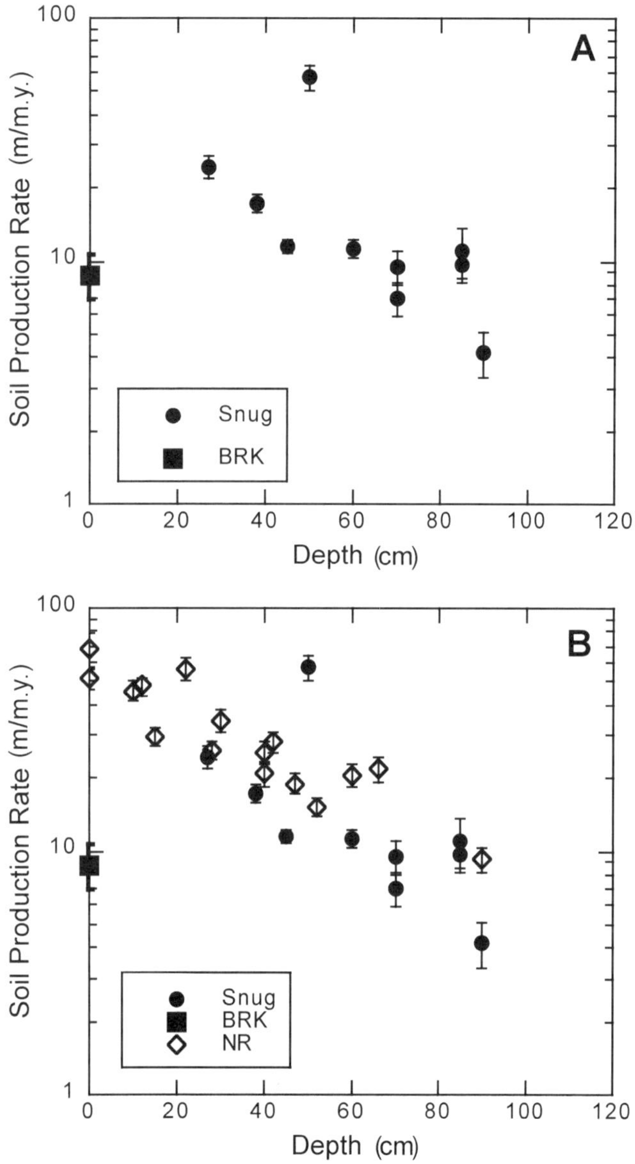

Figure 6. Soil production (m/m.y.), plotted as a function of overlying soil depth (cm). Solid black circles in A are rates derived from ^{10}Be concentrations (Table 1) in samples from the lowland site (Snug) between the Nunnock River (NR) site of Heimsath et al. (2000) and the town of Bega. Comparison with the soil production function reported in Heimsath et al. (2000) from both ^{10}Be and ^{26}Al is shown in B, where the open diamonds are from Nunnock River. Calculation details are the same as above; see text equation 5 for the variance-weighted best fit to the soil production data of A, while the fit to the Nunnock River data is in the text as equation 6. BRK—bedrock.

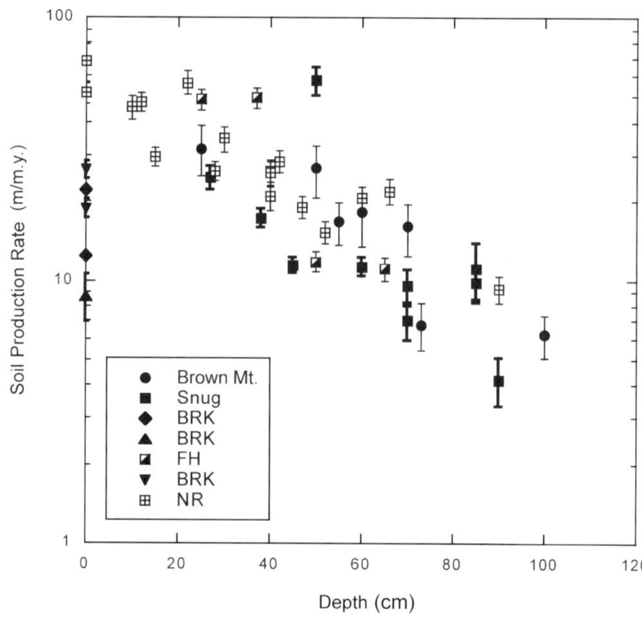

Figure 7. Soil production rates plotted for all four field sites discussed here. The two new sites, Brown Mountain and Snug, are the same data as Figures 5A and 6A, respectively, while the data from the Heimsath et al. (2000, 2001) studies are the same as plotted with Brown Mountain and Snug data in Figures 5B and 6B, respectively. Erosion rates of bedrock exposed (BRK) at the soil surface are plotted for all sites with different symbols. Variance-weighted best fit to all the soil production data plotted here is reported in the text as equation 7. NR—Nunnock River; FH— Frogs Hollow.

That landscapes with such different climates and tectonic histories show such similar soil production functions is surprising, but may support the notion of a rapid achievement of steady-state to current conditions. If biota are the dominant force in producing and transporting soil, then the differences between the sites are not dramatic. All sites have burrowing organisms, trees, and receive roughly similar rainfall rates, though there is a nearly twofold decrease in precipitation with elevation, as documented in the site descriptions. Frost conditions have not been quantified, but are likely to be minimal on the highland sites, as temperatures below freezing are not common.

That exposed bedrock at the highland and lowland sites appears to be eroding more slowly than the maximum soil production rates determined under relatively thin depths of soil raises an important distinction between soil production functions (Carson and Kirkby, 1972; Cox, 1980; Dietrich et al., 1995). Specifically, if lithology is uniform beneath the actively eroding soil layer, then lower rates of erosion for exposed bedrock document the hypothesis articulated by Gilbert (1877) from his observations that weathering processes attributed to frost and solute processes would occur most rapidly beneath a thin soil cover. A thin soil cover would enable moisture retention at the soil-rock interface, such that chemical and physical weathering processes could occur. Dietrich et al.'s (1995) modeling yielded an unreasonable modeling prediction of the spatial variation of soil thickness using such a humped soil production function. Specifically, they used both humped and exponentially declining soil production functions in a model predicting spatial patterns of soil thickness at the same soil-mantled field site in northern California, which was used subsequently by Heimsath et al. (1997, 1999). Using the humped function predicted widespread bedrock exposure not observed in the field, while using the exponential decline predicted soil depths consistent with observations. Carson and Kirkby (1972) explained that having maximum soil production rates beneath a finite soil thickness would lead to landscapes with either exposed bedrock, or soil thicknesses greater than the finite depth, with no intermediate soil depths. This suggestion was documented for the highland site of Heimsath et al. (2001), where modeling spatial variation of soil depth led to a better prediction of soil thickness when the exponential decline function was used, and Small et al. (1999) presented similar findings for an alpine landscape. Ahnert's (1987) modeling presented a different perspective, which was that variations in the underlying bedrock could lead to soil depth variations across a hypothetical landscape eroding at the same rate everywhere. Importantly, variations in lithology documented at the highland site may help to explain why the soil production functions from the highlands and lowlands are different from those presented from the escarpment base and northern California (Heimsath et al., 1997, 1999), emphasizing the importance of knowing the chemical weathering story by quantifying both the extents and rates of chemical weathering across all sites following, for example, methods of Riebe et al. (2001, 2003).

Across the Escarpment

Eight samples define catchment-averaged erosion rates from small, first- and zero-ordered streams draining the escarpment to Nunnock River, while ten samples from a main spur ridge define point-specific erosion rates for an altitudinal transect along a main spur ridge from coastal lowland to the Brown Mountain site (BM; Table 1). We include tops of three granite tors from the escarpment crest, near Brown Mountain, as well as three samples from the tops of outcropping granite faces along the steep escarpment face, where the only cliff exposures are found for this region of the escarpment (plotted as Tor Top). Together these 24 samples, along with average soil production rates from the scarp base site and the lowland site (plotted as separate ridge crest points), yield a striking trend of erosion rates as a function of elevation (Fig. 8), from 3.26 ± 0.28 m/m.y. at the scarp crest to an apparent peak of 36.3 ± 4.8 m/m.y. near the scarp base and declining to 24.5 ± 2.5 m/m.y. across the coastal lowlands. Local agreement between the rates determined for different processes

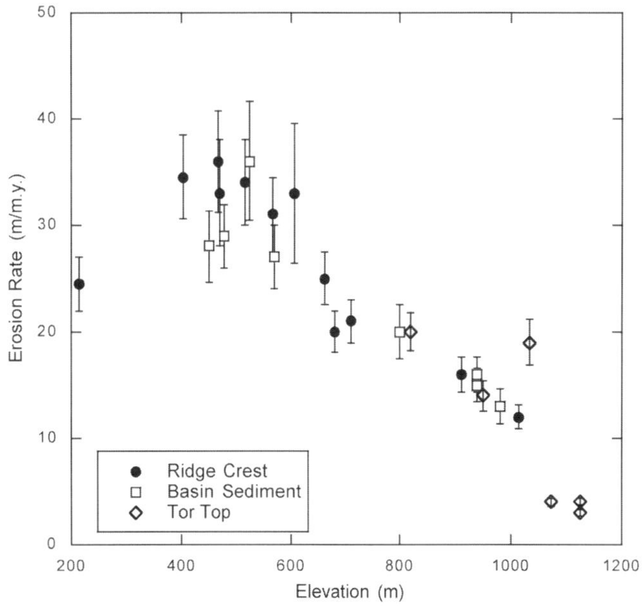

Figure 8. Erosion rate data from the ridge crest (filled black circles), small catchment averages (open squares), and tor tops (open diamonds) plotted as a function of elevation across the escarpment face. For most of the samples, the ridge crest and small catchment samples were adjacent and relatively nearby (within one to two hundred meters) for similar elevations. Ridge crest samples were from a single spur ridge leading directly up the escarpment from the study area of Heimsath et al. (2000), except for the NR-50 sample at 1015 m, which was from the ridge crest draining to the highest elevation small catchment sample, NR-38, with an average elevation of 980 m. The three tor tops from the highest elevation are from ridge crests draining to and near the headwaters of Nunnock River, while the tor top at 1033 m is from the top of a small granite cliff near the escarpment lip. Average values for the soil production rates at the scarp base (NR) and the coastal lowland (Snug) sites are included as additional two ridge crest samples.

suggests local steady-state between soil production, bedrock erosion, and small channel incision.

There are at least two mechanistic reasons to predict this order of magnitude decline in erosion rates, which are not entirely intuitive, because conventional wisdom might suggest increases in erosion rate with elevation. Firstly, observations that precipitation decreases with elevation across this escarpment would help explain the decline in erosion rates with elevation if there were a clear process link. At this stage we can simply suggest that higher precipitation at the escarpment base may support higher densities of biota, which are, in turn, likely to be more active across the landscape, potentially leading to higher erosion rates. While this suggestion is extremely speculative and needs field testing, it is critical that a process-based explanation be determined. Interestingly, the order of magnitude decrease of denudation with a <1000 m rise in elevation reported here is nearly the opposite of the less than a factor of 1.4 difference in rates reported by Riebe et al. (2004) across a <700 m altitudinal transect used as a steep climatic gradient in the Santa Rosa Mountains of Nevada. Their observations suggest a sharp decline in chemical weathering with altitude by coupling precipitation with total denudation, which may play a role here. We are currently testing this hypothesis. Secondly, both increasing stream power with distance downslope and increasing relief at the escarpment base would predict higher erosion rates with lower elevations across our field areas. Morphometrically, the correlation shown in Figure 8 fits with the low relief and gently convex-up ridges near and upslope of the scarp crest, as well as across the coastal lowlands, in comparison with the much higher relief and sharply convex ridge crests downslope on the scarp and onto the escarpment base site of Heimsath et al. (2000; see inset topographic map of Fig. 1). That is, lower erosion rates would be expected at the top of the escarpment and out on the coastal lowland based on the morphology, which is what the data show (note Fig. 6A, the soil production and bedrock erosion rates from Snug at an elevation of just over 200 m). The highest rates shown by Figure 8 are from the sharply convex-up ridge crest across and just upslope of the Heimsath et al. (2000) study area and from small catchments draining that ridge. Local relief, i.e., elevation difference between main spur ridge and the second- and third-order tributary draining the escarpment to Nunnock River, is greater at these sample sites than it is anywhere else on the escarpment, on the order of 50–60 m, and we observed debris-flow deposits in the main tributaries.

Increases in both slope and drainage area (a proxy for discharge) to the main tributaries draining the escarpment, of which Nunnock River is one example, with distance from the headwaters would predict the highest stream incision rates near the base of the escarpment. While the higher relief and over-steepened slopes adjacent to the streams at lower elevations at the escarpment base would suggest that stream incision rates are greater than ridge crest lowering rates, the lower elevation samples plotted in Figure 8 show similarity between the rates, within error. Importantly, the open boxes are average rates of first- and zero-order catchments draining from the spur ridges into the main tributaries, and are therefore not measures of the incision of Nunnock River or of the other main incision dissecting tributaries. We specifically chose the small catchments to avoid input of sediments or bedrock eroded by debris flows or landslides, such that the nuclide concentrations in the sediments would reflect average denudation rates of the upslope contributing area (Granger et al., 1996; Bierman and Steig, 1996). That ridge crest and basin averaged rates are similar suggests that the biogenic processes driving hillslope erosion are keeping up with the increased fluvial incision at the escarpment base, such that there is the suggestion for local steady-state where these rates are nearly equal.

Recent studies using low-temperature thermochronology and cosmogenic nuclides inferred denudational histories for passive margin escarpments, including a transect across the same field area that we are using here, and concluded that rift escarpments experience most of their erosion and retreat soon after rift development (Cockburn et al., 2000; Matmon et al., 2002; Persano et al., 2002). These studies provide important new constraints for the denudational histories of escarpments, which are used in modeling efforts that had previously predicted the same finding (Gilchrist et al., 1994; van der Beek and Braun, 1998). The mechanism suggested by Cockburn et al. (2000) and predicted by the above modeling studies, is that a major inland drainage divide fed rivers that rapidly eroded the newly formed escarpment shortly after formation. Studies from the southeastern Australia escarpment further north of our study area focused on the mechanisms of escarpment propagation, but could not resolve the greater than two orders of magnitude difference between inferred retreat rates of ~3 km/m.y. and denudation rates of 0.04–12 m/m.y. estimated since the time of rifting (Seidl et al., 1996; Weissel and Seidl, 1997, 1998). Rates of surface processes reported here and in Heimsath et al. (2000) are more than an order of magnitude lower than the inferred local escarpment retreat rate of ~500 m/m.y. (dividing the roughly 50 km distance between the coast and scarp lip by 100 m.y., the rough time since rifting), highlighting the need to resolve the discrepancy between inferred and observed rates of escarpment retreat. If, instead of using the gross estimate of horizontal retreat obtained by dividing distance from the coast by time since rifting, we use the geometric relationship between lowering rates and retreat rates (where retreat rate equals lowering rate divided by the tangent of the mean escarpment angle), then a contemporary retreat can be estimated. Specifically, lowering rates range from ~12 to ~36 m/m.y. (excluding tors) across the scarp with an average gradient of 0.25, suggesting an average retreat rate between ~40 and 140 m/m.y., which is an order of magnitude slower than the rate required to retreat the escarpment from the coastline at a continuous pace since rifting. If escarpments such as the one studied here did indeed retreat rapidly soon after rifting and then settle into a period of relative erosional stability, as our measurements and the above thermochronology studies suggest, then a geomorphic mechanism needs to be discovered that can accomplish the initial rapid retreat.

CONCLUSIONS

Our study used 44 new measurements of the in situ–produced cosmogenic nuclide, ^{10}Be, combined with two previously published studies using 46 measurements of both ^{10}Be and ^{26}Al, all measured from bedrock, saprolite, and stream sediments, to enable a comprehensive quantification of the erosion rates and processes across the passive margin escarpment of southeastern Australia. These rates further our understanding of the surface processes shaping landscapes with different climatic and tectonic histories. Specifically, we have documented two new relationships between soil production and the overlying soil thickness, termed the soil production function. These two new functions, from the coastal lowlands and from the highlands near the study areas used for previous studies, agree remarkably well with the earlier findings. We then connect the lowlands with the highlands by reporting erosion rates from samples taken from two different transects across the escarpment face. The first transect samples the convex-up ridge crest leading upslope from the study area at the escarpment base to the highlands above the escarpment lip. The second transect begins near the headwaters of the major escarpment-dissecting river and samples small catchments draining the spur ridges from the highlands to just upslope of the scarp base study area. Rates from these transects define a linear decrease in erosion rate with elevation across the face of the escarpment, with an apparent peak in denudation rate at the scarp base, and agree with the rates determined by our soil production studies. We conclude that biogenic processes of soil production and removal are keeping up with other observed processes of debris flows and bedrock landsliding (that we did not quantify here) across the escarpment to set the overall rate of escarpment retreat. This rate is at least an order of magnitude less than previously inferred, long-term scarp retreat rates long suggested by workers focusing on the problem and therefore calls for the need to resolve the discrepancy between observed rates of landscape denudation and inferred rates of scarp retreat.

ACKNOWLEDGMENTS

We thank Rishi Satyadharma (Steve Bateman) and his family for continued access to and unbounded hospitality at the Nunnock River field area. Bill Dietrich, Ian Prosser, and Kuni Nishiizumi helped initiate this work across the escarpment, and continued discussions have helped our thinking immensely. Paul Rustomji sampled and helped to process the nuclide samples from the coastal lowland site, and we thank him. Comments by Mike Summerfield and an anonymous reviewer helped improve the original manuscript. Cosmogenic nuclide measurements and field support were funded by Dartmouth College and National Science Foundation grants DEB-0128995 and EAR-0239655 to Heimsath. Thirteen nuclide measurements for this study and all the previous analyses were partially performed under the auspices of the U.S. Department of Energy by Lawrence Livermore National Laboratory (LLNL) under contract W-7405-Eng-48, while the research schools of Earth Sciences and Physical Sciences and Engineering at the Australian National University supported the rest.

REFERENCES CITED

Ahnert, F., 1987, Approaches to dynamic equilibrium in theoretical simulations of slope development: Earth Surface Processes and Landforms, v. 12, p. 3–15.

Banfield, J.F., and Eggleton, R.A., 1989, Apatite replacement and rare-earth mobilization, fractionation, and fixation during weathering: Clays and Clay Minerals, v. 37, no. 2, p. 113–127.

Bierman, P.R., 1994, Using in situ produced cosmogenic isotopes to estimate rates of landscape evolution; a review from the geomorphic perspective: Journal of Geophysical Research, ser. B, Solid Earth and Planets, v. 99, no. B7, p. 13,885–13,896, doi: 10.1029/94JB00459.

Bierman, P., and Caffee, M., 2001, Slow rates of rock surface erosion and sediment production across the Namib Desert and escarpment, southern Africa: American Journal of Science, v. 301, no. April/May, p. 326–358.

Bierman, P., and Steig, E.J., 1996, Estimating rates of denudation using cosmogenic isotope abundances in sediment: Earth Surface Processes and Landforms, v. 21, p. 125–139, doi: 10.1002/(SICI)1096-9837(199602)21:23.0.CO;2-8.

Bishop, P., 1985, Southeast Australian late Mesozoic and Cenozoic denudation rates: A test for late Tertiary increases in continental denudation: Geology, v. 13, p. 479–482, doi: 10.1130/0091-7613(1985)13<479:SALMAC>2.0.CO;2.

Bishop, P., 1986, Horizontal stability of the Australian continental drainage divide in south central New South Wales during the Cainozoic: Australian Journal of Earth Sciences, v. 33, p. 295–307.

Bishop, P., 1988, The eastern highlands of Australia: The evolution of an intraplate highland belt: Progress in Physical Geography, v. 12, p. 159–182.

Bishop, P., and Brown, R., 1992, Denudational isostatic rebound of intraplate highlands: The Lachlan River valley, Australia: Earth Surface Processes and Landforms, v. 17, no. 4, p. 345–360.

Bishop, P., and Goldrick, G., 2000, Geomorphological evolution of the East Australian continental margin, *in* Summerfield, M.A., ed., Geomorphology and global tectonics: Chichester, Wiley, p. 227–255.

Bishop, P., Young, R.W., and McDougall, I., 1985, Stream profile change and long-term landscape evolution; early Miocene and modern rivers of the East Australian highland crest, central New South Wales, Australia: Journal of Geology, v. 93, no. 4, p. 455–474.

Brown, E.T., Stallard, R.F., Larsen, M.C., Raisbeck, G.M., and Yiou, F., 1995, Denudation rates determined from the accumulation of in situ–produced ^{10}Be in the Luquillo Experimental Forest, Puerto Rico: Earth and Planetary Science Letters, v. 129, p. 193–202, doi: 10.1016/0012-821X(94)00249-X.

Brown, R.W., Summerfield, M.A., and Gleadow, A.J.W., 2002, Denudational history along a transect across the Drakensberg Escarpment of southern Africa derived from apatite fission-track thermochronology: Journal of Geophysical Research, v. 107, no. B12, 2350, doi: 10.1029/2001JB000745.

Bryant, E.A., Young, R.W., Price, D.M., and Short, S.A., 1994, Late Pleistocene dune chronology; near-coastal New South Wales and eastern Australia: Quaternary Science Reviews, v. 13, no. 3, p. 209–223, doi: 10.1016/0277-3791(94)90026-4.

Burke, B.C., and Heimsath, A.M., 2003, Variations in saprolite chemical weathering signatures: Geological Society of America Abstracts with Programs, v. 35, no. 6, p. 258.

Carson, M.A., and Kirkby, M.J., 1972, Hillslope form and process: New York, Cambridge University Press, 475 p.

Chappell, J., 1991, Late Quaternary environmental changes in eastern and central Australia, and their climatic interpretation: Quaternary Science Reviews, v. 10, no. 5, p. 377–390, doi: 10.1016/0277-3791(91)90002-C.

Clark, D.H., Bierman, P.R., and Larsen, P., 1995, Improving in situ cosmogenic chronometers: Quaternary Research, v. 44, p. 367–377, doi: 10.1006/qres.1995.1081.

Cockburn, H.A.P., and Summerfield, M.A., 2004, Geomorphological applications of cosmogenic isotope analysis: Progress in Physical Geography, v. 28, p. 1–42, doi: 10.1191/0309133304pp395oa.

Cockburn, H.A.P., Brown, R.W., Summerfield, M.A., and Seidl, M.A., 2000, Quantifying passive margin denudation and landscape development using a combined fission-track thermochronology and cosmogenic isotope analysis approach: Earth and Planetary Science Letters, v. 179, no. 3–4, p. 429–435, doi: 10.1016/S0012-821X(00)00144-8.

Cox, N.J., 1980, On the relationship between bedrock lowering and regolith thickness: Earth Surface Processes and Landforms, v. 5, p. 271–274.

Culling, W.E.H., 1960, Analytical theory of erosion: Journal of Geology, v. 68, p. 336–344.

Culling, W.E.H., 1965, Theory of erosion on soil-covered slopes: Journal of Geology, v. 73, p. 230–254.

Davis, J.C., Proctor, I.D., Southon, J.R., Caffee, M.W., Heikkinen, D.W., Roberts, M.L., Moore, T.L., Turteltaub, K.W., Nelson, D.E., Loyd, D.H., and Vogel, J.S., 1990, LLNL/UC AMS facility and research program: Nuclear Instruments Methods in Physics Research, Section B, no. B52, p. 269–272.

Dietrich, W.E., Reiss, R., Hsu, M.-L., and Montgomery, D.R., 1995, A process-based model for colluvial soil depth and shallow landsliding using digital elevation data: Hydrological Processes, v. 9, p. 383–400.

Dietrich, W.E., Bellugi, D., Heimsath, A.M., Roering, J.J., Sklar, L., and Stock, J.D., 2003, Geomorphic transport laws for predicting landscape form and dynamics, *in* Wilcock, P., and Iverson, R., eds., Prediction in geomorphology: Washington, D.C., American Geophysical Union, p. 103–132.

Dodson, J., 1977, Late Quaternary palaeoecology of Wyrie Swamp, southeastern South Australia: Quaternary Research (Washington, University, Quaternary Research Center), v. 8, no. 1, Richard Foster Flint memorial volume, p. 97–114.

Dodson, J.R., 1987, Mire development and environmental change, Barrington Tops, New South Wales, Australia: Quaternary Research, v. 27, no. 1, p. 73–81, doi: 10.1016/0033-5894(87)90050-0.

Dodson, J.R., and Wright, R.V.S., 1989, Humid to arid to subhumid vegetation shift on Pilliga Sandstone, Ulungra Springs, New South Wales: Quaternary Research, v. 32, no. 2, p. 182–192.

Dumitru, T.A., Hill, K.C., Coyle, D.A., Duddy, I.R., Foster, D.A., Gleadow, A.J.W., Green, P.F., Kohn, B.P., Laslett, G.M., and O'Sullivan, A.J., 1991, Fission track thermochronology; application to continental rifting of south-eastern Australia, *in* 1991 APEA conference; New concepts for old areas: Victoria, Australia, APEA Melbourne, p. 131–142.

Dumitru, T.A., Miller, E.L., O'Sullivan, P.B., Amato, J.M., Hannula, K.A., Calvert, A.T., and Gans, P.B., 1995, Cretaceous to Recent extension in the Bering Strait region, Alaska: Tectonics, v. 14, no. 3, p. 549–563, doi: 10.1029/95TC00206.

Dunai, T.J., 2000, Scaling factors for production rates of in situ produced cosmogenic nuclides: A critical reevaluation: Earth and Planetary Science Letters, v. 176, p. 157–169, doi: 10.1016/S0012-821X(99)00310-6.

Dunne, J., Elmore, D., and Muzikar, P., 1999, Scaling factors for the rates of production of cosmogenic nuclides for geometric shielding and attenuation at depth on sloped surfaces: Geomorphology, v. 27, no. 1–2, p. 3–11, doi: 10.1016/S0169-555X(98)00086-5.

Foster, D.A., and Gleadow, A.J., 1991, The architecture of Gondwana rifting in southeastern Australia; evidence from apatite fission track thermochronology, *in* Unrug, R., Banks, M.R., and Veevers, J.J., eds., Gondwana eight; Assembly, evolution and dispersal: (Conference Proceedings)Hobart, Tasmania, Australia, p. 597–603.

Gabet, E.J., 2000, Gopher bioturbation: Field evidence for nonlinear hillslope diffusion: Earth Surface Processes and Landforms, v. 25, p. 1419–1428, doi: 10.1002/1096-9837(200012)25:133.0.CO;2-1.

Gilbert, G.K., 1877, Report on the geology of the Henry Mountains (Utah): Washington, D.C., U.S. Government Printing Office, U.S. Geological Geographical and Survey of the Rocky Mountains Region, 170 p.

Gilchrist, A.R., Kooi, H., and Beaumont, C., 1994, Post-Gondwana geomorphic evolution of southwestern Africa: Implications for the controls on landscape developments from observations and numerical experiments: Journal of Geophysical Research, Solid Earth and Planets, v. 99, no. B6, p. 12,211–12,228, doi: 10.1029/94JB00046.

Gleadow, A.J.W., Kohn, B.P., Brown, R.W., O'Sullivan, P.B., and Raza, A., 2002, Fission track thermotectonic imaging of the Australian continent: Tectonophysics, v. 349, p. 5–21, doi: 10.1016/S0040-1951(02)00043-4.

Gosse, J.C., and Phillips, F.M., 2001, Terrestrial in situ cosmogenic nuclides: Theory and application: Quaternary Science Reviews, v. 20, p. 1475–1560, doi: 10.1016/S0277-3791(00)00171-2.

Granger, D.E., and Smith, A.L., 2000, Dating buried sediments using radioactive decay and muogenic production of ^{26}Al and ^{10}Be: Nuclear Instruments & Methods in Physics Research: Section B, Beam Interactions with Materials and Atoms, v. 172, no. 1–4, p. 824–828.

Granger, D.E., Kirchner, J.W., and Finkel, R., 1996, Spatially averaged long-term erosion rates measured from in situ–produced cosmogenic nuclides in alluvial sediment: Journal of Geology, v. 104, no. 3, p. 249–257.

Granger, D.E., Riebe, C.S., Kirchner, J.W., and Finkel, R.C., 2001, Modulation of erosion on steep granitic slopes by boulder armoring, as revealed by cosmogenic ^{26}Al and ^{10}Be: Earth and Planetary Science Letters, v. 186, p. 269–281, doi: 10.1016/S0012-821X(01)00236-9.

Harrison, S.P., 1993, Late Quaternary lake-level changes and climates of Australia: Quaternary Science Reviews, v. 12, no. 4, p. 211–231, doi: 10.1016/0277-3791(93)90078-Z.

Hayes, D.E., and Ringis, J., 1973, Seafloor spreading in the Tasman Sea: Nature, v. 243, p. 454–458.

Heimsath, A.M., Dietrich, W.E., Nishiizumi, K., and Finkel, R.C., 1997, The soil production function and landscape equilibrium: Nature, v. 388, p. 358–361, doi: 10.1038/41056.

Heimsath, A.M., Dietrich, W.E., Nishiizumi, K., and Finkel, R.C., 1999, Cosmogenic nuclides, topography, and the spatial variation of soil depth: Geomorphology, v. 27, no. 1–2, p. 151–172, doi: 10.1016/S0169-555X(98)00095-6.

Heimsath, A.M., Chappell, J., Dietrich, W.E., Nishiizumi, K., and Finkel, R.C., 2000, Soil production on a retreating escarpment in southeastern Australia: Geology, v. 28, no. 9, p. 787–790, doi: 10.1130/0091-7613(2000)028<787:SPOARE>2.3.CO;2.

Heimsath, A.M., Chappell, J., Dietrich, W.E., Nishiizumi, K., and Finkel, R.C., 2001, Late Quaternary erosion in southeastern Australia: A field example using cosmogenic nuclides: Quaternary International, v. 83–5, p. 169–185, doi: 10.1016/S1040-6182(01)00038-6.

Kershaw, A.P., D'Costa, D.M., McEwen Mason, J.R.C., and Wagstaff, B.E., 1991, Palynological evidence for Quaternary vegetation and environments of mainland southeastern Australia: Quaternary Science Reviews, v. 10, p. 391–401, doi: 10.1016/0277-3791(91)90003-D.

King, L.C., 1962, The morphology of the Earth: White Plains, New York, Oliver and Boyd, 726 p.

Kohl, C.P., and Nishiizumi, K., 1992, Chemical isolation of quartz for measurement of in-situ–produced cosmogenic nuclides: Geochimica et Cosmochimica Acta, v. 56, p. 3583–3587, doi: 10.1016/0016-7037(92)90401-4.

Lal, D., 1988, In situ–produced cosmogenic isotopes in terrestrial rocks: Annual Review of Earth and Planetary Sciences v. 16, p. 355–388.

Lal, D., 1991, Cosmic ray labeling of erosion surfaces: In situ nuclide production rates and erosion models: Earth and Planetary Science Letters, v. 104, p. 424–439, doi: 10.1016/0012-821X(91)90220-C.

Lal, D., and Arnold, J.R., 1985, Tracing quartz through the environment: Proceedings of Indian Academic Science (Earth and Planetary Sciences), v. 94, no. 1, p. 1–5.

Lal, D., and Peters, B., 1967, Cosmic ray produced radioactivity on the Earth: New York, Springer-Verlag, Encyclopedia of Physics, v. XLVI/2, p. 551–612.

Lambeck, K., and Stephenson, R., 1985, Post-orogenic evolution of a mountain range; south-eastern Australian highlands: Geophysical Research Letters, v. 12, no. 12, p. 801–804.

Lewis, P.C., and Glen, R.A., 1995, Bega-Mallacoota geological sheet, SJ/55-4, SJ/55-8, second edition: Sydney, Australia, Geological Survey of New South Wales, scale 1:250,000.

Linton, D.L., 1955, The problem of tors: The Geographical Journal, v. 121, p. 470–487.

Masarik, J., Kollar, D., and Vanya, S., 2000, Numerical simulation of in situ production of cosmogenic nuclides: Effects of irradiation geometry: Nuclear Instruments and Methods in Physics Research Section B-BEAM Interactions with Materials and Atoms, v. 172, p. 786–789.

Matmon, A., Bierman, P., and Enzel, Y., 2002, Pattern and tempo of great escarpment erosion: Geology, v. 30, no. 12, p. 1135–1138, doi: 10.1130/0091-7613(2002)030<1135:PATOGE>2.0.CO;2.

McAlpine, J.R., and Yapp, G.A., 1969, Climate of the Queanbeyan-Shoalhaven area; lands of the Queanbeyan-Shoalhaven area, ACT and NSW: Canberra, Australia, Land Resources Service, Commonwealth Science and Industrial Research Organization, p. 57–75.

Moore, M.E., Gleadow, A.J.W., and Lovering, J.F., 1986, Thermal evolution of rifted continental margins; new evidence from fission tracks in basement apatites from southeastern Australia: Earth and Planetary Science Letters, v. 78, no. 2–3, p. 255–270, doi: 10.1016/0012-821X(86)90066-X.

Nishiizumi, K., Winterer, E.L., et al., 1989, Cosmic ray production rates of ^{10}Be and ^{26}Al in quartz from glacially polished rocks: Journal of Geophysical Research, v. 94, p. 17, 907–17, 915.

Nishiizumi, K., Winterer, E.L., Kohl, C.P., Klein, J., Middleton, R., Lal, D., and Arnold, J.R., 1989, Cosmic ray production rates of ^{10}Be and ^{26}Al in quartz from glacially polished rocks: Journal of Geophysical Research, v. 94, no. B12, p. 17,907–17,915.

Nishiizumi, K., Kohl, C.P., Arnold, J.R., Klein, J., Fink, D., and Middleton, R., 1991, Cosmic ray produced ^{10}Be and ^{26}Al in Antarctic rocks; exposure and erosion history: Earth and Planetary Science Letters, v. 104, no. 2–4, p. 440–454, doi: 10.1016/0012-821X(91)90221-3.

Nishiizumi, K., Kohl, C.P., Arnold, J.R., Dorn, R., Klein, J., Fink, D., Middleton, R., and Lal, D., 1993, Role of in situ cosmogenic nuclides ^{10}Be and ^{26}Al in the study of diverse geomorphic processes: Earth Surface Processes and Landforms, v. 18, p. 407–425.

Nishiizumi, K., Finkel, R.C., Klein, J., and Kohl, C.P., 1996, Cosmogenic production of ^{7}Be and ^{10}Be in water targets: Journal of Geophysical Research, v. 101, no. B10, p. 22,225–22,232, doi: 10.1029/96JB02270.

Nott, J.F., 1992, Long-term drainage evolution in the Shoalhaven catchment, Southeast Highlands, Australia: Earth Surface Processes and Landforms, v. 17, no. 4, p. 361–374.

Nott, J.F., Idnurm, M., and Young, R.W., 1991, Sedimentology, weathering, age and geomorphological significance of Tertiary sediments on the far south coast of New South Wales: Australian Journal of Earth Sciences, v. 38, no. 3, p. 357–373.

O'Sullivan, P.B., Kohn, B.P., Foster, D.A., and Gleadow, A.J.W., 1995, Fission track data from the Bathurst Batholith—Evidence for rapid mid-Cretaceous uplift and erosion within the eastern Highlands of Australia: Australian Journal of Earth Sciences, v. 42, no. 6, p. 597–607.

O'Sullivan, P.B., Foster, D.A., Kohn, B.P., and Gleadow, A.J.W., 1996, Multiple postorogenic denudation events—An example from the eastern Lachlan fold belt, Australia: Geology, v. 24, no. 6, p. 563–566, doi: 10.1130/0091-7613(1996)024<0563:MPDEAE>2.3.CO;2.

Ollier, C.D., 1978, Tectonics and geomorphology of the eastern Highlands, *in* Davies, J.L., and Williams, A.J., eds., Landform evolution in Australasia: Canberra, Australian National University, p. 5–47.

Ollier, C.D., 1982, The Great Escarpment of eastern Australia: Tectonic and geomorphic significance: Journal of the Geological Society of Australia, v. 29, p. 13–23.

Ollier, C.D., 1995, Tectonics and landscape evolution in southeast Australia: Geomorphology, v. 12, no. 1, p. 37–44, doi: 10.1016/0169-555X(94)00075-3.

Ollier, C.D., and Marker, M.E., 1985, The Great Escarpment of southern Africa: Zeitschrift für Geomorphologie, Supplementband, v. 54, p. 37–56.

Pain, C.F., 1985, Morphotectonics of the continental margins of Australia: Zeitschrift für Geomorphologie, Supplement Band, v. 54, p. 23–35.

Persano, C., Stuart, F.M., Bishop, P., and Barfod, D.N., 2002, Apatite (U-Th)/He age constraints on the development of the Great Escarpment on the southeastern Australian passive margin: Earth and Planetary Science Letters, v. 200, no. 1–2, p. 79–90, doi: 10.1016/S0012-821X(02)00614-3.

Richardson, S.J., 1976, Geology of the Michelago 1:100,000 sheet 8726: Canberra, Australia, Geological Survey of New South Wales, Department of Mineral Resources and Development, 253 p.

Riebe, C.S., Kirchner, J.W., and Finkel, R.C., 2001, Strong tectonic and weak climatic control on long-term chemical weathering rates: Geology, v. 29, p. 511–514, doi: 10.1130/0091-7613(2001)029<0511:STAWCC>2.0.CO;2.

Riebe, C.S., Kirchner, J.W., and Finkel, R.C., 2003, Long term rates of chemical weathering and physical erosion from cosmogenic nuclides and geochemical mass balance: Geochimica et Cosmochimica Acta, v. 67, p. 4411–4427, doi: 10.1016/S0016-7037(03)00382-X.

Riebe, C.S., Kirchner, J.W., and Finkel, R.C., 2004, Sharp decrease in long term chemical weathering rates along an altitudinal transect: Earth and Planetary Science Letters, v. 218, p. 421–434, doi: 10.1016/S0012-821X(03)00673-3.

Seidl, M.A., Weissel, J.K., and Pratson, L.F., 1996, The kinematics and pattern of escarpment retreat across the rifted continental margin of SE Australia: Basin Research, v. 8, no. 3, p. 301–316, doi: 10.1046/j.1365-2117.1996.00266.x.

Small, E.E., Anderson, R.S., and Hancock, G.S., 1999, Estimates of the rate of regolith production using ^{10}Be and ^{26}Al from an alpine hillslope: Geomorphology, v. 27, no. 1–2, p. 131–150, doi: 10.1016/S0169-555X(98)00094-4.

Stephenson, R., and Lambeck, K., 1985, Erosion-isostatic rebound models for uplift; an application to south-eastern Australia: The Geophysical Journal of the Royal Astronomical Society, v. 82, no. 1, p. 31–55.

Stone, J.O., 2000, Air pressure and cosmogenic isotope production: Journal of Geophysical Research, v. 105, p. 23,753–23,759, doi: 10.1029/2000JB900181.

Stone, J.O., Ballantyne, C.K., and Fifield, L.K., 1998a, Exposure dating and validation of periglacial weathering limits, northwest Scotland: Geology, v. 26, no. 7, p. 587–590, doi: 10.1130/0091-7613(1998)026<0587:EDAVOP>2.3.CO;2.

Stone, J.O., Evans, J.M., Fifield, L.K., Allan, G.L., and Cresswell, R.G., 1998b, Cosmogenic chlorine-36 production in calcite by muons: Geochimica et Cosmochimica Acta, v. 62, no. 3, p. 433–454, doi: 10.1016/S0016-7037(97)00369-4.

Summerfield, M.A., 1999, Geomorphology and global tectonics: Chichester, Wiley, 357 p.

Taunton, A.E., Welch, S.A., and Banfield, J.F., 2000a, Geomicrobiological controls on light rare earth element, Y and Ba distributions during granite weathering and soil formation: Journal of Alloys and Compounds, v. 303, p. 30–36, doi: 10.1016/S0925-8388(00)00597-1.

Taunton, A.E., Welch, S.A., and Banfield, J.F., 2000b, Microbial controls on phosphate and lanthanide distributions during granite weathering and soil formation: Chemical Geology, v. 169, no. 3–4, p. 371–382, doi: 10.1016/S0009-2541(00)00215-1.

Thomas, M.F., 1989a, The Role of etch processes in landform development. I: Etching concepts and their applications: Zeitschrift für Geomorphologie, v. 33, no. 2, p. 129–142.

Thomas, M.F., 1989b, The role of etch processes in landform development. II: Etching and the formation of relief: Zeitschrift für Geomorphologie, v. 33, no. 3, p. 257–274.

Thomas, M.F., 1995, Models for landform development on passive margins; some implications for relief development in glaciated areas: Geomorphology, v. 12, no. 1, p. 3–15, doi: 10.1016/0169-555X(94)00082-3.

Tucker, G.E., and Slingerland, R.L., 1994, Erosional dynamics, flexural isostasy, and long-lived escarpments: A numerical modeling study: Journal of Geophysical Research, ser. B, Solid Earth and Planets, v. 99, no. B6, p. 12,229–12,243, doi: 10.1029/94JB00320.

Tucker, G.E., and Slingerland, R., 1997, Drainage basin responses to climate change: Water Resources Research, v. 33, no. 8, p. 2031–2047, doi: 10.1029/97WR00409.

Twidale, C.R., 1971, Structural landforms; landforms associated with granitic rocks, faults, and folded strata: Cambridge, The Massachusetts Institute of Technology Press, 247 p.

Twidale, C.R., 1985, Granite landform evolution; factors and implications, *in* Hans Cloos Kolloquim: Bonn, Germany, Geologische Rundschau, p. 769–779.

Twidale, C.R., 1993, The research frontier and beyond; granitic terrains: Geomorphology, v. 7, p. 187–223.

Twidale, C.R., and Vidal Romani, J.R., 1994, On the multistage development of etch forms: Geomorphology, v. 11, p. 107–124, doi: 10.1016/0169-555X(94)90076-0.

van der Beek, P., and Braun, J., 1998, Numerical modelling of landscape evolution on geological time-scales: A parameter analysis and comparison with the south-eastern highlands of Australia: Basin Research, v. 10, p. 49–68, doi: 10.1046/j.1365-2117.1998.00056.x.

van der Beek, P., and Braun, J., 1999, Controls on post-mid-Cretaceous landscape evolution in the southeastern highlands of Australia: Insights from numerical surface process models: Journal of Geophysical Research, v. 104, no. B3, p. 4945–4966.

van der Beek, P., Summerfield, M.A., Braun, J., Brown, R.W., and Fleming, A., 2002, Modeling post-break-up landscape development and denudational history across the southeast African (Drakensberg Escarpment) margin: Journal of Geophysical Research, v. 107, no. B12, 2351, doi: 10.1029/2001JB000744.

Weissel, J.K., and Hayes, D.E., 1977, Evolution of the Tasman Sea reappraised: Earth and Planetary Science Letters, v. 36, no. 1, p. 77–84, doi: 10.1016/0012-821X(77)90189-3.

Weissel, J.K., and Seidl, M.A., 1997, Influence of rock strength properties on escarpment retreat across passive continental margins: Geology, v. 25, no. 7, p. 631–634, doi: 10.1130/0091-7613(1997)025<0631:IORSPO>2.3.CO;2.

Weissel, J.K., and Seidl, M.A., 1998, Inland propagation of erosional escarpments and river profile evolution across the southeast Australian passive continental margin, *in* Tinkler, K.J., and Wohl, E.E., eds., Rivers over rock: Fluvial processes in bedrock channels: Washington, D.C., American Geophysical Union Geophysical Monograph 107, p. 189–206.

Wellman, P., 1979, On the Cainozoic uplift of the southeastern Australian highland: Journal of the Geological Society of Australia, v. 26, p. 1–9.

Wellman, P., 1987, Eastern highlands of Australia; their uplift and erosion: Brown Mountain: Journal of Australian Geology and Geophysics, v. 10, p. 277–286.

Wellman, P., and McDougall, I., 1974, Potassium-argon ages on the Cainozoic volcanic rocks of New South Wales: Journal of the Geological Society of Australia, v. 21, no. 3, p. 247–272.

Williams, A.G., Ternan, L., and Kent, M., 1986, Some observations on the chemical weathering of the Dartmoor Granite: Earth Surface Processes and Landforms, v. 11, p. 557–574.

Young, R.W., 1977, Landscape development in the Shoalhaven River catchment of southeastern New South Wales: Zeitschrift für Geomorphologie, Supplement Band, v. 21, no. 3, p. 262–283.

Young, R.W., 1983, The tempo of geomorphological change; evidence from southeastern Australia: Journal of Geology, v. 91, no. 2, p. 221–230.

Young, R.W., and McDougall, I., 1982, Basalts and silcretes on the coast near Ulladulla, southern New South Wales: Journal of the Geological Society of Australia, v. 29, no. 4, p. 425–430.

Young, R.W., and McDougall, I., 1993, Long-term landscape evolution—Early Miocene and modern rivers in southern New South Wales, Australia: Journal of Geology, v. 101, no. 1, p. 35–49.

Manuscript Accepted by the Society 23 June 2005

Printed in the USA

Geological Society of America
Special Paper 398
2006

A parametric study of soil transport mechanisms

Frédéric Herman[†]
Jean Braun[‡]
Research School of Earth Sciences, The Australian National University, Canberra, ACT 0200, Australia

ABSTRACT

Numerical models are used to help constrain empirical parameterizations of soil production and transport mechanisms on soil-mantled hillslopes. The neighbourhood algorithm is used to invert soil thickness versus surface curvature data to provide not only more rigorous estimates of model parameter values but determine which of the model parameters are constrained by cosmogenic exposure ages (Heimsath et al., 2000). We show that linear and depth-dependent creep constants can be constrained by simple geomorphometric measurements, such as the distribution of soil thickness on the landform and its relationship to surface curvature. We also show that this unique data set cannot be used on its own to constrain the parameterization of overland flow, another transport mechanism that is thought to play an important role on soil-mantled hillslopes, or to determine if a soil distribution has reached local steady state. We also demonstrate that, to explain the data, soil production must be a function of soil thickness. These conclusions have important implications for our understanding of landscape evolution on medium to long time scales.

Keywords: geomorphology, hillslopes, soil production, soil dynamics, landscape evolution, neighbourhood algorithm.

INTRODUCTION

Among the processes that govern the rate at which landform evolves through time, the movement of soil is known to play an important role, especially in areas characterized by low to medium relief under vegetated or semi-arid conditions. It is accepted that soil thickness is determined by the balance between soil production and soil transport (Gilbert, 1877; Dietrich et al., 1995; Howard, 1997; Heimsath et al., 1997). The relationship between soil thickness and surface curvature, or, more generally, the local geometry of a given landscape, has been used to constrain the importance of several transport mechanisms on soil-mantled hillslopes (Fig. 1) (Heimsath et al., 1997, 1999, 2000; Braun et al., 2001).

Although it has been demonstrated that landscape evolution strongly depends on the nature of the soil production process and the rate at which it takes place (Rosenbloom and Anderson, 1994), few attempts have been carried out to offer a well-defined parametric representation of a soil production function (Ahnert, 1967; Carson and Kirkby, 1972; Cox, 1980; Dietrich et al., 1995). Recently, an exponential soil production law has been proposed from measurements of the in situ–produced cosmogenic ^{10}Be and ^{26}Al concentrations in bedrock sampled under soils of different depths in the Bega Valley (Fig. 2), southeastern Australia (Heimsath et al., 2000). Similar results were found in northern California (Heimsath et al., 1997, 1999), confirming the potential general applicability of the relationship between soil production and soil thickness.

[†]E-mail: frederic.herman@anu.edu.au.
[‡]E-mail: jean.braun@anu.edu.au.

Herman, F., and Braun, J., 2006, A parametric study of soil transport mechanisms, *in* Willett, S.D., Hovius, N., Brandon, M.T., and Fisher, D.M., eds., Tectonics, Climate, and Landscape Evolution: Geological Society of America Special Paper 398, Penrose Conference Series, p. 191–200, doi: 10.1130/2006.2398(11).

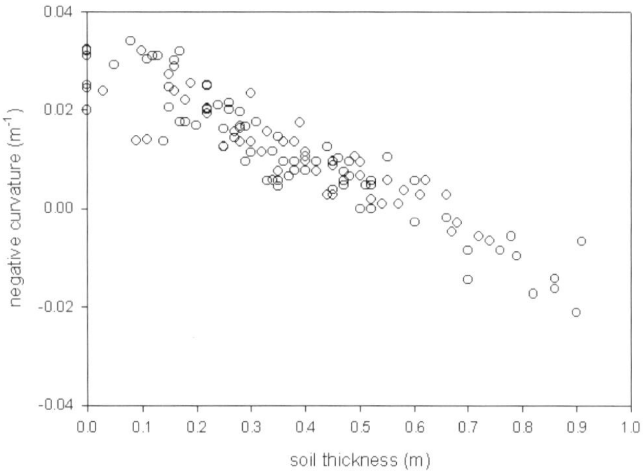

Figure 1. Observed soil thickness–curvature relationship: hillslope curvature (m^{-1}) versus soil depth (m) at Bega Valley site (Heimsath et al., 2000).

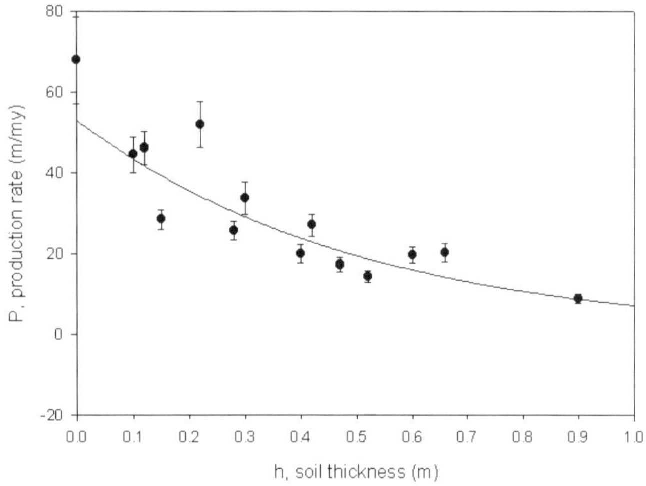

Figure 2. Soil production rates (m/m.y.), calculated from in situ–produced cosmogenic ^{10}Be and ^{26}Al concentrations in bedrock samples, versus normal soil thickness (m) at Bega Valley site (Heimsath et al., 2000), and assumed soil production law ($P=P_0 \exp -h/h_0$, where h_0 = 0.5 m and P_0 = 53 m/m.y.).

Soil is transported downslope by a number of processes, the importance and the rate of which are dictated by many factors, such as local slope, soil type, climate, vegetation, etc., which are commonly described by empirical relationships (Davis, 1892). These mathematical formulations have, in turn, been incorporated in numerical landscape-evolution models (Culling, 1960; Carson and Kirkby, 1972; McKean et al., 1993; Selby, 1993; Kooi and Beaumont, 1994; Dietrich et al., 1995; Howard, 1997; Braun and Sambridge, 1997; Heimsath et al., 1999, 2000; Braun et al., 2001). The simplest law is the linear creep rule, in which soil flux is proportional to slope (Davis, 1892). Several studies have been carried out that support its use (Ahnert, 1967; McKean et al., 1993; Kooi and Beaumont, 1994; Dietrich et al., 1995; Braun and Sambridge, 1997; Hovius et al., 1997; Howard, 1997; Heimsath et al., 1999, 2000; Small et al., 1999). Nevertheless, many observations (morphometric analysis and optically stimulated luminescence analysis) have revealed that nonlinear creep (Anderson, 1994; Howard, 1994; Carson and Kirkby, 1972; Selby, 1993; Montgomery and Dietrich, 1995; Hovius et al., 1997; Roering et al., 1999, 2001; Braun et al., 2001) and overland soil transport (Heimsath et al., 2002) must be taken into account to understand and simulate hillslope evolution.

Recently, a parameterization has been proposed (Braun et al., 2001) that potentially combines the effects of each of the above-mentioned mechanisms. Using this parameterization, the linear relationship between thickness and curvature observed at several sites has been successfully reproduced (Braun et al., 2001). However, the weight of the overland flux appears to be insignificant compared to the creep flux, a conclusion which is not supported by recent work based on optically simulated luminescence data (Heimsath et al., 2002), which suggests that overland flow plays an equally important role in transporting soil along hillslopes.

In response to this apparent contradiction concerning the soil transport mechanisms acting on hillslopes, we have carried out a more rigorous parametric study of the model proposed by Braun et al. (2001), constrained by the observed soil thickness–curvature relationship (Fig. 1). In this way, we hope to offer a theoretical support to better understand surface hillslope evolution and, by doing so, provide better constraints on the rate at which surface processes operate and interact with tectonic processes.

OVERVIEW OF THE HILLSLOPE PROBLEM

Soil Production Law

For many years, difficulties have been encountered in determining an appropriate mathematical formulation to represent the conversion of bedrock to soil (Gilbert, 1877; Culling, 1960; Ahnert, 1967; Carson and Kirkby, 1972; Cox, 1980; Dietrich et al., 1995). In recent years, the emergence of cosmogenic radionuclide analysis methods has helped us to understand the rate at which, in a given geological context, soil is produced from the underlying bedrock. Heimsath et al. (2000) used concentrations of cosmogenic ^{10}Be and ^{26}Al in bedrock at a Bega Valley site (Fig. 2) to derive a soil production law in which the soil production rate, P (m/m.y.), decreases with soil depth as an exponential function:

$$P = P_0 \exp\frac{-h}{h_0}, \tag{1}$$

where h (m) is local soil thickness, and P_0 (m/m.y.) and h_0 (m) are constants.

Heimsath et al. (2000) derived this relationship by adapting the cosmogenic nuclide method so that they could determine a depth-dependant soil production rate. This soil production function was defined by measuring nuclide concentrations in bedrock at different depths of soil to calculate soil production rates, which were then plotted against soil thicknesses (Fig. 2). However, this law should be treated with caution. Indeed, as shown in Figure 2, this parameterization has been derived from relatively few samples. It therefore requires further testing, even though this equation has been incorporated in numerical models (Heimsath et al., 1999, 2000; Braun et al., 2001) that successfully reproduced a range of observations, including the variation of soil with local topographic curvature.

Transport Mechanisms

So far, few models are capable of fully explaining soil-mantled landscape evolution. Assuming that processes such as eolian deposition and dissolution loss are negligible, Braun et al. (2001) proposed a model that reproduces observed soil thickness–curvature relationships by assuming three transport processes. We will follow their approach. Furthermore, we also consider the possibility of a nonlinear, diffusive sediment transport (Roering et al., 1999).

At any point of the hill, soil transport is seen as a combination of three processes (nonlinear, linear, and depth-dependent creep) acting within the soil mantle and one (overland flux) acting at the surface. All four mechanisms assume that transport is in the direction of maximum local slope.

The first assumed mechanism is linear creep (Davis, 1892; Gilbert, 1909), in which soil flux, Q_1, is commonly parameterized by the following expression:

$$Q_1 = K_d \nabla z, \tag{2}$$

where K_d ($m^2\ yr^{-1}$) is a constant governing linear creep efficiency, and ∇z is the local topographic gradient. The second assumed mechanism is depth-dependent creep (Selby, 1993) where soil flux, Q_2, is given by:

$$Q_2 = K_v h^m (\nabla z)^n, \tag{3}$$

where K_v ($m^{2-m}\ yr^{-1}$), and m and n are constants controlling the process efficiency. (K_v is a transport coefficient.) Note that the expression $(\nabla z)^n$ is short-hand for $\|\nabla z\|^{n-1}\nabla z$. The third assumed mechanism is nonlinear transport (Roering et al., 1999):

$$Q_3 = \frac{K_{nl}\nabla z}{1-\left(\frac{\|\nabla z\|}{S_c}\right)^2}, \tag{4}$$

where K_{nl} ($m^2\ yr^{-1}$) is a transport coefficient, and S_c the effective coefficient of friction. Finally, the fourth assumed mechanism is overland soil flux in which soil flux, Q_4, can be expressed as (Moore and Burch, 1986):

$$Q_4 = K_w A^k \nabla z^p, \tag{5}$$

where A is drainage area taken as a proxy for overland water discharge, and K_w ($m^{2-3k}\ yr^{k-1}$), k, and p are constants (Prosser and Rustomji, 2000). It is assumed that water flows down the hill as a sheet flow (i.e., we neglect the contribution of channelized water transport).

Numerical Model

The main purpose of this work is to test parameterizations developed to simulate hillslope landscape evolution. The transient mass balance, or the local rate of soil thickness change ($\partial h/\partial t$), is controlled by soil production and soil transport, and can be expressed as:

$$\frac{\partial h}{\partial t} = -\nabla \cdot Q + \frac{P}{\kappa}, \tag{6}$$

where κ is the ratio of soil density to rock density and $\nabla\cdot$ is the spatial divergence operator.

The rate of surface elevation change, $\partial z/\partial t$, is related to soil production and soil distribution change by:

$$\frac{\partial z}{\partial t} = -\left(P - \frac{\partial z}{\partial t}\right)\sec\alpha\ , \tag{7}$$

where α is the local bedrock slope angle.

By substituting equations 2, 3, 4, and 5 into 6, we obtain:

$$\frac{\partial h}{\partial t} = -\nabla \cdot (K_d \nabla z + K_v (\nabla z)^n h^m + \frac{K_{nl}\nabla z}{1-\left(\frac{\|\nabla z\|}{S_c}\right)^2} + K_w (\nabla z)^p A^k) + \frac{p}{\kappa} \tag{8}$$

and then using equation 7, we obtain:

$$\frac{\partial z}{\partial t} = [(\frac{1}{\kappa} - 1)P - \nabla \cdot (K_d \nabla z + K_v (\nabla z)^n h^m + \frac{K_{nl}\nabla z}{1-\left(\frac{\|\nabla z\|}{S_c}\right)^2} + K_w (\nabla z)^p A^k)]\sec\alpha \tag{9}$$

This partial differential equation must be solved for a set of initial and boundary conditions. We assume the bedrock elevation at $t = 0$ to be a symmetrical cosine function, in the x direction, on which a smaller perturbation has been added, in the y direction. The hill is 200 m high and extends 300 m in the y direction and 600 m in the x direction (Fig. 3). Such a geometry was chosen to simulate geometric properties comparable to those observed in the Bega Valley site by Heimsath et al. The initial soil thickness distribution is uniform ($h = h_0$). Along both lower boundaries, i.e., at $x = 0$ and $x = 600$ m, the boundary conditions are chosen to simulate the presence of a stream removing soil, such that, for all t:

$$z(x = 0, y) \equiv z(x = 600m, y) \equiv h \equiv 0, \tag{10}$$

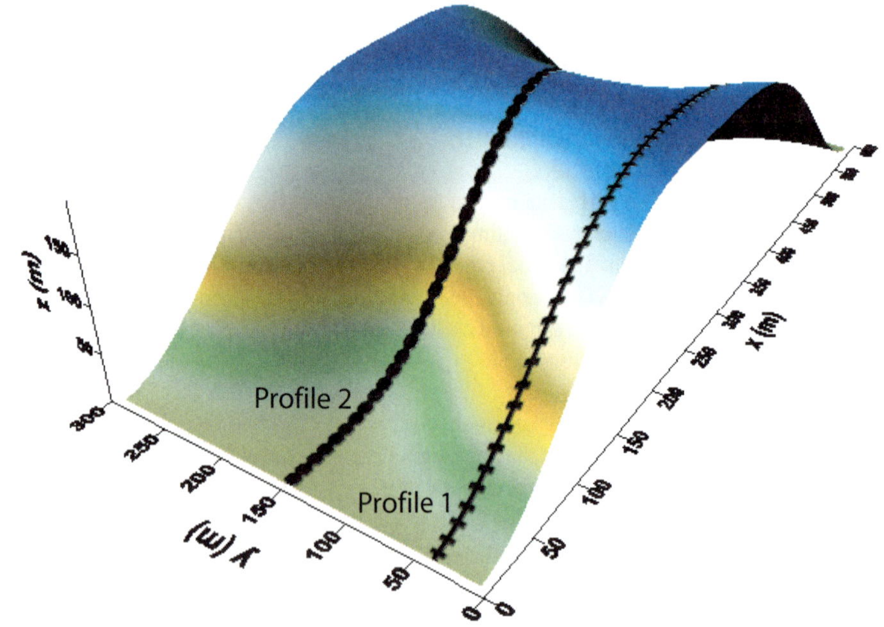

Figure 3. Ground surface elevation simulated. The hill is 200 m high and extends 300 m in y direction and 600 m in x direction.

ditions are used, which means that soil reaching the boundary on one side of the hill is automatically transported to the opposite side at the same x location.

A finite difference scheme is used to solve the equation on a regular, square grid in which $\Delta x = \Delta y = 10$ m. The time dependency is treated explicitly:

$$z(t+\Delta t) = z(t) + \frac{\partial z}{\partial t}\Delta t\,. \tag{11}$$

The time step, Δt, is taken small enough (10 yr) to ensure numerical stability.

Landscape evolution is simulated until a local steady state is reached, i.e., when $\partial h/\partial t \approx 0$ at any point of the landscape. Even though this state is not an actual geomorphologic steady state ($\partial z/\partial t \neq 0$), variations in the soil thickness distribution are small enough in comparison to local relief that we can assume that landform does not drastically change over a short time scale (~4–5 k.y.).

OBSERVATIONAL CONSTRAINTS AND METHODOLOGY

In this study, we make use of data collected on a small hill near the Nunnock River, a tributary of the Bega River, in southeastern Australia (Heimsath et al., 2000). In this region, the lithology mainly consists of intrusive rocks (granite and granodiorite of the Bega Batholith; Lewis and Glen, 1995). Soil mantle varies in thickness across the landscape. Measurements of soil thickness, up to 1 m depth, and surface curvature were gathered (Heimsath et al., 2000; Fig. 1). Furthermore, cosmogenic radionuclide analyses have been performed and used to determine the soil production function for the site (Heimsath et al., 2000; Fig. 2).

As in previous studies (Heimsath et al., 1999; Braun et al., 2001), our purpose here is to constrain which mechanisms govern landscape evolution on hillslopes by reproducing an observed curvature–soil thickness relationship for a given site by simulating soil production and transport on a synthetic hill and assuming that a local steady state is rapidly reached. We propose to improve previously published results in the following manner:

1. Firstly, we explore the significance of the assumed soil production law (equation 1) on the predicted soil thickness–curvature relationship. To assess the significance of the form of the soil production law on the computed

Figure 4. On the left-hand side, we compare different soil production functions to average rates (solid black circles) derived from both ^{10}Be and ^{26}Al measurements at the Bega Valley site (Heimsath et al., 2000). On the right-hand side, observed and synthetic curvature–soil thickness relationships are compared (circles represent the observations at the Bega Valley site, and crosses are from the numerical model). (A) An exponential law is assumed ($P = P_0 \exp -hh_0$, where $P_0 = 53$ m m.y.$^{-1}$ and $h_0 = 0.5$ m). (B) A linear rate is assumed ($P = P_0(1 - h/2h_0)$, where $P_0 = 45$ m m.y.$^{-1}$ and $h_0 = 0.5$ m). (C) The soil production rate is set equal to a constant, $P = 35$ m m.y.$^{-1}$. (D) The soil production rate follows an exponential law for 30 k.y. and then is set to zero. For all models, other parameter values are set as follows: $K_d = 2.08 \times 10^{-3}$ m^2 yr^{-1}, $K_y = 4.68 \times 10^{-3}$ m^{2-m} yr^{-1}, $K_w = 4.7 \times 10^{-7}$ m^{2-3k} yr^{k-1}, $m = 1.12$, $n = 0.66$, $k = 1.07$, $p = 1.60$, and $K_{nl} = 0.0$.

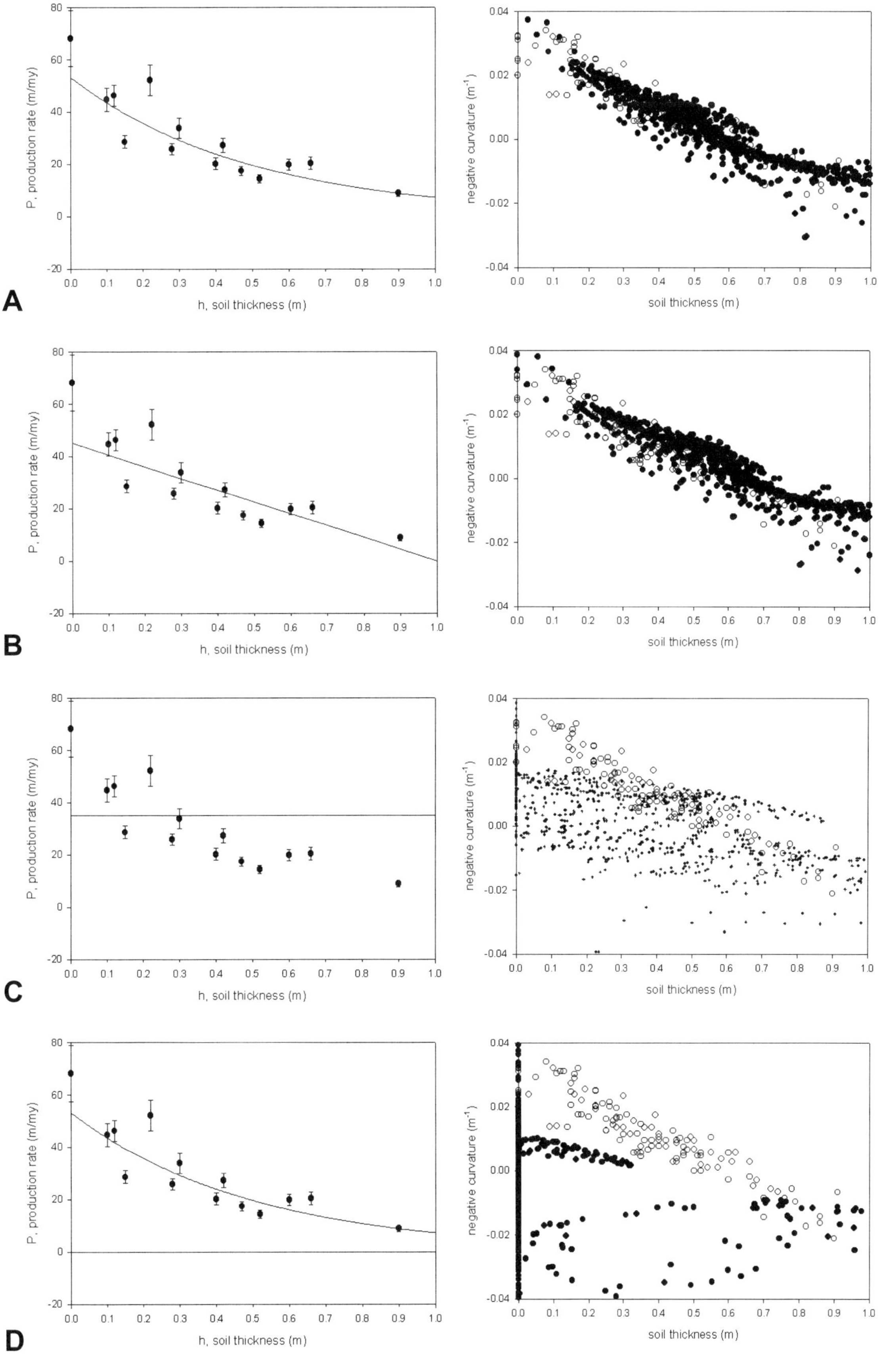
P, production rate (m/my)
h, soil thickness (m)
negative curvature (m^{-1})
soil thickness (m)
A
B
C
D

curvature–soil thickness relationship, we assume and test three soil production laws that reproduce with more or less accuracy the soil thickness–soil production relationship derived from the cosmogenic radionuclide data (Heimsath et al., 2000).

2. Secondly, we test whether the relationship between observed soil thickness and local curvature can be used to determine whether a local steady state has been reached. Local steady state corresponds to a stage at which soil production is balanced by transport (at any point $\partial h/\partial t \approx 0$).
3. Finally, we use a rigorous inversion method, the neighbourhood algorithm (NA) (Sambridge, 1999a, 1999b), to extract better constraints on the relative importance of the transport mechanisms from the observations. The objective of this mathematical procedure is to find models, or parameters, which minimize the misfit to the data. In our study, the parameters are the constants controlling the transport mechanisms (K_d, K_v, K_w, m, n, k, and p), and the data to be fitted are the observed curvature–soil thickness relationship. The misfit function, M, is:

$$M = \frac{(a_m - q_d)^2}{a_d^2} + \frac{(b_m - b_d)^2}{b_d^2} + \frac{(r_m - r_d)^2}{r_d^2} \tag{12}$$

where a, b, and r are the intercept, the slope, and the correlation coefficients issued from linear regressions performed on data (d) and model (m) curvature–soil thickness relationships, respectively. NA is an iterative method making use of simple geometrical concepts to search the multidimensional parameter space. At each stage, the entire parameter space is divided into a set of Voronoi cells, one cell about each previously sampled model. Voronoi cells represent the nearest neighborhood about each point. These cells are then used to guide the next sampling step, in a randomized fashion. As iterations proceed, the algorithm concentrates sampling in regions of the parameter space where the data misfit, through the misfit function, is minimized. Full details of NA can be found in Sambridge (1999a, 1999b).

RESULTS

Soil Production Law

Results are summarized in Figure 4. Firstly, the exponential law proposed by Heimsath et al. (2000) is used (Fig. 4A). Secondly, a linear soil production law is assumed (Fig. 4B). Thirdly, a constant soil production rate is investigated (Fig. 4C). A fourth, hybrid case is also investigated, in which the soil production law is taken equal to the exponential function at the outset of the numerical simulation ($t \leq 30{,}000$ yr, so that the soil geometry imposed is independent from its initial conditions) and is later set equal to zero (Fig. 4D).

The results of our computation clearly demonstrate that to reproduce the observed linear relationship between soil thickness and curvature, soil production must be finite at all times. Moreover, to maximize fit, soil production must decrease as soil depth increases. A constant soil production rate cannot reproduce the soil thickness–curvature relationship, regardless of the assumed value for the production rate. Note, however, that we can reproduce the observed relationship using a linear law as well as an exponential law.

Steady-State Assumption

For the set of parameters used in the previous examples shown in Figure 4, the numerical model reaches steady state after ~50,000 yr. To test whether the observed soil thickness–surface curvature relationship is predicted by the model before it reaches steady state, the landscape evolution has been simulated until t = 10,000 yr, t = 25,000 yr, and t = 50,000 yr, and the results are shown in Figure 5.

For this set of model parameters, a good fit to the data is obtained at all three times of the simulation. The linear relationship between thickness and curvature for soil thickness smaller than 1 m is always predicted. At large soil thicknesses, the relationship breaks down as time proceeds and the system approaches steady state.

This result suggests that soil thickness versus local curvature relationships do not provide good constraints on whether a soil distribution has reached steady state. The linear relationship observed at several sites (Heimsath et al., 1999, 2000) cannot be used as an indication that a local steady state has been reached.

Constraining the Parameterization

Models generated by NA are shown in Figure 6, in which each point, or model, corresponds to a set of parameters (equations 2–4). Results are summarized in Table 1.

The progression of NA in parameter space depends on the defined misfit function. As NA proceeds, sampling is mainly performed where the misfit function is minimum. This is shown in Figure 6, where each dot represents a forward model. The adopted color scheme is related to the percentile value of the misfit function,

$$M,\left(\frac{M - M_{min}}{M_{max} - M_{min}} \times 100\right),$$

so that smaller percentiles (red dots) correspond to minimum misfit values. Therefore, if NA converges toward a single value for a given parameter, small percentiles are concentrated around a unique point in parameter space (e.g., K_d, K_v, n). This means that the value of that given parameter is well constrained by the observations. In contrast, in cases where many models produce similar small percentiles, there exists no local minimum for the misfit function, and many values can be taken, for a given parameter, to reproduce the observed curvature–soil thickness relationship. In this situation, we can state that the observations cannot be used to constrain the given model parameters.

From the results shown in Figure 6, we can conclude the following:

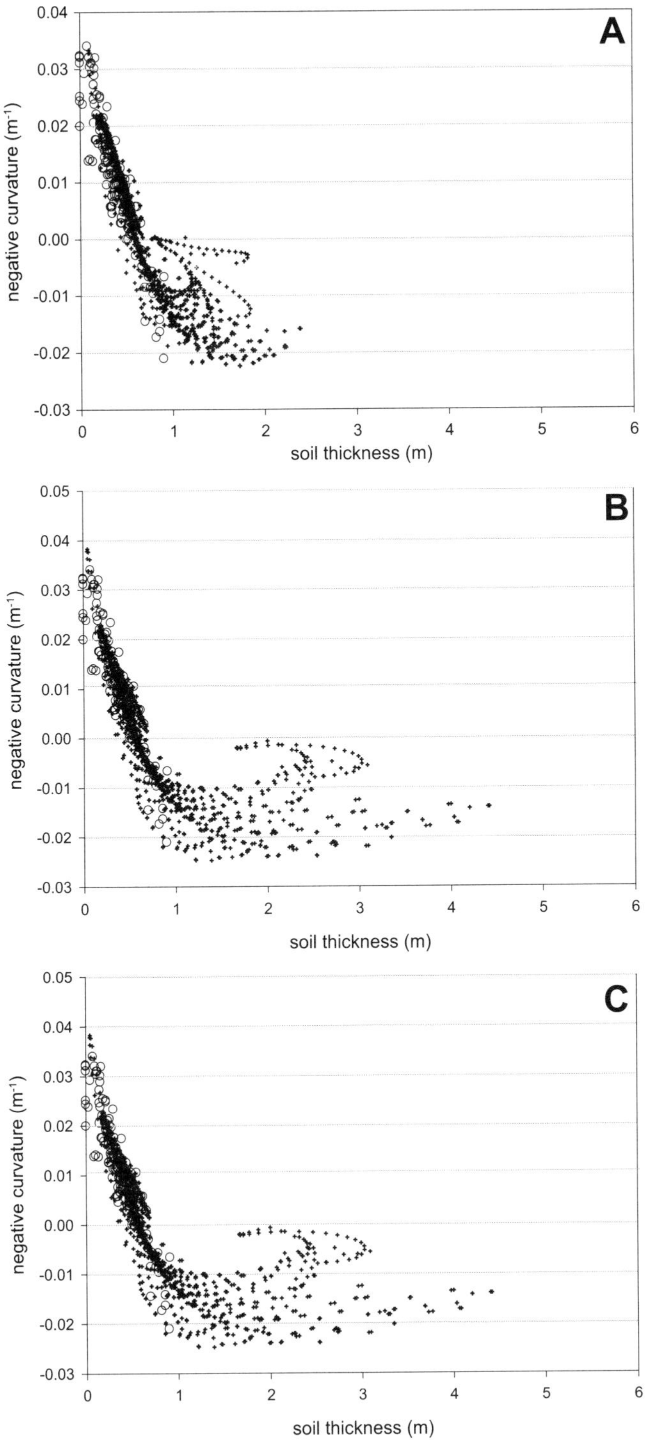

Figure 5. Predicted soil thickness–curvature relationships at different times compared to observations (Bega Valley site) for soil depths between 0 and 5m (note that no data is available for depths greater than 1 m), under steady-state assumption. (A) Represents observed and computed values after 10,000 yr, (B) after 25,000 yr, and C) after 50,000 yr (t = 0 corresponds to the initial conditions).

1. Firstly, the observed curvature–soil thickness relationship (Fig. 2) can be used to constrain the linear creep law parameter, K_d.
2. Secondly, most of the depth-dependent creep parameters are well constrained by the observations, and the optimum values obtained for the slope and thickness exponents are roughly consistent with mechanistically derived values (Olson, 1973), i.e., assuming that soil behaves according to Manning's equation for liquid flow.
3. Thirdly, the nonlinear mass transport (Roering et al., 1999) must be minimized so that the contribution to the total soil flux is insignificant in this part of the landscape.
4. Finally, K_w and p, the overland flow parameters, cannot be constrained by soil thickness and curvature observations alone. In contrast, k is minimized.

DISCUSSION

The work presented here offers a stronger theoretical support for the use of multiprocess parameterizations for modeling soil transport mechanisms on soil-mantled hill slopes. In particular, we have demonstrated the need for the soil production to depend on soil thickness (either linearly or exponentially) for the model predictions to reproduce the observed distribution of soil on the landscape. This implies that if rock to soil conversion is enhanced or inhibited by changes in climate or the exhumation of different lithologies, the distribution of soil on the landform and, consequently, the rate of topographic change by soil transport will be strongly affected. This implies that more observations on the exact nature (and general applicability) of the soil production function are needed to constrain the effect of soil transport on the rate of landform evolution.

The steady-state assumption has been broadly discussed previously (Heimsath et al., 1999). However, this synthetic model clearly shows that the steady-state assumption cannot be tested based on the curvature–soil thickness relationship.

Using an inversion method (NA), optimum values can be determined for K_d, K_v, m, and n. NA tells us that the values of K_w and p cannot be constrained by simple geomorphometric observations, such as soil thickness and local surface curvature. In contrast, k can be determined. This strongly suggests that the parameter controlling the overland flow is the size of the catchments. Therefore, we conclude that soil-mantled hillslopes are only affected where the catchments are sufficiently large. We have used the numerical model to try to determine the type of observations that could be used to constrain the value of the overland flow parameters. The results are shown in Figure 7, where the differences between two soil thickness distributions obtained by setting different values for K_w along two different profiles are shown. All other parameter values are moderately adjusted in order to keep a good fit to the observed curvature–soil thickness relationship (Fig. 7). It can be seen that differences in soil depths are larger where surface is concave, i.e., in the region of maximum topographic convergence. However, the gradients might

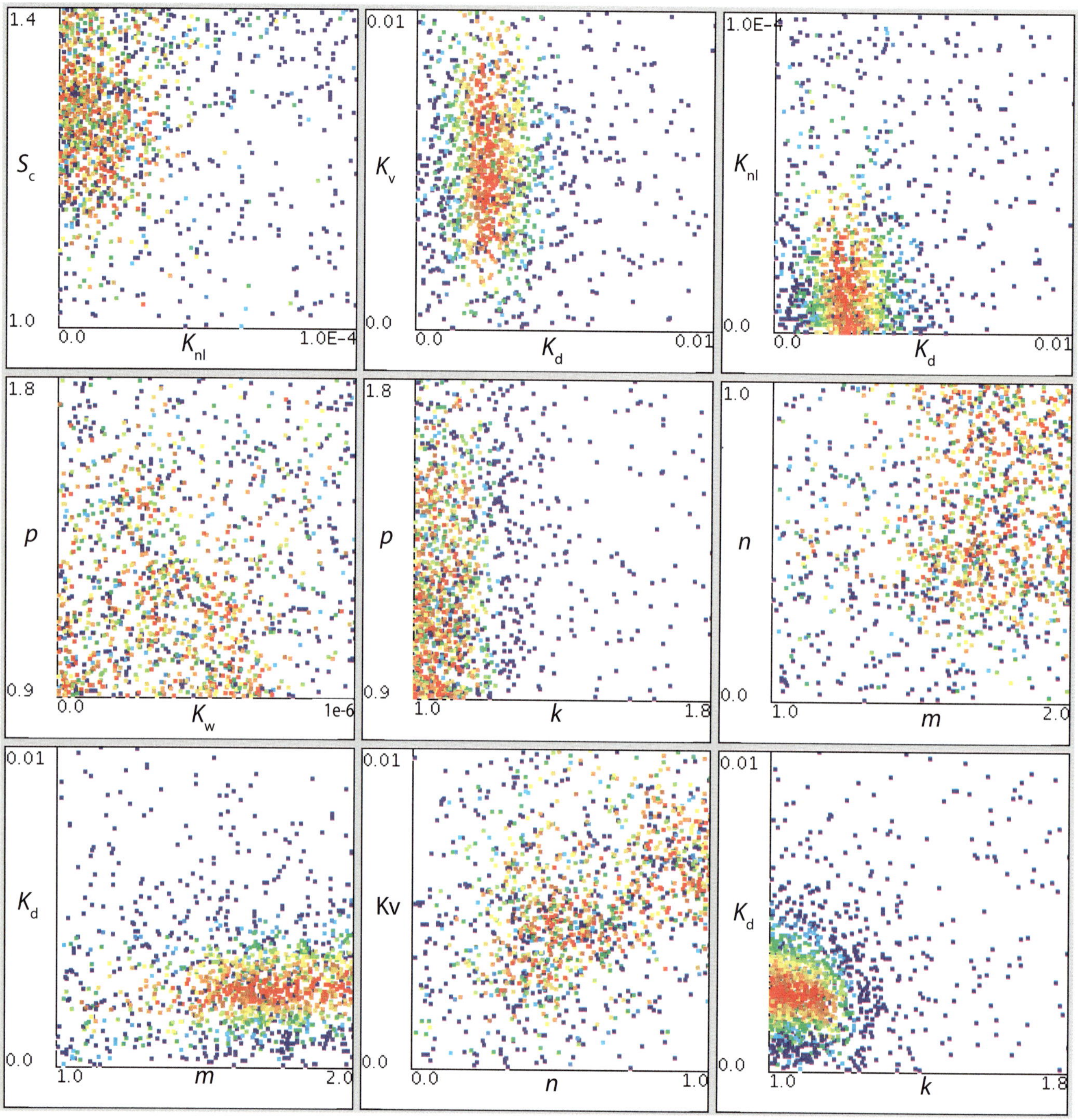

Figure 6. Results from neighborhood algorithm (NA): two-dimensional cross sections in multidimensional parameter space. Each dot represents a forward model run. The color scheme is proportional to data misfit (forward models that produce high misfit are represented in blue, low misfit in red). As the misfit function is minimized, percentiles (M–M_{min} / M_{max}–M_{min}) tend to zero. Results show that a local misfit minimum exists for K_d, K_v, K_{nl}, m, and n, but not for K_w, k, and p (see text for explanation). 6520 models were performed by the algorithm.

TABLE 1: NEIGHBOURHOOD ALGORITHM RESULTS

Transport mechanism	Parameter	Range	Convergence value
Simple creep	K_d	0.0;1e^{-2} ($m^2 \bullet yr^{-1}$)	2e^{-3}
Depth-dependent creep	K_v	0.0;1e^{-2} ($m^{2-m} \bullet yr^{-1}$)	5e^{-3}
	m	1.0;2.0	1.2
	n	0.0;1.0	0.7
Nonlinear	K_{nl}	0.0;1e^{-4} ($m^2 \bullet yr^{-1}$)	0.0
	S_c	1.0;1.4	no convergence
Overland flow	K_w	0.0;1e^{-6} ($m^{2-3k} \bullet yr^{k-1}$)	no convergence
	k	1.0;1.8	1.0
	p	0.9;1.8	no convergence

Note: Acceptable range values for K_d, K_v, K_{nl}, S_c, K_w, m, and n ranges are shown in Braun et al. (2001) and Roering et al. (1999), and k and p ranges are defined from Prosser and Rustomji (2000).

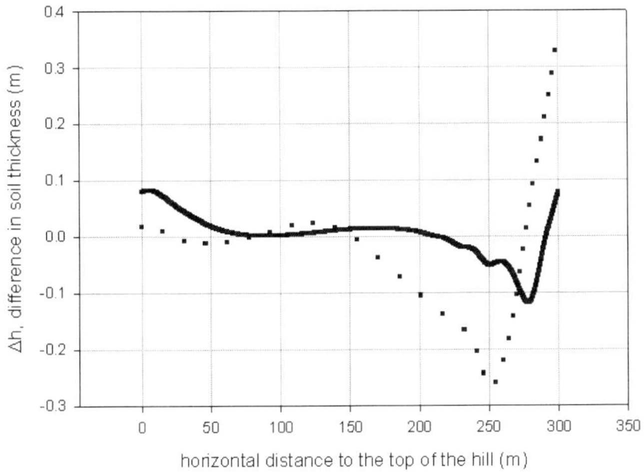

Figure 7. Sensitivity to K_w: The solid line represents the difference (Δh) in computed soil thickness with respect to the best-fitting model along a profile from the top of the hill to a lower boundary at $y = 40$ m (profile 1). The dashed line represents Δh along a profile at $y = 150$ m (central region of the synthetic hill, profile 2). The first parameter set is: $K_d = 2.2 \times 10^{-3}$ (m^2 yr^{-1}), $K_v = 6.27 \times 10^{-3}$ (m^{2-m} yr^{-1}), $K_w = 2.27 \times 10^{-9}$ (m^{2-3k} yr^{k-1}), $m = 1.29$, $n = 0.8$, $k = 1.0$, $p = 1.18$, $K_{nl} = 0.0$; and the second is: $K_d = 2.08 \times 10^{-3}$ (m^2 yr^{-1}), $K_v = 4.68 \times 10^{-3}$ (m^{2-m} yr^{-1}), $K_w = 4.7 \times 10^{-7}$ (m^{2-3k} yr^{k-1}), $m = 1.12$, $n = 0.66$, $k = 1.07$, $p = 1.60$, $K_{nl} = 0.0$. For both sets a good fit to the observed soil thickness–curvature relationship is observed (misfit is smaller than 1×10^{-3}).

get too large in this area, so that the nature of soil depth is highly dependent on landsliding as well as overland flow. Therefore, it is less likely that steady-state conditions are applicable here, as evacuation and infilling can occur at very short time scales for both synthetic and real hills. Our parametric study consequently suggests that to better constrain transport mechanism parameters, further field investigations on mantled hill slopes are required, with a particular emphasis on sites which would offer larger drainage areas with small gradients.

ACKNOWLEDGMENTS

The authors would like to thank M. Sambridge for his useful suggestions concerning the use of the neighbourhood algorithm (NA), as well as Arjun Heimsath, Joshua Roering, and Mark Brandon for their constructive reviews.

REFERENCES CITED

Ahnert, F., 1967, The role of the equilibrium concept in the interpretation of landforms of fluvial erosion and deposition, *in* Macar, P., ed., L'Evolution des versants: Belgium, Université de Liège, p. 22–41.

Anderson, R.S., 1994, Evolution of the Santa Cruz Mountains, California, through tectonic growth and geomorphic decay: Journal of Geophysical Research, v. 99, p. 20,161–20,179, doi: 10.1029/94JB00713.

Braun, J., and Sambridge, M., 1997, Modelling landscape evolution on geological time scales: A new method based on irregular spatial discretization: Basin Research, v. 9, p. 27–52, doi: 10.1046/j.1365-2117.1997.00030.x.

Braun, J., Chapell, J., and Heimsath, A.M., 2001, Sediment transport mechanisms on soil-mantled hillslopes: Geology, v. 29, p. 683–686, doi: 10.1130/0091-7613(2001)029<0683:STMOSM>2.0.CO;2.

Carson, M.A., and Kirkby, M.J., 1972, Hillslope form and process: Cambridge, Cambridge University Press, p. 272–291.

Cox, N., 1980, On the relationship between bedrock lowering and regolith thickness: Earth Surface Processes and Landforms, v. 5, p. 271–274.

Culling, W.H.E., 1960, Theory of erosion on soil-covered slopes: Journal of Geology, v. 73, p. 230–254.

Davis, W.M., 1892, The convex profile of badland divides: Science, v. 20, p. 245.

Dietrich, W.E., Reiss, R., Hsu, M.-L., and Montgomery, D.R., 1995, A process-based model for colluvial soil depth and shallow landsliding using digital elevation data: Hydrological Processes, v. 9, p. 383–400.

Gilbert, G.K., 1877, Report on the geology of the Henry Mountains (Utah): U.S. Geological Survey Professional Paper, 170 p.

Gilbert, G.K., 1909, The convexity of hilltops: Journal of Geology, v. 17, p. 344–350.

Heimsath, A.M., Dietrich, W.E., Nishiizumi, K., and Finkel, R.C., 1997, The soil production function and landscape equilibrium: Nature, v. 388, p. 358–361, doi: 10.1038/41056.

Heimsath, A.M., Dietrich, W.E., Nishiizumi, K., and Finkel, R.C., 1999, Cosmogenic nuclides, topography, and the spatial variation of soil depth: Geomorphology, v. 27, p. 151–172, doi: 10.1016/S0169-555X(98)00095-6.

Heimsath, A.M., Chappell, J., Dietrich, W.E., Nishiizumi, K., and Finkel, R.C., 2000, Soil production on a retreating escarpment in southeastern Australia: Geology, v. 28, p. 787–790, doi: 10.1130/0091-7613(2000)028<787:SPOARE>2.3.CO;2.

Heimsath, A.M., Chappell, J., Spooner, N.A., and Questiaux, D.G., 2002, Creep soil: Geology, v. 30, no. 2, p. 111–114, doi: 10.1130/0091-7613(2002)030<0111:CS>2.0.CO;2.

Hovius, N., Stark, C.P., and Allen, P.A., 1997, Sediment flux from a mountain belt derived from landslide mapping: Geology, v. 25, no. 3, p. 231–234, doi: 10.1130/0091-7613(1997)025<0231:SFFAMB>2.3.CO;2.

Howard, A.D., 1994, A detachment-limited model of drainage basin evolution: Water Resources Research, v. 30, no. 7, p. 2261–2285, doi: 10.1029/94WR00757.

Howard, A.D., 1997, Badland morphology and evolution: Interpretation using a simulation model: Earth Surface Processes and Landforms, v. 22, p. 211–227.

Kooi, H., and Beaumont, C., 1994, Escarpment evolution on high-elevation rifted margins: Insights derived from a surface processes model that combines diffusion, advection and reaction: Journal of Geophysical Research, v. 99, p. 12,191–12,209, doi: 10.1029/94JB00047.

Lewis, P., and Glen, R.A., 1995, Bega-Mallacoota 1:250,000 geological sheet, SJ/55–4, SJ55-8 (second edition): Sydney, Australia, Geological Survey of New South Wales, explanatory notes, 144 p.

McKean, J.A., Dietrich, W.E., Finkel, R.C., Southon, J.R., and Caffee, M.W., 1993, Quantification of soil production and downslope creep rates from cosmogenic ^{10}Be accumulations on a hillslope profile: Geology, v. 21, no. 4, p. 343–346, doi: 10.1130/0091-7613(1993)021<0343:QOSPAD>2.3.CO;2.

Montgomery, D.R., and Dietrich, W.E., 1995, Hydrologic processes in a low-gradient source area: Water Resources Research, v. 31, no. 1, p. 1–10, doi: 10.1029/94WR02270.

Moore, I.D., and Burch, G.J., 1986, Sediment transport capacity of sheet and rill flow: Application of unit stream power theory: Water Resources Research, v. 22, p. 1350–1360.

Olson, R.M., 1973, Essentials of engineering fluid mechanics (third edition): New York, Feefer and Simons International University Editions, 637 p.

Prosser, I., and Rustomji, P., 2000, Sediment transport capacity relations for overland flow: Physical Geography, v. 24, p. 179–193, doi: 10.1191/030913300669852483.

Roering, J.J., Kirchner, J., and Dietrich, W.E., 1999, Evidence for a non-linear, diffusive sediment transport on hillslopes and implications for landscape evolution: Water Resources Research, v. 35, no. 3, p. 853–870, doi: 10.1029/1998WR900090.

Roering, J.J., Kirchner J., Sklar, L.S., and Dietrich, W.E., 2001, Hillslope evolution by nonlinear creep and landsliding: An experimental study: Geology, v. 29, p. 143–146, doi: 10.1130/0091-7613(2001)029<0143:HEBNCA>2.0.CO;2.

Rosenbloom, N., and Anderson, R.S., 1994, Hillslope and channel evolution in a marine terraced landscape, Santa Cruz, California: Journal of Geophysical Research, v. 99, no. B7, p. 14,013–14,029, doi: 10.1029/94JB00048.

Sambridge, M., 1999a, Geophysical inversion with a neighbourhood algorithm, I. Searching a parameter space: Geophysical Journal International, v. 138, p. 479–494, doi: 10.1046/j.1365-246X.1999.00876.x.

Sambridge, M., 1999b, Geophysical inversion with a neighbourhood algorithm, II. Appraising the ensemble: Geophysical Journal International, v. 138, p. 727–746, doi: 10.1046/j.1365-246x.1999.00900.x.

Selby, M., 1993, Hillslope material and processes: Oxford, Oxford University Press, p. 252–258.

Small, E.E., Anderson, R.S., Hancock, G.S., and Finkel, R.C., 1999, Estimates of regolith production from ^{10}Be and ^{26}Al: Evidence for steady state alpine hillslopes: Geomorphology, v. 27, no. 1–2, p. 131–150, doi: 10.1016/S0169-555X(98)00094-4.

Manuscript Accepted by the Society 23 June 2005

Printed in the USA

Geological Society of America
Special Paper 398
2006

Influence of erosion on the kinematics of bivergent orogens: Results from scaled sandbox simulations

S. Hoth[†]
J. Adam[‡]
N. Kukowski
O. Oncken
GeoForschungsZentrum Potsdam, Telegrafenberg, 14473 Potsdam, Germany

ABSTRACT

Scaled two-dimensional sandbox experiments are used to investigate the effect of (1) the location of erosion with respect to the convergence geometry and (2) two erosion modes, distributed or focused, which are thought to represent end members, on the distribution and propagation of deformation within bivergent orogens. This study applies to the brittle part of medium-sized, natural orogens. Particle image velocimetry (PIV) is used to analyze the experiments with respect to surface uplift, thrust displacement, and finite strain.

The experiments suggest that deformation responds immediately to erosion. Retro-wedge erosion amplifies the displacement of the basally accreted material, whereas pro-wedge erosion accelerates and additionally redirects the particle flux of the frontally accreted material. Pro- and retro-wedge erosion retards the propagation of deformation within the pro-wedge. This effect is stronger for pro-wedge erosion. Retro-wedge erosion amplifies vertical growth and leads to strain accumulation along the retro shear-zone and the mid-level detachment. Thus, during retro-wedge erosion, cause (erosion) and response (deformation) are significantly offset in space. Since pro-wedge erosion evokes a complete decoupling of the retro-wedge from the pro-wedge, cause and response are spatially more closely related. With respect to the erosion mode, we found that a more focused erosion leads to a more focused strain accumulation. Similarly, focused erosion applied at the pro-wedge deformation front prohibits accumulation of out-of-sequence displacement. In contrast, distributed pro-wedge erosion amplifies out-of-sequence displacement. Thus, a forward breaking or "piggyback" sequence of thrusting might involve considerable out-of-sequence displacement, which is strongly controlled by the erosion mode.

Keywords: Bivergent orogen, erosion, thrust displacement rates, sandbox simulations, PIV.

[†]E-mail: shoth@gfz-potsdam.de.
[‡]Present address: Department of Earth Sciences, Dalhousie University, Halifax, Nova Scotia B3J 3J5, Canada.

Hoth, S., Adam, J., Kukowski, N., and Oncken, O., 2006, Influence of erosion on the kinematics of bivergent orogens: Results from scaled sandbox simulations, *in* Willett, S.D., Hovius, N., Brandon, M.T., and Fisher, D.M., eds., Tectonics, Climate, and Landscape Evolution: Geological Society of America Special Paper 398, Penrose Conference Series, p. 201–225, doi: 10.1130/2006.2398(12). For permission to copy, contact editing@geosociety.org.

INTRODUCTION

Deformation in collisional orogens is a key factor in controlling the magnitude and location of erosion. Orogenic growth leads to topography, which may enhance monsoonal circulation, provide a rain shadow, and guide or even deflect the drainage pattern or the course of glaciers. Changes in topography may induce the rise or fall of the regional base level to which rivers and glaciers adjust (Kutzbach et al., 1993; Hay, 1996). Following this view, erosion responds primarily to deformation. In contrast, recent field studies in the Southern Alps of New Zealand (Koons, 1990, 1995; Tippett and Hovius, 2000), in the European Alps (Schlunegger and Willett, 1999), in the Chugach/St. Elias Range of southern Alaska (Meigs and Sauber, 2000; Sheaf et al., 2003), in the Andes (Horton, 1999; Montgomery et al., 2001), in Taiwan (Lin, 2000), in the Tien Shan (Pavlis et al., 1997; Strecker et al., 2003), and in the Himalayas (Zeitler, 1985; Zeitler et al., 2001; Wobus et al., 2003; Thiede et al., 2004) have shown that the concentration of erosional energy leads to a similar concentration of mechanical energy (Koons, 1990). These regional-scaled studies indicate that erosion in turn influences the distribution and propagation of deformation within orogenic wedges.

Numerical simulations (Beaumont et al., 1992; Willett et al., 1993; Avouac and Burov, 1996; Willett, 1999; Beaumont et al., 2001) have tested and verified the observations that emerge from the above field studies. They further identify the location of erosion with respect to the convergence geometry as a first-order parameter controlling orogen-scale deformation. Accordingly, the asymmetry of erosion would have a pronounced influence on the distribution of deformation as well as the amount and location of exhumation of rocks within either the pro- or the retro-wedge.

Based on a minimum work approach, Masek and Duncan (1998) and Hardy et al. (1998) highlighted the importance of surface processes on orogenic growth. In this perspective, erosion decreases the load upon an inactive thrust and thus reduces gravitational work and normal stress (and thus frictional work), which favors its reactivation instead of the formation of a new thrust. Similarly, predictions derived from the critical taper theory, which links topography and tectonics in a convergent wedge, show that deformation is very sensitive to mass redistributions by surface processes such as erosion or sedimentation (Davis et al., 1983; Dahlen, 1984; Dahlen et al., 1984; Dahlen and Suppe, 1988; Dahlen and Barr, 1989).

This agrees well with results derived from scaled sandbox simulations, which have also suggested that synkinematic erosion promotes out-of-sequence thrusting in the axial zone and that retroverging thrusts are more persistent through time (Merle and Abidi, 1995; Mugnier et al., 1997; Storti et al., 2000; Marques and Cobbold, 2002; Del Castello et al., 2003; Persson et al., 2004).

However, the influence of the location of erosion with respect to the convergence geometry on orogen kinematics at the scale of individual thrusts has not been well understood. In addition, the interaction between the pro- and the retro-wedge in response to erosion has not yet been addressed. We further recognize that erosion of mountain belts by rivers, glaciers, or bedrock landslides can either be distributed over a wide area or can be very focused in space. Given the sensitivity of crustal wedges to erosion, as outlined above, both erosion modes, which are considered as end members, should evoke a characteristic deformation style, which has not yet been investigated. Therefore, the purpose of this study is to document the influence of the location of erosion with respect to the convergence geometry and the mode of erosion, either distributed or focused, on the kinematic evolution of bivergent orogens, i.e., the propagation and distribution of deformation and the temporal variability of displacement accumulated by individual thrusts. We investigate the resulting spatio-temporal distribution of surface uplift as well. Emphasis is laid upon the interaction of individual structures, but also on the communication between the pro- and the retro-wedge in response to erosion. In order to address both purposes, scaled two-dimensional sandbox simulations were used. In this study, they provided an intermediate spatial scale in terms of their structural resolution and may thus help to link the results from lithosphere-scaled numerical simulations with those from local to regional field studies.

Concepts of Continental Collision Zones

Geodynamic concepts of continental collision zones are based on a kinematic and a dynamic assumption. The fundamental kinematic assumption is that collisional orogens result from the partial subduction of continental lithosphere and accretion of crustal material. According to Willett et al. (1993), a velocity discontinuity (singularity) separates the subducting plate from the overriding plate and is thus the lower limit of accretion (Fig. 1). The asymmetry of the subduction process evokes a polarity in the respective crustal mass transfer, i.e.,

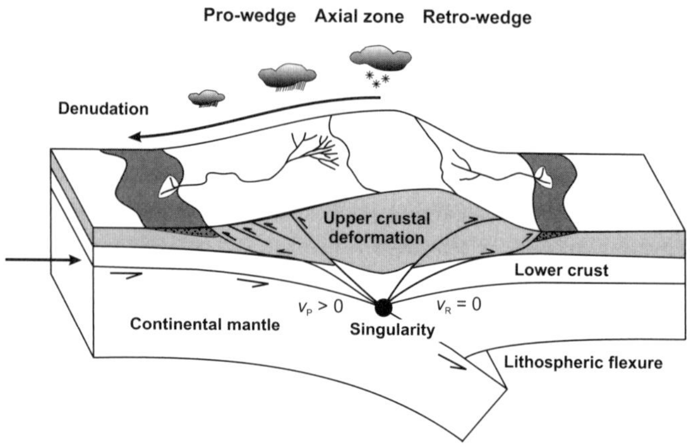

Figure 1. Primary processes controlling the topographic and kinematic evolution of medium-sized continental collision zones (modified from Beaumont et al., 2000a). V_P and V_R are velocities of the downgoing (pro) and the overriding (retro) plate, respectively.

all crustal mass moving into an orogen is derived from the subducting plate and moves toward the overriding plate. This in turn leads to the formation of two crustal sub-wedges. The pro-wedge, located upstream of the singularity, grows by frontal and basal accretion of lower plate material. In contrast, the retro-wedge, located downstream of the singularity, develops by translation of pro-wedge material toward the upper plate. The axial zone is defined as the topographic culmination of the bivergent wedge, which changes its position with respect to the singularity through time.

The transition from subduction to continental collision is coeval with: (1) the progressive deformation and detachment of the pro-wedge from its substratum, and (2) the emergence of frontal accretion within the retro-wedge. In the subsequent continental collision phase, a fully developed doubly vergent orogen grows and propagates toward its respective forelands (Beaumont et al., 1996).

The fundamental dynamic assumption is that a subduction-accretion process leads to the formation of an orogenic wedge with a geometry controlled by the basal and internal mechanical properties (Davis et al., 1983; Willett et al., 2001). A Coulomb-type rheology, which is pressure-dependent but strain-rate–independent, is thought to satisfactorily describe the mechanical behavior of upper crustal rocks and is therefore the stress-limiting factor (Davis et al., 1983; Dahlen, 1984; Dahlen and Suppe, 1988). The key physical assumption implicitly made in the critical wedge concept is that the wedge is at failure everywhere and thus in a critical mechanical state. According to this concept, orogenic wedges evolve toward a state of dynamic force equilibrium between (1) gravitational forces, which arise from the surface slope, (2) a basal shear force, which results from underthrusting of material beneath the wedge, and (3) a compressional force, derived from the convergent motion of the involved continental plates (Willett, 1999). It follows that if the gravitational forces are modified by erosion or sedimentation, the others have to adjust accordingly. A critically tapered wedge slides stably over its base without internal deformation and grows self similar (Davis et al., 1983; Dahlen, 1990).

Surface processes tend to perturb the critical topographic form and may thus control the propagation and distribution of deformation within an orogenic wedge. Outward propagation of deformation toward the foreland is more likely to occur in regions where erosion cannot maintain the surface gradient at, or below, a critical taper, or where enhanced sedimentation within intra-montane basins adds to the thickness of the orogenic wedge (Schlunegger, 1999; Hovius, 2000). In contrast, erosion decreases the thickness of the wedge, which is equivalent to a decrease in the strength of crustal faults and thus drives continued internal deformation until the critical taper is restored (Davis et al., 1983; Dahlen and Suppe, 1988; Willett, 1999). The correction of the perturbed topographic shape back to its critical state may also lead to changes in the pre-existing morphology, the pre-existing drainage pattern, and the erosion rates, which may finally trigger different modes of deformation.

Orogen-Scale Erosion

One implication of the subduction-accretion process is the tectonic advection of material toward the surface of a collisional orogen. How the resulting topography is counteracted by rivers, glaciers, and bedrock landslides depends on a number of factors. Among the most important ones are latitude and altitude of an orogen, its position with respect to global atmospheric and oceanic circulation systems, the erosivity of antecedent rivers, channel slope, orientation of topographic slope with respect to geologic structures, water discharge, lithology, and rate of uplift (Selby, 1982; Seidl and Dietrich, 1992; Howard et al., 1994; Sugai et al., 1994; Hay, 1996; Schmidt and Montgomery, 1995; Hovius, 1998; Meigs and Sauber, 2000; Zeitler et al., 2001).

Given sufficient precipitation, a drainage network will cover the entire orogen up to the channel heads or the snow line. Thereby, the spacing of the transverse streams, which transport most of the material eroded within an orogen, is very regular and is linearly correlated with the half-width of the orogen (Hovius, 1996). This spacing ratio does not differ significantly between young and old orogens and is thus independent of the age or the degree of maturity of an orogen, i.e., the collision stage. Nearly all orogens show, at least in parts, such a drainage network pattern, among them: the Southern Alps, New Zealand; Finisterre Range, Papua New Guinea; Central Range, Taiwan; the Apennines; the Peruvian Andes; and the European Alps (Hovius, 1996). This regular organization of drainage networks and the associated erosion is thought to be characteristic for a distributed erosion pattern. In contrast, monsoon-derived precipitation may focus erosion as demonstrated by Wobus et al. (2003), Hodges et al. (2004), and Thiede et al. (2004). According to Thiede et al. (2004), more than 80% of the annual precipitation (>2000 mm/yr) of the Indian summer monsoon is forced out along the High Himalayan mountain front in NW India at elevations between ~2000 and 3500 m. This altitudinal band corresponds with a ~50–70-km-wide zone of high exhumation rates. Furthermore, Bookhagen et al. (2003) and Thiede et al. (2004) showed that where monsoonal circulation reaches major N-S–oriented valleys in the High Himalayas, moisture is channeled further into the orogen and may lead to higher erosion rates along these valleys. Antecedent rivers with high erosivity, which cross an orogen, might also focus erosion. Both the Indus and the Tsangpo Rivers were capable of cutting steep gorges with several thousand meters of relief into the massifs of Nanga Parbat and Namche Barwa, respectively (Zeitler et al., 2001; Koons et al., 2002). This focused erosion led to the rapid exhumation of Quaternary metamorphic rocks and granites (Zeitler et al., 2001) and may also have attracted channel flow within the middle to lower crust (Zeitler et al., 2001; Beaumont et al., 2001). Finally, orogen-scale glacial erosion can also be considered as focused. Glacial erosion is at its maximum along the mean position of the equilibrium line altitude (ELA), where the net mass balance of glaciers is equal to zero (i.e., accumulation equals ablation), the sliding velocity of glaciers is highest, and so is the erosion rate. Below the ELA, glacial erosion decreases and fluvial processes start to dominate. Due to the transition from warm-based to cold-based glaciers, which are frozen to their beds at higher altitudes, glacial erosion decreases above the ELA as well (Meigs and Sauber, 2000; Burbank, 2002). One prominent example for this scenario is the Chugach/St. Elias Range in southern Alaska, where high deformation and exhumation rates are confined to the altitudinal band spanned by the ELA (Meigs and Sauber, 2000; Sheaf et al., 2003).

Based on the above observations we suggest that orogen-scale erosion can either be distributed or focused. These modes are considered as two alternatives of the spatial distribution of erosion. Given the sensitivity of crustal wedges to erosion as outlined in the preceding section herein, both erosion modes should evoke a characteristic deformation pattern.

EXPERIMENTAL METHOD

We used scaled two-dimensional sandbox simulations to analyze the influence of the location of erosion with respect to the convergence geometry and the erosion mode on the kinematics and the associated tectonic displacement field in bivergent sand wedges. In order to decipher the influence of each parameter, we carried out five experiments: a reference experiment without and four experiments with erosion, where we changed one parameter at a time. We have chosen scaled two-dimensional sandbox simulations because they enable us to simulate large magnitudes of convergence. Also, they allow a direct observation of strain localization and fault propagation with a high temporal and spatial resolution, especially if monitoring techniques such as particle image velocimetry (PIV) are applied. Furthermore, granular flow of sieved, dry quartz sand, which is characterized by a strain-dependent deformation behavior with prefailure strain hardening and postfailure strain softening (stress-strain curve), is similar to the nonlinear deformation behavior of crustal rocks in the brittle field (Mandl et al., 1977; Lohrmann et al., 2003) and can thus be considered as a good analogue for the latter. According to Byerlee (1978), both sand and rocks deformed under brittle conditions tend to localize deformation, i.e., the frictional strength along faults is reduced with respect to the undeformed material. The respective strength contrast for the sand used in this study is on the order of 20% (Table 1).

The observed similarity between the mechanical properties of sand and the brittle upper crust justifies the application of laboratory sand wedges to investigate the kinematics and dynamics of accretionary wedges, fold-and-thrust belts, as well as bivergent orogens. It further supports the application of the critical wedge concept, which assumes a Mohr-Coulomb–type rheology to laboratory sand wedges (Davis et al., 1983). This approach has been followed by a variety of studies (e.g., Malavieille, 1984; Byrne et al., 1993; Kukowski et al., 1994; Storti and McClay, 1995; Wang and Davis, 1996; Gutscher et al., 1998; Nieuwland et al., 2000; Storti et al., 2000; Cobbold et al., 2001) and ensures comparability between results.

The geometric scaling factor (~10^5) between the analogue experiments and nature was derived from the mechanical properties and the density of the sand (Table 1), which were measured with a ring shear apparatus (Hubbert, 1937; Schellart, 2000). Thus, 1 cm in the models corresponds to ~1 km in nature. This scaling approach, which is commonly used for sandbox simulations (Storti et al., 2000; Gutscher et al., 2001; Kukowski et al., 2002; Lohrmann et al., 2003; Hampel et al., 2004), requires that all experiments be performed under normal gravity conditions.

The sandbox simulations were run in a glass-sided deformation apparatus with internal dimensions (length, width, height) of 340 × 21 × 60 cm (Fig. 2A). It consisted of an upper and a lower elastic wooden plate, which overlapped in the center and were only fixed at their respective outward sides to allow for load driven flexure as recognized in natural orogens. The lower plate was supported by a spring to adjust its stiffness. As a first-order approximation, a plate-spring combination was chosen to ensure a linear increase of the flexural amplitude as a function of added load. The corresponding dip angle, β, changed from ~5° to ~6° during experimental runs. Thus, this setup allowed a vertical but not a horizontal migration of the singularity. In order to minimize the influence of the boundary conditions imposed by the experimental device, the singularity was located sufficiently away from the rigid back wall.

A key model assumption was that the mechanical energy delivered into the orogen is derived from the basal pull or drag of the subducting plate (Silver and Reed, 1988). Lower plate subduction and basal shear was simulated by a conveyor belt, which covered only the lower plate and was drawn by a

TABLE 1. PHYSICAL PROPERTIES OF ANALOGUE MATERIALS USED IN EXPERIMENTS

Material	Type	Grain size (μm)	Preparation	Density (g/cm³)	Φ_{peak}† (°)	C_{peak} (Pa)	Φ_{static}‡ (°)	C_{static} (Pa)	$\Phi_{dynamic}$§ (°)	$C_{dynamic}$ (Pa)	Strain softening (%)
Sand	Internal	20–630	Sieved	1.74 ± 0.01	35.36 ± 0.34	43.47 ± 44.34	29.82 ± 0.19	176.38 ± 24.80	28.20 ± 0.05	94.66 ± 6.53	20.24
Glass beads	Internal	300–400	Sieved	1.59 ± 0.01	29.14 ± 0.14	–6.13 ± 18.82	25.91 ± 0.09	78.89 ± 11.56	22.51 ± 0.04	42.71 ± 5.15	13.63
Sandpaper	Basal	<400	Relative to sieved sand	–	32.02 ± 0.32	–63.09 ± 42.49	30.34 ± 0.07	61.37 ± 8.73	28.62 ± 0.15	65.98 ± 19.77	–

Note: Data are derived from ring shear tests. Interpolation of the Mohr envelopes was used to calculate cohesion.
†Peak friction provides the strength of undeformed material.
‡Static-stable friction provides the strength of previously deformed material.
§Dynamic-stable friction provides the strength of active shear zones (Byerlee, 1978).

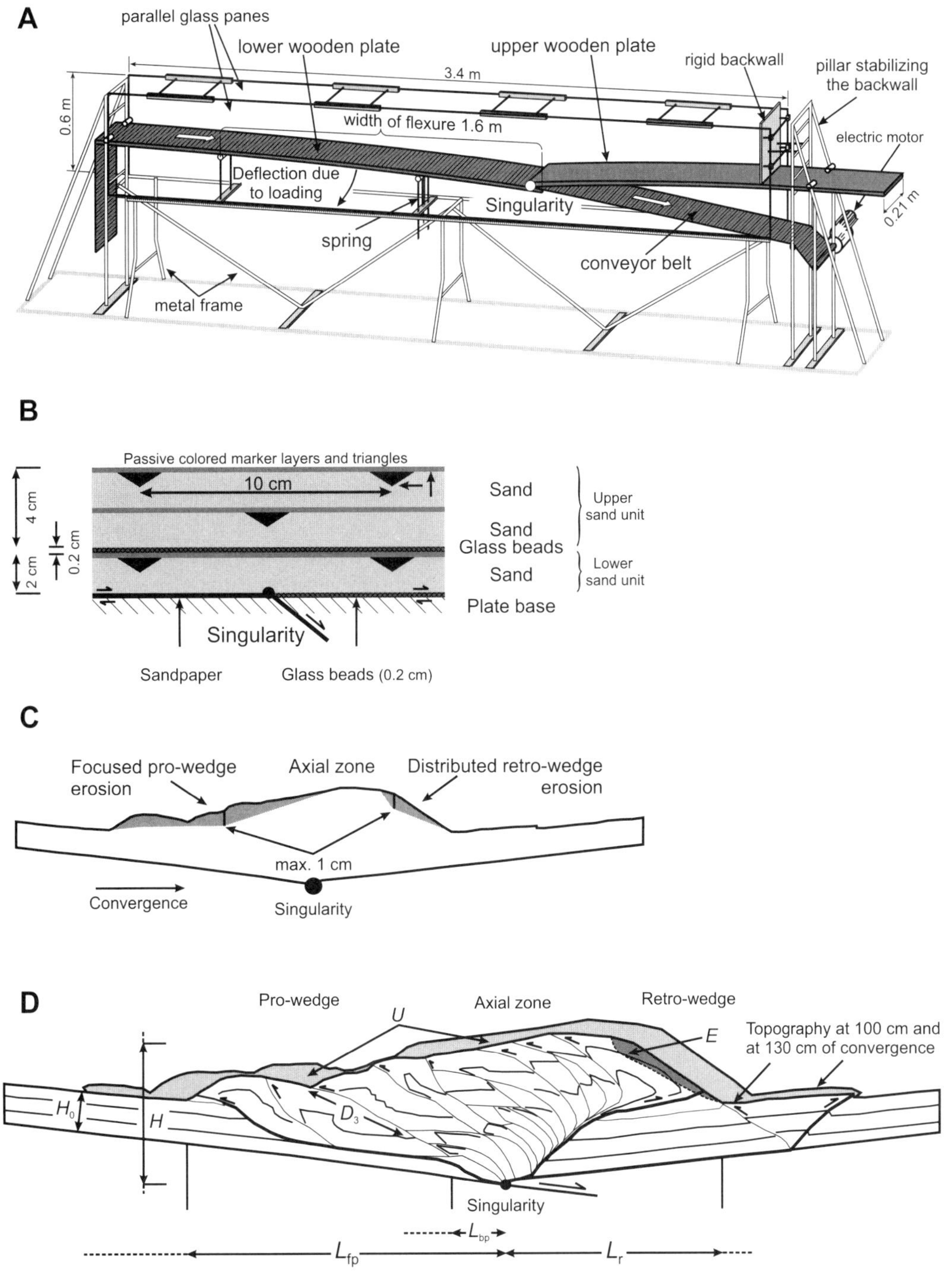

Figure 2. General setup of sandbox simulations. (A) The sandbox that allows load-driven flexure; (B) the mechanic stratigraphy, where the upper (lower) sand unit is frontally (basally) accreted, and the glass-bead layer acts as a possible detachment; (C) geometries of incremental erosion, either distributed or focused; and (D) measured parameters that describe bivergent wedge evolution: (1) propagation of the deformation front of frontal accretion in the pro-wedge (L_{fp}), (2) propagation of the deformation front of basal accretion in the pro-wedge (L_{bp}), (3) topographic height above the singularity (H), (4) propagation of the deformation front of the retro-wedge (L_r), (5) length of individual thrusts ($D_{1...8}$), (6) erosion (E), and (7) incremental surface uplift (U). The subscript p relates to the pro-wedge, the subscript r to the retro-wedge, the subscript f to frontal accretion and the subscript b to basal accretion. H_0 is the height of the undeformed multilayer. All geometric features of individual experiments are normalized with respect to H_0 to account for slight variations in the thickness of the multilayer.

motor (~9 cm/min) beneath the tip of the upper plate. In order to ensure a strong mechanical coupling between the conveyor belt and the sand layer, a sandpaper with a grain size <0.4 cm was chosen. Its friction with respect to the used sand was measured with a ring shear tester (Table 1) and was in the same range as published values for similar experimental devices (Wang and Davis, 1996; Kukowski et al., 2002). A lower basal friction would have only resulted in lower and wider bivergent sand wedges. Excess pore pressures would have played a similar role, since the dip of thrusts decreases with elevated fluid pressures (Cobbold et al., 2001).

The sand layer was 6 cm thick and covered the upper and lower plate base equally. Thus, as a simplifying assumption, we simulated collision of materials that have the same mechanical properties on both sides of the singularity (e.g., Koons, 1990; Storti et al., 2000). We did not account for either the progressive forelandward decrease in the thickness of the sedimentary succession or inherited zones of weakness commonly found in orogenic forelands. Both would have led to a change in the spacing of the thrust ramps only (Bombolakis, 1986; Boyer, 1995; Hardy et al., 1998; Macedo and Marshak, 1999).

Most orogens, e.g., Pyrenees and European Alps, show a variety of possible detachment horizons, which can lead to orogen-scale frontal and basal accretion (e.g., Beaumont et al., 2000b; Pfiffner et al., 2000). Therefore, dry glass beads, which are characterized by a Mohr-Coulomb–type rheology but show significantly lower frictional properties than the used sand (Table 1), were taken to simulate such a weak layer. The glass beads were positioned on top of the lower sand unit (2 cm) followed by the 4-cm-thick upper sand unit (Fig. 2B). Since an internal glass-bead layer only serves as a detachment level if its internal friction is lower than the basal friction (Kukowski et al., 2002), a relatively high basal friction had to be chosen as outlined above.

In order to promote frontal accretion within the upper plate, a basal glass-bead layer was also introduced as a potential detachment horizon between the sand and the upper plate base. Since we intended to study the most general and simplest scenario, we did not account for the variety of materials with differing frictional properties found in continental collision zones. Convergence was on the order of 150 cm (~150 km in nature) and was thus similar to medium-sized orogens like the European Alps or the Pyrenees, but imposed a limitation as well. Thus, the applied kinematic boundary conditions resembled those in numerical and other physical simulations (Willett, 1999; Beaumont et al., 2000b; Storti et al., 2000) and mimicked the geometry of numerous published sections through small orogens (e.g., Beaumont et al., 2000b; Pfiffner et al., 2000; Schaefer et al., 2000).

In the erosion experiments, the sand wedge was allowed to produce an initial topography during the first 40 cm of convergence. This amount of convergence was chosen to ensure that the bivergent sand wedge had been fully established. Erosion was subsequently simulated at every 10 cm of convergence with a vacuum cleaner. In order to detect the influence of the location and the mode of erosion on the kinematics and the associated tectonic displacement field in bivergent sand wedges, we made five simplifying assumptions: First, only the pro- or the retro-wedge was subject to erosion in any given model. Second, we used the same convergence interval (10 cm of convergence) for both erosion modes, to ensure comparison between experiments. Third, both erosion modes were investigated separately, despite the observation that most natural bivergent orogens undergo distributed and focused erosion at the same time. Fourth, we assumed that all eroded material leaves the model system. Fifth, we assumed the erosion rate to be linearly dependent on elevation. This first-order approach was based on the observation that the rate of erosion is positively correlated with local relief (Ahnert, 1970; Milliman and Syvitski, 1992; Summerfield and Hulton, 1994) and that local relief is in turn positively correlated with elevation (Summerfield and Hulton, 1994). Although this causal relationship might not be valid everywhere, e.g., Tibetan Plateau, and is incompatible with fluvial incision laws, it would provide a first-order estimate. An elevation-dependent erosion law has been successfully applied in numerical simulations of continental collision zones as well (Pfiffner et al., 2000). Accordingly, we simulated distributed erosion during sandbox experiments as follows: erosion increased linearly from the toe of either the pro- or the retro-wedge toward the top of the bivergent sand wedge, where it reached its maximum of 1 cm/10 cm of convergence (Fig. 2C). Erosion in the focused mode reached a maximum of 1 cm/10 cm of convergence at the middle of either the pro- or the retro-wedge slope and decreased toward zero at the respective tops and toes.

In order to scale the magnitude of erosion and to compare it with natural examples, we calculated the erosion efficiency parameter $r = v_e/v_c$, where v_e is the erosion rate, i.e., 1 cm of erosion per 10 cm of convergence, and v_c is the convergence velocity, i.e., 10 cm of convergence. It is highlighted here that the velocity with which the conveyor belt is drawn (~9 cm/min) is of no importance for the above calculation, since deformation of a Mohr-Coulomb material like sand is not time-dependent. We further assume that the instantaneous removal of sand as simulated reflects a continuous removal of 1 cm per 10 cm of convergence. The resulting value $r = 0.1$ is thus well in the range of natural values obtained from currently deforming orogenic wedges, such as Taiwan: 0.03–0.18 (Dadson et al., 2003), New Zealand: 0.5–1.3 (Hovius et al., 1997; Koons et al., 2003), southern Alaska: 0.02–0.94 (Meigs and Sauber, 2000), and the Himalayas: 0.07–0.25 (Bilham et al., 1997; Zeitler et al., 2001; Pratt-Sitaula, 2004; Thiede et al., 2004).

All experiments were monitored in cross section with two, high-speed, high-resolution digital cameras to image the particle flow in the sand wedge (Adam et al., 2005; Hampel et al., 2004). Images were taken at every 2 s. Based on an adaptive cross correlation algorithm, successive images were compared, and the incremental and cumulative particle flow field, its horizontal and vertical components, and all its derivatives, such as the incremental and cumulative strain, were calculated. The visualization of finite strain allows the comparison between individual structures in one experiment and the evaluation of the manner in which tested parameters influence finite strain accumulation at similar structures in all experiments. The accuracy of the displacement measurement is ~0.5 mm (equal to 50 m in nature). From these digital images, time series data of several geometric features, such as the propagation of deformation, the spatio-temporal distribution of surface uplift and individual particles paths were derived (Fig. 2D). These geometric features are thought to be indicative of the expected kinematic evolution of a bivergent sand wedge and allow a quantitative comparison between experiments. They may also provide a link to data from natural bivergent orogens. All geometric data derived from PIV analysis were normalized with respect to H_0, which is the height of the undeformed multilayer, to account for slight variations in its thickness (Fig. 2D). If not stated otherwise, all measured geometric data were taken at every fifth image, which corresponds to a convergence interval of 1.5 cm. Examination of both cross sections during and after each experiment revealed that the plane strain assumption is valid. Reproducibility is justified by comparison of the trends and mutual relations of different data sets (propagation of deformation, particle paths, amount of erosion) with the ones derived from similar experimental series or repeated experiments.

Finally, two conventions were made: The term pro-shear is used for all shear zones that dip toward the upper plate. Accordingly, retro–shear zones dip toward the lower plate. The term retro–shear zone is taken only for the shear zone bounding the initial pop up (see "Results"). The term incremental refers to a convergence interval of 10 cm.

RESULTS

Reference Experiment without Erosion

Kinematic Evolution of a Bivergent Sand Wedge

Based on visual inspection of the reference experiment, the kinematic evolution of a bivergent sand wedge can be divided into four stages (Table 2). During stage I, initial layer-parallel shortening leads to the formation of two conjugate shear zones, which nucleate at the singularity. They define a symmetric pop-up (Fig. 3A). Further convergence leads to a rapid uplift associated with progressive back tilting of the pop-up toward the upper plate. In stage II, three narrowly spaced thrust faults are formed within the pro-layer (Fig. 3B–C) and result in an increasing asymmetry of wedge topography and kinematics. At this stage, rates of thrusting along the retro–shear zone and uplift rates of the axial zone and the retro-wedge are high but start to decrease (Fig. 4A). Stage III commences after ~40 cm of convergence (Fig. 3D–E). The initial pro-wedge now provides sufficient load to activate the internal glass bead layer, which facilitates the coeval formation of thrust imbricates above and duplexes below it. The former is referred to as frontal, the latter as basal accretion. Lateral growth of the pro-wedge is now attained by cyclic formation of flat-topped box anticlines at its toe. Each cycle consists of a thrust initiation phase, during which deformation propagates outward along the internal glass bead layer and finally steps up to form a flat-topped box anticline. Thereby, the respective back thrust terminates at the tip of the previous deformation front. During the underthrusting phase, accreted material is transferred toward the axial zone, back rotated, and finally passively uplifted in the hanging wall of the retro–shear zone. All frontal accretion cycles are equally spaced and thus suggest stable mechanical conditions. Additionally, three footwall shortcuts, which occurred late within an accretion cycle, were observed (Fig. 4A).

Coeval to frontal accretion, the sand layer beneath the glass bead layer detached from the one above and was transferred toward the axial zone. Beneath the base of the pro-wedge, duplexes were formed, stacked, and finally uplifted in the hanging wall of the retro–shear zone. Thus, compared to the pro-wedge, which grew by discrete steps, the axial zone and the retro-wedge were continuously fed with pro-wedge–derived material. Thereby, the axial zone and the retro-wedge grew in width and height. These two modes of addition of new material to the respective sub-wedges further amplified the existing topographic and kinematic asymmetry of the bivergent sand wedge.

TABLE 2. SUMMARY OF KEY CHARACTERISTICS OF BIVERGENT WEDGE EVOLUTION

Stage	Distribution of deformation	Thrusting at the retro shear-zone, uplift of axial zone and retro-wedge	Degree of symmetry of bivergent sand wedge
I[†]	Pop up	Fast	Symmetric
II[‡]	Pro-wedge: high-frequency thrust fault, nucleation close to *S*	Fast, but starts to decrease	Asymmetric
III[§]	Pro-wedge: low frequency nucleation of thrust imbricates far from *S*, duplex formation	Slow	Further increase of asymmetry
IV[#]	Pro-wedge: lower-frequency nucleation of thrust imbricates far from *S*, duplex formation; retro-wedge: thrust initiation within upper plate	Slow	Asymmetric, but symmetry increases

Note: Inspired by Storti et al. (2000, their Fig. 6). Bivergent wedge evolution has been described with different sequences of stages as indicated below.

[†]Corresponds to stage 1 in Storti et al. (2000) and stage 1 in Willett et al. (1993).
[‡]Corresponds to stage 1 in Storti et al. (2000) and stage 2 in Willett et al. (1993).
[§]Corresponds to stage 2 in Storti et al. (2000) and stage 2 in Willett et al. (1993).
[#]Corresponds to stage 3 in Willett et al. (1993).

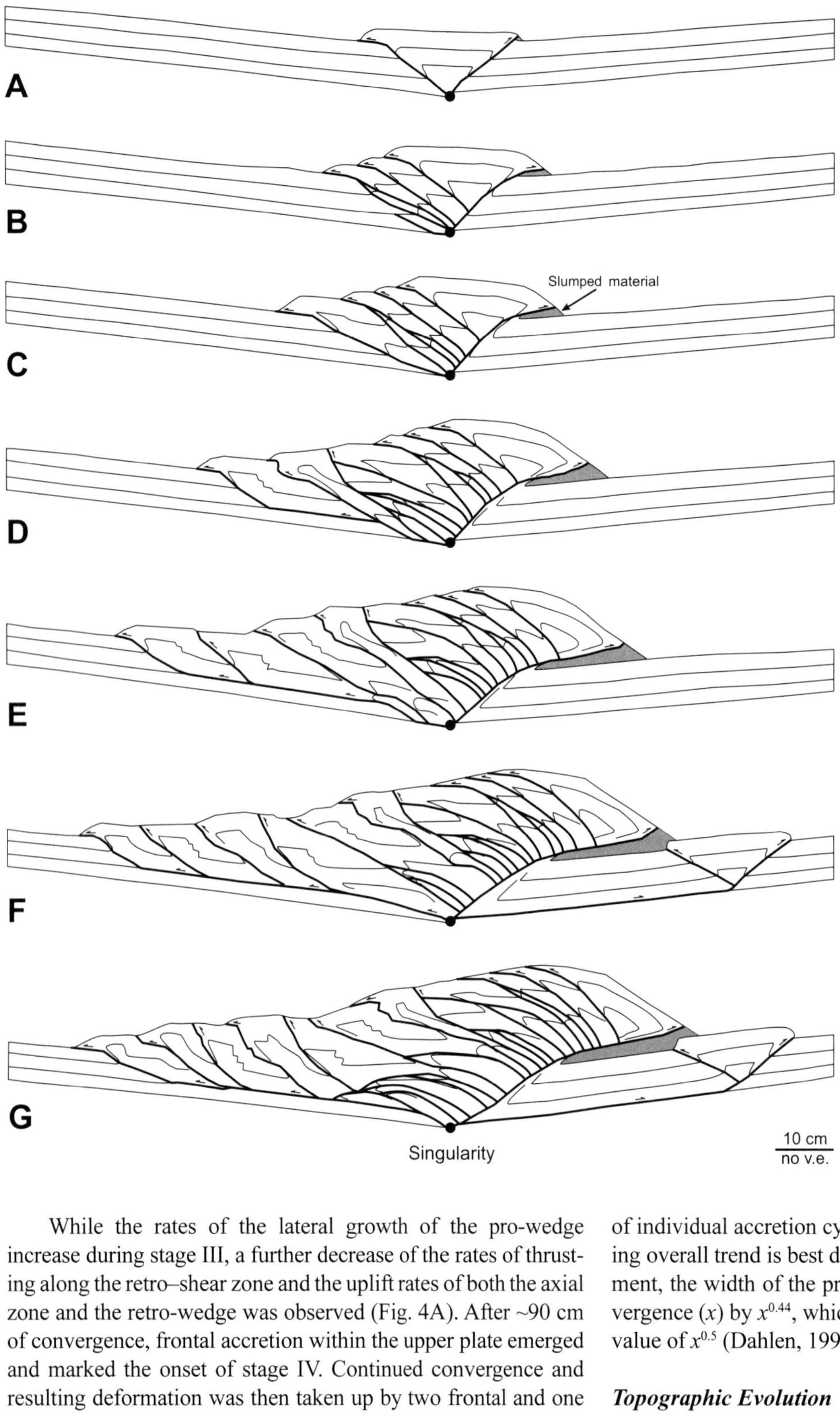

Figure 3. Line drawings of sequential evolutionary stages of the reference experiment: (A) 7 cm convergence; (B) 20 cm convergence; (C) 30 cm convergence; (D) 60 cm convergence; (E) 90 cm convergence; (F) 120 cm convergence; and (G) 140 cm convergence. A footwall shortcut can be observed in (C) before deformation propagates far into the undeformed pro-layer. Frontal accretion within the upper plate emerges after ~90 cm of convergence; v.e. is vertical exaggeration.

While the rates of the lateral growth of the pro-wedge increase during stage III, a further decrease of the rates of thrusting along the retro–shear zone and the uplift rates of both the axial zone and the retro-wedge was observed (Fig. 4A). After ~90 cm of convergence, frontal accretion within the upper plate emerged and marked the onset of stage IV. Continued convergence and resulting deformation was then taken up by two frontal and one basal accretion system (Fig. 3F–G, Fig. 4B). This resulted in a slowdown of the rate with which the pro-wedge grew laterally (Fig. 4A).

It follows from the above description that the propagation of deformation, i.e., the propagation of frontal accretion, consisted of individual accretion cycles and footwall shortcuts. The resulting overall trend is best described by a power law. In this experiment, the width of the pro-wedge grew proportional to the convergence (x) by $x^{0.44}$, which is close to the theoretically predicted value of $x^{0.5}$ (Dahlen, 1990).

Topographic Evolution

The topography of the reference experiment can be divided into two domains (Fig. 4C), one that comprises the pro-wedge and one that includes the axial zone and the retro-wedge. Topography in the former was highly segmented and grew by discrete steps, whereas topography in the latter was smooth and grew

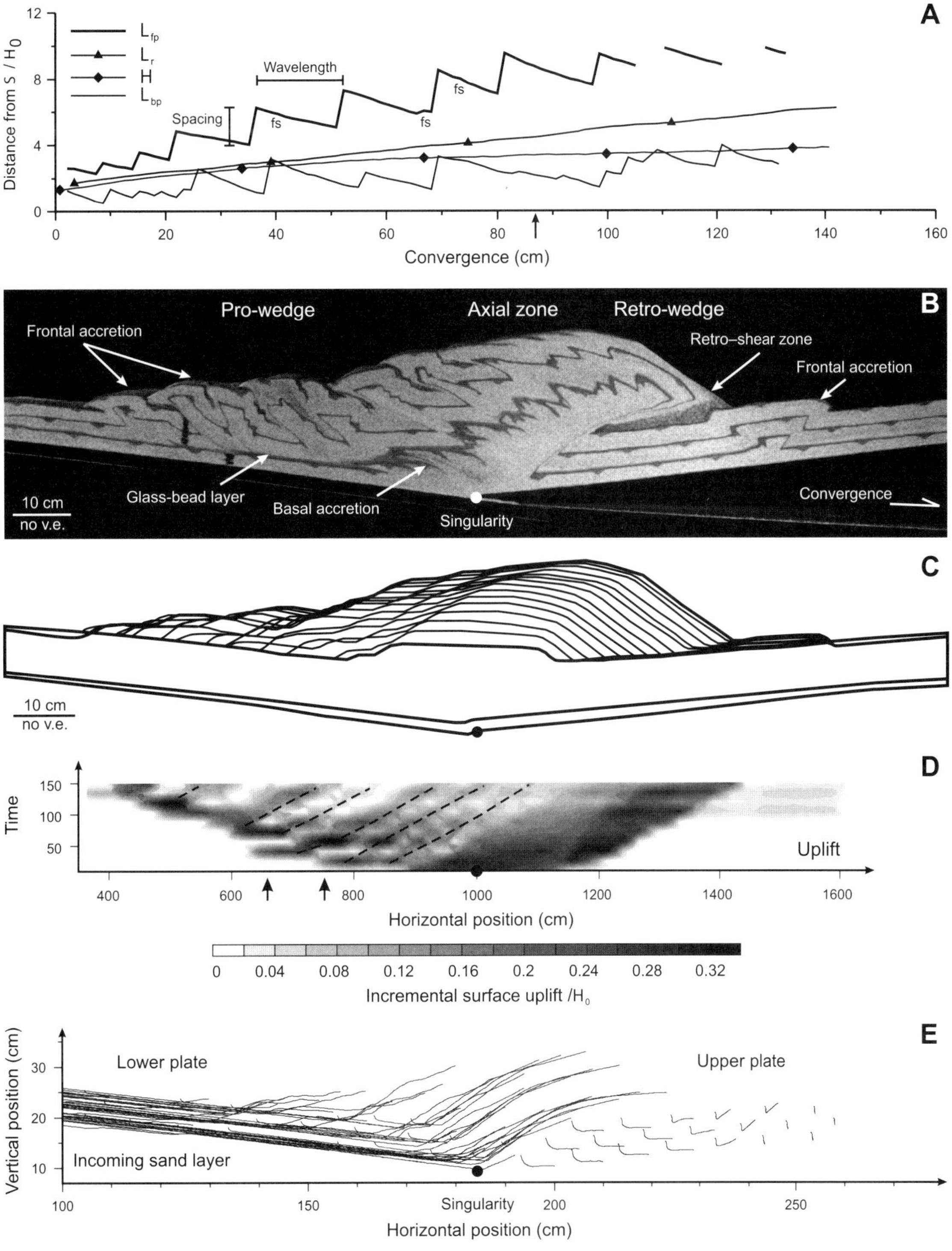

Figure 4. Synopsis of reference experiment. (A) Evolution of geometric parameters as defined in Figure 2D, taken from the digital images at every 1.5 cm of convergence. Arrow indicates onset of upper plate accretion; fs shows footwall shortcuts. (B) Photograph after 140 cm of convergence; v.e. is vertical exaggeration. (C) Topographic evolution. Outlines were taken at every 10 cm of convergence. The first, i.e., after 10 cm, and the last, i.e., after 150 cm of convergence, outline are given in complete form to indicate the magnitude of flexure. Two growth modes can be distinguished: Cyclic accretion within the pro-wedge results in distinct steps in topography, whereas continuous addition of pro-wedge–derived material to the axial zone and the retro-wedge leads to near-concentric growth pattern. (D) Incremental surface uplift with high spatial and temporal variability. Arrows indicate positions where incremental surface uplift changes by 10 during the evolution of the bivergent sand wedge. Dashed lines trace activity of ramp segments. For a more detailed description and a color image see Figure 8. (E) From particle image velocimetry (PIV)-extracted particle paths after 140 cm of convergence.

almost concentrically. The corresponding distribution of incremental surface uplift mirrors these two domains (Fig. 4D).

Maximum incremental surface uplift was associated with the initiation of thrusts within the pro-layer and was located above the respective ramp segment. While convergence proceeded, new thrust faults were formed and older ones were abandoned. This was coeval with a decay of incremental surface uplift for a given ramp segment. However, almost all ramp segments identified in Figure 4D showed a renewed increase in incremental surface uplift, which suggests a reactivation of the respective thrust. The respective increase may reach one order of magnitude, without any modifications of the imposed kinematic boundary conditions (Fig. 4D). Also, the onset of frontal accretion within the retro-wedge resulted in a significant reduction of incremental surface uplift within the pro-wedge.

Experiments with Erosion

In experiments with erosion, the evolving bivergent sand wedges showed remarkable differences as well as similarities to the reference experiment. A summary of the similarities is given first, followed by an analysis of the differences between the experiments.

One of the key observations that emerged from the analysis of the PIV images is that despite the differences in the location and the mode of erosion, all sand wedges evolved into a bivergent state with a pro-wedge, an axial zone, and a retro-wedge (Fig. 5). All four bivergent sand wedges subject to erosion showed at least the first three phases out of the four-stage evolutionary model proposed for the reference experiment. This similarity in bivergent wedge evolution can also be observed in the time series data obtained for several geometric features (Fig. 6). However, only the experiment with distributed retro-wedge erosion evolved into the fourth phase. Finally, all experiments showed a simultaneous occurrence of frontal and basal accretion and no activation of the internal upper plate glass-bead layer.

Particle Paths

For each experiment, two triangular markers with a similar initial position were traced. One marker was located above and one below the internal glass-bead layer, as a representation for frontal and basal accretion. The spatial distribution of these two markers indicates that: (1) different magnitudes of displacement between both accretion modes must have occurred, since the distance between both markers varied between experiments (Fig. 5A–E). They are closest in the distributed pro-wedge erosion experiment and farthest in the reference experiment (Figs. 5A and 5C). (2) The markers for frontal accretion reached a similar position in the reference as well as in the retro-wedge erosion experiments (Figs. 5A, 5B, and 5E). However, the markers for basal accretion show a larger displacement, but they are located within the same region as the respective marker in the reference experiment (Figs. 5A, 5B, and 5E). (3) During pro-wedge erosion, the markers (representative of frontal accretion) indicate a higher displacement and a change in the direction of the particle flow. The markers for basal accretion attained a similar position as observed in the reference experiment (Figs. 5A, 5C, and 5D).

Visual inspection of the photographs (Fig. 5A–E) indicates that the internal lower plate glass-bead layer was exposed at the surface of the pro-wedge during focused pro-wedge erosion, which did not occur in any of the other experiments. Distributed pro-wedge erosion, however, removed nearly the entire upper sand unit and thus led to a dominance of stacked duplexes within the pro-wedge. Only four out of eight thrust imbricates were completely preserved at the final stage of this experiment (Fig. 5A–E). This agrees well with the observation that distributed pro-wedge erosion leads to the maximum amount of mass denuded in any of the four erosion experiments (Fig. 6, Table 3). Also, a mass flux steady-state (tectonic advection equals erosion) was reached at two convergence intervals only during this experiment (Fig. 6). This is consistent with the observation that the experiment with distributed pro-wedge erosion shows the least flexural deflection (Fig. 7). The more general observation, which is derived from Figure 6, is that the amount of incrementally denuded material increases through time as a consequence of the increase in length of either the pro- or the retro-wedge. It has to be emphasized that the simulated erosion modes assume an ideal-shaped bivergent wedge with a smooth topographic gradient. This, however, is perturbed by frontal and basal accretion. Thus, the amount of material taken away at any increment of convergence is dependent on the state within the accretion cycle, i.e., during thrust initiation phases, topography is built up far away from the former pro-wedge toe, which results in an overestimation of the material to be denuded. In contrast, during underthrusting phases, the envelope of the pro-wedge is much smoother and adjustment of the simulated erosion law to this envelope is more accurate.

Frontal and Basal Accretion

Similar to the reference experiment, time series data provided by PIV images show that the propagation of frontal accretion is composed of individual accretion cycles and to a lesser degree of footwall shortcuts (Fig. 6). The resulting overall trend is again best described by a power law (Table 3). At the final stage of each experiment, eight thrusts could be identified within the frontal accretion complex. The experiment with focused pro-wedge erosion, however, shows only six thrust imbricates.

The spacing of frontal accretion, sensu Marshak and Wilkerson (1992), which describes how far deformation propagates into the foreland (Fig. 6), remained fairly constant throughout the experiments. However, the wavelength, which is the time expressed in convergence between two consecutive thrust initiation events, varied within and between experiments and was thus more sensitive to the parameters tested. As a measure of this variability, the sample standard deviation of the wavelength of frontal accretion (s_f) was calculated for all experiments. It was highest for the experiment with distributed and lowest for the experiment with focused pro-wedge erosion, respectively (Table 3).

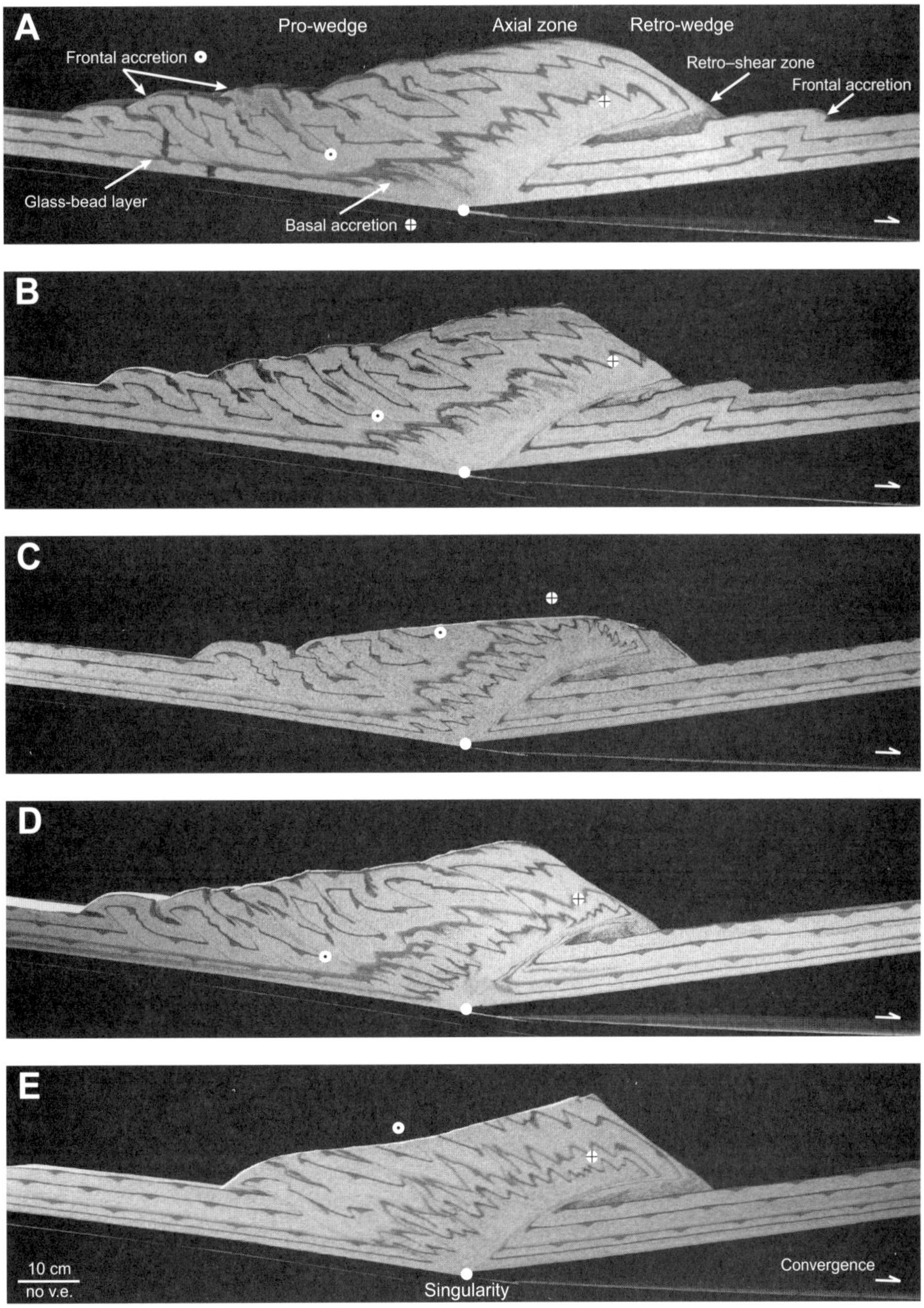

Figure 5. Photographic images of each experiment after 140 cm of convergence. The main structural elements are highlighted. Two markers, one being indicative of the frontal accretion (black dot), the other being indicative of the basal accretion (black cross) are given to show different amounts of exhumation; v.e. is vertical exaggeration.

Basal accretion is, in contrast to the very regular propagation of frontal accretion, more irregular in terms of its wavelength and its spacing (Fig. 6). The respective variability expressed in the sample standard deviation of the wavelength (s_b) was as well calculated for all experiments. It emerged that s_b is always higher than s_f. Also, the range of the latter (0.3) is nearly half the value of the former (0.72) and suggests that basal accretion is more sensitive to the parameters tested (Table 3).

Convergence intervals, where both accretion modes are either in or out of phase, were always too short to be correlated over longer distances. There is, however, a prominent exception. In the experiment with focused pro-wedge erosion, both accre-

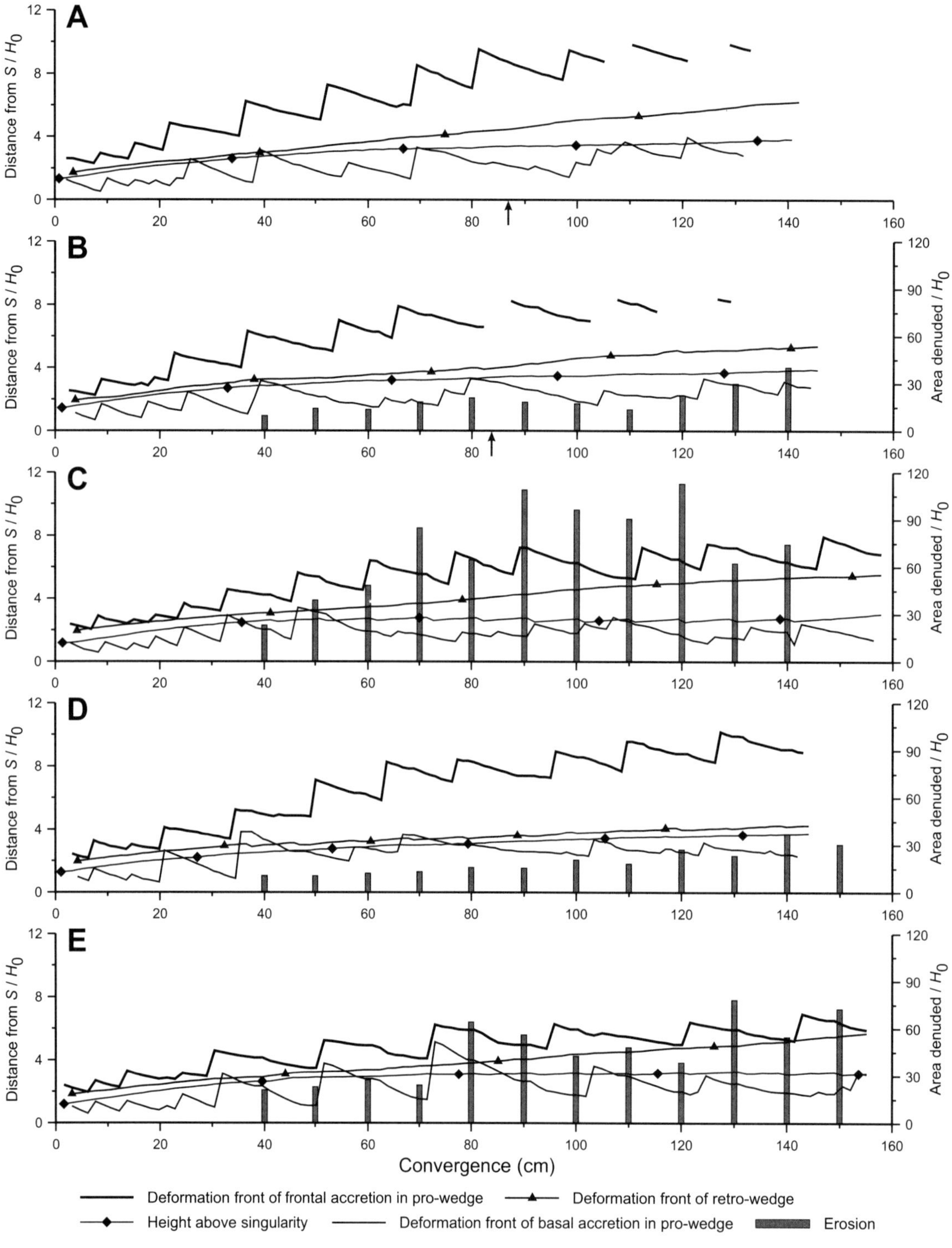

Figure 6. Evolution of geometric parameters taken from the digital images at every 1.5 cm of convergence. The amount of material eroded at every 10 cm of convergence is given as well. (A) Reference experiment; (B) experiment with distributed retro-wedge erosion; (C) experiment with distributed pro-wedge erosion; (D) experiment with focused retro-wedge erosion; and (E) experiment with focused pro-wedge erosion. The terms spacing and wavelength as used in this study are graphically defined. Arrows indicate onset of frontal accretion within the upper plate.

TABLE 3. SUMMARY OF EXPERIMENTAL RESULTS

Observations	No denudation	Pro-wedge denudation		Retro-wedge denudation	
		Distributed	Focused	Distributed	Focused
Frontal accretion in upper plate	Y	N	N	Y	N
Number of thrust imbricates in pro-wedge after 140 cm of convergence	8	8	6	8	8
Cumulative amount of denudation / H_0	0	804	549	217	233
Sample standard deviation of wavelength of basal accretion, S_b	0.47	0.48	0.39	0.78	1.11
Sample standard deviation of wavelength of frontal accretion, S_f	0.15	0.45	0.18	0.29	0.24
Power law equation $y =$	$1.06x^{0.44}$	$1.08x^{0.37}$	$1.34x^{0.30}$	$1.17x^{0.40}$	$0.90x^{0.47}$
$R^2 =$	0.89	0.87	0.83	0.89	0.90
Aspect ratio of retro-wedge $[H/H_0]/[L_r/H_0]$	0.77	0.58	0.70	0.85	0.87
Out-of-sequence displacement index	3.73	4.16	0.97	3.42	3.16
Propagation of frontal accretion	0†	—	–	-	-
Propagation of basal accretion		–	0	-	-
Height above singularity		-	-	+	+
Exhumation of		+	+	0	0
frontally		0	0	+	+
basally accreted material					
Finite strain e_{xy}		-	0	0	+
Retro shear zone		–	–	-	+
Mid-level detachment					

Note: The calculation of the out-of-sequence index is based on the cumulative sum of displacement along a thrust that was accumulated out-of-sequence. The amount of slip taken up by each thrust within an experiment was added and finally normalized with respect to the thickness of the incoming layer, H_0. Propagation of deformation (Fig. 9), exhumation (Fig. 5), and strain accumulation (Fig. 8) are only given qualitatively here, in order to highlight the trends that emerged from the experiments.

†Reference niveau derived from experiment without denudation; – less, + more than reference niveau surfaces.

tion modes were in phase throughout the entire experiment, which suggests that basal and frontal accretion are not decoupled in time by the internal glass-bead layer (Fig. 6). The respective wavelength of basal accretion remained fairly constant, which is evidenced by the lowest s_b value among all the experiments.

After the first erosion increment, basal accretion propagated toward the foreland in pro-wedge erosion experiments, whereas it stepped back in retro-wedge erosion experiments (Fig. 6). The further evolution of basal accretion showed that retro-wedge erosion tends to reduce shorter wavelength activity as compared to the reference or the distributed pro-wedge erosion experiment. During the latter experiment, basal accretion remained nearly stationary with respect to the singularity. It is further pointed out that the height above the singularity can be considered as an envelope for basal accretion (Fig. 6). Finally, it can be shown from Figure 6 that the rate of lateral growth of the pro-wedge during stage III and stage IV, where present, is higher for the reference and both retro-wedge erosion experiments when compared with both pro-wedge erosion experiments. Lateral pro-wedge growth within the latter is nearly stationary. A summary of the relative magnitudes of the propagation of frontal and basal accretion is given in Table 3.

Topographic Evolution

Similar to the reference experiment, the topography of all erosion experiments consisted of two domains: one that comprised the pro-wedge and grew by discrete steps and one that encompassed the axial zone and the retro-wedge, which grew more concentrically (Fig. 7). In both retro-wedge erosion experiments, this equidistant concentric growth pattern as observed in the reference experiment was perturbed. Lines representing incremental stages of wedge evolution merged at an earlier stage, i.e., after the first erosion increment at 40 cm of convergence, than was the case for the reference experiment, i.e., after 120 cm of convergence. Parts of either the axial zone or the retro-wedge, which were not subject to the backstepping pro-wedge erosion, tended to retain their equidistant concentric growth pattern in both pro-wedge erosion experiments. The respective pro-wedge slopes were stationary after ~60 cm of convergence (Fig. 7).

Both identified domains of topographic growth are mirrored, similar to the reference experiment, in the corresponding distribution of incremental surface uplift (Fig. 8), which is characterized by a high spatio-temporal variability. However, the overall evolution of the incremental surface uplift in the retro-wedge erosion experiments and the reference experiment is similar. Maximum incremental surface uplift was confined to either initiated or reactivated ramp segments and to the retro-wedge. A change of incremental surface uplift by one order of magnitude due to reactivation of a thrust ramp as explained in "Reference Experiment without Erosion" was recognized in both retro-wedge erosion experiments (Fig. 8). The experiment with distributed retro-wedge erosion showed a peak in incremental surface uplift at

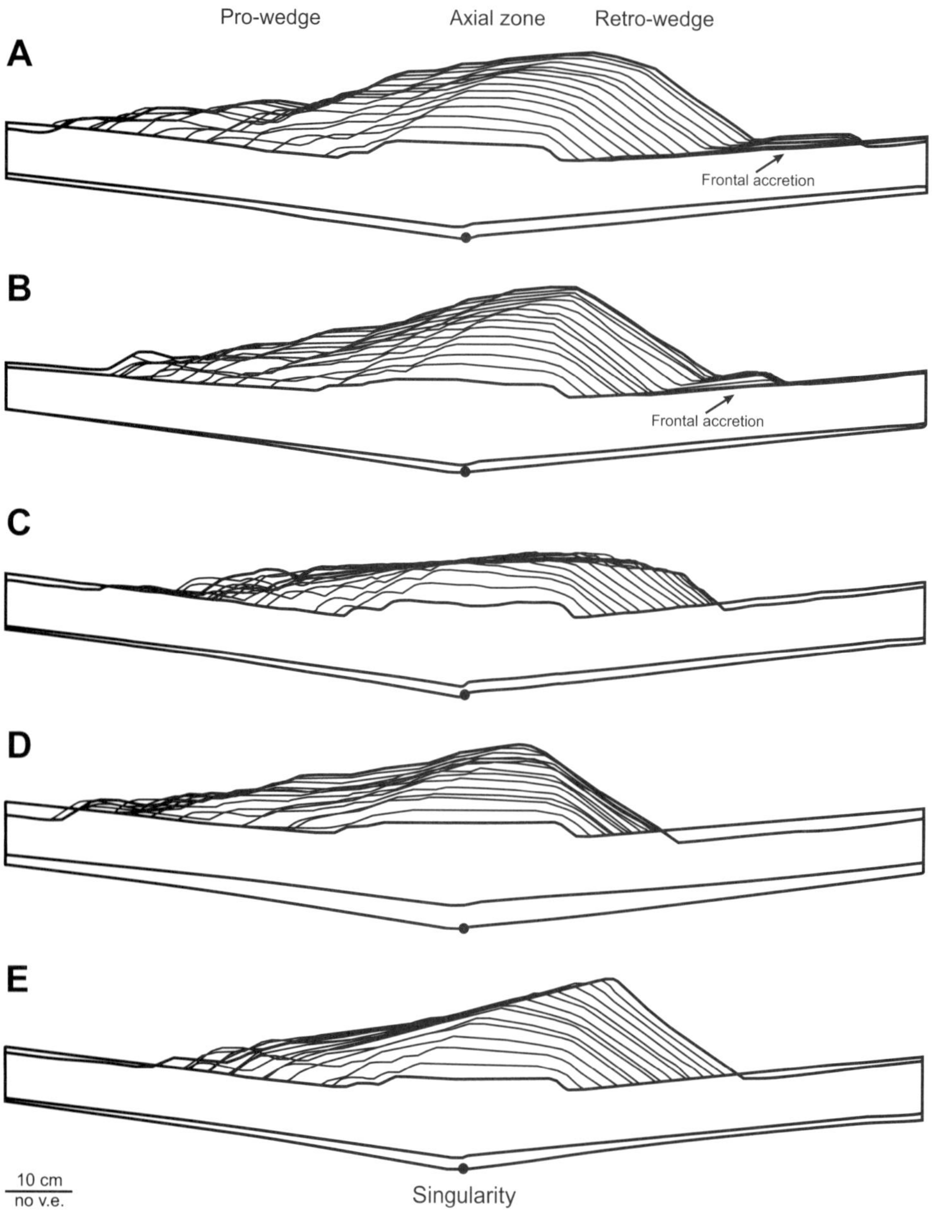

Figure 7. Topographic evolution of each bivergent sand wedge. (A) Reference experiment; (B) experiment with distributed retro-wedge erosion; (C) experiment with distributed pro-wedge erosion; (D) experiment with focused retro-wedge erosion; and (E) experiment with focused pro-wedge erosion. See Figure 4C for explanation. Pro-wedge slopes in both pro-wedge erosion experiments (C and E) are nearly stationary; v.e. is vertical exaggeration.

~90 cm of convergence (Fig. 8). After frontal accretion of upper plate material had set in, a significant decrease of incremental surface uplift within the pro-wedge was observed. Thus, this evolutionary pattern compares well with the reference experiment. A similar feature, but with a lower magnitude and without frontal accretion within the upper plate, was recognized in the experiment with focused retro-wedge erosion.

Both pro-wedge erosion experiments differed significantly with respect to the spatio-temporal distribution of incremental surface uplift, from the reference and both retro-wedge erosion experiments. In the distributed erosion scenario, nearly the entire pro-wedge was heavily denuded, which corresponded to an equally sized area of low-magnitude incremental surface uplift within the pro-wedge. Only minor variations of its respective magnitudes were observed. This general pattern was also recognized in the focused pro-wedge erosion experiment. However, maximum incremental erosion and surface uplift was highest at the toe of the pro-wedge and decreased toward the axial zone (Fig. 8).

Out-of-Sequence Displacement

Frontal accretion in all experiments showed a pure forward-breaking or "piggyback thrust" sequence, although the corresponding displacement along each individual thrust accumulated at several stages during wedge evolution. The result of this reactivation of older thrusts is termed out-of-sequence displacement. Based on the cumulative length evolution of individual thrusts, two end-member kinematic scenarios can be identified, between

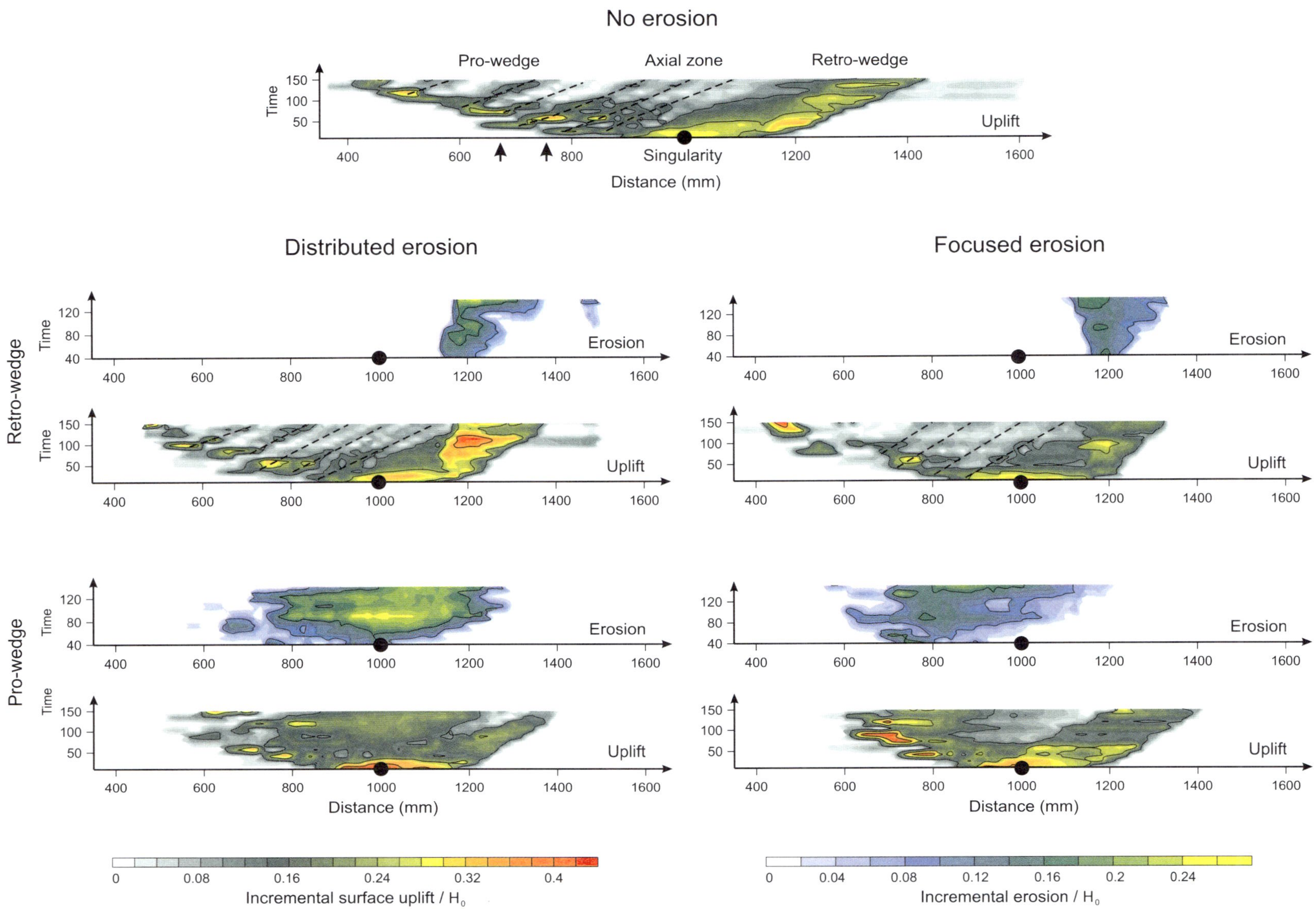

Figure 8. Incremental surface uplift and incremental erosion are mapped at every 0.5 mm along the experiment (abscissa) and are displayed as a function of time (ordinate). Time is expressed in cm of convergence. Data were taken at every 10 cm of convergence. Thus, horizontal lines show the distribution of incremental surface uplift and erosion along the entire bivergent sand wedge for a given time. Lines parallel to the ordinate indicate the variability of incremental surface uplift and erosion for a given lateral position through time. Dashed lines trace the initiation and reactivation of thrust ramps. Reactivation can lead to a temporary increase of incremental surface uplift by one order of magnitude for a given position. Two examples are highlighted by arrows.

which all transitions are possible (Fig. 9). Thrusts, which only show one phase of activity, were found in the early stages of each experiment. These thrusts tended to accumulate only minor amounts of out-of-sequence displacement. In contrast, thrusts that formed at a later stage during wedge evolution were commonly characterized by a phase of activation followed by a period of quiescence and then were reactivated again. In all experiments, both types of thrusts occurred, with one exception. During the experiment with focused pro-wedge erosion, all thrusts within the pro-wedge were only activated once (Fig. 9).

From the cumulative thrust lengths, an out-of-sequence displacement index was derived, which was based on the magnitude of slip accumulated in out-of-sequence. It is highest for the distributed pro-wedge erosion experiment and lowest for the focused pro-wedge erosion experiment (Table 3). The latter experiment is also characterized by the highest cumulative magnitudes of individual thrust length (Fig. 9).

Strain Accumulation

Based on the displacement fields derived from PIV analysis, finite strain after 140 cm of convergence was calculated for each experiment (Fig. 10). The main structural elements, such as the basal and the internal detachment, the retro–shear zone, and each thrust imbricate, as well as the duplexes within the pro-wedge, can be clearly identified for each experiment. A systematic distribution of finite strain accumulated by each individual thrust imbricate was observed as well. Finite strain was lowest within imbricates closest to the toe of the pro-wedge; it was highest in the central part and decreased toward the top. Finally, we point out that the magnitude of retro shear was highest at the retro–shear zone, which separates lower plate from upper plate material and is thus interpreted as long-lived.

However, if similar structures are compared between experiments, it emerges that different magnitudes of finite strain were accommodated in dependence on the location and the mode of erosion. In the experiment with focused retro-wedge erosion, the long-lived, retro–shear zone accommodated the most finite strain relative to all other experiments. In contrast, the same structure accommodated the least magnitude of finite strain in the experiment with distributed pro-wedge erosion. The long-lived, retro–shear zones within the remaining three experiments accumulated similar magnitudes of finite strain (Table 3). A similar pattern is found with respect to the magnitude of finite strain accumulated

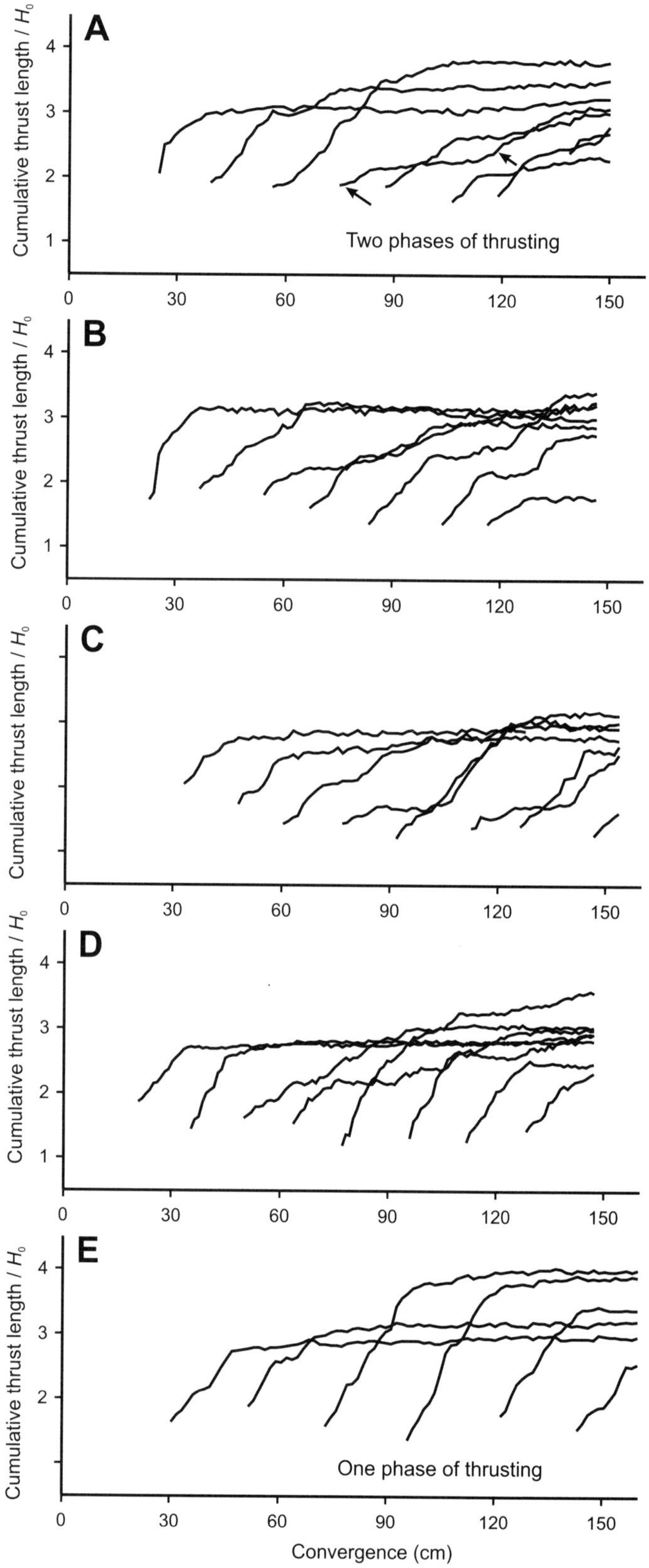

Figure 9. Cumulative thrust lengths of each thrust within the pro-wedge, taken at every 1.5 cm of convergence. (A) Reference experiment, where arrows indicate first and second phase of thrusting; (B) experiment with distributed retro-wedge erosion; (C) experiment with distributed pro-wedge erosion; (D) experiment with focused retro-wedge erosion; and (E) experiment with focused pro-wedge erosion. Experiments D and E have the least out-of-sequence displacement index. Shape of the cumulative thrust lengths curve (elbow to sigmoidal) is strongly controlled by location and mode of erosion.

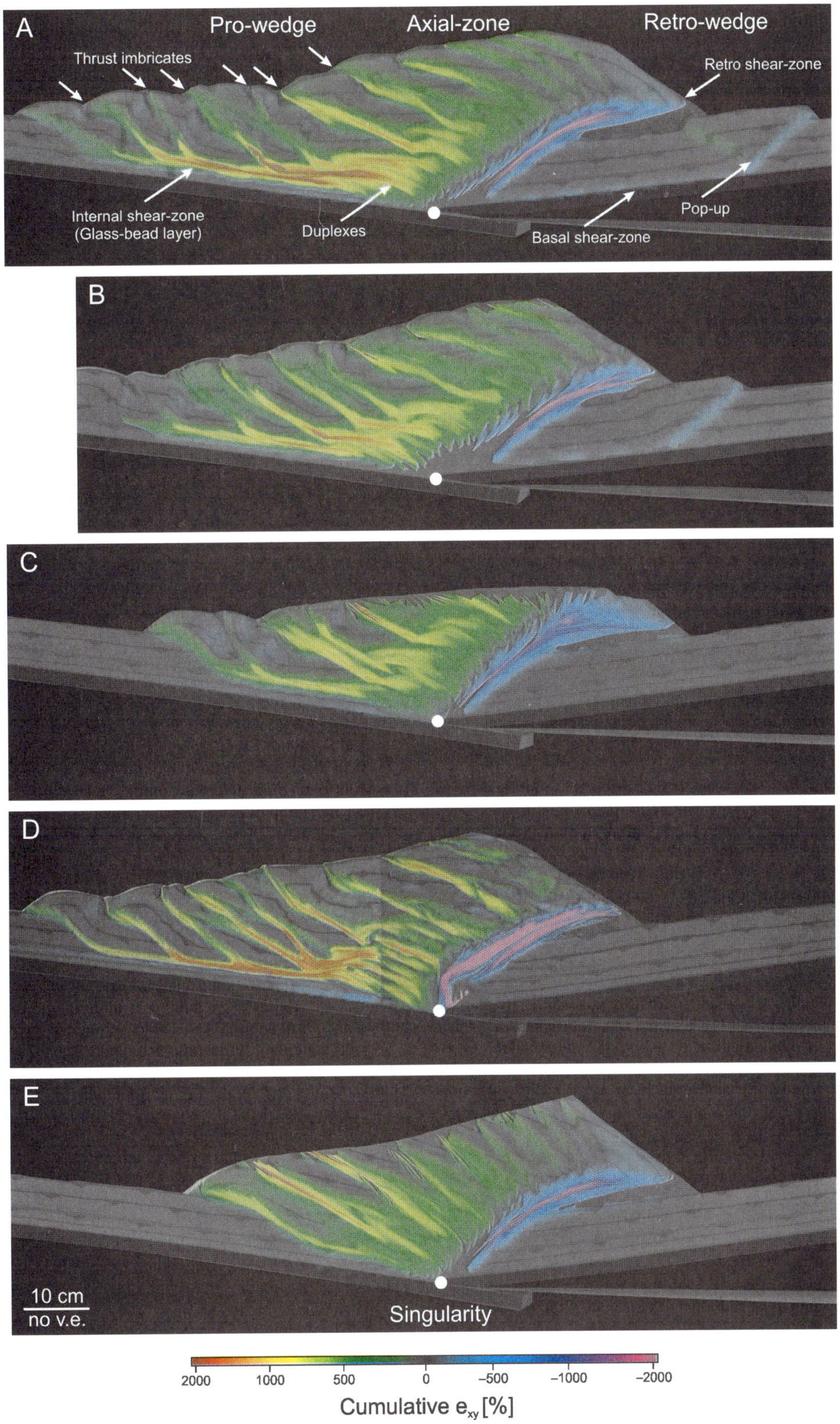

Figure 10. Distribution of finite shear strain, e_{xy}, after 140 cm of convergence, calculated from the cumulative displacement field; e_{xy} describes the local change of the x component of the velocity field in the y direction; v.e. is vertical exaggeration. (A) Reference experiment; (B) experiment with distributed retro-wedge erosion; (C) experiment with distributed pro-wedge erosion; (D) experiment with focused retro-wedge erosion; and (E) experiment with focused pro-wedge erosion. Note the differing magnitudes of finite strain along the retro–shear zone and the glass bead layer between experiments.

by the internal glass-bead layer. There, finite strain is highest in the experiment with focused retro-wedge erosion and lowest in both pro-wedge erosion experiments (Table 3).

DISCUSSION

The purpose of this study has been to demonstrate the influence of the location of erosion with respect to the convergence geometry and the mode of erosion, i.e., distributed or focused, on the upper crustal kinematics of bivergent orogenic wedges. Scaled sandbox simulations were chosen to address this issue. Furthermore, we intended to reduce the number of kinematic boundary conditions and to study the most simple and general case, in order to allow a more self-organized growth of the bivergent sand wedge. It is highlighted here that this study was not aimed at reproducing the geometry of a certain structure or geomorphologic feature. We therefore discuss the significance of the proposed four-stage model of bivergent wedge evolution first, followed by an account of the sensitivity of the model results with respect to the simulated erosion intervals, and finally we analyze the influence of both tested parameters on bivergent wedge evolution.

Four-Stage Evolution Model for Bivergent Sand Wedges

Based on the experiments with and without erosion, we propose a four-stage evolution model for bivergent wedge growth. The experiments shown in this study indicate that the initial asymmetry of the convergence geometry leads to a polarity in the mass transfer, since all material delivered into the evolving bivergent sand wedge is derived from the lower plate. This in turn results in a first-order kinematic segmentation of the bivergent sand wedge into a pro-wedge, an axial zone, and a retro-wedge. The associated evolution of thrusting and uplift rates within each sub-wedge is summarized in Table 2 and agrees well with results from other sandbox and numerical simulations (Malavieille, 1984; Wang and Davis, 1996; Storti et al., 2000; Del Castello et al., 2004; Willett et al., 1993; Beaumont et al., 1996) and suggests reproducibility despite different setups and methodological approaches. There is, however, considerable debate about which boundary conditions are needed to initiate upper plate failure (stage IV), and a variety of factors have been considered: Storti et al. (2000) pointed out that a transition from a vertical backstop (Davis et al., 1983), to a moderate (Lallemand et al., 1994), to a shallow (Byrne et al., 1988), up to a flat lying backstop (Wang and Davis, 1996) is associated with an increase in the development of retro-verging structures, from pure pro-wedges in the first case to fully developed bivergent wedges in the last case. Based on numerical simulations, Byrne et al. (1993) and Pfiffner et al. (2000) showed that either a transition from rigid to deformable backstop–upper plate material or a weak basal detachment produces an increase in the structural complexity of the modeled wedges. Furthermore, Del Castello et al. (2004) suggested that strong upper plate deformation might be a consequence of a subduction reversal.

In contrast, based on the experiments presented in this study, we propose a kinematic model for upper plate failure that tries to integrate the above observations. Given an upper plate basal shear strength, we suggest that for upper plate deformation to occur, a threshold of the height above the singularity (H/H_0) and the width of the retro-wedge (L_r/H_0) must be reached. During the reference experiment, upper plate failure occurred while H/H_0 was 3.4 and L_r/H_0 was 4.4. Similar values were obtained from the experiment with distributed retro-wedge erosion (H/H_0 3.4; L_r/H_0 4.0). An experiment without a mid-level detachment, which also showed upper plate failure, but is not presented here, provided similar values (H/H_0 3.4; L_r/H_0 4.4). Thus, the aspect ratio of the retro-wedge ($[H/H_0]/[L_r/H_0]$) needed for upper plate deformation to occur would be between 0.77 and 0.85 (Table 3). Experiments with either distributed or focused pro-wedge erosion showed that the respective retro-wedges grow as wide but not as high as the ones in the three experiments discussed above (Table 3). Furthermore, the distribution of finite strain (Fig. 10) in both pro-wedge erosion experiments suggests that the retro-wedge is completely decoupled from the pro-wedge, which prohibits stress transfer into the upper plate. This is also evident from calculated particle paths, which are shown schematically in Figure 11. If compared with the reference experiment, particle paths of both pro-wedge erosion experiments are reoriented toward the pro-wedge and dip at a higher angle than their counterparts in the reference experiment. During the experiment with focused retro-wedge erosion, both "thresholds" are met (Table 3). But, this erosion scenario leads to a continuous unloading of the retro–shear zone, which in turn promotes increased slip along it. This interpretation is supported by the observation that the respective retro–shear zone and the mid-level detachment accumulated the highest finite strain magnitudes with respect to all other experiments.

The surface slope of all retro-wedges remained rather uniform throughout the entire wedge evolution and was always at or close to the angle of repose. A lower-tapered retro-wedge sensu Willett et al. (1993) and Wang and Davis (1996) could not be observed, since upper plate accretion started late during the experiments. Nevertheless, it can be concluded that retro-vergent translation and uplift of lower plate–derived material in the hanging wall of the retro–shear zone leads to a continuous increase of the load upon it. Thereby, the strength of the retro–shear zone increases as well. As soon as the strength of the upper plate basal detachment is lower than the one of the retro–shear zone, upper plate deformation will commence. Since we studied the most general case, i.e., material properties did not change across the singularity (dynamic backstop sensu Kopp and Kukowski, 2003), we speculate that the strength contrast between the strength of the retro–shear zone and the strength of the upper plate determines the timing of upper plate deformation. The former depends on the aspect ratio of the retro-wedge and the orientation of stresses within the pro-wedge and both can be modified by erosion. Upper plate strength in turn depends on the thickness of the upper plate and its internal and basal shear strength. This interpretation is supported by numerical simulations (Beaumont et al., 2000b),

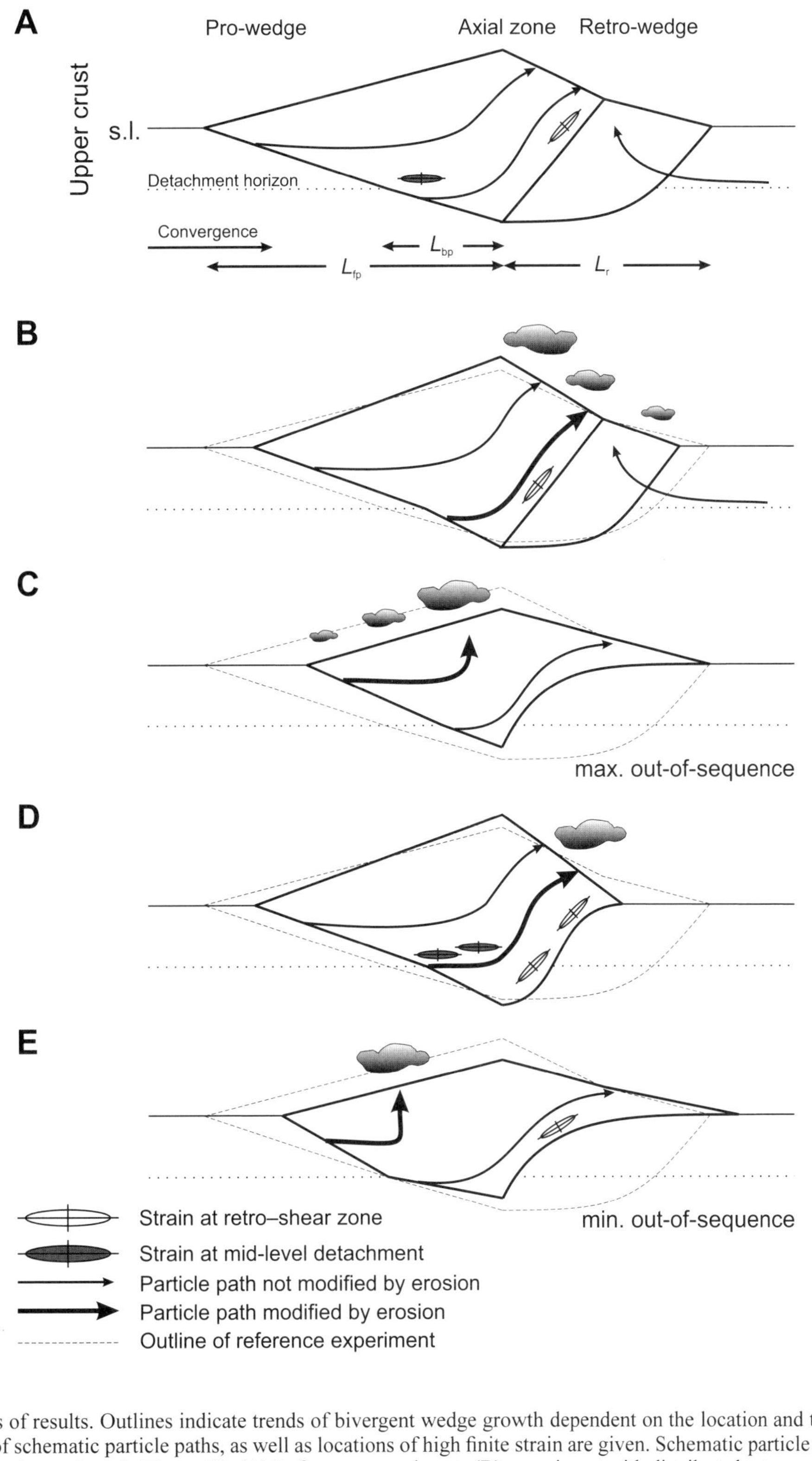

Figure 11. Synopsis of results. Outlines indicate trends of bivergent wedge growth dependent on the location and the mode of erosion. Magnitude and direction of schematic particle paths, as well as locations of high finite strain are given. Schematic particle paths are derived from ~100 calculated particle paths as given in Figure 4D. (A) Reference experiment; (B) experiment with distributed retro-wedge erosion; (C) experiment with distributed pro-wedge erosion; (D) experiment with focused retro-wedge erosion; and (E) experiment with focused pro-wedge erosion. The subscript p relates to the pro-wedge, the subscript r to the retro-wedge, the subscript f to frontal accretion and the subscript b to basal accretion, and s.l. is sea level.

which show that an increased retro-crust strength reduces retro-thrusting. Results from Wang and Davis (1996) indicate a coincidence between a backstop covered with a plastic sheet (μ_b = 0.43) and frontal accretion within the upper plate. In contrast, a backstop covered with sandpaper (μ_b > 0.65) prohibited upper plate deformation.

The proposed four-stage evolution model is supported by observations from natural orogens. Uplift, folding, and retro translation of the initial pop-up might be envisaged as a crustal-scaled backfold. In the case of the Swiss Alps, this backfold is marked by the Austroalpine nappes and is coeval with the onset of frontal accretion but has remained active throughout the collisional history (Pfiffner et al., 2000). Furthermore, based on balanced cross sections, Pfiffner et al. (2000) showed that frontal accretion preceded basal accretion, after collision started during the late Eocene. This has also been documented for the Pyrenees (Beaumont et al., 2000b) and for the Taiwan fold-and-thrust belt (Lacombe and Mouthereau 2002). Therefore, similar to the transition from stage II to stage III in our experiments, a certain load was needed to facilitate slip along possible deeper-lying detachments. A late-stage frontal accretion within the upper plate (transition from stage III to stage IV) has been observed in the Pyrenees (Beaumont et al., 2000b), the European Alps (Pfiffner et al., 2000), the German Variscides (Schaefer et al., 2000), and the Sunda Arc (Silver and Reed, 1988).

Discrete Erosion versus Continuous Deformation

The simulation of erosion within sandbox experiments remains one of the key challenges to be addressed in the future. At present it is only possible to simulate the effect of erosion, i.e., the distribution of unloading across an orogen and not the process of erosion, either by rivers, glaciers, or bedrock landslides. Due to practical limitations, erosion can only be simulated at discrete time intervals that might be considered artificial with respect to a continuously growing bivergent sand wedge. If the applied erosion interval controls the evolution of the bivergent sand wedge, one should expect an immediate response of the wedge with a frequency of the erosion interval. This, however, was not recognized in the time series data, excluding the lowering of the height above the singularity. In addition, experiments from Konstantinovskaia and Malavieille (2004), which involve a 1 cm of convergence interval of erosion, revealed similar results, with respect to the final structural geometries, as those presented in this study. We therefore assume that the interval with which the sand wedge is denuded is of less importance than the mode and location of erosion, as shown below. Second, and most importantly, there is growing evidence that erosion is far more episodic than is often implicitly assumed. Storms, floods, and landslides show a power-law distribution, which means that the bulk of erosional work is done by high-magnitude, low-frequency events (Fujii, 1969; Noever, 1993; Sugai et al., 1994; Hovius et al., 1997; Stark and Hovius, 2001; Guzzetti et al., 2002). The control of periodic (deglaciation, intense monsoon years, severe El Niños) and episodic (high-intensity rainstorms, earthquakes) processes on erosion by rivers, glaciers, and bedrock landslides has been observed in the Himalayas, in Taiwan, in New Zealand, and in the Finisterre Mountains, Papua New Guinea (Harbor and Warburton, 1993; Hovius et al., 1998; Densmore and Hovius, 2000; Guzzetti et al., 2002; Dadson et al., 2003; Keefer et al., 2003; Korup et al., 2004; Barnard et al., 2004; Thiede et al., 2004). Finally, thermochronometric methods can only bracket the time span of an erosion episode. Following this view, Burbank and Beck (1991) speculated that 90% of erosion may have been accomplished in 10% of the time. Taken all together, we suggest that the erosion approach followed in this study is justified by observations that arise from field studies.

Influence of Erosion on Bivergent Wedge Kinematics

Crustal-scaled processes of mountain building have been successfully described numerically by minimum work theory, which is implicitly assumed in the critical taper theory as well. Based on a mechanical analysis of results obtained from sandbox experiments, the minimum work theory has been successfully applied to predict the spacing of thrusts in accretionary wedges (Gutscher et al., 1998). Therefore, we propose that the results shown in this study might be explained in terms of the minimum work concept. Following this view, a bivergent sand wedge subject to erosion has several possible responses to continued convergence by: (1) initiation of a new thrust, either within the upper or the lower plate, (2) continued slip along the deformation front, (3) reactivation of one or more internal thrusts, (4) slip along the retro–shear zone, (5) footwall or hanging wall shortcuts, and (6) internal deformation. Which of these possibilities is activated by the sand wedge depends on the respective gravitational and frictional work. Whereas the former is controlled by the lateral distribution of erosion, the latter is determined by the strength of the undeformed material, which in turn depends on its thickness, as well as the degree of strain softening and the orientation of thrusts.

This interpretation is exemplified by the propagation of frontal accretion within the pro-wedge. It is evident from Figure 6 that the spacing of thrusts is fairly constant throughout all experiments. This agrees well with theoretical considerations and minimum work calculations from scaled sandbox experiments, which predict that the spacing of thrusts, which defines how far deformation propagates into the foreland, is controlled by the depth to the detachment. Given a basal friction that does not change laterally and a constant thickness of the incoming layer, the spacing of thrusts should remain constant (Bombolakis, 1986; Boyer, 1995; Gutscher et al., 1998; Macedo and Marshak, 1999).

In contrast, the wavelength of frontal accretion, which is defined as the time expressed in convergence between two consecutive thrust initiation events, depends on the mode and location of erosion (Fig. 6, Table 3). A permanent unloading of the deformation front, as observed in the focused pro-wedge erosion experiment, leads to an increase of slip along successive thrust imbricates, which in turn retards the propagation of deformation into the foreland (Figs. 6 and 7). Evidence for increased slip during the focused pro-wedge erosion experiment is derived from the cumulative thrust lengths (Fig. 9) and the resulting out-of-sequence displacement index, which was lowest among all experiments. Also, only the experiment with focused pro-wedge erosion exposed the glass-bead layer at the toe of the pro-wedge (Fig. 5). Further support for increased slip is derived from the fact that only six imbricate thrusts were needed to accommodate 140 cm of convergence, whereas all other experiments showed eight imbricates. Focused erosion of the deformation front leads also to a lack of sufficient overthrust length needed to activate the internal glass-bead layer (Kukowski et al., 2002). Thus, frontal accretion by thrust imbrication and basal accretion by duplex formation are in phase throughout the experiment and form one dynamical system. This might explain the observed lowest variability with respect to the wavelength of both accretion modes.

Conversely, the wavelength of frontal accretion during distributed pro-wedge erosion showed the highest variability (Table 3). During this experiment, the highest erosion rates occurred at the central and at the rearward part of the pro-wedge. Also, erosion rates at the respective toeward part showed a higher degree of variability through time than the corresponding part in the focused pro-wedge erosion experiment (Fig. 8). It follows that unloading of the deformation front is more variable through time as well, which in turn influences the wavelength of frontal accretion. Furthermore, in order to restore its critical taper, the pro-wedge deforms internally, i.e., through reactivation of older thrusts. It follows that the propagation of deformation into the foreland is retarded, which agrees with the results from Schlunegger (1999) and Hovius (2000). The preference of internal deformation is mirrored in the diffuse pattern of finite strain and the highest out-of-sequence displacement index among all experiments (Fig. 10, Table 3) and is thus in accordance with Willett et al. (1993). Depending on the magnitude of internal deformation, propagation of deformation into the foreland is retarded and thus adds to its variability.

Retro-wedge erosion unloads the axial zone and especially the retro–shear zone, which eases translation of pro-wedge–derived material toward the upper plate. It follows that in favor of internal deformation, the propagation of frontal accretion can be retarded. The resultant variability of the respective wavelength is similar for both retro-wedge erosion experiments but higher than the one from the reference experiment (Table 3).

Basal accretion is more sensitive to changes of the load gradient determined by erosion, since the respective thrust spacing and the corresponding wavelengths differ significantly within and between experiments (Fig. 6, Table 3). The formation of duplexes depends on the load imposed by the frontally accreted material, the rotation of the glass-bead layer resulting from the stacking and backward translation of the duplexes, and the load upon the retro–shear zone. As noted above, retro-wedge erosion unloads the axial zone and especially the retro–shear zone. It follows that duplexes can be more easily stacked and transferred toward the upper plate. Two consequences emerge: first, at the expense of lateral growth of the pro-wedge, vertical growth of the axial zone and the retro-wedge is promoted. Despite the fact that material is removed, both experiments with retro-wedge erosion have a higher elevation above the singularity than the reference or the pro-wedge erosion experiments. This observation, in conjunction with particle paths, which indicate an acceleration of basal accretion (Fig. 5) lead us to suggest that retro-wedge erosion enhances particle flow of basal accretion. Note that during pro-wedge erosion frontal accretion is accelerated (Fig. 5). A further consequence of the unloading of the axial zone and the retro-wedge is that longer undeformed sand sheets can be drawn beneath the internal glass-bead layer toward the singularity until failure occurs and thus reduces shorter wavelength activity. This is consistent with the predictions from minimum work calculations by Gutscher et al. (1998). Taken all together, our results suggest that the propagation and distribution of deformation within a bivergent sand wedge depend on (1) the location and magnitude of erosion, (2) the internal and basal strength of the sand wedge and of the undeformed foreland, as well as (3) the depth to the detachment within the foreland. Analytical considerations of the growth and erosion of fold-and-thrust belts (Hilley and Strecker, 2004) support our interpretation.

CONCLUDING REMARKS: IMPLICATIONS AND PREDICTIONS FOR NATURAL OROGENS

In this study, we investigated the influence of the location of erosion with respect to the convergence geometry and the mode of erosion, i.e., distributed or focused, which are thought to represent end members, on the kinematics of bivergent sand wedges. Since this study is focused on lower-temperature orogens where brittle behavior prevails, some caution must be taken while transferring and applying our results to natural examples. High exhumation rates may finally lead to the removal of the highest-strength part of the continental crust, which significantly reduces its integrated strength. At this stage, ductile processes might start to dominate (Beaumont et al., 2001; Zeitler et al., 2001; Koons et al., 2002).

Given that one continental plate descends beneath the other during orthogonal continent-continent collision, we propose a four-stage evolution model for the growth of bivergent orogenic wedges. An initial symmetric, crustal-scaled pop-up (stage I) or backfold is followed by the formation of a proto–pro-wedge, where frontal accretion dominates (stage II). Further convergence leads to an increase of the kinematic and topographic asymmetry. Basal accretion commences (stage III). At stage IV, frontal accretion within the upper plate occurs, which may drive the orogen back to more symmetric conditions. This model can be tested against balanced orogen-scale cross sections, and we suggest that the transition from stage II to stage III, i.e., onset of basal accretion, as well as the transition from stage III to stage IV, i.e., onset of frontal accretion within the upper plate, should be identifiable.

Upper plate failure is explained in terms of the strength contrast between the upper plate and the retro–shear zone. Thus, for frontal accretion within the upper plate to occur, no reversal of the convergence geometry or specific backstop geometries must be necessarily invoked.

Propagation of the deformation front is composed of three processes, i.e., initiation of individual thrust imbricates, footwall shortcuts, and the growth pattern of the pro-wedge, which depends on the basal and internal strength of the pro-wedge and the foreland, as well as the magnitude and distribution of erosion. These three processes operate at different timescales (10^4 to 10^6 yr) and may be associated with different seismicity patterns.

We found also that the height of an orogen above the singularity is similar to the distance between the singularity and the deformation front of the basal accretion complex at any given time. This geometric relation might provide an independent measure of topography while balancing orogen-scale cross sections.

The distribution of surface uplift within orogens should indicate at least two domains: one that comprises the pro-wedge and one that includes the axial zone and the retro-wedge. Topography in the former is highly segmented and grows by discrete steps, whereas topography in the latter is smooth and grows almost concentrically. Thus, care should be taken while interpreting surface uplift from different parts of an orogen. Furthermore, surface uplift at a given position may change by one order of magnitude without changes in the boundary conditions. These changes can result from the reactivation of a buried crustal ramp.

Similar to Koons (1990) and Willett (1999), we predict that the location of erosion with respect to the convergence geometry determines the outcrop pattern of metamorphic facies.

We further propose that deformation responds immediately to erosion. Retro-wedge erosion amplifies the displacement of the basally accreted material, whereas pro-wedge erosion accelerates and additionally redirects the particle flow of the frontally accreted material. Pro- and retro-wedge erosion retard the propagation of deformation within the pro-wedge. This effect is stronger for pro-wedge erosion. The evolution of Borneo is consistent with the last prediction. Strong synkinematic erosion under tropical conditions prohibited the growth of a wide, thin-skinned fold-and-thrust belt. The resulting crustal load was not sufficient to generate a foreland basin (Hall and Nichols, 2002). Similarly, Schlunegger and Simpson (2002) demonstrated for the European Alps that a significant decrease in the erosional efficiency during the early Miocene led to a change from vertical extrusion associated with rapid vertical exhumation during the late Oligocene to a mainly horizontally directed extrusion, i.e., the formation of the Jura fold-and-thrust belt and the southern Alps, during middle to late Miocene times.

Also, we suggest that retro-wedge erosion amplifies vertical growth and leads to strain accumulation along the retro–shear zone and the mid-level detachment. This indicates that for retro-wedge erosion, cause (erosion) and response (deformation) are significantly offset in space. Thus, the cause for a certain seismicity pattern observed in the pro-wedge might be the eroding retro-wedge, hundreds of kilometers away. Conversely, pro-wedge erosion evokes a complete decoupling of the retro-wedge from the pro-wedge. Here, cause and response are spatially more closely related. Both results highlight the need for orogen-wide climate-tectonics studies.

We found that a more focused erosion is associated with a more focused tectonic response. This interpretation is supported by observations from the Chugach/St. Elias Range in southern Alaska. There, the windward position of the ELA coincides with a narrow zone of active upper plate deformation associated with high rates of rock uplift (Meigs and Sauber, 2000; Sheaf et al., 2003). Also, strong fluvial erosion in both Himalayan syntaxes has led to high exhumation and deformation rates (Zeitler et al., 2001).

A change from distributed to focused erosion may lead to a continued activation of a certain structure, e.g., deformation front, and may thus retard the propagation of deformation into the foreland of either the upper or lower plate, and may also determine which detachment layer is favored.

Although all of our sandbox simulations exhibit a clear forward-breaking or "piggyback" sequence of thrusting, considerable displacement is accumulated out-of-sequence by reactivation of older thrusts. The magnitude of the latter is strongly controlled by the location and mode of erosion. It is interesting to note that Mouthereau et al. (2001) showed that the increase of erosion rates was associated with an increase of the number of reactivated or out-of-sequence thrusts in the Taiwan fold-and-thrust belt. Similarly, Hodges et al. (2004) demonstrated that strong orographic forcing of precipitation led to concentrated erosion and out-of-sequence thrusting in the higher Himalayan ranges.

Finally, we speculate that a bivergent wedge is very robust with respect to its boundary conditions, e.g., mechanic stratigraphy and erosion. Erosion modifies but does not inhibit the segmentation of a bivergent wedge or the simultaneous propagation of frontal and basal accretion. This postulation gains support from a recent study in the Himalayas where the tectonic displacement field is focused by erosion but does not mimic its asymmetric distribution (Burbank et al., 2003). If, however, a certain threshold is either exceeded or reached, the kinematics might change their mode.

ACKNOWLEDGMENTS

We are grateful to George Hilley and Adrian Pfiffner for encouraging discussion. Very stimulating reviews by Fabrizio Storti and Peter O. Koons and comments from Sean Willett are greatly appreciated. Ulrich Riller is thanked for suggestions on an earlier version of the manuscript. All experiments were carried out in the Geodynamics Laboratory of the GeoForschungsZentrum Potsdam, Germany, with the patient assistance of the Analogue-Lab-Crew. Arne Hoffmann-Rothe is thanked for his help while extracting some of the data from the particle image velocimetry (PIV) software. Günther Tauscher built the experimental setup. Funding to Hoth and Adam was kindly provided by Deutsche Forschungsgemeinschaft (DFG Leibniz-Program) and SonderForschungsBereich 267 project F1.

REFERENCES CITED

Adam, J., Urai, J.L., Wienecke, B., Oncken, O., Pfeiffer, K., Kukowski, N., Lohrmann, J., Hoth, S., van der Zee, W., and Schmatz, J., 2005, Shear localisation and strain distribution during tectonic faulting—New insights from granular-flow experiments and high-resolution optical image correlation techniques: Journal of Structural Geology, v. 27, no. 2, p. 283–301, doi: 10.1016/j.jsg.2004.08.008.

Ahnert, F., 1970, Functional relationships between denudation, relief, and uplift in large, mid-latitude drainage basins: American Journal of Science, v. 268, no. 3, p. 243–263.

Avouac, J.P., and Burov, E.B., 1996, Erosion as a driving mechanism of intracontinental mountain growth: Journal of Geophysical Research, ser. B, Solid Earth and Planets, v. 101, no. 8, p. 17,747–17,769, doi: 10.1029/96JB01344.

Barnard, P.L., Owen, L.A., Sharma, M.C., and Finkel, R.C., 2004, Late Quaternary (Holocene) landscape evolution of a monsoon-influenced high Himalayan valley, Gori Ganga, Nanda Devi: NE Garhwal: Geomorphology, v. 61, no. 1–2, p. 91–110, doi: 10.1016/j.geomorph.2003.12.002.

Beaumont, C., Fullsack, P., and Hamilton, J., 1992, Erosional control of active compressional orogens, *in* McClay, K., ed., Thrust tectonics: London, UK, Chapman and Hall, p. 1–18.

Beaumont, C., Ellis, S., Hamilton, J., and Fullsack, P., 1996, Mechanical model for subduction-collision tectonics of alpine-type compressional orogens: Geology, v. 24, no. 8, p. 675–678, doi: 10.1130/0091-7613(1996)024<0675:MMFSCT>2.3.CO;2.

Beaumont, C., Kooi, H., and Willett, S., 2000a, Coupled tectonic-surface process models with applications to rifted margins and collisional orogens, *in* Summerfield, M.R., ed., Geomorphology and global tectonics: Chichester, UK, John Wiley and Sons, p. 29–55.

Beaumont, C., Munoz, J.A., Hamilton, J., and Fullsack, P., 2000b, Factors controlling the alpine evolution of the central Pyrenees inferred from a comparison of observations and geodynamical models: Journal of Geophysical Research, ser. B, Solid Earth and Planets, v. 105, no. 4, p. 8121–8145, doi: 10.1029/1999JB900390.

Beaumont, C., Jamieson, R.A., Nguyen, M.H., and Lee, B., 2001, Himalayan tectonics explained by extrusion of a low-viscosity crustal channel coupled to focused surface denudation: Nature, v. 414, no. 6865, p. 738–742, doi: 10.1038/414738a.

Bilham, R., Larson, K.M., Freymueller, J.T., Jouanne, F., Le, F.P., Leturmy, P., Mugnier, J.L., Gamond, J.F., Glot, J.P., Martinod, J., Chaudury, N.L., Chitrakar, G.R., Gautam, U.P., Koirala, B.P., Pandey, M.R., Ranabhat, R., Sapkota, S.N., Shrestha, P.L., Thakuri, M.C., Timilsina, U.R., Tiwari, D.R., Vidal, G., Vigny, C., Galy, A., and de Voogd, B., 1997, GPS measurements of present-day convergence across the Nepal Himalaya: Nature, v. 386, no. 6620, p. 61–64.

Bombolakis, E.G., 1986, Thrust-fault mechanics and origin of a frontal ramp, *in* Thrusting and deformation: Oxford-New York, Pergamon International, p. 281–290.

Bookhagen, B., Thiede, R.C., and Strecker, M.R., 2003, Late Quaternary intensified monsoon phases control landscape evolution in the NW Himalaya, *in* AGU fall meeting, San Francisco: Washington, D.C., American Geophysical Union, v. 84, p. F695.

Boyer, S.E., 1995, Sedimentary basin taper as a factor controlling the geometry and advance of thrust belts: American Journal of Science, v. 295, no. 10, p. 1220–1254.

Burbank, D.W., 2002, Rates of erosion and their implications for exhumation: Mineralogical Magazine, v. 66, no. 1, p. 25–52.

Burbank, D.W., and Beck, R.A., 1991, Rapid, long-term rates of denudation: Geology, v. 19, no. 12, p. 1169–1172, doi: 10.1130/0091-7613(1991)019<1169:RLTROD>2.3.CO;2.

Burbank, D.W., Blythe, A.E., Putkonen, J., Pratt, S.B., Gabet, E., Oskin, M., Barros, A., and Ojha, T.P., 2003, Decoupling of erosion and precipitation in the Himalayas: Nature, v. 426, no. 6967, p. 652–655, doi: 10.1038/nature02187.

Byerlee, J., 1978, Friction of rocks: Pure Applied geophysics, v. 116, p. 615–626.

Byrne, D.E., Davis, D.M., and Sykes, L.R., 1988, Loci and maximum size of thrust earthquakes and the mechanics of the shallow region of subduction zones: Tectonics, v. 7, no. 4, p. 833–857.

Byrne, D.E., Wang, W.H., and Davis, D.M., 1993, Mechanical role of backstops in the growth of forearcs: Tectonics, v. 12, no. 1, p. 123–144.

Cobbold, P.R., Durand, S., and Mourgues, R., 2001, Sandbox modelling of thrust wedges with fluid-assisted detachments: Tectonophysics, v. 334, no. 3–4, p. 245–258, doi: 10.1016/S0040-1951(01)00070-1.

Dadson, S.J., Hovius, N., Chen, H., Dade, W.B., Hsieh, M.L., Willett, S.D., Hu, J.C., Horng, M.J., Chen, M.C., Stark, C.P., Lague, D., and Lin, J.C., 2003, Links between erosion, runoff variability and seismicity in the Taiwan orogen: Nature, v. 426, no. 6967, p. 648–651, doi: 10.1038/nature02150.

Dahlen, F.A., 1984, Noncohesive critical Coulomb wedges; an exact solution: Journal of Geophysical Research, ser. B, Solid Earth and Planets, v. 89, no. 12, p. 10,125–10,133.

Dahlen, F.A., 1990, Critical taper model of fold-and-thrust belts and accretionary wedges: Annual Review of Earth and Planetary Sciences, v. 88, p. 55–99.

Dahlen, F.A., and Barr, T.D., 1989, Brittle frictional mountain building; 1, Deformation and mechanical energy budget: Journal of Geophysical Research, ser. B, Solid Earth and Planets, v. 94, no. 4, p. 3906–3922.

Dahlen, F.A., and Suppe, J., 1988, Mechanics, growth, and erosion of mountain belts, *in* Clark, S.P., ed., Processes in continental lithospheric deformation: Geological Society of America Special Paper 218, p. 161–178.

Dahlen, F.A., Suppe, J., and Davis, D., 1984, Mechanics of fold-and-thrust belts and accretionary wedges; cohesive Coulomb theory: Journal of Geophysical Research, ser. B, Solid Earth and Planets, v. 89, no. 12, p. 10,087–10,101.

Davis, D., Suppe, J., and Dahlen, F.A., 1983, Mechanics of fold-and-thrust belts and accretionary wedges: Journal of Geophysical Research, v. 88, no. 2, p. 1153–1172.

Del Castello, M., Pini, G.A., and McClay, K.R., 2003, Unbalanced topography and overloading affecting Coulomb wedge kinematics: Insights from sandbox modeling, *in* European Geophysical Society–American Geophysical Union–European Union of Geosciences Joint Assembly, Nice, France, v. 5, p. 2.

Del Castello, M., Pini, G.A., and McClay, K.R., 2004, Effect of unbalanced topography and overloading on Coulomb wedge kinematics: Insights from sandbox modelling: Journal of Geophysical Research, v. 109, no. B05405, p. 1–17, doi: 10.1029/2003JB002709.

Densmore, A.L., and Hovius, N., 2000, Topographic fingerprints of bedrock landslides: Geology, v. 28, no. 4, p. 371–374, doi: 10.1130/0091-7613(2000)028 <371:TFOBL>2.3.CO;2.

Fujii, Y., 1969, Frequency distribution of the magnitude of the landslides caused by heavy rain-fall: Zisin = Jishin, v. 22, no. 2, p. 244–247.

Gutscher, M.A., Kukowski, N., Malavieille, J., and Lallemand, S.E., 1998, Episodic imbricate thrusting and underthrusting; analog experiments and mechanical analysis applied to the Alaskan accretionary wedge: Journal of Geophysical Research, v. 103, no. 5, p. 10,161–10,176.

Gutscher, M.A., Klaeschen, D., Flueh, E., and Malavieille, J., 2001, Non-Coulomb wedges, wrong-way thrusting, and natural hazards in Cascadia: Geology, v. 29, no. 5, p. 379–382, doi: 10.1130/0091-7613(2001)029<0379:NCWWWT>2.0.CO;2.

Guzzetti, F., Malamud, B.D., Turcotte, D.L., and Reichenbach, P., 2002, Power-law correlations of landslide areas in central Italy: Earth and Planetary Science Letters, v. 195, no. 3–4, p. 169–183, doi: 10.1016/S0012-821X(01)00589-1.

Hall, R., and Nichols, G., 2002, Cenozoic sedimentation and tectonics in Borneo; climatic influences on orogenesis, *in* Jones, S., and Frostick, L., eds., Sediment flux to basins; causes, controls and consequences: London, UK, Geological Society of London Special publication 191, p. 5–22.

Hampel, A., Adam, J., and Kukowski, N., 2004, Response of the tectonically erosive South Peruvian forearc to subduction of the Nazca Ridge: Analysis of 3D-analogue experiments: Tectonics, v. 5, p. 23, doi: 10.1029/2003TC001585.

Harbor, J., and Warburton, J., 1993, Relative rates of glacial and nonglacial erosion in alpine environments: Arctic and Alpine Research, v. 25, no. 1, p. 1–7.

Hardy, S., Duncan, C., Masek, J., and Brown, D., 1998, Minimum work, fault activity and the growth of critical wedges in fold and thrust belts: Basin Research, v. 10, p. 365–373, doi: 10.1046/j.1365-2117.1998.00073.x.

Hay, W.W., 1996, Tectonics and climate: International Journal of Earth Science, v. 85, p. 409–437.

Hilley, G.E., and Strecker, M.R., 2004, Steady state erosion of critical Coulomb wedges with applications to Taiwan and the Himalaya: Journal of Geophysical Research, v. 109, no. B01411, p. 1–17, doi: 10.1029/2002JB002284.

Hodges, K.V., Wobus, C., Ruhl, K., Schildgen, T., and Whipple, K., 2004, Quaternary deformation, river steepening, and heavy precipitation at the front of the higher Himalayan ranges: Earth and Planetary Science Letters, v. 220, no. 3–4, p. 379–389, doi: 10.1016/S0012-821X(04)00063-9.

Horton, B.K., 1999, Erosional control on the geometry and kinematics of thrust belt development in the central Andes: Tectonics, v. 18, no. 6, p. 1292–1304, doi: 10.1029/1999TC900051.

Hovius, N., 1996, Regular spacing of drainage outlets from linear mountain belts: Basin Research, v. 8, no. 1, p. 29–44.

Hovius, N., 1998, Controls on sediment supply by large rivers, *in* Relative role of eustasy, climate, and tectonism in continental rocks: Tulsa, Oklahoma, Society for Sedimentary Geology, p. 3–16.

Hovius, N., 2000, Macroscale process systems of mountain belt erosion, *in* Summerfield, M.R., ed., Geomorphology and global tectonics: Chichester, UK, John Wiley and Sons, p. 77–205.

Hovius, N., Stark, C.P., and Allen, P.A., 1997, Sediment flux from a mountain belt derived by landslide mapping: Geology, v. 25, no. 3, p. 231–234, doi: 10.1130/0091-7613(1997)025 <0231:SFFAMB>2.3.CO;2.

Hovius, N., Stark, C. P., Tutton, M. A., and Abbott, L. D., 1998, Landslide-driven drainage network evolution in a pre-steady-state mountain belt; Finisterre Mountains, Papua New Guinea: Geology, v. 26, p. 1071–1074.

Howard, A.D., Dietrich, W.E., and Seidl, M.A., 1994, Modeling fluvial erosion on regional to continental scales: Journal of Geophysical Research, ser. B, Solid Earth and Planets, v. 99, no. 7, p. 13,971–13,986, doi: 10.1029/94JB00744.

Hubbert, M.K., 1937, Theory of scale models as applied to the study of geological structures: Geological Society of America Bulletin, v. 48, p. 1459–1520.

Keefer, D.K., Moseley, M.E., and deFrance, S.D., 2003, A 38,000-year record of floods and debris flows in the Ilo region of southern Peru and its relation to El Niño events and great earthquakes: Palaeogeography, Palaeoclimatology, Palaeoecology, v. 194, no. 1–3, p. 41–77, doi: 10.1016/S0031-0182(03)00271-2.

Konstantinovskaia, E., and Malavieille, J., 2004, Erosion and exhumation in accretionary orogens: Experimental and geological approaches: Bolletino di Geofisica, v. 45, no. 1, p. 37.

Koons, P.O., 1990, Two-sided orogen; collision and erosion from the sandbox to the Southern Alps, New Zealand: Geology, v. 18, no. 8, p. 679–682, doi: 10.1130/0091-7613(1990)018 <0679:TSOCAE>2.3.CO;2.

Koons, P.O., 1995, Modelling the topographic evolution of collisional belts: Annual Review of Earth and Planetary Sciences, v. 23, p. 375–408, doi: 10.1146/annurev.ea.23.050195.002111.

Koons, P.O., Zeitler, P.K., Chamberlain, C.P., Craw, D., and Meltzer, A.S., 2002, Mechanical links between erosion and metamorphism in Nanga Parbat, Pakistan Himalaya: American Journal of Science, v. 302, no. 9, p. 749–773.

Koons, P.O., Norris, R.J., Craw, D., and Cooper, A.F., 2003, Influence of exhumation on the structural evolution of transpressional plate boundaries; an example from the Southern Alps, New Zealand: Geology, v. 31, no. 1, p. 3–6, doi: 10.1130/0091-7613(2003)031 <0003:IOEOTS>2.0.CO;2.

Kopp, H., and Kukowski, N., 2003, Backstop geometry and accretionary mechanics of the Sunda margin: Tectonics, v. 22, no. 6, p. 11-1–11-16.

Korup, O., McSaveney, M.J., and Davies, T.R.H., 2004, Sediment generation and delivery from large historic landslides in the Southern Alps, New Zealand: Geomorphology, v. 61, no. 1–2, p. 189–207, doi: 10.1016/j.geomorph.2004.01.001.

Kukowski, N., von Huene, R., Malavieille, J., and Lallemand, S.E., 1994, Sediment accretion against a buttress beneath the Peruvian continental margin at 12 degrees S as simulated with sandbox modeling: Geologische Rundschau, v. 83, no. 4, p. 822–831.

Kukowski, N., Lallemand, S.E., Malavieille, J., Gutscher, M.A., and Reston, T.J., 2002, Mechanical decoupling and basal duplex formation observed in sandbox experiments with application to the western Mediterranean Ridge accretionary complex: Marine Geology, v. 186, p. 29–42.

Kutzbach, J.E., Prell, W.L., and Ruddiman, W.F., 1993, Sensitivity of Eurasian climate to surface uplift of the Tibetan Plateau: Journal of Geology, v. 101, no. 2, p. 177–190.

Lacombe, O., and Mouthereau, F., 2002, Basement-involved shortening and deep detachment tectonics in forelands of orogens; insights from recent collisional belts; Taiwan, Western Alps, Pyrenees: Tectonics, v. 21, no. 4.

Lallemand, S.E., Schnuerle, P., and Malavieille, J., 1994, Coulomb theory applied to accretionary and nonaccretionary wedges; possible causes for tectonic erosion and/or frontal accretion: Journal of Geophysical Research, ser. B, Solid Earth and Planets, v. 99, no. 6, p. 12,033–12,055, doi: 10.1029/94JB00124.

Lin, J.C., 2000, Morphotectonic evolution of Taiwan, *in* Summerfield, M.R., ed., Geomorphology and global tectonics: Chichester, UK, John Wiley and Sons, p. 135–146.

Lohrmann, J., Kukowski, N., Adam, J., and Oncken, O., 2003, The impact of analogue material properties on the geometry, kinematics, and dynamics of convergent sand wedges: Journal of Structural Geology, v. 25, no. 10, p. 1691–1711, doi: 10.1016/S0191-8141(03)00005-1.

Macedo, J., and Marshak, S., 1999, Controls on the geometry of fold-thrust belt salients: Geological Society of America Bulletin, v. 111, no. 12, p. 1808–1822, doi: 10.1130/0016-7606(1999)111 <1808:COTGOF>2.3.CO;2.

Malavieille, J., 1984, Modélisation expérimentale des chevauchements imbriqués: Application aux chaines de montagnes: Societé Géologique de France, Bulletin, v. 7, p. 129–138.

Mandl, G., de Jong, L.N.J., and Maltha, A., 1977, Shear zones in granular material; an experimental study of their structure and mechanical genesis: Rock Mechanics, Supplement 9, no. 2–3, p. 95–144.

Marques, F.O., and Cobbold, P.R., 2002, Topography as a major factor in the development of arcuate thrust belts; insights from sandbox experiments: Tectonophysics, v. 348, no. 4, p. 247–268, doi: 10.1016/S0040-1951(02)00077-X.

Marshak, S., and Wilkerson, M.S., 1992, Effect of overburden thickness on thrust belt geometry and development: Tectonics, v. 11, no. 3, p. 560–566.

Masek, J.G., and Duncan, C.C., 1998, Minimum-work mountain building: Journal of Geophysical Research, ser. B, Solid Earth and Planets, v. 103, no. 1, p. 907–917, doi: 10.1029/97JB03213.

Meigs, A., and Sauber, J., 2000, Southern Alaska as an example of the long-term consequences of mountain building under the influence of glaciers: Quaternary Science Reviews, v. 19, p. 543–562.

Merle, O., and Abidi, N., 1995, Approche experimentale du fonctionnement des ramps emergentes: Bulletin de la Société Géologique de France, v. 166, no. 5, p. 439–450.

Milliman, J.D., and Syvitski, J.P.M., 1992, Geomorphic/tectonic control of sediment discharge to the ocean; the importance of small mountainous rivers: Journal of Geology, v. 100, no. 5, p. 525–544.

Montgomery, D.R., Balco, G., and Willett, S.D., 2001, Climate, tectonics, and the morphology of the Andes: Geology, v. 29, no. 7, p. 579–582, doi: 10.1130/0091-7613(2001)029 <0579:CTATMO>2.0.CO;2.

Mouthereau, F., Lacombe, O., Deffontaines, B., Angelier, J., and Brusset, S., 2001, Deformation history of the southwestern Taiwan foreland thrust belt; insights from tectono-sedimentary analyses and balanced cross-sections: Earth and Planetary Science Letters, v. 333, no. 1–2, p. 293–322.

Mugnier, J.L., Baby, P., Colletta, B., Vinour, P., Bale, P., and Leturmy, P., 1997, Thrust geometry controlled by erosion and sedimentation; a view from analogue models: Geology, v. 25, no. 5, p. 427–430, doi: 10.1130/0091-7613(1997)025<0427:TGCBEA>2.3.CO;2.

Nieuwland, D.A., Urai, J.L., and Knoop, M., 2000, In-situ stress measurements in model experiments of tectonic faulting, *in* Lehner, F., and Urai, J., eds., Aspects of tectonic faulting: Berlin, Germany, Springer, p. 155–166.

Noever, D.A., 1993, Himalayan sandpiles: Physical Review, v. 47, no. 1, p. 724–725.

Pavlis, T.L., Hamburger, M.W., and Pavlis, G.L., 1997, Erosional processes as a control on the structural evolution of an actively deforming fold and thrust belt; an example from the Pamir–Tien Shan region, central Asia: Tectonics, v. 16, no. 5, p. 810–822, doi: 10.1029/97TC01414.

Persson, K.S., Garcia-Castellanos, D., and Sokoutis, D., 2004, River transport effects on compressional belts: First results from an integrated analogue-numerical model: Journal of Geophysical Research, v. 109, no. B01409, p. 1–11, doi: 10.1029/2002JB002274.

Pfiffner, O.A., Ellis, S., and Beaumont, C., 2000, Collision tectonics in the Swiss Alps; insights from geodynamic modeling: Tectonics, v. 19, no. 6, p. 1065–1094, doi: 10.1029/2000TC900019.

Pratt-Sitaula, B., Burbank, D.W., Heimsath, A., and Ojha, T., 2004, Landscape disequilibrium on 1000–10,000 year scales Marsyandi River, Nepal, central Himalaya: Geomorphology, v. 58, no. 1–4, p. 223–241, doi: 10.1016/j.geomorph.2003.07.002.

Schaefer, F., Oncken, O., Kemnitz, H., and Romer, R.L., 2000, Upper-plate deformation during collisional orogeny; a case study from the German Variscides (Saxo-Thuringian Zone), *in* Franke, W., et al., eds., Orogenic processes; quantification and modelling in the Variscan Belt: London, UK, Geological Society of London, p. 281–302.

Schellart, W.P., 2000, Shear test results for cohesion and friction coefficients for different granular materials; scaling implications for their usage in analogue modelling: Tectonophysics, v. 324, no. 1–2, p. 1–16, doi: 10.1016/S0040-1951(00)00111-6.

Schlunegger, F., 1999, Controls of surface erosion on the evolution of the Alps; constraints from the stratigraphies of the adjacent foreland basins: International Journal of Earth Science, v. 88, p. 283–304.

Schlunegger, F., and Simpson, G., 2002, Possible erosional control on lateral growth of the European central Alps: Geology, v. 30, no. 10, p. 907–910, doi: 10.1130/0091-7613(2002)030<0907:PECOLG>2.0.CO;2.

Schlunegger, F., and Willett, S.D., 1999, Spatial and temporal variations in exhumation of the central Swiss Alps and implications for exhumation mechanisms, *in* Ring, et al., eds., Exhumation processes; normal faulting, ductile flow and erosion: London, UK, Geological Society of London, p. 157–179.

Schmidt, K.M., and Montgomery, D.R., 1995, Limits to relief: Science, v. 270, no. 5236, p. 617–620.

Seidl, M.A., and Dietrich, W.E., 1992, The problem of channel erosion into bedrock, *in* Functional geomorphology; landform analysis and models: Cremlingen-Destedt, Germany, Catena-Verlag Rohdenburg, p. 101–124.

Selby, M.J., 1982, Controls on the stability and inclinations of hillslopes formed on hard rock: Earth Surface Processes and Landforms, v. 7, no. 5, p. 449–467.

Sheaf, M.A., Serpa, L., and Pavlis, T.L., 2003, Exhumation rates in the St. Elias Mountains, Alaska: Tectonophysics, v. 367, no. 1–2, p. 1–11, doi: 10.1016/S0040-1951(03)00124-0.

Silver, E.A., and Reed, D.L., 1988, Backthrusting in accretionary wedges: Journal of Geophysical Research, v. 93, p. 3116–3126.

Stark, C.P., and Hovius, N., 2001, The characterization of landslide size distributions: Geophysical Research Letters, v. 28, no. 6, p. 1091–1094, doi: 10.1029/2000GL008527.

Storti, F., and McClay, K., 1995, Influence of syntectonic sedimentation on thrust wedges in analogue models: Geology, v. 23, no. 11, p. 999–1002, doi: 10.1130/0091-7613(1995)023 <0999:IOSSOT>2.3.CO;2.

Storti, F., Salvini, F., and McClay, K., 2000, Synchronous and velocity-partitioned thrusting and thrust polarity reversal in experimentally produced, doubly-vergent thrust wedges; implications for natural orogens: Tectonics, v. 19, no. 2, p. 378–396, doi: 10.1029/1998TC001079.

Strecker, M.R., Hilley, G.E., Arrowsmith, J.R., and Coutand, I., 2003, Differential structural and geomorphic mountain-front evolution in an active continental collision zone; the Northwest Pamir, southern Kyrgyzstan: Geological Society of America Bulletin, v. 115, no. 2, p. 166–181, doi: 10.1130/0016-7606(2003)115<0166:DSAGMF>2.0.CO;2.

Sugai, T., Ohmori, H., and Hirano, M., 1994, Rock control on magnitude-frequency distribution of landslide: Transactions of the Japanese Geomorphological Union, v. 15, no. 3, p. 233–251.

Summerfield, M.A., and Hulton, N.J., 1994, Natural controls of fluvial denudation rates in major world drainage basins: Journal of Geophysical Research, ser. B, Solid Earth and Planets, v. 99, no. 7, p. 13,871–13,883, doi: 10.1029/94JB00715.

Thiede, R.C., Bookhagen, B., Arrowsmith, J.R., Sobel, E.R., and Strecker, M.R., 2004, Climatic control on rapid exhumation along the Southern Himalayan Front: Earth and Planetary Science Letters, v. 222, no. 3–4, p. 791–806, doi: 10.1016/j.epsl.2004.03.015.

Tippett, J.M., and Hovius, N., 2000, Geodynamic processes in the Southern Alps, New Zealand, *in* Summerfield, M.A., ed., Geomorphology and global tectonics: Chichester, UK, John Wiley and Sons, p. 109–134.

Wang, W.H., and Davis, D.M., 1996, Sandbox model simulation of forearc evolution and noncritical wedges: Journal of Geophysical Research, ser. B, Solid Earth and Planets, v. 101, no. 5, p. 11,329–11,339, doi: 10.1029/96JB00101.

Willett, S.D., 1999, Orogeny and orography; the effects of erosion on the structure of mountain belts: Journal of Geophysical Research, ser. B, Solid Earth and Planets, v. 104, no. 12, p. 28,957–28,982, doi: 10.1029/1999JB900248.

Willett, S.D., Beaumont, C., and Fullsack, P., 1993, Mechanical model for the tectonics of doubly vergent compressional orogens: Geology, v. 21, p. 371–374, doi: 10.1130/0091-7613(1993)021 <0371:MMFTTO>2.3.CO;2.

Willett, S.D., Slingerland, R., and Hovius, N., 2001, Uplift, shortening, and steady state topography in active mountain belts: American Journal of Science, v. 301, p. 455–485.

Wobus, C.W., Hodges, K.V., and Whipple, K.X., 2003, Has focused denudation sustained active thrusting at the Himalayan topographic front?: Geology, v. 31, no. 10, p. 861–864, doi: 10.1130/G19730.1.

Zeitler, P.K., 1985, Cooling history of the NW Himalaya, Pakistan: Tectonics, v. 4, no. 1, p. 127–151.

Zeitler, P.K., Koons, P.O., Bishop, M.P., Chamberlain, C.P., Craw, D., Edwards, M.A., Hamidullah, S., Jan, M.Q., Khan, M.A., Khattak, M.U.K., Kidd, W.S.F., Mackie, R.L., Meltzer, A.S., Park, S.K., Pecher, A., Poage, M.A., Sarker, G., Schneider, D.A., Seeber, L., and Shroder, J.F., 2001, Crustal reworking at Nanga Parbat, Pakistan; metamorphic consequences of thermal-mechanical coupling facilitated by erosion: Tectonics, v. 20, no. 5, p. 712–728, doi: 10.1029/2000TC001243.

MANUSCRIPT ACCEPTED BY THE SOCIETY 23 JUNE 2005

Printed in the USA

Geological Society of America
Special Paper 398
2006

Response of a steady-state critical wedge orogen to changes in climate and tectonic forcing

Gerard H. Roe†
Drew B. Stolar
Sean D. Willett
Department of Earth and Space Sciences, University of Washington, Seattle, Washington 98195, USA

ABSTRACT

The theories of critical orogenic wedges and fluvial erosion are combined to explore the interactions between tectonics, erosion, and climate. A model framework is developed which allows the derivation of an exact analytical scaling relationship for how orogen width, height, and rock uplift rate vary as a function of accretionary flux and precipitation rate. Compared to a model with prescribed uplift rate, incorporating the tectonic response introduces a powerful negative feedback on the orogen, which strongly damps the system's equilibrium response to changes in forcing. Furthermore, for the most commonly assumed forms of the fluvial erosion law, the orogen is more sensitive to changes in the accretionary flux than in the precipitation rate. And while increases in accretionary flux and precipitation rate both cause an increase in exhumation rate, they have opposite tendencies on the orogen relief. Further analysis shows that the pattern of rock uplift does not affect the scaling relationship and that it is only weakly dependent on the hillslope condition.

Keywords: climate, erosion, tectonics, orography, precipitation.

1 INTRODUCTION

That the mountainous terrains on Earth are shaped by the triumvirate of climate, erosion, and tectonics is one of the central concepts of the earth sciences. At some level it is simply a truism: convergent plate boundaries lead to crustal thickening and drive rock uplift; a host of processes on the land surface (physical and chemical weathering, soil production, hillslope processes, fluvial and glacial erosion, etc.), all dependent to some degree on climate, act in combination to erode material from the landscape; and the local (and indeed global) climate is, in turn, partly controlled by the presence and form of the underlying terrain (Fig. 1). To state, however, that all three are related does not address the strength of the feedbacks, the details of the underlying mechanisms, or their relative importance in different climatic and tectonic settings.

A growing body of observational evidence has pointed to the connection between tectonics and erosion in, for example, the Southern Alps of New Zealand (e.g., Adams, 1980; Koons, 1989), the Alps (e.g., Bernet et al., 2003, Anders et al., 2002), Taiwan (e.g., Suppe, 1980), the Andes (e.g., Montgomery et al., 2001), the Cascades (e.g., Reiners et al., 2002, 2003) and the Olympics (e.g., Brandon et al., 1998) in Washington State, and the Himalayas (e.g., Finlayson et al., 2002). Taken together, such data suggest that at a fundamental level the basic processes of crustal deformation are inextricable from the erosional, and thus climatic, mechanisms driving exhumation (Beaumont et al., 2000).

As the understanding of each of the three components of the system has advanced, research has begun to investigate the coupling mechanisms and strengths of the interactions.

†E-mail: gerard@ess.washington.edu.

Roe G.H., Stolar, D.B., and Willett, S.D., 2006, Response of a steady-state critical wedge orogen to changes in climate and tectonic forcing, *in* Willett, S.D., Hovius, N., Brandon, M.T., and Fisher, D.M., eds., Tectonics, Climate, and Landscape Evolution: Geological Society of America Special Paper 398, Penrose Conference Series, p. 227–239, doi: 10.1130/2005.2398(13). For permission to copy, contact editing@geosociety.org.

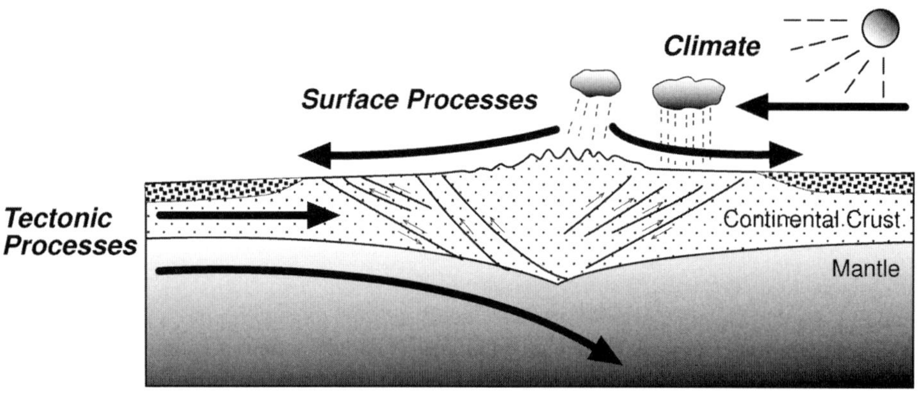

Figure 1. Schematic illustration of the process involved in interplay between climate, tectonics, and erosion (after Willett, 1999).

A handful of modeling studies (e.g., Koons, 1989; Beaumount et al., 1992; Masek et al., 1994; Willett, 1999) have demonstrated the effectiveness of the interactions by incorporating simple representations of orographic precipitation into coupled tectonic-surface process models. Willett (1999), for example, showed the tremendous potential for the direction of the orographic rain-shadow to determine the pattern of exhumation and stress within the orogen.

Another series of papers has investigated in more detail how the relief of orogens is affected by the dominant erosional process and climate feedbacks, focusing on fluvial erosion in longitudinal river profiles (e.g., Howard et al., 1994; Whipple et al., 1999; Whipple and Tucker, 1999; Tucker and Whipple, 2002; Roe et al., 2002, 2003). However, these papers prescribe a fixed rock uplift rate over a fixed length and therefore omit any tectonic response to the modeled patterns of erosion. Without understanding the nature of this tectonic response, it is unclear whether the results reflect the true sensitivity of the real system.

This paper presents a steady-state analytical solution for a simple framework which represents the three components of the system explicitly. It is similar in spirit and form to that of Whipple and Meade (2004) and Hilley et al., (2004), but because of a different formulation reaches somewhat different conclusions. Such analytical solutions provide a conceptual understanding of the coupled system's behavior, but also make important predictions that can be evaluated against observations or compared to more complete numerical models. The tectonic setting considered is that of a critical Coulomb wedge (Davis et al., 1983; Dahlen, 1984, 1990; Willett et al., 1993). The crust is assumed to behave as a plastic, frictional material. Driven by convergence and crustal accretion, surface slopes steepen until gravitational stress and basal traction are in balance while the crust is everywhere at its yield stress. The result is an orogen whose mean cross-sectional profile is wedge-shaped, maintaining a critical taper angle and changing self-similarly in response to changes in tectonic and climate forcing. We apply a model of erosion to this critical orogen that assumes that fluvial processes dominate, except near the divide where hillslope failure—or channel-head processes—maintain a critical hillslope angle.

It is shown in Section 2 that this model framework yields a simple and exact analytical expression for the scaling relationship for the width of the orogen as a function of climate (i.e., the precipitation rate) and tectonics (i.e., the accretionary flux), which is dependent on the parameters in the fluvial erosion law. After examining the sensitivity of the width to plausible combinations of the parameters, further analysis shows the scaling relationship is remarkably robust to relaxation of the model assumptions. We argue that this is a major reason why integrations of a numerical coupled tectonic-surface process model, presented in Stolar et al. (this volume), show close agreement with the scaling relationship derived here. We end with a discussion of other aspects of the system that have been omitted and what might be the consequences on the behavior of the system.

2. AN ANALYTICAL SOLUTION

The framework we use is quasi one-dimensional (Fig. 2). It represents only one side of a critical wedge orogen and is the simplest one which represents both fluvial erosion on a landscape and the tectonic response associated with critical wedge dynamics. We consider the longitudinal profile of a major trunk river,

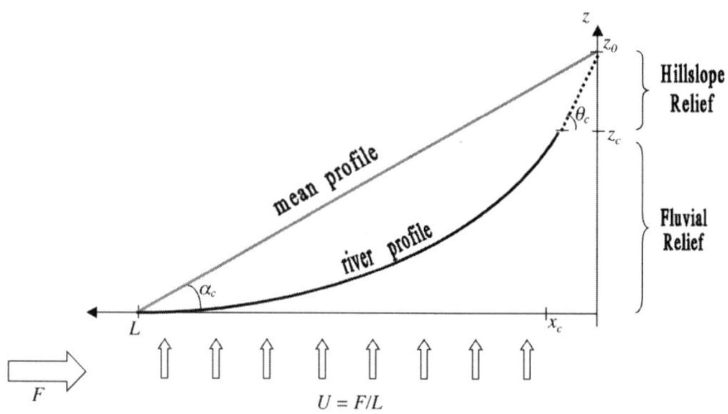

Figure 2. The model framework. The accretionary flux is distributed uniformly as rock uplift over the domain. Together the fluvial and hillslope relief equal the total relief which is constrained to maintain a critical taper angle.

running from the foot of the orogen (at $x = L$) to a channel head (at $x = x_c$). The river channel head is assumed to be connected to the drainage divide (at $x = 0$) by a hillslope which maintains a fixed critical slope, θ_c. Since the orogen is a critical wedge, the mean topographic profile must follow a critical taper angle α_c. This is the key to the coupling of the erosion and tectonics: the total relief (i.e., fluvial plus hillslope), R, divided by the half-width, L, is constrained such that

$$\frac{R}{L} = \tan\alpha_c. \tag{1}$$

Tectonic forcing is in the form of a steady accretionary flux, F, which is assumed to be distributed uniformly under the wedge (i.e., underplating, e.g., Dahlen and Barr, 1989; Willett et al., 2001) so that in steady-state the local rock uplift is uniform and given by

$$U = F/L. \tag{2}$$

We assume that most of the orogenic wedge is dominated by a series of transverse rivers (i.e., oriented perpendicularly to the orogen divide). Erosion rate within the river channel, E, obeys a simple fluvial erosion law:

$$E = KQ^m\left(\frac{dz}{dx}\right)^n, \tag{3}$$

where $Q(x)$ is the fluvial discharge (i.e., the streamflow), and dz/dx is the along-profile river channel slope. K is the erosivity, and m and n are exponents that depend on the physical process governing the erosion (e.g., Whipple and Tucker, 1999). Other functional forms for erosion laws have been postulated which, for example, include an incision threshold (e.g., Sklar and Dietrich, 1998; Snyder et al., 2003), or regard erosion as a transport-limited process (Tucker and Whipple, 2002). Given the myriad processes acting to cause erosion in mountain rivers (e.g., Whipple et al., 2000), the stream power law is probably best regarded as quasi-empirical, justified in part by the characteristic shape of most bedrock river channels, and only in part from basic physical principles. However, as elaborated in the Discussion, the fundamental nature of the feedback studied here is unlikely to change even if erosion operates via a different process. Given the tractability of using the stream power erosion law in this framework, and the current developmental status of alternative representations of fluvial erosion, the incorporation of a more complex erosion law would be premature.

Such a model framework obviously makes some fairly restrictive simplifying assumptions. By construction, the modeled system has far fewer degrees of freedom by which to adjust than the real one has. First, we consider only one side of the wedge. The framework for a two-sided wedge can be constructed by apportioning the accretionary flux so that each side maintains its own critical taper (Whipple and Meade, 2004). Doing so does not affect the sensitivity of the width to changes in tectonic or climate forcing, provided that the erosivity is uniform. In seeking only steady-state solutions, the mass balance is between the accretionary flux and the integrated erosional flux: we do not worry about material accreted into the root or an isostatic response, both of which will affect the temporal evolution of the system. Next, in Sections 3 and 4 the assumptions of spatially uniform rock uplift and the fixed hillslope condition are shown to have only a small impact on the solution's sensitivity. A further factor relates to the quasi one-dimensional framework which effectively fixes the sinuosity of the major trunk rivers. We assume that the tributaries in the drainage network and the interfluve hillslopes somehow adjust so that the major ridge profiles complement the trunk river profiles in order to maintain a critical taper. That such adjustments do in fact happen is supported to some extent by the observation that many orogens have a mean-elevation profile with an approximately constant slope (e.g., Davis et al., 1983). It is further supported by results of integrations of a numerical model (Stolar et al., this volume) that represents some of these extra degrees of freedom, and which yields a width and uplift response in close agreement with that presented below.

The river profile governed by Equation (3) is assumed to be anchored to the foot of the wedge, such that $z = 0$ at $x = L$. As discharge decreases toward the divide the channel slope steepens. When this slope reaches a critical value, α_c, we assume a transition from a fluvial regime governed by Equation (3) to a hillslope dominated regime, which maintains this critical slope. Letting x_c represent this transition point between erosion regimes, we can write

$$\left.\frac{dz}{dx}\right|_{fluvial} = \tan\theta_c \qquad \text{at } x = x_c \tag{4}$$

$$\frac{z_0 - z_c}{x_c - 0} = \tan\theta_c \qquad \text{for } 0 <= x < x_c, \tag{5}$$

where z_c denotes the elevation of the profile at x_c (see Fig. 2).

Lastly, we need to specify river discharge as a function of position on the river. In building an analytical solution, we make the assumption that climatological precipitation rate, P, is uniform. This allows the discharge at a point, $Q(x)$, to be specified as directly proportional to upstream drainage area, A. We can then take advantage of the empirical observation that A can be represented by an expression of the form $A = k_a x^h$ (e.g., Hack, 1957; Montgomery and Dietrich, 1992). Discharge is therefore given by

$$Q(x) = Pk_a x^h. \tag{6}$$

Obviously precipitation is not uniform in mountainous regions, and moreover, it is generally a function of mountain width and relief. It has been shown that incorporating a precipitation feedback can have a substantial impact on the modeled relief of mountain ranges (Roe et al., 2002; 2003). For now we treat it as a prescribed forcing, and we will return to this issue in Section 5 and in the Discussion. Together with an assumption of steady state, (1) to (6) form a closed set of equations which, for prescribed precipitation rate and accretionary flux (P and L), and specified erosion law parameters (K, h, m, n), can be solved to obtain the orogen width, L.

In steady-state there must be a pointwise balance between rock uplift rate and the erosion rate. So from Equations (2), (3), and (6) we get:

$$\frac{dz}{dx} = \left(\frac{F}{LK} \cdot \frac{1}{P^m k_a^m} \right)^{\frac{1}{n}} \cdot x^{-\frac{hm}{n}} . \tag{7}$$

The above expression can be applied at the transition point, x_c:

$$\tan\theta_c = \left(\frac{F}{LK} \cdot \frac{1}{P^m k_a^m} \right)^{\frac{1}{n}} \cdot x_c^{-\frac{hm}{n}} \tag{8}$$

Next, the elevation of the transition point, z_c, can be found by combining Equations (5) and (1):

$$z_c = L\tan\alpha_c - x_c \tan\theta_c . \tag{9}$$

A second expression for z_c can be derived by integrating Equation (7), subject to the lower boundary condition that $z = 0$ at $x = L$:

$$z_c = \left(\frac{F}{LK} \cdot \frac{1}{P^m k_a^m} \right)^{\frac{1}{n}} \cdot \frac{1}{1-\frac{hm}{n}} \left(L^{1-\frac{hm}{n}} - x_c^{1-\frac{hm}{n}} \right) \quad \text{for} \ \frac{hm}{n} \neq 1$$

$$z_c = \left(\frac{F}{LK} \cdot \frac{1}{P^m k_a^m} \right)^{\frac{1}{n}} \cdot \ln\left(L / x_c \right) \quad \text{for} \ \frac{hm}{n} \neq 1 \tag{10}$$

For the case of $hm/n \neq 1$, and setting the right hand sides of Equations (9) and $(10)_1$ equal to each other gives:

$$\left(\frac{F}{LK} \cdot \frac{1}{P^m k_a^m} \right)^{\frac{1}{n}} \cdot \frac{x_c^{1-\frac{hm}{n}}}{1-\frac{hm}{n}} \left\{ \left(\frac{L}{x_c} \right)^{1-\frac{hm}{n}} - 1 \right\} = L\tan\alpha_c - x_c \tan\theta_c \tag{11}$$

Finally, on substitution from Equation (8) and rearranging:

$$\frac{1}{1-\frac{hm}{n}} \left\{ \left(\frac{L}{x_c} \right)^{1-\frac{hm}{n}} - 1 \right\} = \frac{\tan\alpha_c}{\tan\theta_c} \left(\frac{L}{x_c} \right) - 1 \tag{12}$$

For $hm/n = 1$ the equivalent expression is:

$$\ln\left(\frac{L}{x_c} \right) = \frac{\tan\alpha_c}{\tan\theta_c} \left(\frac{L}{x_c} \right) - 1 \tag{13}$$

Equations (12) and (13) are implicit equations in the variable L/x_c. Therefore the solution can be found from the intersection of two curves plotting the left and right hand sides of the respective equations as a function of L/x_c. The graphical solution for the case of $hm/n \neq 1$ [Equation (12)] is shown in Figure 3. For $\alpha_c > \theta_c$ (i.e., taper angle exceeding critical slope), the two curves do not intersect. For $\theta_c > \alpha_c$ however, the curves intersect twice and there are therefore two values of L/x_c which satisfy the equations.

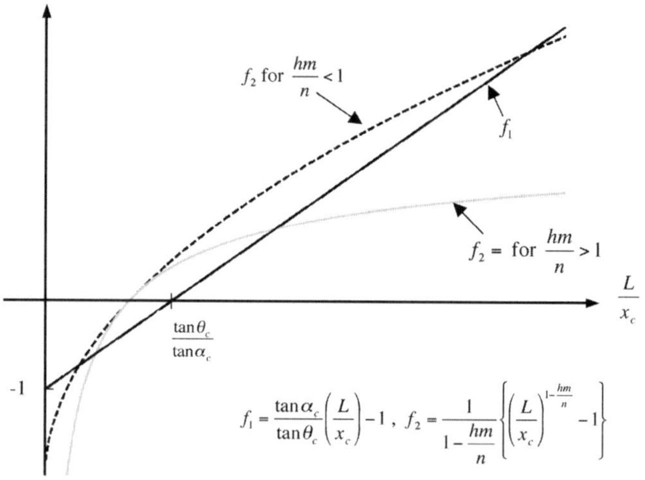

Figure 3. Graphical solution to Equation (12). f_1 and f_2 are the left- and right-hand sides of Equation (12) respectively, plotted as a function of L/x_c. The solution to Equation (12) is given by the intercept of the two functions. Note that two curves for f_2 are plotted, due to the different behavior of the function depending on the sign of $(hm/n - 1)$.

Only one value is physically meaningful however, since by construction $x_c \leq L$. Denoting the acceptable value of L/x_c as λ, and substituting from Equation (8) gives the solution directly:

$$\left[K k_a^m \tan^n \theta_c \right] L^{1+hm} \cdot F^{-1} \cdot P^m = \lambda^{hm} \tag{14}$$

which applies for both the $hm/n \neq 1$ and $hm/n = 1$ cases. Assuming α_c, θ_c, K, and the erosion law exponents remain constant as F and P vary, we can write:

$$L \propto F^{\frac{1}{1+hm}} P^{-\frac{m}{1+hm}} \tag{15}$$

This is an exact analytical scaling relationship for the model framework and approximations, and reflects the fundamental interplay between tectonics, orography, and climate (i.e., F, L, and P), mediated through critical wedge dynamics and fluvial erosion.

As is to be expected, L increases in response to an increase in accretionary flux, and L decreases for an increase in precipitation rate. Since maintaining a critical taper requires relief and width to co-vary, the scaling relationship for the total relief, R, is also given by Equation (15).

Taking log derivatives of Equation (15) shows how fractional changes in L are related to fractional changes in F and L:

$$\frac{\Delta L}{L} = \frac{1}{hm+1} \left(\frac{\Delta F}{F} - m\frac{\Delta P}{P} \right) \tag{16}$$

Therefore, for all values of the discharge exponent $m < 1$ [which most data and theoretical predictions call for (e.g., Whipple et al., 2000)], the orogen width (and total relief) are more sensitive to accretionary flux than to precipitation rate.

Interestingly, the slope exponent n does not enter into Equation (16). This is because the integral of the slope along the profile (in other words, the total relief) is strongly constrained by the imposed critical taper angle. While the lack of dependence on n is only strictly true for the formulation here (see for example Whipple and Meade, 2004), further analysis in Section 4 shows that other plausible formulations of the upper boundary condition give rise to, at most, only a weak dependence on n. Although n does not matter for the functional sensitivity of the system, we note that it does affect the value of λ. Also, while the solution is formally independent of the value of n, a shear stress model of river incision would require that changes in n would also change m, although changes in m could also occur independently of n via changes in channel geometry (e.g., Whipple and Tucker, 1999).

The sensitivity of the rock uplift rate (which in steady-state equals the exhumation rate) to tectonic and climatic forcing can also be determined. Using $U = F/L$ and Equation (15) gives:

$$U \propto F^{\frac{hm}{1+hm}} P^{\frac{m}{1+hm}} \tag{17}$$

Again n does not enter, and since $h \approx 2$ (e.g., Montgomery and Dietrich, 1992) the exhumation rate, like the orogen width, is more sensitive to variations in accretionary flux than to variations in the precipitation rate. Whipple and Meade (2004) do not retain precipitation rate explicitly in their formulation; it is effectively combined into their rock erosivity parameter, K'. They conclude that rock uplift rate is more sensitive to erosivity than to accretionary flux for most parameter combinations. By substituting $K' = KP^m$ into their scaling relationships, the above results can be recovered.

It is interesting to note from Equations (15) and (17) that an increase in tectonic flux causes an increase in both rock uplift rate and relief (or width). And while an increase in precipitation rate also results in an increase in rock uplift rate, it leads to a decrease in relief. These contrasting effects might provide a way of differentiating the causes of relief changes, via the analysis of records of sediment production rates, or of thermochronometry-based estimates of erosion rates.

It is worth comparing the results of the calculations presented here to other simple models which have not included the tectonic response of the critical wedge. Several previous studies have investigated the response of relief to changes in climate and uplift by considering a longitudinal river profile of a fixed length, driven by a uniform rock uplift rate U (e.g., Whipple et al., 1999; Roe et al., 2003). By imposing a fixed length for the river profile (which is taken to represent the major trunk river in a drainage basin extending from the foot of the orogen to the divide), it is implicitly assumed that the orogen width is fixed. Thus the imposed rock uplift rate is always proportional to the implied accretionary flux, and U should be regarded as the equivalent tectonic forcing for this framework. In this "fixed-width" case, Whipple and Tucker (1999) for example, show that the fluvial relief varies as:

$$R \propto U^{\frac{1}{n}} P^{-\frac{m}{n}}. \tag{18}$$

where we have again retained the precipitation rate explicitly. And so taking log derivatives we obtain the relief sensitivity relationship:

$$\frac{\Delta R}{R} = \frac{1}{n}\left(\frac{\Delta U}{U} - m\frac{\Delta P}{P}\right) \tag{19}$$

Comparing Equation (19) to Equation (16), it is seen that the relative importance of tectonic forcing and precipitation forcing is the same as for the critical wedge. Importantly though, the absolute sensitivity is quite different and does depend on the value of n.

To contrast these different sensitivities, Table 1 shows the relief response for both the critical wedge and the fixed-width cases, for three combinations of erosion law exponents representing three different proposed mechanisms for bedrock incision (e.g., Whipple and Tucker, 1999): (a) stream power ($m = 1$, $n = 1$, erosion proportional to rate of release of potential energy), (b) unit stream power ($m = 1/2$, $n = 1$, stream power per unit channel

TABLE 1: SENSITIVITY OF RELIEF OF AN OROGEN FOR CRITICAL WEDGE MODEL, AND FIXED-WIDTH MODEL, DESCRIBED IN THE TEXT, FOR THREE DIFFERENT PROPOSED EROSION MECHANISMS.

Mechanism	m	n	h	Critical wedge		Fixed-width	
				Tectonics	Precipitation	Tectonics	Precipitation
				$F^{\frac{1}{1+hm}}$	$P^{-\frac{m}{1+hm}}$	$U^{\frac{1}{n}}$	$P^{-\frac{m}{n}}$
Stream Power	1	1	2	$\frac{1}{3}$ (23)	$-\frac{1}{3}$ (23)	1 (100)	–1 (100)
Unit Stream Power	$\frac{1}{2}$	1	2	$\frac{1}{2}$ (41)	$-\frac{1}{4}$ (19)	1 (100)	$-\frac{1}{2}$ (41)
Unit Shear Stress	$\frac{1}{3}$	$\frac{2}{3}$	2	$\frac{3}{5}$ (60)	$-\frac{1}{5}$ (15)	$\frac{3}{2}$ (183)	$-\frac{1}{2}$ (41)

Note: m and n are discharge and slope exponents, respectively, in the erosion law, and h is the Hack's law exponent. The ratios in the last four columns are the values of the exponents for the tectonics and precipitation sensitivities, for the appropriate combinations of h, m, and n. To illustrate these sensitivities, the figures in parentheses are the percentage increases in relief due to either a doubling of the tectonic forcing, or a halving of the precipitation rate.

width), and (*c*) unit shear stress ($m = 1/3$, $n = 2/3$, erosion proportional to basal shear stress per unit channel width). For the critical wedge model and using unit shear stress, Table 1 shows that a halving of accretionary flux has the same effect as a quadrupling of the precipitation rate.

What is clear from Table 1 is that, compared to the case of a fixed width, including critical wedge dynamics provides a strong negative feedback on relief changes, thereby damping the response of the system. As also noted by Whipple and Meade (2004), this is because changes in the rock uplift rate tend to oppose changes in the width. This can be seen clearly by imagining that, from an initial equilibrium state, there is a decrease in the accretionary flux. This will tend to cause a decrease in the width of the orogen and so the accreting flux, F, will thus be distributed over a smaller width. The rock uplift rate (i.e., F/L) will therefore be larger than if L had been held fixed. The impact of this can be quite striking. Taking unit shear stress again as an example, the fixed-width model predicts that relief varies inversely with the square root of the precipitation rate. For a critical wedge however, the tectonic feedback means that the relief varies inversely with only the one-fifth power of the precipitation rate. In this case then, the relief is strongly insensitive to climate in the form of precipitation rates.

2.1 Sensitivity to Model Parameters

The analytical solution presented above gives the sensitivity of orogen width (and total relief) to changes in accretionary flux and precipitation rate, but makes the important assumption that everything else is held fixed as the climate or tectonic forcing changes. This is a useful idealization, but as the climate changes over a particular mountain range (perhaps to one that is stormier), it is entirely possible that the dominant process of erosion changes, resulting in different effective values of m and n, or with the role of thresholds becoming more important (e.g., Tucker and Whipple, 2002)

A second issue centers on whether these feedbacks are able to be evaluated in nature. It is difficult, if not impossible, to know the detailed history of the evolution of any particular mountain range. For example, it is unlikely that regional climate (in the form of precipitation rates) will ever be known to a necessary level of accuracy over a several million year time scale. One way to circumvent this would be to compare present-day mountain ranges in different climate and tectonic settings, and to evaluate whether the differences in form between them are consistent with the feedbacks described here. This space-for-time "swap" is possible only if other differences (e.g., lithological) do not swamp the tectonic and climatic signals, or if their effects can be understood well enough to be controlled for. In this context, it is useful to look at the predicted sensitivity of the mountain width to changes in the erosivity, and the critical taper and slope angles.

From field studies, a wide range of erosivities have been inferred (e.g., Stock and Montgomery, 1999). Critical taper and hillslope angles also show some significant variation (e.g., Davis et al., 1983; Montgomery and Dietrich, 1992). Figure 4 shows the sensitivity of the halfwidth to changes in these model parameters, centered on a control case. In this control case, the precipitation rate is prescribed ($P_0 = 1$ m yr^{-1}), as is the accretionary flux ($F_0 = 40$ m^{2}yr^{-1}). Then taking $\alpha_c = 4°$ and $\theta_c = 40°$, the erosivity, K_0 is selected so that for specified erosion exponents (h, m, and n), the halfwidth L_0 is 40 km. This gives a rock uplift rate of 1 mm yr^{-1}. The model parameters are then varied independently to produce the sensitivity plots in Figure 4. In reality, it is likely that all three parameters are controlled at least in part by lithology, and so are not strictly independent. But in the absence of quantitative understanding about how they co-vary, it is worthwhile to examine their separate effects.

Figure 4 shows the sensitivity of orogen width for the three different choices for the erosion law. The width appears least sensitive to variations in the critical slope angle (the reason for this is demonstrated in Section 4). For the critical taper angle, a doubling from 3° to 6° causes a reduction of ~33% in the width. Interestingly, since $R = L\tan\alpha_c$ this means that the total relief is not changing by very much as the taper angle varies. Lastly, variations in the erosivity K can have a substantial impact on the orogen width. Studies on observed river profiles have suggested that erosivity can vary spatially by several orders of magnitude (Stock and Montgomery, 1999). Because of the strong sensitivity of the width to the value of K, Figure 4A suggests caution in comparing orogens with different lithologies, in trying to evaluate the feedbacks presented here.

3. INSENSITIVITY OF SOLUTION TO ROCK UPLIFT PATTERN

The simplifying assumptions made in Section 2 allowed for an exact analytical solution and a demonstration of the strength of the feedback created by incorporating a tectonic response. In the next two sections, the analysis is pursued further and we show that the magnitude of the feedback is surprisingly insensitive to alternative assumptions and model frameworks.

To this point we have assumed the rock uplift is spatially uniform. This is, however, quite a strong constraint on the system: the precise spatial patterns of horizontal and vertical rock velocities are a function of where material is accreted and how it deforms within the orogenic wedge (e.g., Willett et al., 2001). Underplating and frontal-accretion (both of which can be consistent with spatially uniform rock uplift as we have assumed (Masek et al., 1994; Willett, 1999) are useful concepts, but if viscous deformation and thermal coupling, or spatially varying erosion rates, are included, uniform rock uplift is not guaranteed, or expected. For example, the non-uniform temperature field within an orogen is set by topographic form and exhumation rate. This temperature field feeds back on the deformation field via the temperature-dependent viscosity (e.g., Willett, 1999).

The model framework presented here can be extended to incorporate a spatially varying rock uplift function, for which we take the following general form:

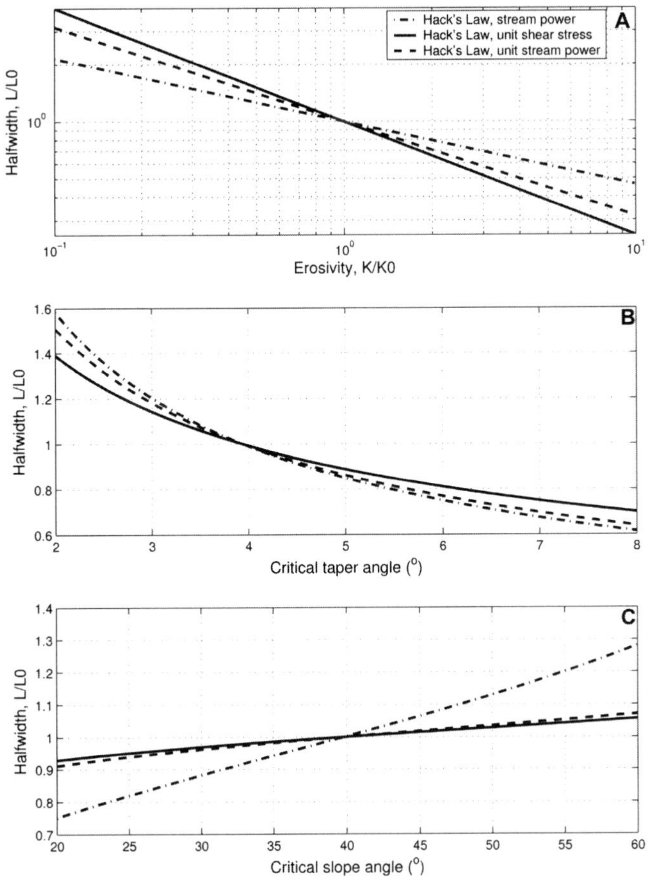

Figure 4. Sensitivity of the orogen width to model parameters, (A) erosivity K, (B) critical taper angle, α_c, and (C) critical slope angle, θ_c. The details of the parameters used in these results are given in the text.

$$U = \frac{F}{L}\phi(x/L), \tag{20}$$

where ϕ is a normalized function (i.e., ${}^{0}\!\int_L \phi dx = 1$). Rock uplift governed by Equation (20) scales with the accretionary flux and has been constructed such that it changes self-similarly as the wedge changes size. In Appendix A it is shown that as long as ϕ is a positive, simply integrable function, the solution can be constructed in a way that is very similar to that for uniform rock uplift, and that furthermore it gives exactly the same sensitivity of the system as Equation (15). The equivalent expression is:

$$\left[Kk_a^m \tan^n\theta_c\right]L^{1+hm}\cdot F^{-1}\cdot P^m = \lambda^{hm}\phi(\lambda^{-1}) \tag{21}$$

This result is an important extension of the analysis: it indicates that, provided that the pattern of rock uplift varies self-similarly, the details of the deformation may not be of primary importance in determining the sensitivity of a critical wedge to a change in forcing. We emphasize that this result applies only to the steady-state form, and that the nature of the deformation will certainly affect how the system achieves a new steady-state in response to a change in forcing. The function ϕ will be determined by the rheological properties of the deforming wedge. Stolar et al. (this volume) presents the rock uplift pattern obtained from an integration of a coupled tectonics-erosion numerical model.

4. ALTERNATIVE FORMULATIONS OF THE MODEL

We obtained the analytical solution in Section 2 by tying the major trunk rivers to the orogen divide via a critical slope. This is possible because of the essentially one-dimensional nature of the model framework. There are other possibilities. Whipple and Meade (2004), for example, obtained their solution by arguing (based on data and two-dimensional land surface models with prescribed uplift) that fluvial relief is approximately proportional to total relief. While critical wedge dynamics requires that the mean topographic profile is constrained to follow the taper angle, real mountain ranges have meandering divides and complicated basin geometries. Therefore the major rivers may be only loosely tied to the average divide position. In this section, the solution is generalized to show that the manner in which the major rivers are tied to the average divide position is of only secondary importance to the sensitivity of the system to changes in forcing. This result accounts for the similarity of the scaling relationship found here to that of Whipple and Meade (2004), and it also shows why the value of the slope exponent n does not contribute significantly to the sensitivity of the system.

From the schematic illustration of Figure 2, we can express the total relief generally as:

$$\text{total relief} = \text{fluvial relief} + \text{hillslope relief}. \tag{22}$$

Even if the hillslope relief depends in complex ways on other parameters in the system, it can always be written in the form $x_c\overline{\tan\theta_c}$, where x_c is still the transition point between fluvial and hillslope regimes, and $\overline{\tan\theta_c}$ denotes the average gradient of the hillslopes. We can evaluate the sensitivity of the solution to changes in x_c and $\overline{\tan\theta_c}$. Using Equation $(10)_1$ for the fluvial relief, Equation (22) can be written as:

$$L\tan\alpha_c = \left(\frac{F}{LKk_a^m P^m}\right)^{\frac{1}{n}} \frac{1}{1-\frac{hm}{n}}\left(L^{1-\frac{hm}{n}} - x_c^{1-\frac{hm}{n}}\right) + x_c\overline{\tan\theta_c} \tag{23}$$

After some manipulation, this can be written as follows:

$$F\cdot L^{-(1+hm)}\cdot P^{-m} = Kk_a^m\left\{\frac{(\tan\alpha_c - \frac{x_c}{L}\overline{\tan\theta_c})(1-\frac{hm}{n})}{1-(\frac{x_c}{L})^{1-\frac{hm}{n}}}\right\}^n \tag{24}$$

Taking log derivatives of the above equation again gives the sensitivities for fractional changes in the various parameters. For simplicity we assume that α_c and K are constant as the forcing changes. This is not necessary, but makes the solution more easily relatable to Section 2 and Whipple and Meade (2004):

$$\frac{\Delta F}{F}-(1+hm)\frac{\Delta L}{L}-m\frac{\Delta P}{P}=n\frac{x_c\overline{\tan\theta_c}}{L\tan\alpha_c-x_c\overline{\tan\theta_c}}\left(\frac{\Delta x_c}{x_c}-\frac{\Delta L}{L}+\frac{\Delta\overline{\tan\theta_c}}{\overline{\tan\theta_c}}\right)+n\frac{(1-\frac{hm}{m})(\frac{x_c}{L})^{1-\frac{h}{mn}}(\frac{\Delta x_c}{x_c}-\frac{\Delta L}{L})}{1-(\frac{x_c}{L})^{1-\frac{h}{mn}}} \quad (25)$$

Equation (25) can be thought of as a perturbation balance equation: in response to an imposed change in forcing (i.e., ΔP, ΔF), it shows how the combination of the other variables in the system must adjust to achieve a new equilibrium state. The left hand side of Equation (25) gives the same sensitivity as the critical slope upper boundary condition [viz, Equation (15)]. All of the dependency on the slope exponent, n, is tied up in the right-hand side of the Equation (25), which reflects changes in the upper boundary condition (i.e., the transition between fluvial and hillslope regimes).

The first term on the right-hand side of Equation (25) can be considered small. The numerator in the fractional factor is the hillslope relief, and the denominator is the fluvial relief. The factor is therefore small since the latter is in general much larger than the former. This explains the relative insensitivity of the solution to changes in θ_c in Figure 4, and the factor of n explains why using simple stream power erosion (which has the highest value for n) gave the greatest sensitivity. The second term on the right hand side contains a factor x_c/L, and so might also be considered small. However it is slightly more problematic, since it is possible that $hm \approx n$, in which case both the numerator and the denominator can get close to zero. However using l'Hôpital's rule, it can be shown that in the limit of hm/n tending to one, the second term on the right hand side tends to

$$\left[\frac{n}{\ln(\frac{L}{x_c})-1}\right]\left(\frac{\Delta x_c}{x_c}-\frac{\Delta L}{L}\right)$$

and so the term is indeed always small for $x_c \ll L$. There are two important consequences of Equation (25). The first is that changes in the upper boundary condition have only a weak effect on the scaling relationship and the second is that, since both terms on the right-hand side are small, the dependency on the slope exponent n is also weak for other formulations of the model framework.

4.1 Alternative Upper Boundary Condition: Critical Discharge

The robustness of the scaling relationship to other model formulations can be illustrated concretely by choosing an alternative upper boundary condition for the hillslope processes. Instead of a critical slope, it can be assumed that a critical discharge, Q_c, is required for the formation of fluvial channels. For constant precipitation this is the same as requiring a critical upstream drainage area for channel formation. This assumption has been used (either implicitly or explicitly) in a number of studies (e.g., Whipple et al., 1999; Roe et al., 2003).

In this case then, from Equation (6), the transition point x_c is given by:

$$x_c=\left(\frac{Q_c}{Pk_a}\right)^{\frac{1}{h}}. \quad (26)$$

While Equation (26) gives the transition point between the fluvial and slope-process dominated portions of the profile, one more constraint is required to tie the fluvial profile to the wedge profile at the drainage divide. To do this we assume constant slope above x_c:

$$\left.\frac{dz}{dx}\right|_{x\geq x_c}=\left.\frac{dz}{dx}\right|_{x=x_c} \quad (27)$$

where dz/dx at $x=x_c$ is given by Equation (7). There is no particular justification for Equation (27): it is not obvious what physical process would fix the slope to the value at the channel head where the critical discharge is exceeded. It is used here simply to compare the sensitivity of an alternative framework with that of the critical slope case in Section 2.We note also that Montgomery and Dietrich (1992) have suggested that channel initiation occurs where a combination of slope and discharge conditions are reached. Such a condition would also be readily incorporable into this framework.

Substituting Equation (27) into Equation (23) gives the solution for the critical discharge upper boundary condition.

$$L\tan\alpha_c=\left(\frac{F}{LK}\cdot\frac{1}{P^m k_a^m}\right)^{\frac{1}{n}}\left\{\frac{L^{1-\frac{hm}{n}}}{1-\frac{hm}{n}}-\frac{\frac{hm}{n}x_c^{1-\frac{hm}{n}}}{1-\frac{hm}{n}}\right\} \quad (28)$$

$$\text{where } x_c=\left(\frac{Q_c}{Pk_a}\right)^{\frac{1}{h}}$$

Although the scaling relationship for this framework cannot be written down explicitly, the solution is readily obtained numerically. Figure 5 shows, for the three different erosion laws considered earlier, there is very little difference in model sensitivity using this alternative upper boundary condition. These calculations highlight the robust nature of the scaling relationship and suggest that despite the simplifying assumptions of the model, its implications may be quite broad and generally applicable.

5. SENSITIVITY TO PRECIPITATION PATTERN

To this point the precipitation rate has been assumed to be uniform. Doing so allowed for an exact analytical solution, but in reality precipitation is highly variable in mountainous regions (e.g., Smith, 1979). The pattern of precipitation affects the pattern of river discharge, and so influences the erosion pattern via

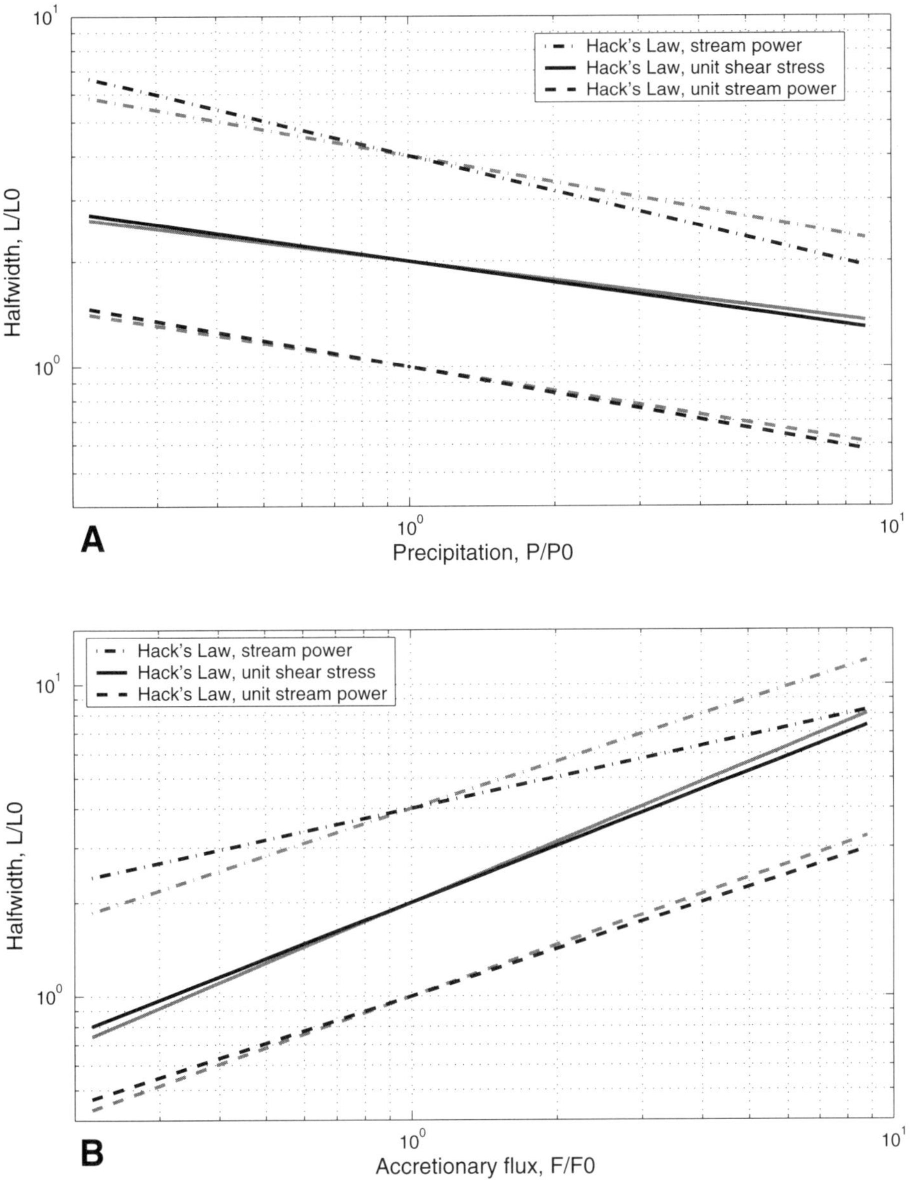

Figure 5. Sensitivity of the solution to boundary conditions. (A) width as a function of precipitation, (B) width as a function of accretionary flux. The black lines are for a critical-slope upper boundary condition, and the gray lines are for a critical-discharge upper boundary condition. The line style indicated in the legend denotes the three different erosion laws tested. For clarity on the plot, the dashed lines have been shifted on the *y*-axis by a factor of two. The similarity of the black and gray lines demonstrates the robustness of the feedback sensitivity to different model formulations.

Equation (3). Imposing plausible patterns in precipitation has been shown to have an important impact on the fluvial relief in the fixed-width model reviewed in Section 2 (e.g., Roe et al., 2003). By allowing the relief to influence the precipitation (both the amount and the pattern), Roe et al. (2003) also showed that the sensitivity of the coupled system to changes in uplift rate was also affected.

We do not pursue a full-coupling with a precipitation feedback here, but we can begin to address what the impacts might be. The results of Section 2 showed that, compared to the fixed-width model, the critical wedge solution was much less sensitive to climate in the form of precipitation rates. In this section, we explore the sensitivity of the two frameworks to changing the precipitation pattern. The sensitivity can be examined by combining a uniform precipitation rate with a δ-function "spike" in the precipitation, located somewhere along the river profile:

$$P(x) = p_0 + p_1\Delta \cdot \delta(x - x_0) \tag{29}$$

where x_0 is the location of the spike along the profile. Roe et al. (2003) discuss the interpretation of this δ-function in terms of expected precipitation patterns, and show that it translates into a discharge pattern given by:

$$Q(x) = \begin{cases} p_0 k_a x^h & \text{for } x_0 \le x \le x_{0^-} \\ p_0 k_a x^h \left[1 + \frac{hp_1\Delta}{p_0 x_0}\left(\frac{x_0}{x}\right)^h\right] & \text{for } x_{0^-} < x \end{cases} \tag{30}$$

It is straightforward to numerically integrate Equation (30) with the critical slope upper boundary condition [i.e., Equations (4) and (5)], and subject to the critical taper constraint. For the curves shown in Figure 6 we take $\alpha_c = 4°$, $\theta_c = 40°$, $p_0 = 1$ m yr^{-1}, and $p_1\Delta = 2000$ m^2 yr^{-1} (equivalent to an increase of precipitation by 2 m yr^{-1} over a 1 km portion of the channel). To facilitate comparison between curves, K is selected so that in the case of uniform precipitation (i.e., $P(x) = p_0$), L equals 40 km. This gives a total relief of ~2800 m. The total relief can then be calculated as a function of the location of the precipitation spike, x_0, for the three different erosion laws. For comparison we also show the same curves for a fixed-width model. The details of this are in Roe et al. (2003). Here we take $L = 40$ km, $U = 1$ mm yr^{-1}, $x_c = 400$ m, and in the same way select K so the fluvial relief equals 2800 m for $P = p_0$.

The response of the critical wedge model is similar to that of the fixed-width model in that the sensitivity of the relief to precipitation increases with the proximity of the anomaly to the divide. As noted in Roe et al. (2003), in the fixed-width case, fractional changes in discharge cause fractional changes in slope of the opposite sign, and so the absolute changes in the slope are largest where the slopes are largest. Figure 6 shows that this also applies to the critical wedge model. An exception is seen for the stream power case for $x_0 <\sim 1500$ m. When x_0 becomes less than the transition point x_c, the δ-function moves divide-wards of the river channel, and the fractional change in the discharge that the river channel must accommodate decreases. This effect is an artifact of having used a δ-function to test the sensitivity, and is likely unimportant in reality.

The overall impression from Figure 6 is that, like the case for the uniform precipitation, the critical wedge model is less sensitive to the pattern of precipitation than the fixed-width model. Nonetheless, the precipitation anomaly chosen here is quite moderate compared to observations, so Figure 6 suggests that the feedback between the precipitation pattern and relief may well be important.

These results should be interpreted cautiously. Importantly, the model assumes that the rock uplift is uniform. Willett (1999) for example showed that imposing a rain shadow in an orogen model incorporating lithospheric deformation causes exhumation to become focused on the wet side of the orogen. The results in this paper show that the sensitivity of the relief changes significantly in going from a fixed-width model to a critical taper model, and it is equally possible that if the physics of the lithospheric deformation can accommodate a localized response to a precipitation pattern, the sensitivity of the total relief may be quite different again.

DISCUSSION AND CONCLUSIONS

Critical wedge dynamics provide a powerful negative feedback on the response of an orogen to tectonic and climate forcing. The self-similar growth implicit in critical wedge theory couples orogen height to width and, for constant mass flux, directly

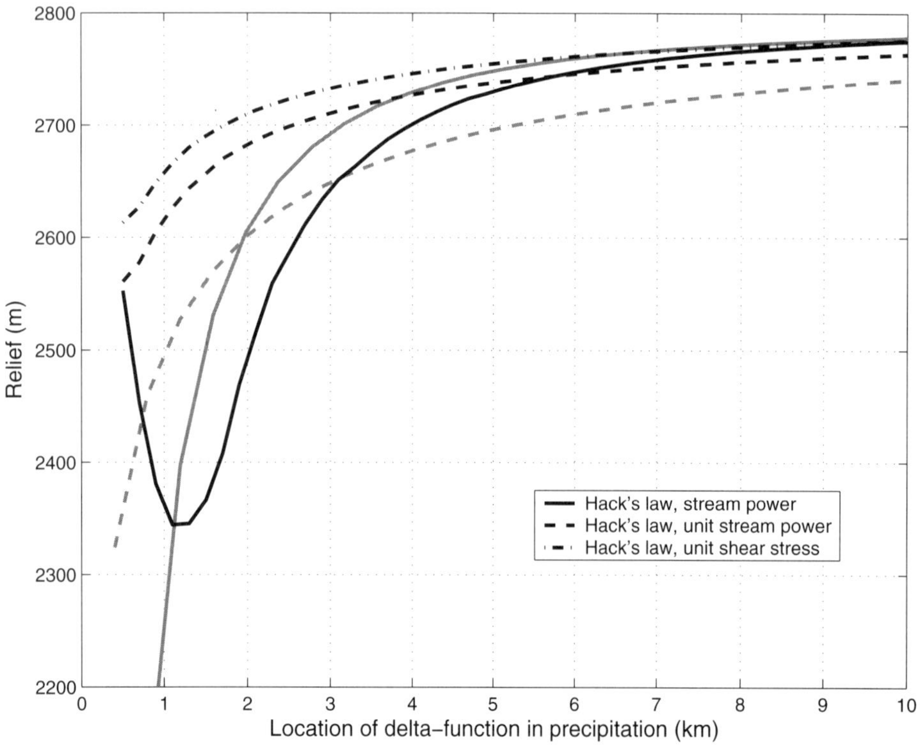

Figure 6. Sensitivity of relief to imposed δ-function in precipitation. The graph plots relief as a function of the location of the δ-function. $x = 0$ is the drainage divide. Lines plotted show the total relief from the critical wedge model for the three different choices of erosion law (black lines), and for comparison, the fluvial relief for the fixed-width model (gray lines). Note for the fixed-width model, unit stream power and unit shear stress have the same sensitivity. For the critical wedge model and the stream power erosion law, the minimum in relief occurs when the transition point x_c becomes greater than the location of the δ-function, x_0. For model details see text.

links rock uplift rate and orogen width. Therefore, changes in tectonic or climatic forcing produce a subdued response in the relief since the system requires concomitant changes in orogen width and rock uplift rate. This paper derived an analytical solution for the response of the orogen width to changes in tectonic forcing (as represented by an accretionary flux), and climate (as represented by the precipitation rate). The strength of the feedback depends on the parameters in the fluvial erosion law. For choices most commonly suggested by theory and observations, the orogen width and exhumation rate are both more sensitive to tectonic forcing than to climate. And in contrast to the case where the width of the orogen is prescribed, the sensitivity of the solution does not depend on the value of the slope parameter n in the erosion law.

While an analytical solution can only be obtained for a critical-gradient hillslope condition, we also showed that the scaling relationship is at most weakly dependent on both the value of n and the way that the major trunk-river profiles connect to the drainage divide. It was also shown that the pattern of rock uplift does not matter for the sensitivity provided that it is integrable and that it changes self-similarly as the orogen size changes. The analyses are important in that they suggest that the results may be quite generally applicable and not just a consequence of the model assumptions. These results are further supported by integrations of a coupled tectonics-surface process model (Stolar et al., this volume), which, although it is a much more complex system with many ways to adjust to changes in forcing, has a sensitivity in close agreement with that predicted by the analytical results.

It is important to emphasize the limitations and caveats of the solution. It implicitly assumes, for example, that as the forcing changes, the dominant physical process governing erosion remains unchanged. The relative importance of different erosional processes may well change with climate, particularly with changes in storminess (e.g., Costa and O'Conner, 1995; Tucker and Bras, 2000; Snyder et al., 2003).

Another complication would be the presence of active glaciers, although the same feedback ought to exist. Existing models of glaciation at the landscape scale assume erosion which scales with the rate of glacier sliding (e.g., Braun et al., 1999; MacGregor et al., 2000; Tomkin and Braun, 2002), both of which tend to increase with precipitation rate. However, in an orogen with significant glaciated areas, it is quite possible that because of the very different effectiveness of glacial and fluvial erosion (e.g., Hallet et al., 1996), the integrated erosion rate over the orogen may be more sensitive to summertime temperature (to the degree that it dictates the permanent snow line) than to precipitation.

A further possibility not addressed in the calculations presented here is that the response of the system depends on the strength of the rain shadow (e.g., Willett, 1999; Whipple and Meade, 2004). As an orogen grows it is to be expected that, all else being equal, the strength of the rain shadow increases, both because the average ascent and descent on the windward and leeward flanks of the orogen, respectively, would be expected to increase, but also because more storms would be blocked on the windward side (e.g., Smith, 1979). An additional factor is that the results in Section 5 and in Roe et al. (2003) showed that the pattern of precipitation and in particular the near-divide precipitation is important.

A long-term objective of this work is to be able to reconcile observed topographic forms with the climatic and tectonic settings which have given rise to them. Ongoing studies into each of these three components of Earth's system continually yields new insights, a result of which is that new physical mechanisms and uncertainties emerge. A major challenge for the field therefore, is to understand whether there are aspects of this enormously complex system which behave in robust and predictable ways, or the myriad feedback pathways and nonlinear interactions preclude a quantitative understanding of the system. All of the caveats noted above are reasons why any particular orogen might not behave according to Equation (15), or why different orogens might behave differently from each other. On the other hand, the fundamental nature of the feedback does not depend on the system being governed by the strict definitions of critical wedge dynamics. Provided the steady-state orogen width has a tendency to increase in response to an increase in accretionary flux (or a decrease in precipitation rate), the feedback will tend to operate to damp the system response to changes in forcing. Moreover taking one particular framework, we have shown that the feedback is strong, that its magnitude is insensitive to some of the details of the model formulation and furthermore, that it captures the behavior of a more complex numerical model (Stolar et al., this volume). The results lend confidence that aspects of the system are indeed robust and comprehensible, and that it therefore remains a worthwhile, if challenging, research goal to pursue an understanding of how climate, erosion, and tectonics have combined to sculpt Earth's landscapes.

ACKNOWLEDGMENTS

The authors are grateful to Kelin Whipple, David Montgomery, and Bernard Hallet for stimulating and instructive conversations, to Jonathon Tomkin and Nicole Gasparini for considered and helpful reviews, and to Mark Brandon, the editor.

APPENDIX A: SOLUTION FOR NON-UNIFORM UPLIFT RATE

For the framework of Section 2 the total relief in the wedge, R, can be written as

$$R = L \tan\alpha_c = \int_L^{x_c} \frac{dz}{dx} dx + x_c \tan\theta_c. \tag{31}$$

Taking the rock uplift function given by Equation (20), requiring a pointwise balance between erosion and this uplift rate, and rearranging for dz/dx, gives an equation similar to Equation (7):

$$\frac{dz}{dx} = \left(\frac{F}{LK} \cdot \frac{1}{P^m k_a^m} \right)^{\frac{1}{n}} \cdot \left[\phi(x/L) \right]^{\frac{1}{n}} \cdot x^{-\frac{hm}{n}}, \quad (32)$$

which, upon substitution into Equation (31), gives:

$$L \tan \alpha_c = x_c \tan \theta_c + \int_L^{x_c} \left(\frac{F}{LKP^m k_a^m} \right)^{\frac{1}{n}} \left[\phi(x/L) \right]^{\frac{1}{n}} x^{-\frac{hm}{n}} dx. \quad (33)$$

Next we make a substitution of variables $y = x/L$:

$$L \tan \alpha_c = x_c \tan \theta_c + \left(\frac{F}{LKP^m k_a^m} \right)^{\frac{1}{n}} \times \int_1^{\frac{x_c}{L}} L^{1-\frac{hm}{n}} \phi(y)^{\frac{1}{n}} y^{-\frac{hm}{n}} dy \quad (34)$$

Finally, using Equation (32) with $x = x_c$, and rearranging gives:

$$\left(\frac{L}{x_c} \right) \frac{\tan \alpha_c}{\tan \theta_c} - 1 = \left[\phi(x_c/L) \right]^{-\frac{1}{n}} \left(\frac{x_c}{L} \right)^{\frac{hm}{n}-1} \int_1^{\frac{x_c}{L}} \phi(y)^{\frac{1}{n}} y^{-\frac{hm}{n}} dy \quad (35)$$

Provided $\phi(y)$ is an integrable function, then like the uniform rock uplift solution, Equation (35) is an implicit equation in L/x_c. The solution can be written as $L/x_c = \lambda$, with the value of λ to be found. Using Equation (32) with $x = x_c$, and rearranging gives:

$$\lambda^{hm} \phi(\lambda^{-1}) = \left[k_a^{\ m} \tan^n \theta_c K \frac{h}{n} \right] L^{1+hm} \cdot F^{-1} \cdot P^m. \quad (36)$$

REFERENCES CITED

Adams, J., 1980, Contemporary uplift and erosion of the Southern Alps, New Zealand: Geological Society of America Bulletin Part I, v. 91, p. 1–114.

Anders, A.M., Montgomery, D.R., and Hallet, B., 2002, Comparing spatial variations in precipitation and erosion index to differences in long term exhumation along the European Alps: Geological Society of America Abstracts with Programs, v. 34, no. 6, p. 108.

Beaumont, C., Fullsack, and Hamilton, J., 1992, Erosional control of active compressional orogens, *in* McClay, K.R., ed., Thrust tectonics: New York, Chapman and Hall, p. 1–18

Beaumont, C., Kooi, H., and Willett, S.D., 2000, Couple tectonic-surface process models with applications to rifted margins and collisional orogens, *in* Summerfield, M., ed, Geomorphology and Global Tectonics: Chichester, UK, Wiley, p. 29–56.

Bernet, M., Brandon, M.T., Gaver, J.I., and Molitor, B.R., 2003, Fundamentals of detrital zircon fission-track analysis for provenance and exhumation studies with examples from the European Alps, *in* Bernet, M., and Spiegel, C., eds. Detrital thermochronology—Provenance analysis, exhumation, and landscape evolution of mountain belts: Geological Society of America Special Paper 378, p. 25–36.

Brandon, M.T., Roden-Tice, M.K., and Garver, J.I., 1998, Late Cenozoic exhumation of the Cascadia accretionary wedge in the Olympic Mountains, NW Washington State: Geological Society of America Bulletin, v. 110, p. 985–1009.

Braun, J., Zwartz, D., and Tomkin, J.H., 1999, A new surface processes model combining glacial and fluvial erosion: Annals of Glaciology, v. 28, p. 282–290.

Costa, J.E., and O'Conner, J.E., 1995: Geomorphically effective floods, *in* Costa, J.E., et al., eds., Natural and anthropogenic influences in fluvial geomorphology: Washington, D.C., American Geophysical Union Geophysical Monograph 89, p. 45–56.

Dahlen, F.A., 1984, Noncohesive critical Coulomb wedges: An exact solution: Journal of Geophysical Research, v. 89, p. 10125–10133.

Dahlen, F.A., and Barr, T.D., 1989, Brittle functioning mountain building 1. Deformation and mechanical energy budget: Journal of Geophysical Research, Solid Earth and Planets, v. 94, p. 3906–3922.

Dahlen, F.A., 1990, Critical taper model of fold-and-thrust belts and accretionary wedges: Annual Review of Earth and Planetary Sciences, v. 18, p. 55–99, doi: 10.1146/annurev.ea.18.050190.000415.

Davis, D., Suppe, J., and Dahlen, F.A., 1983, Mechanics of fold-and-thrust belts and accretionary wedges: Journal of Geophysical Research, v. 88, p. 1153–1172.

Finlayson, D.P., Montgomery, D.R., and Hallet, B., 2002, Spatial coincidence of rapid inferred erosion with young metamorphic massifs in the Himalayas: Geology, v. 30, p. 219–222.

Hack, J.T., 1957, Studies of longitudinal stream profiles in Virginia and Maryland: U.S. Geological Survey Professional Paper 294-B, 97 p.

Hallet, B., Hunter, L., and Bogen, J., 1996, Rates of erosion and sediment yield by glaciers: A review of field data and their implication: Global Planetary Change, v. 12, p. 213–235.

Hilley, G.E., Strecker, M.R., and Ramos, V.A., 2004, Growth and erosion of fold-and-thrust belts with an application to the Aconcagua fold-and-thrust belt, Argentina: Journal of Geophysical Research, v. 109, B01410, doi: 10.1029/2002JB002282.

Howard, A.D., Dietrich, W.E., and Seidl, M.A., 1994, Modeling fluvial erosion on regional to continental scales: Journal of Geophysical Research, v. 99, p. 13,971–13,986, doi: 10.1029/94JB00744.

Koons, P.O. 1989, The tomographic evolution of collisional mountain belts: A numerical look at New Zealand: American journal of Science, v. 289, p. 1041–1069.

MacGregor, K.R., Anderson, R.S., Anderson, S.P., and Waddington, E.D., 2000, Numerical Simulations of glacial-valley longitudinal profile evolution: Geology, v. 28, p. 1031–1034, doi: 10.1130/0091-7613(2000)028<1031:NSOGVL>2.3.CO;2.

Masek, J.G., Isacks, B.L., Gubbels, T.L., and Fielding, E.J., 1994, Erosion and tectonics at the margins of continental plateaus: Journal of Geophysical Research, v. 99, p. 13,941–13,956, doi: 10.1029/94JB00461.

Montgomery, D.R., and Dietrich, W.E., 1992, Channel initiation and the problem of landscape scale: Science, v. 255, p. 826–830.

Montgomery, D.R., Balco, G., and Willett, S.D., 2001, Climate, tectonics, and the morphology of the Andes: Geology, v. 29, p. 579–582.

Reiners, P.W., Elhers, T.A., Garver, J.I., Mitchell, S.G., Montgomery, D.R., Vance, J.A., and Nicolescu, S., 2002, Late Miocene exhumation and uplift of the Washington Cascade Range: Geology, v. 30, p. 767–770, doi: 10.1130/0091-7613(2002)030<0767:LMEAUO>2.0.CO;2.

Reiners, P.W., Ehlers, T.A., Mitchell, S.G., and Montgomery, D.R., 2003, Coupled spatial variations in precipitation and long-term erosion rates across the Washington Cascades: Nature, v. 426, p. 645–657, doi: 10.1038/nature02111.

Roe, G.H., Montgomery, D.R., and Hallet, B., 2002, Effects of orographic precipitation variations on the concavity of steady-state river profiles: Geology, v. 30, p. 143–146, doi: 10.1130/0091-7613(2002)030<0143:EOOPVO>2.0.CO;2.

Roe, G.H., Montgomery, D.R., and Hallet, B., 2003, Orographic precipitation and the relief of mountain ranges: Journal of Geophysical Research, v. 108, B62315, doi: 10.1029/2001JB001521.

Sklar, L., and Dietrich, W.E., 1998, River longitudinal profiles and bedrock incision models: Stream power and the influence of sediment supply, *in* Tinkler, T.J., and Wohl, E.E., eds., Rivers over Rock: Fluvial processes in bedrock channels: Washington, D.C., American Geophysical Union Geophysical Monograph 107, p. 237–260.

Smith, R.B., 1979, The influence of mountains on the atmosphere, *in* Saltzman, B., ed., Advances in Geophysics: Academic Press, New York, v. 21, p. 87–230.

Snyder, N.P., Whipple, K.X., Tucker, G.E., and Merritts, D.M., 2003, Importance of a stochastic distribution of floods and erosion thresholds in the bedrock river incision problem: Journal of Geophysical Research, v. 108, no. B2, p. 2117, doi: 10.1029/2001JB001655.

Stock, J.D., and Montgomery, D.R., 1999, Geologic constraints on bedrock river incision using the stream power law: Journal of Geophysical Research, v. 104, p. 4983–4993, doi: 10.1029/98JB02139.

Suppe, J., 1980, A retrodeformable cross section of northern Taiwan: Geologi-

cal Society of China Proceedings, v. 23, p. 46–55.
Tomkin, J.H., and Braun, J., 2002, The influence of alpine glaciation on the relief of tectonically active mountain belts: American Journal of Science, v. 302, p. 169–190.
Tucker, G.E., and Bras, R.L., 2000, A stochastic approach to modeling the role of rainfall variability in drainage basin evolution: Water Resources Research v. 36, p. 1953–1964, doi: 10.1029/2000WR900065.
Tucker, G.E., and Whipple, K.X., 2002, Topographic outcomes predicted by the stream erosion models: sensitivity analysis and intermodel comparison: Journal of Geophysical Research, v. 107, doi: 10.1029/2001JB000162.
Whipple, K.X., Kirby, E., and Brocklehurst, S.H., 1999, Geomorphic limits to climate-induced increases in topographic relief: Nature, v. 401, p. 39–43, doi: 10.1038/43375.
Whipple, K.X., and Tucker, G.E., 1999, Dynamics of the stream-power river incision model: implications for height limits of mountain ranges, landscape response time scales, and research needs: Journal of Geophysical Research, v. 104, p. 17,661–17,674, doi: 10.1029/1999JB900120.
Whipple, K.X., Hancock, G.S., and Anderson, R.S., 2000, River incision into bedrock: mechanics and the relative efficacy of plucking, abrasion, and cavitation: Geological Society of America Bulletin, v. 112, p. 490–503, doi: 10.1130/0016-7606(2000)112<0490:RIIBMA>2.3.CO;2.
Whipple, K.X., and Meade, B.J., 2004, Dynamic coupling between erosion, rock uplift, and strain partitioning in two-sided, frictional orogenic wedges at steady-state: an approximate analytical solution: Journal of Geophysical Research, v. 109, doi: 10.1029/2003JF000019.
Willett, S.D., 1999, Orogeny and orography: The effects of erosion on the structure of mountain belts: Journal of Geophysical Research, v. 104, p. 28,957–28,981, doi: 10.1029/1999JB900248.
Willett, S.D., Beaumont, C., and Fullsack, P., 1993, Mechanical model for the tectonics of doubly-vergent compressional orogens: Geology, v. 21, p. 371–374, doi: 10.1130/0091-7613(1993)021<0371:MMFTTO>2.3.CO;2.
Willett, S.D., Slingerland, R., and Hovius, N., 2001, Uplift, shortening, and steady-state topography in active mountain belts: American journal of Science, v. 301, p. 455–485.

Manuscript Accepted by the Society 23 June 2005

Geological Society of America
Special Paper 398
2006

Climatic and tectonic forcing of a critical orogen

Drew B. Stolar†
Sean D. Willett
Gerard H. Roe
Department of Earth and Space Sciences, University of Washington, Seattle, Washington 98195, USA

ABSTRACT

Critical wedge theory provides a direct link between the form of an orogen, the rate of orogen evolution, and the accretionary and erosional fluxes that promote orogen growth and decay, respectively. We explore several fundamental characteristics of an eroding critical orogen: (1) the sensitivity of steady-state orogen size to tectonic and climatic forcing, (2) the response time of a critical orogen to perturbations in forcing, and (3) the behavior of surface topography and the rock uplift field in a system in which they are not prescribed. To do this, we develop a numerical model that couples a two-dimensional, planform surface erosion model with a two-dimensional, plane-strain finite element model of deformation. We first present a base model in which a critical orogen evolves to a steady-state under boundary conditions similar to those of analog sandbox experiments. We find that mean topography and tectonic uplift reach steady states, whereas planform topography remains dynamic throughout the simulation. From a suite of simulations, we determine the steady-state scaling relationship between orogen size and tectonic and climatic forcing and find good agreement with predictions from one-dimensional models. In addition, we examine the response of the steady-state orogen to climatic and tectonic perturbation with four simulations in which changes in tectonic and climatic conditions lead to either growth or contraction of the orogen to a new steady state. We show that the response time to perturbation agrees well with predictions from a one-dimensional semi-analytical model. We find that the transient evolution of erosion rate and erosional flux is potentially useful for distinguishing between tectonic and climatic forcing mechanisms.

Keywords: tectonics, erosion, critical wedge, numerical modeling.

INTRODUCTION

The evolution of convergent mountain belts has long been a topic of interest in tectonics. Increasingly over the last two decades, geomorphology has added to this study by investigating how erosion of topography by surface processes can alter the form and rate of orogen evolution. In addition, linkages over a wide range of spatial and temporal scales have been proposed to exist between climate and tectonics, through the influence of surface topography, which is both product and player in the competition between climate and tectonics. The coupling among tectonics, erosion, and climate appears to be rich in complexity, and a broad community of scientists is now discovering and examining first-order linkages in the coupled system.

Coupling between tectonics and climate exists through various mechanisms that operate in both directions. That is, tectonics and climate both affect each other, resulting in feedback between processes. Tectonics influences local and regional climate primarily by raising high mountains and plateaus leading to orographic enhancement of precipitation (Smith, 1979; Koons,

†E-mail: dstolar@u.washington.edu.

Stolar, D.B., Willett, S.D., and Roe, G.H., 2006, Climatic and tectonic forcing of a critical orogen, *in* Willett, S.D., Hovius, N., Brandon, M.T., and Fisher, D.M., eds., Tectonics, Climate, and Landscape Evolution: Geological Society of America Special Paper 398, Penrose Conference Series, p. 241–250, doi: 10.1130/2006.2398(14). For permission to copy, contact editing@geosociety.org.

1990; Molnar, 1997; Willett, 1999a; Beaumont et al., 2001; Roe et al., 2003). Climate affects tectonics through enhanced erosion of tectonic structures or entire mountain belts, altering the gravitational stresses and thus internal deformation (Beaumont et al., 1992, 2001; Willett, 1999a).

Active convergent orogens are particularly sensitive to mass redistribution by surface processes, as crustal thickening can result in orogens with critical topography, such that regional topographic gradients are in balance with the strength of the crustal rock. In the case where deformation is frictional plastic and crustal accretion occurs by shortening above a frictional detachment, topographic gradients attain a constant slope and the orogen takes the cross-sectional form of a wedge (Davis et al., 1983). This tectonic system, in spite of its seeming complexity, lends itself well to analytical treatment (e.g., Dahlen, 1984; Davis et al., 1983). If an orogenic wedge is in a critical state, any perturbations to the cross-sectional form of the orogen, such as those that arise from surface erosion, lead to compensatory deformation to restore the wedge to its critical state. Thus, there is a direct coupling between erosion and tectonic deformation.

The effect of erosion on a critical wedge, including feedback to the erosion process, has also been treated extensively by analytical and numerical methods. Dahlen and Barr (1989) and Barr and Dahlen (1989) considered the effects of erosion on the internal kinematics and heat budget of a wedge with an assumed erosion rate, but did not incorporate any feedback from tectonics on erosion rates. Koons (1990) and Willett et al. (1993) also discussed the effects of erosion on orogen structure and kinematics but did not address the strength of the feedback between topography and erosion rates. This feedback is included in a class of numerical models that couple plane-strain finite element models of crustal deformation to one-dimensional (1-D) erosion models (e.g., Batt and Braun, 1997; Willett, 1999a), but these models include complexities in the tectonic model, such as isostatic compensation and temperature-dependent viscous deformation, that obscure the connections with critical wedge theory.

The most explicit examination of the linkages between critical wedge theory and erosional processes has been conducted in the studies of Hilley et al. (2004), Whipple and Meade (2004), Roe et al. (this volume), and K.X Whipple and B.J. Meade (2004, personal commun.)These models are based on the coupling of a critical taper angle, which defines the mean elevation profile, and a one-dimensional hillslope-fluvial channel profile. As a group, the studies examine how the coupling of tectonics and erosion inherent in a critical orogen affects the steady-state orogen size as a function of tectonic and erosional forcing and the response time of a critical orogen to climatic and tectonic perturbation. Because of the assumptions required for solution, these models do not address how several important degrees of freedom (e.g., spatial variability in rock uplift, spatial and temporal variability in the channel network) influence the behavior of a coupled orogen.

In this paper, we develop a numerical model that allows for these degrees of freedom and contains more complete descriptions of the tectonic and erosional processes relevant to eroding critical orogens. We use the model to (1) examine the time evolution of a critical orogen to a steady state, (2) quantify the scaling relationship between steady-state orogen size and tectonic and erosional forcing, (3) examine the behavior of additional degrees of freedom allowed in our model, and (4) quantify the controls on the response time of a critical orogen to perturbations in forcing. Where appropriate, we make direct comparisons of the results from this model to equivalent analytical models.

COUPLED MODEL

The coupled model consists of a finite element mechanical model, which describes deformation of a plastic material in cross section, and a surface process model (SPM), which simulates fluvial and hillslope erosion in planform.

The mechanical model is a variant of models presented in several studies of orogenic evolution (Beaumont et al. 1992, 2001; Willett, 1999a, 1999b); deformation of the crust occurs in response to convergence and accretion of new material. At orogen scale, this occurs under circumstances similar to Figure 1, in which accretion represents the transfer of crustal material from a downgoing or subducting plate to an overriding plate. In our numerical experiments, which match the basic design of analog sandbox experiments (Malavieille, 1993; Wang and Davis, 1996), accretion is driven by the side and basal velocity boundary conditions shown in Figure 2. Deformation of the subducted, downgoing plate is not considered further in this model. As in the analog experiments, deformation occurs first at the point *S*, where the basal velocity decreases from the convergence velocity V_c to zero, and later outboard of point S along shear zones that propagate upwards from the basal surface.

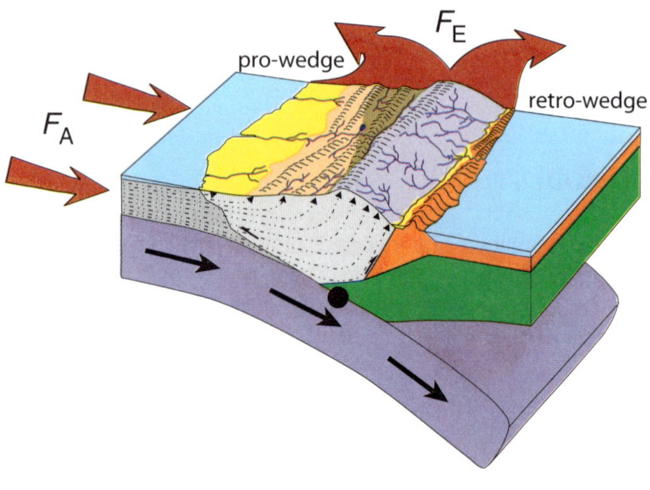

Figure 1. Schematic diagram of a doubly vergent wedge (pro- and retro-wedges are labeled). In this model, material enters and exits the wedge via accretionary (F_A) and erosional fluxes (F_E), respectively. Large black circle is equivalent to point S in Figure 2 and denotes where the upper crust detaches from the downgoing plate (after Yeh, 2003).

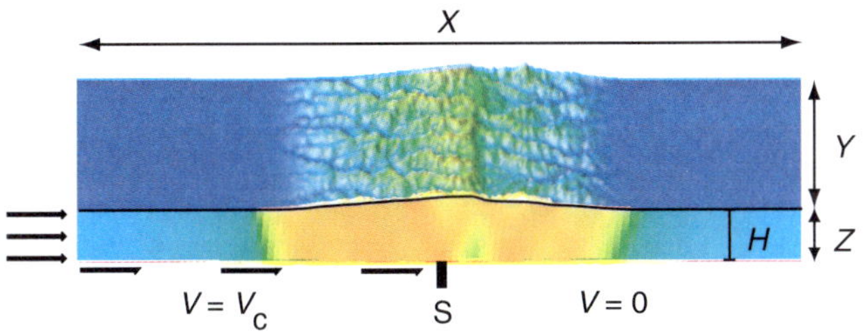

Figure 2. Numerical model with boundary conditions. Mechanical model is shown in the *x-z* plane, and the surface process model is shown in the *x-y* plane. Material to the left of point S moves laterally at a velocity V_c. To the right of point S, the velocity is zero. The upper surface of the mechanical model (black line) is equivalent to the mean elevation profile of the planform topography and is assumed to be a stress-free surface. H is the thickness of the incoming material.

Decoupling of the incoming material from its base and accretion into the wedge is affected by the mechanically weak décollement. The physical interpretation of the basal surface is dependent on both the setting and scale of the specific application of the numerical model. In reference to orogenic, crustal-scale wedges, the basal surface represents a décollement either within the upper crust, at the upper-lower crust boundary, or at the Moho. In analog sandbox experiments, it represents the boundary between sand and an underlying material, typically Mylar (Davis et al., 1983; Wang and Davis, 1996). In the models of this paper, we assume that the basal surface remains horizontal throughout the simulation, such that isostatic compensation of excess topography is not allowed. This permits more careful comparison with the analytical critical wedge theory.

The finite element formulation solves the quasi-static form of the Navier-Stokes equation to calculate deformation of a material obeying a nonlinear viscous flow law. Plastic behavior is simulated by requiring that (1) material behave rigidly (highly viscous) for stresses below the Mohr-Coulomb yield strength of the material and (2) that strain occur at a rate such that the stress never exceeds the yield strength (Willett, 1992). Numerical solution with this formulation yields horizontal and vertical velocities within the cross-sectional domain that, when linked to the SPM, provide one of the mechanisms by which the two models are coupled.

We simulate planform surface erosion with a variant of the Cascade surface process model (Braun and Sambridge, 1997). A particular strength of Cascade is its representation of topography by an irregular network of nodes. This grid allows for accurate tracking of material in systems with significant amounts of lateral strain, such as the wedge investigated in this study.

Erosion of the model topography occurs by fluvial incision and bedrock landsliding. The rate of fluvial incision, E, is calculated by the stream power law:

$$E = KQ^mS^n, \qquad (1)$$

where K is a measure of the substrate erodibility, Q is the discharge at a point along the channel network, S is the downstream slope (e.g., Whipple and Tucker, 1999), and m and n are exponents related to the physical processes of bedrock erosion (e.g., Whipple et al., 2000). Here, we use the unit stream power law (m = 1/2, n = 1). The discharge is equal to the product of the precipitation rate, P, which is assumed to be constant over the landscape, and the upstream contributing area, A. Combining equation 1 with tectonic forcing terms yields equations that describe the horizontal and vertical velocities of a representative channel node:

$$dx_i/dt = v_i, \qquad (2a)$$

$$dz_i/dt = u_i - KQ_i^mS_i^n, \qquad (2b)$$

where v_i and u_i are the horizontal and vertical velocities at the ith node, respectively. We assume completely detachment-limited conditions, such that transported sediment and sediment eroded from the bed do not influence the rate of erosion (e.g., Howard, 1994). Water is allowed to exit the model domain only through the left and right boundaries (Fig. 2).

To simulate hillslope erosion, we use a simple threshold-slope model of bedrock landsliding, which is consistent with results from large-scale topographic analyses (e.g., Burbank et al., 1996; Montgomery, 2001). This is implemented, when necessary, by lowering nodal connections until all slopes are at or below the critical value, which is set in our models to 30°. The landslide mass is assumed to not influence fluvial incision.

The coupling of the mechanical and surface process models is achieved through a sequence of four steps: (1) the mechanical model calculates the velocity field within the entire cross-sectional domain, (2) the rates of uplift and lateral advection from the surface of the mechanical model are interpolated to each SPM node, (3) within the SPM and at much smaller time steps, the surface topography is uplifted, advected, and eroded for the length of the mechanical model time step, and (4) a mean elevation profile is calculated for the SPM topography and set to be the upper boundary of the mechanical model. This sequence is repeated for the length of the simulation.

Nondimensionalization of the coupled model follows naturally from the scale-independence of purely frictional critical wedges (Dahlen, 1984; Willett, 1999a). By using the height of the incoming material, H, and the convergence velocity, V_c, as characteristic length and rate scales, respectively, this approach allows for generalization of the numerical results to both sandbox wedges, where $H \sim 10^{-1}$ m and $V_c \sim 10^4$ m/yr, and fold-and-thrust belts, where $H \sim 10^3$ m and $V_c \sim 10^{-2}$ m/yr. For instance, nondimensionalization of equations 2a and 2b yields:

$$dx_i^*/dt^* = v_i^*, \qquad (3a)$$

$$dz_i^*/dt^* = u_i^* - N_e(Q_i^*)^m S_i^n, \qquad (3b)$$

where N_e, the erosion number, is $KH^{2m}V_c^{m-1}$.

All equations and model quantities presented in the following sections are given in nondimensional form. The model dimensions (Fig. 2) are $\Delta X^* = 1000$, $\Delta Y^* = 8$, and $\Delta Z^* = 1$. The along-profile width, ΔX^*, and strike-parallel width, ΔY^*, are large enough such that the boundaries do not influence the evolution of the mean elevation profile. Note that one unit of nondimensional time represents the amount of time required to accrete one crustal thickness of material into the domain. For example, if $H = 10^3$ m and $V_c = 10^{-2}$ m/yr, $T^* = 1$ would be equivalent to 0.1 m.y.

BASE MODEL EVOLUTION

We first examine the evolution of the coupled wedge as it grows to a steady-state and then characterize the topography and deformation field at steady state. The tectonic boundary conditions and the requirement of purely plastic deformation are chosen to imitate the analog sandbox experiments of Malavieille (1993) and Wang and Davis (1996) and the numerical experiments of Willett (1999b), noting that none of these included erosion. We present here the time evolution of the planform topography, strain rate field, mean elevation profile, erosional flux, and wedge volume.

Two of the most significant characteristics of a mountain belt are the spatial extent over which erosion and deformation occur. Within the coupled model, these can be measured with the planform topography and strain rate field, respectively (Fig. 3). In addition, because the mechanical model responds directly to the gravitational stresses imposed by the mean elevation profile, the surface evolution (Fig. 4) is intimately tied to the pattern and magnitude of deformation. Results from pre-steady-state stages are shown to illustrate the period of rapid change in the early evolution of the wedge.

Beginning from a flat initial topography, wedge growth initiates as a single pop-up structure centered above the S point. Deformation occurs within conjugate shear zones; the width of each shear zone is dictated by the resolution of the mechanical model mesh. Growth of the pop-up structure continues until T^* = 1/4. By T^* = 1/2, shear zones form on either side of the pop-up structure, and deformation within these shear zones results in uplift and widening of the pro- and retro-wedges. This occurs until T^* = 2, when the mean elevation of the divide reaches the approximate steady-state elevation. Between T^* = 2 and T^* = 4, the deforming region expands from three to five shear zones as the wedge widens and the flanks are uplifted. At T^* = 4, both the deformation field and the mean elevation profile are very similar to those of T^* = 20, indicating that the wedge has achieved a large-scale steady state.

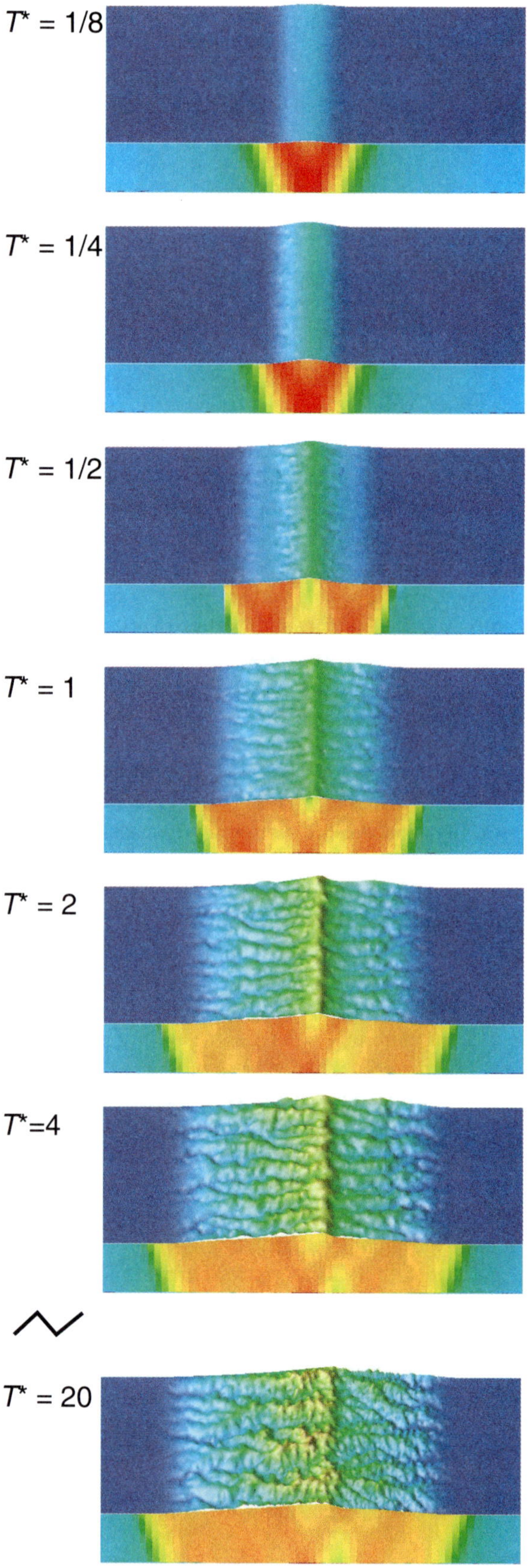

Figure 3. Planform topography and *xx*-component of strain rate through time for the base model. Warmer colors represent higher elevations and higher rates of strain.

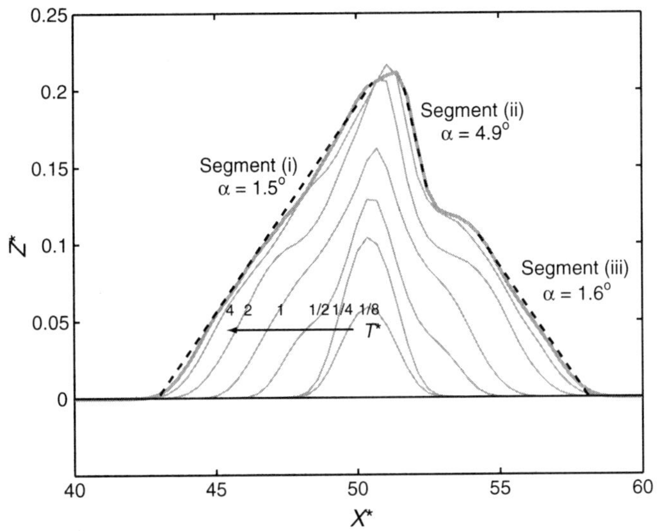

Figure 4. The mean elevation profile through time, from initiation of a pop-up structure to steady state. Regression lines and calculated slopes, α, also shown for segments of the steady-state profile.

At steady state, the mean elevation profile can be divided into three segments: (1) the pro-wedge segment, which has a slope of 1.5°, (2) the upper retro-wedge segment, which has a slope of 4.9°, and (3) the lower retro-wedge segment, which has a slope of 1.6° (Fig. 4). This compares well with critical wedge theory (Dahlen, 1984) and the analog sandbox experiments of Wang and Davis (1996), which reveal that the slopes of segments 1 and 3 are equal to the minimum taper angle (1.5°), and that the slope of segment 2 is less than the maximum taper angle (6.0°).

Next, we examine predictions for the time evolution of erosional flux and wedge volume. These are important quantities in natural systems, because they can be related to observations from sedimentary basins such as provenance, deposition rates,

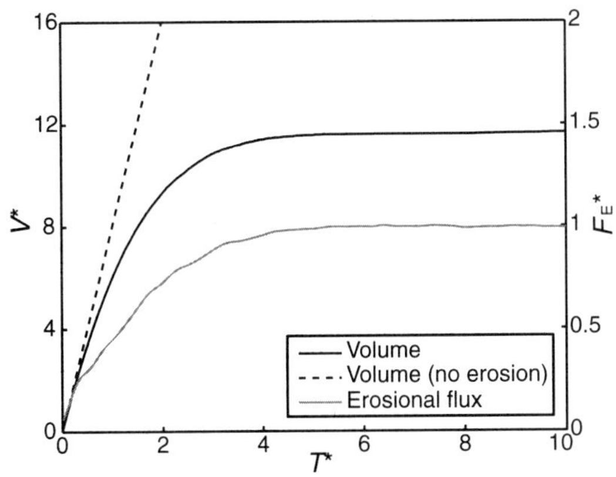

Figure 5. Evolution of wedge volume and erosional flux of the base model. Also shown is the volume of a noneroding wedge forced by the same accretionary flux.

and sediment volumes. In Figure 5, we plot the evolution in time of the erosional flux, which is the integral of the erosion rate over the SPM domain, and wedge volume, which is defined as the volume of the excess topography above the original base level. Also plotted in Figure 5 is a line representing the volume of an equivalent noneroding wedge, which increases with a slope of HV_c.

Erosional flux increases rapidly from $T^* = 0$ and asymptotically approaches a steady value, which is achieved at $T^* \sim 4$. We note that the volume of the noneroding wedge departs from the volume of the eroding wedge at a rate equal to the erosional flux. At $T^* = 4$, the noneroding wedge is more than 2.5 times larger than the eroding wedge.

Steady-State Width Scaling Relationship

The size of the wedge at steady state is a fundamental measure of the response of the coupled system to tectonic and climatic forcing. Roe et al. (this volume) analyzed the coupling between a one-dimensional river profile, which erodes according to equation 1, and a critical wedge in which the mean elevation profile is defined by a constant taper angle. A scaling relationship was derived between steady-state wedge width, W, accretionary flux, F_A, and precipitation rate, P:

$$W \propto F_A^{1/hm+1} P^{-m/hm+1}, \tag{4}$$

where h is the Hack's Law exponent, which relates along-profile distance x to contributing area A ($A \propto x^h$). Equation 4 is very similar to the scaling relationship derived by Whipple and Meade (2004). As demonstrated by Roe et al. (this volume), equation 4 is valid for systems in which topography is advected laterally and in which rock uplift is allowed to be nonuniform but is constrained to change self-similarly. It can easily be shown that equation 4 also applies to doubly vergent wedges (i.e., wedges composed of pro- and retro-wedges) (Whipple and Meade, 2004). Thus, the remaining differences between the analytical and numerical models are that a non-self-similar rock uplift pattern and an emergent and evolving two-dimensional (2-D) channel network are allowed for in the numerical model.

In comparing the steady-state predictions of the analytical and numerical models, we use results from a suite of simulations with different tectonic and climatic forcing for three different fluvial erosion laws. Each erosion law has the general form of equation 1 but with different exponents: the unit stream power law ($m = 1/2$, $n = 1$), the unit shear stress law ($m = 1/3$, $n = 2/3$), and the stream power law ($m = 1$, $n = 1$) (e.g., Whipple and Tucker, 1999). The ratio of the steady-state wedge widths and a reference width are plotted versus the ratio of the climatic and tectonic forcing and reference forcing in Figure 6. Also plotted in Figure 6 are the predictions of equation 4 (Roe et al., this volume), using calculations of the Hack's law exponent h from the model topography. There is very close agreement between the analytical and numerical predictions, indicating that the nonuniform rock uplift

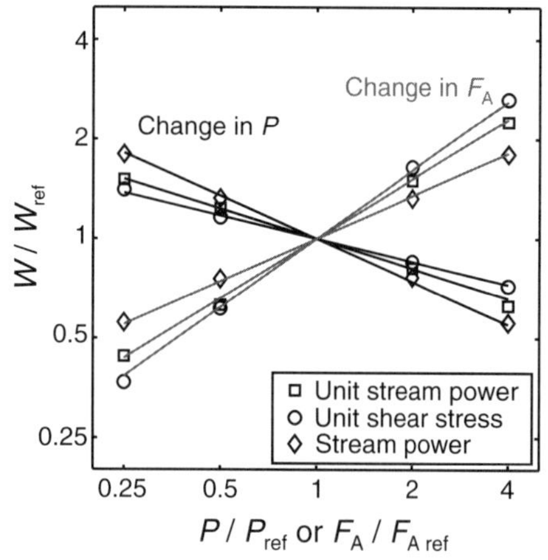

Figure 6. Comparison of the steady-state wedge width determined by numerical model with predictions from width-scaling law (equation 4) (see Roe et al. [this volume] and Whipple and Meade [2004]). Numerical results are denoted by symbols, which indicate the particular erosion law used in the simulation. Analytical predictions from Roe et al. (this volume) for precipitation and accretionary flux scaling are shown with black and gray lines, respectively.

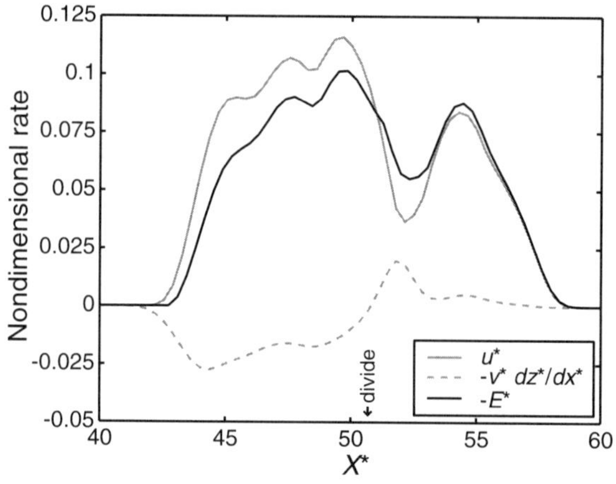

Figure 7. Profiles of selected terms in equation 5. The erosion rate profile is the sum of the rock uplift rate and the contribution from the lateral advection of topography. Position of main drainage divide is noted.

and evolving two-dimensional channel network do not change the basic relationship between wedge size and tectonic and climatic forcing.

Spatial Distribution of Exhumation Rates

A steady-state mean elevation profile requires a wedge-scale balance between accretionary and erosional fluxes and a local balance between rock uplift and erosion rates. For a steady-state tectonic system in which rock moves with vertical and horizontal velocities $u^*(x^*)$ and $v^*(x^*)$, respectively, the two scales can be related by an expression for the flux through the mean elevation profile $z^*(x^*)$:

$$F_A^* = \int_W \left(u^*(x^*) - v^*(x^*)\frac{dz^*}{dx^*} \right) dx^* = -\int_W E^*(x^*)dx^* = -F_E^*, \quad (5)$$

where $E^*(x^*)$ is the erosion rate. The $v^*(x^*)\ dz^*/dx^*$ term in equation 5 is included to account for vertical motion of the mean elevation profile caused by lateral advection of topography (e.g., Willett et al., 2001).

The steady-state profiles of selected terms from equation 5 are shown in Figure 7. The rock uplift rate $u^*(x^*)$ is greater on the pro-wedge than on the retro-wedge. Because the topographic slope of the mean elevation profile switches sign at the divide, the lateral advection term is negative on the pro-wedge and positive on the retro-wedge. The result is that the erosion rate is more uniform than the rock uplift rate. The most significant deviation from uniformity occurs near $X^* = 52.5$, which coincides with the retro-wedge segment of the shear zone initiating at the S point, which is at $X^* = 50$. The maximum erosion rate at $X^* = 49$ is offset from the main divide at $X^* = 51.5$. The segment of the mean elevation profile defined by the maximum taper angle lies above the region of lowest erosion. Overall, we find that the coupled tectonic-erosional system evolves to a steady-state in which non-uniform erosion exists without spatial gradients in precipitation rate and without temporal and spatial changes in the mechanism of erosion.

Planform Topography

Planform topography and the mean elevation profile both vary as the wedge evolves to a large-scale steady state. However, while the mean elevation profile eventually achieves a steady form (Fig. 4), planform topography never reaches a true steady state (Fig. 3). In the approach to a large-scale steady state, erosion progressively shapes the planform topography, resulting in creation of interfluvial relief and a change in the geometry of drainage basins. Initially, however, erosion does not significantly alter the planform topography. Not until outward propagation of deformation at $T^* = 1/2$ does erosion in transverse rivers etch the sides of the wedge, thereby narrowing and roughening the divide. Distinct drainages develop at $T^* = 2$ to form a series of narrow transverse drainages, and are deeply incised by $T^* = 4$. Between $T^* = 4$ and $T^* = 20$, the main divide becomes more sinuous, and basins draining both the pro- and retro-wedges increase in along-strike width.

The reorganization of drainage basins at large-scale steady-state is driven by lateral advection of topography. As with the models of Willett et al. (2001), the incorporation of lateral advection leads to a highly transient topography. Although the SPM used here has a simple representation of ridge migration by erosion (Tucker and Slingerland, 1994), it does allow for the generation of new topographic features at the toe of the wedge and their progressive uplift and erosion as they are advected across the orogen. The direction of advection leads to an interesting asymmetry in the mechanisms that promote basin reorganization on the pro- and retro-wedges. On the pro-wedge, advection of river junctions from the toe toward the divide and truncation of basins at the divide lead to an increase in the along-strike width of basins over time. Reorganization of retro-wedge drainages occurs when the headwaters of pro-wedge rivers are advected over the divide and decapitated by retro-wedge rivers. Addition of drainage area (and thus discharge) by this process leads to growth of some retro-wedge basins at the expense of others. This also leads to changes in the pro-wedge channel network, because the decapitated basins, having less drainage area, are at a competitive disadvantage. Overall, these mechanisms favor a more sinuous main divide and drainage basins with larger along-strike widths.

Advection of topography and basin reorganization lead to a highly transient topography, even at large-scale steady state. We quantify the variability of the elevation field in a fixed reference frame by translating the planform topography at selected time steps to a regular grid and then calculating the standard deviation of the elevation over time at each grid point. In Figure 8, elevation contours of the mean elevation surface are superimposed on the map of topographic variability. Both the mean elevation surface and topographic variability are calculated over a period of $\Delta T^* = 86$, during which the mean elevation profile and deformation field are steady. Over most of the wedge, the standard deviation of elevation is a significant fraction of the mean elevation (1/5–1/3). The magnitude of the variability is much larger on the pro-wedge, where horizontal velocities are greater.

RESPONSE TO CLIMATIC AND TECTONIC PERTURBATION

Geologic time series, such as basin deposition rates and thermochronometric age series, testify to the time-dependent response of landscapes to unsteady tectonic and climatic forcing. By providing predictions for this response, theoretical models are important tools for interpreting the geologic record. In the preceding section of this paper, we examined how several char-

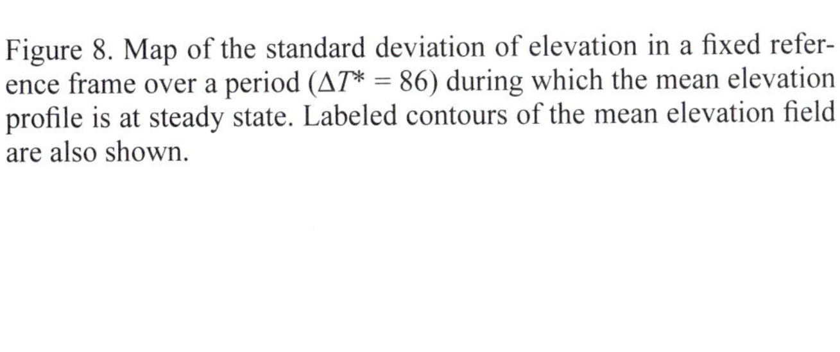

Figure 8. Map of the standard deviation of elevation in a fixed reference frame over a period ($\Delta T^* = 86$) during which the mean elevation profile is at steady state. Labeled contours of the mean elevation field are also shown.

acteristics of the coupled tectonic-erosional system (e.g., wedge volume, erosional flux, and exhumation rate) vary as functions of time and space under steady tectonic and climatic forcing. In this section, we impose perturbations in climatic and tectonic forcing on the steady-state base model and examine the time-dependent response as expressed by three important properties: wedge width, erosional flux, and mean erosion rate.

Because of the strong negative feedback between erosion rate and topographic slope, the perturbations in forcing are ultimately expected to lead to new equilibrium states. The relationship between the initial and final steady states can be determined for each of the three properties. The final steady value of the erosional flux is simply defined, because, at steady state, it must equal the accretionary flux. The sensitivity of wedge width and mean erosion rate to accretionary flux and precipitation rate follow from the results of the 1-D analyses of Whipple and Meade (2004) and Roe et al. (this volume). Equation 4 relates the steady-state wedge width to the accretionary flux and precipitation rate, and a similar expression can be written for the mean erosion rate, $\bar{E}$, which is the erosional flux divided by the width of the wedge:

$$\bar{E} \propto F_A^{\,hm/(hm+1)} P^{\,m/(hm+1)}. \tag{6}$$

Unlike the wedge width, the mean erosion rate is positively dependent on both accretionary flux and precipitation rate. What is not given from equations 4 and 6, however, is the time-dependent behavior of the coupled system following a change in tectonic or climatic forcing.

We present four simulations in which step-function tectonic and climatic perturbations are imposed on the steady-state base model presented in the previous section. We chose perturbations in accretionary flux and precipitation rate such that, for the four simulations, the wedge grows and contracts to only two different steady-state widths. For example, given $m = 1/2$, equation 4 implies that doubling the precipitation rate leads to the same, smaller steady-state width as a decrease in accretionary flux by a factor of $1/\sqrt{2}$; halving the precipitation rate results in the same, larger steady-state width as an increase in accretionary flux by a factor of $\sqrt{2}$. In the left and right panels of Figure 9, we plot the time series of wedge width, mean erosion rate, and erosional flux for growth and contraction of the wedge, respectively. Because time is nondimensionalized with the convergence rate, and the convergence rate is adjusted to change the accretionary flux, we use the initial convergence velocity to nondimensionalize time in simulations with variable accretionary flux.

Climatic Perturbation

The evolution of wedge width in response to perturbations in precipitation rate is shown in Figures 9A and 9B (black lines). An increase in precipitation rate leads to a decrease in wedge width until a new steady width is achieved (Fig. 9B); conversely, a decrease in precipitation rate leads to growth of the wedge to

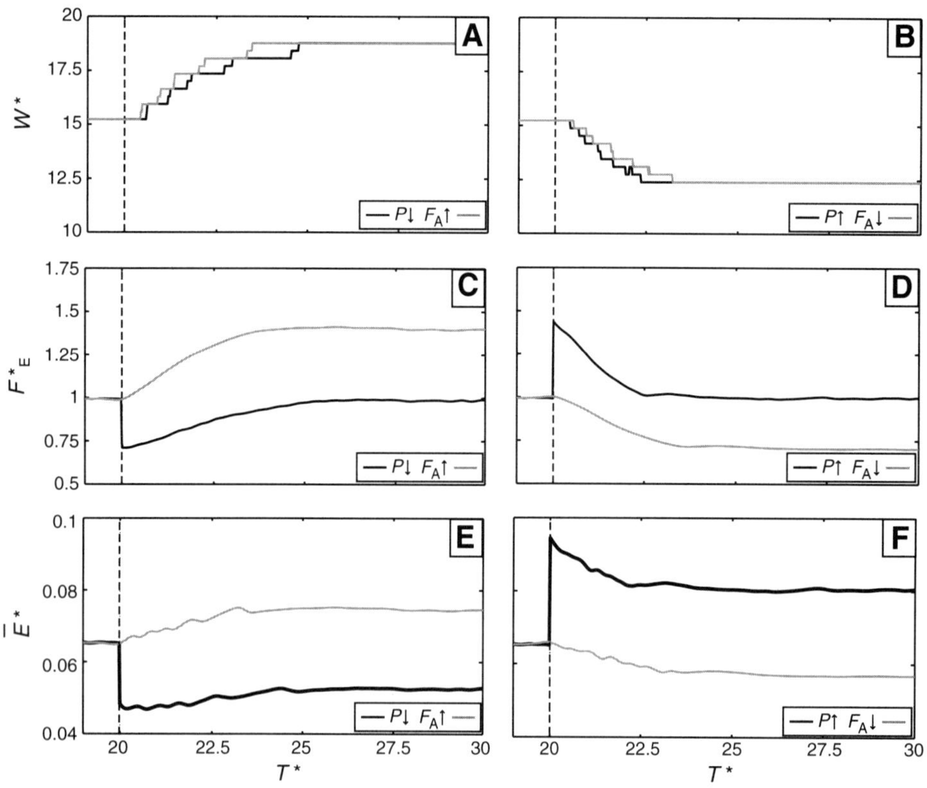

Figure 9. Response of wedge width, erosional flux, and mean erosion rate to step-function changes in tectonic and climatic forcing. Results for wedge growth and wedge contraction are shown in the left and right panels, respectively. Response to change in precipitation rate is shown in black. Response to change in accretionary flux is shown in gray.

a new steady form (Fig. 9A). The stair-step nature of the curves is due to the finite resolution of the model mesh. The response of the erosional flux due to a change in precipitation is shown in Figures 9C and 9D (black lines). An instantaneous increase in precipitation rate leads to an instantaneous increase in erosional flux. However, because the accretionary flux is unchanged, the erosional flux gradually returns to its original value as the system tends to its new equilibrium state (Fig. 9D). Following from equation 4, the value of the erosional flux immediately after the change in precipitation rate is equal to the accretionary flux required to maintain the original width under the new rate of precipitation: $F_E = F_A \times (P_{new}/P_{old})^m$. Because the tectonic flux is unaffected by the change in precipitation rate, the wedge cannot maintain its width at this erosional flux and the wedge contracts. The converse is true for a decrease in precipitation rate. The erosional flux decreases to a value consistent with the wedge topography under the lower rate of precipitation. Erosional flux then gradually increases to return to the level of the accretionary flux as the wedge grows to a new steady state (Fig. 9C).

Mean erosion rate responds to a precipitation change in an interesting manner; it exhibits an instantaneous change, which overshoots the final steady value, followed by a gradual return to the new equilibrium at which the mean erosion rate matches the mean rock uplift rate (Figs. 9E and 9F). An increase in precipitation leads to a new steady state that is characterized by a smaller width, a higher rock uplift rate, and consequently a higher erosion rate; equivalently, a decrease in precipitation leads to a steady state with a lower erosion rate and a larger width.

Tectonic Perturbation

Changes in tectonic flux result in a new steady-state wedge width that matches the width produced by equivalent changes in precipitation rate (Figs. 9A and 9B), showing consistency with equation 4. As is to be expected, increased accretionary flux leads to an increase in wedge width (Fig. 9A), and conversely, decreased accretionary flux leads to a decrease in width (Fig. 9B). A step-function change in accretionary flux produces a gradual change in mean erosion rate. An increase in accretionary flux leads to a permanent increase in erosion rate; a decrease in flux leads to a decrease in erosion rate. Finally, a change in accretionary flux leads to a permanent change in erosional flux (Figs. 9C and 9D).

Comparison to One-Dimensional Transient Model

K.X Whipple and B.J. Meade (2004, personal commun.) also investigated the transient response of an orogenic wedge to perturbation. They employed a steady-state width-forcing law, similar to equation 4, and incorporated an assumption that growth and decay of the wedge occur self-similarly, which implies a quasi-equilibrium throughout the response. The result is a differential equation for wedge width that can be solved analytically for certain combinations of h and m, numerically in all cases.

In Table 1, we list the e-folding times predicted by our numerical model and the model of K.X Whipple and B.J. Meade (2004, personal commun.). There is a very close agreement between the predictions, indicating that the assumption of self-similar growth in the K.X Whipple and B.J. Meade (2004, personal commun.) model is also valid for the numerical model. In the numerical simulations, self-similarity of wedge growth is evident from the evolution of the mean elevation profile (Fig. 4) and follows from two assumptions of the mechanical model: existence of a quasi-equilibrium at each time step and maintenance of critical topography by plastic deformation.

The numerical results support an important implication of the K.X Whipple and B.J. Meade (2004, personal commun.) model: the orogen response time is more strongly dependent on the precipitation rate than on the accretionary flux. In the climate change simulations, the ratio of the larger and smaller final precipitation rates is 4; the difference in response times is $\Delta T^* = 1.4$ (Table 1). In comparison, the ratio of larger and smaller accretionary fluxes is 2, but the response times are identical.

DISCUSSION AND CONCLUSIONS

Under steady forcing conditions, the coupled model presented here evolves to a flux steady state in which the erosional flux balances the accretionary flux (Willett and Brandon, 2002). Associated with this balance are a steady deformation field, a steady mean elevation profile, and a planform topography in which temporal variability is driven by lateral advection. When perturbed by changes in climatic and tectonic forcing, the coupled model evolves to a new steady state via characteristic and, in some cases diagnostic, responses of orogen width, mean erosion rate, and erosional flux. These responses will be more complicated in natural orogenic systems in comparison to the simplified numerical models shown here but can still be indicative of specific processes.

The width evolution in Figures 9A and 9B illustrates that perfect knowledge of wedge width through time is not sufficient to assign causation to changes in either tectonic or climatic forcing. Mean erosion rate and erosional flux, however, have diagnostic responses to perturbations in forcing. Permanent changes in accretionary flux lead to permanent changes in erosional flux, whereas changes in precipitation rate only temporarily influence the erosional flux (Figs. 9E and 9F) (Bonnet and Crave, 2003). This differs from the response of mean erosion rate, because the erosion rate is permanently changed by changes in either accretionary flux or precipitation rate (Figs. 9C and 9D). Mean erosion rate gradually achieves a new steady value in response to a change in accretionary flux, whereas it responds abruptly to a variation in precipitation rate. Given a step-function change in forcing, the utility of erosion rate as a diagnostic tool, however, is dependent on the time-resolution of the record. Coarse records might not resolve the difference between abrupt and gradual changes in erosion rate if the response time is short and may only show permanent changes in erosion rate, which would not be diagnostic.

Isostatic compensation has a significant influence on orogen size and, therefore, on the response time. Local isostasy requires formation of a crustal root and, for typical crustal and mantle densities, the thickness of the root is 4–6 times greater than the height of excess topography. Thus, the response time of a compensated orogen is much greater than that of an uncompensated orogen, such as the one presented in this study. However, isostatic compensation does not change the form of the response of wedge width, mean erosion rate, and erosional flux; hence, these records and the simulations in this paper characterize the response of both compensated and uncompensated wedges.

In addition to isostasy and the magnitude of the perturbation, the system response time is intimately tied to the rheology of the orogen. Frictional material is expected to respond locally and immediately to any perturbation to the mean elevation profile (Dahlen and Barr, 1989). However, in most orogenic wedges, for example, Taiwan, there is abundant evidence for nonfrictional deformation (i.e., viscous flow), including ubiquitous ductile fabrics in the core of the range (Clark et al., 1993). Importantly, because a viscous orogen has an inherent response time scale determined by the viscosity of the material, the response of a viscous orogen is expected to be longer than that of a purely frictional wedge.

The close agreement between the numerical model and 1-D steady-state models (Roe et al., this volume; Whipple and Meade, 2004) and transient models (K.X Whipple and B.J. Meade, 2004, personal commun.) indicates that many first-order aspects of the coupled system can be described by coupling of relatively simple erosional and tectonic models. One important degree of freedom not available in the simpler models is the ability of the spatial distribution of rock uplift and topographic slope to adjust, such that local and orogen-wide balances are achieved between tectonic and erosional forcing. Our results demonstrate that nonuniform rock uplift can exist in a system with uniform precipitation and a steady accretionary flux (Fig. 7). We are currently expanding upon these results by examining how the pattern of rock uplift is influenced by the distribution of precipitation, the erosion law, and the crustal rheology, and, more fundamentally, how the erosional and tectonic systems

TABLE 1. COMPARISON OF PREDICTED RESPONSE TIMES

Perturbation	τ^*	
	K.X. Whipple and B.J. Meade (2004, personal commun.)	This paper
$P_f = 2P_i$	1.3	1.4
$P_f = (1/2)P_i$	2.8	2.8
$F_{Af} = \sqrt{1/2}F_{Ai}$	1.8	2.0
$F_{Af} = \sqrt{2}F_{Ai}$	2.0	2.0

achieve a balance through adjustment of internal parameters, such as rock uplift rate and topographic slope.

In summary, we have presented a suite of numerical simulations that examine the steady-state and transient behavior of a coupled orogenic wedge. In this system, steady state implies a balance between accretionary and erosional fluxes and between uplift and erosion rates, but not a static planform topography. We find that the existence of nonuniform rock uplift does not require spatial variations in precipitation or erosional processes. Finally, we predict that records of erosion rate and erosional flux can be used to distinguish between tectonic and climatic perturbations.

ACKNOWLEDGMENTS

We thank Jon Tomkin and Nicole Gasparini for constructive reviews. This work was supported by a National Science Foundation (NSF) Graduate Research Fellowship to Stolar.

REFERENCES CITED

Barr, T.D., and Dahlen, F.A., 1989, Brittle frictional mountain building—2. Thermal structure and heat budget: Journal of Geophysical Research, v. 94, p. 3923–3947.

Batt, G. and Braun, J., 1997, On the thermo-mechanical evolution of compressional orogens: Geophysical Journal International, v. 128, p. 364–382.

Beaumont, C., Fullsack, P., and Hamilton, J., 1992, Erosional control of active compressional orogens, *in* McClay, K., ed., Thrust tectonics: London, Chapman and Hall, p. 1–18..

Beaumont, C., Jamieson, R.A., Nguyen, M.H., and Lee, B., 2001, Himalayan tectonics explained by extrusion of a low-viscosity crustal channel coupled to focused surface denudation: Nature, v. 414, p. 738–742, doi: 10.1038/414738a.

Bonnet, S., and Crave, A., 2003, Landscape response to climate change: Insights from experimental modeling and implications for tectonic versus climatic uplift of topography: Geology, v. 31, p. 123–126.

Braun, J., and Sambridge, M., 1997, Modelling landscape evolution on geological time scales: A new method based on irregular spatial discretization: Basin Research, v. 9, p. 27–52, doi: 10.1046/j.1365-2117.1997.00030.x.

Burbank, D.W., Leland, J., Fielding, E., Anderson, R.S., Brozovic, N., Reid, M.R., and Duncan, C., 1996, Bedrock incision, rock uplift and threshold hillslopes in the northwestern Himalayas: Nature, v. 379, p. 505–510, doi: 10.1038/379505a0.

Clark, M.B., Fisher, D.M., Lu, C.Y., and Chen, C.H., 1993, Kinematic analyses of the Hsuehshan Range, Taiwan—A large-scale pop-up structure: Tectonics, v. 12, p. 205–217.

Dahlen, F.A., 1984, Noncohesive critical coulomb wedges—An exact solution: Journal of Geophysical Research, v. 89, p. 125–133.

Dahlen, F.A., and Barr, T.D., 1989, Brittle frictional mountain building. 1. Deformation and mechanical energy budget: Journal of Geophysical Research, v. 94, p. 3906–3922.

Davis, D., Suppe, J., and Dahlen, F.A., 1983, Mechanics of fold-and-thrust belts and accretionary wedges: Journal of Geophysical Research, v. 88, p. 1153–1172.

Hilley, G.E., Strecker, M.R., and Ramos, V.A., 2004, Growth and erosion of fold-and-thrust belts with an application to the Aconcagua fold-and-thrust belt, Argentina: Journal of Geophysical Research, v. 109, doi: 10.1029/2002JB002282.

Howard, A.D., 1994, A detachment-limited model of drainage-basin evolution: Water Resources Research, v. 30, p. 2261–2285, doi: 10.1029/94WR00757.

Koons, P.O., 1990, A 2-sided orogen—Collision and erosion from the sandbox to the Southern Alps, New Zealand: Geology, v. 18, p. 679–682, doi: 10.1130/0091-7613(1990)0182.3.CO;2.

Malavieille, J., 1993, Late orogenic extension in mountain belts—Insights from the Basin and Range and the late Paleozoic Variscan Belt: Tectonics, v. 12, p. 1115–1130.

Molnar, P., 1997, The rise of the Tibetan Plateau: From mantle dynamics to the Indian monsoon: Astronomy & Geophysics, v. 38, p. 10–15.

Montgomery, D.R., 2001, Slope distributions, threshold hillslopes, and steady-state topography: American Journal of Science, v. 301, p. 432–454.

Roe, G.H., Montgomery, D.R., and Hallet, B., 2003, Orographic precipitation and the relief of mountain ranges: Journal of Geophysical Research, v. 108, doi: 10.1029/2001JB001521.

Smith, R.B., 1979, The influence of mountains on the atmosphere: Advances in Geophysics, v. 21, p. 132.

Tucker, G.E., and Slingerland, R.L., 1994, Erosional dynamics, flexural isostasy, and long-lived escarpments—A numerical modeling study: Journal of Geophysical Research, v. 99, p. 12,229–12,243, doi: 10.1029/94JB00320.

Wang, W.H., and Davis, D.M., 1996, Sandbox model simulation of forearc evolution and noncritical wedges: Journal of Geophysical Research, v. 101, p. 11,329–11,339, doi: 10.1029/96JB00101.

Whipple, K.X, and Meade, B.J., 2004, Controls on the strength of coupling among climate, erosion, and deformation in two-sided, frictional orogenic wedges at steady state: Journal of Geophysical Research, v. 109, doi: 10.1029/2003JF000065.

Whipple, K.X, and Tucker, G.E., 1999, Dynamics of the stream-power river incision model: Implications for height limits of mountain ranges, landscape response timescales, and research needs: Journal of Geophysical Research, v. 104, p. 17,661–17,674, doi: 10.1029/1999JB900120.

Whipple, K.X, Hancock, G.S., and Anderson, R.S., 2000, River incision into bedrock: Mechanics and relative efficacy of plucking, abrasion, and cavitation: Geological Society of America Bulletin, v. 112, p. 490–503, doi: 10.1130/0016-7606(2000)1122.3.CO;2.

Willett, S.D., 1992, Kinematic and dynamic growth and change of a Coulomb wedge, *in* McClay, K., ed., Thrust tectonics: New York, Chapman and Hall, p. 19–31.

Willett, S.D., 1999a, Orogeny and orography: The effects of erosion on the structure of mountain belts: Journal of Geophysical Research, v. 104, p. 28,957–28,981, doi: 10.1029/1999JB900248.

Willett, S.D., 1999b, Rheological dependence of extension in wedge models of convergent orogens: Tectonophysics, v. 305, p. 419–435, doi: 10.1016/S0040-1951(99)00034-7.

Willett, S.D., and Brandon, M.T., 2002, On steady states in mountain belts: Geology, v. 30, p. 175–178, doi: 10.1130/0091-7613(2002)0302.0.CO;2.

Willett, S.D., Beaumont, C., and Fullsack, P., 1993, Mechanical model for the tectonics of doubly vergent compressional orogens: Geology, v. 21, p. 371–374, doi: 10.1130/0091-7613(1993)0212.3.CO;2.

Willett, S.D., Slingerland, R., and Hovius, N., 2001, Uplift, shortening, and steady-state topography in active mountain belts: American Journal of Science, v. 301, p. 455–485.

Yeh, E.-C., 2003, Structural evolution of slate belts: Examples from Taiwan and eastern Pennsylvania [Ph.D. thesis]: University Park, Pennsylvania, The Pennsylvania State University, 191 p.

Manuscript Accepted by the Society 23 June 2005

Printed in the USA

Geological Society of America
Special Paper 398
2006

First-order topography over blind thrusts

Michael A. Ellis[†]
Center for Earthquake Research and Information, Department of Earth Sciences, University of Memphis, 3876 Central Avenue, Memphis, Tennessee 38152, USA
Alexander L. Densmore
Institute of Geology, Department of Earth Sciences, ETH Zentrum, CH-8092 Zürich, Switzerland

ABSTRACT

The first-order shape of a mountain range over a blind thrust fault is controlled by the competition between the geometry of the underlying thrust fault and local base levels that flank the emerging range. We use an elastic half-space model to show that the surface displacements caused by slip across a blind thrust are sensitive to the geometry and depth of the fault, as well as to its growth history. Accumulation of this displacement field over time creates a tectonic topography; in the absence of erosion, this is the same as the topography of the resulting range. Erosion modifies this tectonic topography by creating local relief on both flanks of the range and forcing divergence between the tectonic and topographic divides. Importantly, erosion is modulated by the relative base levels on either side of the range. We show that in many cases the position of the topographic divide is predicted by relative base level through a simple geometric ratio, such that the average slope of both range flanks is the same. Importantly, attributes of the topography on one flank of an emerging range depend on what happens to base level on the opposite flank. The role of relative base levels in determining the first-order range shape illustrates the importance of regional-scale events (e.g., sediment supply to piggyback basins or the evolution of a regional fluvial network) in the development of mountainous topography in a complex fold-and-thrust belt system.

Keywords: tectonic geomorphology, blind-thrust, neotectonics, topography.

INTRODUCTION

In this paper, we investigate the first-order controls that shape an emerging mountain range over a seismically active blind thrust fault. The development of these structures holds both scientific interest (e.g., they are commonly the leading edge of fold-thrust belts) and societal relevance (these structures are often found in densely populated regions). If we are to use topography as a signal of active tectonics, it is important that we understand the potential variety of first-order shapes that arise from simple and reasonable thrust-fault configurations. Our motivation stems in part, therefore, from the question, to what extent can we determine (or at least infer with some confidence) the geometry and evolution of the underlying fault from the range topography?

In principle, this task is not far removed from the now-common practice of inferring fault slip from coseismic and postseismic surface displacements (Mendez and Anderson, 1991; Molnar, 1987; Stein and Ekström, 1992) or the underlying structure

[†]E-mail: mellis@memphis.edu.

Ellis, M.A., and Densmore, A.L., 2006, First-order topography over blind thrusts, *in* Willett, S.D., Hovius, N., Brandon, M.T., and Fisher, D.M., eds., Tectonics, Climate, and Landscape Evolution: Geological Society of America Special Paper 398, Penrose Conference Series, p. 251–266, doi: 10.1130/2006.2398(15).

from near-surface structural data (Laubscher, 1965; Dahlstrom, 1969; Suppe, 1983; Allmendinger, 1998). In the former case, surface displacements provide a picture of the last earthquake, which is an important but short-term view of the development of the structure. In the latter, the structural geometry of folded beds gives a long-term picture of deformation that is effectively smoothed by both the resolution of the data and the integrating effects of time.

Topography on the other hand is an active recorder of tectonic activity over a range of time scales from 1 yr (i.e., individual earthquakes) to more than 10^6 yr. Blind thrust systems form within this time scale, generating some of the youngest actively deforming topographies in a variety of tectonic settings. Topography has long been recognized as the expression of the interplay or coupling between surface processes (modulated largely by climate and climate-change) and tectonic processes (for a brief review, see Merritts and Ellis, 1994). The topography of blind thrust systems, therefore, provides us with the opportunity to place rigorous limits on our understanding of the coupling or interplay between tectonics, climate, and surface processes.

One of the earlier attempts (and one that dropped below our radar screens for a long time) in using topography to infer tectonics was made by Seeber and Gornitz (1983). These authors showed that increases in steepness on the larger Himalayan rivers corresponded to a previously inferred ramp in the underlying Main Boundary fault, which separates the downgoing Indian plate from the overlying Eurasian plate. In the middle 1990s (as both topographic data and geochronological techniques significantly improved), this approach was revitalized, and a number of investigations have attempted to tie together characteristics of the topography (particularly aspects of the fluvial system) and tectonic forcing (e.g., Champel et al., 2002; Hurtrez et al., 1999; Kirby and Whipple, 2001; Lavé and Avouac, 2000, 2001).

These investigations have contributed a number of challenges largely related to the process rules for an incising river (e.g., when to vertically incise, when to cut laterally, and how much to meander). The behavior of the fluvial system, however, tends to be considered in isolation, without reference to the controlling length scales set by an emerging range. The evolution of a channel on the foreland-sloping flank, for example, is typically considered without reference to what happens on the rear flank (e.g., Kirby and Whipple, 2001).

The rules for tectonic processes used in these investigations, in contrast to those of erosion, have received less critical appraisal. Folds (and hence the pattern of incremental displacements) are commonly related to assumed fault bends (e.g., Lavé and Avouac, 2000) or fault tips, via the kinematics derived from Suppe (1983). Other studies have assumed that the displacement pattern varies uniformly over the length scale of the foreland-sloping topographic flank of an emerging range (Kirby and Whipple, 2001), which implies a consistent relationship between the shape of the range and that of the tectonic forcing.

We effectively step back and examine the first-order profile shape of an emerging range at the length scale of an individual thrust fault. We show that the shape of the range is fundamentally controlled by the competition between the geometry of the underlying thrust fault and the local base levels that flank the emerging range. The first part of our treatment addresses the early topographic development of the range, prior to significant erosion, when the range shape reflects only the distribution of slip across the underlying fault. The second part is concerned with the role played by local base levels, which, in concert with the displacement field, will determine the evolving shape of the range.

There are a number of interesting consequences of the competition between tectonics and local base levels, including the generation of local relief across the range and the potential coupling of the range shape to surface and tectonic processes occurring at regional length scales. These are described and illustrated (with various degrees of speculation) in the third and final section of the paper.

DEVELOPMENT OF TECTONIC TOPOGRAPHY

We use the phrase tectonic topography to describe the cumulative tectonic envelope produced by the distribution of slip across a fault. In the absence of erosion, the tectonic topography and the surface topography are coincident. The conventional model of a thrust-related tectonic topography is an asymmetric range in which the steeper, narrower flank faces the direction of fault vergence and overlooks a footwall depocenter or foredeep (Fig. 1). We show here, however, that the initial shape of the emerging range is sensitive to the geometry and extent of the thrust fault, and that even in the absence of significant erosion, the shape of the range need not correspond simply to this conceptual model.

We present a series of numerical experiments that explore likely variations in thrust fault configurations. What follows is by no means an exhaustive search of parameter space, and in fact we have avoided doing this, because it would miss the point that these experiments are designed as a guide to our intuition, not as a taxonomy of tectonic models.

Our approach is a departure from typical developmental analyses of fault-related folds (e.g., Suppe, 1983; Erslev, 1991;

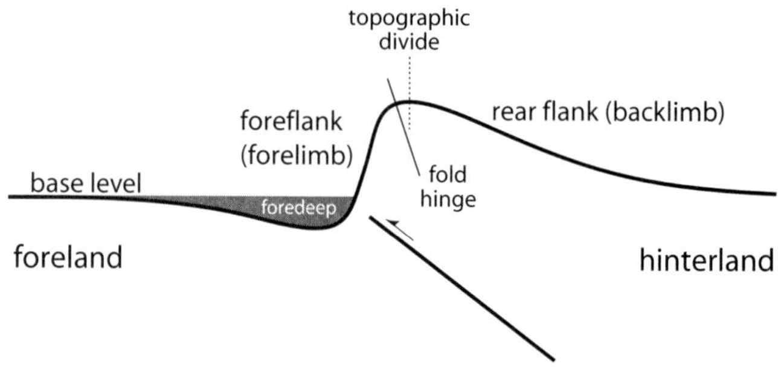

Figure 1. The basic ingredients of a thrust-related mountain range have given rise to this standard view of the first-order shape, which consists of an asymmetric range in which the narrow and steeper foreflank faces the direction of fault vergence.

Wickham, 1995; Allmendinger, 1998) in two ways. First, we adopt an elastic deformation field as the basic tectonic increment (rather than prescribe a set of geometrically derived rules or assume an a priori distribution of strain). Second, we assume that blind thrust faults may initiate as isolated faults within the crust that subsequently propagate and link to other faults, notably basal décollements. We justify each of these aspects in the following section, but the importance from a geomorphic perspective is that this approach allows us to tackle the subtler variations of surface displacement in the vicinity of the fault. Accurate modeling of surface displacement is important, because it sets the directions and magnitudes of fluid and sediment fluxes off the landscape, and so exerts direct control on spatial patterns of erosion and deposition.

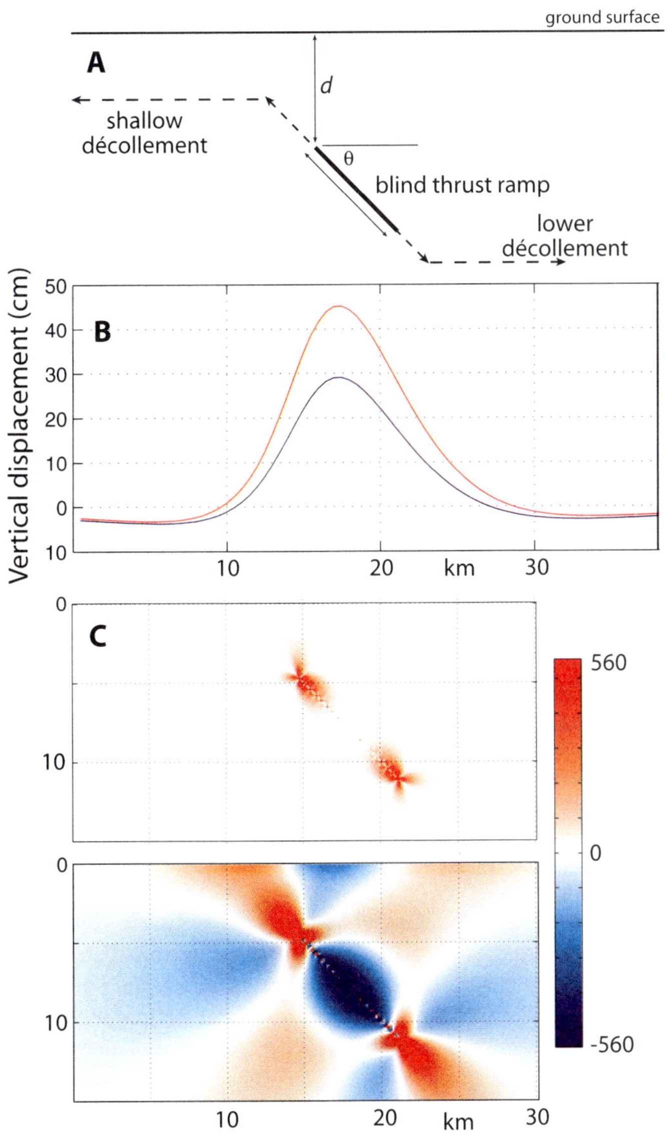

Model Rationale

The model used here is an attempt to simulate the first-order process by which blind thrust folds grow, which happens via earthquakes. We use the three-dimensional boundary-element program 3ddef (Gomberg and Ellis, 1994) to calculate the deformation around dislocations (faults) within an isotropic and homogeneous elastic half-space. This approach is common practice within the seismological and geodetic communities in interpreting coseismic and short time-scale displacement data (e.g., Stein and Ekström, 1992; Mendez and Anderson, 1991). The time scale implicit in this approach (~earthquake repeat times) is appropriate to the response times of geomorphic processes, and in trying to develop a topography incrementally, we need to predict deformation at this time scale.

The basic idea is as follows: An increment of elastic deformation, generated by slip across one or more faults (Fig. 2A) and driven by a homogeneous pure shear, is applied repeatedly to generate a finite displacement envelope, or tectonic topography. Deformation of the ground surface is a direct result of the net distribution of strain within the crust surrounding the fault, the bulk of which is derived from slip across the fault and not (as might be assumed) by residual secular stresses. (Interestingly, distortional strain energy associated with slip across a fault is significantly higher than dilational strain energy, and in general, it is the distribution of distortional strain that gives rise to the shape of the deformed ground surface [Fig. 2B].) Every time we apply a slip or deformation event, we assume that the stress state of the crust can be characterized as homogeneous pure shear. The rupture

Figure 2. (A) Model fault parameters consist of θ (fault dip), w (downdip fault width), d (depth to upper tip), and L (along-strike fault length, not shown). L was constant and equal to 20 km in all models. (B) Vertical displacement of the surface from an increment of slip across a fault with $d = 0.5$ km, $\theta = 45°$, and $w = 9$ km. The two curves show models with values of Poisson's ratio typical of crustal rocks, 0.25 (red), and for an almost incompressible (and from a geological point of view, hypothetical) material, 0.49 (blue). This figure demonstrates that the distribution of volume change does not significantly change the shape of the resultant deformed surface and that if all of the volume change were recovered (which we regard as unlikely), the results derived from the model here remain valid to first order. (C) Distribution in cross section of dilational (upper panel) and distortional (lower panel) strain energy after a slip event on the same fault as in Figure 2B. The input strain energy was 560 Nm per unit volume, highlighted on the scale. The initial strain energy has been subtracted from the plots in order to highlight the effect of slip across the fault. Thus a value of zero (which corresponds to white) is equivalent to the initial or background strain energy, and in these places, the strain energy has not been changed by fault slip. As expected, a significant volume adjacent to the fault is completely relaxed (deviatoric stresses here were used for driving slip). The residual strain energy field is both fairly heterogeneous and in places significantly higher than the initial level. This is the strain field that we assume is effectively homogenized over time by a variety of postseismic processes, probably mainly by aftershocks. Notice that dilational strain energy is negligible in comparison to that caused by distortion.

of a blind thrust, however, is unable to relax deviatoric stresses throughout the crust, and there will be residual secular stresses. Added to these are significantly larger stresses generated by slip across the fault (Fig. 2C). In using this deformation increment as a template, we are assuming that the residual secular stresses plus those generated by slip across the fault are relaxed prior to the next application and that the secular tectonic driving force establishes a homogeneous state of stress within the crust.

Residual stresses are relaxed and elastic strains are made permanent via a variety of postseismic processes. Such processes include aseismic fault slip, pressure solution, grain-boundary sliding, precipitation of mineral species in dilatant regions, and aftershocks. Of these, we suspect that aftershocks relax the bulk of postseismic stresses. A well-established result from seismology is that the locations of aftershocks can be well-modeled by the spatial distribution of slip across fault planes, and that in effect, aftershocks tend to homogenize the heterogeneous strain field that envelops a rupture (e.g., Bakun et al., 1986; Beroza, 1991; Dreger, 1997; Kilb et al., 1997; King et al., 1994; Mendoza and Hartzell, 1988; Stein et al., 1994).

In reality, it is unlikely that all residual stresses will be relaxed. A small proportion of the elastic deformation is recovered and manifest as postseismic deformation. Somewhat serendipitously, the use here of an elastic half-space model, rather than for example a layered model, better predicts the sum of coseismic and postseismic deformation (King et al., 1988), and therefore this effect is likely to be negligible. In other words, the increment of deformation that we impose here is representative of the entire earthquake cycle and not simply the coseismic deformation.

Strains that are not recovered or made permanent by the processes listed above, however, will accumulate over time and contribute to the secular tectonic strain field. The accumulation of strains and associated off-fault damage associated with repeated ruptures allow the fault to propagate (e.g., Maerten et al., 2002; Manighetti et al., 2001, 2004). For this reason, we limit the finite displacement across isolated faults to values that yield a displacement/fault-length ratio of well below 10^{-1}, which is typical of isolated faults in a variety of tectonic settings (e.g., Schlische et al., 1996; Densmore et al., 2004). In practice, we limit the displacement across active faults to less than 1 km, which yields a topographic relief of the same order.

A compelling argument in favor of the model approach used here is that thrust-related mountains appear to be created by the accumulation of earthquakes, and the style of incremental deformation is similar to the readily observed geological structure. There are now numerous examples of earthquakes across blind thrusts (e.g., King and Vita-Finzi, 1981; Lin and Stein, 1989; Yielding et al., 1981; Stein and Ekström, 1992; Bodin and Horton, 2004), all of which have resulted in growth that is directly attributable to slip across the fault plane. Coupled to this fundamental observation is that the finite geological structures associated with such faults (e.g., synclines and depocenters adjacent to hanging-wall anticlines) are readily explained by the model used here. This approach has been used with success in relating the distribution of first-order topography to active faulting along the San Andreas fault (Bilham and King, 1989), in the inversion of topography for regional tectonics along the Ionian plate margin (King et al., 1994), and in linking tectonics and topography in the central Basin and Range (Ellis, 1991). Elastic deformation models have also been used to interpret the finite features of topography generated by reverse faults (King et al., 1988; Stein et al., 1988) and normal faults (King and Ellis, 1990).

Other types of deformation are no doubt important, perhaps to differing degrees in different settings (e.g., aseismic creep, layer-parallel shortening and buckling, salt diapirism). Our treatment here is confined, therefore, to settings in which topography is largely if not wholly due to displacements across active faults.

The second aspect of the model that differs from more established fold-growth models is the assumption that thrust faults may initiate as isolated faults, with tips at both ends. As displacement increases and tip-zone stresses accumulate (as described above), the thrust will propagate both updip and downdip. The rationale for such a scenario can be supported from various points of view. Gretener (1972) provided a mechanical argument in which thrust faults initiate within relatively stiff layers within the crust; Eisenstadt and DePaor (1987) suggested that the growth of isolated faults are better able to explain off-fault structures, and Ellis and Dunlap (1988) showed that displacement gradients along thrust faults are consistent with growth and linkage of initially isolated thrust faults. We take pains to point out, however, that the isolated thrust fault is intended only as an initial and early structure. Thrust faults develop by linking to either other thrust faults, to lower or higher décollements, or they may refract through the crust. Once this sort of linkage has occurred, the fault configurations in the present models do not differ from those in current thrust-fold models.

We do not account for isostatic compensation, assuming instead that the crust is always strong enough to support the emerging range. This is a reasonable assumption, because we are concerned here with the early evolution of topography over blind thrusts, and topographic loads are likely to be small compared to the strength of the crust. In any case, if the effective elastic thickness of the crust is relatively low (less than the typical wavelength of the folds generated in these experiments, which are ~10–30 km), mass redistribution through erosion and deposition will add to the effects that we describe here.

Experimental Setup

A fault is represented by a single dislocation with an along-strike length of 20 km and a downdip width of 8–15 km. The fault is divided into a number of kilometer-scale sub-elements, and stress-free boundary conditions are specified at the center of each sub-element. Displacement occurs across each sub-element in response to an imposed driving force, so that the strain energy is minimized throughout the model space. This means, for example, that in any particular increment of deformation, if there is more than one fault, the amount of slip across any fault will be a function of the amount of slip across every other fault. Any single model, therefore, assumes that all faults are active simultaneously.

We drive model deformation by imposing a uniform pure shear of a magnitude large enough to generate displacement gradients of the order 10^{-4}. A typical strain rate in actively deforming regions is ~10^{-7} yr^{-1}, so that each model deformation increment effectively simulates ~1 k.y. of deformation. The resultant slip across a fault plane is ~10^{-5} of the downdip width and is equivalent to a model earthquake of **M** 5–7, depending on the size of the fault. We effectively model and sum only the largest earthquakes that any specific fault can sustain; smaller earthquakes provide negligible bulk strain (Kostrov, 1974). Both horizontal and vertical components of displacement are applied to a horizontal initial surface to generate the tectonic topography. We confine the models to two dimensions, or plane strain. Permutation of the few model fault parameters (Fig. 2A) provides a sufficiently large number of fault configurations and deformation patterns to think about. Adding three-dimensional components addresses further issues related to lateral fault growth by tip propagation and fault linkage, which will be issues of future research.

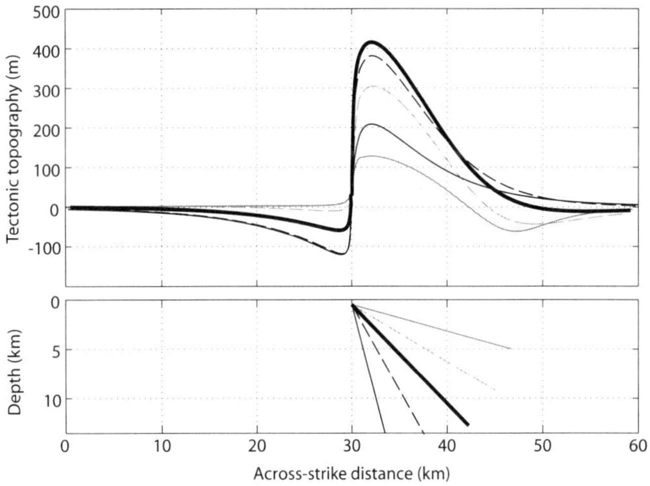

Figure 3. Displacement of the ground surface (top panel) as a function of thrust fault dip (lower panel, no vertical exaggeration). Line types in the lower panel correspond to the displacement patterns shown in the top panel. Displacements of the surface are calculated with both vertical and horizontal components (i.e., these are not simply uplift functions), but due to the necessary vertical exaggeration, the horizontal component of displacement is not as apparent as is the vertical. Each thrust fault is blind, buried below the surface by 0.5 km. Dips vary from 15° to 75° in 15° increments. Notice that all displacement patterns yield an asymmetric fold that is skewed toward the sense of vergence, but that the magnitude and location of subsidence is sensitive to fault dip.

RESULTS

Effect of Fault Dip

Deformation of an initially flat surface by slip on a thrust fault, the leading tip of which is 0.5 km below ground surface is shown in Figure 3. We emphasize at the outset that the important aspect of the tectonic topographies shown in Figure 3 (and subsequent model figures) is the shape of the ranges, not the relative amplitudes. Deformation is driven by the same uniform pure shear, and because the fault dip differs between models, so too will the slip that each sustains. The finite slip in each case shown here (and all subsequent model results) is different and reflects what might accumulate in the same amount of time under the same driving force.

The fault dip controls the position and relative magnitude of a region of subsidence. At shallow dips of less than ~35–40°, subsidence occurs over the rear of the fault, above the buried trailing tip, and there is a notable absence of accommodation generation on the footwall in front of the fault, in contrast to the standard model (Fig. 1). This pattern reverses itself as dip increases, so that at dips greater than 45°, subsidence occurs only on the footwall, and the entire hanging wall experiences rock uplift, as in the standard model. The tectonic topography owes its shape to the relative proximity to the ground surface of the leading and trailing tips of the fault. For example, significant extension occurs immediately above and parallel to the trailing tip of the fault as material is literally pulled away from this region toward the leading tip. Subsidence occurs if this zone of extension is sampled by the ground surface, because as material extends subparallel to the ground surface, it also shortens vertically (and therefore the ground subsides), which is what happens when the fault has a gentle dip and the trailing tip is relatively near the surface. As fault dip increases, the ground surface does not sample the region of extension, and subsidence does not occur. By the same token, significant rock uplift occurs above the fault because material is literally squeezed into the hanging-wall wedge. The region of footwall subsidence grows in width and depth as the fault dip increases, because again increasingly more of the extension under the leading tip is sampled by the surface at steeper dips.

In all cases, prior to significant erosion, the fold, and thus the initial range shape, is strongly asymmetric. Because the topographic divide in the absence of erosion is defined by the fold hinge, it will be skewed toward the vergence direction in each of the examples shown in Figure 3.

Effect of Depth

As the depth to the leading tip of a 15°-dipping thrust fault increases, the asymmetry of the resulting range is reversed. At a depth greater than ~4 km, the steeper side of the range is defined by the back limb and faces downdip, away from the vergence direction (Fig. 4). The combination of burial and shallow fault dip leads to subsidence over the trailing fault tip and relatively uniform rock uplift over the footwall. Thus, a range emerging over a blind, shallowly dipping thrust fault buried to 8 km is superficially similar to a standard model fold over a fault with the

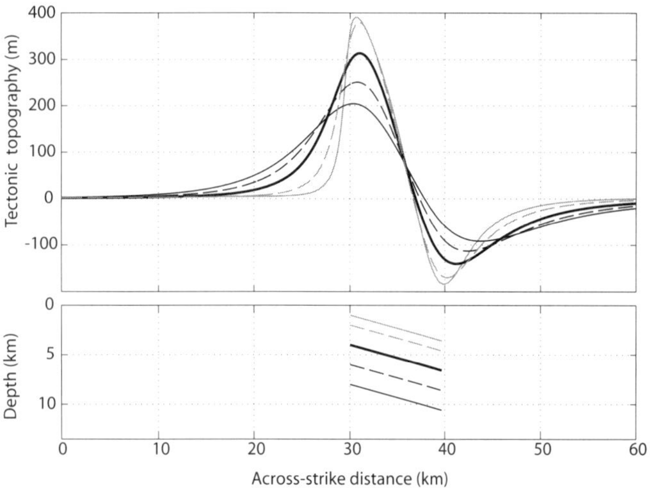

Figure 4. Displacement of the ground surface (top panel) as a function of thrust fault burial depth (lower panel) for a fault that dips 15° and is buried to depths of 1, 2, 4, 6, and 8 km. Line types in the lower panel correspond to the displacement patterns shown in the top panel. Notice that the asymmetry of the thrust-related fold is reversed from that of the standard model and from those generated by faults in which the upper tip is near the surface. All else being equal, a reasonable (and often made) interpretation of a range with this shape and region of subsidence is that the underlying fault verges in a sense opposite to that shown here. This interpretation might be compounded further by the likelihood that the rear range front will be sharply defined along its length because of the adjacent subsidence (expressed by onlapping sediments, or an axially draining river, etc.). By contrast, the foreflank will likely be embayed along its margin, because the range slope is relatively shallow. A similar sense of asymmetry is found for faults that dip less than 45°.

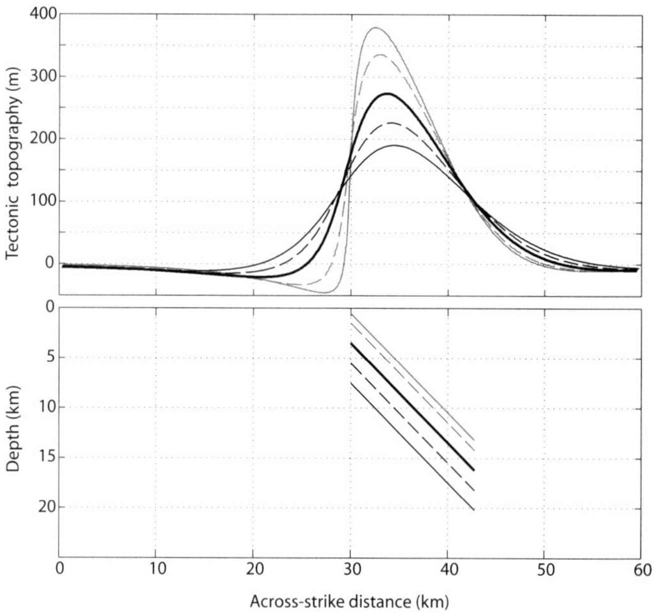

Figure 5. Displacement of the ground surface (top panel) as a function of thrust fault burial depth (lower panel) for a fault that dips 45° and is buried to depths of 1, 2, 4, 6, and 8 km. Line types in the lower panel correspond to the displacement patterns shown in the top panel. The resultant fold (and topography in the absence of erosion) is fairly symmetric if the upper fault tip is deeper than ~4 km, but as the upper tip nears the surface, the fold begins to verge toward the slip direction, closer in style to the standard model of Figure 1. Notice that there is never any subsidence over the hanging wall and only on the footwall when the fault tip is shallower than ~1 km. This means that while the fault is significantly blind, neither margin of the resultant range is well defined along strike and is instead likely to be embayed.

opposite vergence. Inverting from the shape of the range (which in the absence of erosion is equivalent to the shape of the fold) to the underlying fault is not necessarily straightforward. This is an important result, because it is counterintuitive and yet is generated by a fault geometry that is probably fairly common in forelands of fold-and-thrust belts.

The same tendency toward a reversal in asymmetry as a function of depth occurs for all faults that dip less than 45°. For a 45°-dipping fault, the original (and standard) sense of asymmetry is retained within reasonable burial depths (Fig. 5). As burial depths increase, however, the fold becomes more symmetric, and there are no regions of subsidence.

Effect of Fault-Tip Propagation and Linkage to a Lower Décollement

A reasonable fault growth scenario in fold-and-thrust belts is that of a gently dipping (e.g., 30°) thrust that initiates in the middle of the upper crust (~7 km), propagates both updip and downdip and later links to a lower décollement at ~8 km (Gretener, 1972; Eisenstadt and DePaor, 1987; Ellis and Dunlap, 1988). We simulate this scenario by allowing the model fault to propagate in 2 km increments both updip and downdip when its total displacement-to-width ratio is ~0.1, which corresponds to ~1200 model earthquakes on each fault. This is not to suggest that thrust faults propagate in 2 km steps (which is a figure chosen largely for numerical convenience), although there is some inferential evidence that the 1985 M 6.1 Kettleman Hills, California, earthquake may have done something like this (Ekström et al., 1992).

Surface deformation accumulates relatively slowly at first because the early fault is small and sustains proportionately less displacement than later, larger faults (Fig. 6A). The initial asymmetry of the range is again reversed from the standard model because of deep burial and shallow dip (Fig. 6A). A zone of subsidence, which may develop into a piggyback basin, develops on the hanging wall as the fault propagates downdip. The tectonic topography becomes more symmetric over time, although importantly the steepest, most sharply defined range front remains on the back limb of the fold. The forelimb slopes

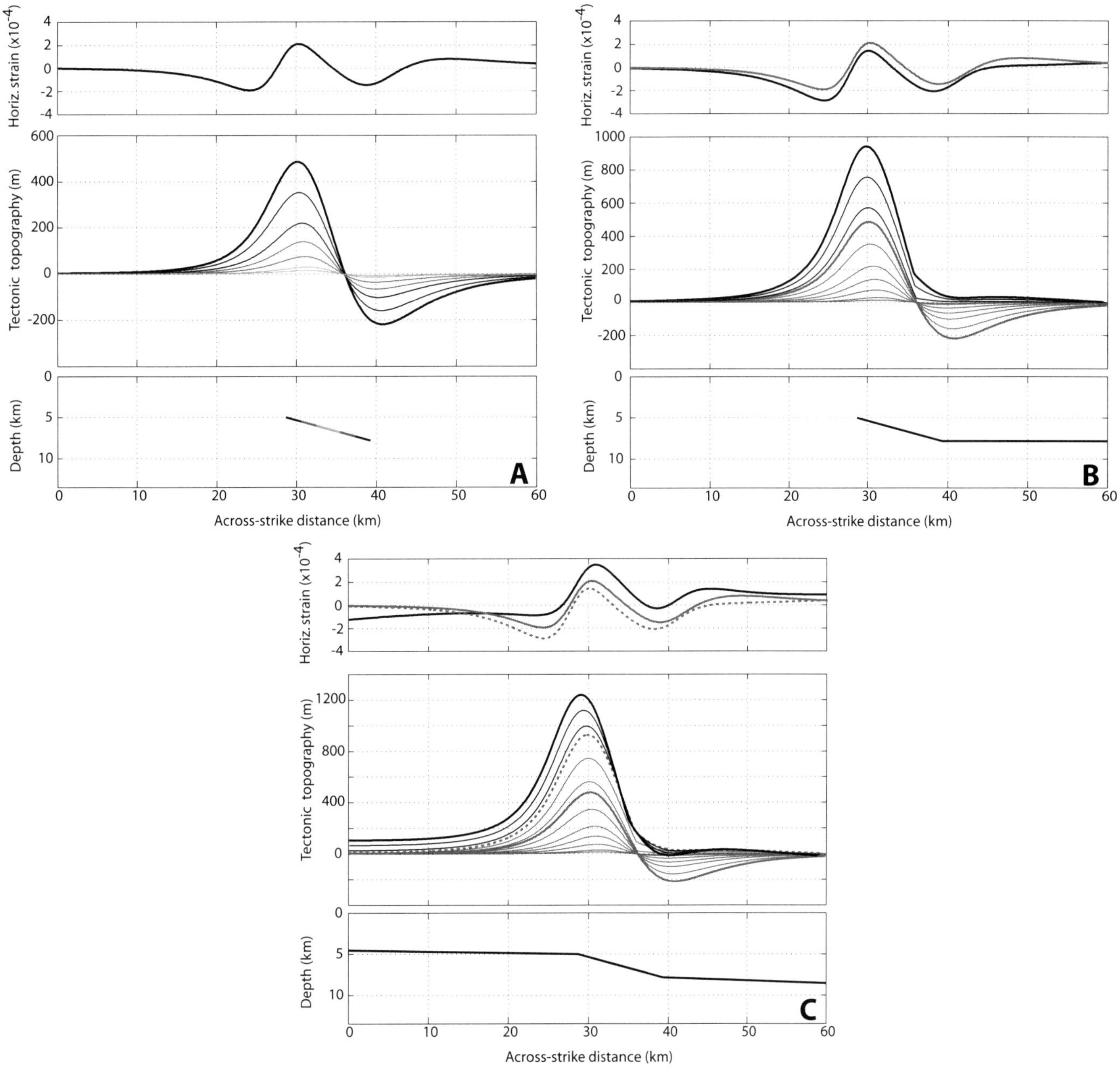

Figure 6. (A) Displacement of the ground surface (middle panel) as a function of a buried thrust fault, dipping at 15° and propagating both updip and downdip (lower panel). The fault acquires slip until the ratio of total slip to downdip width is 0.1, which corresponds to ~1200 model earthquakes on each fault segment. At this point, the fault is allowed to propagate 2 km in both directions. The fold generated by this model is markedly asymmetric toward the hinterland and will likely have a sharp boundary along the rear flank and active subsidence on the hanging wall. The top panel shows the distribution of horizontal strain at the surface; negative values indicate shortening. As expected, extension occurs over the fold hinge, which is likely to be manifest as so-called keystone normal faults at the crest of the range. Shortening occurs along both margins, notably along the foreflank margin. (B) A continuation of the model illustrated in A, showing displacement of the ground surface (middle panel) as a function of a buried thrust fault, dipping at 15° and linked to a gently sloping décollement (lower panel). The most prominent change is observed as relatively uniform rock uplift of the hanging wall as a consequence of removing the tip zone effects from the lower part of the ramp thrust fault. The (geologically) immediate effect is to raise the base level along the rear flank, exhuming possible piggyback sediments deposited during the earlier stages (A). The top panel shows the distribution of horizontal strain at the surface (gray prior to fault linkage, black after linkage); negative values indicate shortening. Notice that the rate of shortening increases along both margins after linkage. (C) A continuation of the model illustrated in B, showing displacement of the ground surface (middle panel) as a function of a buried thrust fault, dipping at 15° and linked to a gently sloping décollement (lower panel). The resultant fault-bend geometry causes a relatively rapid advection of the fold toward the foreland. The extent to which topography is advected depends on the relative rates of advection and erosion, but if there exists a flux steady state, remnants of rear flank topography may be preserved in the upper reaches of foreflank topography, wind gaps will exist, and valleys on the rear flank will be beheaded.

gently in the direction of fault vergence, but because there is no footwall subsidence, the front of the range is not cleanly defined. The combination of a foreland range front that is poorly defined and a rear edge that is well defined may give the misleading impression of an underlying thrust fault verging in the opposite direction (Fig. 6B).

As the isolated thrust fault links to a lower flat at ~8 km, the most significant change in surface deformation occurs in the zone of hanging-wall subsidence. This region now undergoes almost uniform rock uplift, and the rate of local horizontal shortening increases (Fig. 6B). The reversal in the sense of displacement may have significant geological effects. For example, if deposition occurred on the hanging wall during fault propagation, linkage may drive exhumation, minor folding, and removal of these hanging-wall sediments, and base level for the back limb of the fold will be shifted away from the fold toward the hinterland. An example of this scenario might be found in the central Taiwan fold-and-thrust belt, where the Puli piggyback basin is currently undergoing exhumation, and small active folds are observed within the basin sediments (Mueller et al., 2002).

Fault Evolution: A Fault Bend Fold

Following on from the previous model, the ramp now links to a shallow décollement at ~5 km, resulting in the familiar fault-bend geometry (Fig. 6C). The décollement is assumed to develop (geologically) instantaneously, and we do not model its incremental propagation. The continuation of the history of surface deformation shows the importance of the horizontal component of deformation (Fig. 6C). As material moves over the upper bend, it enters a dominantly horizontal deformation field. The fold hinge is advected relatively rapidly toward the foreland and away from the highest rate of surface and rock uplift, which is locked to the ramp section of the linked fault. This means that the tectonic input to the topographic shape of the range will depend on the relative rates of advection and erosion. In the extreme case where the rate of advection is significantly higher than that of erosion, the tectonic input to the shape of the range is the finite structure itself, so that the topographic divide moves at the rate of advection. At the other end of the spectrum, in which erosion rates are significantly higher than advection, the finite structure will be largely removed, and the tectonic input to the range shape is the incremental deformation field, which is locked to the ramp section of the fault.

From a structural perspective, these models make some interesting and testable predictions. The rate of horizontal shortening, relatively high and localized by the leading and trailing fault tips, undergoes systematic changes as the fault geometry evolves. In particular, rates increase in the hanging wall when the ramp segment links to the lower décollement, a direct consequence of the increased slip rate resulting from an increase in fault width (measured downdip). As the leading tip links with the shallow décollement, the rate of tectonic advection increases rapidly, and near-surface strain is wholly extensional, markedly so where the rock uplift rate is highest (Fig. 6C). At a longer length scale, once tectonic advection becomes dominant (i.e., once the ramp has linked to both a lower and higher décollement), the fold hinge is advected toward the foreland, away from the axis of maximum incremental fold growth. Younger folds will be developed over the ramp, where the rate of rock uplift is highest, and so will be displaced toward the hinterland from the longer-term finite fold axis. These younger structures may be recorded by the deformation of fluvial terraces or paleochannel profiles that are draped over the range.

These examples present a series of tectonic topographies that might be expected over a reasonable suite of thrust fault geometries, and that set the initial conditions for the redistribution of mass by surface processes. As the range emerges from a pre-faulting ground level, the initial topographic divide and first-order shape of the range are given by the tectonic topography. But as fluvial erosion becomes increasingly important, the position of the divide and the shape of the range express the competition between rates of surface and tectonic processes. Rates of surface erosion, in turn, depend largely on the position of local base levels, which we address in the following section.

SHAPING THE RANGE VIA SURFACE EROSION AND THE ROLE OF LOCAL BASE LEVELS

The behavior of the topographic divide is more difficult to address quantitatively than is the growth of the underlying structure, because the processes that enable a divide to migrate are not well understood. It seems likely that divide migration occurs by a combination of landslides, which cause abrupt retreat of the catchment margin (Weissel and Seidl, 1998; Hovius et al., 1998; Bishop, 1995), coupled with incision, possibly by debris flows, in the upper reaches of channels (Stock and Dietrich, 2003). In some settings, differential erosion by valley or cirque glaciers may also cause divide migration (Brocklehurst and Whipple, 2002). Because all of these processes are somehow slope-dependent, we propose that, as a first-order simplification, the divide position is set by competition between the flanks of the range, each of which evolves toward a specific range-scale slope. This suggests a simple geometric model for predicting the steady-state position of the topographic divide. If the mean slopes of the opposing flanks of a range are equal, the fractional position of the divide D_r is given by

$$D_r = \frac{F}{F+R}, \tag{1}$$

where F and R are the elevations of the divide above the foreland and hinterland base levels, respectively (Fig. 7A). A useful ratio is defined by $\mu = rF/fR$, where r and f correspond to the fractional horizontal distance from the divide to the hinterland and foreland base levels, respectively (Fig. 7A). If the mean slopes are equal, $\mu = 1$.

If base levels on both sides of the range are the same, the predicted steady-state position of the divide is at the center of the

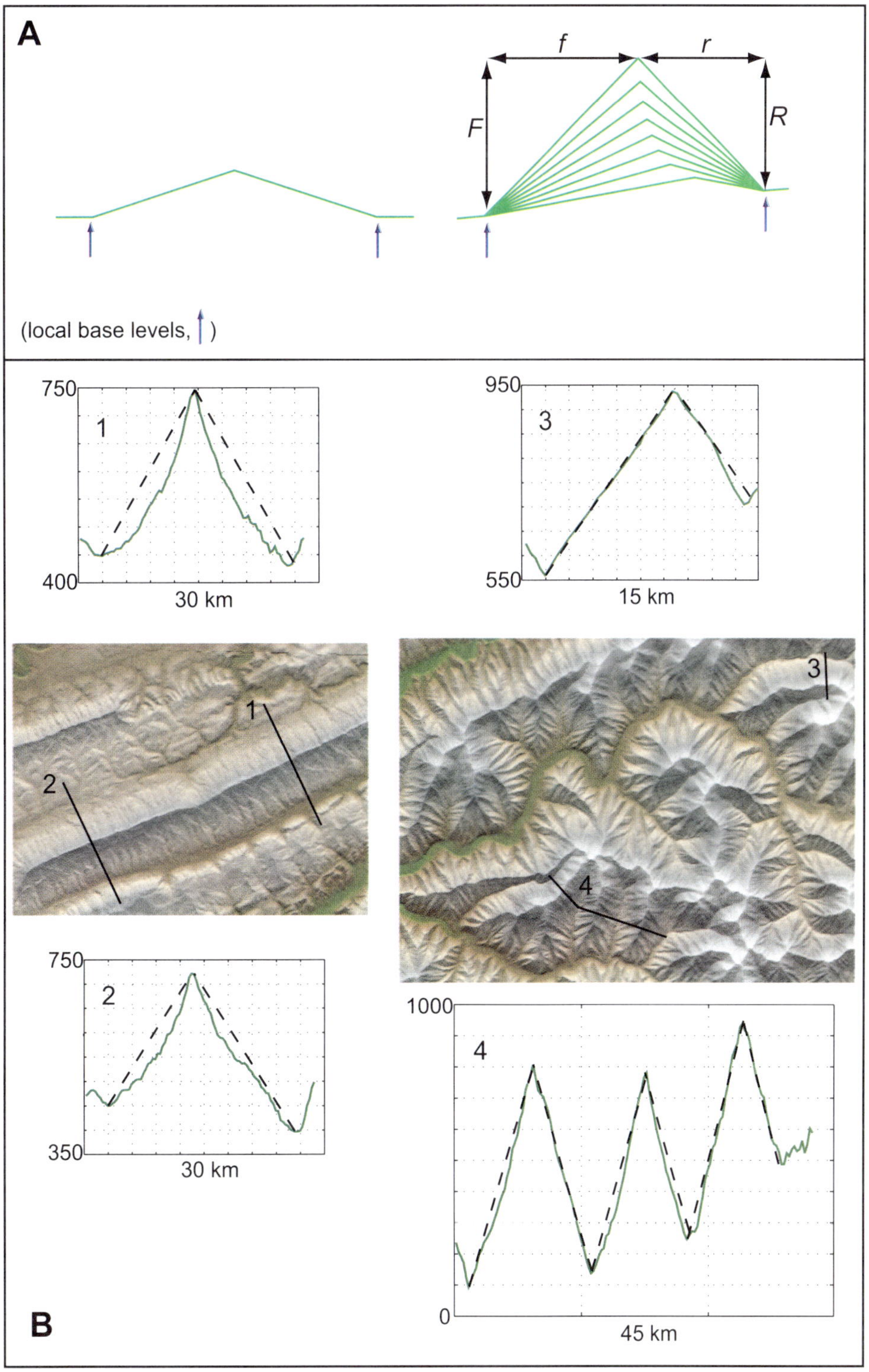

Figure 7. (A) The relative position of the topographic divide is given by a simple geometric relation between parameters that describes the dimensions of a range in settings in which range-scale slopes are assumed to be equal and dictated by the same processes on both flanks of the range (equation 1). Where flanking base levels (shown as blue arrows) are at the same height, the divide will sit in the middle of the range. Where flanking base levels occur at different heights, the position of the divide is a function of relief, but the position is always closer to the higher base level. (B) Divide positions on a number of ridge spurs between well-defined flanking base levels in parts of the Appalachians in southwest Virginia, shown by the shaded relief maps. The terrain has evolved as a result of both debris flows and fluvial processes. Topographic profiles across parts of the topography (numbered lines on the maps) are shown in insets as solid lines. The dashed lines in these insets are of equal slope and consistently connect the divide to the flanking base level. We take this as an empirical demonstration that the topographic divide is predicted by equation 1, where hillslopes are generated by a mixture of erosional processes.

range, an expected and trivial result. If base levels differ across the range, and (importantly) if the processes that control divide migration are the same on both range flanks, the divide position will be shifted toward the side of the higher base level, such that $\mu \rightarrow 1$. These relations are straightforward for single hillslopes, but to make them useful, we need to know if they hold if the process of divide migration is dictated at the range scale by a mixture of fluvial incision, glacial incision, debris flows, and bedrock landsliding.

To test the broader applicability of equation 1, we use two simple illustrations. We first examine the topographic divide positions of ridges with length scales of 10–40 km (similar to those underlain by blind thrusts) from parts of the Appalachian Mountains in southeastern Kentucky and southwestern Virginia. This region is relatively tectonically inactive at the length scale of upper crustal faults, climate is relatively uniform, and regional rates of denudation, derived from a variety of techniques, are relatively low and spatially uniform at ~30 m m.y.$^{-1}$ (Matmon et al., 2003). Hillslopes here are shaped by a combination of debris flows and shallow landslides, as well as by fluvial incision (Eaton et al., 2003). The divide positions predicted by equation 1 match the actual divides remarkably well on a variety of two-dimensional transects across this simple landscape (Fig. 7B).

The second test consists of a numerical experiment in *Zscape*, a landscape evolution model that incorporates both fluvial processes in a mixed bedrock-alluvial channel and bedrock landslides driven by oversteepening of hillslopes (Densmore et al., 1998). In the experiment, we build a tectonic topography from the accumulating displacements of a 30°-dipping thrust fault with its leading tip at a depth of 0.5 km (e.g., Fig. 3), such that the initial divide is positioned close to the foreland side of the range. Base level at the rear of the range is ~175 m higher than at the front of the range. The topographic divide is initially set by the tectonic forcing, but as erosion becomes more important, the divide migrates toward the hinterland and settles into a position dictated by the elevations of the hinterland and foreland base levels, as predicted by equation 1 (Fig. 8). This experiment is by no means conclusive, as the model processes that operate at the channel head (a mixture of fluvial incision and relatively crude bedrock landslides) are far more limited in *Zscape* than they are likely to be in nature, where other erosional agents, such as glaciers or debris flows, may play a role. There are, in addition, a number of other factors that may influence the position of the range divide, including different bulk rock strengths on opposite flanks of the range (a situation that may arise simply from a uniformly dipping sequence of bedded rocks), and orographic

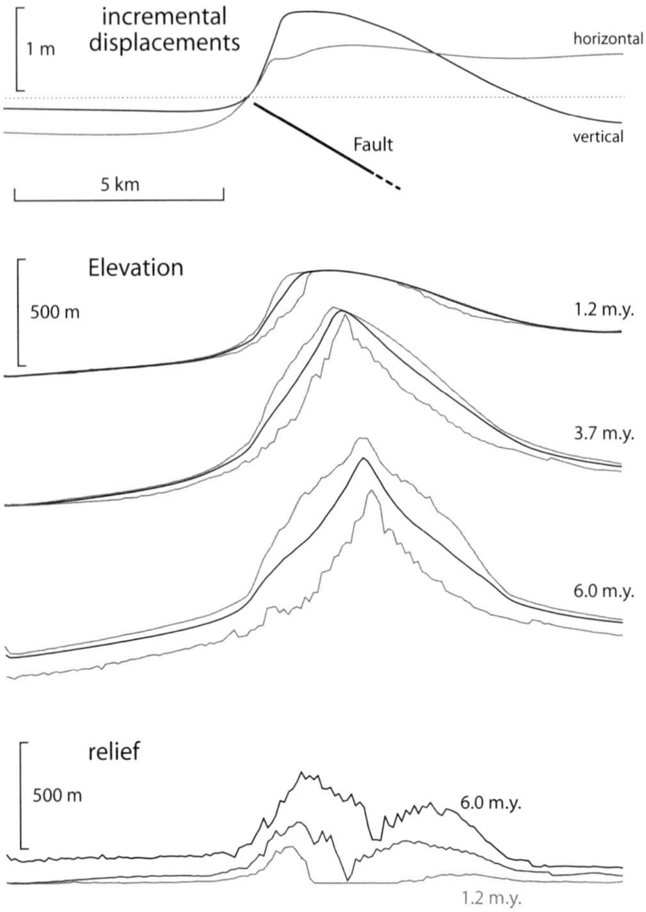

Figure 8. Results of a numerical experiment using *Zscape* (a three-dimensional landscape evolution model; Densmore et al., 1998) designed to further test the influence of flanking base levels on the position of the topographic divide. The model setup (upper panel) in cross section consists of a blind thrust dipping at 30°, and increments of vertical and horizontal surface displacements, calculated with *3ddef* (Gomberg and Ellis, 1994) are shown as dark and gray lines, respectively. The middle three panels are swath profiles across 10 km of an initially gently inclined and slightly rough surface (see Densmore et al., 1998, and Ellis et al., 1999, for details of these types of models) at three stages: after 1.2 m.y., 3.7 m.y., and 6.0 m.y. of model run time. The early range shape at 1.2 m.y. clearly reflects the tectonic displacement field. Flanking base levels subsequently pull the topographic divide toward the left (the local hinterland), but the spatially fixed displacement field maintains the generation of relatively high local relief on the foreflank, as shown by the sequential development of local relief in the lower panel.

precipitation, which may lead to enhanced water and sediment flux on one flank of the range and thus drive migration of the divide away from that flank. Nevertheless, we are encouraged by the empirical evidence from the Appalachians and by the *Zscape* results to suggest that the conceptual model is essentially correct and that the position of the topographic divide is, all else being equal, dictated by the flanking local base levels.

HYPOTHETICAL AND REAL EXAMPLES, AND THE GENERATION OF LOCAL RELIEF

The various tectonic topographies generated over blind thrusts (e.g., Figs. 3–7) when coupled with the influence of local base levels present an array of scenarios in which the evolving first-order shape of the emerging range is likely to be highly transient and a complex reflection of both tectonic and surface processes. To illustrate the importance of base level, we follow the development of a hypothetical range through a simple history in which the tectonic forcing and climate is chosen to be relatively simple and largely constant (Fig. 9). The point of this exercise is to illustrate that the topographic attributes of a simple thrust-related range may be highly transient despite a relatively simple tectonic forcing field.

We start with a 30°-dipping thrust fault that has initiated in the middle upper crust and that is propagating both downdip and updip toward a regionally gently sloping surface (Fig. 9A). Any accommodation generated on the hanging wall is assumed filled by sediments derived from the hinterland. This particular scenario addresses the initial evolution of a blind thrust range in which fault displacement is relatively low and the thrust fault has yet to link to a lower décollement.

In the early stages, prior to significant erosion, the blind thrust will generate a range in which the tectonic and topographic asymmetry is strongly skewed to the hinterland, similar to the

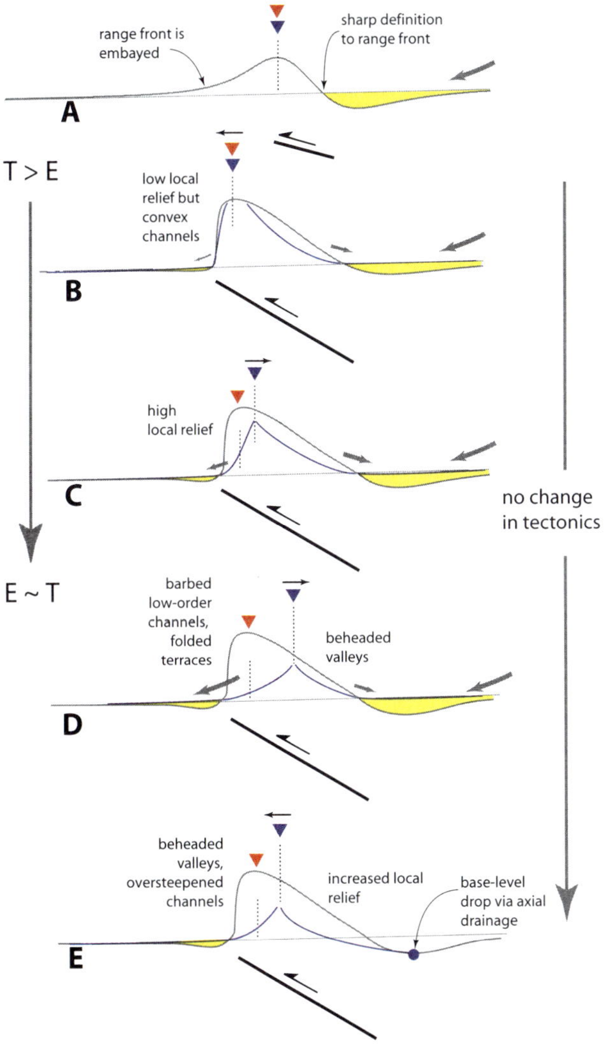

Figure 9. Cross-sectional profiles illustrating the hypothetical sequential (top to bottom) development of a blind thrust faulted range. The conditions have been chosen to emphasize the role played by flanking base levels in the evolution of the first-order shape of the range. (A) The initial fault propagates both updip and downdip and slowly rotates to a steeper dip, as a response to the overall crustal shortening. The initial range shape is the same as the tectonic displacement field, and the topographic divide (blue triangle) is coincident with the fold hinge (red triangle). We assume also that the accommodation generated on the hanging wall is filled by sediments (yellow) derived mainly from the hinterland. Importantly, the shape of the range and the definition of its margins will at this early stage give the impression the underlying fault verges in a sense opposite to that shown. (B) As the fault propagates toward the surface, the shape of the tectonic topography is essentially reversed, and the range shape, still dominated by the tectonics because erosion remains insignificant, is now asymmetric toward the sense of fault vergence. The tectonic and topographic divides are still coincident and have migrated toward the foreland. At this stage, tectonic displacements outstrip rates of erosion (i.e., T > E) and any channels that drain the foreflank will be steep and possibly convex up. (C) When the fluvial network reaches the divide, base levels begin to exert control on the evolving shape of the range. This will initially force the topographic divide to migrate back toward the hinterland. (D) The combination of divide migration and continuing tectonic displacements generates and maintains high local relief on the foreflank. Topography of the rear flank will include beheaded valleys as the divide migrates toward the rear, and remnants of rear flank topography may survive temporarily in the upper reaches of the foreflank network system. (E) The final stage in this scenario shows the tapping and emptying of sediments in the piggyback basin (via, for example, an axial river system). Base-level fall in the piggyback basin promotes migration of the divide back toward the foreland, thereby increasing local relief on the rear flank and oversteepening channels on the foreflank. Thus, attributes of the topography on each flank of the range depend in part on the relative elevations of flanking base levels, and what happens on one side invariably affects the other.

development shown in Figure 4. The resulting steep, narrow, rear-facing flank, which is also better defined along its edge by virtue of onlapping sediments derived from the hinterland, may reasonably (but incorrectly) be interpreted as the result of a thrust fault that verges toward the hinterland (Fig. 9A). The early channel network, if there is one, will reflect the different slopes and watershed areas available to the channels. Channels will be more closely spaced and straighter on the rear-facing flank and more dendritic and widely spaced on the gently sloping foreflank. At this stage, however, the topographic divide is dictated entirely by the tectonics.

As the fault propagates toward the surface, the asymmetry of the range will be reversed, and the steep, narrower flank will now face the foreland, reflecting more simply the sense of fault vergence. Note that the effective foreland migration of the fold hinge and topographic divide is not the result of tectonic advection nor of an increased rate of fault slip but of a changing displacement field as the fault propagates toward the surface. If the rate of erosion on the foreflank is less than that of displacement, as is likely in the early stages of development, foreland draining channels will be convex up (Fig. 9B). Remnants of the upper reaches of the foreflank channel network may be preserved on the rearward flank and recognized as barbed channels and wind gaps. As rates of fluvial erosion increase, local relief will be higher on the foreland flank than on the rear flank (Fig. 9C) and preservation of earlier channels will be concomitantly reduced, with relatively little evidence to indicate the earlier rearward position of the divide.

The topographic divide of our hypothetical range is now sitting near the foreland edge (Fig. 9C), and at this stage, it is likely that the channel network has reached the divide, and thus the role of flanking base levels will be important in determining what happens next in the scenario. The regionally inclined surface through which the range is emerging (plus the filling of any accommodation space generated over the hanging wall) will yield a difference in flanking base levels of one to several hundred meters, depending on the width of the emerging range, which is likely to be ~10–20 km. The position of the divide, if controlled purely by base level, will depend on the absolute relief. At a moderate relief of, for example, 500 m and a difference in flanking base levels of 200 m, the divide is predicted to lie more than two-thirds of the range-width away from the foreland edge. This means that the divide will migrate toward the rear of the range. However, because material advection is toward the foreland, relatively large local relief will be generated on the foreflank (Fig. 9D). As the topographic divide begins to migrate back toward the hinterland, channel valleys on the rear flank will be beheaded, leaving a distinct geomorphic signature in the landscape. Importantly, channels draining the foreflank and attributes of the local topography will be modified as a function purely of base-level–driven divide migration. Neither the rate of slip nor the geometry of the fault has changed during this part of the scenario, and changes in the topography cannot be attributed to changes in the tectonic forcing.

At this stage, the topographic divide is sitting near the rear of the range, a position dictated by flanking base levels. If the piggyback basin is now tapped by a large, axially draining river such that base level begins to drop at the rear of the range, the topographic divide will be driven toward the front of the range, driven by the headward retreat of channels that drain the rear flank (Fig. 9E). Thus, local relief on the rear flank will be increased, foreflank channels will be beheaded, and their profiles will again be modified.

The point we have attempted to make with this hypothetical scenario is that the attributes of topography, whether expressed in the fluvial system or in hillslopes, are sensitive to both tectonic forcing and to the relative elevations of the flanking base levels. Importantly, flanking base levels may be altered by a variety of mechanisms, including local tectonics and the local fluvial system, but also by changes external to the local system. We next illustrate the importance of local base levels with reference to a number of geological examples.

Geological Examples

We know of several real-world examples that may illustrate the general principles described in the preceding sections. Meigs et al. (1999) presciently recognized that the greater dissection and local relief on the southern flank of the Santa Monica Mountains in southern California were driven by a combination of (1) south-vergent displacement on an underlying system of blind thrusts and (2) lower base level on the southern side of the range. Ranges with similar tectonics and base levels are present in a number of fold-and-thrust belts worldwide, including the Siwalik Hills of the frontal Himalayas in northern India and southern Nepal. Displacement on the Himalayan Main Frontal thrust has generated a series of anticlines within the Siwalik Formation, a suite of syntectonic fanglomerates shed from the northern Lesser Himalayas. The anticlines are exposed as a series of en echelon ranges, each tens of kilometers in length along strike and ~10 km in width (Yeats and Lillie, 1991). One of these is the Mohand Anticline south of Dehra Dun, India (Rao, 1986; Powers et al., 1998), which is underlain by a northward-dipping thrust fault, which has been intercepted by a drill hole from the peak of the range (Fig. 10A). The topographic divide lies 2.3 km north of the fold hinge, over the back limb of the fold, resulting in a strongly asymmetric range with long south and short north flanks (Fig. 10A). The expected tectonic displacement field for the thrust fault beneath the Mohand Anticline has the opposite sense of asymmetry, with a short, steep south flank facing the vergence direction. Base level on the northern flank of the range, however, is ~200 m higher than on the southern flank, and the position of the topographic divide is approximately as predicted by equation 1 from the base-level differences (Fig. 10A). Local relief, measured over a length scale of ~1 km, is significantly higher on the southern flank of the range, which indicates one or both of the following: that maximum rock uplift occurs

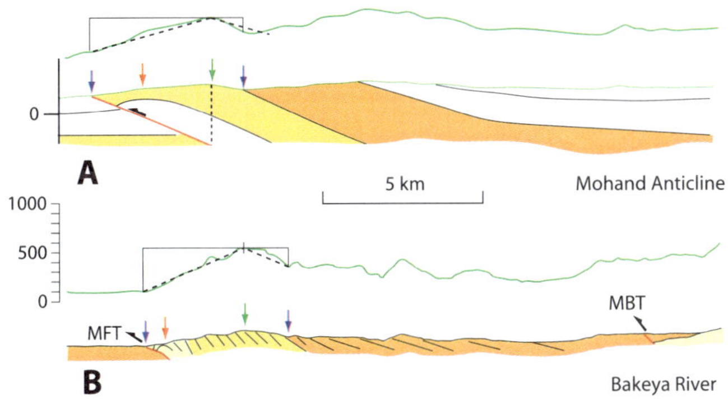

Figure 10. Geological cross sections and topographic profiles across (A) the Mohand Anticline (modified from Yeats and Lillie, 1991), and (B) the Bakeya River section (modified after Lavé and Avouac, 2000). The topographic profile (green line) is raised and exaggerated threefold for clarification. The red arrows mark the surface positions of the fold hinges. The position of the flanking base levels is well defined by axially draining rivers along the rear and by the foreflank margin (shown by blue arrows). The topographic divide is shown the green arrow, and its position is predicted by equation 1 (demonstrated here by the equally sloped dashed lines), which indicates that in all likelihood, flanking base levels control its position. In both cases, the fold hinges are shifted to the foreland side of the topographic divide. MFT—Main Frontal thrust; MBT—Main Boundary thrust.

south of the topographic divide and/or that tectonic advection is helping to maintain higher local relief.

A second example from the Siwalik Hills comes from farther southeast in southern Nepal, where the Bakeya River cuts across a series of actively growing folds (Lavé and Avouac, 2000). One of these folds is illustrated in Figure 10B; again, the fold hinge lies close to the mapped trace of the Main Frontal thrust, and the topographic divide is offset to the north. Base level on the northern flank of the range is ~250 m higher than on the southern flank, and the position of the topographic divide matches that predicted by equation 1 (Fig. 10B).

A third example is offered by the eastern end of the Kyrgyz Range in the northern Tien Shan, Kyrgyzstan (Oskin and Burbank, 2002). The Kyrgyz Range is being generated by an emergent north-vergent thrust. Both tectonic and topographic relief is up to ~4 km, and Oskin and Burbank (2002) described a systematic change in the relative position of the tectonic and topographic divides. At the eastern tip of the fault, the topographic and tectonic divides are coincident for ~15–20 km. Closer to the center of the fault trace, the divides separate, such that the tectonic divide lies in the north of the range (reflecting the northward-vergent thrust fault) and the topographic divide has migrated increasingly farther to the south toward the center of the fault. Again, the position of the topographic divide is consistent with the difference in flanking base level, which is significantly higher in the south than in the north. In this example, the process of divide migration is clearly related to the onset of glacial erosion as the range is pushed up and through the equilibrium line altitude, ELA (Oskin and Burbank, 2002).

A final example comes from preliminary work in the Flaming Mountain, a blind thrust range emerging from the vast south-facing piedmont that flanks the Bogda Shan, the northeastern extension of the Tien Shan, China (Davis et al., 2003). Flaming Mountain is less than 10 km in width, with a steep southern flank. It is crossed by a number of antecedent channels that drain the Bogda Shan to the north, but relatively little development of a fluvial network has occurred over the emerging range. Flaming Mountain is an example of the earliest stages of range development. The topographic divide lies between the two fold hinges, ~3.2 km from the range front and significantly farther toward the rear than the bedrock fold hinge, visible in bedrock at ~1.5 km from the southern range front. Younger, gentle fold hinges are picked out by deformed abandoned antecedent channels, ~4.6 km to the north of the range front. It is possible that the younger folds record the incremental displacement field that is fixed to the underlying fault, while the bedrock hinge is evidence of the near-surface burial depth of the blind thrust fault (e.g., Fig. 3).

DISCUSSION

Can we infer with some level of confidence the underlying tectonic geometry from the shape of a thrust-related range? The potential variety of shapes and means by which these might change over time would suggest that the answer is no, or at least that it is not straightforward.

There are, however, topographic signatures that allow us to make some headway. The first of these is the ratio $\mu = rF/fR$. If a flux steady state does not exist, such that rates of tectonic displacements are greater than those of erosion, the value of μ will depart from 1, and at least part of the range shape will be due to tectonics, rather than strictly due to relative base levels. It is not straightforward, however, to infer the sense of vergence or fault dip from the specific value of μ. This is because the topographic divide may be advected toward the foreland faster than it can be eroded, or its position may reflect a dominantly vertical rock uplift pattern. In the latter case, the topographic divide will tend toward the position set by the displacement field illustrated in Figures 3–6; in some of these examples, the position of the divide is closer to the hinterland than to the foreland. In the former case, where tectonic advection is greater than erosion, fluvial profiles over the foreland flank will be convex upward, reducing local relief. At this stage, therefore, a value of $\mu \neq 1$ tells us only that tectonic forcing is a factor in controlling the range shape, but by itself, it does not uniquely point to the dip or sense of thrust vergence. (The nonuniqueness might be resolved by field mapping, which should reveal whether the topographic divide is being held up by a fold hinge, or other information about the thrust fault.)

While a value of $\mu \neq 1$ tells us that tectonics has a controlling influence on the range shape, the opposite is not true. That is, a value of $\mu = 1$ does not indicate that base-level control is necessarily the controlling factor in shaping the range, nor that tectonic forcing is unimportant. It is possible, for example, that both the tectonic forcing and base-level differences will drive the divide toward the same position.

The picture is actually a little brighter than that painted here, because we have attempted to cover a wide range of possible scenarios rather than those that are most likely. In most of the examples described above, the tectonic asymmetry of the fold is skewed toward the foreland, so that the initial topographic divide will be similarly skewed. Likewise, in most cases base level is lower on the foreland side of the range, such that the topographic divide will migrate toward the hinterland. Therefore, a value of $\mu > 1$ (where the position of the topographic divide is closer to the foreland than predicted by equation 1) most likely indicates a sense of vergence toward the foreland. Along the same lines, a value of $\mu = 1$ probably indicates that the first-order shape of the range is controlled by differences in base level rather than by tectonics.

The second topographic signature that may be useful is that of local relief, measured over a length scale that encompasses the entire hillslope from channel to its neighboring ridge. In flux steady state, we expect local relief to be higher on the foreland flank, assuming that in all likelihood the topographic divide has been driven by base-level control to the rearward side of the site of maximum rate of vertical rock uplift.

The recognition that local base levels play an important role in shaping the emerging range has significant and intriguing consequences. While local base levels are initially set and maintained by local processes, they may also and surely will change as a consequence of external processes (Ellis et al., 1999). The simplest example is of a piggyback basin being filled by sediment derived from farther back in the mountain belt, the production of which is related to slip rate and climatic conditions in an entirely separate range. Sediment supply to the piggyback basin may (geologically) suddenly be cut off by emergence in the hinterland of a more resistant rock type, by climate changes, or by disruption of the transporting river system. These controls on sedimentation in the piggyback basin will drive local base level for the rear flank of the range up or down and will thus affect the geomorphology of the range as a whole. Similarly, base levels of a fold-and-thrust belt that are close enough to sea level will be locked in to sea-level changes, albeit with time lags (that may be measurable) related to the rate at which base-level fall is transmitted to the various ranges.

The inevitable result of these competing factors is that the first-order shape of the range is likely to be a very transient feature. The shape is influenced not only by local tectonic and surface processes but also by the behavior of the regional fluvial network. There is a tremendous and largely unexplored richness in the realization that the regional fluvial system may be an important communications network that couples together the topographic and geologic evolution of an entire orogen.

CONCLUSIONS

We have used a fault-related elastic strain field as an increment of deformation to build tectonic topographies over blind thrust faults of various geometries. The model relies on the assumption that a spatially heterogeneous state of strain induced by faulting can be made largely permanent and that the crust can be returned to a state of homogeneous pure shear prior to the next increment of deformation. The models and topographies explored here, particularly those that involve an isolated thrust fault, are relevant only for the initiation and early development of thrust-related ranges.

Given these caveats, the first-order shape of a mountain range emerging over a blind thrust fault is initially and fundamentally dictated by the dip and depth of the fault plane. Shallowly dipping thrust faults will generate a tectonic topography in which a steep and relatively narrow flank faces away from the vergence direction, and subsidence will be generated over the hanging wall. As the fault propagates toward the surface, this sense of asymmetry changes, so that ultimately the steep and narrow flank faces toward the vergence direction. Changes in the position of the fold hinge may be preserved in features of different ages across the range, so that the fold axis recorded by Holocene terraces may be rearward of the finite or cumulative fold axis.

Initially, in the absence of erosion, the topographic divide is coincident with the fold axis. As deformation and surface erosion increase, the first-order shape of the range reflects a continuing competition between tectonic and fluvial processes, with the latter being controlled largely by base levels on either side of the range. For a range that develops on a regional slope toward the foreland, where base level is lower on the foreland flank, the topographic divide will migrate toward the rear of the range. This divergence between the fold hinge and topographic divide creates a topographic asymmetry that is opposite to the structural asymmetry of the fold.

Base levels on the flanks of the range may be controlled by either the tectonics of the underlying thrust fault or by external conditions related to sediment deposition or erosion by the regional fluvial system. The first-order shape of the range is likely to be a very transient feature of the landscape in regions of active faulting, and will depend on factors such as proximity to global sea-level changes, the degree of fault connectivity, global and regional climate changes, and changes in the fluvial system geometry. In the end, the delivery of a tectonic topography to the surface and its modification by surface processes is not straightforward, and characteristics of the topography (e.g., concavity of fluvial networks, watershed measurements) will reflect more than simply the underlying tectonic geometry.

ACKNOWLEDGMENTS

We are grateful to Sanjeev Gupta for bringing to our attention the Mohand Anticline, to Alex Davis for sharing preliminary observations on the blind thrust and fold south of the Bogdan Shan. Jean Braun, Hugh Sinclair, and Sean Willett provided valuable and insightful reviews that we gratefully acknowledge. Much of this work was done while the first author was a Visiting Professor in the Institute of Geology at ETH in late 2002, and he is grateful to colleagues and friends there for providing a scientifically stimulating and open environment. This research was partially supported by National Science Foundation (NSF) award EAR-980-3484. Center for Earthquake Research and Information contribution no. 488.

REFERENCES CITED

Allmendinger, R.W., 1998, Inverse and forward numerical modeling of trishear fault-propagation folds: Tectonics, v. 17, p. 640–656, doi: 10.1029/98TC01907.

Bakun, W.H., King, G.C.P., and Cockerham, R.S., 1986, Seismic slip, aseismic slip, and the mechanics of repeating earthquakes on the Calaveras fault, California, *in* Das, S., Boatwright, J., and Scholz, C.H., eds., Earthquake source mechanics: Washington, D.C., American Geophysical Union Geophysical Monograph 37, p. 195–207.

Beroza, G.C., 1991, Near source modeling of the Loma Prieta earthquake: Evidence for heterogeneous slip and implications for earthquake hazards: Bulletin of the Seismological Society of America, v. 81, p. 1603–1621.

Bilham, R., and King, G., 1989, The morphology of strike-slip faults: Examples from the San Andreas fault, California: Journal of Geophysical Research, v. 94, p. 10,204–10,216.

Bishop, P., 1995, Drainage rearrangement by river capture, beheading and diversion: Progress in Physical Geography, v. 19, p. 469–493.

Bodin, P., and Horton, S., 2004, Source parameters and tectonic implications of aftershocks of the M_w 7.6 Bhuj earthquake of 26 January 2001: Bulletin of the Seismological Society of America, v. 94, p. 818–827.

Brocklehurst, S.H., and Whipple, K.X., 2002. Glacial Erosion and Relief Production in the Eastern Sierra Nevada, California: Geomorphology, v. 42, p. 1–24.

Champel, B., van der Beek, P., Mugnier, J.L., and Leturmy, P., 2002, Growth and lateral propagation of fault-related folds in the Siwaliks of western Nepal: Rates, mechanisms, and geomorphic signature: Journal of Geophysical Research, v. 107, doi: 10.1029/2001JB000578.

Dahlstrom, C.O.A., 1969, Balanced cross section: Canadian Journal of Earth Sciences, v. 6, p. 743–757.

Davis, A.M., Liu, J.G., and Brown, M.A.N., 2003, Drainage development related to blind thrust evolution along the Flaming Mountain, Turpan Basin, Xinjiang, NW China: Geological Remote Sensing Group Annual Conference and AGM: Geology Unveiled, [abs.], IEEE Symposium.

Densmore, A.L., Ellis, M.A., and Anderson, R.S., 1998, Landsliding and the evolution of normal fault-bounded mountains: Journal of Geophysical Research, v. 103, p. 15,203–15,219, doi: 10.1029/98JB00510.

Densmore, A.L., Dawers, N.D., Gupta, S., Guidon, R., and Goldin, T., 2004, Footwall topographic development during continental extension: Journal of Geophysical Research, v. 109, doi: 10.1029/2003JF000115.

Dreger, D.S., 1997, The large aftershocks of the Northridge earthquake and their relationship to mainshock slip and fault zone complexity: Bulletin of the Seismological Society of America, v. 87, p. 1259–1266.

Eaton, L.S., Morgan, B.A., Kochel, R.C., and Howard, A.D., 2003, Role of debris flows in long-term landscape denudation in the central Appalachians of Virginia: Geology, v. 31, p. 339–342, doi: 10.1130/0091-7613(2003)0312.0.CO;2.

Eisenstadt, G., and DePaor, D.G., 1987, Alternative model of thrust-fault propagation: Geology, v. 15, p. 630–633, doi: 10.1130/0091-7613(1987)152.0.CO;2.

Ekström, G., Stein, R.S., Eaton, J.P., and Eberhart-Phillips, D., 1992, Seismicity and geometry of a 110 km-long blind thrust fault. 1. The 1985 Kettleman Hills, California, earthquake: Journal of Geophysical Research, v. 97, p. 4843–4864.

Ellis, M.A., 1991, Tectonics and topography in regions of active extension: Eos (Transactions, American Geophysical Union), v. 72, p. 461.

Ellis, M.A., and Dunlap, W.J., 1988, Displacement variation along thrust faults: Implications for the development of large faults: Journal of Structural Geology, v. 10, p. 183–192, doi: 10.1016/0191-8141(88)90115-0.

Ellis, M.A., Densmore, A.L., and Anderson, R.S., 1999, Development of mountainous topography in the Basin Ranges, USA: Basin Research, v. 11, p. 21–41, doi: 10.1046/j.1365-2117.1999.00087.x.

Erslev, E.A., 1991, Trishear fault-propagation folding: Geology, v. 19, p. 617–620, doi: 10.1130/0091-7613(1991)0192.3.CO;2.

Gomberg, J., and Ellis, M.A., 1994, Topography and tectonics of the central New Madrid seismic zone: Results of numerical experiments using a three-dimensional boundary element program: Journal of Geophysical Research, v. 99, p. 20,299–20,310, doi: 10.1029/94JB00039.

Gretener, P.E., 1972, Thoughts on overthrust faulting in a layered sequence: Bulletin of Canadian Petroleum Geology, v. 20, p. 583–607.

Hovius, N., Stark, C.P., Tutton, M.A., and Abbot, L.D., 1998, Landslide-driven drainage network evolution in a pre-steady-state mountain belt: Finisterre Mountains, Papua New Guinea: Geology, v. 26, p. 1071–1074, doi: 10.1130/0091-7613(1998)0262.3.CO;2.

Hurtrez, J.E., Lucazeau, J., Lavé, J., and Avouac, J.P., 1999, Investigation of the relationships between basin morphology, tectonic uplift and denudation from the study of an active fold belt in the Siwalik Hills (central Nepal): Journal of Geophysical Research, v. 104, p. 12,779–12,796, doi: 10.1029/1998JB900098.

Kilb, D., Ellis, M., Gomberg, J., and Davis, S., 1997, On the origin of diverse aftershock mechanisms following the 1989 Loma Prieta earthquake: Geophysical Journal International, v. 128, p. 557–570.

King, G., and Ellis, M., 1990, The origin of large local uplift in extensional regions: Nature, v. 348, p. 689–693, doi: 10.1038/348689a0.

King, G.C.P., and Vita-Finzi, C., 1981, Active folding in the Algerian earthquake of 10 October 1980: Nature, v. 292, p. 22–26, doi: 10.1038/292022a0.

King, G.C.P., Stein, R.S., and Rundle, J.B., 1988, The growth of geological structures by repeated earthquakes, 1: Conceptual framework: Journal of Geophysical Research, v. 93, p. 13,307–13,318.

King, G.C.P., Stein, R.S., and Lin, J., 1994, Static stress changes and the triggering of earthquakes: Bulletin of the Seismological Society of America, v. 84, p. 935–953.

Kirby, E., and Whipple, K.X., 2001, Quantifying differential rock-uplift rates via stream profile analysis: Geology, v. 29, p. 415–418, doi: 10.1130/0091-7613(2001)0292.0.CO;2.

Kostrov, V.V., 1974, Seismic moment and energy of earthquakes, and seismic flow of rocks: Izvestiya Akademii Nauk, USSR Physical Solid Earth, v. 1, English Translation, p. 23–44.

Laubscher, H.P., 1965, Ein kinematisches modell der Jurafaltung: Eclogae Geologicae Helvetiae, v. 58, p. 231–318.

Lavé, J., and Avouac, J.P., 2000, Active folding of fluvial terraces across the Siwalik Hills, Himalayas of central Nepal: Journal of Geophysical Research, v. 105, p. 5735–5770, doi: 10.1029/1999JB900292.

Lavé, J., and Avouac, J.P., 2001, Fluvial incision and tectonic uplift across the Himalayas of central Nepal: Journal of Geophysical Research, v. 106, p. 26,561–26,591, doi: 10.1029/2001JB000359.

Lin, J., and Stein, R.S., 1989, Coseismic folding, earthquake recurrence, and the 1987 source mechanism at Whittier Narrows, Los Angeles Basin, California: Journal of Geophysical Research, v. 94, p. 9614–9632.

Maerten, L., Gillespie, P., and Pollard, D.D., 2002, Effects of local stress perturbation on secondary fault development: Journal of Structural Geology, v. 24, p. 145–153, doi: 10.1016/S0191-8141(01)00054-2.

Manighetti, I., King, G.C.P., Gaudemer, Y., Scholz, C.H., and Doubre, C., 2001, Slip accumulation and lateral propagation of active normal faults in Afar: Journal of Geophysical Research, v. 106, p. 13,667–13,696, doi: 10.1029/2000JB900471.

Manighetti, I., King, G., and Sammis, C.G., 2004, The role of off-fault damage in the evolution of normal faults: Earth and Planetary Science Letters, v. 217, p. 399–408, doi: 10.1016/S0012-821X(03)00601-0.

Matmon, A., Bierman, P.R., Larsen, J., Southworth, S., Pavich, M., and Caffee, M., 2003, Temporally and spatially uniform rates of erosion in the southern Appalachian Great Smoky Mountains: Geology, v. 31, p. 155–158.

Meigs, A., Brozovic, N., and Johnson, M.L., 1999, Steady, balanced rates of uplift and erosion of the Santa Monica Mountains, California: Basin Research, v. 11, p. 59–73, doi: 10.1046/j.1365-2117.1999.00083.x.

Mendez, A.J., and Anderson, J.G., 1991, The temporal and spatial evolution of the 19 September 1985 Michoacan earthquake as inferred from near-source ground-motion records: Bulletin of the Seismological Society of America, v. 81, p. 844–861.

Mendoza, C., and Hartzell, S.H., 1988, Aftershock patterns and mainshock faulting: Bulletin of the Seismological Society of America, v. 78, p. 1438–1449.

Merritts, D., and Ellis, M., 1994, Introduction to special section on tectonics and topography: Journal of Geophysical Research, v. 99, p. 12,135–12,141, doi: 10.1029/94JB00810.

Molnar, P., 1987, Inversion of profiles of uplift rates for the geometry of dip-slip faults at depth, with examples from the Alps and the Himalaya: Annales Geophysicae, v. 5, p. 663–670.

Mueller, K., Chen, Y., and Powell, L., 2002, Modern strain and structural architecture of the central Taiwanese orogen—Evidence of active backstepping in response to erosion: Eos (Transaction s, American Geophysical Union), Fall Meeting Supplement, v. 83, T61B-1279.

Oskin, M., and Burbank, D., 2002, Geomorphic evolution of steady-state in a glaciated mountain range: Kyrgyz Range, Western Tien Shan: Eos (Transactions, American Geophysical Union), v. 83, Fall meeting supplement, abs. T72B-10.

Powers, P.M., Lillie, R.J., and Yeats, R.S., 1998, Structure and shortening of the Kangra and Dehra Dun reentrants, Sub-Himalaya, India: Geological Society of America Bulletin, v. 110, p. 1010–1027.

Rao, D.P., 1986, An approach to the quantitative estimates of rates of recent movements: The Sub-Himalayan example: Bulletin of the Royal Society of New Zealand, v. 24, p. 507–517.

Schlische, R.W., Young, S.S., Ackermann, R.V., and Gupta, A., 1996, Geometry and scaling relations of a population of very small rift-related normal faults: Geology, v. 24, p. 683–686, doi: 10.1130/0091-7613(1996)0242.3.CO;2.

Seeber, L., and Gornitz, V., 1983, River profiles along the Himalayan arc as indicators of active tectonics: Tectonophysics, v. 92, p. 335–367, doi: 10.1016/0040-1951(83)90201-9.

Stein, R.S., and Ekström, G., 1992, Seismicity and geometry of a 110 km-long blind thrust fault, 2. Synthesis of the 1982–1985 earthquake sequence: Journal of Geophysical Research, v. 97, p. 4865–4883.

Stein, R., King, G.C.P., and Rundle, J.B., 1988, The growth of geological structures by repeated earthquakes. 2: Field examples of continental dip-slip faults: Journal of Geophysical Research, v. 93, p. 13,319–13,331.

Stein, R.S., King, G.C.P., and Lin, J., 1994, Stress triggering of earthquakes: Evidence for the 1994 M = 6.7 Northridge, California, shock: Annali di Geofisica, v. 37, p. 1799–1805.

Stock, J., and Dietrich, W.E., 2003, Valley incision by debris flows: Evidence of a topographic signature: Water Resources Research, v. 39, doi: 10.1029/2001WR001057.

Suppe, J., 1983, Geometry and kinematics of fault-bend folding: American Journal of Science, v. 283, p. 684–721.

Weissel, J.K., and Seidl, M.A., 1998, Inland propagation of erosional escarpments and river profile evolution across the southeast Australian passive continental margin, *in* Tinkler, K.J., and Wohl, E.E., eds., Rivers over rock: Fluvial processes in bedrock channels: Washington D.C., American Geophysical Union Geophysical Monograph 107, p. 189–206.

Wickham, J., 1995, Fault displacement-gradient folds and the structure at Lost Hills, California (U.S.A.): Journal of Structural Geology, v. 17, p. 1293–1302, doi: 10.1016/0191-8141(95)00029-D.

Yeats, R.S., and Lillie, R.J., 1991, Contemporary tectonics of the Himalayan frontal fault system: Folds, blind thrusts, and the 1905 Kangra earthquake: Journal of Structural Geology, v. 13, p. 215–225, doi: 10.1016/0191-8141(91)90068-T.

Yielding, G., Jackson, J.A., King, G.C.P., Sinvhal, H., Vita-Finzi, C., and Wood, R.M., 1981, Relations between surface deformation, fault geometry, seismicity, and rupture characteristics during the El Asnam (Algeria) earthquake of 10 October 1980: Earth and Planetary Science Letters, v. 56, p. 287–304, doi: 10.1016/0012-821X(81)90135-7.

Manuscript Accepted by the Society 23 June 2005

Printed in the USA

Geological Society of America
Special Paper 398
2006

Influence of erosion and deposition on deformation in fold belts

Guy Simpson
Geology Institute, Sonneggstrasse 5, ETH Zürich, Switzerland

ABSTRACT

This paper investigates whether surface processes such as erosion and deposition have any influence on folding of the upper crust at small to intermediate spatial scales (i.e., <200 km). This problem is studied using a simple mathematical model comprising a thin elastic-plastic beam overlying a viscous substrate. A critical nondimensional parameter that controls whether surface processes influence folding is the ratio of deformation to fluvial erosion characteristic time scales, denoted as R. When $R \ll 1$ (i.e., the rates of surface processes are slow relative to the rate of deformation), numerical results show that surface processes have a negligible influence on folding. Deformation in this regime is characterized by the formation of a train of low-amplitude, short-wavelength folds, which develop serially outward with time. When $R \gg 1$, a new folding mode is entered whereby erosion and deposition dramatically amplify the folding instability leading to localization of rock uplift and exhumation on a single, long-wavelength mega-antiform with relatively high topography. Thus, relatively rapid erosion and deposition at the scale of individual folds can significantly modify the nature, amplitude, and wavelength of folding and can lead to higher topography than in the absence of surface processes. Surface processes influence folding mainly through reducing the influence of gravity but also by modifying the strength distribution. Results indicate that the response of folding to surface processes is dynamic in origin and threshold-like in behavior.

Keywords: folding, deformation, erosion, deposition, surface processes, FEM, Finite Element Method, numerical modeling.

INTRODUCTION

It is now generally accepted that erosion and deposition can have a major influence on the mechanical (and thermal) evolution of the lithosphere at the scale of natural orogens (Koons, 1990; Beaumont et al., 1992; Willett, 1999; Avouac and Burov, 1996). Surface processes, by redistributing mass, modify the stress state of the lithosphere, which may induce both internal deformation and an overall isostatic adjustment (Beaumont et al., 2000). In addition, prolonged erosion in rapidly deforming regions can lead to advection of hot weak rock close to the surface, leading to positive feedback mechanisms including enhanced deformation, weakening, localization, and further exhumation (Zeitler et al., 2001; Koons et al., 2003). Regardless of which mechanism is dominant, both modeling (Beaumont et al., 2000) and field studies (e.g., Norris and Cooper, 1997; Pavlis et al., 1997) indicate that erosion tends to localize deformation and exhumation within relatively narrow belts.

The ability of erosion and deposition to modify deformation becomes less clear at scales <100 km, where the rock strength and rigidity is such that isostasy is no longer compensated locally. Nevertheless, some field observations indicate that erosion and/or deposition may also be important at scales on the order of several kilometers (Norris and Cooper, 1997; Pavlis et al., 1997). The critical factor at such scales is the local stress state and rheology of the deforming rocks (Simpson, 2004c). If stresses are relatively low, such that the upper crust behaves elastically, the rigidity is likely to be such that local erosion and deposition have

Simpson, G., 2006, Influence of erosion and deposition on deformation in fold belts, *in* Willett, S.D., Hovius, N., Brandon, M.T., and Fisher, D.M., eds., Tectonics, Climate, and Landscape Evolution: Geological Society of America Special Paper 398, Penrose Conference Series, p. 267–281, doi: 10.1130/2006.2398(16). For permission to copy, contact editing@geosociety.org.

little influence on deformation. In this case, the lithosphere is envisaged as a rigid block onto which surface processes etch the landscape. If, however, the crust is stressed to the failure limit (such as is envisaged in critical wedge models), stress changes due to erosion and/or deposition may drive the local crust into a stress state where plastic failure is satisfied. Deformation must then take place in order to reduce stresses and maintain a stable stress state (Chapple, 1978; Stockmal, 1983; Davis et al., 1983; Koons, 1994). In this case, the response length scale is controlled by the length scale of erosion-deposition rather than the characteristic length scale related to elastic flexure, the former being more than two orders of magnitude smaller than the latter. This point implies, in some cases, a strong direct coupling between erosion/deposition and the local deformation field, even at relatively small scale. This point also emphasizes the importance of understanding the stress state and rheological behavior of the crust when considering how strongly surface processes influence deformation.

Understanding the extent and nature of the coupling between erosion/deposition and deformation is important for addressing the long-standing problem of how climate change modifies the height of mountain ranges. If surface processes are envisaged to be dynamically uncoupled from deformation (as is assumed in most surface evolution modeling), enhanced erosion due to increased precipitation will lead to a decrease in mean elevation (e.g., Bonnet and Crave, 2003), although mountain peaks may actually rise due to isostatic rebound (England and Molnar, 1990). If, on the other hand, deformation is dynamically coupled to surface processes, enhanced erosion under some circumstances may localize deformation leading to actual growth of topography (Avouac and Burov, 1996; Zeitler et al., 2001). Once again, which of these two very different scenarios occurs in nature will depend on stress state, rheology, and the spatial scale of interest.

The purpose of this study is to investigate the strength and nature of coupling between surface processes and folding at small to intermediate spatial scales. This scale (<200 km) is small enough such that individual folds can be resolved and also small enough such that the rigidity due to elasticity must be considered finite. The approach taken here is to study a simple model coupling erosion and deposition on the surface with folding of an elastic-plastic layer, representing the upper crust, overlying a viscous layer. The response of this system to compression in the absence of surface processes is known: a series of buckle folds are initiated and develop serially outward from a small initial perturbation (Zhang et al., 1996) before being disrupted by faults at large strain (Dixon and Liu, 1992). This study addresses how erosion and deposition influence the spatial pattern, amplitude, and timing of this folding.

The mechanical part of the model presented is based on thin-beam assumptions where the original two-dimensional (2-D) problem of interest is reduced to one dimension (1-D) by simplifying the depth variation of deformation (Timoshenko and Woinowsky-Krieger, 1959). A thin-plate version of this model has also recently been developed to investigate three-dimensional (3-D) coupling between surface processes and deformation (Simpson, 2004a, 2004b, and 2004c). Note that a beam is simply a plate that is assumed to have infinite extent in one direction (here assumed the y direction) and to be loaded and supported by conditions independent of this direction. The main advantage of the thin-beam approach compared to the thin-plate is that it is simpler and therefore easier to extract physical insight. Elastic thin-plate or thin-beam models are mostly used within the geological literature to determine the response of the lithosphere to vertical loads (Jordan, 1981; Karner and Watts, 1983). Some studies have shown that in-plane (horizontal) loads acting on preexisting structures may generate significant additional vertical deflections (e.g., Lambeck, 1983) even when the applied stresses are well below the critical elastic buckling stress (van Balen et al., 1998; van Balen and Podladchikov, 1998). In this study, I consider the possibility that the applied in-plane loads actually generate buckling. This does not require excessive stresses, because a plastic rheology (in addition to elasticity) is considered. This type of continuous deformation is known as plastic buckling and is especially well known in the engineering literature (e.g., Bloom and Coffin, 2001; see also Zhang et al., 1996; Gerbault et al., 1999; Gerbault, 2000). Unlike plane-strain models, the thin-beam formulation doesn't permit development of discontinuous deformation (e.g., faulting). This limitation means that the model is applicable to deformation characteristic of fold belts (e.g., Zagros) at relatively small strains, before faulting becomes the dominant mode of shortening. Several studies have shown that viscoelastic folding may be important in nature (e.g., see Biot, 1961; Schmalholz and Podladchikov, 1999). This work neglects viscous deformation of the beam in order to keep the model simple and the number of free parameters to a minimum.

MODEL FORMULATION

The model developed here consists of a thin elastic-plastic beam restrained underneath by local viscous dashpots and subject to surface erosion and deposition (Fig. 1). A coordinate system is

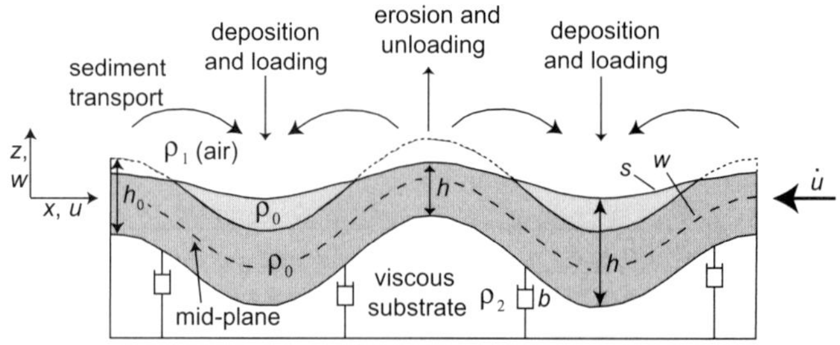

Figure 1. Schematic diagram of model consisting of an elastic-plastic layer (thin beam) overlying a viscous substrate consisting of local dashpots. A horizontal displacement is applied to one end of the beam, which causes folding. Erosion and deposition on the surface of the beam modify the distribution of vertical loads and the behavior of folding.

adopted such that x coincides with the mid-plane of the beam, whereas z is perpendicular to the beam and positive upward. The beam adopted is of Euler-Bernoulli type, which is derived assuming that (1) sections normal to the mid-plane remain normal during deformation (implying transverse shear deformation is neglected) and (2) strains normal to the mid-plane are neglected implying that direct stresses in the normal direction are small or of the same order as applied loads in this direction (Timoshenko and Woinowsky-Krieger, 1959). With these two assumptions, the state of deformation can be described by two displacements $u(x)$ and $w(x)$, the in-plane displacement of the mid-plane and the normal deflection of the mid-plane, respectively (Table 1). The force equilibrium equations computed by integration of the local equilibrium equations (e.g., see Zienkiewicz and Taylor, 2000) are

$$\frac{\partial^2 M_x}{\partial x^2} = \rho_1 g(s - h_0/2) - \rho_2 g w - \rho_0 g(h - h_0) - b\frac{\partial w}{\partial t} \qquad (1)$$

and

$$\frac{\partial P_x}{\partial x} = 0, \qquad (2)$$

where ρ_1 is the density of the overlying medium (air), ρ_2 is the density of the viscous material underlying the beam, b is the effective viscosity of the dashpots, ρ_0 is the density of eroded/deposited material, s is the surface topography, h_0 is the initial uniform beam thickness, h is the actual beam thickness ($h = h_0/2 + s - w$), M_x is the bending moment and P_x is the in-plane stress resultant. The first two terms on the right hand side of equation 1 account for loads created by replacement of material on either side of the beam as it deforms (i.e., hydrostatic restoring pressure), the third term accounts for loads due to erosion and deposition, and the fourth term is the restraining force created by the underlying viscous dashpots. Note that when there is no erosion and deposition, $s = w + h_0/2$, $h = h_0$ (i.e., the topography mimics the underlying deforming beam), and the load terms in 1 reduce to $(\rho_2 - \rho_1)gw$, as anticipated (e.g., see Turcotte and Schubert, 2002). The behavior introduced by the viscous dashpots is not strictly analogous to that of a true viscous continuum, since the dashpots are assumed to be spatially independent (i.e., they act locally), whereas in the continuum case the viscous response depends on the wavelength of the disturbance (see Biot, 1961).

TABLE 1. NOTATION

Symbol	Units	Definition
w	m	Beam deflection
u	m	In-plane displacement
s	m	Topographic elevation
q	$m^2 s^{-1}$	Fluid flux accumulated from rainfall
x	m	Spatial coordinate parallel to beam
z	m	Spatial coordinate perpendicular to beam
L	m	Horizontal dimension
t	s	Time
P_x	Pa m	In-plane stress resultant
M_x	Pa m^2	Bending moment
σ_x	Pa	In-plane stress (negative in compression)
σ_0	Pa	Plastic yield stress
ε_x	–	Total strain $(= \varepsilon_x^e + \varepsilon_x^p)$
$\varepsilon_x^e, \varepsilon_x^p$	–	Elastic and plastic strains
λ	–	Plastic multiplier
$\dot{u}$	m s^{-1}	Imposed deformation rate
h	m	Beam thickness (= $h_0/2 + s - w$)
h_0	m	Initial uniform beam thickness
v	–	Poisson's ratio
E	Pa	Young's Modulus
$\Delta\rho$	kg m^{-3}	Density difference
ρ_1	kg m^{-3}	Density of overlying medium (air)
ρ_2	kg m^{-3}	Density of underlying medium
ρ_0	kg m^{-3}	Density of material eroded/deposited
κ	$m^2 s^{-1}$	Erosional diffusivity
c	$(m^2 s^{-1})/(m^2 s^{-1})^n$	Fluvial erosion coefficient
n	–	Erosion exponent
α	m s^{-1}	Rainfall rate
g	m s^{-2}	Acceleration of gravity

This difference is reflected in the effective viscosity parameter b, which has dimensions Pa s m^{-1} and which should be interpreted as the standard dynamic viscosity divided by the characteristic wavelength of deformation in the overlying beam. Note that the effective substrate viscosity will also be influenced by the thickness of the weak layer if the décollement becomes thin relative to the thickness of the overlying layer (see Goff et al., 1996). The bending moment and in-plane stress resultant are defined as

$$M_x = -\int_{-h/2}^{h/2} z\sigma_x \, dz \tag{3}$$

and

$$P_x = \int_{-h/2}^{h/2} \sigma_x \, dz \quad . \tag{4}$$

Integration of stresses through the beam thickness is carried out numerically due to the nonlinear nature of the constitutive relation. The rheology of the beam is assumed to be elastic–pure plastic. The relationship between stress and elastic strain ε_x^e is given by

$$\sigma_x = \frac{E}{(1-v^2)}\varepsilon_x^e \tag{5}$$

where E is Young's modulus and v is Poisson's ratio. Plasticity is assumed to be governed by a 1-D von Mises–type yield criteria

$$F = \left|\sigma_x\right| - \sigma_0 , \tag{6}$$

where σ_0 is the yield stress. Plastic yielding occurs when $F > 0$ by an associative flow law

$$\delta\varepsilon_x^p = \lambda \frac{\partial F}{\partial \sigma_x} , \tag{7}$$

where λ is the plastic multiplier and $\delta\varepsilon_x^p$ is the plastic strain increment. Stresses of yielding points are required to remain on the yield surface, as specified by the consistency condition

$$\frac{\partial F}{\partial \sigma_x}\Delta\sigma_x = 0 . \tag{8}$$

Additionally, it is assumed that elastic and plastic strains are linearly additive:

$$\varepsilon_x = \varepsilon_x^e + \varepsilon_x^p , \tag{9}$$

where superscripts e and p refer to elastic and plastic strains, respectively. Finally, the kinematic relationship between total strain and the displacements u and w is

$$\varepsilon_x = \frac{\partial u}{\partial x} + \frac{1}{2}\left(\left(\frac{\partial u}{\partial x}\right)^2 + \left(\frac{\partial w}{\partial x}\right)^2\right) - z\frac{\partial^2 w}{\partial x^2} . \tag{10}$$

Note that finite deformation is included, whereas rotations and strains are assumed to be small. The fact that strains are small means that the linear addition of elastic and plastic strains (i.e., equation 9) is an acceptable approximation even though displacements are finite (i.e., equation 10).

Erosion and deposition influence the deforming beam by modifying the distribution of vertical loads (via right-hand side of equation 1). The surface evolution model considered here is a simple transport-limited model, which incorporates two main sediment transport processes: concentrative fluvial transport dominant in channels and dispersive transport dominant on hillslopes (Simpson and Schlunegger, 2003). The model is governed by two equations for two unknown functions, the topographic elevation, s, and the fluid flux, q, due to runoff from rainfall. In one-dimension, the equations can be written as

$$\frac{\partial s}{\partial t} = \frac{\partial}{\partial x}\left(\left(\kappa + cq^n\right)\frac{\partial s}{\partial x}\right) + \frac{\partial w}{\partial t} , \tag{11}$$

where κ is the hillslope erosion diffusivity, c is a fluvial erosion coefficient, n is a fluvial erosion power exponent, and q is the surface fluid flux due to rainfall, which can be determined by integrating the mass-balance condition

$$\frac{\partial q}{\partial x} = \alpha , \tag{12}$$

where α is the rainfall rate. The first bracketed term in equation 11 represents erosion/deposition, whereas the second term accounts for deformation of the underlying beam. Hillslope sediment transport tends to be dominant near drainage divides, whereas fluvial sediment transport becomes increasingly more important further away from ridge crests, due to the increase in accumulated fluid flux from rainfall. The combined topographic evolution occurs by nonlinear diffusion. Note that this simple surface evolution model differs from more common stream-power formulations (e.g., see Tucker and Whipple, 2002, and references therein). One advantage of the formulation presented here is that mass is automatically conserved within the model domain, and thus all material eroded off anticlines is deposited in synclines.

To nondimensionalize the governing equations, the following scales are adopted

$$\begin{gathered} x = \tilde{x}L;\ z = \tilde{z}L;\ w = \tilde{w}L \\ u = \tilde{u}L;\ h = \tilde{h}L;\ s = \tilde{s}L \\ t = \tilde{t}\frac{L}{\dot{u}};\ q = \tilde{q}\alpha L; \\ \sigma_x = \tilde{\sigma}_x E;\ F = \tilde{F}E; \\ M_x = \tilde{M}_x L^2 E;\ P_x = \tilde{P}_x LE. \end{gathered}$$

where tildes indicate nondimensional quantities, L is the initial length of the spatial domain, E is Young's Modulus, α is the mean rainfall rate, and $\dot{u}$ is the imposed boundary velocity. Substituting these relations into the governing equations and rearranging the terms leads to the following nondimensional equations

$$\frac{\partial^2 \tilde{M}_x}{\partial \tilde{x}^2} = \frac{\rho_1 g}{E}\left(\tilde{s}L - \frac{h_0}{2}\right) - \frac{\rho_2 gL}{E}\tilde{w} - \frac{\rho_0 g}{E}\left(\tilde{h}L - h_0\right) - \frac{b\dot{u}}{E}\frac{\partial \tilde{w}}{\partial \tilde{t}} , \tag{13}$$

$$\frac{\partial \tilde{P}_x}{\partial \tilde{x}} = 0 , \tag{14}$$

$$\tilde{M}_x = -\int_{-\tilde{h}/2}^{\tilde{h}/2} \tilde{z}\tilde{\sigma}_x d\tilde{z}\,, \quad (15)$$

$$\tilde{P}_x = \int_{-\tilde{h}/2}^{\tilde{h}/2} \tilde{\sigma}_x d\tilde{z}\,, \quad (16)$$

$$\tilde{\sigma}_x = \frac{1}{\left(1-\nu^2\right)}\tilde{\varepsilon}_x^e\,, \quad (17)$$

$$\tilde{F} = \left|\tilde{\sigma}_x\right| - \frac{\sigma_0}{E}\,, \quad (18)$$

$$\delta\tilde{\varepsilon}_x^p = \lambda\frac{\partial\tilde{F}}{\partial\tilde{\sigma}_x}\,, \quad (19)$$

$$\frac{\partial\tilde{F}}{\partial\tilde{\sigma}_x}\Delta\tilde{\sigma}_x = 0\,, \quad (20)$$

$$\tilde{\varepsilon}_x = \tilde{\varepsilon}_x^e + \tilde{\varepsilon}_x^p\,, \quad (21)$$

$$\tilde{\varepsilon}_x = \frac{\partial\tilde{u}}{\partial\tilde{x}} + \frac{1}{2}\left(\left(\frac{\partial\tilde{u}}{\partial\tilde{x}}\right)^2 + \left(\frac{\partial\tilde{w}}{\partial\tilde{x}}\right)^2\right) - \tilde{z}\frac{\partial^2\tilde{w}}{\partial\tilde{x}^2}\,, \quad (22)$$

$$\frac{\partial\tilde{s}}{\partial\tilde{t}} = \frac{\partial}{\partial\tilde{x}}\left(\left(\frac{\kappa}{\dot{u}L} + \frac{c\alpha^n L^{n-1}}{\dot{u}}\tilde{q}^n\right)\frac{\partial\tilde{s}}{\partial\tilde{x}}\right) + \frac{\partial\tilde{w}}{\partial\tilde{t}}\,, \quad (23)$$

and

$$\frac{\partial\tilde{q}}{\partial\tilde{x}} = 1\,. \quad (24)$$

The most important nondimensional parameter relevant to this study is the ratio of the deformation to fluvial erosion characteristic time scales appearing in equation 23, i.e.,

$$R = \frac{c\alpha^n L^{n-1}}{\dot{u}} = \frac{\tau_d}{\tau_e}\,, \quad (25)$$

where the deformation time scale is $\tau_d = L/\dot{u}$ and the fluvial erosion time scale is $\tau_e = L^{2-n}/(c\alpha^n)$. Bear in mind that a process with a large characteristic time scale is one that takes place slowly. Thus, a large value of R corresponds to a situation where the rates of fluvial processes are fast relative to the rate of deformation, whereas the reverse is true for a small value of R. The other erosion parameter appearing in equation 23 [i.e., $\kappa/(\dot{u}L)$, the ratio of the deformation to hillslope erosion time scales] is generally small (because the hillslope diffusivity is so small) and is not important for deformation.

INSTABILITY ANALYSIS FOR INFLUENCE OF EROSION/DEPOSITION ON ELASTIC BEAM

A linear instability analysis (see Biot, 1961) of a simplified model helps demonstrate the potential influence of surface processes on folding. Consider a purely elastic beam of nondimensional unit length lying on an inviscid (very weak) substrate (i.e., $b\dot{u}/E = 0$ in equation 13) that is pinned at both ends and subject to an applied compressive horizontal force $\tilde{P}_x$. If the applied force is sufficiently large, the beam will become unstable and buckle into a series of folds. We wish to determine how surface processes such as erosion and deposition influence the critical buckling force and the wavelength of buckling.

The equation for the deflection of the beam is obtained by adding the horizontal load $\tilde{P}_x\partial^2\tilde{w}/\partial\tilde{x}^2$ resulting from the applied force $\tilde{P}_x$ to the left-hand side of equation 13 and then using relations 15, 16, 17, and 22. Assuming the normalized beam thickness $\tilde{h}$ is constant (i.e., $\tilde{h} = h_0/L$) for integration (see equation 15) and dropping quadratic terms in 22 (to eliminate nonlinearities) yields

$$\tilde{D}\frac{\partial^4\tilde{w}}{\partial\tilde{x}^4} + \tilde{P}_x\frac{\partial^2\tilde{w}}{\partial\tilde{x}^2} + \frac{\rho_2 gL}{E}\tilde{w} - \frac{\rho_1 g}{E}\left(\tilde{s}L - \frac{h_0}{2}\right) + \frac{\rho_0 g}{E}(\tilde{h}L - h_0) = 0\,, \quad (26)$$

where $\tilde{D}$ is the dimensionless flexural rigidity defined as

$$\tilde{D} = \frac{\tilde{h}^3}{12(1-\nu^2)}\,. \quad (27)$$

The terms ρ_0, ρ_1, and ρ_2 are the densities of the beam material eroded and deposited, the overlying material (air), and the underlying inviscid substrate, respectively ($\rho_2 > \rho_1 > \rho_0$). Two simple limiting cases for the topographic surface are considered. In the first case, it is assumed that there are no surface processes, so that the surface mimics the deformation of the underlying beam (termed "no redistribution" by Biot and Odé, 1965). Substituting $\tilde{s} = \tilde{w} + h_0/(2L)$ and $\tilde{h} = h_0/(2L) + \tilde{s} - \tilde{w}$ into equation 26 yields

$$\tilde{D}\frac{\partial^4\tilde{w}}{\partial\tilde{x}^4} + \tilde{P}_x\frac{\partial^2\tilde{w}}{\partial\tilde{x}^2} + \frac{(\rho_2-\rho_1)gL}{E}\tilde{w} = 0\,. \quad (28)$$

In the second case, it is assumed that the surface is maintained at its initial horizontal level, implying that uplifted material is efficiently eroded from anticlines and deposited in adjacent synclines (similar to the "perfect redistribution" case studied by Biot and Odé, 1965). Substituting $\tilde{s} = h_0/(2L)$ and $\tilde{h} = h_0/(2L) + \tilde{s} - \tilde{w}$ into equation 26 yields

$$\tilde{D}\frac{\partial^4\tilde{w}}{\partial\tilde{x}^4} + \tilde{P}\frac{\partial^2\tilde{w}}{\partial\tilde{x}^2} + \frac{(\rho_2-\rho_0)gL}{E}\tilde{w} = 0\,. \quad (29)$$

Note that equations 28 and 29 only differ in the nondimensional gravity term, which is of the form $\tilde{w}\Delta\rho gL/E$. This term is significantly smaller in the case where erosion and deposition are considered, since ρ_1 is the density of air, whereas ρ_0 is the density of the material eroded/deposited. Thus, one can already anticipate that surface processes should modify the buckling behavior of the elastic beam (see also Biot and Odé, 1965).

The general solution of equations 28 and 29, consistent with $\tilde{w}(0)=0$ and $\tilde{w}(1)=0$, is

$$\tilde{w}(\tilde{x}) = c_1 \sin(\tilde{x}k) + c_2 \sin\left(\frac{\tilde{x}}{k}\sqrt{\frac{\Delta\rho g L}{\tilde{D}E}}\right), \tag{30}$$

where

$$k = \frac{\sqrt{-2\tilde{D}e(\tilde{P}_x E + \sqrt{\tilde{P}_x^2 E^2 - 4\tilde{D}E\Delta\rho g L})}}{2\tilde{D}E}$$

and c_1 and c_2 are arbitrary constants. This solution satisfies the original differential equations when

$$\tilde{P}_x = \tilde{P}_c = m^2\pi^2\tilde{D} + \frac{\Delta\rho g L}{m^2\pi^2 E}, \tag{31}$$

where m (= 1,2,3,...) is the number of sinusoidal buckle folds in the model domain of unit length. This equation defines a series applied horizontal forces for which nonzero values for $\tilde{w}$ exist. The value of m for which $\tilde{P}_e$ is smallest is given by

$$m_c = \frac{(\Delta\rho g L\tilde{D}^3 E^3)^{1/4}}{\pi\tilde{D}E}. \tag{32}$$

If $\tilde{P}_x$ is smaller than the critical value $\tilde{P}_c$, the beam will not deflect in response to the applied horizontal load, whereas if $\tilde{P}_x$ is equal to $\tilde{P}_c$, the beam buckles. The expression for the critical buckling force, $\tilde{P}_c$, contains contributions from two terms. The first term is due to the rigidity of the beam and is the classical Euler buckling load. In the simple model considered, this term is not influenced by erosion and deposition. However, variations in the beam thickness due to surface processes will actually modify $\tilde{D}$ and may eventually become important (see next section). The second term is due to gravity, which is directly influenced by erosion/deposition by modifying $\Delta\rho$. The effect of erosion and deposition on deformation of the elastic beam (Fig. 2) is to reduce both the critical buckling force, $\tilde{P}_c$, and the number of folds per unit length (i.e., increase in fold wavelength) by ~50% relative to the case when no erosion or deposition occurs. In synclines, light rocks are depressed to replace relatively dense underlying rocks, whereas with anticlines, the opposite occurs. In both cases, the folding creates forces that attempt to restore the surface to its original flat configuration. Deposition in synclines counteracts the upward-directed restoring force by adding an additional downward-directed load. Similarly, erosion from anticlines counteracts the downward-directed restoring force by removing a downward-directed load. Together these surface effects tend to amplify the folding instability.

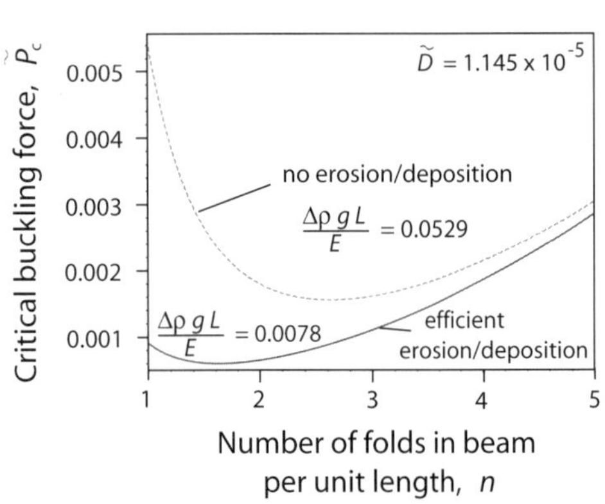

Figure 2. Influence of erosion and deposition on the buckling of an elastic beam over an inviscid substrate (calculated from equation 31). The effect of efficient erosion and deposition is to reduce the influence of gravity on folding, which reduces both the critical buckling force and the number of folds per unit length (i.e., increase in fold wavelength).

Although this example demonstrates the general way in which erosion and deposition influence the initial folding instability, the model is too simple to be applied further to nature. The greatest deficiency of the model is the assumed elastic behavior, which means that exceedingly large compressive stresses are required, even including erosion and deposition, in order to initiate buckling (see Fig. 2). In reality, the beam is likely to deform plastically before the elastic buckling stress is achieved. This possibility in considered is the next section.

INFLUENCE OF EROSION AND DEPOSITION ON FOLDING OF AN ELASTIC-PLASTIC BEAM

The aim of this section is to investigate how erosion and deposition modify folding of elastic-plastic material beyond the initial buckling instability. However, in order to demonstrate the importance of erosion and deposition, we begin by illustrating the nature of deformation in the absence of surface processes. The equations investigated are 13, 14, 23, and 24, in conjunction with the constitutive (i.e., equations 17, 18, 19, 20) and kinematic relations (i.e., equations 21 and 22) and the definitions for $\tilde{M}_x$ and $\tilde{P}_x$ (equations 15 and 16). This system of equations is solved for the four unknowns w, u, s, and q, which are subject to the following boundary conditions: (1) simple support conditions (i.e., $\tilde{w}=0$) at $\tilde{x}=0$ and $\tilde{x}=1$, (2) imposed displacement for $\tilde{u}$ at $\tilde{x}=0$ and $\tilde{x}=1$ (i.e., boundary at $\tilde{x}=0$ is fixed, boundary at $\tilde{x}=1$ is moved at a constant rate $\dot{u}$ toward the left), (3) fixed elevation for $\tilde{s}$ at $\tilde{x}=0$ and $\tilde{x}=1$, and (4) $\tilde{q}=0$ at ridge tops (i.e., where $\partial\tilde{s}/\partial\tilde{x}=0$ and $\partial^2\tilde{s}/\partial\tilde{x}^2<0$). Initial conditions include an initial localized deflection on $\tilde{w}$ with an amplitude 0.001 h_0/L located at $\tilde{x}$ = 1/2 and $\tilde{s}=\tilde{w}+h_0/(2L)$. The equations were solved numerically by the finite element method using modified Newton-Raphson iterations.

For ease of illustration, topography in the following figures is superimposed on the mid-plate deflection. The difference between the two curves then represents the amount of material eroded/deposited. Note however, that the topography is actually located a distance $h_0/(2L)$ above the mid-plate deflection.

Nondimensional parameters used in calculations are listed in Table 2. Note that the ratio L/h_0 is only 20, which is generally considered to fall at the limit for applicability of thin-beam theory (where L/h_0 is more typically ~100).

TABLE 2. NONDIMENSIONAL PARAMETER VALUES USED IN CALCULATIONS

Parameter	Value
$\frac{\rho_1 g L}{E}$	1.96×10^{-5}
$\frac{\rho_0 g L}{E}$	4.51×10^{-2}
$\frac{\rho_2 g L}{E}$	5.29×10^{-2}
$\frac{1}{(1-\nu^2)}$	1.0989
$\frac{h_0}{L}$	0.05
$\frac{b\dot{u}}{E}$	3.17–31.70
$\frac{\sigma_0}{E}$	0.001
n	2
$\frac{\kappa}{\dot{u}L}$	2.5×10^{-8}–2.5×10^{-7}
$R = \frac{c\alpha^n L^{n-1}}{\dot{u}}$	0.0032–3170

Deformation in the Absence of Surface Processes

Deformation in the absence of erosion and deposition is characterized by distributed folding (Fig. 3). Folding commences when compressive stresses reach the plastic yield stress, which first occurs at the site of the initial perturbation located in the center of the model domain. This results in the formation of a central anticline flanked on either side by two synclines. These structures continue to grow until approximately ~2% of bulk shortening is accommodated before deformation is transferred externally to generate a new set of folds. The mechanism responsible for this shift in the location of deformation is related to creation of plastic fold hinges, which concentrates deformation in hinge regions while leaving fold limbs to rotate passively. Subsequent deformation tends to be divided into two distinct zones. External zones are dominated by rapid vertical

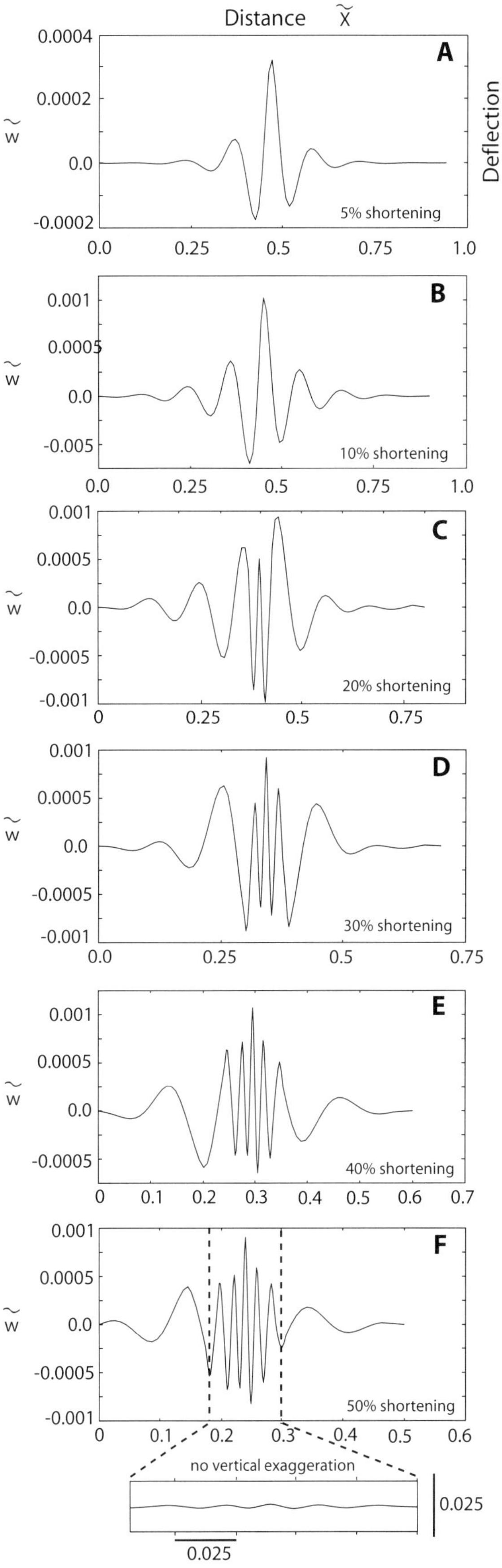

Figure 3. Folding of a compressed elastic-plastic beam overlying a viscous substrate in the absence of surface erosion and deposition. The beam contained a localized initial deflection (imperfection) with a normalized amplitude of 5×10^{-5} at $\tilde{x} = 0.5$. Note that the vertical scale has been exaggerated by varying amounts in order to optimize visualization of the folds. The bottom diagram shows a series of folds with no vertical exaggeration.

motion related to active growth of new buckle folds, whereas the internal zone involves relatively minor vertical motion but continues to accommodate horizontal shortening as folds are flattened between plastic fold hinges. The result of this deformation partitioning is a progressive decrease in fold wavelength from external to internal zones.

Folding in the absence of surface processes is controlled by five dimensionless parameters $(\rho_2 - \rho_1)\, gL/E$, σ_0, E, h_0/L, and $1/(1 - \nu^2)$. The parameter $(\rho_2 - \rho_1)\, gL/E$, resulting from the restoring forces created by displacing material on either side of the beam as it folds, has a strong influence on the fold wavelength, with large values favoring shorter wavelength folding. The ratio h_0/L is the normalized beam thickness, with large values favoring long fold wavelengths. The parameter $b\dot{u}/E$ is the ratio of the characteristic time scale for deformation of the viscous substrate (= bL/E) to the characteristic time scale for beam deformation (= $L/\dot{u}$). Increasing this ratio tends to decrease fold growth rates, while the fold wavelength is decreased and begins to show a dependency on the applied load (see Biot, 1961; Zuber and Parmentier, 1996). The ratio σ_0/E is an indication of how close the behavior of the beam corresponds to the elastic limiting case. Large values of σ_0/E (more elastic behavior) favor relatively large fold growth rates and long fold wavelengths. Note that the fold wavelength for plastic buckling is significantly reduced relative to elastic buckling, due to lower buckling stresses (Johnson, 1980; Martinod and Davy, 1992). Had the simulations presented in Figure 3 been performed with a purely elastic beam, only one buckle fold would have formed at an unrealistic normalized critical stress of ~0.03580 (as determined from equations 31 and 32). This illustrates the potential importance of plastic buckling (as opposed to elastic buckling) in explaining natural fold belts (see Johnson, 1980; Martinod and Davy, 1992; Gerbault et al., 1999).

Influence of Surface Processes on Folding

The influence of surface processes on folding depends on the rate of surface processes relative to the rate of deformation. To demonstrate this, attention is focused solely on the parameter R, defined in equation 25 as the ratio of the beam deformation to fluvial erosion characteristic time scales. This parameter is singled out mainly because it is anticipated to show a large range of variation from region to region (given that it will vary depending on climate and rate of deformation), with small values reflecting deformation-dominated conditions and large values reflecting erosion-dominated conditions. Although not investigated in detail, two other parameters are also potentially relevant. The first is the ratio of the viscous deformation to beam deformation characteristic time scales ($b\dot{u}/E$), the second is the ratio of the beam deformation to hillslope erosion characteristic time scales $[\kappa/(\dot{u}L)]$. Within this study, both of these parameters have relatively small values (see Table 2) and thus do not significantly influence results. However, one could imagine situations (e.g., deformation is very rapid or the effective viscosity of the viscous substrate is very high) where $b\dot{u}/E$ could become large relative to unity (meaning that the viscous substrate becomes important in influencing deformation in the beam).

Small Rate of Fluvial Surface Processes Relative to Rate of Deformation

Folding is not significantly influenced by surface processes when the rates of fluvial erosion and deposition are slow or of the same order as the rate of imposed deformation (i.e., relatively small R values; Fig. 4). During initial fold growth, rock uplift rates are so high relative to erosion rates that topography within the internal zone is advected vertically with the underlying deforming crust almost without modification by surface processes (Fig. 5). Only when activity on internal folds wanes (i.e., due to creation of plastic fold hinges) does the rock uplift rate decrease sufficiently to enable erosion to remove mass from anticlinal highs and deposit sediment in basins, leading to decay of the surface topography. Although this mass redistribution creates a load on the underlying crust, this load is imposed after the active buckling stage during the period when folding is largely passive. Thus, the load created by erosion and deposition does not participate dynamically to any folding instability. Moreover, since the load is created after active folding, when stresses have been somewhat relieved, the crust where the load is imposed is largely elastic (with the exception of plastic fold hinges) and is relatively rigid. Together, these features combine to ensure that erosion and deposition have a negligible influence on folding (compare Fig. 3 with Fig. 4, and Fig. 5A with Fig. 5B).

Although the case when the rates of fluvial surface processes are small relative to rate of deformation is of little interest for studying the influence of surface processes on folding, some features are worthy of additional comment. One interesting point is that although the surface topography closely mimics the underlying structure within the relatively young external structures, this relationship breaks down progressively toward the internal zone, where the age of the structures is such that the surface topography reflects the time- and space-integrated deformation rather than deformation on individual structures (Fig. 4). Even in the youngest structures, one typically observes an offset between the structural divide and the topographic (drainage) divide, with the topographic divide tending to be displaced toward the side of the structure that has the highest local base level. This observation

Figure 4. Folding when the rates of surface processes are slow (or of the same order) compared to the rate of imposed deformation (i.e., R = 3.171, see equation 25). In this case, surface processes have no influence on folding. Note that in the youngest and outermost (external) structures, topography closely mimics the deflecting beam, whereas in the older, internal structures, erosion and deposition is so advanced that the topography bears little or no relation to the underlying structure. Note also the slight offset between topographic and structural divides in external structures, with topographic divides displaced toward the side of the structure that has the highest base level.

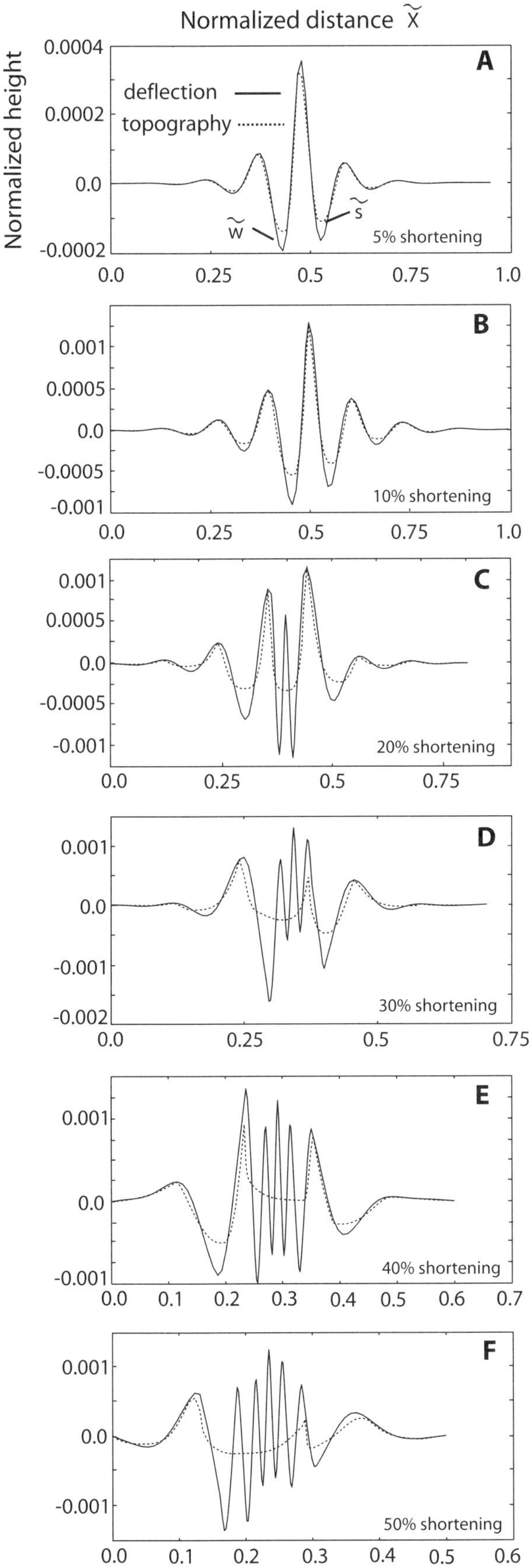

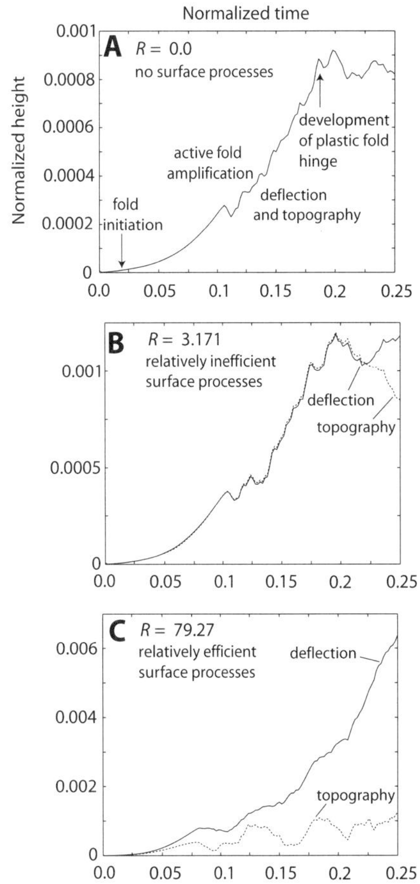

Figure 5. Variation in normalized deflection and topography through nondimensional time for a single anticline structure (taken from near the center of the model domain) using three different simulations considering (A) no surface processes, (B) relatively inefficient surface processes (i.e., R = 3.171), and (C) relatively efficient surface processes (i.e., R = 79.27). Note that the vertical scale is different in each case. When surface processes are absent or slow relative to the rate of imposed deformation, the topography is advected with the underlying structure during the active amplification phase of buckling. Only after active fold growth wanes (due to the creation of plastic fold hinges) do rock uplift rates drop enough to enable erosion to decrease the topography on anticlines (B). In the case where surface processes are relatively efficient, erosion takes place even during the most rapid phase of fold growth. This causes the amplitude of folding to be enhanced. Note that the topography when surface processes are inefficient is characterized by growth followed by decay, whereas the topography actually grows when surface processes are relatively efficient.

indicates the importance of local base level in controlling the drainage divide on actively growing anticlines.

Large Rate of Fluvial Surface Processes Relative to Rate of Deformation

As the rates of fluvial surface processes are increased relative to the rate of deformation (i.e., R is increased), anticlines can be eroded and synclines can be infilled during the rapid active buckling phase of deformation (Fig. 5C). The resulting mass redistribution has three major effects on subsequent folding. First, it amplifies the initial buckling instability by reducing the influence of gravity (see instability analysis section), which tends to enhance further deformation leading to increased rates of rock uplift and subsidence (Figs. 6 and 5C). Second, it increases the wavelength of the buckling instability (also by reducing the influence of gravity), such that initially formed structures become refolded with a longer wavelength (Fig. 6E). Third, the mass redistribution modifies the integrated strength distribution of the deforming layer by reducing the strength (or beam thickness) of exhuming structures and increasing the strength (or beam thickness) of infilling structures. The enhancement of deformation leads to a major reorganization in the structure of surface topography and in the distribution of surface mass transport. Topography formed before the onset of fold-amplification, which is no longer related to active deformation, gets progressively destroyed by erosion and deposition relative to topography related to active localized deformation. This reorganization of topography increases the length scale of mass distribution and further amplifies deformation, since mass can be more efficiently moved from sites undergoing rock uplift to sites undergoing subsidence.

The initial response of the surface topography to folding in the presence of relatively rapid fluvial surface processes is decayed relative to the case when surface processes had been less efficient. However, the enhancement of rock uplift rates induced by mass redistribution eventually increases the topographic heights of anticlines and the depth of basins relative to the case when no surface processes are considered (Fig. 7). This effect tends to be greatest when surface processes are rapid enough to enhance deformation but not fast enough to keep pace with rock uplift rates once localization has been initiated (Fig. 8). Within this intermediate regime, the topographic relief is typically a factor of 2–10 times greater than when surface processes are absent. As the relative rate of fluvial surface processes is increased further, the topographic relief is reduced, because erosion rates can keep pace with the rapid rates of rock uplift experienced during localized deformation (Fig. 8).

The close correspondence observed between topography and underlying structure observed when surface processes are slow relative to deformation (low R) tends to decrease as rates of erosion and deposition are increased (e.g., see Fig. 6C). This difference is due mainly to the tendency of efficient surface processes to rapidly decay the topography above inactive structures while only preserving topography related to the most active structures.

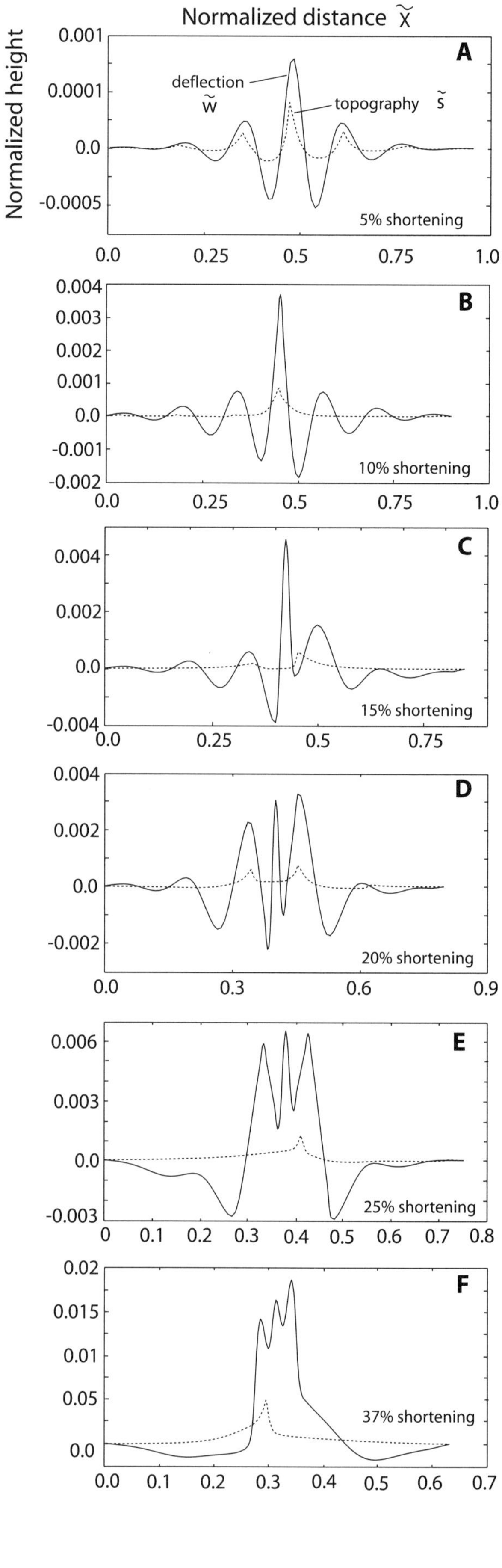

Figure 6. Folding when the rates of surface processes are fast compared to the rate of imposed deformation (i.e., $R = 79.27$, see equation 25). Note the marked difference in the behavior of folding compared to the case when surface processes are relatively slow or absent (i.e., Figs. 3 and 4).

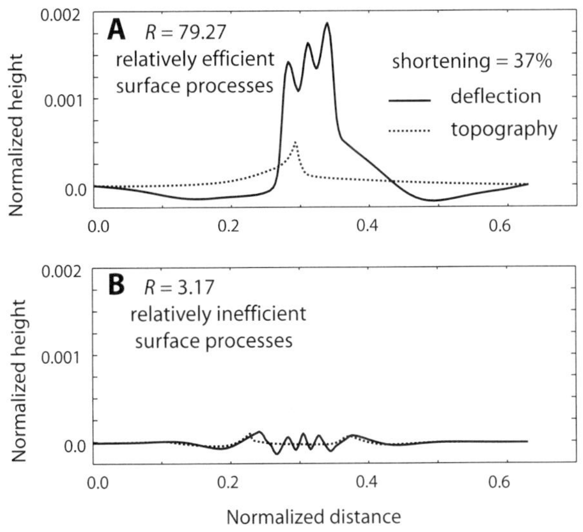

Figure 7. Comparison of deflection and topography resulting from deformation taking place in the presence of (A) relatively efficient surface processes and (B) relatively inefficient surface processes. Both A and B are plotted at the same scale and the same amount of shortening (= 37%). Note that erosion and deposition in A (relative to B) have drastically amplified the deflection and have even led to an increase in maximum elevation due to enhanced rock uplift rates.

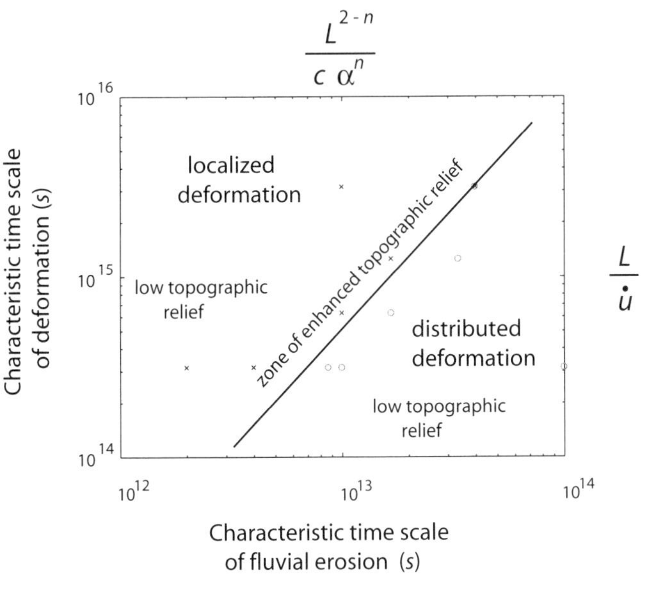

Figure 8. Phase diagram showing the approximate position of the boundary separating distributed deformation (where surface processes are insignificant with respect to influencing deformation) and localized folding (where surface processes strongly influence folding). The ratio of the characteristic time scales plotted on each axis form an important nondimensional parameter *R*, discussed in the text (see equation 25). The circles and crosses represent results from simulations that showed distributed and localized deformation, respectively. Topography is the highest (enhanced by a factor of 2–10) in an intermediate window where erosion and deposition are efficient enough to amplify folding but inefficient enough to be out-paced by rapid rates of rock uplift and subsidence.

In this sense, the surface topography in the face of rapid erosion and deposition more closely mimics the instantaneous rock uplift rate rather than the total finite strain, which is the case when no erosion and deposition take place. The discrepancy between topography and structure is even more accentuated by the tendency of erosion and deposition to increase the wavelength of deformation during the localization phase. For example, an early formed series of anticlines and synclines is often refolded later at a longer wavelength. In this case, the surface topography will not record the early short wavelength structure but will only preserve the latest long-wavelength deformation (e.g., see Fig. 6E).

DISCUSSION AND IMPLICATIONS

The purpose of this paper has been to present a simplified mathematical model to investigate how surface processes influence folding of the upper crust at small to intermediate spatial scales.

Timing of Erosion and Deposition Relative to Timing of Deformation Instability

A critical factor controlling the importance of surface processes in influencing folding is related to the rate of erosion and deposition relative to the rate of rock uplift during the active buckling phase of folding. An important property of folding is that individual structures have a finite lifetime and that rock uplift rates vary strongly throughout this lifetime (Fig. 5). If surface processes are rapid enough, such that material can be eroded as anticlines are actively amplified (i.e., large *R*), the load created by mass redistribution can participate in the dynamic buckling instability, causing it to be amplified. If, on the other hand, the rate of surface processes are relatively slow (i.e., low *R*), significant mass redistribution only occurs after active fold growth ceases, when rock uplift rates have decreased sufficiently to enable erosion to remove mass from anticlinal highs and deposit sediment in basins, leading to a decay of the surface topography. However, in this latter case, the load created by mass redistribution does not contribute to the dynamic folding instability, since it is imposed after the active buckling stage during the period when folding is largely passive. Thus, a major conclusion of this paper is that erosion and deposition have an important influence on deformation if they occur contemporaneously with the development of the folding instability.

Relative Roles of Gravity and Weakening in Influencing Folding

Erosion and deposition influence deformation of the elastic-plastic beam by two main mechanisms. The first is by modifying the spatial distribution of gravitational loads [i.e., mainly due to the term $\rho_0 g(h - h_0)$ in equation 1] and the second is by changing the beam thickness and therefore the integrated strength (i.e., equations 3 and 4). In the modeling carried out here, both mechanisms lead to positive feedback, which amplifies deformation. Initially, the dominating mechanism is that caused by modifying the gravitational stresses. However, once long-wavelength fold amplification has been initiated, the thinning (and therefore weakening) of the crust under the mega-antiform while the adjacent basins are thickened (and strengthened) tends to further amplify deformation (see Zeitler et al., 2001; Finlayson et al., 2002). Note that this may not necessarily be true in nature if the mantle contains significant rigidity, in which case thinning of the upper crust may even cause strengthening (e.g., see Braun and Beaumont, 1989). One could argue that rheological effects related to variations in the beam thickness would be absent if deformation was slow enough to maintain horizontal isotherms and therefore a constant elastic thickness during rock uplift and exhumation. However, simulations here reveal that the behavior remains unchanged when variations in strength (caused by variations in thickness) are neglected (see dashed curve in Fig. 9). This implies that erosion and deposition exert the most influence on folding through modifying gravitational stresses and not through weakening.

It is important to note that the main response of the surface topography to erosion and deposition in the models investigated is not related to a static (i.e., isostatic) readjustment of the lithosphere to erosion and deposition (e.g., see England and Molnar, 1990; Masek et al., 1994) but to dynamic amplification of the background tectonic deformation. Had the response been purely isostatic in origin, erosion would have generated a small component of rock uplift due to unloading, but the overall effect would have been a lowering of mean elevation. Instead, erosion, through modifying the load, amplifies background deformation, which generates a rock uplift far exceeding the isostatic component. Thus, it is entirely possible that aggressive erosion could significantly increase the mean elevation (as indicated by the large-scale numerical results of Avouac and Burov, 1996), especially if erosion is spatially localized (i.e., in narrow river systems), as is often the case.

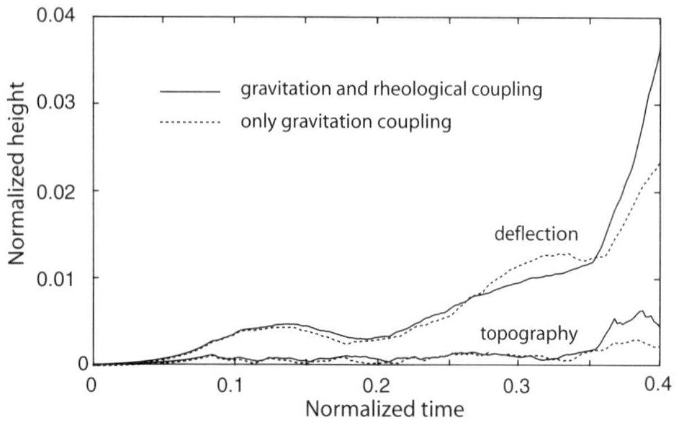

Figure 9. Normalized maximum deflection and maximum topography plotted as a function of nondimensional time for the case when erosion and deposition are relatively efficient (i.e., $R = 79.27$). Note that the rate of rock uplift (change in deflection with time) becomes increasingly more rapid with time, which eventually results in an increase in maximum topography. The difference between the deflection and topography curves represents the amount of exhumation which has taken place. A simulation carried out where the variation in strength resulting from variations in beam thickness was deactivated (dashed curved) shows similar behavior, indicating that the main factor causing the amplified folding is related to gravity and not weakening.

Importance of Mechanical Behavior

Recognition that the crust contains significant elastic rigidity has created uncertainty regarding whether erosion and deposition can influence deformation at small to intermediate spatial scales (e.g., see Gilchrist et al., 1994; Montgomery, 1994). On the one hand, the crust is viewed as being elastic and relatively rigid, in which case one anticipates very little influence of surface processes on deformation, especially at relatively small scale (e.g., Small and Anderson, 1998). On the other hand, the crust is envisaged to be always on the verge of failure (i.e., has no elastic rigidity), in which case even minor erosion and deposition on the surface is expected to generate internal deformation (e.g., Dahlen, 1984) to maintain local isostasy. It is important to emphasize that the main difference between these two end-member cases is the mechanical behavior that depends partially on the magnitude of stresses. When stresses are below the fracture strength, the response will be elastic (neglecting viscous creep), whereas the response will be plastic if stresses equal or exceed the fracture strength. Therefore, neither the elastic nor the plastic model will always be applicable. Rather, it is necessary to include the possibility for both responses and let the model choose which is currently applicable based on calculated stress levels. Thus, uncertainty regarding the influence of surface processes in deformation stems largely from the use of oversimplified rheologies in models, which essentially eliminates alternative modes of deformation. Elastic flexure (e.g., Turcotte and Schubert, 2002) and critical wedges (e.g., Dahlen, 1984), while providing us with insight, are two examples of models that suffer from such rheological oversimplification.

Some caution must be exercised when using inferred elastic thicknesses to argue about how the upper crust responds to erosion and deposition. This is because estimates of elastic thickness reveal nothing about the depth where the strength resides (e.g., see Watts, 2001). In the simple model considered within this study, all strength resides in the upper elastic-plastic layer. However, in reality, it is likely that the rheological profile for continental crust consists of two major elastic-plastic plates, one related to the upper crust and a second related to the upper mantle (e.g., Brace and Kohlstedt, 1980; however, see also Maggi et al., 2000). Most strength resides in the deeper upper mantle and not in the upper crust. Moreover, the typical means that we use to estimate the elastic thickness (e.g., wavelength of bending at trench) probably reflects the stronger and deeper of these two beams. Thus, elastic thickness estimates may bear little relationship to the elastic strength of the upper crust. It is therefore possible to severely overestimate the strength of the upper crust, and therefore underestimate the influence of surface processes on deformation in cases where significant strength resides in the upper mantle.

Implications for Natural Orogens

A prediction based on the modeling carried out within this study is that the behavior of folding may be markedly different depending on the relative rates of surface processes and deformation (see Fig. 8). If the rates of surface processes are slow relative to the rate of deformation, folding is expected to be distributed, of short wavelength, and completely uninfluenced by surface processes. This situation is probably realized in many natural fold belts (e.g., Fars zone of Zagros orogen). If, however, the rate of surface processes is rapid relative to the rate of deformation, the model predicts initially distributed short-wavelength folding followed by relatively rapid localized folding at long wavelengths. In this latter stage, whole series of anticlines and synclines may be up-warped as erosional unroofing begins to amplify folding at longer wavelengths, whereas adjacent structures may be swamped and buried in broad down-warps (e.g., see Fig. 6E). This type of behavior shares similarities to structures in the central Zagros mountains of southwest Iran (i.e., northeast of Dezful), where two major wavelengths of deformation have been described: the oldest related to short-wavelength (~10–20 km), low amplitude (~1 km) buckle folding, and a recent second long-wavelength (~100 km) flexing, which has a maximum amplitude of close to 10 km (see Falcon, 1961; Oberlander, 1965). If this interpretation is correct, it implies that deformation in the central Zagros is currently being strongly amplified at long wavelengths by very localized but efficient unroofing, which occurs along deep canyons (which cut gorges with relief of ~2 km). Moreover, this unroofing may have contributed to the surface uplift of the Bakhtiari Mountains, which have maximum elevations of ~4000 m. At more advanced stages of fold amplification, it is anticipated that fold-belt style deformation formed earlier in the history would be completely disrupted and exhumed in a mega-antiform flanked by relatively shallow basins (which later may themselves be partially or completely exhumed). At this stage, advanced stage of fold amplification, the structure may bear no similarity to the fold-belt from which it initially developed, the difference being due to the amplifying influence exerted by erosion and deposition.

The final point here is the relatively sharp transition (e.g., see Fig. 8) separating regimes of normal distributed folding and localized folding, which is strongly influenced by surface processes. The implication is that relatively minor changes in climate or tectonic deformation rates may profoundly influence the nature of folding and surface topography (if the boundary is crossed). For example, a change to a relatively erosive climate may cause a shift in folding toward longer wavelengths, large amplification of rock uplift and exhumation rates, and an increase in topography, especially if erosion is localized in narrow river systems. This type of behavior indicates that the response of folding to changes in climate and/or rate of deformation is dynamic in origin and threshold-like in behavior.

CONCLUSIONS

The most important conclusions of this study can be summarized as follows:

1. A critical parameter controlling whether erosion and deformation are important in influencing folding is R, the ratio of the deformation to fluvial erosion characteristic time scales (see equation 25). A large value of R corresponds to a situation where the rates of fluvial processes are fast relative to the rate of deformation, whereas the reverse is true for a small value of R.
2. If $R \gg 1$, erosion and deposition can have a major influence on folding at small to intermediate spatial scales (i.e., ~200 km). The main effects are enhanced rates of rock uplift and subsidence and a shift toward longer-wavelength deformation, such that initially formed, short-wavelength folds tend to become refolded to form a central mega-antiform flanked by relatively shallow basins. When $R \ll 1$, surface processes have negligible influence on folding.
3. Erosion and deposition influence folding both through reducing the influence of gravity and by modifying the rheology. Within this study, the most important of these effects is gravitational in origin, especially initially, when the amount of mass redistribution is small and results in only minor variations in thickness (and strength).
4. The surface topography resulting from folding is low when $R \ll 1$ and for $R \gg 1$. However, for intermediate values of R, the maximum topography is greater than for the case when no surface processes are present. In this case, surface processes are rapid enough to amplify deformation, but not fast enough to keep pace with rock uplift rates once localization has been initiated. Thus, within the model investigated here, surface processes are capable of increasing the surface topography. This height increase is, however, not caused by static isostatic rebound but results from amplifying the dynamic folding instability.

ACKNOWLEDGMENTS

I thank Yuri Podladchikov, Neil Mancktelow, Alex Densmore, and Mike Ellis for providing encouragement and for stimulating discussions during the course of this work. I am also grateful to Fritz Schlunegger and the Swiss National Science Foundation (grant No. 620-57863) with whom this research was initiated. Very thorough reviews by Raymond Fletcher and Todd

Ehlers, and comments from Sean Willett greatly improved the final content of this manuscript.

REFERENCES CITED

Avouac, J.P., and Burov, E.B., 1996, Erosion as a driving mechanism of intracontinental mountain growth: Journal of Geophysical Research, v. 101, p. 17,747–17,769, doi: 10.1029/96JB01344.

Beaumont, C., Fullsack, P., and Hamilton, J., 1992, Erosional control of active compressional orogens, *in* McClay, K.R., ed., Thrust tectonics: New York, Chapman and Hall, p. 1–31.

Beaumont, C., Kooi, H., and Willett, S., 2000, Coupled tectonic-surface process models with applications to rifted margins and collisional orogens, *in* Summerfield, M.A., ed., Geomorphology and global tectonics: Chichester, John Wiley and Sons Ltd., p. 29–55.

Biot, M.A., 1961, Theory of folding of stratified viscoelastic media and its implications in tectonics and orogenesis: Geological Society of America Bulletin, v. 72, p. 1595–1620.

Biot, M.A., and Odé, H., 1965, Theory of gravity instability with variable overburden and compaction: Geophysics, v. 30, p. 213–227, doi: 10.1190/1.1439558.

Bloom, F., and Coffin, D., 2001, Handbook of thin-plate buckling and postbuckling: New York, Chapman and Hall, 786 p.

Bonnet, S., and Crave, A., 2003, Landscape response to climate change: Insights from experimental modelling and implications for tectonic versus climatic uplift of topography: Geology, v. 31, p. 123–126.

Brace, W.F., and Kohlstedt, D.L., 1980, Limits on lithospheric stress imposed by laboratory experiments: Journal of Geophysical Research, v. 85, p. 6248–6252.

Braun, J., and Beaumont, C., 1989, Dynamic models of the role of crustal shear zones in asymmetric continental extension: Earth and Planetary Science Letters, v. 93, p. 405–423, doi: 10.1016/0012-821X(89)90039-3.

Chapple, W.M., 1978, Mechanics of thin-skinned fold and thrust belts: Geological Society of America Bulletin, v. 89, p. 1189–1198, doi: 10.1130/0016-7606(1978)89<1189:MOTFB>2.0.CO;2.

Dahlen, F.A., 1984, Noncohesive critical Coulomb wedges: An exact solution: Journal of Geophysical Research, v. 89, p. 10,125–10,133.

Davis, D., Suppe, J., and Dahlen, F.A., 1983, Mechanics of fold-and-thrust belts and accretionary wedges: Journal of Geophysical Research, v. 89, p. 1153–1172.

Dixon, J.M., and Liu, S., 1992, Centrifuge modeling of the propagation of thrust faults, *in* McClay, K.R., ed., Thrust tectonics: New York, Chapman and Hall, p. 53–69.

England, P., and Molnar, P., 1990, Surface uplift, uplift of rocks, and exhumation of rocks: Geology, v. 18, p. 1173–1177, doi: 10.1130/0091-7613(1990)018<1173:SUUORA>2.3.CO;2.

Falcon, N.L., 1961, Major earth-flexuring in the Zagros Mountains of south west Iran: Quaternary Journal of the Geological Society of London, v. 117, p. 367–376.

Finlayson, D.P., Montgomery, D.R., and Hallet, B., 2002, Spatial coincidence of rapid inferred erosion with young metamorphic massifs in the Himalayas: Geology, v. 30, p. 219–222.

Gerbault, M., 2000, At what stress level is the central Indian Ocean lithosphere buckling?: Earth and Planetary Science Letters, v. 178, p. 165–181, doi: 10.1016/S0012-821X(00)00054-6.

Gerbault, M., Burov, E., Poliakov, A., and Daignières, M., 1999, Do faults trigger folding in the lithosphere?: Geophysical Research Letters, v. 26, p. 271–274, doi: 10.1029/1998GL900293.

Gilchrist, A.R., Summerfield, M.A., and Cockburn, H.A.P., 1994, Landscape dissection, isostatic uplift, and the morphological development of orogens: Geology, v. 22, p. 963–966, doi: 10.1130/0091-7613(1994)022<0963:LDIUAT>2.3.CO;2.

Goff, D.F., Wiltschko, D.V., and Fletcher, R.C., 1996, Décollement folding as a mechanism for thrust-ramp spacing: Journal of Geophysical Research, v. 101, p. 11,341–11,352, doi: 10.1029/96JB00172.

Johnson, A.M., 1980, Folding and faulting of strain-hardening sedimentary rocks: Tectonophysics, v. 62, p. 251–278, doi: 10.1016/0040-1951(80)90196-1.

Jordan, T.E., 1981, Thrust loads and foreland basin evolution, Cretaceous, western United States: American Association of Petroleum Geologists Bulletin, v. 65, p. 2506–2520.

Karner, G.D., and Watts, A.B., 1983, Gravity anomalies and flexure of the lithosphere at mountain ranges: Journal of Geophysical Research, v. 88, p. 10,449–10,477.

Koons, P.O., 1990, Two-sided orogen: Collision and erosion from the sandbox to the Southern Alps, New Zealand: Geology, v. 18, p. 679–682, doi: 10.1130/0091-7613(1990)018<0679:TSOCAE>2.3.CO;2.

Koons, P.O., 1994, Three-dimensional critical wedges: Tectonics and topography in oblique collisional orogens: Journal of Geophysical Research, v. 99, p. 12,301–12,315, doi: 10.1029/94JB00611.

Koons, P.O., Norris, R.J., Craw, D., and Cooper, A.F., 2003, Influence of exhumation on the structural evolution of transpressional plate boundaries: An example from the Southern Alps, New Zealand: Geology, v. 31, p. 3–6, doi: 10.1130/0091-7613(2003)031<0003:IOEOTS>2.0.CO;2.

Lambeck, K., 1983, The role of compressive forces in intracratonic basin formation and mid-plate orogenies: Geophysical Research Letters, v. 10, p. 845–848.

Maggi, A., Jackson, J.A., McKenzie, D., and Priestley, K., 2000, Earthquake focal depths, effective elastic thickness, and the strength of the continental lithosphere: Geology, v. 28, p. 495–498, doi: 10.1130/0091-7613(2000)028<0495:EFDEET>2.3.CO;2.

Masek, J.G., Isacks, B.L., Gubbels, T.L., and Fielding, E.J., 1994, Erosion and tectonics at the margins of continental plateaus: Journal of Geophysical Research, v. 99, p. 13,941–13,956, doi: 10.1029/94JB00461.

Martinod, J., and Davy, P., 1992, Periodic instabilities during compression or extension of the lithosphere. 1. Deformation modes from an analytical perturbation method: Journal of Geophysical Research, v. 97, p. 1999–2014.

Montgomery, D.R., 1994, Valley incision and the uplift of mountain peaks: Journal of Geophysical Research, v. 99, p. 13,913–13,921, doi: 10.1029/94JB00122.

Norris, R.J., and Cooper, A.F., 1997, Erosional control on the structural evolution of a transpressional thrust complex on the Alpine fault, New Zealand: Journal of Structural Geology, v. 19, p. 1323–1342, doi: 10.1016/S0191-8141(97)00036-9.

Oberlander, T., 1965, The Zagros streams: A new interpretation of transverse drainage in an orogenic zone: New York, University Press, Syracuse Geographical Series, no. 1, 168 p.

Pavlis, T.L., Hamburger, M.W., and Pavlis, G.L., 1997, Erosional processes as a control on the structural evolution of an actively deforming fold and thrust belt: An example from the Pamir-Tien Shan region, central Asia: Tectonics, v. 16, p. 810–822, doi: 10.1029/97TC01414.

Schmalholz, S., and Podladchikov, Y.Y., 1999, Buckling versus folding: Importance of viscoelasticity: Geophysical Research Letters, v. 26, p. 2641–2644, doi: 10.1029/1999GL900412.

Simpson, G.D.H., 2004a, A dynamic model to investigate coupling between erosion, deposition and three-dimensional (thin-plate) deformation: Journal of Geophysical Research, Earth Surface, v. 109, F02006, doi: 10.1029/2003JF000078.

Simpson, G.D.H., 2004b, Dynamic interactions between erosion, deposition and three-dimensional deformation in compressional fold-belt settings: Journal of Geophysical Research, Earth Surface, v. 109, F03007, doi: 10.1029/2003JF000111.

Simpson, G.D.H., 2004c, The role of river incision in enhancing deformation: Geology, v. 32, p. 341–344, doi: 10.1130/G20190.2.

Simpson, G.D.H., and Schlunegger, F., 2003, Topographic evolution and morphology of surfaces evolving in response to coupled fluvial and hillslope sediment transport: Journal of Geophysical Research, v. 108, no. B6, 2300, doi: 10.1029/2002JB002162.

Small, E.E., and Anderson, R.S., 1998, Pleistocene relief production in Laramide mountain ranges, western United States: Geology, v. 26, p. 123–126, doi: 10.1130/0091-7613(1998)026<0123:PRPILM>2.3.CO;2.

Stockmal, G.S., 1983, Modelling of large-scale accretionary wedge deformation: Journal of Geophysical Research, v. 88, p. 8271–8287.

Timoshenko, S.P., and Woinowsky-Krieger, S., 1959, Theory of plates and shells: New York, McGraw-Hill, 580 p.

Tucker, G., and Whipple, K.X., 2002, Topographic outcomes predicted by stream erosion models: Sensitivity analysis and intermodel comparison: Journal of Geophysical Research, v. 107, doi: 10.1029/2001JB000162.

Turcotte, D.L., and Schubert, G., 2002, Geodynamics: Applications of continuum physics to geological problems (2nd edition): Cambridge, Cambridge University Press, 528 p.

van Balen, R.T., and Podladchikov, Y.Y., 1998, The effect of inplane force variations on a faulted elastic thin-plate; implications for rifted sedimentary basins: Geophysical Research Letters, v. 25, p. 3903–3906, doi: 10.1029/1998GL900005.

van Balen, R.T., Podladchikov, Y.Y., and Cloetingh, S.A.P.L., 1998, A new multilayered model for intraplate stress-induced differential subsidence of faulted lithosphere, applied to rifted basins: Tectonics, v. 17, p. 938–954, doi: 10.1029/1998TC900003.

Watts, A.B., 2001, Isostasy and flexure of the lithosphere: Cambridge, Cambridge University Press, 478 p.

Willett, S.D., 1999, Orogeny and orography: The effects of erosion on the structure of mountain belts: Journal of Geophysical Research, v. 104, p. 28,957–28,981, doi: 10.1029/1999JB900248.

Zeitler, P.K., Koons, P.O., Bishop, M.P., Chamberlain, C.P., Craw, D., Edwards, M.A., Hamidullah, S., Jan, M.Q., Khan, M.A., Khattak, M.U.K., Kidd, W.S.F., Mackie, R.L., Meltzer, A.S., Park, S.K., Pecher, A., Poage, M.A., Sarker, G., Schneider, D.A., Seeber, L., and Shroder, J.F., 2001, Crustal reworking at Nanga Parbat, Pakistan: Metamorphic consequences of thermal-mechanical coupling facilitated by erosion: Tectonics, v. 20, p. 712–728, doi: 10.1029/2000TC001243.

Zhang, Y., Hobbs, B.E., Ord, A., and Mühlhaus, H.B., 1996, Computer simulations of single-layer buckling: Journal of Structural Geology, v. 18, p. 643–655, doi: 10.1016/S0191-8141(96)80030-7.

Zienkiewicz, O.C., and Taylor, R.L., 2000, The finite element method, Volume 2: Solid Mechanics: Boston, Butterworth-Heinemann, 459 p.

Zuber, M.T., and Parmentier, E.M., 1996, Finite amplitude folding of a continuously viscosity-stratified lithosphere: Journal of Geophysical Research, v. 101, p. 5489–5498, doi: 10.1029/95JB02514.

Manuscript Accepted by the Society 23 June 2005

Printed in the USA

Geological Society of America
Special Paper 398
2006

Long-term evolution of tectonic lakes: Climatic controls on the development of internally drained basins

Daniel Garcia-Castellanos[†]
Netherlands Centre for Integrated Solid Earth Science, Faculty of Earth and Life Sciences, Vrije Universiteit Amsterdam, De Boelelaan 1085, 1081HV Amsterdam, Netherlands, and *Institute of Earth Sciences "Jaume Almera" (CSIC), Solé i Sabarís s/n, 08028 Barcelona, Spain*

ABSTRACT

A quantitative model is proposed for the long-term evolution of lakes and internally drained basins resulting from tectonic vertical motions, sediment infill, outlet erosion, and climatic regime. The model accounts for the formation of a water body in the topographic basin created by tectonic uplift across a river, where incision capability is calculated using a stream power-law. The model also addresses the notion that, after cessation of tectonic forcing, lakes are transitory phenomena over geological time scales. High uplift rates across an antecedent river, in combination with low upstream precipitation, result in river defeat and lake formation. In addition to geometrical, lithological, and tectonic parameters, the evaporation rate at the lake surface is revealed as a key factor triggering drainage closure (endorheism) and significant lake life extension by preventing outlet erosion. Post-tectonic lake extinction is ensured by sediment overfill and/or outlet erosion. Once uplift comes to an end and drainage reopens (lake capture), shallow lakes at high altitudes undergo a faster reintegration into the drainage network and extinction. Vertical isostatic movements of the lithosphere significantly delay this process in lakes larger than 50–200 km. The development of an internally drained basin out of an open lacustrine basin requires that uplift across the outlet persists until the lake is large enough to evaporate all collected water. The evolution of tectonic lakes has, therefore, similar dependency on geometrical constraints (initial relief, length of the river, hypsometry of the catchment), lithological parameters (rock erodibility), tectonics (uplift rate, duration, and its spatial distribution), and climate (precipitation and evaporation rates).

Keywords: endorheic, surface processes, landscape evolution, lithospheric flexure, numerical model.

INTRODUCTION

Lakes develop in local topographic minima formed by processes as diverse as crustal tectonic deformation, volcanism, glacial excavation, or meteoritic impacts. On geological time scales, lakes are ephemeral systems that, once the forcing mechanism ceases, disappear by outlet erosion and/or sediment overfill (e.g., Ollier, 1981). Most lakes on Earth are related to the last glaciation and are therefore relatively recent. In contrast, most long-living lakes lasting beyond glacial-interglacial cycles are maintained by very active tectonic processes, such as continental rifting (e.g., Lake Baikal [Williams et al., 1997], 31,500 km^2 in surface and Pliocene-Quaternary in age) or continental compression (e.g., Lake Titicaca in the Andean Altiplano, 8,000 km^2, at

[†]E-mail: danielgc@ija.csic.es.

Garcia-Castellanos, D., 2006, Long-term evolution of tectonic lakes: Climatic controls on the development of internally drained basins, *in* Willett, S.D., Hovius, N., Brandon, M.T., and Fisher, D.M., eds., Tectonics, Climate, and Landscape Evolution: Geological Society of America Special Paper 398, Penrose Conference Series, p. 283–294, doi: 10.1130/2006.2398(17). For permission to copy, contact editing@geosociety.org.

least Pliocene in age) interfering with an existing river network. Independently of the tectonic mechanism generating the tectonic barrier, it is often interpreted that the formation of a tectonic lake is coeval with fast vertical movements (Fig. 1A) and that its extinction coincides with slower or negligible tectonism and/or increased erosion power across the lake outlet. An example for this is Lake Bungunnia, a 33,000 km^2 water body existing in southern Australia between 2.5 and 0.7 Ma, thought to be a result of the rise of a tectonic dam (Stephenson, 1986) between the Murray Basin and the ocean. Dry climate (low precipitation/evaporation ratios), sometimes in combination with active tectonism, can neutralize the lake outlet, closing its drainage (endorheic basins such as the present Andean Altiplano, the Dead Sea, the Caspian Sea, the Aral Sea) and presumably contributing to the extension of the life of the lacustrine system.

A quantitative understanding of the long-term evolution of lakes and internally drained basins is crucial because these systems are climate-sensitive discontinuities in the sediment flow from orogens to basins, and therefore their sedimentary infill constitutes a record of the climatic and tectonic evolution (e.g., Tiercelin, 2002; Yan et al., 2002). Intramountain basins, often presenting long lacustrine periods (e.g., Burbank, 1983), are conspicuous examples of interplay between climate and tectonics. In an open lake, the amount of output water discharge is reduced by evaporation from its surface (Fig. 1A). If evaporation becomes larger than the collected precipitation, the lake becomes the center of a closed (endorheic) drainage basin, under which circumstances the lake's level is climatically controlled (e.g., Lake Issyk-Kul; De Batist et al., 2002). An extreme example of endorheism is the Dead Sea, where evaporation reaches values of 1600 mm/yr, whereas the mean water runoff collected by the Jordan River (mean discharge of 17.1 m^3/s) is only 0.013 mm/yr. In Lake Titicaca, the evaporation rate of 1100 mm/yr in the lake's surface nearly matches the water it collects from the Altiplano, where precipitation rates range between 600 and 800 mm/yr. In the Minchin paleolake (Pleistocene highstand of the intramountain Altiplano Basin [Fornari et al., 2001], with a water surface larger than 40,000 km^2), the mean precipitation rate must have been higher than 300 mm/yr to compensate an evaporation rate similar to the present one. In such cases, due to the absence of river incision along an outlet, the lake's life expectancy is likely to increase. In closed drainage basins, either a climate change to wetter climatic conditions, or the sediment overfill of the lake (both inducing lake level rise) or a basin capture undertaken by an adjacent river can trigger drainage opening (Garcia-Castellanos et al., 2003), abruptly activating basin-wide incision and

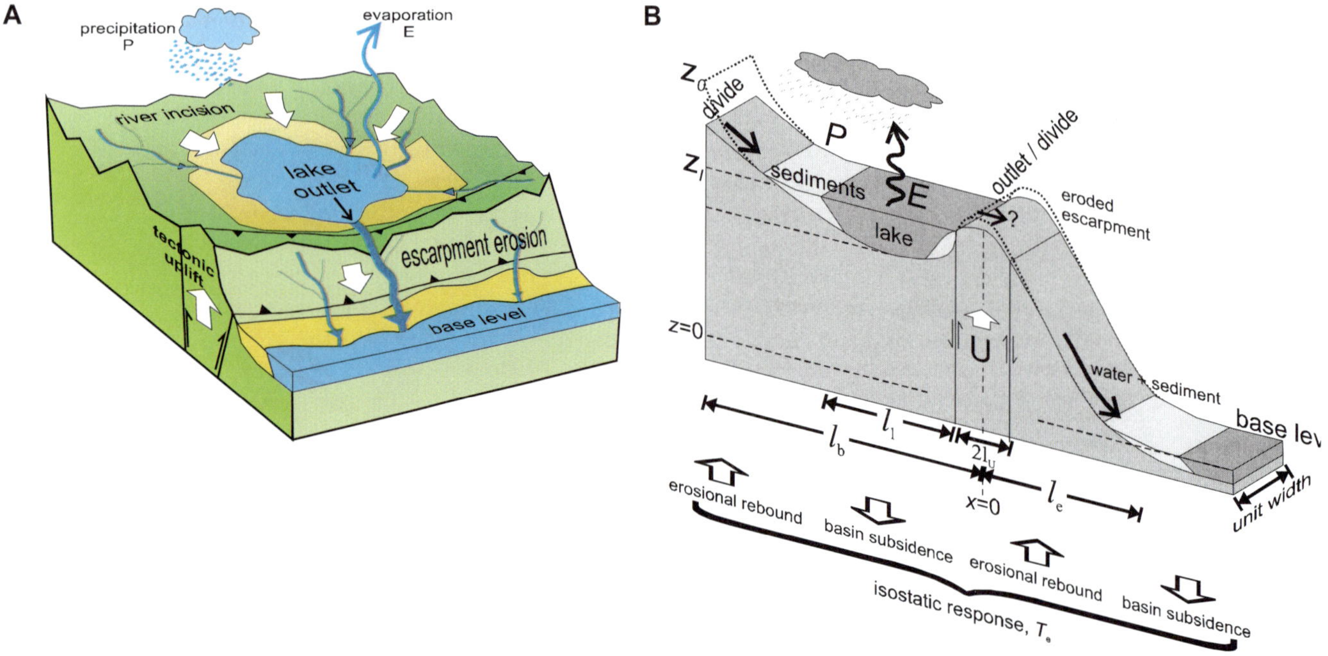

Figure 1. (A) Cartoon showing the main processes involved in the evolution of tectonic lakes, including the growth of a tectonic topographic barrier, river transport of sediments from the catchment toward the lake, erosion and river incision of the barrier, and the hydrological balance between precipitation, lake evaporation, and outlet discharge. (B) One-dimensional (1D, along-stream) approach and main tunable parameters of the cross-section model proposed here: *P*—collected precipitation (runoff, in mm/yr); *E*—evaporation rate (mm/yr); *U*—uplift rate (mm/yr); l_u—length of the uplifted area; l_b—basin length; l_e—escarpment length. As a result of the numerical simulation, lake altitude z_l, lake length along the modeling plane l_l, and other properties are predicted through time; thin arrows indicate the flow of water. Low *P*/*E* ratios can trigger drainage closure and endorheism, inhibiting water flow and erosion at the uplifting barrier. Large white arrows indicate fluvial and tectonic mass transport and vertical tectonic movements.

lake extinction, as reported in a diversity of geological scenarios including the Pleistocene Lake Bonneville (O'Connor, 1993) and the Ma'adim Vallis in Mars (Irwin et al., 2002). Again, though these processes are often cited in the literature as governing the extinction of lacustrine basins, little is known about the relative importance of the parameters involved, and a quantitative study of the process interplay between hydrology, erosion, and tectonics is needed.

Studies of river incision and sediment transport have provided quantitative models of landscape evolution driven by surface water flow (e.g., Howard, 1994). These models have been useful to quantitatively understand the dynamics of tectonic defeat of antecedent rivers (Kühni and Pfiffner, 2001; van der Beek et al., 2002; Sobel et al., 2003). These studies conclude that the main controlling factors are the mean precipitation in the catchment, the rate of tectonic uplift across the river path, and the rock strength of the uplifting barrier. However, these models do not explicitly account for the formation of a lake in front of the uplifted area and its effects on the water balance between collected runoff and evaporation. This is of key relevance for internally drained basins, because by definition the transition from an intramountain basin to an internally drained basin requires the evaporation at the lake surface to compensate the collected water runoff. Even before drainage closure is reached by tectonic uplift, evaporation reduces water discharge across the uplifting area, limiting the erosion rates and presumably accelerating the drainage closure process.

Coupling well-established quantitative approaches to both tectonic and surface processes, this paper focuses on the interplay between these two scales in controlling the long-term evolution of lakes. A cross-sectional model of lake evolution is designed to account for four processes: tectonic uplift of a structural barrier, hydrological balance, surface sediment transport, and flexural isostasy. The model permits parameterization of the potential formation of lakes and closed (internally drained) basins, their sediment infill, and their extinction as a function of the timing of tectonic uplift.

FUNDAMENTALS AND FORMULATION

The scenario of a river undergoing tectonic uplift along part of its path is addressed via a one-dimensional (1D, along-stream cross section) numerical model of river incision and transport. Although a 1D approach might seem an oversimplification, it will be later shown that other assumptions inherent to the adopted transport model impede a direct quantitative comparison with geological scenarios. Thus, a 1D model is chosen as the easiest way to simulate the process interplay while allowing an easy reproducibility of the results. Note also that using a 2D (planform) model would introduce infinite further arbitrary choices for the initial topographic setting. The present approach is similar to the along-stream approach used by Sobel et al. (2003; based on Howard, 1994, among others) to calculate the defeat of rivers by tectonic uplift, but here lakes are explicitly considered and assumed to have the same width (in the direction perpendicular to the modeling plane) as the hydrographic basin.

The lake's history is predicted as a result of the competition between tectonic uplift versus river incision and transport. Due to the 1D nature of the model, hereafter the length of a feature measured along the modeling plane will be assumed to be equivalent to the area of that feature in nature. The origin of the along-stream coordinate x is chosen to be at the center of the uplifting barrier, which has a half-length l_u and is assumed to be invariable along the strike (Fig. 1B). Tectonic uplift is incorporated to the model as a rectangle function $U(x) = U$ for $x = [-l_u, l_u]$ and $U(x) = 0$ elsewhere, and is applied from the initial time $t = 0$ until $t = t_u$ (thus, the period after t_u is referred to as post-tectonic). In accord with the cross-section nature of the model, the lake of length l_l forming in front of this uplifting barrier has a unit width along strike. The adopted initial topography (Figs. 1B and 2A) corresponds to an equilibrium river profile (i.e., one with erosion rate constant along the river) descending from an altitude z_0 at $x = -l_b$ (left boundary of the model) to zero at $x = l_e$. At $x > l_e$, a basin is defined with a constant water level at $z = 0$, which plays the role of a geomorphological base level.

Erosion is modeled via a combination of fluvial incision and diffusive rock flow. These processes are usually represented via a stream power law and a diffusive transport law, respectively (Howard, 1994; Tucker and Slingerland, 1997; Whipple and Tucker, 1999), giving the erosion rate e at any given location x of the topographic profile $z(x)$ as:

$$e(x) = -Kd\frac{\partial^2 z}{\partial x^2} + K_f Q_w^{\,m}\left|\frac{\partial z}{\partial x}\right|^n, \tag{1}$$

where K_d is the constant of diffusive transport, K_f is the constant of fluvial incision or erodibility, and Q_w is the water discharge. Note the different units used here for K_f (Table 1) with respect to the previous references. The exponents m and n determine the relative importance of water discharge against slope in driving fluvial incision, and a value of $n/m \approx 2$ is generally accepted based on field data (e.g., van der Beek and Bishop, 2003). The values $m = 1/3$ and $n = 2/3$ are adopted here, corresponding to the assumption that incision rate is proportional to basal shear stress in the flowing water (Howard, 1994; Whipple and Tucker, 1999). The dependence of K_f on lithology and other parameters is not well understood, and the values reported in the literature are highly variable. Besides that, this model overlooks the role of the sediment load and its granulometry in defining the incision rates. These uncertainties indicate that the results from the model cannot be directly compared with a particular geological scenario without performing previously a detailed calibration. Finally, note that the first term in equation 1 is incorporated to produce erosion in areas where slope is zero, such as the outlet saddle of a lake. Rather than representing hillslope processes, it is here intended to capture any other nonfluvial erosion and transport mechanism, such as water-flow shear stresses, vegetation, frost, etc.

In the absence of significant underground water flow, the water discharge leaving a lake is determined by the water balance

TABLE 1. INPUT PARAMETER VALUES OF THE REFERENCE MODEL SHOWN IN FIGURE 2

Parameter	Variable	Value
Tectonics		
Uplift rate	U	2 mm/yr
Uplift duration	t_u	1 m.y.
Uplift area half-length	l_u	5 km
Elastic thickness (km)	T_e	∞
Climate		
Precipitation rate	P	500 mm/yr
Evaporation rate	E	800 mm/yr
Lithology		
Erodibility	K_f	
bedrock		$1.3\ 10^{-8}$ m/s $(m^2/s)^{1/3}$
sediment		$1.3\ 10^{-7}$ m/s $(m^2/s)^{1/3}$
Diffusive transp. coef	K_d	0.01 m^2/yr
Geometry		
Max. initial topography	z_0	2000 m
Basin length	l_b	50 km
Escarpment length	l_e	10 km

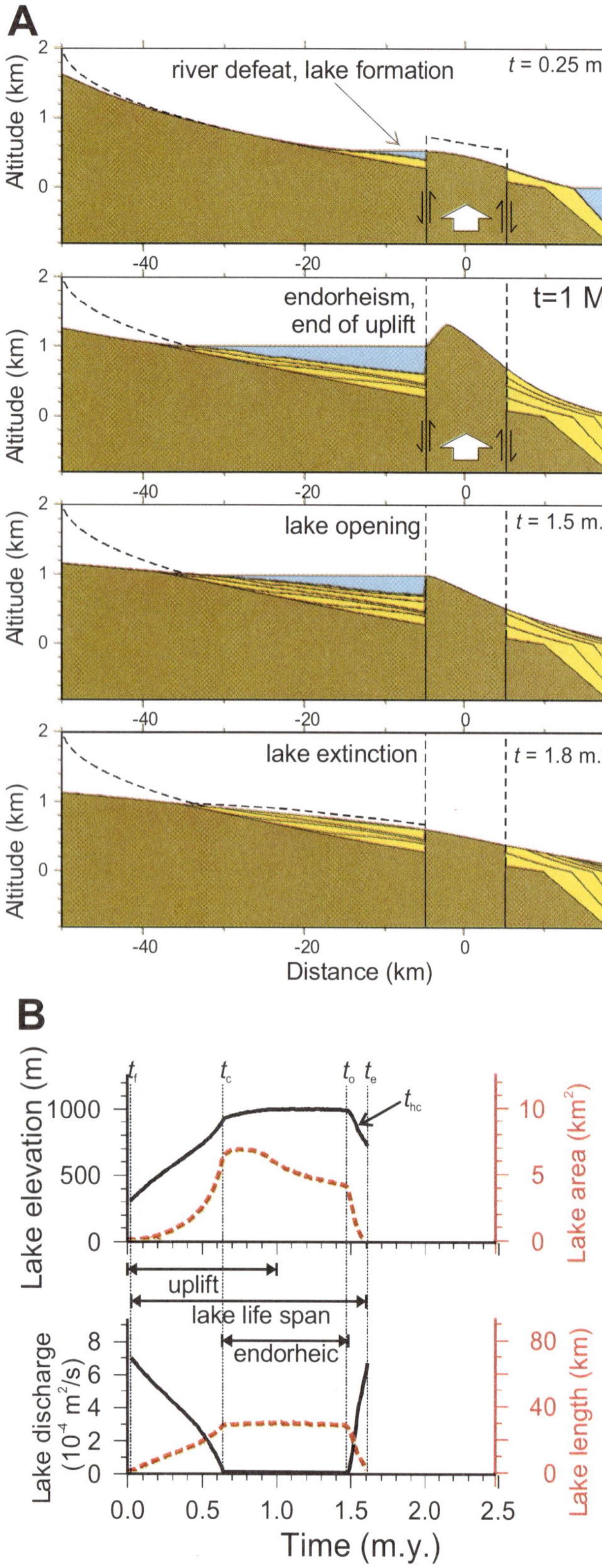

between precipitation and evaporation, which for a steady-state water flow can be written as

$$Q_w(x = 0) = P \bullet l_b - E \bullet l_l, \quad (2)$$

where $Q_w(x = 0)$ is the water discharge at the outlet, P is the mean precipitation rate in the lake's catchment basin of length l_b (measured along the modeling plane and assumed to be invariable along the strike), E is the evaporation rate at the lake's surface, and l_l is the length of the lake (representing its area in real three-dimensional nature). Note that here the term precipitation is used to designate the component of precipitated water that is collected as surface runoff, this to avoid addressing local infiltration and evapotranspiration processes that are out of the scope of this paper. By definition, for a closed lake $Q_w(x = 0) = 0$ and therefore $P/E = l_l/l_b < 0$, whereas for an open lake (in which the lake's water level reaches the outlet), the output discharge is $P \cdot l_b - E \cdot l_l$, which

Figure 2. (A) Time evolution of the reference model (Table 1). Eroded rock is indicated with a dashed line. Sediment horizons are every 0.25 m.y.; vertical exaggeration ×8. (B) Evolution of lake altitude, lake area across the modeling plane (representing volume in the 3D nature, dashed line in top panel), lake length l_l (representing surface, dashed line in bottom panel), and outlet water discharge of the lake. A lake forms in this example between $t_f = 0.02$ m.y. until $t_e = 1.62$ m.y. Lake length and elevation increase until evaporation on its surface equals the precipitation collected in the lake. At that stage (time of closure, t_c), the output water discharge becomes zero and the lake becomes endorheic. After the cessation of uplift at $t_u = 1$ m.y., the altitude of the drainage divide between the lake and the ocean side decreases until it reaches the lake level (time of drainage opening t_o). The lake extinguishes at t_e, both by sediment infill and outlet erosion. The parameter t_{hc} is the time of half capture (time at which closing the drainage again would require a reduction in the precipitation/evaporation ratio of a factor 2, see text).

produces outlet incision leading to the integration of the lake in the drainage network (drainage capture).

The area of a lake is related to its water level, z_l, through the hypsometric curve of its floor topography. Within the adopted 1D approach, this hypsometric curve is expressed in terms of lake length $l_l(z)$ (representing lake surface in nature) below an altitude z, and coincides with the lake bathymetric profile. At the time of drainage opening, by definition, $l_l(z_{min}) = 0$ and $l_l(z_{max}) = l_b \cdot P/E$, where z_{min} corresponds to the minimum elevation of the lake floor, and z_{max} corresponds to the outlet level at the time of opening. During opening, the excess water leaving the lake (i.e., the water that would have been evaporated in the lake portion lying now above z_l) is related to the lake level z_l, and the water discharge along the outlet river follows the equation

$$Q_w(x) = Pl_b - El_l(z_l) + K_H x^h, \tag{3}$$

where x is measured from the outlet in opposite direction to the lake (i.e., toward the escarpment) and the second term corresponds to Hack's Law of river water discharge. For simplicity, $h = 1$ is assumed (which implies that the planform geometry of the catchment, including the lake, is rectangular), and then K_H coincides with the precipitation P, which is assumed to be constant over the entire region. Measured values of h often range between 1.55 and 1.65 in the higher parts of a catchment (e.g., Rodriguez-Iturbe and Rinaldo, 1997), but are <1 in the lower parts. For intramountain lakes, our experiments overestimate the amount of water reaching the lake relative to that leaving it. For the purpose of the present work, it is sufficient to keep in mind that higher values of the Hack's Law exponent would facilitate the drainage closure of the basin, not affecting the qualitative role played by the mean precipitation P and other parameters. From equations 1 and 3, the changes in elevation along the outlet river profile $z(x)$ are:

$$\frac{dz}{dt} = U(x) - e(x) = U_I(x) + U_I(x) + K_d \frac{\partial^2 z}{\partial x^2} - K_f(\mathrm{P\,I_b} - \mathrm{E\,I_l(z_l)} + P \cdot x^h)^m \left|\frac{\partial z}{\partial x}\right|^n \tag{4}$$

where $U(x)$ is the rock uplift rate resulting from tectonic deformation $U_T(x)$ and flexural isostatic readjustments $U_l(x)$. Equation 4 reflects that the reduction of $l_l(z_l)$ as the outlet is eroded produces a feedback effect that initially accelerates the capture process. Note that slope is zero by definition in the outlet, and therefore outlet erosion corresponds only to the third (diffusive) term in the right side of equation 4. However, since curvature in the outlet area is proportional to the slope imposed by fluvial incision in the escarpment (Fig. 1), erosion rate of the outlet is also indirectly controlled by water discharge. This implies that equation 4 cannot be reduced to a simple version restricted to the outlet point, and numerical solutions of the entire domain are required. On the other hand, reasonable values of K_d between 10^{-3} and 10^{-1} m^2/yr (e.g., Flemings and Jordan, 1989; Braun et al., 2001) have a negligible effect on the numerically predicted lake evolutions, because the discretization of topography applied in the numerical method implies nonzero slopes also in an outlet or a drainage divide, allowing for the erosion of such points in the same way as diffusion does. In conclusion, though the diffusive term is analytically required, it has no influence on the numerical results.

Equations 1 and 4 are solved numerically with an improved version of the code *tAo* (Garcia-Castellanos et al., 1997, 2002; http://cuba.ija.csic.es/~danielgc/lakes/), which uses the finite difference technique with a regular grid (dx = 100 m in this work), in three steps: (1) determination of the drainage flow following a maximum slope criteria and identifying the local topographic minima; (2) determination of water discharges in every location, and the extension of lakes accounting for evaporation; and (3) determination of the amount of material eroded/deposited at each location. Iterating these at time steps of 20 yr yields to the topographic and drainage evolutions described later in this paper.

To check the possible influence of isostasy on lake evolution, surface mass transport and changes in water volume in the lake are isostatically compensated via a 1D thin-plate flexural model with laterally constant elastic thickness T_e. Large T_e values imply a more rigid behavior of the lithosphere under surface mass transport, resulting in longer wavelength distribution of vertical movements $U_l(x)$ in equation 4 (e.g., Turcotte and Schubert, 2002). In contrast, small T_e values imply that erosional unloading and sediment loading are locally compensated by vertical rebound and subsidence, respectively.

LAKE FORMATION BY TECTONIC UPLIFT ACROSS A RIVER

A reference model setting is designed to check the role of tectonic uplift, climate, and rock strength on the initiation and evolution of lakes (Fig. 2). For the sake of simplicity, precipitation rate P, lake evaporation rate E, and uplift rate $U(x)$, are set constant through time. Uplift $U(x)$ stops at t_u = 2 m.y., and is localized along a 10-km-long area (l_u = 5 km) centered at x = 0, next to the base level. Other parameters are listed in Table 1.

For fast tectonic uplift rates and relatively low precipitation (P) and initial topographic gradients (low z_0), incision along the river is defeated by uplift, and a local topographic minimum develops in front of the uplifting area. In the model, this coincides with the onset of river aggradation in the upstream side of the tectonic barrier. If river aggradation cannot pace with uplift (e.g., if sediment yield is low due to low topography gradient), then a lake develops. For the parameter values used for the reference model (Table 1; model evolution in Fig. 2), the lake formation time is t_f = 0.02 m.y. Double precipitation rate (1000 mm/yr) results in more efficient river incision and later lake formation (t_f = 0.06 m.y.). A slower uplift rate of U_T = 0.5 mm/yr (compared to the 2.0 mm/yr adopted for the reference model) results also in a longer alluvial aggradation period, delaying the lake formation until t = 0.21 m.y. Other parameters, such as low bedrock erodibility and/or low transport coefficient, facilitate river defeat, similar to the predictions by Sobel et al. (2003).

Water accumulates in the new basin forming a lake with the outlet located in the topographic maximum of the uplifted area. Whereas in Sobel et al. (2003) the initiation of internal drainage is considered coincident with the defeat of the river, in the present model river defeat implies only the formation of a lake, and internal drainage only develops if the lake surface becomes large enough to evaporate the water discharge collected by the lake. Before this condition is accomplished, the lake remains open, and as the outlet is uplifted and incised by the outlet river, so changes the lake level. The difference between river defeat and internal drainage is quantitatively important. The minimum uplift rate required to defeat the river (keeping all other parameter values as in Table 1) is 0.35 mm/yr, whereas in order to create a closed drainage basin, the uplift rate must be at least 1.12 mm/yr. Similarly, the topographic difference between the mountain-side boundary of the model (x =) and the base level also influences the timing of lake formation and drainage closure, since higher initial slopes induce a more efficient erosion of the uplifting barrier. Whereas z_0 > 3810 m is required to inhibit river defeat (all other parameters being as in Table 1), the development of internal drainage is impeded for z_0 > 2780 m.

If the increasing lake length reaches the critical value at which the potential evaporation at its surface becomes larger than the water collected (equation 2), then the lake becomes closed (t_c = 0.62 m.y. in the reference model in Fig. 2). This situation is similar to that of the present Lake Issyk-Kul (De Batist et al., 2002), which has gradually become endorheic since the late Pleistocene. Another example is the closure of the Ebro Basin (NE Iberia) 35 m.y. ago as a result of tectonic shortening in the Pyrenees and other surrounding mountain belts (Garcia-Castellanos et al., 2003). In the model, the occurrence of closure depends on the uplift rate, U, and its duration, t_u, on the climatic ration P/E and on the ratio between l_b and l_l. As expected, long uplift periods (large t_u) promote long closed-drainage periods (Fig. 3), and low uplift rates delay the time of closure t_c (Fig. 4).

Once the lake becomes closed, the lake level (altitude of the lake surface) stops increasing, because it becomes controlled by the hydrological balance, rather than by the altitude of the outlet. The small increase in lake level during this period (Figs. 3B and 4A) is related only to the progressive sedimentation near the lake's shore (alluvial aggradation above the lake level; lacustrine progradation within the lake): the sediment delivered to the lake displaces its shore, so the lake's altitude must increase to preserve a length sufficient to evaporate the collected water (equation 2). The lake keeps a nearly constant length (l_l) as long as it remains closed, because the collected water remains nearly invariable (the internal catchment is only slightly reduced by divide retreat in the topographic barrier). Drainage closure promotes additional sediment supply to the lake basin coming from the uplifted area,

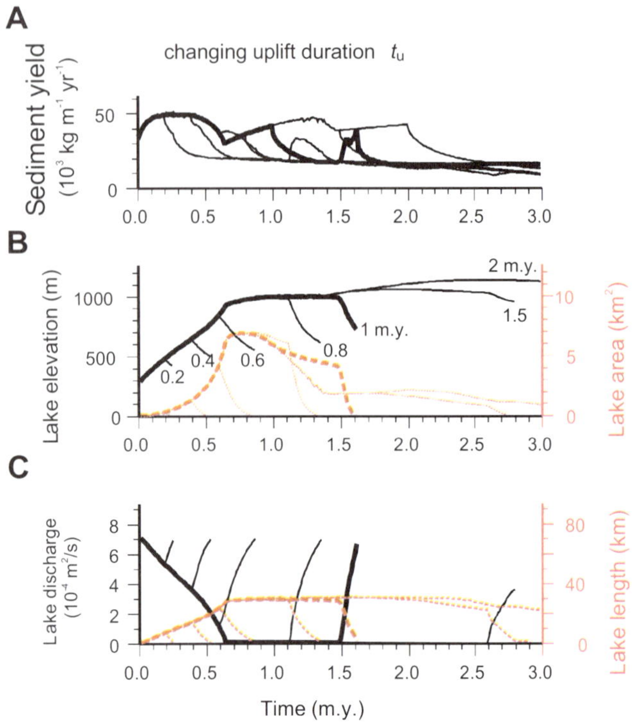

Figure 3. Evolution of the reference model in Figure 2 and Table 1 (bold lines) and a set of models differing only in uplift duration t_u (thin lines; labels indicate t_u). (A) Sediment yield toward the lake and the base level (identical to the erosion rate in the whole model domain); (B) lake level and lake volume; (C) output water discharge and lake area. High uplift rates and/or high uplift durations increase the duration of the lake, particularly if drainage closure (endorheism) is attained. Other legend as in Figure 2B.

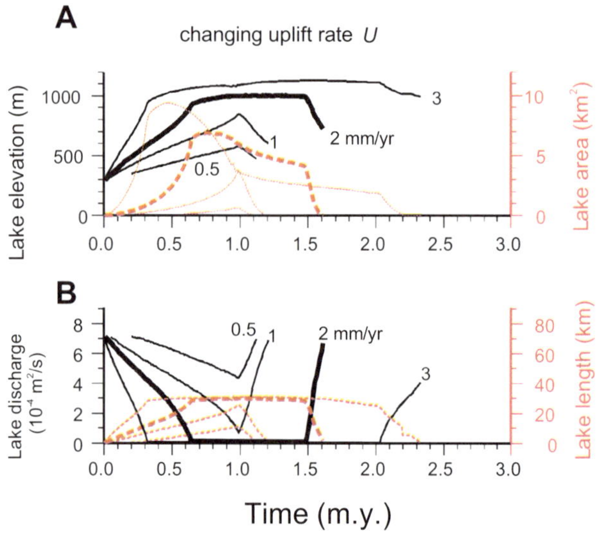

Figure 4. (A) Evolution of lake elevation (bold line) and lake length (dashed line) along the modeling plane; and (B) evolution of lake area (dashed line) and output discharge (bold line) of the reference model shown in Figure 2 and Table 1 (bold lines), and a set of models differing only in uplift rate U (thin lines; labels indicate U in mm/yr). Note the abrupt increase in lake life if the uplift rate is enough to close the drainage (i.e., if evaporation reduces output discharge to zero).

shifting the highest deposition rates toward the outlet side of the lake (Fig. 2A, t = 1 m.y.). Nevertheless, the total erosion rate occurring in the whole model domain reaches a minimum at t_c as a result of the reduced water flow at that time (Fig. 3A).

According to the model predictions, lake evaporation exerts a key control on lake evolution and on the development of internally drained basins. In order to become a closed drainage basin, the lake must attain a sufficient area (length in the present 1D model) so that evaporation in its surface equals the collected water. The dependency of the timing of lake development on evaporation is shown in Figure 5. The time of lake formation is independent of the evaporation rate (t_f = 0.02 m.y.), because evaporation only takes place at the lake surface (evapotranspiration in the rest of the model is not implemented in the model for simplicity). Lake closure (endorheism) occurs only for evaporation rates higher than the assumed precipitation (Fig. 5; Table 1), and it significantly increases the duration of the lake t_e–t_f.

As tectonic uplift comes to an end, the barrier separating the closed lake from the base level is eroded until its maximum topography becomes lower than the lake's level. This corresponds to the time of drainage opening t_o (1.48 m.y. in the reference model), initiating the process of lake capture. During capture, maximum erosion rates in the outlet are reached as a result of the increased water discharge in that area (Fig. 3A). Erosion of the barrier decreases the lake elevation z_l and length l_l, inducing a progressive increase in the water discharge delivered to the escarpment. This increase in water discharge induces an increase of erosion at the escarpment and thus a faster lake-level fall. As

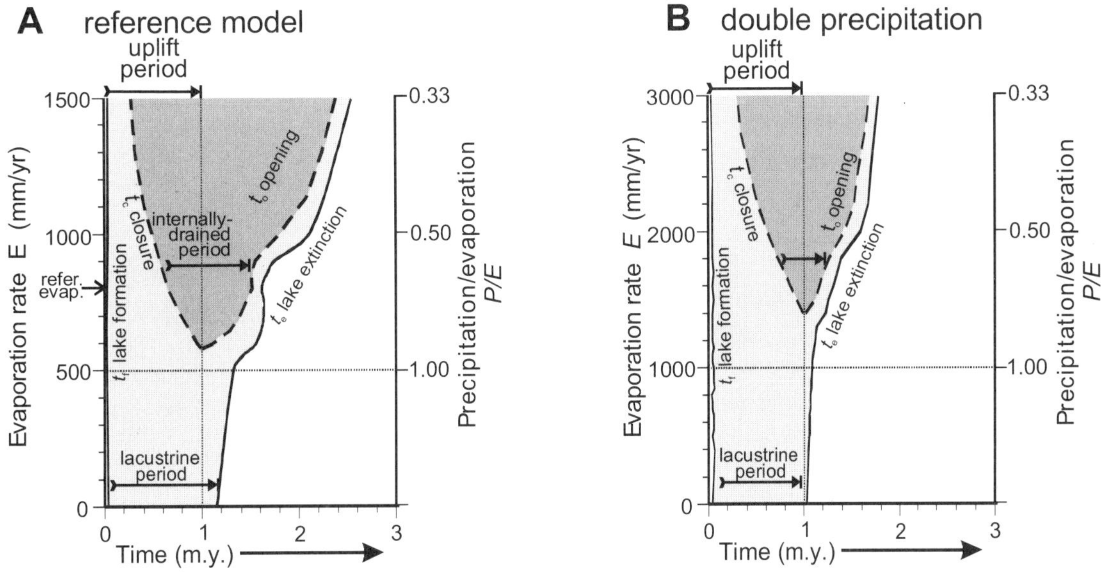

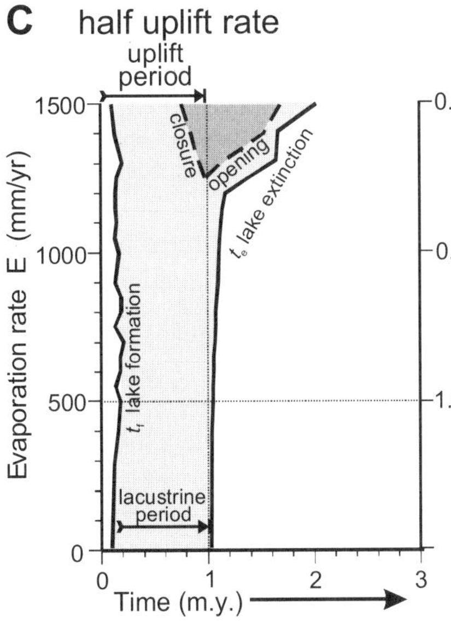

Figure 5. Effect of the climatic parameters P, E (precipitation and evaporation rate) on the time of lake formation t_f, time of drainage closure t_c, time of opening t_o, and time of lake extinction t_e. The light shade indicates the period during which a lake is formed; the dark shade indicates endorheism (closed drainage), lasting from t_c to t_o. (A) Results of the reference model for different evaporation rates (note that for the reference model in Figure 2, a precipitation rate of P = 500 mm/yr was used); (B) same as A for double the precipitation (1000 mm/yr); (C) same as A for half uplift rate (1 mm/yr). The time of formation (river defeat) is not dependent on E, but on precipitation (t_f = 0.2 m.y. for P = 500 mm/yr; t_f = 0.6 m.y. for P = 1000 mm/yr) and uplift rate. For values of $E > P$, the lake can undergo an endorheic period, which duration is proportional to E. This closed-drainage period significantly prolongs the life of the lake relative to lower E values. Though the general trends are similar, the different timings calculated in A and B indicate that the timing of the lake is dependent not only on the ratio P/E, but also on the absolute precipitation itself. Higher precipitation rates and lower uplift rates shorten the life of the lake by increasing erosion versus uplift in the outlet.

the capture of the entire lake progresses, this feedback effect is counteracted by the decreasing topographic difference between lake level and base level. The lake's bathymetry attained at the time of drainage opening also influences the changes in outlet erosion (equation 4). Closed lakes undergoing sediment overfilling disappear shortly after capture, because their floor is nearly flat and easy to integrate into the drainage network. The effect of these parameters in the post-tectonic lake opening and extinction deserves attention in a separate section.

OPENING AND EXTINCTION OF A LAKE

To show more conspicuously the effect of each parameter on the acceleration of the opening, a measure is necessary of the time at which closing the lake again (e.g., by a climatic change to dry conditions) becomes improbable. The time of half capture, t_{hc}, is here defined as the time at which the lake is reduced by a factor 2 in length, as a result of the ongoing capture. This is equivalent to define t_{hc} as the time at which a reduction by a factor 2 of the P/E ratio is required to close again the drainage (i.e., to decrease the lake level below the outlet level). A time of half capture (t_{hc}) closely after opening (t_o) implies that the drainage opening soon becomes irreversible because water discharge and incision rate in the outlet have quickly increased. Both t_{hc} and the time required for lake extinction t_e increase for large values of depth and length of the lake or for low precipitation values. Note that by definition, $t_o < t_{hc} < t_e$. A t_{hc} value closer to t_o than to t_e indicates that capture is faster at the beginning than at the end. In particular, lake extinction will not occur if the bottom of the lake is below the base level. Also, t_{hc} and t_e decrease nearly linearly with the ratio z_l/l_e (lake elevation divided by escarpment length). The model predicts that maximum incision rates in the outlet area and maximum sediment delivery at the base level occur (Fig. 3A) between t_{hc} and t_e. The time of lake extinction t_e is proportional to the evaporation rate, since endorheism impedes water discharge and erosion in the outlet.

The velocity of lake capture and its reintegration in the fluvial network is also strongly dependent on the ratio z_l/l_e (Fig. 6A). Lakes slightly above the base level undergo very little incision in the outlet, due to the small slope along the outlet river, and extinguish very slowly, mostly by sediment overfilling. High-altitude lakes, in contrast, disappear soon after opening, depending on the other geometrical parameters (e.g., slower if lake depth and length are large). For fixed values of lake depth and length at the time of opening, the lake hypsometry inherited at that stage has two opposing effects on the timing of capture. On one side, it controls the reduction in lake surface, as erosion in the outlet reduces its level. At t_o, a high lake-floor concavity (wide and flat bottom in the center of the lake and steep floor near the shore) implies a smaller reduction in lake length for a given outlet incision. In turn, this implies a smaller increase in water discharge and incision at the outlet, and therefore a slower capture. On the other hand, a high floor concavity also implies smaller distance between the lake's bottom and the base level, requiring shorter time to propagate erosion into the lake's center. The model results indicate that this latter effect dominates over the first and that lakes with high-concavity bathymetry suffer slower captures. Overall, however, the effect of the hypsometric concavity on the lake capture is secondary relative to the effects of lake altitude, evaporation, precipitation, and tectonic uplift, described above. Note also that the time of half capture, t_{hc} (Fig. 6), is always similar or smaller than half of t_e–t_o, showing that the first half of the capture moves at similar speed or faster than the second.

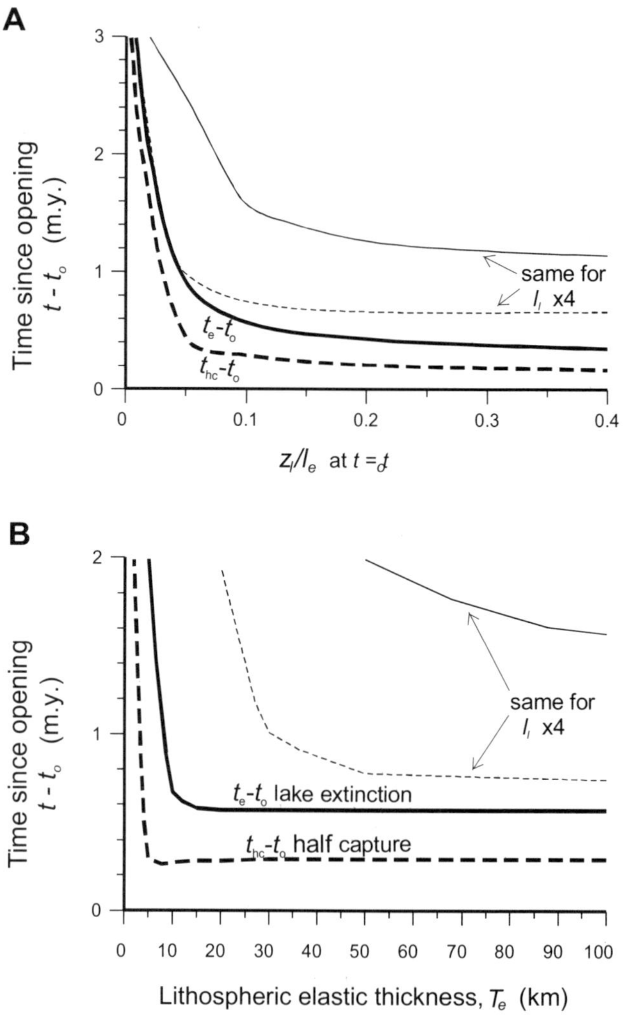

Figure 6. Timing of lake extinction after drainage opening. (A) Effect of lake elevation (normalized with the escarpment length l_e) on the time of half capture (dashed line) and time of extinction (plain line), both measured from the time of opening (t_{hc}–t_o and t_e–t_o, respectively). Thick lines show the results for a lake length of l_l = 50 km; thin lines are calculated for l_l = 200 km. Lakes elevated high above the base level and/or separated from the base level by a high-slope topographic barrier undergo a faster post-tectonic extinction. (B) Same as A, but varying the lithospheric elastic thickness T_e. Large T_e values reduce the isostatic rebound in the lake outlet in response to erosion, accelerating lake extinction.

Because the formation and closure of the lake depends strongly on the uplift rate and initial slope, whereas its opening and extinction depend rather on the accumulated uplift, it is not possible to find a simple parameterization of the lake's life based on the results of the model. Figure 7 shows an attempt to parameterize the lake history as a function of the adimensional quotient between the total uplift $U{\cdot}t_u$ and the maximum initial topography z_0. Note that the reference model was arbitrarily chosen so that $U{\cdot}t_u/z_0 = 1$ (Table 1). Comparison between Figures 7A and 7B shows that the regions of the parameter space where the lake forms and closes are similar, but significant differences remain, particularly for low values of $U{\cdot}t_u/z_0$. For example: a total uplift 4 times smaller than the reference model (i.e., $U{\cdot}t_u/z_0 = 0.25$) will result in no lake if the reduction is performed on U, whereas if t_u is decreased, then a lake develops through nearly 1 m.y. Figure 7C shows the same results as in Figure 7A, but uses double the uplift duration than that of the reference model. Although the lake formation and closure remain invariant (relative to Fig. 7A), the adimensional times of opening and extinction of the lake are significantly delayed. However, it is interesting to remark that

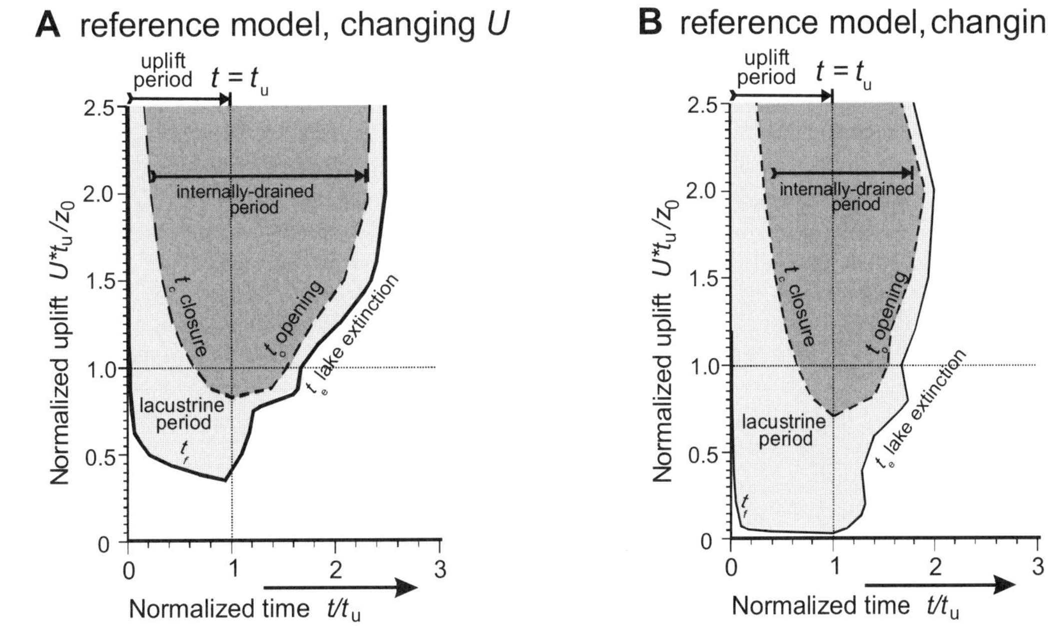

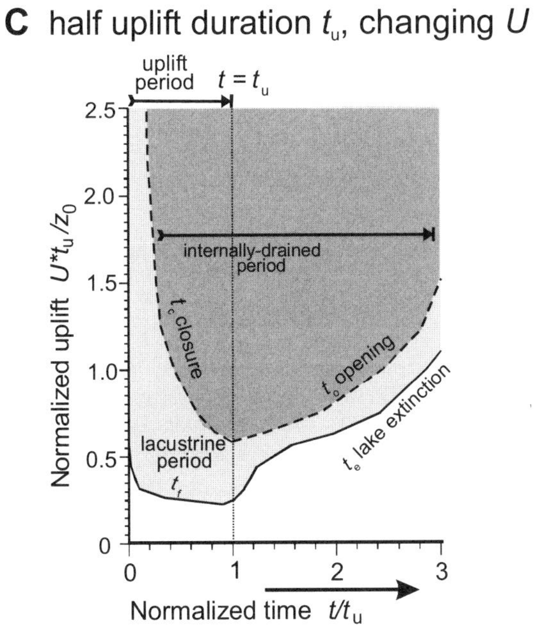

Figure 7. Effect of U, t_u, z_0 (uplift rate, uplift duration, and initial maximum topography) on lake evolution. The light shade indicates the existence of a lake; the dark shade indicates the lake is endorheic (closed drainage). The graphs show the lake evolution in adimensionalized time as a function of an adimensional total uplift. (A) Results of the reference model when changing the uplift rate; (B) same as A, but changing the duration of the uplift; (C) same as A for half uplift duraiion (0.5 m.y.). The differences between the plots indicate that it is not possible to find a precise, simple parameterization of the lake evolution as a function of these parameters. The relative similarity between them shows that, keeping the climatic and lithological parameters fixed, and only as a first approach, lake evolution is mostly controlled by the ratio between the uplift-generated topography $U{\cdot}t_u$ and the initial topography z_0. $U{\cdot}t_u/z_0 = 1$ (corresponding to the reference model) is plotted for reference.

the numerical experiments predict that, generally, an open lake survives for only a short time after uplift ceases, whereas an endorheic basin frequently lasts more than twice the duration of the uplift (see also Fig. 4).

Finally, the effects on this drainage evolution of the vertical motions related to lithospheric flexure were investigated, which were not incorporated in the models above. For this, the load/unload at every column related to sedimentation, erosion, and/or changes in water column were considered, and the lithospheric vertical motions $U_l(x)$ (equation 4) related to this surface mass redistribution were calculated. The reference model (Table 1) was then recalculated for different values of lithospheric elastic thickness (T_e). Large T_e values imply longer-wavelength and smaller-amplitude vertical motions. The results show (Fig. 6B) that post-tectonic uplift at the outlet, related to erosional rebound of the escarpment, induces a relevant delay on t_{hc} and t_e only for large lakes and low T_e. For a lake length of l_l = 50 km, relevant delay (larger than 10%) of t_{hc} and t_e (relative to $T_e = \infty$, i.e., no isostatic vertical movements) are predicted only for T_e values smaller than 7 and 11 km, respectively. In contrast, for l_l = 200 km, t_{hc} and t_e are significantly delayed for T_e < 40 km and T_e < 100 km, respectively (Fig. 3C). T_e values smaller than 3 km (for l_l = 50 km) and 20 km (for l_l = 200 km) produce an outlet rebound related to escarpment erosion that exceeds the erosion rate of the divide, thus delaying the opening of the basin during much larger periods of time.

INTERPRETATION AND DISCUSSION

The lake evolution model presented here assumes a 1D (along-stream) approach to water and sediment flow. In nature, water and sediment are collected in catchments with little symmetry across the river strike. Two-dimensional (planform) modeling was discarded for this study, not only for being extremely expensive in terms of computation time, but also to facilitate the reproducibility and interpretation of the results and because of the difficulty of designing a simple and universal setup (e.g., initial 3D topography). Besides, results from a planform model predict lake capture evolutions that are appreciably dependent on the resolution of the discretization grid (Garcia-Castellanos et al., 2003). A preliminary comparison with the results by Garcia-Castellanos et al. (2003) indicates that out-of-plane effects accelerate the capture process by a factor of ~2–4 (though this depends strongly on the initial 3D topography adopted for the model), maintaining the same qualitative effects of each parameter. Therefore, although the models shown here do provide a reliable quantification of the relative importance of each of the processes involved, the absolute values of lake timing obtained here must be viewed with caution.

A second and more fundamental limitation that must be kept in mind comes from the power-law model adopted for river incision and transport. The ongoing research on the appropriate equations describing these processes (e.g., Whipple and Tucker, 1999; van der Beek and Braun, 1999; Sklar and Dietrich, 2001) has not yet provided an approach with true prediction capability. The different forms proposed for the stream power law are often more sophisticated than that used here, incorporating to equation 1 a dependence on sediment load, granulometry, channel width, or climatic episodicity, for example. The laws proposed remain sometimes grounded on empirical relationships rather than on the physics of the involved processes, or are subjected to poorly constrained parameters such as lithology, vegetation, climatic episodicity, etc. Sets of m and n values (equation 1) different from those used here but still agreeing with field measurements would predict different velocities of upstream migration of erosion (e.g., Whipple and Tucker, 1999) and therefore lake capture. For the purposes of this paper, the simplified form adopted is sufficient, but the results presented here exemplify the extraordinary potential of the stream power law to improve our understanding of landscape evolution, and encourages the search for improved process-based models with increased predictability. Conversely, they suggest that a broad database giving constraints on the evolution of a large number of lakes may help us to refine the available surface process models using the techniques described in this paper.

According to the modeling results, the evolution of lakes is sensitive to the initial geometrical configuration, lithology, tectonic uplift, and climate in a similar degree, and all these factors can change the timing of lake evolution by several orders of magnitude. The most remarkable result is that evaporation, by reducing outlet water discharge in lakes, can substantially reduce erosion rates and extend the lake's life. Evaporation is therefore a key factor triggering and controlling the formation of internally drained basins. A good example for the stage of drainage closure is Lake Issyk-Kul (northern Tien-Shan, Kyrgyzstan), an ~180 × 50 km closed lake trapped at 1600 m above sea level in the forefront deformation of the Himalayan collision. Lake Issyk-Kul, with a P/E coefficient of 0.28 remains internally drained since the late Pleistocene (De Batist et al., 2002), although short periods of historical reopening have been reported. Together with the intense seismic activity in the region and the shortening velocities between the Indian and Asian domains, this suggests that, in geological time scales, Lake Issyk-Kul is presently being incorporated into the internally drained central Asia.

The topography upstream from the tectonic barrier is also a key factor controlling the reduction of water output through the lake's outlet, and it cannot be neglected in simulating the development of internally drained basins. All other parameters being constant, a relatively flat upstream topography implies that the lake attains a large area with small amounts of uplift, favoring evaporation and drainage closure, whereas a steep topography implies smaller lakes and less water evaporation for the same amount of uplift. This means that the formation of internally drained basins requires slower uplift rates in flat regions (e.g., Lake Bungunnia, S. Australia) than in mountainous regions (e.g., Lake Issyk-Kul, Himalayan collision zone).

The model successfully reproduces the notion that, after tectonic forcing ceases, lakes open and extinguish by a combination of sediment fill and outlet erosion (Fig. 2A). Lake elevation is the most relevant geometrical parameter controlling the timing of this process and dominates over the lake's length and depth (Fig. 6A). Thus, decreasing simultaneously by one order of magnitude all three geometrical parameters, capture slows down by a similar amount as a result of lower lake altitude and potential energy. In turn, if only length and depth are reduced and lake elevation is kept unchanged, capture is accelerated.

The flexural response of the lithosphere to surface mass redistribution (isostatic rebound and subsidence) also induces a scale-dependent effect. For the range of elastic thickness most frequently reported in continental lithosphere ($5 < T_e < 20$ km; Watts, 1992), lakes larger than 50–200 km have a post-tectonic evolution sensitive to lithospheric flexure. An important implication is that surface transport in large, closed lacustrine basins interplays with isostasy and erosion in the surrounding areas to control the long-term landscape evolution, and that the water budget in internal basins might have a key effect on the topographic and drainage evolution of continental margins (in addition to other factors such as isostatic rebound or escarpment retreat; Tucker and Slingerland, 1994). In particular, the peneplanation of Gondwana prior to breakup facilitated, together with the flexural uplift of the new continental margins, the formation of large, closed internal basins in southern Africa (Summerfield, 1991), triggering a period of intracontinental deposition lasting for most of the Cenozoic. The increase in altitude of these basins by sediment infill, possibly in combination with a wetter climatic phase, might have eventually triggered their drainage opening in late Tertiary times, in a similar way to that proposed by Garcia-Castellanos et al. (2003) for the Ebro Basin (NE Iberia).

In the last decade, numerical modeling techniques have revealed that the spatial distribution of erosion and surface transport controls the internal structure and tectonic evolution of orogens (e.g., Beaumont et al., 1992; Avouac and Burov, 1996; Willett, 1999; Persson et al., 2004). These studies have not explicitly addressed the effects of internally drained, endorheic basins on surface transport during orogenesis, despite the fact that these occupy large areas of many mountain belts on Earth (e.g., Andean Altiplano, Tibetan Plateau, and Tarim Basin). The results obtained in this work indicate that the hydrologic balance in intramountain basins determines their duration, possible closure, and the volume of their sedimentary infill. Long-lasting lacustrine basins are promoted and prolonged by low initial topographic gradients (low z_0), large and durable uplift rates (high U, t_u) acting over large areas (large l_u), dry climate (low P/E ratio), hard lithology (low K_f, K_d), and in the case of lakes larger than ~50 km, low lithospheric rigidities.

Orographic effects on precipitation (not explicitly incorporated to the model) block the inflow of humid air into intramountain areas (e.g., Roe et al., 2003), lowering the precipitation/evaporation ratio and thus facilitating endorheism, as shown in Figure 5. Accounting for this would in fact prolong the endorheic periods obtained with the model above, since the growing tectonic barrier would result in drier climate upstream from the uplift region. A longer endorheic period would imply larger sediment accumulation in the closed basin and lower erosion of its flanks, which in turn would induce a propagation of shortening toward the external parts of the orogen (e.g., Persson et al., 2004), reinforcing the closure of the basin by localizing precipitation further away from it. The ongoing drainage closure of Lake Issyk-Kul mentioned above, for example, may be not just a consequence but also a cause of the further propagation of tectonic deformation into Asia. Similarly, it is a remarkable concurrence of drainage closure in the Ebro Basin (NE Iberia) at 35 Ma and the dramatic reduction of tectonic deformation along the Pyrenees at ca. 30 Ma (Garcia-Castellanos et al., 2003, and references therein). This suggests that the coupling between orographic precipitation and tectonic deformation might be sufficient to explain the formation of long-lasting intramountain basins, such as the Neogene Ebro Basin, the Lake Issyk-Kul Basin, or the Andean Altiplano, and an inherited tectonic structure, such as a crustal-scale weakening, is not required. From the perspective of this feedback effect, the formation of an endorheic intramountain basin during the early stages of orogenesis, favored by a dry climatic setting and a low-relief inherited topography, might be reinforced and persist in the later tectonic and drainage evolution of the orogen. Future numerical models coupling the dynamics of tectonic, surface, and climatic processes should allow further investigation of the relevance of such effects.

CONCLUSIONS

Within the approaches inherent to the model described above, the following conclusions can be made:

1. The defeat of a river by tectonic uplift and the development of an internally drained basin are dependent on geometrical constraints (initial relief, length of the river), lithological parameters (rock erodibility), tectonics (uplift rate, duration, and its spatial distribution), and climate (precipitation and evaporation rates). For example, quantitatively determining the climatic conditions and uplift rate under which a particular lake and/or internally drained basin was formed requires information on the paleotopography upstream from the tectonic barrier (a flatter upstream topography would require less uplift to close drainage and vice versa). This ultimate goal, however, remains limited by our capability to accurately predict river incision and transport.
2. River defeat (uplift rate $U_T >$ river incision rate e) results in the formation of a lake, which only develops into an internally drained basin if (a) the ratio P/E (precipitation/evaporation) is significantly lower than 1; and (b) uplift persists until the lake is large enough to evaporate all collected water. In the notation used above, the necessary condition to generate an internally drained basin out of a defeated river is $l_l > l_b \cdot P/E$.
3. The drainage closure of a lake induces an important prolongation of the lake's life beyond the end of tectonism by

inhibiting erosion along the drainage outlet. The duration of the lake is increased by a time comparable to the duration of the endorheic period. This leads to a prolongation of the lake duration by a factor generally larger than 2, relative to a lake formed under similar conditions but not attaining endorheism.

4. The post-tectonic extinction of lakes larger than 50–200 km is significantly delayed by flexural isostatic uplift occurring in response to erosion at the topographic barrier.

ACKNOWLEDGMENTS

Sean Willett, Mark Brandon, and two anonymous reviewers are acknowledged for their constructive comments and criticisms; their interest and effort helped to clarify and improve this paper. The manuscript benefited also from illuminating discussions with Mike Summerfield and Gerard Hérail. Funding was provided by the Netherlands Centre for Integrated Solid Earth Science (ISES).

REFERENCES CITED

Avouac, J.P., and Burov, E.B., 1996, Erosion as a driving mechanism of intracontinental mountain growth: Journal of Geophysical Research, v. 101, no. B8, p. 17,747–17,769, doi: 10.1029/96JB01344.

Beaumont, C., Fullsack, P., and Hamilton, J., 1992, Erosion control of active compressional orogens, *in* McClay, K.R., ed., Thrust tectonics: London, Chapman and Hall, p. 1–18.

Braun, J., Heimsath, A.M., and Chappell, J., 2001, Sediment transport mechanisms on soil-mantled hillslopes: Geology, v. 29, p. 683–686, doi: 10.1130/0091-7613(2001)0292.0.CO;2.

Burbank, D.W., 1983, The chronology of intermontane-basin development in the northwestern Himalaya and the evolution of the northwest syntaxis: Earth and Planetary Science Letters, v. 64, p. 77–92, doi: 10.1016/0012-821X(83)90054-7.

De Batist, M., Klerkx, J., Imbo, Y., Giralt, S., Lignier, V., Beck, C., Delvaux, D., Vermeesch, P., Kalugin, I., and Abdrachmatov, K., 2002, Bathymetry and sedimentary environments of a large, high-altitude, tectonic lake: Lake Issyk-Kul, Kyrgyz Republic (Central Asia), *in* Klerkx, J., and Imanackunov, B., eds., The Issyk-Kul Lake: Evaluation of environmental state and its remediation: Dordrecht, Netherlands, Kluwer Academic Publishers, North Atlantic Treaty Organization Science Series IV, Earth and Environmental Series, v. 13, p. 101–123., p. 101–124.

Flemings, P.B., and Jordan, T.E., 1989, A synthetic stratigraphic model of foreland basins development: Journal of Geophysical Research, v. 94, p. 3851–3866.

Fornari, M., Risacher, F., and Feraud, G., 2001, Dating of paleolakes in the central Altiplano of Bolivia: Palaeogeography, Palaeoclimatology, Palaeoecology, v. 172, p. 269–282, doi: 10.1016/S0031-0182(01)00301-7.

Garcia-Castellanos, D., Fernàndez, M., and Torné, M., 1997, Numerical modeling of foreland basin formation: A program relating thrusting, flexure, sediment geometry and lithosphere rheology: Computers & Geosciences, v. 23, no. 9, p. 993–1003, doi: 10.1016/S0098-3004(97)00057-5.

Garcia-Castellanos, D., Fernàndez, M. and Torné, M., 2002, Modelling the evolution of the Guadalquivir foreland basin (South Spain): Tectonics, v. 21, p. 9-1–9-17, doi: 10.1029/2001TC001339.

Garcia-Castellanos, D., Vergés, J., Gaspar-Escribano, J.M., and Cloetingh, S., 2003, Interplay between tectonics, climate and fluvial transport during the Cenozoic evolution of the Ebro Basin (NE Iberia): Journal of Geophysical Research, v. 108, no. B7, p. 2347, doi: 10.1029/2002JB002073.

Howard, A.D., 1994, A detachment-limited model of drainage basin evolution: Water Resources Research, v. 30, p. 2261–2285.

Irwin, R.P., Maxwell, T.A., Howard, A.D., Craddock, R.A., and Leverington, D.W., 2002, A large paleolake basin at the head of Ma'adimVallis, Mars: Science, v. 296, p. 2209–2212, doi: 10.1126/science.1071143.

Kühni, A., and Pfiffner, O.A., 2001, Drainage patterns and tectonic forcing: A model study for the Swiss Alps: Basin Research, v. 13, no. 2, p. 169–197, doi: 10.1046/j.1365-2117.2001.00146.x.

O'Connor, J.E., 1993, Hydrology, hydraulics, and geomorphology of the Bonneville Flood: Geological Society of America Special Paper 274, 90 p.

Ollier, C., 1981, Tectonics and landforms: Longman, UK, Harlow, v. 324, p. 1981.

Persson, K.S., Garcia-Castellanos, D., and Sokoutis, D., 2004, River transport effects on compressional belts: First results from an integrated analogue-numerical model: Journal of Geophysical Research, v. 109, no. B1, p. B01409, doi: 10.1029/2002JB002274.

Rodriguez-Iturbe, I., and Rinaldo, A., 1997, Fractal river basins: Chance and self-organization: New York, Cambridge University Press, 547 p.

Roe, G.H., Montgomery, D.R., and Hallet, B., 2003, Orographic precipitation and the relief of mountain ranges: Journal of Geophysical Research, v. 108, p. 2315, doi: 10.1029/2001JB001521.

Sklar, L.S., and Dietrich, W.E., 2001, Sediment and rock strength controls on river incision into bedrock: Geology, v. 29, no. 12, p. 1087–1090, doi: 10.1130/0091-7613(2001)0292.0.CO;2.

Sobel, E.R., Hilley, G.E., and Strecker, M., 2003, Formation of internally-drained contractional basins by aridity-limited bedrock incision: Journal of Geophysical Research, B, Solid Earth and Planets, v. 108, 25 p.

Stephenson, A.E., 1986, Lake Bungunnia; a Plio-Pleistocene megalake in southern Australia: Palaeogeography, Palaeoclimatology, Palaeoecology, v. 57, p. 137–156, doi: 10.1016/0031-0182(86)90011-8.

Summerfield, M.A., 1991, Global geomorphology: Harlow, UK, Longman, 536 p.

Tiercelin, J.L., 2002, Lake sediments; archives of active tectonics and climate changes: Palaeogeography, Palaeoclimatology, Palaeoecology, v. 187, p. 209–212, doi: 10.1016/S0031-0182(02)00476-5.

Tucker, G.E., and Slingerland, R.L., 1994, Erosional dynamics, flexural isostasy, and long-lived escarpments: A numerical modeling study: Journal of Geophysical Research, v. 99, p. 12,229–12,243, doi: 10.1029/94JB00320.

Tucker, G.E., and Slingerland, R.L., 1997, Drainage responses to climate change: Water Resources Research, v. 33, no.8, p. 2031–2047, doi: 10.1029/97WR00409.

Turcotte, D.L., and Schubert, G., 2002, Geodynamics: Cambridge, UK, Cambridge University Press, 456 p.

van der Beek, P., and Bishop, P., 2003, Cenozoic river profile development in the Upper Lachlan catchment (SE Australia) as a test of quantitative fluvial incision models: Journal of Geophysical Research, v. 108, no. B6, p. 2309, doi: 10.1029/2002JB002125.

van der Beek, P., and Braun, J., 1999, Controls on post-Mid-Cretaceous landscape evolution in the southeastern highlands of Australia; insights from numerical surface process models: Journal of Geophysical Research, v. 104, no. B3, p. 4945–4966, doi: 10.1029/1998JB900060.

van der Beek, P., Champel, B., and Mugnier, J.-L., 2002, Control of detachment dip on drainage development in regions of active fault-propagation folding: Geology, v. 30, p. 471–474, doi: 10.1130/0091-7613(2002)0302.0.CO;2.

Watts, A.B., 1992, The effective thickness of the lithosphere and the evolution of foreland basins: Basin Research, v. 4, p. 169–178.

Whipple, K.X., and Tucker, G.E., 1999, Dynamics of the stream-power river incision model; implications for height limits of mountain ranges, landscape response timescales, and research needs: Journal of Geophysical Research, Solid Earth and Planets, v. 104, no. 8, p. 17,661–17,674, doi: 10.1029/1999JB900120.

Willett, S.D., 1999, Orogeny and orography: The effects of erosion on the structure of mountain belts: Journal of Geophysical Research, v. 104, no. B12, p. 28,957–28,981, doi: 10.1029/1999JB900248.

Williams, D.F., Peck, J., Karabanov, E.B., Prokopenko, A.A., Kravchinsky, V., King, J., and Kuzmin, M.L., 1997, Lake Baikal record of continental climate response to orbital insolation during the past 5 million years: Science, v. 278, p. 1114–1117, doi: 10.1126/science.278.5340.1114.

Yan, J.P., Hinderer, M., and Einsele, G., 2002, Geochemical evolution of closed-basin lakes: General model and application to lakes Qinghai and Turkana: Sedimentary Geology, v. 148, p. 105–122, doi: 10.1016/S0037-0738(01)00212-3.

Manuscript Accepted by the Society 23 June 2005

Printed in the USA

Geological Society of America
Special Paper 398
2006

Knickpoints and hillslope failures: Interactions in a steady-state experimental landscape

Alessandro Bigi[†]
Department of Agricultural Economics and Engineering, University of Bologna, Viale Fanin 50, I-40127 Bologna, Italy
Leslie E. Hasbargen[‡]
Department of Geology, University of Delaware, Newark, Delaware 19713, USA
Alberto Montanari[§]
Faculty of Engineering, University of Bologna, Via del Risorgimento 2, I-40136 Bologna, Italy
Chris Paola[#]
Department of Geology and Geophysics and St. Anthony Falls Lab, University of Minnesota, Minneapolis, Minnesota 55455, USA

ABSTRACT

Hillslope stability depends strongly on local conditions, such as lithology and rock strength, degree of saturation, and critical slope angle. Common triggers for slope failure include severe storms, earthquakes, and removal of material from the toe of the hillslope. In this paper, we focus on the latter, in a model in which streams incise the toe and destabilize the hillslope. We investigate possible interactions between migrating knickpoints and hillslope failures in a small-scale, steadily eroding experimental landscape that experiences steady rainfall and base-level fall conditions. We monitored knickpoint propagation and hillslope failure activity with time lapse photography over a time period in which numerous knickpoints migrated through the drainage basin. We then investigated temporal and spatial relationships between hillslope failures and knickpoints and compared these results to Monte Carlo simulations of hillslope failure distributions. When focusing along a single channel, we found that, statistically (significant at the 98% confidence level), a greater number of failures occur downstream from a migrating knickpoint. These results highlight both the organized and random nature of hillslope and knickpoint interactions.

Keywords: knickpoints, landslide triggering, evolution, hillslope failure.

INTRODUCTION

Landslides have a profound impact on societal structures and cause billions of dollars of damage in the United States alone each year (Schuster and Highland, 2001). They also play a significant role in denudational processes in eroding drainage basins. Hence, there is significant motivation to understand the controls of landslide initiation, as well as the timing, location, and size of landslides. In this paper, we study stream and hillslope interactions in a controlled experimental drainage basin. We test the hypothesis that stream incision driven by migrating knickpoints can impose a spatial and temporal pattern on landslide

[†]Corresponding author e-mail: abigi@agrsci.unibo.it.
[‡]E-mail: lesh@udel.edu.
[§]E-mail: alberto.montanari@unibo.it.
[#]E-mail: cpaola@umn.edu.

Bigi, A., Hasbargen, L.E., Montanari, A., and Paola, C., 2006, Knickpoints and hillslope failures: Interactions in a steady-state experimental landscape, *in* Willett, S.D., Hovius, N., Brandon, M.T., and Fisher, D.M., eds., Tectonics, Climate, and Landscape Evolution: Geological Society of America Special Paper 398, Penrose Conference Series, p. 295–307, doi: 10.1130/2006.2398(18). For permission to copy, contact editing@geosociety.org.

distributions. Before we begin our analysis, we provide a brief overview of landslide studies.

Local controls on landsliding involve several important variables, including lithologic structure, soil development and structure, pore fluid pressure, topographic setting, and vegetative cover. A Mohr-Coulomb relation between stress and failure (Terzaghi et al., 1996) forms the basis for estimating the critical stress at which a hillslope will fail. There are several ways for the motion threshold to be exceeded. Common triggers include severe rain storms, where increased pore fluid pressure facilitates failure (Selby, 1982; Burton et al., 1998; Dietrich et al., 1995; Iverson, 2000), earthquakes, where ground shaking induces failure (Keefer, 1994; Harp and Jibson, 1995; Crozier et al., 1995; Havenith et al., 2003), and removal of material and support from the toe of the hillslope (Schumm, 1956; Densmore et al., 1997; Larsen and Parks, 1997). We investigate this last mechanism in our study. Vegetation exerts control on substrate resistance, and this adds further complexity to the susceptibility of hillslope failure (Sidle, 1992; Duan, 1996; Montgomery et al., 2000; Gabet and Dunne, 2002).

Spatial and temporal distributions provide an estimate of the contribution of landsliding to the erosional budget for landscapes, and offer probabilistic tools for predicting landslide events in a given area (van Asch and van Steijn, 1991; Hovius et al., 1997; Larsen and Torres-Sanchez, 1998; Miller and Sias, 1998; Hermanns et al., 2000; Densmore and Hovius, 2000; Trauth et al., 2000; Stark and Hovius, 2001; Guzzetti et al., 2002; Martin et al., 2002; Dadson et al., 2003; Brardinoni and Church, 2004). Increasingly, physics-based modeling has coupled topographic and weather information with stability criteria, and this approach offers predictions for the location and timing of landsliding (Iida, 1984; Casadei et al., 2003). Landslide size-frequency studies have also employed physical experiments (Densmore et al., 1997) in which a simulated granular hillslope composed of beans responded to base-level fall.

Due to the local conditions that control hillslope stability, one might expect that landslide occurrences are stochastic in time and space (e.g., Benda and Dunne, 1997). Some spatial patterning results from the origination of shallow landslides in hollows in low-order drainages (Campbell, 1975; Montgomery and Dietrich, 1994; Benda and Dunne, 1997). In this case, the spatial pattern is governed by drainage basin structure. Temporal patterns of landslides are largely related to triggering by storm or earthquake events. However, time series of shallow landslides generated from experimental (Densmore et al., 1997) and physics-based models (Benda and Dunne, 1997) are highly stochastic, emphasizing the sensitivity to local conditions.

We turn now to an additional means of destabilizing hillslopes in natural settings: removal of material from the toe of the hillslope by streams. This mechanism has received less attention, largely due to the longer time scales involved in stream incision. While bedrock incision by streams involves a variety of mechanisms (Wohl, 2003; Hancock et al., 2003; Sklar and Dietrich, 2003), we focus on two styles of stream incision in this paper: a steady profile lowering due to uniform erosion of the bed, and knickpoint migration.

Knickpoints can play a significant role in river incision and valley development (Ahnert, 1998). Knickpoints are step changes in bed surface elevation where intense, localized erosion takes place (Brush and Wolman, 1960; Gardner, 1983; Bennett et al., 2000). Formation of knickpoints and their upstream migration have been linked to concentration of overland flow (Mosley, 1974; Merritt, 1984), and rill and gully erosion (Bryan, 1990; Slattery and Bryan, 1992). They are an important erosional process in bedrock channels (Miller, 1991), and landscape evolution (Dietrich and Dunne, 1993; Zaprowski et al., 2001). The mechanism of upstream knickpoint migration has been the object of several experimental, theoretical, and field studies (Brush and Wolman, 1960; Holland and Pickup, 1976; Gardner, 1983; Bryan and Rockwell, 1998; Bennett et al., 2000; Parker and Izumi, 2000; Alonso et al., 2002; Crosby and Whipple, 2002), where the effects of varying bed material, water table height, slope, and flow discharge on migration rate have been explored.

While numerous studies of hillslope stability, river incision, and knickpoint behavior have been conducted, these studies have not looked into a systematic relationship between knickpoint propagation and hillslope failure activity. There are good reasons for this omission. Knickpoints in bedrock channels require a significant amount of time to propagate up through a drainage basin, and so field studies of stream incision and hillslope response are limited by the length of time required to observe this kind of behavior in natural landscapes. Hillslope adjustment to a lowering river bed via landsliding has been documented as a significant response by Burbank et al. (1996) in the northwestern Himalaya, but the connection between stream incision and knickpoint propagation is not known for this case.

We pursue the effect of stream incision on hillslope stability in this paper by postulating that hillslope failures should exhibit spatial and temporal patterns controlled by knickpoint propagation. Namely, if a stream is actively downcutting at the toe of the hillslope, and stream incision is due to the migration of a knickpoint, then hillslope failures will follow in the wake of the knickpoint. According to this assumption, knickpoint location and migration should have an influence on the spatial and temporal patterns of landslides. A small experimental drainage basin provides a convenient setting to test the strength of coupling between stream incision and hillslope failures. Recent research (Hasbargen and Paola 2000, 2003; Lague et al., 2003) has demonstrated the utility of monitoring experimental eroding landscapes under steady uplift and rainfall conditions. These experiments have documented hillslope failures and knickpoints as common erosional processes within laboratory drainage basins (Hasbargen, 2003). Hence, basic interactions between stream incision and failures can be studied in a controlled environment, and they provide a suitable setting to test the idea that knickpoints impose a spatial and temporal pattern of failures in their wake. As a side note, hillslope failures come in a variety of sizes and styles of movement. While we recognize the diverse character of mass movements on hillslopes in natural settings, we will use the terms "hillslope failure" and "landslide" interchangeably throughout this paper. Because weathering and soil development are absent in our experimental landscape, shallow landslides of soil are absent in our experiment. Landslides in our erosional facility are analogous to deep-seated bedrock landslides.

DESCRIPTION OF THE EXPERIMENTAL APPARATUS

A small-scale physical experimental apparatus was set up at St. Anthony Falls Laboratory of the University of Minnesota (Hasbargen and Paola, 2000; Hasbargen, 2003). The apparatus consisted of a nearly circular steel tank ~1 m in diameter and 1 m deep with a single outlet dammed by a motor-controlled gate (Fig. 1). The outlet was 1 cm wide. A motor was attached to the sliding gate via a cable. We ran the motor continuously during the experiment, dropping the outlet at a slow, constant rate. The effect of dropping base level is equivalent to uniform block uplift of the basin relative to base level. We took numerous measurements of the outlet height during the experiment to verify steady base-level fall. A set of 8 greenhouse misters placed 70 cm above the upper level of the tank sprinkled rain (droplet size <200 μm)

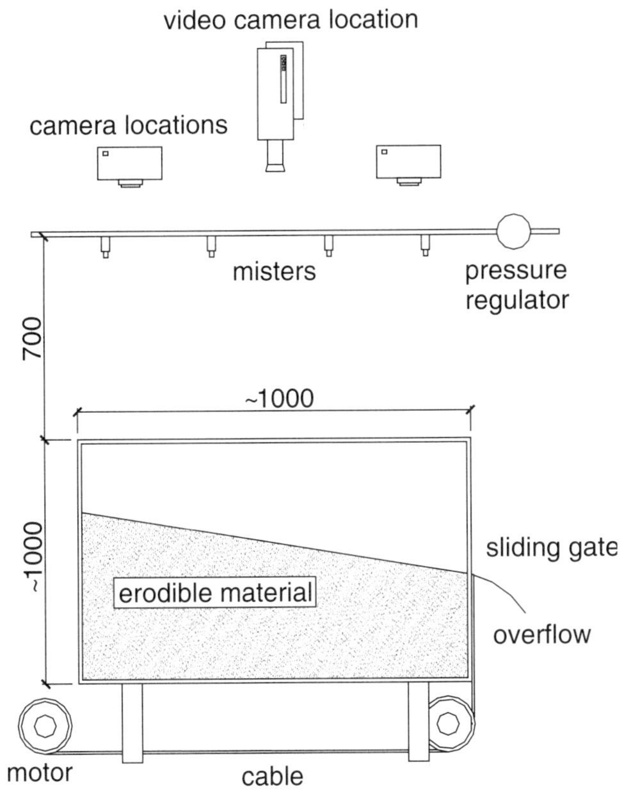

Figure 1. Schematic cross section of the experimental erosion facility. A steel cable drops the outlet; it is wrapped on a pulley and driven by an electric motor mounted on the floor. Dimensions are in millimeters.

over the basin, generating runoff. A pressure regulator maintained a constant pressure to the mist apparatus.

The general behavior of this experimental system under variety of conditions has been reported in Hasbargen and Paola (2000, 2003). For the present study, we carried out detailed measurements of landslide distribution and knickpoint migration in one experiment. The experimental conditions were: rainfall rate = 10.83 μm/s, base-level fall rate = 3.1 μm/s. Before beginning the experiment, we calibrated the rainfall spatial distribution by collecting rainfall in pans at the top of the basin. The spatial coefficient of variation was <15% over a measurement interval of 12 min ($n = 27$). At the beginning of the experiment, the basin was filled with erodible material consisting of a well-sorted silica silt (D_{50} = 45 μm) mixed with kaolinite (1% by weight). These were mixed with water to the consistency of a medium cement in a cement mixer, poured into the basin, and allowed to settle overnight.

After preliminary calibration of rainfall and base-level fall rate, we initiated the experiment by turning on the motor-controlled outlet and rainfall. The run lasted for 63 h, at which time the sliding gate outlet reached the bottom of the basin. Initially, the surface was flat, and runoff patterns were essentially random (Fig. 2). Upon complete dissection of the initial surface, a 3–5 order drainage network had developed (Fig. 3). The dissection of the initial flat surface required ~15 h. Digital stereo-images (1280 × 960 pixel resolution) of the basin were collected at 7–10 min intervals. Time-lapse video recorded snapshots every 100 s. Sediment flux at the basin outlet was measured by weighing sediment and water in a handheld graduated cylinder, and recording the time required to fill it. We episodically applied a thin veil of colored sand over the surface as a tool for the identification of topographic features, such as terraces (Fig. 4) and hillslope failures (Fig. 5). The erosion of the colored sand provided an indication of both the length of time required to erode the surface, and a visual record of spatially variable erosion.

We collected 330 photographs, of which 96 were selected for mapping the position of knickpoints and the location and size of hillslope failures. These 96 photographs covered the last part of the experiment, an interval of 13 h, during which time the landscape eroded at a flux steady-state condition (Willett and Brandon, 2002), an amount of time sufficient for the landscape to erode through 14 cm of the substrate, or ~1 unit of local relief (11 cm).

EXPERIMENTAL RESULTS

Qualitative Observations of Drainage Basin Dynamics

Despite temporally steady external forcing and flux steady-state conditions, the drainage network exhibited interesting non-steady dynamic behavior, as in the experiments reported in Hasbargen and Paola (2000, 2003). We noted short-term processes, such as hillslope failures and runoff-based erosion; intermediate time-scale processes, such as knickpoint development and

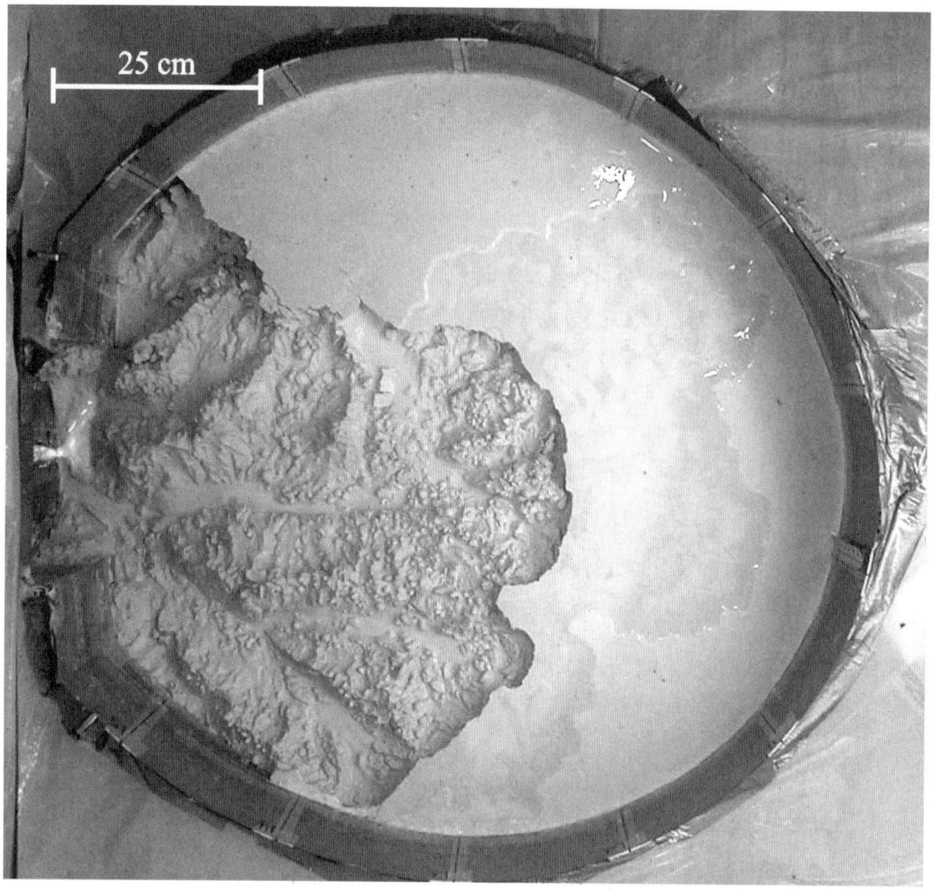

Figure 2. Vertical view of the experimental basin at the beginning of the run during dissection of the initial flat surface.

Figure 3. Vertical view of the dendritic pattern of the drainage network. The photo was taken long after complete dissection of the initial surface (note rind near the top of the basin walls) at run time 62:13 h. Outlet is at left center. Basin length (left to right) is 98 cm; basin width (top to bottom) is 86 cm.

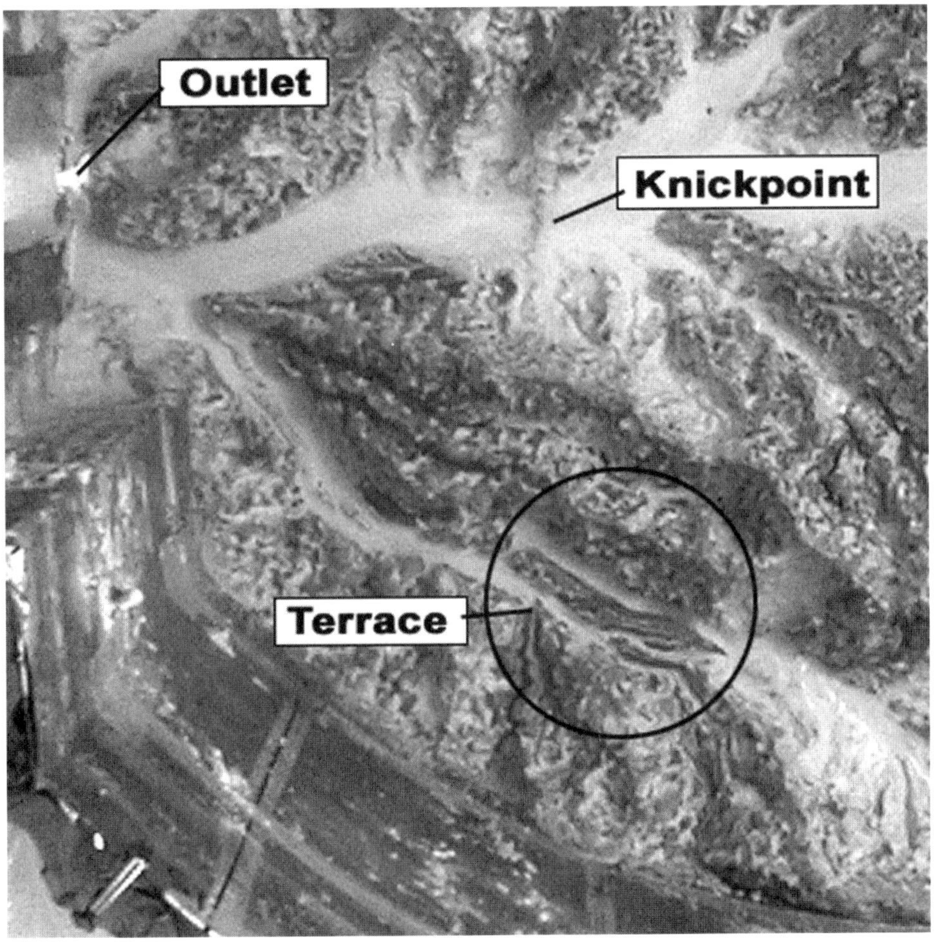

Figure 4. Image of the experimental basin showing wide valley of trunk stream, diagnostic of temporary sediment storage; knickpoint incising into trunk stream; and island and terrace-like structures in the tributary channel in the lower portion of image.

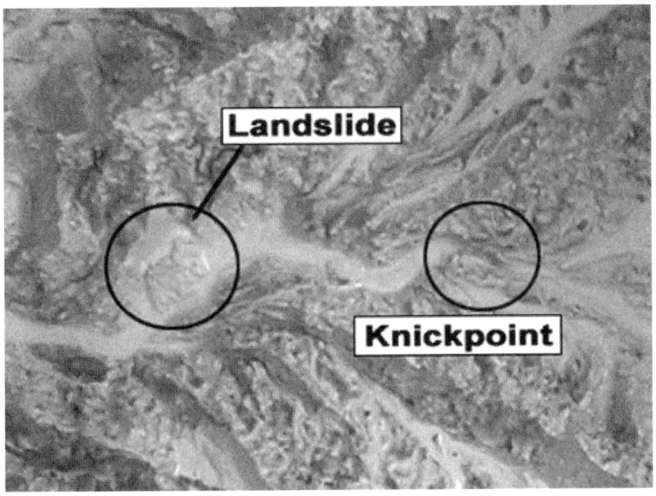

Figure 5. Close-up view of a hillslope failure in the trunk stream after the passage of a knickpoint. Approximate width of image is ~25 cm.

migration, and temporary sediment storage in trunk valleys; and long-term behavior marked by lateral ridge crest movements. Runoff-based erosion was the dominant mechanism for removing material from the basin; it was also the most difficult to detect visually, because flows were rarely more than a few mm deep. We observed knickpoints that typically initiated in the lower reaches of the basin and propagated upstream, apparently triggering hillslope failures.

Rainfall was necessarily stopped before collecting photographs. We took advantage of these occasions to determine how long surface runoff persisted after cessation of rainfall. Flow at the outlet approached nil within a few seconds, implying rapid draining of groundwater from hillslopes and/or very low transmissivity of groundwater in the substrate. Overall, we could not detect obvious effects on landscape development from rainfall cessation and recommencement. Hillslope failures were easier to detect than surface runoff erosion and appeared as abrupt shifts in ridge crest location and as concave hillslopes with a slump at the hillslope toe (Fig. 5).

Episodic cycling between temporary sediment storage in trunk streams and excavation of stored sediment by knickpoints marked the evolution of the landscape throughout the entire

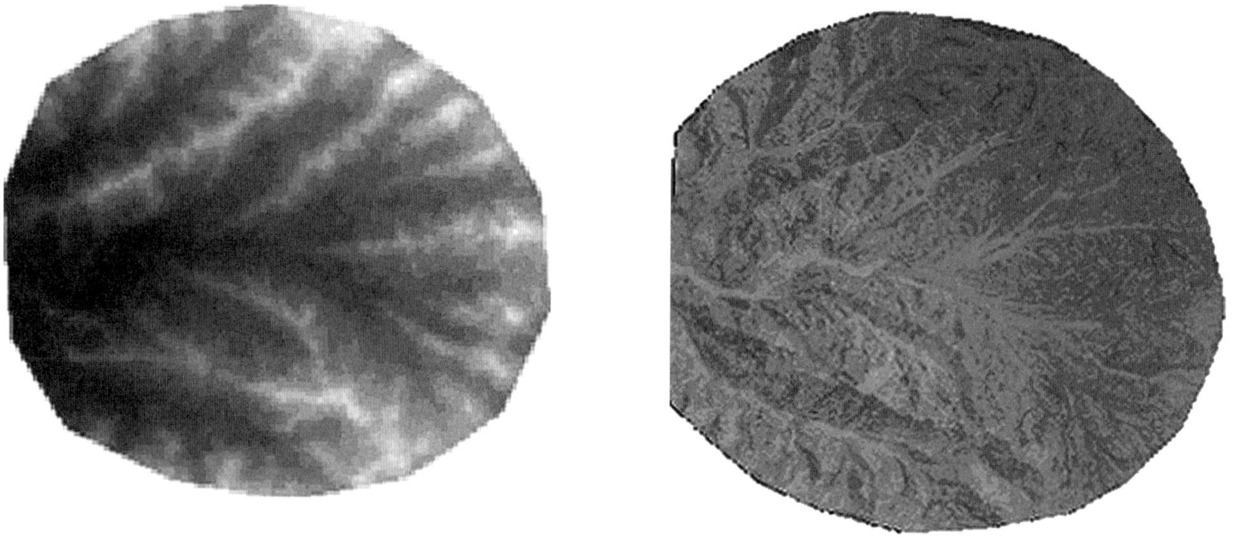

Figure 6. Grayscale digital elevation map (DEM) with pits removed (left), and vertical photograph (right, July 18, photo 17, run time 54:16 h). Long dimension of each image is 90 cm, and width is 86 cm. Grid spacing for DEM is 0.5 cm. Total relief is 22 cm; local relief is 11 cm.

experiment. During times of temporary sediment storage, trunk streams exhibited wide, smooth valley floors with no clearly defined channel. Sediment storage occupied up to 15% of the basin surface area at any one time. Depositional episodes ended when a knickpoint initiated and migrated upstream. Streams deepened and narrowed and rapidly flushed out stored sediment. Similar types of behavior have been noted for experimental channels following an abrupt base-level drop (Cantelli et al., 2004). After the passage of the knickpoint, sediment storage began again, and the sequence was repeated. In some cases, a smooth valley floor developed more than one channel as a knickpoint propagated upstream. Some channels developed marked sinuosity, though most reaches were straight. We did not observe braided channels, though around knickpoints some islands developed. The time scale for a cut and fill cycle ranged from 1 to 1.5 h, equivalent to 1.1–1.7 cm of base-level fall. When this fall distance is scaled to the typical knickpoint height, the cut/fill cycle time is equivalent to the time needed for ~2 knickpoint heights of erosion.

Measurements of Hillslope Failure, Knickpoint, and Temporary Sediment Storage Area

We offer the following measures to give an approximate scale of our experimental landscape, and to facilitate comparison to natural drainage basins. We extracted a few digital elevation models (DEMs) from stereo photographic snapshots of the basin during the course of the experiment to provide basic topographic information (see Fig. 6). Maximum relief was 24 cm in our experiments, with local ridge-to-valley relief of 10–12 cm. Maximum relief divided by basin length (98 cm) yielded a regional slope value of 0.24. Mean steepest descent slope computed at 5 mm grid spacing was 0.90. Hillslope lengths (distance from divide to channel) ranged from 1 to 10 cm. Low-order channel widths were 0.1–0.2 cm, and trunk valley widths during times of sediment storage ranged up to 5 cm. Upon dissection by a knickpoint, channel widths of trunk streams decreased to ~1 cm. Experimental hillslope failures commonly extended from ridge crest to valley floor, and ranged in size from 1 to 50 cm^2. As noted earlier, shallow landslides associated with soil development were absent in the experimental basin. The hillslope failures in our experiment are more appropriately thought of as deep-seated landslides. We present topographic profiles of a few failures and a lower-order channel in Figures 7 and 8. Relief of most knickpoints was ~0.5 cm, or 5% of the local ridge-to-valley relief. However, some

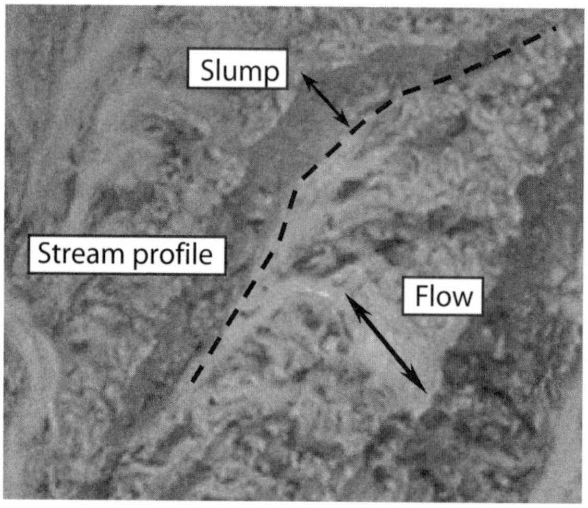

Figure 7. Vertical photograph of two mass movements on hillslopes, showing the location of slump, flow, and stream profile for which topographic profiles were extracted from a digital elevation model (run time, h:min = 54:16).

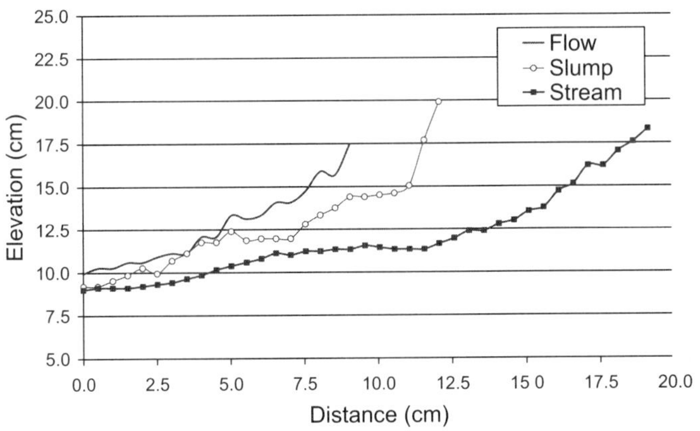

Figure 8. Example of hillslope and channel profiles in the experimental landscape depicted on Figure 7. The feature designated flow appears to be a mass flow, and the feature designated slump exhibits signs of a scarp and irregular deposit at the toe of the hillslope. A stream profile for a second-order channel is plotted to provide a lower limit for relief.

knickpoints exhibited heights of 1–2 cm (10%–20% of local relief). Flat surfaces resembling terraces were noted during this experiment (Fig. 4). They were small, with surface area of 10–20 cm^2. They typically persisted for relatively long times (about 1 h on average, which was the time required to erode through ~1 cm, or 10% of local relief in the basin). Interestingly, all of these features, when scaled to appropriate vertical and horizontal length scales, such as the hillslope length and local relief, are of the same order as natural mountainous landscapes.

In order to analyze the spatial distribution of hillslope failures, we examined a total of 96 images taken during the last 13 h of the experiment. During this period, base level fell 14 cm, a distance roughly 50% of the total relief of basin. We mapped 505 hillslope failures, and we noticed that the landslide trace remained visible for about two minutes after collapse. Since the time step between photographs was about seven minutes, it is likely that not all of the hillslope failures that occurred between photographic snapshots were mapped.

The surface area and centroid location of each landslide were manually determined on scaled photographs using the image processing software Scion Image. Table 1 reports statistics computed on the mapped failures. The hillslope failures we detected had a mean area of 11.1 cm^2 and a standard deviation of 9.7 cm^2. The smallest recorded failure had an area of 1.4 cm^2; the largest had an area of 62 cm^2. Smaller events were more numerous than large ones. The size-frequency distribution of landslides can be approximated by a power law with decreasing numbers of larger events (Fig. 9). This result is comparable to those of studies of hillslope failures in natural settings (e.g., Hovius, et al. 1997; Stark and Hovius, 2001; Guzzetti et al., 2002; Brardinoni and Church, 2004), although ideally a much larger number of events over a broader size range would better define such a relation.

TABLE 1. MAIN CHARACTERISTICS OF THE EXPERIMENTAL RUN

Study interval	13 h
Rainfall rate	10.8 μm/s
Base-level fall rate	3.1 μm/s
Average time step between photographs	7 min
Number of photographs evaluated	96
Number of hillslope failures mapped	505
Smallest failure detected	1.4 cm^2
Largest failure detected	62.8 cm^2
Mean failure area	11.1 cm^2
Basin area	6400 cm^2

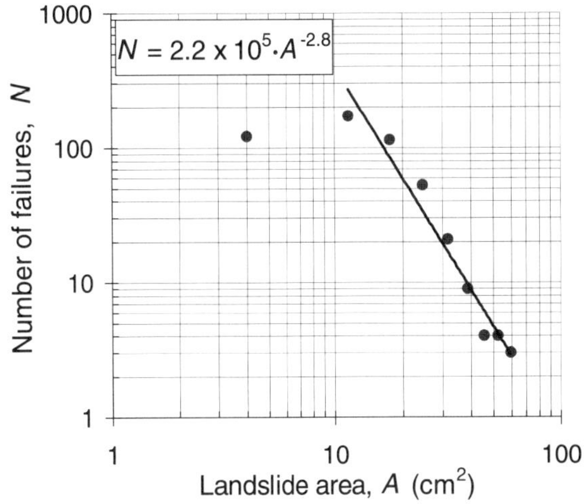

Figure 9. Frequency distribution of hillslope failures during a 13 h period under steady erosion conditions. Typical of landslide distributions in natural settings, the distribution exhibits power-law scaling, though the range in failure size is smaller than for natural settings. The power law on the diagram yields the number of failures (N) of a given area (A). Decreasing numbers of small events could result from difficulty in detection, or represent a lower limit to mass movement size.

We also measured knickpoint locations from the same set of photographs. Three major knickpoint events initiated within 20 cm of the outlet and propagated throughout the basin. Nine other knickpoints initiated 20–50 cm from the outlet. Of these, some propagated to the edges of the basin, while others dissipated. Such behavior has been observed in other studies (Parker, 1977; Hancock, 1997). Migration speed for knickpoints in trunk streams was ~0.6 cm/min; in the upper part of the experimental basin, we measured decreased velocities (~0.4 cm/min) as the knickpoint migrated into smaller tributaries.

Spatial Patterns and Relationships Between Hillslope Failures, Knickpoints, and Temporary Sediment Storage

The main purpose of this study was to identify and characterize the possible presence of patterns in the spatial distribution of landslides, and determine what, if any, relationship exists between landslides and knickpoints. Before we search for relationships between processes, it is helpful to know if any pattern is detectable in the spatial distribution of hillslope failures. Hence, a null hypothesis might be that hillslope failures are randomly distributed in space across the landscape. To test this hypothesis, we applied a Monte Carlo simulation technique (Metropolis, 1987). We numerically generated 100 sets of 505 randomly located landslides over the experimental basin, so that observed and synthetic hillslope failures were equal in number. We then analyzed the clustering tendency in experimental and random distributions of landslides. A qualitative visual inspection of failure locations for a 4 h period mapped onto a single image (Fig. 10) suggests that some clustering exists in the experimental data.

Clustering tendency was determined by measuring for each failure, i, the distance w_i to the closest landslide j, for both randomly distributed and experimental data sets. We indicate the distance between the centroids of landslide i and j with $d_{i,j}$; w_i is computed accordingly to the relationship

$$w_i = \min_{\substack{j=1 \\ j \neq i}}^{N} d_{i,j} \tag{1}$$

for $i = 1, \ldots, N$, where $N = 505$ is the number of observed landslides in the considered time interval. The sample frequency distribution of w_i computed for the experimental data is shown in Figure 11, along with the 90% confidence envelope of the frequency distribution of w_i computed on the 100 random sets of 505 landslides. It can be seen that we cannot reject the hypothesis that the minimum distance among landslides is slightly lower in the experimental basin at the 90% confidence level. Even at a 98% confidence limit, some clustering is still apparent in the experimental data. This latter outcome confirms that both spatial ordering as well as randomness contribute to the stochastic nature of slope failures in our experimental landform.

Our analysis above does not take into account the relation of failures to knickpoint locations. We know that knickpoints account for a significant amount of incision in our basin, and hence they should contribute to destabilization of hillslopes. We quantify their effect by focusing on failures along a single channel and counting the number of landslides that occurred upstream and downstream from knickpoint locations. This resulted in 47 failures downstream and 13 failures upstream from knickpoint locations. Clearly, the greater number of landslides downstream could be evidence of knickpoint triggering. However, this could also result from knickpoint location. For instance, a concentration of knickpoints in the upper part of the trunk stream could result in a high number of downstream hillslope failures even if landslides were randomly distributed.

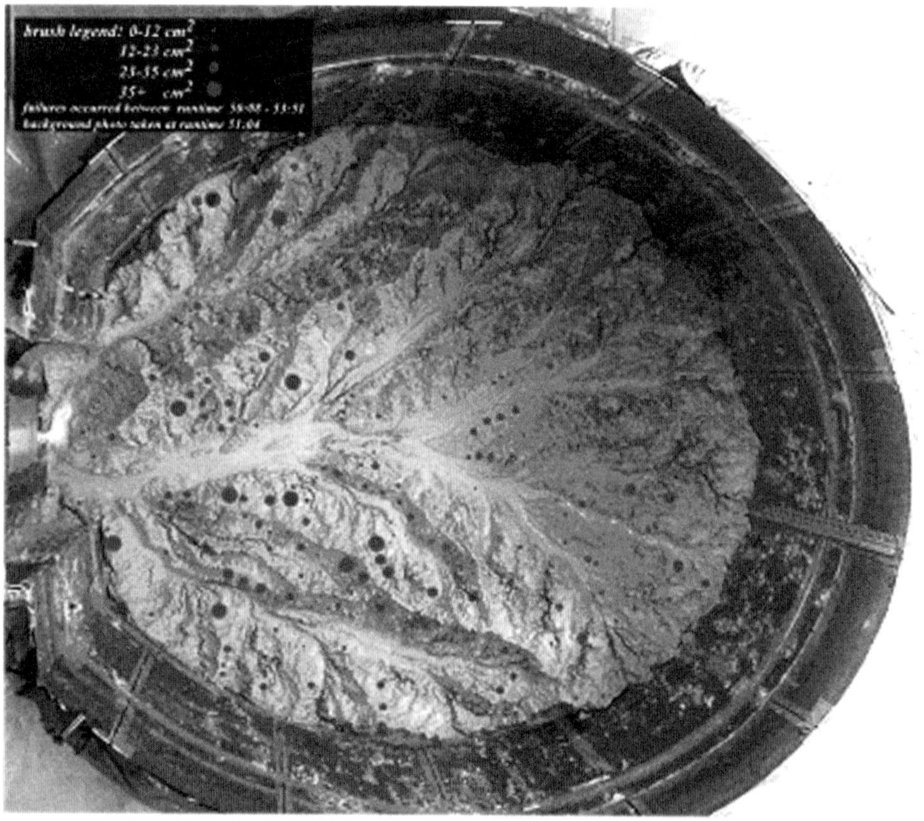

Figure 10. Hillslope failures that occurred during a 4 h period (from 50:08 h to 53:51 h), plotted as gray circles onto a photograph taken at the beginning of the 4 h period, with area proportional to actual failure area. Some clustering of failures is apparent in the photograph; see text for analysis of the significance of this clustering.

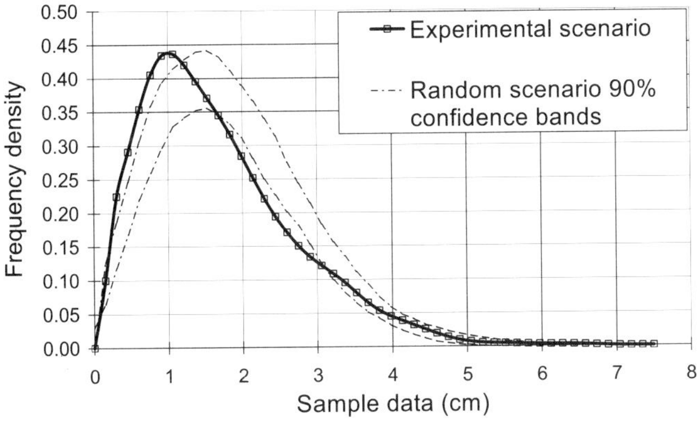

Figure 11. Comparison between the sample frequency distributions of the minimum distance between landslides for experimental and Monte Carlo simulated landslides. Random and experimental distributions are similar. However, the peak frequency at 1 cm separation distance suggests tighter clustering for experimental landslides.

We performed an initial test on the size of failures upstream and downstream from a knickpoint to test for any bias in the actual volume of material removed by failures. Are failures downstream from knickpoints larger than upstream? In order to check from a statistical point of view whether or not significant differences exist, the probability density functions of upstream and downstream landslides were estimated. Figure 12 shows the obtained sample probability density functions. A quantitative comparison was carried out between the mean values and the standard deviations of the two distributions. The 90% confidence limits of the estimated mean values of upstream and downstream hillslope failure areas are 11.81–20.91 cm^2 and 13.98–19.24 cm^2, respectively. The 90% confidence limits of the estimated standard deviations of upstream and downstream hillslope failure areas are 7.56–15.16 cm^2 and 9.42–13.32 cm^2, respectively. Since mean and standard deviations between the two populations overlap substantially, one cannot reject (at the 90% confidence level) the hypothesis that mean and standard deviation of upstream and downstream landslide areas are identical. The similarity of the two probability density functions is also clearly shown in Figure 12. This analysis allows us to conclude that no significant link exists between landslide size and knickpoint location.

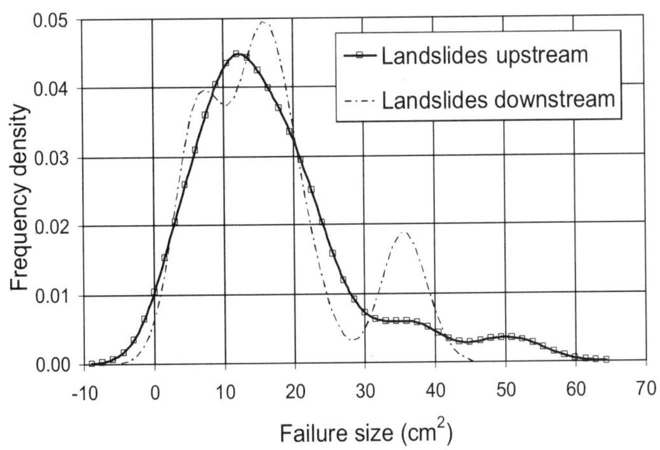

Figure 12. Comparison between the sample frequency distributions of the landslide area (that is, size) that occurred in the main channel upstream and downstream of knickpoints.

We then turned to analyzing the significance of increased numbers of landslides downstream from knickpoints by means of another Monte Carlo simulation. Since landslide area statistics are invariant upstream and downstream from knickpoints, the number of landslides is directly related to total landslide area. We generated 100 random data sets of an equal number ($N = 60$) of hillslope failures along the trunk stream and computed the mean and 90% confidence bands for downstream events. The results are reported in Table 2. The mean number of downstream landslides in the Monte Carlo experiments is 40, with a 90% upper confidence limit of 44. This is slightly lower than the experimental result ($n = 47$). Therefore, one cannot reject the hypothesis, at the 90% confidence level, that downstream landslides occur less frequently in the random scenario than in the experimental basin. Indeed, the above hypothesis cannot be rejected for a confidence level up to 98%.

So, we further focused our efforts by considering the along-stream distance x_t between a knickpoint at time t (the time each photograph was taken) and the experimental landslides at the same time t that occurred downstream on the trunk stream. We then repeated the Monte Carlo analysis by developing 100 sets of 47 downstream randomly located events, while maintaining the same knickpoint positions as observed in the experimental run. This allowed us to get 100 sets of 47 random distances, which we designate with the symbol x^*_t.

A comparison between the experimental probability density functions of x_t and the 90% confidence envelope of the Monte Carlo–generated probability density functions of x^*_t is shown in Figure 13. Clearly, the probability density function of experimental landslides is significantly different in this scenario from that obtained with a random distribution. In the Monte Carlo simulations, the probability density function is unimodal. The mode

TABLE 2. NUMBER OF LANDSLIDES LOCATED DOWNSTREAM AND UPSTREAM OF KNICKPOINTS IN THE RANDOM AND EXPERIMENTAL LANDSLIDE DISPLACEMENT MODELS

Landslide displacement	Number of landslides downstream from a knickpoint	Number of landslides upstream from a knickpoint
Random (100 simulations)	40 (36–44)	20 (24–16)
Experimental	47	13

Note: For the random location, the mean values computed in 100 simulations are given, along with the respective 90% confidence limits (shown in parentheses).

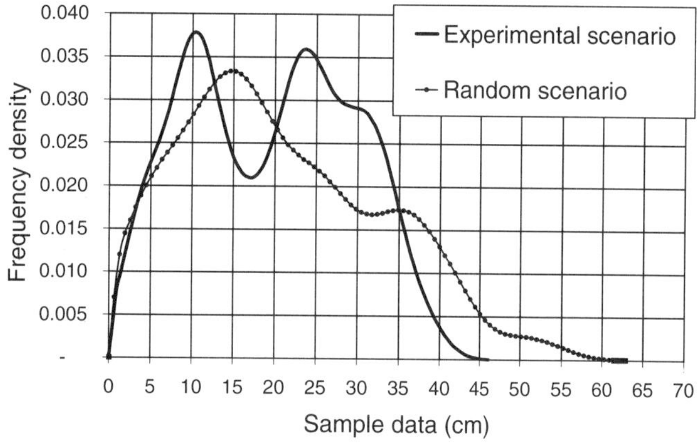

Figure 13. Comparison between the sample frequency distributions of the distances between landslides and closest upstream knickpoint in the experiment and in the random displacements. Experimental frequency distribution is bimodal, with peaks at 10 and 24 cm. A random distribution exhibits a single peak at 10 cm.

of the distribution, x^*_m, indicates the most probable distance between knickpoints and downstream landslides if the locations were random. The magnitude of x^*_t is determined by the geometry of the system and the statistical behaviors of the above stochastic processes. For instance, a knickpoint moving with lower velocity in the upper part of the basin would lead to a smaller value of x^*_m.

The experimental data are characterized by a bimodal probability density function. The value of the first mode, x_1, is close to the value of x^*_m, which is suggestive of significant landslide triggering unrelated to knickpoints. However, the presence of a second mode, x_2, suggests the possible presence of a landslide triggering effect induced by knickpoint migration. The parameter x_2 is a characteristic distance between knickpoints and downstream hillslope failures induced by them. The value of x_2 is ~24 cm and represents a delay distance between the passage of a knickpoint and landslide activation. The hypothesis of the presence of statistically significant differences between the probability density functions of observed and simulated x^*_t cannot be rejected for a confidence level up to 99%.

We performed one final analysis on the relationship between failures and knickpoint locations. For the *i*th landslide in the main channel that collapsed during the last 13 h of the experiment downstream from a knickpoint, for *i* ranging from 1 to 47, the distance h_i from the basin outlet was recorded, as well as the distance k_i from the basin outlet to the closest upstream knickpoint. Therefore, 47 couples (h_i,k_i) were collected. If a relationship exists between h_i and k_i, one would expect to see a significant cross-correlation between them. Indeed, the cross-correlation coefficient between the couples (h_i,k_i) is 0.65. We performed a Monte Carlo simulation as a null hypothesis by computing the cross-correlation coefficient between the vector of the k_i values and 1000 vectors obtained by randomly generating the h_i distances, from a uniform distribution, in the range $[0, k_i]$, thereby breaking the connection between the locations of landslides and the corresponding knickpoint closest upstream. We generated a sample of 1000 values of cross-correlation coefficients between knickpoint positions and randomly distributed downstream landslides. The mean value of the sample is equal to 0.45 and the upper 90% confidence limit is 0.59. The observed value of the cross-correlation coefficient was exceeded in 6 out of 1000 synthetic random realizations. Therefore, one cannot reject the hypothesis of the presence of a significant relationship between knickpoints and downstream landslides positions at the 99% confidence level. This analysis confirms the findings obtained with the previous investigations.

DISCUSSION

We conducted our experiments under steady rainfall and base-level fall conditions. This effectively removed landslide triggering events, such as severe storms and earthquakes, from consideration, and allowed us to focus on stream incision as the dominant triggering mechanism. In a landscape where streams incise continuously at a constant rate, one might expect that hillslope failures would occur as local conditions for instability are met, and a spatial pattern of landslides independent of knickpoint position would result. If stream incision varies in a systematic way, such as that which occurs during upstream migration of a knickpoint, failure patterns should exhibit clustering downstream from the knickpoint. Interestingly, a synoptic basin-scale view of failures in our experiment revealed a statistically significant tendency to clustering at the 98% confidence level. In particular, by selecting failures along a stream and correlating failure locations with knickpoint locations, we observed a statistically significant link. In particular, we find that hillslope failures are more likely to occur downstream from a knickpoint.

In fact, the results of the three quantitative analyses we carried out to test the hypothesis that stream incision by migrating knickpoints can impose a spatial and temporal pattern on landslide distributions are consistent. They show that one cannot reject the above hypothesis with a confidence level ranging from 98% and 99%. Therefore, a high probability exists that a significant relation between knickpoint and landslide positions is indeed present, even if this result must be interpreted in a statistical sense, i.e., the relation is likely to be present but not certain. The significance of this conclusion, which comes from three different analyses that ended up with consistent results, can be evaluated in view of its confidence level, which is an indication of how strong the relation is. Of course, the spatial pattern of landslides is also affected by the presence of significant hillslope triggering factors that are not related to knickpoint migration.

We offer the following exercise to get some sense of the knickpoint contribution to overall stream incision. Recall that we conducted our experiment at steady base-level fall under flux steady-state conditions. This implies that, on average, stream incision is equal to the base-level fall rate, in this case

3.1 µm/s. Under these conditions, we can estimate the contribution of knickpoints to overall stream incision. We observed 12 knickpoint events during the time that we measured knickpoint and hillslope failure locations. During this observational time (13 h), base level dropped 14.5 cm. While knickpoint heights varied, and most knickpoints initiated at a location upstream from the outlet, we can nonetheless set limits on knickpoint contribution by assuming that the average knickpoint height was 0.5 cm, and that this height was maintained while the knickpoint propagated through the entire stream network. Given these assumptions, knickpoints account for 6 cm of incision, or nearly half of the total stream incision during the observation period. If stream incision by knickpoints is the only destabilizing factor, then this implies that knickpoints are responsible for ~50% of the failures. When we compare this estimate to the number of failures downstream (n = 47) from a knickpoint to the total failures along the stream (n = 60), we find a higher percentage (78%). Note that this calculation likely includes some failures that result from steady stream incision downstream from the knickpoint. The point we make here is that streams incise in our experiment via both knickpoint propagation and steady bed lowering. Given these two mechanisms of incision, it is not surprising that the relationship between knickpoint and failure locations is statistically significant. We can also conclude that knickpoint incision plays a significant role in destabilizing hillslopes.

The rather large mean separation distance (24 cm) we observed between knickpoints and landslides triggered by them is somewhat surprising. This could represent the time required for base-level lowering in the channel to propagate laterally to the toe of the hillslope (recall that channels typically narrow and deepen around a knickpoint). Hillslopes adjacent to channels in v-shaped valleys that lack a floodplain would respond more rapidly to stream incision than hillslopes bordering wide valleys. As far as landslide area is concerned, we did not find any significant difference depending on knickpoint position. As a matter of fact, the probability density functions of hillslope failure size were not significantly different for events downstream and upstream from knickpoints. This outcome seems to indicate that the landslide size in our experiment was not significantly different for events triggered by knickpoint migration than by other causes.

Do natural eroding drainage basins behave like our model? This, of course, we cannot answer definitively. Our experiment was not a scaled model of a natural prototype. In fact, it is not possible to construct a scale model with the linear dimensions of our tank that captures the flow conditions of a larger drainage basin (see Peakall et al., 1996, for an overview of scaling issues in experimental geomorphology). Our experimental streams are dominated by laminar flow (maximum Reynolds number is ~750), whereas natural mountain streams are always turbulent. We do observe significant overlap in Froude numbers between experimental and natural drainages (both are dominantly subcritical, with some stream sections at or above critical flow). Sediment concentration for the experiment we present here approached 25%, and such concentrations are unusual in natural rivers (though not unheard of). Feedback between local deposition and erosion may be enhanced at this concentration and could be a source of instability within model streams (Hasbargen and Paola, 2000). Flow depth to grain size ratios vary downstream in our experiment, but for flow depths on the order of 1 mm, this ratio was ~20. Natural streams typically have much higher ratios. A gravel-bedded stream with 1 cm grains and 1 m depth, for instance, has a depth to grain size ratio of 100. Our goal was not to build an exact replica of a natural landscape, but to represent the major processes in a setting in which we could isolate and analyze landscape interactions that would take an extremely long time in nature. We do not think that any of the scale effects noted above would cause qualitatively different dynamics between knickpoints and landslides. The presence of additional features and processes in a landscape, such as vegetation, gophers, tree throw, creep, weathering and soil development, storm and earthquake events, road construction, etc., would add greater complexity to natural systems. One would expect such effects to complicate and obscure stochastic interactions between streams and hillslopes. While there are significant differences between our model landscape and natural settings, the approximate similarity of landscape features and of process activity (failure sizes on the order of the hillslope length; hillslopes of ~90% slope; knickpoint heights on the order of several flow depths) suggests that natural landscapes experiencing stream incision and knickpoint propagation might also exhibit similar dynamics. Indeed, increased numbers of failures downstream from knickpoint have been noted in natural settings (see Stock and Dietrich, 2003, their Figure 5A; K.X Whipple, 2003, personal commun.).

Our results shed light on the role of stream incision in destabilizing hillslopes. In the absence of external stochastic triggers, such as storm events and earthquakes, we still find that hillslope failures are strongly stochastic. Part of the variability can be ascribed to local conditions, such as slope steepness, pore fluid pressure, past history of failures, etc. Part of the control on landslides, however, is exerted by conditions at the toe of the hillslope. Stream incision, either via steady downcutting or by episodic knickpoint propagation, can introduce a weak structure to spatial patterns of landslides.

Our results suggest that spatial patterns of hillslope failures could be a result of recent incision by streams at the toe of the hillslopes. This observation, which must now be checked against natural settings, could allow for rapid identification of regions in a drainage basin prone to increased landscape instability. For anyone concerned with landslide forecasting in mountainous landscapes, locations downstream from knickpoints merit special attention.

CONCLUSION

We have presented the results from a small-scale erosion experiment under steady forcing conditions where we focused our attention on the relationship between hillslope failure activity and knickpoint migration. We measured landslide sizes and locations and knickpoint locations over time under steady forcing conditions. We tested the possible existence of spatial organization of hillslope failures, such as clustering, using randomly generated spatial distributions of landslides as a reference. We observed a statistically significant relation between knickpoint positions and downstream landslide location. The detection of a significant link between migrating knickpoints and landslides confirms our intuition concerning the effectiveness of knickpoints in triggering failures. We note, however, that stream downcutting in the absence of knickpoint propagation, and local effects like slope and pore pressure, also control hillslope stability. These effects superimpose a random spatial component onto the spatial patterns associated with knickpoint migration.

ACKNOWLEDGMENTS

We thank Robert S. Anderson, Kelin Whipple, and an anonymous reviewer for helpful criticism that substantially improved this paper. Bigi was partially supported by a scholarship provided by the Faculty of Engineering of the University of Bologna. This work was supported by the Science and Technology Centers program of the National Science Foundation via the National Center for Earth-Surface Dynamics under the agreement number EAR-0120914 and by the Italian National Research Council Group for the Prevention of Hydrogeological Disasters (contract n.01.01014.PF42). The experimental device and all the experimental equipment was provided by St. Anthony Falls Laboratory, and the support of all the staff has been fundamental to our work.

REFERENCES CITED

Ahnert, F., 1998, Introduction to geomorphology: London, Arnold, 360 p.

Alonso, C.V., Bennett, S.J., and Stein, O.R., 2002, Predicting headcut erosion and migration in concentrated flows typical of upland areas: Water Resources Research, v. 38, p. 39-1–39-15.

Benda, L., and Dunne, T., 1997, Stochastic forcing of sediment supply to channel networks from landsliding and debris flow: Water Resources Research, v. 33, p. 2849–2863, doi: 10.1029/97WR02388.

Bennett, S.J., Alonso, C.V., Prasad, S.N., and Roemkens, M.J.M., 2000, Experiments on headcut growth and migration in concentrated flows typical of upland areas: Water Resources Research, v. 36, p. 1911–1922, doi: 10.1029/2000WR900067.

Brardinoni, F., and Church, M., 2004, Representing the landslide magnitude-frequency relation: Capilano River basin, British Columbia: Earth Surface Processes and Landforms, v. 29, p. 115–124, doi: 10.1002/esp.1029.

Brush, L.M., Jr., and Wolman, G., 1960, Knickpoint behavior in noncohesive material—A laboratory study: Geological Society of America Bulletin, v. 71, p. 59–73.

Bryan, R.B., 1990, Knickpoint evolution in rillwash, *in* Bryan, R.B., ed., Soil erosion—Experiments and models: Catena Supplement, v. 17, p. 111–132.

Bryan, R.B., and Rockwell, D.L., 1998, Water table control on rill initiation and implications for erosional response: Geomorphology, v. 23, p. 151–169, doi: 10.1016/S0169-555X(97)00110-4.

Burbank, D.W., Leland, J., Felding, E., Anderson, R.S., Brozovic, N., Reid, M.R., and Duncan, C., 1996, Bedrock incision, rock uplift and threshold hillslopes in the northwestern Himalayas: Nature, v. 379, p. 505–510, doi: 10.1038/379505a0.

Burton, A., Arkell, T.J., and Bathurst, J.C., 1998, Field variability of landslide model parameters: Environmental Geology, v. 35, p. 100–114, doi: 10.1007/s002540050297.

Campbell, R.H., 1975, Soil slips, debris flows and rainstorms in the Santa Monica Mountains and vicinity, Southern California: U.S. Geological Survey Professional Paper 851, 51 p.

Cantelli, A., Paola, C., and Parker, G., 2004, Experiments on upstream-migrating erosional narrowing and widening of an incisional channel caused by dam removal: Water Resources Research, v. 40, p. W03304, doi: 10.1029/2003WR002940.

Casadei, M., Dietrich, W.E., and Miller, N.L., 2003, Testing a model for predicting the timing and location of shallow landslide initiation in soil-mantled landscapes: Earth Surface Processes and Landforms, v. 28, p. 925–950, doi: 10.1002/esp.470.

Crosby, B.T., and Whipple, K.X., 2002, Knickpoint migration in the Waipaoa River: An examination of the rate and form of transient behavior in fluvial networks: Eos (Transactions, American Geophysical Union), v. 83(47), Fall meeting supplement, abstract H21G-10.

Crozier, M.J., Deimel, M.S., and Simon, J.S., 1995, Investigation of earthquake triggering for deep-seated landslides, Taranaki, New Zealand: Quaternary International, v. 25, p. 65–73, doi: 10.1016/1040-6182(94)00036-5.

Dadson, S.J., Hovius, N., Chen, Hongey, Dade, W.B., Hsieh, Meng-Long, Willett, S.D., Hu, Jyr-Ching, Horng, Ming-Jame, Chen, Meng-Chiang, Stark, C.P., Lague, D., and Lin, Jiun-Chuan, 2003, Links between erosion, runoff variability and seismicity in the Taiwan orogen: Nature, v. 426, p. 648–651.

Densmore, A.L., and Hovius, N., 2000, Topographic fingerprints of bedrock landslides: Geology, v. 28, p. 371–374, doi: 10.1130/0091-7613(2000)028<371:TFOBL>2.3.CO;2.

Densmore, A.L., Anderson, R.S., McAdoo, B.G., and Ellis, M.A., 1997, Hillslope evolution by bedrock landslides: Science, v. 275, p. 369–372, doi: 10.1126/science.275.5298.369.

Dietrich, W.E., and Dunne, T., 1993, The channel head, *in* Beven, K., and Kirkby, M.J., eds., Channel network hydrology: New York, John Wiley, p. 175–219.

Dietrich, W.E., Reiss, R., Hsu, M., and Montgomery, D.R., 1995, A process-based model for colluvial soil depth and shallow landsliding using digital elevation data: Hydrological Processes, v. 9, p. 383–400.

Duan, J., 1996, A coupled hydrologic-geomorphic model for evaluating effects of vegetation change on watersheds, [Ph.D. thesis]: Corvallis, Oregon, Oregon State University, 133 p.

Gabet, E.M., and Dunne, T., 2002, Landslides on coastal sage-scrub and grassland hillslopes in a severe El Niño winter: The effects of vegetation conversion on sediment delivery: Geological Society of America Bulletin, v. 114, p. 983–990, doi: 10.1130/0016-7606(2002)114<0983:LOCSSA>2.0.CO;2.

Gardner, T.W., 1983, Experimental study of knickpoint and longitudinal profile evolution in cohesive, homogeneous material: Geological Society of America Bulletin, v. 94, p. 644–672, doi: 10.1130/0016-7606(1983)94<664:ESOKAL>2.0.CO;2

Guzzetti, F., Malamud, B.D., Turcotte, D.L., and Reichenbach, P., 2002, Power-law correlations of landslide areas in central Italy: Earth and Planetary Science Letters, v. 195, p. 169–183, doi: 10.1016/S0012-821X(01)00589-1.

Hancock, G., 1997, Experimental testing of the Siberia landscape evolution model [Ph.D. thesis]: Newcastle, New South Wales, Australia, University of Newcastle, 467 p.

Hancock, G.S., Anderson, R.S., and Whipple, K.X., 2003, Beyond power: Bedrock river incision process and form, *in* Tinkler, K., and Wohl, E., eds., Rivers over rock: Fluvial processes in bedrock channels: American Geophysical Union Geophysical Monograph 107, p. 35–60.

Harp, E.L., and Jibson, R.L., 1995, Inventory of landslides triggered by the 1994 Northridge, California, earthquake: U.S. Geological Survey Open-File Report 95-213, 18 p.

Hasbargen, L.E., and Paola, C., 2003, How predictable is local erosion rate in erosional landscapes?, *in* Wilcox, P.R., and Iverson, R.M., eds., Prediction in geomorphology: American Geophysical Union Geophysical Monograph 135, 256 p.

Hasbargen, L.E., 2003, Erosion in steady state drainage basins [Ph.D. thesis]: Minneapolis, Minnesota, University of Minnesota, 228 p.

Hasbargen, L.E., and Paola, C., 2000, Landscape instability in an experimental drainage basin: Geology, v. 28, p. 1067–1070, doi: 10.1130/0091-

7613(2000)028<1067:LIIAED>2.3.CO;2.

Havenith, H.B., Strom, A., Jongmans, D., Abdrakhmatov, K., Delvaux, D., and Tréfois, P., 2003, Seismic triggering of landslides, Part A: Field evidence from the northern Tien Shan: Natural Hazards and Earth System Science, v. 3, p. 135–149.

Hermanns, R.L., Trauth, M.H., Niedermann, S., McWilliams, M., and Strecker, M.R., 2000, Tephrochronologic constraints on temporal distribution of large landslides in northwest Argentina: Journal of Geology, v. 108, p. 35–52, doi: 10.1086/314383.

Holland, W.N., and Pickup, G., 1976, Flume study of knickpoint development in stratified sediment: Geological Society of America Bulletin, v. 87, p. 76–82, doi: 10.1130/0016-7606(1976)87 <76:FSOKDI> 2.0.CO;2.

Hovius, N., Stark, C.P., and Allen, P.A., 1997, Sediment flux from a mountain belt derived by landslide mapping: Geology, v. 25, p. 231–234, doi: 10.1130/0091-7613(1997)025<0231:SFFAMB>2.3.CO;2.

Iida, T., 1984, A hydrological method of estimation of the topographic effect on the saturated through flow: Transactions, Japanese Geomorphologic Union, v. 5, p. 1–12.

Iverson, R.M., 2000, Landslide triggering by rain infiltration: Water Resources Research, v. 36, p. 1897–1910, doi: 10.1029/2000WR900090.

Keefer, D., 1994, The importance of earthquake-induced landslides to long-term slope erosion and slope-failure hazards in seismically active regions: Geomorphology, v. 10, p. 265–284, doi: 10.1016/0169-555X(94)90021-3.

Lague, D., Crave, A., and Davy, P, 2003, Laboratory experiments simulating the geomorphic response to tectonic uplift: Journal of Geophysical Research, v. 108(B1), ETG 3-1–3-20.

Larsen, M.C., and Parks, J.E., 1997, How wide is a road? The association of roads and mass-wasting in a forested montane environment: Earth Surface Processes and Landforms, v. 22, p. 835–848, doi: 10.1002/(SICI)1096-9837(199709)22:93.3.CO;2-3.

Larsen, M.C., and Torres-Sanchez, A.J., 1998, The frequency and distribution of recent landslides in three montane tropical regions of Puerto Rico: Geomorphology, v. 24, no. 4, p. 309–331, doi: 10.1016/S0169-555X(98)00023-3.

Martin, Y., Rood, K., Schwab, J.W., and Church, M., 2002, Sediment transfer by shallow landsliding in the Queen Charlotte Islands, British Columbia: Canadian Journal of Earth Sciences, v. 39, p. 189–205, doi: 10.1139/e01-068.

Merritt, E., 1984, The identification of four stages during micro-rill development: Earth Surface Processes and Landforms, v. 9, p. 493–496.

Metropolis, N., 1987, The beginning of the Monte Carlo method: Los Alamos Science, v. 15, p. 125–130.

Miller, D.J., and Sias, J., 1998, Deciphering large landslides: Linking hydrological, groundwater and slope stability models through GIS: Hydrological Processes, v. 12, p. 923–941, doi: 10.1002/(SICI)1099-1085(199805)12:63.0.CO;2-3.

Miller, J.R., 1991, The influence of bedrock geology on knickpoint development and channel-bed degradation along downcutting streams in south-central Indiana: Journal of Geology, v. 99, p. 591–605.

Montgomery, D.R., and Dietrich, W.E., 1994, A physically-based model for the topographic control on shallow landsliding: Water Resources Research, v. 30, p. 1153–1171, doi: 10.1029/93WR02979.

Montgomery, D.R., Schmidt, K.M., Greenberg, H.M., and Dietrich, W.E., 2000, Forest clearing and regional landsliding: Geology, v. 28, p. 311–314, doi: 10.1130/0091-7613(2000)028<311:FCARL>2.3.CO;2.

Mosley, M.P., 1974, Experimental study of rill erosion: Transactions of American Society of Agricultural Engineering, v. 17, p. 909–913.

Parker, G., and Izumi, N., 2000, Purely erosional cyclic and solitary steps created by flow over cohesive bed: Journal of Fluid Mechanics, v. 419, p. 203–238, doi: 10.1017/S0022112000001403.

Parker, R.S., 1977, Experimental study of basin evolution and its hydrologic implications [Ph.D. thesis]: Fort Collins, Colorado, Colorado State University, 331 p.

Peakall, J., Ashworth, P.J., and Best, J.L., 1996, Physical modelling in fluvial geomorphology; principles, applications and unresolved issues, *in* Rhoads, B.L., and Thorn, C.E., eds., The scientific nature of geomorphology: Chichester, Wiley, p. 221–253.

Roering, J.J., Kirchner, J.W., Sklar, L.S., and Dietrich, W.E., 2001, Hillslope evolution by nonlinear creep and landsliding: An experimental study: Geology, v. 29, p. 143–146, doi: 10.1130/0091-7613(2001)029<0143:HEBNCA>2.0.CO;2.

Schumm, S.A., 1956, Evolution of drainage systems and slopes in badlands at Perth Amboy, New Jersey: Geological Society of America Bulletin, v. 67, p. 597–646.

Schuster, R.L., and Highland, L.M., 2001, Socioeconomic and environmental impacts of landslides in the Western Hemisphere: U.S. Geological Survey Open-File Report 01-0276.

Selby, M.J., 1982, Hillslope materials and processes: Oxford, Oxford University Press, 466 p.

Sidle, R.C., 1992, A theoretical model of the effects of timber harvesting on slope stability: Water Resources Research, v. 28, p. 1897–1910, doi: 10.1029/92WR00804.

Sklar, L., and Dietrich, W.E., 2003, River longitudinal profiles and bedrock incision models: Stream power and the influence of sediment supply, *in* Tinkler, K., and Wohl, E., eds., Rivers over rock: Fluvial processes in bedrock channels: American Geophysical Union Geophysical Monograph 107, p. 237–260.

Slattery, M.C., and Bryan, R.B., 1992, Hydraulic conditions for rill incision under simulated rainfall: A laboratory experiment: Earth Surface Processes and Landforms, v. 17, p. 127–146.

Stark, C.P., and Hovius, N., 2001, The characterization of landslide size distributions: Geophysical Research Letters, v. 28, p. 1091–1094, doi: 10.1029/2000GL008527.

Stock, J., and Dietrich, W.E., 2003, Valley incision by debris flows: Evidence of a topographic signature: Water Resources Research, v. 39, p. ESG 1-1-1-25.

Terzaghi, K., Peck, R.B., and Mesri, G., 1996, Soil mechanics in engineering practice: New York, Wiley, 549 p.

Trauth, M.H., Alonso, R.A., Haselton, K.R., Hermanns, R.L., and Strecker, M.R., 2000, Climate change and mass movements in the NW Argentine Andes: Earth and Planetary Science Letters, v. 179, p. 243–256, doi: 10.1016/S0012-821X(00)00127-8.

van Asch, T.W.J., and van Steijn, H., 1991, Temporal patterns of mass movements in the French Alps: Catena, v. 18, p. 515–527, doi: 10.1016/0341-8162(91)90052-Y.

Willett, S.D., and Brandon, M.T., 2002, On steady state mountain belts: Geology, v. 30, p. 175–178, doi: 10.1130/0091-7613(2002)030<0175:OSSIMB>2.0.CO;2.

Wohl, E., 2003, Bedrock channel morphology in relation to erosional processes, *in* Tinkler, K., and Wohl, E., eds., Rivers over rock: Fluvial processes in bedrock channels: American Geophysical Union Geophysical Monograph 107, p. 133–151.

Zaprowski, B.J., Evenson, E.B., Pazzaglia, F.J., and Epstein, J.B., 2001, Knickzone propagation in the Black Hills: A different perspective on the late Cenozoic exhumation of the Laramide Rocky Mountains: Geology, v. 29, p. 547–550, doi: 10.1130/0091-7613(2001)029<0547:KPITBH>2.0.CO;2.

Manuscript Accepted by the Society 23 June 2005

Printed in the USA

Geological Society of America
Special Paper 398
2006

Stable isotopic evidence for a pre–late Miocene elevation gradient in the Great Plains–Rocky Mountain region, USA

Derek J. Sjostrom†
Geology Department and The Environmental and Natural Resources Institute, University of Alaska Anchorage, Anchorage, Alaska 99508, USA
Michael T. Hren
Travis W. Horton
Jacob R. Waldbauer
C. Page Chamberlain
Department of Earth and Environmental Sciences, Stanford University, Stanford, California 94305, USA

ABSTRACT

In order to investigate if high elevations existed in the Rocky Mountains before the late Miocene, we examined oxygen isotope ratios of 63 Tertiary smectite samples as a proxy for the isotopic composition of precipitation. Of these samples, 51 were also analyzed for hydrogen isotope ratios. These smectites were formed as a result of the weathering of volcanic air-fall deposits that blanketed much of western North America during the Tertiary. Smectite-bearing ashfall samples were collected from Eocene, Oligocene, and Miocene deposits along a transect that extends from the western Great Plains to Yellowstone National Park at modern elevations from ~900 to ~2800 m. In general, oxygen and hydrogen δ values of smectite lie along a line parallel to the meteoric water line, which suggests that the isotopic composition of these ash-derived smectites records the meteoric water composition during its formation. There is little evidence for postdepositional exchange with basinal brine fluids, evaporative effects, or diagenesis of these smectites. The $\delta^{18}O$ values of Oligocene and Miocene samples increase ~6‰ linearly from sample sites located at the crest of the Rocky Mountains to sites in western Nebraska and South Dakota. These results mimic the distribution and values of calculated oxygen isotope ratios of theoretical modern smectite over this same geographic traverse of decreasing elevation. This result suggests modern atmospheric circulation patterns and that the resulting distribution of $\delta^{18}O_{precipitation}$ has persisted since the Oligocene. The $\delta^{18}O$ values of Eocene samples increase ~8‰ between the Yellowstone region and central Wyoming, a result that does not correlate with modern $\delta^{18}O_{precipitation}$ trends. Our Eocene results may be explained by climate conditions extant at that time, but tectonic modification in the region between 50 Ma and 37 Ma cannot be excluded as the cause of our results. Because the modern climate system requires interaction with and modification by high-elevation areas, our results suggest that the Rocky Mountains have been at high elevation since at least 50 Ma.

Keywords: stable isotopes, smectite, Rocky Mountains, paleoclimate, paleotopography.

†E-mail: sjostrom@alum.dartmouth.org.

Sjostrom, D.J., Hren, M.T., Horton, T.W., Waldbauer, J.R., and Chamberlain, C.P., 2006, Stable isotopic evidence for a pre–late Miocene elevation gradient in the Great Plains–Rocky Mountain region, USA, *in* Willett, S.D., Hovius, N., Brandon, M.T., and Fisher, D.M., eds., Tectonics, Climate, and Landscape Evolution: Geological Society of America Special Paper 398, Penrose Conference Series, p. 309–319, doi: 10.1130/2006.2398(19). For permission to copy, contact editing@geosociety.org.

INTRODUCTION

There is considerable debate as to when the Rocky Mountains formed as a significant topographic feature. The traditional view holds that the modern relief that exists throughout the area is a recent feature that formed as a result of late- and post-Miocene uplift. According to this interpretation, the low-relief topography and shallow-dipping Tertiary sedimentary deposits that are a common feature at high elevations throughout the Rocky Mountain region are interpreted as evidence for Early and Middle Tertiary deposition on a broad, flat, low-elevation plain (Bradley, 1987; Burchfiel et al., 1992). The present topography resulted from Late Tertiary tectonic activity, uplift, and incision (Epis and Chapin, 1975; Steven et al., 1997; McMillan et al., 2002). In contrast, recent paleofloral and stable isotopic studies suggest that elevations as high or higher than present may have existed in the Rocky Mountains since, at the latest, the end of the Laramide orogeny, and much of the present relief is the result of increased erosion and exhumation due to Late Tertiary climate change (Molnar and England, 1990; Gregory and Chase, 1992; Drummond et al., 1993; Gregory and McIntosh, 1996; Norris et al., 1996; Chase et al., 1998; Wolfe et al., 1998; Dettman and Lohmann, 2000; Peizhen et al., 2001).

To address this debate, we conducted an oxygen and hydrogen isotopic study of authigenic smectites from Eocene (50–49 Ma), Oligocene (37–32 Ma), and Miocene (20–12 Ma) volcanic ashes that are exposed along a transect from the Continental Divide to the Great Plains (Fig. 1). Our study focuses on smectite because its oxygen and hydrogen isotopic values can be used to reconstruct past climate and topographic histories of mountain belts (Lawrence and Taylor, 1971; Stern et al., 1997; Chamberlain and Poage, 2000; Poage and Chamberlain, 2002). In addition, Tertiary volcanic ashes containing abundant smectite are exposed along the eastern flanks of the Rocky Mountains and extend into the western Great Plains. Because these ash layers were initially deposited as volcanic air-fall, time-equivalent deposits blanket a large geographic area. Thus, it is possible to determine isotopic gradients in paleoprecipitation over a broad area, which can be diagnostic of topography.

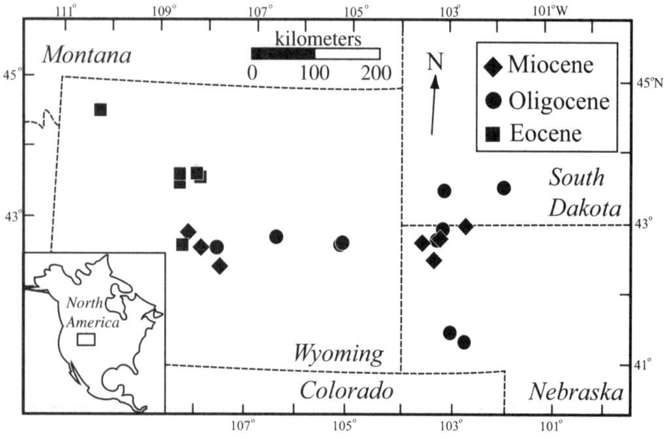

Figure 1. Locations of Tertiary smectite sample sites in the western Great Plains and eastern Rocky Mountain region. All samples were collected from air-fall ashes of Miocene, Oligocene, and Eocene age.

MODERN CLIMATE OF THE WESTERN GREAT PLAINS AND EASTERN ROCKY MOUNTAINS

In contrast to the relatively simple climate systems used in previous isotopic paleoelevation studies, such as Poage and Chamberlain (2001) and Rowley et al. (2001), the modern distribution of the isotopic composition of precipitation in the eastern Rocky Mountains and western Great Plains is fairly complex. The annual average isotopic values of precipitation depends on the annual timing of precipitation, average annual temperature, and the interaction among three isotopically discrete air masses that originate from three different areas (Bryson and Hare, 1974; Nativ and Riggio, 1990; Thordsen et al., 1992; Harvey and Welker, 2000). The modern climate dynamics of the region as described in the following tend to make the change in isotopic composition with elevation difference appear much greater in the western Great Plains and eastern Rocky Mountains than those predicted by the models of Poage and Chamberlain (2001) and Rowley et al. (2001).

The modern climate of the Rocky Mountain region and the western Great Plains is influenced by the direct competition of three distinct air masses that originate over the Arctic, the Gulf of Mexico, and the Pacific Ocean, and by the mountains that create rain shadows for these air masses (Bryson and Hare, 1974). The interplay of these three air masses causes strong seasonal variations in the amount of precipitation and temperature. In the high elevations of the northern Rocky Mountains, dry summer, wet winter conditions dominate as a result of the influence of the eastern Pacific subtropical high-pressure system. During the summer, the westerlies are relatively weak and displaced to the north, causing the summer months to be relatively dry at high elevations. Heating of the continental interior creates a thermal low-pressure system that draws moisture north from the Gulfs of Mexico and California (Bryson and Hare, 1974). This summer monsoon system dominates the low-elevation areas of the northern Rocky Mountains, which are characterized by relatively wet summers and dry winters compared to high-elevation areas (Despain, 1987; Whitlock and Bartlein, 1993). Areas of high elevation in the Rocky Mountains interact with Pacific-sourced air masses, which causes these areas to receive relatively abundant precipitation in the winter (Despain, 1987). These air masses pass over low-elevation areas unaffected, and the lack of orographic precipitation results in lower precipitation amounts in the winter (Despain, 1987). These local climate trends influence the pattern of seasonal precipitation over relatively small geographic areas, in some cases on the scale of only a few kilometers (Despain, 1987; Whitlock and Bartlein, 1993).

East of the Rocky Mountains, the effects of local topography are much reduced. The climate is dominated by interaction of warm, moist air masses that originate over the Gulfs of Mexico and California, cold, dry arctic air masses that originate over the Arctic Ocean, and mild, dry westerlies from the Pacific (Bryson and Hare, 1974). The seasonal dominance of the southern air masses in the summer results in relatively wet, humid summers, with increasing amounts of moisture to the east. In the winter, the Arctic and Pacific air masses dominate, which results in cold, dry winters, especially in the rain shadow of the Rocky Mountains.

The climate variability of the Yellowstone–Rocky Mountain region is reflected in the stable isotope composition of precipitation from the area; $\delta^{18}O_{precipitation}$ ranges from −17‰ to −22‰ $\{\delta^{18}O = 1000[(^{18}O/^{16}O)_{sample}/(^{18}O/^{16}O)_{standard}] - 1\}$; we report all $\delta^{18}O$ and δD values relative to Vienna standard mean ocean water (V-SMOW) (Thordsen et al., 1992; Horton et al., 1999). The dominant control on isotopic composition of precipitation in the Yellowstone area is elevation; lighter isotopic values occur in areas of high elevation (Thordsen et al., 1992). To the east of the Rocky Mountains, oxygen isotopic values of precipitation are more enriched in ^{18}O. In the Wind River Basin of central Wyoming, oxygen isotope values of soil water in equilibrium with modern carbonates range from ~−18‰ to ~−12‰ (Amundson et al., 1996). In the western Great Plains of western Nebraska and South Dakota, oxygen isotope values of precipitation average ~−9.8‰ (Harvey and Welker, 2000). East of the Rocky Mountains, the average oxygen isotope value of precipitation is highly dependent on the annual timing of precipitation. Areas in the eastern part of the study area (western Great Plains) receive more summer precipitation than western areas, and therefore have higher average $\delta^{18}O_{precipitation}$ values (Amundson et al., 1996; Harvey and Welker, 2000). This results in a present-day average $\delta^{18}O_{precipitation}$ difference of ~10‰ between Yellowstone National Park (elevation ~2500 m) and North Platte, Nebraska (elevation, ~900 m) or $\Delta\delta^{18}O_{precipitation}$ = −6.25‰/km elevation difference. This is a much more dramatic $\delta^{18}O_{precipitation}$ gradient than the global average (~−2.8‰/km elevation difference) reported by Poage and Chamberlain (2001). Similarly high isotopic gradients (−4.2‰/km) have also been observed in nearby areas of the eastern Rocky Mountains (Drummond et al., 1993). These greater than average isotopic gradients are the result of elevation increase to the west in our study area, which results in relatively light $\delta^{18}O_{precipitation}$. This is coupled with a much greater average annual amount of isotopically light winter precipitation in the western part of the study area than to the east.

METHODS

Approximately 160 samples were collected from previously described and age-dated Eocene, Oligocene, and Miocene ash layers on an east-west transect from western Nebraska and South Dakota to western Wyoming (Fig. 1; Table 1). Samples were ground with an iron mortar and pestle and dispersed in distilled water using a commercial blender. The samples were centrifuged to a particle size of <0.5μm to remove nonsmectite clay and other minerals with a larger grain size. X-ray analysis was conducted on a Siemens D-500 diffractometer.

Of the 160 samples that were collected, 63 consisted of pure smectite; these pure smectite samples were selected for isotopic analysis. None of the 160 diffractograms showed the presence of significant mixed layering with other clay minerals by inspection and/or by using the $\Delta 2\theta$ (002/003) – $\Delta 2\theta$ (001/002) method of Moore and Reynolds (1997). This includes mixed layering of illite and smectite that would be indicative of diagenetic alteration. The diffractogram of the glycolated sample shown in Figure 2 is typical of the smectite samples selected for isotopic analysis.

Oxygen extraction procedures followed those outlined in Poage and Chamberlain (2002). An internal quartz standard (BX-88) was analyzed with each loading of the extraction line, and an internal smectite standard (01DS069) was analyzed in several of the batch runs. Extraction line runs for which ratios of the standards were outside of the range of ±0.2‰ for quartz or ±0.4‰ for smectite were discarded, and the samples were reanalyzed.

We also collected hydrogen isotopic values of 51 of the smectite samples using a continuous flow online combustion method similar to that described in Sharp et al. (2001). Our analysis of hydrogen isotope standards gave a precision and accuracy of the following: (1) on NBS 22 (oil) the accepted δD value is −118.5‰; our value was −118.6 ± 1.7‰; (2) Clay Mineralogy Source clays were: kga-1 (kaolinite) −64.1 ± 1.5‰; kga-2 (kaolinite) −65.9 ± 0.8‰; STx-1 (smectite) −99.2 ± 1.0‰; and SAz-1 (smectite) −109.2 ± 2.2‰. The δD values of these clays have not been previously reported.

RESULTS

Five results follow from our analyses:

1. With the exception of a few samples, the combined oxygen and hydrogen isotope values of smectite cluster along a line parallel to and displaced from the meteoric water line (Fig. 3; Table 1). Only a small number of the samples (notably samples: 01DS39, 01DS16, 01DS129, and 01DS124) are significantly displaced from this line. This result is similar to results originally described by Lawrence and Taylor (1971), who found that the isotopic composition of kaolinite, montmorillonite, and gibbsite from Quaternary soils throughout North America correlate with the present-day meteoric waters. In fact, the oxygen and hydrogen isotopic values of smectite presented in this paper are similar to the isotopic values of montmorillonite in modern soils from Montana and Idaho (Lawrence and Taylor, 1971).
2. There is a wide range of both hydrogen and oxygen isotope ratios of smectite, and the isotopic values increase toward the east. The δD values of smectite range from −126‰ to −88‰, and $\delta^{18}O$ values of smectites range from 8.7‰ to 20.2‰. In general, both δD and $\delta^{18}O$ values of smectite from eastern sections are ~40‰ and ~6‰

TABLE 1. LOCATION, GEOLOGIC GROUP OR FORMATION, AGE, ISOTOPE VALUE, MODERN ELEVATION, FIELD LOCATION, AND AGE REFERENCE FOR SMECTITE SAMPLES

Sample	Latitude (°N)	Longitude (°W)	Formation	Age (Ma)	$\delta^{18}O$	δD	Elevation (m)	Reference
Miocene samples								
01DS37	42.4373	103.4184	Arikaree	21.9	19.7	–127	1340	2
01DS28	42.769	103.5354	Arikaree	22	18.8	–93.0	1300	2
01DS36	42.4271	103.5419	Arikaree	21.9	16.2	–121	1340	2
01DS44	42.9492	103.6854	Arikaree	22	18.8	–97	1100	2
01DS35	42.4234	103.7525	Arikaree	21.9	15.0	–94.0	1340	2
01DS33	42.7425	103.8113	Arikaree	22	16.7	–93.0	1100	2
01DS34	42.7154	103.8828	Eagle Crag Ash	19.2	16.8	–95.0	1100	2
01DS104	42.446	107.5316	Split Rock	Middle Miocene	15.7	–107	1770	6
01DS102	42.4446	107.5588	Split Rock	Middle Miocene	12.7	–110	1770	6
01DS103	42.435	107.5628	Split Rock	Middle Miocene	12.8	–118	1770	6
01DS116	42.7288	107.5826	Split Rock	Middle Miocene	12.8	–112	2250	6
01DS115	42.7308	107.5845	Split Rock	Middle Miocene	11.9	–113	2250	6
01DS121	42.6817	108.1239	Split Rock	Middle Miocene	14.5	–116	1800	6
Oligocene samples								
01DS55	43.8288	101.6981	Peanut Peak	35	17.7	–107	760	4
01DS69	43.863	102.8605	Chadron	37–34	18.6	–103	850	5
01DS70	43.8614	102.8631	Chadron	37–34	20.2	–	850	5
01DS72	43.8614	102.8631	Chadron	37–34	18.4	–101	850	5
01DS74	43.8614	102.8631	Chadron	37–34	20.0	–101	850	5
01DS75	43.8608	102.8636	Chadron	37	19.6	–99	850	5
01DS12	41.5968	103.1131	White River	31.9–30.6	18.3	–94.0	1170	1
01DS14	41.5968	103.1131	White River	31.9–30.6	18.0	–88.0	1170	1
01DS15	41.5968	103.1131	White River	31.9–30.6	18.0	–92.0	1170	1
01DS16	41.5968	103.1131	White River	31.9–30.6	20.2	–110	1170	1
01DS17	41.5968	103.1131	White River	31.9–30.6	17.6	–102	1170	1
01DS18	41.5968	103.1131	White River	31.9–30.6	17.3	–123	1170	1
01DS19	41.6991	103.3415	White River	31.9–30.6	19.2	–100	1290	1
01DS20	41.6991	103.3415	White River	31.9–30.6	18.6	–90.0	1290	1
01DS39	42.9194	103.4913	Peanut Peak	34.5	16.9	–123	1160	4
01DS31	42.7791	103.5621	Brule	30.6–30	17.2	–93.0	1170	3
01DS30	42.7688	103.5835	Brule	30.6–30	16.8	–107	1170	3
01DS81	42.7625	105.0128	Chadron	37–34	16.6	–101	1600	1
01DS84	42.7125	105.371	White River	31.9–30.6	14.7	–112	1520	1
01DS89	42.6386	106.7545	White River	35.9–35.4	13.9	–107	1830	1
01DS90a	42.6386	106.7545	White River	35.9–35.4	15.5	–122	1830	1
01DS91	42.6386	106.7545	White River	35.9–35.4	14.4	–118	1830	1
01DS92	42.6386	106.7545	White River	35.9–35.4	14.2	–109	1830	1
01DS93	42.6386	106.7545	White River	35.9–35.4	14.1	–122	1830	1
01DS94	42.6386	106.7594	White River	35.9–35.4	13.9	–121	1830	1
01DS95	42.6386	106.7594	White River	35.9–35.4	13.7	–124	1830	1
01DS97	42.6386	106.7594	White River	35.9–35.4	14.3	–117	1830	1
01DS98	42.639	106.765	White River	35.9–35.4	13.4	–126	1830	1
01DS99	42.639	106.765	White River	35.9–35.4	14.3	–124	1830	1
01DS100	42.639	106.765	White River	35.9–35.4	13.6	–120	1830	1
01DS101	42.639	106.765	White River	35.9–35.4	14.0	–113	1830	1
01DS113	42.7574	107.5635	White River	ca. 32	10.9	–126	2200	6
01DS129	42.5848	108.287	White River	ca. 32	10.6	–104	2500	6
01DS129	42.5848	108.287	White River	ca. 32	10.6	–104	2500	6
Eocene samples								
01DS110	42.7575	107.5874	Wagon Bed	ca. 49	13.6	–122	2150	6
01DS111	42.7608	107.59	Wagon Bed	ca. 49	14.7	–122	2150	6
01DS123	42.7091	108.1809	Wagon Bed	ca. 49	16.1	–107	1900	6
01DS124	42.7091	108.1809	Wagon Bed	ca. 49	12.7	–101	1900	6
01DS128	42.5964	108.2933	Wagon Bed	ca. 49	15.7	–113	2500	6
6C-1	43.83	108.55	Lower Aycross	ca. 50	15.8	–	2150	7
6C-3	43.83	108.55	Lower Aycross	ca. 50	15.8	–	2152	7
6A-1	43.501	108.58	Upper Aycross	47.9 ± 0.5	14.5	–	2193	7
6A-2	43.501	108.58	Upper Aycross	47.9 ± 0.5	15.4	–	2195	7
6A-3	43.501	108.58	Upper Aycross	47.9 ± 0.5	14.8	–	2197	7
Mnau	44.8542	110.1567	Lamar River	49	9.4	–	2160	8
MNI	44.8542	110.1567	Lamar River	49	8.7	–	2160	8
CC2	44.8655	111.0488	Lamar River	49	10.2	–	2651	8
CC3	44.8655	111.0488	Lamar River	49	10.1	–	2651	8
SR1	44.8946	111.3345	Lamar River	49	9.2	–	2329	8
SR2	44.8946	111.3345	Lamar River	49	9.7	–	2329	8

Note: 1—MacFadden and Hunt (1998); 2—Love (1970); 3—Larson and Evanoff (1998); 4—Terry (1998); 5—Evans and Welzenbach (1998); 6—LaGarry (1998); 7—Sundell et al. (1984); 8—Fritz (1980).

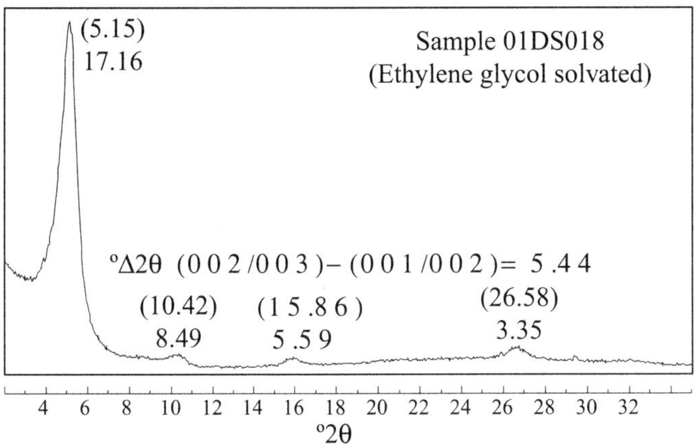

Figure 2. Representative X-ray pattern of glycolated smectite.

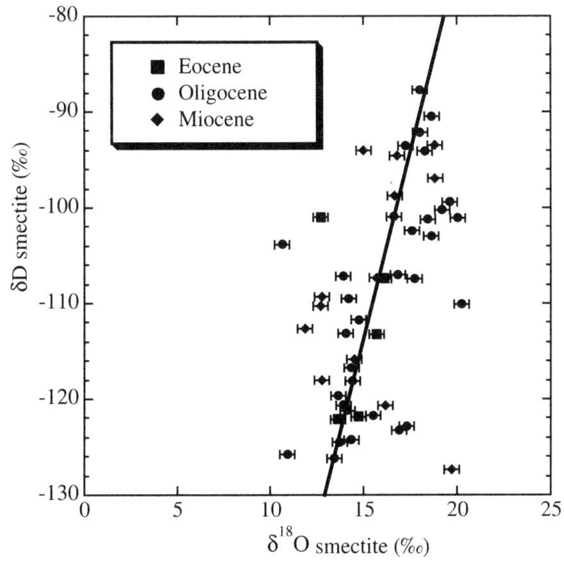

Figure 3. Combined hydrogen and oxygen isotope analyses of smectites examined in this study. Error bars represent analytical uncertainty of oxygen isotope analysis; analytical uncertainty for hydrogen isotopes falls within the range covered by each symbol. Trend of isotopic data is approximately parallel to the meteoric water line, which suggests these smectites formed in equilibrium with meteoric water and have not been modified isotopically by postformation events.

higher, respectively, than δD and $\delta^{18}O$ values of smectite from western sections, irrespective of geologic age.

3. The $\delta^{18}O$ values of smectites from Oligocene and Miocene samples show a strong correlation with modern altitude, with lower values at higher altitudes (Figs. 4 and 5). However, there is essentially no correlation between $\delta^{18}O_{smectite}$ and modern elevation for the Eocene samples (Fig. 6). The δD data from smectite of Oligocene and Miocene samples show a similar correlation with altitude, however, the correlation is not as good as the correlation of oxygen isotope ratios and altitude. Presumably the lower correlation results from the increased scatter in hydrogen isotopes of smectite; scatter which may reflect some degree of post-depositional exchange, evaporative effects, and exchange with basinal fluids, all of which affect hydrogen isotopes to a greater degree than oxygen isotopes. For this reason, we largely restrict our analysis to oxygen isotopes, and simply use the hydrogen isotopes of smectite to assess the role of evaporation and exchange with basinal brines.
4. The $\delta^{18}O_{smectite}$ values for Oligocene and Miocene samples are similar to $\delta^{18}O$ for modern hypothetical smectites calculated to be in isotopic equilibrium with modern precipitation. The dashed lines in Figures 4, 5, and 6 represent the theoretical oxygen isotope value of smectite formed in equilibrium with modern precipitation and was calculated with a smectite-water fractionation factor (Savin and Lee, 1988), modern meteoric precipitation isotope values from North Platte, Nebraska (Harvey and Welker, 2000), central and eastern Wyoming (Coplen and Kendall, 2000), and Yellowstone National Park (Thordsen et al., 1992), and modern annual average temperature at these three sites.
5. No temporal trends in isotopic values were evident where multiple samples were collected throughout a stratigraphic section. The variation in $\delta^{18}O_{smectite}$ was typically much less than 3‰ at these locations. Considering our analytical error of ±0.4‰, there is essentially no difference in oxygen isotope ratios temporally between Miocene and Oligocene samples. Eocene samples have slightly lower oxygen isotope values than Miocene and Oligocene samples overall (Fig. 7).

DISCUSSION

There are several possible explanations for the geographic variation of $\delta^{18}O_{smectite}$ from Tertiary volcanic airfall deposits of the eastern Rocky Mountains and western Great Plains, which include: (1) the modern elevation gradient and climate system on the east side of the Rocky Mountains has been in place since the Oligocene; (2) the smectites do not record paleoprecipitation, but instead have equilibrated with modern waters; (3) the smectites formed relatively recently in these volcanic air-fall ashes; and/or (4) the oxygen isotope values of smectites have been altered by later diagenesis or exchange with basinal fluids. For the following reasons, we believe that the possible effects of points 2 through 4 did not significantly influence the results presented in this paper. Instead, we suggest that the oxygen isotope values of smectites record the paleoprecipitation and elevation gradient on the east side of the Rocky Mountains that has existed since at least Oligocene times.

First, it could be argued that the similarity between the spatial distribution of $\delta^{18}O_{smectite}$ from Early Tertiary air-fall deposits and the $\delta^{18}O$ values of modern precipitation is the result of

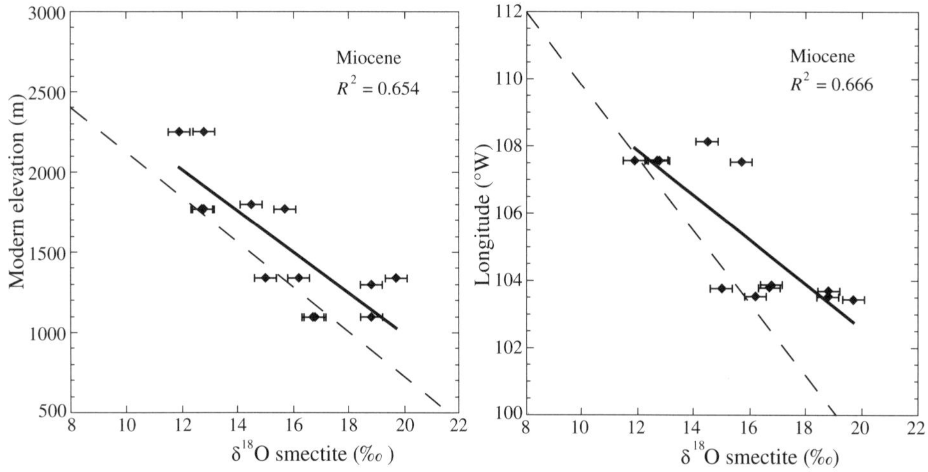

Figure 4. Results of oxygen isotope analysis of Miocene smectite versus modern elevation and longitude. Error bars represent analytical uncertainty of ±0.4‰. The dashed line represents the theoretical oxygen isotope value of smectite formed in equilibrium with modern precipitation (see text). Heavy solid line represents the linear regression best fit to the measured smectite isotopic data. Miocene isotopic results correlate well with modeled, modern smectites.

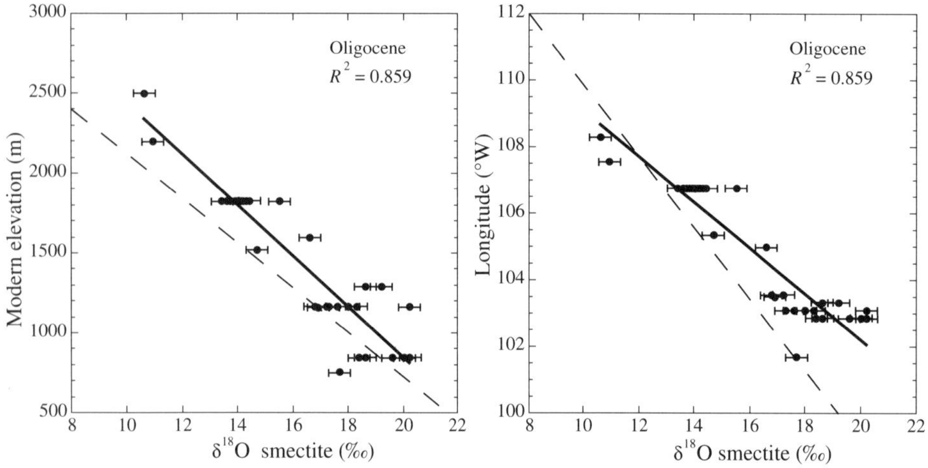

Figure 5. Results of oxygen isotope analysis of Oligocene smectite versus modern elevation and longitude. Error bars represent analytical uncertainty of ±0.4‰. The dashed line represents the theoretical oxygen isotope value of smectite formed in equilibrium with modern precipitation (see text). Heavy solid line represents the linear regression best fit to the measured smectite isotopic data. Oligocene isotopic results correlate well with modeled, modern smectites. Together with the Miocene results from Figure 4, these data suggest that the oxygen isotope gradient in the Rocky Mountain region has remained relatively unchanged since at least 37 Ma.

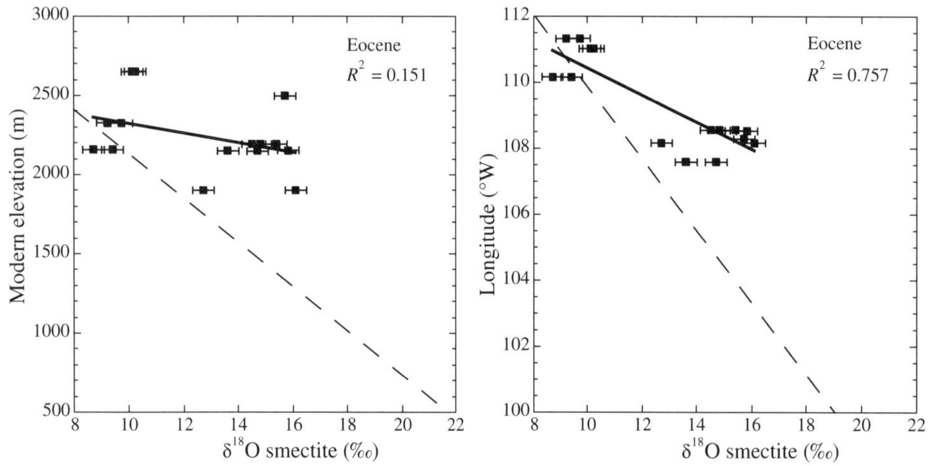

Figure 6. Results of oxygen isotope analysis of Eocene smectite versus modern elevation and longitude. Error bars represent analytical uncertainty of ±0.4‰. The dashed line represents the theoretical oxygen isotope value of smectite formed in equilibrium with modern precipitation (see text). Heavy solid line represents the linear regression best fit to the measured smectite isotopic data. Eastern, low-elevation samples are heavier than our modeled modern smectites, which is likely due to a warmer climate in the Eocene. The western, high-elevation samples, which correlate well with modeled, modern smectites, cannot be explained by global climate change alone.

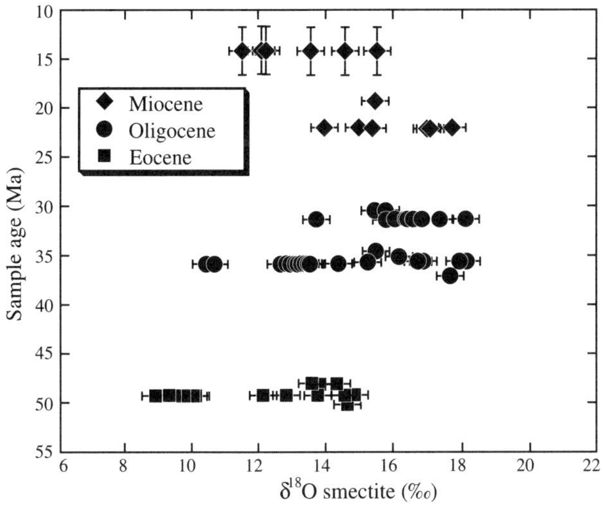

Figure 7. Oxygen isotope results versus age of sample. No correlation exists between age and isotope ratio (the R^2 for this relationship is <0.01). Error bars represent analytical uncertainty of ±0.4‰. Error for the ages is reported within the cited reference listed in Table 1 and falls within the range covered by the symbol where absent. Eocene-aged samples are slightly lighter than younger samples; these smectites formed in locations of relatively high modern elevation and low average modern $\delta^{18}O_{precipitation}$. It appears that geographic location rather than age is a much more important factor for $\delta^{18}O_{smectite}$, which suggests long-term stability of atmospheric circulation patterns in the sample area.

isotopic equilibrium with modern precipitation, and may not be representative of Early Tertiary isotopic composition. It has been shown, however, that pedogenic clay minerals retain their original structural oxygen isotopic composition and do not reequilibrate isotopically after formation (Lawrence and Taylor, 1971; Yeh and Savin, 1977; Stern et al., 1997; Savin and Hsieh, 1998). This work is corroborated by studies showing that smectite and calcite (which is extremely resistant to postdepositional isotopic exchange; Cerling and Quade, 1993) record similar isotopic trends in rocks where both minerals have been examined (Stern et al., 1997; Poage and Chamberlain, 2002).

Second, it is possible that the smectite we analyzed may be much younger than the age of the sedimentary deposits, and the isotope values preserved in these minerals therefore record relatively recent formation. We consider this possibility unlikely, because there is abundant evidence that smectite, as an alteration product, forms relatively rapidly in the shallow subsurface. Strontium isotopic studies and petrographic analysis of calcite, smectite, and silica cements of sandstones from the White River Group, Arikaree Group, and Miocene-Pliocene Ogallala Group suggest that these authigenic minerals formed in the Tertiary soon after deposition of the sediments (Stanley and Benson, 1979; Stanley and Faure, 1979). Additionally, studies of glacial moraines as young as 10^4 yr old in the Rocky Mountains contain smectite as a weathering product (Blum and Erel, 1997), Quaternary soils in this region contain authigenic smectite (Lawrence and Taylor, 1971), and ash from the 1980 Mt. St. Helens eruption has already converted to clay minerals in many locations (G. Thompson, 2003, personal commun.).

Third, it is possible that the oxygen isotopes of the smectites we measured record some later period of diagenesis or exchange with basinal brines. However, during burial diagenesis, smectite converts to mixed-layer illite/smectite to pure illite with increasing burial depth. The conversion of smectite to illite begins to take place at temperatures above 35 °C, and most of the conversion takes place between 55 °C and 95 °C (Perry and Hower, 1970). X-ray diffraction analysis conducted on the air-fall ashes examined in this study showed only a small percentage of mixed layering of illite/smectite (<5% illite) in a small number of samples. Thus, it is unlikely that the ashes reached burial temperatures of ~50 °C, whereas temperatures greater than 100 °C are necessary to cause isotopic exchange between smectite and water (O'Neil and Kharaka, 1976; Yeh and Savin, 1977). Assuming the modern geothermal gradient of the western Great Plains, which is ~28 °C/km, and an average annual surface temperature of ~10 °C, temperatures of 35 °C would be achieved at a burial depth of ~900 m. However, estimations of the original thickness or preserved thickness of the Tertiary and Quaternary deposits do not exceed 800 m (Love, 1970; Hunt, 1990; Larson and Evanoff, 1998). Moreover, there is little isotopic evidence for later exchange with basinal brines, as the combined δD and $\delta^{18}O$ values of smectite cluster along a line parallel to the meteoric water line (Fig. 3) rather than forming a field with a significantly shallower slope due to water-rock interaction (see Longstaffe, 1987).

We, therefore, suggest the observed geographic trends in smectite oxygen isotope values record the long-term precipitation history of the eastern Rocky Mountains. Our data suggest that the Oligocene and Miocene paleoprecipitation patterns were similar to that of today; and there was likely a significant difference in the climate of this region in the Eocene. As mentioned already, the isotopic gradient on the east side of the Rocky Mountains is primarily controlled by the interaction of three distinct air masses, each with a different isotopic signature. Climate models show that this unique climate regime is related to the topography of the region (Kutzbach et al., 1989; Huber and Sloan, 1999; Sewall et al., 2000). The fact that the Miocene and Oligocene isotopic gradients are similar to modern isotopic gradients, therefore, suggests that the Rocky Mountains as a significant topographic feature have existed since at least that time.

However, since the Oligocene, there has been a continuous ~2–4 °C decrease in global temperature (Wolfe, 1978; Prothero and Heaton, 1996; Terry, 2001) and an increase in the oxygen isotope composition of ocean waters due to development of significant ice sheets since ca. 25 Ma (Zachos et al., 2001). Both of these effects could also influence the isotopic composition of precipitation. To assess these effects, we calculated the isotopic composition of smectite, assuming that the isotopic changes in ocean water (Zachos et al., 2001) cause a direct effect on $\delta^{18}O_{precipitation}$, and the temperature changes affect both the $\delta^{18}O_{precipitation}$ (Savin and Hsieh, 1998) and the isotopic fractionation between smectite and water (Savin and Lee, 1988). The results of these calculations show that the smectite in the Oligocene and Miocene samples would have essentially the same $\delta^{18}O$ values as modern samples (our results for these calculations were less than our analytical error of ±4 for our oxygen isotope analysis), and they plot essentially on top of the calculated modern smectite values represented in Figures 4–6 by the dashed lines. Therefore, we suggest that magnitude of change in annual temperature and oxygen isotope composition of ocean water as reported in the literature does not significantly influence the isotopic gradient observed in our Oligocene and Miocene samples when compared to present values, and the stability of atmospheric circulation patterns is a much more important consideration. This hypothesis is in agreement with the conclusions of Amundson et al. (1996), who suggested that atmospheric circulation, not average temperature, determined the $\delta^{18}O$ of precipitation in the eastern Rocky Mountains of Wyoming during the Pleistocene.

Unlike the Oligocene and Miocene samples, Eocene smectites from central Wyoming have $\delta^{18}O$ values that are higher than would be expected for their current elevation. There are several possible explanations for the trends observed in the Eocene data from central Wyoming: (1) these smectites formed at lower elevations and have been uplifted prior to Oligocene time; (2) the central Wyoming smectites formed from evaporative waters during the Eocene; (3) these smectites formed at warmer temperatures; and/or (4) the precipitation in central Wyoming during the Eocene had higher $\delta^{18}O$ values than modern precipitation. Our hydrogen and oxygen isotopic data show no evidence for increased evaporation in the basins of central Wyoming during the Eocene, and there is no compelling evidence for uplift of the basins during the Oligocene. We prefer the interpretation that the relatively high $\delta^{18}O_{smectite}$ values are directly linked to the climatic history, because there were significant climatic changes during the Eocene that would influence the isotopic composition of precipitation in this region. We cannot, however, exclude tectonic modification between the time of deposition of our Eocene and Oligocene samples as a possible explanation of our results.

Previous paleoclimate studies suggest the average temperatures in the Eocene in North America were 10–20 °C warmer than present before a rapid, global cooling event in the early Oligocene (33.2 Ma), when average temperatures in North America fell to 2–4 °C warmer than present (e.g., Wolfe, 1978; Prothero and Heaton, 1996; Terry, 2001). In addition to temperature changes and the lack of significant ice sheets, differences in past oceanic circulation patterns could potentially have affected the stable isotopic composition of precipitation. During the Eocene, oceanic basins were not as restricted in the equatorial regions as they are presently. Because no land mass was extant between North and South America prior to the late Miocene–early Pliocene, seawater exchange took place between the Pacific and Atlantic basins. Additionally, prior to the Oligocene closing of the Tethys Seaway, the northern proto-Indian and northern Atlantic Oceans were open to circulation through the Mediterranean basin. In spite of these circulation differences, modeling of oceanic circulation for the early Eocene Pacific Ocean suggests circulation patterns that closely resembled present-day circulation (Bice et al., 2000). Mean annual temperatures for surface waters in the early Eocene North Pacific (30°–60°N latitude) are thought to have been ~1–2 °C warmer than present, and the global averaged mean annual atmosphere-ocean moisture flux has been calculated to be 10% greater than present, with most of the increase in tropical latitudes (Bice et al., 2000). Based on modeled early Eocene salinity, Bice et al. (2000) predicted the $\delta^{18}O$ value of early Eocene surface water in the North Pacific to be −1.0‰ to −1.2‰, or ~0.5‰ to 1.5‰ lighter than present-day values. Using this difference in $\delta^{18}O$ values for surface waters in the North Pacific for the early Eocene, annual average temperatures for the interior of North America 10–20 °C warmer in the Eocene, and the relationship between average annual temperature and $\delta^{18}O_{precipitation}$ (Gat, 1980), we predict the Eocene $\delta^{18}O_{precipitation}$ in western Nebraska to be approximately −5‰ to 0‰, or 5‰ to 10‰ heavier than present $\delta^{18}O_{precipitation}$ values. Assuming this value for $\delta^{18}O_{precipitation}$ and the fractionation factor for smectite of Savin and Lee (1988) ($1000\ln \alpha_{smectite\text{-}water} = 2.58 \times 10^6 T^{-2} - 4.19$), we would expect that smectite formed in the Eocene throughout our study area would be 4‰–6‰ higher than modern smectites. In addition, general circulation models for the Late Tertiary suggest that as a result of a relatively warm climate, the monsoonal effects were intensified along the east side of the Rocky Mountains (Huber and Sloan, 1999; Sewall et al., 2000). This would add to the increase in the $\delta^{18}O$ values of precipitation in central Wyoming relative to today, but the magnitude of this effect is difficult to quantify. Since these effects can cause the isotopic shift observed in our Eocene samples relative to modern from the eastern, low-elevation part of our study area, it is not necessary to invoke tectonics as the cause of the relatively high $\delta^{18}O$ values of these samples, although we do not rule out Eocene uplift as a possible explanation for our results. The preceding discussion does not explain our results for samples from western Wyoming, because these Eocene smectites are not isotopically heavier than our modeled modern smectites. We propose these relatively light isotopic values in western Wyoming may be the result of cooler than global average temperatures in this area (rather unlikely without high elevations), a relatively large amount of isotopically light winter precipitation in the early Cenozoic Rocky Mountain region, which has been suggested by Norris et al. (1996) and Dettman and Lohmann (2000), uplift or topographically high elevations in western Wyoming in the Eocene, or isotopic rainout effects due to a topographic barrier to the west of western Wyoming. All of these possibilities suggest high elevations or topography in the Early Tertiary Rocky Mountains, which is consistent with other isotopic studies from the region (Norris et al., 1996; Dettman and Lohmann, 2000; Fricke, 2003). We cannot, however, determine from the isotopic data from our Eocene samples if tectonic or climate modification caused the shift in isotopic gradient between our 50–49 Ma and 37 Ma samples.

CONCLUSIONS

We conclude that an isotopic gradient, which suggests areas of high elevation, has existed in the Rocky Mountain–Great Plains region since at least 37 Ma. This conclusion is similar to those of previous isotopic studies of older rocks in the area (e.g., Dettman and Lohmann, 2000; Fricke, 2003). Because this isotopic gradient reflects the unique climatic regime that results from the interaction of three air masses with the high topography of the Rocky Mountains, we suggest it is very unlikely that high elevations in the Rocky Mountains could have formed in the late Miocene without significant modification of the atmospheric circulation patterns of western North America and the resulting distribution of pre–late Miocene $\delta^{18}O_{precipitation}$ values compared to the modern distribution of $\delta^{18}O_{precipitation}$. If there were not a topographic barrier before the late Miocene, we predict unobstructed precipitation originating over the Pacific Ocean would have extended deep into the continental interior of North America. Additionally, with no orographic rainout effects due to interaction with a topographic barrier, the oxygen isotope composition of precipitation in the western part of our study area would have been relatively enriched in ^{18}O. Therefore, we would expect the west-to-east gradient of $\delta^{18}O_{smectite}$ to be shallower and $\delta^{18}O_{precipitation}$ to decrease from west to east in western Wyoming if there were no pre–late Miocene Rocky Mountains. Our $\delta^{18}O_{smectite}$ results, however, increase from west to east, which suggests that pre–late Miocene high-elevation areas existed in the Rocky Mountain region, and for the Oligocene and Miocene samples, the values mimic the modern distribution and values of $\delta^{18}O_{precipitation}$. We propose that this indicates Oligocene and later high elevation in the Rocky Mountains (in other words, an elevation gradient between the Rocky Mountains and the western Great Plains) approximately equal to modern elevations.

Results from our Eocene samples can be explained by climate and/or tectonic modification since the time of deposition for these samples, but both of these explanations suggest high elevations in the Rocky Mountain region ca. 50 Ma. Our results do not agree with the traditional view that the surface uplift of the Rocky Mountains occurred in the late Miocene and early Pliocene. There is additional evidence that the Basin and Range and Sierra Nevada Mountains were at high elevations during much of the Tertiary as well (Wolfe et al., 1997; Wolfe et al., 1998; Poage and Chamberlain, 2001). The idea of late Cenozoic surface uplift in many areas of the American west should be reconsidered, because evidence used to support this view may be the result of changes in global temperature rather than changes in topography, a concept originally proposed by Molnar and England (1990).

ACKNOWLEDGMENTS

We thank R. Reynolds for the use of his X-ray diffraction laboratory and for assistance with the interpretation of our diffraction patterns. We are indebted to M. LeBeau, the rangers and staff at Yellowstone National Park, and the National Park Service for allowing us to collect samples within Yellowstone. We thank P. Ryan, J. Aronson, M. Zhao, J. Spencer, and P. Koch for very helpful reviews of early versions of this manuscript. National Science Foundation (NSF) grants EAR-0001208 and EAR-0309011 provided funding to Chamberlain in support of this project.

REFERENCES CITED

Amundson, R., Chadwick, O., Kendall, C., Wang, Y., and DeNiro, M., 1996, Isotopic evidence for shifts in atmospheric circulation patterns during the late Quaternary in mid–North America: Geology, v. 24, p. 23–26, doi: 10.1130/0091-7613(1996)024 <0023:IEFSIA>2.3.CO;2.

Bice, K.L., Sloan, L.C., and Barron, E.J., 2000, Comparison of early Eocene isotopic paleotemperatures and the three-dimensional OGCM temperature field: The potential for use of model-derived surface water $\delta^{18}O$, *in* Huber, B.T., MacLeod, K.G., and Wing, S.L., eds., Warm climates in earth history: Cambridge, Cambridge University Press, p. 79–131.

Blum, J.D., and Erel, Y., 1997, Rb-Sr isotope systematics of a granitic soil chronosequence: The importance of biotite weathering: Geochimica et Cosmochimica Acta, v. 61, p. 3193–3204, doi: 10.1016/S0016-7037(97)00148-8.

Bradley, W.C., 1987, Erosion surfaces of the Colorado Front Range: A review, *in* Graf, W.L., ed., Geomorphic systems of North America: Boulder, Colorado, Geological Society of America, Geology of North America, Centennial Special Volume 2, p. 215–220.

Bryson, R.A., and Hare, F.K., 1974, The climates of North America, *in* Bryson, R.A., and Hare, F.K., eds., Climates of North America (world survey of climatology, volume 11): New York, Elsevier, p. 1–47.

Burchfiel, B.C., Cowan, D.S., and Davis, G.A., 1992, Tectonic overview of the Cordilleran orogen in the western United States, *in* Burchfiel, B.C., et al., eds., The Cordilleran orogen: Conterminous United States: Boulder, Colorado, Geological Society of America, Geology of North America, v. G-3, p. 407–479.

Cerling, T.E., and Quade, J., 1993, Stable carbon and oxygen isotopes in soil carbonates, *in* Swart, P.K., Lohmann, K.C., McKenzie, J., and Savin, S., eds., Climate change in continental isotopic records: American Geophysical Union Geophysical Monograph 78, p. 217–232.

Chamberlain, C.P., and Poage, M.A., 2000, Reconstructing the paleotopography of mountain belts from the isotopic composition of authigenic minerals: Geology, v. 28, p. 115–118, doi: 10.1130/0091-7613(2000)028 <115: RTPOMB>2.3.CO;2.

Chase, C.G., Gregory-Wodzicki, K.M., Parrish, J.T., and DeCelles, P.G., 1998, Topographic history of the western Cordillera of North America and controls on climate, *in* Crowley, T.J., and Burke, K.C., eds., Tectonic boundary conditions for climate reconstructions: Oxford, Oxford University Press, p. 73–99.

Coplen, T.B., and Kendall, C., 2000, Stable hydrogen and oxygen isotope ratios for selected sites of the U.S. Geological Survey's NASQAN and Benchmark Surface-Water Networks: U.S. Geological Survey Open-File Report 00-160.

Despain, D.G., 1987, The two climates of Yellowstone National Park: Proceedings of the Montana Academy of Science, v. 47, p. 11–19.

Dettman, D.L., and Lohmann, K.C., 2000, Oxygen isotope evidence for high-altitude snow in the Laramide Rocky Mountains of North America during the Late Cretaceous and Paleogene: Geology, v. 28, p. 243–246, doi: 10.1130/0091-7613(2000)028 <243:OIEFHS>2.3.CO;2.

Drummond, C.N., Wilkinson, B.H., Lohmann, K.C., and Smith, G.R., 1993, Effect of regional topography and hydrology on the lacustrine isotopic record of Miocene paleoclimate in the Rocky Mountains: Palaeogeography, Palaeoclimatology, Palaeoecology, v. 101, p. 67–79, doi: 10.1016/0031-0182(93)90152-9.

Epis, R.C., and Chapin, C.E., 1975, Geomorphic and tectonic implications of the post-Laramide, late Eocene erosion surface in the southern Rocky Mountains, *in* Curtis, B.F., ed., Cenozoic history of the southern Rocky Mountains: Geological Society of America Memoir 144, p. 45–74.

Evans, J.E., and Welzenbach, L.C., 1998, Episodes of carbonate deposition in a siliciclastic-dominated fluvial sequence, Eocene-Oligocene White River Group, South Dakota and Nebraska, *in* Terry, D.O., et al., eds., Depositional environments, lithostratigraphy, and biostratigraphy of the White River and Arikaree Groups (late Eocene to early Miocene, North America): Geological Society of America Special Paper 325, p. 93–116.

Fricke, H.C., 2003, Investigation of early Eocene water-vapor transport and paleoelevation using oxygen isotope data from geographically widespread mammal remains: Geological Society of America Bulletin, v. 115, p. 1088–1096, doi: 10.1130/B25249.1.

Fritz, W.J., 1980, Reinterpretation of the depositional environment of the Yellowstone "fossil forest": Geology, v. 8, p. 309–313, doi: 10.1130/0091-7613(1980)8 <309:ROTDEO>2.0.CO;2.

Gat, J.R., 1980, The isotopes of hydrogen and oxygen in precipitation, *in* Fritz, P., and Fontes, J., eds., Handbook of environmental isotope geochemistry: New York, Elsevier, p. 21–24.

Gregory, K.M., and Chase, C.G., 1992, Tectonic significance of paleobotanically estimated climate and altitude of the late Eocene erosion surface, Colorado: Geology, v. 20, p. 581–585, doi: 10.1130/0091-7613(1992)020 <0581:TSOPEC>2.3.CO;2.

Gregory, K.M., and McIntosh, W.C., 1996, Paleoclimate and paleoelevation of the Oligocene Pitch-Pinnacle flora, Sawatch Range, Colorado: Geological Society of America Bulletin, v. 108, p. 545–561, doi: 10.1130/0016-7606(1996)108<0545:PAPOTO>2.3.CO;2.

Harvey, F.E., and Welker, J.M., 2000, Stable isotopic composition of precipitation in the semi-arid north-central portion of the U.S. Great Plains: Journal of Hydrology, v. 238, p. 90–109, doi: 10.1016/S0022-1694(00)00316-4.

Horton, T.W., Chamberlain, C.P., Fantle, M., and Blum, J.D., 1999, Chemical weathering and lithologic controls of water chemistry in a high-elevation river system: Clark's Fork of the Yellowstone River: Wyoming and Montana: Water Resources Research, v. 35, p. 1643–1655, doi: 10.1029/1998WR900103.

Huber, M., and Sloan, C., 1999, Warm climate transitions: A general circulation modeling study of the late Paleocene thermal maximum: Journal of Geophysical Research, v. 104, p. 16,633–16,655, doi: 10.1029/1999JD900272.

Hunt, R.M., Jr., 1990, Taphonomy and sedimentology of Arikaree (lower Miocene) fluvial, eolian, and lacustrine paleoenvironments, Nebraska and Wyoming: A paleobiota entombed in fine-grained volcaniclastic rocks, *in* Lockley, M.G., and Rice, A., eds., Volcanism and fossil biotas: Geological Society of America Special Paper 244, p. 69–111.

Kutzbach, J.E., Guetter, P.J., Ruddiman, W.F., and Prell, W.L., 1989, Sensitivity of climate to late Cenozoic uplift in southern Asia and the American West: Numerical experiments: Journal of Geophysical Research, v. 94, p. 18,393–18,407.

LaGarry, H.E., 1998, Lithostratigraphic revision and redescription of the Brule Formation (White River Group) of northwestern Nebraska, *in* Terry, D.O., et al., eds., Depositional environments, lithostratigraphy, and biostratigraphy of the White River and Arikaree Groups (late Eocene to early Miocene, North America): Geological Society of America Special Paper 325, p. 63–91.

Larson, E.E., and Evanoff, E., 1998, Tephrostratigraphy and source of the tuffs of the White River sequence, *in* Terry, D.O., et al., eds., Depositional environments, lithostratigraphy, and biostratigraphy of the White River and Arikaree Groups (late Eocene to early Miocene, North America): Geological Society of America Special Paper 325, p. 1–14.

Lawrence, J.R., and Taylor, H.P., 1971, Deuterium and oxygen-18 correlation: Clay minerals and hydroxides in Quaternary soils compared to meteoric waters: Geochimica et Cosmochimica Acta, v. 35, p. 993–1003, doi: 10.1016/0016-7037(71)90017-2.

Longstaffe, F.J., 1987, Stable isotopic studies of diagenetic processes, *in* Kyser, T.K., ed., Stable isotope geochemistry of low temperature fluids: Mineralogical Association of Canada, v. 13, p. 187–257.

Love, J.D., 1970, Cenozoic geology of the Granite Mountains area, central Wyoming: U.S. Geological Survey Professional Paper 495-C, p. 1–154.

MacFadden, B.J., and Hunt, R.M., Jr., 1998, Magnetic polarity stratigraphy and correlation of the Arikaree Group, Arikareean (late Oligocene–early Miocene) of northwestern Nebraska, *in* Terry, D.O., et al., eds., Depositional environments, lithostratigraphy, and biostratigraphy of the White River and Arikaree Groups (late Eocene to early Miocene, North America): Geological Society of America Special Paper 325, p. 143–165.

McMillan, M.E., Angevine, C.L., and Heller, P.L., 2002, Postdepositional tilt of the Miocene-Pliocene Ogallala Group on the western Great Plains: Evidence of late Cenozoic uplift of the Rocky Mountains: Geology, v. 30, p. 63–66, doi: 10.1130/0091-7613(2002)030 <0063:PTOTMP>2.0.CO;2.

Molnar, P., and England, P., 1990, Late Cenozoic uplift of mountain ranges and global climate change: Chicken or egg?: Nature, v. 346, p. 29–34, doi: 10.1038/346029a0.

Moore, D.M., and Reynolds, R.C., 1997, X-ray diffraction and the identification and analysis of clay minerals: New York, Oxford University Press, 378 p.

Nativ, R., and Riggio, R., 1990, Precipitation in the southern High Plains: Meteorological and isotopic features: Journal of Geophysical Research, v. 95D, p. 22,559–22,564.

Norris, R.D., Jones, L.S., Corfield, R.M., and Cartlidge, J.E., 1996, Skiing in the Eocene Uinta Mountains? Isotopic evidence in the Green River Formation for snowmelt and large mountains: Geology, v. 24, p. 403–406, doi: 10.1130/0091-7613(1996)024 <0403:SITEUM>2.3.CO;2.

O'Neil, J.J.R., and Kharaka, Y.K., 1976, Hydrogen and oxygen isotope exchange reactions between clay minerals and water: Geochimica et Cosmochimica Acta, v. 40, p. 241–246, doi: 10.1016/0016-7037(76)90181-2.

Peizhen, Z., Molnar, P., and Downs, W.R., 2001, Increased sedimentation rates and grain sizes 2–4 Myr ago due to the influence of climate change in erosion rates: Nature, v. 410, p. 891–897, doi: 10.1038/35073504.

Perry, E., and Hower, J., 1970, Burial diagenesis in Gulf Coast pelitic sediments: Clays and Clay Minerals, v. 18, p. 165–177.

Poage, M.A., and Chamberlain, C.P., 2001, Empirical relationships between elevation and the stable isotope composition of precipitation and surface waters: Considerations for studies of paleoelevation change: American Journal of Science, v. 301, p. 1–15.

Poage, M.A., and Chamberlain, C.P., 2002, Stable isotope evidence for a pre–Middle Miocene rainshadow in the western Basin and Range: Implications for the paleotopography of the Sierra Nevada: Tectonics, v. 21, p. 1–10, doi: 10.1029/2001TC001303.

Prothero, D.R., and Heaton, T.H., 1996, Faunal stability during the early Oligocene climate crash: Palaeogeography, Palaeoclimatology, Palaeoecology, v. 127, p. 257–283, doi: 10.1016/S0031-0182(96)00099-5.

Rowley, D.B., Pierrehumbert, R.T., and Currie, B.S., 2001, A new approach to stable isotope paleoaltimetry: Implications for paleoaltimetry and paleohypsometry of the High Himalaya since the late Miocene: Earth and Planetary Science Letters, v. 188, p. 253–268, doi: 10.1016/S0012-821X(01)00324-7.

Savin, S.M., and Hsieh, J.C.C., 1998, The hydrogen and oxygen isotope geochemistry of pedogenic clay minerals: Principles and theoretical background: Geoderma, v. 82, p. 227–253, doi: 10.1016/S0016-7061(97)00103-1.

Savin, S.M., and Lee, M., 1988, Isotopic studies of phyllosilicates, *in* Bailey, S.W., ed., Hydrous phyllosilicates: Reviews in Mineralogy 19, p. 189–223.

Sewall, J.O., Sloan, L.C., Huber, M., and Wing, S., 2000, Climate sensitivity to changes in land surface characteristics: Global and Planetary Changes, v. 26, p. 445–465, doi: 10.1016/S0921-8181(00)00056-4.

Sharp, Z.D., Atudorei, V., and Durakiewicz, T., 2001, A rapid method for determination of hydrogen and oxygen isotope ratios from water and hydrous minerals: Chemical Geology, v. 178, p. 197–210, doi: 10.1016/S0009-2541(01)00262-5.

Stanley, K.O., and Benson, L.V., 1979, Early diagenesis of High Plains Tertiary vitric and arkosic sandstone, Wyoming and Nebraska, *in* Scholle, P.A., and Schluger, P.R., eds., Aspects of diagenesis: Society of Economic Paleontologists and Mineralogists Special Publication 26, p. 401–423.

Stanley, K.O., and Faure, G., 1979, Isotopic composition and sources of strontium in sandstone cements: The High Plains sequence of Wyoming and Nebraska: Journal of Sedimentary Petrology, v. 49, p. 45–54.

Stern, L.A., Chamberlain, C.P., Reynolds, R.C., and Johnson, G.D., 1997, Oxygen isotope evidence of climate change from pedogenic clay minerals in the Himalayan molasse: Geochimica et Cosmochimica Acta, v. 61, p. 731–744, doi: 10.1016/S0016-7037(96)00367-5.

Steven, T.A., Evanoff, E., and Yuhas, R.H., 1997, Middle and late Cenozoic tectonic and geomorphic development of the Front Range of Colorado, *in* Bolyard, D.W., and Sonnenberg, S.A., eds., Geologic history of the Colorado Front Range: Denver, Colorado, Rocky Mountain Association of Geologists, Colorado Front Range Guidebook, p. 115–124.

Sundell, K.A., Shive, P.N., and Eaton, J.G., 1984, Measured sections, magnetic polarity, and biostratigraphy of the Eocene Wiggins, Tepee Trail, and Aycross Formations within the southeastern Absaroka Range, Wyoming: Earth Science Bulletin, v. 17, p. 1–48.

Terry, D.O., 1998, Lithostratigraphic revision and correlation of the lower part of the White River Group: South Dakota to Nebraska, *in* Terry, D.O., et al., eds., Depositional environments, lithostratigraphy, and biostratigraphy of the White River and Arikaree Groups (late Eocene to early Miocene, North America): Geological Society of America Special Paper 325, p. 15–37.

Terry, D.O., 2001, Paleopedology of the Chadron Formation of northwest Nebraska: Implications for paleoclimatic change in the North American Midcontinent across the Eocene-Oligocene boundary: Palaeogeography, Palaeoclimatology, Palaeoecology, v. 168, p. 1–38, doi: 10.1016/S0031-0182(00)00248-0.

Thordsen, J.J., Kharaka, Y.K., Mariner, R., and White, L.D., 1992, Controls on the distribution of stable isotopes of meteoric water and snow in the greater Yellowstone National Park, USA, *in* Kharaka, Y.K., and Maest, D., eds., Water-rock interaction: Rotterdam, Balkema, p. 591–595.

Whitlock, C., and Bartlein, P.J., 1993, Spatial variations of Holocene climatic change in the Yellowstone region: Quaternary Research, v. 39, p. 231–238, doi: 10.1006/qres.1993.1026.

Wolfe, J.A., 1978, A paleobotanical interpretation of Tertiary climates in the Northern Hemisphere: American Scientist, v. 66, p. 694–703.

Wolfe, J.A., Schorn, H.E., Forest, C.H., and Molnar, P., 1997, Paleobotanical evidence for high altitudes in Nevada during the Miocene: Science, v. 276, p. 1672–1675, doi: 10.1126/science.276.5319.1672.

Wolfe, J.A., Forest, C.E., and Molnar, P., 1998, Paleobotanical evidence of Eocene and Oligocene paleoaltitudes in midlatitude western North America: Geological Society of America Bulletin, v. 110, p. 664–678, doi: 10.1130/0016-7606(1998)110<0664:PEOEAO>2.3.CO;2.

Yeh, H.W., and Savin, S.M., 1977, Mechanism of burial metamorphism of argillaceous sediments: O-isotope evidence: Geological Society of America Bulletin, v. 88, p. 1321–1330, doi: 10.1130/0016-7606(1977)88<1321:MOBMOA>2.0.CO;2.

Zachos, J.C., Pagani, M., Sloan, L., Thomas, E., and Bilups, K., 2001, Trends, rhythms, and aberrations in global climate 65 Ma to present: Science, v. 292, p. 686–693, doi: 10.1126/science.1059412.

Manuscript Accepted by the Society 23 June 2005

Printed in the USA

Geological Society of America
Special Paper 398
2006

Downstream development of a detrital cooling-age signal: Insights from $^{40}Ar/^{39}Ar$ muscovite thermochronology in the Nepalese Himalaya

I.D. Brewer
Department of Geosciences, Pennsylvania State University, University Park, Pennsylvania 16802, USA
D.W. Burbank
Department of Earth Science, University of California, Santa Barbara, California 93106, USA
K.V. Hodges
Department of Earth, Atmospheric, and Planetary Sciences, Massachusetts Institute of Technology, Cambridge, Massachusetts 02139, USA

ABSTRACT

The character and distribution of cooling ages in modern river sediment provide useful constraints on rates and patterns of uplift and erosion within actively deforming mountain ranges. Such sediment effectively samples all locations within the catchment area, irrespective of remoteness. We evaluate how successfully detrital cooling ages may be used to constrain hinterland erosion rates by examining the modern catchment of the Marsyandi River in central Nepal. Over the 100–200-km-length scale of the catchment, laser fusion $^{40}Ar/^{39}Ar$ data for detrital muscovite collected from 12 separate sites illustrate the downstream development of a detrital cooling-age signal that is both systematic and representative of the contributing area. Comparisons of paired samples indicate that, at short spatial (tens of meters) and temporal (hundreds of years) scales, the detrital cooling-age signal is consistent. The distribution of bedrock cooling ages in a subcatchment and the resulting detrital signal at the catchment mouth can be modeled as a function of the erosion rate, relief, hypsometry, catchment area, and muscovite distribution. Given that independent constraints are available for most of these variables, the detrital age signal should be a robust indication of the spatially averaged erosion rate. In the Marsyandi, our model predicts erosion rate differences of approximately twofold, with higher rates (>2 mm/yr) along the southern topographic front of the Himalaya.

Keywords: erosion rate, Himalaya, detrital ages, argon dating, Nepal.

INTRODUCTION

In the absence of reliable temporal controls, rates of erosion, topographic evolution, and deformation during orogenic growth are poorly defined. Bedrock thermochronology is typically restricted to rocks currently exposed at the surface, where it provides a limited temporal record, because high erosion rates in active orogens rapidly remove older rocks. Detrital mineral thermochronology, however, combines the antiquity of the stratigraphic record with the quantitative analysis of thermochronology (Bernet et al., 2001; Carrapa et al., 2003; Cerveny et al., 1988; White et al., 2002). Because grab samples in a foreland basin contain millions of grains, each from a different point within the contributing catchment, detrital samples

Brewer, I.D., Burbank, D.W., and Hodges, K.V., 2006, Downstream development of a detrital cooling-age signal: Insights from $^{40}Ar/^{39}Ar$ muscovite thermochronology in the Nepalese Himalaya, *in* Willett, S.D., Hovius, N., Brandon, M.T., and Fisher, D.M., eds., Tectonics, Climate, and Landscape Evolution: Geological Society of America Special Paper 398, Penrose Conference Series, p. 321–338, doi: 10.1130/2006.2398(20). For permission to copy, contact editing@geosociety.org.

provide a potent integration of orogenic data. When samples can be extracted from stratigraphic successions, detrital thermochronology provides the potential to document successive changes in the frequency of cooling ages and rates in an orogenic hinterland. Amid current controversy regarding thermal, flux, and topographic steady state in orogens (Whipple, 2001; Willett and Brandon, 2002; Willett et al., 2001), detrital ages serve to define changes in lag times between mineral cooling and deposition—a key indicator of changing erosion rates or thermal state (Bernet et al., 2001; Bullen et al., 2003; Carrapa et al., 2003).

Before detrital thermochronological data can be reliably interpreted, however, the processes controlling the cooling-age signal need to be explored. Because a detrital sample represents an integration across a catchment, it is critical to know how the observed spectrum of detrital ages is influenced by patterns of erosion rate, variations in lithology, mechanical breakdown of grains during transport, and differences in sediment production and transport processes and rates (Spiegel et al., 2004). Due to practical limitations on the number of grain ages that can be measured, commonly only one sample is analyzed at a given stratigraphic level or point on a river system. Hence, the consistency of the detrital age signal should also be assessed: does it vary at spatial and temporal scales of tens of meters and hundreds of years?

Although the widest applicability of detrital age analyses may be to stratigraphic successions, many of the key controls are impossible to define in ancient catchments, e.g., spatial and temporal variations in erosion rate, lithology, source-area boundaries, topographic relief, and geomorphic processes. The hinterland may have been eroded away, and the boundaries of former catchments are almost always unknown. Consequently, modern catchments provide the best opportunity to explore the multiple variables that influence the detrital age signal carried to the foreland. Armed with such knowledge, and a good understanding of the possible uncertainties, interpretations of data from the rock record can be improved (Spiegel et al., 2004).

The distribution of bedrock cooling ages in a catchment can be modeled as a function of topographic relief, erosion rate, and the subsequent geothermal gradient (Mancktelow and Grasemann, 1997; Stüwe et al., 1994). For $^{40}Ar/^{39}Ar$ dating of muscovite when the erosion rate is ≤3 mm/yr, the detrital cooling-age signal is determined by the hypsometry (the distribution of area with elevation) of the catchment (Brewer et al., 2003). These simple models, however, are applicable to catchments with spatially uniform erosion rates and homogeneous distributions of the target thermochronometer (in this case muscovite). In most active orogens, such restrictions would preclude the study of larger, orogenic-scale catchments and their foreland deposits.

In this study, we focus on the detrital system of the complete Marsyandi River, a trans-Himalayan river in central Nepal. We examine how the cooling-age signals of tributary catchments combine to produce a modern trunk-stream signal. We use $^{40}Ar/^{39}Ar$ analysis of individual muscovite grains to examine how the lithology, erosion rate, and hypsometry of individual catchment areas vary, and we investigate how these parameters control the evolution of the trunk-stream cooling-age signal as it travels downstream, from the headwaters to the foreland basin. This allows an examination of how well the cooling-age signal at the catchment mouth represents the contributing area upstream, and hence provides a baseline study for the interpretation of detrital ages preserved in the geological record.

The wide range in closure temperatures for different thermochronometers allows their application to many types of temperature-dependent geological problems. High-temperature thermochronometers are typically used to date the crystallization of minerals, whereas low-temperature thermochronometers are typically used to investigate late-stage cooling. The ages of detrital minerals have been used previously to assess provenance and source-area character (e.g., Adams et al., 1998; Garver and Brandon, 1994; Gehrels and Kapp, 1998), to estimate the age of the enclosing strata (e.g., Bullen et al., 2003; Carrapa et al., 2003; Garver et al., 1999; Najman et al., 2001), and to reconstruct orogenic cooling and erosion histories (e.g., Brandon and Vance, 1992; Copeland and Harrison, 1990; Spiegel et al., 2004; White et al., 2002).

Most of these previous investigations used sediments preserved in foreland basins, and despite the obvious value of the studies, interpretations can be ambiguous given the unknown variability in parameters, such as erosion rate, relief, and lithology, which modulate the detrital signal. Stock and Montgomery (1996) suggested that the range of detrital cooling ages in a particular catchment, in combination with a specified geothermal gradient, allows a theoretical calculation of catchment relief. Brewer et al. (2003) investigated how the interactions between geothermal gradient, erosion rate, and relief can be used in conjunction with the catchment hypsometry to predict the distribution of detrital cooling ages.

Although these studies (Stock and Montgomery, 1996; Brewer et al., 2003) reinforce the contention that detrital cooling-age data can provide useful tectonic-geomorphologic insights, they required uniform erosion rates across the catchment(s) of interest. Studies of detrital signals at the orogenic scale, however, should consider variable erosion rates, as well as variations in topography and lithology, because a large catchment produces a complex signal with detrital grains from multiple tributaries. In this paper, we build on previous work to investigate the parameters controlling the cooling-age signal from an entire orogenic-scale catchment. Field data from the Marsyandi Valley in central Nepal are used to examine the spatial pattern of erosion in the modern Himalaya.

GEOLOGICAL BACKGROUND

The Himalaya represent a favorable study area for detrital age dating because the orogen encompasses stark contrasts in tectonic rates, erosion rates, lithology, and depth of exhumation. North of the Himalayan topographic axis, the Indus-Tsangpo suture zone marks the surface boundary between lithologic units of Eurasian plate affinity to the north and Indian plate affinity to the south. After the initial collision at ca. 50–54 Ma (Rowley, 1996; Searle et al., 1997), the downgoing Indian plate was imbricated along a series of south-vergent thrust-fault systems, and the Himalaya formed as a result of the subsequent crustal thickening. Deformation continues today and global positioning system (GPS) data (Wang et al., 2001; Bilham et al., 1997) indicate that ~40% of the convergence between the Indian and Eurasian plates occurs across the Himalaya.

Thrust-fault systems mark many of the principal boundaries between tectonostratigraphic zones in the Himalaya. The Main Central thrust system juxtaposes high-grade metamorphic rocks and leucogranites of the Greater Himalayan sequence against lower-grade metasedimentary rocks of the Lesser Himalayan zone. Farther south, the Lesser Himalayan zone is separated from the foreland basin of the Himalaya by the Main Boundary thrust system. The topographic expression of the Indo-Asian collision closest to the foreland, the Siwalik Hills, corresponds to the Main Frontal thrust system. The initiation ages of these major thrust systems are progressively younger from north (22–28 Ma for the Main Central thrust system; Searle and Godin, 2003) to south (Pleistocene to Holocene for the Main Frontal thrust system), although evidence exists for out-of-sequence thrusting along these and other less significant fault systems in the Himalayan realm over the Miocene-Holocene interval (Hodges, 2000; Wobus et al., 2003).

The surface trace of the Main Central thrust system marks the approximate physiographic transition from the Lesser to the Greater Himalaya, although in places this transition occurs >20 km south of the Main Central thrust (Wobus et al., 2003). The transition is marked by an abrupt increase in mean elevation and relief. As a consequence, much of the steep southern front of the Greater Himalaya, where erosion rates are likely to be more rapid, is developed on the metamorphic and igneous rocks of the Greater Himalayan sequence.

A fourth important fault system bounds the top of the Greater Himalayan sequence: the South Tibetan fault system. Carrying primarily unmetamorphosed, Neoproterozoic-Paleozoic clastic and carbonate sedimentary rocks of the Tibetan zone in its hanging wall, the South Tibetan fault system incorporates a variety of structures, but chief among them are low-angle, north-dipping detachments with normal-sense displacement (Burchfiel et al., 1992). In the study area, the South Tibetan fault system is composed of two primary splays: the Chame detachment and Machhapuchhare-Phu detachment (Fig. 1), which are commonly

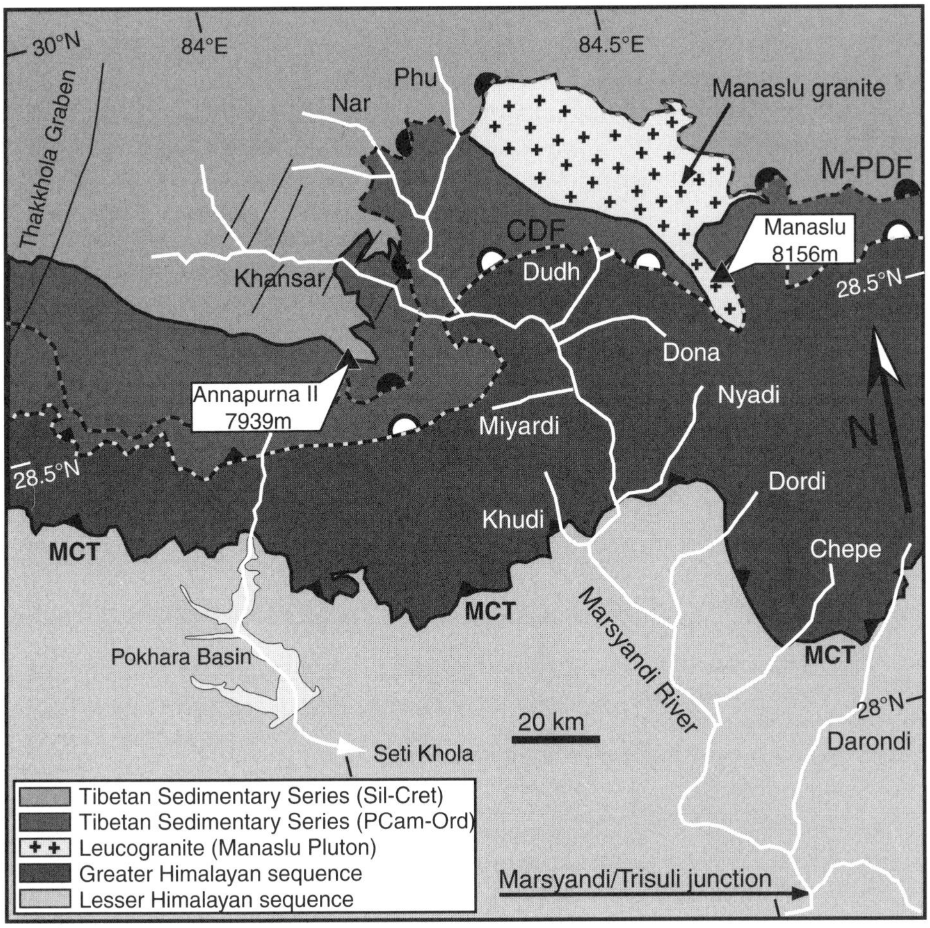

Figure 1. Simplified geological map of the Marsyandi region adapted from Searle and Godin (2003), Hodges et al. (1996), and Colchen (1986). The south-verging Main Central thrust (MCT) separates the Greater Himalaya sequence from the Lesser Himalaya sequence. Other major south-verging thrust faults, the Main Boundary thrust and the Main Frontal thrust, are to the south of this diagram. The South Tibetan fault system, with normal displacement, forms two splays in this region: the Chame detachment fault (CDF) and the Machhapuchhare-Phu detachment fault (MDF). Sil—Silurian; Cret—Cretaceous; PCam—Precambrian; Ord—Ordovician.

separated by the greenschist-to-amphibolite grade marble of the Annapurna Yellow Formation (Coleman, 1996; Hodges et al., 1996), although in the northeastern part of the study area, the Phu Detachment is interpreted to cut the top of the Manaslu pluton (Searle and Godin, 2003).

The Marsyandi River system of central Nepal (Fig. 1) has its headwaters north of the South Tibetan fault system and drains portions of the Tibetan, Greater Himalayan, and Lesser Himalayan zones over an area of ~4760 km^2. Its major tributaries flow over subsets of these tectonostratigraphic zones, and thus sediments in these tributaries sample different zones of bedrock. As they flow into the main Marsyandi trunk stream, individual tributary signals are progressively mixed downstream. The Khansar Khola ("khola" is the Nepali word for river) and Nar Khola predominately drain Tibetan zone sedimentary bedrock. The Dudh Khola drains Tibetan zone and Greater Himalayan rocks, as well as a major Miocene leucogranite, the Manaslu pluton (Le Fort, 1981). The Dona Khola drains this pluton, as well as a variety of metamorphic rocks of the Greater Himalayan sequence. The Miyardi and Nyadi Kholas have headwaters in the Greater Himalayan zone and exclusively sample this bedrock before emptying into the Marsyandi. The Khudi, Dordi, Chepe, and Darondi Kholas bridge the Main Central thrust system and thus have sediments with provenances in both the Greater and Lesser Himalayan sequences.

METHODOLOGY

Sampling Strategy

Detritus shed from a major mountain belt is primarily transported to the foreland basin by fluvial systems. En route, the development of the detrital age signal is influenced by the flux and composition of sediment contributed from tributaries. Thus, the foreland signal represents a complex integration of the dynamics of erosion and sediment transport in each tributary catchment. The sampling strategy in this investigation was designed to: (1) maximize the statistical robustness of cooling-age signals from individual tributaries; and (2) investigate the downstream development of the trunk-stream cooling-age signal; while (3) using small enough catchment areas to define adequately the spatial variation in cooling ages across this part of the orogen. With these objectives and given a finite number of laboratory age analyses, tradeoffs are unavoidable between obtaining the optimal representation of any single tributary, and reliably reconstructing the evolution of the detrital age signal along the course of the entire river. For example, a focus on small catchments will increase the resolution of spatial variation in bedrock cooling ages, but leaves fewer analyses to constrain the downstream evolution of the trunk stream. Conversely, more analyses on an individual sample will provide a more reliable characterization of the specific cooling-age signal, but limit the spatial resolution of the study.

Detrital sand samples were collected within the Marsyandi catchment from sites ranging from the Tibetan zone to the junction with the Trisuli River in the Lesser Himalayan zone (Fig. 2). At each sample site, large-grained sand was collected from bars within the modern river channel. Care was taken to avoid the influence of small tributaries and fill deposits, such as terraces, and to collect samples either upstream of sediment-mixing zones at river junctions or sufficiently below junctions (>1 km) such that sediments could be considered well mixed by these turbulent rivers.

Although 27 samples were collected for point counting (to characterize the mineralogical constitution of the sediment), only 14 samples from twelve separate locations were selected for $^{40}Ar/^{39}Ar$ dating. Six samples were chosen from the main stem of the Marsyandi River, and five samples were taken at

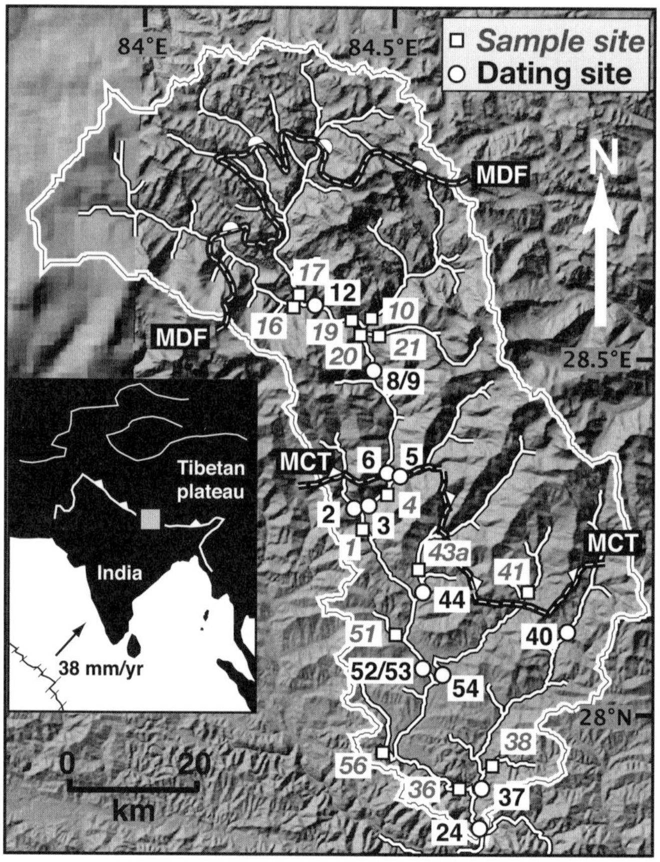

Figure 2. Map of Marsyandi catchment based on a 90 m digital elevation model (DEM). Sample locations are displayed with squares (and gray sample numbers) for point-counting sites and circles (and black sample numbers) for $^{40}Ar/^{39}Ar$ analysis/point-count sites. The Marsyandi catchment (upstream of site 24) is outlined in white. The Main Central thrust (MCT) has triangles indicating south vergence, whereas the Machhapuchhare-Phu detachment fault (MDF) has half circles indicating extension down-to-the-north. The inset shows the approximate position of the sampling area within Central Asia and the modern geodetic Indo-Asian convergence rate (Wang et al., 2001).

the mouths of major tributaries (Fig. 2). One sample (S-40) represents a subcatchment within the overall Darondi Khola (S-37), taken to assess the relative input of ages from the Greater Himalayan sequence portion of the catchment in comparison to the entire catchment. To examine the temporal variability of the detrital signal, two samples were collected from the same location; one from the modern river bed (S-8), and one from a fill terrace elevated 2 m above it (S-9). To examine the natural spatial variability of the detrital signal, samples were collected 45 m apart on the downstream (S-53) and upstream (S-52) ends of the same sandbar. Samples S-8 and S-44 were reported in Brewer et al. (2003).

$^{40}Ar/^{39}Ar$ Analytical Protocols

This investigation focused on detrital muscovite, which has been widely used in detrital mineral geochronology and appears to have fewer problems with excess argon than biotite (Roddick et al., 1980). Individual muscovite grains between 500–2000 µm were analyzed at the $^{40}Ar/^{39}Ar$ laser microprobe facility at the Massachusetts Institute of Technology (Hodges, 1998). This grain size was used because, despite very rapid Himalayan erosion, it typically yields grains with sufficient radiogenic ^{40}Ar for reliable analysis. Hodges and Bowring (1995) provide additional details on the analytical techniques. Apparent ages calculated for each muscovite with an estimated 2σ uncertainty, obtained by propagating all analytical uncertainties, are available in the Data Repository.[1]

Detrital cooling-age signals are commonly represented as probability density (PD) distributions, which represent the probability of finding a grain of a particular age, as a function of the age (Hurford et al., 1984; Silverman, 1986). Assuming that a Gaussian kernel represents the distribution of error (Brandon, 1996), a probability density function can be calculated for each grain, given the age (t_c) and analytical uncertainty (σ). For a sample of N grains collected from a specific locality, the probability density functions of individual grains (n) can be combined:

$$P(t) = \sum_{n=1}^{n=N} \frac{1}{\sigma(n).\sqrt{2\pi}} e^{\left(\frac{-(t-t_c(n))^2}{2\sigma(n)^2}\right)}. \quad (1)$$

By normalizing the area under the resulting curve to unity, a summed PD distribution is generated that represents the distribution of probability, as a function of age, within the sample.

Point Counting

Point counting of detrital samples serves two key purposes in this study. First, the abundance of muscovite in each tributary is a key ingredient in calculating how much the tributary contributes to the trunk-stream detrital age signal. Second, when detrital minerals are used as conservative tracers, the relative abundance of a particular mineral species can define the relative contribution from individual tributaries. Although the latter technique has much lower resolution than the thermochronological approach, it provides an additional constraint on relative erosion rates.

Because of our emphasis on thermochronology, traditional point counting was primarily used to quantify the abundance of muscovite within the same 500–2000 µm fraction that was used for dating. Of the 600–900 grains counted in each sample, quartz, plagioclase, alkali feldspar, and micas were the major constituents counted, with additional minerals grouped together, and rock fragments considered to be an additional species. Crystalline carbonate was considered to be a mineral, whereas granular carbonate was considered a rock fragment. Rather than produce a very detailed description of the exact mineralogy in each sample (outside of the scope of this study), this broad approach provided an approximate quantification of major constituents (Table 1), which was sufficient for our modeling purposes. All reported errors from the point-counting results are taken from the statistical analysis of Van der Plas and Tobi (1965). To investigate the consistency of individual counts, repeat counts on three samples (S-1, S-2, S-3; Table 1) showed that the results were indistinguishable at the 2σ confidence level.

$^{40}Ar/^{39}Ar$ RESULTS

Before attempting to interpret the $^{40}Ar/^{39}Ar$ results using a modeling approach, the broad trends in the data illustrate a systematic downstream pattern in the trunk-stream age signal (Fig. 3). The catchment of the sample farthest upstream (S-12) drains the headwaters of the Marsyandi from the edge of the Tibetan Plateau to the crest of the Annapurna massif. The signal is dominated by an age population concentrated between 12 and 16 Ma. One source of these ages may be hydrothermal veins with $^{40}Ar/^{39}Ar$ muscovite plateau dates of ca. 14 Ma that occur within Tibetan zone sedimentary rocks in this area (Coleman and Hodges, 1995). Alternatively, the catchment contains a small portion of the Annapurna Yellow Formation and the upper Greater Himalayan sequence, which may additionally contribute muscovite. The next sample downstream (S-8/S-9) was collected well south of the South Tibetan fault system and is influenced by two additional major tributaries, the Dudh and Dona Kholas, which drain the top of the Greater Himalaya sequence and Manaslu granite, and the Tibetan zone. Whereas the population of ages observed upstream (S-12) is still represented, the cooling-age signal is dominated by a major age population from 15 to 20 Ma. The weak expression of the younger than 15 Ma age population in the downstream sample and the paucity of muscovite (<1%: Table 1) in the upper reaches of the catchment suggest that the area upstream of sample S-12 (mostly Tibetan zone) makes

[1]GSA Data Repository item 2006027, Table DR1, Detrital $^{39}Ar/^{40}Ar$ ages of muscovites from the Marsyandi River catchment, is available online at www.geosociety.org/pubs/ft2006.htm, or on request from editing@geosociety.org or Documents Secretary, GSA, P.O. Box 9140, Boulder, CO 80301, USA.

TABLE 1. POINT-COUNTING DATA FROM THE MARSYANDI VALLEY

Sample	Catchment	*n*	Qtz	Pl	Kfs	Ms	Bt	Frag	Opq	Other
S-1	Marsyandi	893	43.3 ±3.3	15.4 ±2.4	3.6 ±1.2	5.3 ±1.5	3.7 ±1.3	17.9 ±2.6	0.3 ±0.4	10.4 ±2.0
S-1.2	Marsyandi	798	45.6 ±3.5	14.7 ±2.5	4.6 ±1.5	6.5 ±1.7	4.0 ±1.4	18.5 ±2.8	0.1 ±0.2	5.9 ±1.7
S-2	Khudi	790	37.7 ±3.4	16.2 ±2.6	2.3 ±1.1	10.1 ±2.1	9.6 ±2.1	18.5 ±2.8	0.0 ±0.0	5.6 ±1.6
S-2.2	Khudi	765	42.7 ±3.6	19.6 ±2.9	2.1 ±1.0	11.9 ±2.3	11.2 ±2.3	7.4 ±1.9	0.2 ±0.3	4.3 ±1.5
S-3	Marsyandi	802	26.4 ±3.1	19.6 ±2.8	8.4 ±2.0	2.0 ±1.0	1.6 ±0.9	33.9 ±3.3	0.0 ±0.0	7.9 ±1.9
S-3.2	Marsyandi	800	30.3 ±3.2	15.8 ±2.6	10.0 ±2.1	2.6 ±1.1	2.4 ±1.1	35.3 ±3.4	0.1 ±0.2	3.4 ±1.3
S-4	Marsyandi	691	29.1 ±3.5	13.7 ±2.6	8.7 ±2.1	1.9 ±1.0	3.2 ±1.3	36.2 ±3.7	0.4 ±0.5	6.8 ±1.9
S-5	Nyadi	745	41.2 ±3.6	20.4 ±3.0	8.6 ±2.1	6.3 ±1.8	4.6 ±1.5	12.6 ±2.4	0.3 ±0.4	5.8 ±1.7
S-6	Marsyandi	779	27.3 ±3.2	17.1 ±2.7	9.6 ±2.1	3.5 ±1.3	2.6 ±1.1	35.6 ±3.4	0.8 ±0.6	3.6 ±1.3
S-8	Marsyandi	747	28.1 ±3.3	21.8 ±3.0	15.4 ±2.6	1.9 ±1.0	0.7 ±0.6	30.7 ±3.4	0.4 ±0.5	1.1 ±0.8
S-10	Dudh	804	40.2 ±3.5	29.9 ±3.2	16.0 ±2.6	1.7 ±0.9	0.0 ±0.0	9.3 ±2.1	0.1 ±0.2	2.7 ±1.1
S-16	Khansar	786	5.6 ±1.6	1.8 ±0.9	1.4 ±0.8	0.1 ±0.2	0.8 ±0.6	79.1 ±2.9	0.9 ±0.7	10.2 ±2.2
S-17	Nar	769	11.8 ±2.3	2.3 ±1.1	1.4 ±0.9	0.5 ±0.5	0.4 ±0.4	79.2 ±2.9	0.7 ±0.6	3.6 ±1.4
S-19	Marsyandi	784	15.9 ±2.6	14.0 ±2.5	7.5 ±1.9	1.8 ±0.9	2.0 ±1.0	53.6 ±3.6	0.6 ±0.6	4.5 ±1.5
S-20	Marsyandi	808	27.2 ±3.1	21.9 ±2.9	10.3 ±2.1	1.1 ±0.7	1.5 ±0.8	34.9 ±3.4	0.2 ±0.3	2.8 ±1.2
S-21	Dona	777	29.6 ±3.3	21.6 ±3.0	17.4 ±2.7	0.5 ±0.5	2.8 ±1.2	11.1 ±2.3	0.5 ±0.5	16.3 ±2.7
S-24	Marsyandi	768	49.1 ±3.6	9.1 ±2.1	3.1 ±1.3	6.8 ±1.8	3.9 ±1.4	23.4 ±3.1	0.5 ±0.5	3.9 ±1.4
S-36	Marsyandi	842	43.7 ±3.4	13.2 ±2.3	4.5 ±1.4	5.1 ±1.5	3.7 ±1.3	25.5 ±3.0	0.6 ±0.5	3.7 ±1.3
S-37	Darondi	869	62.7 ±3.3	13.1 ±2.3	1.2 ±0.7	4.6 ±1.4	2.9 ±1.1	12.5 ±2.2	1.0 ±0.7	1.8 ±0.9
S-38	LH trib	524	44.1 ±4.3	1.1 ±0.9	0.6 ±0.7	0.0 ±0.0	5.2 ±1.9	47.5 ±4.4	0.8 ±0.8	0.8 ±0.8
S-40	Darondi	624	55.8 ±4.0	13.3 ±2.7	0.6 ±0.6	7.2 ±2.1	3.8 ±1.5	13.8 ±2.8	0.5 ±0.6	4.5 ±1.7
S-41	Chepe	633	28.6 ±3.6	8.7 ±2.2	0.8 ±0.7	29.7 ±3.6	23.5 ±3.4	2.2 ±1.2	0.2 ±0.3	6.3 ±1.9
S-43a	Dordi	821	35.7 ±3.3	16.8 ±2.6	10.0 ±2.1	7.9 ±1.9	8.3 ±1.9	16.7 ±2.6	0.6 ±0.5	3.8 ±1.3
S-44	Dordi	777	43.8 ±3.6	14.3 ±2.5	2.7 ±1.2	9.7 ±2.1	9.1 ±2.1	12.6 ±2.4	1.4 ±0.8	6.4 ±1.8
S-51	LH trib	614	43.3 ±4.0	7.3 ±2.1	4.9 ±1.7	3.1 ±1.4	2.3 ±1.2	34.5 ±3.8	0.2 ±0.3	4.4 ±1.7
S-52	Marsyandi	858	31.6 ±3.2	13.6 ±2.3	5.5 ±1.6	10.6 ±2.1	11.0 ±2.1	22.5 ±2.9	0.5 ±0.5	4.8 ±1.5
S-53	Marsyandi	897	29.5 ±3.0	10.9 ±2.1	3.7 ±1.3	10.7 ±2.1	16.1 ±2.5	24.2 ±2.9	0.8 ±0.6	4.0 ±1.3
S-54	Chepe	884	40.4 ±3.3	9.3 ±2.0	2.1 ±1.0	23.2 ±2.8	11.9 ±2.2	7.3 ±1.8	0.3 ±0.4	5.4 ±1.5
S-56	LH trib	598	52.0 ±4.1	9.2 ±2.4	6.9 ±2.1	2.2 ±1.2	3.8 ±1.6	23.2 ±3.5	1.0 ±0.8	1.7 ±1.0

Note: Abbreviations: LH trib—Lesser Himalayan tributary; *n*—number of grains, Qtz—quartz, Pl—plagioclase, Kfs—potassium feldspar, Ms—muscovite, Bt—biotite, Frag—rock fragments, Opq—opaques, other—any other minerals. All errors are 2σ and are calculated with the statistical analysis of Van der Plas and Tobi (1965).

a minor volumetric contribution of muscovite when compared to other tributaries upstream of sample S-8/S-9.

Another 20 km farther downstream, the next trunk-stream sample (S-6) displays the same 15–20 Ma population seen in the upstream sample, but contains additional 5–15 Ma ages. After the Marsyandi crosses the Main Central thrust zone (S-3), the 0–10 Ma population becomes more dominant and the 15–20 Ma peak represents less than half the total probability. The growing downstream proportion of young cooling ages is clearly influenced by tributaries like Nyadi Khola (S-5), which drains the lower Greater Himalayan sequence and contains exclusively the 0–10 Ma age population. Additional downstream catchments contribute primarily 3–14 Ma ages, such that the lower Marsyandi River (S-52/S-53) shows a dominance of the 5–10 Ma age population after the influx of these tributaries.

In the Darondi Khola, a sample (S-40) collected above the Main Central thrust zone displays a clear 0–12 Ma population, whereas the sample (S-37) at the catchment mouth is similar, but includes a single older age component. Prior to merging with the Trisuli River, the trunk-stream Marsyandi sample (S-24) comprises a prominent 5–10 Ma signal, a lesser 10–15 Ma signal, and a weak 15–20 Ma signal.

POINT-COUNTING RESULTS

The point-counting results (Table 1) indicate that the sediment composition changes systematically with the downstream addition of tributary material. Samples from the areally extensive Khansar and Nar catchments are rich in rock fragments containing up to 80%, mainly limestone, clasts from the Tibetan Sedimentary Series. The percentage of rock fragments in the tributaries rapidly decreases downstream, with some addition of granitic rock fragments, to between 20% and 30% toward the catchment mouth. The Dudh and Dona Kholas, draining the Manaslu granite, are quartz- and feldspar-rich and somewhat deficient in micas. The percentage of muscovite in the tributaries increases rapidly southward: Nyadi, Khudi, and Dordi Kholas all contain 5%–10% muscovite. Chepe Khola (S-54) has the largest fraction of muscovite, with nearly 30% (S-41) near the Main Central thrust (Fig. 2). This becomes diluted with the addition of quartz and rock fragments through the Lesser Himalaya, but remains over 20% at the catchment mouth. The Darondi Khola samples (S-37, S-40) show a similar trend of decreasing muscovite and increasing quartz through the Lesser Himalaya. The percentage of muscovite at the mouth of the Chepe, however, is ~4 times

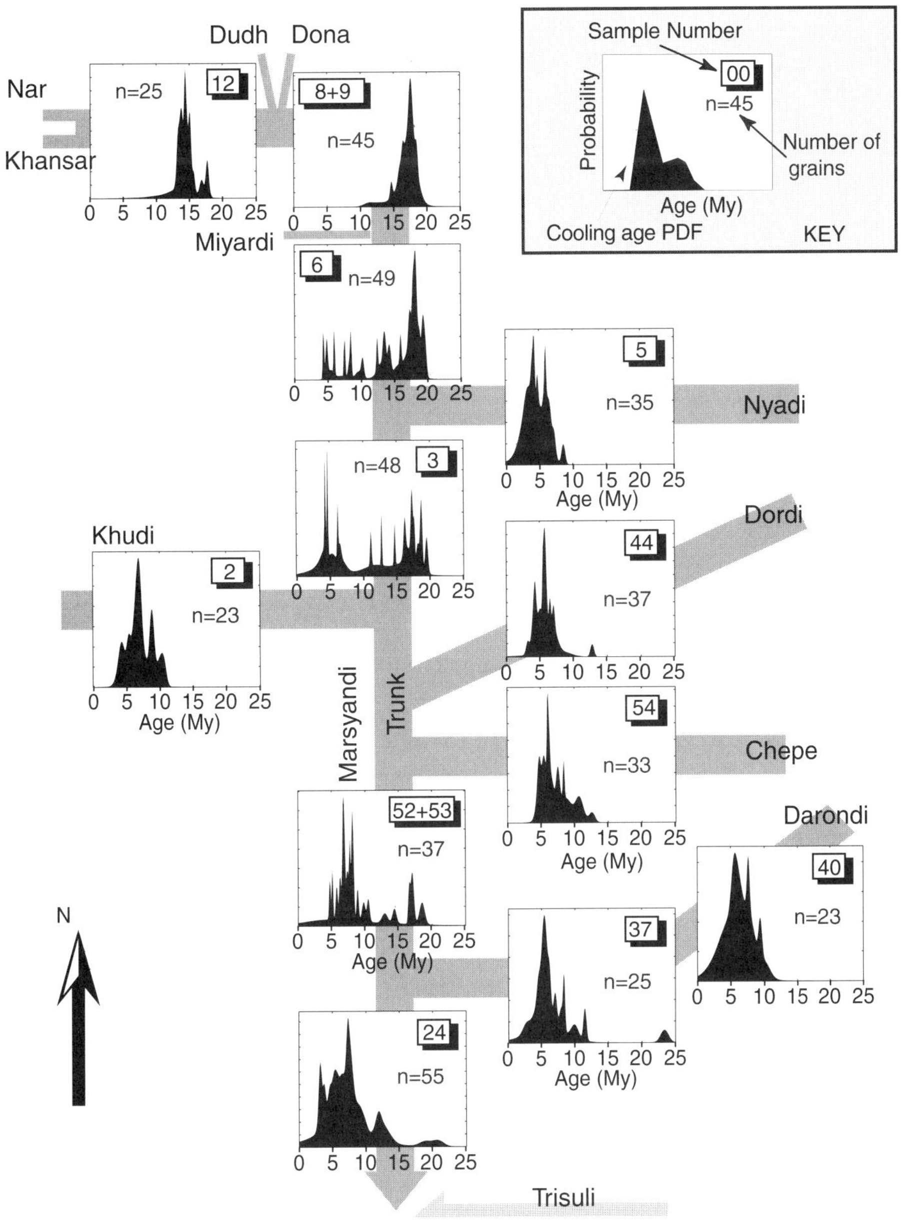

Figure 3. Detrital cooling-age probability density (PD) plots for samples from the Marsyandi catchment. All axes range from 0 to 25 m.y. on the x axis. Areas under the PD plot (shaded black) represent a total probability of one in each case, so that the probability scale on the y axis is relative. The individual plots have been arranged topologically to indicate their position within the Marsyandi River system: geographic locations are shown in Figure 2. Additional grain ages in sample 24 (229 ± 18 Ma) and sample 54 (27 ± 0.9 Ma) are not illustrated. PDF—probability density function.

greater than that at the mouth of the Darondi. Catchments contained wholly within Lesser Himalayan rocks (S-51, S-56, S-38) all show quartz contents of over 40%, high fractions of rock fragments, and muscovite abundances of 0%–3%. The high proportion of rock fragments may be a function of the immaturity of the sediment in relatively small catchments (S-38, S-51).

MODELING OF EROSION RATES AND SEDIMENT MIXING

Although the $^{40}Ar/^{39}Ar$ analyses display systematic changes within the Marsyandi catchment, further insights into spatial variations in erosion (and perhaps tectonic) rates can be gained with numerical modeling. Such modeling allows us to examine the impact of individual parameters (e.g., relief, hypsometry, lithology, and erosion rate) and discriminate between those characteristics of the data that can be explained by the model and those that cannot, forcing us to understand the limitations of initial assumptions. Thus, in combination with the $^{40}Ar/^{39}Ar$ data, we use a numerical model to: (1) assess the spatial variation of parameters that control the hinterland cooling-age distributions; (2) understand how these parameters are manifested in the detrital cooling-age signal observed at the catchment mouth, and; (3) examine the reliability and resilience of the cooling-age signal.

To do this, we construct a theoretical PD distribution of cooling ages for each catchment within the Marsyandi drainage (Brewer et al., 2003). These PD distributions were generated by: (1) inputting the real topographic characteristics of each catchment; and then (2) finding the optimal match between modeled and observed data PD distributions by varying the catchment erosion rate. Once theoretical PD distributions had been generated for each tributary, we modeled the relative contribution from each tributary to the trunk stream. Hence, we examined the systematic mixing of age populations in order to understand and predict the downstream evolution of the cooling-age signal within the Marsyandi Valley.

Modeling the Detrital Cooling-Age Signal

Theoretical PD distributions had to be generated for each tributary. Firstly, to predict bedrock cooling ages within a tributary catchment, the predicted depth of the closure isotherm (at the time when muscovite passed through its closure temperature) was divided by the rate of erosion. Given a crust of predetermined thermal characteristics, the depth of the closure temperature is a function of the topographic relief and the rate of erosion. For erosion rates ≤3 mm/yr and topographic relief ≤6 km, the 350 °C closure isotherm of muscovite experiences negligible deflection due to surface topography (Mancktelow and Grasemann, 1997; Stüwe et al., 1994; Brewer et al., 2003). We used a simple thermal model (Brewer et al., 2003) to predict the depth of the closure isotherm as a function of catchment relief and erosion rate. The thermal model assumes vertical movement of rock toward the surface (given uniform surface heat flow [57 × 10^{-3} Wm^{-2}], heat production [1.0 × 10^{-6} Wm^{-3}: Fowler, 1990], and conductivity) through a steady-state landscape containing hillslopes at a threshold for landsliding (~30° angle: Burbank et al., 1996). These assumptions produce a linear distribution of muscovite cooling ages with elevation. Given this, and with a uniformly eroding catchment, the cooling-age PD distribution is controlled by the distribution of land area with elevation, i.e., the hypsometry. Thus, the likelihood of sampling a particular age at the mouth of a catchment is predicted from the proportion of land containing that age.

In addition to the thermal and topographic parameters described in the preceding, the model required two further inputs: hypsometry and erosion rate. With this basic model, the larger Marsyandi Valley can be broken into individual catchments to represent the contribution from each of our tributary age samples. Given the hypsometry of a catchment, we modeled the cooling-age distribution as a function of the erosion rate. Because hypsometry can be extracted from a 90 m digital elevation model (DEM) for each catchment, the erosion rate was the only unknown parameter. Hence, the optimal model (with lowest mismatch between theoretical and observed PD distributions: Brewer et al., 2003) was found by varying the erosion rate within an individual catchment. Once the best model erosion rate for a tributary catchment had been found, we fixed it for subsequent analyses.

Ultimately, using all the tributaries, we wanted to predict how sediments from tributary catchments coalesce to form the trunk-stream signal. As a consequence of steady-state assumptions (that mean elevation and hypsometry are statistically invariant over time scales exceeding 0.1 m.y.), the flux of material out of a catchment will balance the volume of rock moving into it. With vertical denudation for a unit time, the eroded volume is the product of the catchment area and the erosion rate. The progressive downstream summation of these products should model the evolving trunk-stream detrital signal.

If a mixing model were to rely solely on tributary area and erosion rate to predict relative contribution of muscovite to the trunk-stream detrital signal, a uniform distribution of muscovite must be assumed across the study area. In this study area, however, muscovite is heterogeneously distributed: it is common in high-grade metamorphic rocks, for example, but often absent in carbonates. Even within the high-grade rocks, muscovite shows 5-fold differences in abundance. Hence, the contribution from a carbonate catchment to the detrital muscovite age signal will be negligible, almost irrespective of its erosion rate. We used the percentage of muscovite at the mouth of each tributary, from the point-counting results (Table 1), to calculate a correction factor for the amount of muscovite per unit area present in each catchment (Fig. 4). Note that the percentage of muscovite varies over two orders of magnitude within the Marsyandi river system: any model that ignores a lithological correction factor here would yield profoundly biased results.

To model the trunk-stream signal at a particular location, we considered only the predetermined tributary PD distributions upstream of the sample. Our standard procedure (1) determined the erosion rate (*dz/dt*) that optimized the fit with the observed data for each catchment given known relief and hypsometry (Table 2), and then (2) used the tributary area (A) and abundance of muscovite (M_{msc}) to calculate the relative contribution of the detrital signal from each tributary (PDF_A and PDF_B) to the trunk-stream detrital signal (PDF_C) downstream:

$$PDF_c = \frac{V_{msc_A}}{V_{msc_A} + V_{msc_B}} PDF_A = \frac{V_{msc_B}}{V_{msc_A} + V_{msc_B}} PDF_B, \quad (2)$$

where V_{msc} is the volume of muscovite supplied per unit time from a given tributary:

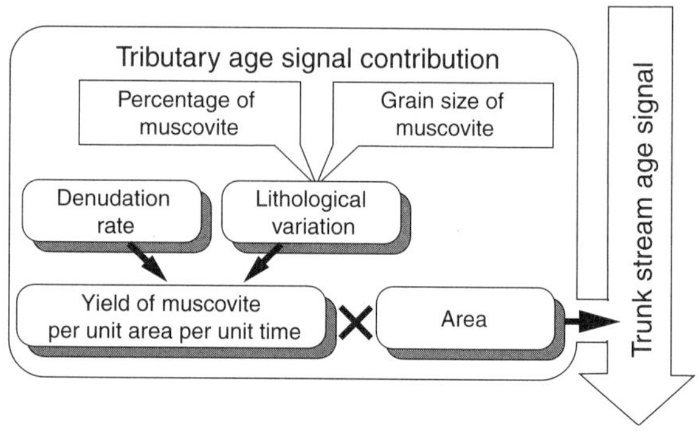

Figure 4. Parameters controlling the contribution of an individual tributary to a trunk-stream cooling-age signal. The foreland signal can be modeled as a specified mix of several such tributaries.

TABLE 2. TOPOGRAPHIC CHARACTERISTICS OF THE MARSYANDI VALLEY AND ITS ASSOCIATED TRIBUTARIES

Basin	Area (km^2)	Elevation (m)				Lithology		
		Mean	Min	Max	Relief	~%TSS	~%GH	~%LH
Marsyandi	4760	3332	244	8152	7908			
Khansar	713	4794	2634	7824	5190	80	20	–
Nar	884	5209	2634	7097	4463	80	20	–
Dudh	392	4694	1958	7669	5711	10	90	–
Dona	129	4851	1895	8152	6257		100	–
Miyardi	60	4050	1496	5842	4346	–	100	–
Nyadi	215	3440	926	7495	6569	–	80	20
Khudi	136	2565	796	4914	4118	–	90	10
Dordi	351	2885	553	7756	7203	–	70	30
Chepe	309	1808	440	4872	4432	–	50	50
Darondi	609	1470	277	5787	5510	–	40	60

Note: The approximate litho-tectonic division of the basins are given in percentage area containing: Tethyan Sedimentary Series (TSS) structurally above the Machhapuchhare detachment fault; Greater Himalayan sequence (GH) complex, leucogranites, and Annapurna Yellow Formation; and Lesser Himalayan sequence (LH).

$$V_{msc} = \frac{dz}{dt} A\, M_{msc}\,. \tag{3}$$

Although equation 2 describes mixing of two tributaries, when assessing the relative contribution of a tributary to the trunk stream (consisting of areas with different lithologies and erosion rates), we had to work progressively downstream, substituting the calculation of V_{msc} in a single tributary (equation 3) for the total volume of muscovite contributed per unit time from t tributaries upstream:

$$V_{msc} = \sum_{i=1}^{n} \frac{dz}{dt_i} A_i M_{msc_i}\,. \tag{4}$$

Where tributary additions were unconstrained by ages at the tributary mouth (for example, the inaccessible area represented by the Miyardi Khola: Fig. 2), an erosion rate was assigned that was consistent with the surrounding tributaries and that minimized the mismatch of the trunk-stream signal in the sample directly below the junction. For the purposes of the modeling, the observed PD distributions were smoothed using a 2 m.y. scrolling window. This reduced the "peakedness" of the PD plots caused by individual grains, meaning that the calculated mismatch was less affected by individual grain peaks, but instead reflected the overall pattern of the entire signal.

Modeling Results of PD Distributions

We previously described the results of modeling samples S-8 and S-44 (Brewer et al., 2003), and here we integrate these with the entire data set. The overall pattern of theoretical PD distributions that emerges from combining individual catchments (upstream Marsyandi, Nyadi, Khudi, Dordi, Chepe, and Darondi: Fig. 5) to produce the trunk-stream signal captures the primary age populations in the observed data. The main peaks of the theoretical plots generated for each tributary will align closely with those of the data, because we minimized the mismatch by varying the erosion rate. The tails observed on either side of the younger peaks, however, are harder to match with our approach (i.e., see the Khudi Khola theoretical PD plot for illustration). The trunk stream displays a systematic pattern of change downstream as tributary signals are added. The >15 m.y. peak, prominent in the upstream samples, becomes diluted downstream as the 5–10 m.y. peak becomes increasingly important; although in comparison to the data, the predicted catchment mouth PD plot (as compared to S-24) is relatively depleted in the 10–15 m.y. age range.

In the upstream reaches of the Marsyandi, we had to modify our modeling procedure in order to find the best match to the observations. For our most upstream sample (S-12), draining the Nar and Khansar catchments (equal to one-third of the total Marsyandi catchment), the optimal erosion rate was modeled as ~1 mm/yr. By applying this rate to the entire upstream catchment area, the predicted flux to the upstream sample (S-12) produced a signal that, in comparison to the next downstream sample (S-8/S-9), was too dominant in the 10–15 m.y. age range. In order to reproduce the downstream observations, the relative contribution from the Nar and Khansar had to be reduced by ~50% compared to that indicated by the predicted erosion rate and the raw data for the percentage of muscovite (Table 1).

The reason for the initial mismatch is uncertain, but is likely a result in part of model assumptions that are violated in this catchment. First, we assumed that muscovite is uniformly distributed, whereas the lithologies of the catchment suggest that muscovite is more abundant in the southernmost part of the catchment that includes Greater Himalayan rocks (Searle and Godin, 2003). Second, we applied a uniform erosion rate to the entire catchment, whereas the lower relief and arid terrain in southern Tibet (north of the South Tibetan fault system) would be expected to erode more slowly than the wetter, higher relief part of the catchment underlain by Greater Himalayan rocks. Third, the point-count data from this catchment indicate a high fraction of rock fragments (~80%: Table 1), most of which are carbonates. This fraction rapidly decreases downstream through a reach with no major tributaries: a behavior suggesting that some combination of dissolution and mechanical breakdown is removing the carbonates from the sand fraction. Thus, the dominance of rock fragments makes estimates of muscovite abundance uncertain. Given these uncertainties in the appropriate flux, our decision to reduce the upstream contribution by 50% in order to match the observed downstream cooling ages seems permissible.

Although the model results mimic age distributions from tributaries with older cooling ages (i.e., the upstream catchments), they fail to reproduce the full range of ages in tributaries that yield younger cooling ages. The strong asymmetry observed in the older age tails is, in particular, difficult to replicate (the 8–13 Ma ages observed in the Chepe Khola, for example: Fig. 5). This mismatch may result from variations in erosion rate in the catchments to the south of the range crest, where rock uplift and erosion rates are expected to be locally controlled by the geometry of the active subsurface structure (e.g., Burbank et al.,

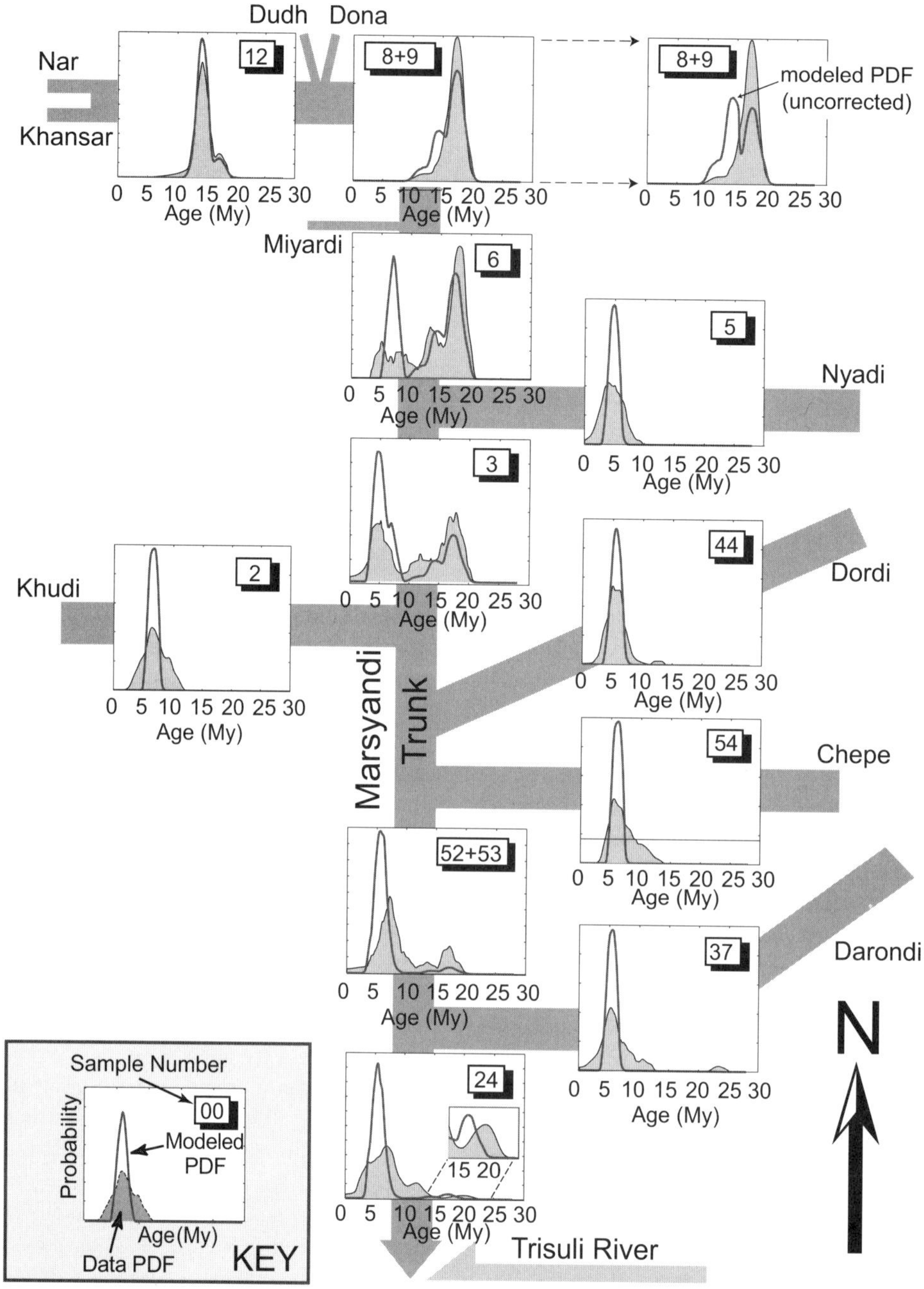

Figure 5. Model results compared to $^{40}Ar/^{39}Ar$ analyses. Real-data probability density (PD) plots (identical to data used in Fig. 3) are shaded gray and smoothed using a 2 m.y. window. Solid black lines are model PD plots generated using the methodology described in the text and are also smoothed with a 2 m.y. window. The *x* axis of each plot ranges from 0 to 30 Ma, with probability on the *y* axis, and the area under each curve represents a probability of one. Inset for site S24 shows the 15–25 Ma observed age distributions (and theoretical curve) with a vertically exaggerated probability scale. Uncorrected theoretical PD plot for site S-8/S-9 is shown in the upper left. PDF—probability density function.

2003; Lavé and Avouac, 2001; Pandey et al., 1995; Seeber and Gornitz, 1983). In the absence of detailed subtributary data or bedrock ages from these regions, however, we have limited the modeling to the same catchment resolution as our data. Another cause of mismatches would be areas not included in the modeling (stippled in Fig. 6), yet contributing to the detrital cooling-age signal. There were insufficient thermochronological data from these areas to constrain their erosion rate. Most of these areas lie in the Lesser Himalaya, and if they have intermediate erosion rates, as their low relief and topography would suggest, then they

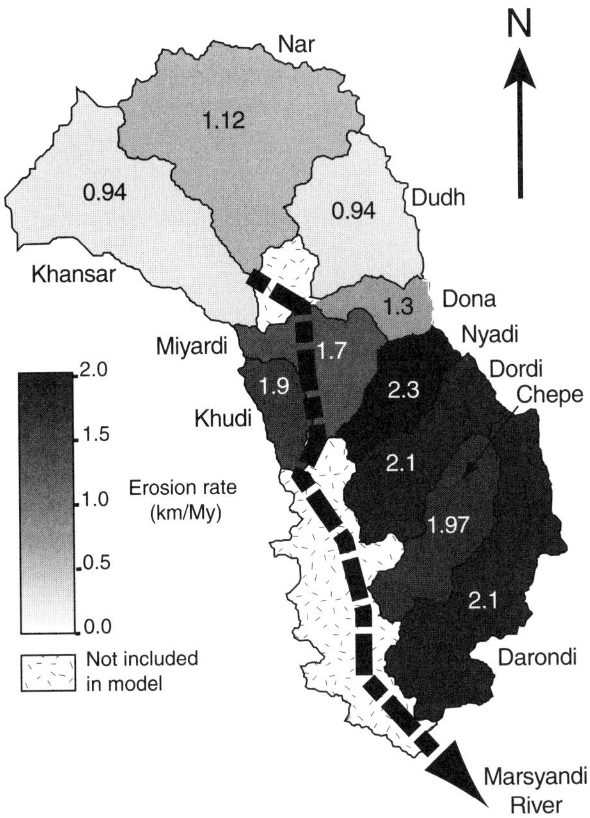

Figure 6. Spatial variation in erosion rates at the catchment scale. Erosion rates are taken from the results of modeling the detrital cooling-age probability density (PD) distributions for individual tributaries. The shaded areas indicate zones not included in the calculations and the dashed black line indicates the approximate path of the trunk stream. Highest erosion rates are predicted in the middle areas of the Marsyandi catchment, where rivers drain the steep front of the High Himalayas. Slowest erosion rates occur to the north of the topographic axis, in the rain shadow.

may be an additional source of 10–15 m.y. ages not represented in the model. Finally, our model assumes a topographic and thermal steady state since the time when rocks passed through the muscovite closure temperature, thereby implying steady erosion since closure. Whereas the model is insensitive to changes in erosion rate, such changes may have occurred. If rates have accelerated in the past 5 m.y. (Harrison et al., 1997; Wobus et al., 2003, Hodges et al., 2004), then older ages reflecting rocks that cooled prior to the onset of acceleration could be preserved at high elevations in a catchment, and hence preserve indications of slower erosion rates.

DISCUSSION

The foregoing analysis makes numerous assumptions about controls on the observed cooling ages. For example, cooling is attributed simply to geomorphic erosion, and tectonic erosion by extensional faulting is not considered. It is assumed that magmatic and fluid heating does not affect the observed cooling ages. Finally, folding or faulting since closure is assumed to have left the basic structure of the cooling ages unaffected.

Given the presence of a suite of normal faults, Miocene plutons, and active deformation in the study area, these assumptions need justification. Most previous studies of the South Tibetan fault system have concluded that the majority of slip occurred in the early Miocene and was completed by ca. 20 Ma (Burchfiel et al., 1992). More recently, Searle and Godin (2003) concluded that slip in the Marsyandi region was more recent than 19 Ma, but that most rapid cooling was complete by 15 Ma. In our study, the preponderance of observed ages are younger than 10 Ma, suggesting that extensional faulting has not been an important control on the muscovite cooling history. Similarly, emplacement of the Manaslu pluton at ca. 22–19 Ma (Harrison et al., 1995; Hodges et al., 1996) undoubtedly caused a thermal perturbation. Indeed, those detrital samples (S8, S9) that are most proximal to the Manaslu granite are dominated by 15–20 Ma muscovite ages (Fig. 3). Nonetheless, both upstream and downstream, cooling ages are much younger than this magmatic event and are deemed unaffected by it. We consider the possible effects of neoformed muscovites and hydrothermal heating to be small. In the vicinity of the Main Central thrust, brittle faulting (Hodges et al., 2004) and young cooling ages (Harrison et al., 1997; Wobus et al., 2003) suggest that deformation is ongoing. Quaternary displacements have been observed on at least one strand of the South Tibetan fault, but the associated rates of slip are low (Hurtado et al., 2001). Inevitably, some deformation has occurred since these rocks passed through their closure isotherm. The consistent pattern of young ages that is apparent south of the range crest (Figs. 3 and 5), however, suggests that deformation has not significantly offset the cooling ages when compared to the spatial scale of the Marsyandi Valley.

Resilience of the Detrital Signal

One key concern with the application of detrital dating is the survivability of the target mineral(s) in the sediment routing system. If the river network causes rapid comminution of grains, a sample will only represent contributions of a small upstream area that is dependent on the rate of attrition. In our study such comminution would strongly influence the muscovite cooling-age signal that reaches the foreland, as information becomes progressively lost downstream. The ideal target thermochronometer is neither destroyed nor altered during the weathering and transportation history of the sediment, and should not be affected by postdepositional weathering or diagenesis.

Both chemical and physical processes may cause the breakdown of muscovite. Previous studies in Himachal Pradesh of Oligocene to lower Miocene Himalayan foreland strata (Najman et al., 1997), however, have indicated that chemical weathering has a negligible effect. Because we are studying a much younger Himalayan fluvial system, and because the Himalaya continues to undergo rapid erosion with little storage of sediments within the hinterland, we assume that chemical alteration of muscovite does not significantly affect our results.

The process of physical attrition of muscovite grains during their passage through the fluvial system is more difficult to assess. With a hardness of 1–2 on Mohs scale and a well-developed basal cleavage, muscovite is susceptible to physical breakdown. However, muscovites do survive transport from the Himalaya to the distal Bengal Fan, a distance of >2000 km, with little disturbance of their $^{40}Ar/^{39}Ar$ systematics (Copeland and Harrison, 1990). Given that the length scale of the Marsyandi catchment is an order of magnitude smaller, physical breakdown of muscovite is less likely to be significant. The persistence of the 15–20 Ma age signal from the upper Greater Himalaya sequence, through all our trunk-stream samples, lends support to this conclusion.

Rather than being a result of mechanical breakdown, we interpret the downstream decrease in the observed 15–20 Ma age signal to be an effect of dilution. Our modeling assumes a sediment flux proportional to the erosion rate and catchment size, but contains no function for the downstream loss of cooling-age signal with distance. Therefore, if comminution were significant over this length scale, the model would be expected to overrepresent the 15–20 m.y. age fraction, particularly in the lower reaches of the river as muscovite grains from upstream are increasingly destroyed. Instead, the model shows the opposite effect and points to dilution by detritus from more southerly catchments. We suggest that mica probably travels in the turbulent wash load of Himalayan rivers, where it experiences few grain-to-grain impacts and little downstream comminution.

Predicted cooling-age distributions are a function of erosion rate and the associated volume of eroded sediment per unit

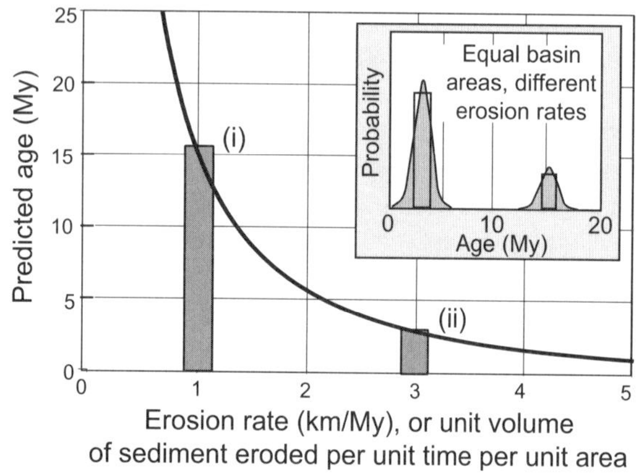

Figure 7. Predicted cooling ages from our model for specified erosion rates. The nonlinearity of the erosion rate–cooling age relation results from changes in the depth of the closure isotherm at different erosion rates. Relative abundances of detrital ages can be predicted based on erosion rates. For two catchments of equal area, but eroding at different rates [(i) 1 km/m.y.; (ii) 3 km/m.y.], 16 Ma cooling ages will constitute only 25% of the amalgamated age spectrum (inset).

area (Fig. 7). Thus, to investigate the relative proportion of an entire catchment that is producing a signal of a specific age, we need to correct for the volumetric contribution of that age. For example, although the 15–20 Ma age fraction in sample S-24 is relatively minor (Fig. 5), because it includes older cooling ages, it derives from an aerially significant upstream area. Note that the relationship between predicted age and erosion rate (Fig. 7) is not linear in our model, because the depth of the closure isotherm varies as a function of denudation rate (Brewer et al., 2003). Reducing the erosion rate by a factor of two, from 2 to 1 mm/yr for example, has a large effect and changes the predicted cooling age threefold, from 5 to 15 m.y. (Fig. 7). In contrast, the exponential form of this curve means that, when comparing terrains producing cooling ages exceeding ~30 m.y., the difference between the volumetric contribution per unit area as a function of cooling age will be minor. In such cases, the probability distribution of age populations in the sediment will more closely reflect the size of the contributing areas.

Reliability of the Cooling-Age Signal

Two important assumptions in detrital thermochronology studies are: (1) the sample is efficiently mixed during transportation; and (2) the signal is not prone to events that cause strong temporal variations in observed detrital ages. Landslides, rock falls, and localized storms could influence the latter. If these assumptions are correct, a grab sample should provide a reliable representation of the entire signal from the river and upstream catchment. Two pairs of samples were collected to investigate these assumptions.

The first pair addressed the homogeneity of fluvial mixing by examining the degree to which two samples varied within the modern river. Comparable results were produced by samples S-53 and S-52, which were collected ~45 m apart on the same sand bar in the trunk stream. To test if the age distributions from these two samples were statistically differentiable, we applied the Monte Carlo methodology as described by Brewer et al. (2003) to each observed population of ages to generate random grain ages, and then summed these ages to generate theoretical PD distributions. To represent the large range in errors seen in both our single-grain analyses and the summed PD distributions (Fig. 3), a more complex method of specifying the 1σ analytical uncertainty was used for this approach. Because no consistent relationship of age uncertainty with grain age was observed (Fig. 8A), a PD plot was constructed of measured uncertainties (Fig. 8B). This error PD plot illustrates that, whereas most 1σ analytical uncertainties are <1.5 m.y., larger errors are not uncommon. Therefore, for each modeled grain, the distribution of error ages (Fig. 8B) was randomly sampled (using the same Monte Carlo method), and the resultant 1σ error was applied to the model-derived grain age.

Based on 500 individual grab samples randomly drawn from the modeled theoretical PD distribution for samples S-52/S-53 (Fig. 5), the calculated mismatch between the grab-sample

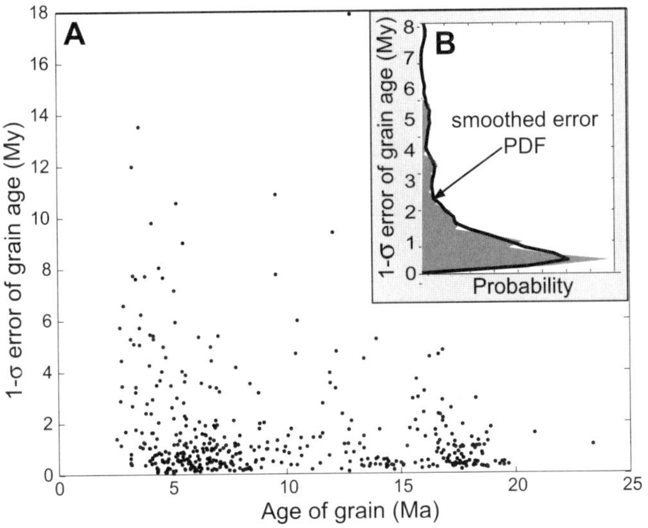

Figure 8. (A) Age versus error (1σ) for analyses with greater than 40% radiogenic ^{40}Ar. No clear relationship between age and error can be seen. Inset (B) shows a probability density function (PDF) plot generated from the 1σ errors. This plot is smoothed (solid line) and used in the error determination in the numerical model. Based on this plot, most assigned errors will be between 0 and 1.5 m.y.

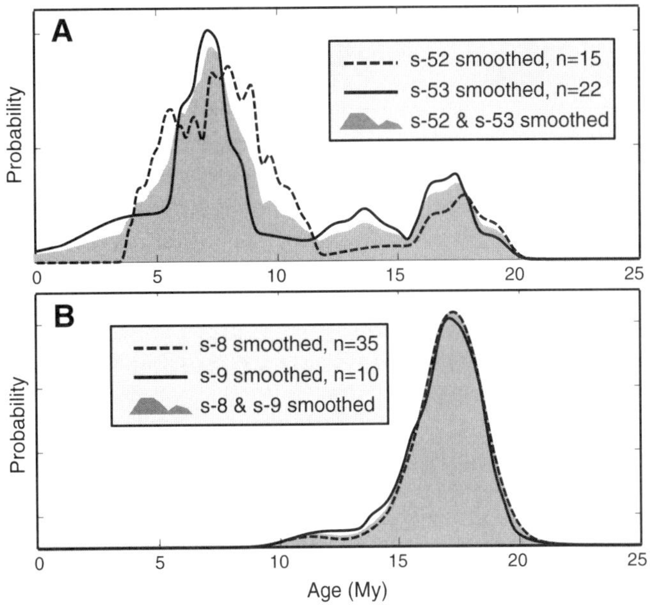

Figure 9. Results of repeat sampling to test (A) the spatial variability of the detrital cooling-age signal, and (B) the temporal variation of the signal. For each plot, the shaded probability density (PD) plot is the combined data used in the construction of Figure 5, while the PD plots represented by solid and dashed black lines are the contributing data. All curves are smoothed with a 2 m.y. scrolling window.

summed PD distribution and the theoretical PD distribution (Brewer et al., 2003) was compared against the observed mismatch between S-52 and S-53. A 2 m.y. smoothing window was again used before comparing curves to reduce the influence of individual grains. The theoretical modeling predicted a mismatch of ~20 ± 14% (2σ), whereas the observed S-52/S-53 mismatch was 32% (Fig. 9). Hence, the observed mismatch of the two actual samples lies within the expected range of variability, and they cannot be proven to be statistically different from each other. Furthermore, in comparison to the observed sample mismatch, the theoretical mismatch is a minimum estimate, because it is drawn from a population defined by the observed samples, whereas the samples themselves were drawn from a broader age population.

The second pair of samples was collected from the modern river bed (S-8) and an adjacent terrace fill (S-9) to evaluate the temporal consistency of the detrital cooling-age signal at a particular location (Fig. 9B). Would samples that are separated by hundreds to thousands of years show the same age distribution? If not, the question of sediment storage and production becomes important. Large bedrock landslides, for example, might produce a pulse of sediment containing a restricted suite of ages. The same comparative procedure was applied to these samples, resulting in a 5% mismatch between S-8 and S-9 and an expected mismatch of 21 ± 15% derived from the modeling. Thus the two PD distributions are very similar and cannot be considered statistically different. One caveat to this analysis, however, is that the almost unimodal cooling signal may mask temporal variability in sediment supply from this catchment area.

Spatial Variations of Erosion Rate

Spatial variations in cooling ages derived from low-temperature thermochronology in orogenic belts are commonly used as proxies for analogous variations in erosion rates (e.g., Blythe et al., 2000; House et al., 2002). The results of our PD distribution modeling indicate that, within the Marsyandi catchment, erosion rates vary by more than twofold: from 0.9 to 2.3 km/m.y. (Fig. 6). The highest rates of 1.9–2.3 km/m.y. are found in catchments draining the topographic front of the Himalaya. Rates decrease to the north, with the Tibetan Sedimentary Series eroding at rates of 0.9–1.1 km/m.y. Areas to the south of the Main Central thrust probably have intermediate rates, but it is difficult to constrain the signal from solely the Lesser Himalaya, because most of the rivers also drain the Greater Himalaya sequence.

Consider the results from the Darondi Khola (Fig. 3). Approximately 40% of the catchment area lies above the Main Central thrust zone, as represented by sample S-40. At the catchment mouth, sample S-37 yielded similar ages to S-40, but drains the additional area of the Lesser Himalaya, which is 1.5 times the size of the Greater Himalaya part of the catchment. The similarity suggests that either: (1) the similar erosion rates prevail throughout the catchment; or (2) the Greater Himalaya sequence is producing most of the muscovite found at the catchment mouth. The latter could be explained either by no effective erosion of the Lesser Himalaya sequence, or low percentages of muscovite in the Lesser Himalayan lithologies. Point counting suggests that the Greater Himalaya sequence is probably dominating the signal, because Lesser Himalayan catchments have low abundances of muscovite compared to those draining the Greater Himalayan sequence (Table 1). In addition, lower rates of erosion in the Lesser Himalaya, compared to higher rates on the topographic front, could be an explanation for the older PD distribution tails that are difficult to fit with the model for catchments draining both regions.

Estimates of Relative Erosion Rates from Point-Counting Data

As defined by the point counting, the proportion of muscovite in the tributaries is used to scale each catchment flux in the PD distribution modeling described in the preceding sections. The complete mineral proportions of trunk stream and tributaries, however, may also be used for an additional calculation of erosion rate. Mixing the sediment of two rivers together should produce a resulting downstream sample that is representative of the relative contributions from each of the rivers. In reality, the natural variability of the fluvial system and point-counting errors mean that the data are not well resolved. Because of this uncertainty, we use a basic model to examine the general pattern of contributions from each of the inputs and resulting mixing within the trunk stream.

The model uses point-counting data from each of the upstream samples and mixes them to produce a resulting downstream signal. The contribution of each upstream sample is varied from 0% to 100%, and the best solution is picked by finding the minimum residual to the downstream sample. This procedure assumes ideal mixing and no selective deposition or attrition of individual mineral species. We have already suggested that the carbonate rock fragments are susceptible to comminution and dissolution, and hence in some areas, they will bias the ratios. Therefore, we found that, instead of (1) mixing all the species at once; or (2) dropping out problematic individual mineral species and recalculating the mixing ratio (with or without renormalizing), the most consistent results were obtained by solving for the optimal mixing ratio (ϕ_i) for the observed concentration (M) of each mineral species (j) individually in the upstream trunk (a) and catchment (b) samples and downstream (c) samples:

$$[M_{j_c}] = \phi_j [M_{j_a}] + (1-\phi_j)[M_{j_b}] \tag{5}$$

and then calculating the mean mixing ratio ($\bar{\phi}$) for the number of species point counted (n) in the sample:

$$\bar{\phi} = \frac{1}{n}\sum_{j=1}^{n}\phi_i . \tag{6}$$

This procedure weighted each mineral species equally, eliminating instances where individual mineral species controlled the mixing ratio due to their large volume (e.g., quartz).

Given mean mixing ratios ($\bar{\phi}$) and catchment areas (A), relative erosion rates (dz/dt_{rel}) for individual tributaries within the Marsyandi were then calculated (Fig. 10):

$$\frac{dz}{dt_{relb}} = \frac{\bar{\phi} V_a}{(1-\bar{\phi}) A_b} . \tag{7}$$

To use this approach to examine the relative mixing of two tributaries, the volume (V_a) of material contributed per unit time in tributary a needs to be known. Thus, initially the relative erosion rates in the two most upstream tributaries were determined, assuming an arbitrary erosion rate of 1 unit per unit time in tributary a. Subsequently, as a function of area, V_a was calculated:

$$V_a = A_a \frac{dz}{dt_{rel_a}} . \tag{8}$$

Moving systematically downstream, relative erosion rates (compared to that of tributary a) of downstream tributaries can be calculated, such that the relative rate of tributary b is calculated with tributary a representing the upstream trunk-stream sample (equation 7), and V_a is calculated assuming that the total contributed volume per unit time of the trunk stream is a summation of t tributary inputs upstream:

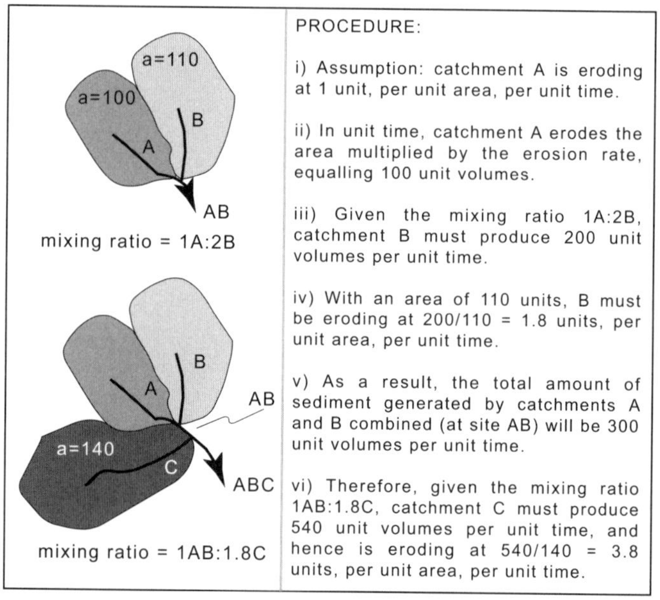

Figure 10. The procedure used to convert point-counting results into relative erosion rates. Point counting of sediment samples from the mouth of catchment A and the mouth of catchment B, in conjunction with a downstream sample AB, can be used to calculate a mixing ratio for the two catchments. When combined with catchment area, extracted from a digital elevation model, this ratio is used to calculate a relative erosion rate.

$$V_a = \sum_{i=1}^{t} A_i \frac{dz}{dt_{rel_i}} . \quad (9)$$

Finally, we rescaled the relative erosion rate in the Dordi catchment to match the erosion rate determined from cooling ages in the same catchment, and then applied this same rescaling factor to all other catchments. This was solely done to provide relative erosion-rate estimates (Fig. 11) in magnitudes that could be easily compared to those calculated using thermochronology (Fig. 6).

Although clear differences appear between the erosion rates calculated from thermochronology (Fig. 6) and those calculated from the point-counting analysis (Fig. 11), the general pattern of low erosion in the north of the region, high erosion rates in catchments on the southern flank of the main topographic axis, and intermediate erosion rates to the south, is the same. Whereas the PD distribution modeling suggested variations in erosion rate of up to 2.5-fold, the point-counting model yielded variations in predicted rates more than tenfold. These discrepancies are not surprising due to the expected variability of sediment within the river: (1) species such as rock fragments will become comminuted downstream; (2) carbonate rock fragments and minerals will experience chemical, as well as physical, erosion; and (3) hydraulic sorting will be important for minerals with different densities, although sorting appears to have little impact on muscovite in the 500–2000 μm size range, as shown by samples S-52 and S-53 (Fig. 9).

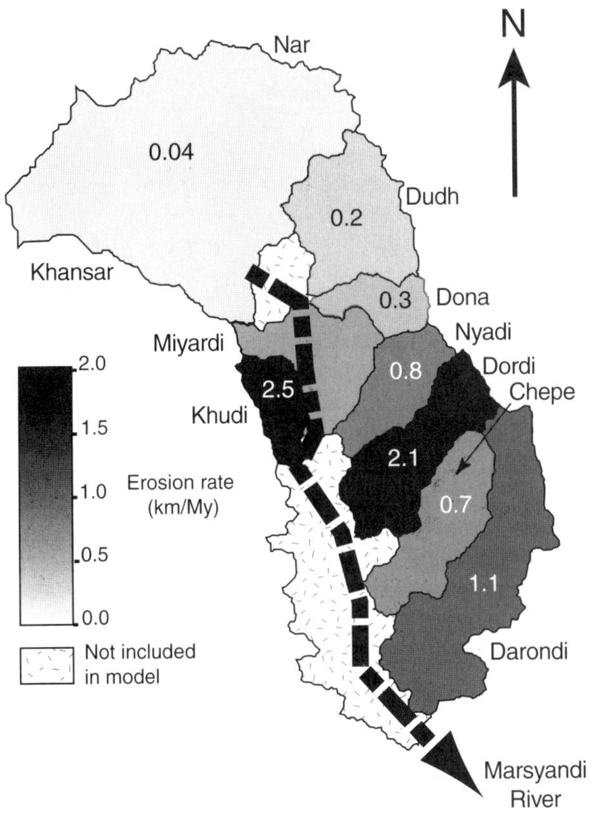

Figure 11. Spatial variation in erosion rates at the catchment scale based on point-counting data (Table 1) using the methodology illustrated in Figure 10. Relative erosion rates are normalized by the Dordi Khola to allow direct comparison with those derived from the cooling ages (Fig. 6). The stippled areas indicate zones not included in the calculations, and the dashed black line indicates the approximate path of the trunk stream. Note that the overall pattern of erosion is similar to that predicted from cooling ages, with high erosion rates in the middle areas of the Marsyandi catchment and slowest erosion rates to the north of the topographic axis.

CONCLUSIONS

Through examination of the downstream development of a detrital mineral cooling-age signal, this investigation provides insights into long-standing questions concerning the interpretation of detrital mineral ages. In the past, a sample collected from the foreland basin might be used to interpret erosion rates in the upstream catchment area, but where the erosion was occurring and how much of the catchment it represented remained unknown. Here, we show that there are systematic and predictable changes in the detrital cooling-age signal of a large, transverse Himalayan river. For example, large regions of a catchment may contribute little to the foreland signal if they are either eroding slowly or have a low abundance of the mineral, such as muscovite, that is being dated.

Our analysis of the Marsyandi Valley catchment detrital system yields insights on how the inputs from tributary catchments combine to form the signal at the catchment mouth. By implementing a thermal-erosional numerical model, we can use the cooling-age distributions from individual tributaries and observable characteristics of the catchment (relief and hypsometry) to determine a best-fit erosion rate. More typically, erosion rates have been estimated using individual bedrock cooling ages or suites of samples in vertical relief profiles (e.g., Blythe et al., 2000; Bullen et al., 2003; House et al., 2002; Ehlers and Farley, 2003). Because detrital samples can yield dozens to hundreds of cooling ages from throughout a catchment, when combined with catchment hypsometry and a thermal model, they provide a potentially potent tool for estimating erosion rates (Brewer et al., 2003; Bernet et al., 2004a).

The numerical modeling indicates that the detrital cooling-age signal changes systematically in the trunk stream, whereby the input of an age population from an individual tributary is a function of the abundance per unit area of the thermochronometer, the area of the catchment, and the rate at which it is eroding. As seen in other studies (Bernet et al., 2004a, 2004b), our mixing model for detrital ages was more successful in matching the central ages of observed cooling-age populations than in reproducing the relative abundance of those populations (Fig. 5). Previous studies have suggested that the failure to match relative abundances may be unavoidable because of the natural unsteadiness in fluvial systems (Bernet et al., 2004a). We are more optimistic, but suggest that the use of larger numbers of ages (>80 for complex age distributions), smaller catchments (across which erosion rates are more likely to be constant), thorough sampling (all important tributary catchments), and better statistics on mineral abundances should underpin significant improvement in this approach.

In our analysis, the distribution of muscovite is a critical factor in determining the representation of a particular area of the catchment in the foreland sample. For example, ~30% of the Marsyandi catchment is composed of the Tethyan Sedimentary Series, and yet these strata produce only a small fraction of the foreland detrital cooling-age signal because of the paucity of muscovite within these strata. Most previous detrital dating studies that examined downstream mixing (e.g., Bernet et al., 2004b) assumed that the mineral being dated had a uniform areal distribution at the scale of the catchments being compared. The ~100-fold variation in muscovite abundance documented here suggests that this assumption needs to be evaluated in almost all studies where heterolithic distributions of rocks can be expected. Whereas examination of geologic maps may provide insights on the spatial distributions of rock types (Bernet et al., 2004b), the comparative fraction of the target mineral needs to be measured for calibration when relative fluxes are calculated.

Comparison of the detrital age data versus modeling results indicates that comminution of muscovite by fluvial processes seems to be insignificant at the 100–200 km scale (within the 500–2000 μm analytical size range used). The 15–20 Ma signal derived from the Marsyandi headwaters is persistent downstream, although it decreases in significance. We argue that the calculated volumes of muscovite added to the trunk stream are consistent with the downstream dilution of this signal by younger age populations, rather than mechanical breakdown of muscovite. The absence of significant mica loss may be attributable to muscovite moving in suspended load during the monsoon season, thus encountering less abrasion than during bedload transportation.

In detrital studies, the reproducibility of detrital age samples needs to be assessed. Here, we evaluated samples from (1) different sites on the same sandbar and (2) from nearby deposits of different ages and showed that the detrital cooling-age signal appeared spatially and temporally reliable in this area. A recent study of the consistency of fission-track ages in detrital samples from the Rhone delta (Bernet et al., 2004b) also concluded that, within a 2σ uncertainty, the same component age peaks were represented in all samples. Such tests are recommended for other modern detrital studies.

The results of the thermochronometry suggest that the topographic front of the Himalaya is eroding faster than regions north and south of it. In particular, those catchments draining the Greater Himalayan sequence are generating abundant ca. 4–10 Ma cooling ages that appear to reflect erosion rates exceeding ~2 mm/yr. Areas to the north of the Himalayan topographic axis, on the edge of the Tibetan Plateau, experience the lowest erosion rates and produce grain ages between 10 and 15 Ma for those catchments sourced in the Tibetan zone, and 15–20 Ma for those catchments draining the top of the Greater Himalayan sequence and the Manaslu granite. Intermediate rates are probably found in the Lesser Himalaya, although the large tributaries draining this zone have headwaters in, and cooling-age signals dominated by, the Greater Himalaya. The pattern of erosion from thermochronology is broadly consistent with the erosion rates calculated from point-counting data. The latter predicts a greater contrast in the erosion rates that should be interpreted cautiously, however, because of the intrinsic variability of sediments within the river system.

Several Himalayan tectonic interpretations can be compared with the south-to-north increase in erosion rates predicted by this study. First, if the Main Central thrust and South Tibetan fault system are active, high erosion rates of the Greater Himalaya might be expected with the southward extrusion of the Greater Himalaya sequence (Beaumont et al., 2001; Hodges et al., 2001). Such extrusion predicts highest erosion rates toward the middle part of the Greater Himalaya, about halfway between the South Tibetan fault and the Main Central thrust (Beaumont et al., 2001). Our data cannot rule this out, but appear more consistent with rapid erosion in the vicinity of the Main Central thrust. Similarly, our study's gradient in erosion across the Greater Himalaya is inconsistent with the interpretation of uniform erosion across the Greater Himalaya that emerges from apatite fission-track studies, which show uniformly young ages spanning the Greater Himalaya (Burbank et al., 2003). Alternatively, the gradient of increased erosion rates toward the south is consistent with either an active Main Central thrust or a shift of the deformation front a few tens of kilometers into the Main Central thrust footwall (Harrison et al., 1997; Wobus et al., 2003).

Any of these tectonic interpretations introduces the complicating factor of lateral advection. In most investigations to date, cooling ages are converted into rock exhumation rates using an assumption of one-dimensional thermal and kinematic processes. For instance, in our model, the depth of the closure isotherm varies is a function of erosion rate and topographic relief, but rock particles still move vertically toward the surface. The thermal structure of an active convergent orogen is not particularly well represented by a simple model of horizontal isotherms, because it is subject to the lateral advection of heat into the orogen (e.g., Batt and Braun, 1997; Jamieson and Beaumont, 1988; Jamieson et al., 1998; Willett, 1999), and rock particles move laterally, not solely vertically, during erosion. Although beyond the scope of this paper, this complicating factor needs to be considered when evaluating our results.

This baseline investigation illustrates that, when assessed with the array of samples from within the Marsyandi catchment, the downstream evolution of the detrital cooling-age signal in the trunk stream is consistent. It can be numerically modeled as a function of contributions from tributaries with variable erosion rates, areas, relief, hypsometries, and muscovite abundances. The detailed interpretation of the most downstream (or foreland) sample by itself, however, is complex because of the patterns of tributary mixing and differential erosion. For example, the foreland sample is dominated by a 4–10 Ma population of grain ages, and one might argue that the southern catchments (the Nyadi Khola and downstream), which represent ~40% of the area sampled, control the signal. On closer inspection, however, cooling ages from these catchments cannot explain all the intricacies of the older signal: the upstream catchments that drain the upper Greater Himalayan rocks and Tibetan Sedimentary Series, representing ~55% of the catchment area, are needed to account for the 15–20 Ma age population (Figs. 2 and 5). Our modeling suggests that, despite a larger contributing area, this upstream signal is so much less dominant in the foreland due to the effects of volumetric contribution: catchment lithology and erosion rate.

In this study, we have dated ~500 muscovite grains from 12 locations, providing new insights into how lithology and catchment characteristics, both easily observable in other modern locations, influence the detrital age signal of the Marsyandi River. Although we agree that the distribution of cooling ages within the foreland provides useful information about the range of erosion rates in the hinterland, considerable caution is needed when interpreting the source and importance of detrital age populations from stratigraphic samples. The ability to confidently relate modern detrital age samples to hinterland cooling and erosion patterns (Bernet et al., 2001) is not easily duplicated in the past (Spiegel et al., 2004). Yet in order to extract the maximum amount of information from the stratigraphic record, an understanding of analogous but commonly unknown parameters (hypsometry, catchment area, lithology, erosion rate) that control the cooling-age signal is vital. An appreciation for both for the complexity of interactions in modern detrital systems, and for the primary controls on the detrital age signal, will support improved interpretations of the spatial and temporal evolution of orogens based on detrital age studies in the ancient stratigraphic record.

ACKNOWLEDMENTS

We thank P. DeCelles, M. Brandon, M. Bernet, and an anonymous reviewer for their incisive comments. Advice and assistance from R. Slingerland, K. Furlong, M. Bullen, W. Olszewski, J. Hurtado, A. White, and M. Krol significantly improved this analysis. Logistical assistance in Nepal was provided by Himalayan Experience. This research was supported by National Science Foundation grants to Burbank (EAR-9627865, 9896048, 9909647) and by grants from the Chevron Corporation and the Krynine Fund to Brewer.

REFERENCES CITED

Adams, C.J., Barley, M.E., Fletcher, I.R., and Pickard, A.L., 1998, Evidence from U-Pb zircon and $^{40}Ar/^{39}Ar$ muscovite detrital mineral ages in metasandstones for movement of the Torlesse suspect terrane around the eastern margin of Gondwanaland: Terra Nova, v. 10, p. 183–189, doi: 10.1046/j.1365-3121.1998.00186.x.

Batt, G., and Braun, J., 1997, On the thermomechanical evolution of compressional orogens: Geophysical Journal International, v. 128, p. 364–382.

Beaumont, C., Jamieson, R.A., Nguyen, M.H., and Lee, B., 2001, Himalayan tectonics explained by extrusion of a low-viscosity crustal channel coupled to focused surface denudation: Nature, v. 414, p. 738–742, doi: 10.1038/414738a.

Bernet, M., Zattin, M., Garver, J.I., Brandon, M.T., and Vance, J.A., 2001, Steady-state exhumation of the European Alps: Geology, v. 29, no. 1, p. 35–38, doi: 10.1130/0091-7613(2001)029<0035:SSEOTE>2.0.CO;2.

Bernet, M., Brandon, M.T., Garver, J.I., and Molitor, B.R., 2004a, Fundamentals of detrital zircon fission-track analysis for provenance and exhumation studies with examples from the European Alps, *in* Bernet, M., and Spiegel, C., eds., Detrital thermochronology: Geological Society of America Special Paper 378, p. 25–36.

Bernet, M., Brandon, M.T., Garver, J.I., and Molitor, B., 2004b, Downstream changes of Alpine zircon fission-track ages in the Rhône and Rhine Rivers: Journal of Sedimentary Research, v. 74, p. 82–94.

Bilham, R., Larson, K., Freymueller, J., and Project Idylhim members, 1997, GPS measurements of present-day convergence across the Nepal Himalaya: Nature, v. 386, p. 61–64.

Blythe, A.E., Burbank, D.W., Farley, K.A., and Fielding, E.J., 2000, Structural and topographic evolution of the central Transverse Ranges, California, from apatite fission-track, (U-Th)/He and digital elevation model analysis: Basin Research, v. 12, p. 97–114, doi: 10.1046/j.1365-2117.2000.00116.x.

Brandon, M.T., 1996, Probability density plot for fission-track grain-age samples: Radiation Measurements, v. 26, p. 663–676, doi: 10.1016/S1350-4487(97)82880-6.

Brandon, M.T., and Vance, J.A., 1992, Fission-track ages of detrital zircon grains: Implications for the tectonic evolution of the Cenozoic Olympic subduction complex: American Journal of Science, v. 292, p. 565–636.

Brewer, I.D., Burbank, D.W., and Hodges, K.V., 2003, Modelling detrital cooling-age populations: Insights from two Himalayan catchments: Basin Research, v. 15, p. 305–320, doi: 10.1046/j.1365-2117.2003.00211.x.

Bullen, M.E., Burbank, D.W., and Garver, J.I., 2003, Building the northern Tien Shan: Integrated thermal, structural, and topographic constraints: Journal of Geology, v. 111, p. 149–165, doi: 10.1086/345840.

Burbank, D.W., Leland, J., Fielding, E., Anderson, R.S., Brozovic, N., Reid, M.R., and Duncan, C., 1996, Bedrock incision, rock uplift and threshold hillslopes in the northwestern Himalayas: Nature, v. 379, p. 505–510, doi: 10.1038/379505a0.

Burbank, D.W., Blythe, A.E., Putkonen, J., Pratt-Sitaula, B., Gabet, E., Oskin, M., Barros, A., and Ojha, T.P., 2003, Decoupling of erosion and precipitation in the Himalayas: Nature, v. 426, p. 652–655, doi: 10.1038/nature02187.

Burchfiel, B.D., Zhileng, C., Hodges, K.V., Yuping, L., Royden, L.H., Changrong, D., and Jiene, X., 1992, The South Tibetan detachment system, Himalayan orogen: Extension contemporaneous with and parallel to shortening in a collisional mountain belt: Geological Society of America Special Paper 269, 41 p.

Carrapa, B., Wijbrans, J., and Bertotti, G., 2003, Episodic exhumation in the Western Alps: Geology, v. 31, no. 7, p. 601–604, doi: 10.1130/0091-7613(2003)031<0601:EEITWA>2.0.CO;2.

Cerveny, P.F., Naeser, N.D., Zeitler, P.K., Naeser, C.W., and Johnson, N.M., 1988, History of uplift and relief of the Himalaya during the past 18 million years; evidence from sandstones of the Siwalik Group, *in* Kleinspehn, K.L., and Paola, C., eds., New perspectives in basin analysis: New York, Springer-Verlag, p. 43–61.

Colchen, M., LeFort, P., and Pêcher, A., 1986, Annapurna–Manaslu–Ganesh Himal: Paris, Centre National de la Recherche Scientifique, 136 p.

Coleman, M.E., 1996, The tectonic evolution of the central Himalaya, Marsyandi Valley, Nepal [Ph.D. thesis]: Boston, Massachusetts Institute of Technology, 221 p.

Coleman, M.E., and Hodges, K.V., 1995, Evidence for Tibetan Plateau uplift before 14 Myr ago from a new minimum age for east-west extension: Nature, v. 374, p. 49–52, doi: 10.1038/374049a0.

Copeland, P., and Harrison, M.T., 1990, Episodic rapid uplift in the Himalaya revealed by $^{40}Ar/^{39}Ar$ analysis of detrital K-feldspar and muscovite, Bengal Fan: Geology, v. 18, p. 354–359, doi: 10.1130/0091-7613(1990)018<0354:ERUITH>2.3.CO;2.

Ehlers, T.A., and Farley, K.A., 2003, Apatite (U-Th)/He thermochronometry: Methods and applications to problems in tectonic and surface processes: Earth and Planetary Science Letters, v. 206, p. 1–14, doi: 10.1016/S0012-821X(02)01069-5.

Fowler, C.M.R., 1990, The solid Earth: An introduction to geophysics: Cambridge, Cambridge University Press, 490 p.

Garver, J.I., and Brandon, M.T., 1994, Fission-track ages of detrital zircons from Cretaceous strata, southern British Columbia: Implications for the Baja BC hypothesis: Tectonics, v. 13, no. 2, p. 401–420, doi: 10.1029/93TC02939.

Garver, J.I., Brandon, M.T., Roden-Tice, M., and Kamp, P.J.J., 1999, Exhumation history of orogenic highlands determined by detrital fission-track thermochronology, *in* Ring, U., Brandon, M.T., Willett, S., and Lister, G., eds., Exhumation processes: Normal faulting, ductile flow, and erosion: Geological Society Special Publication 154, p. 283–304.

Gehrels, G.E., and Kapp, P.A., 1998, Detrital zircon geochronology and regional correlation of metasedimentary rocks in the Coast Mountains, southeastern Alaska: Canadian Journal of Earth Sciences, v. 35, p. 269–279, doi: 10.1139/cjes-35-3-269.

Harrison, T.M., McKeegan, K.D., and LeFort, P., 1995, Detection of inherited monazite in the Manaslu leucogranite by ^{208}Pb/^{232}Th ion microprobe dating; crystallization age and tectonic implications: Earth and Planetary Science Letters, v. 133, p. 271–282, doi: 10.1016/0012-821X(95)00091-P.

Harrison, T.M., Ryerson, F.J., Le Fort, P., Yin, A., Lovera, O.M., and Catlos, E.J., 1997, A late Miocene-Pliocene origin for the central Himalayan inverted metamorphism: Earth and Planetary Science Letters, v. 146, p. E1–E7, doi: 10.1016/S0012-821X(96)00215-4.

Hodges, K.V., 1998, ^{40}Ar/^{39}Ar thermochronology using the laser microprobe, *in* McKibben, M.A., and Shanks, W.C., eds., Applications of microanalytical techniques to understanding mineralizing processes: Society of Economic Geologists Reviews in Economic Geology, v. 7, p. 53–72.

Hodges, K.V., 2000, Tectonics of the Himalaya and southern Tibet from two perspectives: Geological Society of America Bulletin, v. 112, p. 324–350, doi: 10.1130/0016-7606(2000)112<324:TOTHAS>2.3.CO;2.

Hodges, K.V., and Bowring, S., 1995, ^{40}Ar/^{39}Ar thermochronology of isotopically zoned micas: Insights from the southwestern USA Proterozoic orogen: Geochimica et Cosmochimica Acta, v. 59, p. 3205–3220, doi: 10.1016/0016-7037(95)00209-I.

Hodges, K.V., Parrish, R.R., and Searle, M.P., 1996, Tectonic evolution of the central Annapurna Range: Nepalese Himalayas: Tectonics, v. 15, p. 1264–1291, doi: 10.1029/96TC01791.

Hodges, K.V., Hurtado, J.M., and Whipple, K.X., 2001, Southward extrusion of Tibetan crust and its effect on Himalayan tectonics: Tectonics, v. 20, p. 799–809, doi: 10.1029/2001TC001281.

Hodges, K.V., Wobus, C., Ruhl, K., Schildgen, T., and Whipple, K., 2004, Quaternary deformation, river steepening, and heavy precipitation at the front of the Higher Himalayan ranges: Earth and Planetary Science Letters, v. 220, p. 379–389, doi: 10.1016/S0012-821X(04)00063-9.

House, M.A., Gurnis, M., Kamp, P.J.J., and Sutherland, R., 2002, Uplift in the Fiordland region, New Zealand: Implications for incipient subduction: Science, v. 297, p. 2038–2041, doi: 10.1126/science.1075328.

Hurford, A.J., Fitch, F.J., and Clarke, A., 1984, Resolution of the age structure of the detrital zircon populations of two Lower Cretaceous sandstones from the Weald of England by fission track dating: Geological Magazine, v. 121, p. 269–277.

Hurtado, J.M., Jr., Hodges, K.V., and Whipple, K., 2001, Neotectonics of the Thakkhola Graben and implications for recent activity on the South Tibetan fault system in the central Nepal Himalaya: Geological Society of America Bulletin, v. 113, no. 2, p. 222–240, doi: 10.1130/0016-7606(2001)113<0222:NOTTGA>2.0.CO;2.

Jamieson, R.A., and Beaumont, C., 1988, Orogeny and metamorphism; a model for deformation and pressure-temperature-time paths with applications to the Central and Southern Appalachians: Tectonics, v. 7, p. 417–445.

Jamieson, R.A., Beaumont, C., Fullsack, P., and Lee, B., 1998, Barrovian regional metamorphism: Where's the heat, *in* Treloar, P.J., and O'Brien, P.J., eds., What drives metamorphism and metamorphic reactions: Geological Society [London] Special Publication 138, p. 23–51.

Lavé, J., and Avouac, J.P., 2001, Fluvial incision and tectonic uplift across the Himalaya of Central Nepal: Journal of Geophysical Research, v. 106, p. 26,561–26,591, doi: 10.1029/2001JB000359.

Le Fort, P., 1981, Manaslu leucogranite: A collision signature in the Himalaya, a model for its genesis and emplacement: Journal of Geophysical Research, v. 86, p. 10,545–10,568.

Mancktelow, N.S., and Grasemann, B., 1997, Time-dependent effects of heat advection and topography on cooling histories during erosion: Tectonophysics, v. 270, p. 167–195, doi: 10.1016/S0040-1951(96)00279-X.

Najman, Y., Pringle, M., Godin, L., and Oliver, G., 2001, Dating of the oldest continental sediments from the Himalayan foreland basin: Nature, v. 410, p. 194–197, doi: 10.1038/35065577.

Najman, Y.M.R., Pringle, M.S., Johnson, M.R.W., Robertson, A.H.F., and Wijbrans, J.R., 1997, Laser ^{40}Ar/^{39}Ar dating of single detrital muscovite grains from early foreland-basin sedimentary deposits in India: Implications for early Himalayan erosion: Geology, v. 25, p. 535–538, doi: 10.1130/0091-7613(1997)025<0535:LAADOS>2.3.CO;2.

Pandey, M.R., Tandukar, R.P., Avouac, J.P., Lavé, J., and Massot, J.P., 1995, Interseismic strain accumulation on the Himalayan crustal ramp (Nepal): Geophysical Research Letters, v. 22, no. 7, p. 751–754, doi: 10.1029/94GL02971.

Roddick, J., Cliff, R., and Rex, D., 1980, The evolution of excess argon in Alpine biotites—A ^{40}Ar/^{39}Ar analysis: Earth and Planetary Science Letters, v. 48, p. 185–208, doi: 10.1016/0012-821X(80)90181-8.

Rowley, D.B., 1996, Age of initiation of collision between India and Asia: A review of stratigraphic data: Earth and Planetary Science Letters, v. 145, p. 1–13, doi: 10.1016/S0012-821X(96)00201-4.

Searle, M.P., and Godin, L., 2003, The South Tibetan detachment and the Manaslu leucogranite: A structural reinterpretation and restoration of the Annapurna-Manaslu Himalaya, Nepal: Journal of Geology, v. 111, p. 505–523, doi: 10.1086/376763.

Searle, M.P., Corfield, C., Stephenson, B., and McCarron, J., 1997, Structure of the North Indian continental margin in the Ladakh-Zanskar Himalayas: Implications for the timing of obduction of the Spontang ophiolite, India-Asia collision and deformation events in the Himalaya: Geological Magazine, v. 134, p. 297–316, doi: 10.1017/S0016756897006857.

Seeber, L., and Gornitz, V., 1983, River profiles along the Himalayan arc as indicators of active tectonics: Tectonophysics, v. 92, p. 335–367, doi: 10.1016/0040-1951(83)90201-9.

Silverman, B.W., 1986, Density estimation for statistics and data analysis: London, Chapman and Hall, 175 p.

Spiegel, C., Siebel, W., Kuhlemann, J., and Frisch, W., 2004, Toward a comprehensive provenance analysis: A multi-method approach and its implications for the evolution of the Central Alps, *in* Bernet, M., and Spiegel, C., ed., Detrital thermochronology—Provenance analysis, exhumation, and landscape evolution of mountain belts: Geological Society of America Special Paper 378, p. 37–50.

Stock, J.D., and Montgomery, D.R., 1996, Estimating palaeorelief from detrital mineral age ranges: Basin Research, v. 8, no. 3, p. 317–328.

Stüwe, K., White, L., and Brown, R., 1994, The influence of eroding topography on steady-state isotherms: Application to fission track analysis: Earth and Planetary Science Letters, v. 124, p. 63–74, doi: 10.1016/0012-821X(94)00068-9.

Van der Plas, L., and Tobi, A.C., 1965, A chart for judging the reliability of point counting results: American Journal of Science, v. 263, p. 87–90.

Wang, Q., Zhang, P.-Z., Freymueller, J.T., Bilham, R., Larson, K.M., Lai, X.A., You, X., Niu, Z., Jianchun, W., Li, Y., Liu, J., and Yang, Z., 2001, Present-day crustal deformation in China constrained by global positioning system measurements: Science, v. 294, p. 574–577.

Whipple, K.X., 2001, Fluvial landscape response time; how plausible is steady-state denudation?: American Journal of Science, v. 301, p. 313–325.

White, N.M., Pringle, M., Garzanti, E., Bickle, M., Najman, Y., Chapman, H., and Friend, P., 2002, Constraints on the exhumation and erosion of the High Himalayan Slab, NW India, from foreland basin deposits: Earth and Planetary Science Letters, v. 195, p. 29–44, doi: 10.1016/S0012-821X(01)00565-9.

Willett, S.D., 1999, Orogeny and orography: The effects of erosion on the structure of mountain belts: Journal of Geophysical Research, v. 104, p. 28,957–28,982, doi: 10.1029/1999JB900248.

Willett, S.D., and Brandon, M.T., 2002, On steady states in mountain belts: Geology, v. 30, no. 2, p. 175–178, doi: 10.1130/0091-7613(2002)030<0175:OSSIMB>2.0.CO;2.

Willett, S.D., Slingerland, R., and Hovius, N., 2001, Uplift, shortening, and steady state topography in active mountain belts: American Journal of Science, v. 301, no. 4–5, p. 455–485.

Wobus, C.W., Hodges, K.V., and Whipple, K.X., 2003, Has focused denudation sustained active thrusting at the Himalayan topographic front?: Geology, v. 31, p. 861–864, doi: 10.1130/G19730.1.

Manuscript Accepted by the Society 23 June 2005

Printed in the USA

Geological Society of America
Special Paper 398
2006

The geochemistry of rivers in tectonically active areas of Taiwan and New Zealand

Anne E. Carey
Department of Geological Sciences, The Ohio State University, Columbus, Ohio 43210-1398, USA
Shuh-Ji Kao
Research Center for Environmental Changes, Academia Sinica, Taipei, Taiwan
D. Murray Hicks
National Institute of Water and Atmospheric Research, Christchurch, New Zealand
Carmen A. Nezat
Department of Geological Sciences, University of Michigan, Ann Arbor, Michigan 48109-1063, USA
W. Berry Lyons
Byrd Polar Research Center, The Ohio State University, Columbus, Ohio 43210-1002, USA

ABSTRACT

The geochemistry of river water, river sediments, and suspended matter in three mountainous watersheds in New Zealand and Taiwan is used to determine chemical erosion yields in regions of rapid tectonic uplift. Suspended matter from all three rivers is depleted in soluble alkali metals and alkaline earths compared to upper continental crust material and marine clays, reflecting the bedrocks' origin as marine sediments that had undergone previous weathering cycles prior to uplift and subjection to the current chemical weathering regime. The New Zealand rivers are depleted in Mg^{2+} and enriched in Ca^{2+} and $Na^+ + K^+$ compared to global average river water, but the Taiwan river is enriched only in Mg^{2+} compared to global average. The Haast River, draining the Southern Alps of New Zealand, is depleted in $Cl + SO_4$ compared to the global average, but has higher alkalinity and slightly higher H_4SiO_4. The chemical weathering yields determined here compose only a small portion (1%–5%) of the total weathering in these systems, but are still among the highest chemical yields ever reported. Our new data, in comparison to previously determined physical erosion yields in these watersheds, show that physical erosion strongly enhances chemical erosion. This work demonstrates the importance of chemical erosion as a process denuding the landscape, especially in high-standing, tectonically active mountainous landscapes.

Keywords: weathering, Taiwan, New Zealand, streams, sediments.

INTRODUCTION

The influence of tectonic uplift on erosion rates has been qualified and quantified over the past few years (Hovius et al., 1997; Willett, 1999; Montgomery and Brandon, 2002; Willett and Brandon, 2002; Finlayson et al., 2002). These investigations and those of other investigators have focused on the role of physical erosion as the primary process leading to mass loss and topographic reduction. The relationship between physical erosion and chemical erosion also has been a topic of much discussion over the past decade. The linkages between chemical and mechanical weathering have been discussed in

Carey, A.E., Kao, S.-J., Hicks, D.M., Nezat, C.A., and Lyons, W.B., 2006, The geochemistry of rivers in tectonically active areas of Taiwan and New Zealand, *in* Willett, S.D., Hovius, N., Brandon, M.T., and Fisher, D.M., eds., Tectonics, Climate, and Landscape Evolution: Geological Society of America Special Paper 398, Penrose Conference Series, p. 339–351, doi: 10.1130/2006.2398(21). For permission to copy, contact editing@geosociety.org.

numerous recent works, which have argued that the control of chemical weathering is directly related to the rates of physical or mechanical weathering, or occurs primarily in regimes of the world where mechanical denudation (i.e., erosion) is high (Louvat and Allègre, 1997; Gaillardet et al., 1999; Millot et al., 2002). Recently Riebe et al. (2001) argued that weathering is strongly controlled by tectonics and only weakly controlled by climate, and that physical and chemical weathering rates are strongly correlated.

Most of the riverine studies comparing physical to chemical erosion yields have been done on large river systems (e.g., Gaillardet et al., 1999) and most recently on mountainous, smaller systems (Anderson et al., 2002). However, it is now clear that smaller watersheds in active orogenic regions, such as island arcs and oceanic-continental convergent plate boundaries, supply very large amounts of sediment to the ocean. The seminal work of Milliman and Syvitski (1992) was the first to document this, but more recent research has added credence to their original ideas (Milliman, 1995; Hicks et al., 1996; Maneux et al., 1999; Milliman et al., 1999; Restrepo and Kjerfve, 2000; Mertes and Warrick, 2001; Campbell and Church, 2003).

Herein we present the geochemical characteristics of three rivers, in New Zealand and Taiwan, in order to qualify and quantify the nature of chemical weathering within the watersheds and to assess the role of climate and physical erosion yields on chemical weathering processes. The role of chemical weathering as a process denuding the landscape is not frequently considered in regions of rapid tectonic uplift where the high physical erosion rates result in rapid landscape denudation.

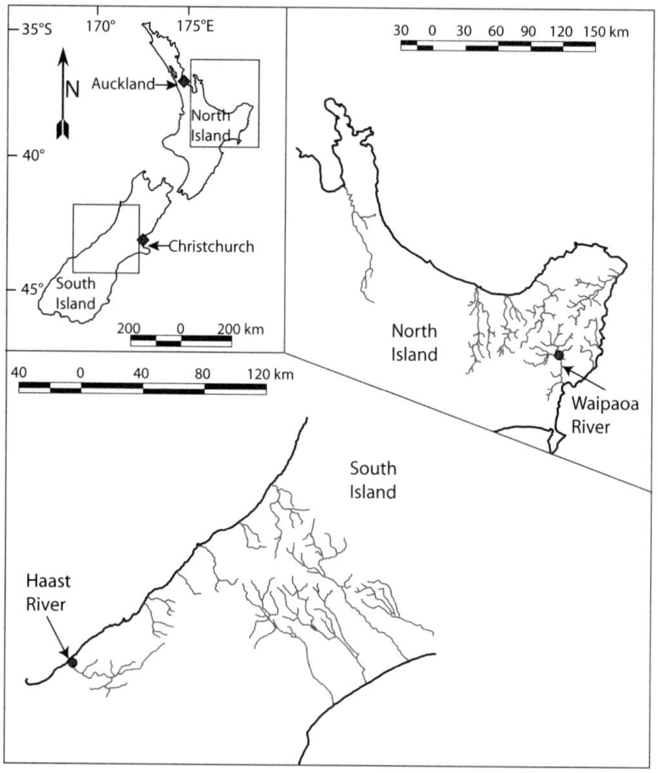

Figure 1. Map of New Zealand showing location of the Haast River on the west coast of South Island and of the Waipaoa River on the East Cape of North Island. Circles indicate gauge locations where samples were collected for this study.

METHODS

Site Descriptions

Haast River

The Haast River (Fig. 1; Table 1) is a temperate river on South Island, New Zealand, hosted in the quartzo-feldspathic Haast schist, a regionally metamorphosed rock of Carboniferous to Early Cretaceous age, which is the metamorphic equivalent of the graywacke found to the east and southeast on South Island (Batt et al., 2000). The river's headwaters are found in the chlorite zone of the schist, while lower reaches of the river flow through the higher-grade biotite and garnet-oligoclase zones of the schist (Mason, 1962; Riddolls, 1987). Lamprophyric dikes are exposed in the Haast River region, and granite pegmatites occur in the northern portion of the drainage (Batt et al., 2000). The Haast watershed's schist lithologies are diverse, ranging from weakly fissile, semischist to garnet schist (Hicks, 1999). Soils in the alpine region to the northeast range from 0 to 22,000 yr old, and contain mica, chlorite, quartz, albite, and K-feldspar (Mokma et al., 1973). Apparently mica is rapidly converted to clay minerals (Mokma et al., 1973). On slopes, the soils are thin, and mass wasting is a major surficial process that accounts for a regional denudation rate of 9 ± 4mm yr^{-1} (Hovius et al., 1997).

The watershed drains the Main Divide of the Southern Alps in the western portion of South Island, New Zealand, where the Alpine fault separates the Australian plate from the Pacific plate. The Alpine fault has undergone ~500 km of strike slip motion and >20 km of vertical motion (Norris et al., 1990). Modern uplift rates near the Main Divide are as high as 12 mm yr^{-1} (Basher et al., 1988). Haast River discharge has been gauged at Roaring Billy since June 1969. From 1970 to 2004, the mean discharge was 187,819 L s^{-1} (Table 1). At the gauge, the average annual rainfall is 7400 mm (Hicks and Griffiths, 1992), but within the watershed the rainfall is highly influenced by orographic effects. A Manning autosampler deployed there from May 1999 to November 1999 collected 41 runoff samples from 4 storm events. During the time that the autosampler was deployed, 10 water samples were collected by hand during base-flow conditions at that same location. Based on the ratings curve, mean annual suspended loads are estimated to be 4.18 Mt yr^{-1}. Based on 45 depth-integrated suspended sediment gaugings collected over the period of August 1967 to December 1995, the maximum gauged sediment concentration during the time of autosampler deployment was 6234 mg L^{-1}. For samples used to develop the suspended sediment ratings curve, water discharges ranged from 53 m^3 s^{-1} to 3745 m^3 s^{-1}, and

TABLE 1. WATERSHED ENVIRONMENTAL DATA AND STREAM DISCHARGE DATA

	Haast River	Waipaoa River	Lanyang Hsi River
Watershed area (km^2)	1350	2200	980
Watershed area above gauge (km^2)	1080	1580	820
Average catchment elevation (m)	1040	374	938
Maximum catchment elevation (m)	2738	1213	3535
Average slope	32.4°	24.3°	25.3°
Temperature (mean annual, °C)	9 (water)	17 (water)	18 (air)
Rainfall, average annual (mm)	7412	1471	3000
Runoff, average annual (mm)	6002	668	2100
Erosion rate (t km^{-2} yr^{-1} ± standard error)	4500 ± 29%	6753 ± 10%	12,800
Land use	44% alpine grass tussock 47% forest (native) 9% bare rock or permanent snow cover	84% grassland (pasture) 7% forest (exotic) 10% scrubland	95% forest 5% agriculture (vegetable gardening and fruit trees)
Mean discharge[†] (m^3 s^{-1})	215	28	87.5
Mean annual discharge[‡] (m^3 s^{-1})	194	33	63
Discharge on day of sampling[§] (m^3 s^{-1})	361	32	112
Number of samples in sediment rating curve	45	301	1435

[†]Mean discharge for 1999 for Haast; for 2000 for Waipaoa and Lanyang Hsi.
[‡]Period of record 1970–2004 for Haast; 1960–2002 for Waipaoa; 1950–2000 for Lanyang Hsi.
[§]2 April 1999 for Haast; March 2000 for Waipaoa; 15 December 2000 for Lanyang Hsi.

the highest sediment concentration gauged was 5128 mg L^{-1}. In addition to the suspended sediment samples, 5 grab samples were collected by hand in March 1998 for dissolved chemistry (major anions and major, minor, and trace cations) analyses.

Waipaoa River

The Waipaoa watershed (Table 1) is 2200 km^2, and the highest elevation in the catchment is 1213 m above mean sea level. The watershed is located on the eastern side of Raukumarara Range in the northeast region of North Island, New Zealand, within the active deformation zone of the Hikurangi subduction margin (Gomez et al., 1999). Uplift rates in the headwater are as high as 4 mm yr^{-1} (Pillans, 1986). There is a broad regional uplift of ~1 mm yr^{-1}, and downcutting by streams at rates higher than the uplift have been observed (Berryman et al., 2000). The watershed consists of mudstones, argillites, and graywacke with a cover of poorly consolidated Miocene-Pliocene sandstone, siltstone, and mudstone (Gomez et al., 2001). Deforestation of the region began with Polynesian settlement ~800 yr ago (Orpin et al., 2002). Further conversion to pasture land during the nineteenth century led to intense erosion and increased sediment yield (Gomez et al., 1999). Both landslide-induced mass transport and gully erosion are important mechanisms, with the latter leading to higher and less variable suspended load concentrations (Hicks et al., 2000). Post-1960, reforestation of some of the most severely eroded areas has helped to reduce the gully erosion and decrease the sediment yield (Orpin et al., 2002). The Waipaoa River discharge has been gauged since 1978 in two locations by the National Institute of Water and Atmospheric Research (NIWA).

The western watersheds of the Southern Alps and the eastern drainages of the northeast portion of North Island are the highest sediment yielding regions of New Zealand and are some of the highest in the world (Hicks et al., 1996). Very high rainfall in the Southern Alps due to the orographic effect and the soft, readily erodible surface landscape in the northeastern portion of North Island are important reasons for these very high physical erosion yields (Hicks et al., 1996). Physical erosion yields for the Haast and Waipaoa watersheds are 4500 and 11,540 t km^{-2} yr^{-1}, respectively.

Lanyang Hsi

The Lanyang Hsi watershed is 980 km^2 and lies in the Central Range of northeastern Taiwan (Fig. 2; Table 1). The Central Range has many peaks greater than 3000 m elevation, and it forms the backbone of the island. The Lanyang Hsi lies in a region of active tectonism, with the main stem of the river following a strike-slip fault (Kao and Liu, 1996). Uplift rates in the eastern Central Range of ~5 mm yr^{-1} are produced by the oblique collision between a portion of the Philippine Sea plate and the Asian continent (Teng, 1990).

The watershed lies in Tertiary-age metasandstones, argillites, phyllites, and slate (Kao and Liu, 1996). These metamorphosed rocks are primary Cenozoic cover of an accretionary wedge, and are derived from continental crust materials (Chen et al., 1990). Erosion rates in the eastern Central Range are 2–8 mm yr^{-1} (Fuller et al., 2003). Hillslope mass wasting is the major process supplying sediments to streams (Hovius et al., 2000). Previous work has indicated that anthropogenic activities, such as road construction, can increase sediment yields by a factor of 2.7 (Kao and Liu, 1996). The suspended sediment fluxes prior to road construction were ~3100 t km^{-2} yr^{-1} (Kao and Liu, 1996). Li (1976) estimated the total denudation rate of the Lanyang Hsi area to be 8000 t km^{-2} yr^{-1}.

The maximum elevation in the watershed is 3535 m (Kao and Liu, 2001). Its total watershed area is 980 km^2, with an area

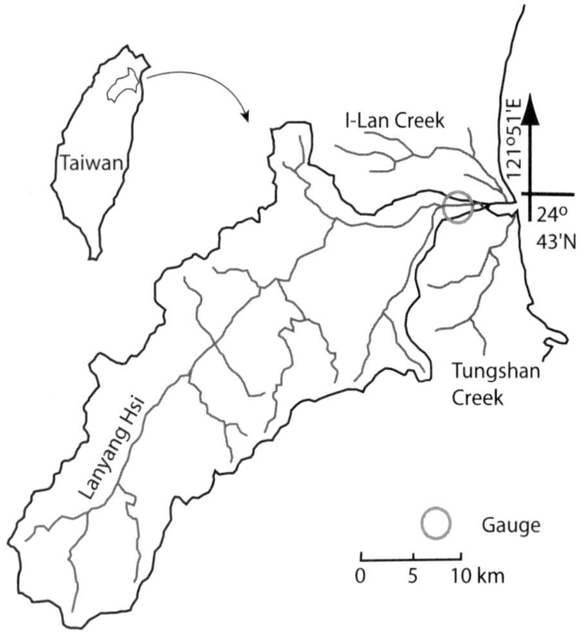

Figure 2. Map of Taiwan showing location of the Lanyang Hsi watershed. Inset map shows location of the stream gauging station where samples were collected for this study.

of 820 km^2 above the gauge location sampled for this study (Kao and Liu, 2001) (Fig. 2). Mean annual precipitation ranges between 2000 and 5000 mm and averages 3000 mm, and ~70% of that runs off. The stream has been gauged at two locations since 1950 for suspended sediment concentrations (sampled ~30 times per year), and the gauge continuously records water discharge (Kao and Liu, 2001). Much of the floodplain area is cultivated in fruit trees, but the primary agricultural area is the headwaters at elevations greater than 850 m (Kao et al., 2005), where agriculture along the stream banks at elevations as high as 1250 m has been practiced since ca. 1980 (Kao and Liu, 2002).

The watershed has been subjected to two major episodes of anthropogenic disturbance: the construction of the Central Cross-Island Highway during 1957–1960 and a county road construction project during 1975–1980. For several years after the completion of each project (1960–1963 and 1980–1982), the suspended sediment load was higher than the average of 4.1–7.4 Mt yr^{-1} (Kao and Liu, 2001). Preconstruction (1950–1957) suspended sediment loads were 0.6–4.4 Mt yr^{-1} (Kao and Liu, 2002).

Analytical Methods

Water samples for major ion analyses were collected in deionized water (18 MΩ)–soaked linear polyethylene bottles (Welch et al., 1996), which were rinsed 3 times with river water before sample collection. Samples for ion analyses were stored at room temperature in the dark until return to the laboratory for filtration and analysis.

For all river samples, major anions were determined by ion chromatography (IC) using a Dionex DX-300 or DX-120 instrument following the methods of Welch et al. (1996). For the Lanyang Hsi samples, major cations were determined by ion chromatography (Welch et al., 1996), and H_4SiO_4 was determined as reactive silicate by colorimetric methods (Mullin and Riley, 1955). For the New Zealand samples, major cations and Si were determined using a Leeman Labs Plasma Spec III inductively coupled plasma–optical emission spectrometer (ICP-OES). The precision of the IC measurements were ±2% or better, while the ICP-OES measurements were ±8% or better, with most being better than ±3%. Minor and trace elements were analyzed using a Perkin-Elmer Elan 6000 inductively coupled plasma–mass spectrometer (ICP-MS). Precision of all the ICP-MS measurements was better than ±10%. Analytical data for major ions, water discharge data, and date of sampling for all water samples are given in Appendix A1.

In addition to the solute measurements, a series of river sediment samples and suspended sediments were collected and analyzed. Major and minor elements were analyzed using X-ray fluorescence spectrometry (XRF), after manually crushing samples with a deionized water–cleaned mortar and pestle to <150-µm-sized particles. All the New Zealand samples were analyzed in our laboratory as pressed pellets with polyvinyl alcohol as binder using a Phillips PW 2450 XRF spectrometer. The Taiwan samples were analyzed by XRAL Laboratories, Don Mills, Ontario, Canada, as fused discs. Rare earth elements, Th, and U were analyzed by ICP-MS after Na_2O_2 fusion and acid dissolution by XRAL. The precision of duplicate analyses of the ICP-MS measurements were all 4% or better. Analytical data for stream sediments and suspended matter are given in Appendix A2.

For our pressed powder XRF measurements, U.S. Geological Survey standards AGV-1 (n = 15), BIR-1 (n = 12), G-2 (n = 12), and SDC-1 (n = 9) were measured within ±12% of the reported values, except for Na_2O (49%), Cr (82%), and Ni (82%). Because the pressed power method may not be as accurate for oxide analysis due to heterogeneity of the sample mix, replicate samples (n = 4) were also analyzed by fused disc technique by XRAL. The results indicated that the pressed power results overestimated some of the oxides. The overestimate of oxides was: 11% (Na_2O); 21% (Fe_2O_3); 29% (K_2O); and 52% (CaO).

RESULTS AND DISCUSSION

Water Geochemistry

The water chemistry of the three rivers is shown in Table 2 and Appendix A1. The total dissolved solids (TDS) of the rivers were 52, 155, and 370 mg L^{-1} for the Haast, Lanyang His, and Waipaoa, respectively. In general, the New Zealand rivers were more dilute than the global average due to high rainfall and high runoff (Close and Davies-Colley, 1990). These TDS values are well within the range for mountain belts, with watersheds of non-sedimentary rocks having lower values (Stallard, 1995). Before

TABLE 2. SOLUTE DATA FOR THE HAAST, WAIPAOA, AND LANYANG HSI RIVERS

	Na^+ (μmol L^{-1})	K^+ (μmol L^{-1})	Ca^{2+} (μmol L^{-1})	Mg^{2+} (μmol L^{-1})	H_4SiO_4 (μmol L^{-1})	Alkalinity (μeq L^{-1})
Haast (*n* = 7)	71	31	244	29	57	531
Waipaoa	1282	80	1485	266	187	2775
Lanyang Hsi	329	44	1000	798	114	2267
	Cl^- (μmol L^{-1})	SO_4^{2-} (μmol L^{-1})	Rb^+ (nmol L^{-1})	Sr^{2+} (nmol L^{-1})	Ba^{2+} (nmol L^{-1})	U^{6+} (nmol L^{-1})
Haast (*n* = 7)	36.1	37.6	12	740	29	0.22
Waipaoa	266	911	20	6430	377	2.3
Lanyang Hsi	35.5	824	21	3790	45	0.48

utilizing any of the solute data for weathering calculations, we corrected the solution chemistry for rainfall input by assuming that all of the chloride was derived from precipitation. The other solutes were then corrected by subtracting from the concentration of each major ion its concentration in proportion to its ratio to chloride in seawater (Bruland, 1983). No corrections were made to the H_4SiO_4 data, because we assumed that there was no H_4SiO_4 in the precipitation.

The precipitation-corrected Si:Na+K molar ratios were 0.17, 0.33, and 0.86 for the Waipaoa, the Lanyang His, and Haast, respectively, indicating significant chemical fractionation in the river waters compared to the rocks undergoing weathering, and that current chemical weathering is superficial, with cation-rich secondary minerals, rather than kaolinite, being the weathering products. This is supported by soil investigations (Mokma et al., 1973; Huh et al., 1998). The Haast waters were more dilute and more Ca^{2+}-rich than "pristine" waters draining metamorphic rocks, and the Waipaoa had higher Ca^{2+}:Mg^{2+} than pristine waters draining shale watersheds (Meybeck, 1987). The Haast water data are well within the range of values reported by others (Kim et al., 1999; Jacobson and Blum, 2003), confirming that it is a dilute Ca^{2+}-HCO_3^- water. Most of the rivers draining the Southern Alps are Ca^{2+}-HCO_3^- systems (Close and Davies-Colley, 1990). The Lanyang Hsi is also considered a Ca^{2+}-HCO_3^- water, while the Waipaoa has higher Na^+ concentrations than the other two rivers. Both the Waipaoa and the Lanyang Hsi have an order of magnitude higher SO_4^{2-} concentration than the Haast. The Waipaoa also has a much higher Cl^- concentration than the others (Table 2).

The geochemistry of these three rivers is well demonstrated by ternary plots of the major ions, alkalinity, and Si (Fig. 3). The

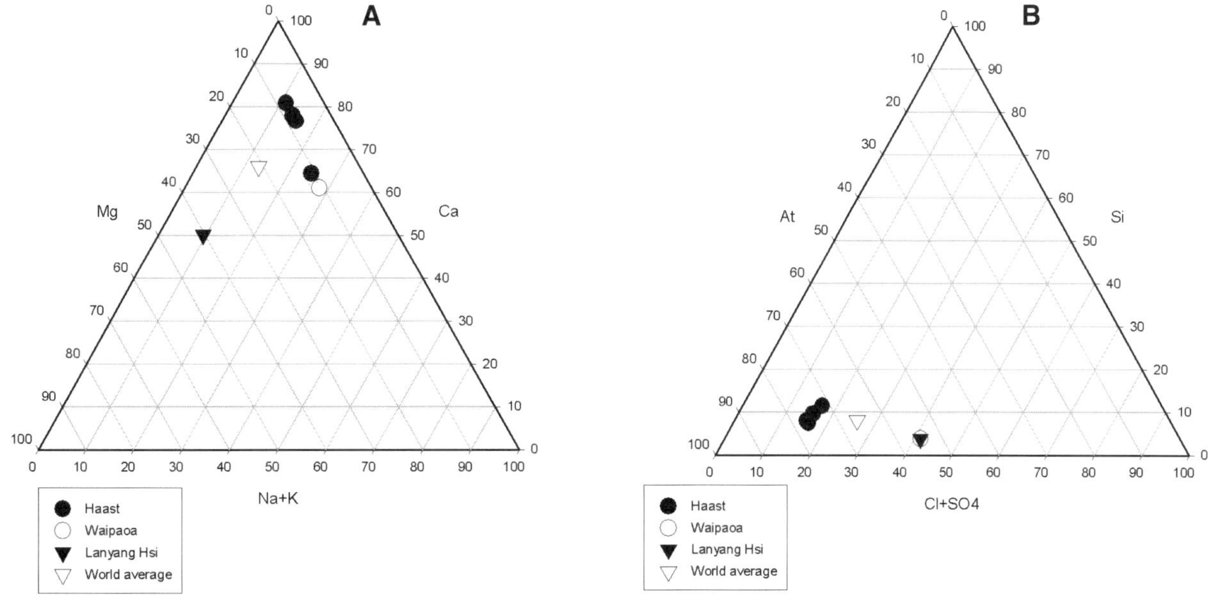

Figure 3. (A) Trilinear plot of major cations (data in meq L^{-1}) from the three rivers studied here. Four samples from the Haast River are shown as filled circles; the sample from the Waipaoa River is shown as an open circle; and the Lanyang Hsi sample shown is shown as a closed, inverted triangle. For reference, the world average river water (data from Hu et al., 1982) is shown as an open inverted triangle. (B) Trilinear plot of anions and silica (in meq L^{-1}) plotted in the manner of Hu et al. (1982). Shown are titration alkalinity (At), dissolved reactive silica (Si) and chloride plus sulfate (Cl + SO_4) from the three rivers studied here. Four samples from the Haast River are shown as filled circles; the sample from the Waipaoa River is shown as an open circle; and the Lanyang Hsi sample is shown as a closed, inverted triangle. For reference, the world average river water (data from Hu et al., 1982) is shown as an open inverted triangle.

global average river waters are enriched in Mg^{2+} relative to our New Zealand rivers but depleted in Ca^{2+} and $Na^{+} + K^{+}$. On the other hand, the Lanyang Hsi is enriched in Mg^{2+} relative to the global average. The Lanyang Hsi and Waipaoa have similar relative compositions of Cl^{-}, SO_4^{2-}, alkalinity, and H_4SiO_4, while the Haast river waters are depleted in $Cl^{-} + SO_4^{2+}$, and have higher alkalinity and slightly higher H_4SiO_4. The differences in the major element concentrations of these rivers potentially reflect: (1) the differences in the geochemistry/mineralogy of the geologic materials undergoing weathering; (2) differences in climatic regimes; and/or (3) differences in the impact of anthropogenic activities either within the watershed itself, or in the chemistry of atmospheric input.

Geologic Influences

As noted above, the watershed rock types are very different (i.e., sedimentary versus metamorphic), but all rocks were originally marine sediments derived from continental sources. However, the higher relative Mg^{2+} concentration in the Lanyang Hsi waters and the Waipaoa, and the Ca^{2+} and $K^{+} + Na^{+}$ in the New Zealand waters reflect differences in the bedrock mineralogy within the watersheds. For example, it has been proposed that the higher Ca^{2+} concentration in the Haast waters is a result of small amounts of calcite within the Haast schist in the Southern Alps, which weathers ~350× faster than plagioclase in this environment (Jacobson et al., 2003). Numerous other studies have recently documented the importance of $CaCO_3$ dissolution on solute geochemistry of streams, even though the $CaCO_3$ is present as a minor phase in the bedrock (White et al., 1999; Anderson and Dietrich, 2001).

The Ca:Sr molar ratios of the three rivers are 230, 264, and 330 for the Waipaoa, Lanyang Hsi, and Haast, respectively, while the ratios of those elements in the sediments are 395, 158, and 148, respectively. The higher Ca:Sr ratio in the Haast river water, compared to the solid weathering product, supports the conclusions of Kim et al. (1999) and Jacobson and Blum (2003) that much of the Ca^{2+} and Sr^{2+} in rivers in the Southern Alps is derived from $CaCO_3$, and not from silicate mineral weathering, as calcite in these drainages has much higher Ca:Sr ratios than the silicate mineral fraction does. Of these three rivers, only the Waipaoa has riverine water ratios lower than the sediment ratios. Low Na:Ca ratios and high alkalinity to Si ratios in both the Haast and the Lanyang Hsi also support the notion that the dissolution of hydrothermal calcite in these metamorphic rocks, rather than plagioclase weathering, is the source of Ca^{2+} to these waters. The high alkalinity to Si ratios also support this. Using the silicate end-member technique of Gaillardet et al. (1999), we can estimate the amount of cations generated from silicate weathering compared to the Ca^{2+} produced from the dissolution of $CaCO_3$. We have calculated the amount of Ca^{2+} generated by silicate weathering by using the Ca:Na molar ratio of 0.35 from Gaillardet et al. (1999) and our precipitation-corrected Na^{+} data. This Ca^{2+} is then subtracted from our precipitation-corrected Ca^{2+} value to obtain the Ca^{2+} from $CaCO_3$ dissolution. This was done after correcting for precipitation input of cations by assuming that all the Cl^{-} present in the river water is from precipitation, then using the seawater cation:Cl ratios to establish cation input. When this is done, the silicate weathering:carbonate weathering ratios are 31%, 58%, and 61% for the Haast, Lanyang Hsi, and Waipaoa, respectively. Although these figures should be taken only as estimates, they do suggest that in all three watersheds, the dissolution of $CaCO_3$ is an important process. In addition, our results strongly support the notion that in rapidly uplifting, young orogenic mountain belts, carbonate weathering plays a significant role in denudation (Jacobson and Blum, 2003). This in turn may be reflected in the age of the landscape, for the reason that in rapidly uplifting areas, such as these three regions of our study, continued exhumation may favor carbonate weathering over silicate weathering due to the continued presence of $CaCO_3$ within the bedrock (Jacobson et al., 2002).

The Waipaoa and Lanyang Hsi waters have higher concentrations of minor elements (Rb, Sr, Ba) than does the Haast (Table 2). This, too, may reflect a greater influence of silicate weathering within the Waipaoa and Lanyang Hsi watersheds (Beaucaire and Michaud, 1982). The higher Sr^{2+} and Ba^{2+} concentrations in the Waipaoa suggest that some authigenic mineral phase, such as barite, may be important, because it is unlikely that plagioclase or K-feldspar exists in sufficient quantity to produce these solutes.

The relatively high K:Mg ratio in the Haast River, compared to the other two rivers (Table 2), suggests that biotite and/or K-feldspar is an important weathering reactant in the Southern Alps. Jacobson et al. (2003) recently argued that biotite is the main source of K^{+} to streams and rivers that pass through the Haast schist. Jacobson et al.'s (2003) K:Rb ratios range from 0.86×10^3 to 1.7×10^3, very similar to those we determined for both the Haast (2.6×10^3) and the Lanyang Hsi (2.1×10^3), suggesting that biotite is an important source of K^{+} in both watersheds, even though there is a much more significant source of Mg^{2+} in the Lanyang Hsi. The K:Rb ratio in the Waipaoa is 4×10^3, suggesting another possible source besides biotite weathering of K^{+} to these waters.

The highest dissolved U value, 2.3 nmol L^{-1}, was observed in the Waipaoa River (Table 2). This value is above the global mean value of 1.3 nmol kg^{-1} for rivers. The two rivers draining crystalline bedrock have values below the global mean (Table 2). This fits with the previous work that the highest riverine U concentrations are found in rivers draining sedimentary rock systems, especially carbonates and black shales (Palmer and Edmond, 1993). Although the high U in the Waipaoa is associated with the highest alkalinity of the three rivers, the Waipaoa also has the lowest Ca:Na, suggesting that the U may not be derived from the dissolution of $CaCO_3$ alone. The Waipaoa River also has the highest Sr and Ba values, indicating the U source may be a trace authigenic mineral phase.

Climatic Influences

The mean annual temperatures of these three watersheds vary dramatically—ranging from 11 °C in the Haast to 18 °C in the Lanyang Hsi. The high precipitation rates, especially in the higher elevations of the Southern Alps and on the east coast of Taiwan, lead to very high physical erosion yields (Milliman and Syvitski, 1992; Hicks et al., 1996), a process which exposes fresh mineral surfaces to chemical weathering. These weathering-limited systems provide very short contact times for soils and floodplain sediments to react. Selective dissolution of the most soluble mineral phases should occur under these circumstances, leading to low TDS waters that are greatly undersaturated with respect to most minerals. Minerals such as calcite and biotite, as described above, should weather faster than many of the other primary minerals in silicate rocks (Blum et al., 1994). In addition, the higher elevations in the Haast River watershed have undergone glaciation in the recent past, and glaciated landscapes are known to produce higher yields of Ca^{2+} and K^{+}, and lower Si yields, than most nonglaciated landscapes (Anderson et al., 1997). The relatively higher amounts of Ca^{2+} and K^{+} in the Haast River probably reflect not only the abundance of calcite and biotite in the bedrock but also the effect of glaciation on production of fresh mineral surfaces.

The annual rainfall in the East Cape region of North Island, New Zealand, is not as high as in the other two drainages, but the nature of the lithology along with the rapid uplift rates are the primary factors controlling the physical erosion rates (Hicks et al., 1996). The poorly consolidated sediments there result in a difference of 36 in the sediment yields for this lithology compared to sediment yields from the schists of the Southern Alps for a given rainfall amount (Hicks et al., 1996).

Anthropogenic Influence

There is strong evidence that precipitation chemistry in the Lanyang Hsi watershed is greatly affect by anthropogenic inputs of nitrogen compounds and SO_4^{2-} (Lin et al., 2000). This acidic deposition may be enhancing chemical weathering rates, as the annual volume-weighted SO_4:Cl ratios in precipitation at a nearby long-term ecological research site are a factor of ~6 times greater than the marine salt ratio. The high SO_4^{2-} in the Waipaoa River water is unlikely to be due to anthropogenic input, but may actually reflect a relatively high S content of the bedrock.

As noted above, both the Lanyang Hsi and the Waipaoa watersheds have been affected by landscape change, which in turn has affected their physical erosion yields. Road construction in the Lanyang Hsi drainage and deforestation in the Waipaoa have increased the sediment fluxes in these rivers (Kao and Liu, 2002; Hicks et al., 2000).

Sedimentary Geochemistry

The geochemistry of solid particles being eroded from a landscape provides important insights into the nature and intensity of the weathering process taking place there. In order to examine the overall composition of river and sediment geochemistry, element abundances from the riverine sediments from the three rivers investigated here were first normalized to the upper continental crust values of Taylor and McLennan (1985). Plots of these normalized ratios, the so-called spidergrams, have been extremely useful in depicting element fractionation in larger river systems (e.g., Gaillardet et al., 1999). Previous work by the group at the Université de Paris 7 has demonstrated that the spidergrams of most riverine suspended loads and sediments have a regular pattern where some elements have low variabilities, while other display significant variations (Gaillardet et al., 1999). These variations are due to the depletion and/or enrichment of elements during the chemical weathering and physical transport processes. In most cases, the alkali metals and the alkaline earths show depletions relative to the upper continental crust, because they are the most easily solubilized during the incongruent weathering of alumino-silicate minerals (Nesbitt et al., 1980; Gaillardet et al., 1999). The extent of the fractionation of these elements depends on weathering intensity (i.e., climate related) and the residence time of the particles both in the soil zone and in the riverine floodplain. Indices of weathering intensity have been determined using the degree of fraction of the alkali metals and the alkaline earths as compared to other elements within the sediments that are more resistant to weathering, such as Al, Ti, Th, or the rare earth elements (Gaillardet et al., 1999).

The riverine sediments from the three watersheds normalized to upper continental crust are shown in Figure 4. In general, the soluble alkali metals and alkaline earths (Na, K, Rb, Ca, Mg, and Ba) are depleted relative to the upper continental crust. The exceptions to this generalization are Rb for the Lanyang Hsi and Ca for the Waipaoa (Fig. 4). Some of the major rivers investigated by Gaillardet et al. (1999) also have positive (i.e., >1) Ca and Rb fractionations, so these patterns are not unusual. The elements that repeatedly demonstrate negative fractionations in the larger rivers (i.e., Na, K, Zr) also show negative fractionations (<1) in our samples (Fig. 4). The Haast and the Lanyang Hsi samples have either positive or no fractionation for many of the elements most resistant to chemical erosion. These include Th, the rare earth elements (REEs), Al, Ti, and Fe. On the other hand, the Waipaoa sediments, with the exceptions of Ca, Si, Cu, and Lu, are depleted relative to the upper continental crust in all of the elements. This almost total elemental depletion in the Waipaoa sediments probably reflects the fact that they represent marine sediments that have recently undergone a previous weathering cycle, prior to the current chemical weathering regime. The previous weathering of this material during the Miocene may have left it geochemically depleted, so that its composition was not equivalent to upper continental crust.

The intensity of the weathering regime should be reflected in the peaks and valleys of the spidergrams. For example, for the Congo, Amazon, and Niger Rivers, all tropical systems where intense chemical weathering occurs, the upper continental crust–normalized Na values approach 0.1 (Gaillardet et al., 1999), while the loss of Na in the New Zealand and Taiwan sediments is much less extensive than in the tropical rivers (Fig. 4). In the Waipaoa and the Lanyang Hsi, greater loss of upper

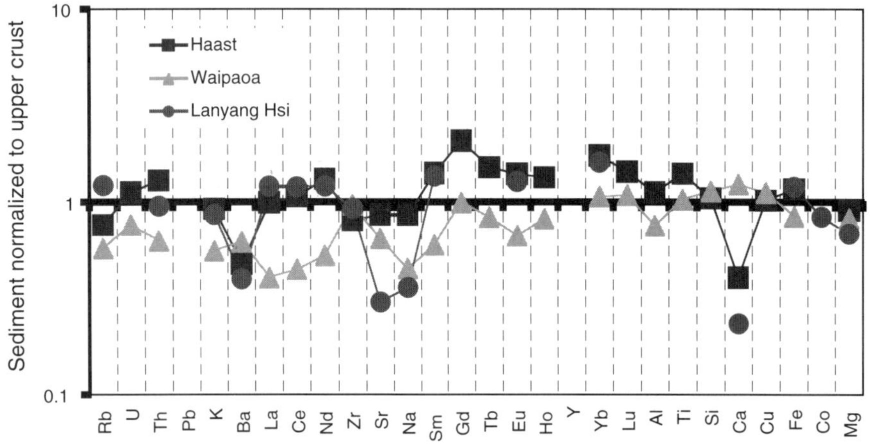

Figure 4. Plot of fine-grained channel sediments from the Haast River (squares), Waipaoa River (triangles) and the Lanyang Hsi (circles) normalized to the upper continental crust (UCC) values of Taylor and McLennan (1985). Elements plotting above the line are enriched relative to UCC, while elements plotting below the line are depleted relative to UCC.

continental crust–normalized Na compared to the Haast sediments may be related not only to the higher mean annual temperatures, and hence weathering intensities in these watersheds, but also to the possible longer residence times of the sediments within the catchments.

Yet because the source of sediments to these rivers is recycled continental materials (i.e., either marine sediments or metamorphosed marine sediment), there is a strong likelihood that upper continental crust may not appropriately represent the bulk chemistry of the weathering bedrock. The more traditional approach of Nesbitt and Young (1984) can also be used to evaluate the impact that the recycling of these sediments has had on the potential soluble products that can be generated from the chemical erosion of these rocks. We have used the chemical index of alteration (CIA*) as modified by Colin et al. (1999), where

$$CIA^* = \left[\frac{Al_2O_3}{Al_2O_3 + Na_2O + K_2O} \right] \times 100 \ .$$

As rocks weather, they lose Na and K relative to Al, and the CIA value increases with weathering. The upper continental crust yields a CIA* value of 67.6%. Slates and argillites found south and west of the Lanyang Hsi watershed, rocks with compositions that closely approximate the bedrock, have values of 85% (Chen et al., 1990). Phyllite from central Taiwan has a CIA* value of 80% (Chen et al., 1990). Our own analyses of five samples of the Haast schist yield a CIA* of 73.8 ± 6.2%, while three samples of unweathered Miocene mudstone/sandstone collected near the Waipaoa drainage have a mean CIA* value of 78.1%. Clearly, the rocks being weathered in each of these watersheds have been previously weathered and have lost a portion of their most soluble cations prior to the current weathering episode. Haast river sediments have a CIA* of 71.1 ± 0.8%, while the suspended sediments from the Haast had a CIA* of 74.8 ± 0.2%. The Waipaoa river sediment had a CIA* of 72.6%, and the mean CIA* from Waipaoa suspended load (n = 6) was 81.8 ± 0.9%. Lanyang Hsi sediment yielded a CIA* of 78.5%, a value very similar to the source bedrock. Therefore, the river sediment and suspended load values of CIA* for these systems are essentially the same as their bedrock source, suggesting rapid production and transport of these particles from landscape to the ocean.

Finally, we have also calculated the $(La/Yb)_N$ values for the sediments from these three rivers in order to evaluate the impact of weathering and erosion on the rare earth element (REE) chemistry (Table 3). The Haast River sediments have a $(La/Yb)_N$ value similar to that of the global average riverine suspended matter, and between those of the deep-sea clay and the North American shale composite (NASC) (Table 3), indicating also that the REE geochemistry of the Haast schist has been little affected by the current weathering episode. The $(La/Yb)_N$ value for the Lanyang Hsi is slightly more heavy REE–enriched than the deep-sea clay value, while the Waipaoa sediment value is depleted in light REE. These two values, outside the range of world river, deep-sea clay, and NASC, might also suggest that these rocks have undergone more intense chemical weather than the Haast sediments have. All the data presented in this section indicate that the bulk chemistry values of the solids being trans-

TABLE 3. La AND Yb NORMALIZED TO CHONDRITE VALUES [I.E., $(La/Yb)_N$] FOR HAAST, WAIPAOA, AND LANYANG HSI RIVER SEDIMENTS

	$(La/Yb)_N$	Data source
Haast River sediments (n = 4)	8.2	This study
Waipaoa River sediments	5.5	This study
Lanyang Hsi sediments	10.8	This study
Global suspended load mean	8.7	Martin and Whitfield (1983)
North American shale composite (NASC)	7.0	Taylor and McLennan (1985)
Deep-sea clay	10.1	Chester (2003)

ported in these systems are not very different from that of the bedrock material. These data strongly support the notion that production, supply, and transport of the physical weathering products from these watersheds is rather rapid, allowing for little chemical change of the weathered solids.

Comparison of Suspended Load to River Sediment Geochemistry

In the cases of the Haast and the Waipaoa, we have analyzed both riverine sediments and suspended load samples collected directly from the water column of the rivers. The suspended loads are enriched in Al and Fe but depleted in Si compared to the channel sediments (Fig. 5). A similar relationship has been observed between suspended load and bedload in the Ganga-Brahmaputra Rivers (Galy and France-Lanord, 2001). This relationship has been attributed to particle sorting of minerals during transport, as clays are separated from the coarse fraction, which is enriched in quartz (Galy and France-Lanord, 2001). The sediments from the Waipaoa are more depleted than either those of the Haast or the Lanyang Hsi, suggesting that the bedrock source also greatly influences the particles. The majority of both the sediment and the suspended load values are enriched in Fe and Al relative to upper continental crust. Because of the very high sediment fluxes from these watersheds, the residence times of particles in the soils and floodplains of these rivers are thought to be very short (Hicks et al., 1996, 2000; Kao and Liu, 2001).

Denudation Rates

The physical erosion yields for these rivers have been previously determined (Hicks et al., 1996, 2000; Kao and Liu, 2001). We have recently calculated the chemical erosion yields for the two New Zealand systems (Lyons et al., 2005). All these data are shown in Table 4. In addition, we estimate the chemical erosion yield of the Lanyang Hsi using our data presented herein, and the flow data from Kao and Liu (2000). Clearly physical erosion dominates the mass loss in all of these watersheds, with the largest chemical erosion yield percentage from the Haast at 5% of the total loss (Table 4). Even though these chemical erosion yields are a small portion of the total loss, these yields are very high and demonstrate that chemical weathering is an important process in these locations, especially compared to other types of landscapes globally. The warmer climate in Taiwan may explain the higher chemical erosion rates there, compared to the Haast, even though there are similar rock types in both watersheds and the rainfall is higher in the Haast watershed. These chemical erosion yields are noticeably higher than in larger, continental-scale rivers like the Congo, and are similar to those draining continental mountainous regions such as the Himalaya (Table 4).

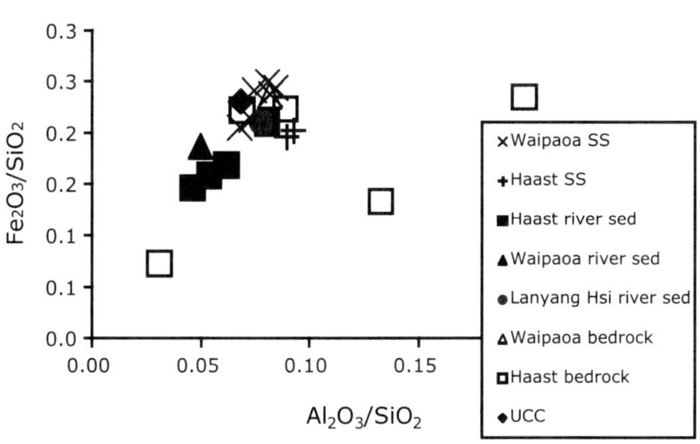

Figure 5. Plot of Fe_2O_3/SiO_2 versus Al_2O_3/SiO_2 in suspended matter (SS), riverine sediments, and bedrock for the Haast and Waipaoa Rivers. Suspended matter is depleted in quartz compared to the bed sediments and source rock, as indicated by higher ratios for the suspended matter. Upper continental crust (UCC) is shown for comparison and is indicated as a filled diamond.

CONCLUSIONS

In this paper, we have presented data on the geochemistry of river water and river sediments from three regions of rapid tectonic uplift where physical erosion yields are some of the highest in the world. Our data suggest that although the absolute chemical erosion rates are high, the physical erosion yields are still dominating the loss of mass in these watersheds. The most soluble elements, alkali metals and alkaline earths, are rapidly lost from the aluminosilicate minerals in the bedrock. The dissolution of $CaCO_3$, even though it is a minor mineral phase, appears to be a very important process in all three watersheds, but among these watersheds, is most important in the Southern Alps of New Zealand. This work suggests that chemical weathering and chemical erosion yields are substantial, even though these bedrocks have previously undergone an erosional cycle. Spidergrams indicate that the chemical erosion patterns in these three small watersheds are similar to ones from larger, continental-scale rivers. A comparison of the suspended load geochemistry and the river sediments geochemistry indicates that the sediment is more intensely weathered. This in turn suggests that the residence time of soil and floodplain sediment in the watershed is a critical factor in the extent of chemical weathering and erosion that occurs in rapidly uplifting water-

TABLE 4. PHYSICAL AND CHEMICAL EROSION YIELDS FROM THE HAAST, WAIPAOA, AND LANYANG HSI WATERSHEDS

	Physical ($t\ km^{-2}\ yr^{-1}$)	Chemical ($t\ km^{-2}\ yr^{-1}$)	Chemical (%)
Haast	4500	218	5
Waipaoa	11,540	163	1
Lanyang Hsi	8160	274	3
Congo[†]	5	8	62
Brahmaputra[‡]	3450	289	8
Ganga[‡]	2500	245	9

[†]Data from Gaillardet et al. (1999).
[‡]Data from Galy and France-Lanord (2001).

sheds. All the data presented herein point to the conclusion that physical erosion strongly enhances chemical erosion and that tectonics is an important consideration in long-term chemical weathering rates. In addition, this work demonstrates that more emphasis should be placed on high-standing, tectonically active landscapes, such as island arcs and oceanic-continental plate boundaries, in order to constrain better the loci of CO_2 uptake during chemical weathering.

ACKNOWLEDGMENTS

The authors acknowledge John Porteous of the National Institute of Water and Atmospheric Research (NIWA) for his efforts in deployment and servicing of the automatic sampler on the Haast River. Hannah Brackley from Landcare Research, New Zealand, provided the six Waipaoa River suspended load samples, and we thank her for that. We thank E.Y. Graham for the analysis of the New Zealand solute samples, and K. Neumann for helping to collect the New Zealand samples. Carey and Lyons would like to thank National Taiwan University and Academia Sinica, Taiwan, for their kind and generous hospitality during a visit to Taiwan. We are extremely grateful to Michael Bickle and Lou Derry for their thoughtful and helpful reviews. Their insights greatly improved the paper. Partial funding for this study was provided by the National Science Foundation grant EAR-9725543.

TABLE A1. ANALYTICAL DATA FOR WATER SAMPLES

River	Sample number and location	Date	Discharge ($m^3 s^{-1}$)	Na	K	Ca	Mg (μmol/kg)	Cl	Si	NO_3	SO_4
Haast	2.1 at Highway 6 bridge	2-Apr-99	361	56	32	220	23	26	61	0.0	37
Haast	2.2 Between Thomas River and Roaring Billy	2-Apr-99	361	48	22	254	25	18	54	1.2	43
Haast	2.3 At Roaring Billy	2-Apr-99	361	215	30	320	54	102	80	1.0	34
Haast	2.4 Just upstream of the confluence of the Landsborough River	2-Apr-99	361	36	13	136	14	14	45	0.0	26
Haast	NZ 3G At Roaring Billy	21-Jul-99	67.2	60		343	38	31	68	0	53
Haast	NZ 3I At Roaring Billy	21-Sep-99	116.3	51	44	301	35	31	59	0	18
Haast	NZ 3J At Roaring Billy	17-Nov-99	1889	29	43	134	14	31	35	0	52
Waipaoa	Te Karaka gauge	Mar-00	32	1282	80	1485	266	266	187	0	911
Lanyang Hsi	At gauge	Dec-00		329	44	1000	798	36	140	21	824

TABLE A2. ANALYTICAL DATA FOR SOLID SAMPLES

Sample name	Si (g/kg)	Ti (g/kg)	Al (g/kg)	Fe (g/kg)	Mn (g/kg)	Mg (g/kg)	Ca (g/kg)	Na (g/kg)	K (g/kg)	P (g/kg)	V (mg/kg)	Cr (mg/kg)	Ni (mg/kg)	Cu (mg/kg)	Zn (mg/kg)	Rb (mg/kg)	Sr (mg/kg)	Zr (mg/kg)	Nb (mg/kg)	Ba (mg/kg)
River sediments																				
Haast 1	335	3.6	78.8	29.1	0.48	10.5	9.2	20.3	22.2	0.6	81	25	11	11	53	78	241	169	9	462
Haast 2	324	4.7	82.5	32.9	0.59	11.0	12.3	22.0	22.5	0.8	92	36	15	12	59	79	278	252	11	481
Haast 3	316	4.5	85.6	36.8	0.61	13.2	13.2	23.7	21.7	0.8	101	39	17	15	63	71	259	158	11	474
Haast 4	295	4.6	99.2	44.8	0.76	14.3	13.0	21.6	28.1	0.9	116	59	23	19	86	105	308	194	9	616
Waipaoa	347	3.2	58.5	26.1	0.36	11.1	36.4	11.6	14.4	0.5	69	88	30	16	68	63	204	232	4	665
Suspended load																				
Haast 3/26–33	290	5.4	91.3	49.0	0.8	19.7	15.5	20.5	34.1	0.9	133	73	32	25	93	128	295	234	11	661
Haast 3/34–41	287	5.5	93.2	50.2	0.8	20.3	15.5	19.4	34.9	0.9	136	70	31	25	93	126	295	236	12	663

	Y (mg/kg)	La (mg/kg)	Ce (mg/kg)	Pr (mg/kg)	Nd (mg/kg)	Sm (mg/kg)	Eu (mg/kg)	Gd (mg/kg)	Tb (mg/kg)	Dy (mg/kg)	Ho (mg/kg)	Er (mg/kg)	Tm (mg/kg)	Yb (mg/kg)	Lu (mg/kg)	Th (mg/kg)	U (mg/kg)
River sediments																	
Haast 1	16	28.8	62.1	6.7	29.7	5.4	1.1	4.9	0.6	3.9	0.69	2.3	0.3	2.2	0.26	12.4	2.7
Haast 2	20	40.7	87.3	9.3	40.6	7.3	1.48	6.4	0.8	5.1	0.89	3	0.4	2.9	0.36	14.5	3.2
Haast 3	18	26.4	58.2	6.5	30.1	5.8	1.3	5.5	0.7	4.7	0.79	2.6	0.3	2.5	0.3	12.3	2.4
Haast 4	21	31.3	69.8	7.8	36	6.8	1.37	6.4	0.8	5.3	0.94	3	0.4	2.8	0.4	12.5	2.9
Waipaoa	13	13	29.1	3	13.7	2.7	0.63	2.8	0.4	2.5	0.51	1.6	0.2	1.6	0.25	6.3	1.9
Suspended load																	
Haast 3/1, ashed[†]	26	30.0	64.2	7.3	30.5	6.2	1.36	5.3	0.7	4.8	1.02	2.8	0.4	2.3	0.3	14.4	2.5
Haast 3/4–9, ashed	27	33.0	73.0	8.0	33.0	6.3	1.42	5.8	0.8	5.1	1.09	2.8	0.4	2.7	0.4	15.1	2.6
Haast 3/18–25, ashed	23	25.4	56.1	6.2	25.5	5.2	1.25	4.6	0.6	4.1	0.94	2.4	0.4	2.3	0.3	16.0	2.4
Haast 3/26–33, ashed	26	32.1	70.5	7.8	32.6	6.6	1.50	5.4	0.8	5.1	1.01	2.7	0.3	2.6	0.4	16.7	2.6
Haast 3/34–41	22	25.1	55.7	6.2	26.2	5.3	1.18	4.6	0.6	4.0	0.85	2.2	0.3	2.0	0.3	12.5	2.2

[†]Ashed samples were combusted at 550 °C for 2 hours before pressing pellets for X-ray fluorescence spectrometry analysis.

REFERENCES CITED

Anderson, S.P., and Dietrich, W.E., 2001, Chemical weathering and runoff chemistry in a steep headwater catchment: Hydrological Processes, v. 15, p. 1791–1815, doi: 10.1002/hyp.240.

Anderson, S.P., Drever, J.I., and Humphrey, N.F., 1997, Chemical weathering in glacial environments: Geology, v. 25, p. 399–402, doi: 10.1130/0091-7613(1997)025 <0399:CWIGE>2.3.CO;2.

Anderson, S.P., Dietrich, W.E., and Brimhall, G.H., Jr., 2002, Weathering profiles, mass-balance analysis, and rates of solute loss: Linkages between weathering and erosion in a small, steep catchment: Geological Society of America Bulletin, v. 114, p. 1143–1158, doi: 10.1130/0016-7606(2002)114<1143:WPMBAA>2.0.CO;2.

Basher, L.R., Tonkin, P.J., and McSaveney, M.J., 1988, Geomorphic history of a rapidly uplifting area in a compressional plate boundary: Cropp River, New Zealand: Zeitschrift für Geomorphologie Neue Folge: Supplementbände, v. 69, p. 117–131.

Batt, G.E., Braun, J., Kohn, B.P., and McDougall, I., 2000, Thermochronological analysis of the dynamics of the Southern Alps, New Zealand: Geological Society of America Bulletin, v. 112, p. 250–266, doi: 10.1130/0016-7606(2000)112<250:TAOTDO>2.3.CO;2.

Beaucaire, C., and Michaud, G., 1982, Origin of dissolved minor elements (Li, Rb, Sr, Ba) in superficial waters in a granitic area: Geochemical Journal, v. 16, p. 247–248.

Berryman, K., Marden, M., Eden, D., Mazengarb, C., Ota, Y., and Moriya, I., 2000, Tectonic and paleoclimatic significance of Quaternary river terraces of the Waipaoa River, east coast, North Island, New Zealand: New Zealand Journal of Geology and Geophysics, v. 43, p. 229–245.

Blum, J.D., Erel, Y., and Brown, K., 1994, $^{87}Sr/^{86}Sr$ ratios of Sierra Nevada stream waters: Implications for relative mineral weathering rates: Geochimica et Cosmochimica Acta, v. 58, p. 5019–5025.

Bruland, K.W., 1983, Trace elements in sea-water, *in* Riley, J.P., and Chester, R., eds., Chemical oceanography: London, Academic Press, v. 8, p. 158–220.

Campbell, D., and Church, M., 2003, Reconnaissance sediment budgets for Lynn Valley, British Columbia: Holocene and contemporary time scales: Canadian Journal of Earth Sciences, v. 40, p. 701–713, doi: 10.1139/e03-012.

Chen, C.-H., Jahn, B.-M., Lee, T., Chen, C.-H., and Cornichet, J., 1990, Sm-Nd isotopic geochemistry of sediments from Taiwan and implications for the tectonic evolution of southeast China: Chemical Geology, v. 88, p. 317–332, doi: 10.1016/0009-2541(90)90096-P.

Chester, R., 2003, Marine geochemistry, 2nd edition: Oxford, Blackwell Science Ltd., 506 p.

Close, M.E., and Davies-Colley, R.J., 1990, Baseflow water chemistry in New Zealand rivers 1: Characterisation: New Zealand Journal of Marine and Freshwater Research, v. 24, p. 319–341.

Colin, C., Turpin, L., Bertaux, J., Desprairies, A., and Kissel, C., 1999, Erosional history of the Himalayan and Burman ranges during the last two glacial-interglacial cycles: Earth and Planetary Science Letters, v. 171, p. 647–660, doi: 10.1016/S0012-821X(99)00184-3.

Finlayson, D.P., Montgomery, D.R., and Hallet, B., 2002, Spatial coincidence of rapid inferred erosion with young metamorphic massifs in the Himalayas: Geology, v. 30, p. 219–222, doi: 10.1130/0091-7613(2002)030<0219:SCORIE>2.0.CO;2.

Fuller, C.W., Willett, S.D., Hovius, N., and Slingerland, R., 2003, Erosion rates for Taiwan mountain basins: New determinations from suspended sediment records and a stochastic model of their temporal variation: Journal of Geology, v. 111, p. 71–87, doi: 10.1086/344665.

Gaillardet, J., Dupré, B., and Allègre, C.J., 1999, Geochemistry of large river suspended sediments: Silicate weathering or recycling tracer?: Geochimica et Cosmochimica Acta, v. 63, p. 4037–4051, doi: 10.1016/S0016-7037(99)00307-5.

Galy, A., and France-Lanord, C., 2001, Higher erosion rates in the Himalaya: Geochemical constraints on riverine fluxes: Geology, v. 29, p. 23–26, doi: 10.1130/0091-7613(2001)029 <0023:HERITH>2.0.CO;2.

Gomez, B., Eden, D.N., Hicks, D.M., Trustrum, N.A., Peacock, D.H., and Wilmshurst, J., 1999, Contribution of floodplain sequestration to the sediment budget of the Waipaoa River, New Zealand, *in* Marriott, S.B., and Alexander, J., eds., Floodplains: Interdisciplinary approaches: Geological Society [London] Special Publication 163, p. 69–88.

Gomez, B., Rosser, B.J., Peacock, D.H., Hicks, D.M., and Palmer, J.A., 2001, Downstream fining in a rapidly aggrading gravel bed river: Water Resources Research, v. 37, p. 1813–1823, doi: 10.1029/2001WR900007.

Hicks, D.M., 1999, A regionally applicable suspended sediment rating for the schist terrain of Otago: Journal of Hydrology, New Zealand, v. 38, p. 77–96.

Hicks, D.M., and Griffiths, G.A., 1992, Sediment load, *in* Mosley, M.P., ed., Waters of New Zealand: New Zealand Hydrological Society, p. 229–248.

Hicks, D.M., Hill, J., and Shankar, U., 1996, Variation of suspended sediment yields around New Zealand: The relative importance of rainfall and geology: International Association of Hydrogeological Sciences Publication 236, p. 149–156.

Hicks, D.M., Gomez, B., and Trustrum, N.A., 2000, Erosion thresholds and suspended sediment yields, Waipaoa River basin, New Zealand: Water Resources Research, v. 36, p. 1129–1142, doi: 10.1029/1999WR900340.

Hovius, N., Stark, C.P., and Allen, P.A., 1997, Sediment flux from a mountain belt derived by landslide mapping: Geology, v. 25, p. 231–234, doi: 10.1130/0091-7613(1997)025 <0231:SFFAMB>2.3.CO;2.

Hovius, N., Stark, C.P., Chu, H.-T., and Lin, J.-C., 2000, Supply and removal of sediment in a landslide-dominated mountain belt: Central Range, Taiwan: Journal of Geology, v. 108, p. 73–89, doi: 10.1086/314387.

Hu, M.-H., Edmond, J.M., and Stallard, R.F., 1982, Major ion chemistry of some large Chinese rivers: Nature, v. 298, p. 550–553.

Huh, Y., Panteleyev, G., Babich, D., Zaitsev, A., and Edmond, J.M., 1998, The fluvial geochemistry of the rivers of Eastern Siberia: II. Tributaries of the Lena, Omoloy, Yana, Indigirka, Kolyma and Anadyr draining the collisional/accretionary zone of the Verkhoyansk and Cherskiy ranges: Geochimica et Cosmochimica Acta, v. 62, p. 2053–2075, doi: 10.1016/S0016-7037(98)00127-6.

Jacobson, A.D., and Blum, J.D., 2003, Relationship between mechanical erosion and atmospheric CO_2 consumption in the New Zealand Southern Alps: Geology, v. 31, p. 865–868, doi: 10.1130/G19662.1.

Jacobson, A.D., Blum, J.D., Chamberlain, C.P., Poage, M.A., and Sloan, V.F., 2002, Ca/Sr and Sr isotope systematics of a Himalayan glacial chronosequence: Carbonate versus silicate weathering rates as a function of landscape surface age: Geochimica et Cosmochimica Acta, v. 66, p. 13–27, doi: 10.1016/S0016-7037(01)00755-4.

Jacobson, A.D., Blum, J.D., Chamberlain, C.P., Craw, D., and Koons, P.O., 2003, Climatic and tectonic controls on chemical weathering in the New Zealand Southern Alps: Geochimica et Cosmochimica Acta, v. 67, p. 29–46, doi: 10.1016/S0016-7037(02)01053-0.

Kao, S.-J., and Liu, K.-K., 1996, Particulate organic carbon export from a subtropical mountain river (Lanyang Hsi) in Taiwan: Limnology and Oceanography, v. 41, p. 1749–1757.

Kao, S.-J., and Liu, K.-K., 2000, Stable carbon and nitrogen isotope systematics in a human-disturbed watershed (Lanyang-Hsi) in Taiwan and the estimation of biogenic particulate organic carbon and nitrogen fluxes: Global Biogeochemical Cycles, v. 14, p. 189–198, doi: 10.1029/1999GB900079.

Kao, S.-J., and Liu, K.-K., 2001, Estimating the suspended sediment load by using the historical hydrometric record from the Lanyang-Hsi watershed: Terrestrial, Atmospheric and Oceanic Science, v. 12, p. 401–414.

Kao, S.-J., and Liu, K.-K., 2002, Exacerbation of erosion induced by human perturbation in a typical Oceania watershed: Insight from 45 years of hydrological records from the Lanyang-Hsi River, northeastern Taiwan: Global Biogeochemical Cycles, v. 16, no. 1, doi: 10.1029/2000GB001334.

Kao, S.-J., Shiah, F.-K., and Owen, J.S., 2005, Export of dissolved inorganic nitrogen in a partially cultivated subtropical mountainous watershed in Taiwan: Water, Air, and Soil Pollution, v. 156, p. 211–228.

Kim, J.P., Hunter, K.A., and Reid, M.R., 1999, Geochemical processes affecting the major ion composition in the South Island, New Zealand: Marine & Freshwater Research, v. 50, p. 699–709.

Li, Y.-H., 1976, Denudation of Taiwan Island since the Pliocene Epoch: Geology, v. 4, p. 105–107, doi: 10.1130/0091-7613(1976)4 <105:DOTIST>2.0.CO;2.

Lin, T.-C., Hamburg, S.P., King, H.-B., and Hsia, Y.-J., 2000, Throughfall patterns in a subtropical rain forest of northeastern Taiwan: Journal of Environmental Quality, v. 29, p. 1186–1193.

Louvat, P., and Allègre, C.J., 1997, Present denudation rates on the island of Réunion determined by river geochemistry: Basalt weathering and mass budget between chemical and mechanical erosions: Geochimica et Cosmochimica Acta, v. 61, p. 3645–3669, doi: 10.1016/S0016-7037(97)00180-4.

Lyons, W.B., Carey, A.E., Hicks, D.M., and Nezat, C.A., 2005, Chemical weathering in high sediment yielding watersheds, New Zealand: Journal of Geophysical Research, v. 110, F01008, doi: 10.1029/2003JF00088.

Maneux, E., Dumas, J., Clément, O., Etcheber, H., Charritton, X., Etchart, J., Veyssy, E., and Rimmelin, P., 1999, Assessment of suspended matter input into the oceans by small mountainous coastal rivers, the case of the Bay of Biscay: Compte Réndu, Academy of Science, Paris, v. 329, p. 413–420.

Martin, J.-M., and Whitfield, M., 1983, The significance of the river input of chemical elements to the ocean, *in* Wong, C.S., et al., eds., Trace metals in sea water: NATO Conference Series IV, v. 9, p. 265–296.

Mason, B., 1962, Metamorphism in the Southern Alps of New Zealand: American Museum of Natural History Bulletin, v. 123, p. 217–248.

Mertes, L.A.K., and Warrick, J.A., 2001, Measuring flood output from 110 coastal watersheds in California with field measurements and SeaWiFS: Geology, v. 29, p. 659–662, doi: 10.1130/0091-7613(2001)029 <0659:MFOFCW>2.0.CO;2.

Meybeck, M., 1987, Global chemical weathering of surficial rocks estimated from river dissolved loads: American Journal of Science, v. 287, p. 401–428.

Milliman, J.D., 1995, Sediment discharge to the ocean from small mountainous rivers: The New Guinea example: Geo-Marine Letters, v. 15, p. 127–133, doi: 10.1007/BF01204453.

Milliman, J.D., and Syvitski, J.P.M., 1992, Geomorphic/tectonic control of sediment discharge to the ocean: The importance of small mountainous rivers: Journal of Geology, v. 100, p. 525–544.

Milliman, J.D., Farnsworth, K.L., and Albertin, C.S., 1999, Flux and fate of fluvial sediments leaving large islands in the East Indies: Journal of Sea Research, v. 41, p. 97–107, doi: 10.1016/S1385-1101(98)00040-9.

Millot, R., Gaillardet, J., Dupré, B., and Allègre, C.J., 2002, The global control of silicate weathering rates and the coupling with physical erosion: New insights from rivers of the Canadian Shield: Earth and Planetary Science Letters, v. 196, p. 83–98, doi: 10.1016/S0012-821X(01)00599-4.

Mokma, D.L., Jackson, M.L., Syers, J.K., and Stevens, P.R., 1973, Mineralogy of a chronosequence of soils from greywacke and mica-schist alluvium, Westland, New Zealand: New Zealand Journal of Science, v. 16, p. 769–797.

Montgomery, D.R., and Brandon, M.T., 2002, Topographic controls on erosion rates in tectonically active mountain ranges: Earth and Planetary Science Letters, v. 201, p. 481–489, doi: 10.1016/S0012-821X(02)00725-2.

Mullin, J.B., and Riley, J.P., 1955, The spectrophotometric determination of silicate-silicon in natural waters with special reference to sea water: Analytica Chimica Acta, v. 12, p. 162–170, doi: 10.1016/S0003-2670(00)87825-3.

Nesbitt, H.W., and Young, G.M., 1984, Prediction of some weathering trends of plutonic and volcanic rocks based on thermodynamic and kinetic considerations: Geochimica et Cosmochimica Acta, v. 48, p. 1523–1534, doi: 10.1016/0016-7037(84)90408-3.

Nesbitt, H.W., Markovics, G., and Price, R.C., 1980, Chemical processes affecting alkalis and alkaline earths during continental weathering: Geochimica et Cosmochimica Acta, v. 44, p. 1659–1666, doi: 10.1016/0016-7037(80)90218-5.

Norris, R.J., Koons, P.O., and Cooper, A.F., 1990, The obliquely-convergent plate boundary in the South Island of New Zealand—Implications for ancient collision zones: Journal of Structural Geology, v. 12, p. 715–725, doi: 10.1016/0191-8141(90)90084-C.

Orpin, A.R., Carter, L., Kuehl, S.A., Trustrum, N.A., Lewis, K.B., Alexander, C.R., and Gomez, B., 2002, Deposition from very high sediment yield New Zealand rivers is captured in upper margin basins: MARGINS Newsletter, v. 9, p. 1–4.

Palmer, M.R., and Edmond, J.M., 1993, Uranium in river water: Geochimica et Cosmochimica Acta, v. 57, p. 4947–4955, doi: 10.1016/0016-7037(93)90131-F.

Pillans, B., 1986, A late Quaternary uplift map for North Island, New Zealand: Royal Society of New Zealand Bulletin, v. 24, p. 409–417.

Restrepo, J.D., and Kjerfve, B., 2000, Water discharge and sediment load from the western slopes of the Colombian Andes with focus on Rio San Juan: Journal of Geology, v. 108, p. 17–33, doi: 10.1086/314390.

Riddolls, P.M., 1987, New Zealand geology: Wellington, New Zealand Science Information Publishing Centre, Department of Scientific and Industrial Research, geological map of New Zealand, scale 1:2,000,000.

Riebe, C.W., Kirchner, J.W., Granger, D.E., and Finkel, R.C., 2001, Strong tectonic and weak climatic control of long-term chemical weathering rates: Geology, v. 29, p. 511–514, doi: 10.1130/0091-7613(2001)029 <0511:STAWCC> 2.0.CO;2.

Stallard, R.F., 1995, Tectonic, environmental and human aspects of weathering and erosion: A global review using a steady-state perspective: Annual Review of Earth and Planetary Science, v. 23, p. 11–39.

Taylor, S.R., and McLennan, S.M., 1985, The continental crust: Its composition and evolution: London, Blackwell, 312 p.

Teng, L.S., 1990, Geotectonic evolution of late Cenozoic arc-continental collision in Taiwan: Tectonophysics, v. 183, p. 57–76, doi: 10.1016/0040-1951(90)90188-E.

Welch, K.A., Lyons, W.B., Graham, E., Neumann, K., Thomas, J.M., and Mikesell, D., 1996, Determination of major element chemistry in terrestrial waters from Antarctica by ion chromatography: Journal of Chromatography A, v. 739, p. 257–263, doi: 10.1016/0021-9673(96)00044-1.

White, A.F., Bullen, T.D., Vivit, D.V., Schulz, M.S., and Clow, D.W., 1999, The role of disseminated calcite in the chemical weathering of granitoid rocks: Geochimica et Cosmochimica Acta, v. 63, p. 1939–1953, doi: 10.1016/S0016-7037(99)00082-4.

Willett, S.D., 1999, Orogeny and orography: The effects of erosion on the structure of mountain belts: Journal of Geophysical Research, v. 104, p. 28,957–28,981, doi: 10.1029/1999JB900248.

Willett, S.D., and Brandon, M.T., 2002, On steady states in mountain belts: Geology, v. 30, p. 175–178, doi: 10.1130/0091-7613(2002)030<0175:OSSIMB>2.0.CO;2.

Manuscript Accepted by the Society 23 June 2005

Geological Society of America
Special Paper 398
2006

Active tectonics in the Subandean belt inferred from the morphology of the Rio Pilcomayo (Bolivia)

Jean-Louis Mugnier[†]
Centre National de la Recherche Scientifique, and University J. Fourier, Maison des Geosciences, BP 53-38041, Grenoble, Cedex 9, France
David Becel
Didier Granjeon
Institut Français du Pétrole, BP 311-92500, Rueil Malmaison, France

ABSTRACT

Recent thrust tectonic mechanisms of southern Bolivia are inferred from the incision of the Rio Pilcomayo, the main river that crosses the thin-skinned Subandean thrust belt. The fluvial shear stress model, and more specifically its adaptation to calculate a nondimensional form of the excess critical shear stress, is used to relate channel river properties to the rate of fluvial incision into bedrock. The following parameters have been compiled to apply this model: (1) grain size of the bedload from field work; (2) drainage area from a digital elevation model; (3) slope of the river from 1:50,000 topographical maps and global positioning system measurements; (4) width of the river from field work and topographical maps; and (5) water discharge from published data. A comparison between the nondimensional shear stress (excess critical shear stress) of the Rio Pilcomayo and a structural cross section through the southern Subandean zone shows that the excess critical shear stress increases in the hanging wall of three thrust faults. These faults show field evidence of recent thrust activity, and we estimate the relative hanging-wall vertical motion for the thrusts based on the excess critical shear stress within the Tertiary sediments at their hanging wall. Our analysis indicates that the frontal structure and the third thrust at the boundary between the eastern and western zones of the Subandean belt are active, but the shortening on the second thrust (Aguarague) is more than half of the total thrust motion. We use and adaptation of the Suppe (1983) equations to estimate the variation of the uplift rate through a folded structure. The ratio between uplift rate and horizontal shortening varies from 1.1 in the western flank to 0.25 in the eastern flank of the Aguarague structure. A nondimensional index of erodibility (ratio between the excess critical shear stress and the horizontal shortening) is estimated from the relative uplift rate and the excess critical shear stress. This index of erodibility varies weakly, from 10 ± 2 for the Tertiary sandstones to 6 ± 2 for upper Devonian rocks. The excess critical shear stress reaches its maximum at the transition between the Subandean zone and Interandean zone. This high value cannot only be driven by tectonics, and a lithological effect (low erodibility) is inferred.

Keywords: Bolivia, active tectonics, Andes, fluvial morphology, incision, thrust.

[†]E-mail: mugnier@ujf-grenoble.fr.

Mugnier, J.-L., Becel, D., and Granjeon, D., 2006, Active tectonics in the Subandean belt inferred from the morphology of the Rio Pilcomayo (Bolivia), *in* Willett, S.D., Hovius, N., Brandon, M.T., and Fisher, D.M., eds., Tectonics, Climate, and Landscape Evolution: Geological Society of America Special Paper 398, Penrose Conference Series, p. 353–369, doi: 10.1130/2006.2398(22). For permission to copy, contact editing@geosociety.org.

INTRODUCTION

The structural geometry of the Bolivian Subandean belt is now well known from field, petroleum, and geophysical studies, (e.g., Giraudo and Limachi, 2001; Zubieta-Rossetti, 2001). However the few neotectonic analyses that have been performed in this area are either based on satellite image analysis (Lamb, 2000) or on a compilation of structural observations (Lavenu et al., 2000). In this paper, we use the morphology of a major river that crosses the Subandean belt (Rio Pilcomayo) to systematically detect the zones of recent uplift.

The analysis of the fluvial characteristics is based on a methodology already used on the Himalayan front (Lavé and Avouac, 2001) to estimate the incision rate from a nondimensional form of the shear stress. Structures are assumed to be fault bend folds (Suppe, 1983) in order to relate rock uplift to the amount of thrust motion and geometry at depth. The motion along the active tectonic faults in the Subandean area is deduced from a comparison of the fluvial incision and the predicted uplift pattern. At the scale of the hanging wall of a single active structure, a method is developed to estimate the contrast in erodibility between Tertiary sandstones and older (Mesozoic to Devonian) rocks.

METHOD

The method of fluvial analysis involves a comparison between tectonic uplift and the fluvial incision pattern. The basis for the estimation of the river incision, for the calculation of uplift, and for their comparison is presented below. Some specific components of the analysis are detailed in Appendices 1 and 2.

River Incision as Derived From the Geometry of the Modern Rivers

River channel geometry expresses the channel adjustment to various external factors related to climate, lithology, tectonics, and also sediment supply from surrounding hillslopes. Nevertheless, simple rationales have generally been used to derive a resultant macroscopic relationship and to relate channel river properties to the rate of fluvial incision into bedrock. Most of these relationships can be reduced to a stream power function (Howard et al., 1994) of the drainage area, A, and the local river gradient, S:

$$E = K_p A^m S^n, \tag{1}$$

where E is the incision rate and K_p is an erodibility coefficient depending on lithology, flow regime (Howard et al., 1994), and sediment supply (Sklar and Dietrich, 1998). The m and n parameters are exponents that vary with different erosional processes (Kirby and Whipple, 2001).

This stream power law is appealing because it has a minimum number of parameters, and because it can be empirically calibrated from topographic data (e.g., Stock and Montgomery, 1999). But its simplicity comes at the price of obscuring the richness of the mechanical processes causing incision (Hancock et al., 1998; Sklar and Dietrich, 1998). The physical basis for this relation is complex, and field investigations are still too scarce to definitively discriminate between the various proposed physical models (Stock and Montgomery, 1999; Snyder et al., 2000; Tomkin et al., 2003; Van der Beek and Bishop, 2003). Nevertheless, the fluvial shear stress model and its adaptation to calculate the nondimensional form of the shear stress excess (excess critical shear stress) seems a robust approach (Lavé and Avouac, 2001) that provides a physical framework to incorporate various morphological parameters within an empirically based law.

Hydrodynamics Model

It is inferred that bedrock incision primarily relates to fluvial shear stress τ (Howard et al., 1994). The fluvial shear stress τ represents the forces exerted by the flowing water on the underlying bedrock, and it follows, from a simple hydrodynamic model developed in Appendix 1:

$$\tau = \rho g \left(\frac{1}{K_s} \frac{Q}{W} \right)^{\frac{3}{5}} S^{\frac{7}{10}}, \tag{2}$$

where ρ is the fluid density, g the acceleration due to gravity, S the slope of the bottom of the river channel, Q the flow discharge, and K_s the Manning-Strickler coefficient.

The Manning-Strickler coefficient (Strickler, 1923) is defined for a variable grain size as:

$$K_s = \frac{26}{(D_{90})^{\frac{1}{6}}}, \tag{3}$$

when D_{90} is the grain diameter (meters) not exceeded by 90% of the bedload.

Bedload Motion

The empirical law developed by Lavé and Avouac (2001) infers that incision occurs because the bedload at the base of the river moves above the top of the bedrock. The critical shear stress to move a pebble of diameter D is (Shields, 1936):

$$\tau^* = \frac{\tau}{[(\rho 1 - \rho) g D]}, \tag{4}$$

where $(\rho_1 - \rho)g$ is the density contrast between water and gravel, and τ^* has a nondimensional form and is called the Shields stress when $D = D_{50}$ (Einstein, 1950), where D_{50} is the grain diameter not exceeded by 50% of the bedload.

Incision of the Bedload

Many workers (e.g., Snyder et al., 2003) have postulated that the bedrock-river incision rate is a power-law function of shear stress above a threshold. In the Siwaliks of central Nepal, a comparison of the river morphology and the incision of the neighboring uplifted terraces (Lavé and Avouac, 2000, 2001) suggests that:

$$E \approx K\,(* - \tau_b^*), \tag{5}$$

where K is an erodibility coefficient, and τ_b^* may be seen as a threshold for incipient motion, and τ_b^* can be called the critical Shields stress (Einstein, 1950). Here, $(\tau^* - \tau_b^*)$ is an adimensional form of a shear stress excess.

If a linear relationship is inferred between Q and A (with p the linear coefficient), a succession of substitutions (see Appendix 1) gives the following form for the excess critical shear stress:

$$\mathrm{E} = K\frac{\rho}{\rho_1 - \rho}\frac{1}{D_{50}}\left(\frac{D_{90}^{1/6}}{26}\frac{p}{W}\right)^{\frac{3}{5}} A^{\frac{3}{5}} S^{\frac{7}{10}} - \mathrm{K}\tau_{\mathrm{b}}^{*}. \tag{6}$$

Equation 6 only differs from equation 1 by assuming a threshold for the beginning of incision, and the excess critical shear stress can be considered as a stream power function that takes explicitly into account the width of the river and grain-size distribution. The power dependence of area and slope (m and n in equation 1) are respectively 3/5 and 7/10 in this stream power function.

Although the excess critical shear stress is based on physical equations, it cannot be considered as yielding quantitative estimates of fluvial shear stress for at least four reasons:

1. Q has a constant value in the excess critical shear stress function. For rivers characterized by great seasonal hydrological conditions, most of the bedload motion and incision occur during peak flow, and according to Lavé (1997), the decadal daily peak flow (Q_{10}) would be the reference to calculate the shear stress. Nonetheless, the concept of a reference flood event is a simplification that does not take into account the crucial role of the frequency-magnitude distribution of flooding (Turcotte and Green, 1993; Enzel et al., 1993; Molnar, 2001). When coupled with a threshold shear stress for initiation of sediment motion, erosional efficiency becomes more strongly dependent on the distribution of floods (Snyder et al., 2003). Therefore, choosing a single reference flood that has quantitative predictive capability for a real fluvial system seems highly unlikely.
2. In our analysis, we have assumed a rectangular channel, where there is no dependence between the width of the river and the flow discharge. In narrow gorges, this assumption is correct, but it is not correct in zones where the low-flow river is bounded by alluvium and are flooded every year during the seasonal peak of flow (see the paragraph concerning the width of the Rio Pilcomayo). Because peak flow corresponds to the most efficient flood in carrying bedload, we consider that the width of the active channel during the peak of flow is the width of the zone of pebbly alluvium not recovered by soils or vegetation. This method neglects the variation in the depth of the river through a transverse section and underestimates the maximum shear stress reached in the deepest parts.
3. In numerous cases, the preservation of abandoned terraces indicates that the active channel moves laterally, and the width of the valley floor is greater than the width of the active channel. It has been suggested (Brocard, 2002; Tomkin et al., 2003) that the vertical channel incision is responsible for downcutting of the entire valley floor, and the long-term rate of valley incision would be in this case slower than the rate of channel incision. For simplicity, we do not consider the width of the valley, a simplification that could lead us to overestimate the incision rate.
4. Quantitative testing of bedrock incision models has failed in most cases to successfully account for the observed geological incision rates (Tomkin et al., 2003; Van der Beek and Bishop, 2003), a result that leads us to be cautious concerning the quantitative prediction of incision models. If Lavé and Avouac (2000) successfully calibrated the excess critical shear stress model in Nepal, this calibration does not mean that the model is applicable everywhere. In this model, K is not only characteristic of the erodibility of the bedrock, but also of other parameters like the bedload supply. These hidden parameters could affect the general applicability of the model.

Uplift Rates as Derived from the Geometry of the Structures

The time scale considered in recent tectonics studies is typically in the order of a few thousands of years and incorporates several seismic cycles in zones of intense deformation (Sibson, 1989). The strain to be elastically stored during interseismic periods is released coseismically, and the succession of reversible deformation does not affect bed length conservation or area conservation. Therefore the geometric description of the structures proposed by Suppe (1983) is valid for the kinematic description of recent tectonics.

Forward kinematic modeling (Endignoux and Mugnier, 1990) has shown that when the fault trajectories and the mode of folding above these faults are known, it is possible to predict the whole kinematic field, and therefore the vertical component of displacement of the rocks close to the surface and their uplift relative to a datum fixed in the crust of the foreland.

For an increment of deformation, a description of the velocity field was proposed by Hardy and Poblet (1995) for simple geometries. For a kink-band style of deformation, the moving thrust sheet can be divided in regions of constant velocity bounded by kinematic discontinuities that are either faults or active axial surfaces (Mosar and Suppe, 1991) or layers affected by layer-parallel shearing (see Appendix 2).

In each region, the velocity vector is parallel to the underlying fault segment and the magnitude of the vertical velocity Vv_i is defined (Hardy and Poblet, 1995) by:

$$Vv_i = -S_i \sin \phi_i, \quad (7)$$

where the slip rate and the dip of this fault segment are S_i and ϕ_i, respectively.

Slip rate on a fault characterized by a complex trajectory is not constant, and the ratio Rt_i (Suppe, 1983), reduction in slip magnitude is:

$$Rt_i = S_i/So, \quad (8)$$

where So is the slip rate along the lowermost segment of the ramp.

The estimation of the slip ratio Rt for complex geometries is detailed in Appendix 2. The calculation is adapted from the works of Suppe (1983) and Suppe and Medwedeff (1990).

From a combination of equations 7 and 8, the uplift for the segment *i*, is:

$$Vv_i = Rt_i\, So \sin \phi_i. \quad (9)$$

This approach would suggest that uplift rate is estimated unequivocally from slip rate and structure geometry. Nonetheless, some cases are equivocal, and different uplift patterns can be inferred from the same geometry. This occurs when there is a difficulty in distinguishing active axial surfaces from inactive axial surfaces and from the subjective distribution of the layer-parallel shearing between the two ends of a layer in a thrust sheet.

Uplift Rate, Incision Rate, and Fluvial Steady State

The topography of an active mountain belt results from the combination of crustal deformation and erosion. Growth of topography generally coincides with increases in erosion rate. Thus with time, we might expect that erosion will become fast enough to balance rock uplift, resulting in a steady-state balance between incision and uplift. Nonetheless, tectonics and climate evolve with time, and the question is under what conditions can we expect river profile to be in equilibrium with the imposed tectonic, lithologic, and climatic setting (Whipple and Tucker, 1999)? Recent studies suggest that river incision appears to be relatively constant when averaged over time scales longer than the glacial climate cycle (~100 k.y.), but varies at shorter time scales (e.g., Pazzaglia and Brandon et al., 2001; Brocard et al., 2003).

Nevertheless, the transient response to tectonic forcing (Whipple and Tucker, 1999) or climatic forcing (Whipple et al., 1999) has been analyzed, and it is suggested that the transient stage decreases when the drainage area increases. Furthermore, geomorphic studies of the frontal active thin-skinned thrust belt of Nepal (Hurtrez et al., 1999; Kirby and Whipple, 2001; Champel et al., 2002) suggest that relief and rivers reach steady state in less than 10 k.y. The short transient stage in the Nepal fold-and-thrust belt could reflect the very rapid rates of uplift of relatively soft rocks (F. Pazzaglia, 2004, personal commun.).

In summary, we are aware that the steady-state assumption for a river profile has to be used with caution, and use it in this paper to perform a close comparison of uplift rate and fluvial characteristics (shear stress, nondimensional shear stress: excess critical shear stress). When fluvial steady state is assumed, the profile of the river remains constant through time. Because the horizontal component of motion can be neglected when channel gradient is small (Kirby and Whipple, 2001), the fluvial incision rate equals the vertical rate of motion:

$$E = Vv. \quad (10)$$

In summary, the large drainage area of the Rio Pilcomayo upstream of the Subandean belt and the rapid rate of deformation in the Subandean belt (Lamb, 2000) favor short transient stages, and we consider a steady-state balance between uplift and incision in this zone.

THE RIO PILCOMAYO THROUGH THE SUBANDEAN BELT

Geological Setting

Mountain building in the Andes is associated with subduction beneath western South America. During the Cenozoic, the central Andes grew progressively eastward and continental crust thickened by regional horizontal shortening (McQuarrie, 2002). Crustal thickening and eastward propagation of the central Andes has developed a topographic load on the South American lithosphere, producing a retroarc foreland basin system (Kley, 1996). The eastern slope of the Bolivian Andes consists of a belt of generally east-vergent fold and thrust faults in the Eastern Cordillera, Interandean zone, and Subandean zone (Fig. 1). This fold-and-thrust belt is flanked on the east by the Chaco Plain, interpreted as the aggradational surface of the foreland basin. The Eastern Cordillera exposes mostly Ordovician rocks overlain unconformably by Mesozoic rocks. The Interandean zone includes tightly folded and thrust-faulted Silurian to Tertiary rocks, but mainly consists of Devonian rocks.

The Subandean zone of Bolivia is a thin-skinned thrust belt (Roeder, 1988; Baby et al., 1992), which forms the external border of the Andean chain (Fig. 2). The geology of the thrust system is well known from seismic profiling and drilling, due to intensive petroleum exploration in the Subandean zone (Zubieta-Rossetti, 2001; Giraudo and Limachi, 2001). The thrust tectonics includes a Silurian to Mesozoic series, overlain by Neogene sediments. The basal detachment is located at the base of the Silurian rocks, and several other detachments are located in Devonian shales (Baby et al., 1992). Duplexes develop between these different detachment levels. They presumably develop as passive roof duplexes (Baby et al., 1992; Belotti et al., 1995). Nonetheless some cross sections have been balanced (Leturmy et al., 2000) by assuming forward propagation of the roof thrust.

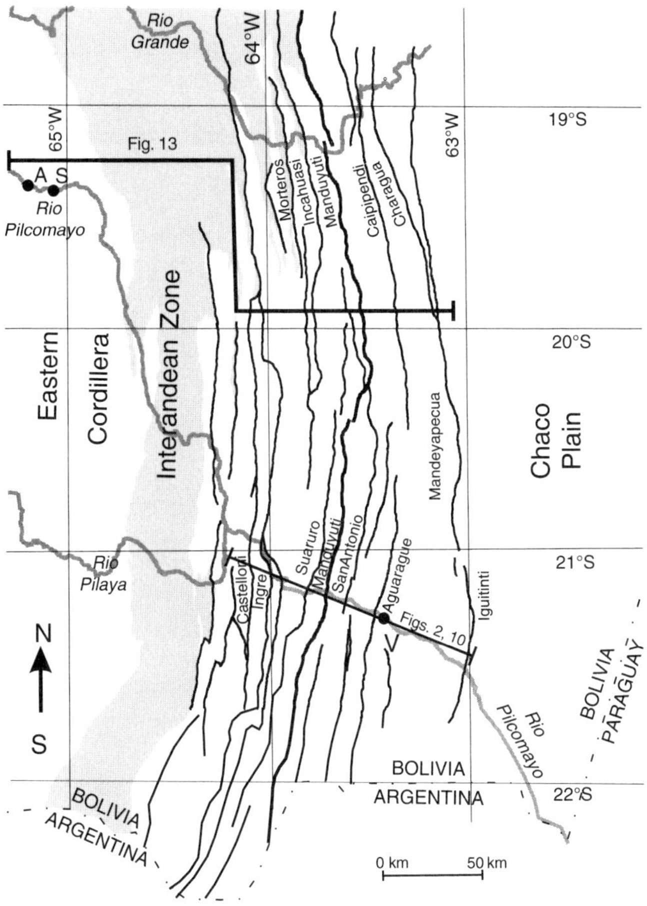

Figure 1. Structural map of the study area. A and V: gauge stations of Abapo and Villamontes (from Guyot, 1993). S refers to Sotomayor (Fig. 6). The gray zone is underlain by Silurian and Devonian rocks and generally coincides with the Interandean zone.

Thrust tectonics mainly occurred in the Eastern Cordillera from 21 Ma to 9 Ma (Gubbels et al., 1993). In contrast, the Subandean zone was relatively undeformed prior to 6 Ma (Baby, 1995), and the deformation propagated to this zone after 6 Ma. The mean shortening direction is nearly E-W in southern Bolivia (Baby, 1995; Lamb, 2000), and the shortening rate calculated from balanced cross sections is between 7 mm/yr (Leturmy et al., 2000) and 13 mm/yr (Rochat, 2003) in southern Bolivia.

Neotectonics and Uplift Rate Estimation

There are very few detailed studies of active faults in the Bolivian Andes. In the southern Subandean zone, analysis of Landsat images (Lamb, 2000) indicates highly eroded terraces at the hanging wall of the Aguarague structure and the Sayurenda anticline (Fig. 2). The erosion of these terraces is presumably driven by uplift induced by tectonic activity. For the more external structure, the seismo-tectonic map suggests that the Iguitinti fault (Fig. 2), located to the east of the Sayurenda anticline, is not active along the Pilcomayo cross section, but is only active northward (Mandeyapecua fault of Lavenu et al., 2000). The tectonic and seismo-tectonic maps of the Bolivian Andes (GEOBLO and YPFB, 1978; Sergeomin and YPFB, 1996) suggest that the Manduyuti structure, which is the major fault in the median part of the Subandean zone thrust, has been active during the Pliocene-Pleistocene (Fig. 2).

During our field work along the Rio Pilcomayo, we have found (Fig. 3) uplifted and tilted strath terraces above the Aguarague and Suaruro thrusts. The strath surfaces are typically covered by ~4 m of pebbly sediments and ~3 m of silt. They are rather undeformed, and record mainly several tens meters of uplift and a weak (less than 2%) tilting. The Suaruro fault breaches back the Manduyuti thrust, and our observations confirm the recent activity of the previously mapped active faults (GEOBLO and YPFB, 1978; Sergeomin and YPFB, 1996; Lavenu et al., 2000).

Total shortening rate has been estimated through the Subandean belt for very short periods (few yr) by GPS measurements

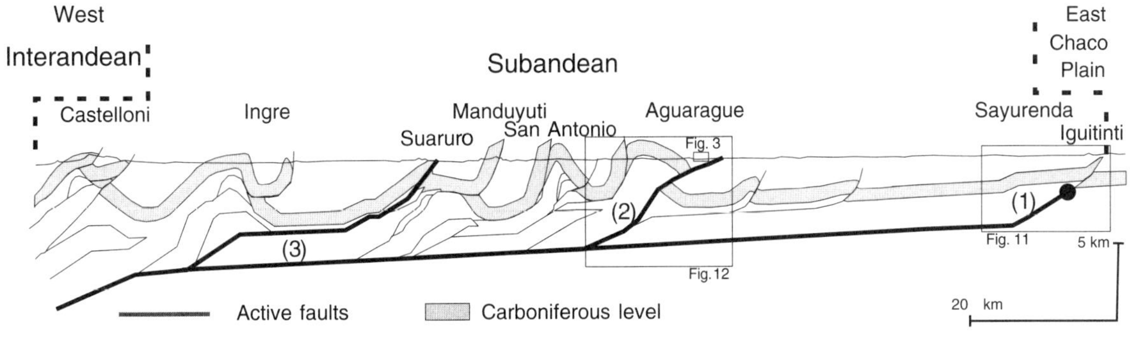

Figure 2. Structural cross section of the Subandean zone (adapted from Giraudo and Limachi, 2001). This cross section is close to the Rio Pilcomayo (location on Fig. 1). It is based on numerous drill holes and seismic profiles acquired by petroleum companies during the exploration and production of gas fields. (1), (2), (3) refer to the Sayurenda, A Aguarague, and Suaruro structures, respectively.

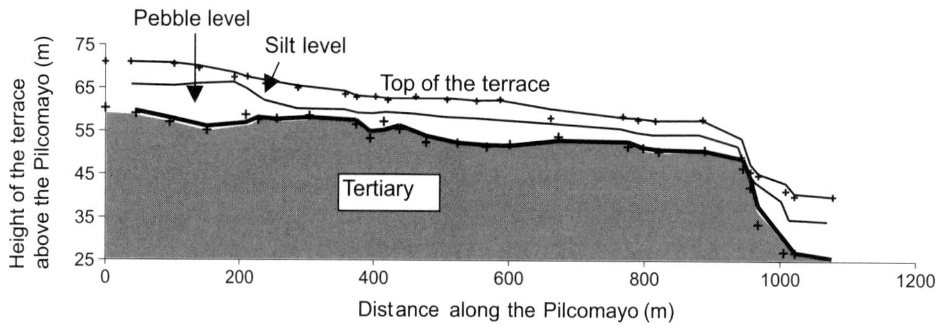

Figure 3. A profile of the uplifted strath terrace located above the eastern flank of the Aguarague structure (location on Fig. 2). Each cross refers to a measure of the height with a laser range distance meter. The strath is covered by ~4 m of pebbles and ~3 m of silt.

(~7–9 mm/yr from Kendrick et al., 1999) or for very long periods (several m.y.) by balanced cross sections (~7–13 mm/yr from Leturmy et al., 2000; Rochat, 2003). Whatever the time step considered, these methods give a shortening rate between 7 and 13 mm/yr. Therefore a constant velocity is assumed, and the shortening rate through the Subandean belt would be 10 ± 3 mm/yr for recent time.

Remnants of morphological surfaces (Kennan et al., 1995) show that the long-term incision is ~0.4 mm/yr (1200 m for ~3 m.y.) in the Eastern Cordillera. Estimates of rock uplift rate have been made by Lamb (2000) from a compilation of geological data and deformation-field modeling. It is suggested that the uplift rate is more than ten times greater in the Subandean and Interandean zones than in the Eastern Cordillera.

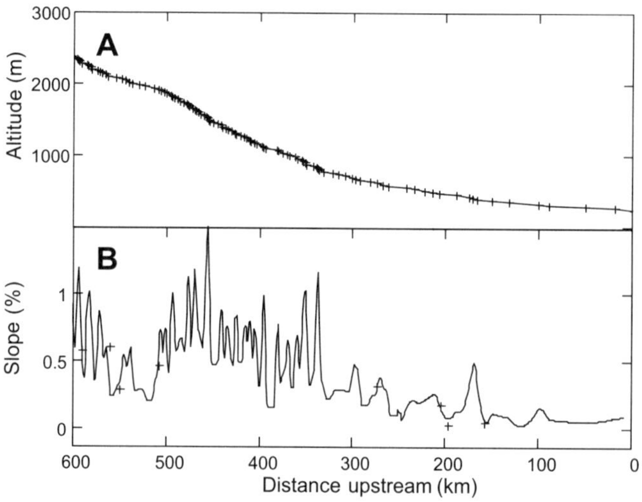

Figure 4. (A) Long profile of the Rio Pilcomayo. The *x* axis is the curvilinear length of the low-flow channel measured on the topographic maps (Instituto Geographico Militar, 1979) at a 1:50,000 scale, with an origin at the border between Bolivia and Argentina. (B) Slope versus distance upstream for the Rio Pilcomayo. The solid line refers to the gradient deduced from a 4-km-large sliding window through the linear interpolation of the profile extracted from the 1:50,000 maps, and the crosses refer to the local slope measured by global positioning system (GPS). The sliding window smoothes the transition between two constant slope values in the linear interpolation, but the maximum and minimum slope values are preserved, because the width of the sliding window is shorter than the distance between two points.

The Rio Pilcomayo

The Rio Pilcomayo is located in southern Bolivia and crosses all the structural zones located to the east of the Altiplano (Fig. 1). The data characteristics of the river are presented in the following sections.

Profile, Gradient, and Drainage Area of the Rio Pilcomayo

The river profile of the Rio Pilcomayo (Fig. 4A) has been deduced from 1:50,000 scale topographic maps with elevation contour lines every 20 m (Instituto Geographico Militar, 1979). The river slope varies from ~1% in the gorges of the Interandean zone to <0.1% in the Chaco Plain (Fig. 5).

The precision of the river slope deduced from the topographic maps has been checked through a comparison with the local gradient measured by satellite geodesy. Local gradients have been deduced from rapid-static (crosses on Fig. 4B) and cinematic global positioning system (GPS) measurements (with a 3 s time-cadenced acquisition) along 8 segments of nearly 1 km length each. These segments of the river profile are formed by a succession of pool and riff configurations (Fig. 6B) on the kilometer scale. Nonetheless, the mean slope deduced from a linear interpolation through the GPS data is close to the gradient deduced from the topographic map. In detail, this comparison suggests that the map-deduced slope gives a good estimation of the river slope when the distance between two elevation contour lines is smaller than 13 km, but smoothes the slope variations for greater distances. Because the river slope is a prime parameter that influences the fluvial shear stress (see equation 2), a multipolynomial regression has also been performed between the abscissa 100 and 210 km to estimate the river profile simultaneously from the gradients measured by GPS and from the topographic maps.

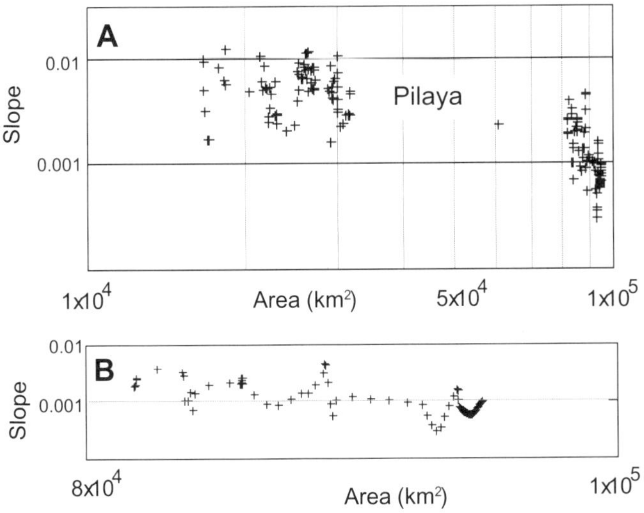

Figure 5. Slope-area plot for the Rio Pilcomayo: (A) for the whole Rio Pilcomayo basin; and (B) for the Rio Pilcomayo through the Subandean belt.

The drainage area of Rio Pilcomayo has been calculated from the digital elevation model (DEM) GTOPO 30 after elimination of local sinks with the help of software developed by Becel (2004). The total drainage area at the mountain front is 94,000 km^2 for the Rio Pilcomayo. The slope-area plot of the Rio Pilcomayo (Fig. 5A) outlines the influence of the large basin of the Pilaya, a tributary of the Pilcomayo characterized by an area close to the area of the Rio Pilcomayo itself, but with smaller flow. Although the slopes are much dispersed, two domains of different concavity (slope of linear regression of the logarithm of local channel gradient and drainage area data) are clearly evidenced: the domain of the Subandean zone and the upstream domain.

Channel Width

The width of the low-flow channel and the width of the pebbly-sandy alluvial plain (including the low-flow channel) have been measured along the Rio Pilcomayo. The measurements were done on topographic maps (Instituto Geographico Militar, 1979) at every 5 m elevation change (Fig. 7). The width of the sandy alluvial zone varies from ~2000 m in the Chaco Plain to ~25 m in the gorges of the Interandean belt.

These measurements have been compared to 12 transverse profiles of the Rio Pilcomayo studied in the field by kinematic GPS (Fig. 6A) or by laser distance-meter. We found that the misfit between the field and map-deduced measurements was less than the precision of the measurement (25 m for the map measurement and 0.05 m in the field), a result suggesting that the topographic maps are of good quality and form a robust database to estimate the width of the river.

The transverse profiles performed with kinematic GPS (Fig. 6A) show that the alluvial plain between the high-flow and low-flow boundaries is nearly horizontal (elevation changes on

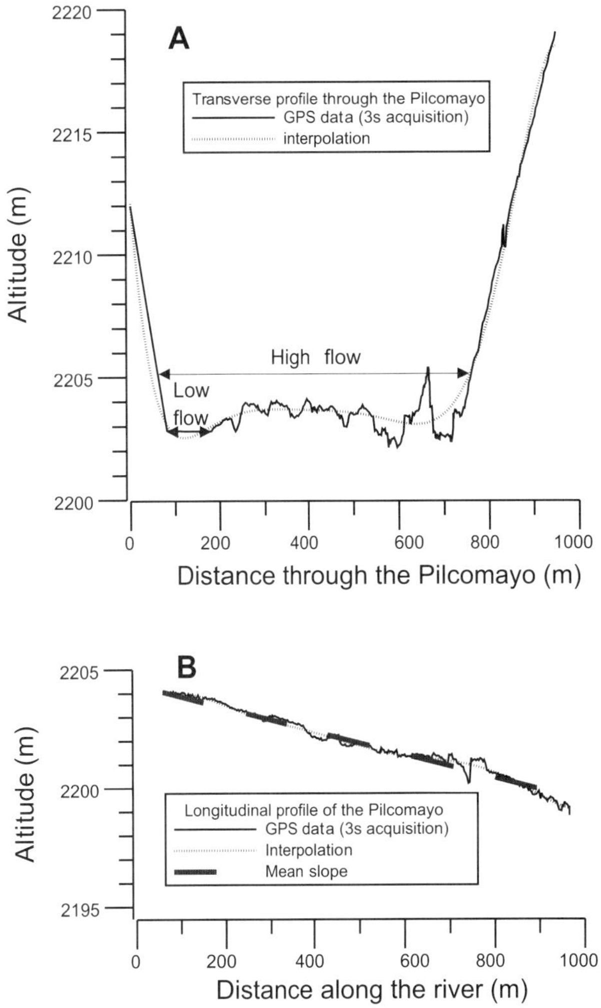

Figure 6. (A) Transverse and (B) longitudinal profiles of the Rio Pilcomayo measured with kinematics global positioning system (GPS; acquisition frequency 3 s) at Sotomayor (distance 560 km upstream on Fig. 4)

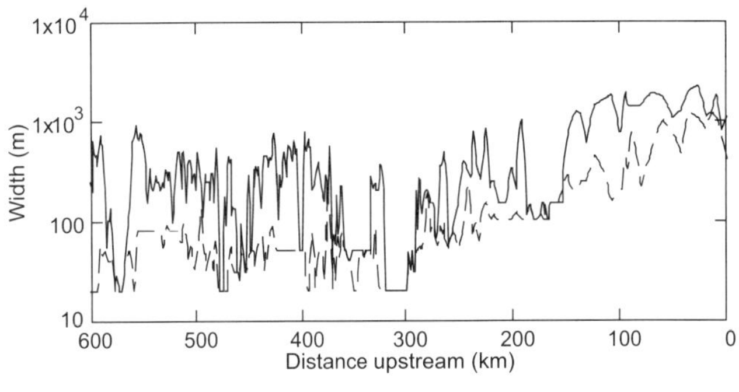

Figure 7. Width of low flow (dashed line) and active channel (continuous line). Vertical scale is logarithmic.

the order of one meter), except for locally incised channels. Therefore, for any flow greater than that restricted to the narrow channels, the width of the river is the width of the alluvial zone, and we assume that this width defines the width of the active channel.

Water Discharge

Annual precipitation measured at meteorological stations on the Rio Pilcomayo basin ranges from 800 mm/yr to 600 mm/yr (Roche et al., 1990), whereas precipitation in the basin of the large tributary of the Pilaya is only 300 mm/yr. The mean rainfall at 2 gauge stations (A. Talula and Villamontes, A and V on Fig. 1) on the river is, respectively, 480 mm/yr and 478 mm/yr (Filizola et al., 2002).

Daily estimates of the water discharge have been reconstructed for the Rio Pilcomayo at the Villamontes gauge station. The low flow has been estimated in October 2001 to be 50 m^3/s. The daily peak discharge has been estimated from the daily record for the period 1977–1982 (Guyot, 1993). The maximum daily discharge recorded during this period was 2000 m^3/s. Close to Villamontes, a catastrophic flood has been inferred from a high flooding surface. This surface is nearer and at a higher altitude than the city of Villamontes and forms a pair of planar terraces 11 m above the low-flow level of the Rio Pilcomayo. A flood of 7000 m^3/s has been estimated by using the simple hydrologic model explored in Appendix 1 and assuming that the channel has not changed since this catastrophic event. This latter hypothesis may overestimate the flow, because the width could have been smaller at the time. The silty sediment of the flooding level has been carbon dated on a large charcoal fragment (>10 cm long) found 40 cm below the surface. We obtained a radiocarbon age of 465 ± 35 yr B.P. (Sample Lyon 11316), which gives a calibrated age (Stuiver et al., 1998) of 1410–1475 A.D. This age is only suggestive, because the charcoal could be inherited, and because the last flood event may postdate the deposition of the charcoal.

Grain Size of the Bedload

We sampled fifteen sites along the Rio Pilcomayo to estimate the grain size of the bedload of the river. The sampling was performed on the pebbly bars of the river, where we dug a hole with an area of 1 m^2 and a depth of 2 times the maximum size of the pebbles. We collected 200–500 kg of material and sieved it with 1, 2, and 4 cm square mesh sieves. Material coarser than 4 cm was weighted separately. D_{90} and D_{50} were deduced graphically from the granulometric curves.

We find (Fig. 8) that the D_{90} and D_{50} slightly increase from the upper catchment in the Eastern Cordillera to the Interandean boundary, are rather stable in the Interandean zone and in the western part of the Subandean zone, and decrease rapidly in the eastern part of the Subandean zone. The mean D_{90} in the Subandean zone is 0.16 m, and the mean D_{50} is 0.06 m. This leads (from equation 3) to a Manning-Stricker coefficient close to 36.

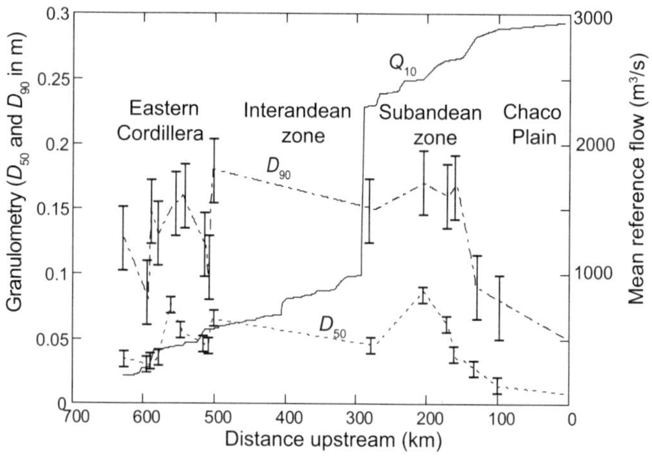

Figure 8. Grain size (dashed line is D_{50}; dotted-dashed line is D_{90}) and mean reference flow (continuous line) along the Rio Pilcomayo (distance upstream).

COMPARISON BETWEEN THE NONDIMENSIONAL SHEAR STRESS AND THE GEOLOGY

Calculation of the Nondimensional Shear Stress (Excess Critical Shear Stress)

We calculated the excess critical shear stress (equation 5) every kilometer along the river profile. For this calculation, the parameters (width, slope, grain size, drainage area) were estimated every kilometer from an interpolation of the database. The interpolation was performed using a 4 km sliding window.

The estimation of the reference flow and the critical Shields stress is not straightforward, and is presented below.

Estimation of a Reference Flow

The comparison of the gauge data of A. Talula and Villamontes suggests a linear relationship between area and discharge (Filizola et al., 2002). Nonetheless, the weak precipitation in the basin of the tributary of the Pilaya has to be taken into account. The peak flow, Q, at each point along the river profile (Fig. 8) was computed from the drainage area, A, the mean annual precipitation over the catchment, P, deduced from the precipitation map (Roche et al., 1990), and the reference flow estimated at the Villamontes gauge station (index g):

$$Q = Q_g (P A)/(P_g A_g). \tag{11}$$

In the absence of a long-term gauge monitoring at Villamontes, and because there is no theoretical basis to use the Q_{10} event as the reference flood event, we rather defined an upper and lower bound for the reference flood event Q_g: the lower bound is the highest measured peak flow during the 5 yr of record, and the upper bound is the catastrophic flow associated with the high flooding terrace of Villamontes. The difference between the upper and lower bound shear stresses (Fig. 9) illustrates the uncertainty of the characteristics of the reference flood event.

In the following, the uncertainty of the reference flow is incorporated in the calculation, but a single curve calculated with a reference flow of 3000 m^3/s is plotted for clarity of the illustration.

Estimation of the Critical Shields Stress

The value of the critical Shields stress τ_b^* has been found to range between 0.03 and 0.08 (Buffington and Montgomery, 1997) and is 0.045 for coarse sand and gravel (Shields, 1936). A value of 0.03 was used by Lavé and Avouac (2001), because it gave the best fit between their incision model and the incision deduced from terraces.

For the bedload of the Rio Pilcomayo, characterized by a D_{50} of 0.06 m and a complex distribution of grain size, τ_b^* has been empirically estimated at Villamontes (Fig. 10). In this portion of the river, field criteria suggest a transition between sedimentation and erosion, because human installations are recovered during high flow by pebbly sediments, whereas upstream they are eroded. Assuming $\tau^* - \tau_b^* = 0$, and taking into account the parameter values measured at this place, we find that τ_b^* is close to 0.03–0.04.

Evolution of the Nondimensional Shear Stress through the Subandean Belt

Qualitative Comparison

To compare the fluvial excess critical shear stress and the structures of the Subandean belt, we generated a projection of the morphological parameters of the rivers on a structural cross section. A N100° cross section was chosen for the lower part of the Rio Pilcomayo. Its location minimizes the distance of projection, and its orientation is very close to the thrust-transport direction determined by Baby (1995). The morphological parameters of the river and the intersections between the major structural features (faults, anticline and syncline axes) and the river have been projected parallel to the main structural trend of the Subandean belt (i.e., N-S). For each segment located between two major structural features, the structural cross section interpreted from seismic profiles (Giraudo and Limachi, 2001) has been horizontally anamorphosed to be superposed with the studied cross section.

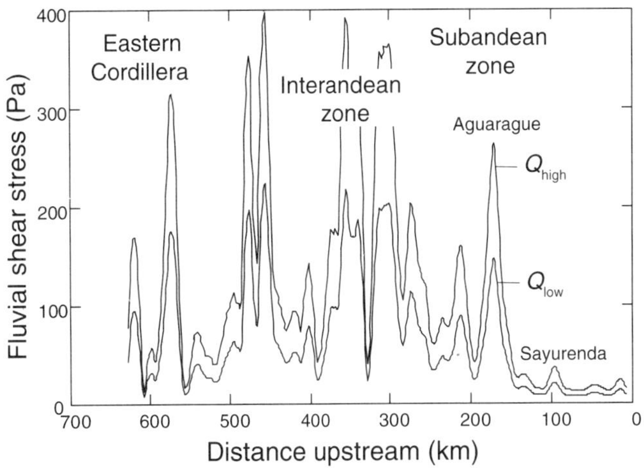

Figure 9. The fluvial shear stress along the Rio Pilcomayo (distance upstream) for the highest daily peak measured between 1977 and 1982 and a catastrophic daily peak discharge. The mean reference event is located between these two extreme limits.

This comparison (Fig. 10) outlines two types of relationships between structures and excess critical shear stress. No increase of the excess critical shear stress occurs for some anticline structures (Castelloni, Frontal Manduyuti, San Antonio, and Ingre anticlines on Fig. 2), whereas three others structures are outlined by localized zones of increased excess critical shear stress (Sayurenda, Aguarague, Suaruro). These three structures have previously (GEOBLO and YPFB, 1978; Lamb, 2000) been recognized as active.

We do not infer a major lithological control for the excess critical shear stress peaks, because the rivers cross the same

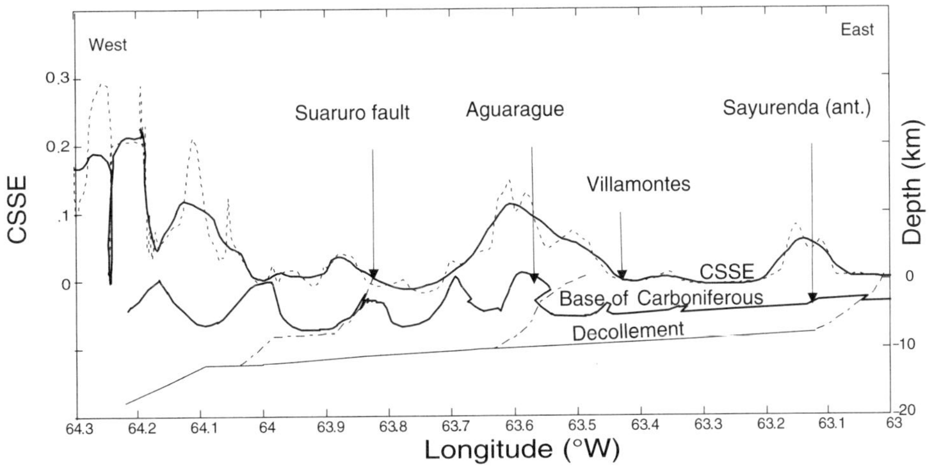

Figure 10. Projection of the Rio Pilcomayo nondimensional shear stress (CSSE) on a structural cross section (modified from cross-section E-E′ of Giraudo and Limachi, 2001).The excess critical shear stress is shown by a dashed line (calculated every km) and a thin continuous line (smoothed with a 4 km sliding window), and the excess critical shear stress scale is on the left side. The scale of the structural cross section is shown on the right side. The décollement is shown by a thin continuous line, the ramps by dashed-dotted lines, and the base of the Carboniferous level by a thick black line.

lithological succession in the two types of structure–excess critical shear stress relationships. Therefore, we infer that the peaks of excess critical shear stress along the river are mainly controlled by the distribution of the tectonic activity. Nonetheless, a close inspection of the relationship between excess critical shear stress and lithology (see the detailed analysis of the Aguaragüe structure for details) also indicates a variation in the erodibility with the lithology.

Estimation of the Shortening Rate Partitioning through the Subandean Belt

We performed a quantitative comparison of the excess critical shear stress (CSSE) and of the uplift rate at the hanging wall of the three active thrusts. To remove the influence of lithology, the comparison only involves the zones where bedrock consists of Tertiary sandstone. Assuming a constant erodibility for the Tertiary sandstones, the subscript referring to the different thrusts (Fig. 2), we find that

$$K = E_1/\text{CSSE}_1 = E_2/\text{CSSE}_2 = E_3/\text{CSSE}_3. \quad (12)$$

By substituting the uplift rate for the incision and for the uplift rate its value as a function of the structure geometry, we find:

$$So_1R_1\sin(\phi_1)/\text{CSSE}_1 = So_2R_2\sin(\phi_2)/\text{CSSE}_2 = So_3R_3\sin(\phi_3)/\text{CSSE}_3. \quad (13)$$

Assuming that the total shortening through the Subandean zone, *St*, is the sum of the shortening through the three structures ($So_1 + So_2 + So_3 = St$), the proportion of the total shortening that occurs on the different thrusts is deduced from the equation 13. We find, for example in the case of the second structure, that:

$$\frac{So_2}{St} = \left[\frac{\text{CSSE}_3}{\text{CSSE}_2}\cdot\frac{\sin\theta_2}{\sin\theta_3}\cdot\frac{R_2}{R_3} + \frac{\text{CSSE}_1}{\text{CSSE}_2}\cdot\frac{\sin\theta_2}{\sin\theta_1}\cdot\frac{R_2}{R_1} + 1\right]^{-1}. \quad (14)$$

This equation indicates that the ratio of partitioning So_i/St can be deduced from the excess critical shear stress and the geometry of the structure, and the value of the total deformation rate is not necessary to perform the estimation.

The uncertainty on the ratio of partitioning So_i/St has been estimated, taking into account the uncertainty on the reference flood value (2000 m^3 to 7000 m^3) and the uncertainty on the dip of the faults (±5°). We find that the partitioning ratio So_i/St is respectively 10 ± 8%, 67 ± 17%, and 23 ± 9% for the structures of the Iguitinti, Aguaragüe, and Suaruro.

These results indicate that the most external structure of the Subandean belt is weakly active, and more than half of the thrust motion occurs along the piggyback structure of Aguaragüe. By using the above calculated value of So_i/St and a total shortening rate *St* of 7–13 mm/yr, an absolute value of the shortening rate through the structure can be estimated. We find a shortening rate (So_i) of 1 ± 1 mm/yr, 7 ± 3.5 mm/yr, and 2.5 ± 1.5 mm/yr for the Iguitinti, Aguaragüe, and Suaruro structures, respectively.

Evolution of the Nondimensional Shear Stress through the Active Structures

The Excess Critical Shear Stress through the Easternmost Structure

Due to the poor definition of the seismic data at depth, the importance and the kinematics of the duplexes that affect the Silurian and Devonian rocks are still in discussion. For example, Labaume and Moretti (2001) suggested that the frontal structure (Fig. 11B) has developed above a passive roof duplex, whereas

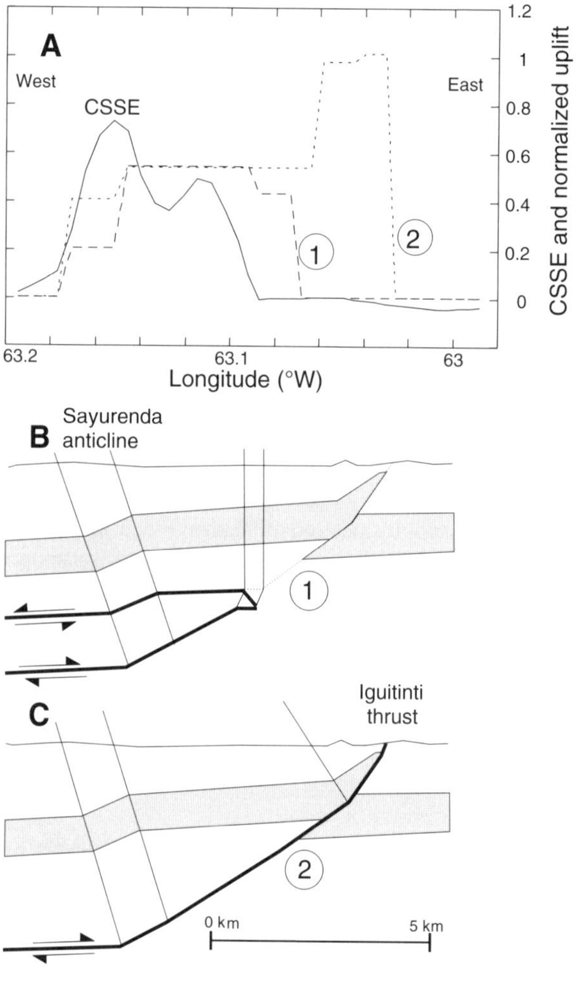

Figure 11. Uplift and nondimensional shear stress (CSSE) along the Rio Pilcomayo through the easternmost structure. (A) Comparison of the nondimensional shear stress (CSSE) and the uplift rates predicted by the two structural models (1) passive roof interpretation, and (2) simple flat-ramp geometry. The uplift rate is normalized to the slip on the flat (V_v/So) and is shown on the vertical axis. (B) Passive roof interpretation (modified from Labaume and Moretti, 2001). (C) Simple flat-ramp geometry (from Giraudo and Limachi, 2001).

Giraudo and Limachi (2001) suggested simple flat-ramp geometry (Fig. 11C). Therefore, we have tested the two structural interpretations by comparing the excess critical shear stress pattern with the uplift pattern predicted by these two structural interpretations. We consider that the influence of the lithology on the incision pattern is negligible, because the Rio Pilcomayo incises only Tertiary sediment in this frontal structure.

A passive roof duplex at depth only induces the uplift of the Sayurenda anticline, whereas a ramp would induce a peak of uplift at the hanging wall of the Iguitinti thrust. The comparison of the shape of the two uplift curves to the shape of the excess critical shear stress (Fig. 11A) shows that the passive roof duplex hypothesis is more compatible.

The Contrast of Erodibility Involved in the Aguarague Structure

We also performed a close inspection of the relationship between excess critical shear stress and lithology for the Aguarague structure, because the distance of projection is small for this portion of the river, the lithologies are mixed, and the river profile is most nearly in steady state due to its proximity to the plain.

This structure (Fig. 12B) is mainly formed by an anticline above a complex thrust. Additionally, another thrust in the western flank of the structure has been reactivated to absorb the layer-parallel shearing induced by the motion of the western syncline above the ramp (Zoetemeijer and Sassi, 1991). The uplift rate has been calculated from the horizontal shortening rate So_2 by using the geometric analysis detailed in Appendix 2. We find that the uplift rate locally exceeds the displacement rate along the ramp beneath the Aguarague structure (uplift rate/displacement rate = 1.1). This high uplift rate is the sum of the effects of the motion along the lower ramp and of the out-of-the-syncline-thrust reactivation.

The shape of the excess critical shear stress and uplift curves are similar (Fig. 12A), suggesting that excess critical shear stress is proportional to the uplift. In the absence of a direct measure of the uplift rate for this structure, and due to the high uncertainty on this estimation of So_2, we use a nondimensional index of erodibility (erodibility of the lithology/shortening rate, K_r/So_2) to estimate the influence of the lithology independently from the absolute rate of the tectonics.

Assuming $E = V$v all along the profile, and from equations 5 and 9, it is found for any segment j of the river:

$$K_r\,(\tau^*_j - \tau_b^*) = So_2\,R_j \sin\phi_j, \qquad (15)$$

where the subscript r refers to the lithology (2 for Mesozoic, 3 for the Tertiary rocks, 11 for the Carboniferous rocks, and 12 for the Devonian).

A mean value of the nondimensional index of erodibility (K_r/So_2) is calculated for each lithology:

$$K_r/So_2 = \frac{1}{n}\sum\nolimits_n^1 R_j \cdot \sin\phi_j/\,(\tau^*_j - \tau_b^*), \qquad (16)$$

where n is the number of 500-m-length river segments that cross the lithology r.

The residual R (Fig. 12A) for the mean value of the K_r/So_2 ratio has been estimated across the structure, with the following estimation of R:

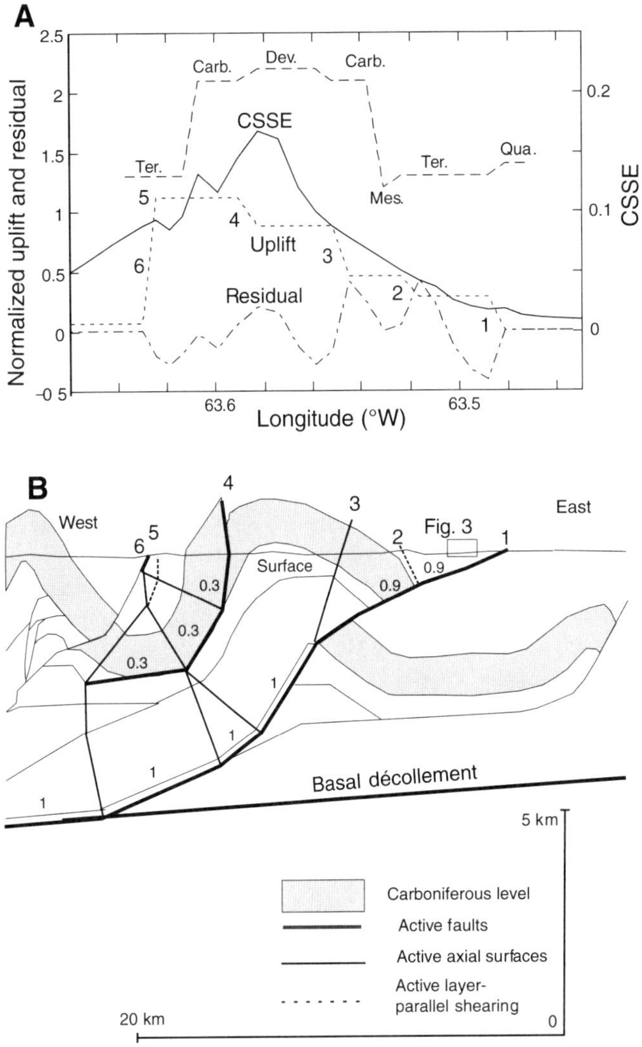

Figure 12. (A) Comparison of the uplift rate predicted by the structural model, the nondimensional shear stress, and the lithology along the Rio Pilcomayo and through the Aguarague structure. The continuous line is excess critical shear stress (CSSE; scale on the right side). The lithology is represented by a curve (long-dashed line) where every plateau is linked to a stratigraphic level. The uplift rate (short-dashed line) is normalized to the slip on the flat (V_v/So) and is shown on the left vertical axis. Residual is related to the calculation of the ratio erodibility/slip rate (see equation 17) and is shown on the left vertical axis. The italic numbers refer to the axial surface or to faults that control the location of the steps on the uplift rate curve. (B) Aguarague structural model. The boundaries of the domains of constant velocity (see method in the text) are: (1) the active fault; (2) a layer-parallel shearing; (3) an active axial surface; (4) a fault reactivated in its upper part due to layer-parallel shearing in its hanging wall; (5) a layer-parallel shearing linked to the merging of active axial surfaces; and (6) a reactivated fault. The numbers along the fault refer to the total slip ratio Rt (see text).

$$R = \frac{\dfrac{R_j \sin\phi_j}{\tau_j^* - \tau_b^*}}{\dfrac{K_r}{So_2}} - 1 \quad (17)$$

The absolute value of the residual is slightly smaller in the western part of the structure, where uplift rate and excess critical shear stress have high values. The small scale spatial fluctuations are clearly linked to the changes of lithology or to the steps in the uplift pattern distribution. This residual is always less than 0.4, and 25% of residual values are greater than 0.1. We therefore consider that the relative uncertainty for K_r is 25% at a 90% confidence level.

The result for the Mesozoic rocks is $K_2/So_2 = 7.9 \pm 2$, for the Tertiary rocks $K_3/So_2 = 9.5 \pm 2$, for the Carboniferous $K_{11}/So_2 = 8.2 \pm 2$, and for the Devonian $K_{12}/So_2 = 6.3 \pm 2$. Although the uncertainty is great, it appears that the relative erodibility of the different units varies by less than a factor 2, and Tertiary rocks are more erodible than Devonian rocks.

The value of the erodibility coefficient (K) can be estimated from the previous paragraphs concerning the estimate of St, the partitioning of the deformation So_2/St, and the nondimensional index of erodibility K/So_2. Nonetheless, the relative uncertainties accumulate in the calculation procedure and reach a total of ~75%. We find that the erodibility is 67 ± 50 mm/yr for the Tertiary, 57 ± 42 mm/yr for the Carboniferous, and 44 ± 33 mm/yr for the Devonian, and the contrast of erodibility is hidden in the uncertainty.

The Nondimensional Shear Stress through the Cordillera Oriental and Interandean Zone

In the upper part of the river, the excess critical shear stress is smaller than in the frontal part of the Interandean zone, and the excess critical shear stress is close to zero in the zones where the river flows above sediment of fill terraces, probably due to the high erodibility of the fill terraces. The highest critical shear stress excess along the Rio Pilcomayo is located in the Interandean zone, more specifically in the lower Devonian rocks at the boundary between the Subandean zone and the Interandean zone (Fig. 13B).

The hypothesis of an intense uplift in this area is consistent with the high relief characteristic of the Interandean zone (Masek et al., 1994), the slight increase of the regional topographic slope (Fig. 13A), and structural model predictions (Williams et al., 1994; Lamb, 2000). Nonetheless, if we take into account the maximum of the excess critical shear stress (0.26) and the erodibility of the upper Devonian estimated in the Aguarague structure (44 mm/yr), the incision rate would be ~11 mm/yr. This value seems too high in respect to the shortening rate (~10 mm/yr) to be only related to the uplift above a deep crustal ramp. Therefore, a lithological effect needs to be considered in addition to the uplift effect. The low erodibility is presumably linked to the strong burial (~10 km) and weak metamorphism that affect the lower Devonian rocks.

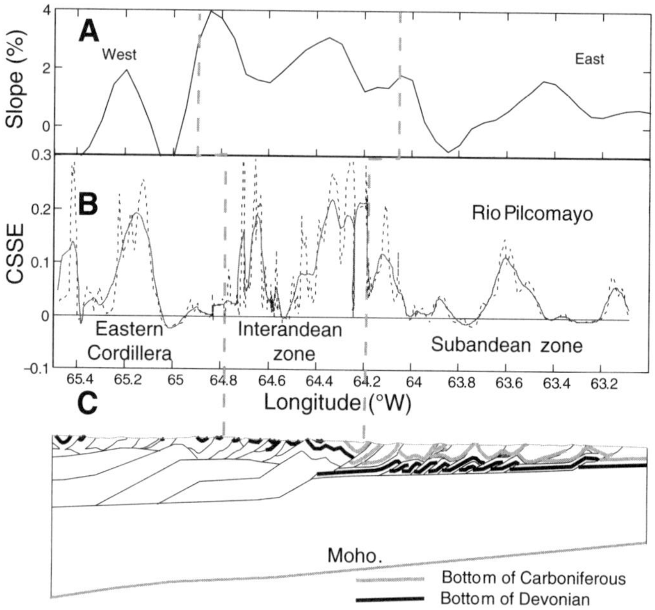

Figure 13. (A) Regional topographic slope averaged between 20°S and 21°S. Regional topographic swath widths, parallel to the strike, are 100 km. Sliding windows, parallel to the profile, are 14 km. The boundaries between Interandean and Subandean zones and between Interandean and eastern Cordillera (vertical hachured lines) are defined by the eastern large outcrop of Devonian layers. (B) Nondimensional shear stress versus longitude for the Rio Pilcomayo. The excess critical shear stress (CSSE) is shown by a dashed line and a thin continuous line (smoothing using a 4 km sliding window), and the excess critical shear stress scale is on the left side. (C) Structural cross section through the eastern side of the Andean belt of southern Bolivia (adapted from McQuarrie, 2002, and Giraudo and Limachi, 2001).

DISCUSSION

In this study, the excess critical shear stress has been used as a hydrological indicator of incision rate, because it integrates a great number of morphological and sedimentological parameters. The incorporation of these parameters into the model embedded a series of assumptions, each of which adds uncertainty to quantitative predictions. For example, the erodibility coefficients of the rocks of the Subandean belt are between 67 ± 50 mm/yr and 44 ± 33 mm/yr, and the uncertainty is so great that the contrast of erodibility is not clear. Furthermore, a simpler approach, such as using the fluvial shear stress, would mimic the variations of the excess critical shear stress. The question arises: is this complexity useful, and what is the gain with respect to a simpler strategy? We argue that it is useful for the following reasons.

If the excess critical shear stress estimation is affected by a strong uncertainty, a nondimensional strategy of analysis furnishes some quantitative information. For example, the absolute erodibility of the rocks of the Subandean belt cannot be considered meaningful, but the relative index of erodibility (coefficient erodibility/shortening rate) seems able to detect an erodibility contrast of less than a factor two.

The mean grain size in the excess critical shear stress estimation is of prime importance in equation 6. When mean grain size is constant, its role is implicitly incorporated in the constant term of any stream power function, and when mean grain size evolves like a power law along the river profile, its role is incorporated in the *m* and *n* exponents of a stream power function. Our measurements show, however, that the mean grain size of the Rio Pilcomayo is not constant and does not evolve regularly. The abrupt decrease in the mean grain size from 7 cm to less than 1 cm in less than 100 km along the lowermost part of the river influences the estimate of the excess critical shear stress. Its explicit incorporation in this study allows detection of the weak tectonic activity of the most external structure in the morphology of the river: the excess critical shear stress of this structure (Iguitinti) is more than 20% of the maximum value reached by the excess critical shear stress along the river (Fig. 10), whereas the value of the fluvial shear stress of the Iguitinti structure is less than 10% of the maximum (Fig. 9): the increase in shear stress across the structure is not greater than the noise, whereas the excess critical shear stress is meaningful. The abrupt decrease of mean grain size also influences the width of the active channel. If we consider that the critical Shields stress remains constant, an order of magnitude decrease in grain size implies a 45× increase in width, a dependence that partly explains why the width varies from 25 m to 2 km along the river profile. The great importance of the grain size would presumably require a more systematic measure of the sediment caliber in the future and further theoretical studies to understand the cause of this grain-size variation.

River width is of great importance in the excess critical shear stress calculation to estimate the uplift in restricted zones, because river width response can be as sensitive as slope to tectonics. For example, river slope increases five times, whereas the width decreases five times through the active Aguarague structure. The width is frequently estimated by using a power-law scaling with the drainage area (e.g., Dunne and Leopold, 1978), but this method smoothes local variations and would induce an underestimation by a factor 4 of the critical shear stress in the uplifted zone of the Aguarague, leading to a strong attenuation of the tectonic signal. Because localized tectonics is found in most of the active thin-skinned thrust belts, the measurement of the river width is of great importance for morphological studies of these tectonics settings.

Structural studies show that the uplift pattern through an active structure of a thrust belt varies dramatically from ~0 to several mm/yr in a few kilometers. The example of the Subandean belt shows that the excess critical shear stress is a powerful approach to better determine the partitioning of deformation in an active thrust belt or to test structural assumptions concerning the geometry at depth and the related uplift patterns.

In the case of the Subandean belt, the morphological study of the Rio Pilcomayo confirms the presence of at least three active structures, with very little activity on the most external one and more than half of the activity taken up through the Aguarague structure. It is also corroborated that the structural assumption of a passive duplex is appropriate to the external part of the Subandean belt. A direct estimate of the uplift rate from dating of strath surfaces is nonetheless needed to better quantify the tectonic activity of this zone.

APPENDIX 1

Calculation of the Excess Nondimensional Shear Stress (Excess Critical Shear Stress)

The fluvial shear stress, τ, represents the forces exerted by the flowing water on the underlying bedrock and is defined as:

$$\tau = \rho g R S_e, \qquad \text{(A1)}$$

where ρ is the fluid density, g the acceleration of gravity, S_e the energy slope, and R the hydraulic radius.

Assuming a uniform (variation along the profile $d/dx = 0$) and steady-state (variation during time $d/dt = 0$) flow, the energy slope is equal to the slope S of the bottom of the river channel:

$$S = S_e. \qquad \text{(A2)}$$

The hydraulic radius R for a rectangular channel is expressed by:

$$R = \frac{Wh}{W + 2h} \qquad \text{(A3)}$$

where W is the width of the river and h the fluid depth. For a large river $W >> h$ and $R = h$.

According equations A1 and A2, we can express the shear stress for a large river as:

$$\tau = \rho g h S. \qquad \text{(A4)}$$

Fluvial shear stress can be related to channel gradient and discharge by considering the continuity of mass relationship:

$$Q = WhU, \qquad \text{(A5)}$$

and the Manning equation: $U = K_s h^{\frac{2}{3}} S^{\frac{1}{2}}$, (A6)

where Q is the flow discharge and U the mean flow velocity. The Manning-Strickler coefficient K_s (Strickler, 1923) is defined for a variable grain size as:

$$K_s = \frac{26}{\left(D_{90}\right)^{\frac{1}{6}}}, \qquad \text{(A7)}$$

with D_{90} the grain diameter (expressed in meters) is not exceeded by 90% of the bedload.

Combining equations A4–A6 gives

$$\tau = \rho . g \left(\frac{1}{K_s} \frac{Q}{W} \right)^{\frac{3}{5}} S^{\frac{7}{10}}. \qquad \text{(A8)}$$

The critical shear stress needed to move a pebble of diameter D is (Shields, 1936):

$$\tau^* = \frac{\tau}{[(\rho 1 - \rho) g D]}, \quad \text{(A9)}$$

where $(\rho_1 - \rho)g$ is the density contrast between water and gravels.

Here, τ^* has a nondimensional form and is called the Shields stress when $D = D_{50}$ (Einstein, 1950), where D_{50} is the grain diameter not exceeded by 50% of the bedload.

Motion occurs if

$$\tau^* > \tau_c^*, \quad \text{(A10)}$$

where τ_c^* is the critical Shields stress.

In the Siwaliks of central Nepal, a comparison of the river morphology and the incision of the neighboring uplifted terraces (Lavé and Avouac, 2000, 2001) suggests that:

$$E \approx K(\tau^* - \tau_b^*), \quad \text{(A11)}$$

where K is an erodibility coefficient, and τ_b^* is a value empirically deduced from the incision of a dated terrace. Because τ_b^* may be seen as a threshold for incipient motion, τ_b^* is also termed a critical Shields stress in the following. Also, $(\tau^* - \tau_b^*)$ is a nondimensional form of the shear stress excess and is called the CSSE for excess of critical shields stress.

By substituting in equation A9 the value of τ deduced from equation A8 and the value of K_s deduced from equation A7, we find the value of τ^*. By substituting this value of τ^*in equation A11, we find:

$$E = K \frac{\rho}{\rho_1 - \rho} \frac{1}{D_{50}} \left(\frac{D_{90}^{1/6}}{26} \frac{1}{W} \right)^{\frac{3}{5}} Q^{\frac{3}{5}} S^{\frac{7}{10}} - K\tau_b^*. \quad \text{(A12)}$$

Assuming a linear relationship between Q and A:

$$Q = pA \quad \text{(A13)}$$

and

$$E = K \frac{\rho}{\rho_1 - \rho} \frac{1}{D_{50}} \left(\frac{D_{90}^{1/6}}{26} \frac{p}{W} \right)^{\frac{3}{5}} A^{\frac{3}{5}} S^{\frac{7}{10}} - K\tau_b^*. \quad \text{(A14)}$$

APPENDIX 2

Uplift Induced by Slip-Rate Variation and Layer-Parallel Shearing above a Complex Thrust Trajectory

For a kink-band style of deformation, the moving thrust sheet can be divided into regions of constant velocity bounded by kinematic discontinuities that are either faults or active axial surfaces (Mosar and Suppe, 1991). In each region, the velocity vector is parallel to the underlying fault segment (Fig. A1A), and its magnitude is defined by the slip rate on this fault segment (Hardy and Poblet, 1995):

$$Vh = -S \cos \phi, \quad \text{(A15)}$$

$$Vv = S \sin \phi, \quad \text{(A16)}$$

where Vh, Vv S, and ϕ are the horizontal velocity, the vertical velocity, the slip rate along the fault, and the dip of the fault segment, respectively.

Slip rate on a fault is generally not constant, and the ratio R (Suppe, 1983), reduction in slip magnitude across the active axial surface, is:

$$R = S/So = \sin(\gamma - \theta)/\sin(\Delta\phi + \gamma - \theta), \quad \text{(A17)}$$

where S and So are the slip rate along the upper segment of the ramp and the lower segment of the ramp, respectively, $\Delta\phi$ is the change of dip of the fault, γ is the axial angle, and θ is the cutoff angle (Fig. A1B).

For a succession of segments (Fig. A1D), the total slip ratio Rt is calculated by:

$$Rt_i = R_1 R_2 ... R_{i-1}. \quad \text{(A18)}$$

Where R_{i-1} is the reduction in slip rate, at a change of dip of the fault, between the segment i of the ramp and the lower segment $i - 1$.

For the segment i, the uplift is:

$$Vv_i = Rt_i \; So \sin \phi. \quad \text{(A19)}$$

Layer-parallel shear frequently occurs in fault-propagation folding (Mosar and Suppe, 1991) or in complex fault systems (Suppe, 1983). The layer-parallel shearing is a third type of kinematic discontinuity that is not a fault or an axial surface and is parallel to the layers. For an increment of deformation, this discontinuity is located at the boundary between unsheared layers and sheared layers (Fig. A1C). The velocity jump between the two sides of the discontinuity is:

$$\Delta S = S_{top} - So \quad \text{(A20)}$$

and

$$\Delta S = \alpha \bullet v_\alpha, \quad \text{(A21)}$$

with α angular shear and v_α the migration velocity of the shear zone perpendicular to the layer.

The calculation of α is described in detail by Mosar and Suppe (1991), and v_α is linked to the propagation rate (V_p) of the

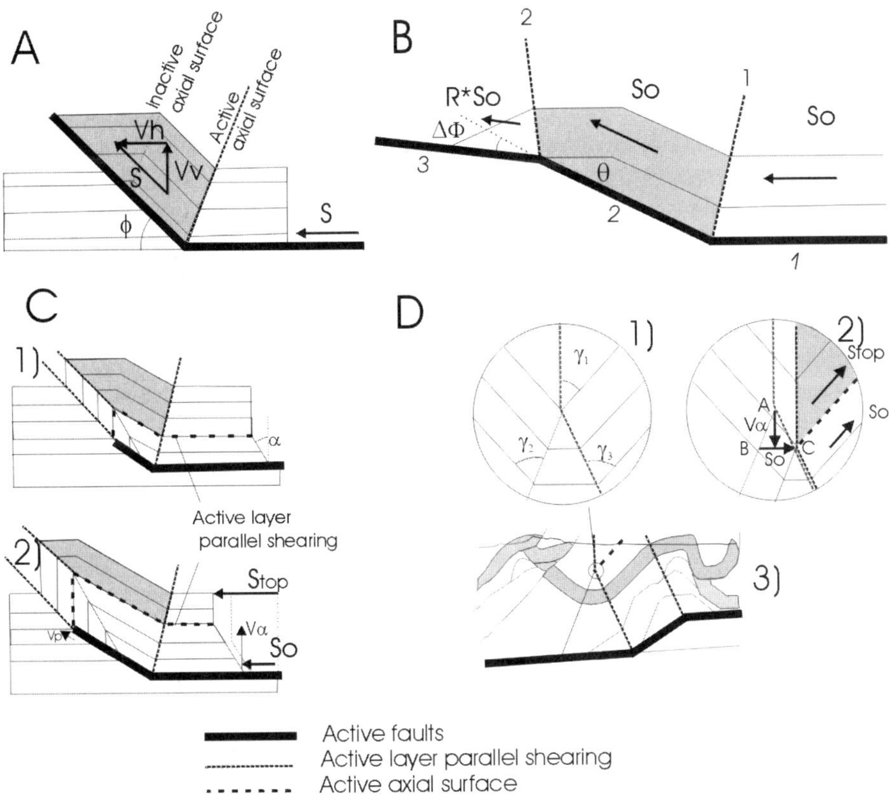

Figure A1. Relationships between slip rate and uplift rate: (A) geometric boundary of a zone of constant velocity; (B) slip variation induced by a change of the cutoff angle through an active axial surface; (C) velocity variation induced by layer-parallel shearing: the case of a fault propagation fold (adapted from Mosar and Suppe, 1991) (1 and 2 refer to two stages of the deformation; (D) velocity variation induced by layer-parallel shearing: the case of merging of one active axial surface with one inactive surface. 1 refers to the definition of the axial surface angles; 2 refers to the zones of distinct velocity; and 3 is the whole geometry of the structure.

fault through the beds for self-similar growth of a simple-step fault-propagation fold:

$$v_\alpha = V_p \sin\theta. \tag{A22}$$

In more general cases, layer-parallel shearing is related to any change of dip of the bedding, and Suppe (1983) has shown that the study of the axial surfaces and of their branching provides a way to estimate the total shearing through a structure. To estimate velocity, the only complement is the introduction of v_α in the analysis of Suppe (1983). In the case of the branching of two axial surfaces in a complex example from the Subandean belt, the migration velocity v_α of the shear zone is the height of the triangle ABC (Fig. A1D):

$$V_\alpha = \frac{\tan(\gamma 3)\cdot\tan(\gamma 2)}{\tan(\gamma 3)+\tan(\gamma 2)}\cdot So, \tag{A23}$$

where γ_2 and γ_3 are the axial angles of the axial surfaces perpendicular to its direction.

S_{top} is the sum of the velocity of the lower-right domain (Fig. A1D) and of the jump of velocity between the two sides of the discontinuity:

$$S_{top} = So\left(1+\Delta\alpha\cdot\frac{\tan(\gamma 3)\cdot\tan(\gamma 2)}{\tan(\gamma 3)+\tan(\gamma 2)}\right), \tag{A24}$$

where $\Delta\alpha$ is the change of angular shear between the upper-right domain and the lower-right domain (Fig. A1D, part 3) calculated from the work of Suppe (1983).

ACKNOWLEDGMENTS

We thank Frank Pazzaglia and Eric Kirby for thorough and constructive reviews. François Jouanne processed the global positioning system (GPS) data, and Patrice Baby introduced us to the geology of the Interandean belt. Alex Whittaker is acknowledged for improvement of the language. Peter van der Beek and Donald Fisher performed a careful editorial reading. This work has been supported by a grant from Institut Français du Pétrole.

REFERENCES CITED

Baby, P., 1995, Importance du raccourcissement et de la sédimentation synorogénique dans la structuration des zones d'arrière arc des Andes centrales (Orocline Bolivien) [Mémoire d'habilitation à diriger les recherches]: Grenoble, Université de Grenoble, 430 p.

Baby, P., Hérail, G., Salinas, R., and Sempere, T., 1992, Geometry and kinematic evolution of passive roof duplexes deduced from cross-section balancing: Example from the foreland thrust system of the southern Bolivian Subandean zone: Tectonics, v. 11, p. 523–536.

Becel, D., 2004, Modèlisation numérique de l'érosion et de la sedimentation le long de la rivière Pilcomayo (Bolivie) [Thèse]: Grenoble, Université Joseph Fourier, 270 p.

Belotti, H.J., Saccavino, L.L., and Schachner, G.A., 1995, Structural style and petroleum occurrence in the Sub-Andean thrust belt of northern Argentina, *in* Tankard, A.J., et al., eds., Petroleum basin of South America: American Association of Petroleum Geologists Memoir 62, p. 545–555.

Brocard, G., 2002, Origine, variabilité spatio-temporelle et signature morphologique de l'incision fluviale dans les Alpes Dauphinoises (SE France): Mémoire Hors Série Géologie Alpine, v. 43, 165 p.

Brocard, G., Van der Beek, P., Bourles, D., Siame, L., and Mugnier, J.L., 2003, ^{10}Be dating of alluvial terraces in the French western Alps: Assessing inheritance, soil development and wind ablation effects and implications for long-term fluvial incision rates and post-glacial river relaxation time: Earth and Planetary Science Letters, v. 209, p. 197–214, doi: 10.1016/S0012-821X(03)00031-1.

Buffington, J.M., and Montgomery, D.R., 1997, A systematic analysis of eight decades of incipient motion studies, with special reference to gravel-bedded rivers: Water Resources Research, v. 33, p. 1993–2029, doi: 10.1029/97WR03190.

Champel, B., Van der Beek, P.A., Mugnier, J.L., and Leturmy, P., 2002, Growth and lateral propagation of fault-related folds in the Siwaliks of western Nepal: Rates, mechanisms and geomorphic signature: Journal of Geophysical Research, v. 107, p. 107, doi: 10.1029/2001JB000578.

Dunne, T., and Leopold, L.B., 1978, Water in environmental planning: New York, W.H. Freeman, 818 p.

Einstein, H.A., 1950, The bedload function for sediment transportation in open channel flow: U.S. Department of Agriculture Technical Bulletin 1026, 71 p.

Endignoux, L., and Mugnier, J.-L., 1990, The use of a forward kinematic model in the construction of balanced cross-section: Tectonics, v. 9, p. 1249–1262.

Enzel, Y., Ely, L., House, P., Baker, R., and Webb, R., 1993, Paleoflood evidence for a natural upper bound to flood magnitudes in the Colorado River Basin: Water Resources Research, v. 29, p. 2287–2297.

Filizola, N., Fraizy, P., Guyot, J.L., Seyler, F., Baby, P., and Hérail, G., 2002, Actual erosion by rivers in the Bolivian Andes, *in* Proceedings, 5th International Symposium on the Andean Geodynamics: Toulouse, France, International Symposium on Andean Geodynamics, p. 211–214.

GEOBOL (Servicio Geologia, Bolivia) and YPFB (Yacimentos Petroliferos Fiscales Bolivianos), 1978, Mapa geologico de Bolivia: La Paz, Bolivia, 1 sheet, scale 1:1000,000.

Giraudo, R., and Limachi, R., 2001, Pre-Silurian control in the genesis of the central and southern Bolivian fold belt: Journal of South American Earth Sciences, v. 14, p. 665–680, doi: 10.1016/S0895-9811(01)00068-2.

Gubbels, T., Isacks, B., and Farrar, E., 1993, High-level surfaces, plateau uplift and foreland development, Bolivian central Andes: Geology, v. 21, p. 695–698, doi: 10.1130/0091-7613(1993)021 <0695:HLSPUA>2.3.CO;2.

Guyot, J.L., 1993, Hydrogéochimie des fleuves de l'Amazonie Bolivienne: ORSTOM (Office de Recherche Scientifique des territoires d'Outre Mer) Ed., Etudes et Thèses no. 1157-4, 264 p.

Hancock, G.S., Anderson, R.S., and Whipple, K.X., 1998, Beyond power: Bedrock river incision process and form, *in* Wohl, E., and Tinkler, K., eds., Rivers over rock: Fluvial processes in bedrock channels: American Geophysical Union Geophysical Monograph 107, p. 35–60.

Hardy, S., and Poblet, J., 1995, The velocity description of deformation, 2: Sediment geometries associated with fault-bend and fault-propagation folds: Marine and Petroleum Geology, v. 12, p. 165–176.

Howard, A.D., Dietrich, W.E., and Seidl, M.A., 1994, Modeling fluvial erosion on regional to continental scales: Journal of Geophysical Research, v. 99, p. 13,971–13,986, doi: 10.1029/94JB00744.

Hurtrez, J.E., Lucazeau, F., Lavé, J., and Avouac, J.P., 1999, Investigation of the relationships between basin morphology, tectonic uplift and denudation from the study of an active fold belt in the Siwalik Hills (central Nepal): Journal of Geophysical Research, v. 104, p. 12,779–12,796, doi: 10.1029/1998JB900098.

Instituto Geographico Militar, 1979, Topographical maps of Bolivia: La Paz, I.G.M. (Instituto Geographico Militar) publisher, scale 1:50,000, 25 sheets.

Kendrick, E.C., Bevis, R.F., Smalley, J.R., Cifuentes, O., and Galban, F., 1999, Current rates of convergence across the central Andes: Estimates from continuous GPS observations: Geophysical Research Letters, v. 26, p. 541–544, doi: 10.1029/1999GL900040.

Kennan, L., Lamb, S.H., and Rundle, C., 1995, K-Ar dates from the Altiplano and Cordillera Oriental of southern Bolivia: Implications for Cenozoic stratigraphy and tectonics: Journal of South American Earth Science, v. 28, p. 163–186.

Kirby, E., and Whipple, K., 2001, Quantifying differential rock-uplift rates via stream profile analysis: Geology, v. 29, p. 415–418, doi: 10.1130/0091-7613(2001)029 <0415:QDRURV>2.0.CO;2.

Kley, J., 1996, Transition from basement-involved to thin-skinned thrusting in the Cordillera Oriental of southern Bolivia: Tectonics, v. 15, p. 763–775, doi: 10.1029/95TC03868.

Labaume, P., and Moretti, I., 2001, Diagenesis-dependence of cataclastic thrust fault zone sealing in sandstone: Example from the Bolivian Sub-Andean zone: Journal of Structural Geology, v. 23, p. 1659–1675, doi: 10.1016/S0191-8141(01)00024-4.

Lamb, S., 2000, Active deformation in the Bolivian Andes, South America: Journal of Geophysical Research, v. 105, p. 25,627–25,653.

Lavé, J., 1997, Tectonique et érosion: L'apport de la dynamique fluviale à l'étude sismotectonique de l'Himalaya du Népal central [thèse]: Paris, University of Paris VII, 226 p.

Lavé, J., and Avouac, J.P., 2000, Active folding of fluvial terraces across the Siwalik Hills (Himalaya of central Nepal): Journal of Geophysical Research, v. 105, p. 5735–5770, doi: 10.1029/1999JB900292.

Lavé, J., and Avouac, J.P., 2001, Fluvial incision and tectonic uplift across the Himalaya of central Nepal: Journal of Geophysical Research, v. 106, p. 26,561–26,593, doi: 10.1029/2001JB000359.

Lavenu, A., Thiele, R., Machette, M., Dart, R., Bradley, L., and Haller, K., 2000, Maps and database of Quaternary faults in Bolivia and Chile: U.S. Geological Survey Open-File Report 00-283, 45 p.

Leturmy, P., Mugnier, J.L., Baby, P., Coletta, B., and Vinour, P., 2000, Interactions between thin skinned thrust tectonics and sedimentation: A view from numeric and analogue models: Tectonophysics, v. 320, p. 45–67, doi: 10.1016/S0040-1951(00)00023-8.

Masek, J.G., Isacks, B.L., Timothy, L., Gubbels, E., and Fielding, J., 1994, Erosion and tectonics at the margins of continental plateaus: Journal of Geophysical Research, v. 99, p. 13,941–13,956, doi: 10.1029/94JB00461.

McQuarrie, N., 2002, The kinematic history of the central Andean fold-thrust belt, Bolivia: Implications for building a high plateau: Geological Society of America Bulletin, v. 114, p. 950–963, doi: 10.1130/0016-7606(2002)114 <0950:TKHOTC>2.0.CO;2.

Molnar, P., 2001, Climate change, flooding in arid environments, and erosion rates: Geology, v. 29, p. 1071–1074, doi: 10.1130/0091-7613(2001)029 <1071:CCFIAE>2.0.CO;2.

Mosar, J., and Suppe, J., 1991, Role of shear in fault-propagation folding, *in* McClay, K.R., ed., Thrust tectonics: London, Chapman & Hall, p. 123–132.

Pazzaglia, F., and Brandon, M., 2001, A fluvial record of long-term steady-state uplift and erosion across the Cascadia forearc high, western Washington State: American Journal of Science, v. 301, p. 385–431.

Rochat, P., 2003, Structure et cinématique de l'Altiplano nord-Bolivien au sein des Andes centrales [Thèse]: Grenoble, Université Joseph Fourier, Mémoire Hors Série Géologie Alpine 38, 193 p.

Roche, M.A., Aliaga, A., Campos, J., Pena, J., Cortes, J., and Rocha, N., 1990, Hétérogénéité des précipitations sur la cordillère des Andes Boliviennes, *in* Lang, H., and Musy, A., eds., Hydrology in mountainous regions: Toulouse, France, International Association of Hydrology Publishing, v. 193, p. 381–388.

Roeder, D., 1988, Andean age structure of Eastern Cordillera (province of La Paz, Bolivia): Tectonics, v. 5, p. 23–29.

Sergeomin (Servicio de Geologia y Minas) and YPFB (Yacimentos Petroliferos Fiscales bolivianos), 1996, Mapa geologico de Bolivia: La Paz, Bolivia, scale 1:1,000,000, 1 sheet.

Shields, I.A., 1936, Anwendung der ähnlichkeitsmechanik und der Turbulenz forshung auf Geschiebewegung: Berlin, Mitteilungen der Preussichen Versuchsansalt für Wasserbau und Schiffau, Heft 26, 123 p.

Sibson, R.H., 1989, Earthquake faulting as a structural process: Journal of Structural Geology, v. 11, p. 1–14, doi: 10.1016/0191-8141(89)90032-1.

Sklar, L., and Dietrich, W.E., 1998, River longitudinal profiles and bedrock incision models: Stream power and the influence of sediment supply, *in* Wohl, E., and Tinkler, K., eds., Rivers over rock: Fluvial processes in bedrock channels: American Geophysical Union Geophysical Monograph 107, p. 237–260.

Snyder, N.P., Whipple, K., Tucker, G.E., and Merrits, D.J., 2000, Landscape response to tectonic forcing: Digital elevation model analysis of stream profiles in the Mendocino triple junction region, northern California: Geological Society of America Bulletin, v. 112, p. 1250–1263, doi: 10.1130/0016-7606(2000)112<1250:LRTTFD>2.0.CO;2.

Snyder, N.P., Whipple K., Tucker G.E., and Merritts, D.J., 2003, Importance of a stochastic distribution of floods and erosion thresholds in the bedrock river incision problem: Journal of Geophysical Research, v. 108, no. B2, 2117, doi: 10.1029/2001JB0016552117.

Stock, J.D., and Montgomery, D.R., 1999, Geologic constraints on bedrock river incision using the stream power law: Journal of Geophysical Research, v. 104, p. 4983–4993, doi: 10.1029/98JB02139.

Strickler, A., 1923, Beiträge zur Frage der Geschwindigkeitsformeln: Bern, Switzerland, Mitteilung no. 16, Amt f. Wasserwirtschaft.

Stuiver, M., Reimer, P.J., Bard, E., Beck, J.W., Burr, G.S., Hughen, K.A., Kromer, B., McCormac, F.G., van der Plicht, J., and Spurk, M., 1998, INTCAL98 radiocarbon age calibration 24,000–0 cal B.P.: Radiocarbon, v. 40, p. 1041–1083.

Suppe, J., 1983, Geometry and kinematics of fault-bend folding: American Journal of Science, v. 283, p. 684–721.

Suppe, J., and Medwedeff, D.A., 1990, Geometry and kinematics of fault-propagation folding: Eclogae Geologicae Helvetiae, v. 83, p. 409–454.

Tomkin, J., Brandon, M., Pazzaglia, F., Barbour, J., and Willet, S., 2003, Quantitative testing of bedrock incision models for the Clearwater River, NW Washington State: Journal of Geophysical Research, v. 108, B6, 2308, doi: 10.1029/2001JB 000862.

Turcotte, D., and Green, L., 1993, A scale-invariant approach to flood-frequency analysis: Stochastic Hydrology and Hydraulics, v. 7, p. 34–40.

Van der Beek, P.A., and Bishop, P., 2003, Cenozoic river profile in the upper Lachlan catchment (S.E. Australia) as a test of quantitative fluvial incision models: Journal of Geophysical Research, v. 108, p. 2309, doi: 10.1029/2002JB002125.

Whipple, K., and Tucker, G., 1999, Dynamics of the stream-power river incision model: Implications for height limits of mountain ranges, landscape response timescales, and research needs: Journal of Geophysical Research, v. 104, p. 17,661–17,674, doi: 10.1029/1999JB900120.

Whipple, K., Kirby, E., and Brocklehurst, S., 1999, Geomorphic limits to climatically induced increases in topographic relief: Nature, v. 401, p. 39–43, doi: 10.1038/43375.

Williams, C.A., Conners, C., Dahlen, F.A., Price, E.J., and Suppe, J., 1994, Effect of the brittle-ductile transition on the topography of compressive mountain belts on Earth and Venus: Journal of Geophysical Research, v. 99, p. 19,947–19,974, doi: 10.1029/94JB01407.

Zoetemeijer, R., and Sassi, W., 1991, 2D-reconstruction of thrust evolution using fault-bend fold method, *in* McClay, K.R., ed., Thrust tectonics: London, Chapman & Hall, p. 133–140.

Zubieta-Rossetti, D., 2001, Les sédiments syn-orogéniques du Subandin et de l'avant pays de Bolivie [thèse]: Grenoble, Université Joseph Fourier, Mémoire Hors Série Géologie Alpine 37, 109 p.

Manuscript Accepted by the Society 23 June 2005

Printed in the USA

Geological Society of America
Special Paper 398
2006

Tectonically driven exhumation of a young orogen: An example from the southern Apennines, Italy

Marcello Schiattarella[†]
Department of Geological Sciences, Basilicata University, I-85100, Potenza, Italy
Paola Di Leo
Institute of Methodologies for Environmental Analysis, National Research Council, I-85050, Tito Scalo (Potenza), Italy
Paolo Beneduce
Department of Geological Sciences, Basilicata University, I-85100, Potenza, Italy
Salvatore Ivo Giano
Department of Geological Sciences, Basilicata University, I-85100, Potenza, Italy
Claudio Martino
Department of Geological Sciences, Basilicata University, I-85100, Potenza, Italy

ABSTRACT

In many young orogens of the Mediterranean region, Quaternary tectonics and regional uplift are traditionally considered strictly correlated, but few data are actually available for precise calculations of uplift rates. Such quantitative studies are needed to define the relationships between local faulting and large-scale uplift, and to formulate correct hypotheses on the geodynamic scenario in which the regional raising occurred. In addition, data about burial depths of sediments or tectonic loadings suffered by sedimentary and low-grade metamorphic rocks are essential for comparisons with the uplift rates obtained in the same areas. Such comparisons between quite different data sources improve the comprehension and choice of the most reliable mechanism responsible for the regional uplift and exhumation.

Uplift rates have been calculated for a large sector of the Lucanian Apennine (axial zone of the southern Italian Apennines), and also reviewed for the Calabrian arc (southern termination of the Italian peninsula), using both geomorphological observations (elevation values, ages, and arrangement of depositional and erosional land surfaces and other morphotectonic indicators) and stratigraphical and structural data (sea-level related facies, base levels, fault kinematics, and offset estimations). Such data have been compared with those derived from clay mineralogy of Mesozoic pelagic deposits (Lagonegro units), outcropping in the same sector of the chain, which give information on burial depths.

The values of the Quaternary uplift rates of the southern Apennines axial zone vary from a minimum of 0.2 mm/yr near the town of Potenza to a maximum of ~1.2–1.3 mm/yr in Agri high valley, a severely deformed Quaternary intermontane basin, still tectonically active, in the Pollino Mountains, a carbonate ridge with elevation up to 2200 m above sea level (asl); intermediate values (0.5–0.7 mm/yr) have been calculated for the other studied areas. The erosion rate from a key area of the Lucanian Apennine, obtained from both quantitative geomorphic analysis and missing volumes calculations on a catchment basin as wide as 150 km^2, has been estimated at 0.2 mm/

[†]E-mail: schiattarella@unibas.it.

Schiattarella, M., Di Leo, P., Beneduce, P., Giano, S.I., and Martino, C., 2006, Tectonically driven exhumation of a young orogen: An example from the southern Apennines, Italy, *in* Willett, S.D., Hovius, N., Brandon, M.T., and Fisher, D.M., eds., Tectonics, Climate, and Landscape Evolution: Geological Society of America Special Paper 398, Penrose Conference Series, p. 371–385, doi: 10.1130/2006.2398(23). For permission to copy, contact editing@geosociety.org.

yr for the middle Pleistocene to Holocene time span. Since in the upper part of that chronological interval erosion and uplift rates match well, the axial-zone landscape could have reached a flux steady-state during the late Pleistocene.

Using geomorphological features and late Pliocene to Pleistocene deposits involved in the genesis of erosional and depositional land surfaces, similar rates (≈0.6 mm/yr) have been obtained for a quite large time span (~2 m.y.) in the Melandro basin and adjacent Maddalena Mountains. Therefore, during the last 2 m.y., the total uplift amount of the axial zone of the Lucanian Apennine is ~1.2–1.3 km, with local peaks of 1.5 km. On the other hand, the Mesozoic pelagic units experienced tectonic loading of 4–5 km, as estimated by means of illite crystallinity (in the range 0.6–1.1 $\Delta°2\theta$), percentage of illitic layers in illite/smectite mixed layers (60%–90%) and white mica polytypes (in the range of 15%–35%), and confirmed by other independent data.

The Quaternary uplift and the related erosion rates of the southern Apennines are unquestionably due to extensional tectonics coupled with thermal/isostatic regional raising and, in minor extent, to strike-slip faulting acting in the earlier deformational stage of the south Apennines chain. The gap of several kilometers deriving from the comparison between uplift rates and tectonic loading values may be explained only with different exhumation modalities starting from late Miocene times. This age can be obtained assuming a fixed rate of 0.6 mm/yr, which represents the best long-term estimate for the axial zone of the chain. Going back in time using such a conservative rate to get a denudation value of ~4–5 km, Tortonian age is reached. At that time, contractional tectonics were still active in this sector of the southern Apennines. Tectonic denudation may be a reliable explanation for the discrepancy between the different sources of data. Such phenomena led to the exhumation of the Mesozoic core of the chain (Lagonegro units), causing low-angle extension on its Tyrrhenian side by reactivation and inversion of older thrusts, and stacking on its eastern margin of Cretaceous to Miocene pelagic and flysch units (i.e., frontal imbricate fan units) by gravity megasliding.

Keywords: tectonic geomorphology, clay mineralogy, southern Apennines, Italy.

INTRODUCTION

In the last decade, research on uplift and/or erosion rates for different chronological intervals has been produced worldwide. The studies mainly demonstrate that vertical motion of the crust can be combined with different tectonic regimes and act during the entire evolution of an orogen, although it is characterized by nonsteady behavior. Nevertheless, in many young circum-Mediterranean chains (e.g., Italian Apennines and Greece), Quaternary (postorogen) extensional tectonics and regional uplift are traditionally believed to be strictly correlated, in spite of the paucity of data for precise calculations of uplift rates from those regions. Such quantitative studies are in fact needed to define the relationships between local faulting and large-scale uplift due to different geodynamic mechanisms (e.g., thermal raising or isostatic compensation, lithosphere delamination or lithospheric thinning and necking, lithospheric flexure, footwall uplift, slab detachment, and so on). Also data about burial experienced by sedimentary and low-grade metamorphic rocks (Di Leo, 2001, 2003, and references therein) are useful for comparisons with the uplift rates obtained in the same areas (England and Molnar, 1990). Such comparisons between quite different data sources improve the comprehension and choice of the most reliable mechanism responsible for the regional uplift and exhumation of the ancient core of an orogenic chain (Schiattarella et al., 2003).

Uplift and/or erosion rates have been calculated for a large sector of the Lucanian Apennine (southern Italian Apennines), and also reviewed for the Calabrian arc (southern termination of the Italian peninsula), using geomorphological, stratigraphical, and structural data. Geomorphic data consist essentially of elevation values, ages and arrangement of depositional and erosional gently dipping land surfaces, and other morphotectonic indicators (see Widdowson, 1997, and Watchman and Twidale, 2002, for a wide review). Stratigraphic observations have been performed on both marine and continental deposits in order to identify sea-level related facies or local base levels and marker beds, whereas structural analysis has been used to define fault kinematics and offset estimations, so as to calculate slip rates in an appropriate way.

The study area covers the axial zone of the southern Apennines (Ortolani et al., 1992) and includes—from north to south—the Potenza Pliocene piggyback basin, the Mt. Li Foi area and catchment basin of Tito-Picerno River, the Pignola Quaternary basin, the Melandro River basin and surrounding uplands, the

Agri River high valley, and the Mercure mid-Pleistocene basin and adjacent Pollino Ridge (Fig. 1).

Morphotectonic data have been compared with those derived from the analysis of Mesozoic pelagic deposits outcropping in the Lucanian Apennine (Lagonegro units; Pescatore et al., 1999, and references therein), which give information on depths of burial using geothermometers based on clay minerals. The analysis of this composite data set allowed us to infer the maximum

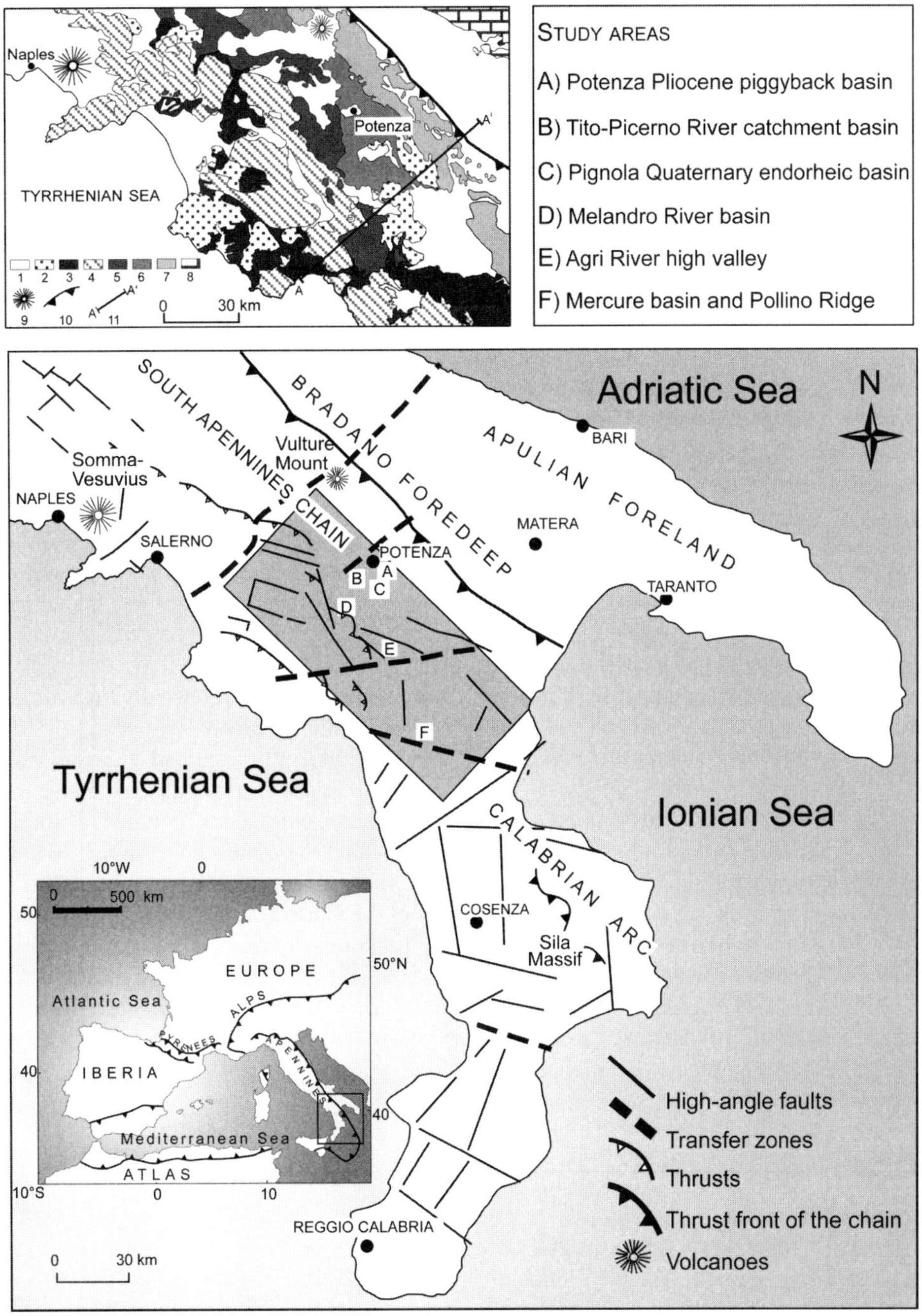

Figure 1. Tectonic sketch map of southern Italy, shaded box shows the study area (i.e., the axial zone of the southern Apennines), its location in the western Mediterranean framework (bottom inset), and geological scheme of the study area (top). Legend (after Pescatore et al., 1999): 1. Pliocene-Quaternary clastics and Quaternary volcanics. 2. Miocene syntectonic deposits. 3. Cretaceous to Oligocene ophiolite-bearing internal units (Ligurian units). 4. Mesozoic-Cenozoic shallow-water carbonates of the Apenninic platform. 5. Lower-middle Triassic to Miocene shallow-water and deep-sea successions of the Lagonegro-type Monte Arioso unit. 6. Lower-middle Triassic to Miocene shallow-water and deep-sea successions of the Lagonegro-type Groppa d'Anzi unit. 7. Cretaceous to Miocene deep-sea successions of the Lagonegro-type Campomaggiore unit. 8. Mesozoic-Cenozoic shallow-water carbonates of the Apulian platform. 9. Volcanoes. 10. Thrust front of the chain. 11. Trace of the regional cross section shown in Figure 10.

Quaternary uplift and the exhumation age of the south Apennines axial zone, including indications about the geodynamic setting in which these phenomena occurred.

GEOLOGICAL AND GEOMORPHOLOGICAL SETTINGS

The geological backbone of south Italy is formed by the orogenic chain of the southern Apennines, its flat-topped Apulian foreland, and the internal nappes of the Calabrian arc (Fig. 1). The southern Apennines are a northeast-verging fold-and-thrust belt, built on the western border of the Apulian plate from late Oligocene to Pleistocene time (Pescatore et al., 1999). It is mainly composed of both shallow- and deep-water sedimentary deposits, derived from Mesozoic-Cenozoic circum-Tethyan sources and from the Neogene-Pleistocene foredeep deposits. Thrusting in the frontal eastern part of the accretionary wedge was followed by backarc extension to the rear (Malinverno and Ryan, 1986). Contractional tectonics continued into the early-middle Pleistocene in the external zone (Pieri et al., 1997).

The Calabrian arc, which from a geographic point of view represents the southern continuation of the Apennines, is made of crystalline metamorphic units thrust on the Mesozoic African-Apulian carbonate domains, locally outcropping in tectonic windows. Its Neogene tectonic transport is toward the southeast, and the thrust front is offshore (i.e., in the Ionian Sea). The arc is strongly fragmented by Pliocene-Quaternary neotectonics and therefore articulated in longitudinal and transversal horst and graben structures (Ghisetti and Vezzani, 1981).

The southern Apennines are characterized by an asymmetric topographic profile (Fig. 2). The summit line of the mountain belt is markedly shifted toward the inner (i.e., Tyrrhenian) margin and does not correspond to the regional water divide. Consequently, the eastern flank of the chain has a greater length and a lower mean gradient than the western flank. The highest summits exceed 2000 m above sea level (asl), whereas the mean elevation of the whole belt is ~650 m asl (Amato and Cinque, 1999). The mountain belt tops are often characterized by a gentle topography mainly represented by relics of an ancient erosional land surface, which unconformably cuts lithological contacts, high-angle faults, and other tectonic structures (Brancaccio et al., 1991; Russo and Schiattarella, 1992; Amato and Cinque, 1999; Ascione and Romano, 1999, among others). The formation of the summit surface can be assigned to the late Pliocene–early Pleistocene, as inferred by the presence of lower-middle Pliocene clastic marine deposits on the top of the Maddalena Ridge (Schiattarella et al., 2003), which were involved in the planation of the 1200 m asl paleosurface (S1 in the cited paper and hereafter). Adopting a counting-from-the-top criterion and regional-scale basin correlations (Brancaccio et al., 1991; Santangelo, 1991; Amato and Cinque, 1999; Schiattarella et al., 2003), the S2 and S3 land surfaces ages can be set up respectively at 1.2 and 0.8–0.7 Ma. A lower surface (S4), both depositional and erosional, forming the depositional top of middle Pleistocene continental sediments and cutting pre-Quaternary siliciclastic deposits, is localized at 500–600 m asl in different basins of the axial zone and can be dated at 0.125 Ma. The uppermost part of the continental clastic deposits filling the intermontane Quaternary basins of the southern Apennines is upper Pleistocene in age, as in the case of the Agri Valley (Giano and Schiattarella, 2002) and the Mercure basin (Schiattarella et al., 1994). Their depositional top is dissected and terraced by the fluvial net, forming the S4 surface. The beginning of the fluvial incision is due to a regional tectonic event, occurring at ca. 0.125 Ma, similar to that observed in coastal areas of southern Italy, where the Tyrrhenian marine terraces are severely uplifted (Westaway, 1993; Bordoni and Valensise, 1998).

It is generally assumed that regional Quaternary uplift suspended a pre-Quaternary erosional base level to which this gentle paleolandscape was graded. The erosional land surfaces occur along the entire orogenic belt in several inset levels. In fact, they are located at different elevations both around the top of the mountains and along the flanks of the valleys. These land surfaces are formed of many relics isolated by fluvial dissection and faulting, reaching a maximum size of some tens of square kilometers, and are elevated 500–1500 m above the valley floors from the Tyrrhenian flank of the chain to the foredeep (Amato and Cinque, 1999). The Calabrian arc is partly characterized by flat-topped mountains as well, especially

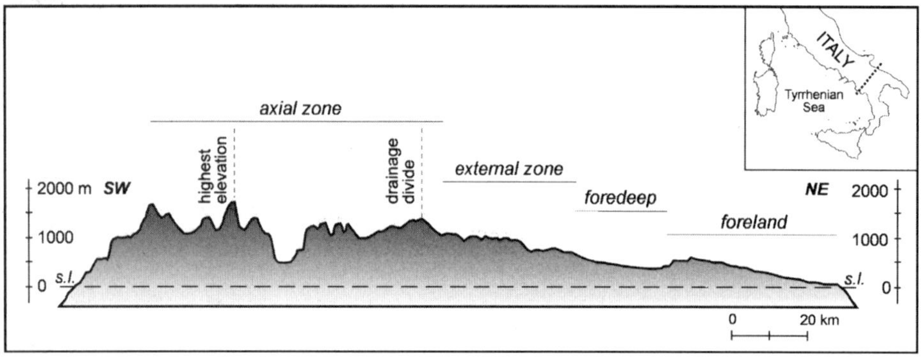

Figure 2. Topographic profile of southern Italy, from Tyrrhenian to Adriatic coastlines (modified after Cinque, 1992); s.l. is sea level.

in the Sila Massif, where the top paleolandscapes are deeply affected by tropical-type weathering and show tens-of-meters-thick regolith and/or altered rock horizons (Guzzetta, 1974).

In the axial zone of the chain, both Mesozoic to Tertiary shallow-water carbonates and coeval pelagic deposits crop out (Fig. 1). The platform limestone and dolomite mainly constitute the western flanks of the Melandro and Agri Valleys, whereas the deep-sea carbonate and siliceous successions (Lagonegro units) form the eastern slopes. The Tito-Picerno and Potenza basins are located inside the outcrop area of the Lagonegro units, unconformably covered by Pliocene marine and Pleistocene continental clastics. The Pollino Ridge represents the junction zone between the southern Apennines and Calabrian arc and is constituted by Mesozoic-Cenozoic platform carbonates overthrust by Ligurian (i.e., internal ophiolite-bearing) units. This ridge is sharply bounded on the northwest by the Mercure River structural depression, a lacustrine basin filled by mid-Pleistocene sediments. In all these areas the upland erosional land surfaces are present. The axial zone was affected by strike-slip faulting during late Pliocene–early Pleistocene times, followed by extensional tectonics from middle Pleistocene to present (Schiattarella, 1998; Giano et al., 2000).

The representative Lagonegro-type deposits, analyzed in this work to estimate burial depths, outcrop in the Lucanian Apennine, between Potenza and Sellata mountain pass and near Sasso di Castalda (Pignola-Abriola facies and Lagonegro–Sasso di Castalda facies, respectively, after Scandone, 1975).

UPLIFT AND EROSION RATES

The arrangement of erosional and depositional surfaces at different heights along the axial zone of the Lucanian Apennine has been detected on the basis of both field survey and map analysis. The investigations have been focused along several transects with different orientation that cross the valleys of the Pergola-Melandro and Tito-Picerno streams (Schiattarella et al., 2003), the high valley of the Agri River (Giano and Schiattarella, 2002), and the Mercure River basin (Schiattarella et al., 1994), because of the capability of these kind of intermontane basins to record changes in base level during the past 2 m.y. Other geological and geomorphological observations have been performed in the adjoining areas, such as the Potenza piggyback basin, where the erosional surfaces cut prevalently Pliocene clastics, the Pignola endorheic basins, where the surfaces are represented by the depositional top of Quaternary lacustrine deposits, and the Pollino Ridge, where the highest paleosurfaces of the whole study area are located. Such surfaces have been used as chronological markers to reconstruct tectonic and morphogenetic events, starting from

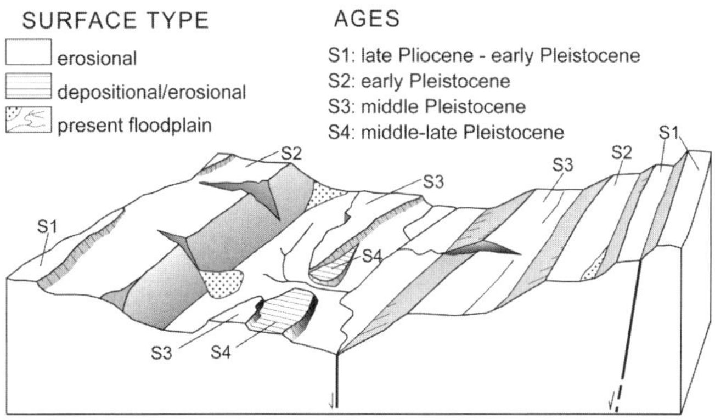

Figure 3. Block diagram showing the morphostructural set of the intermontane valleys of the Lucanian Apennine: S1 to S4 represent the different surfaces (adapted from the Tito-Picerno River basin).

TABLE 1. LAND SURFACES MORPHOMETRIC PARAMETERS, AGES, AND UPLIFT RATES

Erosional and depositional land surfaces	Elevation range (m)	Age (Ma)	Local uplift rate (mm/yr)	Partitioned uplift rate (mm/yr)	Regional uplift rate (mm/yr)
Tito-Picerno and Pergola-Melandro valleys					
S1	1250–1150	1.8	0.40	0.50	0.66
S2	1025–750	1.2	0.35	0.38	0.75
S3	750–710	0.73	0.34	0.21	0.99
S4	575–610	0.125	0.52	0.52	
Agri high valley					
S1	>1300	1.8	0.48	0.33	0.78
S2	1250–1000	1.2	0.56	0.74	1.0
S3	950–750	0.73	0.44	0.38	1.16
S4	650–580	0.125	0.72	0.72	
Mercure basin and Pollino Ridge					
S1	2100–1800	1.8	0.87	0.75	1.05
S2	1600–1400	1.2	0.93	1.06	1.25
S3	1200–800	0.73	0.84	0.84	1.37
S4	480–500	0.125	0.88	0.88	

the late Pliocene, and to calculate regional and local uplift rates of the axial zone of the chain.

Generally, four levels of polygenic land surfaces, well preserved along divides and suspended on the present valley floors, are recognizable in the study area (Fig. 3). Three of them are erosional (S1 to S3), whereas the lowest (S4) is depositional (Table 1). According to previous regional interpretations (Brancaccio et al., 1991; Santangelo, 1991; Amato and Cinque, 1999; Schiattarella et al., 2003), the ages of these land surfaces are included in a time span ranging from 1.8 to 0.125 Ma. A more ancient land surface, probably sculptured in late Pliocene times, can be recognized in the Campania-Lucania Apennines as well.

The uplift rates have been calculated using the difference in height between the absolute (i.e., sea level) or local (i.e., present-day talweg) erosion base levels and the several generations of land surfaces. In some cases, vertical erosion (i.e., incision) rates have been also calculated and converted into local uplift rate (bedrock incision rate, after Burbank and Anderson, 2001), assuming that eustatic changes did not produce relevant effects in this sector of the orogen. In order to illustrate the uplift trend for every single time span (Fig. 4; Table 1), stage (or partitioned) uplift rates have also been calculated using the elevation and age differences between the average level of a given order of land surfaces and that immediately below (example in Fig. 5). In any case, after a decrease between 1.2 and 0.7 Ma, from the beginning of the middle Pleistocene an increasing (local or stage) rate can be observed, similar to what occurred in the northern Apennines (Bartolini, 1999). In contrast, a quasi-linear growing regional uplift rate is clearly recorded by land surfaces of the southern Apennines since the end of the Pliocene (Fig. 4).

The values of the local uplift often indicate the amount of uplift due to faulting. Based on both field and cartographic cross-cutting criteria, we establish the relationships between planation surfaces and faults. This allows suitable discriminations of different generations of morphological and structural elements. In addition, the morphological features of the different surfaces and their bordering slopes (i.e., surface morphofacies, concave-convex slopes, and fluvial channel geometry) also contribute to correlate the base levels. Therefore, it is possible to separate the uplift due to the regional raising from the effects of local tectonics. The estimation of the total offsets produced by faulting thus allows calculation of the slip rates along the fault planes and the comparison of these data with the local uplift rates.

Detailed morphometric profiles intercepting basin-border faults (Schiattarella et al., 2003) show that slip rates vary between 0.3 and 0.5 mm/yr (from 1.8 to 1.2 Ma) and 0.5–0.8 mm/yr (from 1.2 to 0.7 Ma). The last kinematics of the analyzed faults is expressed by normal slip responsible for major Quaternary offsets and basin opening during middle to late Pleistocene times (Giano et al., 2000). Former neotectonic deformational stages affected the study areas and the whole south Apennines chain starting from the late Pliocene (Schiattarella, 1998), but their strike-slip kinematics with horizontal offsets of few kilometers did not favor the relief growth and the creation of large intermontane basins like those generated during middle-late Pleistocene times. This is also documented

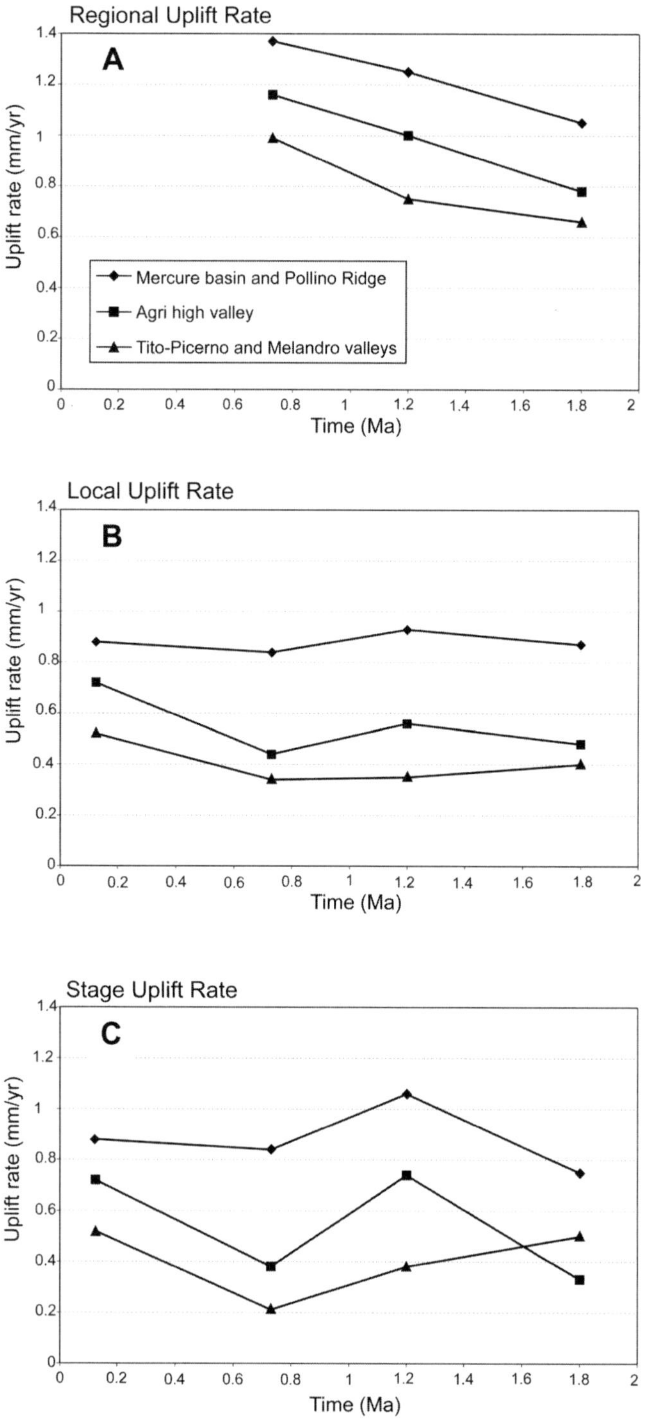

Figure 4. Diagrams showing the variation of the regional (A), local (B), and stage or partitioned (C) uplift rates from the axial zone of the southern Apennines during the Quaternary.

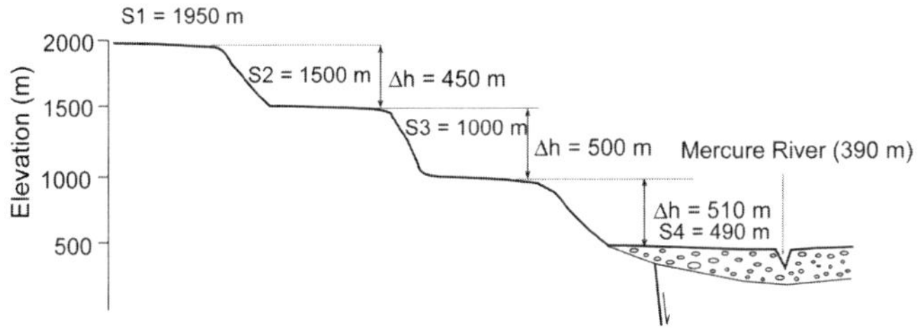

Figure 5. Schematic morphometric profile across the Pollino Ridge and adjacent Mercure basin used for calculating stage uplift rates (S1 to S4 represent the different surface levels from this area).

by low rates of a normal component of faulting during the late Pliocene–early Pleistocene time span (Table 1; Fig. 4).

Quaternary regional and local rates from the study area and those obtained by a wide review of the literature of the past 15 yr on morphotectonics, paleosurfaces, and marine terraces of southern Italy (see, for example, Ascione and Romano, 1999, for Campania region; Amato, 2000, and Schiattarella et al., 2003, for Lucania region; Westaway, 1993, and Cucci and Cinti, 1998, for Calabria region; for a complete list of papers on these topics see also Bordoni and Valensise, 1998) have been used to compile the map shown in Figure 6 (the technique adopted for contouring is the Thissen net). Although the uplift rates refer to different Quaternary chronological ranges (in general, more recent for the terraces along the present-day coastlines), a clear pattern can be deduced. As a matter of fact, the trend of the iso-lines follows the strike of the chain, and a longitudinal and transversal segmentation of the orogen in homogeneously uplifted zones can be discerned. In addition, the boundaries between uplifted and downthrown blocks match the major morphotectonic features of the southern Apennines and Calabrian arc (Fig. 6, sketch map in the frame).

In order to establish the dynamic state of the orogen (Willett and Brandon, 2002), erosion rates from a key area of the Lucanian Apennine have been obtained from both quantitative geomorphic analysis and missing volumes calculations on a catchment basin as wide as 150 km^2 (Tito-Picerno River basin, Fig. 7). Postulating a linear correlation between the suspended

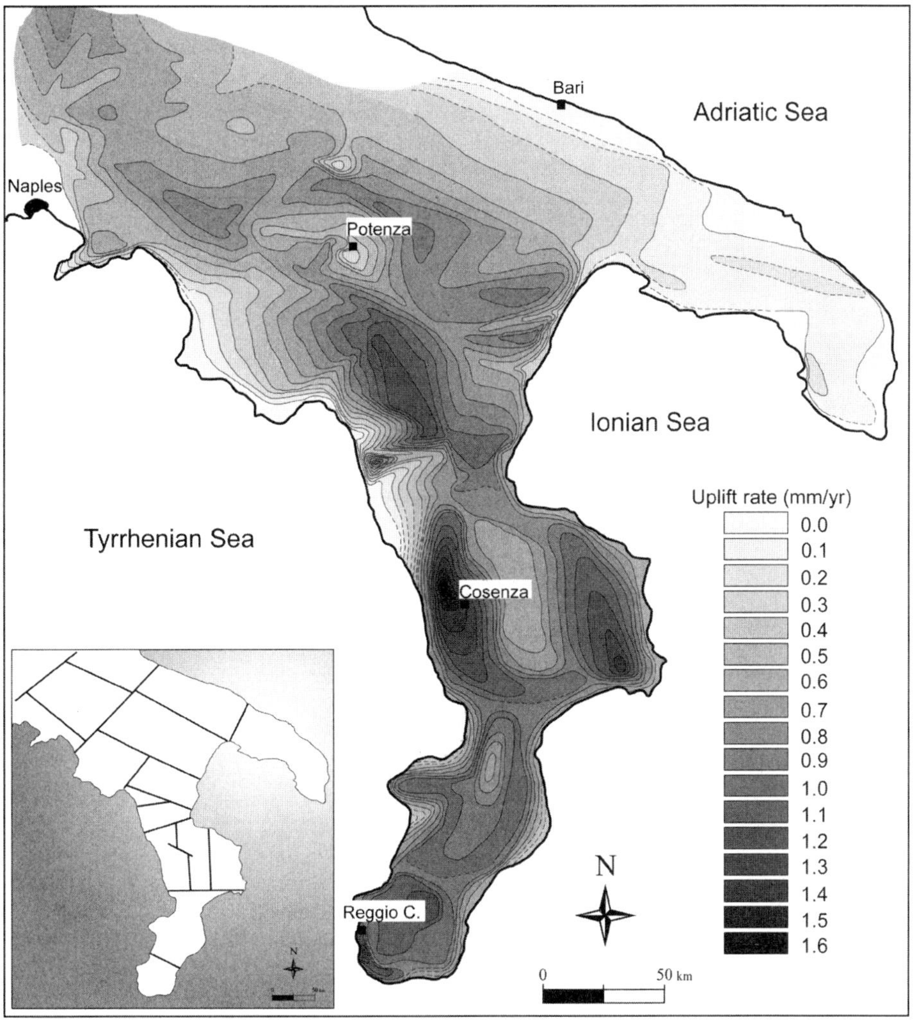

Figure 6. Contour map of the late Pliocene to late Pleistocene uplift rates (inset: segmentation of differently uplifted blocks).

sediment yield and the volumes eroded from the whole hydrographic basin (Tu and V respectively, Fig. 7 inset), as effectively shown by several subbasins of the major catchment area, the erosion rate has been estimated at 0.2 mm/yr for the middle Pleistocene to present time span (Pascale, 2003; Schiattarella et al., 2004). Missing volumes have been calculated by a graphical method based on the morphological features of the basins (i.e., slopes, surfaces, and divides related to different Pleistocene base levels) to evaluate eroded rock volumes for different time intervals defined by the age of the S1 to S3 erosion surfaces. The obtained value is in good agreement with the long-term denudation rates (0.22–0.3 mm/yr) estimated by Amato et al. (2003) for an area fairly close to the investigated catchment basin. Since erosion and uplift rates match well in the upper part of that chronological interval, the axial zone landscape could have reached a flux steady state (Brandon et al., 1998; Willett and Brandon, 2002) during the late Pleistocene.

Finally, it is worth noting that in the high Agri Valley—one of the most active seismic zone of Italy—during the last 30 k.y., the same fault system was characterized by a slip rate decreasing to 0.1 mm/yr (data from Giano et al., 2000).

BURIAL DEPTH ESTIMATES

Shale and clay-rich beds intercalated in the Mesozoic pelagic successions outcropping in the Lucanian Apennine (Lagonegro units) constitute an excellent set of materials to study thermal maturity and estimate maximum burial of these deposits using inorganic thermal indicators such as the Kübler index of illite (KI; Kübler, 1967), percentage of illitic layers in illite/smectite mixed layers (%I in I/S), and percentage of $2M_1$ polytype (%$2M_1$). It is worth noting that in young orogens, as in the case of the Lucanian Apennine, burial depth can be considered synonymous with tectonic loading, because it is the thrust emplacement that mainly leads to thermal maturity of deposits.

Triassic to Cretaceous deposits (Lagonegro units) were sampled and analyzed for clay mineralogy along the Pignola-Abriola and Sasso di Castalda sections. The Monte Facito, Cal-

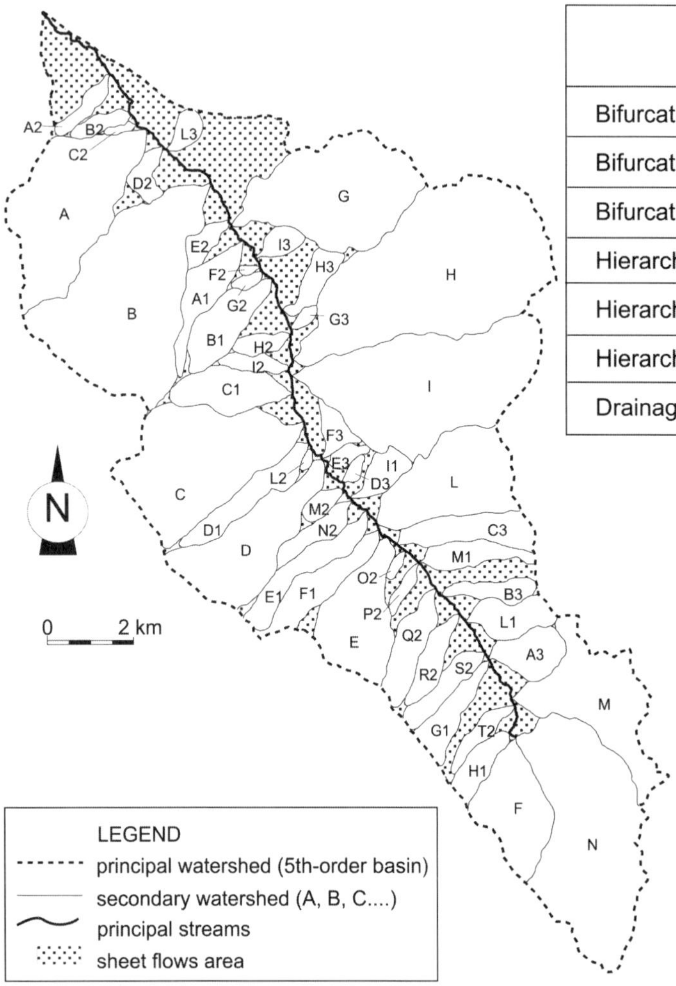

	Tito-Picerno Valley	Subbasins
Bifurcation ratio (Rb)	6	2 < Rb < 7
Bifurcation direct ratio (Rbd)	5.35	2 < Rbd < 7
Bifurcation index (R)	0.66	0 < R < 1.3
Hierarchic anomaly number (Ga)	418	2 < Ga < 54
Hierarchic anomaly index (Δa)	0.50	0.28 < Δa < 1
Hierarchic anomaly density (ga)	2.76	1.14 < ga < 8
Drainage basin area (A)	151 km²	2<A<14.25

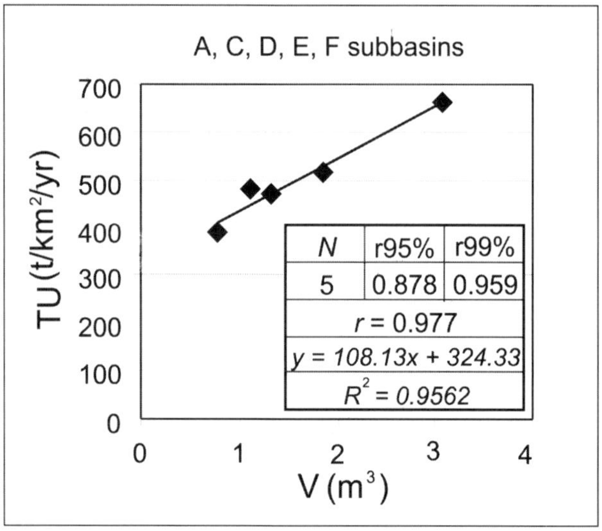

Figure 7. Tito-Picerno River catchment basin and its geomorphic parameters (data after Pascale, 2003, and Schiattarella et al., 2004). Inset: suspended sediment yield (TU) and the volume (V) eroded from the whole hydrographic basin.

cari con selce, Scisti silicei, and Galestri Formations were chosen because of their regional extent and the high frequency of pelitic layers rich in clay minerals. Fifty-eight clay-rich and shale samples were collected from the entire Pignola-Abriola succession: 3 samples came from the Monte Facito Formation, 18 samples came from the Calcari con selce Formation, 31 samples came from the Scisti silicei Formation, and 6 samples came from the Galestri Formation. Ten clay-rich and shale samples were collected along the Sasso di Castalda section: 4 samples came from the Calcari con selce Formation and 6 samples came from the Scisti silicei Formation.

The mineralogy of the shale and clay-rich beds, determined by X-ray diffractometry (Moore and Reynolds, 1997) using both Rigaku miniflex and Siemens D5000 diffractometers, is mainly composed of illite and quartz. Calcite is observed in the lower part of the successions (i.e., in the Monte Facito and Calcari con selce formations), and plagioclase is only observed in the pelitic and cherty members from the Scisti silicei Formation (for member definition, see Di Leo et al., 2002) and in the Galestri Formation. Clay minerals are not homogeneously distributed along the successions: in the Monte Facito and Calcari con selce Formations illite prevails, in the Scisti silicei Formation illite/smectite (I/S) mixed layers are largely present, and in the Galestri Formation kaolinite as well as I/S mixed layers are abundant.

In the Pignola-Abriola succession, starting from the lowermost Monte Facito Formation to the uppermost Galestri Formation (Fig. 8), Kübler index values, determined following Kisch's (1991) recommendations and calibrated with crystallinity index standards (CIS; Warr and Rice, 1994), show a clear increasing trend ranging between 0.6 and 1.1 Δ°2θ. Small amounts of coarse-grained illite with KI values <0.4 Δ°2θ were also found in some samples from the upper part of the Pignola-Abriola succession, i.e., in the cherty member of Scisti silicei and in the Galestri Formations. The stratigraphic position of the samples where the coarse-grained illite were observed (samples containing illite with KI values typical of anchizone were found in stratigraphically higher positions with respect to those containing illite with KI typical of a diagenetic zone) suggests this mineralogical phase is not diagenetic in origin. This illite was not taken into account in defining maximum temperatures and burial depths. The KI values observed all along the Pignola-Abriola section suggest maximum temperatures in the range of 100–180 °C. Conodont color alteration index (CAI) values (Di Leo et al., 2003) determined in both Calcari con selce and Scisti silicei Formations (Fig. 8) also suggest these sediments reached maximum temperatures in the range of 100–180 °C (Garcia-Lopez et al., 1997).

Relationships between temperatures and KI values have been defined by many authors in different geotectonic settings (e.g., Jaboyedoff and Thélin, 1996; see also Merriman and Frey, 1999, and Di Leo, 2001, 2003, for a review). However, the correlation of reaction progress in the smectite–I/S mixed layers–illite series with temperature proposed in Aldega et al. (2003) seems to be the most exhaustive, since it was defined on data sets from a wide number of thermal indicators other than clay mineralogy, such as vitrinite reflectance, fluid inclusions, and fission tracks.

In the Sasso di Castalda section, KI values are in the range of 0.7–0.8 Δ°2θ, suggesting that the sampled sediments reached maximum temperature of ~120 °C (Fig. 9). The portion of the Pignola-Abriola section corresponding to the same stratigraphic level as sampled in the Sasso di Castalda section, i.e., at the passage from the upper calcareous-pelitic member from the Calcari con selce Formation to the basal-pelitic member from the Scisti silicei Formation, also show KI values of ~0.8 Δ°2θ, suggesting that paleotemperatures reached by these formations at this stratigraphic level also reached values of ~120 °C (see Fig. 8).

The percentage of illitic layers in I/S mixed layers, here used as another comparative geothermometer jointly with the Kübler index, estimated following Moore and Reynolds' (1997) suggestions, is in the range of 75%–90% (R = 1 and R = 3 ordering, Reickeweite number) for the Scisti silicei Formation in both sections and decreases to 60%–70% (R = 0.5–1 ordering) upsection, i.e., in the samples from the Galestri Formation (Figs. 8 and 9).

Another indicator of diagenesis and low-grade metamorphism commonly used in studies dealing with reconstruction of geological processes is based on the definition of the white mica polytypes, although polytypic sequence should not be used other than as indicator of reaction progress, since accurate correlation with temperature is not possible (Merriman and Peacor, 1999). The existence of an apparent prograde sequence of polytypism for white micas from $1M_d$ to 1M to $2M_1$ has been well demonstrated (cf. Velde and Hower, 1963; Hoffman and Hower, 1979). The abundances of white mica polytype $2M_1$ (in the range of 15%–35%) estimated in the Monte Facito, Calcari con selce, and Scisti silicei Formations from the Pignola-Abriola succession are consistent with maximum temperatures in the range of 100–180 °C.

Starting from the temperature estimates by clay mineralogy–based geothermometers and considering an average geothermal gradient of 20–30 °C/km (Mongelli et al., 1996), a tectonic loading corresponding to ~4–5 km of burial depth can be reconstructed for both the Pignola-Abriola and Sasso di Castalda sections. Paleotemperatures estimated in the Pignola-Abriola section by means of clay mineralogy were also used in Aldega et al. (2003) as thermal constraints to check the goodness of the thermal modeling and find the set of unique conditions that accounts for the thermal maturity pattern. This set of conditions includes a variable geothermal gradient ranging from 30 °C/km (during most of the passive margin and orogenic stages) to 60 °C/km (during rifting stage), and a burial depth of ~4–5 km.

A comparable range of temperatures and amount of burial were estimated by means of other thermal indicators, such as vitrinite reflectance, fluid inclusions, and fission tracks, for the same units outcropping in other sectors of the Lucanian Apennine (Aldega et al., 2003). Correlation and interpretation of these data also suggest that the maximum temperatures reached by these sediments, typical of a late diagenetic zone, were caused by burial reaching a maximum of 4–5 km. Apatite fission-track data beside temperatures comparable with those furnished by the other thermal indicators also give information on cooling ages. Apatite fission-track data from deposits sampled in the axial zone of the southern Apennines provided concordant final cooling ages of ca. 2.5 Ma, indicating that cooling was ongoing during the late Pliocene. However, other apatite fission-track data relative to Miocene siliciclastic deposits unconformably lying on the Lagonegro units and tectonically covered by Mesozoic carbonate platform units indicate a quite older final cooling of ca. 9.2 Ma, suggesting that cooling (i.e., exhumation) of the axial zone of the chain started even in mid-Tortonian times.

DISCUSSION AND CONCLUSIONS

The values of the Quaternary uplift rates of the south Apennines axial zone (Fig. 1) vary from a minimum of 0.2 mm/yr (Pignola basin) to a maximum of ~1.2–1.3 mm/yr (Agri Valley and Pollino Ridge). Intermediate values (0.5–0.7 mm/yr relative to Mt. Li Foi area, Tito-Picerno and Melandro valleys) have been calculated for the other studied areas. An increasing trend of uplift rate values toward the south can be also observed. On a regional scale, the segmentation of the peninsula in uniform sectors appears quite clear (Fig. 6). It is worthy to note that the

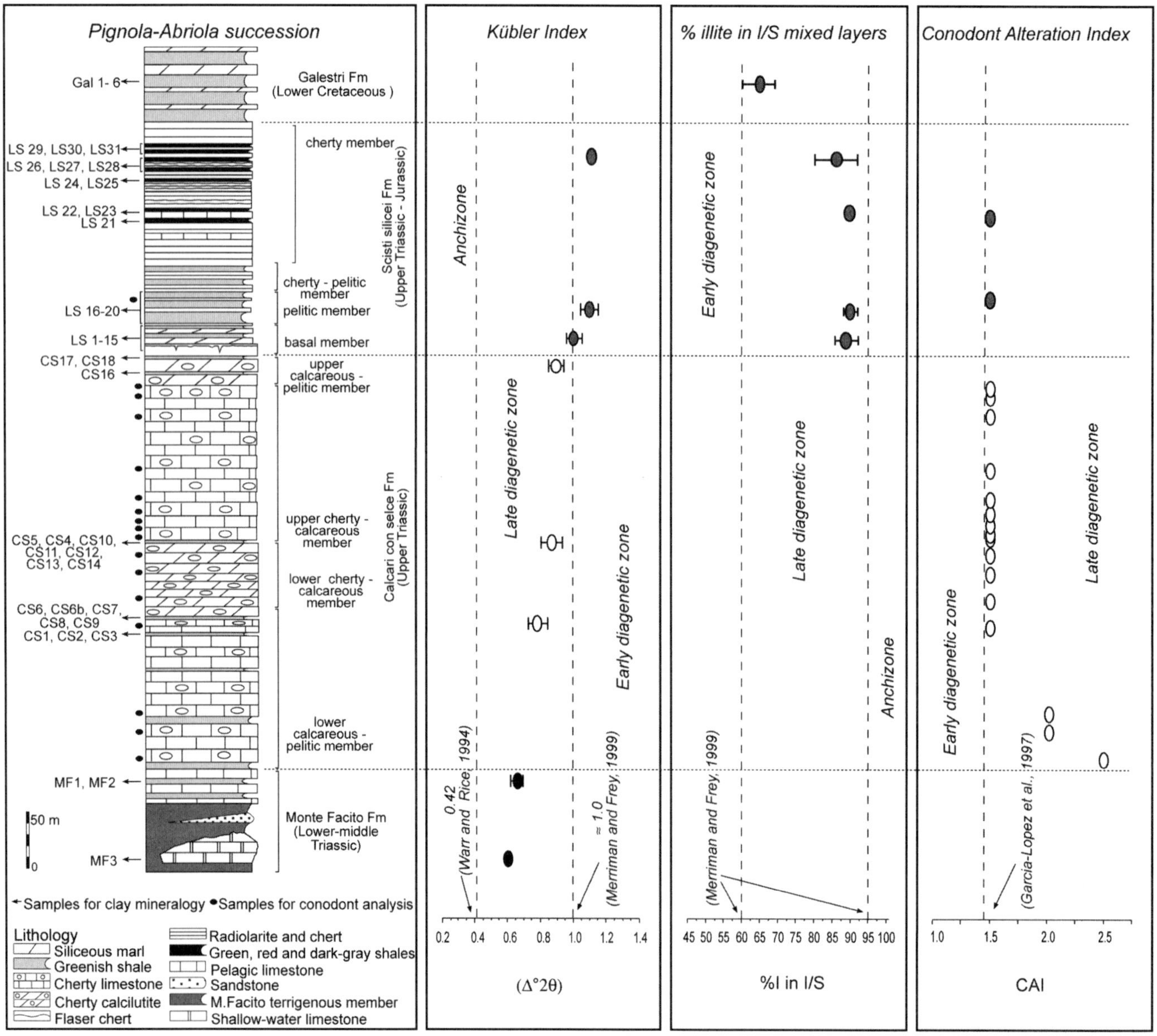

Figure 8. Variation of Kübler index (KI), percentage of illite in the illite/smectite (I/S) mixed layers, and conodont color alteration index (CAI) through the Pignola-Abriola section. Depth-related sample clusters are represented as mean values with error bars (CAI data after Di Leo et al., 2003).

boundaries between differentially uplifted blocks match the major morphotectonic features of the southern Apennines. Maximum rates can be observed along the axial zone of the chain. An acceleration of the local component of vertical motion starting from early-middle Pleistocene times and slowing of the uplift rates since the late Pleistocene have been deduced as well.

Erosion rates from the Lucanian Apennine have been estimated at 0.2–0.3 mm/yr for the middle Pleistocene to Holocene time span (Amato et al., 2003; Pascale, 2003; Schiattarella et al., 2004). Since erosion and uplift rates match well in the late Pleistocene, the axial zone landscape could have reached a flux steady state (Brandon et al., 1998; Willett and Brandon, 2002) just in that stage.

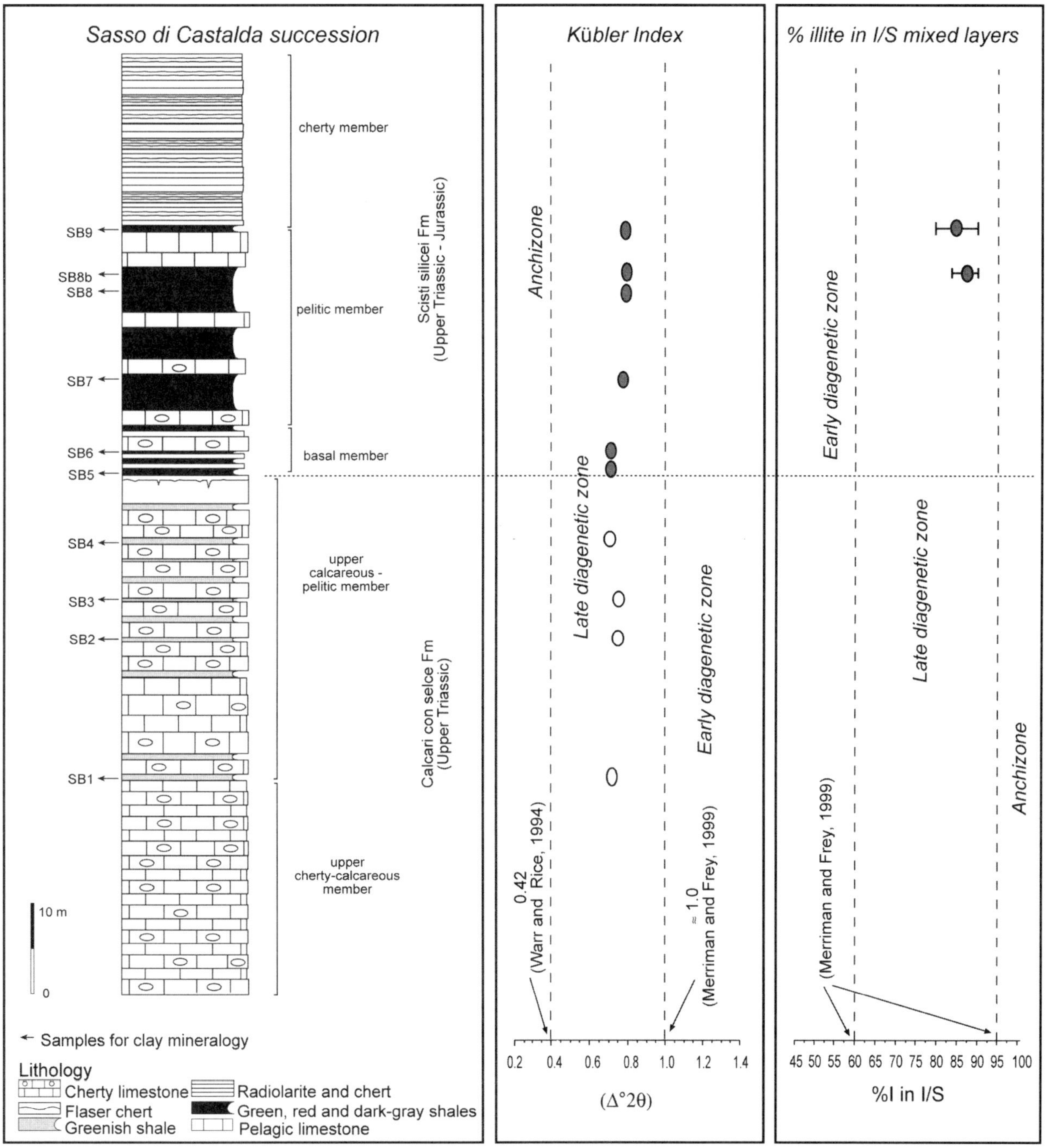

Figure 9. Variation of Kübler index (KI) and percentage of illite in the illite/smectite (I/S) mixed layers through the Sasso di Castalda section in the Melandro basin.

In the Melandro basin and adjacent Maddalena Mountains, similar rates (≈0.6 mm/yr) have been obtained for a quite large time span (ca. 2 m.y.), using geomorphological features and late Pliocene to Quaternary deposits involved in land-surface planation (see also Amato and Cinque, 1999), which is in good agreement with data from other parts of southern Italy (Cucci and Cinti, 1998; Amato, 2000). An exception is the paper by Corrado et al. (2002) on the exhumation of the Monte Alpi area—not far from the Pollino Ridge (Fig. 1)—where a rather high rate (4 mm/yr, as a result of extensional tectonics in the last 0.7 m.y.) is given.

The slip-rate value of 0.3 mm/yr from the Pergola-Melandro basin relative to 1.8 Ma land surfaces is lower than those calculated by means of more recent features (Schiattarella et al., 2003). Such a value suggests low activity of faults and small vertical displacements at that time, in agreement with strike-slip faulting acting during the early Pleistocene (Schiattarella, 1998). The extensional tectonics responsible for major vertical displacements may have initiated at ca. 0.8–0.7 Ma in the axial zone of the chain. The highest slip-rate value of 0.8 mm/yr is in fact recorded in the Agri River high valley at that time. In the same area, the slip rate slowed down to 0.1 mm/yr in the last 30 k.y. (data from Giano et al., 2000). It is remarkable that in the external sectors of the chain, the tectonic regime changed from transcurrent to extensional during the time span between 0.7 and 0.5 Ma (Pieri et al., 1997), as a consequence of a diachronous activation of normal faulting from inner to outer areas.

On this basis, it can be affirmed that in the Melandro Valley, a comparable partitioning between local (i.e., tectonic) and regional uplift has to be invoked to justify the present elevation of the relict morphological features, whereas the major part of the relief energy in the Agri River high valley can be ascribed to the activity of basin-border faults. Yet, also in this area, the local morphostructural offsets have to be coupled with regional raising of the orogen to reach the total amount of Quaternary uplift. Maximum uplift rates along the axial zone of the chain have been recorded in the Mercure basin–Pollino Ridge area. The southward increasing uplift rate trend may be related to the higher faulting activity in the southernmost part of the investigated area. In any case, during the last 2 m.y., the total uplift of the axial zone of the Lucanian Apennine is ~1.2–1.3 km, with local peaks of 1.5 km.

On the other hand, in the pelitic beds from Lagonegro units, the mineral assemblage of the <2 μm fraction, mainly composed of illite and I/S mixed layers, and the thermal indicators as the percentage of illitic layers in I/S mixed layers (60% to 90%, R ordering in the range of 0.5–3), the illite crystallinity expressed as Kübler index (KI mode = 0.88 $\Delta°2\theta$), the percentage of white mica polytype $2M_1$ (in the range of 15%–35%), and the CAI (values in the range of 1.5–2.5) are consistent with a tectonic loading corresponding to 4–5 km burial depth.

The Quaternary uplift and the related erosion rates of the southern Apennines are unquestionably due to extensional tectonics coupled with thermal/isostatic regional raising and, in minor extent, to strike-slip faulting acting in the earlier deformational stage of the south Apennines chain (Schiattarella, 1998; Schiattarella et al., 2003). The gap of several kilometers deriving from the comparison between uplift rates and tectonic loading values may be explained only with different exhumation modalities starting from late Miocene times. This age can be obtained assuming a fixed rate of 0.6 mm/yr, which represents the best long-term estimate for the axial zone of the chain. Going back in time, using such a conservative rate to get a denudation value (i.e., rock uplift) of ~4–5 km, Tortonian age is reached, which is in concordance with the older exhumation age furnished by apatite fission-track analysis (Aldega et al., 2003). At that time, contractional tectonics were still active in this part of the southern Apennines, while extensional faulting acted along the internal (i.e., Tyrrhenian) side of the chain as a consequence of the spreading of a backarc basin (Malinverno and Ryan, 1986; Rehault et al., 1987). A lot of low-angle normal or transtensional contacts—diversely interpreted as due to pre-Quaternary orogen-parallel extension of the belt (Ferranti et al., 1996) or as structures associated with Pliocene-Pleistocene strike-slip faulting (Schiattarella, 1996)—are present in the Mesozoic units outcropping in the western belt of the south Apennines chain. As a rule, Miocene siliciclastic formations tectonically overlie different older successions in the axial zone. Younger-on-older tectonic contacts are also recognizable in the external sectors of the orogen, where Miocene synorogenic units are often detached from their Cretaceous to Oligocene pre-orogenic successions and moved forward.

In the light of all the field evidences, tectonic exhumation ("tectonic erosion" sensu Mancktelow, 2000, and references therein; see also "tectonic denudation" in Burbank and Anderson, 2001) may be a reliable explanation for the discrepancy between the different data sources examined here (i.e., between the amounts of rock uplift and surface uplift, sensu England and Molnar, 1990). Such phenomena (see Ring et al., 1999, for a wide review on these processes) led to the exhumation of the Mesozoic core of the chain (Lagonegro units), causing low-angle extension on its Tyrrhenian side by reactivation and inversion of older thrusts, and stacking on its eastern margin (i.e., frontal sector of the orogen) of Cretaceous to Miocene pre- and synorogenic units by gravity megasliding (Fig. 10). From late Pliocene times, as corroborated by younger apatite fission-track cooling ages (Aldega et al., 2003), this structure was first eroded and largely planed, then cut off by transcurrent and normal high-angle faults responsible for the recent uplift of the orogen in combination with several mechanisms acting on a regional scale, such as deep stacking of crustal elements in the axial portion of the chain and footwall uplift of its western sector by backarc extension.

ACKNOWLEDGMENTS

This study was supported by Italian Ministry of University and Research Cofin 2000 (M. Schiattarella) and National

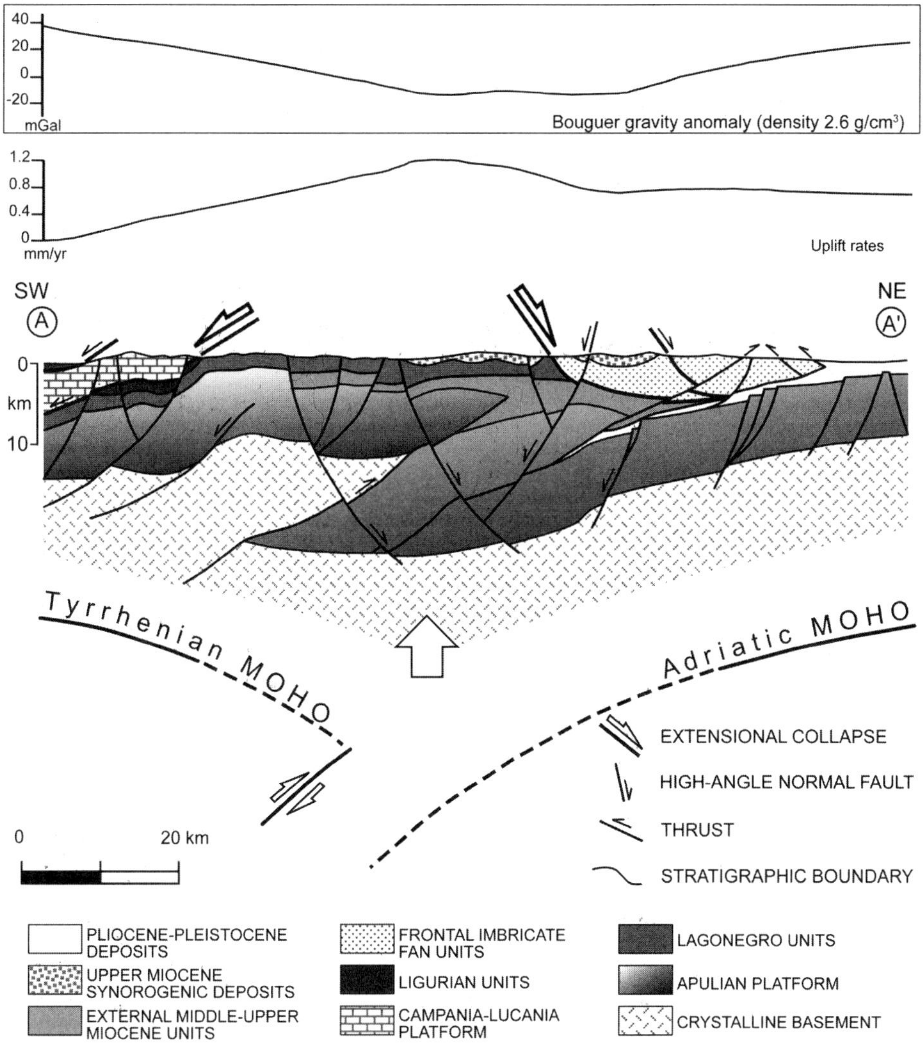

Figure 10. Regional cross section of the southern Apennines with deep structural interpretation and its relationships with uplift rate (from Fig. 6) and gravimetric (after Bernabini et al., 2002) profiles. The trace of the cross section is shown in Figure 1. Note the strong correspondence among the minimum in Bouguer gravity anomaly curve, the maximum in the uplift rate profile, crustal thickening, and the exhumed Mesozoic core of the chain formed by the Lagonegro units.

Research Council Agenzia 2000 (P. Di Leo) funds. We wish to thank the Mineralogy Department of the Natural History Museum of London for the use of the Nonius PDS 120 X-ray powder diffraction system and John S. Loring for his permission to use the LINKFIT program for the estimation of mica polytypes. We also thank Mark Brandon, Frank Pazzaglia, and Massimiliano Zattin for their useful comments and suggestions in reviewing the manuscript.

REFERENCES CITED

Aldega, L., Cello, G., Corrado, S., Cuadros, J., Di Leo, P., Giampaolo, C., Invernizzi, C., Martino, C., Mazzoli, S., Schiattarella, M., Zattin, M., and Zuffa, G., 2003, Tectono-sedimentary evolution of the southern Apennines (Italy): Thermal constraints and modelling: Atti Ticinensi di Scienze della Terra, Serie speciale, v. 9, p. 135–140.

Amato, A., 2000, Estimating Pleistocene tectonic uplift rates in the southeastern Apennines (Italy) from erosional land surfaces and marine terraces, *in* Slaymaker, O., ed., Geomorphology, human activity and global environmental change: New York, Wiley & Sons, p. 67–87.

Amato, A., and Cinque, A., 1999, Erosional land surfaces of the Campano-Lucano Apennines (S. Italy): Genesis, evolution, and tectonic implication: Tectonophysics, v. 315, p. 251–267, doi: 10.1016/S0040-1951(99)00288-7.

Amato, A., Aucelli, P.P.C., and Cinque, A., 2003, The long-term denudation rate in the southern Apennines Chain (Italy): A GIS-aided estimation of the rock volumes eroded since middle Pleistocene time: Quaternary International, v. 101–102, p. 3–11, doi: 10.1016/S1040-6182(02)00087-3.

Ascione, A., and Romano, P., 1999, Vertical movements on the eastern margin of the Tyrrhenian extensional basin: New data from Mt. Bulgheria (southern Apennines, Italy): Tectonophysics, v. 315, p. 337–356, doi: 10.1016/S0040-1951(99)00279-6.

Bartolini, C., 1999, An overview of Pliocene to present-day uplift and denudation rates in the northern Apennine, *in* Smith, B.J., Whalley, W.B., and Warke, P.A., eds., Uplift, erosion and stability: Perspectives on long-term landscape development: Geological Society [London] Special Publication 162, p. 119–125.

Bernabini, M., Di Bucci, D., Orlando, L., Parotto, M., Tiberti, M.M., and Tozzi, M., 2002, Nuovi vincoli gravimetrici 3D in Italia centro-meridionale per un modello della crosta profonda: Memorie della Società Geologica Italiana, v. 57, p. 335–342.

Bordoni, P., and Valensise, G., 1998, Deformation of the 125 ka marine terrace in Italy: Tectonic implications, *in* Stewart, I., and Vita-Finzi, C., eds., Late Quaternary coastal tectonics: Geological Society [London] Special Publication 146, p. 71–110.

Brancaccio, L., Cinque, A., Romano, P., Rosskopf, C., Russo, F., Santangelo, N., and Santo, A., 1991, Geomorphology and neotectonic evolution of a sector of the Tyrrhenian flank of the southern Apennines (region of

Naples, Italy): Zeitschrift für Geomorphologie Supplementbände, v. 82, p. 47–58.

Brandon, M.T., Roden-Tice, M.K., and Garver, J.I., 1998, Late Cenozoic exhumation of the Cascadia accretionary wedge in the Olympic Mountains, northwest Washington State: Geological Society of America Bulletin, v. 110, p. 985–1009, doi: 10.1130/0016-7606(1998)110<0985:LCEOTC>2.3.CO;2.

Burbank, D.W., and Anderson, R.S., 2001, Tectonic geomorphology: Oxford, Blackwell Science, 274 p.

Cinque, A., 1992, Distribuzione spazio-temporale dei movimenti tettonici verticali nell'Appennino campano-lucano: Alcune riflessioni: Studi Geologici Camerti Special Volume 1992/1, p. 33–38.

Corrado, S., Invernizzi, C., and Mazzoli, S., 2002, Tectonic burial and exhumation in a foreland fold and thrust belt: The Monte Alpi case history (southern Apennines, Italy): Geodinamica Acta, v. 15, p. 159–177, doi: 10.1016/S0985-3111(02)01086-0.

Cucci, L., and Cinti, F.R., 1998, Regional uplift and local tectonic deformation recorded by the Quaternary marine terraces on the Ionian coast of the northern Calabria (southern Italy): Tectonophysics, v. 292, p. 67–83, doi: 10.1016/S0040-1951(98)00061-4.

Di Leo, P., 2001, Phase equilibria and geothermobarometry in very low-grade metapelites: Plinius, v. 26, p. 168–186.

Di Leo, P., 2003, Use of clay mineralogy in reconstructing geological processes: Thermal constraints from clay minerals: Atti Ticinensi di Scienze della Terra, Serie speciale, v. 9, p. 55–68.

Di Leo, P., Dinelli, E., Mongelli, G., and Schiattarella, M., 2002, Geology and geochemistry of Jurassic pelagic sediments, Scisti silicei Formation, southern Apennines, Italy: Sedimentary Geology, v. 150, p. 229–246, doi: 10.1016/S0037-0738(01)00181-6.

Di Leo, P., Martino, C., Mastandrea, G., and Perri, E., 2003, Kübler index and conodont color alteration index as thermal indicators from lower Triassic to Jurassic pelagic successions of the Lucanian Apennine, southern Italy: Late Triassic–Early Jurassic events in the framework of the Pangea break-up: Capri, Italy, Proceedings, International Geological Correlation Programme (IGCP) 458 Workshop, September 30–October 1, 2003, p. 34–35.

England, P., and Molnar, P., 1990, Surface uplift, uplift of rocks, and exhumation of rocks: Geology, v. 18, p. 1173–1177, doi: 10.1130/0091-7613(1990)018<1173:SUUORA>2.3.CO;2.

Ferranti, L., Oldow, J.S., and Sacchi, M., 1996, Pre-Quaternary orogen-parallel extension in the southern Apennine belt, Italy: Tectonophysics, v. 260, p. 325–347, doi: 10.1016/0040-1951(95)00209-X.

Garcia-Lopez, S., Brime, C., Bastida, F., and Sarmiento, G.N., 1997, Simultaneous use of thermal indicators to analyze the transition from diagenesis to metamorphism: An example from Variscan Belt of northwest Spain: Geological Magazine, v. 134, p. 323–334, doi: 10.1017/S0016756897006882.

Ghisetti, F., and Vezzani, L., 1981, Contribution of structural analysis to understanding the geodynamic evolution of the Calabrian arc (southern Italy): Journal of Structural Geology, v. 3, p. 371–381, doi: 10.1016/0191-8141(81)90037-7.

Giano, S.I., and Schiattarella, M., 2002, Geomorfologia e neotettonica dell'alta Val d'Agri, *in* Boenzi, F., and Schiattarella M., eds., Guida all'Escursione Geomorfologica: Dalla Val d'Agri a Matera: AIGeo Assembly, Excursion Guidebook, Potenza, TecnoStampa, p. 22–34.

Giano, S.I., Maschio, L., Alessio, M., Ferranti, L., Improta, S., and Schiattarella, M., 2000, Radiocarbon dating of active faulting in the Agri high valley, southern Italy: Journal of Geodynamics, v. 29, p. 371–386, doi: 10.1016/S0264-3707(99)00058-7.

Guzzetta, G., 1974, Ancient tropical weathering in Calabria: Nature, v. 251, p. 302–303.

Hoffman, J., and Hower, J., 1979, Clay mineral assemblages as low grade metamorphic geothermometers: Application to the thrust faulted Disturbed Belt of Montana, U.S.A: Society of Economics and Mineralogist Special Publication 26, p. 55–79.

Jaboyedoff, M., and Thélin, P., 1996, New data on low-grade metamorphism in the Briançonnais domain of the pre-Alps, western Switzerland: European Journal of Mineralogy, v. 8, p. 577–592.

Kisch, H.J., 1991, Illite crystallinity: Recommendations on samples preparation, X-ray diffraction settings, and interlaboratory samples: Journal of Metamorphic Geology, v. 9, p. 665–670.

Kübler, B., 1967, La cristallinité de l'illite et les zones tout à fait supérieurs du métamorphisme, *in* Kübler, B., ed., Etages tectoniques, Colloques de Neuchâtel: Neuchâtel, Université Neuchâtel à la Baconnière, p. 105–121.

Malinverno, A., and Ryan, W.B.F., 1986, Extension in the Tyrrhenian Sea and shortening in the Apennines as result of arc migration driven by sinking of the lithosphere: Tectonics, v. 5, p. 227–245.

Mancktelow, N., 2000, Exhumation of high-pressure rocks, *in* Ranalli, G., Ricci, C.A., and Trommsdorff, V., eds., Crust-mantle interactions: Siena, Proceedings of the International School Earth and Planetary Sciences, p. 41–52.

Merriman, R.J., and Frey, M., 1999, Very low-grade metapelites: Mineralogy, microfabric and measuring reaction progress, *in* Frey, M., and Robinson, D., eds., Low-grade metamorphism: Oxford, Blackwell Science, p. 61–107.

Merriman, R.J., and Peacor, D.R., 1999, Very low-grade metapelites: Mineralogy, microfabric and measuring reaction progress, *in* Frey, M., and Robinson, D., eds., Low-grade metamorphism: Oxford, Blackwell Science, p. 10–58.

Mongelli, F., Harabaglia, P., Martinelli, G., Squarci, P., and Zito, G., 1996, Nuove misure di flusso geotermico in Italia meridionale: Possibili implicazioni sismotettoniche: Proceedings, 14th Gruppo Nazionale Geofisica della Terra Solida (GNGTS) Conference, Rome, October 23–25, 1995.

Moore, D.M., and Reynolds, R.C., 1997, X-ray diffraction and identification of clay minerals: Oxford, Oxford University Press, 375 p.

Ortolani, F., Pagliuca, S., Pepe, E., Schiattarella, M., and Toccaceli, R.M., 1992, Active tectonics in the southern Apennines: Relationships between cover geometries and basement structure: A hypothesis for a geodynamic model: International Geological Correlation Programme (IGCP) no. 276 Newsletter, v. 5, p. 413–419.

Pascale, P., 2003, Analisi geomorfica quantitativa e morfostrutturale del bacino idrografico della Fiumara di Tito-Picerno (Appennino lucano) [M.S. thesis]: Potenza, Italy, Basilicata University, 119 p.

Pescatore, T., Renda, P., Schiattarella, M., and Tramutoli, M., 1999, Stratigraphic and structural relationships between Meso-Cenozoic Lagonegro basin and coeval carbonate platforms in southern Apennines, Italy: Tectonophysics, v. 315, p. 269–286, doi: 10.1016/S0040-1951(99)00278-4.

Pieri, P., Vitale, G., Beneduce, P., Doglioni, C., Gallicchio, S., Giano, S.I., Loizzo, R., Moretti, M., Prosser, G., Sabato, L., Schiattarella, M., Tramutoli, M., and Tropeano, M., 1997, Tettonica Quaternaria nell'area Bradanico-Ionica: Il Quaternario, v. 10, p. 535–542.

Rehault, J.P., Moussat, E., and Fabbri, A., 1987, Structural evolution of the Tyrrhenian back-arc basin: Marine Geology, v. 74, p. 123–150, doi: 10.1016/0025-3227(87)90010-7.

Ring, U., Brandon, M.T., Willett, S.D., and Lister, G.S., 1999, Exhumation processes, *in* Ring, U., Brandon, M.T., Lister, G.S., and Willett, S.D., eds., Exhumation processes: Normal faulting, ductile flow and erosion: Geological Society [London] Special Publication 154, p. 1–27.

Russo, F., and Schiattarella, M., 1992, Osservazioni preliminari sull'evoluzione morfostrutturale del bacino di Castrovillari (Calabria settentrionale): Studi Geologici Camerti, Special Volume 1992/1, p. 271–278.

Santangelo, N., 1991, Evoluzione stratigrafica, geomorfologica e neotettonica di alcuni bacini lacustri del confine Campano Lucano (Italia meridionale) [Ph.D. thesis]: Naples, Italy, University of Naples "Federico II", 109 p.

Scandone, P., 1975, The preorogenic history of the Lagonegro basin (southern Apennines), *in* Squyres, C., ed., Geology of Italy: Tripoli, The Earth Sciences Society of the Libyan Arabian Republic, p. 305–315.

Schiattarella, M., 1996, Tettonica della Catena del Pollino (Confine Calabro-Lucano): Memorie della Società Geologica Italiana, v. 51, p. 543–566.

Schiattarella, M., 1998, Quaternary tectonics of the Pollino Ridge, Calabria-Lucania boundary, southern Italy, *in* Holdsworth, R.E., Strachan, R.A., and Dewey, J.F., eds., Continental transpressional and transtensional tectonics: Geological Society [London] Special Publication 135, p. 341–354.

Schiattarella, M., Torrente, M.M., and Russo, F., 1994, Analisi strutturale ed osservazioni morfostratigrafiche nel bacino del Mercure (Confine Calabro-Lucano): Il Quaternario, v. 7, p. 613–626.

Schiattarella, M., Di Leo, P., Beneduce, P., and Giano, S.I., 2003, Quaternary uplift vs. tectonic loading: A case-study from the Lucanian Apennine, southern Italy: Quaternary International, v. 101–102, p. 239–251, doi:

10.1016/S1040-6182(02)00126-X.
Schiattarella, M., Beneduce, P., and Pascale, S., 2004, Comparazione tra i tassi di erosione e sollevamento dell'Appennino lucano: L'esempio della Fiumara di Tito e Picerno: Bollettino dell'Associazione Italiana di Cartografia, v. 121-122, p. 367–385.
Velde, B., and Hower, J., 1963, Petrological significance of illite polytypism in Paleocene sedimentary rocks: American Mineralogist, v. 48, p. 1239–1254.
Warr, L.N., and Rice, A.H.N., 1994, Interlaboratory standardization and calibration of clay mineral crystallinity and crystal size data: Journal of Metamorphic Geology, v. 12, p. 141–152.
Watchman, A.L., and Twidale, C.R., 2002, Relative and 'absolute' dating of land surfaces: Earth-Science Reviews, v. 58, p. 1–49, doi: 10.1016/S0012-8252(01)00080-0.
Westaway, R., 1993, Quaternary uplift of southern Italy: Journal of Geophysical Research, v. 98, p. 21,741–21,772.
Widdowson, M., 1997, The geomorphological and geological importance of palaeosurfaces, *in* Widdowson, M., ed., Palaeosurfaces: Recognition, reconstruction and palaeoenvironmental interpretation: Geological Society [London] Special Publication 120, p. 1–12.
Willett, S.D., and Brandon, M.T., 2002, On steady states in mountain belts: Geology, v. 30, p. 175–178, doi: 10.1130/0091-7613(2002)030<0175:OSSIMB>2.0.CO;2.

Manuscript Accepted by the Society 23 June 2005

Printed in the USA

Geological Society of America
Special Paper 398
2006

Slow, patchy landscape evolution in northern Sweden despite repeated ice-sheet glaciation

A.P. Stroeven[†]
Department of Physical Geography and Quaternary Geology, Stockholm University, S-106 91 Stockholm, Sweden
J. Harbor[‡]
Department of Earth and Atmospheric Sciences, Purdue University, West Lafayette, Indiana 47907, USA
D. Fabel[§]
Department of Geographical and Earth Sciences, Glasgow University, Glasgow, G12 8QQ, UK
J. Kleman
C. Hättestrand
Department of Physical Geography and Quaternary Geology, Stockholm University, S-106 91 Stockholm, Sweden
D. Elmore[#]
Purdue Rare Isotope Measurement Laboratory, Purdue University, West Lafayette, Indiana 47907, USA
D. Fink[††]
Australian National Tandem for Accelerator Research, Australian Nuclear Science and Technology Organisation, PMB 1, Menai, New South Wales, 2234, Australia
O. Fredin
Geological Survey of Norway/University of Bergen, Leiv Eirikssons vei 39, N-7491 Trondheim, Norway

ABSTRACT

The conventional assumption that erosion by ice sheets is pervasive and effective in landscape evolution is tested in northern Sweden using geomorphic mapping and cosmogenic nuclide analyses of formerly glaciated surfaces. The following evidence indicates that recent glaciations in this region have produced only slow and patchy landscape evolution: (1) Geomorphic mapping shows that at least 20% of the repeatedly glaciated study region in the northern Swedish mountains has landforms that are relict, i.e., clearly nonglacial in origin. (2) The contrast between cosmogenic apparent exposure ages from relict landforms in the northern Swedish mountains and from overlying glacial erratics and juxtaposed glacially eroded bedrock surfaces, which are consistent with last deglaciation, implies that the relict landforms have been preserved through multiple glacial cycles. (3) Apparent ^{10}Be and ^{26}Al exposure ages for tor summit bedrock surfaces in the northern Swedish lowlands reveal that these relict landforms have survived at least eleven exposure and ten burial events with little or no erosion over the past ~1 m.y. (4) The northern Swedish lowland and mountains are primarily covered by glacial landforms. However geomorphic mapping suggests that even these landforms may have undergone limited erosion during the last glacial cycle. Cosmogenic ^{10}Be and ^{36}Cl data from what appear to be heavily scoured areas

[†]E-mail: arjen@geo.su.se.
[‡]E-mail: jharbor@purdue.edu.
[§]E-mail: dfabel@ges.gla.ac.uk.
[#]E-mail: elmore@purdue.edu.
[††]E-mail: fink@ansto.gov.au.

Stroeven, A.P., Harbor, J., Fabel, D., Kleman, J., Hättestrand, C., Elmore, D., Fink, D., and Fredin, O., 2006, Slow, patchy landscape evolution in northern Sweden despite repeated ice-sheet glaciation, *in* Willett, S.D., Hovius, N., Brandon, M.T., and Fisher, D.M., eds., Tectonics, Climate, and Landscape Evolution: Geological Society of America Special Paper 398, Penrose Conference Series, p. 387–396, doi: 10.1130/2006.2398(24). For permission to copy, contact editing@geosociety.org.

in one glacial corridor indicate erosion of only ~2 ± 0.4 m of bedrock during the last glaciation. These results suggest that in some areas the overall modification produced by ice sheets may be more restricted than previously thought, or it has occurred preferentially during earlier Quaternary glacial periods.

Keywords: Fennoscandian ice sheet, landscape evolution, terrestrial cosmogenic isotopes, glacial erosion.

INTRODUCTION

In the late nineteenth century, Ramsay (1864, 1876), Judd (1876), Bonney (1871, 1874), and Irving (1883), among others, had a protracted debate about whether landscape modification by ice was purely superficial, or if it was in fact largely responsible for the great depth and unique form of glaciated valleys. Although the significance of glacial erosion was still doubted by some influential geologists in the early twentieth century (e.g., Bonney, 1910; Garwood, 1910), since the 1880's it has generally been accepted that glaciers and ice sheets are capable of significant erosion (Davies, 1969). This notion that ice sheets are pervasive agents of erosion has been fostered by:

1. The visual impression of huge amounts of erosion in glaciated mountain landscapes; i.e., steep-sided glacial troughs, overdeepened basins, which, particularly in coastal settings, are impressive because they have been inundated by the sea (fjords);
2. the ubiquity of lakes and lake basins in formerly glaciated regions, and the virtual absence of these in regions that have never been glaciated;
3. the large volumes and ubiquitous occurrence of glacial and glacifluvial deposits; and
4. measurements of subglacial erosion by mountain glaciers (the primary source of field knowledge).

Hence, it is the overwhelming presence of a diverse array of evidence of the actions of ice that has led to the popular notion of pervasive ice-sheet modification of underlying landscapes. In addition, there is abundant offshore evidence for the erosion, transport, and deposition of large amounts of debris by glaciers. Typically, these sediments occur in packages called fans, a word that links to the idea that these deposits relate to the location of the grounding line of former ice streams (e.g., Solheim et al., 1996).

Notwithstanding the indisputable evidence for local large volumes of glacial erosion (carving of fjords, U-shaped valleys, and piedmont lakes) and deposition (offshore fans), the pervasiveness of ice-sheet erosion has continued to be questioned (Glasser and Hall, 1997; Hall and Sugden, 1987; Kleman and Stroeven, 1997; Lidmar-Bergström, 1997; Sugden, 1968, 1974, 1976). Alternative models include the possibility that individual ice sheets may have a patchy impact on the landscape, but that the combined impact of consecutive ice sheets affect most of the landscape at some point in time. Several types of observations support this alternative model. First, the large-scale geomorphology of many mountainous regions in formerly glaciated areas indicates large spatial differences in glacial erosion (Sugden 1968, 1974). For example, the typical geomorphology of these mountains is a patchwork of glaciated terrain and relict upland surfaces (Ahlmann, 1919; Gjessing, 1967; Kleman and Stroeven, 1997; Peulvast, 1988; Reusch, 1901; Sugden and Watts, 1977; Wråk, 1908). Glacial terrain is recognized through the presence of U-shaped valleys, cirques, horns and arrêtes, hanging valleys, mountain asymmetry, and the ubiquitous presence of lakes (e.g., Sugden and John, 1976). Relict (or remnant) upland surfaces, on the other hand, are recognized through the presence of winding V-shaped valleys, mountain symmetry, tors, weathering mantles, and an absence of (water-filled) rock basins (Kleman and Stroeven, 1997; Stroeven et al., 2002b; Sugden and Watts, 1977). This patchwork pattern of glacial erosion and preservation could either result from restricted ice extents (glaciers limited to valleys) or, where we know that large ice sheets covered the mountains, such as in northern Sweden, from complex patterns of ice-sheet basal thermal regimes.

Second, from the cross-cutting patterns of sets of glacial lineations (glacial morphology formed in a flow-parallel direction), it was concluded that glaciers have only very gently remolded pre-existing morphological features during subsequent glacial stages (Kleman et al., 1997). In these situations, neither erosion nor deposition was pervasive enough to remove the pre-existing glacial morphology. One reason for this apparent inability to change the landscape surface could be that glacial erosion is a very slow process. This explanation is presumably at odds with abundant observations on contemporary glaciers regarding the relative efficacy of glacial evacuation of loose debris and even erosion into competent bedrock. Another reason could be that the conditions appropriate for landscape modification by ice sheets only occurred during short periods of time (a transient feature). Because this explanation appears more likely, there now is a growing awareness that ice sheets can in fact protect the landscape for extensive periods of time (André, 2004; Clarhäll and Kleman, 1999; Dyke, 1993; Hättestrand and Stroeven, 2002; Kleman, 1992, 1994; Kleman and Borgström, 1990; Kleman and Hättestrand, 1999).

Third, the existence of interstadial and interglacial sediments in formerly glaciated regions signifies the relative inability of subsequent ice sheets to erode these sediments (Lagerbäck and Robertsson, 1988; Mangerud et al., 1979). In comparison to the preservation of "undated" glacial morphology, the presence of interstadial or interglacial sediments helps define the length of the time during which these sediments were subglacially preserved. Similarly significant, but harder to date, are occurrences of periglacial features or glacial deposits underneath more recent tills.

Finally, an abundance of direct observations and remotely sensed data indicate that contemporary ice sheets, ice caps, and polar glaciers are able to preserve older geomorphological features, sediments, or vegetation. For example, Arctic ice caps have retreated across older geomorphological features such as raised shorelines (e.g., Jonsson, 1983), the East Antarctic Ice Sheet retreated across Pliocene marine sediments during the Holocene (Fabel et al., 1997; Pickard, 1986), and the melting Greenland Ice Sheet and Arctic ice cap margins have exposed vegetation in growth position (Bergsma et al., 1984; Falconer, 1966; Goldthwait, 1960; Holmgren et al., 1984). The implication of all these observations is that although the ice masses once covered these locations, they were unable to erode their substrate.

Even though it is now becoming apparent that large areas underneath former ice sheets were either preserved or only slightly modified during the last glacial cycle, we have much less evidence for the behavior of earlier ice sheets. Based on the nonglacial nature of some mountain upland surfaces (Kleman and Stroeven, 1997; Sugden, 1968, 1974; Sugden and Watts, 1977) and extensive lowland surfaces (Lidmar-Bergström, 1997; Rudberg, 1954; Wråk, 1908), however, it can be argued that these have remained unaffected by glacial erosion for the entire duration of the Cenozoic ice ages (Hättestrand and Stroeven, 2002; Kleman and Stroeven, 1997). Moreover, relict glacial depositional morphology may be of a variety of ages, and glacial erosional features may be relict features. These questions are pertinent to the issue of whether the long-term evolution of landscapes in some areas of repeated glaciation is slow and patchy, and are addressed in this paper by means of cosmogenic nuclide studies performed in northern Sweden.

METHODOLOGY

The primary method used in our effort to estimate amounts, rates, and patterns of glacial erosion is the interpretation of cosmogenic radionuclide concentrations in samples from rock surfaces. The cosmogenic radionuclides ^{10}Be (half-life = 1.51×10^6 yr), ^{26}Al (half-life = 7.1×10^5 yr), and ^{36}Cl (half-life = 3.01×10^5 yr) are produced in rocks near the ground surface by reactions with secondary and tertiary cosmic ray neutrons and muons (Lal and Peters, 1967). These radionuclides are commonly used in studies of landscape evolution because they only form within a few meters of Earth's surface in quartz (^{10}Be,^{26}Al) and feldspar (^{36}Cl), which are ubiquitous minerals in crustal rocks and sediments (cf. Bierman, 1994; Cerling and Craig, 1994; Fabel and Harbor, 1999; Gosse and Phillips, 2001; Nishiizumi et al., 1993). The in situ production provides a means of determining the amount of time the mineral has been at or near the ground surface (apparent exposure age; Lal, 1991). Because nuclide production decreases with depth, removal of two or more meters of irradiated rock during a glacial event will create a zero age surface (Fabel et al., 1997). In this context, areas known to have been ice covered should have exposure ages equivalent to deglaciation if they were significantly eroded by ice, and older exposure ages if they suffered limited erosion or were completely protected.

Depending on the level of sophistication of the data set for the different subregions studied, we employed one or more of the following three interpretations. First, single radionuclide evidence was used to acquire minimum limiting ages and maximum limiting erosion rates for these sites. This represents the simplest interpretation of the data, as it assumes continuous exposure (i.e., no burial by ice). Second, when both ^{26}Al and ^{10}Be were measured on the same samples, we used the ^{26}Al/^{10}Be ratio and the individual radionuclide concentrations to calculate minimum limiting total exposure and shielding (burial) durations, closely following the Bierman et al. (1999) approach. Finally, we calculated the minimum total history required to produce the exposure and shielding duration for samples with multiple isotopes, using the Deep Sea Drilling Project (DSDP) site 607 marine benthic foraminifera oxygen isotope record (Lazarus et al., 1995; Raymo et al., 1989; Ruddiman et al., 1989) as a proxy for the approximate duration of periods of ice-sheet cover versus ice-free conditions (Fabel et al., 2002).

The sampling strategy reflects the questions posed in the introduction in that we collected (1) samples of relict upland bedrock surfaces (upland tors), (2) samples of bedrock surfaces created by significant glacial erosion ("glacial surfaces": on uplands and in the valleys), and (3) samples of bedrock or boulders from deglaciation features (i.e., meltwater channels, rock slope failures) and of erratics (ice-transported boulders). Field sampling procedures and laboratory procedures followed standards set by others, and have been detailed elsewhere (e.g., Fabel et al., 2002; Stroeven et al., 2002a, 2002c).

RESULTS

The cosmogenic radionuclide results given in Table 1 refer to the three geomorphological settings of interest: (1) individual mountain blocks, (2) individual glacial valleys, and (3) the northern Swedish lowland (Fig. 1). The first region is represented by eight bedrock samples for which cosmogenic radionuclides in a simple approach would indicate that they have been preserved through at least the last glacial cycle, because their apparent exposure ages are older than local deglaciation ages (32.7 ± 2.0 to 72.6 ± 4.4 ka). Local deglaciation ages of the relict bedrock surfaces are represented by four dates on erratic boulders (7.4 ± 0.5 to 12.7 ± 0.8 ka) and two dates from glacial surfaces (cut transverse to ice flow), formed when the ice sheet plucked bedrock to considerable (>2 m) depth (9.7 ± 0.7 and 10.8 ± 0.7 ka; cf. Fabel et al., 2002).

In the deep glacial valleys, bare bedrock outcrops are a common feature (Fig. 2). Because these valleys appear to have been deepened locally by hundreds of meters of glacial erosion, the expectation is that bedrock samples reflect deglaciation ages.

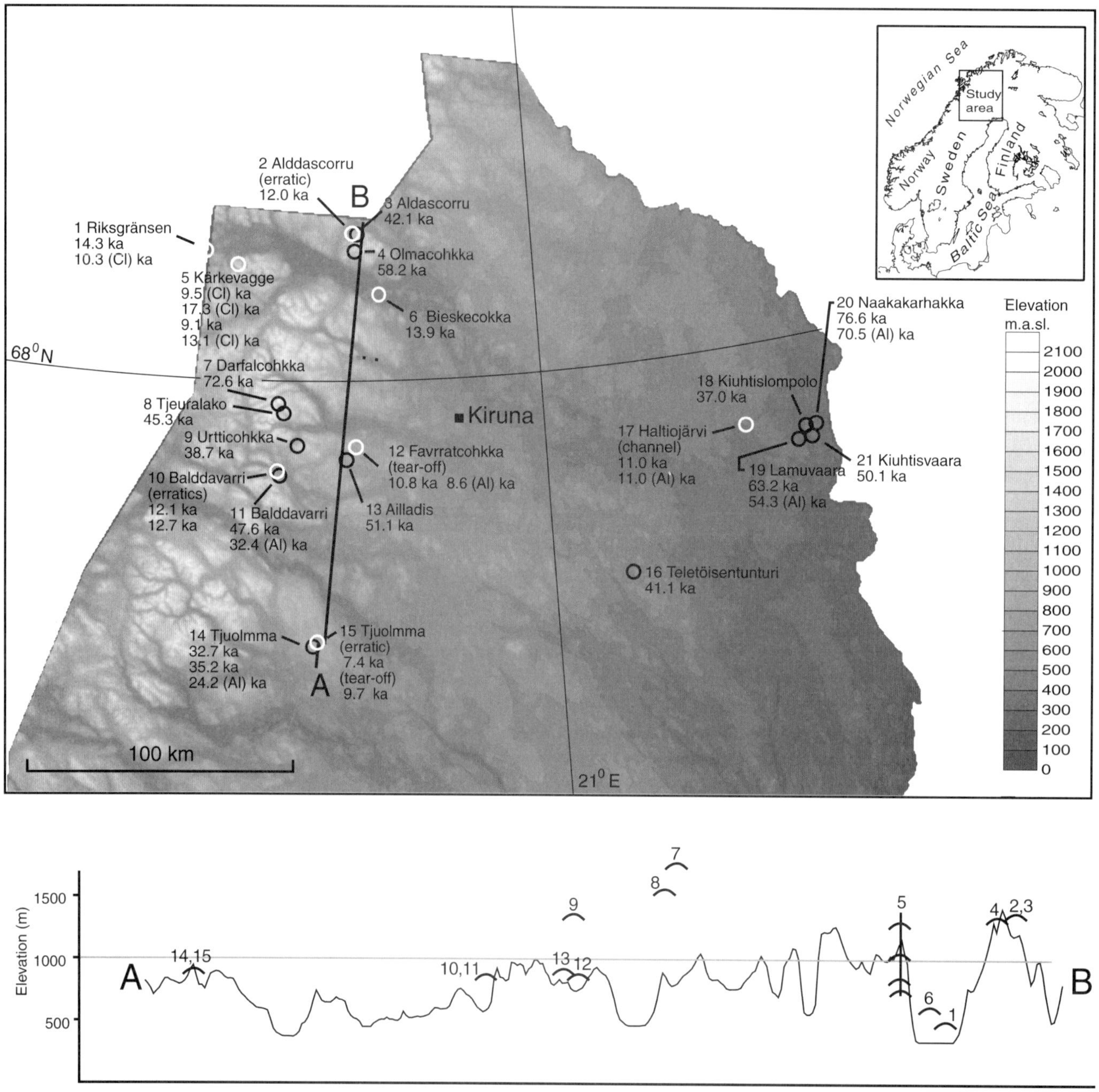

Figure 1. Map of northern Sweden study area with sample locations, location of profile A–B, and inset map showing the study region. Location numbering system follows that given in Table 1. Samples denoted by white circles were expected to yield deglaciation ages (between 8 and 13 ka), whereas samples denoted by black circles were expected to contain an inheritance signal. Unless noted otherwise, apparent ages are given in ^{10}Be ka and refer to bedrock samples. Digital elevation model of 50 m resolution (courtesy of the National Land Survey [Lantmäteriverket], 2002. Excerpt from GSD elevation data, case no. L2000/646). Bottom panel is profile A–B, which runs east of, and parallel to, the axis of the mountain range. The profile intersects a number of surfaces of intermediate elevation (800–1000 m above sea level [asl]), which exhibit an array of nonglacial characteristics (cf. Fig. 2), and prominent glacial valleys. Sample locations are projected onto the transect (except for the northern Swedish lowland locations, sites 16–21).

Two samples derived from the bottom (10.3 ± 1.7 and 14.3 ± 0.9 ka), slopes (9.5 ± 1.5 and 17.3 ± 2.8 ka), and rock bastions within these valleys (13.9 ± 0.9 and 47.6 ± 2.9 ka) show a broad range of ages. They can be compared to local deglaciation ages derived from erratics and the upland glacial surfaces mentioned before and the ages of two boulders in a rock slope failure (9.1 ± 0.6 and

TABLE 1. COSMOGENIC NUCLIDE DATA AND APPARENT EXPOSURE AGES

Sample location and elevation (m asl)[†]	Sample type	Correction factor[‡]	^{10}Be[§] ($\times10^5$ atom g^{-1})	^{26}Al ($\times10^5$ atom g^{-1})	^{36}Cl ($\times10^5$ atom g^{-1})	Apparent exposure age[§] ^{10}Be (ka)	^{26}Al (ka)	^{36}Cl (ka)
(i) Relict bedrock surfaces								
3. Alddasčorru (1380)	bedrock relict	0.98	7.43 ± 0.04			42.1 ± 2.5		
4. Olmáčohkka (1355)	bedrock relict	0.98	10.03 ± 0.05			58.2 ± 3.5		
7. Dárfalcohkka (1790)	bedrock	0.98	17.57 ± 0.12			72.6 ± 4.4[#]		
8. Tjeuralako Plateau (1560)	bedrock	0.96	9.03 ± 0.06			45.3 ± 2.7[#]		
9. Urttičohkka (1360)	bedrock relict	0.95	6.55 ± 0.06			38.7 ± 2.3		
13. Áilladis (910)	bedrock relict	0.98	5.96 ± 0.04			51.1 ± 3.1		
14. Tjuolmma (920)	bedrock relict	0.96	3.85 ± 0.03	17.31 ± 0.84		32.7 ± 2.0	24.2 ± 1.9	
14. Tjuolmma-R (920)	bedrock relict	0.96	4.14 ± 0.03			35.2 ± 2.1		
Erratics on relict surfaces								
2. Alddasčorru (1380)	erratic	0.98	2.14 ± 0.07			12.0 ± 0.8		
10. Bálddavárri (900)	erratic	0.97	1.38 ± 0.02			12.1 ± 0.7		
10. Bálddavárri (810)	erratic	0.94	1.34 ± 0.03			12.7 ± 0.8		
15. Tjuolmma (945)	erratic	0.99	0.92 ± 0.02			7.4 ± 0.5		
Glacial erosion of relict surfaces								
12. Fávrratčohkka (850)	Bedrock lee-side scarp	0.96	1.20 ± 0.04	5.83 ± 0.98		10.8 ± 0.7	8.6 ± 1.6	
15. Tjuolmma (930)	Bedrock lee-side scarp	0.98	1.15 ± 0.04			9.7 ± 0.7		
(ii) Glacial erosion in troughs								
1. Riksgränsen (510)	bedrock	0.98	1.20 ± 0.02			14.3 ± 0.9		
1. Riksgränsen (490)	bedrock	0.91			0.46 ± 0.02			10.3 ± 1.7
5. Kärkevagge (1050)	bedrock	0.98			1.14 ± 0.08			9.5 ± 1.5
5. Kärkevagge (1320)	bedrock	0.98			3.75 ± 0.16			17.3 ± 2.8
6. Bieskecohka (600)	bedrock	0.97	1.22 ± 0.03			13.9 ± 0.9		
11. Bálddavárri (890)	bedrock	0.97	5.50 ± 0.03	22.74 ± 0.60		47.6 ± 2.9	32.4 ± 2.2[#]	
Deglacial slope failure								
5. Kärkevagge (840)	boulder	0.98	1.01 ± 0.02			9.1 ± 0.6		
5. Kärkevagge (CM1) (730)	boulder	0.91			1.08 ± 0.06			13.1 ± 2.1
(iii) Northern Swedish lowland/tors								
18. Kiuhtislompolo (350)	bedrock	0.96	2.59 ± 0.03			37.0 ± 2.3		
19. Lamuvaara (300)	bedrock	0.96	4.18 ± 0.05	21.69 ± 1.72		63.2 ± 3.9	54.3 ± 5.6	
20. Naakakarhakka (360)	bedrock	0.96	5.37 ± 0.05	29.66 ± 1.76		76.6 ± 4.7	70.5 ± 6.2	
21. Kiuhtisvaara (350)	bedrock	0.97	3.53 ± 0.05			50.1 ± 3.1		
Northern Swedish lowland/deglaciation								
17. Haltiojärvi (280)	bedrock	0.97	0.73 ± 0.03	3.31 ± 0.49		11.0 ± 0.8	11.0 ± 1.8	
Northern Swedish lowland/drumlin								
16. Teletöisentunturi (575)	bedrock	0.99	3.65 ± 0.04			41.1 ± 2.5		

[†]Site location numbers match those given in Figure 1; elevation is in parentheses, asl—above sea level.

[‡]Correction factor includes geometric shielding (Dunne et al., 1999) and sample thickness using an attenuation coefficient of 160 ± 10 g cm^{-2} and rock density of 2.75 g cm^{-3}.

[§]Nuclide ratios measured by accelerator mass spectrometry at the Purdue Rare Isotope Measurement Laboratory (PRIME Lab) and the Australian National Tandem Accelerator for Applied Research (ANTARES) at the Australian Nuclear Science and Technology Organisation (ANSTO). Procedural blanks were used to correct measured ratios. Total Al measured by inductively coupled plasma–atomic emission spectrometry (ICP-AES) and assigned 2% uncertainty. Apparent ages are calculated assuming zero erosion and sea level and high latitude (>60°) ^{10}Be and ^{26}Al production rates of 5.1 ± 0.3 atom g^{-1} yr^{-1} and 31.1 ± 1.9 atom g^{-1} yr^{-1}, respectively. Production rates were scaled to site-specific latitude and altitude using Stone (2000) and a mean sea-level temperature of 5 °C. Age uncertainties are based on analytical errors only and do not include systematic uncertainties (which are commonly considered to be ~10%; Gosse and Phillips, 2001). ^{36}Cl production rates at sea level and latitude >60°N: $P_{Ca\ spallation}$ = 66.8 ± 6.8 (gCa)$^{-1}$ yr^{-1}, $P_{Ca\ muon\ capture}$ = 4.8 ± 0.5 (gCa)$^{-1}$ yr^{-1}, P_K = 154 ± 10 (gK)$^{-1}$ yr^{-1}. Neutron capture (including thickness corrections) treated according to method of Liu et al. (1994), with $P_{f,\ air}$ = 597 ± 83 g^{-1} yr^{-1}.

[#]Previously unpublished and measured at ANSTO (Fink et al., 2004). All other data from Fabel et al. (2002), Hättestrand et al. (2004), and Stroeven et al. (2002a, 2002c).

Figure 2. (A) View toward the east across the uplands of intermediate elevation. The glacial valley to the right (south) is hanging with respect to the glacial valley to the left (north, with lake Rautasjaure). Surfaces in the middle distance north and south of Rautasjaure have previously been mapped as relict surface with tors. (B) View toward the west across the Leavas valley (of nonglacial character) on one of the intermediate elevation uplands fringing the mountain range (background). (C) Glacially scoured bedrock surfaces at Riksgränsen (sample locations and border in the extreme foreground). (D) Tor at Naakakarhakka (site 20) with typical deep-weathering characteristics (cf. Hättestrand and Stroeven, 2002). (E) Upland surfaces of intermediate elevation are often characterized by occurrences of tors (black bedrock protrusion in the foreground), similar to those found in the northern Swedish lowland (sites 18–21).

13.1 ± 2.1 ka). The rock fall was triggered by, and deposited onto, a decaying ice surface (cf. Rapp, 1960).

The northern Swedish lowland region is characterized by low-relief surfaces that slope toward the Baltic depression. One of these low relief surfaces is characterized by a sudden absence of glacial lineations and the presence of numerous tors (Hättestrand and Stroeven, 2002), four of which were dated (37.0 ± 2.3 ka to 76.6 ± 4.7 ka). They have radionuclide inventories in excess of the local deglaciation inventory, represented by a bedrock sample from a meltwater channel (11.0 ± 0.8 ka; Stroeven et al., 2002c). More typically, however, the northern Swedish lowland region is characterized by extensive sets of (partly cross-cutting) glacial lineations and an associated system of eskers. Drumlins are the most impressive glacial lineations, in terms of amounts of material eroded and deposited. A pilot study including one apparent exposure date concluded that drumlins are complex features that have formed over multiple glacial cycles (41.8 ka; cf. Hättestrand et al., 2004).

DISCUSSION

The cosmogenic radionuclide data gathered in this study strongly support the concept that some ice sheets have a minimal influence on the long-term landscape evolution of formerly glaciated regions. The results from apparent exposure ages are consistent with expectations, based on the patterns of erosion as deduced from geomorphological evidence (Hättestrand and Stroeven, 2002; Stroeven et al., 2002b). The evidence is all counter to the concept that recent ice sheets are pervasive agents of erosion. This view can only be maintained as a possible hypothesis for localities in northern Sweden that have not yet been studied using cosmogenic nuclide approaches.

Relict Uplands (~900–2100 m above sea level)

The evidence from the mountains clearly shows that relict surfaces, as defined by geomorphology (Figs. 2A, 2B, and 2E), are also "relict" in terms of bedrock samples having inherited cosmogenic radionuclide inventories. For the first time, it has been possible to advance beyond the semiqualitative statements that geomorphology can offer, and determine the evolution history (minimum exposure age and burial period) for these surfaces. Fabel et al. (2002) integrated the evidence from Mount Tjuolmma, the best locality in the mountains, and arrived at a conservative youngest age for the relict bedrock surface of 845^{+461}_{-418} ka. This result is quantitatively consistent with recent results from mountain summits in other formerly glaciated regions (Briner et al., 2003; Marsella et al., 2000; Small et al., 1997).

Hence, from a long-term landscape evolution point of view, these relict surfaces have been insignificantly lowered during the Cenozoic ice ages. Processes that would act to lower these surfaces are frost weathering and periglacial slope processes, such as solifluction. Of these options, periglacial processes dominate as indicated by (1) the absence of linear features of fluvial origin, (2) the ubiquitous presence of gentle convex-concave slope profiles, (3) the presence of sporadic permafrost, (4) the ubiquitous presence of active and relict periglacial phenomena across these uplands, and (5) the presence of tors (our sample localities). In conjunction with the sporadic presence of permafrost (Lundqvist, 1962; Sollid et al., 2000), transport of solutes in (melt)water through the active layer probably promoted widespread but low rates of long-term landscape lowering (Stroeven et al., 2002b).

Glacial Valleys (~500–900 m above sea level)

The two samples from Riksgränsen, although sampled from roche moutonnée surfaces at the bottom of a prominent glacial valley that forms the Torneträsk depression (Fig. 2C), (which is the most likely location for severe glacial scouring), can test for the concept of selective linear erosion (Sugden, 1968). Because the glacial valleys occur adjacent to relict surfaces at intermediate and high elevations (Kleman and Stroeven, 1997), the interpretation is that there have been large spatial differences in rates of glacial erosion. The preglacial fluvial drainage pattern and relief of the landscape have dictated these spatial differences in erosion. When ice first started to grow on these mountains, it preferentially deepened existing valleys and depressions. Because pressure-melting conditions at the base of an ice sheet are required for both basal sliding and extensive glacial erosion to occur, it is likely that these conditions were first met where the ice sheet was thickest, such as in pre-existing valleys. The early glaciers and ice sheets thus enhanced the existing relief, which created a positive-feedback effect on subglacial temperature and pressure-melting patterns (Mazo, 1991; Oerlemans, 1984). Pre-existing depressions and valleys aligned to the ice-flow direction were preferentially eroded, and other areas were subject to much less or no erosion, a trademark of the process of selective linear erosion. A comparison of our data from Riksgränsen with older bedrock ages indicates a magnitude of $\sim 2 \pm 0.4$ m of valley-bottom erosion of bedrock during the last glacial cycle (Stroeven et al., 2002a). This rock loss depth yields an erosion rate for the glacial corridor that is at least an order of magnitude higher than maximum erosion rates over relict surfaces, which typically are 1–2 m/m.y., a result that is clearly consistent with the concept of selective linear erosion.

Although two features formed by the process of selective linear erosion are typical ingredients in this landscape, i.e., sharp glacial surface–preglacial surface boundaries and hanging valleys, which are formed by glacial valley widening and deepening, respectively, and although the pattern of erosion determined by cosmogenic nuclide dating is consistent with the concept of selective linear erosion, the magnitude of erosion of the trunk valley is surprisingly lower than expected. Considering that the larger glacial valleys have not been deepened by more than ~400 m (Stroeven et al., 2002b) (although some valleys appear to have experienced as much as almost 900 m of erosion [Kleman and Stroeven, 1997]), 2 m of erosion per glacial cycle (e.g., Riksgränsen) is insufficient to explain the presence of these depressions. These results indicate that the overall modification produced by ice sheets along glacial corridors may either be more restricted than previously thought, or it has occurred preferentially during earlier Quaternary glacial periods. The potentially old age of significant deepening of the glacial valleys is corroborated by the presence of inherited nuclides in a clearly glacially scoured bedrock surface on the Bálddavárri rock bastion in the Kaitum depression (47.6 ± 2.9 ka; Table 1).

Northern Swedish Lowland (~0–500 m above sea level)

The northern Swedish lowland is characterized by a stepped "plains with residual hills" morphology generally at 300–400 and 400–550 m above sea level (e.g., Fredén, 1994, p. 50–51), referred to as the Muddus Plains (Wråk, 1908). The Muddus Plains are regarded as the youngest preglacial fluvial surfaces of the region, the youngest of which, then, was the last base level for the evolution of the preglacial fluvial drainage system in the mountain range. If this interpretation is correct, the implication is that the northern Swedish lowland has not been significantly deepened (Lidmar-Bergström, 1997), save for an array of piedmont lakes that fringe the mountain range and probably relate to the inception stages of successive Fennoscandian ice sheets (Fredin, 2002; Kleman, 1992). The area with tors investigated in this study (Figs. 1 and 2D) is part of the lower Muddus surface, and the results qualitatively support the notion that these surfaces have not been degraded significantly. In fact, when Stroeven et al. (2002c) considered the mean value of the two tors yielding both ^{26}Al and ^{10}Be results, they concluded that the tors have survived for at least 605 k.y., and experienced a maximum erosion rate of 1.6 m/m.y.

These results are also in concert with the cosmogenic radionuclide results from the relict surfaces in the mountains. If the determined rate of erosion proves to be typical for the northern Swedish lowland as a whole, and the apparent exposure age of the Teletöisentunturi drumlin is provisionally consistent with this assumption, then the general subglacial condition of Fennoscandian ice sheets has been one of preservation or deposition. Because lineation systems overwhelmingly relate to the deglaciation of the last ice sheet, the inference is that successive Fennoscandian ice sheets were largely cold-based during inception and growth stages (east of the mountain range). Destruction of the cold-based core occurred during deglaciation and predominantly along fast-flowing corridors (e.g., Kleman and Hättestrand, 1999).

CONCLUSIONS

Geomorphological and cosmogenic evidence indicates that erosion over relict surfaces, which comprises ~20% of the area in the northern Swedish mountains and most of the northern Swedish lowland, has been negligible. These relict areas need to be accounted for as frozen bed patches in basal boundary conditions for ice-sheet and landscape development models, and as potential refuges for biota that can survive long periods of frozen conditions. The geomorphological and cosmogenic nuclide evidence from the northern Swedish mountains also provides strong support for a model of patchy and slow landscape evolution as a result of repeated ice-sheet glaciation in this region. Although this may not hold for all ice sheets, it is a model that should be tested for other areas of repeated ice-sheet glaciation, as it has potentially significant implications for ice-sheet reconstructions, landscape development models, and interpretations of sedimentary sequences in ocean basins.

ACKNOWLEDGMENTS

The authors are indebted to Andrew Meigs for a very helpful review, and to Colin E. Thorn and Robert G. Darmody of the University of Illinois, who supplied the ^{36}Cl results reported in this paper through National Science Foundation grant BCS-9818667. Swedish Natural Sciences Research Council grants G-AA/GU 12034-300 and G-AA/GU 12034-301 to Stroeven, and National Science Foundation grants OPP-9818162 and OPP-0138486 to Harbor financially supported this work.

REFERENCES CITED

Ahlmann, H.W., 1919, Geomorphological studies in Norway: Geografiska Annaler, v. 1, p. 1–148.

André, M.-F., 2004, The geomorphic impact of glaciers as indicated by tors in north Sweden (Aurivaara, 68° N): Geomorphology, v. 57, no. 3-4, p. 403–421, doi:10.1016/S0169-555X(03)00182-X.

Bergsma, B.M., Svoboda, J., and Freedman, B., 1984, Entombed plant communities released by a retreating glacier at central Ellesmere Island, Canada: Arctic, v. 37, no. 1, p. 49–52.

Bierman, P.R., 1994, Using in situ produced cosmogenic isotopes to estimate rates of landscape evolution: A review from the geomorphic perspective: Journal of Geophysical Research, v. 99, no. B7, p. 13,885–13,896, doi: 10.1029/94JB00459.

Bierman, P.R., Marsella, K.A., Patterson, C., Davis, P.T., and Caffee, M., 1999, Mid-Pleistocene cosmogenic minimum-age limits for pre-Wisconsinan glacial surfaces in southwestern Minnesota and southern Baffin Island: A multiple nuclide approach: Geomorphology, v. 27, no. 1–2, p. 25–39, doi: 10.1016/S0169-555X(98)00088-9.

Bonney, T.G., 1871, On the formation of cirques and their bearing upon theories attributing the excavation of alpine valleys mainly to the action of glaciers: Quarterly Journal of the Geological Society, v. 27, p. 312–324.

Bonney, T.G., 1874, Notes on the Upper Engadine and the Italian valleys of Monte Rosa, and their relation to the glacier erosion theory of lake basins: Quarterly Journal of the Geological Society, v. 30, p. 479–489.

Bonney, T.G., 1910, Presidential address to the British Association for the Advancement of Science: Science, v. 32, p. 321–336 and p. 353–363.

Briner, J.P., Miller, G.H., Davis, P.T., Bierman, P.R., and Caffee, M., 2003, Last Glacial Maximum ice sheet dynamics in Arctic Canada inferred from young erratics perched on ancient tors: Quaternary Science Reviews, v. 22, no. 5–7, p. 437–444.

Cerling, T.E., and Craig, H., 1994, Geomorphology and in situ cosmogenic isotopes: Annual Review of Earth and Planetary Sciences, v. 22, p. 273–317, doi: 10.1146/annurev.ea.22.050194.001421.

Clarhäll, A., and Kleman, J., 1999, Distribution and glaciological implications of relict surfaces on the Ultevis plateau, northwestern Sweden: Annals of Glaciology, v. 28, p. 202–208.

Davies, G.L., 1969, The Earth in decay: A history of British geomorphology 1578–1878: New York, Elsevier, 390 p.

Dunne, J., Elmore, D., and Muzikar, P., 1999, Scaling factors for the rates of production of cosmogenic nuclides for geometric shielding and attenuation at depth on sloped surfaces: Geomorphology, v. 27, p. 3–11, doi: 10.1016/S0169-555X(98)00086-5.

Dyke, A.S., 1993, Landscapes of cold-centered late Wisconsinan ice caps, Arctic Canada: Progress in Physical Geography, v. 17, no. 2, p. 223–247.

Fabel, D., and Harbor, J., 1999, The use of in-situ produced cosmogenic radionuclides in glaciology and glacial geomorphology: Annals of Glaciology, v. 28, p. 103–110.

Fabel, D., Stone, J., Fifield, L.K., and Cresswell, R.G., 1997, Deglaciation of the Vestfold Hills, East Antarctica: Preliminary evidence from exposure dating of three subglacial erratics, *in* Ricci, C.A., ed., The Antarctic region: Geological evolution and processes: Siena, Terra Antartica Publication, p. 829–834.

Fabel, D., Stroeven, A.P., Harbor, J., Kleman, J., Elmore, D., and Fink, D., 2002, Landscape preservation under Fennoscandian ice sheets determined from in situ produced ^{10}Be and ^{26}Al: Earth and Planetary Science Letters, v. 201, p. 397–406, doi: 10.1016/S0012-821X(02)00714-8.

Falconer, G., 1966, Preservation of vegetation and patterned ground under a thin ice body in northern Baffin Island, N.W.T: Geographical Bulletin, v. 8, no. 2, p. 194–200.

Fink, D., Hotchkis, M., Hua, Q., Jacobsen, G., Smith, A. M., Zoppi, U., Child, D., Mifsud, C., van der Gaast, H., Williams, A., and Williams, M., 2004, The ANTARES AMS facility at ANSTO: Nuclear Instruments & Methods in Physics Research Section B–Beam Interactions with Materials and Atoms, v. 223-224, p. 109–115.

Fredén, C., ed., 1994, Geology, national atlas of Sweden: Stockholm, SNA Publishing, 208 p.

Fredin, O., 2002, Glacial inception and Quaternary mountain glaciations in Fennoscandia: Quaternary International, v. 95–96, p. 99–112, doi: 10.1016/S1040-6182(02)00031-9.

Garwood, E.J., 1910, Features of alpine scenery due to glacial protection: The Geographical Journal, v. 36, p. 310–339.

Gjessing, J., 1967, Norway's paleic surface: Norsk Geografisk Tidsskrift, Norwegian Journal of Geography, v. 21, no. 2, p. 69–132.

Glasser, N.F., and Hall, A.M., 1997, Calculating Quaternary glacial erosion rates in northeast Scotland: Geomorphology, v. 20, no. 1–2, p. 29–48, doi: 10.1016/S0169-555X(97)00005-6.

Goldthwait, R.P., 1960, Study of ice cliff in Nunatarssuaq, Greenland: U.S. Army Snow Ice and Permafrost Research Establishment, Technical Report, v. 39, p. 1–108.

Gosse, J.C., and Phillips, F.M., 2001, Terrestrial in situ cosmogenic nuclides: Theory and application: Quaternary Science Reviews, v. 20, p. 1475–1560, doi: 10.1016/S0277-3791(00)00171-2.

Hall, A.M., and Sugden, D.E., 1987, Limited modification of mid-latitude landscapes by ice-sheets: The case of northeast Scotland: Earth Surface Processes and Landforms, v. 12, p. 531–542.

Hättestrand, C., and Stroeven, A.P., 2002, A relict landscape in the centre of Fennoscandian glaciation: Geomorphological evidence of minimal Quaternary glacial erosion: Geomorphology, v. 44, no. 1–2, p. 127–143, doi: 10.1016/S0169-555X(01)00149-0.

Hättestrand, C., Kosche, S., Näslund, J.-O., Fabel, D., and Stroeven, A.P., 2004, Drumlin formation time—Evidence from northern and central Sweden: Geografiska Annaler, v. 86A, no. 2, p. 155–167, doi:10.1111/j.0435-3676.2004.00221.x.

Holmgren, B., Olsson, I., Skye, E., and Alm, G., 1984, Climate and glaciation in Kong Karls Land, eastern Svalbard, *in* Mörner, N.-A., and Karlén, W., eds., Climatic changes on a yearly to millennial basis: Dordrecht, D. Reidel Publishing Company, p. 291–302.

Irving, R.A., 1883, On the mechanics of glaciers with especial reference to their supposed power of excavation: Quarterly Journal of the Geological Society, v. 39, p. 62–81.

Jonsson, S., 1983, On the geomorphology and past glaciation of Storöya, Svalbard: Geografiska Annaler, v. 65A, no. 1–2, p. 1–17.

Judd, J.W., 1876, On the origin of Lake Balton in Hungary: Geological Magazine, ser. 2, v. 3, p. 5–15.

Kleman, J., 1992, The palimpsest glacial landscape in northwestern Sweden—Late Weichselian deglaciation landforms and traces of older west-centered ice sheets: Geografiska Annaler, v. 74A, no. 4, p. 305–325.

Kleman, J., 1994, Preservation of landforms under ice sheets and ice caps: Geomorphology, v. 9, p. 19–32, doi: 10.1016/0169-555X(94)90028-0.

Kleman, J., and Borgström, I., 1990, The boulder fields of Mt. Fulufjället, west-central Sweden—Late Weichselian boulder blankets and interstadial periglacial phenomena: Geografiska Annaler, v. 72A, no. 1, p. 63–78.

Kleman, J., and Hättestrand, C., 1999, Frozen-bed Fennoscandian and Laurentide ice sheets during the Last Glacial Maximum: Nature, v. 402, p. 63–66, doi: 10.1038/47005.

Kleman, J., and Stroeven, A.P., 1997, Preglacial surface remnants and Quaternary glacial regimes in northwestern Sweden: Geomorphology, v. 19, no. 1–2, p. 35–54, doi: 10.1016/S0169-555X(96)00046-3.

Kleman, J., Hättestrand, C., Borgström, I., and Stroeven, A.P., 1997, Fennoscandian paleoglaciology reconstructed using a glacial geological inversion model: Journal of Glaciology, v. 43, no. 144, p. 283–299.

Lagerbäck, R., and Robertsson, A.-M., 1988, Kettle holes—Stratigraphical archives for Weichselian geology and palaeoenvironment in northernmost Sweden: Boreas, v. 17, p. 439–468.

Lal, D., 1991, Cosmic ray labeling of erosion surfaces: In situ nuclide production rates and erosion models: Earth and Planetary Science Letters, v. 104, no. 1/2, p. 424–439.

Lal, D., and Peters, B., 1967, Cosmic-ray produced radioactivity on the earth: Handbook of Physics, v. 46, p. 551–612.

Lazarus, D., Spencer-Cervato, C., Pika-Biolzi, M., Beckmann, J.P., Salis, K., Hilbrecht, H., and Thierstein, H., 1995, Revised chronology of Neogene DSDP holes from the world ocean: College Station, Texas, Ocean Drilling Program Technical Report, no. 24.

Lidmar-Bergström, K., 1997, A long-term perspective on glacial erosion: Earth Surface Processes and Landforms, v. 22, p. 297–306, doi: 10.1002/(SICI)1096-9837(199703)22:3<297::AID-ESP758>3.3.CO;2-I.

Liu, B.L., Phillips, F.M., Fabryka-Martin, J.T., Fowler, M.M., and Stone, W.D., 1994, Cosmogenic Cl-36 accumulation in unstable landforms. 1. Effects of the thermal-neutron distribution: Water Resources Research, v. 30, no. 11, p. 3115–3125, doi: 10.1029/94WR00761.

Lundqvist, J., 1962, Patterned ground and related frost phenomena in Sweden: Sveriges Geologiska Undersökning, v. C583, 101 p.

Mangerud, J., Sønstegaard, E., and Sejrup, H.-P., 1979, Correlation of the Eemian (interglacial) stage and the deep-sea oxygen-isotope stratigraphy: Nature, v. 277, p. 189–192.

Marsella, K.A., Bierman, P.R., Davis, P.T., and Caffee, M.W., 2000, Cosmogenic ^{10}Be and ^{26}Al ages for the Last Glacial Maximum, eastern Baffin Island, Arctic Canada: Geological Society of America Bulletin, v. 112, no. 8, p. 1296–1312, doi: 10.1130/0016-7606(2000)112<1296:CBAAAF>2.3.CO;2.

Mazo, V.L., 1991, Interactions of ice sheets: Instability and self-organization: IAHS Publication, no. 208, p. 193–205.

Nishiizumi, K., Kohl, C.P., Arnold, J.R., Dorn, R., Klein, J., Fink, D., Middleton, R., and Lal, D., 1993, Role of in situ cosmogenic nuclides ^{10}Be and ^{26}Al in the study of diverse geomorphic processes: Earth Surface Processes and Landforms, v. 18, p. 407–425.

Oerlemans, J., 1984, Numerical experiments on large-scale glacial erosion: Zeitschrift für Gletscherkunde und Glazialgeologie, v. 20, p. 107–126.

Peulvast, J.-P., 1988, Preglacial landform evolution in two coastal high latitude mountains: Lofoten-Vesterålen (Norway) and Scoresby Sund area (Greenland): Geografiska Annaler, v. 70A, no. 4, p. 351–360.

Pickard, J., 1986, The Vestfold Hills: A window on Antarctica, *in* Pickard, J., ed., Antarctic oasis: Sydney, Academic Press Australia, p. 333–351.

Ramsay, A.C., 1864, On the erosion of valleys and lakes: Philosophical Magazine, v. 28, p. 293–311.

Ramsay, A.C., 1876, The origin of lake basins: Geological Magazine, v. 3, p. 136–138.

Rapp, A., 1960, Recent development of mountain slopes in Kärkevagge and surroundings, northern Scandinavia: Geografiska Annaler, v. 42, p. 73–200.

Raymo, M.E., Ruddiman, W.F., Backman, J., Clement, B.M., and Martinson, D.G., 1989, Late Pliocene variation in Northern Hemisphere ice sheets and North Atlantic Deep Water circulation: Paleoceanography, v. 4, no. 4, p. 413–446.

Reusch, H., 1901, Nogle bidrag til forstaaelsen af hvorledes Norges dale og fjelde er blevne til: Norges Geologiske Undersøkelse, v. 32, p. 124–217.

Rudberg, S., 1954, Västerbottens bergggrundsmorfologi—Ett försök till rekonstruktion av preglaciala erosionsgenerationer i Sverige: Geographica, v. 25, 457 p.

Ruddiman, W.F., Raymo, M.E., Martinson, D.G., Clement, B.M., and Backman, J., 1989, Pleistocene evolution: Northern Hemisphere ice sheets and North Atlantic Ocean: Paleoceanography, v. 4, no. 4, p. 353–412.

Small, E.E., Anderson, R.S., Repka, J.L., and Finkel, R., 1997, Erosion rates of alpine bedrock summit surfaces deduced from in situ ^{10}Be and ^{26}Al: Earth and Planetary Science Letters, v. 150, p. 413–425, doi: 10.1016/S0012-821X(97)00092-7.

Solheim, A., Riis, F., Elverhøi, A., Faleide, J.I., Jensen, L.N., and Cloetingh, S., 1996, Impact of glaciations on basin evolution: Data and models from the Norwegian margin and adjacent areas—Introduction and summary: Global and Planetary Change, v. 12, no. 1–4, p. 1–9, doi: 10.1016/0921-8181(95)00008-9.

Sollid, J.L., Holmlund, P., Isaksen, K., and Harris, C., 2000, Deep permafrost boreholes in western Svalbard, northern Sweden and southern Norway: Norsk Geografisk Tidsskrift, Norwegian Journal of Geography, v. 54, no. 4, p. 186–191.

Stone, J.O., 2000, Air pressure and cosmogenic isotope production: Journal of Geophysical Research, v. 105, no. B10, p. 23,753–23,759, doi: 10.1029/2000JB900181.

Stroeven, A.P., Fabel, D., Harbor, J., Hättestrand, C., and Kleman, J., 2002a, Quantifying the erosional impact of the Fennoscandian ice sheet in the Torneträsk-Narvik corridor, northern Sweden, based on cosmogenic radionuclide data: Geografiska Annaler, v. 84A, no. 3–4, p. 275–287, doi: 10.1111/j.0435-3676.2002.00182.x.

Stroeven, A.P., Fabel, D., Harbor, J., Hättestrand, C., and Kleman, J., 2002b, Reconstructing the erosion history of glaciated passive margins: Applications of in-situ produced cosmogenic nuclide techniques, *in* Doré, A.G., Cartwright, J.A., Stoker, M.S., Turner, J.P., and White, N., eds., Exhumation of the North Atlantic Margin: Timing, mechanisms and implications for petroleum exploration: Geological Society [London] Special Publication 196, p. 153–168.

Stroeven, A.P., Fabel, D., Hättestrand, C., and Harbor, J., 2002c, A relict landscape in the centre of Fennoscandian glaciation: Cosmogenic radionuclide evidence of tors preserved through multiple glacial cycles: Geomorphology, v. 44, no. 1–2, p. 145–154, doi: 10.1016/S0169-555X(01)00150-7.

Sugden, D.E., 1968, The selectivity of glacial erosion in the Cairngorm Mountains, Scotland: Transactions (Institute of British Geographers), v. 45, p. 79–92.

Sugden, D.E., 1974, Landscapes of glacial erosion in Greenland and their relationship to ice, topographic and bedrock conditions: Institute of British Geographers Special Publication, v. 7, p. 177–195.

Sugden, D.E., 1976, A case against deep erosion of shields by ice sheets: Geology, v. 4, no. 10, p. 580–582, doi: 10.1130/0091-7613(1976)4 <580: ACADEO> 2.0.CO;2.

Sugden, D.E., and John, B.S., 1976, Glaciers and landscape: A geomorphological approach: London, Edward Arnold, 376 p.

Sugden, D.E., and Watts, S.H., 1977, Tors, felsenmeer, and glaciation in northern Cumberland Peninsula, Baffin Island: Canadian Journal of Earth Sciences, v. 14, p. 2817–2823.

Wråk, W., 1908, Bidrag till Skandinaviens reliefkronologi: Ymer, v. 28, p. 141–191.

Manuscript Accepted by the Society 23 June 2005

Printed in the USA

Geological Society of America
Special Paper 398
2006

Facing reality: Late Cenozoic evolution of smooth peaks, glacially ornamented valleys, and deep river gorges of Colorado's Front Range

Robert S. Anderson
Department of Geological Sciences and Institute for Arctic and Alpine Research (INSTAAR), Campus Box 450, University of Colorado, Boulder, Colorado 80309, USA
Catherine A. Riihimaki
Department of Earth Sciences, University of California, Santa Cruz, California 95064, USA
Elizabeth B. Safran
Environmental Studies Program, Lewis and Clark College, Portland, Oregon 97219, USA
Kelly R. MacGregor
Geology Department, Macalester College, St. Paul, Minnesota 55105, USA

ABSTRACT

Thirty to forty m.y. of post-Laramide degradation of the southern Rocky Mountains likely produced relatively low-relief topography within the crystalline cores of the ranges, and capped the adjacent sedimentary basins with easily eroded sediments. We focus on the modern, more dissected topography of these ranges, reflecting late Cenozoic evolution driven by fluvial and glacial exhumation, each of which affects different portions of the landscape in characteristic ways. Ongoing exhumation of the adjacent basins, in places by more than 1 km, is effectively lowering base level of streams draining the crystalline range cores. The streams have incised deep bedrock canyons that now cut the flanks of the range. Over the same time scales, glaciation of the headwaters of the major streams has modified the range crests. We utilize the topography of the northern Front Range of Colorado to explore the response of a Laramide range both to the exhumation of the adjacent basin and to glaciation in the high elevations. We break the problem of whole landscape evolution into three related, one-dimensional problems: evolution of the high smooth summit surfaces; evolution of the longitudinal profiles of adjacent glacial troughs; and evolution of the fluvial profiles downstream of the glacial limit. We review work on the high summit surfaces, showing quantitatively that they are steady-state features lowering at rates on the order of 5 μm/yr, and are entirely decoupled from the adjacent glacial troughs. Glaciers not only truncate these high surfaces, but greatly alter the longitudinal profiles of the major streams: major steps occur at tributary junctions, and profiles above the glacial limit are significantly flattened from their original fluvial slopes. We extend existing models of glacial valley evolution by including processes that allow headwall retreat. This serves to enhance the headward retreat of east-facing valleys, and explains the asymmetric truncation of the high smooth surfaces that form the spine of the range. Fluvial profiles downstream of the glacial limit commonly display a prominent convexity inboard of the range edge. Stream-power–based numerical models of profile evolution of specific rivers demonstrate that this reflects a transient response of the streams to base-level lowering. This response varies significantly with drainage basin area. We explore the degree to which this differential response controls the

Anderson, R.S., Riihimaki, C.A., Safran, E.B., and MacGregor, K.R., 2006, Facing reality: Late Cenozoic evolution of smooth peaks, glacially ornamented valleys, and deep river gorges of Colorado's Front Range, *in* Willett, S.D., Hovius, N., Brandon, M.T., and Fisher, D.M., eds., Tectonics, Climate, and Landscape Evolution: Geological Society of America Special Paper 398, Penrose Conference Series, p. 397–418, doi: 10.1130/2006.2398(25). For permission to copy, contact editing@geosociety.org.

location of major remnants of pediments on the edge of the Great Plains, such as the prominent Rocky Flats and adjacent surfaces.

Keywords: glacial valleys, river incision, erosion, landscape evolution.

INTRODUCTION

The Laramide Ranges of the western United States were generated by compressional tectonics in the late Cretaceous and early Tertiary (e.g., Dickinson et al., 1988; Bird, 1998; Erslev, 2001). These crystalline-cored ranges are bounded by vast basins filled with syn- and post-tectonic sediments that partially buried this region tens of millions of years ago. Because the region has been relatively quiescent since the early Tertiary, its post-tectonic history has been inferred largely from qualitative arguments about how a suite of distinctive and characteristic landforms might have developed. These landforms include widespread surfaces of low relief within the ranges, narrow, glaciated "spines" at the range crests, deeply incised fluvial canyons at the range edges, and remnants of former river valley bottoms preserved as benches and surface fragments in the sedimentary basins. To clarify the post-tectonic history of the Laramide region, we consider it essential to couple conceptual models with quantitative models of Laramide landform evolution. In this paper, we use a set of one-dimensional (1D) numerical models to simulate the evolution of several key landforms. The models are simple but incorporate many essential details of real systems. Our efforts aim to complement previous empirical work and to place additional constraints on the timing, magnitude, relative importance, and linkages of the geomorphic processes that have sculpted the Laramide landscape for the last few million years.

The most recent significant geomorphic event affecting the Laramide region involves the exhumation of the range-bounding, sediment-filled basins. Many of these basins have been eroded to depths of 1 km or more (e.g., McMillan et al., 2002; Leonard, 2002). The exhumation began in the late Cenozoic, although the timing is poorly constrained and very likely varies from basin to basin (e.g., Dethier, 2001; Reiners and Heffern, 2002; Riihimaki, 2003). The Laramide landscape also reflects the influences of both regional and local geophysical and tectonic processes. Broad-scale tilting of late Cretaceous seaway deposits, and of gravel sheets emplaced in the mid-Cenozoic is thought to reflect dynamic topography driven by mantle flow (Mitrovika et al., 1989; Heller et al., 2003), while down-to-the-east tilting of paleochannels of mid-Cenozoic age in deposits of exterior basins to the east is claimed to reflect both local tectonics and exhumation of the basin deposits (McMillan et al., 2002; Leonard, 2002). Possible mechanisms driving local tectonic activity include the propagation of the Rio Grande Rift into Colorado from the south (Chapin and Cather, 1994; Erslev, 2001), and the arrival of the Yellowstone hotspot in the NW of the province (e.g., Smith and Braile, 1993).

Despite these local effects, we expect the post-tectonic history of the Laramide ranges to be characterized largely by topographic decay. Low-temperature geochronology reveals that no more than a few kilometers of erosion have occurred within the crystalline cores of the ranges since the Laramide orogeny or shortly thereafter; apatite fission-track ages, documenting exhumation from roughly 120 °C, are early Tertiary (e.g., Crowley et al., 2002; Kelley and Chapin, 1997; Kelley and Duncan, 1986). Calculated mean Tertiary erosion rates are typically less than ~0.1 mm/yr, although there is significant scatter. More recently, the ranges have experienced topographic rejuvenation linked to basin exhumation (e.g., Izett, 1975). The mechanism driving exhumation in the Laramide region is not known, and we do not address that question in this paper. Rather, we focus on the response of the adjacent ranges, in particular the Front Range, to basin exhumation. In effect, the basin exhumation, driven by the incision of the master streams within them, serves to lower base level for the fluvial system draining the adjacent range. We may therefore treat this portion of the southern Rockies as a natural experiment that allows us to probe the response of a partly buried landscape to both lowering of base level and to climatically driven erosion processes (e.g., glaciation).

In part, this paper constitutes a review and assembly of our past work (e.g., on high surfaces [Anderson, 2002] and glacial valleys [MacGregor et al., 2002]). Herein we extend our models and our discussions of these features, we address the linkages among them, and we address for the first time the fluvial system as it responds to the base-level lowering of the South Platte. The features we aim to explain in the Front Range specifically, and for the Laramide ranges in general, are described below.

Introduction to Front Range Physiography

The Front Range is roughly 50 km from the base of the Flatirons at the mountain front to the continental divide (Fig. 1). The range is bordered by the very flat floor of Denver Basin at 1500 m elevation, into which a number of low-gradient smooth surfaces (pediments, pediments buried by alluvium, and more narrow fingers of alluvium) protrude from the Front Range. Collectively, these geomorphic surfaces and deposits define former levels of the Denver Basin as it has been exhumed. Among these surfaces, the most prominent is the Rocky Flats pediment between Boulder and Golden (e.g., Scott, 1960, 1975). The range steps upward abruptly by 800 m from the adjacent plains. If one avoids the canyon floors, a walk from plains to crest would then encounter a rolling surface at 2300–3000 m elevation that stretches toward the spine of the range. This physiographic feature has been called the subsummit surface, or the Rocky Mountain surface (e.g., Epis and Chapin, 1975). The topography then steps up by another 1000 m to the narrow summit spine that forms the drainage divide (Continental

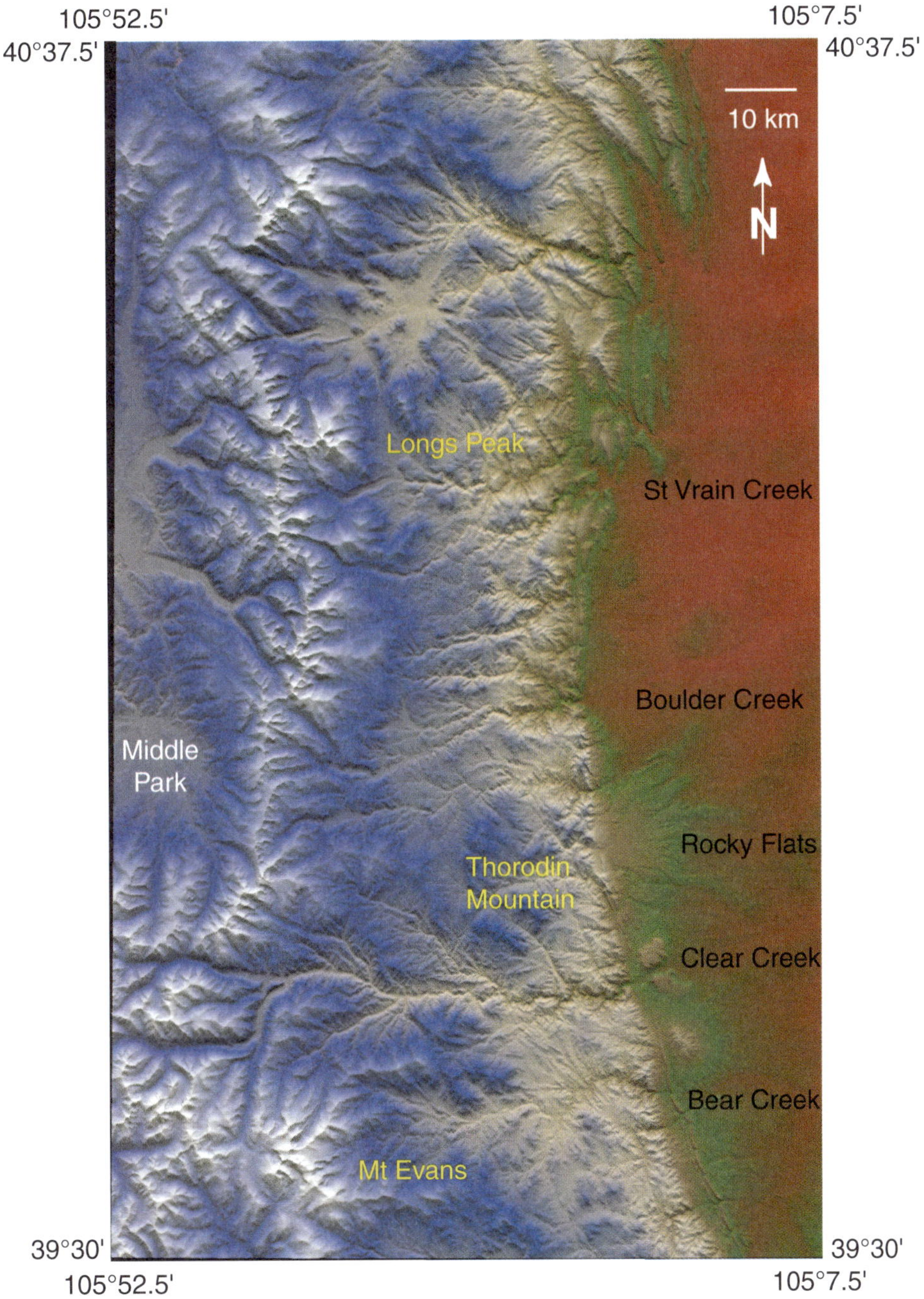

Figure 1. Digital elevation model (DEM) of northern portion of the Colorado Front Range, colored to emphasize the major geomorphic features (outlines in Fig. 2). White of the range crest (Continental Divide) shows strong asymmetry of the glacial ornamentation; most cirques face eastward, while broadest high smooth surfaces round off to the west. This is best seen north of Longs Peak and due west of Thorodin Mountain. Note prominent eastward-trending ridges from the divide that include Mt. Evans and Longs Peak. Several of the prominent streams draining eastward from the Front Range are labeled. The Rocky Flats surface is the best-preserved pediment, more than 100 m above the surrounding floor of the Denver Basin.

Divide) between the Colorado and Mississippi Rivers. The few-kilometer-wide range crest is dominated not by glacial arêtes, but by smooth tundra-covered surfaces, with occasional bedrock knobs poking through shallow regolith. It is easy walking; most of the work of the climb to the summits is done getting to the ridge. A walk up the river canyons is significantly different. Following any of the gently sloped rivers on the plains, all tributaries of the South Platte, into the mountain front, one ascends through a steepening, deep bedrock gorge with walls that at first are quite narrow and steep. Above an inflection in the profile, many of these rivers become less deeply incised into the subsummit surface. Further upstream one encounters the end moraines from the glaciers that have periodically occupied the headwaters. Thereafter, the character of the valley profile alters significantly. Major steps and flats dominate the profile up to the tip of the drainage, where the channel form halts, and one is left standing in a cirque basin or a bench just below the divide.

It is these major features (Fig. 2), held in common among those Laramide ranges that have not experienced late Cenozoic tectonism, for example tectonism associated with the Rio Grande Rift, that we wish to explain quantitatively: the high smooth surfaces of the summit spine, the glacial canyons that bound them, the fluvial canyons downstream of the glacial troughs, and the pediments on the edge of the plains. Below we describe the conceptual framework upon which our numerical modeling is based.

Conceptual Model of Late Cenozoic Landscape Evolution

It is likely that in the roughly 40 m.y. between the end of the Laramide orogeny and the initiation of basin exhumation, the topography of the range had decayed significantly, although perhaps not monotonically (e.g., Evanoff, 1990; Mears, 1993). The relief between the crest of the range and the basin floor has become diminished as the products of the weathering of the range have been transported to the nearby depositional basin. The basin floor used to be significantly higher than it is now, as witnessed by the remnants of older surfaces on the fringe of the plains. The local relief between river and ridgeline was likely small as well. By the late Cenozoic, both the subsummit surface and the summit spine had likely evolved into low-relief surfaces adjusted to transport the products of weathering of the crystalline bedrock of the core of the range.

With the initiation of exhumation of the adjacent Denver Basin driven by incision of the South Platte River, the Front Range landscape was rejuvenated. Tributaries of the South Platte then responded by incising into the crystalline core, and are presently in a state of transient response to the exhumation event. The hillslopes of the subsummit surface remain relatively intact, showing only minor relief where they exist beyond the reach of this transient response. The exceptions are the steep canyons of the front of the range. That the exhumation occurred within the last few m.y. means that the climate was undergoing strong glacial-interglacial cycles. This resulted in major glacial incision of the headwaters of these tributary streams, which in turn both increased the local relief of the spine, and fed the fluvial system downstream with glacial sediment.

In the following sections, we describe in detail our current quantitative understanding of the geomorphic processes that generated key Laramide landforms, beginning with the high surfaces in the montane interior and working basinward to the pediments carved into sediment fill.

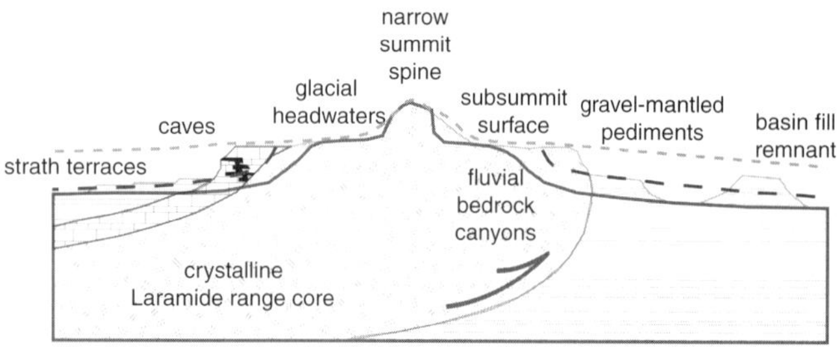

Figure 2. Schematic cross section through a Laramide range. Laramide thrusting of crystalline rock over Mesozoic sedimentary sections set up a lithology contrast that manifests itself as a major topographic break at the range front today. The narrow summit spine is separated from the range front by a broad, low-relief subsummit surface. Glacial canyons etch into the summit spine otherwise characterized by high smooth surfaces, creating high local relief. Rivers are in transient response to lowering of the adjacent basin floor, each showing a broad convexity. Downstream of the convexity, the local relief is again high, as crystalline rock forms canyon walls that have been deepened by recent river incision. In some instances (e.g., Bighorn Mountains, Wind River Mountains, Black Hills), the Paleozoic limestones tilted by Laramide thrusting are exposed and allow cave formation. Timing of the incision event can be derived by dating caves, or by dating strath terraces and pediments at the range edges.

HIGH SURFACES

A large fraction of the spine of the Front Range is a very smooth surface. A walk on the Continental Divide is for the most part a stroll in alpine meadows, only occasionally interrupted by glacially carved arêtes. We first briefly summarize the set of geomorphic features and processes acting on these high surfaces, and the recently published numerical model of them (Anderson, 2002). We then apply this model to demonstrate how morphologic asymmetry of the range crest can be produced by asymmetric (east versus west) headward retreat of glacial canyons in the Front Range.

Features of the High Surfaces

Aerial photos and digital elevation models (DEMs) (Figs. 3 and 4) show that these surfaces are truncated by glacial headwaters, more extensively on their eastern than western sides. This strong glacial asymmetry of the range has long been noted. The high surfaces of the Front Range, like those of the Wind River range described in Anderson (2002), have the following features in common: broad convex-up parabolic profiles, thin (roughly 1 m) but uniform regolith cover, occasional tors, which are commonly found along the crestline, and bedrock edges that separate the surface from bounding glacial headwalls. Some of these surfaces are more than 1 km wide. The smooth quality arises not only from the lack of significant roughness on the surface, but also from the low curvature, typically $0.5–8.0 \times 10^{-4}$ m/m^2 (see Fig. 5). The tors are often surrounded by block fields, in which large blocks on the order of 1 m diameter dominate the surface of the otherwise smooth surface.

At present, these surfaces lie at elevations of roughly 3–4 km, and experience mean annual temperatures that subject them to periodic freezing. They display the activity of periglacial processes. Sorted nets and polygons are occasionally present, and frost heave of individual blocks is common. Both the bedrock of the tors and the blocks derived from them are often riven by frost shattering.

Figure 3. Mid-November photograph of summit spine of the Front Range, looking north from above Berthoud Pass. The highest summit surfaces are white from a skiff of recent snow; the subsummit (Rocky Mountain) surface is dark green with forest cover. The floor of the Denver Basin at the eastern edge of the Great Plains is in the distance on the right.

Figure 4. Photograph of representative high surfaces on the Continental Divide. James Peak is the most prominent peak on the divide in the scene. Note abrupt truncation of surfaces on several sides by glacial valley walls and headwalls. The high surface in the foreground slopes gently toward the viewer, and displays several regolith-mantled fingers that are bounded by tor-studded ridges.

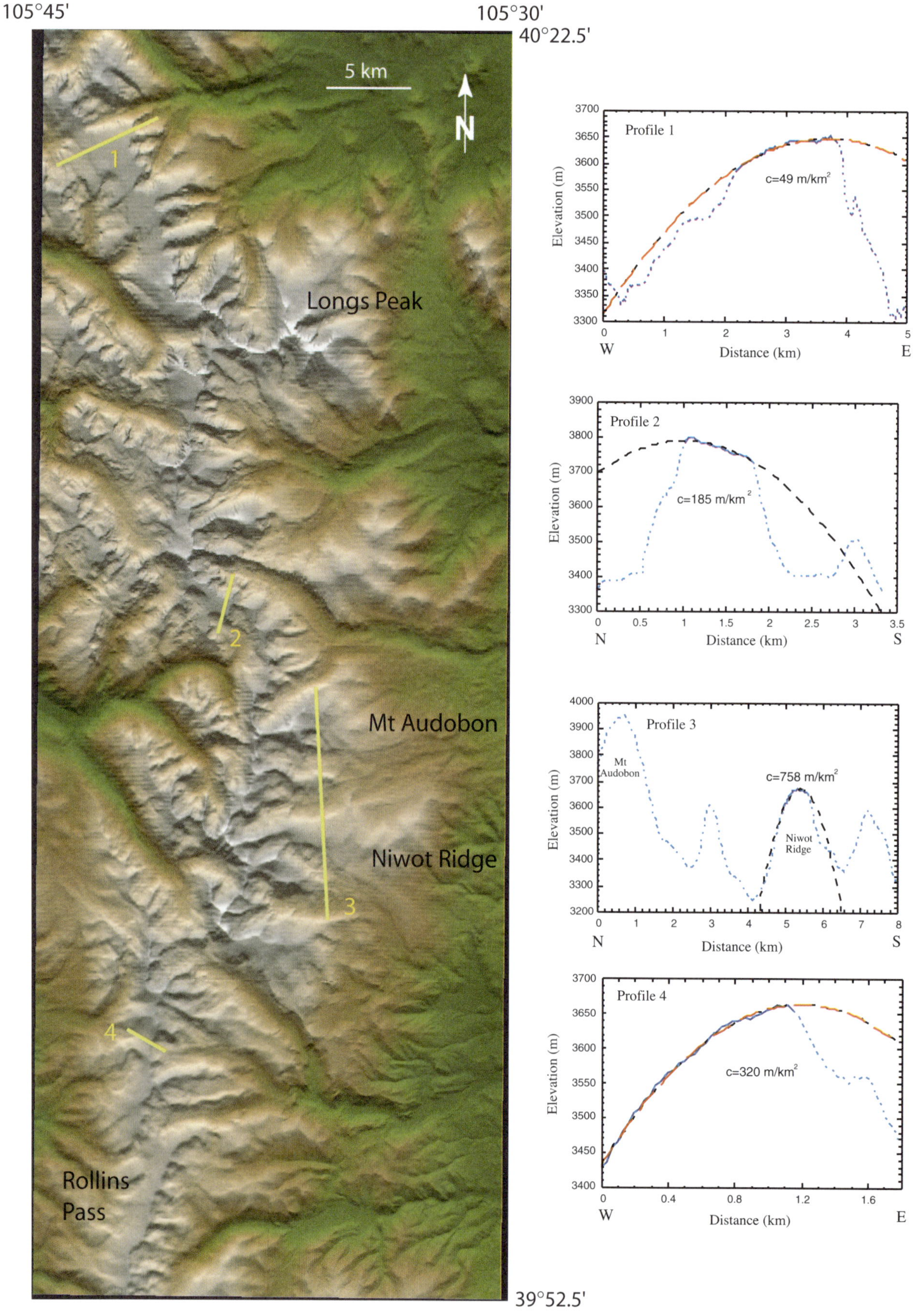

105°45'
105°30'
40°22.5'
39°52.5'
5 km
N
1
2
3
4
Longs Peak
Mt Audobon
Niwot Ridge
Rollins Pass
Profile 1
c=49 m/km²
Elevation (m)
Distance (km)
W
E
Profile 2
c=185 m/km²
N
S
Profile 3
Mt Audobon
c=758 m/km²
Niwot Ridge
Profile 4
c=320 m/km²

The rate of weathering in the landscape is difficult to measure. However, application of cosmogenic radionuclides to this problem has yielded constraints that are consistent in crystalline rocks from the Sierras to the Laramide ranges (Small et al., 1997). Bare bedrock weathering rates are commonly 3–15 µm/yr. Working in the Wind River Mountains on analogous high smooth surfaces, Small et al. (1997) showed that these rates match well with the rates derived from bedrock samples beneath the regolith. Small et al. (1997) argued that this implies that the surfaces in the Wind River Mountains are in steady state; most importantly for the evolution of local relief, they indicate lowering at rates on the order of 5 µm/year (5 m/m.y.). The present profiles of the Front Range crest surfaces (Fig. 5), and the few measured bare bedrock rates from the range crest, imply that these surfaces are eroding at comparably low rates, and have achieved steady-state forms. We review below the quantitative basis for this conclusion.

Model of High Surface Evolution

Conservation of regolith demands that:

$$\frac{dh}{dt} = \frac{\rho_r}{\rho_b} w - \frac{dQ_r}{dx}, \tag{1}$$

where h is the thickness of regolith, t is time, Q_r is flux of regolith, w is the rate of production of regolith by weathering, ρ_r is the density of rock, and ρ_b is the bulk density of the regolith. Here we have made use of the axial symmetry of these surfaces in order to cast the problem in one dimension, in distance from the divide, x. We have also assumed that the deposition of dust is negligible, and that chemical dissolution of regolith is a minor contribution to the evolution of the surface (see discussion in Anderson, 2002). This equation is coupled to the complementary equation for the lowering of the bedrock interface:

$$\frac{dz_b}{dt} = -w, \tag{2}$$

where z_b is the elevation of the regolith-bedrock interface, and the topographic surface is simply $z = z_b + h$. Solution of these equations requires rules for both the generation of regolith by weathering, w, and transport of regolith, Q_r.

In order to explain the existence and location of tors in the landscape, Anderson (2002) argued that the weathering rate must be a function of regolith thickness, which displays a strong maximum at a finite regolith thickness. One such rule is (Fig. 6, inset):

$$w = \frac{dz_b}{dt} = \min\left[A_o + bh, A_1 e^{-\left(\frac{h}{h_*}\right)}\right], \tag{3}$$

where min refers to the minimum of the two terms. The first equation linearly increases from a minimum rate of A_o on bare bedrock ($h = 0$), where b is the rate of increase with regolith thickness, while the second equation decays exponentially with regolith thickness, where A_1 is the bare bedrock weathering rate, and h_* is the length scale for decay. The maximum weathering rate occurs at the regolith thickness at which the two curves cross.

We assume frost creep dominates the transport of regolith downslope (Anderson, 2002), although similar formulation of a regolith transport rule could be carried out for other dominant processes. Following the field and theoretical work of Matsuoka and Moriwaka (1992), and acknowledging that regolith discharge should not occur in the absence of regolith(!), the discharge of regolith per unit width of hillslope becomes

$$Q = f\beta\left[\zeta_*^2 - e^{-h/\zeta_*}(h\zeta_* + \zeta_*^2)\right]\frac{\partial z}{\partial x} = k\frac{\partial z}{\partial x}, \tag{4}$$

where f is the annual number of frost events, β is the soil strain associated with ice lensing, and the probability distribution of frost penetration depth decays exponentially with depth, with a characteristic depth scale of ζ_*. Here k is an effective landscape diffusivity associated with the frost-creep process. Importantly, the discharge smoothly feathers to zero as the regolith approaches zero thickness (Fig. 6, inset), and approaches the maximum discharge appropriate for the slope, dz/dx, as $h \gg \zeta_*$. Note that in the case of uniform regolith, the efficiency of transport (represented by k) is uniform, and discharge becomes linear in slope. This should result in diffusive behavior of the hillslopes.

If the hilltop is at steady state, the regolith thickness must be uniform, so that the weathering rate of the bedrock interface is uniform. Such hilltops should display uniform curvature,

$$c = \frac{\partial^2 z}{\partial x^2} = \frac{\rho_r}{\rho_b}\frac{w}{k}, \tag{5}$$

meaning that their profiles should be parabolic. Therefore, if the hill crest has achieved a steady-state form, one should be able to fit topographic profiles across the hilltops using

$$z = z_{top} - \frac{c}{2}\left(x - x_{top}\right)^2, \tag{6}$$

where the crest of the surface lies at (x_{top}, z_{top}).

High Surfaces of the Continental Divide

In Figure 5, we show four examples of curve fits from the summit surfaces of the Front Range, using 30-m-resolution DEMs available from the U.S. Geological Survey. While the curvature varies by roughly an order of magnitude among these examples, all may indeed be well fit using a parabolic profile. We therefore argue that these surfaces are in steady state. If so, and if the lowering rate associated with weathering is on the

←

Figure 5. Digital elevation model (DEM) of crest between Trail Ridge at the north and Rollins Pass at the south, with traces of topographic profiles labeled. Note strong asymmetry of glacial cirques, and excellent preservation of high smooth surfaces. Topographic profiles 1–4: Best-fitting parabolas to the high surfaces are shown with calculated curvatures (c).

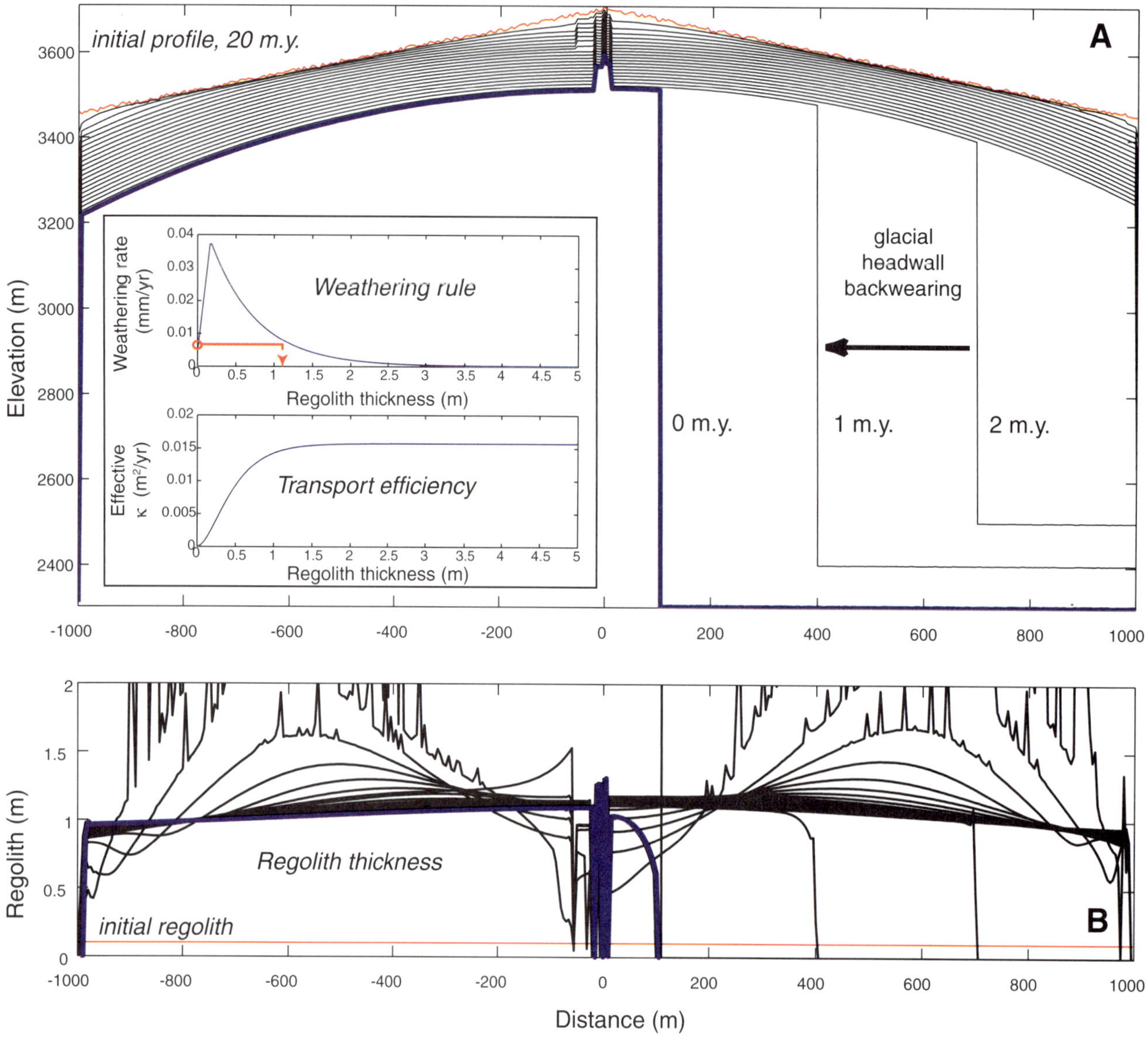

Figure 6. Simulation of high surface profile evolution. Inset: model rules for weathering rate, and for the transport of regolith as a function of regolith thickness. Transport rate feathers to zero for zero regolith thickness. Freeze event frequency = 5/yr; characteristic freeze depth = 0.25 m; slope = 0.10. (A) Profile evolution over 20 m.y.; time steps of 1 m.y. shown. Blue line is final profile. Glacial canyon evolution is crudely mimicked by both deepening and backwearing of valley at the right boundary for the last 3 m.y. of simulation. Smooth parabolic profile is simply truncated by backwearing. A small bare bedrock tor adorns the crest. (B) Evolution of regolith thickness profile. Original uniform thin cover (0.1 m) evolves to a uniform cover of roughly 1 m at end of simulation (blue line). This corresponds to the thickness of regolith needed in the weathering rule such that the bare bedrock edge of the surface and the bedrock beneath the regolith are lowering at the same rate (see arrow in inset in A), as required for a steady-state form.

order of 5 m/m.y., then the topographic diffusivity k may be constrained to lie between 0.2 and 0.02 m^2/yr (assuming $\rho_r/\rho_b \sim 2$). The variation in curvature can be attributed to variation in both the weathering rate, w, and the efficiency of frost creep, k, with microclimate.

That the surfaces may be in steady-state may seem surprising given that the present profiles often display abrupt truncation by headwalls and valley sides that are clearly of glacial origin (Figs. 4 and 5). Anderson (2002) argued that these surfaces are fully decoupled from the glacial valleys beside them. The decoupling occurs at the bare bedrock rims that bound the surfaces. The rims are kept bare by the strong curvature there (regolith is removed down the steep headwall more rapidly than it is delivered to the rim), and therefore they decrease at rates dictated by the (very slow) bare bedrock weathering rate. This is illustrated in Figure 6, in which we numerically simulate the evolution of a high surface adjacent to a glacial valley. Glacial valley deepening and extension (here simply imposed by dictating a boundary retreat rate on the right-hand side of the simulation) are allowed to occur only within the last 3 m.y. of the 20 m.y. simulation. As expected, the surface itself is "ignorant" of the evolution of the glacial valleys adjacent to it. It shows no signs of reacting to the rapidly eroding boundaries.

Tors with Block Fields—A Role for Lightning?

One feature of the high surfaces bears further discussion. Anderson (2002) noted that when they exist, tors (bare bedrock knobs) are largely confined to the crests of these surfaces. He argued that the weathering rule of the sort used here can result in tors, while a rule in which weathering rate simply exponentially decreases with regolith thickness cannot. As can be seen in two of the profiles presented in Figure 5, and in the photograph of Figure 4, knobs do ornament the crests of some of these surfaces. Some of these tors are indeed isolated bedrock outcrops. In many instances, the tors are surrounded by block fields (felsenmeer), consisting of blocks ranging up to >1 m in diameter. The blocks have clearly originated from the outcrop or tor. These block fields display an open framework, with little if any fine-grained regolith in the interstices between blocks, at least in the top meter.

The origin of these block fields is difficult to explain. Repeated frost-cracking is commonly invoked (e.g., Washburn, 1979), which can certainly explain some fraction of the degradation of the outcrop (tor) upslope of the block field. However, in more than one instance, one of us (Anderson) has observed a block of roughly 1 m diameter that is resting on the bedrock surface of a tor at a distance of at least 1 m from a joint-bounded hole from which it must have come. No frost action that we know of could have hoisted the block out of its hole and placed it on the surface so far away. We posit that lightning may be an important process on these surfaces, inspired by the following observations. (1) Every climber of these mountains knows to get off the summits by mid-afternoon in the summer, as thunderstorms pound the mountains, charging the atmosphere with ice-axe buzzing electricity. (2) Evidence of lightning strikes is common. Paleomagnetists who sample rocks from mountainous areas know that summits are places to avoid, as it is difficult to find samples not magnetically reset by lightning (S. Bogue, 1996, personal commun.; R. Coe, 1996, personal commun.). (3) Walking the Continental Divide in this same area in the summer of 1969, one of us (Anderson) witnessed 11 sheep being thrown roughly 2 m into the air by lightning. In addition, the ball lightning dug a trench through the regolith roughly 50 m long, 10–20 cm wide, and 5–10 cm deep (Fig. 7). Many eyewitness accounts exist of heavy objects being thrown into the air by lightning (e.g., Uman, 1969). We therefore hypothesize that lightning strikes on these slowly eroding surfaces may be responsible for both cracking apart outcrops and moving the resulting blocks of rock. The mechanism by which this might operate, propelling rocks this massive such distances, could involve water flashing to steam within joints between blocks. While there is no doubt that enough power exists in lightning strikes to accomplish this work, one might reasonably ask whether lightning occurs with sufficient frequency and density to be worthy of serious consideration. The world map of lightning strike density (http://science.nasa.gov/headlines/images/lightning2/lightningmap_large.gif) shows that lightning strikes roughly 20–30 times per year per km^2 in the Laramide region. As the resolution in these maps is low, we would expect that the enhancement in mountainous areas renders this a minimum estimate. Assuming that this frequency is appropriate over long time scales, assuming it is spatially random, and ignoring the likely enhancement factor, every 1 m^2 of the high surface should be struck about every 50 k.y. If a 1 m block is indeed removed (eroded) from the surface upon impact, then the lowering rate would be 1 m/50,000 yr, or 20 µm/yr. This is clearly in the range of the 5 µm/yr lowering rate measured using cosmogenic radionuclides (Small et al., 1997), implying that this process occurs with sufficient frequency to be a viable candidate for the degradation of tors and the origin of the felsenmeer on these particular surfaces. Clearly, this process will not operate

Figure 7. Photograph taken summer 1969 showing 4 of 11 sheep killed by ball lightning strike. Scene is within 100 m elevation of Continental Divide, just west of the divide on a smooth high surface. Trace of ball lightning is drawn, showing a portion of the 10-cm-wide, 50-m-long trench it dug through vegetation into regolith.

in arctic environments, where other spectacular tors exist, meaning that one must appeal to other processes there. Further study is needed; our hope in including this speculative section is to inspire such study.

Summary of High Surfaces

In summary, the high smooth surfaces that characterize the summit spine of the Front Range are likely steady-state surfaces that operate in a manner not unlike that described in the word-picture of G.K. Gilbert (1909). The processes of weathering and of downslope transport are periglacial and have likely remained so throughout the late Cenozoic. We speculate that lightning plays an important role in detaching blocks on the crests of these surfaces. They are eroding at rates that will likely have lowered them only 10 m over the entire Quaternary, while the neighboring glacial valleys have eroded at rates several orders of magnitude faster, averaged over the many glacial cycles of the Quaternary. That these valleys currently truncate the high surfaces, leaving in some cases less than half of the original surface intact, implies that the glacial valleys have both deepened and migrated headward over the Quaternary. It is therefore inevitable that the local relief in the headwaters of the Front Range has increased dramatically since the inception of late Cenozoic glaciation. The ornamentation, and the aesthetic drama of the high parts of the range are due to the operation of glaciers.

GLACIAL CANYONS

Glaciers leave a distinctive signature in the landscape, both in cross valley (e.g., Harbor et al., 1988; Harbor, 1992) and longitudinal valley (MacGregor et al., 2000; 2002) profiles. The Front Range is particularly instructive, because the glacial limit (Madole et al., 1988) lies roughly halfway between the summit spine and the range front. One can therefore distinguish the signatures between fluvial and glacial erosion on the landscape. In contrast, for example, the glaciers of the Wind River Range were so extensive that they exited the range and gouged deeply into the easily eroded rock of the adjacent basin floors.

Consider the profiles of Clear Creek and its tributaries (Figs. 8 and 9). While the lower portions of the trunk valley profile are smooth, the upper reaches display multiple steps and flats, many of which occur at tributary junctions. In a following section, we discuss the broad convexity of the fluvial profile, but focus here on the steps and flats in the glacially impacted headwaters. MacGregor et al. (2000) argued that this longitudinal valley signature of glacial occupation is dictated by the long-term discharge of ice through the system. Numerical simulations of longitudinal valley profile evolution showed patterns that were robust against the details of the erosion rule used. In the simulations they reported, the first manifestation of glacial erosion is the flattening of the original stream profiles occupied by the glacier. The slope-area relationship so characteristic of fluvial systems is overwritten

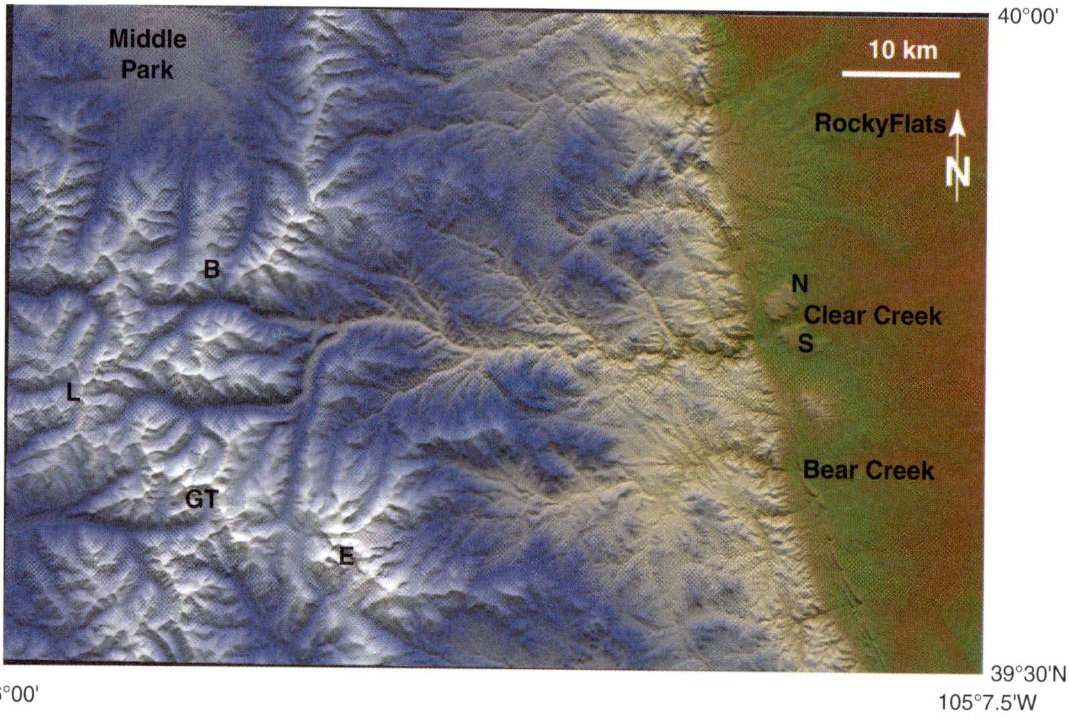

Figure 8. Digital elevation model (DEM) of the Front Range centered on Clear Creek, showing prominent tributaries. See Figure 10 for longitudinal profiles of creeks. B—Berthoud Pass, GT—Grays and Torres peaks, E—Mt. Evans, L—Loveland Pass. Clear Creek exits the mountain front to run between the North (N) and South (S) Table Mountains at Golden.

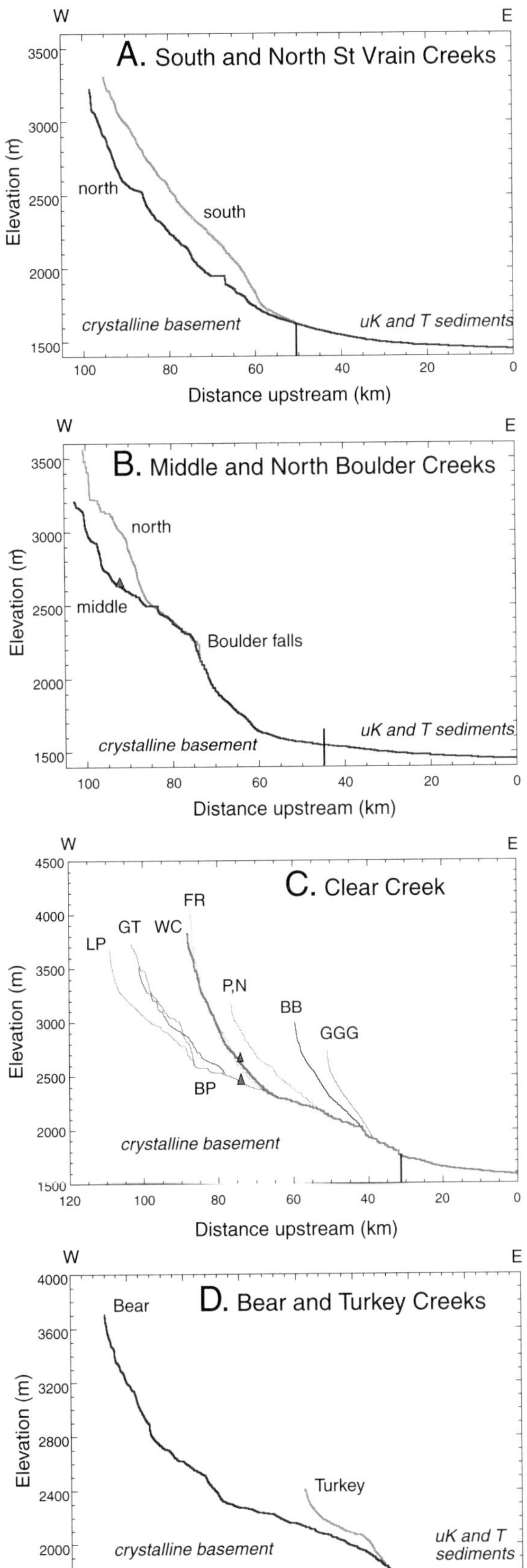

Figure 9. Profiles of several creeks draining the east side of the northern Front Range. Note prominent convexity in each main profile, reflecting capture of the river in the midst of a transient response to lowering of base level on the South Platte. (A) St. Vrain Creek, (B) North and Middle Boulder Creeks, (C) Clear Creek, and (D) Bear and Turkey Creeks. Tributaries reaching high elevations show steps and flats associated with glacial modification (e.g., above join of BP and LP in Clear Creek; and above triangle [glacial limit] in Boulder Creek profiles). uK—upper Cretaceous; T—Tertiary; LP—Loveland Pass; GT—Grays and Torreys; WC—West Chicago Creek; FR—Fall River; BP—Berthoud Pass; P, N—Pine and North Clear Creek; BB—Beaver Brook; GGG—Golden Gate Gulch.

as the more efficient glacial erosion lowers the valley floor. The simplest explanation for effects of glacial occupation of a valley is as follows. Model results incorporating subglacial erosion rules of various types have all shown that erosion rate is roughly proportional to the ice discharge (MacGregor et al., 2000; Fig. 10). Recall that the instantaneous ice discharge in a glacial system reaches a maximum at the equilibrium line elevation (ELA), and feathers to zero at both the headwall and the terminus. The resulting pattern of erosion should mimic the pattern of ice discharge. The erosion should lead to steepening of the original profile between the ELA and the headwall, and flattening of the valley floor between the ELA and the terminus. Over many glacial cycles of varying duration and intensity, this pattern of erosion shrinks and swells. The integration of the erosion pattern over many cycles yields the pattern of total erosion.

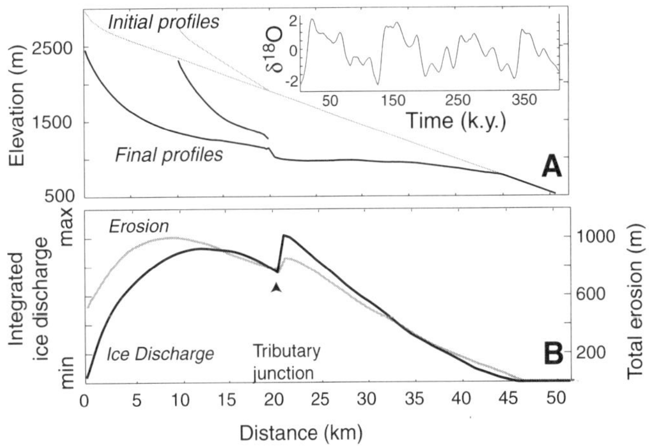

Figure 10. Simulation of glacial longitudinal valley evolution. (A) Evolution of longitudinal profiles from initial fluvial profiles (gray) to final glacial profiles (black) in the face of repeated glacial occupation of the valley. Inset shows history of deep-sea $\delta^{18}O$ that serves as a proxy for the climate history forcing precipitation and melt histories over the 400 k.y. simulation. The step in the valley floor corresponds to a junction with a 10-km-long tributary valley. The tributary valley is left hanging above the trunk valley floor. (B) Profiles of total ice discharge and of total erosion over the simulation. The erosion pattern closely mimics that of ice discharge. The increase in total erosion downstream of the junction reflects the added ice from the tributary (after MacGregor et al., 2000).

The glacial signature is strong at tributaries as well. Long-term ice discharge and hence total erosion jump abruptly at tributary junctions (Fig. 10). In the main valley, the increase in ice discharge leads to greater erosion and a step in the long profile of the valley. In addition, the disparity in ice discharge between small tributaries and the trunk valleys leads to the hanging of tributary valleys (Fig. 10). The height of the hang should be more dramatic the greater the disparity in long-term ice discharge, while the amplitude of the step in the main valley will decrease (MacGregor et al., 2000). We note the correspondence of the major steps in the Clear Creek profiles with tributary junctions.

While useful in addressing first-order issues of glacial longitudinal valley morphology, these early simulations leave several issues unresolved. The simulations were unable to produce overdeepened cirques. As there were no supraglacial erosional processes included in the model, glacial headwall retreat could not occur. Instead, headwalls simply continued to steepen as valley erosion continued. This runs counter to the suggestion that considerable headwall backwearing has occurred in some alpine settings (e.g., Brocklehurst and Whipple, 2002; Whipple et al., 1999). The subglacial erosion rules used were very simple, either invoking the empirically derived relation between sliding and erosion rate (Humphrey and Raymond, 1994), or a theoretical relation between sliding and abrasion (Hallet, 1979).

In an attempt to target these issues, and to make the simulations more relevant to situations akin to those found in the Laramide ranges, MacGregor (2002) updated the model rules. The modifications included the presence of a smooth high surface (or plateau) above the headwall of the glacier and a suite of headwall processes, including blowing of snow off the high surface, avalanching of snow down steep headwalls, frost-cracking and erosion of the bare rock headwall above the glacier, and a more process-specific set of subglacial erosion rules. We merely outline these alterations here, and focus on the model results.

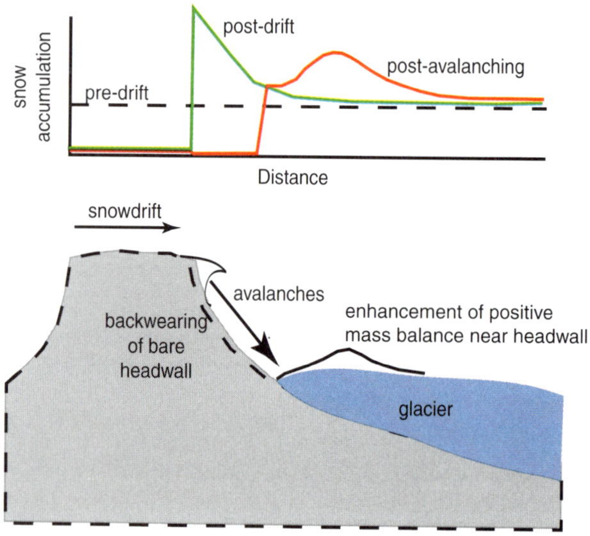

Figure 11. Schematic overview of headwall processes, and modifications to the glacial valley simulation designed to capture them. Snow is blown off high smooth summit surfaces to accumulate in both a major cornice and further downslope. Failures of the cornice accomplish further snow transfer onto the glacier surface. This both adds to the positive mass balance of the glacier, and rids the headwall of protective snow cover. The headwall is then attacked by frost-cracking processes to generate blocks, which ravel to crenulations in the headwall, and are swept downward in avalanches and debris flows. The pre-snow-drift positive mass balance (snowfall) is modified by drift such that all snow blown off the surface is accounted for in an exponential pattern. Avalanching occurs when the accumulation slope is greater than 30°, and all snow is then redistributed on the glacier surface in a Gaussian pattern conserving the snow mass.

New Processes Included in Glacial Valley Profile Simulations

Inclusion of the smooth high surface (plateau in Fig. 11) allows the high surface to act as an additional source of snow for the head of the glacier. This reflects the present-day observation that these high surfaces do not accumulate great depths of snow in the winter (e.g., the February photograph of Fig. 4). Rather, the snow blows off the surfaces to accumulate as large seasonal cornices that overlook the glacial cirques below, largely to the east (downwind). We assume that this same eolian sweeping of the surfaces occurred over the glacial epoch, and that the cornices acted as an additional source of snow. We incorporate avalanching of the snow onto the valley floor simply, by assuming a threshold slope for complete, annual avalanching. In addition, we enforce conservation of cross-sectional area of the snow, and dictate a reasonable run-out pattern onto the glacier. This serves two purposes: to augment the positive mass balance of the glacier near its headwall, and to remove snow cover from the steep headwall, exposing it to subaerial geomorphic processes (Fig. 11). The bare headwall is allowed to erode at a rate dictated by freeze-thaw activity (e.g., Matsuoka and Sakai, 1999). The bedrock particles freed by this process are then delivered either to the ice surface or the bed of the glacier via the bergschrund. The subglacial erosional rules include explicit treatment of both glacial quarrying and abrasion. The processes are linked in that quarried material supplies the tools for the abrasion of bedrock down-valley. The only other source of tools is clasts derived from erosion of the headwall, which encounter the glacial bed either immediately (through the bergschrund) or some distance downstream from the headwall (e.g., Lliboutry, 1994). In addition, the population of clasts is allowed to evolve down-valley. Clasts wear down as they are employed to abrade the bed, for a distance that is dependent on their original size. Finally, the entire valley is allowed to rise isostatically as rock is removed from the valley. This calculation is performed assuming either that the unloading driving the rebound is (1) only local, driven by the glacial erosion pattern, or (2) more regional, principally driven by exhumation of the adjacent sedimentary basin (e.g., Small and Anderson, 1998).

Inclusion of these processes in the longitudinal valley evolution simulations results in more realistic deepening and extension of the glacial valleys (Fig. 12), and can in some instances generate glacial cirques (MacGregor, 2002). Model runs targeting the details at the valley head (as opposed to tributary junctions) show that significant headwall retreat is expected to occur; this could not happen in the earlier model simulations, as no headwall processes were incorporated. In the model run depicted in Figure 12, the climate forcing is taken to be an asymmetric triangular pattern through time (e.g., Harbor, 1992), reflecting slow decay into glacial climates, and rapid emergence from them. During simulations of 400 k.y. periods, headwalls retreat by ~1 km, enough to truncate significantly the high smooth surfaces described above. While isostatic response to the removal of rock from these valleys drives tens of meters of uplift, it is insufficient to counter the erosional downwearing that effectively lowers the valley relative to the climatically determined ELA. This feedback, first appearing in Oerlemans' (1984) simulations, would yield smaller and smaller glaciers through uniform climate swings in the absence of either a deepening of the glacial cycles, or considerably more uplift than the glacial incision alone can generate.

Sediment Output History: The Fluvial Transition

One of the most interesting elements in an alpine geomorphic system is the glacial-fluvial interface. Glacial erosion produces vast amounts of both fine and coarse sediment, which is subsequently handed over to the fluvial system. While the initial impact of this sediment delivery to the fluvial system is immense, the timing of sediment delivery may dictate the timing of river downcutting and terrace formation far downstream of the glacial-fluvial interface (Hancock and Anderson, 2002). The sediment output history from glacial headwaters strongly varies over the glacial cycle. Averaged over a relatively short time scale associated with subglacial sediment storage, sediment output from the glacier must be the integral of the erosion rate over the entire subglacial footprint.

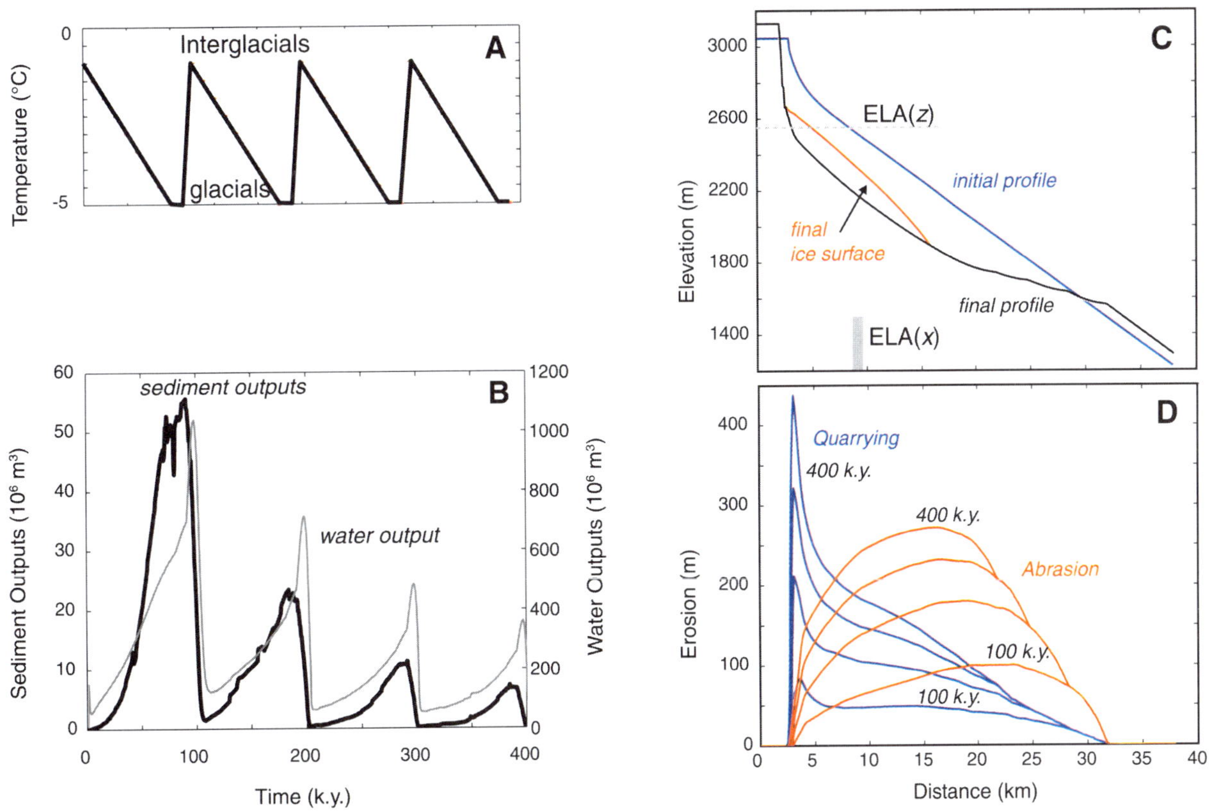

Figure 12. Glacial valley profile evolution, including headwall retreat (after MacGregor, 2002). (A) Climate history used in the 400 k.y. simulation: temperature decreases linearly for 80% of a 100 k.y. cycle, stays cold for 10% (glacial maximum), and warms rapidly during the final 10%. The total amplitude of the temperature change is 4 °C. (B) Time series of sediment and water output from the glacier, showing very strong modulation of sediment output over a glacial cycle. (C) Valley profile evolution after 400 k.y. The headwall is dramatically steepened and lengthened, and retreats by roughly 1 km. The lower 10 km of the valley floor flattens. Successive glaciers become shorter as the erosion lowers the valley into warmer microclimates. Long-term mean location of the equilibrium line altitude (ELA) is shown, along with its down-valley location. (D) The cumulative patterns of abrasion (gray) and quarrying (black). Quarrying is most important near the valley headwall, while abrasion dominates down-valley and is closely tied to the glacier size.

In Figure 12B, we show the sediment output history from the simulation of valley evolution already described. The sediment output varies with the size of the glacier and with climate, as expected. We note that the impact on the fluvial stream will be largely derived from the bedload component of the sediment load, as this affects the channel morphology and the thickness of alluvium in a reach. While the glacial flour (fine silt in suspension) emanating from a glacial terminus stream is quite visible, it is not appropriate to assume, as is commonly done, that the bedload is a small fraction of the suspended load (usually taken to be 10% in nonglacial settings). In recent work on a small valley glacier in Alaska, it has been shown (Riihimaki, 2003; Riihimaki et al., 2005) that when it can be measured in the system, bedload comprises almost 30% of the load. In addition, recent work on a glacially impounded lake into which glacially derived sediments are debouched demonstrated that the volume of coarse load trapped in the delta to the system is approximately equivalent to the volume of fine-grained silt deposited across the basin floor (Loso et al., 2004).

Summary of Glacial Impacts on the Landscape

In summary, glaciation of the headwaters of the east-flowing valleys has left its mark in several ways. Not only do the valleys display glacial signatures in cross valley and longitudinal profiles, but the glaciers likely produced sediment at rates high enough to impact the fluvial system downstream significantly. We note several possible feedbacks. For example, headwall retreat into the high surfaces of the summit spine may result in the diminution of their own snow/avalanche supply, which in turn might stall headward erosion and allow relict high surfaces to be preserved. As headwalls steepen, supraglacial processes become important in the evolution of the summit spine.

FLUVIAL CANYONS

The bedrock fluvial canyons of the Front Range are sandwiched between glacially influenced headwaters and the alluvial reaches of the western portion of the Great Plains. Their profiles reflect this position in the landscape. Steps and flats in the headwaters represent the strong signature of glacial occupation, as discussed above. Below the glacial limit, the profiles are smooth, but commonly display a convexity within the crystalline mountain front that interrupts the profile (Fig. 10). All of these profiles smoothly grade to the South Platte River, which acts as the lower boundary for these streams. In modeling the evolution of these river profiles, we must therefore acknowledge the control exerted by both the headwaters and the main trunk stream.

In this section, we will only attempt to model the portions of the fluvial system in which bedrock erosion dominates. More specifically, we will not count against the model any misfit of the profiles out of this domain. We will specify the lowering rate of the master stream as the lower boundary condition, and refer to ongoing work at the regional scale for understanding of the long-term evolution of these master streams (Riihimaki, 2003). We discuss in a separate section the reality of the headwaters connection, and how we have treated it in the present model.

Bedrock Incision

Stream incision is accomplished by a combination of processes that include quarrying of rock blocks from the bed, abrasion of bedrock, and minor dissolution in susceptible rock types (e.g., Hancock et al., 1998; Whipple et al., 2000). The efficiency of these processes varies strongly with lithology, the pace of incision dictated by the most resistant rock in a reach. In some cases, the switch from dominance of one to another process is frequent, as both the character of the rock or the flow characteristics that are relevant to incision are very local. We acknowledge that there are many formulations of the problem, some casting it as a function of local shear stress, others as a function of local stream power (see reviews in Whipple et al., 2000; Whipple and Tucker, 1999; Tucker and Whipple, 2002). Most researchers will also acknowledge that the incision process has a threshold, either or both because the sediment mantling the bed must be entrained to make way for erosion of the bedrock beneath it (Hancock and Anderson, 2002) or because this sediment is itself the tool of abrasion (e.g., Lavé and Avouac, 2001; Sklar and Dietrich, 2001). This assumption has led researchers to incorporate a probability distribution of discharges into landscape evolution models, as only some fraction of discharge events will incite erosion (Snyder et al., 2003; Tucker and Bras, 2000; Baldwin et al., 2003; Molnar, 2001).

We choose a middle ground for now, in which we appeal to the stream-power formulation of river incision, but incorporate an incision threshold. Importantly, we also explicitly address the drainage topology in that the spatial pattern of water discharge reflects the pattern of drainage area as a function of position in the basin. We also include a first-order approximation of the orographic effects that lead to a nonuniform and evolving pattern of precipitation within the landscape. We do not smooth the drainage area into a power-law distribution, but instead represent the interesting discrete jumps of area at tributary junctions.

Specific stream power (defined as power per unit area of bed) may be written:

$$\omega = \rho_w gQS/W, \tag{7}$$

where ρ_ω is the density of water, g is the acceleration due to gravity, Q is the water discharge, S is the local slope of the channel (dz/dx), and W is the channel width. Vertical incision is then cast as a function of excess stream power over a threshold stream power, ω_c:

$$E = \frac{dz}{dt} = \mathrm{k}(\omega - \omega_c) = \mathrm{k}\,[(\rho_w \mathrm{gQS/W}) - \omega_c]. \tag{8}$$

In this formulation, the constant k reflects primarily the erodibility of the channel. The threshold reflects the need to transport sediment of the caliber that exists on the local bed in order to accomplish the scouring of the local sediment to reach the bedrock in any particular year. In addition, there is likely a threshold involved in both the abrasional and quarrying processes themselves.

Linkage to the Glacial Headwaters

Glaciers impact the downstream fluvial system through their prodigious production of sediment. As discussed at length in Hancock and Anderson (2002), the coarse end member of this sediment load can result in major aggradation of the river downstream, which can reduce or eliminate access of the river to its bedrock bed for large fractions of the glacial cycle. We have already depicted a reasonable time series of sediment delivery from the glacial headwaters. The formal coupling should therefore involve modeling of bed aggradation and degradation through formulas for conservation of sediment and for bedload sediment transport (e.g., Hancock and Anderson, 2002). The upper boundary condition for this transport includes the location of input of sediment (the glacial terminus), and the strength of the source of sediment. This leads to prolonged periods of aggraded conditions that limit vertical incision, but that allow lateral planation. It is in these periods that the river can produce broad strath surfaces in reaches of weak bedrock (e.g., the Wind River as it passes through the Wind River Basin [Chadwick et al., 1997; Hancock and Anderson, 2002; Hancock et al., 1999]). While Hancock and Anderson (2002) showed that wide straths of the Wind River system can be explained by this mechanism, fluvial systems in harder rock will not easily carve laterally. In these cases, the signature of the glacial sediment input variation will likely be fill terraces rather than strath terraces.

Alternatively, one can address this issue by modulation of an armoring function (Sklar and Dietrich, 2001; see also Riihimaki, 2003) through the glacial-interglacial cycle. In simplest terms, this reduces to the inclusion of a factor, which varies from 0 to 1, by which the calculated instantaneous bedrock incision rate is multiplied.

Orographic Response

The precipitation pattern dictates the spatial pattern of stream discharge. Following the development of Roe et al. (2002, 2003), the stream discharge is taken to be

$$Q = \int_0^x P'(x) \frac{dA(x')}{dx'} dx' , \tag{9}$$

where $P(x)$ is the runoff (precipitation-evaporation) pattern, and $A(x)$ is the spatial distribution of drainage area, as derived from a DEM of the drainage basin. The far-field atmospheric speed, V, the local slope of the topography presented to the atmosphere, and the temperature of the atmosphere dictate the precipitation pattern. We follow the work of Roe et al. (2002, 2003) who defined the convergence of air-column moisture flux, C, as

$$-\nabla \cdot F \equiv C = \left(\alpha_o + \alpha_1 \left[V \frac{dz}{dx} \right] \right) e_{sat}(T) , \tag{10}$$

where F is moisture flux, e_{sat} is the saturation vapor pressure at the surface, which is a strong function of temperature, T, and α_0 and α_1 are coefficients. The term in square brackets represents the rate of vertical lift of the air mass. Roe et al. (2002) acknowledged a spatial lag between the location of hydrometeor formation in the air mass at height H above the ground and the location of precipitation on the ground by incorporating a spatial lag that is distributed as a Gaussian. The resulting precipitation pattern is

$$P = \frac{L\sqrt{\pi}}{2} \int_x^\infty C e^{-((x-x')/L)^2} dx' . \tag{11}$$

The length scale, L, may be scaled using the fall speed of the precipitation, w_{sett}, the height of formation of droplets, and the airspeed, V, giving $L = V(H/w_{sett})$. The surface temperature is taken to be $T = T(z_{basin}) - \Gamma[z(x)]$, where Γ is the lapse rate, 6.5 °C/km, and $T(z_{basin})$ is the temperature in the basin adjacent to the range.

Given the axial symmetry of the range, and the relatively narrow valleys incised into it, we assume that gradients in the mean elevation of the range profile, and not those in the stream profile itself, control the pattern of atmospheric flow relevant to generating the precipitation pattern. The important storms that deliver precipitation to the present range are upslope storms, meaning that they are not driven by the mean westerly flow, but back up into the range from the Great Plains to the east. We posit a long-term pattern of precipitation in the Front Range utilizing the records of the Mountain Research Station, which provides a range-normal transect of precipitation.

Initial Condition for the Fluvial Profile Model

Any numerical model of river profile evolution must start with a specified profile: the initial condition. We choose to initiate the model using a steady profile ($dz/dt = 0$ everywhere) in which the incision is perfectly balanced by a uniform but very slow rock uplift rate, U_o. This uplift is meant to signify the regional isostatic response to slow uniform erosion, the rate of which was set by the weathering rate in the landscape, on the order of a few microns per year. Following the work of Whipple and Tucker (1999), we solve the erosion equation for river incision under these circumstances to yield the local slope of the stream. This is then integrated to yield the initial profile. The slope of the channel, S, is

$$S = \frac{W(U_o + \omega_c)}{\rho g k Q} . \tag{12}$$

Note that we must know the pattern of river discharge, $Q(x)$, in order to perform this calculation, meaning that we must also assume an initial precipitation pattern. Here we appeal to a pattern of precipitation, $P(x)$, that is reasonable given the initial topography and the orographic effects described above. The river discharge is then calculated using equation 9, incorporating the real distribution of drainage area, $A(x)$. The initial profile, $z(x)$, is then obtained by integration:

$$z = z_o - \int_0^x S(x)dx \,, \qquad (13)$$

where z_0 is the elevation at the headwaters of the system. (Equivalently, one could integrate this profile from the junction with a master stream, $x = x_{jct}$, upward:

$$z = z_{jct} - \int_{x_{jct}}^x S(x)dx \qquad (14)$$

where z_{jct} is the elevation at the junction with the master stream, x_{jct}.) Given an initial guess at the river profile and using an initial guess of uniform precipitation, P_o, one can assess the slopes, $S(x)$ and then integrate. This profile can then be used to correct for orographic effects on precipitation to generate a new $P(x)$; this alters $Q(x)$, and hence $S(x)$ and $z(x)$, and the process can be repeated until the profile converges on one that accommodates both a more realistic precipitation pattern and the requirements of uniform, steady rock uplift.

Fluvial Model Results

In Figure 13, we present a simulation of the evolution of Middle Boulder Creek over 3 m.y. A lowering rate of 0.15 mm/yr was imposed at the outer edge of the calculation, mimicking incision of the South Platte River (Dethier, 2001). A single hardness contrast of 4-fold was used at the mountain front. Isostatic rebound in response to the unloading of both the soft bedrock of the basin and the crystalline bedrock valleys within the range was employed, using mantle and bedrock densities of 3300 kg/m^3 and 2700 kg/m^3, respectively. We assumed that cross valley profiles of rock removal were very broad in the basin fill and had 20° bounding slopes in the crystalline core, a value derived from averaging slopes on the DEM near the present mountain front. The response to lowering of the boundary was rapid; a subtle convexity in the profile within the rock of the basin migrated 60 km to arrive at the mountain front within 1 m.y. Thereafter, the migration of the convexity in the profile was more prominent, and slowed, progressing 20 km in the remaining 2 m.y. of the simulation. The position and amplitude of the convexity at the end of the model run roughly coincided with that in the modern Middle Boulder Creek profile. The base of the convexity is now at Boulder Falls, essentially at the junction of the North and Middle forks of Boulder Creek.

Reported late Pleistocene and Holocene rates of incision of Boulder Creek (Schildgen et al., 2002) are much higher than those in the model run at these times. This likely reflects the reliance on dates from fill deposits to constrain river incision. As our target is the incision of bedrock channels, the appropriate datum is the bedrock beneath the fill. In addition, incision rates based

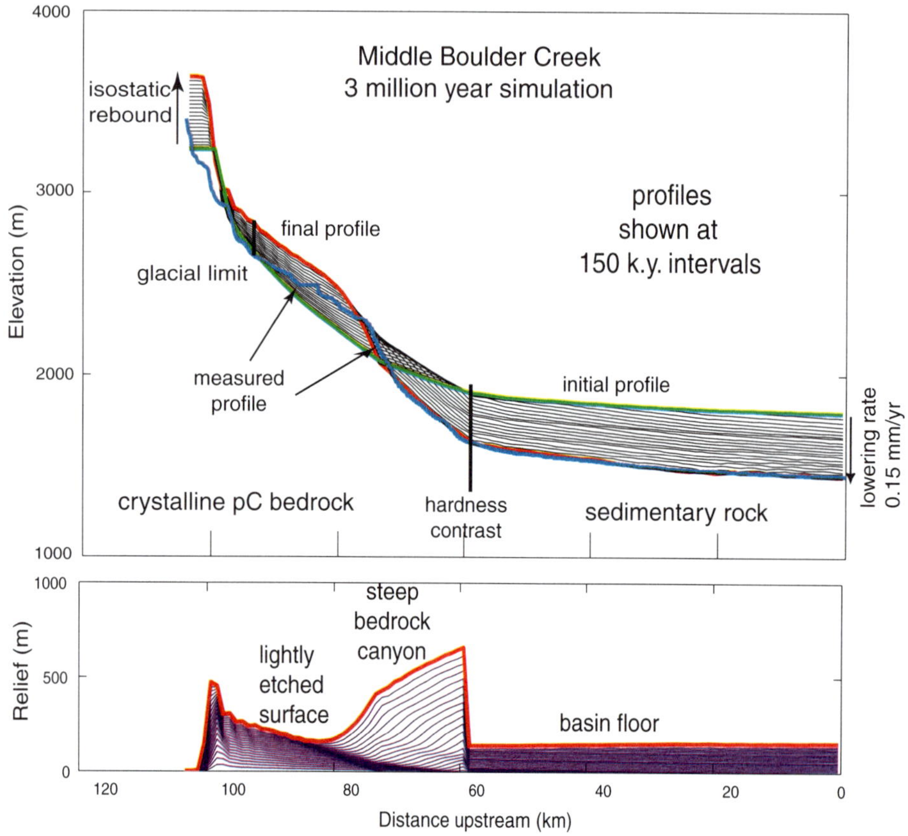

Figure 13. Middle Boulder Creek 3 m.y. model results. Initial condition, shown in green and calculated from assumption of steady very low uplift rate (see text), is subjected to steady lowering of base level on the left side of the calculation space at a rate of 0.15 mm/yr. Lithology contrast at mountain front is assumed to be vertical, and is reflected in persistent break in slope in the profile. Migration of response upriver is shown in 20 profiles evenly spaced at 150 k.y. intervals. Final profile is shown in red, while modern river profile is shown in blue. The break in slope migrates rapidly across Cretaceous bedrock of the basin floor, and slows at the crystalline mountain front, to be found at 3 m.y. roughly 15 km into the front. Isostatic rebound is driven not only by local incision, but by exhumation of the much wider basin at the mountain front. The calculation is expected to fail above the glacial limit, as glacial erosion is not captured in the fluvial model; pC—Precambrian.

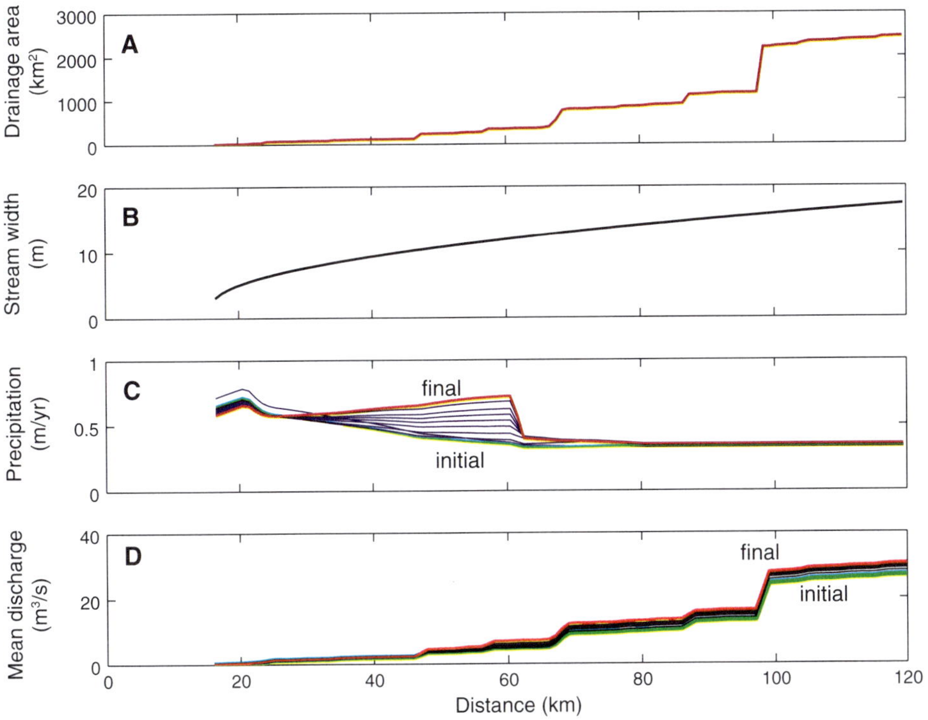

Figure 14. Orographic feedback in Middle Boulder Creek simulation. (A) Drainage area distribution in Middle Boulder Creek, assumed not to change through the simulation. (B) Width distribution of Boulder Creek channel, assumed to go as $x^{0.5}$. (C) Evolution of the pattern of precipitation, calculated according to the orographic precipitation rules discussed in the text. As the topographic step at the mountain front becomes more prominent, the precipitation generated by it increases. The amplitude of the final precipitation pattern is consistent with modern records. (D) Evolution of mean annual river discharge as the precipitation pattern changes. Enhancement in the lower reaches is more than 10%.

upon the time interval since the last glacial can be misleadingly fast (Hancock and Anderson, 2002), as the access of the river to channel bedrock in this interval is likely greater than the mean over a glacial-interglacial cycle (see also Wegmann and Pazzaglia, 2002).

The evolution of the abrupt mountain front enhances orographic precipitation, and hence alters the expected pattern of river discharge through time (Fig. 14). The enhancement of precipitation occurs near the range front, as the mean slope of the topography changes most dramatically there, increasing from a minor step to one on the order of many hundred meters.

The profile of modeled local relief up the creek (Fig. 14) shows the strong increase in relief at the mountain front, and decay of relief into the range. Above the convexity, where little erosion has taken place, the relief remains subdued, reflecting the rolling nature of the subsummit surface. This captures the essence of the DEM-derived cross valley profiles depicted in Figure 15. We note that in reality, the mountain front has become more crenulated near the exit of Middle Boulder Creek, reflecting the launching of incision waves up the smaller tributaries in response to lowering of the trunk stream.

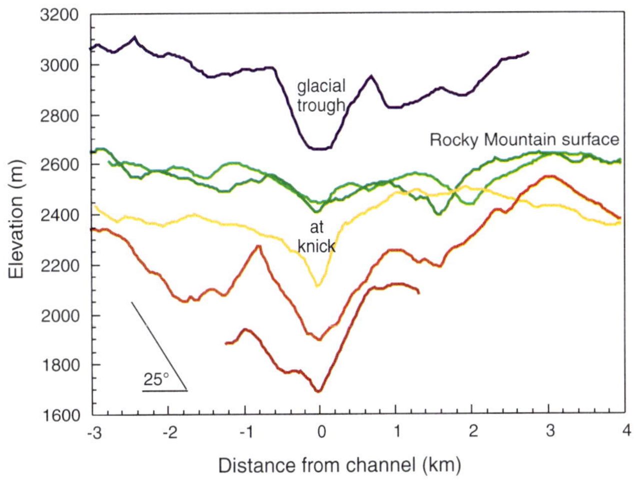

Figure 15. Profiles of hillslopes adjacent to Boulder Creek, aligned such that Boulder Creek is centered at 0. Uppermost profile 1 shows obvious U-shaped cross section. Profiles 2 and 3 display roughness typical of the subsummit surface, with tributaries of Boulder Creek only slightly incised. Profile 4 is centered over the present position of the knickpoint, at Boulder Falls, and shows significant incision into the surface, and steep bounding slopes. Profiles 5 and 6 show very steep canyon walls maintained well after passage of the knickpoint, and more complexity associated with the response of smaller tributaries to the incision of Middle Boulder Creek.

Stream Evolution and the Preservation of the Rocky Flats Surface

We have begun to explore the variability of the response of the various streams draining the Front Range to the lowering of the trunk stream, which in this area is the South Platte.

While we have illustrated the transient nature of the response of Middle Boulder Creek to the lowering of the South Platte River, other nearby streams appear to have been much less effective in their response. We focus on two of these, Coal Creek and Ralston Creek. These streams cross and bound, respectively, the prominent pediment named the Rocky Flats surface (e.g., Scott, 1960; Fig. 16). Importantly, neither of these streams reaches the Continental Divide, and hence they fail to tap glacial headwaters. Their drainage areas at the exit from the mountain front (Ralston Creek, 55 km^2; Coal Creek, 44 km^2) are much smaller than Boulder Creek (300 km^2) to the north, or Clear Creek (1800 km^2) to the south. Their headwaters are limited by the anomalous topography of the 3-km-tall Thorodin Mountain. This interrupts the otherwise low-relief Rocky Mountain, or "subsummit," surface roughly halfway between the range crest and the range front. Coal Creek rises on the east side of this massif, while Ralston Creek snakes around it to the south and taps a portion of its western side. Finally, Coal Creek crosses a very prominent band of strong Precambrian quartzite (Coal Creek quartzite), clasts of which form 90% of the clasts on the Rocky Flats surface (Shroba and Carrara, 1996).

The profiles of Coal Creek and Ralston Creek merge smoothly with the profiles of the adjoining pediments. Rather than becoming less steep outboard of the mountain front, as does Middle Boulder Creek, for example, Coal Creek steepens. Coal Creek is still coupled to the Rocky Flats surface at the mountain front, and becomes progressively more incised with distance into the basin. One should therefore expect that the age of this surface, meaning the time of its abandonment by the stream contributing sediments to it, will depend upon distance from the mountain front, i.e., it is expected to be diachronous (Riihimaki, 2003). Ralston Creek, with greater drainage area, has incised more deeply into both the mountain front and the basin. In the process of this incision, it has left two terraces lower than the Rocky Flats surface (Verdos and Slocum surfaces of Scott, 1960). A slight convexity occurs a few kilometers into the crystalline rock of the range, indicating that the effect of base-level lowering has propagated into the range front. The net result is that Ralston and South Boulder Creeks, at the southern and northern edges of the Rocky Flats surface, respectively, are deeply incised into the surface, while Coal Creek is still un-incised at the exit from the mountain front. In essence, then, the preservation of the Rocky Flats surface has been allowed by the presence of the Thorodin Mountain massif, which most likely represents the decayed remnant of a topographic high within the post-Laramide landscape, and which in turn limits Coal Creek to a small drainage area. The remaining Rocky Flats surface resides in something like an erosional shadow downstream of the topographic high.

Three factors dictate the timing and intensity of the response of a stream to exhumation in the Laramide basins: (1) the position of the stream within the fluvial network, (2) the drainage area of the stream, and (3) the geologic and geomorphic nature of the headwaters (Riihimaki, 2003). We scale this timing as follows. The stream-power formulation of the fluvial incision problem is a kinematic wave equation, in which the celerity of the wave may be thought of as the rate of upstream propagation of a knickzone (or convexity) (e.g., Anderson, 1994; Whipple and Tucker, 1999; Crosby and Whipple, 2002). Ignoring the erosion threshold, the rate is proportional to both the discharge of water, Q, and the erodibility of the rock, k, and is inversely proportional to the channel width:

$$c = k\,[(\rho g Q/W)]\,. \tag{15}$$

In general, both k and Q vary upstream, so that only in uniform rock and in allochthonous rivers (Q uniform) is the wave celerity a constant. The time, T, of arrival of a knickpoint at a position x within the system is then

$$T(x) = \int_{x_o}^{x} \frac{1}{c}\,dx\,. \tag{16}$$

Here x_0 is taken to be the junction with the trunk stream casting off the knickpoint. Alternatively, and more relevant to the Front Range natural experiment, the present position of the knickpoint (convexity) in each drainage will be set by

$$L = \int_{0}^{T} c\,dt\,, \tag{17}$$

where T is the time since the lowering began on the master stream (South Platte) at the base of the system. Because the celerity is related to the erodibility of the rock, the knickpoints should migrate rapidly across the portion of the network dominated by soft Cretaceous sediments, and slow markedly as they enter the mountain front (see Zaprowski et al. [2001] for an example of knickpoint migration in channels beside the Black Hills). In addition, as the knickpoint migrates higher into the network, loss of drainage area will dictate loss of discharge, and hence slow the progress. This will be especially true at tributary junctions, where discharge declines in discrete and sometimes large jumps.

CONCLUSIONS

The Front Range is ideal for separating and understanding the development and functioning of each of its geomorphic components. The range is ornamented at its crest by glacial erosion and along its front by fluvial incision. That the glaciers of the Front Range only reached halfway to the range front allows us to discriminate among the signals of each portion of the system. In addition, at the scale of an individual range, there has been little tectonic activity for the last 40 m.y., providing us the opportunity to investigate the operation of geomorphic processes in the absence of tectonics and associated feedbacks.

The various geomorphic components of the alpine landscape presented by the Front Range can be accounted for in several straightforward 1D models. The high surfaces and their counterpart at mid-elevations, the subsummit surface or Rocky Mountain surface, are dominated by slow weathering and transport rates that lower these portions of the landscape at

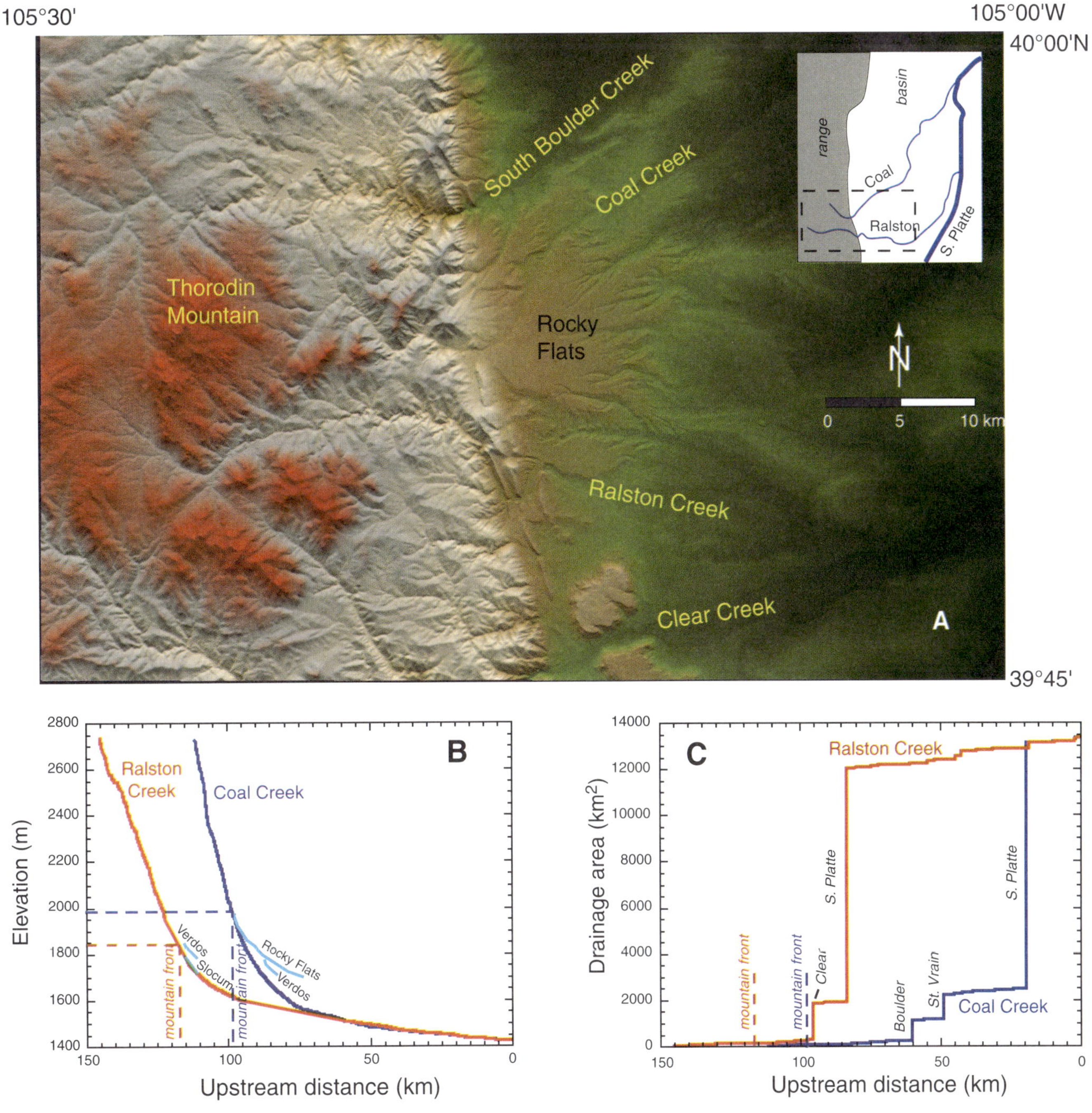

Figure 16. Digital elevation model (DEM) and profiles of the Rocky Flats surface and associated streams (see inset with boxed area for larger fluvial context). (A) DEM of entire drainage area of Coal Creek as it rises on the east side of Thorodin Mountain. Ralston Creek rises to the west of Thorodin Mountain, but does not access the glaciated terrain of the Continental Divide. (B) Coal Creek and Ralston Creek profiles, along with profiles of Rocky Flats, Verdos, and Slocum surfaces. (C) Drainage basin area profiles for each creek. Names of creeks that add significantly to the drainage basin areas are shown vertically. Profiles extend beyond the junction of Coal Creek with the South Platte. Coal Creek is only slightly etched into the Rocky Flats surface at the mountain front. Ralston Creek exits the crystalline range at an elevation 150 m below that at which Coal Creek exits.

rates on the order of 5–10 μm/yr. The glacial troughs adjacent to these within ~10 km of the crest are decoupled from the high surfaces. They have deepened significantly, perhaps by large fractions of a kilometer in places, over the late Cenozoic epoch, producing significant local relief along the spine of the range. In addition, significant headward retreat of glacial headwalls has truncated the high surfaces. Headwall retreat, driven largely by westward growth of glacial valleys east of the divide, results in a highly asymmetric range crest, with intact long, smooth slopes from the west, meeting glacial headwalls at or near the crest of the surfaces. This glacial deepening and extension serves to ornament the crest, and has transformed what was likely a boring alpine landscape in mid-Cenozoic time into one that merits National Park status.

The fluvial system in the Front Range is sandwiched between these glacial headwaters and an incising trunk stream, the South Platte River. The fluvial channel is supplied with both meltwater and sediment from the glacial system, the location of the transition being the glacial terminus. Because both the terminus position and the maximum ice discharge (and along with it, the maximum erosion rate) will co-vary in a glacial cycle, sediment delivery (both fine and coarse) will be highly pulsed over time scales of 10^5 to 10^6 yr. The montane fluvial system is currently caught in the midst of a transient response to base-level lowering on the South Platte River, and its ability to incise is likely highly punctuated by pulses of aggradation associated with glacial sediment delivery. Each stream draining the mountain front is in a different state within this transient response. The large streams draining the glaciated crest (e.g., Boulder Creeks, Clear Creek) display strong convexities well up into the crystalline core of the range, while the knickzones of some smaller streams have yet to reach the mountain front (e.g., Coal Creek). This differential response is dictated by both the location of the stream in the South Platte drainage network, which sets the timing of the initiation of base-level lowering, the distance to the mountain front from this confluence, and the water discharge of the tributary stream. This latter is set largely by drainage area, which is strongly affected by anomalies in the topography (e.g., Thorodin Mountain) within the ancient post-Laramide landscape. Consequently, the streams on the edge of the Denver Basin are presently at different levels relative to the prominent pediments, and will have experienced different incision histories. This raises a cautionary flag for correlation of surfaces within the basin (Riihimaki, 2003). The position of the stream within the fluvial network, and the drainage characteristics of each stream must be taken into account when attempting to deduce a basinwide history of incision from sparsely dated surfaces.

Although we have shown that the glacial troughs and high surfaces are effectively decoupled from one another, the same cannot be said of other components in the system. Future work should address (model) the formal linkage between the glacial and fluvial portions of the landscape, and between channels and hillslopes.

Important and intriguing questions about the evolution of the Front Range remain. The timing of the initiation of exhumation of the Denver Basin (or any other basin in the Laramide province) remains poorly constrained. Identification and dating of scraps of fluvial deposits on the subsummit surface may aid in this. We have not addressed the cause of the incision of the South Platte River, which has produced the clear response within its tributaries on which we have focused. Debate still revolves around the roles of climate change and of tectonics in inciting this broad exhumation of the Laramide basins. While local tectonic events may be invoked in the NW of the Laramide province (arrival of the Yellowstone hotspot [e.g., Smith and Braile, 1993]), or in the south (northward propagation of the Rio Grande Rift [Chapin and Cather, 1994; Erslev, 2001; Formento-Trigilio and Pazzaglia, 1998; Leonard, 2002; MacMillan et al., 2002]), no single regional geophysical event has been identified that would incite regional exhumation in the late Cenozoic. The primary long-wavelength geophysical mechanism to which one might appeal is the dynamic topographic response to the movement of the Farallon slab beneath this region (e.g., Mitrovika et al., 1989; Heller et al., 2003). On the other hand, significant changes in late Cenozoic climate (e.g., Zachos et al., 2001) may have driven changes in the probability distribution of storm sizes and hence in stream discharges to which the fluvial system will have been subjected (Zhang et al., 2001). In either case, the temporal and spatial pattern of exhumation appears to demand that the fluvial signal is propagating up the system. While this alone does not discriminate between tectonic and climatic (discharge of water and sediment) forcing mechanisms, we presently favor a climatic driver (Riihimaki, 2003). The primary data needed to solve this riddle will come from more dating of depositional and erosional surfaces within the basins and down the trunk streams.

ACKNOWLEDGMENTS

This research was supported by two grants from the National Science Foundation (NSF) (EAR-0003604 and OPP-9818251), an NSF graduate fellowship (to Riihimaki), a National Aeronautics and Space Administration (NASA) graduate fellowship (to MacGregor), and a University of California Presidential postdoctoral fellowship award (to Safran). The manuscript was improved significantly by the careful reviews of Frank Pazzaglia and Eric Kirby. Finally, we thank the organizers of the Penrose conference for a spectacular and informative meeting.

REFERENCES CITED

Anderson, R.S., 1994, Evolution of the Santa Cruz Mountains, California, through tectonic growth and geomorphic decay: Journal of Geophysical Research, v. 99, p. 20,161–20,179, doi: 10.1029/94JB00713.

Anderson, R.S., 2002, Modeling the tor-dotted crests, bedrock edges, and parabolic profiles of high alpine surfaces of the Wind River Range, Wyoming: Geomorphology, v. 46, p. 35–58, doi: 10.1016/S0169-555X(02)00053-3.

Baldwin, J.A., Whipple, K.X., Tucker, G.E., 2003, Implications of the shear stress river incision model for the timescale of postorogenic decay of topography: Journal of Geophysical Research, v. 108 no. B3, 2158.

Bird, P., 1998, Kinematic history of the Laramide orogeny in latitudes 35°–49°N, western United States: Tectonics, v. 17, p. 780–801, doi: 10.1029/98TC02698.

Brocklehurst, S.H., and Whipple, K.X., 2002, Glacial erosion and relief production in the Eastern Sierra Nevada, California: Geomorphology, v. 42, no. 1–2, p. 1–24, doi: 10.1016/S0169-555X(01)00069-1.

Chadwick, O.A., Hall, R.D., and Phillips, F.M., 1997, Chronology of Pleistocene glacial advances in the central Rocky Mountains: Geological Society of America Bulletin, v. 109, p. 1443–1452, doi: 10.1130/0016-7606(1997)109<1443:COPGAI>2.3.CO;2.

Chapin, C.E., and Cather, S.M., 1994, Tectonic setting of the axial basins of the northern and central Rio Grande Rift, *in* Keller, G.R., and Cather, S.M., eds., Basins of the Rio Grande Rift; structure, stratigraphy, and tectonic setting: Geological Society of America Special Paper 291, p. 5–25.

Crosby, B.T., and Whipple, K.X., 2002, Knickpoint migration in the Waipaoa River: An examination of the rate and form of transient behavior in fluvial networks: Eos (Transactions, American Geophysical Union), v. 83, no. 47, Fall meeting supplement, p. F582.

Crowley, P.D., Reiners, P.W., Reuter, J.M., and Kaye, G.D., 2002, Laramide exhumation of the Bighorn Mountains, Wyoming: An apatite (U-Th)/He thermochronology study: Geology, v. 30, p. 27–30, doi: 10.1130/0091-7613(2002)0302.0.CO;2.

Dethier, D.P., 2001, Pleistocene incision rates in the western United States calibrated using Lava Creek B tephra: Geology, v. 29, p. 783–786, doi: 10.1130/0091-7613(2001)0292.0.CO;2.

Dickinson, W.R., Klute, M.A., Hayes, M.J., Janecke, S.U., Lundin, E.R., McKittrick, M.A., and Olivares, M.D., 1988, Paleogeographic and paleotectonic setting of Laramide sedimentary basins in the central Rocky-Mountain region: Geological Society of America Bulletin, v. 100, p. 1023–1039, doi: 10.1130/0016-7606(1988)100<1023:PAPSOL>2.3.CO;2.

Epis, R.C., and Chapin, C.E., 1975, Geomorphic and tectonic implications of the post-Laramide, late Eocene erosion surface in the Southern Rocky Mountains, *in* Curtis, B.F., ed., Cenozoic history of the southern Rocky Mountains: Geological Society of America Memoir 144, p. 45–74.

Erslev, E.A., 2001, Multistage, multidirectional tertiary shortening and compression in north-central New Mexico: Geological Society of America Bulletin, v. 113, p. 63–74, doi: 10.1130/0016-7606(2001)113<0063:MMTSAC>2.0.CO;2.

Evanoff, E., 1990, Early Oligocene paleovalleys in southern and central Wyoming: Evidence of high local relief on the late Eocene unconformity: Geology, v. 18, p. 443–446, doi: 10.1130/0091-7613(1990)018<0443:EOPISA>2.3.CO;2.

Gilbert, G.K., 1909, The convexity of hilltops: Journal of Geology, v. 17, p. 344–350.

Hallet, B., 1979, A theoretical model of glacial abrasion: Journal of Glaciology, v. 23, p. 39–50.

Hancock, G.S., and Anderson, R.S., 2002, Numerical modeling of fluvial strath-terrace formation in response to oscillating climate: Geological Society of America Bulletin, v. 114, p. 1131–1142, doi: 10.1130/0016-7606(2002)114<1131:NMOFST>2.0.CO;2.

Hancock, G. S., Anderson, R.S. and Whipple, K. X, 1998, Bedrock erosion by streams: Beyond stream power, *in* Tinkler, K., and Wohl, E., eds., Rivers over rock: American Geophysical Union Geophysical Monograph, p. 35–60.

Hancock, G.S., Anderson, R.S., Chadwick, O.A., and Finkel, R.C., 1999, Dating fluvial terraces with Be-10 and Al-26 profiles: Application to the Wind River, Wyoming: Geomorphology, v. 27, p. 41–60, doi: 10.1016/S0169-555X(98)00089-0.

Harbor, J.M., 1992, Numerical modeling of the development of U-shaped valleys by glacial erosion: Geological Society of America Bulletin, v. 104, p. 1364–1375, doi: 10.1130/0016-7606(1992)104<1364:NMOTDO>2.3.CO;2.

Harbor, J.M., Hallet, B., and Raymond, C.F., 1988, A numerical model of landform development by glacial erosion: Nature, v. 333, p. 347–349, doi: 10.1038/333347a0.

Heller, P.L., Dueker, K., and McMillan, M.E., 2003, Post-Paleozoic alluvial gravel transport as evidence of continental tilting in the U.S. Cordillera: Geological Society of America Bulletin, v. 115, no. 9, p. 1122–1132, doi: 10.1130/B25219.1.

Humphrey, N.F., and Raymond, C.F., 1994, Hydrology, erosion and sediment production in a surging glacier; Variegated Glacier, Alaska, 1982–83: Journal of Glaciology, v. 40, p. 539–552.

Izett, G.A., 1975, Late Cenozoic sedimentation and deformation in northern Colorado and adjoining areas, *in* Curtis, B.F., ed., Cenozoic history of the southern Rocky Mountains: Geological Society of America Memoir 144, p. 179–209.

Kelley, S.A. and Chapin, C.E., 1997, Internal structure of the southern Front Range, Colorado, from an apatite fission-track thermochronology perspective, *in* Bolyard, D.W., and Sonnenberg, S.A., eds., Geologic history of the Colorado Front Range: Rocky Mountain Association of Geologists Symposium Guidebook, p. 19–30.

Kelley, S.A., and Duncan, I.J., 1986, Late Cretaceous to middle Tertiary tectonic history of the northern Rio Grande Rift, New Mexico: Journal of Geophysical Research, v. 91, p. 6246–6262.

Lavé, J., and Avouac, J.P., 2001, Fluvial incision and tectonic uplift across the Himalayas of central Nepal: Journal of Geophysical Research, v. 106, no. B11, p. 26,561–26,591, doi: 10.1029/2001JB000359.

Leonard, E.M., 2002, Geomorphic and tectonic forcing of late Cenozoic warping of the Colorado piedmont: Geology, v. 30, p. 595–598, doi: 10.1130/0091-7613(2002)030<0595:GATFOL>2.0.CO;2.

Lliboutry, L., 1994, Monolithologic erosion of hard beds by temperate glaciers: Journal of Glaciology, v. 40, p. 433–450.

Loso, M.G., Anderson, R.S., and Anderson, S.P., 2004, Post Little Ice Age record of fine and coarse clastic sedimentation in an Alaskan proglacial lake: Geology, v. 32, p. 1065–1068, doi: 10.1130/G20839.1

MacGregor, K.R., 2002, Modeling and field constraints on glacier dynamics, erosion, and alpine landscape evolution [Ph.D. dissertation]: Santa Cruz, University of California, Santa Cruz, 277 p.

MacGregor, K.C., Anderson, R.S., Anderson, S.P., and Waddington, E.D., 2000, Numerical simulations of longitudinal profile evolution of glacial valleys: Geology, v. 28, no. 11, p. 1031–1034, doi: 10.1130/0091-7613(2000)028<1031:NSOGLP>2.3.CO;2.

Madole, R.F., VanSistine, D.P., and Michael, J.A., 1988, Pleistocene glaciation in the upper Platte valley drainage basin, Colorado: Geologic Investigations Series Map I-2644.

Matsuoka, N., and Moriwaka, K., 1992, Frost heave and creep in the Sor Rondane Mountains, Antarctica: Arctic and Alpine Research, v. 24, p. 271–280.

Matsuoka, N., and Sakai, H., 1999, Rockfall activity from an alpine cliff during thawing periods: Geomorphology, v. 28, p. 309–328, doi: 10.1016/S0169-555X(98)00116-0.

McMillan, M.E., Angevine, C.L., and Heller, P.L., 2002, Postdepositional tilt of the Miocene-Pliocene Ogallala Group on the western Great Plains: Evidence of late Cenozoic uplift of the Rocky Mountains: Geology, v. 30, p. 63–66, doi: 10.1130/0091-7613(2002)030<0063:PTOTMP>2.0.CO;2.

Mears, B., Jr., 1993, Geomorphic history of Wyoming and high-level erosion surfaces, *in* Snoke, A.W., Steidtmann, J.R., and Roberts, S.M., eds., Geology of Wyoming: Geological Survey of Wyoming Memoir 5, p. 608–626.

Mitrovica, J.X., Beaumont, C., and Jarvis, G.T., 1989, Tilting of Continental Interiors by the Dynamical Effects of Subduction: Tectonics, v. 8, p. 1079–1094.

Molnar, P., 2001, Climate change, flooding in arid environments, and erosion rates: Geology, v. 29, p. 1071–1074, doi: 10.1130/0091-7613(2001)029<1071:CCFIAE>2.0.CO;2.

Oerlemans, J., 1984, Numerical experiments on large-scale glacial erosion: Zeitschrift für Gletscherkunde und Glazialgeologie, v. 20, p. 107–126.

Reiners, P.W., and Heffern, E.L., 2002, Pleistocene exhumation rates of Wyoming intermontane basins from (U-Th)/He dating of clinker: Geological Society of America Abstracts with Programs, v. 34, no. 6, p. 321.

Riihimaki, C.A., 2003, Quantitative constraints on the glacial and fluvial evolution of alpine landscapes [Ph.D. dissertation]: Santa Cruz, University of California, Santa Cruz, 205 p.

Riihimaki, C.A., MacGregor, K.R., Anderson, R.S., Anderson, S.P., and Loso, M.G., 2005, Sediment evacuation and glacial erosion rates at a small alpine glacier: Journal of Geophysical Research - Earth Surface, v. 110, No. F3, F03003, 10.1029/2004JF000189

Roe, G.H., Montgomery, D.R., and Hallet, B., 2002, Effects of orographic precipitation variations on the concavity of steady-state river profiles: Geology, v. 30, p. 143–146, doi: 10.1130/0091-7613(2002)030<0143:EOOPVO>2.0.CO;2.

Roe, G.H., Montgomery, D.R., and Hallet, B., 2003, Orographic precipitation and the relief of mountain ranges: Journal of Geophysical Research, Solid Earth, v. 108, no. B6, 2315, doi: 10.1029/2001JB001521.

Schildgen, T., Dethier, D.P., and Bierman, P., 2002, Al-26 and Be-10 dating of late Pleistocene and Holocene fill terraces: A record of fluvial deposition and incision, Colorado Front Range: Earth Surface Processes and Landforms, v. 27, no. 7, p. 773–787, doi: 10.1002/esp.352.

Scott, G.R., 1960, Subdivision of the Quaternary alluvium east of the Front Range near Denver, Colorado: Geological Society of America Bulletin, v. 71, p. 1541–1543.

Scott, G.R., 1975, Cenozoic surfaces and deposits in the southern Rocky Mountains, *in* Curtis, B.F., ed., Cenozoic history of the southern Rocky Mountains: Geological Society of America Memoir 144, p. 227–248.

Shroba, R.R., and Carrara, P.E., 1996, Surficial geologic map of the Rocky Flats Environmental Technology Site and vicinity, Jefferson and Boulder Counties, Colorado: U.S. Geological Survey Miscellaneous Investigations Series Map I-2526, scale 1:12,000.

Sklar, L.S., and Dietrich, W.E., 2001, Sediment and rock strength controls on river incision into bedrock: Geology, v. 29, p. 1087–1090, doi: 10.1130/0091-7613(2001)029<1087:SARSCO>2.0.CO;2.

Small, E.E., and Anderson, R.S., 1998, Pleistocene relief production in Laramide mountain ranges, western United States: Geology, v. 26, p. 123–126, doi: 10.1130/0091-7613(1998)026<0123:PRPILM>2.3.CO;2.

Small, E.E., Anderson, R.S., Repka, J.L., and Finkel, R., 1997, Erosion rates of alpine bedrock summit surfaces deduced from in situ Be-10 and Al-26: Earth and Planetary Science Letters, v. 150, p. 413–425, doi: 10.1016/S0012-821X(97)00092-7.

Smith, R.B., and Braile, L.W., 1993, Topographic signature, space-time evolution, and physical properties of the Yellowstone–Snake River plain volcanic system: The Yellowstone hotspot, *in* Snoke, A.W., Steidtmann, J.R., and Roberts, S.M., eds., Geology of Wyoming: Geological Survey of Wyoming Memoir 5, p. 694–754.

Snyder, N.P., Whipple, K.X., Tucker, G.E., and Merritts, D.J., 2003, Importance of a stochastic distribution of floods and erosion thresholds in the bedrock river incision problem: Journal of Geophysical Research, v. 108, no. B2, 2117, doi: 10.1029/2001JB001655.

Tucker, G.E., and Bras, R.L., 2000, A stochastic approach to modeling the role of rainfall variability in drainage basin evolution: Water Resources Research, v. 36, p. 1953–1964, doi: 10.1029/2000WR900065.

Tucker, G.E., and Whipple, K.X., 2002, Topographic outcomes predicted by stream erosion models: Sensitivity analysis and intermodel comparison: Journal of Geophysical Research, Solid Earth, v. 107, 2179, doi: 10.1029/2001JB000162.

Uman, M.A., 1969, Lightning: New York, McGraw-Hill, 264 p.

Washburn, A.L., 1979, Geocryology: A survey of periglacial processes and environments: London, E. Arnold, 406 p.

Wegmann, K.W., and Pazzaglia, F.J., 2002, Holocene strath terraces, climate change and active tectonics: The Clearwater River basin, Olympic Peninsula, Washington State: Geological Society of America Bulletin, v. 114, no. 6, p. 731–744, doi: 10.1130/0016-7606(2002)114<0731:HSTCCA>2.0.CO;2.

Whipple, K.X., and Tucker, G., 1999, Dynamics of the stream-power river incision model: Implications for height limits of mountain ranges, landscape response timescales and research needs: Journal of Geophysical Research, v. 104, p. 17,661–17,674, doi: 10.1029/1999JB900120.

Whipple, K.X., Kirby, E., and Brocklehurst, S.H., 1999, Geomorphic limits to climate-induced increases in topographic relief: Nature, v. 401, p. 39–43, doi: 10.1038/43375.

Whipple, K.X., Hancock, G.S., and Anderson, R.S., 2000, River incision into bedrock: Mechanics and the relative efficacy of plucking, abrasion, and cavitation: Geological Society of America Bulletin, v. 112, no. 3, p. 490–503, doi: 10.1130/0016-7606(2000)112<835:RAPOBI>2.3.CO;2.

Zachos, J., Pagani, M., Sloan, L., Thomas, E., and Billups, K., 2001, Trends, rhythms, and aberrations in global climate 65 Ma to present: Science, v. 292, no. 5517, p. 686–693, doi: 10.1126/science.1059412.

Zaprowski, B.J., Evenson, E.B., and Pazzaglia, F.J., 2001, Knickzone propagation in the Black Hills and northern High Plains: A different perspective on the late Cenozoic exhumation of the Laramide Rocky Mountains: Geology, v. 29, no. 6, p. 547–550, doi: 10.1130/0091-7613(2001)029<0547:KPITBH>2.0.CO;2.

Zhang, P., Molnar, P., and Downs, W.R., 2001, Increased sedimentation rates and grain sizes 2–4 Myr ago due to the influence of climate change on erosion rates: Nature, v. 410, p. 891–897.

Manuscript Accepted by the Society 23 June 2005

Printed in the USA

Geological Society of America
Special Paper 398
2006

Mountain fronts, base-level fall, and landscape evolution: Insights from the southern Rocky Mountains

Kurt L. Frankel[†]
Frank J. Pazzaglia
Department of Earth and Environmental Sciences, Lehigh University, Bethlehem, Pennsylvania 18015, USA

ABSTRACT

Mountain ranges in the southern Rocky Mountains, first uplifted during the early Cenozoic Laramide orogeny, have followed separate landscape evolutionary pathways in the late Cenozoic. We present a model that reconstructs the post-Laramide tectonic and geomorphic history of Sierra Nacimiento and the Taos Range, two nearly adjacent rift-flank ranges in north-central New Mexico that serve to illustrate the various processes shaping landscapes across the southern Rocky Mountains. The Sierra Nacimiento landscape reflects the exhumation of hard Precambrian rocks from beneath a softer Phanerozoic sedimentary cover. The exhumation is continuous, but not steady, being driven by distal base-level fall. Downstream diverging river terraces in the Jemez River valley on the eastern flank of Sierra Nacimiento and late Pliocene to Holocene fluvial deposits on the western Sierra Nacimiento piedmont document the base-level fall. The timing and contemporary rates of incision from these river systems suggest that exhumation is being propagated from south to north as knickzones work their way headward from the Rio Grande. In contrast, the Taos Range landscape reflects alternating active stream incision and aggradation astride, and throttled by, an active range-front normal fault. The distinction between the exhumation-dominated and tectonic-dominated mountain front is best quantified by analyses of first-order stream gradients and a watershed metric we call the drainage basin volume to drainage basin area ratio (R_{va}). Gradients of first-order streams in the exhumation-dominated Sierra Nacimiento have a mode of 6.8 degrees, significantly less than the 17.7 degrees obtained from a comparable data set of Taos Range first-order streams. The distinct stream gradient and R_{va} populations hint at an important change in the processes shaping hillslopes and low-order channels, which is supported by the lack of slope-clearing landslides in the Sierra Nacimiento landscape and the presence of such landslides in the Taos Range. Analogue and numeric models find that steep, rugged, faceted topography associated with tectonically active mountain fronts like the Taos Range can only be produced and maintained by creep and landslides where the sediment flux scales as a power law with respect to average hillslope or low-order channel gradient. Here, the fingerprint of active tectonics is recorded by both high R_{va} values and steep modal channel gradients. By comparison, the Sierra Nacimiento landscape is shaped primarily by creep where the sediment flux has a linear relationship to average hillslope and low-order channel gradient. In this situation, the signatures of distal base-level fall are low R_{va} values and relatively gentle modal channel gradients.

[†]Present address: Department of Earth Sciences, University of Southern California, Los Angeles, California 90089, USA; e-mail: kfrankel@usc.edu.

Frankel, K.L., and Pazzaglia, F.J., 2006, Mountain fronts, base-level fall, and landscape evolution: Insights from the southern Rocky Mountains, *in* Willett, S.D., Hovius, N., Brandon, M.T., and Fisher, D.M., eds., Tectonics, Climate, and Landscape Evolution: Geological Society of America Special Paper 398, Penrose Conference Series, p. 419–434, doi: 10.1130/2006.2398(26).

Keywords: tectonic geomorphology, landscape evolution, Taos Range, Sierra Nacimiento, Rocky Mountains.

INTRODUCTION

The geomorphic evolution of mountain fronts has been used to interpret the processes and rates of tectonic deformation (reviewed in Keller and Pinter, 2001; Burbank and Anderson, 2001). Topographic metrics including mountain front linearity (Bull and McFadden, 1977), slopes, alluvial fault scarps and facets (Menges, 1990a; dePolo and Anderson, 2000), valley excavation (Harbor, 1997), and alluvial fan geometry (Ferrill et al., 1996), among others, are the core data utilized in mountain front studies. In this context, there is a growing appreciation for the confounding effects of processes other than offset along the range-front fault, including changes in climate, influences of variably resistant lithologies, and regional base-level fall, which contribute to the shapes and forms of mountain fronts (Ellis et al., 1999). Additionally, it is now better appreciated that the flux of sediment off of hillslopes does not necessarily scale linearly with respect to slope (cf. Culling, 1960), especially in steep, tectonically active settings (Kirkby, 1984; Anderson, 1994; Howard, 1994; Brozovic et al., 1997; Densmore et al., 1998; Roering et al., 2001; Montgomery and Brandon, 2002), making it difficult to interpret the evolution or isolate the tectonic component of landscapes using topographic metrics alone, even if a multiparameter approach is used (Wells et al., 1988).

This paper explores the landscape evolution of two mountain fronts, the Taos Range and Sierra Nacimiento, in north-central New Mexico that share a common tectonic origin, but then diverge with respect to their tectonic histories (Fig. 1). The distribution of gradients among channels of similar order as well as the rate and spatial distribution of fluvial incision provide the core data set for mountain front landscape evolution. Sierra Nacimiento is a Proterozoic basement-cored, ~80-km-long Laramide uplift, located on the western flank of the Rio Grande Rift adjacent to the (Quaternary) Valles caldera. A distinctly linear west-facing escarpment coincident with a major Laramide high-angle fault zone forms the western boundary of the range (Woodward, 1987). The escarpment is steep, faceted, and terminates against spurs extending from a low-relief upland surface (Fig. 2A). The overall morphology of the range is consistent with accepted ideas about how a tectonically active mountain front is expressed geomorphically (Bull and McFadden, 1977), and it has been proposed that the Nacimiento fault zone has experienced significant post-Laramide offset (Formento-Triglio and Pazzaglia, 1998; Machette et al., 1998). This view is

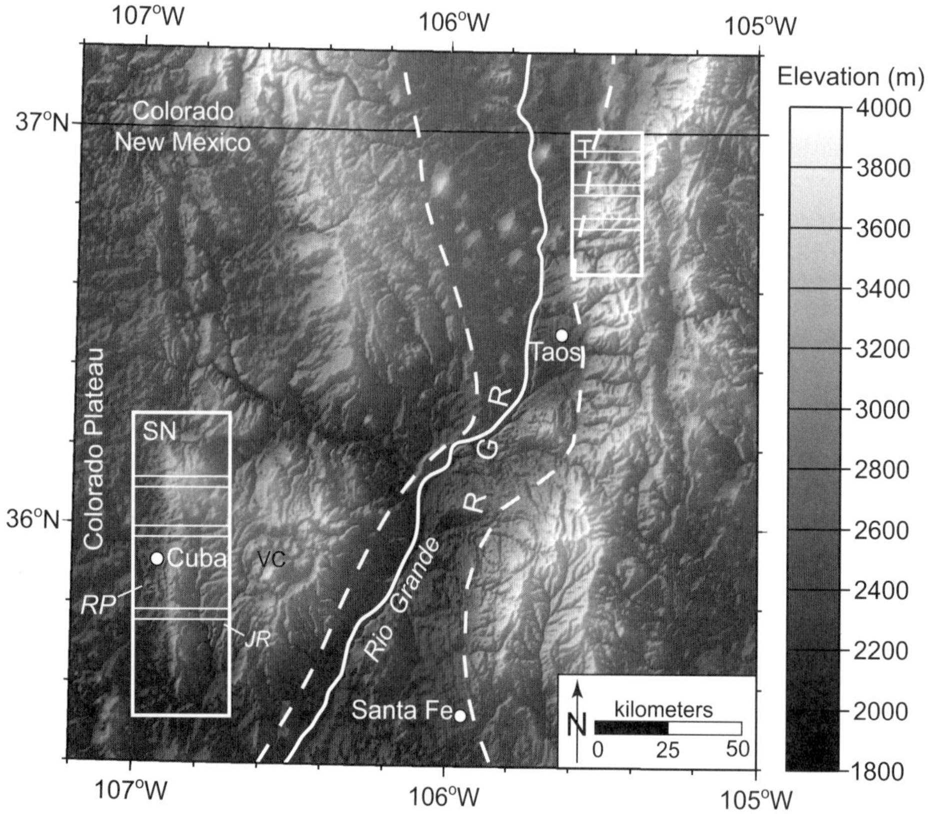

Figure 1. Hill-shaded 90-m-resolution digital elevation model (DEM) of the study areas. Sierra Nacimiento (SN) is located in the southwest portion of the image on the western flank of the Valles caldera and the Rio Grande Rift (RGR; delineated with dashed white lines). The Taos Range (T) is located in the northeast corner of the image, on the eastern flank of the Rio Grande Rift. Boxes enclosing the ranges are the locations of Figures 2A and 2B. Horizontal white lines crossing Sierra Nacimiento and the Taos Range indicate 4 km swaths of topography used to calculate maximum and minimum elevations in Figure 3. VC—Valles caldera, JR—Jemez River, and RP—Rio Puerco.

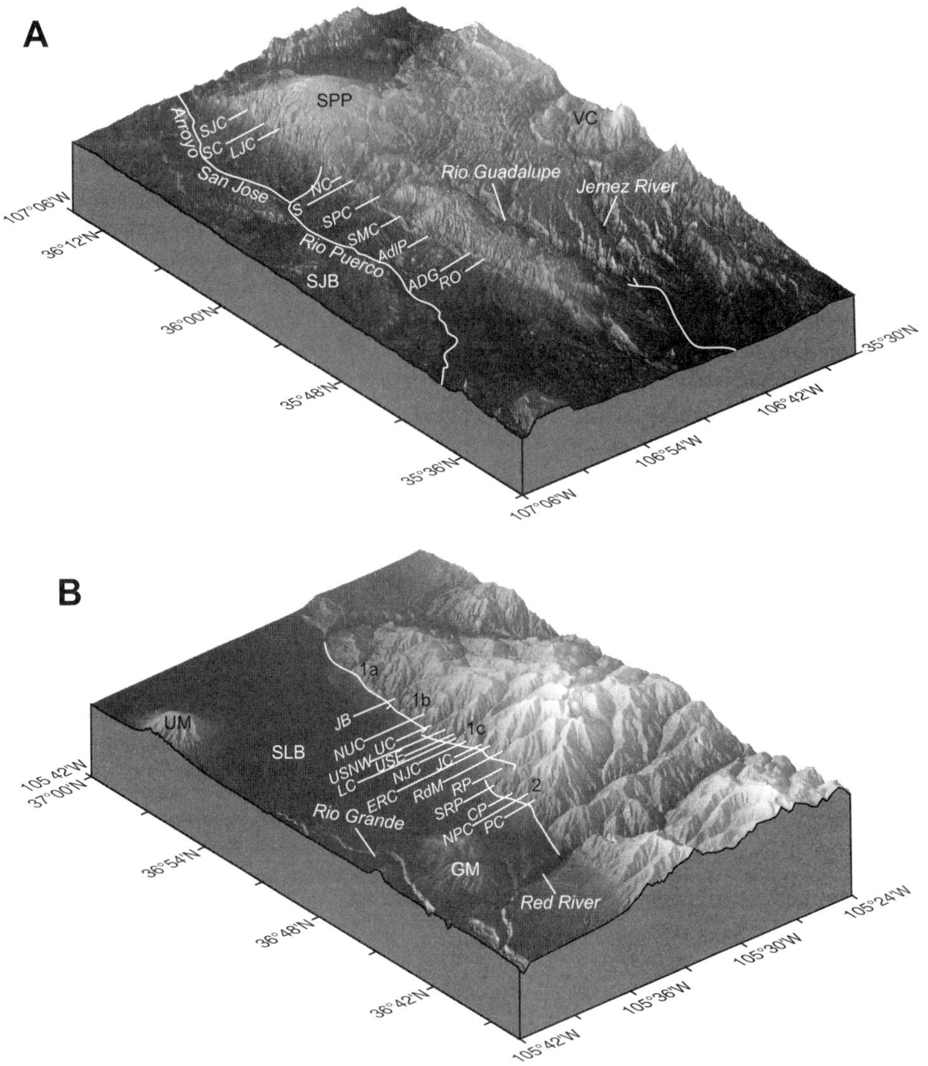

Figure 2. Three-dimensional views of 30-m-resolution hill-shaded digital elevation models (DEM) of Sierra Nacimiento and the Taos Range. Locations of the ranges are shown by white rectangles in Figure 1. (A) Sierra Nacimiento. Note the linear range front, change in the east-west extent of the range from south to north, and the faceted spurs extending from the mountain front westward into the San Juan Basin (SJB). SJC—San Jose Creek, SC—Salado Creek, LJC—La Jara Creek, NC—Nacimiento Creek, S—Senorito, SPC—San Pablo Canyon, SMC—San Miguel Canyon, AdlP—Arroyo de los Pinos, ADG—Arroyo Dedos Gordos, RO—Rito Olguin, SPP—San Pedro Parks, VC—Valles caldera. (B) Taos Range. Note the linear, segmented range front, lack of spurs extending from the mountain front, and increase in elevation from north to south. Fault segments 1a, 1b, 1c, and 2 are as defined by Menges (1990a). JB—Jaroso Basin, NUC—North Urraca Canyon, UC—Urraca Canyon, USNW—Upper Sunshine northwest, USE—Upper Sunshine east, LC—Latir Creek, NJC—North Jaracito Canyon, JC—Jaracito Canyon, ERC—El Rito Canyon, RdM—Rito del Medio, RP—Rito Primero, SRP—South Rito Primero, CP—Canada Pinabete, NPC—North Penasquito Canyon, PC—Penasquito Canyon, SLB—San Luis Basin, GM—Guadalupe Mountain, UM—Ute Mountain.

countered by seismic quiescence (Sanford et al., 1991; House and Hartse, 1995) and restriction of offset Quaternary deposits to the southernmost terminus of the range-front fault (Baltz, 1967; Formento-Trigilio and Pazzaglia, 1998; cf. Machette et al., 1998). This leads to the overall conclusion that the mountain front is instead being exhumed as softer Phanerozoic sedimentary rocks are stripped off a resistant Precambrian core. The Taos Range, in comparison, is located on the eastern flank of the Rio Grande Rift, marking the northern boundary of the uplifted rift flank in northern New Mexico (Fig. 1). It is a rugged footwall uplift bound to the west by a clearly active range-front fault and actively subsiding hanging-wall basin (Menges, 1990a). Mountain front linearity and landforms (Fig. 2B) correlate well with the frequency of late Pleistocene and Holocene ruptures along the range-front fault, which stands as a model for fault segmentation in extending regions of western North America (Menges, 1990a). An early history of Laramide and early Rio Grande Rift footwall uplift has been largely overprinted by a Pliocene-Quaternary pulse of rapid uplift and subsequent footwall incision.

The goal of the comparison between the Taos Range and Sierra Nacimiento is to quantify the topographic metrics and geomorphic processes that distinguish a mountain front that continues to be uplifted along an active range-front fault (Taos Range) from one that is being erosionally exhumed (Sierra Nacimiento). The comparison provides data supporting numeric models that argue for divergent mountain front landforms as a function of a hillslope sediment flux law that scales either linearly or as a power function with respect to mean hillslope gradient (Densmore et al., 1998; Roering et al., 2001). We expand the applicability of such model results to observations of low-order stream channels and suggest a channel gradient threshold consistent with the erosional processes occurring on the hillslopes.

GEOMORPHOLOGY OF TECTONICALLY ACTIVE MOUNTAIN FRONTS

Landforms in tectonically active areas exhibit characteristic topographic metrics (e.g., steeper slopes, greater relief, and erosional dissection) that help distinguish them from tectonically quiescent settings (Bull and McFadden, 1977). Mountain front metrics that tend to be relatively sensitive to broad ranges in the rate of footwall uplift include escarpment sinuosity, the valley floor width to valley height ratio (Bull and McFadden, 1977), hypsometry (Strahler, 1952; Pike and Wilson, 1971; Ohmori, 1993), and mean hillslope gradient.

Linear mountain fronts tend to be correlated with footwall uplifts where offset on the range-front fault is rapid enough to inhibit drainage basin elaboration (widening) and subsequent embayment of the growing escarpment. Such mountain fronts are typically accompanied by seismicity and an exposed fault cutting Quaternary deposits. These mountain fronts exhibit widely spaced, sharp, triangular facets terminating upward to truncated spurs. Sediment eroded from the escarpment accumulates in basins on the subsiding piedmont as steep, short alluvial fans (Bull, 1984).

In contrast, strongly embayed mountain fronts are typically associated with an inactive range-front fault. The mountain front lacks sharp, triangular facets and instead has closely spaced, linear drainages that elaborate considerably into the dissected uplift. Sediment liberated from the escarpment spreads out as thin, broad alluvial fans extending into the axis of the slowly or nonsubsiding basin (Bull, 1984). The inactive range-front fault is commonly buried by sediment filling the mouths of canyons.

Coupled landscape evolution–tectonic models (Densmore et al., 1998), analogue models (Bonnet and Crave, 2003; Lague et al., 2003), and space-for-time substitution field studies (Harbor, 1997) provide insight into the topographic development of mountain front escarpments under the influence of variable rates of tectonic uplift. A threshold or critical hillslope gradient, above which landsliding becomes the dominant process transporting sediment off of hillslopes, is an important component of numeric models that generate realistic-looking landscapes (Tucker and Slingerland, 1994, 1996; Kooi and Beaumont, 1996; Densmore et al., 1998). Parameterizing hillslope sediment flux to both nonlinear hillslope creep and threshold-dominated landsliding in steep terrains represents a major improvement in model depictions of real landscapes over those parameterized to route sediment proportional to a linear transport law (diffusion only). Results from Densmore et al. (1998) are particularly pertinent to our investigation, because their model generates visually distinct, end-member escarpments when hillslope processes are represented as a transport law that scales linearly with mean hillslope gradient or becomes nonlinear for gradients above a predefined threshold. The former case produces a mountain front that resembles, to a first order, Sierra Nacimiento (Fig. 2A), and the latter produces topography that is similar to most mountain fronts bounded by an active normal fault, including the Taos Range (Fig. 2B).

Although the impressive relief of Sierra Nacimiento and the linearity of the mountain front are similar to many other Laramide ranges in the southern Rocky Mountains, neither metric is related to active uplift along a range-bounding fault (Leonard and Langford, 1994; Leonard, 2002). If such landforms can result from nontectonic processes or alternatively, persist as the result of tectonic processes long dead, then a clear problem exists in the application of simple landform metrics or a landscape evolution model to interpret contemporary tectonic activity.

SETTING

The mountains of northern New Mexico are part of the southern Rocky Mountains. They were constructed by crustal shortening and regional uplift associated with the late Cretaceous–Eocene Laramide orogeny, and subsequently by a major north-south–oriented Neogene rift (Rio Grande Rift), which has structurally inverted the former Laramide highland within a broader epeirogenically uplifted topographic swell (Eaton, 1986). Climate in northern New Mexico was semiarid and seasonal throughout the Neogene, much like the modern-day climate (Tedford, 1981; Tedford and Barghoorn, 1999). At the onset of the Pleistocene, high-amplitude, glacial-interglacial climate cycles ensued, and the climatic regime oscillated between wetter, cooler glacial periods and effectively drier and warmer interglacial periods. Glacial climates are thought to be times of depressed seasonality and greater effective precipitation delivered mainly in the form of increased snow pack (Thompson, 1991; Reneau, 2000). Interglacial periods are typically viewed as times when there is a northerly shift in the jet stream, and there is a return to a drier, warmer climate with increased seasonality and delivery of precipitation in the summer as intense convective storms (Pendall et al., 1999; Reneau, 2000; Dethier, 2001).

Sierra Nacimiento

Sierra Nacimiento is cored by Mesoproterozoic crystalline rocks that were uplifted and deformed during the Ancestral Rocky Mountain and Laramide orogenies along a steeply eastward-dipping reverse fault (Woodward, 1987; Pazzaglia et al., 1999). Ash-flow tuffs and associated volcaniclastics of the Abiquiu Formation were deposited across the range during a period of extensive volcanic activity in the Oligocene and early Miocene (Church and Hack, 1939; Vazzana and Ingersoll, 1981; Ingersoll et al., 1990; Moore, 2000), and today cover parts of a low-relief surface at the range's northern crest (Figs. 1 and 2A). The unconformity buried by the Abiquiu Formation is typically considered part of the broader Rocky Mountain erosion surface (Gregory and Chase, 1994; Chapin and Kelley, 1997). It is tilted 3°–7° to the east and lies at 3300 m, an elevation higher than its source, suggesting post-Laramide rift-flank deformation associated with opening of the Rio Grande Rift (Church and Hack, 1939; Eaton, 1986; Woodward, 1987; Formento-Trigilio and Pazzaglia, 1998).

The range is ~10–16 km wide, ~80 km in length, and has ~1 km of relief along its western escarpment, the majority of which is concentrated across an inactive fault separating resistant rocks of the Precambrian core and easily erodible, syn-Laramide sediments of the San Juan Basin (Woodward et al., 1972; Woodward,

1996). The northern half of the range cooled during the Eocene and preserves an apatite fission-track (AFT) partial annealing zone. The total unroofed section in this part of the range is ~4.4 km, with surface uplift and rock uplift of ~2.1 km and ~6.5 km, respectively (Pazzaglia and Kelley, 1998). The block that forms the southern portion of the range, in contrast, is narrower in width, does not preserve the Rocky Mountain erosion surface or the partial annealing zone (Figs. 3A, 3B, and 3C) and has late Eocene to early Oligocene fission-track cooling ages, indicating that total exhumation and rock uplift in the range increases toward the south (Pazzaglia and Kelley, 1998; Formento-Trigilio and Pazzaglia, 1998).

The Sierra Nacimiento piedmont has a cover of late Tertiary–Quaternary alluvium and travertine, related to the axial (along-strike) Rio Puerco drainage system and its range-front tributaries, which unconformably overlie and are inset into tilted early Cenozoic and Mesozoic sedimentary rocks (Woodward, 1987; Formento-Trigilio and Pazzaglia, 1998; Frankel, 2002). The sub-Abiquiu unconformity at the northern end of the range represents the headwaters of an ancestral drainage system called the Rio Chacra, stemming from a time when base level was higher than present, and from which, the modern, master drainage of the eastern Colorado Plateau, the Rio Puerco, has evolved (Bryan and McCann, 1936, 1938). There is no evidence for glaciation of Sierra Nacimiento during the Pleistocene, although some periglacial features are present at the highest elevations (Reneau, 2000).

Taos Range

The Taos Range, like Sierra Nacimiento, has a Proterozic origin, followed by Ancestral Rocky Mountain and Laramide deformation, and late Cenozoic uplift along the Rio Grande Rift flank. The core of the range is composed of Precambrian crystalline rocks that have been intruded and overlain by Oligocene igneous rocks associated with a caldera (Lipman and Reed, 1989). Apatite fission-track data indicate late Cenozoic rift-flank footwall uplift and erosion at rates between 0.14 and 0.28 mm/yr (Pazzaglia and Kelley, 1998).

The range is bound to the west by an active, west-dipping, segmented normal fault (Menges, 1990a). The mountain front rises steeply some 1400 m above the San Luis basin floor (Figs.

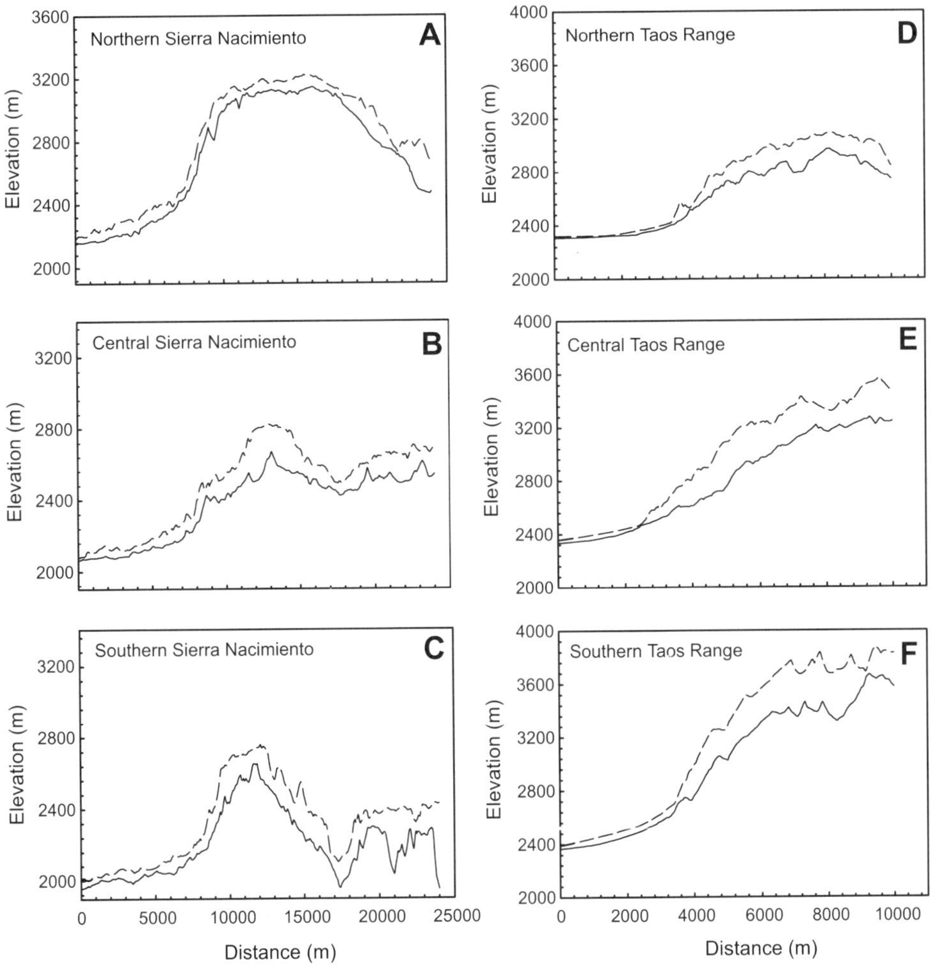

Figure 3. Maximum and minimum elevation profiles for northern (A), central (B), and southern (C) Sierra Nacimiento and for the northern (D), central (E), and southern (F) Taos Range. Profiles were constructed by averaging maximum and minimum elevation values for 4 km swaths of along-strike topography from 100-m-resolution digital elevation model (DEM) data. Swath locations are shown in Figure 1. The difference between the maximum and minimum elevation profiles for each plot is the average amount of incision and relief in that region. The southern and central parts of both ranges have large amounts of incision. In contrast, the northern part of each range is characterized by relatively little incision. The change in range-front relief along Sierra Nacimiento is attributed to a wave of exhumation propagating from south to north. The increase in relief from north to south in the Taos Range mountain front is the result of an increase in fault offset and rock uplift in the same direction.

3D, 3E, and 3F). It is distinctly faceted and dissected by west-flowing, deeply incised drainages that terminate as alluvial fans spilling onto the San Luis basin piedmont. The largest drainages and fans are developed at the segment and subsegment boundaries. Fan metrics, such as area and slope, correlate with the area and slope of their contributing basins (Pazzaglia, 1989). Offset along each segment increases from north to south on a fault beginning as a single feature in the south that splays, steps basinward (west), and decreases in offset to the north. As a result, the range is broken into several blocks, each one tilting to the north. This segmentation is most obvious along the northern 35 km of the range front, designated segment 1 (subsegments 1a, 1b, and 1c; Menges, 1990a), and is the part of the range we use in this study (Fig. 2B). At least 700 m of post–middle Pliocene offset has occurred along subsegment 1c (in the south) in comparison to ~200 m of offset over the same period of time in subsegment 1a (in the north). Surface uplift of the mountain front since the middle Pliocene is known from the correlation of the tops of faceted spurs to a basalt flow uplifted in the northernmost portion of the range and dated at 4.3 Ma. Fault-slip rates based on the offset basalt flows range from 0.12 to 0.23 mm/yr (Menges, 1990a), which correspond well with long-term footwall uplift rates (Pazzaglia and Kelley, 1998).

Regional base level for the Taos mountain front is defined by the Rio Grande, which has incised 250 m into basin-fill basalt and axial-stream sediments in the center of the San Luis basin. Larger tributaries, such as the Red River and Rio Hondo, have integrated with the Rio Grande and are also deeply incised, exposing the piedmont facies of the basin fill. In contrast, the basins draining segment 1 are not yet integrated into the Rio Grande and grade to the elevation of the San Luis basin as local base level.

The Taos Range experienced minor glaciation in the Pleistocene (Richmond, 1962). However, most of the cirque orientations show that it was the eastern rather than western portion of the range crest that was glaciated. There was little to no glaciation in the headwaters of the largest basins draining the west-facing escarpment along segment 1.

TABLE 1. GEOMORPHIC METRICS FOR DRAINAGES IN THE WEST FLANK OF SIERRA NACIMIENTO AND THE TAOS RANGE

Range	Drainage	V/A (m)	Average first-order stream gradient
Sierra Nacimiento	Arroyo de los Pinos	109 ± 8	0.22 ± 0.09
Sierra Nacimiento	Rito Olguin	81 ± 7	0.20 ± 0.16
Sierra Nacimiento	Arroyo Dedos Gordos	75 ± 7	0.20 ± 0.13
Sierra Nacimiento	San Pablo Canyon	80 ± 7	0.15 ± 0.11
Sierra Nacimiento	Nacimiento Creek	77 ± 7	0.11 ± 0.07
Sierra Nacimiento	La Jara Creek	74 ± 7	0.19 ± 0.13
Sierra Nacimiento	Salado Creek	88 ± 7	0.30 ± 0.14
Sierra Nacimiento	San Jose Creek	104 ± 8	0.28 ± 0.12
Sierra Nacimiento	Senorito	102 ± 8	0.15 ± 0.07
Sierra Nacimiento	San Miguel Canyon	93 ± 8	0.17 ± 0.11
Taos	Jaracito Canyon	180 ± 10	0.26 ± 0.15
Taos	Latir Creek	139 ± 9	0.23 ± 0.16
Taos	Rito del Medio	156 ± 9	0.28 ± 0.15
Taos	Upper Sunshine East Basin	133 ± 9	0.25 ± 0.11
Taos	Upper Sunshine NW Basin	122 ± 8	0.20 ± 0.11
Taos	North Jaracito Canyon	141 ± 9	0.32 ± 0.23
Taos	Penasquito Canyon	89 ± 7	0.16 ± 0.09
Taos	Canada Pinabete	148 ± 9	0.25 ± 0.15
Taos	Rito Primero	174 ± 10	0.29 ± 0.15
Taos	South Rito Primero	118 ± 8	0.33 ± 0.04
Taos	North Penasquito Canyon	147 ± 9	0.26 ± 0.05
Taos	North Urraca Canyon	132 ± 9	0.25 ± 0.16
Taos	Urraca Canyon	139 ± 9	0.26 ± 0.14
Taos	Jaroso Basin	104 ± 8	0.16 ± 0.09
Taos	El Rito Canyon	180 ± 10	0.27 ± 0.22

Note: V/A is the drainage basin volume to drainage basin area ratio.

METHODS AND RESULTS

First-Order Stream Gradients

Gradients of first-order channels were measured in the Taos Range and Sierra Nacimiento following a methodology similar to that of Merritts and Vincent (1989), who demonstrated a clear influence of the rate of uplift on low-order channel gradients. First-order channel gradients of both the range block and ~5 km of adjacent piedmont were extracted from 10-m-resolution digital elevation models (DEM) using Arc/Info (Table 1). The channels were defined by a flow accumulation threshold with a drainage area of 0.025 km^2. This resulted in a subsequent ordering of the entire drainage basin stream network, similar to that which would be obtained from standard 1:24,000 scale topographic maps (Pazzaglia, 1989). First-order channel long profiles were extracted from the DEM at 4.5 m increments, and these data were parsed using a FORTRAN code to calculate a gradient based on the minimum and maximum elevations in the long profiles. Each channel gradient value was spatially positioned as a point datum at the location of the channel head. The point gradient data were then contoured by fitting a least-squares polynomial trend surface that allows for a smooth fit through data with a large variance (Figs. 4A and 4B). Repeated analyses using different drainage area thresholds for defining a first-order channel and variable amounts of piedmont versus mountain front areas show that there is little change in the general shape of the resulting trend surfaces.

The channel gradient contour map of Sierra Nacimiento (Fig. 4A) is dominated by rather uniform gradients (~0.1) throughout the central part of the range. There is a sharp decrease in channel gradient to the south coincident with the narrowing of the range and the progressive loss of hard Precambrian rock types. Channel gradient increases to the north, which coincides with the widening of the range, and a greater amount of exposed resistant rock types. Contours in the north bend in response to steeper gradients in the granitic core of the range with respect to those developed directly to the west on the softer piedmont rock types.

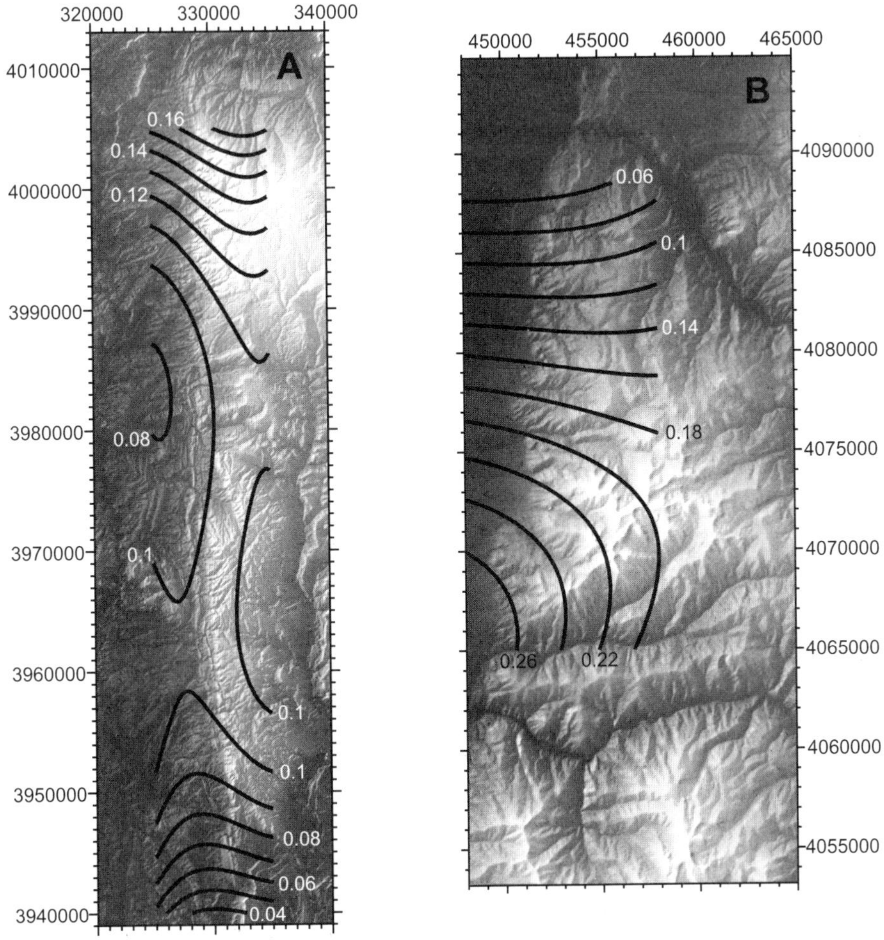

Figure 4. Smoothed first-order stream gradient contour maps for Sierra Nacimiento and the Taos Range. (A) Sierra Nacimiento. The first-order stream gradients increase toward the north. Steeper stream gradients are located further north in the soft, sedimentary rocks of the San Juan Basin compared with gradients in the hard, crystalline core of the range. This is consistent with a northward-migrating wave of exhumation, where the wave has proceeded further through the more easily erodible basin rocks than through the more resistant core of the range. The knickzone exhuming Sierra Nacimiento lies in the northern portion of the range where stream gradients increase and contours are closely spaced in the piedmont. (B) Taos Range. The stream gradients increase from north to south in conjunction with a north-to-south increase in elevation, fault offset, and rock uplift. Sensitivity analyses show that the width of the contouring window does not affect the distribution of stream gradients. Coordinates are in UTM easting and northing for both maps.

Channel gradient contours for the Taos Range (Fig. 4B) are significantly different than what is observed for Sierra Nacimiento. Channel gradients here increase from north to south along the mountain front—consistent with the established notion that displacement of the footwall block increases to the south. However, there is also a notable bend in the contours in subsegment 1c. Here the steepest channel gradients are at the mountain front, with shallower first-order stream gradients in the interior of the block. Channel gradients in the Taos Range are highest where the range is the highest in elevation, widest in east-west extent, and has the largest amount of fault offset.

Overall, the first-order channel gradients in the Taos Range are higher than those found in Sierra Nacimiento drainage basins (Fig. 5). The mode of channel gradients in the Taos Range is 0.32 (17.7°), compared to 0.12 (6.8°) for Sierra Nacimiento. Both data sets have a skewed distribution with more frequent low-gradient, first-order channels; however, the Taos data are nearly bimodal. Modal distributions are in part affected by separate populations of piedmont versus range-front channel gradients. We view this as a small effect on the overall distribution because (1) the contours are generally oriented orthogonal, rather than parallel to the mountain front, and (2) analysis of the range only, excluding the piedmont, reproduced a similar bimodal distribution.

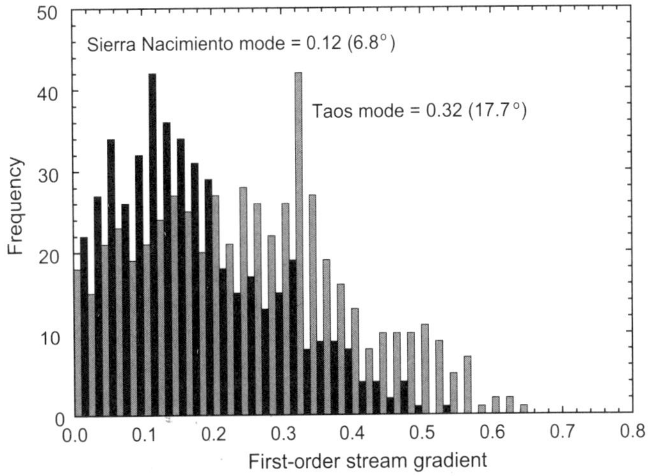

Figure 5. Histogram of first-order stream gradients for Sierra Nacimiento (black) and the Taos Range (gray). The mode of first-order stream gradients in Sierra Nacimiento is 0.12 (6.8°), in contrast to the Taos Range, where the mode of first-order stream gradients is 0.32 (17.7°). The difference in first-order stream gradients is a function of the processes occurring at the respective mountain fronts where low modal gradients are associated with erosional exhumation and tectonically active ranges are characterized by high modal gradients.

Drainage Basin Volume-Area Ratio

A drainage basin volume to planimetric drainage basin area ratio (R_{va}) is proposed as a key metric to quantify the morphologic differences between a landscape that has experienced continuous base-level fall concentrated at a range-front fault as opposed to one that has experienced either base-level stability or base-level fall from a distal source. R_{va} is crudely equivalent to basin hypsometry (Strahler, 1952; Pike and Wilson, 1971), or the valley floor width to valley height ratio (Bull and McFadden, 1977), and its change through time, as a drainage basin initiates and grows, has a physical basis in observed drainage basin evolution analogue models (Schumm and Parker, 1973; Ouchi, 2002; Bonnet and Crave, 2003; Lague et al., 2003). The R_{va} has an advantage over simple drainage basin volume calculations in that it is normalized to basin area, allowing for more direct comparison of different sized basins. R_{va} also removes the subjectivity of defining the distance to a mountain front, as well as valley widths, from topographic maps, as in the traditional valley floor width to valley height ratio (Bull and McFadden, 1977). In addition, R_{va} encapsulates hillslope and fluvial processes acting over the entire drainage basin instead of only a discrete cross-sectional area.

Volume and area are calculated by constructing envelope maps of the topography from high-resolution (10 m) digital elevation data using Arc/Info (Frankel, 2002). First, individual drainage basins are extracted from the DEM by defining the catchment outlet at the range front. The 10-m-resolution DEM of each individual drainage basin is then resampled to 100 m. A maximum-elevation envelope map is produced from the resampled data by interpolating a surface between local maximum elevations in a circular moving window with a 5 km radius. This allows the envelope maps to be pinned to the watershed divide and cover the maximum elevations within each drainage basin. Next, the maximum elevation maps are resampled back to 10 m resolution, so that the original 10 m topography can be subtracted from the envelope surface. Subtracting the original DEM from the maximum elevation envelope map produces a topographic residual map from which area and volume data can then be obtained for individual drainage basins (see GSA Data Repository material for complete methodology[1]).

R_{va} shows quantifiable differences between erosionally exhumed and tectonically active range fronts (Table 1). Sierra Nacimiento R_{va} values are relatively low (74–109 m, mean 88 m) compared to those for basins draining the Taos Range (89–180 m, mean 140 m). R_{va} covaries linearly with channel gradients ($r^2 = 0.73$; Fig. 6), with greater variance in Sierra Nacimiento compared to the Taos Range mountain front. There are four data that fall off the regression line of Figure 6. All of these data come from small, very steep watersheds with trunk channels that do not extend back to the main drainage divide of the respective ranges.

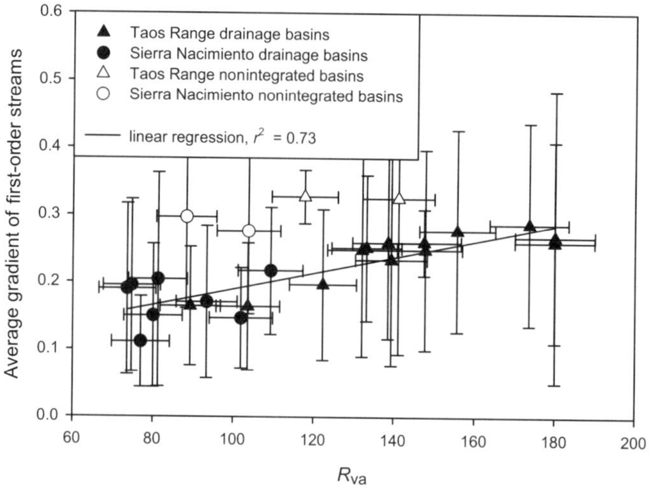

Figure 6. Plot of average first-order stream gradients versus the drainage basin volume to drainage basin area ratio (R_{va}) for 25 drainage basins in Sierra Nacimiento and the Taos Range. The 10 drainage basins from Sierra Nacimiento (circles) have a high variance and generally lower R_{va} values and stream gradients than drainage basins in the Taos Range. A linear relationship exists for all but the smallest drainage basins ($r^2 = 0.73$). The four smallest basins (open symbols) do not extend to the drainage divide, are not fully integrated into the fluvial system, and fall above the general trend for the larger drainage basins. Error bars represent 2σ standard errors for both average gradients of first-order streams and R_{va}.

R_{va} varies systematically with respect to location along the Taos mountain front (Fig. 7A), but shows a high variance along Sierra Nacimiento (Fig. 7B). For the Taos Range, R_{va} increases from north to south, among fault subsegments and the range front as a whole, consistent with the known offset gradient along the range-front fault.

Piedmont Stratigraphy and Mountain Front Deposits

Sierra Nacimiento

Surficial deposits extend west onto the Sierra Nacimiento piedmont, unconformably overlying and inset into sedimentary rocks of the San Juan Basin. These deposits range from thin colluvial hillslope mantles to thick alluvial packages of fining-upward bed sets to travertine deposits, and are not offset by the range-front fault anywhere north of Arroyo Peñasco (Formento-Trigilio and Pazzaglia, 1998; Frankel, 2002). These deposits, which were previously described and mapped as Quaternary and Tertiary terrace and pediment alluvium (QTtp) by Woodward et al. (1970) and Woodward (1987), record the history of mountain front

[1]GSA Data Repository item 2006028, Tables DR1–DR3, is available online at www.geosociety.org/pubs/ft2006.htm, or on request from editing@geosociety.org or Documents Secretary, GSA, P.O. Box 9140, Boulder, CO 80301, USA.

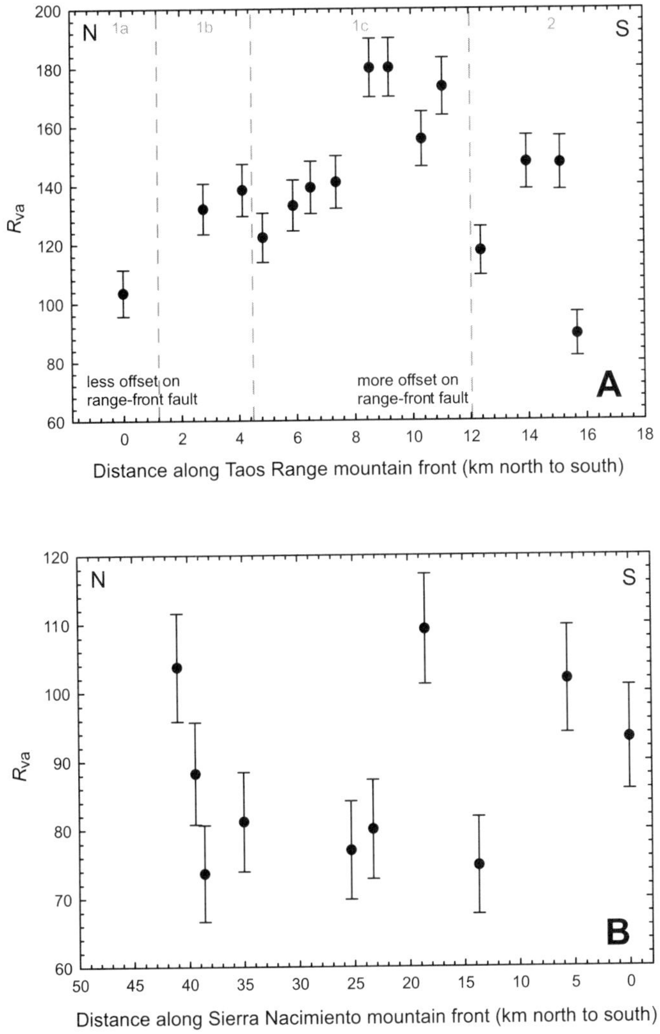

Figure 7. Plot showing the along-strike change in the drainage basin volume to drainage basin area ratio (R_{va}) for the Taos Range and Sierra Nacimiento. (A) Change in R_{va} versus distance from north to south along the Taos Range. R_{va} increases from north to south along fault segment 1 and within fault subsegments 1a, 1b, and 1c (Menges, 1990a) of the range front in conjunction with increasing elevation, fault offset, and rock uplift. (B) R_{va} versus distance from north to south along the Sierra Nacimiento range front. No correlation exists between R_{va} and the distance along the range front for Sierra Nacimiento. Location of drainages is shown in Figure 2. Error bars represent 2σ standard errors for R_{va} values.

exhumation in both their landscape position on the piedmont and in their sedimentology and stratigraphy. Three new formations in order of decreasing age, the Veguita Blanca formation, the Leche formation, and the Laguna Bonita formation were defined (Frankel, 2002; GSA data repository [see footnote one]). The older deposits dominate the southern Sierra Nacimiento piedmont, where they occupy wide interfluves of broadly incised inverted topography. The younger deposits dominate the northern piedmont, where they are vertically stacked and locally incised by deep, narrow valleys. Sierra Nacimiento hillslopes are covered nearly everywhere by a thin (<1 m) grus-rich regolith. Slumps, landslides, and related slope failures are uncommon.

Taos Range

The Taos Range piedmont is dominated by alluvial fans composed of three middle Pleistocene-Holocene lithostratigraphic units (Qf1–Qf3) that conformably overlie a thick, proximal, basin-fill deposit called the Lama formation (Pazzaglia and Wells, 1990; GSA data repository [see footnote one]). The Lama formation is a crude, but thicker, lithostratigraphic equivalent to the Ancha and Tuerto formations of central New Mexico, and the Veguita Blanca formation on the Sierra Nacimiento piedmont. The Taos Range piedmont contrasts strongly with that of Sierra Nacimiento because of the accommodation space produced by subsidence of the hanging-wall basin against the range-front fault. Except for local fan-head trenching, the Taos piedmont is not incised. Alluvial fan deposits exhibit both inset as well as stacked alluvial stratigraphy. Taos Range hillslopes bear a combination of bedrock, regolith, and landslide deposits and landforms (Menges, 1990b).

River Incision

Sierra Nacimiento

River terrace and piedmont deposit elevations have been assembled into a data set indicating the distribution and rate of Pleistocene incision in and around Sierra Nacimiento. Nine Quaternary terraces have been mapped along the Jemez River (Rogers, 1996; Formento-Trigilio, 1997; Formento-Trigilio and Pazzaglia, 1998; Frankel and Pazzaglia, 2002; Ault and Pazzaglia, 2004; Fig. 8), a large perennial drainage flowing from north to south on the eastern flank of the range. These terraces range in age from 1.6 Ma for QTg1, which underlies the lower Bandelier Tuff, to middle Holocene (3160 ± 60 to 4530 ± 40 radiocarbon yr B.P.) for Qt7 (Formento-Trigilio and Pazzaglia, 1998; Frankel and Pazzaglia, 2002). The ages of Qt1 (610 ka) and Qt5 (60 ka) are well known because of the presence of dated tephra and volcanic clasts, respectively.

Jemez River terraces diverge downstream (cf. Formento-Trigilio and Pazzaglia, 1998), consistent with downstream base-level fall and upstream knickpoint retreat along the Jemez channel (Fig. 8). The Jemez valley was more or less uniformly incised following emplacement of the upper Bandelier ash-flow tuff from 1.2 Ma to Qt1 time at a rate of ~0.3 mm/yr. Since Qt1 time, incision has both slowed and been concentrated in the lower and middle portions of the valley below valley distance 40 km, where the rate has been ~0.17 mm/yr.

In comparison, a single dated alluvial terrace on the northern Sierra Nacimiento piedmont indicates a local rate of incision into the Leche formation of 0.22–0.32 mm/yr (Frankel, 2002). Similarly, the Rio Puerco in the same area is argued to be incising at a rate of ~0.2 mm/yr (Slavin, 1991). The southern Sierra Nacimiento piedmont is likely being incised at a comparatively

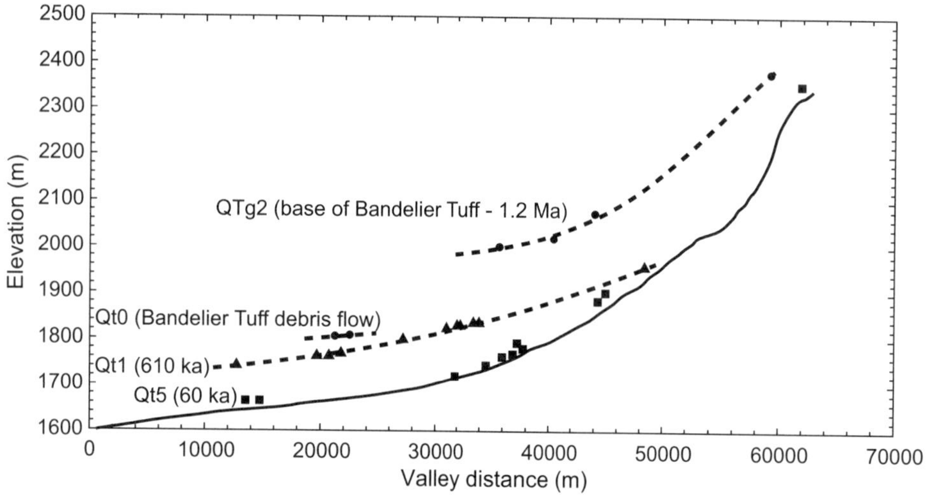

Figure 8. Downstream diverging terraces along the Jemez River on the eastern flank of Sierra Nacimiento. The downstream divergence of terraces indicates the river longitudinal profile is responding to base-level fall resulting from Rio Grande incision. See text for discussion of terrace ages, incision patterns, and incision rates. Terrace data are from Rogers (1996), Formento-Trigilio (1997), Formento-Trigilio and Pazzaglia (1998), Frankel and Pazzaglia (2002), and Ault and Pazzaglia (2004).

slower rate, as suggested by the age and proximity to base level of the Veguita Blanca formation as well as the long-term erosion rate of 0.04 mm/yr (Hallet, 1994; Formento-Trigilio and Pazzaglia, 1998).

Taos Range

River incision in the Taos Range has been reconstructed for the mountain front using the relative elevation of the uplifted middle Pliocene basalt and accordant faceted spurs. The range front is being incised at rates ranging from ~0.05 mm/yr in the northern subsegment (1a) to ~0.2 mm/yr in the southern segment (1c). These rates, which are being driven directly by footwall uplift, are similar, but genetically unrelated to the rate of piedmont incision of the Red River and its tributaries (~0.15 mm/yr), which is being driven by base-level fall and incision of the Rio Grande. Piedmont incision north of the Red River, and distal to the mountain front, is zero or even negative (aggradation), as these streams have not yet been fully integrated to the Rio Grande. The general pattern of erosion for the Taos Range and its piedmont is rapid incision in the uplifting footwall block and locally in the proximal alluvial fans, slow or no incision on the distal piedmont, where streams are not integrated to the Rio Grande, and at about the rate of Rio Grande incision for piedmont regions that are well integrated to large Rio Grande tributaries.

DISCUSSION

Geomorphic Processes on Erosionally Exhumed and Tectonically Active Mountain Fronts

We use the R_{va} metric, first-order channel gradients, piedmont stratigraphy, and mountain front geomorphology to construct a model for landscape evolution that contrasts the geomorphic processes shaping active footwall uplifts (Taos Range) to those shaping erosionally exhumed mountain fronts (Sierra Nacimiento). The Taos Range mountain front experiences coseismic base-level fall associated with active slip on the range-bounding normal fault. Streams draining the mountain front are directly connected to that base-level fall, and they transmit the signal efficiently and rapidly to their tributaries and contributing hillslopes. Streams in the Taos Range are incising at the long-term rate of footwall uplift, a fact strongly supported by the adjustment of first-order channel gradients to fault segments of known, variable slip histories. Together with high R_{va} values and a mean relief (Fig. 3), which also correlates well with the along-strike variation in fault displacement (Fig. 7), we suggest that hillslopes in the Taos Range are also well coupled to the rate of incision and the rate of rock uplift. The presence of landslides in the Taos Range supports the notion of comparatively steeper hillslopes with respect to Sierra Nacimiento, which are largely devoid of landslide scars and deposits.

The base-level fall shaping Sierra Nacimiento is tens of kilometers distant from the mountain front and is likely propagated through the landscape as one or more migrating knickzones. The result of this propagation is to stretch the base-level fall signal along the entire range front and piedmont, leading to overall lower gradients of first-order streams. The steepest first-order channel gradients are located in the northern part of the range where it is the widest, has the lowest relief (Fig. 3), and there has been the least amount of long-term exhumation (Fig. 4). These apparently contradictory data can be reconciled if we make the reasonable assertion that the base-level fall signal first propagates through the fluvial system, but is delayed in its coupling with the adjoining hillslopes. In this case, landscape response times are slow.

Data from numeric landscape evolution models support our qualitative model for divergent landscape evolutionary pathways, even though a specific quantification of the hillslope processes of the studied mountain fronts is beyond the scope of this paper. Densmore et al. (1998) explored the landscape evolution of mountain fronts in the Basin and Range province with the numeric model ZSCAPE. In this model, the sediment flux off of hillslopes was fixed to a simple linear creep (diffusion) equation, or to a linear creep equation plus a separate bedrock landsliding rule, which allowed for advective (nonlinear) removal of material from hillslopes. Model simulations that do not consider landsliding produce a model mountain front (Densmore et al., 1998, their Fig. 7) very similar to the gross topography of Sierra Nacimiento (Fig. 2A). Likewise, the model that allows for landsliding based on a probabilistic rule such that bedrock landslides are possible at any slope angle generates a mountain front morphology (Densmore et al., 1998, their Fig. 5) similar to the Taos Range (Fig. 2B).

The Volume-Area Ratio and Landscape Evolution of Mountain Fronts

Topographic differences between Sierra Nacimiento and the Taos Range are the result of the degree of tectonic activity along the respective range fronts. The erosional volume of material removed from drainage basins on the fault-bounded side of a mountain has been shown to be proportional to the rate and amount of fault offset and the degree of escarpment degradation (Harbor, 1997). In this sense, R_{va} links normalized erosion to the length of time that the mountain front has been exposed to erosion (Fig. 9). Drainage basin evolution in an uplifting mountain front begins with an early phase of incision by steep, short streams (Glock, 1931; Parker and Schumm, 1982). The immature drainage basin will have a small erosional sediment flux out of the system over a small basin area, resulting in a low R_{va} (point 1 on Fig. 9). Continued incision follows as rock uplift persists and the drainage basin elongates (point 2 on Fig. 9). The basin increases in volume, but not much in planimetric area, leading to a relatively high R_{va}. As the drainage basin elaborates, it begins approximating a steady-state condition in terms of relief, mean elevation, and flux of material in and out of the system, such that it is no longer growing in area or volume (point 3 on Fig. 9). At this point, the erosional sediment flux out of the basin is maximized with respect to the drainage area (Horton and DeCelles, 2001), and streams and hillslopes should have reached their maximum gradients. However, if catchment growth is self-similar, R_{va} will not converge to a common steady-state value. When rock uplift ceases, interfluves will start to lower. This results in a decrease in drainage basin volume with little to no change in drainage basin area and marks the beginning of an exponential decay in R_{va} through time (point 4 on Fig. 9).

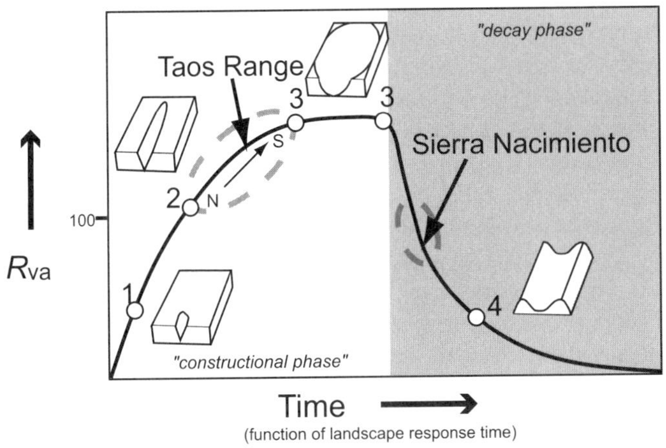

Figure 9. Schematic diagram showing the evolution of the drainage basin volume to drainage basin area ratio (R_{va}) through time. 1—initial uplift and incision. Here, a small volume of material has been removed over a small area from the drainage basin, and R_{va} is relatively low. 2—continued uplift and incision. A large amount of material has been removed from the drainage basin over a relatively small area, and R_{va} is increasing. 3—waning tectonism and elaboration of the drainage network. By this time, a large volume of material has been removed over a relatively large area. This position represents a system in flux steady state where R_{va} will be highest. 4—cessation of tectonic activity and lowering of interfluves in a decaying mountain range. In this case, a moderate volume of material has been removed from the largest possible area, and R_{va} will be low. The R_{va} values in the Taos Range fall between points 2 and 3 on the theoretical curve and increase from north to south. The R_{va} values for Sierra Nacimiento fall between points 3 and 4 on the decay portion of the curve and show no systematic variation with respect to position along the range front. See text for complete discussion.

These hypothesized changes in R_{va} over time are supported by recent studies involving experimental landforms and the effect of landscapes to climate change and tectonic uplift (Bonnet and Crave, 2003; Lague et al., 2003). These analogue models show that mean elevation increases exponentially until it asymptotically approaches a steady-state before undergoing a similar exponential decay as uplift ceases. Bonnet and Crave (2003) also showed that with increased uplift rates comes increased drainage basin denudation. Similarly, the two drainage basins within the Taos Range that are located along the portion of the mountain front with the largest fault offset, Jaracito Canyon and El Rito Canyon, have the highest R_{va} values (180 ± 10 m) in the study area (Table 1). This suggests that the erosional flux of sediment out of the drainage basin is large with respect to the drainage area in this region.

Along-strike variations in R_{va} for the Taos Range may be linked to fault displacements along the range-front fault. Differences in slip rates along strike, across strike, and through time are a function of how normal fault arrays grow. Individual segment displacements and the original length of the segments are key factors in determining the along-strike variation in slip rate. As the fault system grows, faults may be become inactive as fault segments link. However, slip rates and fault displacement will increase toward the center of the fault array due to a positive feedback between strain accumulation and fault displacement. The central segments of a fault zone can accumulate strain faster than the fault tips due to repeated fault rupture and healing, leading to more earthquakes and higher slip rates in the middle of the array (Cowie and Roberts, 2001). The R_{va} data mimics this increase in strain and displacement rate toward the center of the fault array at the boundary between segments 1 and 2 in the Taos Range (Fig. 7A). Thus, we feel that it is not unreasonable to view R_{va} and the evolution of drainage basins as proxies for relative strain rate and fault displacement along a range front.

Base-Level Fall and Exhumation of Sierra Nacimiento

Sierra Nacimiento data are consistent with a regional, distal base-level fall and the continued propagation of that base-level fall signal through the landscape (Garcia et al., 2003). River incision is propagating up-valley, through knickzone retreat, on the Jemez River, as documented by downstream diverging terraces (Fig. 8). The Rio Grande, the regional base-level control for both of these rivers, has been incising through rift basin fill for at least 1 m.y. (Mack et al., 1996). The cause of this axial stream incision is poorly known. It may be related to a change in regional uplift with respect to the Rio Grande Rift, a change in discharge and discharge characteristics related to Quaternary climate, or alternatively, may be driven by the decrease in volcanogenic sediment supply to the axial drainage system.

Sierra Nacimiento is lowest in elevation, narrowest in east-west extent, has the highest relative relief, and has the lowest first-order stream gradients in its southern and central parts (Figs. 3 and 4). Here, tributary piedmont streams of the Rio Puerco have reached, or have nearly reached, their base level of erosion. In essence, the regional base-level fall signal has been effectively transmitted through this part of the landscape. Stream gradients in the central part of the range have reached a relatively uniform value of 0.1 despite a complex juxtaposition of resistant and less-resistant rock types. The landscape position and ages of Quaternary deposits west of the Sierra Nacimiento mountain front indicate that generally, there are higher rates of river incision along the northern half of the piedmont, toward the headwaters of the Rio Puerco and its tributaries. The highest rates of incision along the Jemez River and Rio Puerco should occur where a knickzone persists; the rivers will continue to erode at a slower, long-term rate once the knickzone passes upstream. Piedmont incision began in the southern and central portions of the range in the late Pliocene or early Pleistocene (Frankel, 2002), an age constrained by nearby volcanic necks that were intruded into formerly extant San Juan Basin sedimentary rocks (Hallett, 1994; Hallett et al., 1997), as well as by the estimated age of the Veguita Blanca formation.

In contrast, the northern part of Sierra Nacimiento is just beginning to experience the effects of Rio Grande base-level fall and subsequent range-front exhumation. Stream gradients in this part of the mountain front are relatively high (>0.1; Fig. 4), the range stands high and wide, and very little dissection of the northern part of the range has occurred (Fig. 3). Channel gradients differentiate between a steeper population on the resistant rocks and more gentle population on the erodible rocks in the northern part of the range and piedmont (Fig. 4). Channel gradient contours on the piedmont are shifted northward with respect to the hard rocks of the range. This is precisely the pattern that would be expected from a wave of northward-propagating incision that is currently experiencing resistance to its propagation in the hard Precambrian rocks, but little hindrance in the soft sedimentary rocks of the San Juan Basin.

Sierra Nacimiento exhumation is influenced by the middle Pleistocene base-level fall related to Rio Grande incision, superimposed on a longer-term base-level fall resulting from the location of the range on the flanks of the Rio Grande Rift and Jemez caldera (Karlstrom et al., 2002). The wave of exhumation is a combination of both base-level falls; the latter helping explain the fission-track data and preservation of the Abiquiu Formation, and the former being responsible for more recent incision in the south and steep first-order channel gradients in the north. It is impossible to know the morphology of any individual knickpoint at its place of origin as it began propagating northward. Regardless of the initial shape or rate of base-level fall, the current zone of incision stretches across at least 10 km and spans ~180 m of relief. The rate of knickpoint retreat is poorly constrained and should be a focus of future work in the region.

One consequence of the northward-migrating wave of exhumation is that the axial Rio Puerco is free to slip westward, away from the range front, and progressively incise into the sediments of the San Juan Basin. The axial channel of the Rio Puerco is located further west in the southern part of the basin, where the exhumational wave is thought to have largely passed, but is closer to the mountain front in the north where incision of tributary streams is most active. The westward shift of the Rio Puerco is fostered by the general westward tilt away from the range front, a possible flexural response linked to erosional unloading of the rift flank and Jemez caldera inflation. Entrenchment of the Rio Puerco as a long, nearly linear stream flowing in a strike-parallel valley is controlled by its adjustment to the outcrop belt of soft Cretaceous shale. The result of this westward shift of the rift-flank drainage has been to produce accommodation space for the deposition of Quaternary deposits, such as the Veguita Blanca, Laguna Bonita, and Leche formations. The thickness of these deposits is consistent with the limited accommodation space in the absence of an active range-front fault.

The implications of this regional base-level fall on the landscape evolution of Sierra Nacimiento support an emerging model where much of the rugged topography in the southern Rocky Mountains can be explained by slow exhumation of a Tertiary landscape by streams responding to slow, but persistent base-level fall throughout the late Cenozoic. The inference that Sierra Nacimiento is not being exhumed at the same rate everywhere, but rather in a decidedly south to north pattern, supports our assertion that a base-level fall signal was introduced into the landscape since Jemez River terrace Qt1 time. This landscape has long response times, complicated by local rock type and regional climatic effects. Recent studies in the northern Rockies (Zaprowski et al., 2001), the High Plains (Leonard, 2002; Wisniewski and Pazzaglia, 2002), and Colorado Plateau (Pederson et al., 2002) support both time-transgressive and long landscape response times to regional base-level fall, which are probably locally enhanced by flexural response to erosional unloading (Small and Anderson, 1998; McMillan et al., 2002; Leonard, 2002) or heat-flow–related uplift (Karlstrom et al., 2002). It is not known how the results of these knickzone migration studies scale to the entire orogen, but a reasonable extrapolation would suggest that global processes, such as late Cenozoic eustatic fall, or major rearrangement of drainage patterns due to glaciation would take millions if not tens of millions of years to propagate into the interior of the Rocky Mountains. This view allows for late Cenozoic epeirogeny to have increased the mean elevation of the Rocky Mountains (Pazzaglia and Kelley, 1998; McMillan et al., 2002) and climate change to have increased the erosional efficiency of streams (Dethier, 2001); however it is the propagation of base-level fall signals that ultimately limits the rate and processes of landscape evolution.

CONCLUSIONS

Here, we summarize our model for two divergent landscape evolution pathways, which includes the stream gradient data, the R_{va} data, and the final topographic expression of the mountain front (Fig. 10). The initial landscape in the lower-middle part of the diagram was produced by syn-Laramide uplift and erosion. The primary control on the divergent topographic histories of Sierra Nacimiento and the Taos Range is attributed to the different post-Laramide base-level fall processes. More detailed investigations, such as landslide inventories, hillslope gradients, and measured sediment fluxes, are needed to directly validate our model. However, our current data sets, combined with existing numeric and analogue models provide a measure of mutual validation and quantitative criteria for distinguishing an exhumed, tectonically dead mountain front from one that is truly tectonically active.

This paper distinguishes the landforms and processes by which two different ranges have evolved with respect to their tectonic setting, one that is no longer tectonically active (Sierra Nacimiento) and one that continues to experience offset along a range-bounding fault (the Taos Range). Sierra Nacimiento and the Taos Range share a common origin as Laramide uplifts, but have since diverged in their landscape evolution, leading to different topographic expressions because of the different ways that base level has fallen for each range. We have proposed a conceptual model supported by topographic metrics and field relationships, which explains the diverse aspects of post-Laramide landscape evolution in the southern Rocky Mountains.

Mountain ranges that are not bound by an active range-front fault, such as Sierra Nacimiento, and have experienced a distal base-level fall, tend to have slow landscape response times. In this setting, hillslope and first-order stream gradients will be low, with stream gradient values generally less than 0.2. Sierra Nacimiento, our case example, has a landscape decaying in terms of mean elevation and mean local relief, where the mean R_{va} is 88 m. The Jemez River, on the east flank of Sierra Nacimiento, has incision rates of 0.17 mm/yr over the past 610 k.y. in its southern

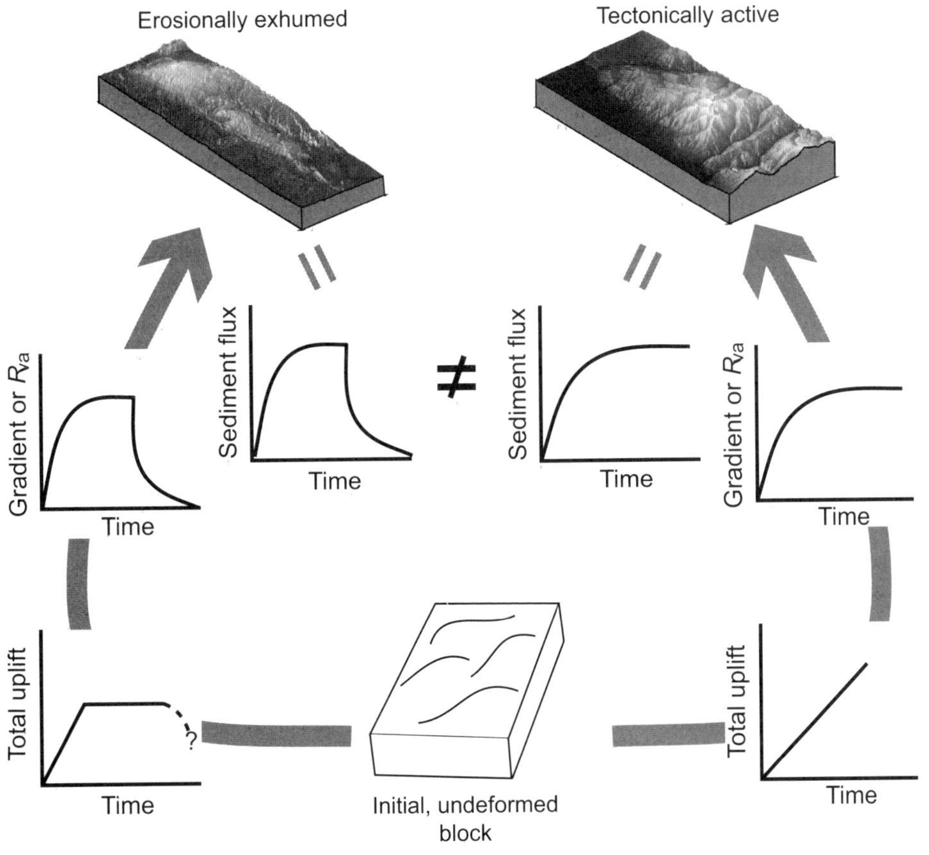

Figure 10. Landscape evolutionary pathways of an erosionally exhumed versus a tectonically active range front. In the erosionally exhumed case (left side of diagram), there is an initial period of uplift followed by tectonic quiescence. Stream gradients and R_{va} values initially increase and then begin to decrease as uplift comes to an end. The reduction in stream gradients will cause a decrease in sediment flux and the response time of the drainage system to be slow. This results in a range front similar in morphology to Sierra Nacimiento. Alternatively, a range front controlled by an active fault (right side of diagram) will experience continued rock uplift through time. Stream gradients and R_{va} will initially increase in response to uplift until they reach a steady-state value, where they will remain as long as uplift and erosion are equal. In this case, stream gradients will remain steep, the sediment flux will remain high, and the response times of the drainage basin will remain fast, resulting in a range front with a similar morphology to the Taos Range.

and middle portions, which decrease upstream. This middle to late Pleistocene incision is driven primarily by distal base-level fall of the Rio Grande. Piedmont incision on the west side of Sierra Nacimiento is consistent with this documented base-level fall signal propagating northward as a wave of exhumation through the range and surrounding piedmont. The knickzone responsible for exhumation is currently positioned in the northern part of the range where stream gradients are highest. Stream gradients in Sierra Nacimiento are lowest in the areas through which the knickzone has already passed. In summary, Sierra Nacimiento is being exhumed by a south-to-north propagating knickzone responding to Pleistocene Rio Grande incision superimposed on late Cenozoic epeirogenic warping of the western flank of the Rio Grande Rift and Valles caldera.

The Taos Range, in contrast, has experienced repeated base-level lowering events directly along its mountain front in response to motion on an active range-bounding normal fault. Hillslope and first-order stream gradients are high, with values consistently greater than 0.2 along the most active fault segments. Consequently, hillslopes have a relatively short response time. Interfluves are high standing and drainage basins have high R_{va} values, averaging ~140 m, which correlate with the increase in fault offset and rock uplift from north to south along the range front. The R_{va} data mimic the increase in strain and displacement rates toward the center of the fault array at the boundary between segments 1 and 2 in the Taos Range. Consequently, the R_{va} and the evolution of drainage basins may be reasonable proxies for strain rate and fault displacement along a range front.

ACKNOWLEDGMENTS

Funding for this project was provided by the U.S. Geological Survey EDMAP program, National Science Foundation grant EAR-9716787, the Colorado Scientific Society, and the Lehigh University Department of Earth and Environmental Sciences. Access to land was kindly granted by the Jicarilla Apache Nation, Navajo Nation, Jemez Pueblo, and Zia Pueblo, along with Evan Andermann, Stanley Crespin, and Cosmos Herrera. We thank Stephanie Briggs, Andrew Drabick, Shari Kelley, and Mike Rampey for their assistance with field work. Jane Selverstone and David Gutzler graciously provided logistical support. Peter Zeitler and David Anastasio gave many helpful suggestions and comments on early versions of the manuscript. Constructive and insightful reviews by Alex Densmore and David Harbor significantly improved the focus of the manuscript.

REFERENCES CITED

Anderson, R.S., 1994, Evolution of the Santa Cruz Mountains, California, through tectonic growth and geomorphic decay: Journal of Geophysical Research, v. 99, p. 20,161–20,179, doi: 10.1029/94JB00713.

Ault, A., and Pazzaglia, F.J., 2004, Quaternary geologic map of the Jemez Springs 7.5′ quadrangle, Sandoval County, New Mexico: New Mexico Bureau of Geology and Mineral Resource, Open-File geologic map OF-XX (number not yet assigned), scale 1:24,000.

Baltz, E.H., 1967, Stratigraphy and regional tectonic implications of part of Upper Cretaceous and Tertiary rocks, east-central San Juan Basin, New Mexico: U.S. Geological Survey Professional Paper 552, 99 p.

Bonnet, S., and Crave, A., 2003, Landscape response to climate change: Insights from experimental modeling and implications for tectonic versus climatic uplift of topography: Geology, v. 31, p. 123–126, doi: 10.1130/0091-7613(2003)031<0123:LRTCCI>2.0.CO;2.

Brozovic, N., Burbank, D.W., and Meigs, A.J., 1997, Climatic limits on landscape development in the northwestern Himalaya: Science, v. 276, p. 571–574, doi: 10.1126/science.276.5312.571.

Bryan, K., and McCann, F.T., 1936, Successive pediments and terraces of the upper Rio Puerco in New Mexico: Journal of Geology, v. 44, p. 145–172.

Bryan, K., and McCann, F.T., 1938, The Ceja del Rio Puerco: A border feature on the Basin and Range Province in New Mexico: Journal of Geology, v. 46, p. 1–16.

Bull, W.B., 1984, Tectonic geomorphology: Journal of Geological Education, v. 32, p. 310–324.

Bull, W.B., and McFadden, L.D., 1977, Tectonic geomorphology north and south of the Garlock fault, California, *in* Doehring, D.O., ed., Geomorphology of arid regions: Binghamton, New York, Proceedings of the Eighth Annual Geomorphology Symposium, State University of New York at Binghamton, p. 115–138.

Burbank, D.W., and Anderson, R.S., 2001, Tectonic geomorphology: Malden, Blackwell Science, 274 p.

Chapin, C.E., and Kelley, S.A., 1997, The Rocky Mountain erosion surface in the Front Range of Colorado, *in* Bolyard, D.W., and Sonnenberg, S.A., eds., Geologic history of the Colorado Front Range: Denver, Rocky Mountain Association of Geologists, p. 101–114.

Church, F.S., and Hack, J.T., 1939, An exhumed erosion surface in the Jemez Mountains, New Mexico: Journal of Geology, v. 47, p. 613–629.

Cowie, P.A., and Roberts, G.P., 2001, Constraining slip rates and spacings for active normal faults: Journal of Structural Geology, v. 23, p. 1901–1915, doi: 10.1016/S0191-8141(01)00036-0.

Culling, W.E.H., 1960, Analytical theory of erosion: Journal of Geology, v. 68, p. 336–344.

Densmore, A.L., Ellis, M.A., and Anderson, R.S., 1998, Landsliding and the evolution of normal-fault-bounded mountains: Journal of Geophysical Research, v. 103, p. 15,203–15,219, doi: 10.1029/98JB00510.

dePolo, C.M., and Anderson, J.G., 2000, Estimating the rates of normal faults in the Great Basin, USA: Basin Research, v. 12, p. 227–240, doi: 10.1046/j.1365-2117.2000.00131.x.

Dethier, D.S., 2001, Pleistocene incision rates in the western United States calibrated using Lava Creek B tephra: Geology, v. 29, p. 783–786, doi: 10.1130/0091-7613(2001)029<0783:PIRITW>2.0.CO;2.

Eaton, G., 1986, A tectonic redefinition of the southern Rocky Mountains: Tectonophysics, v. 132, p. 163–193, doi: 10.1016/0040-1951(86)90031-4.

Ellis, M.A., Densmore, A.L., and Anderson, R.S., 1999, Development of mountainous topography in the Basin Ranges, USA: Basin Research, v. 11, p. 21–41, doi: 10.1046/j.1365-2117.1999.00087.x.

Ferrill, D.A., Stamatakos, J.A., Jones, S.M., Rahe, B., McKague, H.L., Martin, R.H., and Morris, A.P., 1996, Quaternary slip history of the Bare Mountain fault (Nevada) from the morphology and distribution of alluvial fan deposits: Geology, v. 24, p. 559–562, doi: 10.1130/0091-7613(1996)024<0559:QSHOTB>2.3.CO;2.

Formento-Trigilio, M.L., 1997, The tectonic geomorphology and long-term landscape evolution of the southern Sierra Nacimiento, northern New Mexico [M.S. thesis]: Albuquerque, University of New Mexico, 201 p.

Formento-Trigilio, M.L., and Pazzaglia, F.J., 1998, Tectonic geomorphology of the Sierra Nacimiento: Traditional and new techniques in assessing long-term landscape evolution in the southern Rocky Mountains: Journal of Geology, v. 106, p. 433–453.

Frankel, K.L., 2002, Quantitative topographic differences between erosionally exhumed and tectonically active mountain fronts: Implications for late-Cenozoic evolution of the southern Rocky Mountains [M.S. thesis]: Bethlehem, Lehigh University, 138 p.

Frankel, K.L., and Pazzaglia, F.J., 2002, Quaternary geologic map of the Ponderosa 7.5′ quadrangle, Sandoval County, New Mexico: New Mexico Bureau of Geology and Mineral Resources, Open-File geologic map OF-GM 57b, scale 1:24,000.

Garcia, A., Zhu, Z., Ku, T.L., Sanz de Galdeano, C., Chadwick, O.A., and Chacon Montero, J., 2003, Tectonically driven landscape development of the eastern Alpujarran Corridor, Betic Cordillera, SE Spain (Almeria): Geomorphology, v. 50, p. 83–110, doi: 10.1016/S0169-555X(02)00209-X.

Glock, W.S., 1931, The development of drainage systems and the dynamic cycle: The Ohio Journal of Science, v. 31, p. 309–334.

Gregory, K.M., and Chase, C.G., 1994, Tectonic and climatic significance of a late Eocene low-relief, high-level geomorphic surface, Colorado: Journal of Geophysical Research, v. 99, p. 20,141–20,160, doi: 10.1029/94JB00132.

Hallet, R.B., 1994, Volcanic geology, paleomagnetism, geochronology and geochemistry of the Rio Puerco necks, west-central, New Mexico [Ph.D. thesis]: Socorro, New Mexico Institute of Mining and Technology, 340 p.

Hallett, R.B., Kyle, P.R., and McIntosh, W.C., 1997, Paleomagnetic and $^{40}Ar/^{39}Ar$ age constraints on the chronologic evolution of the Rio Puerco volcanic necks and Mesa Prieta, west- central New Mexico: Implications for transition zone magmatism: Geological Society of America Bulletin, v. 109, p. 95–106, doi: 10.1130/0016-7606(1997)109<0095:PAAAAC>2.3.CO;2.

Harbor, D.J., 1997, Landscape evolution at the margin of the Basin and Range: Geology, v. 25, p. 1111–1114, doi: 10.1130/0091-7613(1997)025<1111:LEATMO>2.3.CO;2.

Horton, B.K., and DeCelles, P.G., 2001, Modern and ancient fluvial megafans in the foreland basin system of the central Andes, southern Bolivia: Implications for drainage network evolution in fold-thrust belts: Basin Research, v. 13, p. 43–63, doi: 10.1046/j.1365-2117.2001.00137.x.

House, L., and Hartse, H., 1995, Seismicity and faulting in northern New Mexico, *in* Bauer, P.W., Kues, B.S., Dunbar, N.W., Karlstrom, K.E., and Harrison, B., eds., Geology of the Santa Fe region: New Mexico Geological Society Guidebook, 46th Field Conference, p. 135–137.

Howard, A.D., 1994, A detachment-limited model of drainage basin evolution: Water Resources Research, v. 30, p. 2261–2285, doi: 10.1029/94WR00757.

Ingersoll, R.V., Cavazza, W., Baldridge, W.S., and Shafiqullah, M., 1990, Cenozoic sedimentation and paleotectonics of north-central New Mexico: Implications for initiation and evolution of the Rio Grande Rift: Geological Society of America Bulletin, v. 102, p. 1280–1296, doi: 10.1130/0016-7606(1990)102<1280:CSAPON>2.3.CO;2.

Karlstrom, K.E., Bowering, S.A., Chamberlain, K.R., Dueker, K.G., Eshete, T., Erslev, E.A., Farmer, G.L., Heizler, M., Humphreys, E.D., Johnson, R.A., Keller, G.R., Kelley, S.A., Levander, A., Magnani, M.B., Matzel, J.P., McCoy, A.M., Miller, K.C., Morozova, E.A., Pazzaglia, F.J., Prodehl, C., Rumpel, H.M., Shaw, C.A., Sheehan, A.F., Shoshitaishvili, E., Smithson, S.B., Snelson, C.M., Stevens, L.M., Tyson, A.R., and Williams, M.L., 2002, Structure and evolution of the lithosphere beneath the Rocky Mountains: Initial results from the CD-ROM experiment: GSA Today, v. 12, no. 3, p. 4–10.

Keller, E.A., and Pinter, N., 2001, Active tectonics: Earthquakes, uplift, and landscape: Upper Saddle River, Prentice Hall, 362 p.

Kirkby, M.J., 1984, Modelling cliff development in South Wales: Savigear re-reviewed: Zeitschrift für Geomorphologie, v. 28, p. 405–426.

Kooi, H., and Beaumont, C., 1996, Large-scale geomorphology: Classical concepts reconciled and integrated with contemporary ideas via a surface processes model: Journal of Geophysical Research, v. 101, p. 3361–3386.

Lague, D., Crave, A., and Davy, P., 2003, Laboratory experiments simulating the geomorphic response to tectonic uplift: Journal of Geophysical Research, v. 108, doi: 10.1029/2002JB001785.

Leonard, E.M., 2002, Geomorphic and tectonic forcing of late Cenozoic warping of the Colorado piedmont: Geology, v. 30, p. 595–598, doi: 10.1130/0091-7613(2002)030<0595:GATFOL>2.0.CO;2.

Leonard, E.M., and Langford, R.P., 1994, Post-Laramide deformation along the eastern margin of the Colorado Front Range—A case against significant faulting: The Mountain Geologist, v. 31, p. 45–52.

Lipman, P.W., and Reed, J.C., Jr., 1989, Geologic map of the Latir volcanic field and adjacent areas, northern New Mexico: U.S. Geological Survey Miscellaneous Investigations Series Map I-1907, scale 1:48,000.

Machette, M.N., Personius, S.F., Kelson, K.I., Haller, K.M., and Dart, R.L., 1998, Map and data for Quaternary folds and faults in New Mexico: U.S. Geological Survey Open-File Report 98-521, 443 p.

Mack, G.H., McIntosh, W.C., Leeder, M.R., and Monger, H.C., 1996, Plio-Pleistocene pumice floods in the ancestral Rio Grande, southern Rio Grande Rift, USA: Sedimentary Geology, v. 103, p. 1–8, doi: 10.1016/0037-0738(96)00009-7.

McMillan, M.E., Angevine, C.L., and Heller, P.L., 2002, Postdepositional tilt of the Miocene-Pliocene Ogallala Group on the western Great Plains: Evidence of late Cenozoic uplift of the Rocky Mountains: Geology, v. 30, p. 63–66, doi: 10.1130/0091-7613(2002)030<0063:PTOTMP>2.0.CO;2.

Menges, C.M., 1990a, Late Quaternary fault scarps, mountain front landforms, and Pliocene-Quaternary segmentation on the range-bounding fault zone, Sangre de Cristo Mountains, New Mexico, *in* Krinitzsky, E.L., and Slemmons, D.B., eds., Neotectonics in earthquake evaluation: Boulder, Colorado, Geological Society of America Reviews in Engineering Geology, v. 8, p. 131–156.

Menges, C.M., 1990b, Soils and geomorphic evolution of bedrock facets on a tectonically active mountain front, western Sangre de Cristo Mountains, New Mexico: Geomorphology, v. 3, p. 301–332, doi: 10.1016/0169-555X(90)90009-F.

Merritts, D., and Vincent, K.R., 1989, Geomorphic response of coastal streams to low, intermediate, and high rates of uplift, Mendocino triple junction region, northern California: Geological Society of America Bulletin, v. 101, p. 1373–1388, doi: 10.1130/0016-7606(1989)101<1373:GROCST>2.3.CO;2.

Montgomery, D.R., and Brandon, M.T., 2002, Topographic controls on erosion rates in tectonically active mountain ranges: Earth and Planetary Science Letters, v. 201, p. 481–489, doi: 10.1016/S0012-821X(02)00725-2.

Moore, J.D., 2000, Tectonics and volcanism during deposition of the Oligocene–lower Miocene Abiquiu Formation in northern New Mexico [M.S. thesis]: Albuquerque, University of New Mexico, 147 p.

Ohmori, H., 1993, Changes in the hypsometric curve through mountain building resulting from concurrent tectonics and denudation: Geomorphology, v. 8, p. 263–277, doi: 10.1016/0169-555X(93)90023-U.

Ouchi, S., 2002, Effects of uplift on the development of experimental erosion landform: Geological Society of America Abstracts with Programs, v. 34, no. 6, p. 129.

Parker, R.S., and Schumm, S.A., 1982, Experimental study of drainage networks, *in* Bryan, R., and Yair, A., eds., Badland geomorphology and piping: Norwich, Geobooks, p. 153–168.

Pazzaglia, F.J., 1989, Tectonic and climatic influences on the evolution of Quaternary depositional landforms along a segmented range-front fault, Sangre de Cristo Mountains, north-central New Mexico [M.S. thesis]: Albuquerque, University of New Mexico, 246 p.

Pazzaglia, F.J., and Kelley, S.A., 1998, Large-scale geomorphology and fission-track thermochronology in topographic and exhumation reconstructions of the southern Rocky Mountains: Rocky Mountain Geology, v. 33, p. 229–257.

Pazzaglia, F.J., and Wells, S.G., 1990, Quaternary stratigraphy and geomorphology of the northern Rio Grande Rift, *in* Bauer, P.W., Lucas, S.G., Mawer, C.K., and McIntosh, W.C., eds., Southern Sangre de Cristo Mountains: New Mexico Geological Society Guidebook, 41st Field Conference, p. 423–430.

Pazzaglia, F.J., Woodward, L.A., Lucas, S.G., Anderson, O.J., Wegmann, K.W., and Ester, J.W., 1999, Phanerozoic geologic evolution of the Albuquerque area, *in* Pazzaglia, F.J., and Lucas, S.G., eds., Albuquerque geology: New Mexico Geological Society Guidebook, 50th Field Conference, p. 97–114.

Pederson, J.L., Mackley, R.D., and Eddleman, J.L., 2002, Colorado Plateau uplift and erosion evaluated using GIS: GSA Today, v. 12, no. 8, p. 4–10, doi: 10.1130/1052-5173(2002)012<0004:CPUAEE>2.0.CO;2.

Pendall, E., Betancourt, J.L., and Leavit, S.W., 1999, Palaeoclimatic significance of δD and $\delta^{13}C$ values in piñon pine needles from packrat middens spanning the last 40,000 years: Palaeogeography, Palaeoclimatology, Palaeoecology, v. 147, p. 53–72, doi: 10.1016/S0031-0182(98)00152-7.

Pike, R.J., and Wilson, S.E., 1971, Elevation-relief ratio, hypsometric integral, and geomorphic area-altitude analysis: Geological Society of America Bulletin, v. 82, p. 1079–1084.

Reneau, S.L., 2000, Stream incision and terrace development in Frijoles Canyon, Bandelier National Monument, New Mexico, and the influence of lithology and climate: Geomorphology, v. 32, p. 171–193, doi: 10.1016/S0169-555X(99)00094-X.

Richmond, G.M., 1962, Correlation of some glacial deposits in New Mexico: U.S. Geological Survey Professional Paper, v. 450-E, p. E121–E125.

Roering, J.J., Kirchner, J.W., Sklar, L.S., and Dietrich, W.E., 2001, Hillslope evolution by nonlinear creep and landsliding: An experimental study: Geology, v. 29, p. 143–146, doi: 10.1130/0091-7613(2001)029<0143:HEBNCA>2.0.CO;2.

Rogers, J.B., 1996, The fluvial landscape evolution of San Diego Canyon, Jemez Mountains, New Mexico [M.S. thesis]: Albuquerque, University of New Mexico, 123 p.

Sanford, A.R., Jaksha, L.H., and Cash, D.J., 1991, Seismicity of the Rio Grande

Rift in New Mexico, *in* Slemmons, D.B., Engdahl, E.R., Zoback, M.D., and Blackwell, D.D., eds., Neotectonics of North America: Boulder, Colorado, Geological Society of America, the Geology of North America, v. I, p. 229–244.

Schumm, S.A., and Parker, R.S., 1973, Implications of complex response of drainage systems for Quaternary alluvial stratigraphy: Nature, v. 243, p. 99–100.

Slavin, P.J., 1991, Summary of recent work on the geomorphology of the upper Rio Puerco, *in* Julian, B., and Zidek, J., eds., Field guide to geologic excursions in New Mexico and adjacent areas of Texas and Colorado: New Mexico Bureau of Mines and Mineral Resources Bulletin 137, p. 171–172.

Small, E.E., and Anderson, R.S., 1998, Pleistocene relief production in Laramide mountain ranges, western United States: Geology, v. 26, p. 123–126, doi: 10.1130/0091-7613(1998)026<0123:PRPILM>2.3.CO;2.

Strahler, A.N., 1952, Hypsometric (area-altitude) analysis of erosional topography: Geological Society of America Bulletin, v. 63, p. 1117–1142.

Tedford, R.H., 1981, Mammalian biochronology of the late Cenozoic basins of New Mexico: Geological Society of America Bulletin, v. 92, p. 1008–1022, doi: 10.1130/0016-7606(1981)92<1008:MBOTLC>2.0.CO;2.

Tedford, R.H., and Barghoorn, S., 1999, Santa Fe group (Neogene), Ceja del Rio Puerco northwestern Albuquerque basin, Sandoval county, New Mexico, *in* Pazzaglia, F.J., and Lucas, S.G., eds., Albuquerque geology: New Mexico Geological Society Guidebook, 50th Field Conference, p. 327–335.

Thompson, R.S., 1991, Pliocene environments in the western United States: Quaternary Science Reviews, v. 10, p. 115–132, doi: 10.1016/0277-3791(91)90013-K.

Tucker, G.E., and Slingerland, R.L., 1994, Erosional dynamics, flexural isostasy, and long-lived escarpments: A numerical modeling study: Journal of Geophysical Research, v. 99, p. 12,229–12,243, doi: 10.1029/94JB00320.

Tucker, G.E., and Slingerland, R., 1996, Predicting sediment flux from fold and thrust belts: Basin Research, v. 8, p. 329–349, doi: 10.1046/j.1365-2117.1996.00238.x.

Vazzana, M.E., and Ingersoll, R.V., 1981, Stratigraphy, sedimentology, petrology, and basin evolution of the Abiquiu Formation (Oligo-Miocene), north-central New Mexico: Geological Society of America Bulletin, v. 92, part 2, p. 2401–2483.

Wells, S.G., Bullard, T.F., Menges, C.M., Drake, P.G., Karas, P.A., Kelson, K.I., Ritter, J.B., and Wesling, J.R., 1988, Regional variations in tectonic geomorphology along a segmented convergent plate boundary, Pacific coast of Costa Rica: Geomorphology, v. 1, p. 239–266, doi: 10.1016/0169-555X(88)90016-5.

Wisniewski, P.A., and Pazzaglia, F.J., 2002, Epeirogenic controls on the Canadian River incision and landscape evolution, High Plains of northeastern New Mexico: Journal of Geology, v. 110, p. 437–456, doi: 10.1086/340441.

Woodward, L.A., 1987, Geology and mineral resources of Sierra Nacimiento and vicinity New Mexico: New Mexico Bureau of Mines and Mineral Resources Memoir 42, 84 p.

Woodward, L.A., 1996, Tectonics of Nacimiento uplift and adjacent areas, *in* Goff, F., Kues, B.S., Rogers, M.S., McFadden, L.D., and Gardner, J.N., eds., Jemez Mountains region: New Mexico Geological Society Guidebook, 47th Field Conference, p. 11–12.

Woodward, L.A., McLelland, D., Anderson, J.B., and Kaufman, W.H., 1970, Geologic map and section of Cuba quadrangle, New Mexico: New Mexico Bureau of Mines and Mineral Resources Geologic Map 24, scale 1:24,000.

Woodward, L.A., Kaufmann, W.H., and Anderson, J.B., 1972, Nacimiento fault and related structures, northern New Mexico: Geological Society of America Bulletin, v. 83, p. 2382–2396.

Zaprowski, B.J., Evenson, E.B., Pazzaglia, F.J., and Epstein, J.B., 2001, Knickzone propagation in the Black Hills and northern High Plains: A different perspective on the late Cenozoic exhumation of the Laramide Rocky Mountains: Geology, v. 29, p. 547–550, doi: 10.1130/0091-7613(2001)029<0547:KPITBH>2.0.CO;2.

Manuscript Accepted by the Society 23 June 2005

Printed in the USA

Index

Bold numbers indicate material in figures and tables.

Bold numbers indicate material in figures and tables.

Bold numbers indicate material in figures and tables.

Bold numbers indicate material in figures and tables.

Bold numbers indicate material in figures and tables.

Bold numbers indicate material in figures and tables.

Bold numbers indicate material in figures and tables.